# CLINICAL SIMULATION

# CLINICAL SIMULATION

## OPERATIONS, ENGINEERING, AND MANAGEMENT

**RICHARD R. KYLE, JR.,
AND W. BOSSEAU MURRAY**

AMSTERDAM • BOSTON • HEIDELBERG • LONDON • NEW YORK • OXFORD
PARIS • SAN DIEGO • SAN FRANCISCO • SINGAPORE • SYDNEY • TOKYO

Academic Press is an imprint of Elsevier

Academic Press is an imprint of Elsevier
30 Corporate Drive, Suite 400, Burlington, MA 01803, USA
525 B Street, Suite 1900, San Diego, CA 92101-4495, USA
84 Theobald's Road, London WC1X 8RR, UK

First edition 2008

Notice
No responsibility is assumed by the publisher for any injury and/or damage to persons
or property as a matter of products liability, negligence or otherwise, or from any use
or operation of any methods, products, instructions or ideas contained in the material
herein. Because of rapid advances in the medical sciences, in particular, independent
verification of diagnoses and drug dosages should be made

**Library of Congress Cataloging-in-Publication Data**
A catalog record for this book is available from the Library of Congress

**British Library Cataloguing in Publication Data**
A catalogue record for this book is available from the British Library

ISBN: 978-0-12-372531-8

For information on all Academic Press publications
visit our web site at books.elsevier.com

Transferred to Digital Printing in 2011

Richard R. Kyle, Jr.

*The book is dedicated to Kenneth E. Kinnamon, DVM, PhD: "When I was ready to learn, he appeared and taught."*

W. Bosseau Murray

*This book is dedicated to my wife, Janette, and children, Charlotte, Ian, and Elaine, for their unwavering support over many years, as well as to Art Schneider, MD, for guiding my early steps in educational principles using simulation.*

# Epigraph

All the world's a stage,
And all the men and women merely players.
They have their exits and their entrances;
And one man in his time plays many parts. . .

William Shakespeare (1564–1616), "As You Like It," Act 2, Scene 7.

# Contents

## Topic VI   Functional Forms at the Institutional Size

## Topic VII   Functional Forms at the State and Nation Size

## Topic VIII   The Big Picture: Sum of Many Smaller Views

## Topic IX   Make Your Own

## Topic X   Buy from Others

## Topic XIX    Tricks of the Trade

## Topic XX    Rehearsing is the Basis of All Learning

## Topic XXI    Expect the Unexpected

## Topic XXII    Borrow Success

# Biographies

**Richard R. Kyle**, Jr., BSE, MS
Formally educated in fluid mechanics and physical chemistry, his lifetime fascinations with engineering, theatrical production, and biomedical investigation found the perfect storm in clinical simulation. In 1997, co-founded the Patient Simulation Laboratory, Uniformed Services University, Bethesda MD, with one class-one topic to teach using one clinical instructor, one CAE-Eagle patient simulator, a foot-pumped OR bed, and an empty room 15 × 20 feet. Since then, he's ridden the waves of ever increasing simulation program and facility creation. In doing so he's meet new friends and collaborated on expanding simulation-based learning at home and around the globe. Now he wonders how much more the fundamental characteristics of simulation (intentional, deliberate, and scheduled) can reduce harmful novelty in clinical care and clinical education, and at what scales.

**W. Bosseau Murray**, MBChB, FFA(RCS)(Lon), MD(Anaes) (Stell)
Formally educated in medicine and anesthesia, his interests include physics, statistics, mathematics, and common sense as applied in the Health Care Sciences. His aim is to learn how to better be able to teach these essential concepts using innovative educational modalities. Associate Director of the Simulation Development and Cognitive Science Laboratory (also known as the "Simulation Lab"), Professor of Anesthesiology at Pennsylvania State University College of Medicine at the Milton S Hershey Medical Center in Hershey, Pennsylvania, U.S.A.

**Alice L. Acker**, MPA, CHE
Civilian and military health care management: a passion for start-ups, technology, and e-based learning; longtime interests in acting, playwriting and moulage. Adjunct Professor of Master of Healthcare Administration. 2005, Program Manager, Institute for Surgical and Interventional Simulation (ISIS), University of Washington, Seattle, U.S.A.

**J. Lance Acree**, BMechEng, MSEng, MSAeroEng
First encountered simulation in U.S. Air Force pilot training; served as instructor pilot in the Defense Department's school for the C-130 aircraft, learned that instructing in simulation is different from, more challenging and more powerful than instructing in "the real thing." Pentagon budget boy; commanded the C-17 training squadron; earned Boeing 737 type rating through simulation. Senior Executive Consultant for Aviation Training Consulting, LLC (www.atc-hq.com). Severna Park, Maryland, U.S.A.

**Mark R. Adelman**, PhD
BA in Biology (Princeton, 1963), then realized he would probably make a poor medical doctor, switched paths, earned a PhD in Biophysics (University of Chicago, 1969), and has taught undergraduate medical students and graduate students since 1971, but has *never* taught a course that he has formally taken as a student. Always educated by his failures but has *still* not succeeded in getting students to accept failures as learning experiences. Department of Anatomy, Physiology and Genetics, Uniformed Services University of the Health Sciences, Bethesda, Maryland, U.S.A.

**Amy Guillet Agrawal**, MD
Areas of clinical specialization: Critical Care Medicine and Infectious Diseases. Research: Clinical trials on West Nile virus. Physician and clinical researcher in the Critical Care Medicine Department at the National Institutes of Health Clinical Center. 2004: Founder and Director of the Clinical Simulation Service at the National Institutes of Health Clinical Center, Bethesda, Maryland, U.S.A.

**Riva R. Akerman**, MD
Recently an Assistant Professor of Anesthesiology in the pediatric anesthesia division at the University of Miami Miller School of Medicine, Miami, Florida. Very involved in the simulation program at their Center for Patient Safety. Now an Assistant Professor of Anesthesiology and associate program director of the pediatric anesthesia fellowship in the pediatric anesthesia division at Columbia University, New York, New York, USA.

**Guillaume Alinier**, DUT, MPhys (Hons), PGCE, CPhys, MInstP, ILTM Formal education in Physics. Research in ultrasound altimetry, fish deterrence systems, medical technology. Teaches "Medical Design and Technology" and "Simulation Facilitator Training." University Teaching Fellow to the Center for the Enhancement of Learning and Teaching in 2005. Elected Secretary for the Society in Europe for Simulation Applied to Medicine, SESAM, for 2005–2006. Awarded National Teaching Fellowship from Higher Education Academy of Great Britain in 2006. Simulation Center Manager, University of Hertfordshire, U.K.

**John J. Anton**, BA, MS,
Chief Technology Officer and Vice President, Engineering, Government Systems and Education & Training. Medical Education Technologies Inc. Sarasota, Florida, U.S.A.

**Craig Balbalian**
Prehospital and tactical care training program coordinator, US Army Medical Command, Walter Reed Army Medical Center, Washington DC, U.S.A.

**Lorena Beeman**, RN, MS, CCVT
Twenty-five years as a critical care nurse; 14 years as a critical care nurse educator; Cardiovascular and pulmonary pathophysiology critical care and education. In 2001, her retired pharmacist husband became a simulation coordinator at the hospital at which she worked. Motivated to learn not just the scripting, but the setup and mechanics, and organization of the whole simulation process. University of New Mexico Health Sciences Center Clinical Education Department, Albuquerque, New Mexico, U.S.A.

**Thomas E. Belda**, RRT, BA
A decade as respiratory therapist with pediatric and neonatal patients and their families, interests in computers and technology, degree in Information Technology and web-based software development. Learned concepts and technologies of high-fidelity simulation in aerospace simulation projects. Contributes to the big health care machine with the promise of improving patient safety and new methodologies for teaching and training in medical education. Network Systems Engineer – Mayo Clinic Multidisciplinary Simulation Center, Rochester, Minnesota, U.S.A.

**Betsy Bencken**, MS
Formally studied human anatomy with a focus on neuroanatomy and gross anatomy. Twenty-five years of University experience in medical research, education and administration. 2002, Codeveloper of the Center for Virtual Care, University of California at Davis. Expanded this simulation program to encompass students of prehospital care all the way through advanced practitioners. Active in clinical instruction as well as Center management. Davis, California, U.S.A.

**Haim Berkenstadt**, MD
Graduate of the Hebrew University medical school in Jerusalem, served as a military physician at the Israeli Defense Forces Medical Corps, and trained as anesthesiologist at Sheba Medical Center, Tel Hashomer. Practices neuroanesthesia and clinical research in hemodynamic monitoring, serves on the Israeli Board of Anesthesia, and a clinical senior lecturer of anesthesiology at the Tel Aviv University Medical School. A founding member and now Deputy Director of MSR – the Israel Center for Medical Simulation, Tel Hashomer, Israel.

**Anthony Brand**, LCH, BSc, MA, PhD, DUniv, MARH
Studied Physics and first encountered simulation approaches and techniques as an undergrad in the late 1960s; later, as part of his Doctorate, this was extended using a Monte Carlo technique. Currently his simulation work includes collaborating with a variety of clinical practitioners – nurses, osteopaths, homoeopaths, and pharmacists. Anglia Ruskin University, Cambridge, England, U.K.

**Cathleen K. Brannen**, MBA
Thirty years in higher education administration, including 7 years as CFO of a nursing school. Charter and current member of the HealthPartners Simulation Center Oversight Committee. Experienced consumer of health care services. Gardener, woodworker, photographer. Vice President, Finance and Administration, Metropolitan State University, St. Paul, Minnesota, U.S.A.

**Brian C. Brost**, MD, FACOG, FACS
Formal education in chemistry and medicine, appointments as various types of Medical Student, OB/Gyn Residency and Maternal Fetal Medicine Fellowship director, started developing in his garage low-cost but tactilely accurate training models and simulators, work at Mayo Multidisciplinary Simulation Center, Rochester, Minnesota, U.S.A.

**Ronald G. Carovano, Jr.**, BSEE, MBA
Formal education in Electrical Engineering, Management and Finance; Researcher with the Department of Anesthesiology at the University of Florida College of Medicine; Co-Inventor and U.S. Patent-holder of Human Patient Simulator (HPS™) technology developed at the University of Florida and licensed to Medical Education Technologies, Inc. (METI). Fan of the University of Florida Fightin' Gators and U.S. National Soccer Teams. 1989, Director of Development with Medical Education Technologies, Inc., Sarasota, Florida, U.S.A.

**Daniel Castillo**, MD
Cardio-Thoracic Anesthesia and Critical Care Medicine. Medical Director for Simulation Center for Patient Safety, Department of Anesthesiology University of Miami Miller School of Medicine. Miami, Florida, U.S.A.

**Chris Chin**, MBBS, MRCP, FRCA
Consultant Anesthetist, Guy's and St Thomas NHS Foundation Trust. Associate Director, Barts and the London Medical Simulator. After qualifying as a doctor, several years in internal medicine and intensive care before training in Anesthesia in London, U.K. and Toronto, Canada. Subsequently specializing in Pediatric and Cardiac Anesthesia. 2001, Started instructing in the newly opened Barts and the London Simulator Center and in 2003, appointed as Associate Director. London, England, U.K.

**Roger E. Chow**, RT
Formal education in Fine Arts and in Respiratory Therapy. A lifelong hope to be a full-time artist, but I can't be creative when I'm hungry. Started simulating in 2000, currently the Simulation Coordinator at St. Michael's Hospital and Trauma Center, Toronto, Canada.

**Larry A. Cobb**, RN, BSN, CEN, EMT-P
Has enjoyed a varied Nursing/EMS career. From Sports–Medicine, to Neurosurgery, to Orthopedics, to Emergency Medicine – always an educator. He's utilized a portion of his hobby, resin casting, to produce detailed "props" for training/teaching purposes. Currently a Nurse Educator with the University of New Mexico Health Sciences – B.A.T.C.A.V.E. simulation lab, Albuquerque, New Mexico, U.S.A.

**Neil Coker**, BS, EMT-P
Emergency medical services education program director at Amarillo College, Texas Tech University Health Sciences Center, and Temple College. Texas State Emergency Medical Services Training Coordinator. EMS program specialist for City of Dallas and Texas Department of Health. 2004, Director of Simulation Teaching, Assessment, and Research Programs, Temple College, Temple, Texas, U.S.A.

**Edmundo P. Cortez**, MD
Medical Student Clerkship Director/PICU Resident Education Director Assistant Professor of Pediatrics. Medical Director of Pediatric ICU North. Assistant Pediatric Clerkship Director, Rush Children's Hospital, Chicago Illinois, U.S.A.

**Robert C. Cox**, BS
First encountered simulation in U.S. Air Force pilot training; served as instructor pilot and evaluator pilot in the formal school for the C-141 aircraft. Later served as director of the training systems for the C-17, C-5, C-141, and the T-1A aircraft, where he learned to reduce training costs by applying the systems approach to training and good business sense. Over 2000 hours instructing in aircraft simulators. President and CEO of Aviation Training Consulting, LLC (www.atc-hq.com), Altus, Oklahoma, U.S.A.

**Ian Curran**, BSc, AKC, MBBS, FRCA, Pg Dip Med Ed
Enduring passion for undergraduate and postgraduate clinical education, teaching faculty development, professional behaviors, and assessment. The greatest challenge facing educators: creating effective learning. The greatest danger we face as educators is being seduced by the activity of teaching! Consultant in Pain Medicine and Anesthesia, Clinical Tutor and Assoc. Director of Medical Education, St Bartholomew's and The Royal London Hospitals, Assoc. Chair "Anesthetists as Educator" Program, Royal College of Anaesthetists, United Kingdom. Director, Barts and the London Medical Simulation Center, London, England, U.K.

**Sharon M. Denning**, MS, RN, CNA
Majority of nursing career spent in critical care, with more than 20 years as nurse manager and director. Strong interest in education, patient safety, and helping staff to develop clinical expertise. Director of HealthPartners Simulation Center for Patient Safety at Metropolitan State University in St. Paul, Minnesota, U.S.A.

**Peter Dieckmann**, PhD
Psychologist, inspired by Kurt Lewin's formula: $B = f_{(P,E)}$, with a focus on safety and simulation; became involved with simulation by taking part in an experiment. Now researching human error using simulators, training of simulator instructors, and incident reporting/analysis at the Center for Patient Safety and Simulation (TuPASS), Tuebingen, Germany.

**Harold K. Doerr**, MD
Bachelor of Arts in 1979 from Rutgers University; MD from University of Texas Health Science Center in 1987; Residency in Anesthesiology; fellowships in Cardiac Anesthesiology and Cardiology-Ultrasound; principal investigator for Medical Operations Support Team to use patient simulation for medical training in space; created Anesthesia protocol for astronauts following landing; developed a technique for intubating in zero gravity. Houston, Texas, U.S.A.

**Thomas J. Doyle**, MSN, RN
Over 25 years of experience as a registered nurse, hospital administrator, and nurse educator; 4 years as Patient Simulation Program Coordinator; guides integrating patient simulation-based curriculum into clinical education programs. Medical Education Technologies, Inc. Sarasota, Florida, U.S.A.

**Bonnie Driggers**, R.N., M.S., M.P.A.
Served as Director of Clinical Teaching Systems and Programs for the OHSU School of Nursing and Co-Director of the OHSU Simulation and Clinical Learning Center. A founding member and the past chair of the Oregon Simulation Alliance, serves on the Governing Council of the Simulation Alliance. Co-authored the "Oregon Simulation Readiness Report" which presented a comprehensive look at desire and readiness for simulation education in Oregon. Served as Co-Director of the OCNE Clinical Education Model Project focused on the redesign of clinical education in nursing for the consortium. The model includes the use of simulation as one of several clinical learning activities as an adjunct to clinical experiences that support the competencies required for healthcare practitioners. Currently provides statewide simulation education and consults throughout the United States, Canada and overseas in the area of simulation education, program development and implementation. Professor Emeritus at the Oregon Health & Science University (OHSU) Schools of Nursing and Medicine.

**William F. Dunn**, MD
2008, President, Society for Simulation in Healthcare; Associate Professor of Medicine, Division of Pulmonary and Critical Care Medicine, Past Director of Mayo Multidisciplinary Critical Care Fellowship, Mayo Clinic, Rochester, Minnesota, U.S.A.

**David Erez**, EMT-P, MSc
BSc in Sport and Exercise Science, MSc in Cardiac Rehabilitation from University of Auckland, Diploma as Mobile Intensive Care Paramedic from Magen David Adom in Israel. Assisted establishing Divisional High Fidelity Simulation Education Center at Auckland University of Technology; supervised and tutored students in Bachelor in Health Science – Paramedic Program, AUT. Developed educational program for patients at the Auckland Cardiac Rehab Clinic. Instructor and course developer at Israel Center for Medical Simulation, Tel Hashomer, Israel.

**Mark E. A. Escott**, BMBS, MPH, LP
Founded an Emergency Medical Service at Rice University in Houston, Texas, and an EMS training program in Department of Human Performance and Health Sciences as an Adjunct Professor. Involved in paramedic education and simulation at Rice, further developed skills as a medical student at Flinders University (Adelaide, Australia). Aided design of simulation-based medical student curriculum in Emergency Medicine, and did research involving cricoid pressure trainers. Experience with high-tech simulators during Space Medicine Clerkship at NASA/Baylor College of Medicine, Houston, Texas where he designed scenarios for the training of astronauts/flight surgeons for emergency situations. Currently a consultant to Rice University and a resident in Emergency Medicine, PennState-MS Hershey Medical Center, Hershey, Pennsylvania, U.S.A.

**Margaret Faut-Callahan**, CRNA, DNSc, FAAN
Initial interest in clinical simulation related to the enhanced learning opportunities for students in a nonthreatening but clinically relevant environment. She is also interested in the way we educate health professionals, increasing patient safety across many practice settings. As a founding member of our simulation faculty, the dramatic impact on our students has been rewarding. Chair, Adult Health Nursing and Director, Nurse Anesthesia Program, Rush University Medical Center, Chicago, Illinois, U.S.A.

**Valerie Follows**, RN
Registered nurse with Intensive Care diploma and experience in acute care nursing since 1975. Began in Simulation, educating nurses in Canada, expanded the process with Professor Harry Owen, establishing the Simulation Unit at Flinders University in 2001. Now lectures, plans simulations and teaching sessions for the medical students in Flinders University Graduate Entry program; also assists other educators in using our facility. Coordinator of the Clinical Simulation Unit and Associate Lecturer in Simulation at Flinders University, Adelaide, Australia.

**Michael C. Foss**, MA, RDMS, RVT
Formal education in Diagnostic Medical Sonography (ultrasound), teacher education, educational administration, and leadership. Currently Dean, School of Health housing 16 health programs. Have always dabbled in electronics, photography, woodworking, and trying to make things work. Remain fascinated by how Walter Elias Disney could create an environment that suspended disbelief. He will do whatever it takes to help students learn how to provide high-quality health care. Integrates patient simulation into the health curriculum, with the expectation of improved patient care. After all, everything we do is for the patient. Springfield Technical Community College in Springfield, Massachusetts, U.S.A.

**Flight Lieutenant Denis B. French**, RAAFSR
BSN, BNursPrac(Aviation), GradDipNursing(Perioperative), RN Ex-full-time RAAF Nursing Officer with operational experience in East Timor and Iraq as an OR and recovery room nurse. Now part-time ICU nurse and part-time desk officer coordinating health simulation policy, doctrine and implementation across the Australian Defence Force. 2005, Staff Officer Grade 2 Health Simulation, Defence Health Services Division, Australian Defence Force. Canberra, Australia.

**Christopher J. Gallagher**, MD
Anesthesiologist who has practiced in Sweden, North Carolina, California, Florida, and New York. (He keeps getting run out of town on a rail.) Played with dolls as a young man and found this the highlight of my existence. The leap to simulators was a cinch. Simulation Center for Patient Safety, University of Miami Miller School of Medicine. Miami, Florida, U.S.A.

**John Gillespie**, CAT
Retired from USAirways as a Flight Attendant, taught as a Flight Attendant Instructor and performed Certification for the FAA, a crash site investigator for the NTSB; worked at the University of Louisville-School of Medicine, OR Anesthesia Technician at the University of Louisville Hospital; certified as a National Disaster Life Support Foundation Instructor; coordinated the Education and Training Department at Medical Education Technologies Inc.; now the Simulation Center Coordinator at the Medical College of Georgia, Augusta, Georgia, U.S.A.

**Ronnie J. Glavin**, MB, ChB, MPhil, FRCA
Became very interested in medical education during anesthetic training and embarked upon some formal training – A one year Certificate in Medical Education from the University of Glasgow (1989), received a Master of Philosophy degree in Educational Studies (1993). He saw simulation as a way of extending the possibilities of training and of studying how anesthetists really work. Clinical anesthetist in Glasgow, Scotland.

**Michael S. Goodrow**, BS, MEng
Formal education is in engineering mathematics. Used various simulation techniques to analyze weapon system effectiveness for 10 years, and used discrete event simulation to analyze industrial engineering processes for five years. 2001: Joined

the University to operate the Medical School's clinical simulation center. Doctoral candidate in physiology. University of Louisville, Louisville Kentucky, U.S.A.

**Wolfgang Heinrichs**, MD, PhD
Professor of Anesthesiology, staff member and co-director of the department of anesthesia and intensive care at university hospital Mainz (Johannes Gutenberg University Mainz, Medical School). Founder and director of the academic simulation center at Mainz from 1996. Founder and director of the private simulation center Mainz at AQAI. Special interests: Modeling pharmacology, complex models, interfaces and add-on devices to simulation. Mainz, Germany.

**Charles W. Hilton**, MD
Formal education in Endocrinology; served as Program Director in Internal Medicine; lifelong interest in trying to improve educational processes; employs active involvement in learning through immediate feedback and practice makes perfect. Associate Dean for Academic Affairs at Louisiana State Univeristy School of Medicine, New Orleans, Louisiana, U.S.A.

**Steven K. Howard**, MD
Undergraduate training at University of California, Santa Barbara in pharmacology and pursued postgraduate training in anesthesiology at Stanford after medical school. He was exposed to simulation very early in his residency and was captivated by its potential to make us better clinicians. 1989: Director, Patient Safety Center of Inquiry, VA Palo Alto Health Care System, Palo Alto, California, U.S.A.

**Judith C.F. Hwang**, MD, MBA
Studied Biomedical Sciences and English. Formal training in anesthesiology and critical care medicine. Associate Professor in Anesthesiology at University of California, Davis. Interested in fostering team work, communication, and adult learning. 2002, Faculty and Faculty Coordinator at the Center for Virtual Care, University of California, Davis Medical Center, Sacramento, California, U.S.A.

**Constance M. Jewett Johnson**, MPH, BS, RN
Formal education in nursing and health care administration. Experienced health care administrator with experience in the full continuum of care (clinic, hospital, home care, nursing home, senior housing). Experienced in design and building clinical spaces. Knowledgeable about the real world of care delivery and the needs of providers. Director, Center for Continuing Professional Development, HealthPartners Institute for Medical Education. Charter and current member of the HealthPartners Simulation Center Oversight Committee, Metropolitan State University in St. Paul, Minnesota, U.S.A.

**Dan Johnson**, MA, PT
Formal education includes BS in Physical Therapy with masters and doctoral work in education. Teaching experience at the community college level with academic leadership roles as associate and acting academic dean. Currently working within a large health care system in clinical education program development related to staff development and care improvement. Served as Project Manager for development of the Health-Partners Simulation Center for Patient Safety at Metropolitan State University and founding member, Center Oversight Committee, Metropolitan State University in St. Paul, Minnesota, U.S.A.

**Linn Jones**, RRT
Formal education in Respiratory Therapy. Instructor for the Respiratory Therapy program at NAIT and respiratory therapist at the Royal Alexandra Hospital in the Adult Intensive Care unit, professional passion is teaching in adult education. Owner and mother to two Labrador Retrievers Emmy and Ozzy. Seeking a smoother transition from the classroom to the clinical setting for our students, ease some of the pressures fitting students into clinical rotations. Northern Alberta Institute of Technology in Edmonton, Alberta, Canada.

**Alan D. Kaye**, MD, PhD, DABPM
Bachelors of Science degrees in biology and psychology, PhD in pharmacology, Chairman and Program Director in the Department of Anesthesiology, Professor of Pharmacology; devoted to educating medical students and residents. Louisiana State University School of Medicine, New Orleans, Louisiana, U.S.A.

**Lawrence E. Kass**, MD, FACEP, FAAEM
Vice Chair for Education and Director of Residency Training in the Department of Emergency Medicine. Penn State Hershey Medical Center, Hershey, Pennsylvania, U.S.A.

**Valeriy V. Kozmenko**, MD
Formal training in Anesthesiology, Emergency Medicine, Critical Care, Normal Psychology and Abnormal Psychology. 2002, Director of the Human Patient Simulation Lab at Louisiana State University Health Sciences Center, New Orleans, Louisiana, U.S.A.

**Jane E. Kramer**, MD
Pediatric Residency Program Director. Director, Pediatric Emergency Medicine. Rush University Medical Center, Chicago, Illinois, U.S.A.

**Mary Katherine Krause**, MS, CHE
15-year career in field of physician relations and medical education. Associate Vice President for Medical Affairs Administration at Rush University Medical Center, Chicago, Illinois. Assistant Professor of Health Systems Management, Rush

University. 2004, Administrative leadership liaison to the Rush University Simulation Laboratory. Rush University Medical Center, Chicago, Illinois, U.S.A.

**Derek J. LeBlanc**, BA, MA
12 years of paramedic practice and EMS program design and facilitation prior to becoming in 2001 the Program Manager of the Emergency Health Services Atlantic Health Training and Simulation Center in Halifax, Nova Scotia, Canada.

**Howard Levine**, B.Sc
Formally educated at University of Michigan, developed technology used for an advanced medical simulation and training program involving real time teleconferencing and remote control of human patient simulators. Developing new methods in simulation-based distance medical education and platforms suitable for use in the Third World and in technologically impoverished regions of the globe. Currently Executive Vice President and Director of Operations at Digital Realm, Inc, Ann Arbor, Michigan, U.S.A.

**William E. Lewandowski**, BA, MS
Served as an U.S. Army Officer and later worked as an Instructional Designer and Corporate Manager. Eighteen years of experience using, designing, and managing simulation devices and simulation centers, in the military and commercial aviation, prior to moving into clinical simulation in 2000. Owner, William E. Lewandowski Consulting, Daytona Beach, Florida, U.S.A.

**Marilyn Loen**, PhD, RN
Experience in a variety of nursing staff positions, teaching, and leadership in baccalaureate and graduate nursing education. Cofounder of HealthPartners Simulation Center. Professor and Executive Director of Metropolitan State University's School of Nursing, St. Paul, Minnesota, U.S.A.

**Dag K. J. E. von Lubitz**
Combines tools and concepts from once thought to be disparate domains like simulation, telemedicine and global clinical training into new solutions to previously intractable problems. Chairman and Chief Scientist at MedSMART, Inc, Ann Arbor, Michigan and adjunct professor at H&G Dow College of Health Sciences at Central Michigan University, Mt Pleasant, Michigan, U.S.A.

**Christina M. Matadial**, MBBS, MD, Diplomate, American Board of Anesthesiology
Undergraduate Education – Physics and biochemistry – University of the West Indies, Psychology – Broward Community College, University of Miami – Jackson Memorial Hospital, Miami, Florida. Interests: amateur photography. Now with the Simulation Center for Patient Safety, University of Miami Miller School of Medicine. Miami, Florida, U.S.A.

**William C. McGaghie**, PhD
Formal education in educational and social psychology, research methods, and educational measurement; professional researcher in medical education and preventive medicine. Feinberg School of Medicine, Northwestern University, Chicago, Illinois, U.S.A.

**Christopher A. McNeal**, BS, EMTP
Simulation Laboratory Coordinator, Rush University Simulation Laboratory, Rush University Medical Center, Chicago, Illinois, U.S.A.

**Andreas H Meier**, MD, FAAP, FACS
Pediatric surgeon with interest in computer technology, simulation and surgical education; started with simulation in 1999 while at the Center for Advanced Technology in Surgery at Stanford (CATSS) with Tom Krummel. 2002, participating faculty at the Cognitive Science and Simulation Laboratory at Penn State University, Hershey, Pennsylvania, U.S.A.

**Jane Lindsay Miller**, MA, PhD
Formal education in anthropology and education, medical anthropologist with the United Nations Development Program in maternal-child health and infectious disease, educational researcher in learner outcomes and effectiveness of performance-based teaching and assessment. Endless fascination with human behavior and advocate for health care reform. Minneapolis, Minnesota, U.S.A.

**Stefan Mönk**, MD, DEAA
1992–1998, Specialist training in Anesthesia and Intensive Care Medicine; 1997, Cofounder of the Mainz Simulation Center; 1998, Coauthor of the German requirements for Simulation in Anesthesia of the German Anesthesia Society; 1998, Specialist in Anesthesia and Intensive Care Medicine; 1999, European Diploma in Anaesthesiology and Intensive Care (DEAA); 2000, Specialist for Emergency Medicine; 2000, Host of the SESAM Annual Conference in Mainz, Germany (Simulation in Europe Applied to Medicine); 2000–2006, Chief Emergency Physician of the city of Mainz, Germany; 2003, Consultant in Anesthesia at the Mainz Medical School; 2001–2003, Secretary of SESAM; 2003–2005, President of SESAM; 2003–2005, Program Manager for the DGAI-project; 2003, Abstract Chair, International Meeting on Medical Simulation, IMMS; 2004, Workshop Chair, International Meeting on Medical Simulation, IMMS; 2005, left University to become Vice President Production, Research and Development at AQAI Mainz Simulation center; 2005, Member of METI modeling consortium; 2005, Cohost of HPSN (Human Patient Simulator Network) Europe; 2006, Cohost of HPSN Europe.

**Barbara Morgan**, MD
1973, graduate of Louisiana State University School of Medicine in New Orleans; 1976, completed anesthesiology

residency at Baylor College of Medicine in Houston, Texas, and later Chair of Anesthesiology Departments of two hospitals; After Hurricane Katrina in 2005, she responded to an urgent plea for physicians and was assigned my own 30–40 bed unit within the Louisiana State University athletic facility for several days until the federal government responded. Currently, an Associate Clinical Professor in Anesthesiology at LSU Medical School in New Orleans in charge of preoperative evaluation and screening of surgical patients. She assists Dr Kozmenko with his research and work in Human Simulation at LSU Medical School, New Orleans, Louisiana, U.S.A.

**Viren N. Naik**, MD, MEd, FRCPC
Graduate studies and research in health professional education and evaluation. Many years of instruction (swimming, tennis, sailing, medicine). 2002, Medical Director of the Patient Simulation Center, St. Michael's Hospital, University of Toronto, Toronto, Ontario, Canada.

**Harry Owen**, MBBCh (Bristol), MD, FRCA (UK), FANZCA
Supervises a busy program using simulation to improve medical student learning of important technical and nontechnical skills used in caring for very sick and injured patients. Uses cycling for health maintenance, stress relief, and reducing carbon emissions from transport. Trained in the U.K and moved to Australia 20 years ago. Director of the Flinders Clinical Skills and Simulation Unit and Professor of Anesthesia and Pain medicine at Flinders University, Adelaide, Australia.

**Alfredo Guillermo Pacheco**, MD
Ten years as ICU Physician, 10 years as ICU ambulance physician, 5 years as Medical Coordinator and in charge of education and training of emergency staff of Air and Land Transport of Critical Patients at the Integral Emergency Service of the Ministry of Health of the Province of Buenos Aires, Argentina.

**David Patterson**, BSE
Fourteen years in the office of medical education doing educational support and every sort of simulation lab done at the school. Director, Human Simulation Center, Kirksville College of Osteopathic Medicine, A.T. Still University, Kirksville Missouri, U.S.A.

**Carl Patow**, MD, MPH, FACS
Executive Director of an educational institute in a nonprofit health system, including CME, GME, medical library, simulation, online learning, allied health and nursing education. Head and neck surgeon, experienced in managed care leadership. Cellist, sculptor, tree farmer. 2003, Cofounder, Health-Partners Simulation Center for Patient Safety at Metropolitan State University, St. Paul, Minnesota, U.S.A.

**Frédéric Patricelli**, BSEE, MSEE
At Telecom Italia's Corporate University (SSGRR) since 1986 researching Database Technology applied to Telecommunications. Invited speaker at Academy of Sciences (Moscow), Ecole Supérieure des Télécommunications (Paris) and VTT Electonics/Nokia (Helsinki). Led the International Education Business Unit at SSGRR until 2004. Founded ICTEK Worldwide, an international education & consulting company on ICT, and served as guest professor at university of L'Aquila, Italy. Worked at Motorola's Satellite Communications Division (Phoenix), and currently serves as Training Manager at Mobile Telecommunications Co. (Kuwait City).

**Leonard Pott**, MBBCh
Trained as an anesthesiologist in South Africa, and currently working in the U.S.A. Been interested in simulation and assessment since 1992, particularly airway management and decision making. Currently: Assistant Professor of Anesthesiology; Director, Advanced Airway Management; Director, Medical Student Education; Director, Educational Research; Penn State College of Medicine at the Milton S. Hershey Medical Center, Hershey, Pennsylvania, U.S.A.

**Ramiro Pozzo**, Biomedical Engineer
Formal education in: bioengineering/health and social security systems/computer programing; Experience in: maintenance of biomedical equipment in hospital critical areas (ICU, OR), magnetic resonance image processing, disaster area volunteer for United Nations, member of anesthesia equipment standards commission. Integral Emergency Service of the Ministry of Health of the Province of Buenos Aires, Argentina.

**Carla M. Pugh**, MD, PhD, FACS
2001, Assistant Professor of Surgery with a PhD in education; Inventor and U.S. Patent holder for METI's (Medical Education Technologies, Inc.) "Touch Sensitive®" simulation technology. Broad research interests in the use of technology for medical and surgical education and assessment. 2003, Associate Director of Center for Advanced Surgical Education, Northwestern University Feinberg School of Medicine, Chicago, Illinois, U.S.A.

**Commander Fabian E. Purcell**, RANR, MB BS, FANZCA
Graduate Monash University Medical School and Fellow Australian New Zealand College Anaesthetists. Commissioned Royal Australian Navy Reserve 1988. Currently Reserve Directing Staff at the Australian Command and Staff College. Senior Instructor at St. Vincents Simulation Center, Melbourne, Australia.

**Marcus Rall**, MD
Studied medicine in Tuebingen, Cologne and Wurzburg, Germany, with rotations at Harvard Medical School

(Endocrinology, Emergency Medicine (MGH) and Cardiology (BWH)) and at University of Michigan (Emergency Medicine's St. Joseph Mercy Hospital, Ann Arbor).

Firefighter and Paramedic before and during medical school. Always wondered, "why one was never told how to prevent errors, only how stupid other people are?" 1998, Translated and adapted David Gaba's book: "Crisis Management in Anesthesia" into German. Married with a wonderful wife and 2 kids (trying to practice CRM also at home . . . ). Since 1994, anesthesiologist and prehospital emergency physician and clinical lecturer at the University of Tuebingen Medical School, Department of Anesthesiology and Intensive Care Medicine. Founder and director of the Center for Patient Safety and Simulation Tuebingen (TuPASS), Germany.

**Silke Reddersen**, MD
Anesthetist at Tuebingen University Hospital, Germany, gained experience in the operating theater as well as on intensive care unit. Came in contact with simulation in 2000 by taking part in a research project. Working as an instructor with an emphasis on mobile training and incident reporting. Center for Patient Safety and Simulation Tuebingen (TuPASS), Germany.

**Simon Richir**, MEng, PhD
His domains of education and research include technological innovation, innovative projects driving and engineering design of Virtual reality systems. Simon is scientific chair of Laval Virtual international conference. Professor at ENSAM, high French engineers school, and Director of "Presence & innovation" research lab, France.

**Jill Steiner Sanko**, CRNP
Bachelors degrees in nursing and anthropology, Masters degree in nursing; an acute care nurse practitioner; interested in teaching critical thinking during crisis events and applying anthropology to examine the cultural differences among various health care disciplines. Clinical Center, National Institutes of Health, Rockville, Maryland, U.S.A.

**Diane C. Seibert**, BSN, MS, PhD, CRNP
Wide array of nursing experiences including: ICU, Orthopedics, ENT/Eye, Multiservice Unit, Outpatient Care, Labor & Delivery, Newborn Nursery, Postpartum. Certifications as a Women's Health NP, Adult NP, Lamaze Educator and Menopause Clinician. The "go to" person in the Graduate School of Nursing for all things technology. Program Director Family Nurse Practitioner Program. USUHS, Bethesda, Maryland, U.S.A.

**Michael Seropian**, MD, FRCPC
Pediatric anesthesiologist, started with simulation in 1995 and has designed and helped implement multiple programs. He sits as a member of both the board of the Society for Simulation

in Healthcare and the Oregon Simulation Alliance. He is the founder and past director of the OHSU Simulation Center. He has a background in computer programing; has always been entrepreneurial and interested in creating things that work; strong background in electronics, project management, and collaboration. Has worked a great deal with nursing and considers himself discipline agnostic when it comes to simulation. Oregon Health Sciences University, Portland Oregon, U.S.A.

**Paul N. Severin**, MD, FAAP
BS (Biology); Research Assistant (College of St Francis; State University New York Downstate; Illinois State Police Crime Laboratory); Laboratory research (sepsis); PALS Course Codirector. Rush University Medical Center and John H. Stroger, Jr. Hospital of Cook County, Chicago, Illinois, U.S.A. 2003, Member, Rush University Simulation Laboratory Steering Committee and Assistant Director, Affiliated Programs. Rush University Medical Center, Chicago, Illinois, U.S.A.

**Ilya Shekhter**, MS, MBA
Studied biomedical engineering/signal processing and constructed mathematical models of the auditory system; taught signal and system analysis to engineering students at Boston University. 1997–2004, Simulation Engineer, Rochester Center for Medical Simulation, University of Rochester Medical Center, Rochester, NY, USA. From 2004 till date, Technical Director of Medical Simulation, Center for Patient Safety, University of Miami, Jackson Memorial Hospital, Miami, FL, U.S.A.

**G. Allan Shemanko**, MA, RRT
Very diverse past including ambulance attendant, live theater technician, worked in a large inner-city hospital ICU and ER as a respiratory therapist, instructed in the Northern Alberta Institute of Technology Respiratory Therapy Program for 6 years, and completed a graduate degree in distributed learning. He enjoys opera and live theater, foreign travel, dogs (especially basset hounds – they help me learn patience). 2004, Assistant Manager, Respiratory Services, Royal Alexandra Hospital, Edmonton, Alberta, Canada.

**Cynthia H. Shields**, MD
Participated in behind-the-scene theater arts in high school and college. Physical therapy technician and EMT. Came to simulation after 9 years as a clinical anesthesiologist in 2001. Desired a more effective way to teach critical attitudes, skills, and thought processes to clinicians. Director of Anesthesiology Simulation, Uniformed Services University, Bethesda, Maryland, U.S.A.

**N. Ty Smith**, MD
NIH Career Development award: 1966–1971. Interests have included computers, cardiovascular physiology and pharmacology, EEG analysis and display, closed-loop control, drug interactions, physiologic and pharmacologic mathematical

modeling, simulation, noninvasive monitoring, the human pharmacology of inhaled anesthetic agents, and automated record keeping, including voice recognition. First microcomputer in medicine. Founded Journal of Clinical Monitoring, Society for Technology in Anesthesia (which spawned the SSH) and the ASA EMIT Committee. During this long time, managed to put together about 400 publications. Most importantly, has been blessed to work with some incredibly bright and creative people. His most satisfying achievement is founding PACEM, the Pacific Academy of Ecclesiastical Music, a nonprofit organization whose goal is to advance and preserve church music. 1970: Started physiologic/pharmacologic modeling, on an analog/hybrid computer. 1985: Our first digital simulator, "Sleeper" (Sigmagraphics Iris 2300, with two parallel processors). San Diego, California, U.S.A.

**David H. Stern**, MSEE, MD
Studied Physics and Electrical and Biomedical Engineering. Background in electronics repair, video production, photography, computer programing. Clinical research in hemodynamics. Trained in cardiac anesthesia and echocardiography. 1994, Assoc. Prof. of Anesthesiology and Director of the U of R Center for Medical Simulation, University of Rochester Medical Center, Rochester, NY, U.S.A.

**Kristina Lee Stillsmoking**, RN, BSN, M.ED., CNOR
PeriOperative Registered Nurse, Hospital Educator, served in the Gulf and Iraqi Wars; various life experiences provide creativity in managing the simulation training for a variety of specialties and mentorship of new educators in the field of simulation. Two years as Simulation Education Facilitator-GME/Nursing, Charles A. Andersen Simulation Center, Madigan Army Medical Center, Ft. Lewis, Washington, U.S.A.

**Eric Stricker**, M.Sc.
Studied biomedical-engineering; is working in simulation since 2004, research on different types of training with a simulator, and incident reporting/analysis. Currently employed in the Center for Patient Safety and Simulation (TuPASS), Tuebingen, Germany.

**Claudia Sun**, BSME
BS in Mechanical Engineering (University of New Mexico). Studied artificial muscles, medical imaging, and medical equipment design. Worked as Vehicle Design Engineer at Ford Motor Company. 2003, Simulation Center Manager/Engineer, VA Palo Alto/Stanford Simulation Center, Palo Alto, California, U.S.A.

**Kay M.B. Thiemann**, BS, MBA
Studied business finance and communications at undergraduate and graduate levels. Planned and implemented the Mayo Multidisciplinary Simulation Center, Mayo Clinic. Teach elements of business planning, strategic planning, and process improvement. Education Administrator, Mayo Clinic College of Medicine; Clinical Operations Administrator, Mayo Clinic, Jacksonville, Florida, U.S.A.

**Carol I. Vandrey**, RN, BSN, MS, CCRN
Degrees include Bachelor of Science in Nursing, University of Wisconsin and Master of Science, University of Maryland. Vast experience scrounging teaching materials from dumpsters and an active imagination that incorporates found articles into teaching scenarios that replicate actual patient scenarios with uncanny accuracy. Over 30 years of clinical experience in coronary, medical, surgical, thoracic, and pediatric intensive care. Twenty years experience as Deputy Director of the Department of the Army's graduate level Critical Care Nursing Course. Five years of experience with the Patient Simulation Laboratory, Uniformed Services University. Director of the Advanced Cardiac Life Support Program and the Trauma Nursing Core Course programs for Walter Reed Army Medical Center, Washington DC, U.S.A.

**Jochen Vollmer**, Dipl. Math.
Formal education in maths, computer science, and physics. Main interests in education, ergonomics, and physiologic modeling. Founding member of Simulationszentrum Mainz in 1997, since 2003 manager of METI International Customer Support and 2005 Vice President Technics and Support of AQAI Simulationszentrum Mainz, Germany.

**Diane Bronstein Wayne**, MD
Director of large training program in internal medicine, research interest in medical education; Current interests: developing rigorous training programs for physicians, assessment of these programs and linking their use to quality improvement on the clinical service; Board-certified internist and Program Director, Internal Medicine Residency Northwestern University Feinberg School of Medicine. Chicago, Illinois, U.S.A.

**Eileen R. Wiley**, BS, MS
BS in Psychology and MS in Urban and Policy Sciences earned simultaneously at State University of New York at Stony Brook in 1981. 26 years of experience in all aspects of facilities planning, design and construction. 21 years at academic medical centers. Last 16 years at Penn State College of Medicine at Milton S. Hershey Medical Center. Currently, Assistant Director of Facilities Planning and Construction. Main responsibility is space planning and analysis, and facility planning with emphasis on educational facilities, research laboratories, offices and outpatient clinics.

**Paul Williamson**, BEngMech(Hons) MBA
Qualified in mechanical engineering in manufacturing and computer systems engineering plus an advanced MBA from Adelaide, Australia. Moved through engineering and IT project management to start-up business development and venture capital. Travel too much for amateur theater these days, but

it was a good influence along with paramedic experience and pitching to venture capitalists – you see more blood in venture capital. Currently building and commercializing simulators and courseware under the master of engineering at Flinders University, Adelaide, Australia.

**Jörg Zieger**, MD
Started clinical anesthesiology in 1995 (University Hospital Tuebingen/Germany), instructor in simulation courses, special interest in mobile simulation, working in simulation since 2001 in Tuebingen Center for Patient Safety and Simulation (TUPASS, Marcus Rall), Tuebingen, Germany.

**Amitai Ziv**, MD, MHA,
A veteran combat pilot and instructor in the Israeli Air Force; trained as a Pediatrician in Israel (Hebrew University – Hadassah Medical Center) with subspecialties in Adolescent Medicine (University of Pennsylvania, USA) and in Medical Management, and a Masters degree (Tel-Aviv University) in Health Administration; on the editorial board of the Journal of the Society for Simulation in Healthcare and chair of the Credentialing, Accreditation, Technology, and Standards (CATS) Committee; a clinical senior lecturer at the Department of Behavioral Sciences of the Tel Aviv University Medical School, and Adjunct Associate Professor of Pediatrics at Case Western Reserve University. Responsible for Risk Management, Quality Assurance and Medical Education; founder and Director of MSR – the Israel Center for Medical Simulation and Deputy Director of the Sheba Medical Center at Tel Hashomer, Israel.

# Glossary of Degrees and Awards

| | |
|---|---|
| AKC | Associateship of King's College |
| BA | Bachelor of Arts |
| BEngMech | Bachelor of Mechanical Engineering |
| BMBS | Bachelor of Medicine, Bachelor of Surgery (Medical Doctor, Australia) |
| BMechEng | Bachelor of Mechanical Engineering |
| BNursPrac (Aviation) | Bachelor of Nursing Practice (Aviation) |
| BS | Bachelor Surgery |
| BS | Bachelor of Science |
| BSc | Bachelor of Science |
| BSE | Bachelor of Science, Education |
| BSE | Bachelor of Science, Engineering |
| BSN | Bachelor of Science, Nursing |
| BSEE | Bachelor of Science, Electrical Engineering |
| BSME | Bachelor of Science, Mechanical Engineering |
| CAT | Certified Anesthesia Technologist |
| CCRN | Critical Care Registered Nurse |
| CCVT | Certified Cardiovascular Technologist |
| CEN | Certified Emergency Nurse |
| ChB | "Chirurgiae Baccalaureus" Bachelor of Surgery |
| CHE | Certified Healthcare Executive |
| CNA | Certified Nurse Anesthetist |
| CNOR | Certified PeriOperative Nurse |
| CPhys | Chartered Physicist |
| CRNA | Certified Registered Nurse Anesthetist |
| CRNP | Certified Registered Nurse Practitioner |
| DABPM | Diplomate, American Board of Pain Medicine |
| DEAA | European Diploma in Anaesthesiology and Intensive Care |
| Dipl. Math. | Diplomate, Mathematics |
| DNSc | Doctor of Nursing Science |
| DUniv | Doctor of the University |
| DUT | University Technical Diploma |
| EMT-P | Emergency Medical Technician-Paramedic |
| FAAEM | Fellow of the American Academy of Emergency Medicine |
| FAAN | Fellow of American Academy of Nursing |
| FAAP | Fellow of American Academy of Pediatrics |
| FACEP | Fellow of American College of Emergency Physicians |
| FACOG | Fellow of American College of Obstetricians and Gynecologists |
| FACS | Fellow of American College of Surgeons |
| FANZCA | Fellow of Australian New Zealand College Anaesthetists |
| FCA(SA) | Fellow of the College of Anesthesiologists (South Africa) |
| FFA(RCS)(Lon) | Fellow of the Faculty of Anaesthetists, Royal College of Surgeons (London) |
| FLTLT | Flight Lieutenant |
| FRCA | Fellow of Royal College of Anaesthetists |
| FRCPC | Fellow of Royal College of Physicians of Canada |
| GradDipNursing (Perioperative) | Graduate Diploma of Nursing (Perioperative) |
| ILTM | Member, Institute of Learning and Teaching (UK Higher Education Academy) |
| LCH | Licentiate of the College of Homeopathy |
| LP | Licensed Paramedic |
| MA | Master of Arts |
| MARH | Member of the Association of Registered Homeopaths |

| | | | |
|---|---|---|---|
| MB | Bachelor of Medicine | MSEE | Master of Science in Electrical Engineering |
| MBA | Master of Business Administration | MSEng | Master of Science, Engineering |
| MBBCh | Bachelor of Medicine/Bachelor of Surgery | MSAeroEng | Master of Science, Aerospace Engineering |
| MBBS | Bachelor of Medicine and Bachelor Surgery (Doctor of Medicine, UK) | MSN | Master of Science in Nursing |
| MBChB | Bachelor of Medicine/Bachelor of Surgery | PGCE | Postgraduate Certificate in Education |
| | | Pg Dip Med Ed | Postgraduate Diploma in Medical Education |
| MD | Doctor of Medicine | PhD | Doctor of Philosophy |
| MD(Anaes)(Stell) | Doctor of Medicine, Anaesthesia Stellenbosch | PPS | Plenipotentiary for Patient Simulation |
| MD(Sc) | Doctor of Medicine, Science | PT | Physical Therapist |
| MEd | Master of Education | RAAFSR | Royal Australian Air Force Specialist Reserve |
| MEng | Master of Engineering | | |
| MInstP | Member of the Institute of Physics | RANR | Royal Australian Naval Reserve |
| MPA | Master of Public Administration | RDMS | Registered Diagnostic Medical Sonographer |
| MPH | Master of Public Health | | |
| MPhil | Master of Philosophy | RN | Registered Nurse |
| MPhys | Master of Physics | RRT | Registered Respiratory Therapist |
| MRCP | Member, Royal College of Physicians | RT | Respiratory Therapist |
| MS | Master of Science | RVT | Registered Vascular Technologist |

# Foreword

This monograph "Clinical Simulation: Operations, Engineering and Management" by Richard Kyle and Bosseau Murray fills a rapidly growing need as the science of simulation achieves acceptance by the health care field as an important medical educational tool. This acceptance has been very slow to come, especially considering that aviation and other industries have used simulation for over 50 years. And looking at the history of medical education, it is clear that such an opportunity, in fact a revolution, only occurs once every century or so – our last great medical education revolution was with the Flexner Report in 1910. Whatever will be developed during the coming decade may well be the foundation until the 22nd century.

Just as important is the timeliness of the book. The Residency Review Committee of the Accreditation Council on Graduate Medical Education has begun requiring residency programs to have simulation as an integral part of their training programs. The American College of Surgeons (ACS) has also recognized this transformation, and has taken the bold step to certify training centers to ensure the quality of the training that will be provided – other societies and credentialing bodies are sure to follow. A natural by-product of this will be that curricula will become more standardized, the measures of success more uniform, and the overall quality of education will take a giant step forward. Most important is the fact that students can be trained in a safe environment – an environment in which they have "permission to fail" and in which they will be taught via errors, how to recognize them and to avoid or repair them. And all this without jeopardy to a patient.

It has also been recognized, and repeatedly emphasized in these chapters, that this will require that the training be both multispecialty and interprofessional, including not only all specialties of physicians and medical students, but also nurses and other allied health professionals. The ACS application for certification requires that at least three different categories of students be taught by the simulation center. Deans of medical schools and nursing schools are under pressure; now is the time when all levels of medical education are searching for advice on how to establish their own simulation centers, and many of the answers have been succinctly provided by the contributing authors.

Rather than an academic dissertation on the changes in the educational process and the subsequent impact on the training of students, the authors have chosen a practical approach and address critical issues that lead to successful implementation of simulation into a training program and propose pragmatic solutions to transitioning to clinical utilization. Certainly, the theory and philosophy of education are touched upon, but mainly to illustrate a practical point *or* provide a theoretical basis or provide an underlying structure. The practical focus is critical, and the authors have identified the salient features that epitomize the value added by simulation: the importance of objective assessment using a benchmark performance criterion, and then training students to the criterion rather than continue to use the traditional time-based training. The setting of metrics to be achieved and the reporting of outcomes that are fed back to the student, form a critical tool for both the educator and students – emphasizing the value of the debriefing.

However, some of the most valuable information will be that which addresses the day-to-day decisions needed to establish and then maintain a busy simulation center. With resources scarce and expensive, the experience of the contributors is invaluable in sorting through the numerous options in facility design, curricula development, simulator purchasing, scheduling, etc. The treatment of these important issues is comprehensive and illuminating. The chapters are laced with vignettes and lessons learned to help those charged with starting a new simulation center. Just as simulation provides an opportunity to make mistakes on a simulator before operating on a patient, so too does this book let the reader learn the many mistakes before striking out on their own endeavors. The authors are to be commended for recognizing a critical need and then providing such an eminently practical solution for all levels of readers.

Richard Satava, MD
2007

# How to use this book

*"Experience is a hard teacher because she gives the test first, the lesson afterwards."*

Vernon Sanders

This book consists of 82 chapters from 99 contributors, arranged in 22 topics on the *What* and the *How* of clinical simulation. Integrated throughout these works are messages on the *Why*. These topics are arranged in an order that we like to use in all our teachings: starting at the widest view then zooming inward for examination of numerous fine details. These details are like the stones of a structure: useful to the extent that they link with their neighbors, valuable in how each contributes the overall purpose of the structure. For any topic, we encourage you to first read through all of it looking for the larger perspective, and then return for a closer examination of the finer details. For those topics with more than one chapter, each chapter approaches the common theme from a different vantage point. Thus, one chapter may address issues left unresolved for you by the other chapters, or may present it in a way that is more accessible to you than by the others.

On purpose, our topic introductions are framed as our half of a dialog with you; what we might say in conversation about the value a given topic contributes to a successful clinical simulation program. We assumed that questions and curiosity would be your motivation to open this book, and thus, like any good instructor, we attempted to anticipate your primary question: "what will I gain from investing my time and attention upon this topic?" To the extent that these introductions do not mislead you, we were successful in anticipating your needs. To the extent that the authors' contents enlighten you, they were successful in their efforts to communicate the lessons they learned from their experiences. To the extent that you employ their lessons or that they catalyze your own inventiveness, all of us will have reached our goal of helping you become better at helping others become better clinicians.

Note that the book chapters' *content* are printed and bound, while the chapters' *appendices* are provided via a web site created and hosted by Elsevier, and accessed through this one common URL: www.elsevierdirect.com/9780123725318

Even though the Appendices are only on the web site, we believe that this structure of the book will help you, the reader, rapidly find the *Why*, the *What*, and *How* of clinical simulation needed to enhance the learning of your trainees.

# Introduction

*"How many things are looked upon as quite impossible until they have been actually effected?"*

*Pliney the Elder*

While the words "simulation" and "simulators" may be new for many of us in clinical education, they are not new phenomena in our own personal history of learning, in the history of clinical education, in the history of human learning. At the smallest individual scale, from birth each of us used simulation as a way to master speech, walking, and attaining individual autonomy. One could argue that each of us used simulation in imagining what our lives would be like after we graduated from the schools we applied to, that we and the admissions interviewers used simulation to estimate how compatible we might be. At the largest social scale, from inception all societies used simulation to convey its beliefs to attain group consistency. All culturalization processes engage in active audience participation. All use role models of desired outcomes. All include punishments and rewards for failure and successes. All employ repetition. All consist of many small ingestible bits of newness that once digested, become incorporated into amazing capabilities. All require a person dedicated to helping others become better. You, the clinical educator, *are* that person for your clinical students. You are reading this book to become a better clinical teacher to help your students become better clinicians to help their patients become better. Your goal is to become so proficient with simulation that your students can't see your effort – only the results. This invisibility is the acme of the professional. In a word, you appear a "magician". Consider this book to be a "how to" guide in your efforts to become a better magician.

Clinical simulation is pretend for the purpose of improving behaviors for someone else's benefit. Clinical simulation is not fakery, not a con game where the purpose is to fool others for one's own benefit. Clinical simulation is clinical theatrics with full contact audience participation. Today, entire schools and their libraries are devoted to teaching the arts and crafts of live theater. Courses address writing, dialog, acting, music, props, sound effects, lighting, costumes, stage design, theater construction, finances, patronage, publicity, location, personnel and personalities. The essence that makes theatrical story telling universally accessible to the audience and infinitely malleable in their forms is the very same used in successful clinical simulation. All theatrical productions require competent producers, directors, actors, writers, prop masters, personnel and facility managers. You, in your effort to become a competent simulation professional, must become knowledgeable with all, and expert in many of these trades.

Those of us staging clinical simulations face the same issues and challenges perennial to all live performances: present a compelling event for the audience to fully engage in, while displaying adaptability given the unpredictable. Since all clinical simulations include under-scripted students at the center of the action, the risks and uncertainties to the outcomes are greater than in any other form of live production. Yet, this reality echoes all clinical care, where under-scripted patients are the center of the action. As educators, our task is done when the only source of novelty that our students face is their next patient, not their environment, not their equipment, not their own knowledge, skills, or attitudes

Technology can be defined as "the way we do things" more than just "with what we do". Technologies, like electric lights and amplified sounds, have reduced many of the limitations to theatrical production. Such technologies are used to make accessibility easier for the audience to experience staged events. The better stage professionals understand that some new technologies can be used to improve their theatrical presentations. The best stage professionals have mastered the use of any technology for the purpose of pulling their audiences' consciousness out of their seats and fully engage them in the action on the stage. Yet, technological advances have done little to change the purpose of most theater or the value of the stories told to most audiences. Usually, if the audience is captivated by the production technology employed, let alone even notices it at all, then the intended story-telling purpose of the theatrical event has been lost.

Clinical simulation has recently employed new technologies like robotics and physio-pharmacological computational models. Such technologies are used to make it easier for the student to experience staged clinical events. The better clinical instructors thoughtfully use only those technologies that improve their teachings. The best clinical instructors have mastered the use of any technology for the purpose of making their students' fully engaged in clinical care. Yet, technological advances have done nothing to change the purpose of clinical education or the value of it to students. Always, if clinical students are distracted by the technology employed in their simulated clinical learning experience, then the educational purpose of the class is lost.

A note about manufactured simulators. These are the most recent kind of simulators to join human actors, human cadavers and non-human animals employed by clinical educators. The most distinct feature of manufactured simulators, and also their most significant added value, is their greater predictability. To the extent that the market place offers features you desire, you can select just what you need at the time of placing your order. As your needs mature, so will the commercial offerings. No cadaver, animal or human actor can offer such feature specificity combined with boundless reproducibility and ease of access. However, such extraordinary capability brings extraordinary expectations. Users all too easily come to expect engineering and marketing miracles from the makers and vendors. The payers all too easily come to demand validation and documentation of performance (like we have learned to expect when buying an automobile) unlike they have ever demanded of non-manufactured clinical simulators. This tension pulls both ways. For the first time ever, we all can legitimately challenge ourselves with two fundamental questions: "just *what* is valid clinical education?" and "*how* can we craft and provide valid clinical education?" This book is an attempt to help you create answers to these *what* and *how* questions.

Let's imagine that we are already on the mountaintop, which is the best vantage point from which to consider paths to get there. Clinical education is *behavior modification*: changing knowledge, skills and attitudes. Books, libraries and other repositories of vetted information provide access to facts and figures. Laboratories, with and without the presence of real patients, provide practice opportunities to gain functional aptitude. Instructor qualities such as honesty and respect provide context within which content is ingested and integrated.

Classical clinical education management structure is based upon individualism. Individuals in isolation are accepted *individually* into a clinical education program. They form a crowd of individuals, each individually rewarded and punished in isolation from one another. The reward and punishment reinforce the role of external motivation, to the detriment of internal motivation. Upon completion, each of the graduates independently joins the next crowd.

But clinical care is a team sport: the team is often seen as consisting of just two members, a clinician and a patient. Yet like an iceberg, most of the other members are much less visible but play equally critical roles in achieving success. Patients seek clinical care for their own benefit, not for that of the clinicians. Success, from the patients' point of view, is mostly within themselves, the patient.

Herein lies the rub: if clinicians actually function as members of a team with the team goal of satisfying the needs of the patient, then how can we expect such a result from any educational process based upon individuals, each receiving isolated reinforcement for self-promotion? If the motivations are all external, how will we ever see clinicians who are internally motivated to perform beyond the currently acceptable minimum? In other words, the current education process reinforces attitudes ("individualism") diametrically opposed to the desired end product ("teams").

Back to behavior modification. Training systems, both familiar and emerging, employ simulation to train attitudes as well as skills. Simulation is a tool, and like all tools, produces results no better than the ability and intent of its wielder. Simulation, by its very nature a schedulable event, provides clinical educators a unique tool: an obstacle course/treasure hunt as well matched to the students' needs as the educator can make it, and provided at a time and location far more driven by a deliberate curriculum than real patients can ever be.

To date, most clinical simulation has been employed within the classical education model, including when the goals of the teachers and students directly address teamwork. Yet, even in this application, almost all of the students' rewards and punishments are individualized. Using this new tool in such an old way is not an inherent limitation of the tool itself. Simulation can build and reinforce the team-centric attitudes and behaviors that correlate directly with the desired end performance. It can also reinforce the internal motivation mechanisms in the clinician. Simulation has, can, and will, provide utility within the classical clinical education model. It just may be the lever to shift the world of clinical education.

While reading one single book cannot substitute for a lifetime of learning, our goal is to bring together pragmatic descriptions of the broad range of topics essential to success in many forms of clinical simulation. We chose these topics based upon what we learned from our experiences in creating clinical simulations. We chose these topics' authors based upon their ability to convey their learning experiences. We requested that the authors address perennial issues, challenges and concepts, and only reference today's products as illustrations. For example, recording and replaying sights and sounds of clinical simulation sessions will always be a key contributor, yet the audio and video tools and technologies available have never been in such a flux as we are currently experiencing. Thus, the challenge to the authors was in crafting lessons in which the fundamental principles are elucidated in ways that are both universal and specific enough to be accessible and usable for a broad audience.

Throughout the book you will happen upon terms like "better", "improvement" and "successful". The repetition is intentional. We have no delusions that of us knows the perfect solutions to all clinical education issues as well as ways to perfectly convey these solutions to you. Yet, each and every suggestion offered here is based upon lessons learned by others in their real life as clinical educators in using simulation to help them be better at their task. To the extent that the experiences of others can teach the teachable, we offer them here.

Richard R. Kyle Jr., and W. Bosseau Murray

# I

# Why Simulate?

*It isn't that they can't see the solution. It is that they can't see the problem.*

G. K. Chesterton

Just as the lessons learned from any clinical simulation experience do not exist in isolation, neither does the use of simulation within clinical education. In fact, the greater that simulation is absorbed into the very fabric of clinical curricula, the more successful it will have become. These three chapters are distinct, wide-angle perspectives providing big-picture context for understanding where and how this tool we call simulation can contribute to your students' learning. Each in their own way share the premise that clinical care is information application, thus clinical education should prepare future clinicians to be very competent information appliers. Also, each shares the premise that clinical care is always a collaborative event, thus clinical education should prepare future clinicians to be very competent collaborators. If you agree with these two premises, then you may also agree that the current execution of clinical education should be changed to improve the performance of our future clinicians. For any one person at any one time, changing oneself into a clinician is a significant challenge; yet, changing today's ways in which we provide clinical education may seem far more challenging given the far greater number of participants, each with their own vested interests. However, in both cases, the only way the desired change is possible is through deliberate, intentional practice. Simulation as a teaching method has a rich and successful history as a tool to safely practice change and explore the consequences of change. Participants in activities like aviation and nuclear power turned themselves into high-reliability organizations through simulation. In doing so, they made many of the pioneering developments in simulation-based learning. No one today would ever accuse clinical care or clinical education as being prime examples of high reliability. However, if we want to make it so, many of the principles, if not also the tools already developed and refined by nonclinicians, are well suited for adoption by clinicians. Thus, you can apply well-founded simulation approaches not only to help your clinical students attain your current educational goals, but also to evaluate your teaching methods, as well as investigate alternatives to the very goals and methods themselves.

Allan Shemanko describes today's traditions-based clinical education as in the preagricultural age, best characterized for a time when day-to-day survival depended upon whatever could be found or caught. While massive effort could and did overcome this unpredictability and allowed our ancestors' survival in small tribes, the uncertainty was still too large to allow the development and expansion of civilization and all its rich benefits. Mark Adelman describes the transformation that we intend to generate in our clinical students during their brief exposure in our schools and to us. All clinical simulation programs will succeed or fail to the extent that they help or hinder this change. Robert Cox and Lance Acree describe how simulation contributed to the transformation of a comparable high risk/low reliability enterprise into a high risk/high reliability one. They show which parts of this transformative activity are common with clinical care, which learning methods are suitable for transplantation, and which wheels we don't need to reinvent.

# From Primitive Cultures to Modern Day: Has Clinical Education Really Changed?

G. Allan Shemanko

## 1.1 Chance-based Hunter-Gatherer Culture

Reviewing the roots of clinical care before looking at the future of simulation is an integral step on the journey to learn where we are going by looking at where we have been. Through the process of reviewing the past, we effectively change perspective in navigating our path to the future. My inspiration for this line of thinking came from a very unlikely source – a department coffee-table book stored in my office. *The History of Medicine: An Illustrated History* follows the history of medicine from ancient times of the hunter-gatherer cultures of humankind, progressing through the ages to our present era of technology, complex medical diagnostic, and interventional wizardry that we call modern medicine [1]. It struck me that medicine has not really progressed from primitive times.

In ancient times, there were four specific requirements for any therapy to be effective: magic, ceremony, superstition, and the patient's belief that the therapy will work [1]. These requirements have not changed! In a modern day, hospital patients present themselves to a registration desk where the ceremony begins. Paperwork, commonly followed by blood-letting and other ceremonial rituals disguised as assessments create the perfect environment for a truly mystical experience. The patient is drawn further and further into the belief that the surgery they are about to undergo will cure them. To appease the superstition, patients will wear their "lucky slippers," remove all forms of adornment (lest they offend the diagnostic imagers), put on the ceremonial hair cover, and then lay on a stretcher ready for the long journey to the operating room. Here the anesthetist, appropriately garbed in the ceremonial scrub costume replete with a colorful cap and surgical mask that dangles from his neck, inserts an intravenous catheter in order to administer his favorite secret induction potion. The anesthetist then proceeds to administer his concoction, reassuring the patient that when they awake, they will be cured because the surgical magic will have already happened. The patient awakens completely ignorant of the intraoperative proceedings; the postoperative surgical pain is a crude reminder that the magic experienced in their drug-induced slumber was indeed manifest. Now, the disease is completely cured – truly a miracle!

Our clinical professions require hard evidence that medicines and other interventional therapeutics work – hence the randomized control trial ritual is known as the *gold standard*. Although there are indeed other methods of proving or disproving what I refer to as the *pharmacomagic* benefit of any given drug, there is still the little issue of the "placebo effect." Physicians attend to their loyal patients and prescribe medications to the sick. Very few patients understand why or even care how these medications work, relying on the simple pharmacomagic properties of the drug to make them better. Patients dutifully consume these concoctions, fully expecting they will work. For those people enrolled in randomized control trials, there will be some who will respond positively even though they were taking a placebo sugar pill. Their belief that they are truly receiving the experimental medication and that the drug will cure them is so strong that their symptoms and even their disease process will disappear. This is a perfect example of true pharmacomagic.

Even in the education of health care providers, there is a certain amount of ceremony involved. Ritualistic registration proceedings followed by the massing of students in classrooms establish the required environment for learning. Our belief that the education system will provide us with the very best learning environment is evident. The "sage of the stage" enters in an appropriate costume complete with lab coat and begins to address the minions. Although we seem to have the environmental creation ceremony and individuals willing to submit themselves to the rituals of clinical and allied health care education, are we providing the best learning environment that we can?

Our ancestors used to follow the migrating herds of animals because they were dependent upon these creatures for their food, clothing, and shelter. Indeed, they were ensuring their very survival; however, this was not the most efficient model to ensure the progression of a population. When humans changed their approach to a more intentional and predictable agriculture model, they certainly became more efficient. Now they began to grow their own food, fenced in their animals, and provided better, more permanent shelters. Without having to spend so much of their day simply surviving, they began to better themselves. They began to plan for the future, and so must we. In planning for our next future, we need to move from the hunter-gatherer model, where education happens in a rather chaotic environment, to a more predictable, intentional agriculture model.

## 1.2 Intended/Predictable/Deliberate Agriculture Model

If we are to progress past primitive magic as a tool to learn the practice of medicine, we need to change our *ceremony* in the classroom. There are many examples of medical residents, respiratory therapists, nurses, and other health care practitioners who learn at the mercy of the gods. They *wait* for the rare but critical condition in a patient in order to learn. They wait for the likely but critical events to happen. They wait for the definite and critical. They wait to be chosen to perform the ceremonial intubation in the operating room, assuming they don't have to fight for the privilege of passing an airway on a patient. Instead, the anesthetist believes that this is a difficult airway and that only they should perform the intubation. How is an anesthetic resident, nurse practitioner, respiratory therapist, or paramedic supposed to learn difficult airway management if they are not allowed to manage a difficult airway? A true apprentice needs to be taught, guided, and afforded full access to the patient, for every patient is different, thereby creating an environment of critical thinking. Lave and

Wenger [2] remark on this very subject when contemplating the work of Becker and his concerns regarding full access for apprentices:

> He recognizes the disastrous possibilities that structural constraints in work organizations may curtail or extinguish apprentices' access to the full range of activities of the job, and hence to possibilities for learning what they need to know to master a trade. (p. 86)

Certainly, one way of providing full access to the patient is to provide a synthetic model. This could take the form of basic, high-fidelity, or even virtual reality simulators now widely available, depending on the task at hand. Technology always expands to fill the need. We are not technology limited, but we are ceremony limited.

We need to move away from the Guttenberg era of ever greater amounts of noninteractive media as the panacea for clinical education! We need to change the way students interact. We need to better understand the processes of teaching and learning in order to adapt technological tools and curricula to make sure they fit our needs and the needs of our patients. We must ensure that that we are engaging in the perpetual scholarship that provides for reproducible results in all health care education programing.

How often does a pneumothorax strike a mechanically ventilated patient? If this experience does not happen very often even to those professionals working 12-hour rotating shifts, how are these individuals going to learn to recognize the problem and provide the treatment in a timely and safe manner? Unfortunately, this learning and teaching scenario either does not happen or if it does, it often occurs during a crisis event, the least likely moment to allow for the inexperienced to gain competence. Although this form of random, situated learning has important learning potential, "baptism by fire" cannot and should not be our basis for clinical education [2]. We need to plan ahead for the crisis instead of waiting for the crisis to come to us. We need to be more predictable in what and how we teach to health care providers in order to provide the best care for the patient. This is the reason many of us are in the profession of health care in the first place.

Teaching and learning should be planned and reproducible if we are to move past the random hunter-gatherer stage of old and move into the intended, more predictable agriculture model of the future. In one sense, we are well on our way with our current apprenticeship model of health care education. However, in moving from the unique to the ubiquitous in health care education, we need to map the impact of health care education:

- from enrolment to engagement,
- from the classroom to the real world,
- from the text to critical thinking,

- from exposure to mastery,
- from the procedure to understanding the process.

In order to make this leap, there are many educators and practitioners who would be against change for the sake of change. Dr W. Edwards Deming once said "In God we trust. All others bring data." He believed that a scientific approach, combined with systems thinking and data analysis, will allow us to understand how to create processes that will consistently deliver what our students need.

### 1.2.1 Where Do We Go From Here?

Students need to attain familiarization through simulation before going into a clinical rotation and treating live patients. We can predict what and how health care providers, including physicians, are going to learn using simulation. We can require these learners to practice and demonstrate their skills on simulators rather than on real patients. The feedback that can be provided through high-fidelity simulation can be much more comprehensive to the learner than an instructor saying "good job." Students can see changes in blood pressure and oxygen saturation – this is the profound feedback that they would receive from a real patient. Through simulation, learners are immersed in learning rather than being bystanders. "The effectiveness of the circulation of information among peers suggests, to the contrary, that engaging in practice, rather than being its object, may well be a *condition* for the effectiveness of learning [2]." Students will still be afforded the opportunity to be involved in the care of patients who develop the rare but critical event during a clinical rotation. The only difference is that their learning will not be dependent upon the development of such a scenario; it will be complimented by it. Their simulation experiences will also prime them to gain the maximum value from their live patient experiences.

With the population around the world aging, there may be a shortage of all health care professionals including respiratory therapists, nurses, physicians, and others. This could require the delegation of authority for nonphysicians to perform certain skill sets that have traditionally fallen under the authority of physicians and others in independent practice. Governments are currently planning for a worldwide influenza pandemic. It is anticipated that 15–35% of all health care practitioners could succumb to the flu if a vaccine is not available, effectively removing them from providing health care to the sick and injured [3]. Cross-training of health practitioners will be part of the contingency plan used to address the anticipated impact on human resources. This requires planning for effective education if these individuals are going to provide safe health care in this environment. Simulation could certainly help fill the training gap in teaching the practitioners these additional (new to them) skills as our population ages or should a pandemic scenario come to fruition. There is already talk of using dentists for triage in mass casualty events. A good application of resource management indeed, but when and how are these dentists to gain familiarity in the additional clinical and teamwork skills required? Through simulation.

### 1.2.2 How Do We Get There?

From the perspective of students, they are now becoming much savvier in choosing their educational institutions! Not only are they asking how many hours of didactic versus laboratory the school will provide, they are also now beginning to ask about simulation and whether this form of education is provided. The policy makers need to take heed – simulation is here. Education must dictate policy rather than the reverse. The development of sound education policies needs to include the best that simulation has to offer, and curricula need to reflect developing and reinforcing the lifetime-of-learning cultural needs of learners. We need to change the behavior of our policy makers and our educators similar to the ways advertisers change behaviors of their customers. We have the attitude and we have simulation as the approach to implement solutions.

With the ever-increasing need to practice the critical but infrequent scenarios such as recognizing and treating malignant hyperthermia or tension pneumothorax, simulation can be a lifesaver. There are many examples of the increasing need to practice the likely but critical events that are likely to come across our paths depending on where we work. For example, providing hemodialysis will certainly put the nurse or other health care provider in the way of critical events. These events have been cataloged, and as such becomes a curriculum for these health care providers. Why would we not want to prepare these individuals as much as possible in order to set them up for success in treating their patients?

Even more important than the perennially low frequency, high-acuity events is the fact that for each clinical student, at one time, each and every procedure, process, diagnostic decision, and treatment event is *their* first time – by definition the lowest frequency event they can possibly experience. Simulation can address this very real "fear factor" – not by wishing it way, but by augmenting a curriculum that acknowledges and respects this fundamental law of learning.

Research and product development will eventually produce haptics ("sense of touch") of sufficient performance at an affordable price in order to provide yet another dimension to the apprenticeship model of clinical education. Imagine the surgical resident never having cut into flesh before, praying he/she could have the experience of the surgeon. Now imagine the ability of the resident to stand in the surgeon's virtual shoes, seeing what the surgeon is seeing, and feeling what the surgeon is feeling through the use of virtual reality. Taking that one step even further, imagine the surgeon now being able to feel or sense how hard the resident is pressing on the flesh about to be incised. By placing the resident in the same orientation as the surgeon (please feel free to insert any clinical relationship here) and using tactile feedback afforded by virtual reality tools, desired learning is occurring at a much faster and more efficient rate than was ever before possible.

Similarly, the crisis management in anesthesiology program can also help provide the tools necessary to care for patients. There are similar curricula for perfusionists and heart–lung bypass crisis diagnosis and intervention practice. Simulation is commonplace where there is a need to practice the definite and critical, with complex surgeries being but one example.

Simulation can also help us learn to work together better using crisis resource management (CRM) principles with a simulator, with standardized patients, or a combination of the two. Standardized patients can help us with patient/physician/health care worker interaction. Perhaps the first time a new health care professional comes in contact with a patient should actually be a simulated event. We can learn how to interact with patients more compassionately and combine skilled interview techniques with standardized patients to help us elucidate the problems.

Standardized patients can also help us with interprofessional communication, cooperation, and collaboration. We have all met and have had to deal with an individual who had less than the best bedside or professional manner. Standardized patients can be used to modify inappropriate behaviors so that the student is better prepared for the outside world. Standardized patients can be used to help teach everything from conducting a patient interview to learning to perform a pelvic or rectal examination. When we are dealing with such private areas of the body, it would be prudent to have the student well prepared in conducting these examinations. And, when the practicing is first done on a device simulator, we can use these standardized patients to test competencies. Clinical schooling is difficult – why not set up our future doctors, respiratory therapists, nurses, and other critical health care providers for success rather than failure? If we fail to appropriately use all of the tools that we have at our disposal, then we are guilty of setting up students for failure. If we are truly setting our students up for failure, then we have only succeeded in failing ourselves, for we are all aging and will be at the mercy of those who we have taught.

There are many people out there who may wish to be a health care provider, but just do not have what it takes to be one. To my knowledge, there is really no good tool out there that is sensitive enough to predict the future clinical performance of a student in any health profession. Why not use simulation to develop assessment tools to predict successful or not-so-successful outcomes? A significant portion of attrition in health care-related professions can be attributable to a change of career that "may be related to inadequate knowledge of those fields or due to misconceptions regarding them [4]." Simulation could certainly be used as part of a career investigation. Without simulation, how does a nonhealth care professional really know what it is like to touch a patient, or learn what the job is like? With simulation, an interested person could touch a "patient" and perform some therapeutic

intervention to save a life! Learning that you faint at the sight of blood during your first suturing experience or phlebotomy laboratory is probably not the best time. This learning example is known as the null curriculum. Learning that you faint at the sight of blood was not the intended goal of the suturing experience, yet learning has still occurred. Simulation can provide the directed, predictable learning that needs to happen, whether or not it is part of the curriculum.

## 1.3 Conclusion

How do we use simulation to provide the very best of clinical education to help us provide the very best of clinical care? How do we move from the hunter-gatherers of clinical education to the much more intended and predictable agriculture model? Can we keep the magic and ceremony of medicine without affecting our need to move forward? We need to be proactive instead of reactive. This means we need to plan ahead for crisis instead of waiting for the crisis to come to us. We need to ensure that our clinical apprentices, no matter the field, are afforded complete access to and responsibility for their patients in order to set them up for learning success. We need to look at the data that has been already collected, and we need to move forward.

Imagine a time where patients are not exposed to medical residents who have to spend 36 hours on-call simply to gain experience with what may or may not happen. Imagine a time where all health care providers take calls in a simulated clinical setting where directed learning can be anticipated and patients die only when it furthers the education of our students. Imagine a future where a resident, on their first time ever, is given permission and the time to attend a dying patient, hold their hand and grieve for the loss without also having to attend to another patient. Imagine a future where health care students are allowed to make a mistake and see the consequences of their actions without causing injury or death to a real patient. We no longer have to imagine this future – we can have this right now if we make the decision. We just need to make the decision and then follow through with action.

Learning the art of medicine does not need to be haphazard anymore. We have the tools, the magic, the ceremony, and a patient's belief that we can make them better. Now we also have the tools, the magic, the ceremony, and the learner's belief that we can make better health care practitioners. As our population ages along with the masters of our clinical professions, I can only believe that our very survival could become dependent upon our clinical culture moving from the hunter-gatherer approach to education more toward the intentional, more predictable agriculture model. For, as the population ages, so do our masters of clinical education. I would rather not risk the education of our future health care providers to chance – I prefer better odds than that.

## 1.4 Favorite Problem Solvers

1. The Society for Simulation in Healthcare (http://www.ssih.org/), specifically the listserv and the "Ask the Wizards" section has provided us with several key work-arounds that have helped make our simulation experiences more realistic.
2. Association for Standardized Patient Educators (http://www.aspeducators.org/) helped us to identify how standardized patients could help enhance our program and student success.
3. Loyd, G. E., Lake, C. L., and Greenberg, R. *Practical Health Care Simulations.* Elsevier, Philadelphia, 2004. This book provided us with several answers to questions related to planning and setting up our simulation centre – a good all-round reference book suitable for all sizes of simulation centres.
4. Monsters, Inc. (movie). The initial 3 minutes of this movie is a simulation that provided us with a great example of how NOT to conduct a debriefing session!
5. The Institute for Medical Simulation's "Comprehenisve Workshop in Medical Simulation." This intensive workshop helped me to understand the importance of embracing qualities of emotional intelligence while learning and practicing a variety of debriefing techniques for my facilitation "toolbox."
6. Bates, A. W., and Poole, G. *Effective Teaching with Technology in Higher Education: Foundations for Success.* Jossey-Bass, San Francisco, 2003. From this book we learned the best way to successfully incorporate new technology into the workplace.

## References

1. Lyons, A. S. and Petrucelli, R. J. *Medicine: An Illustrated History.* Abradale Press, New York, 1987.
2. Lave, J. and Wenger, E. *Situated Learning: Legitimate Peripheral Participation.* Cambridge University Press, New York, 2002.
3. Government of Alberta. *Alberta's Plan for Pandemic Influenza.* Retrieved March 20, 2006, from http://www.health.gov.ab.ca/influenza/Pandemic_plan.pdf, 2003.
4. Douce, F. H. and Coates, M. A. Attrition in respiratory therapy education: Causes and relationship to admissions criteria. *Respiratory Care* **29**(8), 823–828 (1984).

# 2

# Undergraduate Medical Education is NOT Rocket Science: But that Does NOT Mean it's Easy!

Mark R. Adelman

## 2.1 The Big Picture

If you want to know what this chapter is doing in this book, ask the editors. They believe (and I agree) that simulation includes more than just a full-sized simulator. If you want to know why this chapter might be worth reading (or maybe not), look at the reference, which is the *only* reference in what is nevertheless a very scholarly article [1].

What follows are a series of opinions of a basic science medical educator who has been trying to help students learn for over 35 years. They are opinions shared by many others who have been working at the same task. And have come to similar conclusions. So it is a sort of status report on an ongoing longitudinal study (with no control group). It is based on extensive observations of thousands of subjects, selected readings of scholarly and lay texts, and a healthy dose of common sense. My goal is to convince you of what you already know. To leave you with the conclusion that everything I said is (and was) obvious, and that you already knew it before we began.

My goal is to have you forget that you even read this chapter, and that you ever even heard of me. But also to modify the way you approach medical education, so that you will be more effective and will reach your "ultimate" goal with less effort than you might have otherwise. That is the "*big picture*," as I've been saying to first year medical students for a *long* time.

It might be informative to interject here that the first draft of this chapter was written without any idea of the contents of the rest of the book (other than that the overall goal of this book is to help those interested in improving clinical education), and that I have essentially no first-hand experience with the use of simulators (especially high-fidelity ones) in clinical education. After completing the first draft, I requested that the editors send me a provisional table of contents. Having skimmed the table of contents, I asked to see draft versions of several of the chapters in the sections on *Purposes and Philosophies of Operations, Engineering and Management* and *Lessons already learned from other disciplines*. On the basis of that reading, I felt it worthwhile to insert this very brief paragraph, to call attention to two points that are woven through my chapter, which – except for this insertion – is otherwise largely unchanged from the draft version. First, it is my assertion that undergraduate medical students – at least in their early years – are *not* the sort of "adult learners" who are prepared to derive maximal benefit from high-quality simulation centers. Secondly, it is our job as educators to help our students to transition from those comfortable with "other-directed" to those now performing "self-directed" learning, and we should be doing so using a variety of techniques, including a continuum of "simulation experiences."

It would be wonderful if I could proceed with a prioritized list of key "things" or "concepts" or "teaching tips." But you wouldn't buy it any more than the "typical" first year medical student. We all know there is no "free lunch," just as we all know there is no such thing as a typical medical student. Those are two *big picture* items and both belong on any short list of *key* items. But what would be *the first*? *Always step back so you can get a better view of the big picture*, because the answer is very rarely in the details, and because the answer can almost never be understood in terms of details unless the big picture is kept in mind. Obvious, right? Most people who do not teach first year medical students cannot understand how difficult it is to get first year medical students to recognize, let alone use that obvious principle.

Another obvious fact is that medical school is not conceptually difficult, it's just that there is *so much* to learn. While that fact is largely true, it is also largely irrelevant to the challenge facing undergraduate medical educators. Especially those of us who meet the first year medical students as they enter the front door, but also those who treat them in their clinical rotations. Because much of what follows is from the perspective of a "basic science educator," rather than the perspective of a "clinical science educator," I think it is *extremely* important to state the obvious: education is a continuum. We all start with the student who comes in the front door and we must have respect for (but realistic expectations and evaluations of) what is/is not achieved by those educators who precede us and those who follow us. As an educator of first year medical students, I frequently work clinical correlations into my teaching. And occasionally, I invite clinicians to give such correlations. I am usually struck by how much more skillfully clinicians can do such a presentation. And how often clinicians do not remember what they knew, as first year medical students, at that particular point in the process. A student in May (near the end of their first year) is very different than the same one in September (near the beginning of their first year). Obvious, right? All of it is obvious, but we must keep it constantly in mind as we work our craft. As a pathologist colleague (who teaches second year medical students) and I worked together on gaps in the curriculum, we rediscovered that when second year medical students report that they were not taught anything about macrophages, this does not necessarily mean that the histologists who taught the students as "first years" neglected

to cover the topic. And the fact that a surgeon can teach much (maybe even all) of gross anatomy does not necessarily mean that the surgeon has the skills (or desire) to help first year medical students take the necessary steps toward becoming a self-directed learner. We all have a job to do; we are all part of the process.

It is perhaps worth noting at this point that while the use of complex simulations has been most common in the clinical years of undergraduate medical education, there is growing awareness of the value of simulators in the preclinical arena. I would argue that it is vitally important for all of us to stress the value of simulations of varying kinds in the continuum of medical education. We need to make students aware that much of what they learn is in the framework of an environment that is *not* the reality of everyday medical practice and that learning in a simulated environment is the norm (rather than the exception), ranging from looking at histology slides (very far removed from reality), to virtual microscopy, to demonstrating complex physiological processes as computer models, to practicing complex diagnostic and surgical techniques on ever-more-realistic simulation devices. Knowing very little about the most current (and complex) simulation devices, I would presume to include them in a continuum of tools that allow us to help our students learn in a relatively safe way, how tricky (yet fascinating) it is to deal with a "real" patient, how easy it is to make mistakes, how necessary and useful it is to make such mistakes (and to learn from them), and how gradual the spectrum is from simulation to reality, both in terms of devices and in terms of practices. I look forward to the day when first year medical students not only enjoy learning from an amazingly realistic patient simulator but also intuitively understand the value of actively learning by simulating reality and challenging themselves: by pretesting themselves, not simply to maximize the chance of passing an examination but to reduce the number of mistakes that they will *inevitably* make as they practice medicine on the living.

## 2.2 Successful Applicants Becoming Successful Graduates: Modifying Attitudes

First year medical students are developing medical professionals. Essentially, all of them matriculate at medical school as very successful young adults and nearly all of them will graduate from medical school as even more successful young physicians, ready for still more development. However, they are a rather heterogeneous group of learners when they enter and they travel diverse paths as they proceed. We all know that there are multiple modes of learning; actually, when we say this, most of us are thinking of multiple modes of acquiring knowledge. Remember the old trilogy: KSA = Knowledge, Skills, and Attitudes? Over the years, I have come to understand that our students arrive with lots of knowledge (and the ability to assimilate *much* more), numerous skills (and they develop *many* more), but a number of attitudes that are not optimal for the kind of self-directed learning that we expect life-long learners to have – or to acquire. That is not surprising, given that they are among the most successful graduates of primary, secondary, and postsecondary education systems which are based upon "other-directed" rewards and punishments.

That last statement about attitudes could well be the basis for an entirely separate chapter, but I want to expand it briefly here, so as to provide context for the specific comments that follow, regarding most (but not necessarily all) entering medical students. Because of the ways in which our society views education (mostly as some sort of commodity), and the processes by which the students who enter medical school have been taught and tested, we matriculate some *very* capable students who – for the most part – have some deficits for which neither they nor we are responsible. But these are deficits of which we must be aware and must help them work to overcome. The most immediate challenge is to get students to recognize that although they have been very successful, i.e., have passed a lot of tests, passing tests is a very small part of what they must do as successful medical students, and as successful clinicians. And that the most important tests will be administered long after they graduate – by their patients. We need to convince them that they *can* multitask (they've been doing it for many years) and that time management and having confidence in their own ability to make "educated guesses" are neither new skills nor ones that they are likely to perfect any time soon. That grades in courses matter, but only a little, and that their patients (for the most part) will not care how much they know, only how well they use what they know. That no one has "the truth" hidden in some secret Personal Digital Assistant (PDA) (or whatever technological wonder is currently threatening to replace textbooks). And so on. All obvious, but all concepts and principles that we *must* keep in mind as we work with the medical students who enter our schools and proceed through the process.

First year medical students are *not* – with rare exceptions – "adult" (i.e., self-directed) learners. If we who work with them do our job correctly, those who teach our students later in the process will be teaching students who are closer to being adult learners; but the process is surprisingly slow and variable. Like a meandering stream that almost always gets to the ocean, eventually. Consider the following facts about first year medical students, but *please* do not regard these as criticisms. I have a very high regard for the vast majority of the students with whom I interact. They are hardworking, intelligent, decent people who have an intense desire to become very good doctors – and for very noble reasons. Please consider all of what follows as scholarly critique, rather than petulant criticism. My comments are meant to apply to most students. I guesstimate that some 10–20% figure all of this out for themselves, so long as we do not "mess them up." About 10% do *not* get it during the first year;

I can only hope that things improve later on. Hence I'm talking about let's say 70% of the entering first year medical students.

1. Very few of them have thought about how they think and learn. Call it meta-cognition if you wish; it is a foreign concept. Can one be a successful self-directed learner without engaging in some meta-cognition? Possibly (perhaps even probably), but that will make the overall task much less efficient.
2. Because they have been so successful, and because most are not used to analyzing how they think (never had to), they are ill-prepared to deal with the new challenges of medical school. They are not inclined to accept the freedom (and its flip side, responsibility) of being in charge of their own education. Nor are they inclined to try different strategies; why change what has always worked? Nor do they regard the faculty as people who are there to help them (not even those of us who really *are* there to help!).
3. A surprisingly large number not only do *not* seek out our help (except immediately before an examination, or after a series of disastrous examination results), they also seem to reject the notion that we want to, and *can*, help them. Faculty are often regarded as "the enemy": people who write trick questions, focus on obscure and irrelevant details, and/or are so involved in research that they regard students as an annoyance. It seems to take many students an exceedingly long time to sort out those faculty who fit one or more of the above categories from those who do not. Even when those faculty who do not fit any of the above undesirable categories try repeatedly to make their availability (to help) quite clear, they are still "not accepted" by some first year medical students.
4. While most students recognize the value of reasoning by analogy, very few actually use analogies in their professional lives (as developing doctors). And when they "finally" seek help from professional educators, such students are usually astounded at what they can learn about their learning styles by thinking in terms of analogies to everyday life.
5. Once they make the decision to take control of the learning process, most are very quick to recognize how they have been making things harder for themselves and how easy it is to fix things. But what is even more revealing is the number of students who, in those quiet moments when educator and student can relate as human beings, will freely (often with a wry grin or a self-deprecating remark) acknowledge that what they are doing doesn't really make sense, and that they would give the same advice (that the educator has been giving) to a friend or to their own children.

So the most important challenge, in dealing with first year medical students (and this probably applies to all would-be clinicians in all their student years – to varying degrees), is the challenge of modifying a number of attitudes. And doing so in a manner that leads the student to appropriate the altered attitude as the logical, reasonable, and obvious one to apply if the student is to make the journey with minimal stress and maximal effectiveness. The task is *not* a simple one. The goal is not easily reached. But most medical students "get it" and they must get it on their own (with our help) so that they *own* it. And, as they begin to own it, a number of specific points can be made. I have listed a number of points in what follows, but they are not listed in any priority order, for a number of reasons. Not all students absorb these points in the same way and in the same sequence. Each *real* learning encounter with *each* student is in essence a new teaching session and cannot be taught using PowerPoint or any other technological tool (although such tools may prove useful). And many of the points seem so simple (because they *are*), that students may reject them because they know there is no such thing as "free lunch" and must develop – for themselves – such concepts as "what is hard work," "what is big picture and what is detail," or "what is an important fact to *me*."

## 2.3 Words Matter

It is not surprising that so many students find it difficult to deal with all the "jargon" of medical school. They are the products of an educational process that does not value precision in language and does not understand the relationship between precise use of words and true understanding of concepts. Our students are the product of a social process that fills the public discourse with "y' know," regards proper grammar as antiquated (or is it antique?), insists that correct spelling is the responsibility of the computer, and operates on the principle that even the most nuanced argument must be presented in a 30-second sound bite. Very few know any foreign language and many function as though their native language (especially if it is English) is a foreign language. So we cannot be surprised that, in addition to not knowing Greek or Latin, they do not see the value of being informed that the stem "reticul" is used in so many words: reticular fiber, reticular cell, endoplasmic reticulum, and so on; and that knowing this simple stem makes it much easier to keep track of so many seemingly mysterious words and terms. They know that words are used to transmit information, know that certain words are condensations of entire sentences, know that acronyms are abbreviations that are immensely useful (but must be used in the appropriate context), but bristle at having to learn a bunch of new words, not to mention being expected to spell them correctly and use them in complete sentences! I routinely tell our students that the medical dictionary they are issued when classes begin is an immensely valuable book. They often look at me as though I was speaking a foreign language. They all know how to "google" something, yet rarely do so when the something is a word that is a stumbling block to their understanding of such

a trivial distinction as the difference between an intralobular duct and an interlobular one, and do not seem to believe me when I repeat for the Nth time that the terms are just used to convey a sense of where the duct is found. Much of my work with individual students involves helping them understand, by using examples and making analogies, how often the proper use of a particular word has *big-picture* implications on how and what they learn.

## 2.4 Make the Verbal-Visual Link

One of the notions we try to stress in our teaching of Histology and Cell Biology is the need to "make the verbal-visual link." Students are aware of the old saying about a picture being worth a thousand words. And they understand, almost immediately that "if you can't picture it, you don't understand it," but most have very little practice in converting words to pictures or in extracting words *from* pictures. It takes some time to convince them that this is a valuable skill; one that they need to develop not just to pass our tests (and those in pathology, radiology, etc.) but also in order to "read charts," make a first level diagnosis from a physical exam, etc. What is even more challenging is the task of getting them to commit to practicing "making the verbal-visual link" as much as is needed to develop the skill of doing it rapidly, effectively, and usefully. As a concrete example, many students balk at "wasting time" looking at glass microscope slides, trying to find "stuff" that an instructor could show them in just a few seconds. Many of us have put lots of wonderful images on websites and are investing tremendous resources in developing "virtual microscopy" so that students do not have to spend too much time with old-fashioned microscopes and loan slide sets. It is absolutely true that most of them will rarely (if ever) use a microscope after they have left medical school. But the skill of observing and converting the observations to words that can transmit information to others is such a vital skill for any physician, that we *must* convince the students of the value of the attitude that this skill *must* be developed, even if it isn't easy, takes a lot of time, and – in the context of a first year Histology course – seems very far from relevant to medical practice. It may be useful to explain an algorithm (one of many) I often trot out in helping students (who have *finally* sought help) use the microscope as a learning tool. The algorithm (call it *four questions*) is based on the premise that most students do not understand that the best way to learn and prepare for a test is to test oneself (and/or one's peers) repeatedly.

## 2.5 The Four Questions Algorithm

The "four questions" algorithm goes something like this: Place any loan slide on the stage of your microscope. Make sure that light is going through the specimen, and that the specimen is in focus with either the 10× objective or the 40× objective. Then switch objectives (to/from 10 or 40), and *without looking in the oculars*, move the stage slightly, but not so far that the specimen is no longer in the light path. Then look in the oculars and, with reference to the pointer (or crosshairs or reticule – all our loan microscopes have one), ask yourself the following questions:

1. What is at the tip of the pointer?
2. What does it do?
3. How does it do that?
4. So what? (that is, what is the significance of its normal function?) Obviously, the use of this algorithm requires some questions and answers (such as an explanation of my assertion that all students have a built-in "my answer is bull-shit" meter). It is in fact the framework for a *lot* of discussion and education, both by the educator and by the self-directed learner. And this leads into the next "item."

## 2.6 Analysis and Diagnosis

All first year medical students know that they will be making diagnoses as part of their professional career. Some even understand the notion of a differential diagnosis. But very few recognize that any diagnosis is, in essence, an analysis leading to an "educated guess." (See following section on probability and uncertainty.) Even fewer have had much practice analyzing the implications of observations – for example, looking at a picture, describing the elements present, making a tentative conclusion as to what the facts imply, and acting on that educated guess. We spend a lot of time explaining to our students that we test them with practical exams (at the microscope), for multiple purposes, including the desire to determine if they have developed the knowledge of what an organ looks like in the microscope (and how that relates to the function(s) of the organ), the skill of observing the visual manifestation of the structure and converting the information to a verbal equivalent, and the attitude of a professional who understands that solving many different puzzles is the only way to develop the ability to solve new puzzles. Medical students are – as I've said – quite intelligent and industrious, but it is often a hard sell on that last point. They seem to think – as do many computer junkies – that skilled diagnosis can be reduced to a list of choices and a simple algorithm, both of which can be rapidly learned and *voila!* We all know that "gifted" diagnosticians get there by working hard and thinking about what they are doing. Pattern recognition (whether it is of histology slides or clinical scenarios) is developed over time and is best applied by someone who understands when certain patterns are truly diagnostic and when they must be used cautiously

in a complex process of differential diagnosis and feedback evaluation of the analysis.

## 2.7 Probability and Uncertainty

Perhaps the most difficult attitudinal issue for first year medical students is the notion that so many things in medicine are *not* certain and that so many decisions must be based on probabilities. It is my perception that students arrive having learned about probabilities as a mathematical subject, but have not actually recognized how much it plays a role in their everyday lives. And the part of their everyday lives that has been centered on education has *not* provided them with the attitude that dealing with probabilities is the reality. Instead, most arrive firmly convinced that medicine is a set of true facts that, if applied, will lead to success. Of course they recognize the reality that medicine frequently fails, but there seems to be a fundamental disconnect between what they understand as the reality in one context (life) and the reality they expect in their education as physicians. Here I am *not* claiming to know what the average student thinks; I doubt that anyone really knows what *anyone* else thinks. I am simply observing how they behave when confronted with uncertainty: they are unwilling to accept and deal with it. Students behave as though the faculty member who does not give them a simple yes/no answer *must* be deliberately hiding some essential fact from them. And this attitude is *extremely* difficult to modify. Every colleague with whom I have spoken about this tells me essentially the same story. Student X asks for an identification of a particular cell. The educator tries to help the student figure it out, using various elements of verbal-visual linkage, analysis, etc. And brings the student to the point of accepting that the cell is, e.g., a fibroblast. Or rather is *probably* a fibroblast. But could be a macrophage. But, given the available information and the probabilities, it is *most likely* a fibroblast. The observed reality is that the "average" student is extremely unsatisfied. How will they be sure that, on the examination, they will be able to identify the cell at the pointer? The truthful answer is that they cannot be certain; but they can increase the probability of reaching the most likely choice if they practice the analytical approach we are trying to teach. The educator must be prepared to go through the process a very large number of times, repeatedly pointing out to the student that they routinely make probabilistic decisions in everyday life and are usually right, but – in those cases where they are wrong – they deal with the consequences and move on to the next decision. Again, in those rare moments when student and educator talk as equals – caring human beings – the students understand and admit that their behavior (demanding certainty) is unrealistic and does not make sense. But modifying attitudes ingrained by the educational system that helped them gain admission to medical school is neither easy, nor rapidly achieved.

## 2.8 Short-term and Long-term Views

Students have become so used to studying for one exam, because that exam is *crucial*, that they have not developed the attitude of constantly thinking about the long-term goals of building knowledge, skills, and attitudes that they will *use* as practicing physicians. It requires considerable effort to get them to stop studying *just* for the test. Equally difficult is the task of getting them to appropriate the reality that some skills are learned in stages, only the last of which is the desired level of skill, but that *all* the stages must be not only mastered but also retained. Repeated examples (clinical scenarios, experiences with simulations, analogies from everyday life, likely test questions from a second-year course exam – or the United States Medical Licensing Examination) must be presented and employed *not* to show them how much they do not know, but rather how what they have learned (or are learning) has relevance to all the upcoming hurdles. They understand this "intellectually," but find it very difficult to modify the behaviors that got them to medical school. If there is one single thing I could do to improve the education our students receive before they walk in the front door of our medical schools, it would be to *stop* the obsessive focus on testing and the resultant delivery of the message that passing specific-staged tests is the single most important thing students must do to become educated!!

## 2.9 Test-taking Strategies and Educational Value of Tests

First year medical students bring with them an astonishing number of learned test-taking strategies, many of which don't help them do well on our examinations, or – more importantly – are not useful strategies for the daily tests that are presented in the practice of medicine. Many of the strategies "work" in the context of "standard" exams. For example, knowing (i) that the longest answer is probably wrong (or right) or (ii) that words like "never" and "always" are red-flags. Such facts may be useful in the game of test-taking, but are not very useful in the game of practicing medicine. Perhaps most astonishing to me (and others) is the number of students who do not understand the value of going over an examination *after* it is done (whether they did well or poorly) so as to *learn* from the examination. Mentioning the notion that doctors *do* in fact learn from mistakes resonates with many students, but they do not seem to see the obvious implications for how to analyze their mistakes so as to learn from them. A few examples (from the long list that any experienced educator can supply):

1.  Students often dismiss an answer option because it is "too easy." Working with each student as an individual, one can help them decide which particular fallacy in thinking (or flawed assumption) made them reject the obvious.

It may be the old theme of not looking for zebras when one hears hooves. Or it may be the notion that teachers are *not* necessarily trying to trick them – and that it is impossible to guess what trick such an evil teacher would be using anyway! Or it may be that the student, having experienced crushing defeat (a "C"!!!) on an examination, is now convinced that he/she doesn't know anything and thus must distrust what seems like the obvious answer.

2.  Perhaps related to the above, but also a reflection of the "lack of big-picture syndrome," is the fact that many students, after being shown that the "correct" answer was not just obvious, but was also one that they probably would have chosen before they came to medical school, still refuse to accept that reading any question from the perspective of general information will often lead them to sorting out the correct answer from amongst a number of choices that seem confusingly equally likely, because the student is ignoring an obvious element of the correct answer that has been obscured by all the "distractors."

3.  Returning briefly to the theme of simulations, I note that most students do not recognize the immense value of simulating the test situation by posing questions to themselves (and small groups of their peers), as a means of studying the material on which they will be tested. (Recall the four questions algorithm.) At our institution, it has long been a practice for the second year medical students to set up a "practice practical" for the first year medical students. This is usually done just before our first examination and is very useful in exposing the students to the mechanics of a practical examination. However, despite multiple discussions with students about how our practical exams are constructed (and why), I have yet to be successful in conveying to them the multiple levels at which such examinations are in fact simulations of the various testing elements to which they will routinely be subjected throughout their careers. The analogies are transparent and obvious, and this perhaps explains why students do not think though the useful implications. (I am working on an article on this, tentatively entitled *The Anatomy of a Practical Exam: Rationale and Logistics of a Simulation Experience.*)

4.  Returning to the themes of *words* and *analysis*, I am often struck by the way in which students categorize their mistakes on an examination in response to my suggestion that they go over the examination and try to sort out what sorts of mistakes they made. One of the most common terms they use is "that was just a dumb mistake." It takes a lot of effort to convince them that the word "dumb" in this context is usually not only inaccurate, but also judgmental (self-deprecating), and – even worse – a "cop-out," because if a mistake is "dumb," one is off the hook since there is little one can do about being "dumb" except to work harder. Which brings me to the last of my "points."

# 2.10 Studying Hard Versus Studying Smart

For many first year medical students who have problems with the material they encounter at the start of undergraduate medical education, their inclination is to assume that they are not working hard enough and to spend more time studying. In my experience, most students in fact spend ample time (perhaps even too much time) studying, but do not understand how to study in a more efficient fashion. Many fall back on time-proven techniques of rereading texts, highlighting notes, making lists of "important" stuff, studying examinations from previous iterations of the course, etc. Despite having been urged to outline big-picture concepts, identify only a limited number of things to memorize, and to work on trying to decide what is most useful for *them* to learn, students persist in trying to memorize minutiae first, then major topics later. I choose to finish this last major "point" section by listing a few of the concepts that students need to *own* in order to make the transition to the self-directed learning style that they *will* in fact achieve if we all do our job and they do theirs. Unfortunately, many of these concepts are neither subtle nor easy to sell.

1. Students overestimate the value of detailed factual knowledge and underestimate the value of simple information, common sense, etc. If you ask the average student whether they expect future patients to care how much they (as physicians) know, and suggest that future patients are much more likely to care about how well the physician can *use* whatever they know, every student (even those who have never been patients) will immediately understand that all the detailed knowledge is worthless if they do not have it in some sort of accessible toolkit and can actually use it. But when you stress the value of restricting memorization to the most useful things (like a list of all the organs in the body and a 25-word definition of what each does), they will balk at the absurdity of this being a useful exercise. And will probably still do so after you point out that the reason they misidentified a section of parathyroid for one of parotid was that they *forgot* about the existence of the parathyroid. (Trust me – I have collected a long list of such "common errors" that are only understandable in the context of a student who has lost sight of the big picture.) Another example might be helpful. We routinely stress that there are four basic tissues in the body (epithelium, connective tissue, muscle, and nerve) and that one can identify (and begin to understand the function of) any organ by observing, identifying the tissues present, describing the amounts and arrangements of those tissues, etc. And that one should *not* attempt to memorize details like the types of collagen present in, nor the substances produced by, a particular organ (say a gland), until one can state a very brief description of the specific versions of the four basic tissue types that make up that organ. Yet, I can assure you it is much easier to find a first year medical student in June (at the end of the first year) who can name the type of collagen in a specific connective tissue of an organ than to find one who can quickly name the four basic tissue types. One very interesting consequence of all this is that students in fact spend more time memorizing than we want them to, do not memorize the most useful things, and then – when examination questions seem harder than they in fact are – the students complain that we were testing on picky little details.

2. Students lack a number of fundamental organizational skills – as applied to studying – that most of us would like to assume they bring in the front door. They do *not* understand (or display skill in) outlining a topic, nor the value (to them) of starting from the major points and working down into the details. They do not recognize that putting material down on paper "from their memory" is an excellent way of combining learning, and reviewing, and self-testing. If you suggest a chart as a way of reviewing material, be sure to suggest the rows and columns but do *not* fill in the chart, because they are more likely to try to memorize its contents than understand the logic of its construction. If you suggest that they make a sketch of something and label it as a means of reviewing material, be sure you do *not* label your sketch and be sure to urge them to discard the labeled diagram immediately after they have prepared it. And do not be surprised if a student asks how they can be sure that they have included "everything" in the chart or correctly labeled the diagram – and seems nonplussed if you respond by suggesting that they look at similar charts and diagrams (in the texts you have assigned) for confirmation or corrections. The current mantra is for us as educators to integrate what we are teaching and avoid wasting student time by redundant presentations. Do *not* buy this! Because many students have not learned and thus do not understand that the only meaningful integration that occurs is what *they* do in their own minds. And that time is only wasted with redundancy if students who have already learned something keep reading about it (or attending optional lectures) because they are unwilling to test out the premise that they have learned the material, by simulating reality and learning from any mistakes they make.

3. We devote a great deal of effort to gathering student evaluations of our courses – as we should. Not because they are our customers, but because they are our students and every educator knows there is much to be learned from students. But not necessarily what is the best way to teach. Because most students do not spend enough time analyzing how they think and learn. Because we know,

better than they are likely to, that there is no *best* way to teach all students and that while "popularity" may be a very useful indicator of effective teaching, it is not an infallible guideline. In fact, for many students, the most effective teacher will be the one they least enjoy. And for most students, there is immense value in being taught the same things by many different educators, because the most important things we are likely to teach them are the things that are so obvious that there is no way of predicting which iteration will be the charmed one, from which they "get it."

4. Related to the above, I would like to comment on the notion (growing in popularity) that we must give the students more time to study and learn on their own. To the extent that we see the problem as excessive reliance on didactic teaching (lectures and the like) and not enough on experiential learning (small group discussions, discovery at the microscope or in the library, experience at a simulator, or with a mock patient), we must correct the imbalance – and are doing so. But anyone who thinks that giving first year medical students more free time and a list of tasks to do, problems to solve or methods to master, without taking pains to make sure they have the skills and attitudes necessary – that educator is flirting with disaster. Work with all students until your are fairly sure that each is indeed an independent learner, with all that entails, and meet with them often so you can be reasonably certain that they are testing themselves and neither kidding themselves nor you about what they are learning.

## 2.11 Broader Perspective

OK ENOUGH! I'll close with a redundant reference to the importance of the *big picture*. Redundant because we all know of the danger of losing sight of the forest for the trees. But we all tend to forget that while it is easy to see the forest from the outside, it is exceedingly difficult to have an overall view of it from the inside. So one last example from my own experience. In countless instances, I have had a student call me over to a microscope and with utter frustration point to the scope and ask me "what is that thing at the pointer and how on earth would I be expected to know that?" And in most such cases, I simply switch the objective lens from $40\times$ (where the student almost always is) to $10\times$ (as an approximation to the bigger picture) and ask him/her to describe what they now see. And – if necessary – work them through the verbal-visual tasking until they have described what is there. And usually (but not always, because it takes time), the student will say something like "You mean its just a . . . ?" And I respond, more or less: "Seems likely to me."

## 2.12 Conclusion

The approach of helping the student see the bigger picture and figure it out for themselves has always seemed – to me – the obvious way to go. And I know it works in many areas. Having had no experience in teaching clinical sciences with space age simulators, I certainly cannot be sure about this. But the next time a student is peering at some display, or strip chart output, or listening intently to some sound produced by a simulation device – and the student seems totally befuddled and is trying to look closer or listen more carefully – ask the student to step back and describe where they are and what they are doing. What is the overall reality of what they are analyzing so intently? Are they trying to deduce something about cardiovascular function from an EKG? Are they trying to figure out what is wrong with a respiratory system by listening with a stethoscope? Is the goal to understand why the simulated patient isn't responding as expected to an anesthetic? Looking closer or thinking smaller *may* get them to an answer, but maybe not. Stepping back may not always make things more obvious. But it frequently will and, even if it doesn't make the answer to the specific question obvious, it will allow the student to recast the question so they see it from a broader perspective and have a better chance of "seeing" what is obvious even when it is not quite as obvious as we would like to think it will be.

## Reference

1. Mark R. Adelman is an Associate Professor of Anatomy, Physiology and Genetics at the Uniformed Services University of the Health Sciences (USUHS), 4301 Jones Bridge Road, Bethesda, MD, 20814-4799. He can be reached at adelman@educationalassistance.org. After receiving a undergraduate Bachelor of Arts degree in Biology (as a pre-medical student) in 1963 from Princeton, and realizing that he would probably make a rotten medical doctor, he switched paths, attended the University of Chicago (for graduate studies) and received a PhD in Biophysics (1969). He then did post-doctoral research in Cell Biology at the Rockefeller University. He has taught undergraduate medical students since 1971, first at Duke University, more recently at USUHS. He has also taught graduate students and undergraduates at Duke, USUHS, and other schools. Most of his teaching has been of Histology and Cell Biology, working with first year medical students; hence most of his 'examples' are based on those experiences. But he has taught many other subjects and there is one common thread to all of his educational efforts: he has NEVER taught a course that he has formally taken as a student. He does not believe ANY of the materials he has presented to students are all that hard, he fails often and is always educated by his failures, tries to get his students to accept that this is the

way it is; but has STILL not succeeded in getting students to accept failures as learning experiences. He recognizes that most of his comments are based on data derived from students and colleagues in the United States. But, given his discussions with other members of IAMSE (International Association of Medical Science Educators), he suspects that much of this chapter will resonate with educators from other countries.

# 3

# Guidance for the Leader-Manager

Robert C. Cox
J. Lance Acree

## 3.1 You as the Reader

Our purpose in writing this chapter was to help the decision maker understand some of the factors we have come across in our experience designing and implementing training systems that employ simulations.[1]

## 3.2 You as the Leader

Integrating simulation into a training program is more than an academic exercise – it can be a major culture change adventure. Culture change is not a job for administrators; it requires leaders. The commitment of a few leaders to bringing constructive change over the long-term will be essential to success in building an integrated and comprehensive training system. Your ability to communicate the purpose and philosophy of this change will be the most critical ingredient in providing leadership throughout the culture change. Since many of the concepts you will be attempting to communicate are somewhat abstract, the following discussion will introduce some specific terminology that you as a leader might find helpful.

Leaders recognize and wrestle with the "trade space" of cost, performance, and risk; they seek to maximize performance output for the least cost and risk input (Figure 3.1). You and your fellow leaders will develop a vision and define goals for your training, so that you can begin the communication work to spread the vision and shape the expectations for the rest of the organization. Prepare yourself to face and overcome the natural organic resistance to change. If the apprentice-training ("Halsted"[2]) model is the only model people have known, they will have difficulty imagining how anything else will work. So you might start by imagining, for yourself, what it would be like to have a world-class performance-driven training system that uses courseware, simulation, and "the real thing" to reduce overall cost and risk while boosting graduate performance.

You may want to write down what you imagine this would look like, especially in terms of the finished product. For example, you might envision your graduates, on graduation day, being proficient with the latest technology and techniques,

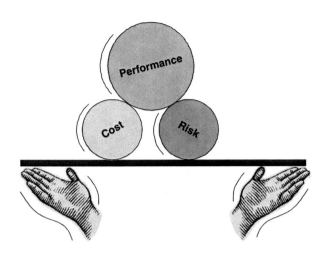

**FIGURE 3.1** The dynamic challenge facing the leader-manager: balancing cost and risk to gain performance.

proficient with your institution's information systems and processes, proficient at teamwork under stress, and so forth. Imagine them having practiced repeatedly in simulations, and having proven their skill against objective standards, so that their first interaction with a real patient is a low-stress event for everyone. Your imagination is going to get a workout in this chapter, so you might as well start stretching out and limbering up.

Leadership is also "the art of the possible." You and your allies will need to identify a *feasible starting solution* – something not too big to accomplish within a year, requiring

**TABLE 3.1**  Two pilot programs

*Examples: starting small but feasible*

*a. Some eye surgeons we work with selected the procedure "peel an epiretinal membrane" for a starting point; this surgical task has distinct beginning and end points, and some simulation is available to help train it. Using a microcosm of a comprehensive training system, we were able to perform experiments to demonstrate quantifiable performance gains in retinal surgery. Of particular interest was the participants' response to the task analysis–based courseware (residents through attending physicians); the case group response was both positive (difference in means >20%) and statistically significant (p value less than 0.05) as compared to the control group response. Appendix 70A.1 contains some of the quantitative results from this example.*

*b. A different community (anesthesiologists) selected a more general, whole-body emergency procedure as a prototype for revising their graduate medical education program. The procedure is known as COVERABCD-A Swift Check, and includes over 20 subalgorithms to address a corresponding number of top life-threatening complications. Starting with this procedure and one of its subalgorithms will enable them to prove the principle; their goal is to expand the program sequentially to the full procedure (all algorithms) while gathering performance gain and cost data.*

maybe 25–50% more resources than you have on hand, to prove to the powers-that-be that the concept works, and begin a spiral development cycle. Perhaps focusing on one or two critical procedures or skills will do the trick (Table 3.1).

This will stretch you as you hold on, with one part of your brain, to the ideal grand vision you imagined in the paragraph above, while simultaneously working on a practical first step with some other part of your melon. Chances are, if you are reading this chapter, you are a visionary. If so, you may have real difficulty identifying a truly feasible starting solution; pick friendly colleagues who are naturally skeptical. Make them help you determine what is feasible. Keep these pragmatists in your inner circle to balance out your idealism; they will force you to think practically.

## 3.3 The Clash of the Titans

If you haven't yet run into the traditional training philosophy in your organization, it won't be long before you do. You must not be surprised when "expert opinion" deems simulation inadequate for training, either because "it's not the real thing" or because "it will hinder training in dealing with real patients" or some similar reaction.

> **Editors' Note:** Most of today's allopathic medical school curriculum is "not the real thing" compared to the daily behaviors and responsibilities of *real* clinicians treating *real* patients. The basic science years are just a continuation of a typical undergraduate science curriculum, with content assumed essential for subsequent learning of clinical thought. The following clinical years, at best, prepare the students to make informed choices in selecting a specific residency topic and program. The residency years transform these extensively prepared individuals into actually functional clinicians. Then come fellowships to master the most rare and arcane specialized skills and abilities. Finally, after 10 years in clinical practice, the former clinical school applicant is now a master craftsman, doing "the real thing."

You are now facing a classic collision between training philosophies, and you will come out better (clothed and in your right mind) if you label this conflict as such every time you encounter it. By proper use of the word *philosophy*, you are grasping a distinct and powerful cognitive tool for achieving lasting culture change. (In contrast, the word *attitude* points to the affective or emotional domain, and *culture* is too broad a term to steer with.) This statesman approach helps keep the collision from degenerating into personality-driven argumentation.[3]

Training philosophies are built on assumptions. To help make allies out of adversaries, you might list the underlying assumptions of the competing philosophies (Table 3.2).

You might recognize the last entry in the table under traditional philosophy, where performance is assumed if sufficient time has passed. The aviation version is "Fly with an instructor pilot for *x* hours; come back alive and don't crash too many aircraft." The academic version is "Sit in the instructor's class for *x* hours; come out sober and don't flunk too many quizzes." The medical version is "Operate with an attending surgeon for *x* years; come out sane and don't kill too many patients." As the smoking wreckage of many an airplane will attest, time spent with a master certainly aids performance, but it is *highly unreliable* as a guarantee of performance. The surviving passengers were the first to complain, followed closely by the people who paid for the airplanes. Eventually, even the pilots recognized that any crash is one crash too many, and aviation adopted an integrated, performance-driven training philosophy. That led aviation to investigate a new approach to training.

**TABLE 3.2** Training philosophies and their underlying assumptions

| Traditional training philosophy | Performance-driven training philosophy |
| --- | --- |
| You can't replace "the real thing." | Simulation augments "the real thing." (By the way, what's your definition of "the real thing?") |
| Unless the simulation is *exactly* like "the real thing," no *real training* can be accomplished. | Simulation does not have to be *exactly* like "the real thing" in order to boost the trainee's performance while reducing overall cost and risk. |
| The trainee should be developing judgment, and that can only be learned when facing the ambiguities of real situations. | Properly constructed simulations include a wide variety of scenarios, some of which are ambiguous, to push the trainee to develop competent judgment. |
| A "war/sea story" told = Lesson Learned. | It's not a Lesson Learned until it's in a lesson. (We'll discuss later how you can be certain the lesson has actually been *learned*.) |
| Probabilistic, random, *ad hoc*: The trainees *might* be trained to perform actual tasks X, Y, and Z to the satisfaction of an expert, during the training period T, *if* the right patient mix appears during T. | Deterministic: The trainees *must* perform *training tasks x, y,* and *z* in period T, under select conditions and to appropriate training standards; they can then be expected to perform *actual tasks* X through Z.[a] |
| Only expert clinicians can determine what should be trained and how it should be trained. | Subject matter experts determine what is to be trained; training system experts help competent authority to determine how it is best trained. |
| Curriculum is a list of medical conditions and procedures; it's a general list I keep in my head. | Curriculum is the blueprint that guides and controls the entire training system; it's highly detailed and kept in a database. |
| Personality-driven | Trainee performance-driven |
| Time spent with a master = performance. | Only performance = performance. |

[a] We're using upper and lower case Xs, Ys, and Zs to distinguish between actual tasks (upper case) and training tasks (lower case).

## 3.4 The Aviation Analogy: Is it Valid?

We occasionally encounter the argument that since engineers design planes, but they don't design humans, treating the patient as if they were an airplane (and the clinicians as if they were flight crew members) does not hold up. But this is a misperception; the aviation analogy, when properly articulated, does not place the patient in the role of the airplane. It is more correct to say that the engineers design aircraft to function *as the interface between the aircrew and the atmosphere*. We can see this by looking at the mission of the aircrew, which is not the simplistic "fly the airplane." This would make a mere device the goal of the activity, which it is not. Likewise, we would not make "operate clinical gadget X" the mission statement of a clinician. Properly stated, the mission of the aircrew is to deliver something *through the atmosphere*. This statement keeps the aircraft and other gadgets in their proper place; they are merely the means to the end. It also identifies the main challenge to be overcome by the aircrew, and that's not the airplane – it's the atmosphere. Figure 3.2 illustrates why this approach makes the aviation analogy work for clinicians.

In this illustration, the airplane functions in the same way as the syringe in the clinician's hand; they both serve merely as interfaces with the primary challenge. Of course, the modern airplane is more complicated than a syringe – and so is the

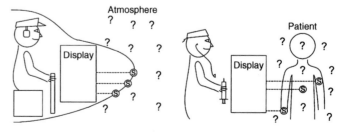

**FIGURE 3.2** The aviation analogy (art courtesy of R. Kyle).

da Vinci Surgical System®. But in the early days, airplanes were simple and crude; take a look at a modern hang glider. These early airplanes and their modern counterparts have hardly any systems, displays, or sensors – no altimeter, no airspeed indicator. Note that these two sensors measure and display *atmospheric* behavior, not airplane system behavior. These sensors were added early on to assist the aircrew in understanding their primary challenge – the atmosphere. Pilots needed to know the airspeed, for example, with greater precision when *the mission* demanded that the gadget (airplane) be operated close to the edge of its *atmospheric behavior* limits, such as the stall speed. Stalling is atmospheric behavior, not airplane system behavior.

The behavior of the atmosphere challenges the aircrew in many other ways: drag, low visibility, wing tip vortices, icing, turbulence, crosswind, tire friction loss. And aeronautical engineers don't design atmospheric behavior. These challenges are highly variable and unpredictable. (How far do *you* trust the weather forecast?) In contrast, the system behavior of the airplane is well understood and thoroughly documented, mainly through systematic engineering and flight-test programs. You will not find a hydraulic system, or a fuel system, or any other kind of system inside a properly certified airplane that does not have reams of charts describing its behavior in excruciating detail. Sometimes, a system failure (such as adding the wrong kind of hydraulic fluid, the equivalent of a drug-delivery error) complicates the mission, and while these complications are serious, they are not the primary challenge. As we move to the outside of the aircraft, we begin to encounter things we can't see and don't thoroughly understand, from boundary layer separations to microbursts. Seen this way, the patient functions like the atmosphere, not like the airplane; the patient generates most of the variability and unpredictability that challenge the clinician.

In aircrew training curriculum, you would expect to see airplane system simulations (that emulate *predictable* hydraulic system behavior, for example), but you would also find a variety of *unpredictable* atmospheric conditions. It is these conditions that call for simulations of dangerous crosswind, turbulence, fog, icing, etc. – things that do not emanate from the airplane itself but from the atmosphere. In its quest to control risk and cost, aviation discovered a reliable way to methodically identify and force these hazardous conditions (and their corresponding simulations) into its training systems. The end result is reliable human performance at acceptable cost and risk.

## 3.5 The Systems Approach to Training

It is a common mistake to start out campaigning for adding simulation to existing training when that is only one part of the solution to risk, performance, and cost. Simulation, whether

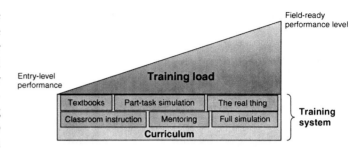

**FIGURE 3.3**   The training system and the training load.

adopted in whole or piecemeal, will ultimately fail to deliver on its promises unless it is understood to be only one element in your *training system*. Unless having a simulation museum is your goal, it would be a mistake to procure simulators and construct simulation facilities without first defining, in a systematic way, what you expect the simulation to help you accomplish. This may sound daunting, but there already exists a process for this; you are going to apply the systems engineering process to the whole training equation. This is called the Systems Approach to Training, and, like systems engineering, it begins with analyzing the training requirement, as opposed to beginning with analyzing glossy sales brochures.

Training exists to build human performance; it seeks to guarantee a minimum level of performance at an acceptable level of risk and at an acceptable cost (see Chapter 8). If we start with the goal of minimum risk, or with the goal of minimum cost, either way we will rapidly conclude that the solution is zero training (= zero risk and zero cost) and consequently zero performance gain. Therefore, we begin by carefully defining that minimum performance level required of the graduate *in the field*, and then work backward toward the entry-level performance, methodically managing risk and cost as we go. It may help to visualize the training required to achieve the minimum performance requirement as a weight, or *training load*, that the training system must lift, or help each trainee learn to lift. The different elements in the training system – the textbooks, the classroom, the lab, the simulation, the instructors, and "the real thing" – all carry part of the training load, and are organized into a system *by the curriculum*. Figure 3.3 shows the training load as an increasing performance requirement, lifted by the various elements of training as they are organized into a system by the curriculum.

## 3.6 Defining the Performance Requirement

World-class human performance, the kind that requires (and deserves) a world-class training system, is quite complex and must be described in multiple variables. This gets at one of the primary arguments used against simulation: "You can't

**FIGURE 3.4** The tent analogy, showing a typical tent, held in place by stakes, poles, and ropes.

define the real performance requirement with a simulator." Before we address this charge, we need to examine how to define complex performance requirements in general. Think of a tent, held in position by a collection of points: the peaks and corners (Figure 3.4).

Without these peaks and corners, the tent collapses; we might say the tent is defined by these points. Simple tents have only a few; elaborate tents have many. But we don't set up tents just for the fun of pounding stakes – we set them up to get what is inside, the nice dry *living space* that exists *between the peaks and corners*. We could say that the living space inside is also defined by these points. If one of these points moves, as when a tent rope slips, the living space inside changes, so an experienced camper pays a lot of attention to the peaks and corner points. You might also have noticed that the concept of *living space* we are using includes not only spatial dimensions but also dampness conditions and probably temperature and ventilation conditions, as well.

Now imagine that your tent is made of a revolutionary fabric that stretches freely in all directions without limit. You can create quite a complex living space by simply adding points. But as you add points to your tent, and the complexity of the living space increases, you will find that at some point you have enough rooms and halls in the tent to meet your needs. You or your significant other will say, "That's good enough – that's all the living space I think we'll need, at least for the near future." By defining the corners and peaks, you will have achieved your target living space requirement.

The analogy we're driving at is this: the living space in the tent is like the performance target our training system is supposed to produce. For the rest of this discussion, we will call this complex performance requirement *the target performance space*. This is not the geometric use of the word *space*, which uses only three dimensions. We're talking about multiple dimensions and variables – procedures, time, instruments, patient status, blood pressure, heart rate. It's closer to the more abstract Operations Research use of the word, as in *solution space*. Operations researchers use this term to describe solutions – in all sorts of combined, nonspatial measures, such as degree-days, ton-miles, unit-cost, and man-hours – that

make up the abstract volume of feasible combinations. Recall that our *living space* concept included temperature, ventilation, and dampness characteristics in addition to length, width, and height.

Someone might object that *performance space* is too abstract to work with, but we already work with abstract concepts without batting an eyelash (try explaining to a child, in words, what *the dollar* and *energy* are; not what they *do* – what they *are*). Once you become familiar with defining points, the corners and peaks, you will find the performance space concept quite powerful and even familiar. Consider the following fairly familiar example of one defining point in a target performance space:

> Given a typical car and a 6′ by 25′ parallel parking space, perform parallel parking, without striking the parking space markers, and placing the curbside tires within 12 inches of the curb in less than 3 minutes.

Many other key points are required to define the "corner points" of the performance space for driving a car, such as the one for merging with heavy expressway traffic, but the parallel parking "corner" has the advantage of evoking painful memories of the driving test. It also is written above as a "Terminal Learning Objective", complete with *task* ("perform parallel parking"), *condition* (car, 6 by 25 ft. parking space), and *standard* (3 minutes, no crashing, 12 inches from the curb).

A collection of terminal learning objectives, written in a form similar to this one, would define all the "corners" of the target performance space for *driving a car*. Remember your tent – the target performance space – held in position by a collection of points, peaks, and corners – the defining terminal learning objectives. If a trainee performs up to standard at each of the terminal learning objectives, you can be reasonably assured that the trainee will perform *acceptably* at *all points in between*. If you think the driving example is too simple to apply to the medical community, remember that we use it to train teenagers – millions of them – to operate potentially dangerous machines in ambiguous and potentially deadly situations, and you still get on the highway with them. The risk has been balanced against the cost and the performance requirement.

You might well wonder how someone came up with 3 minutes, or 12 inches, since no one hacks a stopwatch or whips out a ruler for parking "the real thing." Teenagers certainly wonder, out loud and with enthusiasm. The answer is that these were selected by a competent authority *for training purposes only*. The competent authority knew full well that these metrics are artificial and even arbitrary, but the authority also knew that without them neither the instructors nor the trainees would know what performance goal they should attempt *during training*. Nor would they know if they had achieved success. In the absence of a well-defined terminal learning objective, parallel parking training would become randomized across a wide range of time and distance, bounded only by human imagination. The phrase *for training purposes only* should become part of your hip pocket vocabulary, along with *task, condition, and standards*.

Once the key terminal learning objectives for the target performance space are defined, it is a much simpler task for a manufacturer to build a simulation for it, if the technology is cost-effective. This helps harness the power of market economics to your training goals. In order to minimize risk and cost, the training system designer will first help the competent authority move as much training load as possible out of "the real thing" and into high-fidelity simulation. Then, since high-fidelity simulation is not cheap, the designer begins attempting to move as much training load as possible out of high-fidelity simulation and into some other lower-cost media, such as benchtop models and interactive computer-based training.

This systematic method, in contrast to the traditional *ad hoc* method, will help you manage risk and cost while ensuring that the required performance is achieved consistently. As to the original charge ("You can't define the real performance requirement with a simulator"), you have sidestepped this logic conundrum by defining the *training* performance requirement in great detail, as close to the real performance requirement as possible. Where simulation helps demonstrate this performance, and/or reduces cost and risk, it has been included in the training system along with experience in "the real thing." And you now have a baseline from which to begin updating the entire training system as "the real thing" shifts and changes shape.

## 3.7 Cost Versus Value Added

One of the first questions facing you as a leader contemplating or advocating a simulation-rich training system is how to communicate the cost versus the value added. It is not enough to say that simulation and task analysis are good for training; how good, and how much investment is required to bring that good about need to be quantified as much as possible, so that people will embrace the effort. In a high-level tally, you might spell it out as in Table 3.3.

The arguments summarized in Table 3.3 have the disadvantage of being intuitive and difficult to measure. Measurable increases and decreases will be needed to justify investments of specific man-hours and funds. But for starters, this is not a bad summary; most people easily recognize that the patient is safer if the trainee first practices on a mannequin or some other simulation before attempting to handle their first real case (Figure 3.5). Likewise, they will probably see the intuitive linkage between increasing patient safety and cutting the litigation and settlement costs while improving public image. The "decrease in realism" in the cost column should be kept opposite the increase in patient safety, because while we all

**TABLE 3.3**   Conceptual tally of value added versus cost

| Value added | Cost |
| --- | --- |
| Increase in patient safety | Decrease in realism during training |
| Increase in spectrum of case experience | Increase in training system support cost |
| Increase in standardization | Increase in standardization |
| Decrease in operations cost | |
| Decrease in insurance cost | |
| Decrease in litigation cost | |
| Decrease in out-of-court settlement cost | |
| Increase in public image | |
| Increase in *esprit de corps* | |

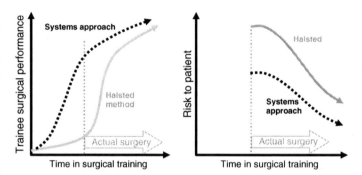

**FIGURE 3.5**   Notional performance and risk-to-patient curves for both the Halsted (traditional) method and the Systems Approach to Training. Note that the systems approach introduces simulation well before the start of actual surgery experience, and continues its use as appropriate afterward.

want the perfect training fidelity that comes from learning on live patients, the patient's best interest must be kept in mind at all times.

## 3.8 Operations Cost

What people may not see, and what you as the leader must continue to highlight, is the potential to reduce *overall* operations cost. Without an integrated training system, training load tends to migrate surreptitiously into the most costly and highest-risk arena – the "real thing." Training that currently occurs in the Operating Room, at hundreds of dollars per minute operating cost, can be transferred into simulation at tens of dollars per minute. Most hospital administrators are examining Operating Room costs, and some have begun to pressure surgeons whose operations take longer than average. But this relatively crude approach to cost reduction does not consider how *training* strongly influences time in the Operating Room. There is the direct time cost of training, as when one clinician takes extra time to show a trainee how to do something. But there is also the indirect cost, as when inefficiency, or avoidable mistakes that should have been caught and corrected during simulation turn short procedures in the Operating Room or Emergency Room (OR/ER) into much longer ones, or worse. How often this happens, and the resulting unplanned costs, are usually hidden from view.

Another indirect cost strongly influenced by training is the cost of turnover. The greater the turnover, the bigger the drain on your resources. Each new person added to the team requires some kind of training, even if it is distributed (and hidden from view) as informal on-the-job training. This training *is not free* – no matter how informal it is, someone has to take extra time to show the trainees how things are done. Throughout their orientation and on-the-job training, newcomers are not fully productive on their own, and they are a steady drain on those who are productive on their own. It might be argued that on-the-job training occurs mainly during "dead time" not used for productive work, but that argument is normally based on data-free analysis. And by the way, doesn't all that "dead time" cost something?

It is highly probable that the most valuable on-the-job training occurs when the instructor is doing productive work, slowed down on purpose to allow the trainee to absorb the knowledge; this is followed by the trainee attempting to do the same procedure under the eye of the instructor, again in slow motion, to allow the instructor to control the risk to the actual patient. But how much does all this "slowing down" cost the institution, if we could sum it all up? More importantly, how well are all the trainees performing? Can anyone describe their performance in objective terms? What on-the-job training have they completed and when? Is their training qualitatively equal? Are they fully productive? What is all this on-the-job training costing us, and can we reduce that cost? The honest answer in most institutions that depend largely on informal, *ad hoc* on-the-job training is *we don't know*. We know that we have intrinsic training costs, but we can't manage them because we can't measure them. The Systems Approach to Training will improve visibility into all of these issues.

One of the intangible benefits of replacing as much as possible of the *ad hoc*, probabilistic, and personality-driven training method with a world-class, curriculum-driven training system will appear as *esprit de corps*. Both the trainees and their instructors, seeing the tangible investment in training, sense that the institution really values their quality performance, and people tend to interpret that as *the institution leaders think I'm important*. This effect internally spurs them on to help you keep the training system strong, and improve it where possible. They will be reluctant to leave an institution where leaders invest so much in their people, and this reduction in turnover will be visible. This attitude percolates beyond the institution and into the community as an increasingly strong public image. Since world-class training systems focus on reducing human error, you can reasonably expect a decrease in the tangible costs associated with human error: insurance premiums, litigation, out-of-court settlements, and negative publicity. As a leader, you will need to keep reminding people about these costs and how a performance-driven training system will help reduce them while building up the morale of your work force.

## 3.9 Standardization: What is it, and who Wants it?

You probably noticed that "increase in standardization" appears in both the "value added" and "cost" columns in Table 3.3. This is not an error; it points out that standardization is viewed by some as value added, while the free spirits among us view it with disdain and possibly horror. Frequently, people associate standardization with regimentation or even a complete loss of creative freedom. Your challenge will be to assure them that an increase in standardization does not mean pervasive regimentation; we still expect instructors to tell their war/sea stories and to pass on their personal techniques and pet peeves. These personal touches have great value, but they do not replace the curriculum. In fact, a well-designed curriculum will make room in the schedule for storytelling and personal technique propagation. You're trying to *standardize what you can* so that it takes less time to build a core competency – this means the instructors will have *more time* for building higher-level skills like adapting and innovating.

What people may not have contemplated is how simulation allows a training system to provide both a wide spectrum of cases and a standardized set of cases and treatments, so that the graduate's performance is a known quantity. As we discussed earlier, the trainee will be trained to handle a representative set of problems (terminal learning objectives) that have been carefully selected to frame the target performance space. Since the *condition* part of the task-condition-standard formula is partly defined by the spectrum of patients, who are in turn described in terms of a large and complex array of variables, it may be appropriate to speak of the *patient space*.

## 3.10 Patients as Training Conditions

In aviation training, conditions described in the terminal learning objectives are frequently weather conditions – the highly variable and somewhat unpredictable challenge of operating in the atmospheric media. In the world of clinician training, the patients may provide the majority of this function all by themselves. Examining the patient is somewhat akin to looking out the windscreen at the weather you're about to fly through. One goal in a systems approach is to avoid artificially limiting the trainee's experience to the random distribution of real patients that appear during his or her clinical training, when a guaranteed distribution of simulated patients can provide uniform experience across a broader sample of the patient space.

As we pointed out earlier, this is "fixed" or "deterministic" training as opposed to the traditional *ad hoc* or "probabilistic" training. Trainees who are struggling can be made to repeat well-defined simulations until their performance is acceptable: they must meet objective performance standards in an objectively defined core competency. The actual patient cases then add real (but highly variable and unpredictable) conditions on which to apply the core competency developed during simulation. In a well-designed curriculum, trainees continually alternate between simulations and "the real thing" so that a powerful coupling effect emerges: real experience internally motivates the trainees to learn as much as they can from simulation, so that they can be confident and self-assured in the next real experience.

By way of example, military aviation seeks to build adaptation/innovation skills for a highly variable and unpredictable spectrum of future combat conditions by first establishing a fixed core competency in the target performance space. Then, the training system adds realistically ambiguous scenarios to help develop adaptation and innovation skills. Military aviators deploying to "the real thing" can then begin from a known starting point, the fixed core competency, and use their adaptation and innovation skills when they encounter their unpredictable adversary on an unfamiliar battlefield.

## 3.11 Equipment as Training Conditions

Another pesky component of the training condition, with its own annoyingly complex set of multiple variables, might be called the *equipment space*. This is the increasingly large array of pharmaceuticals and the hardware and software configurations that plague modern clinical equipment, all of which act together to create a large and growing training load. We're willing to bet you have equipment in your hospital that has untapped capability because of a lack of training. We'd also bet you have equipment in your hospital that no one knows how to use, except as a coat rack; someone used to know, but they forgot or left your institution. This untapped equipment capability costs the institution something. Even for the equipment your people are trained on, the equipment configurations change over time as parts and software "morph" (change) under your feet, and this only increases the already formidable training load. You might also include the training load due to the ever-increasing rate of introducing new pharmaceuticals and their interactions in this "equipment" category. Your training system must evolve with an evolving equipment space.

## 3.12 Increase in Training System Cost

This is where you will brazenly come out in the open with a tangible investment plan for your training system, to challenge the hidden training costs (and costly risks) of the traditional method. As we have seen, the goal of standardizing a core competency translates into a requirement for a process to define that competency (in terms of terminal learning objectives), and define it to a degree of granularity not encountered in the traditional probabilistic model. This new process, curriculum development, will require a tangible investment of time and expense. Once this detailed and comprehensive view of competency has been built, it will pay quantifiable dividends to keep it current as equipment, software, procedures, and techniques continue to change.

These gains come only with some investment of time, energy, and funds; because this cost element is fuzzy, most people assume that the investment will be more than the reduction in operation costs. This is especially true if the actual operating costs are unknown or are obscured by the random on-the-job training, the convoluted billing processes of the typically Byzantine clinical bureaucracy, and the human tendency to hide the true cost of human error. Until an algebraic cost/return model can be built that can project the return on investment, it may suffice to point out that, intuitively, training performed in the OR/ER costs in the hundreds of dollars per minute, while simulation training costs in the tens of dollars per minute. Of course, some training must be accomplished with the real thing, in the highest-cost, highest-risk arena – "on the floor" and on real patients. But extensive and systematic use of simulation means that this live training can be focused on final polishing, because the basic skills and supporting knowledge have been learned and demonstrated in the lower-cost, zero-risk simulations.

The instructor talent (and cost) can be distributed along the training system according to the performance level required; early (basic) training may be performed by general instructors, saving the more specialized instructors for the final, refined training. Closely connected to this concept is the need to train the instructors, especially since the media and methods of simulation-based teaching is significantly different from the media and methods of actual patient-based teaching. The instruction methods traditionally employed to train in the OR/ER will not leverage all the capability of the simulation; simulation instruction has its own skill set. In fact, the desired results of all teaching – the performance gains in the trainee – are strongly constrained by the instructing performance of the instructor. The strong and well-deserved urge of the instructor to protect the live patient from the less-than-competent student is a huge handicap when teaching with simulation. So also is the tendency to mimic the pedagogical instruction techniques from one's past experiences in the traditional method, when the adult learning model is much more effective.

The simulation instructor skill set should therefore be defined with its own terminal learning objectives and trained in its own unique training program.

The cost of change is another element that is usually disguised by Byzantine billing processes. New equipment, new pharmaceuticals, new software and protocols, and new techniques have a way of appearing suddenly and with little training support, and this translates into inefficient use of the highest priced/greatest risk clinical care environment. Estimates range widely, but medical knowledge probably doubles in less than 10 years; this means that demand for skills is also doubling. With proper support processes, simulation can help keep skills up-to-date despite rapid changes in the real world. This leads us to the need to categorize and manage training system support costs.

## 3.13 You as the Leader-Manager

Once your pilot project is beginning to take shape, it will be important to develop the supporting processes and manage them well. This may include (in no particular order):

- Methodically expanding the curriculum development beyond the pilot project.
- Budgeting for expansion beyond the pilot project.
- Procuring courseware specified by the Systems Approach to Training curriculum.
- Procuring training devices (mannequins, benchtop simulators, etc.) specified by the curriculum.
- Building increased use of simulation into future training plans.
- Building a real estate plan – rooms and floor space – to support the training system.
- Developing training for simulation instructors.
- Developing training for terminal learning objectives that require teamwork.
- Developing feedback mechanisms for students and instructors.
- Developing processes to analyze aggregated trainee performance metrics.
- Building a financial model to determine the Return On Investment (ROI) and refining it as you go.
- Developing a system to track clinical equipment configurations and who has been trained to use them.
- Developing change management processes to keep your training system current.

- Soliciting feedback from the field, where your training system graduates end up.

This list represents a sizable effort that will involve some cost to sustain. But all these processes contribute to helping you manage a world-class training system, and will help you understand, quantify, and manage the cost of training for your institution. In the process, you will be materially reducing risk and its costs.

But if all this is too daunting, feel free to call in some training system experts.

## 3.14 Conclusion

Transitioning from the *ad hoc* apprentice training model to a true performance-driven training system is not just a management change – it is a cultural change. That requires leadership and persistence on your part. It requires a good blueprint, a curriculum built on defined training targets. It requires your best communication skill and determination. Expect people to react to culture change; more importantly, expect to win them over, one at a time, with your persistent focus on the value added. And when all the risks and costs are added up, the bottom line is patient safety. It's the right thing to do.

## Endnotes

1. Robert Cox is the CEO of Aviation Training Consulting, LLC (ATC), and Lance Acree is the Senior Executive Consultant. The company provides strategic and operational consulting service and curriculum development for aviation and medical clients. ATC corporate headquarters is located in Altus Oklahoma; the mailing address is: P.O. Box 754, Altus, OK 73522. Phone: (580) 477-1767; Fax: (580) 477-1886; www.atc-hq.com.
2. William Stewart Halsted (1852–1922), usually credited with starting the first formal surgical apprenticeship training program in the United States, as well as initiating the American tradition of requiring junior doctors to stay on the wards around the clock. Hence the term "resident."
3. You may want to obtain a copy of the *Virtual Patient Research Roadmap* from www.FAS.org. This summary is a superb reference document; it cites numerous sources and studies over the years that have examined the value of simulation.

# II

# What's In It For Me

*Progress always involves risk; you can't steal second base and keep your foot on first base.*

Frederick Wilcox

Not until we believe that rewards for us might outweigh risks to us will we engage ourselves and employ our resources toward a new end. Clinical education has three primary job categories: those that provide the face-to-face clinical labor alongside the students, those that provide supporting educational labor, and those who provide program management. The many tasks required in clinical simulation fall under at least one of three roles: the foreground clinical instructor, the background clinical professional, and the academic administrator.

The role of the foreground clinical instructor is the most easily imagined, but the one most often miscast, if not just misperformed of the three. The students must accept the legitimacy of each and every person that they encounter during each and every one of their simulation sessions. Content competency is a necessary but insufficient requirement for these foreground instructors; just as knowing the dialog is a necessary but insufficient requirement for any stage performer. Like any educator, they must not only be competent in conveying ideas, abilities, and attitudes to their students, but also strive to improve how they do so in using simulation. Yet, unlike during real patient clinical teaching, when using simulation, they must redirect their focus from that of their patients' safety to that of their students' betterment. This essential point is often not obvious, even to those who already grasp that using simulation may improve patient safety. For the near future, most foreground instructors will follow the traditional path of developing their teaching using real patients, and only later ever doing so with simulated ones. Many of the learned habits conducive to good patient care and safety actually hinder good teaching. Such habits are hard to change. Lorena Beeman describes the transformation of a competent patient's bedside clinical instructor into an accomplished simulator foreground instructor.

The role of the background instructor is the most unfamiliar of the three, thus is the most difficult to hire or develop with confidence. As manufactured simulators first appear as just another gadget and simulation facilities' technological budgets are noticeable, this role is too often given the job title of simulation technician; yet this title is beyond inadequate, it is incorrect. Everyone directly involved in any clinical simulation production is first and foremost a teacher, a professional occupation. Every thought and action they take should clearly support the students' learning. Thus, a better job title is "Simulation Professional." One necessary characteristic of any successful background instructor is the desire and ability to create and use new tools and new methods. A second necessary characteristic is zero desire to actually perform patient care – clinically trained individuals can become background instructors, but they must forego all responsibility for live patient care, because producing simulated patient care is a full-time activity. Despite their commonality of purpose, the contributions of the background and foreground instructors complement each other. Like our right and left feet, they share the load for the common good, but are neither interchangeable nor individually sufficient for optimal performance. To further this analogy, while we are familiar with stating a given foot is preferred when performing a movement like striking a ball, the other foot is engaged in an equally important task of sustaining a nonmoving link to the ground. Both feet moving to strike a ball at the same time is almost as ineffective as both feet immobile at the same time.

During scenario development, the background instructor is focused upon the "*how* to present" while the foreground instructor is focused upon the "*what* to present." During scenario presentation, the background instructor is focused upon the proper execution of everything under their influence, while the foreground instructor is focused upon the clinical performance of the student. Few enough individuals are masters of either domain; very few are masters of both. Assigning both full-time responsibilities to one single person or several who are all either foreground or all background instructors has never been documented as a recipe for success. Guillaume Alinier shares his adventure in first creating a successful career for himself as a scientist and instructor in a well-defined basic science subject and then taking on the challenge of changing himself into a gifted simulation professional. His successes in his current endeavor are well supported by the professional level of his previous creative accomplishments.

There is a notion that the definition of a good manager is one who can mange any activity in any organization without actually being a subject matter expert in that activity, they just need to know how to manage. While this may be a laudable concept, all good managers first seem to have become experts in managing themselves and then experts in the subject of their organizations' primary activity long before they became expert managers of others. Simulation is very disruptive to clinical education, and most organizations are purposely designed to minimize disruptions. Also, there is little or no "spare" time, people, money, beliefs, or floor space in any clinical education program just waiting for employment. Simulation resources all have to be created out of purposeful enthusiasm or taken away from others. The good simulation manager will go far beyond mere acceptance of these fundamental conditions, making use of whatever resources are available to support the activities of the foreground and background simulation instructors. Alice Acker presents those challenges and their practical solutions common to managing a new clinical simulation program inserted into a well-established, traditional clinical academic ecology.

# 4

# Basing a Clinician's Career on Simulation: Development of a Critical Care Expert into a Clinical Simulation Expert

Lorena Beeman

## 4.1 New Path to Perennial Goal

At the University of New Mexico Health Sciences Center, Albuquerque, New Mexico, U.S.A., our Clinical Education and Human Simulation programs employ the philosophy of the educational theorist Patricia Benner and her phenomenological model of Novice to Expert [1]. The rationale for this decision was multifold, as will be discussed.

This chapter will present how we created a critical care nurse clinician education curriculum utilizing human simulation technology for evaluation that addresses five levels of a professional clinician's career. Curriculum design will accompany this discussion, along with lessons learned and those still being learned. A functional overview of an education model will be presented so that the reader can better appreciate our curriculum design and the use of simulation for documenting the progression of a clinician's professional career.

Historically, in acute care hospitals, the majority of educational time is spent on introductory or core curricula that are mandatory for the area in which the clinician will be practicing. Traditionally, once this introductory phase is completed, educational offerings beyond this level are scarce. What offerings are presented are most commonly associated with establishing and maintaining clinician competency or current concepts in the area of practice.

At our teaching hospital, we decided to establish and develop educational offerings to complement this traditional, mainly lecture-based format, focusing on initiatives to improve patient outcomes, while at the same time providing learning opportunities for clinicians to expand their knowledge and experiential base.

## 4.2 Health Sciences Center Demographics

The Center is the only Level I trauma center in the state of New Mexico, with an approximate population of 1,819,046 (U.S. Census data 2000). The Center has just over 300 staffed beds at the present time. We have three adult critical care units, one pediatric critical care unit, and one newborn intensive care unit. (The adult units will be the focus of this discussion as the curriculum development and use of simulation to evaluate the clinical learner is adult-focused.) The medical-cardiovascular critical care unit has 15 beds and employs 50 nurses (full- and part-time). The neuroscience critical care unit has 10 beds and employs 30 nurses (full- and part-time). The trauma-surgical critical care unit has 18 beds and employs 36 nurses (full- and part-time).

The center is in the process of constructing a new pavilion that will house these critical care units. It is expected to be completed in 2007, and will take each of these critical care units to 24 beds. Obviously, the impact on staffing and educational needs, specifically simulation training, will be dramatic.

Additionally, we have five adult subacute care units. While all of these units have a predominant specific type of patient population focus, all must be able to take patient overflow for each other. One of these units focuses on the cardiovascular patient. It has 20 beds (all with telemetry monitoring), and a staffing goal of approximately 42 full-time nurses. The second unit focuses on the trauma-surgical patient. It has 20 beds as well, all with telemetry monitoring and also has a staffing goal of 42 full-time nurses.

The other three subacute care units were originally medical-surgical units. Owing to the high patient acuity and the concomitant need for telemetry beds, our hospital has gradually brought these units to the subacute level. One of the units focuses on renal patients. It has 20 beds, all of which are now hardwired with monitors. It is staffed with 34 employees (full- and part-time) at present. The fourth focuses on oncology and clinical research inpatients. It has 16 beds, six of which have telemetry monitoring. The unit is staffed for 25 full-time nurses. The fifth subacute care unit is the smallest, and focuses on neuroscience patients. It has six beds (increasing to 10) with telemetry monitoring collectively housed as an old-style ward, within a larger 27-bed unit. All nurses in that unit are expected to be able to staff the subacute care unit as needed.

## 4.3 Simulation Capabilities

The simulation and virtual experience opportunities at our center are referred to as the *B.A.T.C.A.V.E.* It is both a descriptive mnemonic, a physical location, as well as a cost center. The mnemonic stands for *Basic Advanced Trauma Computer Assisted Virtual Experience*, and is cofunded by our center and the University of New Mexico School of Medicine.

Our center has three high-fidelity patient simulators, two adults and one pediatric. At the present time, only one of our high-fidelity simulators is housed in a setting that has piped-in gases (nitrogen, carbon dioxide, and oxygen) and live wall suction (Figure 4.1). It has a control booth with a one-way viewing window, audiovisual capability allowing for recording (Figure 4.2), and audio/video reproduction of the room and its events into a separate breakout room (that can hold 10 clinical learners), or onto a large screen in the main, multipurpose room (Figure 4.3), with space for up to 30 clinical learners.

The other two high-fidelity simulators are currently housed in separate rooms. However, they do not have piped-in gases, live suction, nor a control booth at this time. These simulators require their own air compressors, which are quite noisy (Figure 4.4 and 4.5). (Remodeling is expected to begin for these rooms in the near future, and that, hopefully, will permit the creation of control booths, piped-in gases, and live wall suction.)

We have three mid-fidelity patient simulators that are transportable. One is an adult, one is capable of simulating labor and delivery (Figure 4.6), and one is a neonate (Figure 4.7). We also have one mid-fidelity heart and breath sound simulator (Figure 4.8) that generates signals with the assistance of a computer-based model, and is interactive with multiple, simultaneous clinical learners. We have also recently purchased a high-fidelity cardiac auscultation software package (UMedic®) [2].

We have numerous low-fidelity mannequins that are utilized for skill development (as with intubation technique) or as environmental props in the setting of mid- or high-fidelity simulations for embedded learning opportunities. In addition, numerous partial-task simulators and their attendant clinical devices greatly extend the range of clinical skills experienced prior to first attempts on live patients. These devices are housed within four breakout rooms. These rooms have the space for up to 10 clinical learners each.

Located in separate sites, our center has several high-fidelity, virtual reality simulators. These simulators permit the clinical learner to develop bronchoscopy, endoscopy, colonoscopy, and surgical suturing techniques prior to these procedures being performed on a live patient.

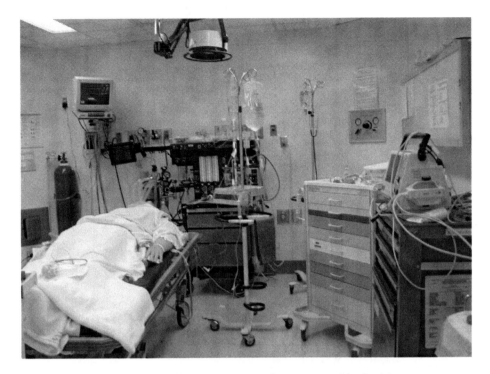

**FIGURE 4.1**   High-fidelity patient simulation room with piped-in gases.

**FIGURE 4.2**   Control booth for the high-fidelity patient simulator referenced in Figure 4.1.

**FIGURE 4.3**  Main room of the B.A.T.C.A.V.E.

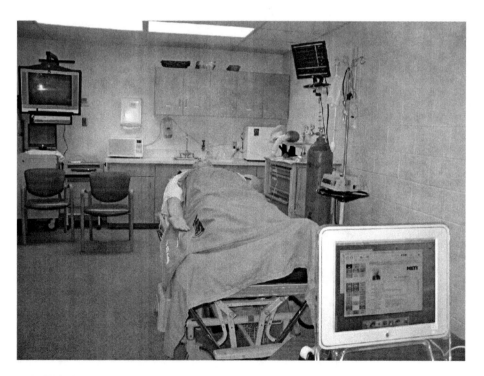

**FIGURE 4.4**  Second high-fidelity patient simulator (same as in Figure 4.1 but run by compressor).

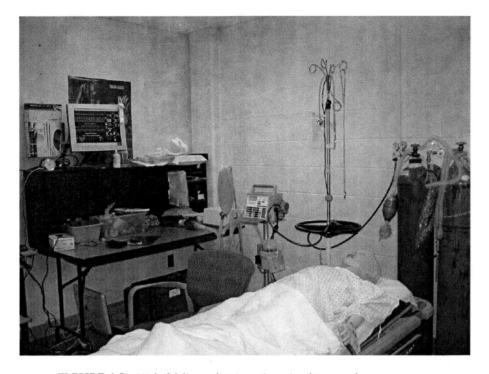

**FIGURE 4.5** High-fidelity pediatric patient simulator run by compressors.

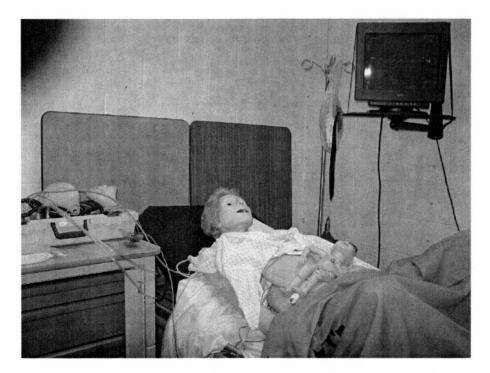

**FIGURE 4.6** Mid-fidelity patient simulator and laboratory for labor and delivery.

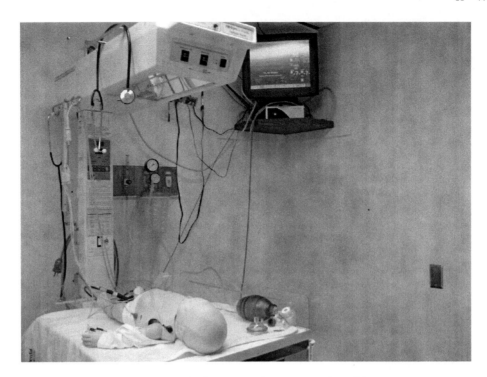

**FIGURE 4.7**   Mid-fidelity neonatal patient simulator laboratory.

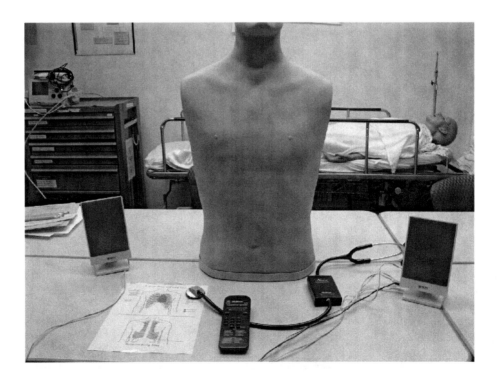

**FIGURE 4.8**   Mid-fidelity cardiac and breath sound auscultation simulator.

## 4.4 Topics of Study

Educators and clinical learners who benefit from this technology at our center are multidisciplinary in health care background, educational experience, and professional experience. Currently, we provide educational opportunities to students from the University of New Mexico Schools of Medicine, Nursing (undergraduate, graduate nursing including family and acute care nurse practitioners), Physician Assistant, and Pharmacy. Additionally, we provide educational opportunities to physical therapy, paramedic, and respiratory therapy students.

Educational opportunities involving simulation are also offered to medical residents, nurses, nurse technicians, respiratory therapists, pharmacists, and paramedics working within the health care sciences center, for the purposes of establishing and maintaining competency, and/or professional development.

### 4.4.1 Magnet Recognition®

Adoption of a new educational philosophy was also congruent with the center's drive toward its application to the Magnet Recognition Program® established by the American Nurses Credentialing Center (a subsidiary of the American Nurses Association) [3], and concomitantly serves as an excellent recruitment and retention tool. This is especially apropos in light of the current professional health care personnel shortage (The Magnet Recognition Program® is described in Table 4A.1).

### 4.4.2 The Reality

While all of this sounds quite appealing and laudatory in its goals, it quickly presented the clinical educators with numerous challenges:

- Using the Benner educational model at a practical level for the bedside clinician.
- Clarity with the use of terminology and appropriate definitions (e.g., Competent as a Benner level versus competency).
- Educating the educator in relation to constructing learning objectives and validation criteria appropriate for the identified Benner level.
- Educating the educator about evaluation roles necessary to achieve learner outcomes (mentor, coach, evaluator).
- Appropriate selection of the level of simulation and environmental fidelity.

The first four of these challenges are referent to the process involved with the adoption and application of the Benner model. The last challenge is referent to the space and fiscal constraints of the B.A.T.C.A.V.E. Because of these constraints, our center's focus is on "sufficient" fidelity to achieve the educational goals, and not "perfect" fidelity just for its own sake.

## 4.5 An Overview of Benner's Novice to Expert

In the early 1980s, Patricia Benner described a skill acquisition model for critical care nurses founded upon clinical experiences acquired over time in conjunction with development of critical thinking in relation to the patient and clinical setting. She described five levels that all clinical learners are expected to progress to over time:

- Novice
- Advanced Beginner
- Competent
- Proficient
- Clinical Expert

Distinctions between the levels are related to how the clinical learner applies six aspects of clinical judgment and skill acquisition (Table 4A.2). How the clinical learner demonstrates these aspects of clinical judgment and skill acquisition within a given level are referred to as *milestones* (Table 4A.3) [1, 4, 5].

The first aspect is performing *reasoning-in-transition*. This is the ability of the clinician to recognize and reason about changes in the patient's baseline physiological, psychological, and/or emotional status and act upon it (thinking in action). The second aspect is applying *skilled know-how*. Essentially, this is the capability of knowing what to do for the patient, and when to do it in the context of the immediate critical care environment (situational relevancy).

The third aspect is demonstrating *response-based practice*. It is one thing to read about clinical manifestations of pathophysiology or responses to medications and successfully complete a written examination, and another to take that knowledge and simultaneously apply it and respond to the patient within the immediacy of the situation at hand (crisis management).

The fourth aspect is presenting *agency*. This involves improving the clinician's engagement with the patient, acceptance of responsibility, and becoming a valued member of the health care team.

The fifth aspect is developing *perceptual acuity and skill of involvement*. In order to problem solve, the clinician must first identify the problem by defining and framing it. This requires perception, and perception requires engagement with the problem, the patient, and the critical care environment.

The sixth aspect is linking *ethical and clinical reasoning*. Good clinical judgments and clinical practice, and optimal patient outcomes must be considered in relation to what the patient or family views as such. It involves the balance between the ethical concepts of beneficence and nonmalfeasance during times when the patient and family are most distressed and vulnerable.

Thus, as one transitions from novice to expert, the clinician will change from a rule-based approach and concrete thinking to the application of abstract principles and evidence-based practice at the bedside. The clinician should transition from a reliance on policy and procedure to application of guidelines, utilizing clinical relevancy and situational thinking. The clinician should ultimately achieve the ability to intuit a situation fraught with multiple complexities, while the involvement of the clinician with the patient and environment of care will transition from detachment to full engagement.

Benner evaluated these six aspects of developing clinical judgment and skill development and their milestone application from Novice to Expert in the context of nine domains in the critical care environment (Table 4A.4). In other words, as trainees progress through each of the levels, their progress can be measured by how they apply the aspects of clinical judgment and skill acquisition relevant for the level (milestones) in the context of the nine domains.

Observation and narrative reports by clinical educators documented the clinical learner's progression from novice to expert. In other words, there is no specific or detailed "measurement scale" to document this progression.

## 4.6 Development of a Tiered Critical Care Education Program

Our center has in place a clinical advancement program for bedside nurse and respiratory therapy clinicians. The program has three levels (or tiers) that are tied into monetary rewards for achievement of excellence in professional development, involvement in research or clinical application of existing research (referred to as *evidence-based practice*), community service, and achievement of hospital- and unit-based goals. In keeping with this program, a three-tiered critical care clinical education program was developed that is based upon Benner's phenomenological model (Table 4A.5) [6]. Tier I takes the clinical learner from Advanced Beginner (graduate nurse) to Competent. (The Novice level of Benner's model was omitted as this is most relevant to student nurses before they enter our institution.) Tier II takes the Competent clinical learner to the level of Proficient. Tier III focuses on the Proficient clinical learner and the progression to that of Clinical Expert. (Note: The milestones for each level remain the same as identified by Benner.)

Milestones at each of Benner's levels (reflecting expected clinical judgment and skill development appropriate for that level) and the domains of critical care drive the development of objectives and validation criteria. Simulation technology and environmental props are incorporated within each level to create the framework for evaluation of in-the-moment situational thinking. Skill development and clinical judgment can then be more effectively practiced, evaluated, and improved, rather than only at the live patient's bedside. Debriefing following each simulated experience joined nicely with Benner's use of narratives to further evaluate these components.

The following sections describe the tiers and how we put into practice Benner's model of Novice to Expert. Each will identify the following:

- Milestones for each level (clinical judgment and skill development).
- Relevant domains of critical care nursing (Table 4A.4).
- Objectives appropriate to the clinical learner's level.
- Validation criteria for the simulation.
- Case scenarios.
- Level of simulation and relevant environmental props.
- Evaluation methods employed.

## 4.7 Tier One: Advanced Beginner to Competent

The goal of the first tier is to facilitate the clinical learner's achievement of Competent. Milestones reflecting this achievement include the initial development and application of critical thinking skills, effectively presenting a report about the patient's status to a colleague or to a physician while the patient is seen as a person with a disease state or condition, not just a disease.

The clinical learner at the Advanced Beginner level is usually a recent nurse graduate or a nurse with no background in critical care or subacute care. At this level, the clinical learner should be focused upon learning about disease processes relevant to the area of practice and the expected progression of the disease (this is referred to as a *disease trajectory*). Assessment still reflects a checklist data collection approach rather than actual interpretation of the data. Usually, technology is viewed as more important than patient assessment, reflecting disengagement and distancing from the patient, family, and circumstances. However, the ability to recognize patient deterioration and the need to initiate a response is essential at this level in order to progress to the level of Competent.

To meet the needs of the Advanced Beginner and to achieve the goal of Competent, a Critical Thinking Approach© tool was developed (Table 4A.6). The tool was developed on the basis of this author's clinical and educational experiences of 24 years, and peer review and input. The tool, in this context, facilitates recognition of patient deterioration (Step 1), necessitating urgent or emergency intervention. The clinical learner is then guided to begin developing anticipatory thinking in relation to the development of complications (Step 2), recognition of clinically significant comorbidities (Step 3), and expected medications, diagnostics, or interventions (Step 4). The tool also fosters consideration for individualization of patient care and engagement with the patient and family (Step 5). The cards are laminated and given to each clinical learner to reference and apply during the simulations.

In 2004, the Essentials of Critical Care Orientation (ECCO)© program developed by the American Association of Critical-care Nurses (AACN) [7] was purchased and implemented. The program is computer-based and self-directed, and replaced the traditional approach of formal, lecture, and classroom-based instruction. The dilemma was how participants could be evaluated for application and demonstration of critical thinking related to the content.

The result was the development of Essentials of Critical Care Orientation Simulation Labs [8]. There are four 3-hour simulation labs that advanced beginners complete once as part of their unit-based, competency-based orientation. These are held monthly. The rationale for this frequency is due to the numbers of clinical learners enrolled into the Essentials of Critical Care Orientation© program, and the effectiveness of the simulation labs in achieving the objectives. On the basis of trial and error (those "lessons learned"), we have determined that the optimal number of clinical learners in the simulation laboratory is eight. Usually, two sequential scenarios are presented in each laboratory session. The group of eight can be subdivided into two groups of four. One group is actively involved in a simulation while the other group observes and subsequently participates in peer review during the debriefing session following each simulation. The groups then reverse the roles. The four Essentials of Critical Care Orientation Simulation Labs that were developed include cardiac, pulmonary, neuroscience, and multisystem case scenarios, and each will be detailed separately.

At Tier I (Advanced Beginner to Competent), cognitive objectives reflect knowledge and comprehension for the Advanced Beginner clinical learner, and comprehension and application for the Competent clinical learner. An overview of each simulation lab, the relevant critical care domains addressed in the lab, and the objectives for each are detailed in Tables 4A.7–4A.10. Validation criteria for each scenario are detailed in Tables 4A.11–4A.14. (The validation criteria provided in these tables were established through relevant literature review and peer review. They are currently undergoing further validity and reliability testing in preparation for future publication.)

During orientation to an actual critical care or subacute care unit, clinicians have either no exposure or limited exposure to crisis recognition and management. The simulated critical care units provide a safe and guided environment in which to identify and perform code roles effectively and safely. Also, the development of clinical judgment to recognize respiratory, hemodynamic, or neurological compromise is of equal or greater importance, and that is repeatedly enforced and discussed in the simulation sessions.

Another feature of the content of the Essentials of Critical Care Orientation Simulation Labs is the nurse's role in preventing "failure to rescue." Failure to rescue is defined as "the death of a patient with one of five life-threatening complications (pneumonia, shock or cardiac arrest, upper gastrointestinal bleeding, sepsis, and deep vein thrombosis) for which early

identification by nurses and nursing interventions can influence the risk of death" [9]. These are reflected in the developed case scenarios.

### 4.7.1 Cardiac Simulation Laboratory

One of the learning needs identified from the Essentials of Critical Care Orientation© program's cardiac content was success in cardiac auscultation. When the clinical learners were asked where to auscultate for the heart sounds S1 and S2, or for the extra heart sounds of S3 or S4 (and with which part of the stethoscope), few could correctly identify the auscultatory sites or demonstrate the correct use of their stethoscope. Fewer still could maintain that referent when patient breath sounds were also present. When asked the clinical significance of the extra sounds, few could correctly state "volume overload" or ventricular stiffness for S3 and S4 respectively, and none could correctly differentiate them (Tables 4A.7 and 4A.11).

For the cardiac auscultation component of the cardiac simulation sessions, a mid-fidelity system is used that is computer controlled and allows for multiple users (Life/form® Auscultation Trainer and Smartscope™ by Nasco [10]). This mid-fidelity heart sound simulator presents the clinical learner with distinguishable S3 and S4 sounds. Unfortunately, this permitted limited achievement of the objectives due to the inability to have both heart and breath sound changes being generated at the same time. The heart and breath sounds generated by our high- and mid-fidelity patient simulators have similar limitations, and further inhibited by the ambient sounds within the simulation laboratories (e.g., compressors).

Thus, the focus of this component of the cardiac simulation laboratory is now on "how to do it in the real world," focusing on intensity changes for evaluating the clinical significance of S1 and S2, and use of intensity changes or lack thereof with the respiratory cycle to help differentiate S3 from S4, and physiologic versus pathologic split S2. Using other expected clinical findings to help support the validity of the clinical learner's interpretation is also stressed, along with anticipated patient interventions. In the future, we hope to use UMedic® Essential Cardiac Auscultation System to more effectively achieve the objectives [2].

For the case scenarios, the cardiac simulation laboratory uses a high-fidelity simulator. The simulator is operated either by an educator, who has completed a Train-the-Trainer course, or a simulation technician. The room resembles an intensive care unit or subacute care patient room, complete with hardwire telemetry, oxygen, wall and air suction, and intravenous infusion pumps with the relevant intravenous drips for the patient scenarios attached.

A standardized hospital crash cart and defibrillator/transcutaneous pacer is available, along with the resuscitation forms for documentation. A "code" button as is found in patient rooms, is present on the wall, and there is a phone in the room to call and initiate the code team response (Figures 4.9 and 10).

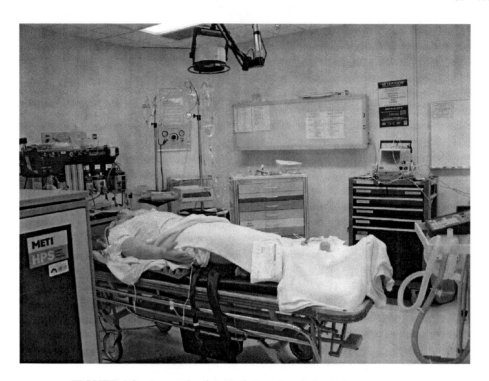

**FIGURE 4.9**   Essentials of Critical Care Simulation Laboratory setup.

**FIGURE 4.10**   Initiation and documentation of a code scenario. (Note: Code Blue button is to the left of the dry erase board.)

The case scenario beginning the simulation is presented to the clinical learners as a transfer or shift report, thus beginning the participants' engagement with the "patient" and circumstantial dynamics. All of the Essentials of Critical Care Orientation Simulation Labs have this basic room setup, and entry into the simulations.

Student-performance evaluation during these simulation sessions optimally requires two educators. One educator has the responsibility of teaching and coaching the clinical learners how to use the defibrillator safely. In preparation for actual live demonstration of defibrillation by each clinical learner, the educator orients each to the features of the device (monitor, defibrillation, cardioversion, and external pacing), electrode and lead placement, paddle versus pad placement, and the mechanics of safe defibrillation technique. Using a handheld battery-operated rhythm generator providing ECG signals for display on the defibrillator, each clinical learner practices recognizing pulseless ventricular tachycardia or ventricular fibrillation without delivering any shock energy.

Following this, each clinical learner is then coached through safe defibrillation with the paddles using "live" shock energy. This is then put into the context of a simulation involving cardiac arrest. The patient's clinical presentation in the peri-arrest session is interactively managed by the same educator, who is physically present in the lab, and actively changes the simulator's variables (e.g., cardiac rhythm, heart rate, blood pressure, and oxygen saturation) on the basis of the clinical learners' assessment and management of the "patient."

The second educator observes the clinical learners participating in code management for safety and achievement of objectives, while documenting the validation criteria, and intervenes as necessary to maintain safety. The participants' performance, based upon these validation criteria, are reviewed during debriefing. Each code "team" member is encouraged to describe his/her "in the moment" perceptions, and critique his/her performance. (This is what is referred to in the Benner model as the clinical learner's narrative understanding of the patient, the environment, interactions with fellow team members, and his/her rationale for action or inaction.)

For all of the simulation labs, the role of the educator during the scenario and during debriefing is that of role model, coach, and mentor. Successes are praised, and mistakes are used as another learning opportunity.

### 4.7.2 Pulmonary Simulation Laboratory

Breath sound auscultation and interpretation is accomplished in this laboratory by using the same mid-fidelity computer-driven program as that used in the cardiac simulation laboratory for heart sounds. Application of this portion of the pulmonary laboratory is incorporated into a pulmonary assessment during the patient scenarios (Tables 4A.8 and 4A.12) and (Figure 4.11).

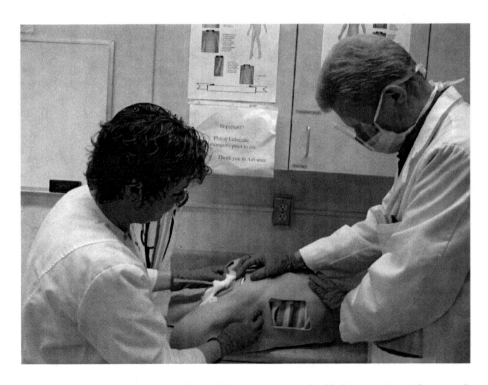

**FIGURE 4.11**  Pulmonary simulation laboratory using imbedded instruction with part-task trainer.

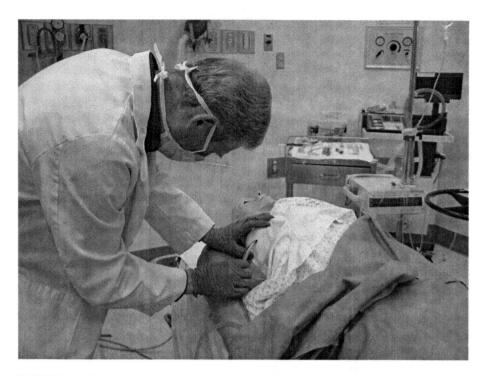

**FIGURE 4.12** Pulmonary simulation laboratory with demonstration of chest tube assessment.

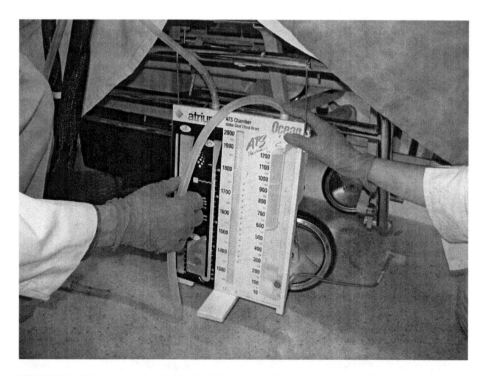

**FIGURE 4.13** Pulmonary simulation laboratory with demonstration of chest tube drainage system troubleshooting using high-fidelity patient simulator.

While the mid-fidelity human simulator has breath sound capabilities such as atelectasis, wheezing, and rhonchi, the high-fidelity simulator (when it is available) is used during the case scenarios, as not only can breath sounds be altered but chest tube drainage system malfunction can also be programed (Table 4A.15).

The second part of this laboratory involves setup and troubleshooting of a chest tube drainage system, which is intimidating to many clinical learners at the Advanced Beginner level. The teaching component of this section involves embedded learning. The educator has part-task trainers available to more visually demonstrate key points, including a chest tube, chest tube drainage system, and passive chest tube insertion mannequin (Life/Form® model 3220-102) to demonstrate chest tube insertion. (Figures 4.12 and 4.13)

The room environment has other additional equipment for this lab (Figure 4.14). While intubation is not the focus at this level of learning and experience, the participants are expected to demonstrate how to set up for intubation, perform cricoid pressure, time the intubation attempt, monitor vital signs while a clinician attempts intubation, and if successful, evaluate for correct placement. An expired $CO_2$ detection device is available for evaluation of correct endotracheal tube placement. If the intubation fails, a laryngeal mask airway is available for placement by the clinical learner. A part-task trainer is available at this point for anatomical demonstration of laryngeal mask airway insertion. The part-task trainer is a head-and-neck model that has a cutaway on one side so that the learner can visualize the placement of the tube (Ambu, Incorporated [11]).

Evaluation of bag-valve-mask technique is achieved by another part-task trainer. The clinical learner attaches the ambu bag to a passive mannequin (Laerdal [12]) that contains a graduated cylinder (made by BEL-ART) containing water (Figure 4.15).

Hyper-insufflation pressure is vividly apparent, based upon the water bubbling within the column. The clinical learner can then teach himself/herself how much to squeeze the bag to avoid the clinically deleterious hyper-insufflation pressures by watching for absence of bubbling in the water column. The device is also applicable for use with the laryngeal mask airway or endotracheal tube.

Each scenario has been scripted and preprogramed into the high-fidelity simulator to facilitate ease of presenting the scenarios. The chest tube drainage system malfunction is the most complex of the programed scenarios in this station, and the programing variables are identified in Table 4A.15.

**FIGURE 4.14** Imbedded instruction for preparation for placement of laryngeal mask airway and assistance with intubation.

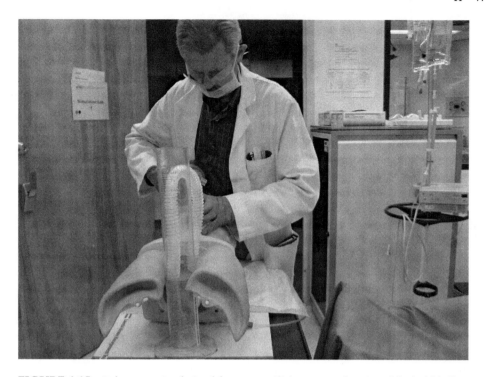

**FIGURE 4.15**   Pulmonary simulation laboratory utilizing part-task trainer "the bubbler" to learn proper bag-valve-mask technique.

### 4.7.3  Neuroscience Simulation Laboratory

This simulation exercise requires the presence of a reactive pupil. Therefore, a high-fidelity simulator is used for this lab (Figure 4.16). The scripted scenario for the laboratory involves a patient with head trauma requiring cervical spine precautions, and having a tibial fracture. The mannequin has trauma features that permit this realism (Tables 4A.9 and 4A.13).

While one might consider incorporating brain death, and request for organ donation into the head trauma scenario, clinical learners at the Advanced Beginner level are often insufficiently prepared on either a personal or professional level to deal with the associated emotional and moral responses that often accompany this situation. Thus, the discussion of brain death and organ donation, and the nurse's role in relation to each is conducted outside of the simulation environment via an interactive case study (Table 4A.13).

Use, setup, and assessment of intracranial pressure monitoring or ventriculostomy is the last component of this lab. This is ICU specific, and subacute clinical learners are permitted to leave at this point.

Only one of our high-fidelity simulation laboratories has the monitor capability to present intracranial pressure values. Unfortunately, as sold, the high-fidelity mannequin in this laboratory cannot realistically support demonstrating placement of the intracranial catheter for pressure monitoring or for ventriculostomy. Also, if this one room with this unique monitoring capability is in use for some other teaching, clinical learners must complete the laboratory in the neuroscience live patient intensive care unit.

### 4.7.4  Multisystem Simulation Laboratory

The multisystem simulation laboratory has the most detailed scripts from an educator's perspective, given the scenarios that are presented with their incumbent plethora of complications. High-fidelity human simulation is used. The goals of this laboratory are to compare and contrast hypovolemic and distributive shock; recognize early presentation signs and symptoms of severe sepsis in the elderly patient; recognize patient deterioration; and identify and demonstrate intervention priorities. Therefore, there are two simulation scenarios for this lab (Tables 4A.10 and 4A.14).

The quality of the environmental props is crucial to the success of the gastrointestinal bleeding and hypovolemic shock simulation. The case scenario involving gastrointestinal bleeding and hypovolemic shock requires the following props (Figure 4.17):

- Arm (or vein boards) for demonstration of placement of large bore intravenous catheters for fluid resuscitation and blood product administration.
- Intravenous infusion pumps for fluid resuscitation.
- Saline-labeled bag for fluid resuscitation and blood component administration.

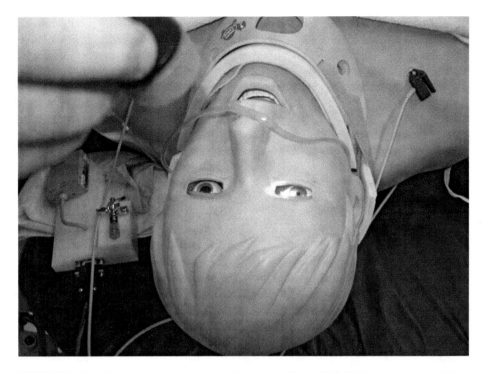

**FIGURE 4.16**  Neuroscience simulation laboratory utilizing high-fidelity patient simulation feature of unequal pupil size and reaction to light.

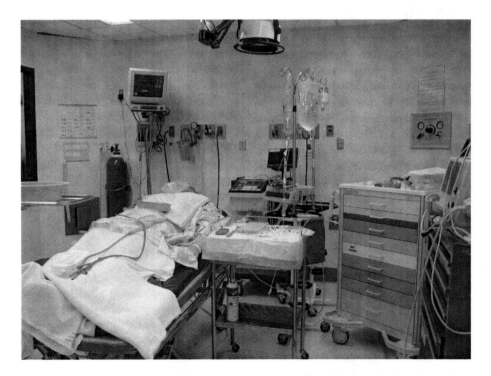

**FIGURE 4.17**  Some of the props utilized for the gastrointestinal bleeding scenario from the multisystem simulation laboratory.

- Blood component–labeled bag with real or faked blood (based upon availability).
- Patient identification band.
- Blood bank forms for demonstration of procedure for verifying blood type and crossmatch to the correct patient.
- Nasogastric tube.
- Iced "slush" and 60-ml syringe for gastric lavage.
- Suction canister containing faked coffee-ground stomach contents.
- Suction tubing connected to live wall suction.
- Tamponade tubes for esophageal varices.
- Crash cart.
- Rapid sequence intubation drug kit.

Following this simulation, there is a detailed debriefing and round robin discussion of the numerous critical thinking aspects required for this scenario. Thus, content is presented twice: once with the simulation and again with guided discussion, questions and answers, and interactive teaching (Table 4A.16).

In contrast, the simulation involving the elderly end-stage renal patient with severe sepsis requires minimal environmental props. This scenario requires a passive mannequin arm to demonstrate the presence of an arterio-venous fistula. The objectives of this simulation are reviewed through a scripted round robin discussion of salient points (Table 4A.17).

### 4.7.5 Incorporating the Family

The right of the family member to be present during procedures and during code resuscitations has been supported by numerous professional health care organizations [13–15].

Combined with the fact that at this level of learning, meeting family needs in addition to trying to meet the needs of the patient and keep on top of the dynamics in the critical care environment is inherently difficult, family member presence is now being built into the scripts for the scenarios. The purpose is to foster the sense that family presence is a normal and acceptable occurrence in today's acute care and critical care hospital environment, for more than visitation purposes.

### 4.7.6 Lessons Learned

The initial labs were striking in their ability to identify problematic areas for participants that required additional or alternative learning modalities, and for identifying mislearning from the actual computer Essentials of Critical Care Orientation© program content. One example is the clinical learner who did not consider heart sounds an important part of an assessment and simply documented what previous clinicians had noted. Another example is the clinical learner who did not realize that one did not need to wait until complete cardiac or respiratory arrest to initiate a code. A third example

is the clinical learner who could differentiate types of abnormal breath sounds, but could not interpret them or identify patient interventions, or else interpreted them incorrectly to the potential for patient harm. Yet another example is the clinical learner who did not understand the concept of reevaluation of the patient following a change in the patient's condition, or following an intervention (such as administration of an opioid analgesic or benzodiazepine). Unfortunately, we have seen these examples repeatedly demonstrated in the labs. The labs therefore allow for timely correction of the mislearning before it is reinforced at the bedside, resulting in less than optimal patient outcomes.

Initially, when our hospital began requiring the simulation labs following completion of the Essentials of Critical Care Orientation© program, we expected rather small numbers of nurses completing the program at the same time, and thus the simulation sessions were scheduled quarterly. Enter the reality of the acute, tertiary care hospital, and the need for telemetry beds. Our hospital, over the past 12 months, has seen three medical-surgical units transition to subacute care status. These nurses, while experienced in medical-surgical patients and environment, were advanced beginners with the subacute care environment. Nurses from each of the three units were enrolled into the Essentials of Critical Care Orientation© program, and concomitantly the simulation labs.

The Essentials of Critical Care Orientation Simulation Labs at our hospital are done in partnership with another city hospital who has purchased the Essentials of Critical Care Orientation© program. They too are experiencing the transition of their medical-surgical units to subacute care status.

It became rapidly apparent that the frequency of the simulation sessions was insufficient to the need and demand for the labs to continue. Hence, we are now producing these sessions approximately every 4–6 weeks, which is a significant time commitment by the educators and by the B.A.T.C.A.V.E.'s simulation technicians. (At the time of this submission, 114 clinical learners have completed the E.C.C.O.© Sim Labs, involving 264 hours of simulation time.) We expect the need for this higher rate to continue due to the high frequency of staff turnover within these units, and the anticipated increase in the number of critical care beds and staff when our critical care pavilion opens in 2007.

### 4.7.7 Lessons Learned by the Clinical Educators

A common theme in these four labs is the recognition of patient deterioration and rapid, prioritized interventions. A common clinical finding when the patient's condition deteriorates is a change in skin color and/or temperature. What this leads to is the necessity of careful orientation of the clinical learners to the simulation environment, and what the mannequin can and cannot simulate for assessment findings, and concomitantly, what part of the assessment the educator must be asked to provide. For example, in the neuroscience simulation lab, the clinical learner can assess pupillary size and response to light, but when performing motor assessment, the educator must be asked the response to grip strength and equality.

While the presence of two educators for each laboratory is optimal and recommended, circumstances may not make this achievable. If only one educator is present to produce the session and to debrief, only one simulation can be completed within the timeframe. For example, in the multisystem simulation lab, the environmental preparation for the gastrointestinal bleeding and hypovolemic shock scenario is quite detailed. If only one educator is present to do this lab, there may not be enough time after presenting one scenario to remove the equipment, debrief, and set up for the next one.

Logically, a scenario should be written so that different responses can unfold depending upon the clinical learners' assessment, interventions, and evaluation of the patient's response (or lack thereof). However, this requires extensive time and skill with the simulator on the part of the educator for mastery of these complex tools. As of yet, not all of the educators have achieved the level of "Clinical Expert" in the area of effective utilization of simulation. Thus, our educators are progressing through Benner's levels of Novice to Expert!

A final lesson learned is that there must always be a backup plan in the event that the mid-fidelity or high-fidelity simulator fails. When this happens during a simulation, the time spent attempting to troubleshoot the problem significantly impairs not only the quality of the simulation, but also the interactions between the clinical learners and the mannequin, and also between the clinical learners and the educator. Our recommendation is to not interrupt the scenario with troubleshooting, if the solution is not immediately apparent.

Rather, be prepared to reorient the clinical learners to the new environment, and redirect assessment questions to the educator. (When this occurs, we try to reinforce with the clinical learners that unexpected changes occur in the "real world" of patient care, and the clinical learner's responses to these changes is yet another learning experience.)

### 4.7.8 Evolving Uses for the Essentials of Critical Care Orientation Simulation Labs

The success and concept of the Essentials of Critical Care Orientation Simulation Labs has also engendered other uses, specifically that of an "educational prescription", which is applied in two differing contexts. The first context involves the clinician entering an ICU or subacute care unit with prior experience. Formerly, a "challenge exam" was given to the clinician, and managerial and education decisions were made on the basis of the score.

Before the Essentials of Critical Care Orientation© program, similar content in our city was provided in eight 6-hour sessions, 2 days per week, as part of a citywide hospital consortium. The challenge exam was developed for this. It was peer-reviewed for readability, accuracy, and reliability. The exam was annually reviewed and questions evaluated for quality by master's-prepared clinicians with test-writing background.

In keeping with the philosophy of advocating and facilitating the development of professional growth, the exam is now used as an educational prescription to identify courses and simulation labs that would serve to provide knowledge, comprehension, and application of the identified learning needs.

The second context involves the clinician who has been identified as having a performance deficit at the time of professional evaluation. The Essentials of Critical Care Orientation Simulation Labs are being used as part of an educational prescription (as relevant to the performance deficit identified). Ultimate evaluation is completed through the development of a targeted "mega simulation" (based upon previously identified problem areas in the performance evaluation). The "mega simulation" involves high-fidelity simulation, an embedded instructor (serving as the clinician's charge nurse or preceptor, for example), video recording (with informed consent obtained), and debriefing.

In this manner, all aspects of the clinician's performance evaluation can be addressed: cognitive domain, psychomotor demonstration, and affective domain (via narrative understanding).

## 4.8 Tier Two: Competent to Proficient

The goal of this tier is to facilitate progression of the Competent clinical learner (staff nurse, acute care nurse practitioner, or respiratory therapist) in relation to clinical judgment and skill development to that of Proficient. Milestones that reflect achievement of the Proficient level include the ability to modify clinical practice by incorporating relevant clinical research into practice (referred to as *evidence-based practice*). The clinical learner is able to focus on not only the "moment" of a situation but also comprehend the contextual variables in play at that moment (content and context).

Ethical decision-making gains relevance at this level along with a growing acceptance of the clinical learner's role as a patient advocate. Professional interaction and collaboration with other disciplines becomes easier as the clinical learner's confidence in his/her clinical judgment and skills develops with each clinical experience.

When constructing content for education at this tier, cognitive objectives should reflect comprehension, application, and analysis. Case scenarios that are developed should involve opportunity for the clinical learner to react to not only the content contained within the scenario but also to the context so as to validate achievement of the Proficient level.

An example of a critical care educational offering written to this tier is that of *Comprehensive Assessment of the Mechanically Ventilated Patient.* Our hospital has three adult critical care units (medical, cardiovascular surgical; trauma-surgical; and neuroscience). The number of patients requiring mechanical ventilation in these units is high, and multidisciplinary cooperation is fundamental to successful weaning from mechanical ventilation, prevention of complications, and optimal patient outcomes.

The ability of the clinical learner to apply critical thinking to the assessment of a mechanically ventilated patient is pivotal to both the prevention of complications and early identification of changes in the patient's condition. However, there is considerable variance among nurses and respiratory therapists when it comes to identifying components of an assessment and correlating it with the phases of respiration.

So as to better address this variance, an assessment algorithm was developed to help standardize how a patient requiring mechanical ventilation is assessed (Tables 4A.18a and b). The Comprehensive Assessment of the Mechanically Ventilated Patient© was developed by this author. Components of the algorithm were developed on the basis of clinical experience, mentoring experience, educational experience, and feedback from clinical learners enrolled in mechanical ventilation courses. The algorithm was also developed from this author's personal experience with being mechanically ventilated.

The algorithm serves as a guideline for assessment of each phase of respiration: ventilation, distribution, perfusion, diffusion, and cellular respiration. Each phase also identifies relevant pathophysiological considerations (respiratory muscle fatigue, oxidative stress, and differentiation of Type I and II acute respiratory failure), and ventilator assessment components. The algorithm is laminated and trifolded, and given to each clinical learner enrolled in a mechanical ventilator course.

To execute this algorithm, a classroom-based, interactive 4-hour course was created that explains each component of the algorithm, presents relevant pathophysiology in detail, discusses the clinical implications for the patient, and identifies ventilator management and treatment modalities for the patient. The content of the course and application of the algorithm are then evaluated using scenarios developed for high-fidelity simulation.

A Puritan-Bennett 7200 mechanical ventilator with graphics capability is attached to a high-fidelity mannequin. The two programmed scenarios involve a patient with acute severe asthma, and acute respiratory distress syndrome, so as to compare and contrast the differences in pulmonary resistance and compliance, and how these are manifested by the ventilator graphics. Each case also has pertinent psychosocial components with which the clinician must cope. The case scenarios and programing variables for each are detailed in Tables 4A.19 and 4A.20. The relevant domains and cognitive objectives are outlined in Table 4A.21, and the validation criteria used for the simulation scenarios are found in Table 4A.22.

The clinical learner is encouraged to use the provided algorithm to assist with patient assessment, identification of pathophysiology, recognition of or prevention of complications (such as auto or intrinsic positive end-expiratory pressure with the acute severe asthma scenario, or increasing pulmonary pressures and concomitant deterioration in ventilatory tidal volume), and relevant interventions and ventilator management.

Evaluation of achievement of the level of Proficient for this course involves an embedded educator within the scenario. This embedded educator is either a respiratory therapist or a clinical expert from one of the adult critical care areas. The role of this educator is to observe and when necessary, foster troubleshooting, decision-making, and interventions by the clinical learner, or correct mislearning expeditiously.

### 4.8.1  Lessons Learned

Both respiratory therapists and staff nurses take this course, which is also one component of a clinical excellence trajectory called Respiratory Advanced Life Support.

The algorithm and its application is included in annual critical care competencies for the three critical care units. The algorithm is included in each unit's competency-based orientation, and is consequently incorporated into clinical practice by the unit preceptors for their orientees. It is also part of the Introduction to Care of the Mechanically Ventilated Patient course to initiate the critical thinking process.

Multiple venues that reinforce comprehensive assessment correlates with better patient outcomes, prevention of complications, or early recognition of compromise – the essential first step in all timely interventions. Interviews involving narrative understanding have been utilized to demonstrate the temporality and practical understanding of the algorithms' concepts, and its incorporation into patient care.

An example of this is a staff nurse who was working in nights in the medical-cardiovascular critical care unit. She had completed the course and simulations 4 weeks before. The patient for whom she was caring was a 72-year-old male with chronic obstructive pulmonary disease, suffering from an acute respiratory infection, for which intubation and mechanical ventilation was required. The staff nurse from the previous shift had not asked the respiratory therapist to view ventilator graphics other than the volume-pressure loop that was currently being displayed. This nurse asked the respiratory therapist to bring up the flow scalar on the ventilator graphic, and subsequently began evaluating it. Both she and the respiratory therapist noted gradual changes in the flow scalar (expiration terminating below the baseline, and the onset of inspiration below the baseline). At this point, the physician was notified and ventilator changes made to correct the worsening of preexisting auto-positive end-expiratory pressure (auto-PEEP). She concluded that she felt that she had helped to prevent the development of tension pneumothorax or hemodynamic consequences related to this condition.

Collectively to date, between the aforementioned venues, 229 clinicians have been introduced to the concepts and utilization of the algorithm, and involving 50 hours of simulation.

The immediate future endeavor is to develop a study that permits more formal demonstration of optimized patient outcomes and prevention of complications through the use of this standardized assessment tool.

## 4.9 Tier Three: Proficient to Expert

At the level of Clinical Expert, the clinical learner has an extensive experiential repertoire from which to draw. The Clinical Expert knows the patient, and can integrate the multiple and complex environmental variables, thereby honing in on the context and crux of the problem. The clinical learner at this level sees the patient as a person, and is strong in patient advocacy. Key milestones of the Clinical Expert reflect:

- Clinical grasp of the patient's complexities
- Resource and evidence-based practice
- Coaching and mentoring of others
- Clinical (as opposed to administrative) leadership

Course content at the Proficient to Clinical Expert tier focuses on incorporation of evidence-based practice whenever possible. Case studies are constructed that involve multiple complexities and the opportunity to demonstrate clinical leadership, mentorship, and patient advocacy. Objectives at this tier reflect analysis, synthesis, and evaluation. An example of clinical education and use of simulation at this level follows.

The goal of this tier is to elicit and reinforce clinical excellence. To enable this, four clinical trajectories were developed that focus on areas of critical care nursing practice common to all adult critical care areas. One program involves interdisciplinary collaboration between nursing and respiratory therapy. One program is broad-based in that anyone involved with cardiac telemetry can complete the trajectory (critical care, subacute care, telemetry monitoring technicians, electrocardiograph technicians, paramedics). Another program is also applicable to adult subacute care settings. The four clinical excellence trajectories that have been established are:

- Hemodynamic Clinical Excellence Trajectory
- Respiratory Advanced Life Support (RALeS)
- Cardiac Electrophysiology Clinical Excellence Trajectory
- Cardiac Pathophysiology Clinical Excellence Trajectory

The clinical excellence trajectories are a combination of preexisting courses and new courses. Only the first three require high-fidelity simulators to evaluate the demonstration of clinical excellence. Each trajectory has specified prerequisites and course content (Table 4A.23). Many of the courses within each trajectory had been offered before for continuing education hours, and are still offered in the traditional, classroom format.

However, bedside clinicians who select a trajectory must select from two tracts during the course of the year. (Of experiential note, we have seen that clinical learners who select a clinical excellence trajectory are usually more committed, experienced, and therefore closer to achievement of Clinical Expert, or they are clinicians who are already practicing at that level and wish to more formally validate this. For those clinicians who do not self-select for such improvement, the real challenge is in convincing them that they should want to do so, and to participate in these sessions, and to learn while doing so, and to actually change their current behaviors and employ their new learning.)

The courses within the tract are taken sequentially over several weeks, culminating in evaluation of clinical excellence. Clinical excellence may be maintained by either repeating the evaluation component every 2 years or by becoming an instructor for a course within the trajectory. Managers are encouraged to use the trajectories as part of the clinician's goals when doing annual evaluations.

The Cardiac Pathophysiology Clinical Excellence Trajectory will be used as an example of curriculum design at the Proficient to Clinical Expert levels. There are six courses within this trajectory that the clinical learner is expected to complete. Each course is 4 hours in length to facilitate ease of scheduling.

- Advanced Cardiac Assessment
- States of the Heart Part I: Atrial Fibrillation and Care of the Post-code Survivor
- States of the Heart Part II: Infective Endocarditis and Aortic Emergencies
- States of the Heart Part III: Care of the Acute Myocardial Infarction Patient and Heart Failure Syndrome
- Coronary Artery Disease Part I: Pathophysiology and Diagnostics
- Coronary Artery Disease Part II: Percutaneous Transluminal Coronary Interventions and Pharmacological Management

The evaluation component of this trajectory utilizes a blend of high-fidelity simulation and a part-task trainer. The part-task trainer used in the simulations is the high-fidelity heart sound simulator. Prior to the start of the scenario, the educator programs the heart sounds relevant to the scenario selected. Thus, when cardiac auscultation is performed as part of cardiac assessment, this simulator is used instead of the mannequin. The Benner domains and clinical objectives for the evaluation components are found in Table 4A.24. Objectives for the evaluation day are derived from the aforementioned, prerequisite course objectives.

Four simulations of cardiac pathophysiology have been developed, two of which are shared here. Case scenarios and high-fidelity programing variables for acute myocardial infarction with cardiogenic shock, and aortic dissection/rupture are detailed in Tables 4A.25 and 4A.27 respectively. Validation criteria for each are found in Tables 4A.26 and 4A.28. The other two programmed states are infective endocarditis and heart failure syndrome.

From these four, the learner selects two by random draw. The learner receives a report on the patient, and is assigned an "orientee" to mentor. The orientee is actually an embedded instructor. The rationale for this is that in order to evaluate clinical judgment, clinical leadership, mentoring, and coaching within the scenario, the learner must be given an environment where an ongoing narrative is natural. The presence of an orientee is an optimal venue to permit this to occur, as such narratives are commonplace when precepting. The learner signs a consent form prior to the evaluation component to permit video recording. Confidentiality is assured. The video recording is used during debriefing to yet again create an environment for narrative understanding. Should sufficient clinical expertise not be demonstrated (based upon inability to demonstrate 80% of the validation criteria), a plan is mutually developed between the learner and clinical educators. (Note: 80% was chosen on the basis of this percentage as being an established score for successful or unsuccessful testing within our facility.)

### 4.9.1  Lessons Learned

Simulation at many levels has become commonplace at our hospital, especially in the critical care and subacute care areas, emergency department, and air transport. It is incorporated into intermediate and advanced life-support courses. It has been routinely used in annual critical care competencies since 2000, and in subacute care competencies since 2001. Orientees to the critical care and subacute care areas are initiated to simulation through their exposure to the Essentials of Critical Care Orientation Simulation Labs.

However, use of simulation technology to establish the progression of a clinician's career from novice to expert is still in its infancy. The curriculum and evaluation using simulation are more robust than when they were initiated. Given that Essentials of Critical Care Orientation Simulation Labs and the clinical excellence trajectories can be tied to performance evaluation changes the dynamics of how learning opportunities were provided in the past. Thus, for this tier, the lessons are still to be learned. The ultimate goal will be to demonstrate optimized patient outcomes and patient safety correlated with the validation of clinical excellence by this means.

## 4.10  Conclusion

Success in using human simulation to promote the development of a clinician's professional career is only as robust and effective as the employment of an educational model that is accepted, understood, and utilized by all involved with clinical education and simulation technology. Additionally, all must share a philosophical belief in fostering clinical excellence.

Identifying an educational framework that identifies criteria for determining the level of the clinical learner, and concomitantly facilitates development of objectives, validation criteria, and evaluation methodology that are congruent with each other is fundamental to successful implementation of the curricula.

On the basis of this framework, the appropriate level of human simulation technology and the props required to optimize the simulation environment may be more easily identified. The next piece is outlining the presentation style of the simulation, such as use of imbedded instructors and part-task trainers, and how they interface with the progression of the simulated case scenario.

The final piece to the puzzle is to educate the educator regarding appropriate, respectful, and sensitive debriefing techniques that are also congruent with the level of the primary learner and the simulation objectives. To this extent, our hospital is creating formalized "Educating the Educator" training sessions to achieve the successful application of the this framework, and minimize variance.

The purpose of this section has been to provide the reader with one example of how simulation can be used to "foster" a clinician's career, and how it can be incorporated into a clinical advancement program and performance evaluation. The trick is not to "base" the clinician's career on simulation, but to use simulation in combination with other teaching and educational opportunities and modalities to foster and validate clinical excellence in patient care.

## References

1. Benner, P. *From Novice to Expert: Excellence and Power in Clinical Nursing Practice*. Menlo Park, California, Addison-Wesley, 1984.
2. UMedic® Essential Cardiac Auscultation System. University of Miami School of Medicine Center for Research in Medical Education. tcoster@crme.med.miami.edu.
3. http://www.nursingworld.org/ancc/magnet/benes.html (American Nurses Credentialing Center Magnet Recognition® Program).
4. Brykczynski, K. A. From novice to expert: excellence and power in clinical nursing practice. Tomey A. M. and Alligood M. R., (eds.), *Nursing Theorists and Their Work*, 5th ed. Mosby, Inc., St. Louis, 2002, pp. 165–85.
5. Benner, P., Hooper-Kyriakidis, P., and Stannard, D. *Clinical Wisdom and Interventions in Critical Care: A Thinking-in-Action Approach*. WB Saunders Co., Philadelphia, 1999.
6. Beeman, L. [Abstract] Development of a tiered (novice to expert) critical care education program. American Association of Critical-Care Nurses National Teaching Institute Program and Proceedings, 2003, p. 491.

7.  http://www.aacn.org/AACN/conteduc.nsf/vwdoc/Ecco Home (American Association of Critical-care Nurses Essentials of Critical Care Orientation© Program).

8.  Beeman, L. (Abstract) The use of human simulation technology to facilitate experiential learning in correlation with "E.C.C.O." American Association of Critical-Care Nurses National Teaching Institute Program and Proceedings, 2005, p. 210.

9.  Needleman, J., Buerhaus, P., Mattke, S., et al. Nurse staffing levels and quality of care in hospitals. New England J. Med. 346(22), 1715–1722 (2002).

10. www.enasco.com (Life/Form® Auscultation Trainer and Smartscope™, LF01142U).

11. www.ambu.com (Laryngeal Mask Airway Head, Ambu, Inc.).

12. www.laerdal.com.

13. http://www.aacn.org/AACN/practiceAlerts.nsf/vwdoc/ PracticeAlertMain (American Associate of Critical-Care Nurses Practice Alert: Family presence during CPR and invasive procedures, November 2004).

14. http://www.ena.org/about/position/PDFs/ (Emergency Nurses Association Position Statement: Family presence at the bedside during invasive procedures and cardiopulmonary resuscitation, April 1994; revised October 2005).

15. Hazinski, M. F., Chameides, L., Ellling, B., et al. (eds.). American Heart Association Guidelines for Cardiopulmonary Resuscitation and Emergency Cardiovascular Care. Circulation 2005; 112(24), IV-9, IV-181.

# Basing a Nonclinician's Career upon Simulation: The Personal Experience of a Physicist

Guillaume Alinier

## 5.1 From Nonclinician to Clinical Simulation Professional

In today's society, workers are very mobile and do not stay in the same company or even in the same field during their entire professional career. This is not only true for the general unskilled or unqualified workforce but also applies more frequently to trained graduates and highly skilled and specialized professionals. Nowadays, an employer sees a degree more as a learning passport than as a field-specific qualification. It is clearly expected that employers will look for specific attributes in prospective candidates, but it also appears that they believe in the transferability of skills of their new recruits. A qualification is seen as someone's ability to learn about a subject, hence it is expected that with the right attributes or aptitude and motivation, one can also acquire the relevant knowledge and skills required to adapt effectively to a different field.

This chapter is about the breed of people who start by selecting a nonclinician path and eventually work in clinical simulation. Their new role requires their grasping a wide range of clinical and educational concepts and skills that should already be second nature to their new colleagues who are trained as an educator, a clinician, or health care professional. There are many different opportunities and responsibilities in the domain of clinical simulation, and nonclinicians could occupy almost any one of them. This ranges from educationalist, communication or team dynamic expert, technician,

or technical manager, center coordinator, administrator, to operations manager. The background of potential simulation center employees will be an important asset in their new role as it could give them a different perspective, additional skills, and innovative ideas that complement those of their clinical colleagues.

## 5.2 Working in a Versatile Environment

Simulation is currently a very exciting field where today's veritable explosion of global simulation activity has come from decades of isolated pioneering efforts. Some of these simulation explorers started developing and testing methods, and technology eventually succeeded in making commercial training products, along with a generalized teaching approach that could be more widely available and affordable [1]. The triggers for this rapid growth have not only been recent technological developments in terms of personal computer power, but primarily clinical educators believing in the potential of simulation training approaches to help others effectively acquire life-saving skills in a safe and controlled environment.

As the appellation "simulation" indicates, everything can and should not be real to allow control of the scenario (patient and environment). The key is determining how much reality is needed, and how to produce it. Fortunately, the realism of most clinical items and settings can be very high, given that most are actual clinical equipment and clinical settings. However, quite often, because of the procedures to be carried out and for ethical, consent, and safety reasons, one of the principal simulated components has to be the patient. For scenarios including invasive procedures, part or the entire patient is substituted by a mannequin, and in the latter case, under computer direction. In other words, it is a machine or nonbiological system designed by a team of engineers to reproduce the physical aspects, as well as the pharmacological and biological behaviors of a human being. Although great discoveries have been made in medicine in general, creating a machine that simulates a whole human being is a particularly challenging task that might never be realized. Fortunately, education is characterized by the instructor illuminating a small fraction of all *reality* as a way to attract the students' attention toward one or two learning objects to the exclusion of all else. This intentional focus and isolation upon a very small segment of reality is the essential basis for simulation's success, since all that is needed to be created is that which is illuminated.

We often make the analogy between simulation in the aviation industry and in health care, with the aviation being ahead of health care by decades. Technically, this is not a very fair comparison as the simulated atmosphere (the patient equivalent) is readily available for testing and has no will of its own while the simulated aircraft (the clinical devices and tools equivalent) has well-defined components that do not change at the whim of frequent equipment purchasing decisions. Even if it is very costly, it is evident that building a device that will simulate the functions of an aircraft strictly obeying the laws of physics is less challenging than building a life-like patient. Socially, this gap is an indictment on the insufficient interest in creating data and using evidence to drive both clinical care and clinical education.

Today's flight simulators can be very realistic. Sitting in the cockpit, one could feel as though you were inside a real airplane as it moves through the air. The technology is now even available to the general public in fun parks as rides which can provoke adrenaline rushes. One key difference between flight and clinical simulation is the perceived return on financial investment. The business linkage between the payers for flight simulation (investment) is very close to the payers for flight (customers). In contrast, the payers for clinical education (investment) are very far from the payers for health care (insurance?) and the payers for the lack of health care (all of us). Thus, the amount of resources expended upon developing and implementing both the technology and the acceptance of flight simulation is orders of magnitude greater than that expended upon clinical simulation.

Even with unlimited finances today, we would not be able to build a patient simulator that really feels and looks like a real patient on which one could connect any type of medical monitoring equipment, provide invasive or noninvasive treatment

and expect the entire range of possible human responses. However, such a tour de force is totally unnecessary, since today's clinical care, to say nothing of clinical education, never sees nor treats the entirety of any one patient. Just as we have very large selection of different types of automobiles and trucks to meet differing needs, we will develop and employ a large selection of different clinical simulation devices, each optimized for particular uses.

A simulation center is a versatile environment in the sense that every aspect of a clinical simulation program is evolving. It needs to follow not only the developments in terms of medical practices across an ever-growing number of disciplines that adopt simulation, but also the development of the simulation technology itself. There is a permanent challenge to stay on top of it all, whether it is the adoption of new resuscitation protocols, hospital policies, new roles of health care practitioners, the release of revolutionary pieces of medical equipment, or even the social and political context. Scenarios and debriefing provided to participants constantly need to be improved just as their clinical competencies need to be improved. Because of recent natural and unnatural catastrophes such as floods, epidemics, and chemical, biological, radioactive, and nuclear terror threats, there has been unprecedented efforts in developing large-scale simulation training programs to review protocols and prepare emergency services and hospitals [2–5]. The consequences after the use of weapons of mass destruction, for example, on our populations would be even more catastrophic if our emergency services have not had the chance to prepare themselves through simulation exercises.

# 5.3 The Personal Experience of a Physicist Working in a Simulation Center

My interests and technical mindset have encouraged me to study Physics at University, both in France and Great Britain. During and after my Master studies in Applied Physics, I started working for two sister companies primarily specializing in underwater acoustics research and in the design and installation of fish deterrence systems. Although I thoroughly enjoyed the scientific research challenges and the technicality of the work, I felt the need to apply my skills and knowledge in a different area where it could directly benefit people. I did not want to carry on working in isolation or always with the same people (as nice as they are), but instead was in search of doing something where I could more directly see the effect of my efforts and meet more people. In 2000, I decided to look for a position in medical physics, but instead I found myself attracted by a university research job in biomedical engineering and clinical education. It consisted on coordinating projects toward the development of a low-cost interactive patient simulator for the training of preregistration nursing and paramedic students. This project was funded by the British Heart Foundation, with the goal of creating a low-cost patient simulator with which real medical pieces of monitoring equipment could be used (blood pressure cuffs, ECG monitors, pulse oxymeters), which at the time was not available in any low-price commercially available simulator. At first, this seemed an intriguing yet interesting project as I had never heard of patient simulators nor knew anything about the training of health care professionals.

Once engaged to take on this job, for 3 years I strived to come up to speed with the overall concepts and the current developments in the field of health care simulation. Very rapidly, the enthusiasm of the pioneers working in this area fully made sense to me. Learning about the patient simulator developments throughout the world, I realized a fair amount of work had already been carried out and employed in high-fidelity patient simulators. It was often overlapping with projects in which the department in which I was now working was engaged through the work of some of their biomedical engineering students. In parallel, I was made aware of the imminent arrival on the market of the intermediate-fidelity patient simulators with similar features. Not intending to reinvent the wheel, and coincidently with the student recruitment difficulties in biomedical engineering, our project's aims or objectives were significantly reduced. Within a couple of years, the University's biomedical engineering teaching program was stopped, which unfortunately almost put an end to our technical developments in patient simulation.

At the same time, I assumed responsibility for the development of the University of Hertfordshire's Intensive Care and Emergency Simulation Center concentrating primarily on delivering simulation-based training to nursing and paramedical students and evaluating the teaching effectiveness of this approach [6]. The center was initiated and inaugurated by my line managers in 1998, but was not very actively used as a simulation center as no one was in charge of the center on a day-to-day basis. Despite this shift in my focus, I believed that remaining active on the engineering development side of patient simulators was still important for me in order not to lose the skills I had previously acquired through my academic qualification and prior work experience. To that effect, alongside the day job in the Faculty of Health and Human Sciences, I have always tried to pursue my research efforts in technically challenging projects [7] to further enhance current patient simulators. To this end, I supervise the regular visits of physics research students from the French University where I graduated in 1997.

An interesting advantage as a technically minded person is to be able to understand how the patient simulator operates at the overall scale and at the subcomponent scale; that is, from the philosophy inherent to its intended use, to the computer interface to the microswitches and pneumatic valves inside the mannequin. It proves particularly useful to rapidly troubleshoot problems with the patient simulators or pieces of medical equipment and quickly identify solutions. There is also certainly a greater awareness in terms of the technology that can be used to further enhance the simulation center, and make it a better learning environment. This includes, for example, choosing appropriate audio and visual equipment to link the observation room or simply providing advice to colleagues regarding the best tool to use to teach a particular skill, concept, or attitude.

Education or pedagogy is a major aspect of anyone working in the control room or on the simulation platform of a simulation center. Very early on, after starting my research job in simulation, I started studying for a Postgraduate Certificate in Teaching and Learning in Higher Education. It has recently become a compulsory qualification for new UK University lecturers. I felt it could be another very valuable asset to become a faculty in a simulation center. This gave me some underpinning knowledge of the theories of learning, teaching, and assessment that I could relate to in my simulation teaching practice, especially during debriefing, or through the examination sessions organized as part of my research on the effectiveness of simulation training [8]. At the end of my 3-year research contract, I was offered a lectureship to carry on operating the Hertfordshire's Intensive Care and Emergency Simulation Center as the center coordinator and take on responsibilities for a broader range of activities such as producing short courses and taking on consultancy work. As the only permanent member of staff of the center, I have a very varied role including scheduling, cleaning and maintaining the patient simulators, controlling them during the simulation sessions, and training other faculty how to facilitate simulation sessions. The latter has even been validated by the University's Postgraduate Medical School as an optional module counting for one-sixth of a Master of Health and Medical Education [9]. This is probably one of the first "simulation faculty training course" attracting academic credits.

Training as a physicist, I had never thought I could ever be involved in teaching or in facilitating learning with such diverse groups of health care trainees and professionals, but simulation opened up these opportunities to me. Hence, I strongly believe that there is a place in simulation centers for physicists who want a career change. I find working in the simulation center very fulfilling and I would not discourage anyone with a nonclinical background trying their chance in health care simulation.

## 5.4 Advice to Engineers and Scientists Aspiring to Work in Health Care Simulation

Engineers and scientists have rarely been renowned for their outstanding communication skills, and I believe that it is an aspect of extreme importance in the field of health care simulation education. It is very relevant and applicable to different aspects of a job in a simulation center, whether it involves interaction with participants, colleagues, and especially the media. Whether you are a faculty or a technical manager, communication should always work both ways. If you are a faculty, you need to work closely with other faculty and actors to make the scenarios work and execute smoothly. Good communication skills will help you in this teamwork activity. Another important part of any simulation is the actual debriefing where, after having listened to the participants discussing their experience, you need to be able to appropriately debrief them. Good communication skills will be an asset to clearly and effectively transmit your teaching points and tips. Similarly, if you intend to become a technician or technical manager, you might have to brief fellow educators about particular features of your simulation room, the audio/visual system, or the patient simulator, and similarly, you will continuously take in their suggestions and requests to improve aspects of the environment or the simulator itself to further their teaching objectives.

I believe that poor communication skills could hinder your career progression in this field. Although one might think that if you have a technical role, you are not very exposed to coming into contact with participants because you can hide in the control room most of the time; accepting this limited role will doom you into never ever extending beyond being a technician. In reality, even if you only have to deal with small numbers of participants at a time, you will certainly be involved in their briefing about the simulation environment, the patient simulator, and maybe the debriefing of scenarios. Your interaction with participants can significantly contribute to the learning success of the simulation session. Alongside your colleagues, you need to be able to make participants feel at ease in this unfamiliar environment to them, and not being a good communicator will not help you or them in this aspect. Similarly, you can greatly contribute to the success of your center's simulation programs by effectively communicating with other educators to refine scenarios provided to your participants, develop the simulation environment and new props.

As an engineer or scientist working in a simulated health care environment places, you are in an ideal position to think "outside the box" and make new ones. It might help you seeing things from a different angle in comparison to your clinical colleagues. This is particularly true and useful when it comes to building simulation props and tricks. Not possessing the physiological knowledge might be an advantage and help you being more innovative and creative. At times, for the development of a simulation component, it is useful to only understand what someone is expecting to see from the outside rather than the full underlying physiological principles. During scenarios, you might have new ideas for future developments that will enhance the simulation experience of the participants. Hence, at any time, it is always useful to take notes before you forget.

Integration of new capabilities required for one new scenario often opens up opportunities for the creation of others. Eventually, exposure to a broader range of scenarios can be very enriching for participants, and helps prevent fatigue and boredom in the educators. Similarly, working collaboratively with clinical colleagues will be something to explore. You might be able to use your skills and technical knowledge to further develop their ideas and improve aspects of the simulation training experience of participants.

The reality of a nonclinician working in a simulation is that you have all chances to become an overall expert of simulation-based training as you will be involved in all aspects of operation of your center. Many people with an engineering background are currently employed by hospitals or universities as "simulation center coordinators" or even "simulation center managers." Taking on educational training might even enable you to become a member of the Faculty team. The more you want to participate in clinical education, the more you will be able to do so through simulation, but only to the extent that you are willing to learn how to be useful.

# 5.5 Conclusion

As we have seen through this chapter, the simulation arena is a versatile environment that offers interesting opportunities to nonclinicians. The brief overview of a physicist's journey in a University simulation center provides a concrete example. The advice provided to engineers and scientists is probably also valid to other professionals wishing to work in the field of health care simulation.

# 5.6 Favorite Problem Solvers

The Society in Europe for Simulation Applied to Medicine, SESAM
   http://www.sesam.ws
   Meeting other people in the field was an eye opener. Because it is Europe-based, there are many participants from the UK with the same problems and in the same situation (institutional/economical/social climate).
   The International Meeting on Simulation in Healthcare, IMSH
   http://ssih.org/
   Much broader in terms of audience and experiences, larger-scale centers, but also valuable things to learn from.
   A Simulation User Website with a lot of useful information:
   http://www.patientsimulation.co.uk
   An open forum of simulation users. Pages with tips and tricks to simply modify simulators, share scenarios. Thank you Neal for your great work at maintaining the pages up-to-date!

# References

1. Cooper, J. B. and Taqueti, V. R. A brief history of the development of mannequin simulators for clinical education and training, *Quality and Safety in Health Care* **13** (Suppl. 1), i11–i18 (2004).

2. Christie, P. M. J. and Levary, R. R. The Use of Simulation in Planning the Transportation of Patients to Hospitals Following a Disaster, *J. Med. Sys.* **22**(5), 289–300 (1998).

3. Abrahamson, S. D., Canzian, S., and Brunet, F. Using simulation for training and to change protocol during the outbreak of severe acute respiratory syndrome, *Crit. Care* **10**, R3 (2006).

4. Kyle, R. R., Via, D. K., Lowy, R. J., et al. A multidisciplinary approach to teach responses to weapons of mass destruction and terrorism using combined simulation modalities, *J. Clin. Anesth.* **16**, 152–158 (2004).

5. Vardi, A., Levin, I., Berkenstadt, H., et al. Simulation-based training of medical teams to manage chemical warfare casualties, *Israeli Medical Association Journal* **4**, 540–544 (2002).

6. Alinier, G., Hunt, B., Gordon, R., and Harwood, C. Effectiveness of intermediate-fidelity simulation training technology in undergraduate nursing education, *J. Adv. Nurs.* **54**(3), 359–369 (2006).

7. Alinier, G. Gordon R, Harwood C, and Hunt B. 12-Lead ECG training: The way forward, *Nurs. Educ. Today* **26**(1), 87–92 (2006).

8. Alinier, G. Nursing students' and lecturers' perspectives of OSCE incorporating simulation, *Nurs. Educ. Today* **23**(6), 419–426 (2003).

9. Alinier, G. Accredited University Module for clinical simulation facilitators at the University of Hertfordshire, *Simulation in Healthcare* **1**(3), 197 (2006).

# 6

# Overcoming Operational Challenges: An Administrator's Perspective

Alice L. Acker

## 6.1 Start-up Management

Managing start-up operations is challenging. Add very rapidly evolving technology like simulation to the equation and things get more interesting. Then, mix in a first-ever collaboration of 15 departments within the School of Medicine with the involvement of the School of Nursing, School of Engineering, Biorobotics Laboratory, and Human Interface Technology Laboratory. Finally, have the process overseen by the President of the University, the Dean of the Medical School, other senior leadership within the University, industry, and professional organizations. Imagine putting this project together with partial funding and space limitations, and you have an idea of the overriding challenges in establishing the Institute for Surgical and Interventional Simulation (Figures 6.1 and 6.2).

This chapter will provide some insight about specific operational obstacles and how to overcome them. It is very important to note that most people are very enthusiastic and supportive of simulation. This passion has enough energy of its own to push many aspects of the project forward. It is the job of the Administrator to harness this positive energy in order to meet operational goals and objectives, given the other real constraints that are present.

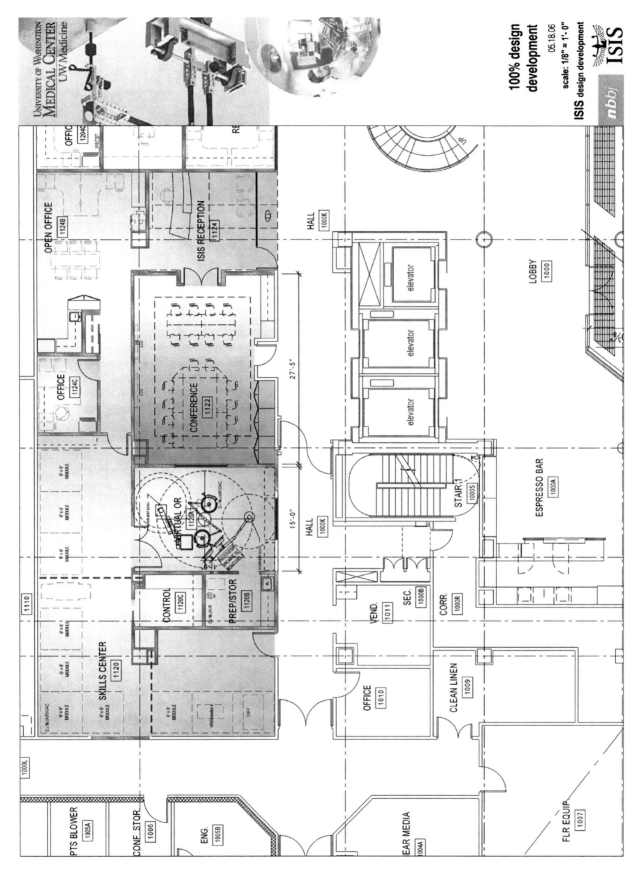

FIGURE 6.1a   Floor plan of our simulation center.

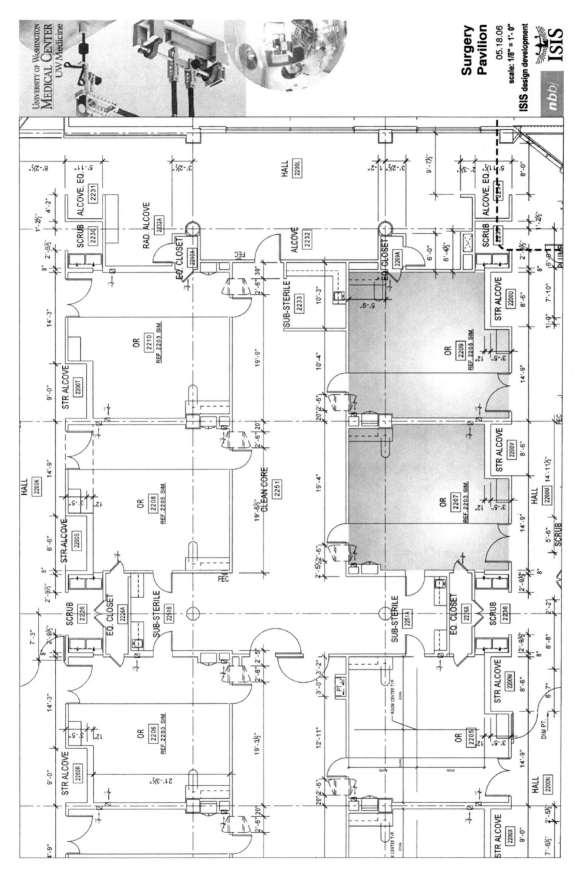

**FIGURE 6.1a** (Continued)

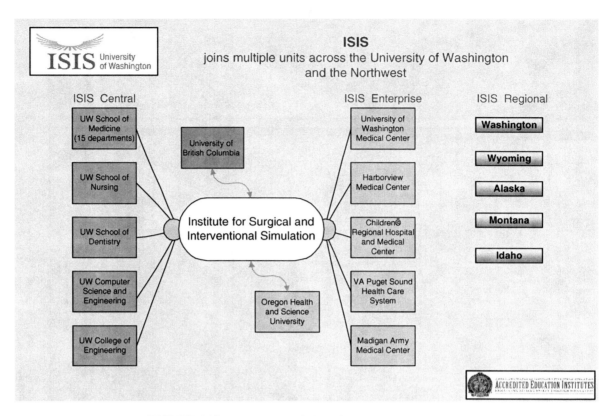

**FIGURE 6.1b**   Organizational plan of our simulation program.

**FIGURE 6.2**   An extract from the webcasts developed by the Community and Media Relations of The University of Washington, advertising our Simulation Center.

# 6.2 Challenge 1: Convincing Others within the Organization, including the People Holding the Purse-strings, that Simulation Is Here to Stay, and Is Worth the Investment of their Resources

Yes, most people are excited about simulation and are quick to draw parallels between flight simulation and clinical simulation. Let someone, layman or clinical professional, control the graspers in a laparoscopic cholecystectomy simulator and they will quickly act like "their" patient's life really depended on whether or not they can place the two clips side by side as close as possible to the gallbladder. However, how do you convince others who have not had this experience? These are some of the challenges.

## 6.2.1 Limited Validation Studies

There is not a lot of data, let alone rigorous and reproduced studies producing data *proving* that clinical simulation shifts behaviors toward laudable goals like improving patient safety. We have all heard the anecdotal evidence that flight simulation reduces aircrew errors. It makes sense, but the single greatest *proof* that flight simulation is *valued* by owners and operators of aircraft is that they do make extensive use of it. The main difference from most medical errors is that "The pilot is the first person at the accident!" and they have a vested interest in safety. But no proof exists that either the tools or techniques used in simulation-based education are valid. But then, all other forms of education, including whatever it was that you and I ever survived through for decades as students and now pass on as educators, was ever validated, either. Simulation is done by, for, and with people who like simulation. Those who don't, don't. And that's your real challenge; to convince enough people to at least try the green eggs and ham.

Simulation is a verb, active learning through interaction that when employed in all its forms can be the tool upon which we can build clinical curricula characterized by deliberate, intentional, scheduled, compounding, and escalating learning process. Chasing validation for its own sake is a distraction. First comes a robust curriculum to power the high expectations of our students. When reproducible results start coming out – that's about all the validation anyone is likely to see.

## 6.2.2 Some Forms of Simulation can be Expensive

You will undoubtedly hear that simulators are expensive (some robotic patient simulators can cost US$250,000 and the environments to make use of their expensive features can cost several times more); competent personnel cost money; space is at a premium; and excusing the Residents from earning income for the clinic while learning in the simulation laboratory within their 80 hours per week costs someone additional money.

## 6.2.3 All Forms of Ineffective Clinical Education and Clinical Care are Even More Expensive

The real question is not what is the cost of disrupting the *status quo* by adding simulation, but what other expenses are reduced to an even greater extent when simulation improves patient care? What are the current costs due to currently accepted but obviously poor practices? Other quotes that pertain to this antiquated concept of education include:

> "If you think education is expensive, try ignorance."
>
> Derek Bok

> "If you think clinical simulation is expensive, you volunteer yourself as your clinician's first-time patient."
>
> Richard Kyle

The claim by simulation proponents that simulation can have a positive impact on patient safety and patient outcomes is important to consider. When money moves, someone is accepting sufficient validity. An article entitled "Grand Rounds: Lessons from the cockpit: How team training can reduce errors on L&D (Labor and Delivery)" in *Contemporary OB/GYN* [1] describes how physicians who participated in teamwork training and completed online courses were offered a 10% reduction in malpractice insurance. In "Make Safety a Priority: Create and Maintain a Culture of Patient Safety" in *Healthcare Executive* [2], the authors emphasize the value of team training in reducing errors. They embrace both high-fidelity and low-fidelity simulators as effective tools to illustrate the components of good teamwork behavior. These are convincing examples in a clinically oriented journal and an administrator-oriented journal, which demonstrate how simulation can ultimately save money through its impact on improved safety and patient outcomes.

## 6.2.4 Simulation Is more than just the Mannequin

Not all simulation curricula need expensive high-fidelity mannequins to be effective. For example, one of our Internal Medicine faculty effectively uses a cardboard cafeteria tray and a baby bottle nipple to teach thoracentesis. First, know what you need to teach, then buy what will enhance your teaching, not the other way around. We recently had a colleague from our health care system ask us how to go about buying a daVinci robot for their pediatric simulation center. Our advice? "Don't." You can use the money to better gain by analyzing what you want to teach, seeing which of these curricula can be used across specialties, and then building what you need to accomplish your learning objectives. Don't confuse "glamorous" technology with technology that will be used to maximum learner benefit. Buying an expensive simulator that will rest and rust unused on a bed or in a dark closet does nothing to convince others that simulation is worth the investment. If you need a high-fidelity simulator to meet your learning objectives, make sure first that a faculty member assumes "ownership" of the simulator by ensuring that there are curricula and classes that will use it. While it is a wonderful crowd-pleaser to wheel out the high-tech devices for open houses and tours, that really does little for our learners than possibly bring an awareness of simulation as well as build the unrealistic public expectation that simulation is all about expensive high-fidelity devices.

### 6.2.5   How do you Prove that Simulation Is not a Passing Trend, Like Total Quality Management, Quality Circles, or Pet Rocks?

Indications that there is growing acceptance of medical simulation abound. The recent inauguration and subsequent explosive growth in The Society for Simulation in Healthcare is an example, as is the incorporation of simulation into the American College of Surgeons accreditation of educational institutes. As more and more applications for simulation are identified and implemented within health care, simulation will become more firmly embedded into how we train clinical professionals. By keeping abreast of the literature, networking with peers (such as reading postings on the listserv by members of The Society for Simulation in Healthcare), you can be aware of many developments and share them with colleagues. Additionally, this offers you a great opportunity to design better research and educational projects for your institution.

### 6.2.6   How do you Encourage Learners and Faculty to Come to your Center?

It is important to get the word out that you exist and to let them know what you do. In response to a faculty member who asked, "I want to use simulation in my teaching, what do I do next," we identified the need for an *Introduction to Institute for Surgical and Interventional Simulation* class. It is offered twice monthly at different times to accommodate schedules. It is listed in the online course schedule. Lasting about 45 minutes, the class provides an overview of the support offered by our center, the simulators, available curricula and options for new ones, information how to schedule a class, and after-hours access. We also take advantage of every opportunity to market our center to learners and faculty at meetings and through "Institute for Surgical and Interventional Simulation Updates," an e-mail to faculty, chairs, community and media

relations contacts, the dean, and other expressing an interest in simulation. Our "Updates" provide information about vendor visits, articles about simulation, new class information, and details about our simulators. The list of people requesting to be on our e-mail list for our "Updates" doubled between the first and second editions.

### 6.2.7   Once People Know about your Simulation Center, How do you make your Facility Inviting?

In our plans for a new center, we incorporated a "Serendipity Room." It is an open area with a kitchen, table, and comfortable chairs where learners and faculty can gather. It will be accessible 24/7, and will be stocked with some beverages and snack foods. We also planned for carrels with Internet-enabled computers and telephones, where people can check e-mail, return pages, and conduct research. Furniture will be modern and comfortable. We are considering iPod© docking stations and speakers in the skills center and virtual OR so that our center creates an environment in which learners will want to learn. An article in *The New York Times* (While in Surgery, Do You Prefer Abba or Verdi, June 10, 2006) states that music, particularly iPods© are commonplace in the Operating Room, with different genres evolving, such as "closing music" or "pancreas music." This might even provide us with some research opportunities, i.e., will Barry Manilow's *Mandy* increase the speed of the procedure as the learners try to escape from it, or will it mellow them out so that they are more cautious? I say this with a sense of humor, obviously, but the environment that we establish for the learners is an important consideration.

### 6.2.8   How do you Demonstrate your Excellence?

Your reputation must be earned through an overall commitment to excellence. Learner surveys are a good indicator of what you are doing well and what you can improve upon. There are good online survey tools such as http://surveymonkey.com/ that you can customize. Keep current in the literature, contribute your knowledge to the simulation community, and take advantage of opportunities to beta test and/or develop new technologies. Get accredited. We received accreditation from the American College of Surgeons (ACS) as a Comprehensive Education Institute. It was a very labor-intensive endeavor, but serves a clear display of our excellence. We display the ACS accreditation logo prominently in our center and in our print and electronic media. Most importantly, you need strong and pertinent curricula since we are all about education. No matter how good your intentions or how nice your simulation center is, it boils down to the excellence of your curricula.

# 6.3 Challenge 2: Getting The Key Players Together to make Strategic and Operational Decisions

This is the biggest challenge that an Administrator faces. At the same time, it offers an amazing amount of latitude and freedom from micromanagement. You must trust your insight about your institution's climate – its priorities and strengths. You have to understand the personalities of the faculty and administrators involved in your governance. Perhaps some are driven more by the need to publish, some are driven solely by the love of simulation, while others are stimulated by their need to be in charge of the next high-visibility program. Since our center is interdisciplinary, I have learned how to work with personnel from many different specialties whose priorities and approaches radically differ from one another. In all cases, the Simulation Center Administrator must wear a multitude of hats – principally, listening and then going with your best judgment. The clinical practiconers also wear many hats. While you might be concerned with budget decision deadlines, they might have a difficult case scheduled in the Operating Room and are prioritizing that – as they should be. It is my experience that our simulation center is guided by some very influential physicians. They are committed to teaching, lecturing, traveling, serving on committees, and also to their clinical practices in addition to our center. I have had to learn how to maximize what limited time I have with them and to rely on my judgment combined with "decision by quorum" to do the right thing. When I was hired, my Supervisor told me, "The great thing about your job is that you can't make any mistakes." I took this to mean that this was such a new and evolving field that there was no history, and therefore no pitfalls to fall into. In retrospect, maybe she was threatening me not to make any mistakes. In any case, I am still employed and still making decisions every day, while our center is flourishing.

## 6.3.1 How do you Make Strategic Decisions?

It is easy to jump on the simulation bandwagon without a clear focus of who you want to be or where you want to go. It's more than buying some simulators, getting some space to put them, and waiting for the learners to come. You need a clear business plan. The classic 5-year plan is a relic since new developments occur so quickly. However, a formal plan to delineate, at the very least, your first year's goals, funding sources, facility, and faculty is essential.

Our Center conducted a day and a half strategic planning session (see Appendix 6A.1 for the Agenda of the strategic planning session). The first day was dedicated to overall objectives, with the second day more focused on operational issues. There was an offsite group dinner the first night, and the event culminated in a keynote speech by a recognized expert in simulation program creation and operation. There were representatives from the dean's office, different clinical specialties, and schools.

The retreat had several important results. Besides the adoption of a mission statement, governance, and measurable goals for the upcoming year, it signaled the University's strong support for the success of our center. It was a perfect launch. The outcomes from the strategic planning session formed the footprint for the center's operations

## 6.3.2 How do you Communicate to your Team Regularly? How do they Communicate with you?

I learned early on that the people involved in our center ran the gamut – from those embracing technology to those who were e-mail- and web-phobic. The e-mail embracers received a disproportionate amount of my attention since it was easy to see what their issues were (for instance, "May we schedule a simulation demonstration this month?") and respond to them. It also made cross-communication difficult since I was getting isolated bits and pieces of information from various individuals. I would spend too much time coordinating schedules to accommodate faculty, vendors, and others seeking out information about our center or receive our expertise. The communication method was fragmented and broken.

The solution was to establish a set time, day, and drop-in place. It is scheduled on the Director's nonclinical day to allow him to be present. We make sure that people know that on Wednesdays at 11:00 AM, they can come to the same conference room with their simulation questions, ideas, or whatever. We make good use of visitor's time by placing them first on the impromptu agenda. This has worked very well. Issues are discussed, opportunities explored, and news is communicated. Although the key players do not make it to every meeting, they do attend several times a month. This is enough to keep the communication flow active and useful.

## 6.3.3 How do you Communicate with Executive Staff?

It is essential to convene the Executive Committee as regularly and as efficiently as possible. Since we receive significant attention from the President of the University, and our initial funding is from the Dean of the School of Medicine (our simulation program is one of his top three initiatives, along with the Institute for Stem Cell and Regenerative Medicine and the Global Health Initiative and Genome Project), the center's senior leadership must know of major milestones promptly.

The Executive Committee meets monthly, starting at 6:30 AM. It is scheduled early in the day to accommodate clinical schedules. Committee members who cannot make the meeting in person are provided with a dial-in number for live audio participation. The agenda is concise, consisting of three subject areas:

1. Schedule of upcoming events
   (This tracking can be used to compile a history of the simulation center's use.)
2. Development
   (Current development efforts are discussed as well as future opportunities.)
3. Operations
   (This includes budget, facility, and personnel.)

The meeting begins promptly at 6:30 AM and ends at 7:30 AM. There are formal agendas and minutes, and the Chairman requires undivided attention. No multitasking or Blackberries are permitted. Reminders and agenda item requests are sent to participants in advance of the meeting. As a result of this structure, the Committee is able to accomplish much in a short time.

## 6.4 Challenge 3: Doing What Needs to be Done with Limited Resources

You have the interest of your institution, you have faculty, and you might even have some simulators and some space. But, you have limited funding, especially to market yourself to potential supporters and to the public. How do you move ahead? It can be a conundrum. This is where being a conservationist comes in handy. Maximizing limited resources by reusing them for many different objectives is a powerful strategy.

### 6.4.1  How do I Get Electronic and Print Design and Production Support?

The Development Office is a tremendous resource. They can provide direct graphic design support for brochures, posters, agendas, PowerPoint presentations, and web design. Ask the graphic artist for the document in a format you can edit. This enables you to make updates or customize when needed. Having a library of PowerPoint slides that you can pick and choose for different audiences is a good idea. Including someone from the Development Office in your meetings will help them understand you, and in turn, cause them to be more effective in getting financial support for your simulation center.

Our center was fortunate to be featured in a large University-wide fund-raising campaign by the University of Washington Foundation. We were able to use print media, a professionally produced 30-second video spot, and web links that the Development Office designed for other uses.

Community and Media Relations is another strong resource. We have used their professional photographs from VIP tours and their *Report to the Community* to feature in our materials. It is important to keep them informed as to what your simulation center is doing. They arranged for two segments featuring our center to be used in a webcast on a Microsoft-hosted web site (www.onten.net). The University of Washington used still photographs and extracts from these webcasts as well as the links to display on the banner of the University's web page (Figure 6.2).

Our Department of Community and Media Relations has facilitated coverage in the special health segment of *The Puget Sound Business Journal* [3]. They promote us heavily during open houses and health fairs, and have funded various printing projects. Using vendor photographs is another good idea. They want to sell their simulators, and have some excellent photographs of their equipment. If you use their photographs, be sure to get their permission in advance and to give them credit. Make sure that the photographs you take of your center represent the different departments that train there – don't limit yourself to mannequin photos when there is so much more going on. Establishing and maintaining strong relationships with different departments throughout your organization will help market your simulation center better than any Madison Avenue advertising company because your simulation program will only be successful when it becomes their simulation program.

## 6.5 The Cliché Conclusion

The role of the Simulation Center Administrator is like a conductor of a symphony – the string section might be a little more sensitive, but no less talented than the percussionists who set the tempo of the selection. Or, there might be a trumpet player who wants to make sure that everyone knows he is a Julliard graduate, and will play over everyone else in the section. The oboe player might want to show off her new, really expensive instrument. Sometimes, the notes in the score don't make sense, or in the case of a new composition, the conductor doesn't really know if the piece really works.

It is the Administrator's job to manage the organization so that the components come together and that the audience (learner) ultimately learns from the selection (curricula). Senior leadership at our center had the foresight to design the organizational structure to include a full-time Administrator. The American College of Surgeons, recognizing the value of a full-time Administrator, requires this for certification as a Comprehensive Education Institute. As a health care executive with many years of experience, this has been the most rewarding and challenging position I have held. Being part of something new is exciting and offers limitless opportunities and challenges!

# References

1. Mann S., Marcus R., and Sachs B. *Lessons From the Cockpit: How team Training can Reduce Errors on L&D*. Grand Rounds, Contemporary OB/GYN. January 2006. pp. 1–7.
2. Leonard M. and Frankel A. Make safety a priority: Create and maintain a culture of patient safety. *Healthcare Executive* **21**(2), 12 (March/April 2006).
3. Sinanan, M. Simulation becoming a better way to train doctors. *Puget Sound Business Journal.* May 12, 2006. http://www.bizjournals.com/seattle/stories/2006/05/15/focus7.html. Last accessed 29 August 2007.

# How to Fit
# in while
# Standing Out

*So many new ideas are at first strange and horrible, though ultimately valuable, that a very heavy responsibility rests upon those who would prevent their dissemination.*

J. B. S. Haldane

We don't engage in clinical simulation like we do hobbies, that is, the doing is the end in itself. We don't buy simulators with our own money like we do for art and jewelry, that is, the appreciation in owning is the end in itself. The resources of others that we expend upon our clinical simulation program, no matter how large or small, are always in support of relationships with organizations surrounding our simulation program. Furthermore, almost all of these organizations and their interrelationships have seniority, if not also priority over your simulation program. Just as many unchallenged beliefs persist by familiarity alone, so do budgets and resource allocations. Usually, the newcomers must overcome far greater hurdles of legitimacy than did any of the programs preexisting at the time their program managers came of age. Yet, the simulation program must honor and respect the mission and calling of the larger community. Thus, having clearly defined reasons for creating a simulation program is an essential foundation for any simulation program's success. Successful simulation programs help their instructors help their learners and their learners' patients:

- Filling in experiential gaps in present training models
- Training to demonstrated competency instead of merely satisfying a schedule
- Patient safety: lessons learned from mistakes without abusing real patients
- Trainee safety: focus upon lessons to learn with minimal unwanted distractions
- Deliberately creating desired lessons of knowledge, skills, and attitudes

- Providing immediate feedback for both student and teacher
- Scheduling "disasters" for solo and team efforts
- Increasing efficient use of scarce learning time
- Improving public relations

Note that proposing simulation as a solution toward solving any one of these problems implies that such ideas are better than the current practices. Don't be surprised if your leadership is less than welcoming to admit that *their* institution is in need of improvement, and by inference that *they* are not already the champions of excellent leadership.

The only loss greater than not using simulation where it may benefit our students is to impose it where or in ways it does not. Unsuccessful simulation programs are founded upon selfish reasons that ignore the passions of the teachers and needs of the learners:

- Novelty, sexiness, just to say "we have one, too"
- Vehicle for self promotion, just for the program manager "to make full Professor"
- Imposed from above "we got it, now you figure out how to make use of it"
- Extra work with no appreciation of the hidden training load on instructors
- Considered as "additional duties as assigned" instead of as core contributions
- "It will teach by itself," not unlike when computers were sold as replacements for staff

A key question about design of any program and its supporting facilities is "how much realism is essential, desired, overkill?" A single fixed answer is impossible, as the question really asks "who are my learners, what do they already know, what do I want them to learn?" At any level of competency and challenge, all students should receive unambiguous go/no go type of instructions about everything they will encounter within the simulation experience. For any either/or choice that the student may take, both choices must have legitimate consequences and effects. If not possible, then the scenario should be constructed and produced so that the student never needs to consider making such a choice. Like any game, the rules for play must be very clear, the boundaries of what is and is not in play made obvious, and who is responsible for what be known and understood. Like learning a new card game, the first round is often played with all cards face up as a prerequisite for the real exercise to assure comprehension of the rules. Since our learners are *supposed* to gain competency while within your educational environments, the answers to the above questions move along with our students' progression. Despite the fluidity of this reality, there are several invariant guideposts for consideration:

- Simulations are stepping stones for the gaps between didactic and real patient-based learning;
- Formal education is amplification through simplification, so too is simulation;
- Like live theater, the *story's* realism is essential, not that of the *art work* on the wall;
- Level of *sensed* realism must match students' comprehension, not just their detection;
- Visual cues are good for static information, audible cues good for changes;
- Teaching requirements come first, purchase orders come last;
- As always, acquire and accumulate more "stuff" in haste, store and repent in leisure.

Ronnie Glavin provides an overview of just what simulation is and is not, and how its strengths can be harnessed to improve existing curricula. Allan Shemanko and Linn Jones add another complementary perspective to this issue. Judith Hwang and Betsy Bencken provide a sense of scale from one point of view, as do Roger Chow and Viren Naik from another. Judith Hwang and Betsy Bencken close by describing the fine art of reaching for the stars while not losing one's footing. In addition, Chapter 40 by Craig Balbalian on prehospital care and tactical environments includes excellent examples of how to think about how much and what kinds of realism are useful, safe, reproducible, and those that are not.

# 7

# When Simulation should and should not be in the Curriculum

Ronnie J. Glavin

## 7.1 What is Simulation?

Simulation in the context of clinical practice is an attempt to recreate one or more aspects of that practice. This ranges from a very simple physical task – using oranges to practice giving intramuscular injections – to recreating a whole clinical environment with interactive, high-fidelity simulators, representing patients in a relevant clinical setting, such as a labor room, or a ward, or an Intensive Care Unit.

Simulation is an interactive event for the participants. Unlike sources of knowledge that are not responsive to the recipient, such as print and prerecorded video, simulators in any of their myriad forms detect some action of the participants, even if it is merely their presence, and modify their output. The orange may not be an active participant but it provokes a reaction in the clinician performing the intramuscular injection. Like two dance partners in a dance, the action of one is detected, interpreted, and reacted to by the other. Like dance instruction, one of the partners needs to both know how to dance well themselves and how to teach others how they can dance better. Likewise, simulation is best employed as a way of teaching and learning those topics that are inherently interactive, and requires that those doing the instruction know both the topic content and how to use interactive teaching. It is this educational role that links simulation with the curriculum.

## 7.2 What is a Curriculum?

John Ruskin, the 19th century English academic, described education in the following way:

> "Education does not mean teaching people to know what they do not know. It means teaching them to behave as they do not behave [1]."

What do we want learners to be able to do at the end of a course? What problems do they have to deal with? What abilities do they need to acquire to equip themselves for these problems? The curriculum is a statement of how that can be achieved.

Lawrence Stenhouse described the curriculum as follows:

"A curriculum is an attempt to communicate the essential principles and features of an educational proposal in such a form that it is open to critical scrutiny and capable of effective translation into practice [2]."

The educational proposal is the end product – what the learner should be able to do at the end of the course. The end product is usually expressed as a set of short statements referred to as the aims of the curriculum. In addition to the aims, a curriculum will usually include teaching methods that will help learners realize these aims and assessment tools that will confirm that the aims have been achieved. The teaching methods for a professional curriculum will include formal methods such as small-group teaching and lectures, but will also include episodes of experience, such as time spent under supervision in the clinical workplace.

## 7.3 Which Components are in a Curriculum?

The educational aims are usually broken down into smaller and more achievable components; variously referred to as objectives, learning outcomes, etc. This can be compared to the Fish Ladder model: to help the learner scale the high dam, educators build fish ladders that allow the new fish to complete the same overall height of accomplishment, but do so in many smaller, attainable steps. These usually reflect the main areas of professional practice. The taxonomy of educational objectives introduced in the 1940s by Benjamin Bloom has had a considerable influence on the classification of these smaller components. Bloom [3] and his colleagues described three domains:

The Cognitive Domain – the field of knowledge
The Psychomotor Domain – the field of practical skills
The Affective Domain – the field of values and beliefs

This has given rise to that well-known triad of "knowledge, skills and attitudes." However, there are other ways of expressing the main dimensions of the health care professional role. For example, in medical practice, the six competencies of the Accreditation Council for Graduate Medical Education [4] or the seven roles of the doctor in the Canadian CanMEDS 2005 project [5] or the seven duties of a doctor listed by the General Medical Council of the UK [6].

Competencies as defined by three different accreditation bodies sharing similar, if not identical, responsibilities are shown in Table 7.1. The terms in this table are reproduced as from their original sources. However, medical is equivalent to clinical and includes nurses, surgeons, in fact all health care workers.

**TABLE 7.1**   Competencies as defined by three different accreditation bodies

| General Medical Council (UK) | Accreditation Council for Graduate Medical Education (USA) | Royal College of Physicians and Surgeons of Canada |
| --- | --- | --- |
| Good clinical care | Patient care | Medical Expert |
| Maintaining good medical practice | Medical knowledge | |
| Working with colleagues | Interprofessional and communication skills | Collaborator |
| Relationship with patients | | Communicator |
| Teaching, training, appraising, and assessing | | Scholar |
| | Practice-based learning and improvement | |
| | Professionalism | Professional |
| | Systems-based practice | Manager |
| | | Health Advocate |
| Probity (integrity, honesty, and "goodness") | | |
| Health | | |

## 7.4 Dimensions of Clinical Practice

These dimensions ensure that all aspects of clinical practice are captured and the full ranges of activities that will be encountered are included. Of course, each of the dimensions will have its sets of theoretical knowledge, procedural knowledge, etc., as well as relevant skills and an underlying value system. A useful way of tying these strands together was published by George Miller in 1990 [7] (Figure 7.1).

The most basic level is KNOWS – which we can take to refer to the acquisition of facts and concepts (e.g., the physiology of the cardiovascular system).

The second level is KNOWS HOW – which deals with the ability to apply different facets of knowledge to a particular clinical situation (e.g., how would one manage a patient presenting at the emergency room with a blood pressure of 70/50, heart rate of 135, Glasgow Coma Score of 13, respiratory rate of 25, and temperature of 36°C? – knowledge of the normal physiology of the cardiovascular system and the pathophysiology of "shock" will come into play. The more specific one makes the case, the greater the degree of application from basic principles to specific management.

The third level is SHOWS HOW – don't tell me what you are going to do; do it! Just because a learner can describe how

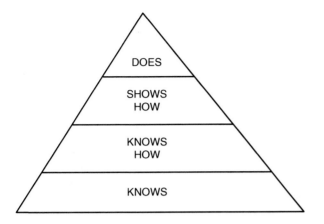

**FIGURE 7.1** Miller's triangle showing the progressive levels of performance.

to do something does not mean that they can actually do it. There can be (often is) a discrepancy between KNOWS HOW and SHOWS HOW.

The final level is DOES – the trainee has demonstrated to the teacher/instructor that he/she can manage a particular case, but what is that trainee's approach to the subsequent management of such cases when the trainer is not physically present?

If we take a very simple clinical task such as scrubbing and gowning to perform a procedure under sterile conditions, then KNOWS refers to the background knowledge about infection, transmission, the need for sterility, etc. KNOWS HOW refers to knowing how (being able to describe in words either written or spoken) to scrub and put on a surgical gown. SHOWS HOW refers to being able to scrub and put on a gown. A trainee may be able to describe the process but may have difficulty operating the tap without use of hands, difficulty in opening out the gown, or putting on surgical gloves without compromising sterility. DOES refers to what happens when the instructor or trainer is not present. Does the learner take short cuts or only pay lip service to the act, or does the learner continue to practice in a way that is consistent with the standard that was taught?

KNOWS and KNOWS HOW fall within the "Cognitive Domain" of Bloom's Taxonomy. SHOWS HOW falls within the "Psychomotor Domain" and DOES falls within the "Affective Domain." Note that everything written above about clinical students in their learning endeavors applies equally well to their clinical educators and their teaching endeavors.

> KNOWS content to be taught
> KNOWS HOW to teach
> SHOWS HOW teaching is done
> DOES teach

**Editors' Note:** Just because someone has learned for themselves (i.e., scaled their own fish ladder) to the level of DOES clinical *care* very well is not the same as DOES clinical *teaching* very well with others who are far less accomplished.

## 7.5 Which of these Curriculum Components are Best Suited to Simulation?

Traditionally, many examinations in medicine concentrated on factual recall (KNOWS and KNOWS HOW), but a recognition in the 1970s that these were not recognizing the SHOWS HOW level brought about the introduction of a new form of assessment – the Objective Structured Clinical Examination. As health care has become more complex, tests of knowledge are no longer sufficient because they cannot demonstrate the ability of the learner to operate at SHOWS HOW level. Skill labs can help with some of these areas but may not include all of the components of SHOWS HOW because they rarely capture any of the context in which the skill is situated and so fail to capture the complexity of the task. As an example, consider the management of a patient suffering from an acute asthmatic attack. The knowledge level (KNOWS) may include some of the underlying theoretical concepts – pathophysiology of the condition, natural history of the disease, pharmacology of the various drugs, etc. The knowledge of how to recognize and manage a patient (KNOWS HOW) involves memorizing a list of activities. However, being able to describe what to do is different from being able to manage the patient (SHOWS HOW). There may be some aspects in which learners can be trained – how to set up and use a nebulizer, how to take blood gases, how to use inhalers, etc., but part-task trainers and standardized patients can only take the management so far. This is where the high-fidelity simulators are most useful as teaching tools. They provide conditions that capture some of the complexity in which a learner can manage a patient because the learner has to integrate practical tasks with knowledge. The part-task components also do little to prepare the learner for dealing with a very worried patient and even-more-concerned relatives. Many times, in our center, senior medical undergraduates in such a scenario do not carry out simple but crucial tasks – oxygen is not applied as early as it should, monitoring is not attached as quickly as it should.

More importantly, we find that the students have difficulty in managing all of the practical tasks while trying to exude an air of confidence and reassurance to the patient (the more you can exude an air of confidence and reassurance to the patient, the more you can charge them!). However, there is an even greater role for simulation, and that is in helping learners reflect on their clinical judgment. Clinical judgment is not easily conveyed in the knowledge, skills, and attitudes triad, although Miller's triangle can easily be applied to it. The level of judgment will also reflect the level of training. At the level of beginner, it may be sufficient to have the learner recognize a problem, initiate simple treatment, and call for help from someone more experienced (the level that we expect from the senior medical undergraduates). However, simulation can be used to create scenarios that are more demanding for more senior doctors (the ones who would be called to help). This is covered in more detail in other chapters (see Chapters 77–80). Once more, the level is at SHOWS HOW but the scenario can be made more challenging by having the patient deteriorate more quickly or having a more complicated course, for example, the acute asthmatic patient could also develop a pneumothorax or could be unresponsive to beta 2 agonist therapy, etc.

The complexities of modern health care are such that the single doctor or other health care professional very seldom works in isolation. Working as a team member features in all three lists in Table 7.1. Collaboration between health care workers, especially in the acute clinical sector, is important for successful outcome, and simulation can help teams reflect on their performance or can help individuals reflect on their performance within a team. Importantly, this is also at the SHOWS HOW level. Other simulations can be carried out to reflect the outcomes specified within health care profession as collaborator. For example, the Clinical Skills Centre in Dundee, Scotland, includes a simulated ward area in which simulated patients are managed by learners from a variety of health care professions [8], so it is not only acute scenarios that lend themselves to simulation, but also stable patients and daily professional interactions. For example, in the simulated ward, there may be patients waiting for discharge medication, investigations such as X-ray have to be organized, new patients admitted, lab results chased up, etc. In the time available, the participants have to prioritize tasks and coordinate activities to work efficiently and effectively.

Simulation can be used to promote those learning outcomes that are more pitched at KNOWS and KNOWS HOW levels, but the types of simulation device better suited are at the desktop type of simulation rather than a high-fidelity patient simulator. Similarly, DOES level is better assessed by asking people who work with the learner on a regular basis to comment on the behavior of the learner. This is normally done through questionnaire techniques in 360° appraisal, etc. This does not mean that simulation, especially high-fidelity-type simulation, cannot be used for the other areas, but for the reasons listed above, it is not usually the most effective way to use simulation.

## 7.6 Linking Simulation to the Curriculum

If you work in a simulation center, how do you attract potential participants? One is more likely to be successful if one can offer a solution to someone else's problems. So you need to have a plan in mind when approaching the individual(s) with control over your target group of learners. Begin a dialog with these key individuals and discuss which aspects of their curriculum are difficult to deliver. It may be difficult to teach or it may be that some crucial learning outcomes have to be "signed off"

as having been completed to a desired standard but cannot be done so in the clinical settings in which the learners work. These outcomes, whether in terms of teaching or assessment, are most likely to be at the SHOWS HOW level. There are some ways to help identify important outcomes that are not being achieved.

Ask newly qualified learners from that course to rate their confidence in dealing with a variety of clinical situations. Some aspects of practice will be difficult to deal with due to lack of exposure or an understandable reluctance on the part of the supervisors to let the learners lead the management of the case.

Enquire about areas of patient complaints or lawsuits (closed claims studies can be helpful) in the institution or in the profession at large. This may be seen as a potentially attractive way of avoiding some problems that may have a sizeable financial impact on the institution.

National reports may have recommendations that can have an impact on certain areas of the curriculum. For example, in the UK, the report from the Confidential Enquiry into Maternal and Child Health [9] for the period 2000–2002 recommended better training in the management of major obstetric hemorrhage. These recommendations have been used by some simulation centers to promote courses for labor room staff.

Findings from a local or national incident-reporting system may reveal a pattern reflecting performance in some areas of a lower standard than general.

Reports from professional bodies conducting quality assurance visits of the institution may have identified some areas of concern that require to be addressed.

The above approaches are ways on exploring potential areas once the dialog has begun.

## 7.7 Example: Emergencies in a General Dental Practice

Our center in Scotland was approached by the national body responsible for dental training. A new system of assessment was being piloted with the intention of national implementation. Working with the project team, we identified some areas that were both of concern to the implementation team but appeared possible to deal with by a simulation-based course – the management of medical emergencies in general dental practice. We chose this area because these do not occur commonly but the trainees were expected to have gained some familiarity with their management.

The course was competency based, thus successful completion of the course required that the learners demonstrate the required competencies at SHOWS HOW level. So, we designed a pilot course that would allow us to teach and assess performance at SHOWS HOW level. We worked with the relevant individuals from the course design team to ensure consistency between the formal teaching and the content of the simulator-based course. We also set a written exam, for use in the pilot study only, to confirm that the candidates had reached the KNOWS HOW level. There were, of course, other areas that we could have considered in our center but rejected them because the opportunities for these areas to be taught and assessed were plentiful in "real" clinical practice. The pilot study confirmed that we could use this course to move the dentists from KNOWS HOW level to SHOWS HOW level. We then carried out some further development by working with experienced dentists who were in clinical practice to ensure that the context was both realistic and appropriate. We then arranged with the course organizers to have the clinical dental trainers attend as faculty on our courses to help with the SHOWS HOW management and to help keep the simulator component in a realistic clinical context.

All newly qualified dental graduates working in general dental practice in Scotland attend this course, which is popular with the participants and the training program organizers, and brings in a steady flow of revenue for our center. We are currently working with the relevant training authority to expand the course and link it with the existing refresher activities, so the story is far from over.

## 7.8 Conclusions

Knowing the educational strengths of simulation can help you attract would-be participants (and hopefully their funding) by engaging in dialog with course organizers and offer solutions to some of their problems.

1. Become familiar with their curricular documents.
2. Identify competencies, objectives, learning outcomes, etc., that fall within the SHOWS HOW level of Miller's triangle.
3. Identify which of those competencies are difficult to achieve in clinical practice.

4. Identify threshold competencies – are there competencies that have to be signed off that allow learners to progress? If these are difficult to achieve or assess in clinical practice, then your simulation-based course may be the solution.
5. Identify which of those competencies are of concern to the course organizers (possibly because of external pressures) – remember that many course organizers will be happier with more conventional areas of teaching – knowledge and basic practical skills – and may not be so happy with the "softer" or nontechnical skills such as communication, decision-making, team working, etc.
6. Work with the course developers to integrate your simulation-based component into the course.

## References

1. Rowntree, D. *Educational Technology in Curriculum Development*. Paul Chapman Publishing, London, 1988.
2. Stenhouse, L. *An introduction to Curriculum Research and Development*. Heinemann Educational Books, London, 1975.
3. Krathwohl, D. R., Bloom, B. S., and Masia, B. B. *Taxonomy of Educational Objectives: Handbook II: Affective Domain*. Longman Group, London, 1956.
4. http://www.acgme.org/outcome/comp/compFull.asp–accessed 14th August, 2007.
5. http://rcpsc.medical.org/canmeds/CanMEDS2005/index.php–accessed 14th August, 2007.
6. http://www.gmc-uk.org/guidance/good_medical_practice/index.asp–accessed 14th August, 2007.
7. Miller, G. E. The assessment of clinical skills/competence/performance. *Acad. Med.* **65**(9), 63–67 (1990).
8. Ker, J., Mole, L., and Bradley, P. Early introduction to interprofessional learning: A simulated ward environment. *Med. Educ.* **37**(3), 248–255 (2003).
9. http://www.cemach.org.uk/publications/WMD2000_2002_ExecSumm.pdf–accessed 14th August, 2007.

# To Simulate or not to Simulate: That is the Question

G. Allan Shemanko

Linn Jones

## 8.1 Our World

Using technology to provide hands-on experiences guided by proven educational principles, we can provide the very best evidence-based learning environment for our future caregivers. In reading this, you have already contemplated simulation – learning from other's experiences is an important consideration. With these tried and true road maps, you can learn why some experiences were more successful than others. Take the time to investigate these experiences and you will save yourself some heartache if you do!

In order to put what you are about to read into perspective, we will provide you with a brief introduction (some would say shameless promotion) so you may compare our situation with your own. Along with this brief introduction, we will describe the reasons for our move into high-fidelity simulation and describe our use of standardized patients for respiratory therapy and paramedic programs at the Northern Alberta Institute of Technology in Edmonton, Alberta, Canada. As we proceed through this chapter, we will describe our experiences with low-fidelity as well as high-fidelity simulation and standardized patients. Our overall objective is to provide you with our experiences in deciding when simulation should and should not be incorporated; many of you will undoubtedly find that these experiences mirror your own situation. We hope that you

can learn from our experiences, and encourage you to share your own with others. The chapter closes with some thoughts on how simulation can be used to save some money for our educational institutions and even our students.

Officially opened in 1963, Northern Alberta Institute of Technology is a board-governed, publicly funded technical educational institute. The Institute offers more than 250 programs, including 34 apprenticeship programs for approximately 17,500 full-time and apprenticeship students plus 50,000 students enrolled in continuing education programs. Four key directions guide the Institute and deliberately assist the leadership in managing all operational issues: "champion student success, excel in teaching and learning, optimize the use of technology, and advance enterprise development" (NAIT Board of Governors, Executive Summary, section 1 [1]). The Institute offers respiratory therapy as well as paramedic programs as an integral part of the School of Health Sciences; these programs are where our expertise lies related to simulation.

## 8.2 High-fidelity Simulation

The first thing that we learned about simulation was that we were already doing simulation. That in itself was an epiphany, strange though that may sound. We knew then that we were not in the dark ages – mind you we really did not know where we were compared to the rest of the world, but at least we hoped we were not in the dark ages! Basic simulation involving CPR dolls, intubation heads, and more recently, advanced cardiac life support trainers had always been part of the training in the two programs we are describing. These experiences helped us to fit high-fidelity simulation into our programs where it made sense, simply because we already knew how we used basic simulation. High-fidelity simulation was integrated into our programs when our need for simulation exceeded the capabilities of basic simulation. Our thinking was much like the artist's analogy of carving a figure from a block of clay – simply remove the bits that do not look like the figure you are creating! For our programs, we found that it was a logical progression to decide where and when high-fidelity simulation should be incorporated.

Our faculty needed to be onboard and prepared for the realm of high-fidelity simulation. Buy-in was an important aspect of simulation – if you don't have the buy-in, you may as well decide not to pursue high-fidelity simulation or any other major change for that matter. We had decided as a group that we would seek outside help in teaching us the necessary intricacies of high-fidelity simulation as quickly as possible, only to realize after the fact that we should have waited somewhat longer. Delays in renovations drove the train-the-trainer dates ever closer. These delays certainly put us at a disadvantage – you need to ensure that your faculty have the necessary tools and time so they may "play" with the simulator prior to any how-to-use-a-sim training. For all the same reasons that we want our students to practice to gain clinical familiarity in a safe setting before they face the real unknowns of real clinical care, we wanted our teachers to gain educational familiarity in a safe setting before they faced the real unknowns of real instruction. With our staff having very little experience in playing with the simulator prior to their own learning, our initial train-the-trainer experience became ineffectual and overwhelming for staff.

One aspect of high-fidelity simulation that we learned, especially related to physiologically based models, was that we needed all our instructors to get their hands on this new technology early and often. The foundation of this idea comes from the text published by Bates and Poole [2]. Simply (but not necessarily inexpensively) providing the simulator's control technology to each and every one of your instructors will advance your simulation program much faster and more efficiently than by limiting their access to only using the one computer attached to the simulator. Having only one simulator limits the access for instructors, especially during the delicate initial phases of first dating. Depending on the workload model your school subscribes to, the "Lone Ranger" instructors will tend to approach the simulator after regular office hours and on weekends in order to learn how to operate it [2]. Lone Rangers are the early adopters, and will do whatever they need to in order to use the technology. Late adopters or stragglers

and butt draggers are those who wait to see what the Lone Rangers do with the new tool and then some of them will choose to go 180 degrees the other way. In order to make high-fidelity simulation more available to the instructors, seriously consider buying the extra laptops and software to allow these so-called late adopters (and even the Lone Ranger types) to take the simulator home with them. We failed to provide extra laptops, only to learn that development of scenarios and instructor comfort in operating the simulator was less than ideal. Then again, what did we expect? It takes time on the simulator in order to be familiar with its various personalities and how each behaves clinically. We should have ensured that instructors were familiar enough with each of the preprogramed patient profiles so that they knew how each behaved clinically and could anticipate how far they could push these simulated patients before causing an irreversible downward spiral toward death. There is nothing quite like showcasing a medical intervention on a physiologically based high-fidelity simulator, only to have the patient die inappropriately in the middle of the demonstration! Instructors have an inherent need to be comfortable and adept with the equipment before they risk embarrassment in front of their students. Today's high-fidelity simulators are incredibly complex; we need to set our faculty up for success. When we do that, everyone wins.

As we progressed into the world of high-fidelity simulation, we believed it would be prudent to decide exactly what simulation in our respective programs would look like. We believed that case studies, standardized patients, and low-fidelity and high-fidelity simulation all had a role to play in educating our respiratory therapists and paramedics of the future.

We found that clear student-performance objectives were crucial in incorporating simulation into our curricula – and we're still learning (see Chapter 3). It is difficult to clearly describe *what* we wish the student to do, *how* we wish them to do it, and *the standard to which we will measure them* so that the student and instructor understands clearly what the objective is. We described the term *student-performance objective* as what the learner was expected to achieve given the specific performance (task), the condition (environment) in which the task was to be completed, and the level of proficiency to which the student was expected to perform. Since the performance objective described what the student would be doing, the statement needed to contain an action verb such as perform, construct, intubate, decannulate, etc. The condition component delineated to the student (and teacher) the situation in which the student would be required to demonstrate the task. Finally, the criterion component described the required level of mastery to which the student's performance would be measured against [3].

One example of our student-performance objectives is: The student will perform a pulmonary assessment on a standardized patient including inspection, percussion, palpation, and auscultation in a simulated Emergency Department according to the procedure outlined in the *Competency-Based Objectives* manual. In this case, the required task is the demonstration of a chest assessment. The condition is that the assessment will occur in the simulated Emergency Department. The level of proficiency is outlined in the *Competency-Based Objectives* manual provided to each student, with each task, subtask, and required background knowledge clearly outlined. The student would understand that there would be noises, hospital staff, patients, smells, and other distractions that may impact their ability to perform the task as the environment is an Emergency Department. As part of this performance objective, the student understands that he/she must introduce themselves to the patient, perform infection-control procedures (wash with alcohol hand rub), and confirm the physician order. As well, he/she will also know that their performance will be evaluated to the standard of an entry-level practitioner.

We found that some instructors were reluctant to use the high-fidelity simulator to its fullest capacity. The cost of neck skins, tension pneumothorax bladders, and the like were perceived to be too expensive for them to use. An analogy would be to purchase a US$50,000 vehicle yet not use the signal lights because they may burn out. Why would a US$12 neck skin prevent faculty and students from using a US$200,000 simulator? Acquisition of "supplies," including everything from laryngeal mask airways to neck skins to syringes, needs to be included in the simulator's operating budget. By encouraging faculty to use the middle- and high-fidelity simulators to their fullest capacity, we got much better return for the money spent on supplies for the high-fidelity simulator. Examples of every item that gets broken or worn out was placed in the Bad Box for unexpected appearances in future scenarios.

Our faculty discussed what we should be using our high-fidelity simulator for in teaching and learning. It did not make sense to us to use the high-end simulator to teach individuals how to perform venipuncture for administration of intravenous medications when a basic IV arm task trainer could certainly suffice. That being said, we found it was still prudent to offer the students the ability to acquire venous access during a resuscitation attempt on the high-fidelity simulator. We have used the simulator to illustrate the physiological response of pharmaceutical agents so that students may have a first-hand experience of a patient's response to a drug. It was truly amazing to see the students as they began to understand the complexities of the patients they will be treating when it comes to pharmacologic agents.

In using the high-fidelity simulator to take students to the next level of airway management, we had fully expected that students would gain more confidence in managing the airway in a difficult intubation. Surprisingly, we found the opposite to be true when we surveyed them – the overall confidence in intubation skills for some students decreased! We believed that this dip in confidence and ability was due to an overconfidence in using the basic intubation task trainers. Students had become complacent with the basic trainers; when presented with a difficult airway patient on the high-fidelity simulator,

their confidence decreased. In analyzing the root causes for this finding, we arrived at the following:

1. When using basic skill simulators, intubation skills are taught in steps.
2. Often, during basic skills practice, patient case scenarios are not used.
3. Some students tend to be less than vigilant in wisely managing their laboratory time appropriately when not immediately supervised by faculty members.
4. In the move to high-fidelity simulation, students are presented with an entire scenario and a full body to contend with, often requiring students to demonstrate multiple skill sets.
5. The conceptual integration of pharmacology, patient assessment, the use of troubleshooting technology, critical thinking, and problem solving certainly raises the level to which students are required to perform.
6. Null curriculum may be an issue
   a. A null curriculum relates to the issues that the instructor does not intentionally plan for, but the student learns.
   b. For example, failure to orient students to a "crash cart" may be a null curriculum issue when they are unable to locate needed equipment in an emergency.
   c. Failure to orient the students to the simulator may also have negative effects. For example, breath sounds are binary – there are either wheezes or not, crackles or not. For our particular simulators, the amplitudes of various chest sounds cannot be dynamically adjusted during a scenario – they are either there, or they are not.
7. Complacency in the students' attitude and ability to intubate a difficult airway.

After discussing the various potential reasons for the drop in student confidence, we have concluded that this simply reflects a healthy respect for intubation. Confidence is a key indicator that the students are ready to progress on to their next challenge. Simulation is the ideal way to provide both the safe development of experiences and the safe evaluation of challenges required for learning. It is indeed a humbling experience not to be able to intubate a live patient the first time!

## 8.3 Low- and Mid-fidelity Simulation

It made sense to us that low-fidelity simulation, the cornerstone of our training in the past, should continue to be used for basic skill acquisition and development. We use these basic task trainers to help students learn the finer points of manual dexterity, such as manual ventilation with a mask, and endotracheal or laryngeal mask airway intubation and suctioning.

These manual skills trainers help the learner to develop the physical skills to perform the procedure and internalize the steps necessary for a successful outcome. Eventually, these learners became very adept at performing these basic skills, even to the point of being cocky. Once they have achieved this level of performance, we begin to introduce more difficulty and more adjunctive equipment incorporated into full scenarios, adding pressures to further prepare these individuals for the clinical practicum experience. Once they have achieved a mastery level of the basic skills required, the student progresses to the high- (or higher-) fidelity simulators to help hone their ever-growing abilities.

We identified that one of the bottlenecks in our student practicum rotations was in the neonatal and paediatric intensive care units, which is a joint practice venture between our clinical partners and the Institute. Our intention was to better prepare students for these clinical rotations. We were able to secure a one-time provincial government grant that allowed us to purchase some much-needed equipment, as well as pay for practicing respiratory therapists to assist us in developing a curriculum to address this issue by offering additional training time for our respiratory therapy students.

We offer this preparatory week of training four times per year to respiratory therapy students about to begin their 8-week neonatal and 3-week paediatric intensive care rotations. Here, industry personnel and the Institute's faculty work side by side to present the curriculum, much of it involving case scenarios using mid-range simulators. The curriculum for this preparatory course focuses on getting the student capable enough to ready an admission bed, secure the patient's endotracheal tube, and place the patient on a mechanical ventilator. The feedback received about student preparedness from other student preceptors has been positive. Clinicians now believe that our students are better prepared and have quantified this amount of benefit. These preceptors estimate that with this additional preparatory training, students entering the neonatal or paediatric intensive care units for the first time are now operating at a higher level compared with students who have not had the preparatory training by an equivalent of one week of clinical experience.

## 8.4 Standardized Patients

We were fortunate in 2004 to have been approved for an educational grant from the Government of Alberta's Ministry of Health and Wellness, and administered by the Ministry of Advanced Education to employ both high-fidelity simulation and standardized patients at the Institute. It was encouraging to know that the Ministry of Health and Wellness was collaborating with the Ministry of Advanced Education to provide the greatly needed resources in the provision of health care education! To our knowledge, we were the first respiratory

therapy program in North America incorporating standardized patients into our curriculum. We have learned much from our standardized patient experiences, and they from us. Students love working with standardized patients, and these "patients" genuinely enjoy working with our students. We have had feedback from these "patients" telling us that our students had more empathy, and acted more professionally than some of their previous health care student encounters.

The course was developed to teach the students basic patient assessment skills that they would be expected to perform on real patients in a few months. Before simulation, the skills were discussed in class, they performed them on each other, and then they were tested didactically. After didactic testing, the students normally went out on clinical rotations for their first "real" contact with patients. Bringing standardized patients into the course seemed to be the perfect fit to help better prepare students for the real world. Our program did not realize the magnitude of our students' gap between their classroom and laboratory learning and their clinical experiences until our use of standardized patients.

Our first event was with 35 second-year respiratory therapy students whose only experience was a 2-week clinical encounter at the end of their first year of training. This pilot was invaluable in the planning and vision for a second attempt at the use of simulation with standardized patients. The second occurrence with standardized patients for respiratory therapy program was with first year, first semester students in a patient assessment class.

First, a little background about these particular students: they were a class of 38 students without any prior "professional" clinical experience or contact with patients in any setting. The students had been in the program for only 3 months and the only experience previous to standardized patients involved didactic lectures, watching video recordings, and practicing on basic simulator mannequins and each other. They had received all didactic information on how assessment skills were to be performed. Their knowledge was tested using short-answer and multiple choice questions prior to seeing the standardized patient.

Once it was decided to incorporate standardized patients into the assessment course, we were challenged with exactly how these "patients" would be employed. There were so many options! The original vision for using standardized patients was to test competency-based performance objectives as described previously. When taking into account a variety of factors such as instructor time, standardized patient budget, and the time allotted for the class, our vision for incorporating standardized patients changed quite drastically from using these highly trained actors for testing performance-based objectives into using simulation to simply "be" the students' first clinical experience in demonstrating their assessment skills.

After agonizing over the objectives of this project, there was still the plain and simple fact that each student would have limited time to spend with the standardized patient. The skills and objectives were narrowed down to the following:

- Perform a patient interview and history taking.
- Obtain a noninvasive blood pressure.
- Determine heart rate.
- Determine respiratory rate.
- Perform the respiratory assessment skills of chest inspection, palpation, percussion, and auscultation of breath sounds.

Even these few objectives seemed overwhelming since taking a history and interview alone could take an hour or more in some situations. Some individual objectives were reduced even further by focusing on some of the most important basic questions to ask a patient from a respiratory therapy perspective. We ensured that there was also some leeway depending on the answers the patients gave, which were determined by the standardized patient's character and history. We decided to use three patient characters: emphysema, congestive heart failure, and a young female asthmatic patient.

## 8.5 Preparing Students for Standardized Patients

This is where the vision for the project was revealed. Once the objectives were identified, the students were required to practice the skills and build confidence before they would see the standardized patients. Students were informed that they would be using their assessment skills on standardized patients. Incredulously, students appeared more focused and passionate than in previous groups of students at this stage of their training without standardized patients. In retrospect, our practice of scheduling students for a short clinical rotation 4 months after an assessment course seemed ridiculous, especially after using standardized patients, as the valuable lessons learned would be lost. Simulations incorporating both task trainers and live standardized patients were excellent ways to keep skills current during the long gaps between when the concepts are introduced and when they are employed on real patients.

Groups of three or four students were set up to rotate every 15 minutes through five skill stations, each with its own objective. The students were preparing to eventually spend an hour putting all of the individual skills together to demonstrate a complete respiratory assessment on a standardized patient. In order for the students to be able to see the entire flow of an assessment, they had the opportunity to watch a video encompassing an entire respiratory assessment and observe a respiratory assessment demonstrated by an instructor.

As the day of standardized patients quickly approached, we felt the nervousness and tension building in the students despite the time they had to prepare. Just the thought of touching and talking to a "stranger" had some students feeling nauseous. Others were nervous to take a history from a "professional patient" and some felt anxious to perform all of the skills at once. The standardized patients were prepared for the limited experience and background of first year students and were made aware that this set of students had never touched or spoken previously to a patient.

## 8.6 Suspending Disbelief

We had outfitted a borrowed diagnostic sonography laboratory so that it looked like an Emergency Department with beds, curtains, and monitors. Each standardized patient was dressed in a gown, sitting on the end of their bed waiting to be assessed. The overall success of this simulation setting was mirrored in the students' faces as they walked into their first practical experience! The same sights, sounds, and smells associated with an Emergency Department were very much present, which supported the suspension of disbelief. The same facial expressions, nervous behaviors, expression of anticipation and trepidation were displayed by each student as all clinicians have expressed during their first patient experiences. For their first time with a real patient, students had to perform infection-control procedures and do their own introductions as they began. Although these standardized patients were unable to assess the procedural skills of the students, they were asked to assess their soft skills and professionalism and to describe how they felt being under their care. They were encouraged to be very forward and honest about their experience as a patient by providing constructive feedback.

Owing to time constraints and the lack of availability of additional instructors for this practical session, only one instructor was in the room for five standardized patients, each having a pair of students attending them. The noise in the room was similar to a normally busy Emergency Department. Even though the student-instructor ratio was less than optimum, the instructor was able to spend time watching and assisting each group when required. The success of this laboratory session was likely due to the appropriate student preparation for this experience. Since the goal of the simulation project was to simulate a student/patient encounter in the Emergency Department, the instructor only intervened in the assessment if absolutely necessary or if a group asked for help.

The instructor participated in the standardized patient feedback process where students were told what it felt like being a patient under their care. The instructor also provided feedback to each student regarding their assessment and interview strengths and weaknesses. General feedback was shared with the entire class once everyone had completed their assessment and had been debriefed.

We soon learned that the biggest reason for using standardized patients in the assessment course was going to be because of an unanticipated gap between clinical rotations and the classroom that we didn't even realize existed until the end of our pilot. The students began participating in the simulation laboratory and classroom lecture more seriously knowing they would have to perform on a standardized patient within a relatively short period of time. Students are now better prepared after touching their first patient in a simulated patient care setting rather than in a real clinical setting. The feedback provided by the standardized patients allowed the students to adjust their technique and attitude before going out to the clinical setting. This resulted in students:

- feeling better prepared for clinical rotations,
- getting over with that first upset feeling in the stomach,
- anticipating the feelings they may have in stressful clinical situations,
- remembering the small yet still very important stuff – close curtains while listening to someone's chest, the importance of introductions, and sitting at patients level, all discussed in class and forgotten.

## 8.7 Summary of Standardized Patients

Incorporating standardized patients into the curriculum is beneficial. Interacting with standardized patients instills confidence in students, ensures a more professional approach to patient care, and allows the students to evaluate their first jitters and patient experience outside the "real" clinical setting. Students enter their first clinical rotation feeling that they have some experience and already have a list of do's and don'ts that they can continue to build upon. The student's first encounter in touching a complete stranger and the nervous anticipation before and during the experience allows the student to conceptualize and identify what emotions they may expect in the clinical environment. Because these first feelings of nervousness have already been dealt with during the standardized patient encounters, students can focus more on learning in the clinical environment rather than dealing with the emotional firsts.

The use of standardized patients during the didactic part of the program inspired and, therefore, motivated the students to better understand and practice their skills. The use of standardized patients also allowed the student to practice communication and other soft skills while providing the opportunity to try different interview methods to make the patient feel more comfortable. By practicing these soft and hard skills in a relatively safer simulated clinical environment, students internalized and reflected on their learning. In other words, students demonstrated these assessment skills not only in the classroom setting but equally well in a stressful patient encounter. Since the student-instructor ratio for this project did not allow for us to test or evaluate competencies, it is difficult to comment on the use of standardized patients for objective testing, but it is something we would like to undertake.

The best example that exemplified a need for simulation was the story of a young male first year, first semester student. He was placed with a female standardized patient of an age similar to his own. The student was extremely nervous, almost to the point of paralysis. He required my assistance to get through the assessment. In the debriefing session, the student expressed shock at his own nervousness and related how the situation took him off guard. The student was very glad that he was able to experience these strong emotions in this safe and familiar educational environment rather than in a more uncontrolled clinical setting. After taking some time to internalize the experience, he expressed through the debriefing process that he had never actually thought about having to perform a chest assessment on a female patient of his own age!

The standardized patient's feedback to the student was paramount for the hard skills part of learning to occur. These patients are able to communicate whether a touch was painful or the patient felt uncomfortable whereas a mannequin cannot. One standardized patient told a student they felt like they were being interrogated during the interview process. That student had no idea how he was being perceived. This feedback motivated the student to focus on improving his soft skills.

It is difficult to comment on when simulation using standardized patients should *not* be used in the curriculum based on the fact it was beneficial in the practical sense for this group of students. We don't think simulation using standardized patients should be used as part of the original learning and practice of a skill. The best fit from our experience is to apply practiced assessment skills on a standardized patient before a clinical experience is scheduled. This was the gap mentioned earlier that, in our opinion, we didn't even realize existed until we used this type of simulation. After using standardized patients in our patient assessment course, it is difficult to imagine these classes without them.

## 8.8 Attrition in Health Sciences Education

From high-fidelity simulation to demonstrate life-saving skills to standardized patients, there are still many opportunities to incorporate simulation into the educational process. Governments and educational institutions are pressured to show positive outcomes where public or private money is concerned. The youth of today is bombarded with an almost limitless abundance of career choices; this makes it exceedingly difficult to choose. Some students enrol in a course of studies and are content. Others register for a program, only to learn that they would rather not be there or that the job market in that particular profession has all but closed. Still others learn that they really are not suited for a particular profession, leaving the program with one less seat filled. There is a cost to government, the educational institutions, and society as a whole due to this brain-drain phenomenon. As our population ages, it will become a necessity for our health care programs to enrol and maintain full placement – we will need these health care workers in the not-too-distant future. We can no longer afford the luxury of attrition in our health care programs!

There are many reasons why students choose to leave the program in which they are enrolled. Changes in career choice and poor academic performance are the leading causes of attrition in our respiratory therapy programs [4]. This is costly to the program whose base funding may be affected because of key performance indicators reported to government. It is also costly to faculty who are required to spend extra time with the poorer performing students, only to have these students leave the program anyway. To predict academic performance, we might use the results from one or more of the various types of standardized examinations such as the Medical College Admissions Test (MCAT), Graduate Record Examination (GRE), or Scholastic Aptitude Tests (SAT); however, none of these tests can predict how well a student will perform in a clinical environment.

In choosing a career path, it is prudent to perform some form of career investigation to fully appreciate the various intricacies of the job. With recent Canadian government legislation aimed at protecting the privacy of the patient, it is becoming increasingly difficult for prospective students interested in health-related careers to make informed career choices. Currently, only clinicians actively treating specific patients or students assigned a patient may have access to those particular patients. The patient's privacy is essential, which translates to the fact that only employees of the hospital or authorized students of an educational institution are allowed to see a patient. This severely limits any clinical access a person interested in a particular clinical field can have. With these limitations, how is a person interested in a career in health care supposed to be able to truly understand what it is like caring for the ill when they are unable to see a patient, or tour a critical care unit to fully appreciate the working environment they aspire to join?

While growing up, children are always stating that they want to be a doctor, a teacher, or a nurse. Invariably, this epistemological concept is not based upon sound facts and individual abilities, but rather on emotion or because of an experience seeing a "nice" nurse, or "kind" doctor who made you feel better when you were sick. This lack of foreknowledge becomes a gap – a costly vortex where students, governments, educational institutions, and in fact society as a whole will lose. For a minimal tuition fee, learners interested in a health care – related program could be provided with a short, 1-week "So You Want to Be a Clinician" program. Students could spend a career day learning some basic physiology, anatomy, and some of the basic equipment used in the field. At the end of the course, these students would be required to care for a robotic patient using all of the tools and knowledge that they have gained over the course. This may serve several purposes, the first being that they get excited about their new skills and about caring for those in need. Secondly, they may have gained a better understanding of the roles and responsibilities of the health care provider. Thirdly, certain behavioral attributes may be identified that could determine the clinical suitability of the candidate for the program. Simply stated, simulation could provide educators and admissions committees with a much needed, legally defensible tool that could provide specific information regarding student selection. Students would also have the necessary tools to make a truly informed decision regarding their career path.

With this concept in mind, we challenge you, the simulation practitioner, to think outside the box. Think outside the normal realm of teaching and learning in order to positively affect the educational system as a whole! Perhaps, learning through play in a setting such as career investigation may help to ensure that adequate numbers of health care providers are there for those who need them most. Who are these needy individuals? Why, us of course! Well, maybe in 10 or 20 years anyway.

## 8.9 Conclusion

If you are reading this book, you are already a simulation convert and are actively engaged in simulation, or at least looking for answers to your own "So You Want to be in the Clinical Simulation Game" questions. Simulation can take all forms, from discussing a case study to the use of standardized patients to using low- and high-fidelity simulators. Essential to their effective use are educational objectives. Without these educational objectives, neither faculty nor students will know what they are supposed to learn using these tools. Further steps to take on the road to simulation include the development of performance-based objectives. These tell the learner what exactly they are supposed to be able to demonstrate and under what conditions. Not only will this aid in learners knowing what skills they are required to demonstrate, it may also help provide students interested in a health-related career an efficient way to conduct a meaningful career investigation. Asking a paramedic to extricate a patient from a collision in the dark, save for a headlight, is completely different from asking him/her to demonstrate this same skill in simulated clinical setting unless that simulated environment is dark as well.

Standardized patients can be used in a number of ways, and do provide the much-needed excuse for students to provide the "human touch" that seems to have been lost along the way in our quest to provide the factual content in clinical education. Go ahead – reward yourselves by thinking outside the box. We did!

# References

1. NAIT Board of Governors. (2004, May 3). *NAIT four year business plan: 2004/05 – 2007/08.* Northern Alberta Institute of Technology, Edmonton, A. B: Author. Retrieved April 19, 2006, from http://www.nait.ab.ca/calendars/ bp2004.pdf

2. Bates, A. W. and Poole, G. *Effective Teaching with Technology in Higher Education: Foundations for Success.* Jossy-Bass, San Francisco, 2003.

3. Hamilton, B., Norton, R. E., and Fardig, G. E., et al. *Module B-2 Develop student performance objectives.* American Association for Vocational Instructional Materials, Winterville, GA, 1983.

4. Douce, F. H. and Coates, M. A. Attrition in respiratory therapy education: Causes and relationship to admissions criteria. *Respiratory Care* **29**(8), 823–828 (1984).

# Simulated Realism: Essential, Desired, Overkill

Judith C.F. Hwang

Betsy Bencken

## 9.1 Realism: What Is it Good for?

The amount of realism required depends on the goal of the educational session:

- There is a difference between teaching a task or a skill versus teaching a concept.
- Different learning objectives call for different teaching tools: You don't need a highly realistic, comprehensive, multicapability US$250,000 patient simulator within its US$500,000 room to teach introductory airway skills to a large number of novice students. You don't want to risk damage to your expensive full-function patient simulator from *any* introductory part-task skills learners.

## 9.2 Dexterity Skills

If the goal is to teach a particular skill and have the students develop their hand–eye coordination, the realism can be limited to the equipment needed for the procedure and the part of the body being affected. For example, when teaching intravenous (IV) catheter insertion, an isolated hand or arm is just as good, as well as being more portable, easier to store, and significantly less costly than any low- to high-fidelity full-body patient simulator (Figure 9.1).

In fact, a simple "IV board" that has multiple veins of varying sizes and at varying depths may be sufficient or even better, though a bit less realistic in appearance overall, introductory teacher of this task or skill (Figure 9.2).

## 9.3 Human with Human Skills

If the goal is to ensure that the student can not only perform a given procedure, but can also simultaneously interact with the patient, it may be useful to combine an isolated hand or arm with a standardized patient (e.g., strap the device on the standardized patient) or to use a low- to mid-fidelity patient simulator whose arm can be cannulated and who can be equipped and instructed to speak with the student. Using high-fidelity patient simulators, however, to teach the skill of IV placement would be considered as an excessive and overpriced realism. On the other hand, one could rationally employ high-fidelity patient simulators when the focus ("objective") of the session includes competency of a skill while dealing with a clinical complication. For example, when the patient experiences an adverse reaction during the IV placement procedure, such as a vasovagal episode, the student must expeditiously gain venous access to treat the patient appropriately, while calling for help ("situational awareness").

## 9.4 Purpose of Realism

Every simulation session must have enough realism for the participants to become fully engaged in the scenario. They must believe and act as if the patient simulator is someone for whom they are responsible and must provide appropriate care. Having a mannequin that blinks and speaks is wonderful, but for many participants that is not enough to overcome the sense of playacting. The clinical content of the scenario itself must be

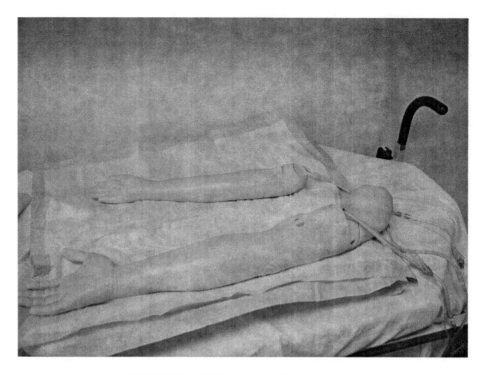

**FIGURE 9.1**   Child and adult IV trainer arms.

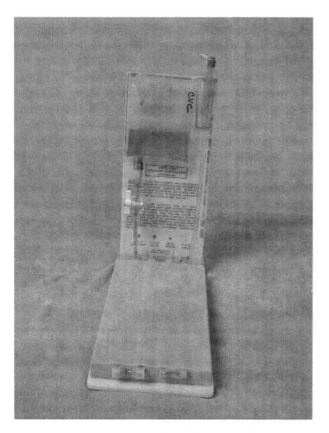

**FIGURE 9.2**   IV board with various-sized veins at various depths.

realistic and the professional attitude of the clinical instructors must be realistic. One simple, no-cost technique for realism: the patient is never "turned off," that is, appearing to be dead when the students are in the room, unless the patient's death supports that session's teaching objective. Often, the "hook" of an adverse event requiring them to take action is needed to fully engage the student.

## 9.5 Learner-centric Realism

For all levels and topics of teaching, the realism that is required depends on the level of the learner. For example, medical students are often sufficiently challenged by the process of integrating the patient's history with the signs and symptoms of their disease and then determining the best course of action. While realism is helpful, their clinical inexperience means that though the patient needs to be realistic, any deficits in their surroundings will be less noticeable. Conversely, residents need to have not only realistic patients, but also surroundings that more closely mimic their usual clinical settings. Otherwise, the differences between their typical clinical setting and the simulation setting tend to distract them from the focus of the simulation, which is providing patient care. In addition, the lack of realism in the surroundings provides the residents with a potential built-in excuse for any behavior or lack of action that occurs during the simulation.

If the goal is team training, the level of realism is again dependent on the level of the learners as well as the need to fully engage them. Often, the fact that the learners are fully immersed in the simulation scenario "blinds" them to their simulated patient's shortcomings and the less than perfectly realistic surroundings.

## 9.6 Conclusion

The amount of realism required depends on the goal of the educational session. It should be just sufficient to convey the specific teaching objectives to the given level of learners, and no more so.

# Realism and the Art of Simulation

Roger E. Chow
Viren N. Naik

Our Patient Simulation Center is located at St. Michael's Hospital in downtown Toronto, Canada. This inner-city teaching hospital is fully affiliated with the University of Toronto, and is one of the busiest trauma and emergency centers in Canada. Our Simulation program services all disciplines and is under the Department of Clinical Education. Our simulation experience has been gained from the 11 years we've been in existence and from interactions with a supportive and sharing simulation community, locally and internationally.

## 10.1 Simulated Realities

In 19th Century Art, painters in the style of realism drew their inspiration from life around them and cherished their interest in visible reality. They accepted the doctrine of imitating nature and striving for realism in their art. In crafting a painting, an artist need only provide so many elements to create the illusion of reality. In other words, not every single detail is needed to create a complete picture for the viewer. Seeing a painting of a head from the neck up exhaling smoke, the viewer will likely conclude that the person is not internally on fire but is smoking, even though objective evidence is not visible in the painting.

Patient simulation offers clinical educators a new type of "canvas" where they can apply the same principles of realism. We can't perfectly duplicate or replicate reality with simulation and we don't need to, but we can present cues that are sufficiently realistic to get buy-in and elicit desired actions and behaviors from the learner. A fake wound on the mannequin's back with a bloody sheet underneath and a low blood pressure should lead the learner to believe there is significant blood loss occurring with their "patient."

Realism is what you make of it, using what you have to work with to convey a scenario. The Oral Tradition of telling tales is a fundamental teaching tool common across all human societies. The mere creation and reception of sounds was all the fidelity employed. The teller and the listener provided all the realism. With limited resources, the creators focused on the intangible aspects of the human condition to convey their tales. When crafting a simulation scenario, the realism worth considering is that which will stimulate the learner's imagination.

While having the latest high-tech robotic patient simulators emplaced within fully functional clinical spaces is great, that alone doesn't provide enough realism. Naïve people are generally impressed by the mannequin and its capabilities, but the "patient" is the most artificial thing about simulation. It's more important to use the technology as a lens to focus on the human aspect of interactions. In our quest for realism, we create context by placing the mannequin in a simulated or real environment, breathe life into it with human interaction, and take advantage of the learner's imagination.

## 10.2 The Orientation

To facilitate learners being more receptive to us, our Center always presents a supportive and nonthreatening atmosphere. Our goal is to provide experiential acquisition of skills and judgment, allowing lessons to be learned from errors made. Before starting, learners are familiarized to the environment and the mannequin, and provided with a set of ground rules. This is done on a need-to-know basis, being careful not to cognitively overload the learners with too many rules. There is no need to share all the capabilities and limitations of simulation or to explain how the props work. Keep it simple and user friendly. The rules will vary with the different levels of learners and with how they utilize simulation. Directing learners on how to approach simulation makes it easier for them to participate fully, once they have a better idea of the playing field and rules of the game.

Our primary rules are:

- Approach the mannequin like he's a real live patient, talk to him, and he will respond realistically appropriately or not.
- Do what you would normally do and treat the situation like it's real.
- Ask for whatever equipment, procedures, or services you want.
- A drug isn't given until it's given, please announce which medication and the dosage.
- There will be a "clinician" in here to help you, so don't worry about what you should or shouldn't do, but this helper usually never initiates, that is your responsibility.

In our early days of simulation, the orientations were done poorly. Everyone usually received a familiarization, along with an inconsistent set of ground rules and presentation of realism. Learners were told any number of the following: pretend to do chest compressions and defibrillation (due to limitations of the MedSim Eagle mannequin), inject the intravenous drugs in this port, and if you can't find the drug you need then just pretend, but call out what and how much is being given, you need to ventilate your "patient" for real or he'll go into a death spiral desaturate and arrest, auscultate for breath sounds at the nipples where the speakers are because there are acoustical dead zones, do not restrain your "patients" arm movements or

you will strip the gears, and the list went on. There were times when we would direct the learners with too many rules, and times when we neglected to inform them with enough. We came to realize the negative influence it had on our simulations and ultimately on the learners.

One particular simulation that brought this to light for us was an operating room scenario: **The OR Pipeline Failure.** The learners are Junior and Senior Anesthesia Residents, and the case is a carotid endarterectomy. After cross-clamp of the carotid artery, an OR pipeline and backup gas system failure occurs, complicated by an empty backup oxygen tank at the rear of the anesthesia gas machine. The objectives are to recognize the gas failure, secure a backup oxygen source, a self-inflating manual resuscitator, ventilate the "patient" with room air until an $O_2$ source arrives, and eventually change the $O_2$ tank at the back of the gas machine. The learner was a Senior Resident who was outstanding in the real operating room, and had handled previous simulent off, she quickly recognized the failure of her anesthetic gas machine and secured the manual resuscitator. She hooked up the resuscitator to an oxygen flowmeter on the wall outlet and proceeded to turn it on. The ball of the flowmeter went up and came down quickly as it depressurized. She watched this happen, hesitated for a moment, shrugged her shoulders, and proceeded to ventilate the patient. Afterward, we discovered that she assumed the $O_2$ flowmeter was one of the pretend elements, accepted it and continued the case. This didn't change the outcome of the scenario as she realized the objectives, but it was revealing to us the errors made on our end. Interesting enough, the residents loved this scenario because they valued highly the lesson of how to change the $O_2$ tank applying the Pin Indexing Safety System. This scenario also provided us with unexpected data implying a gap in the Residents' education.

The lesson learned was that our orientations were setting a confusing tone for the simulation, and as a result, learners were being misled. For learners with previous simulation experiences and performing at high levels, there is less room for mistakes on our part. Following this incident, our orientations have become consistent, clear, and concise with a distinct purpose. We know the objectives before the simulation and focus our orientations toward that. We now use the orientation to put the learner at ease, reduce the gap of unfamiliarity with the mannequin, and to set the simulation off in the right direction.

## 10.3 The Medium of Simulation

In crafting their work, artists will respect the limitations and capabilities of their medium. The medium is the materials which an artist chooses to use, whether it is collage, oil paints, water color, stone, etc. With patient simulation, the medium is a collage of the clinical scenario, the simulation theater, the mannequin, props, and human elements.

## 10.4 Clinical Scenarios: Who are the Scenarios for?

Write the scenarios to address the learning objectives, but design the simulation to the level of the learners, keeping in mind the limitations and capabilities of the medium:

- The best scenarios are taken from real life or clinical events, producing believable simulations and helping to deal with the "That will never happen in real life" issues. The OR Pipeline Failure scenario described earlier was taken and modified from a real-life event. What truly transpired was a construction crew digging right in front of the hospital and cutting into the underground lines, causing simultaneous failures in oxygen and electric power.

- Develop scenarios relevant to your environment. Try to reproduce clinical content unique to your hospital. For example, if your institution is located in a colder climate, create scenarios that encompass hypothermia (ice bags on the simulator's hands and feet for a few minutes before the students come into the room are a great way to provide thermal cues).

   Three years ago, our city suffered an outbreak of Severe Acute Respiratory Syndrome (SARS). As a result, we ran SARS scenarios to help anticipate and troubleshoot potential problems with arrests and infection control. All learners happily participated fully, including the cumbersome gowning and degowning process.

- When fewer resources are available, try presenting the scenario in a different context. As a teaching hospital, many clinical situations have unlimited staff and resources during daytime hours. Simulation can't provide all these resources, but we can manage this limitation by changing the context of the scenario to the setting of a rural community hospital in the middle of the night.

- Explore different scenarios keeping the versatility of the medium in mind. Look at scenarios that can promote interprofessional education, bringing different groups together to manage a clinical problem. Use the scenarios to create situations that will address behavior and communication. Find scenarios that address your institution's and students' future needs, like patient safety. These are all real issues for future clinicians and their patients alike.

An example of versatility is the variety of different groups that scenarios can include for simulation. Our hospital Chaplaincy and Spiritual Care Department utilizes patient simulation in a most simple but effective way. The scenario has a Chaplain Intern on her way to the Neuro-Trauma Intensive Care Unit, to meet with the parents of a 22-year-old fatally wounded gun shot victim. When she arrives at the "ICU," she will be immersed in an unfamiliar environment, seeing the "patient"

intubated, ventilated, and connected to a host of different equipment. The injuries sustained are too extensive and the parents have agreed to withdraw life support. They are by his bedside angry and confused, and have asked for the presence of Chaplaincy for the withdrawal. Some of the main objectives for the Chaplain Interns are to practice and improve their ability to provide spiritual care, and to increase their ability to function as a fully participating member of the health care team.

If scenarios can exploit the capabilities while avoiding the limitations of the medium, then a higher degree of realism should prevail. Realism will be further heightened once we place a scenario into a simulated clinical environment with the props and include the human elements.

## 10.5 Simulation Theater: Design for Effective Story Telling

You can approximate the clinical reality of a physical space by recreating it in your own simulation environment with equipment, props, and a mannequin. Alternatively, you can bring the mannequin into the real clinical environment and produce simulations there. To duplicate clinical settings, visit different areas in the hospital and gather the necessary references. If your resources are unlimited, then duplicate the settings down to every last detail if you want (which isn't necessary). With limited resources, try to approximate just those items essential to establish a sense of place and provide devices with just enough functionality that the human senses directly perceive. Remember that not every physical detail is needed to convince the learner of where they are. If the resources are severely limited, dress up the physical space with whatever equipment is available, and remember not all of it has to be functional. Beg, borrow, or whatever. Some institutions/centers use painted/printed backdrops and curtains, similar to those used in a real theater – these are rapidly changeable. Once you add the mannequin, a few props, and some humanity, there should be enough elements to convince the learner.

- Think the scenario through as you write the script, perform a technical rehearsal, and then a full-dress rehearsal – just like in live theater. This should help establish what equipment and props are needed and reveal possible shortcomings that require revision. Put yourself into the shoes of a learner, act out a specific scenario, and explore the possible routes they may take.

- The theater should be equipped with a public announcement (PA) system and a telephone (the former is easily installed hardware of microphone, cable, and powered speaker). This PA system provides overhead calls for arrests, traumas, pages, etc. A duplicate configuration can be installed under the mannequin's head to provide a quality "patient's" voice.

- The telephone allows a learner to call for X-ray, have someone paged, or order blood from the blood bank. The phone calls are routed into the control room where someone will answer as the appropriate "personnel" and interact with the learner in the scenario. Conversely, the control room can call back into the theater as the service being paged, or as the laboratory with the results.

- The telephone also provides a communication link between an actor in the theater and the operator in the control room. Phone calls are excellent ways for actors to exchange stage directions with the hidden simulator operator without the learner's awareness. This is an important link as simulations are dynamic and encompass many details. Communication between the actor and control room operator is inevitable, as situations will arise during the simulations that require this link to help troubleshoot the problem (malfunctions, missed details, etc.) and preserve the realism.

- A major barrier to realism occurs when the learner walks into a highly realistic "clinical environment" and is confronted with an artificial patient. The most powerful way to instill reality around an obviously unreal mannequin is for the instructors or simulation coordinators providing the orientation to behave toward their simulated patient with all the respect and human-to-human interaction that they (should) do with live patients. Likewise, the simulated patient should respond in character to the extent that it can be made to do so. All people while in student status are Lorenzian ducklings, and will all too blindly accept and respond to their instructor's directions. Make extensive use of this innate ability.

- The second most powerful tool you can take advantage of here is to provide a high-quality human voice for the mannequin. Adding a believable voice to the mannequin brings it to life, and the "patient" can now engage the learner, making a human connection. The mannequin becomes a real "patient" when it can offer feedback and challenge a learner for not observing patient dignity by asking "Hey who are you and what are you doing to me?" It can add to neurological assessments, providing responses for physical examinations. The "patient" can also take on different personalities; scared, aggressive, unstable, offensive, or abusive. Write these strategically into the scenarios to heighten stress levels as well as to create intentional distractions.

- Finally, a voice can be used to add human context to a scenario. A "patient" can say in a worried voice "where is my family?" or "I feel like I'm dying, please call my wife." These are not always scripted into the scenario, and are sometimes improvised depending on the progression and level of buy-in from the learner.

- The mannequin can be further enhanced with a little spontaneous movement, if it has this capability. Our Eagle Medsim mannequin can move his arms and blink his eyes. With our Laerdal SimMan, we expand the use of dynamic functions like trismus and decreased cervical range of motion to mimic normal patient movement. Even though they were not developed for these reasons, it works effectively when timed appropriately. These functions create only subtle movements, but they add another dimension to the mannequin which further humanizes it.

## 10.6 Props: Place and Purpose

Props can be just one of the many elements enhancing the realism or they can have a more significant role influencing the learner. Props provide two essential contributions to simulation success: a sense of place and tools to perform expected actions. They can be as simple as a large printed image representing a wall or more complex like a fluid-feeding "dry" suction system. The following is a short list of examples:

- Patient Charts and Identification: Both of these props will help humanize the mannequin. Some scenarios require a patient chart and others will need operating room records. Obtain blank forms from your hospital and fill them with clinical information relevant to the patient and simulation. Allergy bracelets are simple accessories that can be used to provide vital cues in a scenario.

- Images: Images include X-rays, computerized tomography, electrocardiograph (ECG) tracings, etc. Images will support patient symptoms and provide the learner with a broader context. Acquire real images or get them from the Internet. Use hard copies or electronic ones with their respective viewing devices ready for the learner to see. These props are easy to obtain and can be banked for use in future scenarios. Do remove all real patient–identifying information, and where possible, replace with that of the patient in the scenario.

- An Intravenous System: This prop is both visual and tactile for the learner. If they decide to increase fluids, they must perform the action and confirm administration to the patient. Have functional flowing IV systems that can support fluid drug administration though a cannulation in the mannequin's arm and into a hidden reservoir. Another option is having a closed IV system consisting of a source bag connected in series by two sets of IV tubing to a receiving bag. The IV tubing is taped to the mannequin arm and hidden under gauze, giving the illusion of the cannulated system. With this closed system, the liquid from full bag flows through to the empty bag, and when the IV runs "dry," you can discreetly reverse the system for your "new" IV bag.

- Remember that much like live theater, things don't *actually* have to fully happen for the viewer (in this case, the learner) to believe it's happening. For example, a learner opens an IV valve and sees a solution flow into the patients arm, followed by a change in heart rate and blood pressure. They don't know that the IV tubing is just taped to the mannequin arm and is actually flowing into a hidden collection reservoir. Likewise, hidden pressure bags can propel fake blood and other fluids into the student's field of view.
- Medications: We use containers of expired drugs if they are available. Otherwise, we use prefilled syringes with water. It's a limitation, but usually not a factor affecting our simulations. The drugs are still injected for real through an IV port or Luerlock stopcock. We accept this as one of the "pretend" elements, as most of our current scenarios are not focused on drawing up medications.
- Suction System: Suctioning the mouth of full-function mannequins has always lacked acoustical and visual feedback for the learner since really introducing liquids of (simulated) vomit would most likely destroy the airway and lungs. This can be overcome with a self-feeding suction-fluid system. An internal system feeding the tip of a suction catheter provides fluid ("blood" or "secretions") to be suctioned into a canister. The learner gets tactile feedback and the mannequin is protected. Please see Appendix 10A.1 "The Wet/Dry Suction System" for construction and use instructions.
- Wounds: There are different ways to dress up a mannequin with wounds. One way is to make them with clear craft "window" paint and some acrylic colors. Even if the injuries look a little fake, it's one less element of pretend. The visual cue is better than having the learner ask "what do I see here?" or us saying "Pretend you see this here . . ." Imagine producing a simulation with a plain mannequin who was involved in a bicycle crash, has respiratory distress, tachycardia, and decreasing oxygen saturations. Now, picture the scenario again after dressing up the mannequin with bruising along the ribs, facial abrasions, and cuts on the hand. Reusable wounds are made quickly and easily, and are placed anywhere on the mannequin. Peel them off afterward and store properly for future use. Please see Appendix 10A.2 "Making Wounds" for materials needed and illustrated instructions.
- A Cell Phone: This functional prop adds reality and stress to a scenario. The cell phone can be planted in the pocket of a "trauma patient," and as the scenario unfolds, a timely call is made by a "loved one." The caller is unaware of the trauma and from there you can have the phone call play out in many different ways. It can be used to add context, increase stress, provide distraction, introduce communication issues, etc.

## 10.7 The Human Elements

For patient simulation to mimic reality, human elements need to be prevalent. It is the common thread pulling everything together. We try to define it in terms of anything that provides humanity to the mannequin and the simulation, thereby improving realism.

There is a lot of crossover and many of the elements previously mentioned have humanizing qualities. In this section, we will focus mainly on what actors can contribute to bringing realism into simulation. When we use the term actor here, it refers to one of our own simulation coordinators playing clinical or nonclinical roles. An actor can:

- Simplify the orientation with fewer "ground rules." The learner can be instructed with "treat the patient like he's real, do what you would normally do, there will be a clinical person in the scenario to help and guide you if needed." This makes for a simple and concise set of rules.
- Reduce the gap of unfamiliarity to the environment, allowing the learner to behave more naturally. The presence of an actor, who knows where all the supplies are located, provides this resource. Unfamiliarity with the environment can be used as part of a scenario, forcing the learner to communicate and utilize their resources effectively. The unfamiliarity can also be an unwanted distraction leading the learner away from the objectives, as they become fixated looking for a piece of equipment and their frustration grows.
- Enhance the realism of a scenario or can be used to add stress and distraction.

Sometimes we encounter a learner who is having a hard time buying into the simulation; they go through the motions without commitment even during the crisis. The actor can recognize this situation and intervene naturally. As the "patient" deteriorates in a prearrest state, the actor will stay in character and show urgency or stress with his body language and tone of voice, as he would normally in a real emergency. When reluctant learners see someone else behaving like the patient's outcome matters, they sense the tension in the air, feel the urgency, and the simulation becomes more real for them. This initial intervention is sometimes not enough, necessitating further action. It may be that the learner needs additional context for the scenario before we can achieve buy-in with them. One example was with a hypothermic scenario we ran. The simulation was for Paediatric Emergency Fellows, and the scenario involved a baby being brought into the Emergency Department by a stranger who found him outside in below-freezing temperatures. The objective was to recognize that resuscitation with drugs is ineffective until you warm the "patient" up. When the scenario started, it was clear from the body language of the learner in charge that she was not buying into the simulation. She went through the motions of resuscitation, and kept repeating ". . . it doesn't matter the baby's cold and dead." The other participants were urgently working away with the resuscitation, articulating several times to the learner in charge that the baby should be warmed before declaring death. This comment was met with no response. The actor recognized this impasse and quickly notified the control room to phone into the theater. He made sure everyone heard the phone ringing and proceeded to answer, making up a story that the police had arrived with the parents. They were hysterical and wanted to know what's happening with their baby. Then the question "what should I tell them, they want to come in" was posed to the learner in charge. Again, there was no verbal response, but this clearly caused a physical shift in her appearance and a definite change in her approach, resembling more of a buy-in and commitment to the scenario. She then proceeded to be more aggressive in implementing additional warming techniques, before any further drug resuscitation.

The phone call provided the additional context for the scenario and the baby, and widened the scope of the realism. However, we're not entirely sure why the phone call worked, it may have been the additional context, it may have triggered a parental instinct, or it may have been the threat of the parents coming in. In retrospect, we should have asked what was going through the learner's mind at the time. To gain acceptance from the learners, we should be aware that each and every learner, and all instructors for that matter, perceives "realism" in their own way. Thus, with every session, no matter how rigid the presentation, we need to allow for these variances in perception and be prepared to improvise.

Additionally, the actor can take on roles to heighten the situation with stress and distraction. Actor roles may include family, staff, or student, and can be attributed with a host of different human characteristics. The actor can play an overprotective and emotional family member, an overbearing and challenging staff person, a scared student paralyzed with fear, etc. You can take simple scenarios made for junior learners, add these elements, and produce more complex scenarios suitable for more senior learners. Participating in simulations is inherently stressful for most learners and exhausting for most instructors – use additional stress and distraction judiciously.

- Help modify a learner's approach to the simulation if needed. Some learners come with experience in web-based simulations and will approach patient simulation in the same manner. For example, the learner asks for a medication without the dosage and proceeds to look at the patient monitor for its affect. The actor's response would be to act out what was conveyed in the orientation. He would ask the learner "how much?" or "in what concentration?" and then proceed to draw up and give the drug, followed by closing the loop with an announcement that "x" amount of medication has been given. This illustrates a desired approach for the learner, and can help redirect him toward more realistic actions.
- Offset limited resources (equipment and personnel). For example, during a difficult airway scenario, a Resident may ask for a bronchoscope and for the staff Anesthesiologist to be paged. The actor can place the calls and respond with "the staff Anesthetist has been called down for an emergency caesarean section and is unavailable right now" and "the bronch is being cleaned, it will be ready in 20 minutes." Faced with these plausible responses, the learner can move on to consider other strategies.

The actor's role is indispensable, multifunctional, enhances, and protects the realism we painstakingly create for each simulation. The additional advantage of having a simulation coordinator in the theater as the actor is the protection of the mannequin from inadvertent physical abuse. With heightened adrenaline levels, some learners can become too aggressive, and we need the mannequin alive for a successful simulation program to continue.

## 10.8 Conclusion

When you can incorporate as many humanizing elements as you can into a simulation, it goes to producing a higher caliber of realism. Take advantage of the fact that patient simulation involves the participation of human learners. As you present enough elements of reality for a simulation, their imagination is stimulated and this will facilitate the intention of achieving an interactive buy-in from the learner.

# 11

# Integrating Simulation with Existing Clinical Educational Programs: Dream and Develop while Keeping the Focus on your Vision

Judith C.F. Hwang
Betsy Bencken

## 11.1 Background

The overall objectives of this chapter are to describe the introduction of simulation at the University of California Davis Medical Center and how we have reached our current state of performance. The sections are as follows:

**Have the dream:** Expanding your vision; yet don't be afraid in starting small. Identifying others with the same goal(s). Aiming your attention on your next step while always keeping your goals in sight.

**Teaching with simulation:** Focusing on the educational objective; using simulation is not an excuse for poor lesson planning. Advising instructors on integrating simulation into their teaching and providing technical support but leaving the content development to them. Ensuring that the simulation center is open to all health care providers.

**Focus on your vision:** Simulation is limited only by one's imagination; that means thinking outside the box, creating boxes where none exist, looking at what you have available now, and where you are going.

The Center for Virtual Care (the Center) is located at the University of California Davis Medical Center, the core facility of the University of California Davis Health System.

The Medical Center, in Sacramento, California, U.S.A., is the only level 1 trauma center in Northern California outside of San Francisco and provides adult and pediatric patient care to a region spanning 33 counties, more than 65,000 square miles, and over six million residents. As inland Northern California's only academic medical center, the University of California Davis Health System admits over 32,000 inpatients and provides approximately 864,000 out-patient office visits annually through its entities – UC Davis Medical Center, UC Davis School of Medicine, and UC Davis Medical Group. The University of California Davis School of Medicine has over 660 faculty in all disciplines who teach over 400 medical students. The Medical Center trains over 800 residents and fellows in 40 medical and surgical specialties and has over 6400 full-time staff. In addition, the Medical Center collaborates with other educational institutions within our region to educate and train prehospital/paramedic students, nursing students, and other health care workers.

## 11.2 Have the Dream

The Center began with the vision and leadership of the Chairman of Anesthesiology and Pain Medicine, a department that provides clinical, educational, and research services to the communities of Northern California as part of the University of California Davis Health System. The Chairman drafted a proposal introducing the use of simulation into ongoing clinical care and clinical education at the University of California Davis Medical Center and then funded the original purchase of pediatric and adult high-fidelity patient simulators. From the beginning, the Center was intended to be a multidisciplinary center that would use simulation to improve the training of residents and nurses and hence the delivery of all aspects of patient care in our hospital. Also committed to the concept of using simulation to improve patient care and safety was the Director of Nursing who assigned valuable hospital space to house the patient simulators as well as provided a nursing educator's time to support simulation-based clinical education.

Initially, users of the two patient simulators were the anesthesiology residents and the Medical Center nurses, in part, because those two departments provided resources for the development of the simulation center and simulation was still not well known or understood by the other specialties. Most importantly, the Department of Anesthesiology and Pain Medicine and the Division of Nursing had faculty that were not only aware of, but also interested in utilizing the simulators in their educational efforts. As the anesthesiology faculty and nursing educators became more facile with operating the patient simulators and incorporating them into their respective curriculums, other clinical departments were encouraged to explore the opportunities for using simulation in their educational programs.

## 11.3 Realize the Dream

In 2002–03, the first year of operation, the simulation laboratory consisted solely of one patient room on the third floor of the hospital. This part of the hospital was being remodeled into an endoscopy suite. However, until the new unit opened, we were allowed to use this patient room as it was not going to be remodeled or otherwise used.

Three anesthesiology faculty members taught on an as-needed basis when a specific course was offered, for example, a live televised demonstration of the effects of medications used during induction of anesthesia to first year medical students taking a course in pharmacology (Figure 11.1).

The nursing educator was assigned to teach full time; he divided his time between the Center and the Medical Center's Nursing Skills Lab according to which had the best resources for the class he was teaching. His salary was fully supported by the Division of Nursing and he reported to the manager for the Center of Nursing Education. He gradually incorporated the patient simulator into the nursing skills course taught to nurses newly hired by the Medical Center as well as nurses transferring to a different unit. Throughout this first year, the anesthesiology and nursing faculty were responsible not only for teaching their students but also setting up the patient simulator, the simulation room, and maintaining the equipment. In addition, one of the information technology staff in the Department of Anesthesiology and Pain Medicine served as an ancillary technical support person.

In the fall of 2003, almost 5 months into its second year, the Center expanded as it moved the patient simulators into their current location on the first floor of the Medical Center. With the move, the Center now also incorporated two new entities: a cardiac catheterization simulation suite (Figure 11.2) and a robotic surgery program (Figure 11.3).

Taking over space previously occupied by the Department of Radiology for it CT scanners, the configuration of the rooms was unable to be changed because of asbestos in the walls. Therefore, the rooms were assigned to optimize the operational aspect of each type of simulation. Fortunately, the control rooms for the CT scanners were easily adapted into control rooms for the patient simulators while other rooms were transformed into small and large conference rooms that can be used for teaching, debriefing, etc.

With plans to increase the number of classes presented by anesthesiology and nursing as well as to start classes taught by other departments, two staffing changes occurred. First, the Department of Anesthesiology and Pain Medicine's chairman agreed to assign two of the anesthesiology faculty to the Center on a part-time basis. Their responsibilities included teaching anesthesiology residents and medical students, developing and implementing curriculum for the anesthesiology residents, providing technical support to other interested users as needed, and helping with tours of the Center. Their salaries, however, were still determined by their clinical responsibilities.

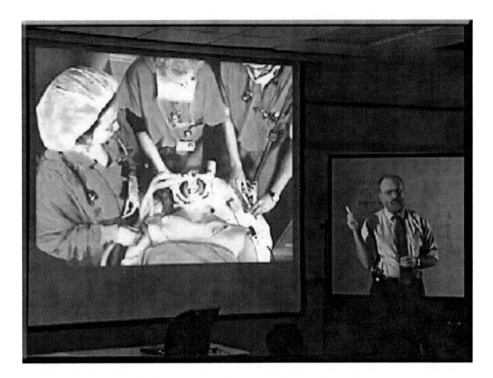

**FIGURE 11.1**  Long-distance teaching with simulation.

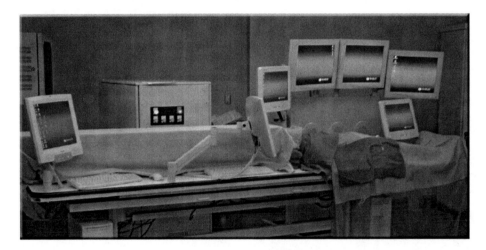

**FIGURE 11.2**  The cardiac catheterization simulation suite.

They reported to the department chairman who also served as the Medical Director for the Center.

Second, the Center hired a full-time "Jill of all trades" early in 2004. Her responsibilities included answering the telephone, managing the overall schedule for the Center (including classes, tours, meetings, etc.), setting up the simulators as well as ancillary clinical equipment (e.g., drugs, laryngoscopes, ventilators, etc.) for the simulation classes, and maintaining and ordering the equipment. Over the last 2 years, her contributions have increased to also include providing technical support for the patient simulator when simulations are simple scenarios that have been predetermined (i.e., only one or two changes need to be made during the scenario) or complex scenarios that have been preprogramed. She also serves as the voice of the patient or as a cast member in the scenario (e.g., a hysterical parent who distracts the paramedic who is trying to provide care to her child).

With an ever-increasing number of classes being scheduled and a widening group of users in our fourth year of operation, the Center is planning to hire a second support person who will

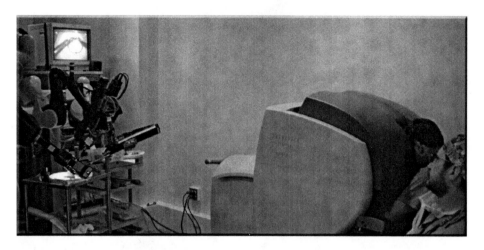

**FIGURE 11.3**   The robotic surgery program.

work part-time. Her duties will overlap those of the full-time support person as well as extend to cover the newly established surgery simulation suite. The surgery simulation suite of the Center will incorporate part-task trainers for minimally invasive surgery as well as the robotic surgery program.

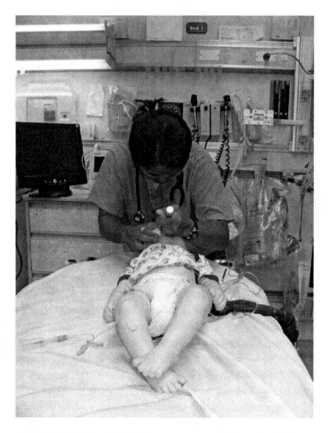

**FIGURE 11.4**   Pediatric emergency medicine simulation.

## 11.4 Teaching with Simulation

During the spring of 2004, two areas within the Medical Center that pioneered incorporating simulation into their curriculum were Pediatric Intensive Care Medicine and Pediatric Emergency Medicine (Figure 11.4).

They appreciated the ability to train their residents to respond to critical incidents in the pediatric intensive care unit and the emergency room, respectively, prior to the residents providing care in those areas. Besides these two areas, local paramedic colleges as well as the University of California Davis Medical School also were relatively quick to implement the use of simulation into selected areas of their curricula.

## 11.5 Pediatric Intensive Care Medicine

The Department of Pediatrics has a pediatric intensivist who wanted to expose residents on their pediatric intensive care unit rotation to potential patient management issues that would arise during the month-long rotation in the pediatric intensive care unit. She developed, on paper, scenarios that ranged from respiratory distress to meningococcemia to cardiac arrest in a patient with cardiomyopathy. Each patient case had its own educational objectives. With the technical help of the full-time nursing educator, whose experience included working in the pediatric intensive care unit, these scenarios were translated into programs on the pediatric simulator.

On the first Thursday morning of the month, the intensivist brings the three or four residents assigned to the pediatric intensive care unit to the Center so that the residents can learn the concepts embedded in the scenarios. While she focuses on the residents and their actions, the nursing educator ensures that the pediatric simulator has the desired clinical signs and symptoms and responds appropriately to the residents' interventions. After each scenario, the intensivist debriefs the residents while the nursing educator prepares for the next clinical scenario. Initially, the sessions lasted for half a day. Over the last year, the sessions have expanded to occupy the entire day.

## 11.6 Pediatric Emergency Medicine

The Department of Pediatric Emergency Medicine also collaborated with the nursing educator, who, once again, provided the expertise in programing and managing the pediatric simulator. In this case, there are two pediatric emergency medicine physicians who are primarily responsible for the teaching sessions. Unlike the pediatric intensive care scenarios, the pediatric emergency medicine scenarios tend to be done "on the fly," that is, the scenarios are not preprogramed. Frequently, the clinical scenario desired is one that has just recently occurred in the emergency room. Consequently, the pediatric emergency medicine physician contacts the nursing educator in the morning of the planned session and discusses what she would like to have happen. This gives the nursing educator a little time to plan different techniques to create the

clinical scenario. Over the last year, the pediatric emergency room physicians have had some of these scenarios preprogramed to provide a backup library of cases for the pediatric emergency room sessions.

The pediatric emergency room simulation sessions are scheduled for an hour and a half each Thursday morning. The sessions are attended by three to five residents at different levels of training as well as two to four fourth year medical students rotating in emergency medicine. However, the classes may be cancelled at the last minute owing to a busy emergency room. On the other hand, when the pediatric section of the emergency room is not overflowing with patients, the portable adult patient simulator is brought to the emergency room as an adolescent patient. The availability of a portable baby simulator in July 2005 has expanded our ability to simulate pediatric cases with an appropriately sized patient simulator in the emergency room itself (Figure 11.5).

Our ability to bring the patient simulator to the emergency room has two benefits. First, it has helped to avoid a potential conflict with the pediatric intensive care unit simulation sessions that occur one Thursday a month. Second, and probably more importantly, the residents are working in their usual environment, which brings a level of realism to the simulation session not otherwise achievable in the Center. This realism means that lack of familiarity with the Center's room setup is no longer a viable excuse for why the residents omitted, or took, certain actions while caring for the patient.

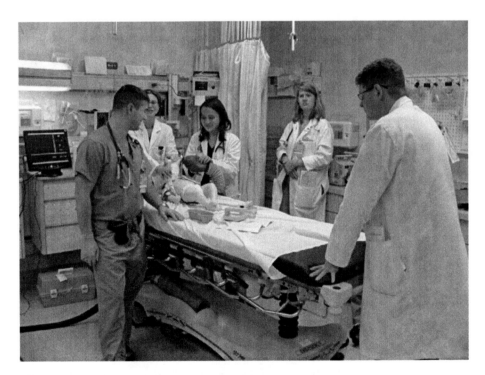

**FIGURE 11.5**  Simulation in the pediatric emergency room.

## 11.7 Paramedics

The need to train the regional and county's paramedic students in airway management led to the development of a simulation-based course to acquire those skills. In the early Spring of 2004, one of the anesthesiology faculty members met with two of the instructors from a local paramedic training program. This initial meeting reviewed the training that had been occurring at the Medical Center for years. Essentially, the paramedic students were assigned to the operating room for a day or two during the clinical portion of their curriculum. During this time, they would approach random anesthesiology faculty to identify patients that were potential candidates for them to intubate. From the perspective of the anesthesiology faculty, there were several concerns. First, the paramedic student's level of skill was unknown, but was frequently nonexistent because the paramedic student had had minimal to no prior practical experience in managing the airway. Second, the paramedic student was unfamiliar with the operating room setup. These concerns, combined with the pressure of the operating room to keep surgical cases moving in a timely fashion, meant that the training quality and quantity each paramedic student received was variable (see Chapter 3).

To standardize the experience for each paramedic student and ensure that they had acquired basic airway management skills prior to providing care to real patients, the paramedic faculty and the anesthesiology faculty member decided to create a half-day course that would focus on mask ventilation and intubation skills. In addition, since the Center had both an anesthesia cart and anesthesia machine, the paramedic students would also familiarize themselves with the operating room setup. After their half day in the Center, the paramedic students would subsequently spend one day in the operating room refining their skills with actual patients.

In May and June of 2004, a Department of Anesthesiology and Pain Medicine faculty began teaching five to six paramedic students mask ventilation on several partial-task trainers for airway management. Next, the paramedic students learn intubation skills with the use of a video laryngoscope, a laryngoscope that has a fiberoptic camera attached to the blade (Figure 11.6).

The video laryngoscope enables the anesthesiology faculty to provide a televised view of the pharynx when he places the laryngoscope into the patient simulator's mouth, allows him to identify the structures of the larynx for all the students, and permits the students to watch as the endotracheal tube is passed through the vocal cords. On their first attempts to intubate a partial-task trainer, the video laryngoscope allows the instructor to confirm the student's knowledge about the anatomical structures as he performs direct laryngoscopy and

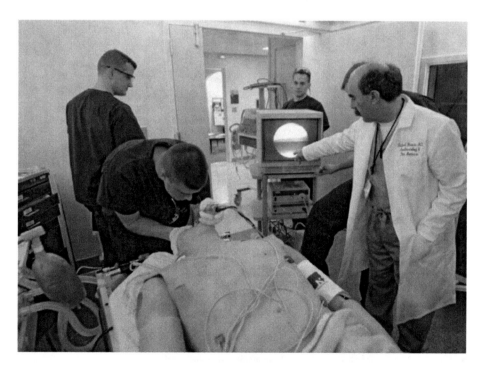

**FIGURE 11.6**   Paramedics learning airway management skills using the video laryngoscope.

intubation, give the student feedback on how to improve the laryngoscopic view, and watch as the student inserts the endotracheal tube.

This course was very well received by the paramedic students and is considered a success by all involved. The paramedic students are much more comfortable with the operating room environment as well as with intubating real patients, both in the operating room and after graduation. At the same time, the anesthesiology faculty report that the paramedic students are technically more competent, as demonstrated by greater facility with mask ventilation and more successful intubations.

The success of the airway management course led to the almost immediate addition of another simulation session for the paramedic students. The second session focuses on using the adult and pediatric human simulators to recreate potential crisis events that the paramedic students might encounter. These scenarios include a person having a cardiac arrest, a person being bitten by a rattlesnake, a pediatric patient with respiratory distress, and a hysterical parent. The educational objectives of these sessions are determined by the paramedic faculty who reinforce the teaching points during the debriefing sessions held after each scenario. The technical support for the patient simulator was initially provided by either the nursing educator or the anesthesiology faculty, but has more recently been provided by the Center's "Jill of all trades."

## 11.8 Third Year Medical Students and Problem-based Learning Discussion

During the first year of the Center's existence, medical student sessions started with a trial integration of the patient simulator into an existing problem-based learning discussion on management of an intensive care unit patient. Over the course of the academic year, all 100 or so third year medical students participate in this problem-based learning discussion that takes place during their internal medicine rotation. In the 2002–03 academic year, two of the anesthesiology faculty members, who were already familiar with the patient discussed in the problem-based learning discussion and the educational goals of that section of the problem-based learning discussion, used the high-fidelity simulator to literally bring the patient to life. Since this was a preliminary integration, half of the medical students on each internal medicine rotation did the traditional problem-based learning discussion (in groups of eight to ten) and half had the patient simulator integrated into their discussion. The patient simulator gave the students the opportunity to not only discuss but also actively treat their patient and see the consequences of their treatment. The introduction of the patient simulator received rave reviews from the third year medical students and has been subsequently integrated into

this problem-based learning discussion for all the third year medical students on their internal medicine rotation.

## 11.9 Medical Students and Anesthesiology

In the summer of 2004, the anesthesiology faculty assigned to the Center began two parallel courses. The first one was for first and second year medical students whose interest in anesthesiology had led to the creation of the Anesthesiology Interest Group. These students would come to the Center during a weekend or during a free period during the week to learn mask ventilation, intubation, and intravenous catheter placement. The sessions were held on a sporadic basis depending on the schedules of those interested in attending and the availability of the anesthesiology faculty.

The second course was for the fourth year medical students who were on an anesthesiology rotation. During their 4-week rotation, the students would come to the Center two to four times for an hour and a half each session. These sessions provided them with the opportunity to act as the resident and assume responsibility for the patient during induction, emergence, and critical events. The students also were able to practice their skills in airway management and intravenous catheter placement.

## 11.10 First Year Medical Students and Doctoring 1

In early 2005, the chief administrative officer of the Center began a new role, that of being the pelvic simulator instructor. She took the pelvic simulator to the Male and Female Genitourinary section of the Doctoring 1 course taught to the first year medical students. Traditionally, the students did their first pelvic examination on a standardized patient after watching a video describing how to perform a pelvic examination. Now, after viewing the video, the students were given an opportunity to practice doing a pelvic examination on the pelvic simulator. Since each 2-hour session had approximately 25 students and there was only one pelvic simulator, half the students used the simulator first while the other half did an examination on the standardized patient first. Feedback from the majority of the students was positive, because the opportunity to practice beforehand helped to alleviate their anxiety about technically performing the examination, gave them a correlation between the anatomy and the examination, and allowed them to focus on their patient as a person (if they did the simulation first). The imperfect tactile realism was an issue for only a minority of the students.

The use of the pelvic simulator in the Doctoring 1 course is now in its third year of use (2006), with equal emphasis being given to the pelvic examination itself as to the student's interaction with his "patient" before and during the examination. The Center recently purchased a second pelvic simulator to allow all the students to do an examination on a simulator prior to the standardized patient. Having the new course director for Doctoring 1 already familiar with simulation from the problem-based learning discussions held during the internal medicine rotation definitely helped to formalize the use of the pelvic simulator in the Doctoring 1 course. In addition, we have started discussions on how to increase the use of simulation in the Doctoring 1 course.

## 11.11 Second Year Medical Students and Doctoring 2

In the Spring of 2005, the patient simulator was introduced into the Doctoring 2 course taught to the second year medical students. In this instance, one of the lead instructors already had experience with the simulator because he supervises the problem-based learning discussion that occurs during the internal medicine rotation and was interested in creating a patient case to reinforce an existing lecture on the management of chest pain. The other instructor – an emergency medicine faculty member – had not used the patient simulator but was interested in exploring its application to teaching. Along with one of the anesthesiology faculty who works part-time in the Center, the educational objectives for the simulation session were determined to be management of a patient with an emergency cardiopulmonary problem, management of the airway with an emphasis on the application of supplemental oxygen, and the use of a defibrillator. To help the students optimize their patient management, two procedure stations were incorporated into simulation sessions: one on airway management and the use of supplemental oxygen and one an introduction to using the defibrillator.

To facilitate the students' familiarity with the patient simulator, the monitors available, and the Center, a video was made that introduced the Center, the patient simulator and its capabilities, and the simulation room itself. The students were shown the video prior to the first simulation session and the video was available for review on the Internet. The students were divided into groups of 15–16 that attended one of six simulation sessions. Each session lasted 2 hours and consisted of four 30-minute stations: airway management, defibrillator use, patient simulation, and video debriefing. At each session, the students were further divided into three groups that started at different stations (initially all stations except the video debriefing) and then rotated to all the stations.

Initial feedback after the first session resulted in some slight modifications of the session. First, the students were provided

with the advanced cardiac life support algorithm for pulseless ventricular tachycardia and ventricular fibrillation prior to their session. Next, the defibrillator station was expanded to include cardiopulmonary resuscitation since many of the students were unsure how to perform cardiopulmonary resuscitation and to include a review of the pulseless ventricular tachycardia and ventricular fibrillation. Finally, except for the first group of students in the patient simulator session, the modified defibrillator session always preceded the patient simulator session, because the students felt that deficiency (i.e., the lack of defibrillation knowledge) had the most impact on their ability to provide optimal patient care.

In 2006, the Doctoring 2 course repeated the sessions with the patient who has an emergency cardiopulmonary issue and added another six sessions with a pediatric patient who has been in a motor vehicle collision. The airway management procedure station for the first patient was modified to focus on application of supplemental oxygen and mask ventilation. Two procedure stations accompany the new pediatric simulation session: intubation (to build on the skills from the first session) and placement of an intravenous catheter. One additional change made this year was to delay the start of the patient simulation session so that all groups would at least have rotated to the advanced cardiac life support station prior to the cardiopulmonary simulation session and to the intravenous placement station prior to the pediatric simulation session.

## 11.12 Other Medical Student Rotations

With each of the above sessions requiring four instructors and with an increase from 6 to 12 sessions in the last 2 years, we have had the opportunity to have more faculty in the Departments of Anesthesiology and Pain Medicine and Emergency Medicine use patient simulations. While not all have been enthusiastic about their experience with the use of simulation in education, we have successfully converted a few. In fact, since July 2005, three of the emergency medicine physicians now regularly bring fourth year medical students rotating on emergency medicine to the Center to learn airway management skills at the start of their rotation. Those who have been unenthusiastic usually focus on the imperfect realism of the patient simulators rather than on the opportunity to provide realistic educational experiences without risk of harm to live patients.

During the 2005–06 academic year, the Department of Pediatrics also began to use the Center and its pediatric simulators (both child and baby) during their orientation day for the third year medical students. As with the other groups, the course directors met with one of the Center's faculty to discuss their existing curriculum, their educational objectives, and ways to incorporate simulation into their teaching. After investigating the resources available at the Center, the director decided to use the baby patient simulator to introduce the students to the differences in examining a pediatric patient versus an adult patient. The sessions are scheduled for half a day every 8 weeks and are taught by the pediatric faculty with technical support from the Center's full-time support person. With the upcoming academic year, the Department of Pediatrics is planning to use simulation during the orientation for its new residents.

## 11.13 Other Simulation Users

With the addition of each new group of learners, we have been able to identify faculty who are interested in incorporating simulation into their teachings and, consequently, new ways to use simulation to teach. At the present time, our user base includes the groups mentioned above as well as respiratory therapy, pharmacists, nursing students, family practice residents, internal medicine residents, pulmonary critical care fellows, and pain medicine fellows. For the majority of each of these groups, one of their principle instructors wanted to provide the students with a risk-free opportunity to review and apply Advanced Cardiac Life Support principles. The one exception is the respiratory therapy group that uses the patient simulator to teach and review the use of different ventilator settings. In this case, the Center provides technical support to change the pulmonary status of the patient simulator while the educator for the respiratory therapists teaches his students, which range from being other respiratory therapists to nurses to residents.

As a certified Advanced Cardiac Life Support instructor, one of the anesthesiology faculty members tailored cardiopulmonary patient scenarios to the learners. If the debriefings with the students afterward are solely by the students' instructor, then the simulation sessions are frequently done by the Center's full-time support person. If the debriefings are to be done by the students' faculty in conjunction with the certified Advanced Cardiac Life Support instructor, then the simulation sessions are jointly produced by the Center's full-time support person and the anesthesiology faculty. When the debriefings are done jointly, the Center's faculty focuses on the application of the Advanced Cardiac Life Support algorithms and the team interactions while the instructor highlights issues that apply to the practice of the specific specialty. The frequency of sessions for each student group varies from annually to semiannually to quarterly.

Beginning in the fall of 2005, the same anesthesiology faculty member also began conducting a 4-hour simulation session for four experienced providers who are recertifying in Advanced Cardiac Life Support. This is done in conjunction with a monthly recertification class sponsored by the Center for

**TABLE 11.1**   Scenarios for Pain Medicine

1. Vasovagal reaction
2. Local anesthetic toxicity resulting in arrhythmias
3. High spinal
4. Opioid overdose
5. Anaphylactic shock

Nursing Education. The experienced providers participate in 4–5 MegaCode scenarios to demonstrate their skills. The overall feedback has been extremely positive; the participants prefer the level of realism both in the scenarios and in their ability to actively apply the algorithms to the traditional methods used in Advanced Cardiac Life Support classes.

In the Spring of 2006, the Pain Medicine faculty were concerned with their fellows' management of potentially critical incidents. The faculty was already aware of the Center since several of their fellows had been sent to learn and practice their airway management and cardiopulmonary resuscitation skills. Consequently, they approached one of the Center's part-time anesthesiology faculty to develop a critical incident course. The Pain Medicine faculty were asked to identify the critical incidents they wanted the fellows to diagnose, to describe an appropriate clinical scenario for the incident, and to delineate how the fellows were to manage each problem. The anesthesiology faculty member then created programs according to the information provided. The five-scenario critical incident course (Table 11.1) was held in one 4-hour session with attendance by both the Pain Medicine fellows and the faculty. The debriefing was done by the Pain Medicine faculty who had helped to develop the scenarios and the associated educational objectives.

## 11.14 Our Approach to Simulation in Education

Our overall approach to using simulation in education is the same for all users. The Center's faculty primarily provides technical support for the various departments that want to integrate patient simulation into a part of their curriculum, but leaves the development of the curriculum and educational objectives to the department itself. Taking into account the number of students to be taught, the Center's faculty helps to identify possible ways to use the Center's simulation resources to best achieve the educational objectives and provides guidance as to the duration of each station and/or the overall session. We believe that having each department's faculty retain control over the educational content helps them to adopt the use of simulation because they realize simulation is not about trying to control what they teach but rather a change in how they teach. It also helps the Center because we do not have enough personnel to teach all the courses or to be the content expert in all topics. Whether the session is held within the Center or elsewhere is again determined by the educational objectives and the patient simulator that will be used. Assuming a portable patient simulator is to be used, the simulation session may be held outside the Center to take advantage of the practitioner's normal clinical environment. On the other hand, if multiple pieces of equipment are required and/or the portable simulator is not adequate to meet the educational objectives, then the simulation session is held in the Center.

## 11.15 Focus on your Vision

Our goal, from the beginning, has been to create a multidisciplinary center that will improve the training of residents and nurses and hence the delivery of all aspects of patient care in the Medical Center. To sustain our growth, we have acquired additional equipment based on the interests of the groups using the Center as well as in anticipation of its potential future users. The equipment has been purchased with a combination of grant money from the Children's Miracle Network and the hospital's Volunteer Services as well as money provided by the School of Medicine's Office of the Dean. The Center's faculty jointly decided with its chief administrative officer what equipment to buy. These decisions then drive the grant applications that are submitted as well as the allocation of funds available

to the Center. Equipment that we purchased to facilitate the teaching of our existing users include part-task trainers for airway management and intravenous catheter placements. We also bought a portable infant mannequin to expand our simulation capabilities. An example of equipment that we purchased in anticipation of its potential future users would be the pelvic simulator that we bought with the expectation that it would be useful for the Department of Obstetrics and Gynecology. Although purchasing the patient simulators before identifying definitive users proved successful over time, it has been less true with the pelvic simulator. While the Department of Obstetrics and Gynecology may find it useful in the future, we are, at the present time, using it to teach medical students as we described above.

We have also started to work in conjunction with the Departments of Emergency Medicine and General Surgery to purchase part-task trainers as well as to develop simulation scenarios. With the help of visiting faculty from the University of Washington in Seattle, WA, U.S.A., and from Southmead Hospital in Bristol, England, who brought their own part-task trainers to help us produce three 3-hour simulation sessions on managing shoulder dystocia, we have been successful in demonstrating the value of simulation to the Department of Obstetrics and Gynecology (Figure 11.7).

Prior to the shoulder dystocia training sessions, we were unable to interest the Department of Obstetrics and Gynecology faculty in using simulation to teach medical students,

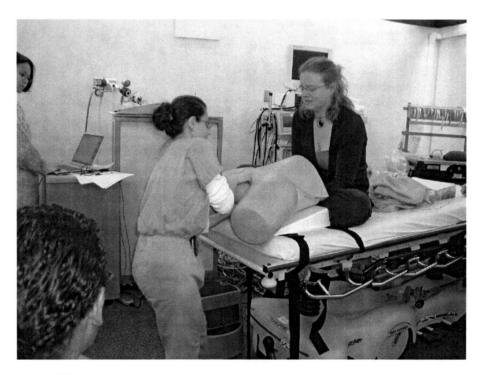

**FIGURE 11.7**   Simulation workshop teaching the management of shoulder dystocia.

residents, or fellows. However, after the sessions, both the residents and the faculty were impressed with the effectiveness of the maternal fetal simulator to improve their skills in managing a difficult delivery. In fact, the chief residents for the upcoming academic year have started to work with one of the Center's faculty to develop several simulation sessions, including orientation for the incoming interns and managing obstetric emergencies for all the residents. In addition, the Chairman of the Department of Obstetrics and Gynecology agreed to fund the purchase of a maternal/fetal simulator that will be kept at the Center to be used by all practitioners who might have to care for a pregnant patient.

As we continue to grow and develop programmatically, we have discovered departments, such as Emergency Medicine, that have purchased equipment independently and that we have invited to join our multidisciplinary center. The Department of Emergency Medicine, like the Center, has purchased some partial-task trainers for central line insertion as well as for learning ultrasound guidance. The tentative plan is to bring the equipment to the Center where it will be stored and maintained for the use of all groups that are interested in using them. The cost of maintaining the equipment will be shared by the Center, the Department of Emergency Medicine, and any other department who might use the equipment to train their residents.

## 11.16 Conclusion

By consolidating the equipment, the clinical educational system as a whole can minimize unnecessary duplication of expenditures; an individual department no longer has to be solely responsible for storage and maintenance, and the equipment is available for use system-wide. In addition, bringing together different departments allows us to explore the implementation of multidisciplinary team scenarios on a more consistent basis.

# IV

# Curriculum: Planning for Success

*In preparing for battle I have always found that plans are useless, but planning is indispensable.*

Dwight D. Eisenhower

"Cur·ric·u·lum; noun; very detailed description of every performance and evaluation task in the transformative process experienced by students for the express purpose of changing their behaviors from those they happen to have when they enter a program of study to those expected of them upon completion; often confused with a simple list of course titles or a calendar listing of topic titles linked with dates, times, and class room locations."

Consider the analogy that an incoming student is like a collection of refined materials: metals, plastics, and glasses, and this collection has decided to come together in such a fashion as to become an automobile, a specific make/brand/type of automobile optimized for a well-defined purpose. But help with this process is needed, our help, as this collection of materials has limited fabrication experience to guide itself and few tools with which to change itself. Curriculum is a very detailed plan of that process to transform this collection of materials into a specific functional entity of predefined performance. Curriculum is a description of those required tools and their performance characteristics. Curriculum is the teaching program for those tool users. Curriculum is the definition of all the myriad fabrication steps, each very well characterized, and each optimally situated in the sequence. Parallel with each fabrication step is an evaluation step, where the actual results are compared with an ideal result, differences are determined, and where they exceed a well-defined range, a decision is taken to either perform rework until an acceptable result occurs, or reject outright. After the final inspection has been passed, the stamp of assurance is bestowed and the brand new creation is sent out to meet the world. It has been written, "that which achieves its effect by accident is not art." We write, "those who graduate without demonstrated performance are not educated."

Once upon a time, all automobiles were rare works of art, handcrafted with loving care by a few master craftsmen over a great deal to time, were very expensive, and thus readily available only to a small fraction of the population. In the age of pervasive horse-pulled mobility, those who didn't have to work for a living often acquired the early automobiles as just another form of conspicuous consumption, as playthings. Most of the people that actually cared for and directly used horses did so for the utility that the beasts provided, not for the sake of ownership of the horse in and of itself.

Then one day, a different process was employed to familiar materials to fabricate a different kind of creation for a different kind of need. What was most remarkable about this difference was how this new approach to creation, this new curriculum if you will, transformed not only the methods of material modifications within one domain of fabrication but subsequently transformed the very nature of almost all other fabrication operations as well. Even more remarkable is that while the initial transformative structure, its products, and their specific fabrication plans have been retired long ago, the overall concept of creating, using, and refining a plan for transforming materials has only gone from success to success. This success is mostly due to reduced randomness, uncertainty, and unnecessary novelty – all causes of poor quality.

Yet, fine craftsmanship and the pride in a difficult job done well are still with us. Proof is that today, while high price still buys conspicuous consumption, it no longer buys proportionally higher performance or higher quality. The fears that systematic approaches to fabrication would destroy quality creations were totally unfounded. What were once only luxuries available to the very few are now standard features available to many. Furthermore, the range of creations for conspicuous consumptions has only grown. Both occurred *because* of the inherent high quality, high reliability, and high predictability due solely of highly detailed plans rigorously applied at each and every step of every fabrication step.

How do lessons learned in manufacturing apply to clinical education? Today, what usually passes for a clinical curriculum looks just like an advertising document, mainly because that is how it is so often used: colorful, exciting, attractive, and very, very short (to prevent boredom). Actual clinical curriculum is colorful, exciting, attractive, and very, very detailed (to prevent ambiguity). Real clinical curriculum starts at the smallest scale: where the instructor and student meet. Functioning clinical curriculum satisfies the needs of the instructors so that they can and will make use of it. Acceptable clinical curriculum honors the passion of the instructors, helps fill in their individual gaps, and yet does not fetter exemplary teaching. Useful clinical curriculum explicitly defines the level of competency in knowledge, skills, and attitudes required for graduation. Curriculum use is very liberating *because* of the rigorous effort applied with great wisdom and forethought.

Lorena Beeman describes her challenges and success in fabricating and employing the concept of a curriculum within her clinical educational "crafts shop" environment. Mark Escott and Lawrence Kass do the same, with special emphasis on how performance standards set by an outside agency can be used to direct curriculum creation and its use within a clinical educational program. Valeriy Kozmenko *et al.* present fundamental rules for designing and using curriculum. Ian Curran places the contributions of simulation-based learning in perspective within a larger clinical education curriculum.

# 12

# Integration of Simulation with Existing Clinical Educational Programs

Lorena Beeman

## 12.1 Setting the Context: Experiential Learning

The traditional approach with many clinical education offerings has typically included classroom instruction, utilizing (at best), interactive lecture allowing for questions and answers, and some discussion. Such instruction is usually augmented with passive audiovisuals. Occasionally, this scope is expanded by utilizing case studies allowing for limited application of content, which is then reviewed through guided instruction. The effectiveness of either interactive lecture or guided instruction using case studies is greatly affected by the number of learners present. There is an inverse relationship between large numbers of learners and the effectiveness of these teaching strategies. This approach seldom reflects any assessment of the learner's ability to synthesize what has been taught or judge the value of the material presented in relation to the clinical setting. The traditional clinical apprentice approach with skill development has been "see one, do one, teach one." This usually involves the use of either live patients, but just for access of their "flesh," or passive mannequins for demonstration of the skill. The problem with these traditional approaches is that they are separated from situational reality that demands critical thinking, problem solving, clinical knowledge, and clinical inquiry within the context of multiple complexities.

Incorporating interactive experiences, that is, the very essence of simulation, into clinical education allows for the limitations of traditional approaches and challenges faced by the clinical educator to be effectively addressed. Of course, a clear educational framework for effectively using simulation is essential. The ownership of simulators and simulation facilities does no more teaching than does having podiums and lecture halls.

The benefits of simulation are that it allows for the demonstration of *experiential learning*. Patricia Benner defines experiential learning as "clinical learning that is accomplished by being open to having one's expectations refined, challenged, or disconfirmed by the unfolding (events)" [1]. Experiential learning occurs in every situation where the clinical learner has preconceptions challenged, clinical enquiry demonstrated, and self-reflection required. Both the clinical learner and the instructor are learning from the experience. The learner is able to reframe key concepts identified within the simulation context and how they can be incorporated into clinical practice. The instructor is able to teach more effectively by identifying the learner's frame of understanding as the simulation unfolded.

Through incorporation of simulation into existing clinical education opportunities, several goals may be actualized:

- The clinical learner's ability to reason-in-transition and apply clinical judgment can be evaluated.
- The clinical learner's reasoning and coping processes may be developed in a safe environment.
- Critical thinking can be developed, demonstrated, and evaluated.
- The clinical learner's application of critical problem solving and creative problem solving may be fostered and subsequently evaluated.
- The clinical learner can develop good habits of applying clinical inquiry, evidence-based practices, and clinical knowledge through exposure to interactive, patient-oriented scenarios.
- The clinical learners can not only discover what lies beyond the point of "gone too far" but by exploring this boundary, they can learn to recognize its presence without having to cross it with live patients.

This last point ("gone too far") is essential and required of all practicing clinicians, is the most costly to learn on the living, and is a key feature that only simulation can provide to clinical education. The Clinical Education Department at the University of New Mexico Health Sciences Center, New Mexico, U.S.A, utilizes the phenomenological model of Patricia Benner to provide the framework for achieving these goals. The model is described in detail in Chapter 4, but components relevant to this discussion will be addressed here.

**The Benner Model of progression from Novice to Expert, and the incorporation of this model into simulation-based clinical education.**

The Benner model identifies five levels that a clinical learner, in context with professional experiences, can progress through to the final level being that of Clinical Expert. Distinctions between these five levels are based on how well the clinical learner demonstrates six aspects of clinical judgment and skill development (Table 12.1). The clinical application of these six aspects of clinical judgment and skill acquisition change across the levels and their application within a level are referred to as milestones (Table 12.2).

**TABLE 12.1**   Benner's six aspects of clinical judgment and skill acquisition

1. Reasoning in transition
2. Skilled know-how
3. Response-based practice
4. Agency
5. Perceptual acuity and skill of involvement
6. Links between ethical and clinical reasoning

Our Clinical Education Department has condensed these five levels into three "tiers" to reflect commitment to fostering continued development of clinical excellence in patient care, and to correlate with the hospital's clinical advancement programs. These tiers are: Advanced Beginner to Competent, Competent to Proficient, and Proficient to Clinical Expert (Refer to Table 12.5 in Chapter 4). (Note: Our three tiers do not include Benner's "Novice" level as the hospital has different programs for students employed by the facility and those transitioning from the student to professional role.)

By being cognizant of Benner's representative milestones for each level, appropriate learning objectives can be developed. On the basis of the objectives, validation criteria can be developed, reflective of the six aspects of clinical judgment and skill development.

Deciding when and where to make use of the most extensive features and the most expensive costs of high-fidelity robotic model-assisted simulators is based upon the answer to the following questions. If the answer to these questions is "No," then other levels of simulation or other methods of teaching are more appropriate.

1. Does the scenario call for some clinical aspect that is unique to the high-fidelity simulator? An example of this is the need to evaluate pupillary response, size, and/or whether the pupils are fixed and dilated as part of a neurological assessment.
2. Does the scenario require the learner's demonstration of validation criteria in an interactive response to a state of pathophysiology? An example of this is the programing of acute severe asthma requiring endotracheal intubation and mechanical ventilation to demonstrate the development of auto-PEEP, its recognition via the ventilator flow scalar/graphic, and subsequent identification of interventions to correct it, or recognition of complications by the clinical learner.
3. Does the scenario require that the learner's demonstration of validation criteria involve the administration of medications? Examples of this include correct application of rapid sequence intubation utilizing induction, sedative, opioid, and neuromuscular blockade agents, or administration of advanced life-support medications.

Whenever possible, we encourage that low-fidelity, passive props be incorporated into a simulation in the context of part-task trainers that are imbedded into a teaching, coaching, or mentoring scenario, rather than separating them from the scenario. Examples of this are provided in detail in Chapter 4. By doing this, learning is kept within the situational context and enhances retention of the clinical knowledge or skill being presented.

**TABLE 12.2** Benner's phenomenological model of Novice to Expert milestones

| Novice | Advanced Beginner | Competent | Proficient | Clinical expert |
|---|---|---|---|---|
| Lack of background experiences to create context | Learning a disease trajectory | Application of critical thinking skills becomes situationally based | Incorporates best practices in routine patient care | Clinical (not administrative) Leadership |
| Need for rules that are context-free | Checklist approach to assessment and interventions | Can present an effective report to a colleague | Ethical decision-making becomes important | "Powered by insight"[a] |
| Need for clear explanations of cause and effect | Technology is more important than the patient | Beginning to individualize the disease trajectory to the patient | Growing acceptance of his/her role as a Patient Advocate | "Head, Hands, Heart"[a] |
| Difficulty with prioritization | Disengagement from the patient, family, and environment | Questions the validity and reliability of technology | Can correlate content+context of the environment | "Bold Voices"[a] |
| Vulnerable | Must develop ability to recognize patient deterioration to progress to Competent level | Begins to look at "best practices" to implement at the bedside | Professional interactions and collaboration become easier | Can handle multiple complexities |
| | Development of critical thinking ability is essential to progression | Goals for educator: prevent disillusionment and reward successes | Numerous experiences to provide "framing" of assessment & interventions | |
| | Beginning to develop effective communication techniques | | | |

[a] These are mottos from former presidents of the American Association of Critical-Care Nurses and reflect the Clinical Practice of the Clinical Expert level in critical care.

---

> **Editors' Note:** In some simulation programs, instead of having students attempt intra-venous access by inserting real needles and real fluids into the US$150,000 simulator's arm and risk corroding its complex interior, they demonstrate their ability to perform the entire process of gaining IV access while consuming real time by using cheap rubber intra-venous training forearms sitting on a nearby counter. Once they see the red flash, an imbedded instructor makes available a permanently mounted tubing port for intra-venous access in the expensive mannequin. This approach works equally well for intra-osseous access for simulated pediatric patients. Alternatively, when the simulated patient is alive and real needle insertion also undesirable, some programs have these actor-patients wear strap-on props – IV boards or entire fake arms.

The milestones of the two levels within each of three tiers also help educators identify their role in relation to the simulated scenario (Refer to Table 12.5). At the tier of Advanced Beginner to Competent, for example, the milestones suggest the need for the educator to be physically present within the simulation laboratory. The roles of the educator may subsequently be those of teacher, coach, role model, guidance counselor, demonstrator of skills, and corrector of mislearning.

At the tier of Competent to Proficient, the educator remains in the simulation laboratory but assumes more of an observer role, and fostering troubleshooting, problem solving, and interdisciplinary collaboration. (Note: As an educator working with clinical learners at these levels, it can be an essential role of the educator to recognize the potential for disillusionment by the clinical learner. You may just salvage a good health care professional and/or prevent burnout.)

Finally, at the Proficient to Clinical Expert Level, the role of the educator often involves testing to determine achievement of clinical excellence. In order to assess the milestones of clinical judgment and skill development at the third tier, it can be beneficial to have one instructor evaluating out of the clinical learner's vision and one instructor serving as the learner's "orientee." The "orientee" is actually an imbedded instructor, but within the context of "preceptor and orientee," it is natural for the preceptor to talk aloud (thereby facilitating the evaluation of clinical judgment and inquiry and the quality of how to teach the orientee skills relevant to the scenario).

Throughout the levels of each of the three tiers, narratives by the learners are encouraged during simulation sessions (i.e., the trainees are encouraged to "think out aloud," say what they are planning, thinking, doing, also NOT doing, and why) as well as during debriefing sessions. This aspect of simulation is congruent with Benner's use of narratives in her research to capture what she references as "aspects of agency" [2]. The aspects of agency that can be captured with the use of simulation include engagement with the patient, engagement with the circumstance, acceptance of responsibility, and becoming a member of the health care team, as well as what the learner was thinking in "real time," and their practical understanding of the situational variables.

Debriefing is an excellent medium for the exemplification of narrative understanding by the learners in a simulation scenario. Learners can practice and develop their competencies in reflection and self-assessment, to the extent that their instructors will make use of this opportunity for them to do so (i.e., instructors keeping their mouths shut most of the time). It also provides another opportunity for the educator to teach, role model, coach, and/or mentor the learner(s).

## 12.2 Applied to Multilevel and Multidisciplinary Learning Groups

Developing educational strategies for a learning group that is homogenous in terms of clinical judgment and skill development is challenging enough. But what about learning groups who are not? There are two examples of this: (i) The learning group that is homogenous in terms of their health care discipline but at different levels of clinical judgment and skill development. (ii) The learning group that is multidisciplinary and not homogenous in relation to Benner's levels.

Teaching a group that is homogeneous in terms of their health care discipline but at different levels of proficiency can be best accomplished by establishing objectives and validation criteria at the Competent level (which our department uses as the median of our educational framework).

If the learning group is multidisciplinary and nonhomogenous, it can be advantageous to identify a single desired outcome that the different members contribute and combine their efforts toward. Thus, objectives and validation criteria for evaluation of competence can be more easily constructed. To expound upon this introductory framework, what follows are three "tales" of clinical education and simulation adventure. One tale will focus on how simulation was incorporated into a "clinical excellence trajectory" involving a learning group that was homogenous as they progressed from Advanced Beginner to Clinical Expert. A second tale will provide an example of a multilevel learning group, but all of the same discipline. The last tale will discuss how simulation was used in the setting of a multilevel and multidisciplinary learning group.

### Example 1: Hemodynamic Monitoring: A Clinical Excellence Trajectory

In our institution, prior to incorporation of simulation, we offered instruction in invasive hemodynamic monitoring using a pulmonary artery catheter, as described in Table 12.3. The

tier for the introductory course was determined to be the second (Competent, moving toward Proficient). Given that, the focus of the content was to provide the clinical learners with mnemonics for application of the determinants of myocardial performance, and a "Fundamentals of Hemodynamic Monitoring Help Sheets" (Tables 12.4a and 12.4b) that had the dual purpose of providing a framework for critical thinking in relation to hemodynamic assessment, and for providing a comprehensive report to a colleague (e.g., a change of shift report), physician, or midlevel practitioner (i.e., Physician Assistant or Acute Care Nurse Practitioner). The limitations to this approach were numerous and are summarized as follows. As a consequence of these limitations, the educators were not meeting the needs of these clinical learners, which subsequently affected their performance at the bedside of real patients.

- Absence of a hands-on approach to helping the clinical learner develop the skills to assist the practitioner with insertion of the pulmonary artery catheter.
- Absence of demonstration/return demonstration of pressure line setup, identification of the phlebostatic axis, zeroing, leveling, and calibration.
- Waveforms were presented via static strips, rather than dynamic (as they would appear on the hardwire monitoring equipment at the patient's bedside).
- Insufficient time for clinical inquiry and clinical judgment related to the case studies.
- Assumption that preceptors were cognizant of and congruent with the content of the American Association of Critical-Care Nurses Procedure Manual for Critical Care in relation to hemodynamic monitoring and assessment [3, 4].

This introductory course was subsequently reframed into a theory portion of the course (4 hours) and an additional simulation lab (4 hours). The simulation lab addresses the aforementioned limitations by placing the teaching components in context with a more realistic learning environment. The simulation lab is divided into two sections. The first reviews waveforms and pressures seen when a pulmonary artery catheter is inserted (central venous, right atrial, right ventricular, pulmonary artery, and pulmonary artery occlusive). The waveforms are presented by using a high-fidelity hemodynamic waveform generator [5] that interfaces with our laboratory's audiovisual system to project on a large screen in a classroom. This permits instruction and discussion of the following: normal pressures, components of waveforms in relation to the cardiac cycle, assessment of pulmonary artery occlusive pressure validity, and clinical correlates with cardiac assessment.

The second section of the introductory hemodynamic simulation lab involves application of the summary assessment table (Tables 12.4a and 4b) in relation to case studies. High-fidelity physical simulation using the manufacturer's preprogramed

**TABLE 12.3** Objectives of introduction to cardiac hemodynamic monitoring and advanced cardiopulmonary hemodynamic monitoring prior to incorporation of simulation

| Introduction to cardiac hemodynamic monditoring (8-hour course; limited hands-on; no simulation) | Advanced cardiopulmonary hemodynamic monitoring |
|---|---|
| 1. Describe each phase of the cardiac cycle and identify the clinical significance of each. | 1. Evaluate indications and underlying principles of cardiopulmonary hemodynamic monitoring. |
| 2. Define the determinants of myocardial performance (CRAS): Contractility, Rate/Rhythm, Afterload, Starling's Law. | 2. Identify and analyze the variables that affect oxygen delivery, demand, and consumption. |
| 3. List and describe factors influencing CRAS. | 3. Assess the significance of oxygen debt in a critically ill patient and recommend interventions. |
| 4. State indications for hemodynamic monitoring with a pulmonary artery or Swan–Ganz cathether. | 4. Summarize the elements and significance of the oxyhemoglobin dissociation curve in relation to arterial oxygen content ($CaO_2$) and oxygen extraction index (OEI). |
| 5. Identify complications associated with insertion of the pulmonary artery catheter, and discuss interventions and/or prevention. | 5. Explain the clinical significance of conditions and interventions that increase oxygen consumption. |
| 6. Describe the insertion process for a pulmonary artery catheter. | 6. Outline strategies to incorporate into the clinical care plan that reflect the concepts of oxygen supply and demand. |
| 7. Identify and discuss the clinical significance related to components of validity and reliability of a pressure line: Pressure line setup; location of the phlebostatic axis and what must be placed there; calibration and zeroing; square wave analysis for dynamic frequency and response. | 7. Assess and evaluate the principles of a cardiopulmonary hemodynamic assessment of the following states of pathophysiology in the patient with Multiple Organ Dysfunction Syndrome (MODS) and preeclampsia. |
| 8. Recognize normal damping, underdamping, and overdamping of a pressure line. | 8. Working in small groups, assess and evaluate, and then present a cardiopulmonary hemodynamic assessment for patient with either systemic lupus erythematosus or a ventricular septal defect. |
| 9. Identify troubleshooting interventions for underdamping and overdamping. | 9. For each case study, identify therapeutic goals and interventions to help achieve them. |
| 10. Recognize normal waveforms of each of the following: Right atrium, right ventricle, pulmonary artery, and pulmonary occlusive. | |
| 11. Identify factors related to an assessment of the validity of the Pulmonary Artery Occlusive Pressure. | |
| 12. Recognize normal pressures and derived calculations, and correlate them with the determinants of myocardial performance. | |
| 13. Discuss interventions for alterations in the determinants of myocardial performance. | |
| 14. Utilizing the Fundamentals of Hemodynamic Help Sheets, perform a hemodynamic assessment in conjunction with interactive case studies. | |

settings for septic, hypovolemic (gastrointestinal bleeding), and cardiogenic shock were run in congruence with case studies developed for each (Table 12.5). The simulator permits visualization of the pulmonary artery catheter waveforms and pressures relevant to each state of shock, and responses to interventions (e.g., fluid resuscitation, positive inotropic medications, vasopressors, and vasodilators).

As an outgrowth of these two courses, and as part of the development of a clinical excellence trajectory for hemodynamic assessment, a separate 4-hour course on normal and abnormal waveforms was developed utilizing the simulator previously described and static strips. The clinical significance of the abnormal waveforms was emphasized. Calculation of

the right atrial and pulmonary occlusive pressures in the presence of respiratory variation is also taught in this course (in accordance with the American Association of Critical-Care Nurses Procedure Manual for Critical Care [3]).

The former Advanced Cardiopulmonary Hemodynamic monitoring course became the fourth component of the clinical excellence trajectory. Like the introductory course, it was reformatted into two, 4-hour courses. The first focuses on advanced cardiac assessment concepts, culminating in assessment of oxygen delivery from the pulmonary hemodynamic perspective. The second course involves assessment and application of the oxyhemoglobin dissociation curve, and the concepts of arterial oxygen content, oxygen extraction

**TABLE 12.4a**   Fundamentals of hemodynamic monitoring help sheets (Developed by Lorena Beeman, RN, MS, CCVT)

| IN the beginning... | Oxygenation: Heart and body | Rate and rhythm | From the outside in... |
|---|---|---|---|
| Perform a clinical assessment and correlate findings with determinants of myocardial performance | What is the $FiO_2$ and method of $O_2$ delivery? | Rate determines the duration of systole and diastole. Rate determines the response to metabolic and oxygen demand. | Mean Arterial Pressure (MAP): Systolic + (Diastolic × 2) ÷ 3<br>• Normal: 70–100 mmHg<br>• <65 mmHg indicates a shock state, and triggering of compensatory mechanisms (SNS, RAA system)<br>• >100 mmHg indicates a hyperdynamic state and increased oxygen consumption (watch for ischemia) |
| Remember the assumptions:<br>• Is the RV normal? If not, LVSWI and SVRI cannot be used.<br>• Is the PAOP valid? If not, what is it assessing? | Assess the work of breathing (SO-BAD):<br>• S = Surgery or trauma (thoracoabdominal)<br>• O = Orthopnea (correlates with PAOP >18 mmHg)<br>• B = Breath sounds<br>• A = Accessory muscles<br>• D = Demand for $O_2$ | Relative rate for hemodynamics. A change in HR +/−20 bpm for more than 3 minutes changes the formula:<br>C.O. = HR × SV<br>• Assess for relative tachycardia<br>• Assess for relative bradycardia | Cardiac Index<br>• Normal heart: $2.5$–$4.0$ L/min/m$^2$<br>• Normal heart: <2.5 indicates a shock state<br>• Cardiac dysfunction:<br>  – $1.9$–$2.2$ = Severe heart failure<br>  – <1.9 = Cardiogenic shock |
| Assess validity and reliability<br>• Pressure line setup correctly?<br>• Reference stopcock leveled to the phlebostatic axis (4th ICS, midway A–P)?<br>• Square wave test normal? | Labwork:<br>• Hemoglobin<br>• $PaO_2$<br>• pH<br>• $PaCO_2$<br>• $PaO_2/FiO_2$ ratio. Normal is >300 | Rhythm determines synchronization of electrical and mechanical events related to the cardiac cycle.<br>• A change in rhythm will alter SVI<br>• A change in rhythm will alter diastole<br>• A change in rhythm alters mechanical contractility | Cardiac Index Increased + SVRI decreased = SIRS or Sepsis |
| Too many bounces or too many boxes: *Underdamped*<br>• Was the pressure line primed under pressure? Change the pressure line.<br>• Is your patient hyperdynamic? Treat the patient, but your numbers are invalid!<br><br>No bounces and hence no boxes: *Overdamped*<br>• Attempt to aspirate blood from the distal port. If unable to aspirate, there is a blood clot at the tip. DO NOT flush!<br>• If blood can be aspirated: Did you let the bag go dry? Did you not maintain 300 mmHg pressure? Do you have large air bubbles in the line? | Calculate the Coronary Perfusion Pressure:<br>Diastolic BP ÷ PAOP or PAD<br>• Normal is 60–80 mmHg<br>• <60 = ischemia<br>• <40 = infarction<br><br>If the PAD is increased + the PVRI is increased, think HYPOXIA! | AV dissociation:<br>• Loss of P waves = Loss of atrial kick<br>• P waves not 1:1 with ventricular complex = development of heart failure signs and symptoms | Stroke Volume Index (SVI) = Amount of blood ejected by the heart per beat (33–47 cc/beat/m$^2$).<br><br>If HR cannot compensate for increased oxygen or metabolic demand, SVI is left to do this. The patient will quickly have demand exceeding supply! |

**TABLE 12.4a**  (Continued)

| IN the beginning... | Oxygenation: Heart and body | Rate and rhythm | From the outside in... |
|---|---|---|---|
| Obtain numbers and calculations, and for each, indicate if decreased or elevated. | Goal is $O_2$ delivery to the cells:<br>• Increase $FiO_2$<br>• Change method of delivery<br>• Decrease $O_2$ demand<br>• Adjust ventilator settings and/or mode<br>• Increase Hemoglobin<br>• Correct pH<br>• Correct temperature<br>• Evaluate for excessive blood administration or sepsis/SIRS<br><br>Goal is coronary perfusion:<br>• Beta-blockers to reduce $O_2$ demand<br>• IV Nitroglycerin<br>• PTCI<br>• CABG<br>• IABP | Ventricular arrhythmias? Perform the DEATHLIR assessment:<br>• D = Drugs (SNS stimulation)<br>• E = Electrolyte abnormalities<br>• A = Acidosis<br>• T = Trauma<br>• H = Hypoxemia<br>• L = LV dysfunction<br>• I = Ischemia/Injury/Infarction<br>• R = Reperfusion<br><br>ID the trigger and treat the cause.<br><br><br><br>Interventions for Rate:<br>• Speed it up ($O_2$, pacing, atropine, vasopressors, relieve ICP)<br>• Slow it down (Antiarrhythmics)<br><br>Interventions for Rhythm:<br>• Convert to NSR<br>• Antiarrhythmics<br>• Dual chamber pacing | |

RV = Right ventricle; LVSWI = Left Ventricular Stroke Work Index; SVRI = Systemic Vascular Resistance Index; PAOP = Pulmonary Artery Occlusive Pressure; ICS = Intercostal space; A–P = Anterior–Posterior; mmHg = millimeters of Mercury; $O_2$ = oxygen; $PaO_2$ = partial pressure of oxygen; pH = hydrogen ion concentration in the blood; $PaCO_2$ = Partial pressure of carbon dioxide; $FiO_2$ = Fraction of inspired oxygen; $PaO_2/FiO_2$ also known as the Hypoxemic Score or Shunt Equation of P/F ratio; BP = Blood Pressure; PAD = Pulmonary Artery Diastolic; PVRI = Pulmonary Vascular Resistance Index; HR = Heart Rate; C.O. = Cardiac Output; SV = Stroke Volume; SVI = Stroke Volume Index; AV = Atrioventricular; SNS = Sympathetic Nervous System; RAA = Renin-Angiotensin-Aldosterone; $L/min/m^2$ = Liters per minute per meter (squared); SIRS = Systemic Inflammatory Response Syndrome; LV = Left ventricle; ID = identify; IV = Intravenous; PTCI = Percutaneous Transluminal Coronary Intervention; CABG = Coronary Artery Bypass Graft; IABP = Intra-aortic Balloon Pump.

**TABLE 12.4b**  Fundamentals of hemodynamic monitoring help sheets (Developed by Lorena Beeman, RN, MS, CCVT)

| Assessment of Starling's Law (Volume, Stretch, Pressure) | Assessment of forward flow (compare to afterload) | Assessment of resistance to forward flow (afterload): compare to forward flow |
|---|---|---|
| Volume<br>• Amount of circulating volume<br>• Intake and output<br>• Atrial contractility<br>• Distribution of volume<br>  – Medications (vasodilators, ACE inhibitors, diuretics)<br>  – Vasoconstriction or vasodilation of venous system<br>  – Triggering of hormones (Aldosterone, ADH) | Reflects effectiveness of diastole, isovolumetric contraction, and systole.<br>Clinical Assessment:<br>• S1 (heard over mitral and tricuspid areas).<br>• Diminished = Decreased contractility.<br>  Accentuated = Hyperdynamic<br>• Hepatojugular reflux (JVD when pressing over liver) = RV failure<br>• Kussmaul's sign (JVD varying with respiratory cycle) = RV failure | The resistance that the ventricles must overcome to eject blood forward.<br>• Valve status (pulmonic and aortic)<br>• Pressure in the receiving circulation (hypertension increases afterload)<br>• Compliance of the receiving circulation (vasoconstriction increases, and vasodilation decreases)<br>• Triggering of the compensatory mechanisms of shock will increase afterload (except SIRS/sepsis) |

*(Continued)*

**TABLE 12.4b**   (Continued)

| Assessment of Starling's Law (Volume, Stretch, Pressure) | Assessment of forward flow (compare to afterload) | Assessment of resistance to forward flow (afterload): compare to forward flow |
|---|---|---|
| | • Pulse pressure<br>• Ischemia/injury/infarction?<br>• Hypoxemia?<br>• Altered mentation (decreased forward flow)<br>• Decreased UOP (decreased forward flow)<br>• Evaluate thyroid function<br>• Evaluate Mg++, Ca++, serum lactate (cellular inotropy) | Clinical assessment:<br>• PaO$_2$<br>• Valve function (insufficient or stenotic)<br>• Systolic blood pressure<br>• Pulse volumes<br>• Core temperature |
| Stretch<br>• Chamber size<br>• Heart sounds (S3, S4)<br>  – S3 = Volume overload (constant sound throughout respiratory cycle). Found at mitral (LV) and tricuspid (RV) areas<br>  – S4 = Stiff, noncompliant ventricle (sound varies between inspiration and expiration). Found at same areas as S3. | Hemodynamic Assessment<br>• LVSWI reflects isovolumetric contraction of the LV: 40–75 gm/m/m$^2$<br>• RVSWI reflects isovolumetric contraction of the RV: 8–12 gm/m/m$^2$<br>• PAS reflects systolic pressure of the RV<br>• Systolic BP does not necessarily represent systolic pressure of LV, however.<br>• Ejection fraction is a better evaluation of forward flow by the LV (normal is 50–70%). <40% EF correlates with heart failure and sudden cardiac death.<br>• Wall motion of LV also is indicative of forward flow (akinesis, dyskinesis, hypokinesis). Dyskinesis is the most concerning because of paradoxical movement by the LV during contraction and increased myocardial O$_2$ consumption. | Hemodynamic Assessment (always correlate with LVSWI or RVSWI)<br>• SVRI = LV afterload (1700–2350 dyne seconds/cm$^5$/m$^2$)<br>• PVRI = RV afterload (250–280 dyne seconds/cm$^5$/m$^2$)<br><br>Note: The SVRI and PVRI do not tell you WHERE the problem is. You must go back and look at the factors affecting afterload. |
| Pressure = Preload<br>• RA (0–8 mmHg) = RV preload pressure<br>• PAOP if valid = LV preload pressure<br>  – Normal heart: PAOP 6–12 mmHg<br>  – Cardiac patient: PAOP up to 18 mmHg<br>  – PAOP >25 mmHg = Pulmonary edema | Interventions to augment contractility:<br>• Oxygen<br>• Coronary perfusion<br>• Volume (for the RV)<br>• Decrease resistance to forward flow (discussed with afterload)<br>• Positive Inotropes<br>  – Dopamine<br>  – Dobutamine (also reduces afterload)<br>  – Inodilators (Amrinone or Milrinone)<br>• IABP<br>• Vasopressors (but these will be at the cost of increased afterload due to vasoconstriction)<br>• Magnesium sulfate (controversial)<br>• Resynchronization therapy (biventricular pacing for patients with end-stage HF) | Interventions to reduce afterload for the LV:<br>• Oxygenate<br>• Vasodilators (Nitroglycerin or Nipride)<br>• Esmolol (for perioperative hypertension)<br>• Labetalol (alpha and beta-adrenergic blocker)<br>• ACE inhibitors<br><br>Interventions to reduce afterload for the RV:<br>• Oxygenate<br>• Improve lung function<br>• Venodilators (Nitroglycerin)<br>• Nitrous oxide (uncommon) |

**TABLE 12.4b** (Continued)

| Assessment of Starling's Law (Volume, Stretch, Pressure) | Assessment of forward flow (compare to afterload) | Assessment of resistance to forward flow (afterload): compare to forward flow |
| --- | --- | --- |
| Interventions for Volume Overload: Pee – Pump – Park <ul><li>Diuretics</li><li>Vasodilators</li><li>Morphine</li><li>NSR</li><li>Improve forward flow</li></ul> Interventions for inadequate volume: <ul><li>Replace volume (crystalloids, colloids, blood products)</li><li>Reduce vasodilators</li><li>Correct neurological defect</li><li>Restore atrial contractility</li></ul> | Interventions to decrease contractility (uncommon): <ul><li>Beta-blockers</li><li>Calcium channel blockers</li><li>Surgical techniques</li></ul> Usually seen with hypertrophic cardiomyopathy | Interventions to increase afterload for the LV (when distributive shock is present: sepsis, neurogenic, anaphylactic): <ul><li>VOLUME, VOLUME, VOLUME!</li><li>Oxygenate</li><li>Squeeze (Vasopressors such as Levophed, Epinephrine, Vasopresin)</li><li>Benadryl and Solu Medrol for anaphylactic shock</li><li>ID and treat the cause</li></ul> |

ACE = Angiotensin Converting Enzyme; ADH = Anti-diuretic Hormone; LV = Left ventricle; RV = Right ventricle; RA = Right atrium/atrial; mmHg = millimeters of Mercury; PAOP = Pulmonary Artery Occlusive Pressure; NSR = Normal Sinus Rhythm; JVD = Jugular Venous Distention; UOP = urine output; LVSWI = Left Ventricular Stroke Work Index; gm/m/m$^2$ = grams per meter per meter (squared); RVSWI = Right Ventricular Stroke Work Index; PAS = Pulmonary Artery Systolic; BP = Blood Pressure; EF = Ejection Fraction; $O_2$ = Oxygen; IABP = Intra-aortic Balloon Pump; SVRI = Systemic Vascular Resistance Index; PVRI = Pulmonary Vascular Resistance Index; cm = centimeter; ID = identify.

**TABLE 12.5** Introduction to cardiac hemodynamic monitoring case studies

| Cardiogenic shock case study | Distributive shock (sepsis) case study | Hypovolemic shock case study |
| --- | --- | --- |
| *History:* 46-year-old male with viral cardiomyopathy. Admitted 3 days ago with progressive dyspnea, lethargy, anorexia, and irregular pulse. Past medical history previously unremarkable. <br><br> *Clinical Assessment Findings:* <ul><li>Neurological: Alert but anxious.</li><li>Cardiovascular: 1–2+peripheral edema; jugular venous distention present at 30°; pulse volumes faint to 1+palpable with pulsus alternans; cool, slightly diaphoretic skin; diminished S1; summation gallop (S3 and S4); displaced point of maximum intensity.</li></ul> | *History:* 39-year-old female Native American. Admitted to the ED with respiratory arrest. Intubated at that time. Chest radiograph revealed a foreign body in the left lung. She was transported to an acute care hospital and admitted to an ICU. The patient was taken to the OR for removal of the foreign body; however, pulmonary infarction around the area resulted, requiring resection. A pulmonary artery catheter is in place. The patient is on benzodiazepine and opioid drips, and Dopamine at 8 mcg/kg/min. <br><br> *Past Medical History:* Significant only for a cholecystectomy 2 years ago. | *History:* 45-year-old male construction worker and part-time truck driver. Found vomiting bright red blood at the construction site, and subsequently collapsed. Coworkers brought him to the ED. In the ED, the patient was hypotensive, tachycardic, pale, diaphoretic. The patient was intubated and given 2 L of 0.9% normal saline by peripheral IV. GI consult was requested. Patient transferred to the ICU, where a pulmonary artery catheter, arterial line (right radial artery), and foley catheter were placed. Patient is still receiving 0.9% normal saline at 200 ml/hour; benzodiazepine and opioid drips. |

*(Continued)*

**TABLE 12.5**   (Continued)

| Cardiogenic shock case study | Distributive shock (sepsis) case study | Hypovolemic shock case study |
|---|---|---|
| • Pulmonary: Scattered coarse crackles throughout all lung fields; tachypneic, worsening with minimal exertion; orthopneic.<br>• Gastrointestinal: Anorexic by report; bowel sounds present in all quadrants but hypoactive.<br>• Urine output: 20 ml per hour per foley catheter.<br><br>*Vital Signs:*<br>BP 74/60 mmHg; MAP 64 mmHg; HR 128 bpm; Sinus tachycardia with frequent PVCs and nonsustained VT; RR 28 breaths/min; Temperature 37.9°C (core per pulmonary artery catheter); SpO$_2$ 88% on a nonrebreather mask.<br><br>*Hemodynamic Findings:*<br>RA 12 mmHg; RVS/RVD 70/18 mmHg; PAS/PAD 70/35 mmHg; PAOP 34 mmHg; C.O. 3.9 L/min; C.I. 2.01 L/min/m$^2$; SVI 16 cc/beat/m$^2$; LVSWI 6.3 g/m$^2$/beat; RVSWI 7.4 g/m$^2$/beat; SVRI 2067 dyne seconds/cm$^5$/m$^2$; PVRI 517 dyne seconds/cm$^5$/m$^2$ | *Clinical Assessment Findings:*<br>• Neurological: Pupils responsive to light. Unable to obtain further assessment results due to sedative and opioid drips.<br>• Cardiovascular: Skin flushed and warm to touch; pulse volumes bounding; accentuated S1; Sinus tachycardia on the monitor.<br>• Pulmonary: Diminished breath sounds throughout all lung fields along with fine crackles in the bases of both lungs. Patient remains intubated and on a mechanical ventilator with the following settings: FiO$_2$ 60%; assist control mode; Pressure control 30 cmH$_2$O; PEEP 8 cm/H$_2$O; rate 12.<br>• Gastrointestinal: Receiving enteral feedings. Bowel sounds present but hypoactive all quadrants.<br>• Urine output: 50 ml/hour per foley catheter.<br><br>*Vital Signs:*<br>BP 74/55 mmHg; MAP 61 mmHg; HR 134 bpm; minute ventilatory rate 16 breaths/min; Temperature 40.8°C (core); SpO$_2$ 70%.<br><br>*Hemodynamic Findings:*<br>RA 20 mmHg; PAS/PAD 75/50 mmHg; PAOP 18 mmHg; C.O. 9.2 L/min; C.I. 4.65 L/min/m$^2$; SVI cc/beat/m$^2$; LVSWI 5.7 g/m$^2$/beat; RVSWI 18 g/m$^2$/beat; SVRI 722 dyne seconds/cm$^5$/m$^2$; PVRI 481 dyne seconds/cm$^5$/m$^2$. | *Past Medical History:* Significant for long-term alcohol abuse (approximately 20 years), smoking (30 years), and previous admission for GI bleeding (approximately 1 year ago). History of esophageal varices and GERD. History of systolic hypertension (noncompliant with medications).<br><br>*Clinical Assessment Findings:*<br>• Neurological: Pupils equal and responsive to light. Arouses easily to voice command.<br>• Cardiovascular: Heart sounds distant; Sinus tachycardia on the monitor; remains pale and diaphoretic; pulse volumes by Doppler.<br>• Pulmonary: Some wheezing in upper lung fields. Patient remains intubated and on a mechanical ventilator with the following settings: FiO$_2$ 50%; SIMV mode; TV 800 ml; rate 16; PEEP 5 cmH$_2$O.<br>• Gastrointestinal: Abdomen is large, with palpable liver. NG tube in place, left nare draining bright red blood (1 liter since admission from the ED and 1 liter in the ED prior to admission).<br>• Urine output: Marginal per foley catheter.<br><br>*Vital Signs:*<br>BP 80/40 mmHg; MAP 53 mmHg; HR 140 bpm; minute ventilatory rate 16 breaths/min; Temperature 35°C (core); SpO$_2$ 84%.<br><br>*Hemodynamic Findings:*<br>RA 1 mmHg; PAS/PAD 30/18 mmHg; PAOP 4 mmHg; C.O. 4.2 L/min; C.I. 1.98 L/min/m$^2$; SVI 14 cc/beat/m$^2$; LVSWI 9.47 g/m$^2$/beat; RVSWI 4 g/m$^2$/beat; SVRI 2101 dyne seconds/cm$^5$/m$^2$; PVRI 727 dyne seconds/cm$^5$/m$^2$. |

BP = Blood Pressure; mmHg = millimeters of Mercury; MAP = Mean Arterial Pressure; HR = Heart Rate; bpm = beats per minute; PVC's = Premature Ventricular Complexes; VT = Ventricular Tachycardia; RR = Respiratory Rate; C = Centigrade; SpO$_2$ = spectrophotometric oxygen concentration; RA = Right atrial; RVS = Right Ventricular Systolic; RVD = Right Ventricular Diastolic; PAOP = Pulmonary Artery Occlusive Pressure; PAS = Pulmonary Artery Systolic; PAD = Pulmonary Artery Diastolic; C.O. = Cardiac Output; C.I. = Cardiac Index; L/min/m$^2$ = Liters per minute per meters (squared); SVI = Stroke Volume Index; LVSWI = Left Ventricular Stroke Work Index; RVSWI = Right Ventricular Stroke Work Index; cm = centimeter; SVRI = Systemic Vascular Resistance Index; PVRI = Pulmonary Vascular Resistance Index; ED = Emergency Department; ICU = Intensive Care Unit; OR = Operating Room; FiO$_2$ = Fraction of Inspired Oxygen; cmH$_2$O = centimeters of water pressure; GI = Gastrointestinal; GERD = Gastro-esophageal Reflux Disease; SIMV = Synchronized Intermittent Mandatory Ventilation; TV = Tidal Volume; PEEP = Positive End-Expiratory Pressure; NG = Nasogastric.

ratio, venous reserve, and oxygen demand. The content is then applied to four case studies: multiple organ dysfunction syndrome, a high-risk obstetric patient with preeclampsia and subsequent complications, a patient with acute respiratory failure secondary to systemic lupus erythematosus, and a patient with complications from an acute myocardial infarction (two of which are shared in Table 12.6). Two of the case studies are done collaboratively between the instructor and the clinical learners. The clinical learners are then divided into two groups, with each assigned one of the two remaining cases. They then apply the assessment criteria, and present to each other with guided instruction by the educator and input from their colleagues.

Realizing that this is insufficient to validate whether clinical learners in this course are truly performing at the Proficient to Clinical Expert levels, a "testing" component was developed to validate achievement of Clinical Expert in cardiopulmonary hemodynamic assessment. The clinical learner is given one or more of the case studies presented in the aforementioned course. The case studies are presented in conjunction with a Laerdal mid-fidelity simulator [6] for the purpose of programing vital signs and cardiac rhythm(s) called for in the case study. As the scenario unfolds, hemodynamic information, laboratory values, and/or other diagnostic results are given to the clinical learner by an educator upon request. The validation criteria for the multiple organ dysfunction case scenario are presented in Table 12.7.

**TABLE 12.6**  Advanced cardiopulmonary hemodynamics case studies

| Multiple organ dysfunction syndrome | Systemic lupus erythematosus (SLE) |
| --- | --- |
| *History*: 68-year-old male involved in a motor vehicle collision with rollover. Multiple rib fractures resulting in pulmonary contusion. Fracture of the left femur. Stabilized at the scene and transported to the ED. At the ED, the patient had a positive DPL, and evolving signs and symptoms of respiratory failure secondary to the pulmonary contusion. The patient was intubated and sent to the OR. Aggressive fluid resuscitation involving 0.9% normal saline, whole blood, FFP, and platelets were required. A splenectomy was also necessary.<br><br>*Past Medical History*: Significant for AMI 3 years ago requiring a CABG, and stenting of the LAD 6 months ago. The patient is also a Type II insulin-dependent diabetic, with long-standing hypertension controlled with a beta-blocker, ACE inhibitor, aldosterone-antagonist, and diuretic. The patient is now in the ICU and is postoperative Day 3.<br><br>*Clinical Assessment Findings*:<br>• Neurological: The patient is receiving Diprivan (Propofol) and an opioid infusion. Neuromuscular blockade is also being utilized affecting neurological assessment.<br>• Cardiovascular: The patient has diminished heart sounds and S3; pulse volumes are by Doppler. The patient is in NSR with a LBBB and occasional PVCS on the monitor. 12-Lead ECG shows an old anterior MI, LBBB, and LVH with significant left axis deviation.<br>• Pulmonary: The patient has coarse crackles in the bases of both lung fields, and displaced bronchial breath sounds over the left lower lobe. The patient's ventilator settings are: $FiO_2$ 80%; SIMV mode with Pressure Control at 35 cmH20; rate 20 breaths/min; PEEP 10 cmH$_2$O.<br>• Gastrointestinal: Bowel sounds have been hypoactive, and are now not present in the left lower quadrant. The patient is receiving enteral feedings. The patient is on a proton-pump inhibitor.<br>• Gentiourinary: Creatinine is increasing. Urine output at 15 ml/hour per foley catheter for the last 2 hours. | *History*: 40-year-old male with history of SLE. History of respiratory arrest approximately 1 year ago. Development of increasing shortness of breath and fatigue over the last 6 days. Presented to the ED with acute respiratory failure secondary to pulmonary hemorrhage as a complication of SLE. The patient was intubated in the ED and transferred to the ICU. An oximetric pulmonary catheter was placed via the right subclavian vein. The patient spent several hours with a systolic BP between 70–80 mmHg. The patient is now in acute renal failure and CVVH has been initiated. The patient is on Dopamine at 20 mcg/kg/hour, and benzodiazepine and opioid drips for sedation and pain control. He is currently receiving 2 units of PRBCs for a hemoglobin of 7 g/dl & hematocrit of 25%. He is also receiving apheresis therapy. Gamma globulin and albumin are to be infused following the blood transfusion. The patient is also receiving a broad spectrum antibiotic.<br><br>*Clinical Assessment Findings*:<br>• Neurological: Pupils equal and reactive to light, and reflexes are intact.<br>• Cardiovascular: Heart sounds distant; no extra sounds appreciated. Pulse volumes are all 1+palpable. 1+peripheral edema. Patient is on the DVT prophylaxis protocol.<br>• Pulmonary: Patient is on a mechanical ventilator (Day 3). Ventilator settings: $FiO_2$ 75%, assist control mode with pressure control at 35; rate 14; aPEEP 10 cmH20. Morning ABGs show the following: pH 7.47, $PaO_2$ 55, $PaCO_2$ 21, $HCO_3$ 30, $SaO_2$ 83%, $PvO_2$ 48, P/F ratio 73. Lungs with scattered fine and coarse crackles throughout both lung fields and also diminished.<br>• Gastrointestinal; Bowel sounds all quads but hypoactive. Enteral tube feedings. Some gastric distention noted (new finding).<br>• Genitourinary: Receiving CVVH.<br>• Skin: Patient is on a specialty bed to maintain skin integrity.<br>• Social: Wife is pregnant with second set of twins. Older twins are 5 years old. |

(*Continued*)

**TABLE 12.6**  (Continued)

| Multiple organ dysfunction syndrome | Systemic lupus erythematosus (SLE) |
|---|---|
| *Vital Signs*: BP 94/60 mmHg; MAP 71 mmHg; HR 105 with runs of nonsustained VT now developing; minute ventilatory rate is 20; Temperature 36°C (core); SpO$_2$ 84%. | *Presenting Vital Signs at this Phase*: BP 94/70 mmHg; MAP 78 mmHg; HR 112 bpm; RR 14 (per vent); Temperature 38.2°C (core), SpO$_2$ 89%. |
| *Hemodynamic Findings*: RA 5 mmHg; PAS/PAD 54/28 mmHg; PAM 37 mmHg; PAOP 19 mmHg; C.I. 4.6 L/min/m$^2$; SVI 43 cc/beat/m$^2$; LVSWI 30 gm-m/m$^2$/beat; RVSWI 19 gm-m/m$^2$/beat; SVRI 1148 dyne seconds/cm$^5$/m$^2$; PVRI 313 dyne seconds/cm$^5$/m$^2$; CPP 41 mmHg; SVO$_2$ 70%; CaO$_2$ 11.8 ml/dl; CvO$_2$ 9.8 ml/dl; DO$_2$I 543 ml/min/m$^2$; VO$_2$I 92 ml/min/m$^2$; OEI 16.7% | *Presenting Hemodynamic Findings at this Phase*: RA 7 mmHg; PAS/PAD 40/28 mmHg, PAM 32 mmHg, PAOP 16 mmHg; C.I. 3.1 L/min/m$^2$; SVI 28 cc/beat/m$^2$; LVSWI 24 gm-m/m$^2$/beat; RVSWI 6.09 gm-m/m$^2$/beat; SVRI 1832 dyne seconds/cm$^5$/m$^2$; PVRI 413 dyne seconds/cm$^5$/m$^2$; CPP 54; SVO$_2$ 48%; CaO$_2$ 8.2 ml/dl; CvO$_2$ 4.7 ml/dl; DO$_2$I 254 ml/min/m$^2$; VO$_2$I 42.7 ml/min/m$^2$; OEI 42%. |
| | *Case Study Progression*: PRBCs have infused. Albumin is infusing slowly with Gamma globulin to follow. Abrupt change in patient's condition. |
| | *Vital Signs*: BP 106/50 mmHg; MAP 69 mmHg; HR 132 bpm; RR 28 (no change in ventilator rate); Temperature 40.2°C; SpO$_2$ 80%. |
| | *Hemodynamic Findings*: RA 16 mmHg; PAS/PAD 72/42 mmHg; PAM 52 mmHg; PAOP 32 mmHg; C.I. 5.8 L/min/m$^2$; SVI 44 cc/beat/m$^2$; LVSWI 22 gm-m/m$^2$/beat; RVSWI 22 gm-m/m$^2$/beat; SVRI 731 dyne seconds/cm$^5$/m$^2$; PVRI 275 dyne seconds/cm$^5$/m$^2$; SVO$_2$ 28%; CaO$_2$ 9.3 ml/dl; CvO$_2$ 3.3 ml/dl; DO$_2$I 539 ml/min/m$^2$; VO$_2$I 348 ml/min/m$^2$; OEI 65%. |

ED = Emergency Department; DPL = Diagnostic Peritoneal Lavage; FFP = Fresh Frozen Plasma; AMI = Acute Myocardial Infarction; CABG = Coronary Artery Bypass Graft; LAD = Left Anterior Descending artery; ACE = Angiotensin Converting Enzyme; ICU = Intensive Care Unit; BP = Blood Pressure; mmHg = millimeters of Mercury; mcg/kg/hour = micrograms per kilogram; PRBCs = Packed Red Blood Cells; g/dl = grams per deciliter; DVT = Deep Vein Thrombosis; NSR = Normal Sinus Rhythm; LBBB = Left Bundle Branch Block; MI = Myocardial Infarction; LVD = Left Ventricular Hypertrophy; SIMV = Synchronized Intermittent Mandatory Ventilation; FiO$_2$ = Fraction of Inspired Oxygen; cmH$_2$O = centimeters of water pressure; PEEP = Positive End-Expiratory Pressure; MAP = Mean Arterial Pressure; HR = Heart Rate; VT = Ventricular Tachycardia; C = Centigrade; SpO$_2$ = spectrophotometric oxygen concentration; RA = Right Atrial; PAS = Pulmonary Artery Systolic; PAD = Pulmonary Artery Diastolic; PAM = Pulmonary Artery Mean; PAOP = Pulmonary Artery Occlusive Pressure; C.I. = Cardiac Index; SVI = Stroke Volume Index; ml/m$^2$ = milliliters per meter (squared); LVSWI = Left Ventricular Stroke Work Index; RVSWI = Right Ventricular Stoke Work Index; gm $-$ m/m$^2$ = grams-meter/meter (squared); SVRI = Systemic Vascular Resistance Index; PVRI = Pulmonary Vascular Resistance Index; dyne seconds/cm$^5$/m$^2$ = dynes per second per centimeter (to the negative fifth power) per meter (squared); ABGs = Arterial Blood Gases; pH = hydrogen ion concentration; PaO$_2$ = partial pressure of oxygen; PaCO$_2$ = partial pressure of carbon dioxide; HCO$_3$ = bicarbonate; SaO$_2$ = saturation of arterial oxygen; PvO$_2$ = partial pressure of venous oxygen; P/F = Partial Pressure of Oxygen (P) divided by the Fraction of Inspired Oxygen (F), also known as the Hypoxemic Score or Shunt Equation; CVVH = Continuous Veno-Venous Hemofiltration; bpm = beats per minute; RR = Respiratory Rate; CPP = Coronary Perfusion Pressure; SVO$_2$ = Venous Oxygen Saturation; CaO$_2$ = arterial oxygen content; CvO$_2$ = venous oxygen content; DO$_2$I = Delivery of Oxygen (Indexed); VO$_2$I = Oxygen Consumption (Indexed).

The "patient" room (simulation laboratory) incorporates environmental props that may be needed for appropriate assessment and interventions during the scenario(s):

- Hemodynamic waveform strips (if requested)
- Laboratory results (if requested)
- Chest radiograph (if requested)
- 12- or 18-lead electrocardiogram (if requested)
- Intravenous drips that are prelabeled (if requested)
- Intra-aortic balloon pump catheter setup (if requested)
- Continuous Renal Replacement Therapy setup (if requested)

The clinical learner being evaluated is assigned an "orientee" to help evaluate narrative understanding of relevant clinical judgment, clinical inquiry, and skill development that are part

of Benner's model. The orientee is an imbedded instructor. Informed consent of the clinical learner is obtained prior to this course, as the scenario is videotaped and then reviewed during debriefing. (Note: The institution's Human Simulation Laboratory maintains any saved videotapes in a secured cabinet that is accessed by the laboratory's simulation technicians. Students may have copies of their videotape, if requested. Retention of the videotapes are generally at the educator's discretion, given the type of evaluation, otherwise they are erased.) Another clinical educator or established clinical expert evaluates the performance according to the established validation criteria as the scenario progresses to its conclusion.

The clinical learner is subsequently debriefed and asked by the instructors to comment on what he/she was thinking to determine the clinical learner's frame of reference in relation to the scenario. The instructors are encouraged to ask questions if

**TABLE 12.7** Validation criteria for the multiple organ dysfunction syndrome simulation scenario

| Validation criteria | Met | Not met | Evaluator comments |
|---|---|---|---|
| 1. The Clinical Learner has reviewed the assumptions associated with invasive hemodynamic monitoring and applied them to the patient. | | | |
| 2. The Clinical Learner has assessed the system for validity and reliability, and corrected any damping problem identified. | | | |
| 3. The Clinical Learner identifies an assessment of oxygenation (systemic and coronary perfusion pressure) as a priority. | | | |
| 4. The Clinical Learner evaluates oxygenation components presented, and identifies intervention strategies. | | | |
| 5. The Clinical Learner correlates clinical assessment findings that correlate with each component of oxygen delivery. | | | |
| 6. The Clinical Learner evaluates each component of Oxygen Delivery Index ($DO_2I$), and identifies intervention strategies. | | | |
| 7. The Clinical Learner performs an assessment of the patient's oxygen demand via clinical assessment and pulmonary hemodynamic calculations. | | | |
| 8. The Clinical Learner identifies whether the oxyhemoglobin dissociation curve has been altered utilizing clinical, laboratory, and pulmonary hemodynamic calculation data. | | | |
| 9. The Clinical Learner interprets the significance of alterations of the oxyhemoglobin dissociation curve and identifies intervention strategies. | | | |
| 10. The Clinical Learner determines whether the patient's venous reserve is being depleted, and correlates the clinical significance of the determination made. | | | |
| 11. The Clinical Learner summarizes the therapeutic goals for the patient and presents them during grand rounds. | | | |

they are unclear as to the clinical learner's intent. If all involved are satisfied that the validation criteria have been met and the milestones for Clinical Expert have been demonstrated, the clinical learner is recognized as a Clinical Expert in Cardiopulmonary Hemodynamic Monitoring. If not, the clinical learner is given time to review the performance and the scenario objectives, study as appropriate, and subsequently allowed to revalidate. Other learning opportunities using different teaching methodologies are explored with the clinical learner such as the Pulmonary Artery Catheter Education Project [7] or demonstration of assessment in the learner's clinical setting.

## Example 2: Critical Care Nursing Competencies

Clinical educators involved with the process of evaluating a clinical learner's competency in relation to demonstration of clinical judgment, advocacy, clinical inquiry, agency, and/or skill performance in a critical care or subacute care area are aware that this process is fraught with difficulties. Criticisms have included the focus on skill performance to the exclusion of evaluation of critical thinking, subjectivity by the evaluator, an artificial environment, and the methodology of the evaluation. Given these legitimate criticisms, our Clinical Education Department, intensive care units, and subacute care units began a reevaluation of the process and framework of critical care performance, application of the Benner model, and how simulation technology could be used to make evaluation of competence more legitimate and robust.

The very definition of the word "competency" is in itself complex. According to Benner, competency involves an "interpretively defined area of skilled performance identified and described by its intent, function, and meanings" [1]. By breaking out the components of this definition, and applying Benner's model, the development of a more robust methodology for evaluation of a learner's competence may be actuated.

Our hospital's critical care and subacute care units have a long history of employing annual competency evaluations and of being familiar with the Benner model of Novice to Expert. The hospital's three intensive care units (medical/cardiac/cardiovascular surgery, trauma/surgical, and neuroscience) collaborate for their annual competency evaluations, as do the five subacute care units (cardiac, medical, oncology/clinical research, neuroscience, and trauma/surgical). Clinical learners in these areas who have completed the Essentials of Critical Care Orientation [7] and Essentials of Critical Care Orientation simulation labs [8, 9] to a critical care or subacute care unit within the last 12 months are considered to have their evaluation of competence met (described in detail in Chapter 4). Therefore, all other clinical learners from these areas are required (per hospital mandate) to attend an annual competence evaluation. These clinical learners are representative of the levels of Competent to Clinical Expert (same discipline, but not homogenous). Note that Benner's level of Competent is distinct from her definition of competency. A clinical learner at the level of Competent reflects clinical judgment and skill performance in the following manner:

- Prefers the environment to be consistent and predictable.
- Task oriented, and has difficulty individualizing to the patient's needs.
- Prefers time management to be structured.
- Hyper-responsible and self-critical.
- Beginning to reframe what variables in an interaction must be addressed and what can be ignored.

Selection of content to be included in evaluation of competence are taken from procedures that are infrequently seen (low volume), but associated with high risk of compromise or complications; National Patient Safety Goals as identified by the Joint Commission on Accreditation of Healthcare Organizations [10]; Centers for Medicare and Medicaid Services core measures [11]; evidence-based practice or practice alerts from the American Association of Critical-Care Nurses [12]; and internal quality-improvement measures. Thus, the selection of content to be included will change on an annual basis.

Congruent with the Benner model, cognitive objectives for the selected content reflect comprehension, application, and analysis. Validation criteria are subsequently constructed from these objectives that drive the determination of competence by the clinical educator/evaluator. The clinical educator evaluating competency at this level is expected to evaluate troubleshooting, problem solving, and foster multidisciplinary and interdisciplinary collaboration among the clinical learners being evaluated. Proficient or Clinical Experts from the units involved are encouraged to become evaluators for the selected content or become an active observer to help facilitate debriefing. For both the critical care and subacute care units, recognition of an impending code, initiation of the code, locating contents of the crash cart, code roles and responsibilities, defibrillator use, and code documentation were identified as an ongoing competence validation need. Examples of how simulation was incorporated as an evaluation methodology for competency follow.

# 12.3 Using Mid-fidelity Simulation for Critical Care and Subacute Care Competency Evaluation

Mid-fidelity simulation manufactured by Laerdal [6] was utilized for this competency topic. The use of this level of simulation (rather than our high-fidelity physical simulators) permits "on-the-spot" programing of vital signs, selection of an arrhythmia, alteration of heart and breath sounds, and intubation capability sufficient to meet the objectives of the scenario(s). (Also, unlike our high-fidelity simulator, the mid-fidelity allows us to program ST segment elevation, not just ST segment depression.)

From the critical care units' perspective, the code scenarios involve evaluation of an "internal" code versus an "external" code. For an internal code, a code button on the wall of each room may be pressed to initiate a code within the unit, but the hospital operator is not called. The personnel within the critical care unit perform all of the code roles: Team leader (nurse and critical care resident), primary nurse, airway and breathing (usually the respiratory therapist), compressions, medication administration, defibrillation/cardioversion/external pacing, documentation, someone to take blood to the laboratory and return test results, and someone to network with the family (if applicable).

With an external code, outside of the critical care unit, a code button is pushed, the hospital operator is called (the number is different for an adult versus pediatric code team) and given the location of the code, who in turn activates the code team's pagers and also pages a "Dr. Heart" overhead. The three critical care units alternate which one carries the nurse code pager. That responding nurse is considered to be the co-team leader along with the critical care resident responding from our medical/cardiac/cardiovascular surgical ICU. A respiratory therapist assigned to one of the critical care areas also carries a code pager. The other code roles are assumed by personnel within the unit calling the code. The emphasis here is on the familiarity with the unit's crash cart (which is standardized) and evaluation of the performance of the other code roles.

From the subacute care perspective, the focus is on recognition of patient deterioration before the patient "codes," how to call a code, managing the patient prior to the code team arrival, and then giving a report to the receiving critical care nurse (if the patient is resuscitated), or how to help the family following the patient's death, and the fostering of the nurse's coping mechanisms following the dying and death experience.

> **Editors' Note:** As potential patients, we are appalled every time one of our clinical instructors demonstrates that their discipline, let alone all the members of the different disciplines that surround a patient, do *not* have a common terminology for declaring/stating/announcing/initiating an emergency.

## 12.4 Using High-fidelity Simulation for Critical Care Competency Evaluation

Another ongoing competency topic for the critical care areas is the evaluation of the recognition of ventilator complications, and/or their prevention. High-fidelity simulation was chosen because of the ability to attach a ventilator to the simulation mannequin, program a state of pathophysiology that reflected the development of auto-PEEP on the ventilator flow graphic (which lower-fidelity simulators cannot do as effectively, if at all), and that would respond to the clinical learner's interventions or lack thereof.

By having the critical care competence evaluation shared by the three units, collaboration is fostered, variance is minimized, and sharing of similarities and differences between patients and experiences is optimized. This has also led to interest by respiratory therapists, pharmacists, and physicians in future multidisciplinary collaboration where competence areas are shared by the disciplines.

In relation to the subacute care units, the sharing of the units in evaluation of competence has involved a "parenting" or "role modeling" experience for those involved. Only

two of the five subacute units (cardiac and trauma/surgical) were "established" units. As with the critical care units, they had an established history of competency collaboration and incorporation of simulation. The other three subacute units were only recently "upgraded" to this status (being formerly medical-surgical in nature). Thus, there has been a learning curve in working together toward common outcomes, and in simulation as an evaluation methodology.

For all of the units discussed in this section, future evaluation of competence will focus on multidisciplinary involvement, work environment issues, prevention of failure to rescue, involvement of the patient's family in code situations and during procedures, medication reconciliation, and communication dilemmas. Simulation and debriefing initiatives will serve to optimize evaluation of competence in these areas.

Because of the effectiveness of simulation in the critical care, subacute care, emergency department, and emergency department areas in our facility, the hospital mandated that all inpatient areas have annual competence evaluation incorporating simulation. Because these areas have either not been required to have annual competency evaluation be mandatory or have not been exposed to the dynamics involved with incorporation of simulation, the clinical education department has recognized the need for formalized train-the-trainer courses involving content like that presented throughout this book. For every area, it is encouraged that pre-readings be assigned to the clinical learners containing each topic's objectives, validation criteria, references for study (including web-based sites), and what is involved with evaluation by simulation. New clinical learners are oriented to the level of fidelity of the simulator being used (what it can and cannot do), participant expectations during the simulation, and what is involved with debriefing so that their performance reflects their state of competence and not just reactions to simulation.

## Example 3: Multidisciplinary Code Team Management

### The "MegaFest"

Each summer, at our hospital, when approximately 120 new interns arrive at our academic facility, they are exposed to a unique methodology to help them rapidly achieve competence in preparing for basic and advanced life support prior to exposure to their clinical rotations. The interns are divided into two groups, because of space limitations and group size, and over the course of a 3-day weekend are exposed to basic life support and advanced life support as applicable to their practice (Table 12.8). This is known as "Intern MegaFest."

Flash back to Tale 2 of this section, and several problematic areas in relation to the code process became apparent, and were congruent between these two tales. These problematic areas included recognition of patient deterioration prior to actual cardiopulmonary arrest, crash cart familiarity, arrhythmia recognition, safe use of the defibrillator, use of

**TABLE 12.8**   Intern MegaFest matrix for skill development rotations (Part I)

Friday: Group I

13:00–13:20: Welcome/introductions/course overview
13:20–14:10: PALS Rapid Cardiopulmonary Assessment
14:10–14:25: Rotate to B.A.T.C.A.V.E. and break into assigned groups
14:30–18:30: Skills stations

| Time | Infant/ child BLS | Adult BLS | DEFIB/TCP | Vascular access (Peds) | Vascular Access (Adults) | BVM/Airway (Peds) | Intubation | SHOCK/ RESP/AED (Peds) |
|---|---|---|---|---|---|---|---|---|
| 14:30–15:00 | AA-1 | AA-2 | AA-3 | AA-4 | AA-5 | AA-6 | AA-7 | Break |
| 15:00–15:30 | AA-2 | AA-3 | AA-4 | AA-5 | AA-6 | AA-7 | Break | AA-1 |
| 15:30–16:00 | AA-3 | AA-4 | AA-5 | AA-6 | AA-7 | Break | AA-1 | AA-2 |
| 16:00–16:30 | AA-4 | AA-5 | AA-6 | AA-7 | Break | AA-1 | AA-2 | AA-3 |
| 16:30–17:00 | AA-5 | AA-6 | AA-7 | Break | AA-1 | AA-2 | AA-3 | AA-4 |
| 17:00–17:30 | AA-6 | AA-7 | Break | AA-1 | AA-2 | AA-3 | AA-4 | AA-5 |
| 17:30–18:00 | AA-7 | Break | AA-1 | AA-2 | AA-3 | AA-4 | AA-5 | AA-6 |
| 18:00–18:30 | Break | AA-1 | AA-2 | AA-3 | AA-4 | AA-5 | AA-6 | AA-7 |

Note: Group II begins the same matrix on Saturday afternoon
(PALS = Pediatric Advanced Life Support; B.A.T.C.A.V.E. = Basic Advanced Trauma Computer Assisted Virtual Experience: BLS = Basic Life Support; TCP = Transcutaneous Pacing; Defib = Defibrillation; BVM = Bag-Valve-Mask; Resp = Respiration; AED = Automatic External Defibrillation; Peds = Pediatric).

**Intern MegaFest matrix for skill development rotations (Part 2)**

Saturday: Group I continues

07:15–08:00: Registration and breakfast
08:00–08:45: Acute Coronary Syndrome lecture
08:45–09:25: PALS Dysrhythmias lecture
09:25–09:30: Breakout into group assignments (same as yesterday)
09:30–12:30: ACLS & PALS rotation preparation

| Time | PALS: Resp. failure | PALS: Shock/ trauma | PALS: Rhythms | PALS: Coping with death | ACLS: VT/VF* | ACLS: PEA/ Resp. Arrest | ACLS: Tachy | ACLS: Brady/ Asystole* |
|---|---|---|---|---|---|---|---|---|
| 09:30–10:15 | AA-1 | AA-2 | AA-3 | AA-4 | AA-5 | AA-6 | AA-7 | Break |
| 10:15–11:00 | AA-2 | AA-3 | AA-4 | AA-1 | AA-6 | AA-7 | Break | AA-5 |
| 11:00–11:45 | AA-3 | AA-4 | AA-1 | AA-2 | AA-7 | Break | AA-5 | AA-6 |
| 11:45–12:30 | AA-4 | AA-1 | AA-2 | AA-3 | Break | AA-5 | AA-6 | AA-7 |

12:30–13:30: Lunch
13:30–16:30: Complete ACLS & PALS rotation preparation

| Time | PALS: Resp. failure | PALS: Shock/ trauma | PALS: Rhythms | PALS: Coping with death | ACLS: VT/VF* | ACLS: PEA/ Resp. Arrest | ACLS: Tachy | ACLS: Brady/ Asystole* |
|---|---|---|---|---|---|---|---|---|
| 13:30–14:15 | AA-5 | AA-6 | AA-7 | Break | AA-1 | AA-2 | AA-3 | AA-4 |
| 14:15–15:00 | AA-6 | AA-7 | Break | AA-5 | AA-2 | AA-3 | AA-4 | AA-1 |
| 15:00–15:45 | AA-7 | Break | AA-5 | AA-6 | AA-3 | AA-4 | AA-1 | AA-2 |
| 15:45–16:30 | Break | AA-5 | AA-6 | AA-7 | AA-4 | AA-1 | AA-2 | AA-3 |

Note: Group II begins the same rotations on Sunday
(PALS = Pediatric Advanced Life Support; ACLS = Advanced Cardiac Life Support; Resp = Respiratory; VT = Ventricular Tachycardia; VF = Ventricular Fibrillation; Tachy = Tachyarrhythmias; Brady = Bradycardias).

the defibrillator to achieve transcutaneous pacing, code roles (as described in Tale 2) and responsibilities and assignment thereof, and code documentation.

Consequently, the code process within the facility was reevaluated for (i) learning opportunities that would be congruent across the disciplines, (ii) opportunities to facilitate multidisciplinary involvement beyond the standard basic and advanced life support, and (iii) opportunities to involve simulation as a means of creating context and evaluation of content and critical thinking. The following describes what has been enacted or is in development at this time.

### 12.4.1 Arrhythmia Recognition Competency Evaluation

An annual arrhythmia competence exam containing 25 normal and abnormal cardiac rhythms is now incorporated as screening criteria for the American Heart Association's Advanced Cardiac Life Support and Advanced Cardiac Life Support for Experienced Providers courses. Clinical learners who are unable to successfully complete basic life-support demonstration and/or testing, and the arrhythmia competence exam (less than 86%), are referred for remediation in these areas and rescheduled into an Advanced Cardiac Life Support course after preliminary requirements are met.

Annual arrhythmia competence for inpatient telemetry units must be demonstrated during the year between advanced life support (therefore, annual arrhythmia evaluation). Nurses new to the hospital are required to demonstrate competence by either completing a basic arrhythmia course and exam or successfully completing the arrhythmia challenge exam (based upon the nurse's previous experience). The annual arrhythmia competence exam serves to "maintain" competence. The rationale for this requirement is based upon the fact that the focus of advanced life support is NOT on routine, daily, shift-to-shift interpretation of the rhythms of telemetried patients who are not in a code or near-code situation. Hence, the requirement of demonstration of annual arrhythmia competency with the intent being prevention of the development of lethal or potentially life-threatening arrhythmias and subsequent prevention for the need for advanced life-support interventions.

### 12.4.2 Intermediate Life Support

Facilitation of crash cart familiarity, technical use of the defibrillator, pad versus paddle placement and use, assignment of code roles and concomitant responsibilities, documentation considerations, and family presence are now part of a course that was developed to address these needs. It is known as "Intermediate Life Support," and is a prerequisite for first-time Advanced Cardiac Life Support and Pediatric Advanced Life Support participants (physician and nursing), and charge nurses on the nontelemetry units. Mid-fidelity simulation technology is used to evaluate competence for each

of the topics. Advanced beginners new to the critical care and subacute care units are also presented with this information, and are required to demonstrate competence as part of the Essentials of Critical Care Orientation cardiac simulation lab (detailed in Chapter 4).

### 12.4.3 Other Initiatives

The trauma/surgical/burn intensive care unit has also recognized the need to maintain these areas of competence following initial ACLS. Multidisciplinary code team management scenarios using mid- or high-fidelity simulation (depending on simulator availability) is now enacted on a monthly basis in our simulation laboratory. Residents who are on this rotation and available nurses and respiratory therapists are the clinical learners. Again, the problematic areas previously mentioned are the foci of competence evaluation.

Use of both mid- and high-fidelity simulation is incorporated into Advanced Cardiac Life Support and Pediatric Advanced Life Support courses. Use of mid- or high-fidelity simulation (dependent upon lab availability) is incorporated into the Advanced Cardiac Life Support "round robin" that is part of the Experienced Provider course content so as to evaluate not only clinical judgment, but also synthesis and evaluation of the core algorithms by the participants.

Through identification of homogenous elements of concern in the basic and advanced life-support programs between the disciplines, the framework for interactive learning and teaching, and differentiation of levels of competence between the clinical learners can be more easily articulated and evaluated by the clinical educator(s).

## 12.5 Conclusion

This chapter has strived to present examples of how simulation technology can be incorporated into existing clinical education programs. The degree of success in this endeavor is directly correlated with the need to have a solid philosophical foundation upon which to create robust educational programs. The phenomenological model of Benner's Novice to Expert has served as a workable framework to improve and/or achieve our educational goals (although this is always an ongoing work in progress as we learn from successes and mistakes).

One of the outstanding benefits that simulation provides is the ability to evaluate experiential learning and see how the clinical learner progresses over time in the evolution of critical thinking, problem solving, clinical judgment, clinical inquiry, and skill acquisition (novice to expert).

This chapter has given examples of not only how simulation was incorporated into existing clinical education programs, but also how simulation technology and the philosophical framework guiding its use changed the context in which teaching was provided. It also has given examples of how the Benner model and simulation can be used with homogenous clinical learners, heterogeneous clinical learners of the same discipline, and heterogeneous multidisciplinary learners.

Along with a philosophical foundation, knowledge of simulation technology levels of fidelity and limitations, a program's success is also correlated with good, upfront preparation, and practice time. Without this commitment, the educator will not recognize the benefits and strengths, and opportunities for experiential learning that simulation technology can provide.

# References

1. Benner, P. *From Novice to Expert: Excellence and Power in Clinical Nursing Practice*. Addison-Wesley, Menlo Park, California, 1984.
2. Brykczynski, K.A. From novice to expert: excellence and power in clinical nursing practice. Tomey, A. M., and Alligood, M. R., (eds.), *Nursing Theorists and Their Work*, 5th ed. Mosby, Inc., St. Louis, 2002, pp. 165–185.
3. Lynn-McHale Wiegand, D. J. and Pruess, T. Pulmonary artery catheter insertion (assist) and pressure monitoring. Lynn-McHale Wiegand, D. J. and Carlson, K. K., (eds.), *American Association of Critical-Care Nurses Procedure Manual for Critical Care*, 5th ed. Elsevier, Inc., St. Louis, 2005, pp. 549–569.
4. Lynn-McHale Wiegand, D. J., and Pruess T. Pulmonary artery catheter and pressure lines, troubleshooting. Lynn-McHale Wiegand, D. J. and Carlson, K. K., (eds.), *American Association of Critical-Care Nurses Procedure Manual for Critical Care*, 5th ed. Elsevier, Inc., St. Louis, 2005, pp. 576–590.
5. Armstrong Medical Industries, Inc. *Rhythm*SIM™ AA–820 and *Rhythm*SIM™ AA-900 TV interface.
6. www.laerdal.com
7. http://pacep.org (Pulmonary Artery Catheter Education Program).
8. http://www.aacn.org/AACN/conteduc.nsf/vwdoc/Ecco Home (American Association of Critical-Care Nurses Essentials of Critical Care Orientation© Program).
9. Beeman, L. (Abstract) The use of human simulation technology to facilitate experiential learning in correlation with "E.C.C.O." American Association of Critical-Care Nurses National Teaching Institute Program and Proceedings, 2005, p. 210.
10. http://www.jointcommission.org/Patientafety/National PatientSafetyGoals/ (National Patient Safety Goals, Joint Commission on Accreditation of Healthcare Organizations, 2007).
11. http://www.cms.hhs.gov/ (U.S. Department of Health and Human Services Centers for Medicare and Medicaid Services Core Measures).
12. http://www.aacn.org (American Association of Critical-Care Nurses Practice Alerts).

# 13

# Incorporating Simulation into Graduate (Resident) Medical Education: With Special Reference to the Emergency Department

Mark E.A. Escott
Lawrence E. Kass

## 13.1 Role of Simulation in Teaching Emergency Medicine

The use of high-fidelity simulation in Emergency Medicine has rapidly expanded over the past decade. For many years, we have utilized low-tech and part-task trainers to teach individual emergency skills to medical students, EMS personnel, and residents. However, complex scenarios required that instructors read aloud from written scripts to describe the patient's changing conditions, forcing the learners to contribute a significant amount of imagination. Where adequate simulators did not exist, students would be taught by first observing one or several procedures, followed by performing the procedure on a patient under supervision, and eventually on their own. Emergency procedures, from blood draws to thoracostomies, were learned in this apprenticeship manner. The integration of part-task trainers and full-body high-fidelity robotic patient simulators into clinical education has dramatically changed our ability to suspend disbelief and utilize preprogramed scenarios to achieve standardization and complexities that were not previously possible. Utilizing this approach for training allows us to minimize the risks to patients, challenging the limitations of the traditional "see one, do one, teach one" axiom. Advanced clinical simulation is also useful to achieve the new Accreditation Council for Graduate Medical Education guidelines (http://www.acgme.org/acWebsite/home/home.asp) established for resident training in patient evaluation and management. In this chapter, we describe our experience with incorporating simulation technology into an emergency medicine residency training curriculum as it relates to teaching and evaluating various patient management skills (Table 13.1).

**TABLE 13.1**  Definitions

*Low-tech trainer:* A device that does not respond to interventions or have the ability to be altered in real time to create a response.

*Part-task trainer:* A device designed to train in performance of a particular task such as cricoid pressure or lumbar puncture.

*High-fidelity trainer:* A device that can be a whole body or part body that is able to respond to treatments or interventions through physiologic and/or pharmacologic modeling.

## 13.2 Procedure Training

Dependence on patients for procedure training is problematic for two major reasons: First, because many of these procedures are relatively rare, a long period of time is required before the clinical exposure necessary for proficiency is achieved. This makes maintenance of proficiency difficult as well. Part-task trainers can be used to achieve initial proficiency prior to attempting a procedure on patients. Instead of watching one or two, the trainee can perform 10 or 20 of a given procedure on different mannequins prior to their first attempt on a live patient. While simulators are limited in the amount of anatomic variation they can present, the process of learning and perfecting a new skill in the simulation laboratory leads to the familiarity with equipment and the development of muscle memory that are vital in the achievement of effective and efficient skill performance. Thus, students can focus their attention upon the uncertainty generated from their unpredictable patients, and not from their invariant equipment and work environments.

Simulation centers have become more common and greater varieties of patient simulators have become more widely available. For instance, in the field of airway management, there are a number of models that can be used in a "difficult airway" training course to challenge the clinician by simulating trismus, edema, and other forms of difficult anatomy. The use of live patients as teaching subjects for Cricoid pressure (Sellick's maneuver) education has been shown to produce uniformly poor results when assessed by important performance descriptors such as the amount of force needed [1]. Part-task training in cricoid pressure application results in impressive improvement in subsequent performance [2]. There is also evidence to support the contention that training in airway management skills in a simulated setting results in improved performance in endotracheal intubation [3]. While we lack specific evidence for many of the various individual skill trainers, there has been no evidence to date that such training is harmful to skill development, so long as the instructors keep their students away from the non-faithful and anti-faithful aspects of these devices.

The second problem with procedure training on patients is that many of the procedures performed in the emergency department are painful and potentially harmful. Learning the procedure in the simulated setting allows for a margin of error without consequence that is not available in the clinical setting. Most importantly, simulation allows students to "go too far," beyond where they would ever be allowed to go with live patients, and thereby learn how to recognize just what "way out of bounds" looks like, and do so safely for everyone (especially the patient). There is the potential for reducing clinical errors, discomfort, and litigation by first learning in the safe setting of the simulation laboratory with feedback from experts in that skill.

The simulation laboratory is particularly useful in those tasks that are complex, rare, or associated with significant risk. It is important to remember that in the life of every student, every procedure is at one time their first attempt. Hence for them, at that moment, it is the most rare, and thereby may very well be complex, with significant risk to them as well as their patients.

## 13.3 Emergency/Resuscitation Skill Integration

As with individual skill training, the integration of skills by incorporating the use of role-playing scenarios has been in use for decades. This began in the form of verbal scenarios that were performed in a sort of oral board fashion. (Note: even a discussion of morbidity and mortality (M&M) where the question is asked: "What if this happened..., What if this was tried....?" can be seen as "simulation.") As part and whole-body mannequins were developed, they began to be used not only for training of individual skills, but also in the integration of multiple skills (Figures 13.1–13.3). A common example is the "mega code" scenarios that were once a part of Advanced Cardiac Life Support training and testing. Inert mannequins that were at first only able to be used for cardiopulmonary resuscitation training received electrocardiogram signal generation capabilities and could then be used for Cardioversion/Defibrillation training as well. However, when tested on these resuscitation algorithms, medication knowledge and dosage was left to a verbal report to the examiner, i.e., "I would then intubate and give x dose of y drug..." In this setting, by not being able to perform the exact skills needed, the evaluator was left to assume that those skills could be performed correctly. In some settings, those skills could be tested individually to compensate and ensure proficiency, but an evaluator would still be left questioning if the knowledge and skills demonstrated could be integrated successfully.

Successful completion of complex tasks such as resuscitation requires not only the proficiency of each individual skill, but also the ability to integrate and prioritize those skills. The Residency Review Committee of Emergency Medicine has recognized the complexity of many patient presentations in emergency medicine and mandated since January 1, 2005, that residents not only be able to demonstrate proficiency in individual skills, but also in resuscitation [4]. Similar guidelines have recently been presented for a national curriculum for fourth year medical student rotations in emergency medicine [5].

The logistics of training and achieving proficiency in these clinical skills relies on a functioning simulation center capable of producing simulated scenarios and proficient in using advanced patient simulators. While individual skills can be developed on part-task trainers, integration of those skills can only be achieved within advanced simulations. The advanced patient simulations can take the common example of the "mega code" scenario and evaluate every portion of the resuscitation. Thus, a simple ventricular fibrillation cardiac

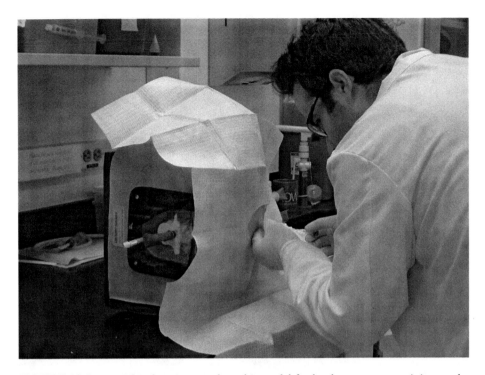

**FIGURE 13.1**   Partial-task trainers such as this model for lumbar puncture training can be utilized for single skill training and assessment. These mannequins serve as a convenient and safe means of skill acquisition.

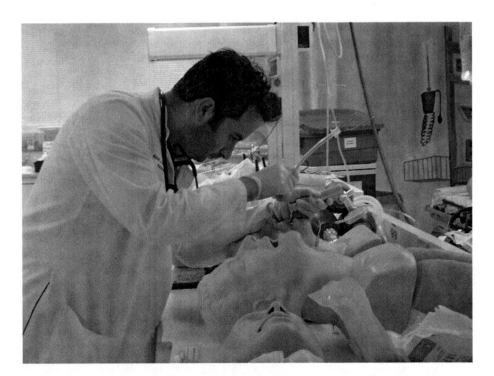

**FIGURE 13.2**   Some partial-task trainers such as the airway head mannequins can be used for integration of a few but crucial airway skills. Trainees may benefit from a number of different mannequins to provide a variety of airway exposure, including some "difficult airway" models.

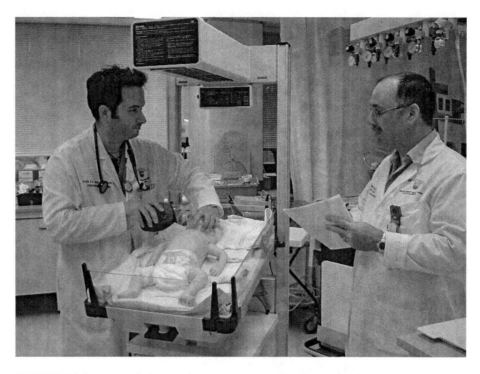

**FIGURE 13.3** Low-tech mannequins require a significant use of imagination, a script, and an evaluator in the room to tell the trainee about changes in the "patient" condition in response to treatment decisions.

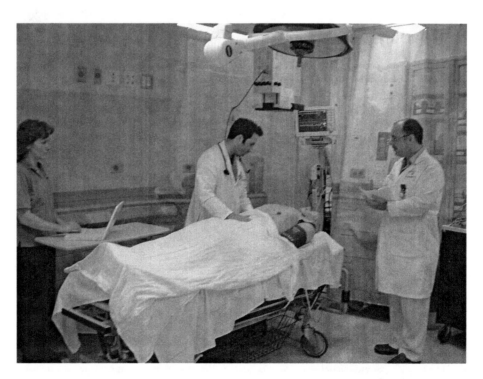

**FIGURE 13.4** Multiple skill integration can be assessed using high-tech mannequins with the evaluators during the "live" scenario. This can be followed by postscenario quizzing of discreet knowledge.

arrest scenario can present not only the basic skills of EKG interpretation, CPR, defibrillation, but also airway management, IV insertion, drug administration and their physiologic effects, and postresuscitation care.

The additional benefit of the high-tech patient simulators is that these scenarios can be preprogramed or parameters can be changed in a location remote from the patient in order to preserve the suspension of disbelief. Some model-assisted simulators will even change on their own, as a result of the treatment given, and display the appropriate physiologic response that would be expected: drugs and their dosages, the types and volumes of intravenous fluids, or the quality of ventilation delivered with a bag-valve-mask. Preprograming is beneficial to the educator, because the scenarios can be performed in a comparable fashion for each resident, medical student, or paramedic, and graded according to a standardized rubric. This increases the reliability, reproducibility, and validity of clinical skill evaluation, which in turn becomes the basis for evaluating the quality of the teaching program itself.

In most circumstances, it is desirable to have a scenario debriefing during which further knowledge can be elicited such as indications, contraindications, and expected complications for particular medications or procedures. For example, one scenario used in our residency training program involves a man in his 50s with an ST segment elevation myocardial infarction who is in cardiogenic shock. This particular scenario is designed to test the clinical management skills involved in treating this acute condition, as well as the relevant medical knowledge as it relates to definitive management (the choice of thrombolysis versus cardiac angioplasty). The residents are then subsequently quizzed about indications and contraindications for thrombolysis or angioplasty as well as medications and management issues of this case to assess the completeness of their factual knowledge and procedural skills (Figure 13.4).

## 13.4 ACGME Core Competencies

Evaluation of performance requires the *a priori* determination of the goals sought and outcomes expected. This can easily be done in a checklist fashion using a motor vehicle trauma case that explicitly describes the expected performances utilizing the various elements of the American College of Graduate Medical Education's General Competencies [6] (Table 13.2). While certain procedures (such as intubation and cardioversion/defibrillation) can be easily incorporated into scenarios such as those described above, it has been our experience that if too many distracters are incorporated, the dramatic events tend to dominate the scenario, often to the detriment of the primary learning objectives of the case. Therefore, we have found it more effective to separate the evaluation of procedural competency from the clinical scenarios for those procedures that are particularly complex or time consuming.

**TABLE 13.2**  American College of Graduate Medical Education Core Competencies

- Patient care: Residents must provide care that is compassionate, appropriate, and effective to the treatment of health problems and the promotion of health.
- Medical knowledge: Residents must demonstrate knowledge about established and evolving biomedical, clinical, and cognate (e.g., epidemiological and social-behavioral) sciences and the application of this knowledge to patient care.
- Practice-based learning and improvement: Residents must be able to investigate and evaluate their patient care practices, appraise and assimilate scientific evidence, and improve their patient care practices.
- Interpersonal and communication skills: Residents must learn to provide effective information exchange and teaming with patients, their families, and professional associates.
- Professionalism: Residents must demonstrate a commitment to carrying out professional responsibilities, adherence to ethical principles, and sensitivity to a diverse patient population.
- Systems-based practice: Residents must demonstrate an awareness of and responsiveness to the larger context and system of health care and the ability to effectively call on system resources to provide care that is of optimal value.

## 13.5 Crisis Resource Management

Emerging as a crucial factor in successful resuscitation is the ability to not only perform and prioritize tasks, but also to be able to successfully orchestrate the resuscitation team. This has been advanced through the development of Crisis Resource Management (CRM) programs to assess and teach the management aspects of critical events. Crew (or Cockpit) Resource Management began in the aviation industry after a 1979 NASA-sponsored symposium of aircraft accidents and near-misses that determined that 85% were due to human error involving poor cockpit communication and discipline. This led to the aviation standard simulation training, in not only their flying skills, but also their Crew Resource Management, or management of the flight "team" (see Chapter 62).

Crew Resource Management and Crisis Resource Management share common grounds in the key principals such as Role Definition, Communication, Global Assessment, Avoiding Fixations, Support, and Resources. The transition to medicine was through Anesthesiology after, similar to aviation, the vast majority of adverse events were found to have an origin in human error. While anesthesiologists were the first developers of the new age of clinical simulation centers, CRM has crossed the boundaries into multiple areas of health care.

Emergency Medicine is well suited to CRM training due to the often critical and chaotic nature of the emergency department, which routinely utilizes a combination of unformed and pre-formed teams to address various critical incidents. Therefore, CRM training is the perfect setting to hold

**TABLE 13.3**   Example simulation performance evaluation checklist

| Trauma/motor vehicle collision | | | | |
|---|---|---|---|---|
| Resident:_____   Evaluator:_____   Date:_____ | | | | |
| *Patient care/Medical knowledge* | | | | |
| Obtains focused history and physical exam for trauma | _U | _NI | _S | _E |
| Initiates appropriate initial treatment(general): | _U | _NI | _S | _E |
|   Obtains intravenous access | _U | _NI | _S | _E |
|   Applies oxygen | _U | _NI | _S | _E |
|   Cardiac monitor | _U | _NI | _S | _E |
|   Fluid bolus | _U | _NI | _S | _E |
| Obtains X-rays in timely fashion | _U | _NI | _S | _E |
| Interprets X-rays correctly | _U | _NI | _S | _E |
| Obtains blood work in timely fashion | _U | _NI | _S | _E |
| Interprets blood work correctly | _U | _NI | _S | _E |
| Does appropriate reassessment after fluids | _U | _NI | _S | _E |
| Initiates appropriate treatment for persistent hypotension (general) | _U | _NI | _S | _E |
|   Repeat fluids | _U | _NI | _S | _E |
|   Prepares for transfusion | _U | _NI | _S | _E |
| Does appropriate assessment for abdominal trauma (FAST or DPL) | _U | _NI | _S | _E |
| Appropriate reassessment for worsening shock | _U | _NI | _S | _E |
| Intubation | | | | |
|   Recognizes need for endotracheal intubation | _U | _NI | _S | _E |
|   Performs endotracheal intubation | _U | _NI | _S | _E |
| Central venous access | _U | _NI | _S | _E |
|   Recognizes need for additional line/central venous access | _U | _NI | _S | _E |
|   Performs central line | _U | _NI | _S | _E |
| *Communication/Professionalism* | | | | |
| Keeps patient and family informed | _U | _NI | _S | _E |
| Directs team appropriately | _U | _NI | _S | _E |
| *System-based Practice* | | | | |
| Consults surgery appropriately | _U | _NI | _S | _E |

Key:   U = unsatisfactory (does not meet minimal expectations)
    NI = needs improvement (minimal deficiencies not expected to have significant effect)
    S = satisfactory (meets expectations for year of training)
    E = exceeds expectations (performance or knowledge beyond expectation for year of training)

interdisciplinary training sessions with emergency department residents, nurses, technicians, and perhaps consultants. These sessions begin as they often do in the Emergency Medicine Department. A nurse or two enter(s) the room and begins a routine task when something starts to go wrong. This requires them to first *recognize* that there is a problem, call for *support*, gather *resources*, establish *roles* of team leader and team member, and begin treatment. This is followed by the entrance of an intern or two, causing a change in leadership followed by the entrance of more senior residents. The first scenario is performed after an orientation to the simulator so that all

involved are aware of the capabilities of the simulator and are familiar with the equipment that is available for their use. Of course, whenever possible, the equipment should be the same equipment available in the usual work environment, unless an objective for the session is dealing with unfamiliar equipment.

The first scenario is performed before any lectures, training, or education in CRM and is then followed by one or two sessions after training. (Note: some simulation centers spend up to 8 hours of preparation and lectures on crisis resource management principles before allowing trainees to participate in a crisis event. Other simulation centers believe that it is important to provide a venue for trainees to experience how difficult it is to manage a crisis with chaos – they believe this turns the trainees into adult learners who "know what they do not know, and know what they cannot do.") This allows the team members to compare the video recorded sessions before and after training in light of their new exposure to the principles of CRM. Having multiple sessions also allows for team members to change roles in order to gain experience as the team leader, team member, first responder, and backup responder and experience the dynamics associated with each. During each of the scenarios, the remainder of the participants can watch a live video display while in a remote room. The remote room can also be used to review the video recording of the session with the scenario participants.

Effective communication skills are vital for the Emergency Medicine physician during critical incidents as well as during everyday interactions with team members. This involves issues such as keeping team members informed of working diagnoses and treatment plans so that they are aware of, and are able to, look for evidence to support or contradict that plan, as well as anticipate future needs. Communication also involves "closing the loop" of requests for tasks. For example, a nurse is given a direct and specific request to start a large bore peripheral IV and, therefore, should report back to the team leader either that the task is completed, there will be a delay in the completion of the task, or the task is not possible to complete. It is useful to keep in mind that physician-nurse communication accounts for 2% of their time but responsible for 37% of medical errors [7]. While the importance of communication skills may be stressed during a crisis (real or simulated), they are still essential for the "routine" communications that occur in the emergency department, particularly when failures in "routine" communications precipitate subsequent crises for the patient.

Following each session, there is facilitated debriefing of the team members, which is largely self-guided by the participants who point out lapses or strengths in their performance (see Chapter 69). The specific goals under the categories of problem recognition, calling for support, gathering resources, and establishing team roles should be identified *a priori*. The participants can then easily be graded (or self-graded) on a simple rubric of Unsatisfactory, Needs Improvement, Satisfactory, or Excellent for each item. This can be utilized to direct future areas of development for individuals and teams, to identify areas of improvement needed in the curriculum, to direct future areas of development for individual teachers, and to identify areas of improvement needed in the curriculum of the teaching program (Table 13.3).

## 13.6 Complex Scenarios

Experienced users of advanced patient simulation facilities have remarkable abilities to construct and reproduce very difficult and realistically flowing scenarios in which students develop and maintain logical approaches to a changing clinical picture. With appropriate and effective use of such recent facilities, we are able to take basic knowledge and skills and challenge students by changing the underlying physiologic parameters.

One important example in emergency medicine is the development of the "can intubate/can ventilate, can't intubate/can ventilate, can intubate/can't ventilate, and can't intubate/can't ventilate" algorithms for airway management. With the advanced simulator, it can be programed to begin with a simple airway that can managed by a bag-valve-mask and oral airway progressing to the need for intubation. The tube can be placed and the "patient" can suffer a pneumothorax, pulmonary edema, bronchospasm, equipment failure, or multiple failures that can challenge the student and refine their abilities.

## 13.7 Conclusion

Resident education requires defining specific objectives in the realms of knowledge, procedural skill, communication skill, proficiency with team work, and systems utilization. Simulation allows educators to design scenarios to explicitly teach and/or test each of these areas. Our experience to date has been very positive. While efforts are ongoing to collect data about its efficacy (i.e., its impact upon subsequent patient care), we believe it is already apparent that with this approach educators can easily identify, and target, specific resident and training program weaknesses.

## References

1. Escott, M. E., Owen, H. and Strahan, A. D. Plummer JL. Cricoid pressure training: how useful are descriptions of force? *Anaesthesia & Intensive Care.* 31, 388–391 (2003).
2. Owen, H., Follows, V., Reynolds, K. J., Burgess, G., and Plummer, J. Learning to apply effective cricoid pressure using a part task trainer. *Anaesthesia.* 57, 1098–1101 (2002).
3. Hall, R. E, Plant, J. R., Bands, C. J, et al. Human patient simulation is effective for teaching paramedic students endotracheal intubation. *Acad Emerg Med.* 12, 850–855 (2005).
4. http://www.acgme.org/acWebsite/RRC_110/110_guidelines. asp#res (accessed June 2, 2006).
5. Manthey, D. E., Coates, W. C., Ander, D. S., et al. Report if the task force on national fourth year medical student emergency medicine curriculum guide. *Ann Emerg Med.* 47, E1–E7 (2006).
6. http://www.acgme.org/outcome/comp/compFull.asp (accessed June 2, 2006).
7. Donchin, Y., Gopher, D., Olin, M., Badihi, Y., Biesky, M. and Sprung, C. L. A look into the nature and causes of human error in the intensive care unit. *Crit Care Med* 23, 294–300 (1995).

# 14

# Theory and Practice of Developing an Effective Simulation-based Clinical Curriculum

Valerity V. Kozmenko
Alan D. Kaye
Barbara Morgan
Charles W. Hilton

## 14.1 Setting the Plan

In 2003, Louisiana State University Health Sciences Center at New Orleans pioneered development and implementation of a mandatory longitudinal simulation curriculum for undergraduate medical students. Developing an effective simulation-based clinical curriculum employing simulation is complex and multifaceted. The content and entire structure of any simulation-based educational program depend on:

- participants' prior knowledge in a given domain;
- participants' learning objectives;
- instructors' expected educational outcomes;
- teaching methods applied to clinical simulation;
- simulated clinical conditions (includes medical and surgical conditions);
- competency of the simulation production team;
- capabilities of the simulator(s).

Current knowledge of the target audience is a critical factor in developing simulation presentations for any curriculum. After matching learning and teaching objectives, specific scenarios are created, and an appropriate simulator or a combination of simulators is selected. Very often, developing learning objectives and matching them with available simulation team capabilities and available technology is a dynamic spiral process that ends only when appropriate balance and compromise are reached. Of course, as your simulation team should be improving its performance and simulator manufacturers should be improving their wares, this dynamic upward spiral never really ends, it just waxes and wanes.

> *The spiral model is a risk-oriented lifecycle model that breaks a project into miniprojects. Each miniproject addresses one or more major risks until all the major risks are addressed. The concept of "risk" is broadly defined in concept of development, and it can refer to poorly understood requirements, poorly understood structure, potential performance problems, problems in underlying technology, and so on [1].*

For example, for a given class of students, the course developer generates a list of learning objectives and examines the capabilities of several simulators to present them. Each of several different simulators is capable of supporting the presentation of different learning objectives, and the course developer needs to select the simulation resources that provide the best examples of the objectives of the training program. In those instances where the available simulation resources are a poor match to the learning objectives, alternate educational methods are employed. At our Center, we have successfully combined a Human Patient Simulator (Medical Education Technologies, Inc, Sarasota Florida, U.S.A[1]), a LAP Mentor laparoscopic surgical simulator (Simbionix, Inc, Cleveland, Ohio, U.S.A), and state-of-the-art Virtual OR Suite (Stryker, Inc, Kalamazoo, Michigan, U.S.A).

The Human Patient Simulator excels at presenting physiological conditions via real clinical vital sign monitors. The LAP Mentor provides a unique opportunity of performing laparoscopic surgery on a virtual patient. The Virtual OR visualization equipment facilitates broadcasting simulation sessions to remote classrooms.

Simulator training and existing technology mutually affect each other: Advances in technology allow more sophisticated methods of teaching whereas educational needs often work as a driving force in new technology development. This process continues *ad infinitum* and a simulation course must evolve accordingly.

Presently, there are two major formats for teaching with a simulator: (i) teacher-focused content-based format, in which the simulator is used solely as a visual aid; and (ii) student-focused simulator-based dynamic methods of teaching with the use of a simulator as an interactive agent.

The first method is often used in a form of minilectures where a simulator is used to demonstrate principles or a concept. For example, a teacher presents a lecture on acute bronchial asthma and uses a simulator to demonstrate the manifestations of the disease: Wheezing, desaturation, change of inspiration/expiration ratio, bronchial constriction, change in the lung volumes and capacities, etc. A major drawback of this method is its failure to actively engage the learners. The students are observers only.

The second method brings a new dimension to traditional clinical education. Here the learners are actively involved in the educational activities, typically in smaller groups. This simulator-based interactive method helps the students develop clinical reasoning and practical skills pertinent to patient care. Because of its focus upon learning via interaction with the simulator, an effective "human-machine" mechanism of interaction must be developed. The "human-machine" interaction is a two-way communication that depends on the skills and attitudes of the simulator operator, the capabilities of the simulator, and skills and attitudes of the learner. Our experience with teaching medical students, residents, nurses, and physicians indicates that the student-focused dynamic method is eagerly accepted by students, and produces better learning than simulator-assisted minilectures (see Appendix 14A.1: A sample session based upon this method). To maximize the likelihood of success, this method should be based upon sound doctrines of adult teaching and learning [2].

## 14.2 Methods of Teaching

Malcolm Knowels in his book *Adult Learner* quotes Eduard Lindeman, a pioneer of adult education, who developed a solid foundation for systematic theory about adult learning:

> The approach to adult education will be via the **route of situations**, not subjects. Our academic system has grown in a reverse order: Subjects and teachers constitute the starting point, students are secondary. In conventional education the student is required to adjust himself to an established curriculum; **in adult education the curriculum is built around the student's needs and interests**. Every adult person finds himself in specific situations with respect to his work, his recreation, his family life ... – situations, which call for adjustments. Adult education begins at this point. **Subject matter is bought into the situation, is put to work, when needed.** Texts and teachers play a new and secondary role in this type of education; **they must give way to the primary importance of the learners.** The resource of highest value in adult education is the learner's experience. If education is life, then life is also education. ... Psychology is teaching us that **we learn what we do, and that therefore all genuine education will keep doing and thinking together. ... Experience is the adult learner's living textbook** [3].[2]

Lindeman describes key assumptions about adult learners [4]:

- Adults are motivated to learn as they experience needs and interests that learning will satisfy.
- Adults' orientation to learning is life-centered.
- Experience is the richest source for adult's learning.
- Adults have a deep need to be self-directing.
- Individual differences among people increase with age.

In clinical simulation, the audience almost exclusively consists of adult learners, or at least those rapidly becoming adult learners. Since teaching with a simulator is based on allowing the audience to educate themselves through the use of a simulator, and the role of the instructor is to only facilitate learning, one needs to know the basic principles of adult learning.

Several methods are useful for teaching adults. Teaching through inquiry is a method that shares several common principles with teaching via simulation: It is student-focused, interactive, and trainee's experience is a driving vehicle of gaining knowledge and skills. Thus, this method is used very often in teaching with simulation.[3]

Teaching through inquiry is both a new and an old method of teaching. It is surprising that regardless of its successful use by the great teachers of the past, it has been for a long time ignored in formal teaching. Confucius and Lao Tse of China, Hebrew prophets and Jesus Christ in Biblical times, Aristotle, Socrates, and Plato in Greece, and many other teachers and philosophers have contributed to and taught with the method of inquiry. Below is the summary of how Postman and Weingartner describe the observable behaviors of teachers who use the method of inquiry [5]:

- *The teacher rarely says what the students ought to know because it preoccupies the students, creates fixations, and deprives them in their search for solution of the problem.*
- *If the teacher's intervention is required, the main method used is questioning.*
- *Generally, the teacher does not accept a single statement as an answer to a question. The teacher is looking for the **reasons**, not **a reason**. The method of finding an answer is as important as is the correct answer.*
- *The teacher encourages student/student (and student/simulator) interaction, and generally avoids acting as a mediator or a judge of the quality of ideas expressed. That is the job of the students' real-time interaction within the simulation.*
- *The teacher-directed path of the scenario develops from the responses of the students and not from the previously determined "logical" structure or end point.*
- *Generally, each lesson poses a challenging, but not impossible problem for the students (see Appendix 14A.1).*

Knowledge of the target audience enables us to determine (i) the scope of interventions that the learners might undertake during the case management, and (ii) equip the simulator operator and the simulated environment with the mechanisms and resources for responding appropriately to the students' interventions. Unless both these goals are met, undesirable outcomes for the learner and/or incapacitating frustrations of both teachers and learners are likely.

Consider the following example:

An instructor plans a simulation session for practicing physicians about a narcotic overdose with respiratory depression, and chooses a high-fidelity patient simulator. Developing the case, the instructor plans the following series of conditions and events to occur:

1. patient is in postoperative pain;
2. learners administer morphine;
3. respiratory depression develops;
4. without appropriate management of the airway, the patient develops hypoxia-mediated myocardial ischemia;
5. appropriate airway management resolves the problem.

When the actual session occurs, the student's responses are different than anticipated. Having recognized the postoperative pain related problem, the trainees administer morphine in incremental fashion to avoid drug overdose. Since the first event in this series has failed to occur, the session reaches a stalemate. To force the case into the scenario as it was planned, the instructor assumes a role of a nurse and administers a bolus of 30 mg of morphine. After that, the patient develops respiratory depression, and the learners have to deal with it.

Question: What do the participants feel about this session and how much do they learn?

Answer: "**They resent and resist situations in which they feel others are imposing their wills on them** [6]."

This example shows fallacies at several major steps in logic:

1. target audience's level of expertise has been ignored;
2. scope of possible interventions has not been defined[4];
3. democratic principle of session has been violated[5];
4. the lesson was designed with rigid logical structure that failed to react to the learners' responses.

This example demonstrates the importance of: (i) a thorough knowledge of the target audience; (ii) determining the scope of possible interventions that could be undertaken during the management of given medical condition; and (iii) providing extensive feedback mechanisms to guide the learners through the case.

## 14.3 Place and Role of Didactic Teaching

It is generally believed that teaching with a simulator serves as a method of "putting all pieces of knowledge together into a case management algorithm," rather than merely a method of gaining new content knowledge. To a point, this is true. Since learning with a simulator is significantly influenced by content knowledge of the learner, the instructor needs to decide how didactic teaching will fit in the simulation curriculum. How much should take place before, during, or after the simulation session? As with every complex issue, the answer to this question is dialectic and it depends on the (i) audience, (ii) learning objectives, (iii) method of teaching, (iv) structure of the course, (v) competence of the simulation production team, and (vi) the simulator's fidelity.

### 14.3.1 Didactic Teaching Before the Simulation Session

Providing didactic teaching relevant to the planned scenario prior to the simulation session is beneficial when students' knowledge across the group is uncertain. Each student must possess a minimum required knowledge base before they can apply it to a specific clinical problem. Didactic teaching before the simulation session also aids the teacher in assessing learner knowledge and allows the teacher to refine the scenarios and interventions to the level of the student. However, the didactics must not be tailored too tightly to the specific scenario, as this may impair recognition of patterns similar but not identical to the planned scenario. To avoid this potential problem, the course developer should ensure that the learners cannot establish an unexamined mental link between a didactic teaching and a simulation session that is going to take place. For example, we accomplish this by meeting with the directors of the major clerkships before the development of the simulation curriculum and determine what subjects the medical students will have been exposed to before they arrive for their simulation session. Also, we decided that the students would not have assigned prior readings and would not know the specific clinical topic of their simulation session. Thus, as stated above, design of the session starts with assessment of the needs and knowledge of the learner.

### 14.3.2 Didactic Teaching During Simulation Session

Given the average cost of teaching with a simulator varies from US $700/hr to US $2400/hr [7], many would say that the session is not the best time to provide content teaching. This might be true if the simulator is used merely as a visual aid during a minilecture. Still, didactic material could be effectively used during the simulation session in conjunction with teaching through inquiry. For example, we encourage our students to use their textbooks and other resources during the case if they need it. This approach serves several purposes. First, it allows us to administer challenging cases and raise the bar of expectations from students' performance without embarrassing them. Second, it helps the students to find a solution without instructors' interrupting the flow of the scenario. Third, it helps maintain a nonjudgmental atmosphere in a stressful environment. Fourth, this best replicates the way students learn in the clinical arena.

### 14.3.3 Didactic Teaching After the Simulation Session

The discovery method is based on posing a problem for the students and letting them find a solution with minimal intervention by the teacher. If the scenario designers have employed excellent needs assessment, the scenarios can be designed for modification and adaptation when in progress, to challenge the students up to and slightly beyond their current capabilities. In such cases, the students are challenged to go beyond what was intended for the session and learn more. For example, we observed this phenomenon in presenting the malignant hyperthermia case to the first year anesthesia nursing students. The complexity of the case was rather high and the students' performance scores gradually improved with each repetition of the case. The students were informed about their progress in mastering their treatment of this pathology. The students reported reviewing material after the simulation sessions. Didactic material could be presented after the simulation session in a form of modeling the terminal (ideal) performance by the instructor. Role modeling is a good supplement to the method of teaching through inquiry, but mostly it should be used at the end of the simulation session, with some exceptions. For example, since our student nurse anesthetists did not have any operating room experience prior to the simulation training course, we demonstrated to them a classic technique of perioperative assessment of the patient and performing an intravenous induction of anesthesia.

## 14.4 Learning Objectives

Houle (1972, pp. 139–312) describes the attributes of learning objectives, "**Objectives lie at the end of actions designed to lead to them.** Objectives are usually pluralistic and require the use of judgment to provide a proper balance in their accomplishment."

As illustrated above in the example of the drug overdose scenario, setting appropriate learning objectives based on painstaking needs assessment determines the educational value of the simulation session. The ultimate goal of teaching with a simulator is: (i) to teach students to perform the best clinical practices based on the generally accepted standard treatment protocols, (ii) develop effective critical thinking skills when forming a differential diagnosis, (iii) correction of common misconceptions, and (iv) reduce time and magnitude of patients' changes required for detection.

**TABLE 14.1**  Decision-making tree of how the learners should perform

| Type of learning objective | Definition | Example |
|---|---|---|
| Primary learning objective | Major learning objective that must be reached at the simulation session. Primary learning objectives are the equivalent of a standard treatment protocol. | According to PALS protocol, hemodynamic support in septic shock includes maintaining airway patency, fluid challenges, vasopressor administration, and IV steroids (if there are symptoms of adrenal insufficiency) |
| Secondary learning objectives | Secondary learning objectives are the components, or subsets, of the primary learning objectives. | Maintaining of airway patency consists of the following procedures: Head tilt, jaw thrust, and an airway device insertion. |
| Variable learning objectives | Variable learning objectives may be relevant issues for some learners and irrelevant for the others. | Teaching about inappropriateness of administration of a beta-blocker to control compensatory tachycardia in a hypovolemic patient could be relevant for inexperienced junior medical students, and it would be irrelevant to the surgery residents. |

Teaching students to follow the standard treatment protocols is the easiest of these three tasks. It is relatively easy to locate valid treatment protocols and to develop a decision-making tree of how the learners should perform (Table 14.1). However, the list of learning objectives should be more detailed than the treatment protocols because the protocols have an outline format rather then a storyboard. As such, protocols cannot anticipate all of the routes the student may take, with consequences as shown in the drug overdose case above.

This classification is based on the scope of the learning objectives rather than on their clinical importance. Provided that an interactive method of session management has been chosen, after the learning objectives have been outlined, the course developer must create a mechanism for simulator reaction based upon a student's successful (or unsuccessful) achievement of the learning objectives. This mechanism builds a backbone structure of the interactive simulation case. These are the *primary learning objectives* because they reflect best practices and should be achieved by all students regardless of other factors that may affect the way the cases progress.

Flawlessly achieving primary learning objectives is not the only goal of the simulation session. Scoring 100% during case management does not necessarily mean that effective learning has occurred. For example, there are several repeating simulation sessions scheduled for a particular day. During each session, a group of several students manages the same clinical topic. Perhaps, a group that has finished their session has passed the information about the session content to a second group of students. In that case, the second group, primed with the content of the case, scores better in reaching fixed learning objectives than the previous groups. Yet, the ultimate goal of the entire training session may have been missed.

There are several ways of dealing with this potential problem. First, the case should have learning objectives that reinforce differential diagnosis and critical thinking. Second, this method in turn is subdivided into several methods of reaching the goal:

1. There should be a set of *secondary learning objectives* that could be reached *only* if critical thinking took place. Secondary learning objectives are primary learning objectives broken down to their basic components. For example, the primary learning objective of diagnosing malignant hyperthermia consists of several secondary learning objectives: Cardiac rhythm analysis, diagnosis of hyper metabolism, ruling out hypoxemia, ruling out light anesthesia, confirmation of the increasing end-tidal carbon dioxide concentration, and confirming metabolic acidosis, etc. The case should be structured in a way that the instructor assesses whether these interventions have been performed. Whether a particular primary learning objective has been reached would depend upon achieving a set of secondary learning objectives.

2. *Case-distracters* are short cases with simple learning objectives and clinical problems that have nothing in common with the pathophysiology of the main case. Their main characteristic is that they have clinical problems or abnormal conditions different from those that the students anticipate. The use of case-distracters is limited to warm-up before the main case of the session, and for offsetting the students' expectations of the session to follow. Although case-distracters are not a substitute for the previously described mechanism of interdependencies between learning objectives, they work extremely well to improve educational value of the simulation sessions. For example, the main case of a session could be one with malignant hyperthermia. When the students arrive in simulation laboratory, they are told that before they proceed to the next case, they need to finish the previous one. An instructor describes the patient's history,

the course of anesthesia during the surgery, previously administered medication, and the stage of surgery. The patient in this hypothetical case-distracter might have a history of bronchial asthma and heavy smoking and will already be intubated. After several minutes of a noncomplicated course at the end of a surgical procedure, the patient suddenly develops extreme hypoxemia with diffusely diminished lung sounds and a series of hypoxemia-induced arrhythmias. The trainees/students are expecting a case of malignant hyperthermia, however, in this particular case, the reason for developing these complications is an obstruction of the endotracheal tube with a big mucous plug. Suctioning the tube resolves the hypoxemia and only then would the patient respond to treatment with antiarrhythmic medications. This "case-distracter" offsets the expectation of malignant hyperthermia. When the group proceeds to the main case with malignant hyperthermia, the students typically perform as if they have never been informed about the session content, as they do not know what to expect any more. A case-distracter could also be a simple case with no complications. Using cases-distracters helps to offset the students' expectations of the course of events in the main case. This method helps to ensure that the performance during the case is driven by critical thinking rather than by fixations and preoccupied mindsets.

3. There is another mechanism of enforcing critical thinking that is based on overlap of pathophysiologic conditions, and we describe this method in section 14.8 Internal Logic.

*Variable learning objectives* cover common misconceptions that the students might have. The importance of correcting common misconceptions and the clear superiority of simulation over traditional training in achieving this goal cannot be overstated. In the traditional live clinical setting, learning to avoid adverse outcomes develops after years of clinical experience. Perhaps the *best and most valid use of simulation training* is to shorten the traditional learning curve and reduce risks, together.

Scenarios can be designed to anticipate and train in an environment that does not harm patients and allows students to learn with the least personal trauma along a deliberate schedule that matches the students' readiness for a lesson with their experience of the same scenarios. The list of variable learning objectives cannot be developed with the use of standard treatment protocols because protocols describe *the correct way to manage the condition* rather than what *should never be done and what can go wrong*. What should the simulator do if the students shock the patient while he is awake and alert? What should be the simulator's response if the students administer a neuromuscular blocking agent before the hypnotic? What should happen if the airway obstruction has not been relieved within 2 minutes? How should the simulator respond if the students initiate blood transfusion before blood typing

and cross-matching has been performed? What should be the consequence of the students' failure to check the airway for presence of foreign bodies before they start mask ventilation? *Attention to the details in setting the variable learning objectives determines the quality of the case.* The relevance of variable learning objectives depends on how well the course developer knows the target audience (see above), on the clinical condition being taught, and on how each particular case contributes to the entire course structure. Most often, variable learning objectives are refined during the spiral development of the case: a developer makes a list of the possible common misconceptions, and this list would be refined during implementation of the course. This cycle needs to be repeated multiple times during the course implementation. Again, initial painstaking needs assessment allows the teacher to anticipate the majority of possible mistakes and thereby preemptively design ways to train to avoid those mistakes.

Using variable learning objectives closely correlates with the main principles of adult learning and teaching. It helps to identify the students' learning needs and provides a mechanism for reaching them. The instructor's nonintrusive role helps the trainees to perceive the session as an interaction with the simulator and makes internalization of a new experience easier. This helps to avoid presenting a threat to the learners' self-confidence due to an overwhelming scenario.

## 14.5 Two Designs: Learning and Engagement

Every high-quality interactive scenario consists of two forms of design: Learning and engaging. *Learning design* is a dynamic mechanism that provides learning experience for the trainees and ensures achieving learning goals. In other words, learning design is a mechanism of reaching learning objectives. *Engaging design* is an interactive framework for the learning experience to occur. There is an overlap in definitions and characteristics of these two designs. The engagement framework is the mechanism that provides interactivity of the simulation sessions and it works side by side with the learning design. As an illustration of a case with a poor engagement design, let us consider a scenario of bronchial asthma attack. If the scenario required only administration of a bronchodilator to improve the patient's state, the engaging potential of such a case would be very low. The ultimate goal of both learning and engagement designs is generating experiences that force people to make a decision and then take action. We want these experiences to require from learners *to apply their knowledge in a given context* rather than to simply test their knowledge. Table 14.2 outlines the differences between knowledge and application of knowledge [8]. *Knowledge application is an **observable** behavior based on understanding specific concepts and principles and executed in a specific context.*

**TABLE 14.2**  Differences between knowledge and application of knowledge

| Knowledge test | Knowledge application |
|---|---|
| Asks for concept | Asks for action |
| Distracters are other concepts | Distracters are common misconceptions |
| States the question | Story sets up need for decision-making |
| Wrong or right | Feedback in context of story |

That is why, when in presenting a simulation session, we do not accept statements from students such as "if the patient does not have pulse, I would shock him; and if the patient has a pulse, I would treat him pharmacologically." Rather than verbalizing treatment protocols, the learners need to assess the virtual patient and initiate treatment. Appropriate engagement design creates a framework for the learning design: The story sets up the need for decision-making that requires action from the trainees, and the simulator provides feedback in the context of the story. Assessment of the simulator's feedback allows the students to decide how right or wrong their decisions were, and to reevaluate their concept of their patient's condition.

# 14.6 Realism of Simulation

Relevance and learner internalization of the simulation experience are the goals of simulator training. With this in mind, the instructor seeks to enhance both *perceptual realism* and *social realism* of the session [9].

*Perceptual realism* is the degree to which the simulated environment is perceived as real. Perceptual realism, in its application to clinical simulation, greatly depends on the competency of the simulation team, the simulated setting, the type of simulator employed, and how the simulator is used. Using real and current clinical equipment and enforcing dress code during the simulation sessions are both means useful for enhancing a sense of reality.

*Social realism* is the degree to which consequences of actions (or of the failure to act) in the simulation mirrors real-life consequences. This is one of the most important factors of simulation that determines transfer of knowledge from the simulation environment to the clinical realm. Social realism depends on how the conventions, or rules of engagement, of the simulation sessions are set on the learning objectives of the sessions and the methods of teaching. It has both technical and moral application to the simulation.

One of the dilemmas facing a course developer is whether the virtual patient should be allowed to die. Some educators believe that simulated patients should not die under any circumstances because the learners may become embarrassed. Others believe that the conventions in the simulated environment should be as close as possible to the conventions of our real world, and we share this view. We believe that the simulation laboratory is the ideal environment for learning REAL consequences of judgment and knowledge[6].

Careful attention to achieving realism assures more successful sessions by allowing the learners to experience what we call "suspending disbelief" or allowing them to be convinced that the patient and disease process is real. Excluding negative outcomes at *appropriate* times in simulator training might spare the trainees' feelings, but it would create a credibility problem later in the course by disrupting the suspension of disbelief. Clinical educators and students, together, are responsible for determining just what is or is not appropriate.

We have collected opinions of the third and fourth year medical students, nurse anesthesia students, and anesthesia residents in the decision to allow simulator patients to die. They confirmed that they experience some emotional stress when the patient died during the session because of their mistakes. However, they have never been overwhelmed or become desperate by these feelings. Moreover, they explicitly state that they would prefer to see realistic consequences of their interventions because it would better prepare them to deal with real world problems. Typically, the trainees comment that they want to practice on the simulator before working on real people.

We believe that not providing realistic responses to potentially disastrous interventions desensitizes the trainees to the possible consequences of their actions. Dietrich Dorner states in his book *The Logic of Failure or Recognizing and Avoiding Error in Complex Situations*:

> **Another likely reason for the violation of safety rules is that people had frequently violated them before.** But as learning theory tells us, breaking safety rules is usually reinforced, which is to say, pays off. Its immediate consequence is only that the violator is rid of the encumbrance the rules impose and can act more freely. Safety rules are usually devised in such a way that a violator will not be instantly blown sky high, injured, or harmed in any other way but will instead find that his life is made easier. And this is precisely what leads people down the primrose path. **The positive consequences of violating safety rules reinforce our tendency to violate them, so the likelihood of disaster increases. And when one does occur, the violator of safety rules may not have another chance to modify his behavior in the future** [10].

Besides realism, the simulation curriculum should provide the students with enhanced educational experiences that improve their learning and which real life cannot provide. Sometimes, enhancement of learning experience must be done at cost of sacrificing some realism. For example, in combat pilot training via simulator, the trainees are provided with the opportunity to view the battlefield from the third-party view rather than from the cockpit of their own aircraft (Figure 14.1).

**FIGURE 14.1** A third-party view within a flight simulator, while a totally nonfaithful representation of real flight, is a powerful learning method for pilot trainees to grasp the importance of situational awareness. (Image courtesy of Evans & Sutherland)

It allows them to assess the three-dimensional position of the opponent's aircraft in relation to the trainee's aircraft and the landscape. Partial-task trainers provide many of the same educational advantages of the "third-party view" for clinical students.

Clinical simulation can provide some enhancements that are not available in real life: Fast forwarding and pausing the case's runtime, reperforming the case, making clinical laboratory test results available immediately, and reviewing video record of performance during the case management. Using these methods should be justified and be very cautiously used, because excessive use of them may set unrealistic expectations in learners. For example, to enhance learning experience, we can provide instant results of bacterial sensitivity to different antibiotics but in real life this test may require several days.

## 14.7 Simulation Scenario as an Instance of Disease

As discussed earlier in this chapter, there are many well-known differences between content-based methods of teaching (textbooks or lectures) and problem-based teaching (standardized patients and robotic simulators). The first method of teaching is mostly static and the learners all too willingly assume the passive role assigned to them. The second method of teaching is predominantly dynamic, and the students play their assigned active role in the learning process.

Let us first introduce the concepts of *class* and *instance of class*, concepts computer programers are well acquainted with.

> *A class is a cohesive package that consists of a particular kind of metadata (for instance, pneumonia can be seen as a class). It describes the rules by which objects behave; these objects are referred to as instances of that class (for instance, Right-sided lobar pneumonia is an instance of the class). A class specifies the structure of data which each instance contains as well as the methods (functions) that manipulate the data of the object and perform tasks [11].*

This concept has clinical application too. For example, to expand the definition, pneumonia can be seen as a "class:" It is an abstract concept that describes an inflammatory process in a lung or lungs and has several distinctive features that places it in its own, distinctive "class." For instance, features of the class might include: Types of pneumonia, location, etiologic agent, morphologic changes, clinical manifestations, complications, etc. An *instance of a class* is a particular occurrence of the class. For example, "right-sided lobar pneumonia in a 38-year-old Caucasian male complicated with pleural effusion and caused by Streptococcus pneumoniae" is an instance of the "pneumonia class."

Very often, the difference between a textbook or lecture and simulation is the same as the difference between a class and a particular instance of that class: The first one (class) provides an abstract description of medical conditions and diseases and the latter one (instance of a class) presents a particular occurrence of the given medical condition. The process of moving from learning about the disease to managing a patient with this particular disease accomplishes a loop in the Whole-Part-Whole learning model [12] and results in understanding of the principles of functioning of the human body on the qualitatively higher level.

Selecting an appropriate "instance of a class," i.e., an instance (or an example) that provides the highest educational value in learning about a class, is an important decision, and it depends on the type of pathology and the learning objectives of the case. Let us consider an example of a patient with multiple trauma complicated by pneumothorax and internal bleeding.

*Martin Graw, a 49 year-old male person drives home from a business lunch (full stomach problem). On the way home, he becomes involved in a motor vehicle collision. His vehicle is hit by a truck from the left side (broken ribs on the left side of the chest, broken left femur, ruptured spleen, profuse internal bleeding). Several minutes after arrival at the Emergency Room, the patient becomes nonresponsive to external stimuli, his skin is cold, and his blood pressure is 75/30 mm Hg, heart rate is 145 beats per minute, urine output is 2 ml/hr.*

One of the learning objectives of this case is performing the Airway-Breathing-Circulation of the ACLS protocol, as well as the consequences of not protecting the airway of a comatose patient with a "full stomach." Airway protection in a comatose patient with a full stomach is one of the most critical decision-making points in the early phase of this case management. How could this message be conveyed with the use of a high-fidelity patient simulator? On the basis of the principles of adult learning, the students need to reach this particular learning objective through their experience with the simulator: If the patient becomes comatose and he has a full stomach, and if his airway remains nonprotected for a defined period of time, the patient should aspirate and potentially die. Someone may argue that there are many instances where comatose patients with a full stomach and no airway protection would not aspirate.

In responding to this arguable question, we need to decide whether we want to teach the students that they may gamble with the patient's life or teach them how to practice safely? Should we let the simulator compute a random number and on the basis of the statistical data decide whether or not the patient aspirates in this particular case? What lesson would the students have learned if they have failed to perform appropriately and nothing bad happened?[7]

The answer is simple and straightforward:

*The positive consequences of violating safety rules reinforce our tendency to violate them, so the likelihood of disaster increases. And when one does occur, the violator of safety rules may not have another chance to modify his behavior in the future [13].*

This example demonstrates that in a simulation session, a curriculum developer needs to create an instance of a given medical condition that delivers the highest educational potential to the students. Developing of an effective simulation case is a spiral process (Figure 14.2). Spiral development of the case begins with identifying educational needs of the learners, setting appropriate learning objectives, developing mechanisms of achieving these learning objectives, and refining the case further (see definition of the spiral cycle development at the beginning of the chapter).

Here are two major forms of curricula: (i) brief intensive course during which the students spend from 6 to 8 hours per day in the simulation room; and (ii) a longitudinal curriculum that provides distributed learning. Both types of curricula have their advantages and disadvantages.

The brief intensive type of curriculum is easier to plan and administer, especially if there is no full-time simulation professional designated to program, operate, and maintain the simulator. One of the major drawbacks of this method is the likelihood that students may be overwhelmed with new information at a rate that exceeds their rate of its internalization. As described above, the learning typically begins upon the presentation of a general concept. Active learning follows, resulting in the trainee's reaching a new level of understanding of the original concept. New knowledge and experience are reduced to their basic elements via methods of analysis, and then these basic elements are being reintegrated into the existing system of knowledge. This process is called internalization. This is not an instantaneous event. Short courses often do not provide enough time for internalization of new experiences.

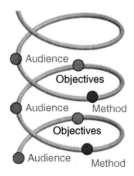

**FIGURE 14.2**  Spiral structure of curriculum development and refinement.

A longitudinal curriculum requires more planning, but this type of curriculum is more beneficial in the long run. First, a longitudinal structure of the curriculum allows multiple repetitions of the same (or similar) case during the course. This creates a unique opportunity for the students to perform several cycles in the Whole-Part-Whole Learning Model. This is how Malcolm Knowels describes the Whole-part-Whole Learning Model:

> Through the "first whole," the model introduces new content for the learners by forming in their minds the organizational framework required to effectively and efficiently absorb the forthcoming concepts into their cognitive capabilities. The supporting cognitive capabilities and component behaviors are then developed in the classical behavioristic style of instruction found in the "part," or several parts, aspects of the Whole-Part-Whole (WPW) Learning Model. After the learner has successfully achieved performance criteria for the individual "parts" or components within the whole, the instructor links these parts together within the second "whole." **The whole-part-whole learning experience provides the learner with complete understanding of the content at various levels of performance and even allows for higher order cognitive development to the levels of improvement and invention** [14].

Alternating use of didactic teaching and simulator training is the key of better comprehension of the problems and concepts at higher cognitive levels. Longitudinal curriculum is capable of engaging all eight types of learning described by Gagne [15] as shown in Table 14.3.

In a longitudinal curriculum, experience of each session builds on the foundation of the internalized experiences from previous sessions. In the case of a brief intensive course, these experiences mostly compete with each other for their place in the short and intermediate term types of memory of the trainees. At first glance, it may seem that scheduling a longitudinal simulation curriculum or curricula could be a complicated issue, but it is not really so. We provide each clinical clerkship with one-day-a-week access to the simulation laboratory[8]. This allows flexible scheduling around other educational activities. To accomplish this, we produce three simulation curricula totaling 700 2-hour long sessions per year with a single high-fidelity simulator.

## 14.8 Internal Logic

A simulation curriculum should be considered a continuum rather than just a collection or battery of simulation scenarios. Experiences of preceding cases have to be viewed as prerequisites for upcoming sessions and reinforce desired behaviors. This approach ensures transfer of experience from one case to another.

Let us consider the following series of cases as an example of a curriculum for the Internal Medicine clerkship: Stable atrial fibrillation, unstable atrial fibrillation with signs of myocardial ischemia and pulmonary edema, left ventricular systolic dysfunction and congestive heart failure complicated with pulmonary edema, and finally, septic shock complicated with Adult Respiratory Distress Syndrome (Figure 14.3).

**TABLE 14.3**   Eight types of learning described by Gagne

| Type | Name | Description |
| --- | --- | --- |
| Type 1 | Signal learning | The individual learns to make a general, diffuse response to a signal. |
| Type 2 | Stimulus-response learning | The learner acquires a precise response to a discriminated stimulus. |
| Type 3 | Chaining | What is acquired is a chain of two or more stimulus-response connections. |
| Type 4 | Verbal association | Verbal associations are the learning of chains that can be expressed with the use of speech. |
| Type 5 | Multiple discrimination | The individual learns to make different identifying responses to as many different stimuli, which may resemble each other in physical appearance to a greater or lesser degree. |
| Type 6 | Concept learning | The learner acquires a capability to make a common response to a class of stimuli that may differ from each other widely in physical appearance. He/she is able to make a response that identifies an entire class of objects or events. |
| Type 7 | Principle learning | In simplest terms, a principle is a chain of two more concepts. |
| Type 8 | Problem solving | Problem solving is a kind of learning that requires the internal events called thinking. Two or more previously acquired principles are somehow combined to produce a new capability that can shown to depend on a "higher-order" principle. |

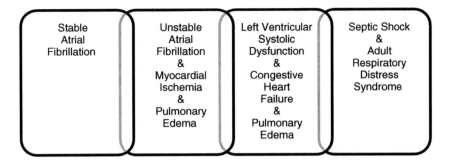

**FIGURE 14.3**  Continuum of signs and symptoms with some shared presentations but with different etiologies in each.

In this example, each case shares some pathophysiologic features with one or more of the other cases:

- Stable atrial fibrillation and unstable atrial fibrillation cases share the same type of arrhythmia but differ in approach to treatment.
- Unstable atrial fibrillation and congestive hear failure cases have the same complication, namely, pulmonary edema, but there are two different mechanisms of developing pulmonary edema and correspondently different methods of treatment.
- Septic shock is similar to all other cases in terms of arterial hypotension but the mechanism of development of arterial hypotension is different in all these cases.
- Fluid accumulation has different mechanisms of development and requires different treatments in cases with septic shock in contrast to (versus) left ventricular systolic dysfunction in contrast to (versus) unstable atrial fibrillation. This structure of a curriculum can help the students improve their ability to create differential diagnoses, critical thinking, and treatment in a dynamic environment. Using an overlap in pathophysiologic mechanisms in a series of simulated cases is a very effective teaching tool because it helps cognitive processes reach higher levels of reasoning.

For example, if the patient complains of severe chest pain and presents with depression of the ST segment of the ECG, the students might apply two separate protocols to the management of congestive heart failure complicated with pulmonary edema: The pulmonary edema and acute myocardial ischemia treatment protocols. If the level of critical thinking that applies to this case is limited to the stimulus-response level (type 2 learning) and the students respond to the key words "myocardial ischemia," "beta-blockers," "increased survival rate" with administration of a beta-blocker, the patient will start deteriorating (i.e., the pulmonary edema will worsen). Experiencing negative outcome as a result of intervention, the students would have to use their critical thinking process to find a way to help the patient (i.e., find a therapy that would treat the cardiac failure without aggravating the pulmonary edema). The instructor will not let them achieve a successful outcome without reaching a higher level of comprehension of the pathophysiologic processes in this case.

## 14.9 Stages of Curriculum Implementation

According to Clark Aldrich, there are four main stages of curriculum implementation: (i) background, (ii) introduction, (iii) engagement, and (iv) practicing [16]. As in Erickson's theory of human psychological development, sequential mastery of an earlier stage is essential for achieving goals in subsequent stages.

**Stage One: Background.** Most of the learners have had no prior simulator-based learning experience. Because the learners are apprehensive about their performance in an unfamiliar environment, it is important to educate them about the underlying educational principles upon which simulator training is based. This teaching should be administered in both written and live forms. Some students might think that they are prepared enough to proceed to full immersion into the simulation and skip the educational component of the simulation curriculum. In that case, providing them with a written material about educational components of the course will help them to successfully resolve the frustration/satisfaction psychological conflict that is going to develop at the Stage Two. During Stage One, the instructor helps the learners to develop appropriate expectation of the course, set the rules governing the structure of the course, and, most importantly, build motivation. Building motivation is the most important component at this stage because it determines how actively the learners are engaged in the simulation activities. Typically, this stage should last from

45 to 60 minutes and be supplemented with a written materials published at the institutional web site.

---

Editors' Note: For a perspective on Web-based learning before their Simulation Session, please see Chapter 63.

---

**Stage Two: Introduction.** At this stage, the learners become familiar with the simulator and its simulated care environment and learn how to manage a relatively simple clinical scenario. During this stage of a simulation course administration, an instructor may expect that "some students balk at the (simulation) elements, . . . and the experts in the subject-matter area start trying to establish their own credibility by loudly disparaging relatively trivial elements [17]." At this point, the learners may experience their first frustration of working with a simulator, and an instructor needs to be ready to deal with that frustration. If the frustration has resulted from the students' misunderstanding of their expected interactions during simulation sessions, then referring them to a written description of educational principles of the course may be a solution to this problem. Immersing the students into a challenging scenario could be another reason for the learners to experience their first frustration. This is a desirable reaction, and this is what an instructor should expect in a successful simulation course implementation. With this regard Clark Aldrich says,

> This first encounter with frustration is a key learning opportunity. It is also a meta-learning opportunity, a chance to learn about something. If you left a gym after two hours without having broken a sweat, without aching muscles, you would suspect that you had wasted you time. Any good trainer knows his or her role is not to eliminate or even reduce the burning in the muscles, but to encourage it (with safety parameters), and to reframe it. Without pain, there is no gain. The ache is the feel of progress. . . . We all need to rethink frustration in the context of learning. If you leave a learning program without having felt waves of frustration, you probably wasted you time." Frustrations in learning activities should be followed by satisfaction of their resolutions: "Frustration during the learning program and then the feeling of resolution afterward is the most reliable sign that learning is going on[9] [18]." There are few things more destructive to the simulation program than easy cases that bring no challenge to the trainees. The introduction stage should normally last from 1 to 2 hours, and its success builds on the successful accomplishment of the first stage.

**Stage Three: Engagement.** The engagement stage is the beginning of the heart of the simulation curriculum where the actual clinical learning occurs. Effective learning during this station is possible ONLY if the trainees have successfully resolved all internal conflicts during first two stages of the course implementation. During this stage, refinement of the course material should occur. The more proficient the learners are in content knowledge, the more they put the simulation sessions to test. When and if deficiencies in the simulation scenarios are revealed, this should be admitted in front of the students. Otherwise, the simulation sessions will be viewed as illegitimate and the engagement factor will be irreparably damaged.

Maintaining a stable level of student engagement is another challenge at this stage of curriculum implementation. Here, the key to success is alternating cases of different stress levels. Monotonous flow of the courses tends to create a gradual desensitization and disengagement of the students. The role of instructor at this stage cannot be overemphasized. The instructor indirectly assists the learners in optimizing the stress level, guides them during self-reflective debriefing, and provides external gratification and role modeling when needed. The "Multiple Exposures" section of this chapter describes practical aspects of how to keep the students engaged during the course.

**Stage Four: Practicing.** The other name for this state is *unchaperoned engagement*. Here the students master their decision-making and procedural skills and depend on the instructor's assistance to a lesser degree. Nevertheless, the vigilance of the instructor is critical at this stage lest an increase of students' self-confidence result in faulty learning arising from the pairing of two events that have no cause-effect relationship whatsoever. Sometimes, at this stage, the trainees' self-confidence may result in superficial analysis of the underlying pathophysiologic mechanisms of medical conditions. Later in this chapter, the "Multiple Exposures" section demonstrates a mechanism of overcoming this deficiency.

### 14.9.1   Assessment Methods

Assessment of educational activities is extremely complex process. Malcolm Knowels, a well-known scholar in adult education, explicitly states:

> *Here is the area of greatest controversy and weakest technology in all education, especially in adult education and training. As Hilgard and Bower (1996) point out regarding educational technology in general, "It has been found enormously difficult to apply laboratory-derived principles of learning to improvement of efficiency in tasks with clear and relatively simple objectives. We may infer that it will be even more difficult to apply laboratory-derived principles of learning to the improvement of efficient learning in tasks with more complex*

*objectives" (p. 542). This observation applies double to evaluation, the primary purpose of which is to improve teaching and learning – not, as is so often misunderstood, to justify what we are doing. One implication of Hilgard and Bower's statement is that difficult as it may be to evaluate training, it is doubly difficult to evaluate education.*

According to Donald Kirkpatrick [19], conceptualization of evaluation process consists of four major steps:

1. *Reaction evaluation*, which is collecting the learners' opinions about the course as it takes place and the changes in their professional self-assessment as a result of the course.
2. *Learning evaluation*, collecting data about the principles, facts and skills acquired during the session. Most often, it has been done via administration of pre- and postactivity multiple choice quizzes.
3. *Behavior evaluation*, collecting data about actual changes in trainees' behavior after the training, which can be done mostly via but not limited to direct observations.
4. *Results evaluation*, collecting data how the training course affected the entire organization: Number of sentinel events, complications, mishaps etc.

*Reaction evaluation* is one of the most controversial methods of assessment that still can be used to monitor the trainees' progress through the course. To illustrate the ambiguity of this method, let us consider the following example.

A group of students with different backgrounds and expertise was presented a case with intraoperative anaphylaxis. The group demonstrated a highly effective case management at all stages: The condition was recognized at the early stages of development, comprehensive differential diagnosis was performed, and correct treatment was administered. After the case, during an interview, the students graded the complexity of the case as easy and their performance as good. To evaluate how the students' self-assessment correlated with their performance we split the group on the basis of their background and expertise into two groups. Then, we reproduced the case separately with more- and less-experienced groups. Less-experienced students failed in recognizing the condition, in establishing a differential diagnosis in initiating treatment, and at the end of the scenario, the patient eventually died. This reproduction of the case took place 2 days later after the first event that was perceived by both groups as easy.

This example demonstrates that reaction evaluation is a highly variable and unreliable method of assessment of the effectiveness of the curriculum if the data is interpreted directly (i.e., direct interpretation the trainee's feedback). When we compared the students' self-assessment through the course with the evaluation of their performance as assessed by the instructor, we found that as the course progressed, the difference between these two assessments gradually decreased. In parallel with the reduction of the difference between self-assessments and external evaluation, the student demonstrated disappearance of immature defense mechanisms such as avoidant behaviors and skepticism. We believe that reaction evaluation data could be nonintrusively obtained during debriefing and needs to be used indirectly (possibly in combination with the other methods of evaluation).

*Learning evaluation:* We evaluated the effectiveness of the course via administration of identical case-specific multiple choice quizzes before and immediately after the session (Figure 14.4). The content of the quizzes was based on the case-specific decision-making trees, and thus the quizzes were paper-based replicas of the simulation sessions. Comparing the results of the pre- and postactivity quizzes, if there is any difference between the scores, we assume that the difference is the result of teaching with the simulator see Appendix 14A.2, for an example Multiple Choice Quiz) [20, 21].

These data show significant procedural knowledge gain right after the session. Six months later, we have readministered the quizzes to assess retention of knowledge. Notable decay of the scores confirmed the well-known postulate that a single exposure to a problem may result in a short-term knowledge and a skill gain but fails to produce a steady long-term gain. A potential drawback of this assessment method is that administration of the case-related multiple choice quizzes may reveal the content of the simulation session and impair the learners' clinical reasoning during management of the virtual patient. In essence, the pretest may give them cues to the clinical case, thus short-circuiting the learning process.

*Behavior evaluation:* This method of evaluation assesses the trainees' behavior during the case management. The behaviors are assessed with the case-specific checklists designed to

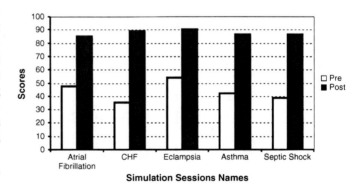

**FIGURE 14.4** On the basis of the large change in the scores on the pre- and postsimulation multiple choice quizzes, we may conclude that at least in the short term, the simulation experiences did help the students' learning.

monitor reaching learning objectives. This type of assessment can be done in several ways:

1. paper-based checklist and direct observation of performance by an evaluator;
2. computer-based checklist in combination with a video asset management system (for example, Studiocode, Camarillo, CA)
3. automated systems capable of tracking all events and all minute details in clinical performance, probably will be used widely in the future when consensus is reached about the scoring criteria.

> **Editors' Note:** For a description of automated pelvic examination with digitized data collection, see Chapter 68.

Development of an assessment tool is extremely difficult due to the complexity of human behaviors in clinical settings. On the basis of the learning objectives and what the assessment tool is supposed to assess, one needs to find an appropriate balance between the "forest" and "trees" types of tool design. For example, a "forest" type of tool may contain a checkbox "endotracheal intubation performed correctly and in a timed manner" with "True" and "False" options. It is obvious that interpretation of the trainee's performance would depend on the evaluator's interpretation of the procedure. Assessing the same intervention with the "tree-"like tool may result in extremely large numbers of subfactors that will determine how the students performed (how soon they recognized indications for intubation, how well the head positioning performed, was the laryngoscope blade insertion atraumatic, have they reported visualization of the vocal cords, how atraumatic was the tube insertion, was tube size appropriate, etc.) Using either extreme in the tool design can make use of the tool less than desirable.

## 14.10 Multiple Experiences

As shown previously, a single experience in management of a clinical problem failed to produce long-term retention of desired outcomes for third year medical students. Unsatisfied with that outcome, we developed a simulation curriculum for student nurse anesthetists that allowed us to achieve more reliable and more precise results. The educational concept was derived from a principle of using multiple repetitions of the cases to produce persistent patterns of performance. We have selected a cluster of cases in which the patients presented similar vital signs but each case had different underlying pathophysiologic mechanisms. The critical pathways and troubleshooting algorithms of these cases had several common decision-making points but required different methods of problem solving. To assess the proficiency of problem solving, we devised each case with a specific checklist that was used to monitor the achievement of learning objectives. We selected a group of four or five cases that comprised to what we called "an active cluster" of the cases. During teaching, we administered the cases that comprised the active cluster and case-distracters in alternating order with the variable ratio of repetitions. This technique has helped to thwart students' expectations that each subject was to be presented only once.

We scored the students' collective performances by using case-specific checklists designed to monitor achieving learning objectives. Each particular case remained in the active cluster until the students demonstrated proficiency in managing that problem in three to four consecutive performances (Figure 14.5). After the students demonstrated stable and

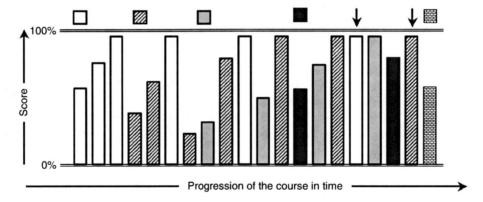

**FIGURE 14.5** Course progression and student performance. Each different shade pattern represents a different clinical case. Arrows point at the simulation scenario that is going to be moved from the active cluster into the case-distracters.

proficient performance, the case was moved to the category of the cases-distracters and replaced in the active cluster with the next one in the curriculum.

Figure 14.5 describes the logistics of the method with a series of five anesthesia-related cases. The arrows show two of the cases that will be moved from the active cluster to the group of case-distracters since the student demonstrated repeatable mastery. To prevent identifying medical conditions based on the patient's profile, we supplied each case with several patient descriptions (name, gender, age, coexisting medical conditions, etc.). Since the students did not know which case had been added to the active cluster and which one had been removed or if the case belonged to the group of distracters, they could not predict the medical condition about to occur in the simulation session.

The effective use of this technique depends upon:

1. learning objectives carefully devised and defined to meet the educational needs of the audience;
2. adequate number of learning objectives per case designed to challenge the learners and provide room for improving their performance;
3. outcomes dependent upon the successful/unsuccessful achievement of learning objectives;
4. multiple repetitions of cases.

# 14.11 Setting the Appropriate Psychological Environment

The importance of providing appropriate emotional environment at the simulation session cannot be overestimated. A valid simulation session must be challenging and exact, while remaining user friendly and nonthreatening to the trainees' self-concept.

"Adults have a self-concept of being responsible for their own decisions," for their own lives. Once they have arrived at that self-concept, they develop a deep psychological need to be seen by the others and treated by the others as being capable of self-direction. **They resent and resist situations in which they feel others are imposing their wills on them**[10].

This presents a serious problem in adult education.

> (Their) typical method of dealing with psychological conflict is to try to flee from the situation causing it, which probably accounts in part for the high dropout rate in much voluntary adult education. As adult educators become aware of this problem, they make efforts to create learning experiences in which adults are helped to make the transition from dependent to self-directing learners [22].

There are several methods of mitigating this normal adult reaction to new, challenging, and unknown experiences.

(1) *An introduction to simulation before the beginning of the course.* To make the simulation course effective and to avoid psychological resistance, an effective introduction to the simulation environment should be given. For our students, we have developed a Web-based pictorial introduction to the simulation learning that explains the technical aspects of working with our patient simulator. Also, we have provided access to the materials that explain educational methods used in simulation training. When students arrive at the simulation laboratory for the hands-on orientation session, we provide a brief demonstration of the features of the simulator and describe the main rules and conventions of the simulation lab: The students are to take charge of case management, and case progression depends only upon the medical condition of their patient and the treatment they administer. Students know in advance that the sessions will be very challenging and that it is possible for the patient to die if treated inappropriately. This Web-based theoretical introduction should not be longer than 10–15 minutes. Using the simulator is indeed the best way to learn about simulation. As stated before, an introduction scenario affects the first impression of simulation. The scenario should be dynamic, interactive, authentic, and challenging.

(2) *Setting the rules and attitudes in the simulation lab.* We have established a set of rules that apply to all simulation sessions:

- The purpose of the sessions is developing skills and expertise, extending the boundaries of their competence.
- The students' freedom in the simulation laboratory accompanies the responsibility for the patient's outcome.
- The students are to learn from their experience.
- Do not be ashamed of less than complete clinical success – the cases are very challenging on purpose, no one learns from their success, just learn from your mistakes and try different methods of treatment at your next opportunity.
- Simulation laboratory is like a confession room – whatever happens here, stays here;
- Feel free to use any source of information if problem solving is needed.
- In most of cases, the instructor plays the role of a cab driver that brought the patient to the student's attention ("I am a cab driver! I know nothing about birthing babies!").
- Try to acquire help from your patient – seek for the simulator's feedback, order clinical laboratory tests or any examination, and do not forget to talk with your patient. Very often, patients can give valuable clues about their conditions, as they have been living with them.

- Please keep the content of the session private so that every student has the same opportunity to learn. (We have only two or three suspected information leaks in the last 4 years. Luckily, we were equipped with the cases-distracters.)

(3) *Debriefing after the session.* We use an instructor-facilitated student/team-centered technique as a method of debriefing after the session. Team members discuss their performance directly with each other. The instructor sets the objectives of debriefing and guides the discussion only when needed, usually by asking questions. The students are expected to verbalize their decision-making and explain their medical interventions for the simulator patient. The instructor facilitates the discussion to ensure that the students correctly analyze their performance. Avoiding judgmental expressions, the instructor guides the discussion to examination of the mishaps or incorrect decisions. This guidance is provided as questions that prompt the students to apply their critical thinking skills to discover their mistakes and help them to find a solution to the problem. If a case had a negative outcome, the instructor's facilitation in the discussion of the incorrect decisions and actions is minimal and limited only to the situations when the students cannot find the answer without external help. It is the instructor's responsibility to monitor trainees' stress level during debriefing and readjust their attitudes if needed. At times, the instructor must be a sort of "cheerleader" intent upon keeping the students interested and not discouraged. At the end of debriefing, the instructor may demonstrate a model of terminal (ideal) performance but it should be done only after the team has finished their self-assessment. The instructor's demonstration of an ideal management is especially effective *after* the students' successful outcome of the case. It helps to reinforce the critical components of the case. Doing so before the students are successful robs them of the great value in discovering the answers for themselves and just reminds them, yet again, that their instructors are more accomplished than they. Both of these outcomes are antiethical to why we do simulation. During the instructor's performance after the session, the students can compare their performance to the expert's level. This serves as a method of external gratification.

(4) Sometimes, the students performed well but the debriefing session shows that the mistake was not consciously avoided. In that case, an instructor may perform management of the virtual patient by himself and purposefully make the mistake to demonstrate possible consequences. After the demonstration, the team may discuss the event, and the instructor may play the devil's advocate role of a person who tries to justify incorrect treatment. This role-playing technique reinforces students' confidence in their own knowledge and teaches

them to apply their critical thinking skills. Also, this method helps to eliminate any overly harsh judgmental atmosphere that might have developed.

# 14.12 Conclusion

Those who employ simulation in clinical education have adopted a great deal of the lessons learned from simulation in aviation and other high-risk, high-reliability organizations. At the same time, medical simulation has very unique challenges. The most important of them is the fact that unlike in flight simulation where the variables and characteristics of the *atmosphere* are relatively few, well characterized and easily measured, the variables and characteristics of *humans* are decidedly not. Presentations, let alone mechanisms of many clinical phenomena are still unknown. Developing any clinical simulation system and curriculum takes courage to reproduce something that we understand only partially. Today's best physiological and pharmacological models only partially simulate even the well-known physiologic and pharmacodynamic processes. In our own developments of simulation-based clinical education, we must aim to do our best for our students while accepting partially realistic physical and physiologic models. In addition to developing increasingly sophisticated models, we are challenged with continuing to learn better ways to learn and teach with simulation, and how to evaluate what has been learned. The future holds a promise of simulation for both learning and evaluation – both the maintenance and proof of competence in various fields. The greatest challenges involve continuing to develop increasingly sophisticated models and enhance the transparency of simulation in both teaching and evaluation and the evaluation itself. If practice makes perfect and deliberate practice achieves this quicker, then simulation with its inherent feature of being able to be "reset" and start over is the perfect modality. The question is – how much practice makes perfect and how often do you need to practice to keep your skills? Ongoing studies are attempting to answer these questions to produce ever more effective simulation-based learning.

# Endnotes

1. http://www.meti.com.
   http://www.simbionix.com/LAP_Mentor.html
   http://www.stryker.com
2. Highlighted by authors.
3. Teaching through inquiry has other names such as discovery method, self-directed learning, and problem-solving learning.
4. NSAID and regional blockade could be an alternative treatment that would prohibit the session from reaching a morphine overdose.

5. The literature describes the democratic practices as following: (i) flat hierarchy classroom management; (ii) cooperative learning environment; (iii) inquiry, discourse, and high student engagement; (iv) student-centered curriculum.

6. The movie "Last Action Hero" vividly illustrates this point. The main character (Arnold Schwarzenegger) has been miraculously transported from the movie realm into real life. At his first encounter with real life he complains, "Wow, it hurts!" when he hits a wall with his fist.

7. Should we teach that pointing a loaded gun at someone is acceptable only because people die from gun shot wounds only once in a while?

8. During an academic year rotating through Medicine, Surgery, Pediatrics and OB/GYN clinical clerkships, each third year medical student participates in eight simulation sessions. Each session is 2-hour long, and typically a team consists of three or four students.

9. At this time, one should revisit the question if a virtual patient should be allowed to die during a simulation session due to student behaviors.

10. Recall the example at the beginning where practicing physicians were forced into treating a narcotic overdose.

# References

1. McConnell, S. *Rapid Development.* Microsoft Press, 2006.
2. Knowels, M. *Adult Learner.* 3rd ed. Elsevier, 2005a, p. 37.
3. Ibid., 2005b, p. 40.
4. Ibid., 2005c, p. 99–100.
5. Ibid., 2005d.
6. Knowels, M. *Self-Directed Learning: A Guide for Learners and Teachers.* Elsevier, 1975a.
7. McIntosh, C., Macario, A., Flanagan, B., and Gaba, D. *Simulation: What Does It Really Cost?* IMMS Proceeding, San Diego, CA, January 2006.
8. Quinn, C. N. *Engaging Learning. Designing e-Learning Simulation Games.* Pfeiffer, 2005.
9. Meigs, T. Ultimate Game Design: Building Game Worlds, McGraw-Hill, 2003.
10. Dorner, D. *The Logic of Failure*, Metropolitan Books, New York, 1996.
11. Wikipedia, *Class in Object-Oriented Programming.*
12. Swanson, R. A. and Law, B. D. *Performance Improvement Quarterly*, **6**(1), 43–53 (1993). this paper was originally presented tot eh European Conference on Educational Research, Enschede, the Netherlands, June 24, 1992.
13. Dorner, D. *Logic of Failure.* Metropolitan Books, New York, 1996.
14. Knowels, M. *Adult Learner*, Elsevier, 2005e, p. 241.
15. Knowels, M. *Adult Learner*, Elsevier, 2005f, p. 80.
16. Aldrich, C. *Learning by Doing, a Comprehensive Guide to Simulation, Computer Games and Pedagogy in e-Learning and Other Educational Experiences*, Pfeiffer Publishing, 2006a, p. 242.
17. Ibid., 2006b.
18. Ibid., 2006c.
19. Knowels, M. *Adult Learner.* Elsevier, 2005g.
20. Kozmenko, V. V. et al., "*Development and Implementation a Human Patient Simulation Curriculum for Junior Medical School Students*" presented at the Swedish-American Workshop on Modeling and Simulation, Cocoa Beach, FL, February 2004.
21. Kozmenko, V. V., Padnos, I., and Kaye, A. *Designing a Simulation Program for the Third Year Medical School Curriculum Using Human Patient Simulation System*, HPSN 2005, Tampa, Florida, March 2005.
22. Knowels, M. *Self-Directed Learning: A Guide for Learners and Teachers*, Elsevier, 1975b.

# 15

# Creating Effective Learning Environments – Key Educational Concepts Applied to Simulation Training

Ian Curran

*Personally I'm always ready to learn, although I do not always like being taught.*

Sir Winston Churchill, British Politician (1874–1965)

## 15.1 Competent Trainers

The purpose of this chapter is to provide a general overview of the educational issues associated with creating effective learning in the simulation center. It will concentrate upon the pivotal role trainers play in creating effective learning. If an educational activity is to be valuable, it is important that trainers understand the fundamental need to develop an environment conducive to learning. A competent training faculty is a prerequisite for the sustained vitality and effectiveness of a simulation center.

## 15.2 The Learning Environment

The learning environment is a complex but critical educational construction borne of essentially three components: the physical environment, the actions and resources of the trainers, as well as the response of the learners – planned and unplanned! The product or consequence of a successful learning environment is that the learner acquires the desired learning objectives. Effective learning can take many forms and needs to be the explicit focus of the trainers. This learning can be described in terms of knowledge, technical skills, and professional behaviors. Professional behaviors are the complex, high-value, and highly valued contribution to any practice that separate professionals from technicians and amateurs.

A simplified diagram conceptualizes the complex relationships between the learning environment, the learner, and the educator/trainer (Figure 15.1). There are many factors that affect the integrity of the learning environment including physical resources such as teaching accommodation, teaching aids, and teaching material. Also of importance are the human resources including the availability and capability of the training faculty and the learners. Finally, there is a need for intellectual, conceptual, and cognitive resources intrinsic to the delivery of the required learning. These factors include the curriculum itself, the program ethos, presentation, assessment, and evaluation of the learning, in addition to concepts such as face and content validity with respect to the curriculum, and feasibility of delivery.

## 15.3 The Trainees and their Physical Environment

Whilst all three components of the learning environment are important and may prove critical, in most adult clinical learning environments, the learner is usually receptive to the educational activity being proposed – they have usually elected to be involved in the educational process by voluntary enrolment or application. However, being a clinical student is often seen as a means to an end, not the end in itself. The trainer, however, would be wise not to presume this receptiveness on the part of the learner; as engaging and retaining the interest of the learner is essential to creating and maintaining an effective learning environment. The trainer should expect to have to continuously work at this aspect of their teaching practice. Note that learners are always drawing in from every stimulus to their senses and their perceptions, comprehension, and prioritization of such perceptions will not automatically coincide with those intended by their trainers. Thus, the trainer has a great responsibility to assure that all the many varied stimuli reinforce their intended teaching objectives and not unintentionally clash, distract, or disorient the learners.

A hostile, uncomfortable, or distracting environment will obviously have a negative impact upon the educational activity. Too hot, too cold, too dark, too noisy all detract from the effectiveness of the physical environment, unless of course, too hot, too cold, too dark, too noisy support the learning objectives of the teaching activity. If the seminar room is too hot (air conditioning failure) or too cold (central heating failure), learners can be distracted and uncomfortable so less inclined to focus on the teaching at hand.

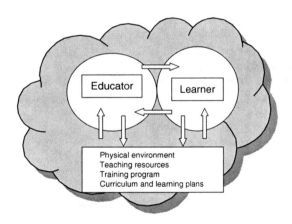

Physical environment
Teaching resources
Training program
Curriculum and learning plans

**FIGURE 15.1** A stylized learning environment.

## 15.4 The Trainer's Role in the Learning Environment

The capabilities and insight of the trainer into the educational process are probably the most important and variable factors leading to the success or failure of a learning environment or teaching activity. For, it is the trainer who, as organizing faculty and "Guider of Learning," interprets the curriculum and tries to guide the learner toward the desired learning outcomes; it is the trainer who generally metaphorically constructs, organizes, and controls the physical environment. For instance, the effects of a workman using pneumatic hammer drills whilst repairing a road outside a teaching or seminar room can undermine the best of teaching activities. I have, on occasion, had teaching activities interrupted or "sabotaged" by birds, flies, bees, sewage, power failure, electric fans, helicopters, fire alarms, emergency vehicles, building contractors, cleaners, surveyors, lost members of the Public, late trainees or faculty members, and a family of ducks! This list is by no means exclusive, so be prepared! Teachers need to be adaptable!

The capability of the trainer as an educator limits or expands the effectiveness of the teaching; the more versatile and competent the trainer, the more likely they are to be effective. I am sure many an undergraduate has sat through an unambitious and uninspiring lecture series acquiring little useful learning owing to the limited ability of the Lecturer. In such circumstances, looking at the lecture schedule, "The Faculty" might believe that the curriculum has been delivered effectively but the necessary learning is rarely achieved by this passive teaching method. Indeed, simulation teaching methods ranging from low- to high-fidelity are often effective at creating the necessary learning as they are engaging, and active learning processes, which under the guidance of an effective educator, offer a better chance of creating "real" learning. Never ever assume that simply because a teaching activity has taken place that the required learning has occurred!

It is the trainer's responsibility to engage the learners, inspire, and motivate them about a subject or content matter. Their task is to bring the subject to life. To be effective educators, they need to be able to adapt their teaching style and methodologies to ensure that the learner acquires the desired learning outcomes. This is why increasing the teaching and educational capabilities of the educators is essential if successful learning is to be created. If the teaching or creation of learning opportunities is effective, then the desired learning outcomes will result; if they are not, then the teaching will have been ineffective and of little value to the learners. The challenge for all educators is to constantly improve their "Teaching Skills Toolkit" through continuing professional development. They must focus upon becoming learner-centered rather than teacher-centered and understand that the key value of teaching is learning.

Many a lecturer or pedagog has delivered what they felt was a sterling piece of teaching, a *tour de force*, only for it to be entirely ineffective for the learners. The invitation for questions at the end to be met with self-satisfying silence could be interpreted that all is well and that the learners have lapped it up unerringly; unfortunately, it could also mean that they are so bewildered that they do not know where to start, are totally confused, and do not wish to appear stupid and so remain silent. The latter situation would also suggest that the trainer has not created an environment conducive to the free flow of ideas or discussion. The easiest way to illuminate this teaching "blind spot" is to actively engage with participants and create an effective dialog with open lines of communication rather than delivering an unstinting monolog without learner feedback. By establishing an early two-way flow of information through effective communication, it will soon become clear to an insightful educator if there are major problems, disconnects, or dislocations between the planned teaching objectives and learner needs. It is pointless trying to run before you can walk!

For instance, a novice educator may diligently prepare a piece of educational work such as an immersion scenario in the simulation center for medical students to explore complex clinical practice, say, a course in clinical medical error. It may quickly become clear to the trainer that the group or subsets of individuals in the group are struggling with some of the basic clinical concepts and that their behavior is of a worryingly low level. If their life-support skills are suspect and their generic professional behaviors are so poor that they struggle to manifest even basic clinical responses appropriately, then there is no point persisting with the planned medical error training as they cannot exhibit basic, safe, clinical practice. With these major clinical deficiencies, it is highly likely that the advanced course-learning objectives will not be achieved, as there is no basic foundation to build upon. The course will effectively be a complete and utter waste of everyone's time unless the trainer can adapt the teaching content and approach to try and salvage some useful learning from the occasion. This example also raises the challenge of how to facilitate learning in groups, where there is a wide range of learner capability.

## 15.5 Illuminating Ignorance or Unconscious Incompetence (What the Educator Is Supposed to Do)

A key role for an educator is as a "Guider of Learning" to shine a spotlight where treasures lie to direct the student's attention to that area of illumination, all the while minimizing inappropriate distracters from the dark areas. A fundamental mistake made by trainers is to assume that the learner is aware of their (the learners') own learning needs, and that as such, they should be able to guide or take responsibility for their own learning. If the learners do not know what they need to know, how can they be expected to illuminate their own ignorance? In reality, learners are often unaware of their own learning needs; they are blissfully unaware or *unconscious of their own incompetence*. The first responsibility of the trainer in such circumstances is to raise the learner's awareness and signpost the learner's learning needs. Once they (the learners) are conscious of their learning needs, effective teaching or self-directed learning can ensure that they dispel their own ignorance through creating or seeking effective learning to resolve their ignorance.

## 15.6 Conscious Incompetence to Unconscious Competence

An important and early task of an educator in any teaching activity is to explore the level of understanding and awareness exhibited by the learners. The findings of this enquiry should then be compared to the intended teaching objectives and learning outcomes for the session. If there is some awareness on the part of the learners, then they are potentially conscious of their own learning needs or incompetence. They are *conscious of their incompetence*, or aware of their own limitations. Such self-awareness or insight is a helpful requisite of any learner. These self-aware learners are usually able to self-start their own learning as they have insight into their own learning needs, and the approach of the educator in this instance can then be very different to that for those who are unaware of their necessary learning needs.

As learners become more proficient, they move from *conscious incompetence* to *conscious competence*, and ultimately, as they practice and refine their knowledge and skills become *unconscious of their competence*; this is the level at which trained professionals and experts work. Paradoxically, this also creates challenges for the expert trainers, for if they are unable to remember or reconnect with the difficulties they experienced on their own learning journey, then they may struggle to facilitate that understanding when faced with new learners. This weakness or "educational Achilles Heel" can be illustrated by the frustrated rants of eminent professors or highly regarded clinicians when dealing with junior clinical colleagues in training. This flaw is usually accompanied by the lines "You just do it like this, its easy" or "Don't you get it, stupid!" or "Trainees aren't what they used to be . . . ." One thing that is certain is that these pillars of clinical excellence were once in a situation of unconscious incompetence themselves, all be it many moons ago, they would do well to remember this and try and reconnect with own learning journey – this may make them more effective educators! Excellent clinicians do not always make excellent clinical educators.

It is therefore vital that if clinicians wish to become effective educators or clinical trainers, then they become aware of fundamental educational principles and concepts. Everyone is embarked upon their own learning journey, and only few have completed it; indeed, I wonder if it is ever possible to be truly unconsciously competent in all that one does! In challenge to the old proverb, and given that most people continue to learn their whole life through, it should be possible "to teach old dogs new tricks" – lets hope so!

## 15.7 A Learning Journey

| Unconscious incompetence | → | Unaware of own inability |
| Conscious incompetence | → | Aware of own inability |
| Conscious competence | → | Able but has to concentrate |
| Unconscious competence | → | Able and automatic (expert) |

## 15.8 Training Faculty – Nature or Nurture?

A training faculty needs to be cultivated and developed. This is a specific but discrete task and will determine whether the simulation center delivers effective educational programs. Because of the flexibility of simulation as a teaching methodology, it requires educators with advanced teaching skills and educational awareness to realize and then harness the potentially enormous learning opportunities on offer. Although considerable initially, investment in the teaching faculty will be amply rewarded over time and is essential for the long-term success and vitality of the simulation center.

The innate personal qualities of the individual trainers form the foundations upon which the advanced teaching skills are built. So, creating an effective training faculty is dependent upon both nature and nurture! The theme of selecting trainers will be discussed later in the chapter.

## 15.9 Trainers – True Value added?

Without effective trainers, a simulation center is simply a room full of expensive equipment that otherwise has great cost but no intrinsic educational value, a "*Sim Warehouse*." The next level up is a "*Sim Museum*," where fancy dioramas of expensive but untouched simulation stuff is shown off to VIPs as proof that the institution is cutting edge based upon all the resources expended on simulation.

It is only by adding to these physical and technological items the key human ingredients of capable educators and receptive learners that there is a transformation in the educational value of all the simulation hardware.

Creating an environment where learning can flourish should be the central challenge and preoccupation for the trainers and simulation center management. Learning, although fundamentally an instinctive human trait, can all too easily be misguided, misdirected, or corrupted. Understanding the blocks or impediments to learning enhances the effectiveness of educators, but this requires that they develop high levels of self-awareness, reflection, and insight into the educational process. Blocks to effective learning include ignorance, prejudice and presumption, creating negative, damaging, or degenerate learning environments; these behaviors undermine the educational process and will not create the desired learning outcomes. Far too often, particularly when giving feedback to learners, clinicians use the expediency of haste or busy schedules as a smoke screen for poor-quality educational behavior. It is often apparently easier to "tell" or be prescriptive toward a trainee about their behavior when there appears to be insufficient time to discuss an issue they may have been struggling with, "Do not do that again or else . . . ." However, simply "telling" people to change their behavior rarely changes deeply engrained behavior in the long-term and so is a largely ineffective intervention. (If "telling" was as effective as we wished it were, then all child rearing would be much easier and faster.) As a consequence, "telling" is unlikely to achieve the desired outcome – a change in behavior.

Only by taking the time to more fully "explore" the individual's real beliefs behind a particular behavior is there any chance of challenging or changing behavior in a sustainable manner. For instance, in an operating theater not a million miles away . . . "I was wondering if we could discuss the incident when you shouted at the theater nurse. That behavior was unacceptable. I would like to explore with you your thoughts on this matter. We need to schedule time to discuss this further as I am concerned about your aggressive behavior." Allowing a proper opportunity to explore the events means there is a better chance of uncovering the real issues, and only if the key issues are revealed is there a chance that new behavior can be established. There may be many reasons why an individual behaves as they do; jumping to incorrect conclusions regarding the behavior of learners is one of the greatest disservices trainers can do to learners.

Therefore, developing effective insight into the adult educational process is essential for educators to discharge their teaching duties adequately.

# 15.10 Learning Through Teaching – The Value Proposition!

The yardstick by which any teaching or training activity is measured should be whether or not effective learning has been created. Confusing teaching activity with effective learning is a common mistake made by novice teachers and too many administrators of clinical education Institutions. Counting lectures given by an Institution's faculty, as outlined above, is a case in point. Delivering an accomplished lecture, although satisfying to the lecturer, counts for naught if it is pitched at the wrong level for the learners in attendance. Understanding that the *real value proposition of teaching is to create effective learning* is the single most important concept for trainers to grasp if they are to become effective educators.

# 15.11 Learner-centered not Trainer-centered

Learning as a process resides with the learner; understanding the importance of making educational activity learner-centered is critical if learning is to be the product of a teaching activity. Appreciating that learning resides with the learner and can be catalyzed or enhanced by the educator is a pivotal concept in education. Many learners learn in spite of their teacher! This might suggest that teachers are not essential for learning to occur and this is indeed true to some extent; good educators, however, act as catalysts to learning, they facilitate learners to learn. Trainer-facilitated learning can be more time-efficient than simply relying upon experiential learning alone. With experiential learning, the process is dependent upon learning by trial and error. Although unguided experiential learning can be very powerful, it is invariably time-inefficient, potentially highly variable, and has associated attendant risks to patients as a training methodology.

# 15.12 For the Learner and the Trainer

Teaching should not been seen as a vehicle for the trainer's own aggrandizement; it is a responsibility not to be taken lightly but one to be discharged responsibly. Guiding learning should be seen as a privilege, and indeed, most educators or trainers value and treasure the very many enjoyable experiences associated with their teaching activities. Teaching activities and responsibilities offer an opportunity for greater self-awareness of the educator or trainer, especially if they remain sensitive to their own learning whilst teaching. The notion of a "trainer as learner" is another key concept for effective trainers to develop. It reflects another key attribute of good educators, in that they continue to learn and grow from being involved in educational activity. The translational impact of the generic skills acquired by competent clinical educators is considerable. Being able to reflect upon and debrief complex professional behavior and also communicate more effectively in the simulation suite readily translates into enhanced clinical practice. Organizations supporting educational activity should value this hidden benefit of developing trainers.

Good educators can be defined by their ability to create new learning for both their trainees and themselves.

# 15.13 Developing the Training Faculty

How then does one go about creating an effective training faculty? Broadly, there are two aspects. Firstly, it is essential to create the right learning environment for the trainers themselves. This can be achieved through role modeling desirable

**TABLE 15.1**   Facilitated debriefing technique

This is a teaching methodology in which the role of the facilitator is to encourage or facilitate the learner or learners to explore through active participation and reflection an event or learning opportunity such as a real clinical or simulated clinical event. As the process encourages active participation, the learning created is often deeper and more profound. The focus of the process is entirely learner-centered and requires considerable advanced communication skills on the part of the facilitator to create a supportive environment in which knowledge, technical, or professional behaviors can be explored and challenged. Using appreciative enquiry, facilitation allows detailed exploration behind the thoughts and beliefs associated with events or behaviors. This methodology is particularly useful in exploring complex professional behaviors or nontechnical skills especially when coupled with audiovisual feedback of complex individual or team behavior. It is the opposite of pedagogy or "telling."

educational behavior ("walking the walk") by the "training the trainers" faculty. Secondly, by creating a specific learning plan and program for the development of the new facilitator/trainers. We insist that our facilitator candidates attend a simulation center course as a learner before being put forward for facilitator training. This allows them to be on the receiving end of the *facilitated debriefing technique* and in so doing they can start to develop an awareness of the potential value of the facilitated teaching methodology to learners (Table 15.1). Once they have experienced the effectiveness/value of the facilitated behavioral debriefing session, they can be considered as "ready" for a formal period of training (see Chapter 65).

## 15.14 Identifying Facilitator Trainers

Identifying which candidates should be put forward for trainer development is a difficult challenge for those operating a simulation center. Some trainer candidates are drawn to teaching activity and self-nominate, others exhibit strong personal behaviors that would make them suitable to be invited to train; others, unfortunately by job title or departmental expectation, are duly delegated (sometimes against their will). I would question the wisdom of forcing unsuitable individuals, who perhaps have limited interest or motivation in teaching, to be allocated significant educational roles. These individuals are perhaps better utilized in spheres of clinical practice other than education. There is potentially nothing more damaging in educational terms than an unwilling or incompetent trainer or facilitator, especially those who resent being required to engage in an activity they do not personally value. They are a threat to the safety and well-being of the learners. Exploring the educational motivation and professional integrity of why people want to become trainer candidates can be a useful exercise in identifying those most suitable for advanced educational training, which will avoid significant problems later. Challenging the

motivation and integrity of those involved in educational activity can reveal dissonant or unhelpful attitudes that are well to be aware of; these include comments such as the following. An individual, when asked what he wanted out of his participation in clinical simulation, his response was, "to make full professor." A questionable motivation from his learners' perspective!

However, when trainers are identified, consideration should be given at an early stage by the "training the trainers" faculty whether each trainer candidate has the "rough diamond" qualities likely to warrant the investment of time, effort, and resources in their development as a facilitator. It is important for the "training the trainers" faculty to understand that not every trainer candidate will have the core personal qualities and motivation required to become a safe and competent facilitator. Particular caution should be exercised where there are concerns regarding underlying personality, interpersonal communication, or their ability to be a role model for the desired behavior of a facilitator. Not everyone will be able to master the learner-centered skills necessary to make an effective facilitator in a simulated training environment.

Generally, those who exhibit self-awareness, an ability to listen and empathize with others, and conduct themselves in a way that is conducive to creating a robust but supportive learning environment should be seen as having the raw qualities necessary to become/develop into a facilitator. There is clearly a judgment to be made by the "training the trainers" faculty in selecting those it feels appropriate to put forward for advanced training. Being aware of some of the personal qualities we find encouraging in a facilitator may help this selection process. The qualities of good and bad educators should be used as a guide to selection and are by no means exhaustive (Table 15.2). Interestingly, many of the qualities of a good

**TABLE 15.2**   Educator qualities

| Good-educator qualities | Bad-educator qualities |
| --- | --- |
| Integrity | Lacking integrity |
| Motivated | Unmotivated |
| Empathic | Critical, judgmental, or opinionated |
| Good communicator | Poor communicator |
| Good listener | Unsympathetic or impatient |
| Supportive | Unsupportive |
| Challenging | Dominating |
| Engaging | Unpleasant, derogatory, or rude |
| Knowledgeable | Ignorant |
| Attentive and diligent | Disinterested |
| Skilled | Unskilled |
| Capable | Incapable |
| Interesting | Boring |
| Enthusiastic | Dull |
| Flexible | Inflexible |
| Adaptable | Limited |
| Enjoyable | Aggressive |
| Welcomes challenging students | Fears challenging students |

or bad educator are similar to the features of a good or bad learning environment; thus, these qualities could just as easily be used to describe training departments!

# 15.15 Creating Effective Health Care Professional Learning

If we accept the main theme of this chapter that the *learning environment* is the critical functional unit in creating learning, then understanding the relationships between the three key components of the learning environment is critical. The three key components are given below.

## 15.15.1 The Physical Environment

Facilities, teaching aids, and resources including simulation center resources (Table 15.3). Compare with Table 15.2.

TABLE 15.3   Learning-environment qualities

| Good learning environment | Bad learning environment |
| --- | --- |
| Honest | Lacking integrity |
| Supportive | Unsupportive |
| Open communication | Closed or poor communication |
| Challenging and constructive | Dominating and unconstructive |
| Motivated | Unmotivated |
| Nonthreatening | Threatening |
| Nonjudgmental | Harsh and judgemental |
| Knowledgeable | Ignorant |
| Attentive | Shoddy |
| Empathic | Critical |
| Functional | Dysfunctional |
| Skilled | Unskilled |
| Capable | Incapable |
| Engaging | Isolating or ignoring |
| Interesting | Disinterested |
| Flexible | Inflexible |
| Inspiring | Frustrating |
| Respectful | Disrespectful |

## 15.15.2 The Human Component

Trainers, learners, program directors, assessors, and managers! We need their support and sanctioned funds!

## 15.15.3 The Intellectual or Cognitive Element

The educational program ethos, faculty motivation, and insight into the educational process of learning.

These three components all play a critical role in creating an effective environment for learning to occur and alone are largely ineffective without the other two elements. If an Institution is to be successful in creating learning, it must be clear about the desired learning outcomes required to create its desired product – Competent Health Care Professionals. I would suggest that the greatest challenge to teaching Institutions and educationalists is the complexity of the educational process by which these professionals are created.

The creation of competent high-quality health care professionals is the true value proposition of expensive health care training. Understanding how to create these high-quality health care professionals should be the preeminent task of all health care training Institutions. However, this complex process is not well understood. A reductionist approach advocates a competency-based microdissection of the curriculum, outlining the necessary knowledge base (even down to the exact facts learners should recall), the necessary technical skills armamentarium, and professional behaviors portfolio. Knowledge and technical skills lend themselves well to being delivered in a competency-based manner; however, the real high-value professional behaviors do not lend themselves to being taught in a similarly reductionist way. The competency-based approach risks underestimating the critical role that is a persistent theme of professional training the world over – that professionals "learn by doing and doing repeatedly" [1].

When one considers how one creates the necessary learning opportunities in professional behaviors, important areas such as communication, team working, crisis management, decision making, conflict management, leadership come to the fore. Simulation techniques are ideal for teaching these domains of professional practice. When health care professionals are dealing with inherently complex systems such as human disease and the infinite variety of biological systems, a reductionist or competency-based view of the process of education is intellectually naïve and unhelpful. Until such time as patients' biology reliably follows the algorithms proposed in the curriculum, it would be better to recognize and understand that the inherent value of any professional is in their ability to manage chaos or uncertainty. The ability to function in this area requires not only competencies, but also, more importantly, qualities or capabilities of judgment that create a capacity to deal with the unknown. This is an essential function of professionals.

# 15.16 Professional Training – An Apprenticeship?

The competency-based approach undervalues a critical feature of the ancient, time-honored method of professional training: that of professional apprenticeship. The apprenticeship model is based upon a total immersion in the professional workplace allowing prolonged and repeated exposure to the inherent vagaries of actual cases and real events attendant with the professional role. Reflection and learning at the "Knee of the

Master" is a tried and tested method of acquisition of learning in the Professions. In modern educational parlance, a professional apprenticeship creates experiential learning and also affords facilitated debriefing or troubleshooting of problems with "The Master" as and when learning opportunities occur. This allows the apprentice to develop professional capabilities in dealing with the diversity of tasks likely to be encountered in his professional career. The impact of experience and the reflective consolidation of learning created professionals "fit for purpose." In modern professional curricula, we have lost much of the intensity and continuity of professional support that typified the apprenticeship model of old; this is particularly of concern with respect to the high-value professional behaviors. There is increasing concern that professionals completing shortened, less intensive, competency-based, time-limited training programs are not "fit for purpose;" this should be cause for concern in Society and the Profession.

## 15.17 Standardized Clinical Learning Opportunities

High-fidelity simulation offers modern day parallels with the key features of professional apprentice-type learning. In these days of structured training programs, shortened time-limited training periods, and limited clinical exposure for vocational learners, simulation offers an opportunity to standardize the learners' experiential exposure to a range of clinical scenarios that may not be deliverable by simply relying on clinical experience alone. In the United Kingdom, higher specialist medical training has been reduced in real terms from over 30,000 hours to less than 8000 hours in under a generation. This has deprived vocational learners (trainees) of vital clinical experiential learning opportunities. If standards of clinical practice are to be maintained, novel high-impact effective clinical-orientated teaching methods need to be developed to compensate for this deficiency in total clinical exposure. The ability of high-fidelity immersion simulation especially when complimented with facilitated debriefing of complex professional behaviors offers an exciting opportunity to deliver these important professional domains of practice in an environment conducive to effective learning.

The clinical environment is not always tolerant of vocational learning needs as clinical priorities rightly take precedence. Immersion simulation also offers the opportunity for vocational learners (trainees) to repeatedly practice drills and skills to improve their proficiency and performance as clinicians. Immersion simulation is also patient-safe with no attendant risk to real patients; indeed, with high-fidelity immersion scenarios, there are opportunities to allow scenarios to progress where in clinical practice early intervention would be mandatory on grounds of patient risk. This potentially affords vocational learners (trainees) new learning opportunities, particularly in critical areas of practice. Critical clinical

events are often infrequent and unplanned, which means that access to the necessary clinical learning opportunities for vocational learners is limited; immersion simulation overcomes this learning-opportunity deficit for a training Institution. Reduced training time inevitably means limited clinical exposure, and training in high-fidelity simulation environments can protect against the vagaries of the clinical caseload and throughput. Simulation when fully integrated and blended into the learning or training program allows training in the clinical scenarios or contexts suggested above. Simulation is an active teaching methodology and offers active learning opportunities.

## 15.18 Teaching Professional Behaviors

Complex professional behaviors are difficult to quantify, and are qualities or capabilities that are more subjective and, as a consequence, less amenable to the reductionist, analytical approach of competency-based training delivery. It is not realistic for an Institution to attempt to deliver complex, behavioral, nontechnical competence on the basis of delivering a lecture or 1-day course. Exposure to facilitated behavioral debriefing and feedback techniques over a prolonged period of time offers excellent new learning opportunities particularly in the professional behaviors domains to those in training. Ideally, the facilitated behavioral debriefing technique should be employed widely including the real clinical environment, but for many reasons, not least the pressure of clinical service, the opportunities for reflective professional training in practice is often not possible due to hectic clinical service departments.

## 15.19 Blended Learning – Using Simulation Techniques

Simulation-based courses, especially if blended into the appropriate part of the training program, allows targeted teaching to be delivered, for example, the pediatric emergency course during the pediatric module of anesthetic training or the advanced airway skills course during the advanced airway module. This also allows relevant, difficult, or rare clinical events to be simulated, such as a "can't intubate, can't ventilate" scenario (fortunately very rare in clinical practice, equally unfortunately very rare in clinical training) at a time when the trainers are available to facilitate the learning. Integrating immersion or other forms of simulation methodology can provide enhanced and consistent learning opportunities to Institutions and training programs. This effectively increases the value added of the training program to the vocational learners.

## 15.20 Simulation – An Educational and Clinical Catalyst?

Simulation centers can also act as centers of teaching excellence within Institutions as the prerequisite skills such as complex facilitated behavioral debriefing are at the high-end of educational practice. Giving focus to and a base for such advanced educational activity potentially enhances the quality of all teaching activity through trickle-down of practice and the impact upon the wider Institutional training community. In addition to the training "ripple-out" effect, trickle-down of professional behaviors awareness can directly enhance clinical practice in many areas of technical and nontechnical practice such as resuscitation, drills and skills, effective team working and communication, leadership, risk management, clinical governance, and many other areas besides.

When an Institution is able to create strong learning environments and supports the ability of its clinical educators to create learning, then an Institution through its educators can challenge and encourage their learners to reflect, explore, and challenge their beliefs and values as considered in the context of their professional role. Then, and only then, is a teaching Institution "fit for purpose" – to create competent Health Care Professionals and worthy of the title "Teaching Hospital." Personally, I would prefer the term "Hospital for Learning" but it is a bit of a mouthful!

teaching to the learners. It is important to understand that the real value of immersion simulation techniques goes way beyond simply "drills and skills" competence. It offers unique learning opportunities when combined with facilitated behavioral debriefs in exploring professional behaviors. Risk management reports can provide useful windows upon the efficacy of the impact of professional behaviors training particularly in the domains of patient safety. The impact of simulation training on health care practice is currently being extensively evaluated in many centers worldwide. This involves detailed, extensive (and expensive) qualitative research to uphold the high face validity that the immersion simulation technique has with patients! Patients cannot understand why they should be the subject matter of the first naïve attempts of health care professionals when they need to be trained. If immersion simulation is valued by other high-risk sectors such as the aviation or nuclear industries, why then is it not good enough for medical training? Cost alone seems a shallow argument against immersion simulation when full consideration is given to the potentially significant benefits outlined above.

> If you think education is expensive, try ignorance
> Derek Bok, Harvard President (1930)

Patients, their families, lawyers, and clinicians would generally subscribe to that sentiment.

## 15.21 A Recipe for Success with Simulation Techniques?

When developing a strategy for the role of simulation in health care education, the following points may be of value. Integrate simulation training and ensure that teaching compliments the relevant training curriculum. Identify the trainers and develop them so that they are "fit for purpose:" to provide facilitated behavioral debriefs, and be aware of professional behavior, and how to challenge people to create new learning in these domains. Clarify the role of simulation in delivering the desired learning outcomes. In particular, integrate simulation training into the greater curriculum and training program. Adopt a blended learning approach where different teaching methods and styles augment one another to produce significant and deeper learning. Communicate the curriculum widely to candidates and educators and clearly establish learning objectives and expected learning outcomes. Provide necessary resources, both people and equipment. Ensure that the faculty is trained so that they can create a strong effective learning environment. Importantly, evaluate teaching delivery in terms of learning created and changes in behavior, not simply measure teaching activity (e.g., number of lectures) or the popularity of the

## 15.22 The Reward of Creating Effective Learning Environments

The consequence of creating effective learning environments populated by effective learner-centered facilitator/trainers is that this then becomes a powerful way of creating effective learning for the learner. This after all is the real value proposition of any educational activity.

Creating learning should be the primary goal of any teaching activity, had this perhaps been achieved in Churchill's day, then his pointed observation at the head of this chapter may well have read differently... "Personally I'm always ready to learn, and I always find it worthwhile being taught."

## Reference

1. Philip E. R. The Expert Mind: Studies of the mental processes of chess grandmasters have revealed clues to how people become experts in other fields as well. *Scientific American*, July 24, 2006.

# The Best Form Follows the Essential Functions

*"Planning without action is futile, action without planning is fatal"*

To repeat: Clinical simulation *is* clinical theatrics with full contact audience participation. Thus, like all theaters, all clinical simulations require some form of venue to house all the participants and furnished to establish the clinical environment and define boundaries of the event. Thus, you should ask yourself where your venue should be located, how big does it need to be, how that space should be partitioned, and what are the essential furnishings? The answer to each is found within your intended functions. If public relations are paramount, then perhaps the best location is in the midst of where your institution usually presents itself to its guests. If you have 10,000 new emergency medical technicians to train each year, then you will need a very large facility. If you have an entire clinical school's worth of different topics to present, then you will need a number of different "stages," each with its own unique and dedicated supporting rooms. In each and every case, you need to furnish every area with at least the minimal items required to support how you intend to use simulation to further your teaching objectives and your students' learning objectives. The simulators themselves, either manufactured or live, are essential yet insufficient resources.

Despite unique needs and resources, there is a common approach to determining location of any simulation activity: close enough to where instructors and students already congregate, yet far enough away from real patient care to prevent cross-contamination and disruption in either direction. Too far way would require using automobiles (find a parking space, again?!) or walking outside (change clothes, again?!). Too close to real patient care risk infection of trainees and visitors (e.g., SARS) and risk fake drugs/equipment/supplies mistakenly used in real patient care. Furthermore, dedicated stage environments allow production of scenarios that should not happen near real patients (e.g., intentional disasters with all their noises, smells, and death). The exception to this advice is when simulation is intentionally produced in real clinical care environments. Known as *in situ* simulation, it is a powerful teaching method, but such power requires great control to prevent harm. Chapter 39 describes the value and challenges of producing **invisible** in situ simulation.

At every scale, for every resource design and application decision to take, the two most important questions to ask are "who are the students?" and "what are the educational objectives of the teachers and the students?" Resources include all those items essential for creating and sustaining any academic program that requires specialized facility for its operation, not the least of which is good will of the neighbors. The first question encompasses "what are the different clinical disciplines?" and "how different are they?" Another way to state the second question is "how do you want your students' knowledge, skills, and attitudes changed by their experience in your educational program?" The more accurately, honestly, and completely you can answer these questions, the more likely you will succeed in your endeavors. Last but not least, don't be surprised that the activity of your simulation program will change its environment, which will in turn change the purposes that it will be asked to address, which will in turn disrupt all the careful analysis and preplanning you made as the basis of your initial program design. Simulation is a dynamic, disruptive activity to both its neighbors and itself. Honor and respect its transformative nature. Expect the unexpected.

Simulation is a stepping-stone for the gap between didactic and patient-based learning. Therefore, look to add simulation to where the gaps are widest and the risks to patients and students greatest. Unlike unguided on-the-job training, the essence of formal education is amplification through simplification, and so too is simulation. We limit the realism of simulated clinical environments and simulated patients *because* real clinical environments and real patients are far too complex, too messy, too noisy, too saturated with too many uncontrollable distracters. Like any live theatrical production, the realism of simulated who, what, where, when, and why of the clinical *story* are the essence of the production, while the clinical "art work" about the stage is merely there to help convey that story. The very robust realism of the ancient Greek and Roman stories is what allowed them to be rewritten for different times and locations by Shakespeare, and his stories subsequently restaged in a myriad of different times and locations.

Since we all detect more than we can comprehend, a key part of any learning is developing the ability to discriminate between what we need to attend to and what we can ignore. Simulation is ideal for creating safe environments and activities within which our students can refine their attention discrimination abilities. Since most students enter simulated environments with a poorly developed sense of discrimination, err on the side of equipping your clinical stages with too little instead of too many unintentional distracters. Each and every item "on stage" should be there for a well-defined teaching purpose aimed at the level of competency of the students on the stage at that moment. At best, having clinical gear just for the sake of showing it off will clutter your environment and who knows how it will clutter your students' minds. Note that having at your disposal less than all the resources that you might desire for your simulation program requires discrimination in yourself. Conveying this attitude and skill to your students is always a valuable lesson for both of you.

Just as we use simulated clinical behaviors as a stepping-stone to greater clinical competency, so too can you use simulation of your desired simulation spaces as a stepping-stone to greater facility utility. Before expending the prodigious amounts of resources required for any all-new simulation facility, consider first constructing and operating a pilot program in renovated spaces. Many decisions about how much space *your* program needs and how *your* space is to be apportioned has no universal right answer. But, we describe some common trade-off issues and their consequences. For example, dividing one large space into two smaller ones with a solid, fixed sight/sound wall greatly reduces the amount of distractions from one side into the other, and allows for scheduling of dissimilar events. In simulation facility design, sound control is a far greater challenge than sight control, and thus will become a key issue in any list of design trade-offs. Yet, this same fixed barrier eliminates the option of easily sharing a common experience with a larger audience. The small carrying capacity of several sports cars can never combine to convey one king-sized mattress like the cargo space in one large truck. Like all conveyances, the more optimized of one performance feature, the less suitable for all others. What *is* universally true for all construction projects, the sooner in the process you *define* your requirements as completely as possible, the more you *use* your requirements as the basis for deciding all the myriad unanticipated questions that arise during construction, the cheaper the final outcome will be. Thus, during your initial designing, periodically ask "if we include this feature to support that function, what other functions will now be precluded?" Do keep in mind that simulation is a stepping-stone to shrink the gap between the shores of two other teaching experiences. As a stepping-stone, it provides the support otherwise missing in the gap, but there is little value in making it more substantive than is necessary for your students' steps.

Irrespective of your intended program type and scope, there are a number of fundamental issues common to all simulation programs and facility design decisions; something as simple sounding as wall placement may influence each of them:

- Managing others – students, instructors, and staff
- Funding – start-up and sustaining incomes and expenditures
- Curriculum – your plans for action to meet your educational goals
- Champions – identified individuals dedicated to your programs' success
- Public relations – guest access, message distribution, and visitor control
- Shifting priorities – from patient care to student learning
- Calls for "proving" the value of simulation in clinical education

All successful simulation programs have been challenged with and successfully addressed all these issues. Lack of success in any one of these will greatly hobble your program. Since individuals least familiar with the invisible essence of simulation can and will count floor space and instructor usage hours, your simulation facility design *will* influence how it is used and by whom, thus how its costs and contributions *will* be ranked within the greater educational effort of your institution. Prepare for such challenges with the best data available. Since there is almost zero data of any consequence currently in use in academic management decision making, by default, your data will be the best in the room.

Jane Miller recognizes that there is a limit to how well anyone can answer such questions about any activity before they ever attempt it, and she provides a framework for thinking that accepts and works within this ambiguity. Michael Seropian presents an algorithmic approach to designing and building based upon actual intended use. Brian Brost *et al.* first show examples of their multiuse facility design as their "answers" to these fundamental design questions, and then follow with how they made their choices in simulators and supporting items as their "answers" to these fundamental furnishing questions. David Stern addresses purchase decision criteria for robotic simulators, and provides explicit examples. Ramiro Pozzo warns that the consequences of location is not limited to just where your facility is sited within your instituion institution, but includes where your insitution institution is sited on the planet. Mike Foss addresses the challenges typical to all renovation projects that you undertake while still living and working within the structure.

# 16

# Thought Thinking Itself Out: Anticipatory Design in Simulation Centers

In linguist Benjamin Lee Whorf's classic essay, *An American Indian Model of the Universe* [1], he underscores the cultural mutability of time and space using the Hopi language as an illustration. As Whorf points out, Hopi has no verb tenses; instead, Hopi speakers use grammar as a means of talking about events that have already occurred and others that may be yet to come. Future events are in the process of "becoming" – actualizing their potential. Little distinction is made between space and time, the two being interconnected. The future "is the realm of expectancy, of desire and purpose, of vitalizing life, of efficient causes, of thought thinking itself out into manifestation".

Perhaps surprisingly, Whorf's observations on this heuristic framework are useful for the process of planning a simulation center. This chapter serves as a case study of one design process that did not employ this anticipatory perspective.[1] Understandably, design team members focused only on the "Newtonian" present. Expectations – the size and configuration of rooms, the placement of walls and equipment, etc. – were predicated on their current, observable reality model of clinical education (e.g., "see one, do one, teach one"), not on how the establishment of a simulation center itself might alter those expectations, necessitating a different kind of design. While the center has exceeded expectations for simulation program development and use, and has benefited from some of the decisions of the initial design team, the facility and the programing resources have required substantial changes in the first 3 years of its existence. This essay offers reflections on this experience and uses an ethnographic approach to the design process. The essay concludes with recommendations for employing an anticipatory design process in the creation of future centers.[2]

## 16.1 Users Defined and Counted

In the fall of 2001, a multiprofessional group of clinicians and educators were assembled to create a 7500-square-foot center for health sciences simulation, the Interprofessional Education and Resource Center (the Center), at the University of Minnesota, Minnesota, U.S.A. The University of Minnesota is one of the nation's largest land grant universities. Founded in 1851, it is actually older than the state of Minnesota itself. With five campuses (in Minneapolis-St. Paul, Duluth, Morris,

**TABLE 16.1**   2005 Enrollment by Academic Health Center College/School

| College/School | Enrollment |
| --- | --- |
| School of Dentistry | 590 |
| Medical School | 1818 |
| School of Nursing | 779 |
| College of Pharmacy | 648 |
| School of Public Health | 640 |
| College of Veterinary Medicine | 447 |
| Total | 4992 |

Note: Enrollment numbers include residents and fellows, as well as matriculated students.

Crookston, and Rochester), the University serves approximately 60,000 students in 370 different fields of study and employs over 18,000 people. The Academic Health Center, established in the early 1970s, serves as an administrative umbrella (and, in some senses, an innovations hub) for the University's six health sciences schools: the Medical School, School of Dentistry, School of Nursing, College of Pharmacy, School of Public Health, and College of Veterinary Medicine. Five of these six schools are served by and fund the Center (the School of Public Health is the exception). As a comprehensive health sciences center serving the upper Midwestern United States, each of the schools has a substantial enrollment. Enrollment for the Medical School alone (which combines students at the University's Twin Cities and Duluth campuses) is 213 students per undergraduate class (Table 16.1). This poses significant challenges for every aspect of the educational experience, and delivery of instruction and assessment using simulation is no exception.

## 16.2 Facility Designers

In its early stages, the simulation center team was led by the Facilities Planning office in the Academic Health Center. The Center's space, which had formerly housed administrative offices for the University's practice plan, was "land-locked" by contiguous spaces used for other purposes ranging from student computing to transplant clinic offices to transplant patient care. This required a complete reconfiguration of the space, including reconfiguration of the Heating Ventilation, and Air Conditioning, electrical, Networking/Information Technologies, and plumbing systems. Although it has proven to be a convenient location for most users (the College of Veterinary Medicine is located on the St. Paul campus and is accessible by campus shuttle), it has required ongoing compromise with clinical and administrative needs of neighboring offices of the independently owned Fairview-University Hospital.

The original design group included faculty from across the health sciences, appointed by their respective deans for their interest in and/or use of live simulations using standardized patients; "virtual" simulations and simulations using human patient mannequins were not considered at all in the construction of the Center. Because few of our schools used objective structured clinical exams (OSCEs) or standardized patients expertise in their use came primarily from the Medical School. Following University policies and protocol, a competitive bid process was used to identify a nationally known architectural firm for the project. Initial perceptions of the project by both Facilities and Medical School staff tended to focus on duplicating primary care clinical spaces (the familiar context for simulations, at the time) rather than on creating a space that could be reconfigured to simulate a variety of clinical contexts. As a result, the architectural firm was selected on the basis of its experience with designing primary care clinics rather than educational spaces for teaching and assessment.

As part of the planning process, a subset of the design group was invited to participate in a site visit to one multiprofessional simulation center with an academic service platform and curricular projects very similar to the expected use of our Center. This was the only formal site visit conducted as part of the design process. The resulting design was, in many ways, an expansion of the site visit center spread over a larger footprint (Figure 16.1). While each planning participant was acting with the best of intentions, no consideration was given to how changing educational paradigms (e.g., the increasing use of performance-based examinations for students and practitioners at various stages of development across the health sciences curriculum) might actually necessitate a different approach to design. Take the Media Center as an example. It is the largest coherent space and is located in the middle of the Center. Equipped with a conference table, projection units, and 18 dedicated monitoring stations (corresponding to the Center's 18 examination rooms), it was the site visit center writ large. However, the location of the monitoring stations (ringing the room) presumed the use of the Center for only one event (i.e., OSCE) at a time, rather than how the Center would come to be used. The design team was not able to think beyond one existing paradigm for teaching and assessment. Team members (the author included) were not capable of allowing "thought to think itself out."

## 16.3 Integrating Infrastructures

At the same time, an audiovisual engineering unit within the University was contracted by Facilities Planning to design an Audio/Video system for monitoring and recording events in the new simulation center. While the A/V system was considered an important part of the project, few of the design team members really understood the implications of decisions made

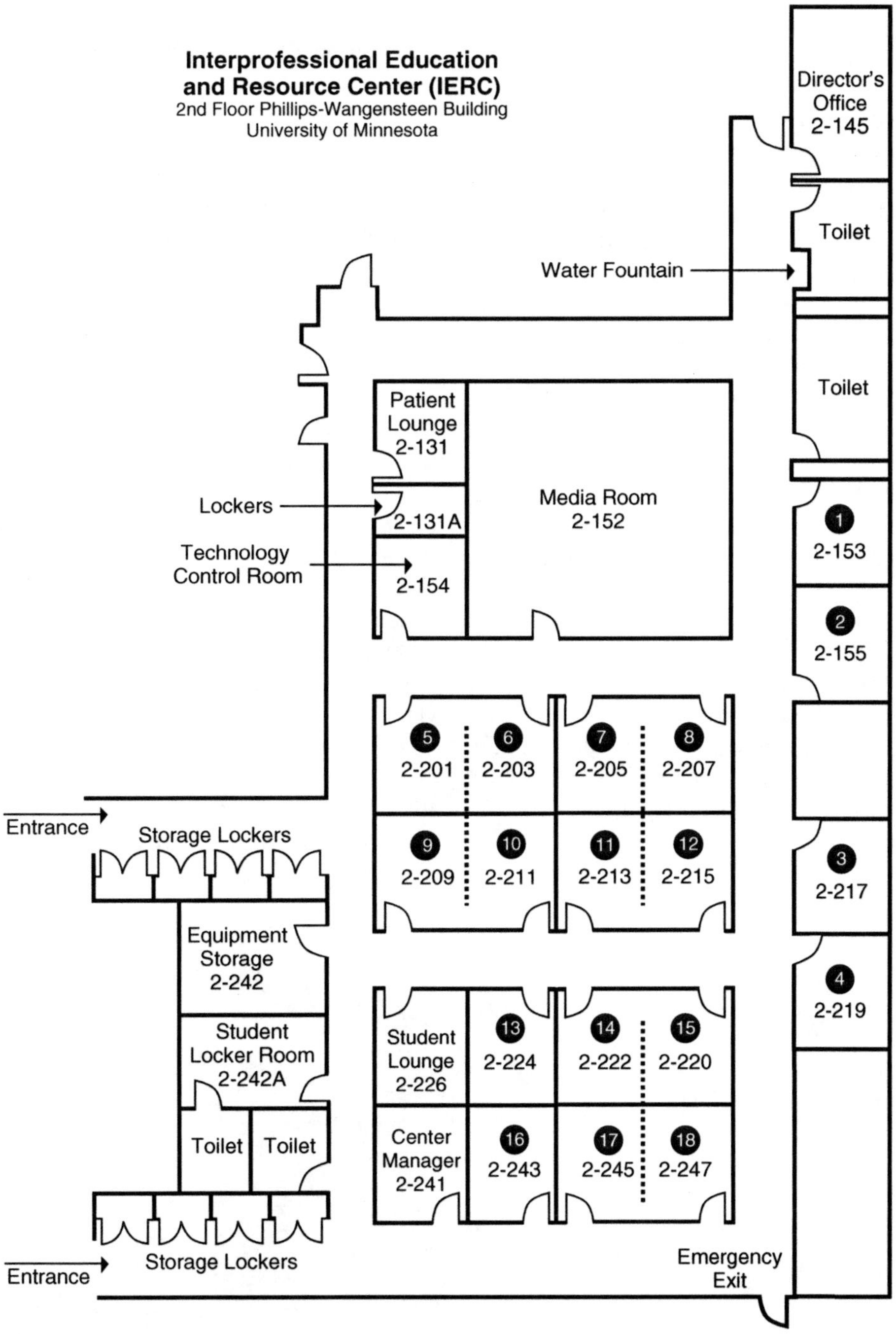

**FIGURE 16.1**   Floor plan of our Center showing the numerous teaching environments and support spaces, their relative sizes, locations, and flow paths between them.

about Networking/Information Technology infrastructure and hardware. The clinical paradigm of deferring to specialists was extended to this process. Almost without exception, decisions about the audiovisual system were made between Facilities and the A/V engineer. As early as the predesign phase, cost and the senior engineer's perception that digital technology was unreliable led to a decision to install analog cameras and recording equipment. This decision resulted in significant costs for retrofitting the space for digital technologies starting in 2005, a little more than 2 years after opening.

## 16.4 Adaptations to Curriculum Constraints

Designation of space within the Center proved to be a similar challenge. As of 2002, the Medical School offered an 18-station OSCE for the Primary Care Clerkship. Every 8 weeks, 30–36 students were expected to interview (and, in a few cases, conduct a minimal physical examination) 18 standardized patients simulating a wide variety of chief complaints and clinical issues. (In some rotations, more than 18 stations were actually included in the examination.) At the time, this examination was considered the prototypical standard for performance-based exams in the Academic Health Center. It was also perceived, to some extent, as a comprehensive examination of third- and fourth-year medical students' clinical skills. Consequently, the decision was made to allocate almost all of the space for simulation of a primary care setting, with individual examination rooms (8′8″ × 11′5″) comprising the vast majority of the square footage (Figure 16.2).

In order to maximize the use of this space, accordion partitions were installed so that 12 of the 18 exam rooms could be combined into six larger rooms (8′8″ × 22′10″) for small group teaching (Figure 16.3). While this would prove to be an essential design decision for maximizing the usefulness of the Center, it would also pose some significant challenges. Sound penetration between these partitions remains a limitation, particularly in testing situations.

Limited spaces for support functions would also prove to be a limitation. Only two offices for a director and a standardized patient trainer were included in the layout and were located at opposite ends of the Center. There was also no consideration given to providing separate space for staging or orienting event participants. Space was also allocated for storage, two small locker rooms, and a small lounge area (presumably for student use), but the spaces were scattered throughout the Center, and were not appropriate, even for the anticipated use of the Center (e.g., the "student" lounge can hold no more than six occupants at a time).

## 16.5 Early Users

Since the simulation facility opened in January of 2003, use has grown in all the health sciences and has become increasingly diverse and complex in content. Within the Medical School alone, usage has nearly doubled, going from just over 5000 room hours (i.e., the number of rooms used multiplied by the number of hours) in 2003 to 8500 in 2005 (Figure 16.4).

With the addition of task trainers (e.g., CathSim) and full-sized mannequin patient simulators (e.g., SimMan) in 2004, these figures reflect an expanding variety of teaching and assessment activities, particularly in the fields of medicine,

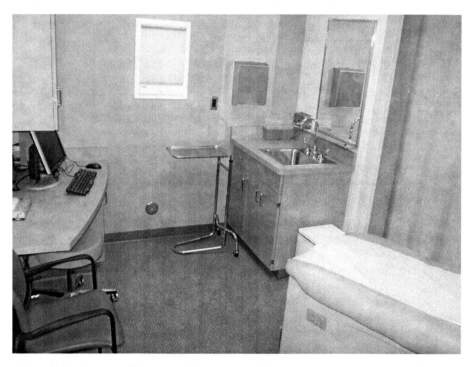

**FIGURE 16.2**   Single exam room.

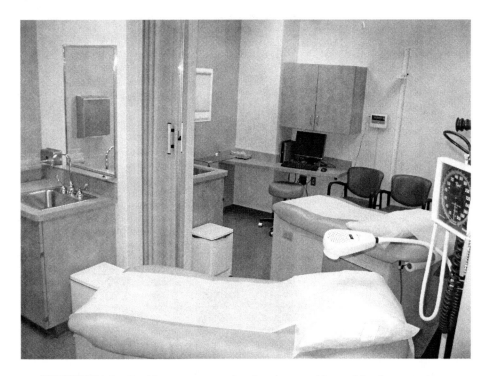

**FIGURE 16.3** Double exam room, showing the movable partition between them.

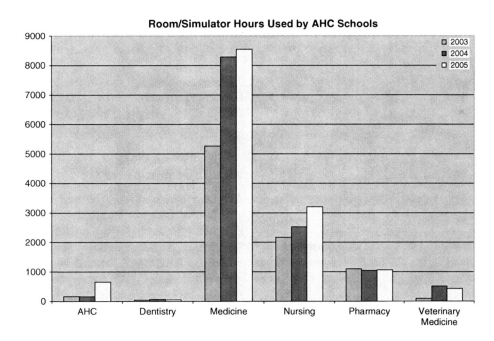

**FIGURE 16.4** Simulation usage over 3 years by six different clinical schools.

nursing, and pharmacy (Figure 16.5). Still, this explosive growth in use represents a fraction of learners' clinical education. A recent study conducted as part of a strategic planning process revealed that almost 1 million hours of clinical training time will be invested in the credentialing of the Medical School class of 2005 [2].

Educational (often course-related) events have also increasingly used audiovisual and simulation technologies (ranging from tabletop static models such as IV arms and birthing pelvises to standardized patients to partial-task trainers and full-body patient simulators) for extending the range and value of simulation experiences. Subsequent investments in

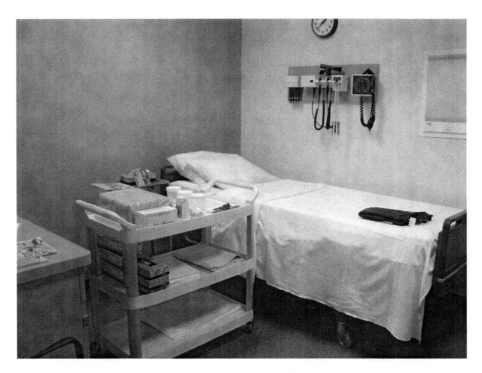

**FIGURE 16.5**   Nursing home staging for First Responder Communication Skills workshop.

standardized patient recruitment and training, digital recording/storage/reproduction devices (e.g., DVD burners with hard drives), desktop computers in each of the exam rooms, and a Web-based electronic exam management system have greatly improved the Interprofessional Education and Resource Center's capacity to develop simulation projects and meet user needs.

# 16.6 "Lessons Learned" Cover each of the Domains Involved in the Design Process

For all of the omissions and missteps that occurred in the planning and development of the Center, it has proved to be a durable and adaptable facility. Three elements have been critical to its current success:

- The support of senior leaders who value competency-based education and see the potential for research and scholarship in simulation;
- the creativity and ingenuity of program staff, faculty, and external clients;
- facilities planners who have continued to respond to the schools' (and the facility's) changing needs. For instance, the addition of an exam management system has set the standard for data collection and management for the new facility. Research projects (most of which were funded

and started in the past year) are flourishing, and staff struggle to keep up with expectations for development of new projects.

Reconceptualizing the intersection of space, time, and learner experience – as "thought thinking itself out" – has been critical to the success of this program [3]. A short list of key Lessons Learned are provided in Table 16.2.

## 16.6.1 Design Team

As noted above, the project was developed under the auspices of the Academic Health Center, with the expectation that the facility would serve at least five of the six health sciences (Medicine, Nursing, Dentistry, Pharmacy, Public Health, and Veterinary Medicine) included in the AHC's academic programs. Yet, many of the early design decisions were made through direct consultation between representatives of the facilities planning office and the Medical School, circumventing collective consultation with all of the planning group participants. As is often the case, time and funding constraints made it difficult to do an extensive review of centers with comparable programs. Similarly, a well-known architectural firm that had experience in designing clinics but no experience in designing educational spaces was selected for the project.

Finally, the design team relied on a local expert for recommendations for the installation of A/V technologies, which, while cost-effective, were limited in their utility and quickly required enhancement. For instance, the initial A/V system

**TABLE 16.2**  Summary of Lessons Learned

| | |
|---|---|
| *Design team* | |
| Lesson #1 | Make decision-making process obvious and representative of all of the stakeholders. |
| Lesson #2 | If stakeholders do not have the necessary expertise, invest in developing that expertise or in hiring consultants appropriately. Do not rely on utilitarian needs assessments alone, but also include prospective planning in the project development process. |
| Lesson #3 | Invest in research on design and equipment through literature review, strategic site visits to centers with comparable programs, then invite their program design experts to visit your environment. |
| *Program* | |
| Lesson #4 | Use an organizational development process – driven by intended use – in the planning of the center to avoid excessive focus on present uses and expectations. |
| Lesson #5 | Consider all of the factors that may have an impact on the frequency and types of future use (such as program accreditation requirements), not only current uses. |
| Lesson #6 | Consider the training and continuing education needs of additional groups (e.g., first responders, hospital personnel, etc.) that may become an important part of the center's mission. |
| *Space* | |
| Lesson #7 | Consider the different types of learning (e.g., in small and large groups) and movement patterns through the proposed design before project confirmation is complete. |
| Lesson #8 | Think generously about access points and spaces for the three Ss (staffing, storage, and staging) in order to create successful simulations. |
| Lesson #9 | Emphasize flexibility and mutability above all other design principles. Invest in mechanical, electrical, and IT infrastructure (i.e., wiring and piping), even if it will not be immediately employed, in order to meet changing technological demands and to maintain the viability of the simulation space for as long as possible. |

was wired through a router that followed a room numbering scheme that did not correspond to the flow of learners through an exam, making it nearly impossible to use the router for recording individual student performances on one tape or DVD. Subsequent investments in a digital A/V and exam management system have dramatically improved the Center's ability to meet educational needs for collection, storage, and review of learner performances, as well as research on learning outcomes and curriculum development. The key limitations of using legacy technologies are listed in Table 16.3.

## 16.6.2 Program Definition

While the facilities planning office uses "charrettes" as part of its planning processes, this project lacked an ongoing discussion of vision and mission with planning group members.[3] As with the design team, planners relied heavily on the Medical School for direction in designing the Center around an educational program to the exclusion of other health science professions. As of 2002, the primary simulation modality used by the Medical School was live simulation using standardized patients. As noted above, the largest of these was an 18-station

OSCE that, up to this point, had been staged in primary care clinics on the weekends or in rented conference rooms at a nearby hotel. Other design team members were most interested in using the Center as a space for teaching physical examination and communication skills, and were initially less interested in its resources (the facility and the staff) for assessment of learner skills. Since these design team members had no comparable experience with live simulations in their respective fields, and the Medical School had no plans to change the design of their exam, accommodating this OSCE became a necessity.

## 16.6.3 Space Allocation and Utilization

Because of the Medical School's prior experience using borrowed spaces to stage simulations, they – and the rest of the design team – tended to approach the project as a classroom rather than as a stage. Little thought was given to learner flow through the space under different conditions (e.g., duplicate clusters of scenarios/stations staged simultaneously; multiple events at the same time; etc.). In fact, when the A/V system was installed, room numbers and cameras were assigned on the basis of the configuration of the rooms on a map, not on

**TABLE 16.3** Limitations of legacy technologies

| Legacy technology | Limitation | Solution |
|---|---|---|
| Recording: VHS tape | • Inability to find replacement parts or units for broken hardware<br>• Inefficient and insecure storage of video exam data<br>• Inability to efficiently scan learner performances for feedback and/or remediation<br>• Inability to integrate performance data from checklists with video<br>• Inability to repurpose student video in other educational tools (e.g., e-portfolios)<br>• Inconsistency with flow of learners during an objective-structured clinical exam or a clinical skills teaching experience | • Convert to CD/DVD as portable recording medium<br>• Convert data management to a digital system<br><br>• Convert to digital technology, allowing for greater flexibility in tracking and storing individual learner performances |
| Exam management: Analog | • Inconsistency of raw data<br>• Insecurity of data<br>• Inability to assess interrater reliability in real time<br>• Inflexibility of long-term data management for curriculum development and research | • Real-time overall exam results<br>• Data aggregation according to performance domains and student competencies<br>• Customized remediation based on specific performance criteria<br>• Curriculum development and instructional strategies based on student outcomes<br>• Real-time assessment of interrater reliability |

how students might move through the space (note the distance between rooms 4 & 5, 8 & 9, and 15 & 16 in Figure 16.1). In retrospect, all members of the planning team should have taken a more active role in interrogating the assumptions behind these design decisions, challenging themselves and their fellow members to think about how learners in their programs would actually move through the space on a daily basis (e.g., Where will they enter? Where will they wait? Where will they exit?), as well as how learners would move through the space during examinations (e.g., How will they receive instructions? Will they need charting stations in the hall? Where will they go when they have completed the exam?).

Similarly, only two offices – at opposite ends of the Center – were included in the plans. Instead of considering the impact that having a dedicated space for simulation might make on the curriculum – seeing the future as "thought thinking itself out" – planners primarily saw the space as an empty box to be filled with walls and doors. One exception to this was the addition of movable partitions that made it possible to pair 12 of the 18 rooms for scenarios requiring more equipment or small group teaching. As noted above, this proved to be an essential (if expensive) addition to Center's resources, but it came with more than substantial financial costs. In spite of additional insulation and the creation of customized baffles between the rooms, sound penetration – especially in exam situations – remains a problem.

## 16.7 Conclusion

Another way of distinguishing between projects that employ anticipatory design and those that do not may be demonstrated by the linearity of the design and implementation process without an anticipatory component. Anticipatory design allows for conceptualization with repeated check-ins not only on issues like materials and equipment to be used, but more importantly, on the vision and mission for the project.

The repeated reassessment of mission and vision are essential to the successful completion of the project. This is not to suggest a project development process devoid of goals and expectations – without goals and timelines, stakeholders could easily become frustrated and disenfranchised. Excessive reevaluation of mission and vision could easily result in a paralyzing cycle of second- guessing. Instead, stakeholders should be driven by some fundamental questions that serve as touchstones for "thought thinking itself out:"

- What's important to you about teaching? What's important to you about assessment?
- What can you do educationally with simulations that you can't without? Why is this important to you?
- What do you want your learners to understand about being a health professional by the time they complete their programs?

- If you currently use simulations, what are the features (educationally) you like about the simulators you use (e.g., data capture)? What are the deficiencies (educationally) you would like to see changed?

These lessons learned are currently being employed in the renovation of an additional simulation center space at the University of Minnesota.

# Endnotes

1. "Anticipatory design" as it is used here draws on complexity theory as a means of attempting to anticipate future flow and programing demands. For an excellent example of this perspective, see SL Brown, KM Eisenhardt [3].
2. There is emerging literature on the design and planning process. Two excellent examples are GE Loyd [4] and MA Seropian [5].
3. The term "charrette" is often used to refer to a process of intense consultation and debate as part of an architectural design process. Its contemporary usage comes from the École des Beaux Arts in Paris during the 19th century, where proctors circulated a cart, or "charrette," to collect final drawings while students struggled to put finishing touches on their work. See Lennertz [6]. For more information, see also www.charretteinstitute.org.

# References

1. Whorf, B. L. An American Indian Model of the Universe. In: John B. C. (ed.), *Language, Thought, and Reality, Selected Writings of Benjamin Lee Whorf.* The MIT Press, Cambridge, MA, 1936/1956.
2. Brandt, B. and Ling, L. *Transforming the University: Final Report of the AHC Task Force on Health Professional Workforce.* University of Minnesota, Minneapolis, Minnesota, 2006.    www1.umn.edu/systemwide/strategic_positioning/ tf_final_reports_060512/ahc_hpw_final.pdf
3. Brown, S. L. and Eisenhardt, K. M. The art of continuous change: linking complexity theory and time-paced evolution in relentlessly shifting organizations, *Administrative Science Quarterly* **42**, 1–34 (1997).
4. Loyd, G. E. Issues in starting a simulation center. In: William F. D. (ed.), *Simulators in Critical Care and Beyond.* Society for Critical Care Medicine, Des Plaines, IL, 2004.
5. Seropian, M. A. General concepts in full scale simulation: Getting started. *Anesth. Analg.* **97**(6), 1695–1705 (2003).
6. Lennertz, B. The Charrette as an Agent for Change. In: *New Urbanism: Comprehensive Report and Best Practices Guide.* New Urban Publications, Ithaca, 2003.

# 17

# Simulation Facility Design 101: The Basics

Michael Seropian

## 17.1 The Virtual Hospital – A Virtual Fantasy?

I once was involved in a role-play. It went something like this. The hospital CEO walks in to announce that a major donor wants the hospital to build a simulation center. The monies are for start-up and construction but cannot be used for ongoing costs. The room is filled with excitement. Surgery has wanted to enter into this domain and sees the opportunity to have the center accredited by the American College of Surgeons. Nursing has long seen simulation as a mainstay for quality, workforce issues, and cost savings. Anesthesiology, Emergency Medicine, Obstetrics, and the Intensive care representatives are all equally enthusiastic to develop simulation programs for team, crisis management, and skills training. The CEO announces that the donor wants the hospital to build a simulation center that is a virtual hospital. It should look and feel like a hospital. Eyes around the table are big and clearly eager and pleased.

This scenario sounds great. Many consider it the utopian solution. Who wouldn't want a simulation facility that looked exactly like a hospital? The reality is that the notion of a virtual

hospital may actually be candy for the mind. It is sweet and desirable. It is however a solution that is likely reserved for those with not only unlimited start-up funds but deep pockets for ongoing costs. The reality is that hospitals are built the way they are to deliver patient care as their primary focus. A hospital environment is filled with distraction and people with cross-purposes. Although dealing with this may be a course in itself, it is just a small part of the totality of clinical education. Education is a secondary or even tertiary focus in a hospital. On the other hand, a simulation center's primary focus must be education. The educational principles and environment must be optimized and must be pedagogically sound. Indeed, classrooms are built the way they are for a reason. The same consideration applies to the construction of a simulation center or facility. Although this point may seem subtle, it has profound implications. Mixing of students and participants undergoing distinct and different simulation experiences can be inherently distracting and artificial. They are in the same relative space for artificial reasons. This is potentially disruptive to the education process. As with most educational activities, the concept of contextual isolation should be a top priority. The notion that people need to learn in an environment that is filled with distractions is fallacious.

## 17.2 Design and Build for the Actual Use

Unless one is able to design a virtual hospital that achieves true separation of activity, the risk of disruption of the immersive experience inherent to simulation is possible if not likely. As people exit a high-fidelity mannequin-based simulation, they are often still "in the moment" as they move to debriefing. Being distracted by passing colleagues having just completed a virtual reality training session is not pedagogically desirable. So what is the answer? There is not one answer. It is important for people developing centers to carefully assess what is desired versus what is needed. They may not be the same thing. The seduction of reproducing the environment to the exacting detail is potent but likely a waste of money. As one considers what and how to build a center, it is important to consider how the activity (especially if it is immersive) allows a certain amount of leeway not possible in a hospital. Once a participant is immersed, do they really look at the wall details? Do the wall details immerse them or does the interactivity of the whole environment achieve that? Some centers have made good attempts at creating hybrid versions of virtual hospitals and educational environments. These centers are worth looking at, not to replicate, but to be able to learn in which direction should the hybridization favor.

Should space be designed so specifically that its utility becomes limited to a specific activity or discipline alone? If a space is created to closely resemble an operating room, then it

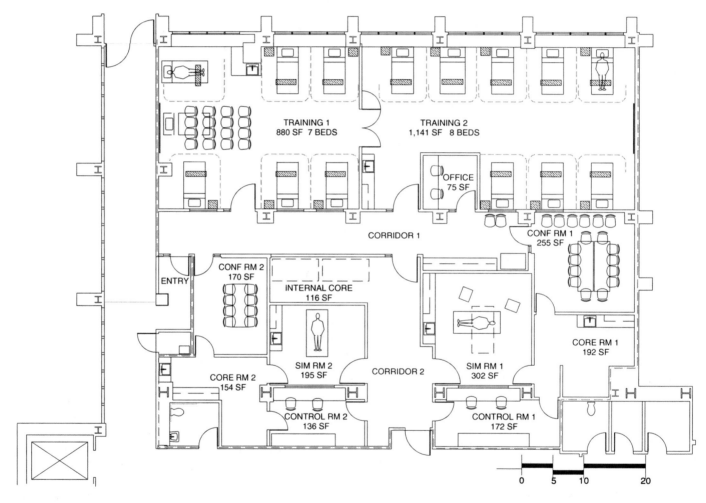

**FIGURE 17.1** Floor plan for a 5000-square-foot simulation center. The apartment concept (also know as a suite) is composed of four distinct functional subsections: simulation, control, core, and conference. There are two such apartments shown, numbered 1 and 2, located in each lower quadrant. Look at # 1. Notice how the conference room is in the far corner, away from noise sources, like conversations in the control room. Trace the typical footpath of the students and their instructors. Locate # 2. Notice how the two control rooms are just a quiet shout away from each other. Note: Permission to use Figure 17.1 has been granted by Dahanukar Brandes Architects, Samuel Merritt College, and SimHealth Consultants LLC. The schematic was developed by Dahanukar Brandes Architects and SimHealth Consultants LLC (facility design consultants) for Samuel Merritt College in Oakland, California. All rights and copyright remain the property of Dahanukar Brandes Architects and SimHealth Consultants LLC.

may become difficult to use that space for anything else. Utilization of the space is therefore limited to the demand created by select specialties. The room lacks flexibility to be appealing to other disciplines and specialties. The common response to such a comment is "we can just put partitions up to hide the elements that label the room." The counter response to this is "why create the detail in the first place if you will ultimately want to hide it?" It is important to recognize that the amount of detail relates to the role it plays in the suspension of disbelief.

This chapter is intended to reveal issues rather that explain what your simulation center should look like. However, one floor plan that does meet most of one program's needs is offered as an example (Figure 17.1). The design of each center is unique but certain guiding principles around function, structure, and need are common in most instances. The topics that will be covered include:

- The center design team
- Function, flow, and utilization
- Simulation type
- Sound
- Gases
- Lighting
- Electrical
- Ventilation

A flow chart provides a graphical representation of the iterative process, that is, creation of a simulation program and its supporting facilities (Figure 17.2).

## 17.3 The Center Design Team

The team that designs a facility should include an architect, the customer, the project manager, the contractor, AV and IT professionals, a simulation facility design consultant, and select faculty who will use the facility. The criterion of what makes a simulation design consultant skilled needs to be further matured and standardized. Some qualities include:

- Prior simulation experience
- Prior simulation facility construction experience
- Good spatial skills and qualities
- Ability to convert ideas into images
- Ability to convert customer vision into functional images
- Ability to preempt and consider consequences and limitations of design decisions
- Ability to understand what talent set needs to be involved and at what point in the process
- Ability to assess and understand construction and architectural drawings
- Ability to understand and guide audiovisual and software professionals in designing audiovisual and storage solutions that are consistent with the intended vision and application.

The team should be assembled or at least considered early on. The criteria for contracting with contractors or architects should not rely on their experience to build medical facilities but their ability to create a studio that will mimic one. The different components of the team will have relative importance at different times. The constant team members, however, are the architect, the customer, and the design consultant. Together, this core team will help the customer realize the vision they set out with.

## 17.4 Simulation Center Design Considerations

The notion that simulation facility design differs from a hospital or classroom is just coming of age. Indeed, we see these concepts appearing on listservs. As alluded to earlier, the design of a simulation facility is not the recreation of a hospital but a studio designed for education that should provide adequate fidelity to immerse individuals to believe they are delivering patient care. In the case of skill trainers, the area should similarly embrace principles that provide an efficient and conducive learning environment. Center design should consider:

- Type of simulation
- Flow
- Budget
- Future plans
- Type, level, and discipline of participants
- Ongoing funding
- Size of available space
- Core values of the institution
- The vision of those proposing the facility

Missing just one of these elements could result in cost to the customer and limitations that were not previously foreseen. There are several examples of this throughout the world.

I often start the design of a facility with a blank sheet of paper. The outline of the space is drafted onto the paper and the interior only includes structures that are immovable or structural in nature (e.g., electrical panels and support columns, respectively). In creating this skeleton outline, the absence of any existing walls (in the case of a remodel) does not bias the creative process. Prior to any design work, detailed program information from the customer needs to be gathered. It is not uncommon that they may not have the program information straight away but this is where the facility design consultant can facilitate by asking contextually relevant questions. The facility is defined by the programs that reside within it. The programs are in turn defined by the curricula and objectives. Having this basic information and knowing what the "must-have's" are allows the designer to create a preliminary sketch. The sketch is not arbitrary and calls on hospital design only at specific junctures.

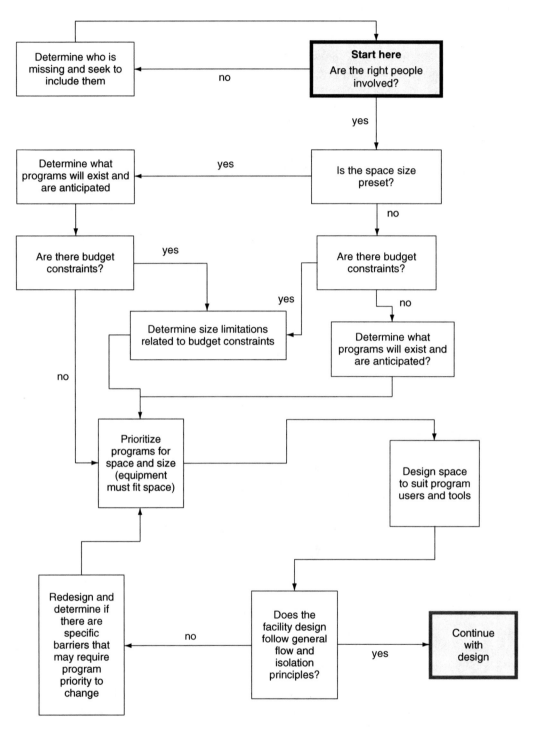

**FIGURE 17.2**   Flow chart: getting started on simulation facility design.

# 17.5 Function and Utilization

The *type of simulation* to be done has a heavy influence on the design of a facility. The concepts for a VR training room versus a computer-based area versus a high-fidelity mannequin-based area are all different and unique. Having determined what programs will exist in a defined space, the design team can move to flow and volume.

- How many people will use each area in any given period of time?
- What concurrent sessions are likely to happen?
- What are the constraints of the users that may weight the use of the facility to a particular time of the year?

Knowing the *intended use* of the facility and the *relative utilization* requirements will help define the space requirements. In some instances, the space is predefined and will therefore be the constraining factor that will influence type and utilization (see flow chart). In either case, it is optimal to aim for a facility that is *functional* and achieves high utilization rates (70–75%) – while keeping future growth in mind.

- *Utilization* – relates to the percentage use of the given space over a defined period of time. A rate of 75% is generally considered capacity. Beyond this, the ability to maintain and flex the space becomes constrained. Similarly, scheduling becomes cumbersome and complex. The utilization rate is not only determined by the number of participants but also by the type of simulation. Scenario-based mannequin simulation may have a lower throughput of people per unit space per unit of time. The exercise of determining how many participants will likely be coming through the center per year (and for what activity), after 3–4 years of operations, is very valuable. In doing this, one can determine how many same use areas are needed and how much overall space is required (if space is not a constraining issue). If a space is to be used with a ratio of one participant per hour, then that would be dramatically different than a space that is used by 3–4 participants per hour. This is important for the design team and will also be beneficial for the customer in justifying space allocation to their respective executive.
- *Functionality* – relates to the ability of the space to comfortably achieve the desired function and to allow for good flow dynamics within the specified space. This includes the ability to give easy access and egress, easy access to washroom facilities, easy access to storage, and all necessary spaces for the defined activity. It also speaks to the ability of one space to not interfere with another. It is important to consider how people enter and leave a defined space. Do they need to traverse another teaching area? In considering functionality, the design team

should consider if the functional integrity of the facility remains intact when in full use. I have often found it useful to draw footpaths on paper to represent groups of participants as they move in and out of the center and how they move within a center.

- *Type of Simulation* – It either determines the use of the space or is defined by the space (virtual hospital example). The type of simulation will define the requirements of the space (large room versus a set of rooms). This is discussed in considerable detail in the next section (Space by Design), as it is likely the most important factor. It can also be the most constraining.

# 17.6 Space by Design

## 17.6.1 Virtual Reality Area

This skill setting may or may not require a single room for a single trainer or may, on the other hand, include many trainers in a single room. In this case, establishing the type of trainer and how it is to be used is of importance. For the purpose of this chapter, we will not touch on virtual reality using "cave" technology. Placing trainers in different environments may affect the objectives of the session. As an example, interventional radiology trainers may be used in a realistic patient environment or in a generic classroom. In the case of the former, the trainee is not only exposed to the immersion of the technology but also to that of the environment as well. The environment is contextually closer to reality than a simple classroom. The objectives of a course can therefore be expanded from just procedural skill training to include cognitive skill training. The design of a more generic setting can range from a large room to a series of rooms with groupings of technology. The program requirements will dictate this. Virtual reality trainers may also have a significant footprint and utility requirements, therefore, will require more than just a cubicle.

## 17.6.2 Skills Training Area

This type of area embraces many of the same issues as with VR training. There is considerably more history with skills centers – especially within nursing. The old metaphor includes beds within a room often aligned along the wall with a central desk area in the middle. This arrangement has it benefits but requires a substantive footprint given the size of the beds and the space required around them. The important question to ask is for what purpose will these beds be used? Do the trainees really need all the beds for their activities? Do the training exercises always require a full-size mannequin or person in the bed? If not, then what percent of the time does it not? If the percent of time that actually truly requires a bed is small, then it is important for the customer to understand the trade-off – loss of functional space. Beds in particular are problematic as they cannot be easily moved and stored (especially with today's more specialized beds). Skill training areas are well suited for large "flexible rooms." In this environment, the instructor can address the entire room but also has the flexibility of setup not being hindered by beds. There are many activities that require tables rather than beds. This is especially true of partial-task trainers. Certain skill areas will require specialized smaller rooms (e.g., patient room). These rooms are useful for physical assessment that would inappropriate in a large room. These rooms can also be utilized for standardized patient encounters.

### 17.6.3 Computer-based Learning Area

Learning that leverages this type of technology has its unique characteristics. It is typically done in a larger area divided into cubicles or self-study areas. The question to be posed is whether a dedicated space for this type of training is required and if one already exists within the institution. The temptation is to make all that is simulation occur in the same facility. The demands of a computer-based environment is much more reliant on self-study that most other forms of simulation. In the case of computer-based training, there may already be sophisticated and well-developed space for this purpose. Careful consideration should therefore be given as to whether this space is required or if this activity can be done at another location. Recall that the simulation facility is premium space for very specialized activities.

### 17.6.4 Full-sized High-fidelity Mannequin-based Area

There have been many variations of design of this type of learning environment. There is no one good answer. The type of simulation, flow, and the participant type (and quantity) are important in the design of this setting.

The high-fidelity area is a special circumstance where more than one room is used for a single activity. The rooms include a conference room, control room, supply area, and simulation room. The challenge is in placing these rooms and areas in proximity to each other, so that flow within the space as a whole is self-contained and undisturbed by other learning activities in the center (revisit Figure 17.1).

This area is much like an apartment. In this case, the apartment contains a control room, a simulation theater, a debrief area, and a core area (where meds, patient supplies, etc., are housed). The apartment should be accessible in multiple ways and for different purposes. In one case, the purpose may be related to the participant and the other specific for the operator (to be able get equipment and deal with things behind the scenes without being seen or heard). If possible, the apartment should be designed in such a way that its rooms could be used together or for multiple activities. In this way, if a full-scale simulation session is not occurring, the debriefing room can be used as a conference room, and the simulation theater could be used for bedside training with the group in the room. This type of arrangement provides maximal flexibility while maintaining contextual isolation for the participant. A center can truly exist with multiple activities and little crossover of participants and faculty. A center that is built with a central corridor presents a specific dilemma when the space on either requires movement through the common corridor for activities excluding entering and leaving the facility. Careful arrangement/placement of rooms is paramount. A design that has its simulation rooms at one end and debriefing rooms at the other opens the center to considerable disruption of different courses as people move in and out of the different learning environments. Learning areas should have little or no overlap and the use of the corridor should be restricted for entrance, egress, bathroom needs, and other non-class-related issues.

## 17.7 Utilities

### 17.7.1 Gases and Suction

As with a hospital, a simulation center will need to provide compressed gases for two purposes: (i) to deliver gases to virtual patients and (ii) to drive equipment. The plumbing for the gases does not need to be hospital grade but must meet the code for the facility. The ventilation must meet specific code requirements before certain gases such as pure oxygen and volatile anesthetics are used. There are several centers that do not use oxygen at all. This is usually satisfactory but may become problematic if the equipment that is in use requires the presence of pure oxygen to function properly (e.g., certain newer ventilators). The cost of decommissioning a ventilator by bypassing its built-in safety features may far outstrip the cost of effective ventilation. Similarly, the cost of ventilation to evacuate anesthetic gases may far outstrip the educational value of having real gases present. Ventilation must also consider heat, which is determined by the number of people in given space and all the heat generating equipment (e.g., an Audio/Video rack, or numerous video displays).

A functioning vacuum/suction system should also exist as participants expect it. This is an example of a detail that adds to a room's fidelity but does not restrict its functionality. The system may be small and self-contained, as are found in dental offices, or may tie into an existing system.

### 17.7.2 Sound

The design and structure of the center should involve the input of a sound specialist. It is important to consider the isolation of sound from one area to another. This includes considerations such as wall treatments, special ceiling materials, wall construction, etc. The ventilation system must also be considered as it can provide considerable background noise. The level of noise is determined by the size of the ducting and the velocity of the air flow. The reverberation and transmission of sound through materials and through hung ceilings can be considerable. These issues degrade sound quality and pose problems for the setup of the Audio/Video system for broadcasting and recording. The size and shape of a room will also change the sound dynamics more than one would expect. Building a room that is a perfect square will produce dead spots (nodes) where sound cancels itself out as it reflects off the walls. This may all sound like minutiae but considering this early will make a large difference in the quality and ease of setup of your AV equipment.

### 17.7.3  Lighting

Lighting affects the experience of the individual, the activity, and the quality of recording if any is to be done. The "temperature" of the lighting should be consistent with the lighting requirements of the video equipment in use. The placement of the lights should also be carefully thought out, lest they reside exactly where the ceiling-mounted LCD projector needs to go. Lighting must also accommodate for function. This is where one considers issues such as OR lights. These lights are costly and require expensive structural modifications to mount them. The design team should carefully evaluate whether they fall into the "candy for the mind" category or if they are truly educational. Beyond expense, they will label your room. This has been discussed earlier but is worthwhile reiterating.

### 17.7.4  Electrical and Information Technologies

Careful attention should be given to the electrical requirements (amount, type, and location) of a room within code limitations. This spans AC outlets to light switches and types of switches (dimmers versus a simple conventional binary switch). Also, consideration must be given to the need to electrically isolate a specific area should a scenario involve an electrical failure. The placement of electrical outlets should be frequent and at multiple heights along the walls. Power outlets for mounted projectors will need to be installed at the appropriate height or hung from the ceiling.

Information Technology (IT) considerations are important and complex. A center may have its own dedicated intranet or may exist on an institutions network. Both have implications. New technologies exist that use IT infrastructure to not only deliver data but to stream audio and video. This may be of some importance for certain centers that wish to transmit to remote locations. An IT professional will be able to quickly determine where junction boxes are required and the amount of conduit and preinstalled dark fiber that would suit the application and its intended growth. In this way, the AV and IT professionals should work closely as much of their cabling may follow common pathways, if not also share common cabling and network equipment.

### 17.7.5  Security

It would be great if this was not a consideration, but in most centers, theft is not unheard of. Not only is this pertinent for equipment, but also for materials that may be sensitive and confidential. It is therefore important to involve your institutional security to consult on what security measures would best suit the environment. This can be as simple as to what side of the door the key lock is on or the use of sophisticated card locks and tracking systems. For confidential information, one should consider redundant security measures as well as methods of encryption in the case of electronic media. Networked video systems originally designed for capturing the behaviors of students during sessions can be used to capture those of intruders after sessions.

### 17.7.6  Storage

This topic does not need much explanation, except to say it is almost always underestimated. All those beds and equipment have to go somewhere lest they line the halls of your new and beautiful center. As the design of a center evolves, storage space is usually the first to be decreased if more space is needed. Storage space is abstract and is not considered to be functional for the purpose of education. The exercise of creating a rough catalog of equipment and supplies and then placing them in the drawn storage areas can quickly make the point of how much storage is really needed.

We are all familiar with the use of storage areas in our own homes such as closets, basements, attics, and garages. The term *storage* connotes hidden and dormant. We are also familiar with the use of utility areas such as laundry, cleaning, and workshops. The term *utility* connotes accessible and active. Perhaps, relabeling all storage space as utility will help focus its essential value added to a simulation facility.

## 17.8  A Walk-through

It is worthwhile taking a deep breath here and to take a stroll through a design example (Figure 17.1). Imagine that four students have come to the center. They are all here for four different courses:

- Pat needs to go to simulation area 1 at 09:00 for a course
- Jeff needs to go to training room 1 at 09:00
- Alex is scheduled for a course at 10:00
- Francis is schedule for a course at 10:15 in sim area 2

Trace the steps of each these individual starting from point X on the schematic. Notice that the ONLY common area is the central hall and core corridor (which is faculty only). As each of these individuals enters separately or together, they do not disturb classes/courses in progress. Similarly, as individuals leave or need to go to the restroom, they cause minimal disruption within the center. All their activities are quarantined to their respective areas. In the event that you want a larger area, Training Rooms 1 and 2 can be opened up and Simulation Rooms 1 and 2 can also have some communication through the core corridor. The center design provides for easy access and egress and meets fire code for exits. Faculty and educators can meet with individuals in the office area if needed. The core corridor acts as a "faculty only" area where preparation for the day can occur and where supplies can be readily available given the natural variation found in full-size high-fidelity mannequin-based simulation. Training Rooms 1 and 2 can have beds or not. This has implications as to its function: a ward-like setting for training multiple participants skills at the bedside or an open area with multiple stations to use task trainers, VR trainers, etc. The key to this design is versatility and function, while maintaining the sanctity of educational isolation when needed.

## 17.9 Conclusion

The design of a simulation facility is deliberate and precise. It requires imagination and forethought. It is not a hospital or clinic; it is space dedicated to immersive interactive education. Attention to the selection of a design and project team is important. Who you include may be less important than who you do not include. The facility design should suit the program needs, which in turn have an obligation to curricular objectives. Flow, utilization, and functionality remain at the forefront of a good design specialist. There is not one single design but a set of guiding principles that have at least partially been covered in this chapter. If nothing else, this chapter should serve to spur the imagination *and* illustrate the complexity of thought that goes into effective design. In ending, summarizing this chapter, I find myself thinking of the virtual hospital again. The idea is truly exciting and hard to resist. I will however remain true to education and fight the desire – for the time being.

# 18

# Creation of Structure-Function Relationships in the Design of a Simulation Center

Brian C. Brost
Kay M.B. Thiemann
Thomas E. Belda
William F. Dunn

## 18.1 More Than the Position of the Walls, Doors, and Windows

This chapter addresses the structural-functional relationships of the various simulation spaces and offers suggestions on how to plan both their physical designs and their uses to create an integrated whole where these diverse relationships complement and reinforce each other. You can design spaces for standardized patient encounters, task-trainer skills development, or full-body robotic patient simulator scenarios. Typically, for in-hospital care teachings, the latter are staged in rooms that look like the real environments that the simulated events are expected to occur in: the labor and delivery suite, operating rooms, emergency rooms, intensive care units, and patient recovery wards. The very presence of any heterogeneous structure inhabited by diverse users both generates and constrains the latter's relationships with, and utility from, the structure. Mindful use of structure-function relationships allows your simulation center to become a living entity that provides the optimal learning environment while being adaptable to a variety of learner-educator experiences and needs.

Structure-function relationships are not intuitively obvious. It is more than the position of the walls, doors, and windows in your simulation space. You need to establish meaningful linkages in your simulation center between the basic scientific principles and clinical practice while being cognizant of what are the educationally important questions. For example, structure-function relationships in the body incorporate the interactions from the cellular level to the movement of the body to the ultimate interaction of the individual with the world around them. In the educational realm, we have to be aware of not only the simulation-based learning principles, but also how the interactions of the learning environment affect the trainee, the actors, etc., during their educational endeavors.

We will share with you some of the thoughts of the designers of our simulation center and wise counsel we received from experienced operators of established simulation centers. These are some of the structure-function relationships of the Mayo Clinic Multidisciplinary Simulation Center, Mayo Clinic, Rochester Minnesota, U.S.A, that allow it to attain its desired function as a highly effective learning environment.

## 18.2 Preliminary Considerations to Designing a Simulation Space

An integral starting point in designing your simulation space lies in two principle thoughts. Determining who your primary customer is and establishing your simulation center's mission and vision. The answer to the question, "Who is your primary customer" is generally initially thought to be the learner. But on an in-depth scrutiny, the primary customer of any simulation center is usually the educator. Our job as a simulation staff is to enhance the educational environment and opportunities available to the educators who have as their primary customer, the learner. Our **mission statement** is *"The Mayo Multidisciplinary Simulation Center will transform clinical education by assisting Mayo Educators in developing, implementing, and evaluating experiential curricula for learners that advance patient care"* and **vision statement** is *"Transformative Learning Toward Demonstrated Excellence."* Your first step is to define your mission and vision statements to reflect the goals and objectives of your center.

The two most crucial resources required for a simulation facility are the bane of all educators' existence: space and funding. We cannot tell you how to acquire these, but we describe factors involved in utilizing these to maximize the results within your educational environment. First, it is essential to consider the learner types and their throughput expected at each session and throughout the year. This defines the overall amount of space and types of rooms required for your center. Second, determine the teaching technologies that will be used in your simulation center (mannequin, task trainer, standardized patients, animal labs, cadaver labs, etc.). This defines the spatial relationship between these various rooms. Incorporating all this information, you can now begin to focus on the interactions between the activities of these rooms.

Most academic simulation centers have been charged by their institutions to provide a simulation environment that is adaptable to a variety of disciplines and learner levels (allied health, nursing, medical students, residents, fellows, staff physicians, etc.). Your simulation center should reflect the particular mission and focus of your department or institution. Another facet that comes into the picture is the degree of realism of the room. How important is it to mimic the actual inpatient and outpatient rooms at your facility? The expectation of realism differs if the room has an educational, team training, safety/quality review, or research purpose. Lastly, but not least, where will the simulation center staff offices reside and that pesky problem of all institutions – storage. These are but a few of the considerations for defining the number of rooms, their relative locations, and inner designs. Using our center as an example, let us address some structure-function relationships to make the simulation center more than a just collection of rooms in the same part of the building (Figure 18.1).

## 18.3 Developing a Conducive and Inviting Educational Environment

The unknown or unexpected is uncomfortable. This is why education is so challenging for both student and teacher alike. Even though our center is covered with signs saying, "We believe that all participants at the Mayo Multidisciplinary Simulation Center are intelligent, well educated, and want to improve so that they can provide high-quality and safe patient care," the process of experiential learning is threatening. Learners coming to your simulation center are always anxious, independent of their student experience level. Learn to work *with* this. Teachers new to simulation are anxious, independent of their clinical experience level. Learn to work *with them* to reduce this. Novices know they do not know, while experienced practitioners know they do not want to be embarrassed by doing or saying the wrong thing. Employ methods to decrease everyone's anxiety that distracts from the explicit teaching and learning objectives the moment they enter through your door.

The most anxious individuals coming to your simulation center are those who have never been there before. Make sure that your greeting/reception area is obvious with someone always available at this desk to greet all center visitors and learners. This receptionist should be situated right in front of the entrance door to your center. This person can also tell them the agenda for the day, where the restrooms/changing rooms are located, and direct them to the waiting area or appropriate simulation room. Note that if you install a receptionist desk, then you are obligated to staff full-time receptionists. An empty receptionist desk is even more disconcerting than the absence of a receptionist desk.

Not knowing where to go or what to do only heightens anxiety. Remember the Disney principle of including helpful signage throughout your center that provides intelligible directions and let them know they have arrived at the right room.

The initial entry spaces should facilitate the change to a learning environment. If you are planning on having larger groups, a centrally located classroom-sized space provides persistent orientation of all the learners. This space can also be used for preparatory briefings, short didactic lectures prior to interactive simulations, use of task trainers or use of models to demonstrate key principles (lung compliance, cardiac rhythm generator, or other systems), or allow direct interaction with scenario participants utilizing AV equipment described later in this chapter.

Often, for simulation scenarios, participants are asked to change into hospital scrubs, wear masks, hats, and gloves. Dressing into clinical "uniforms" is the fastest way to facilitate change from the passive observer into the fully engaged. Be sure to plan for the changing areas that are adequately sized to your largest projected group of learners. Some groups of participants are mostly of the same gender (nurses or allied health staffs are overrepresented by females, surgeons overrepresented by males) and this may increase their changing and toilet room

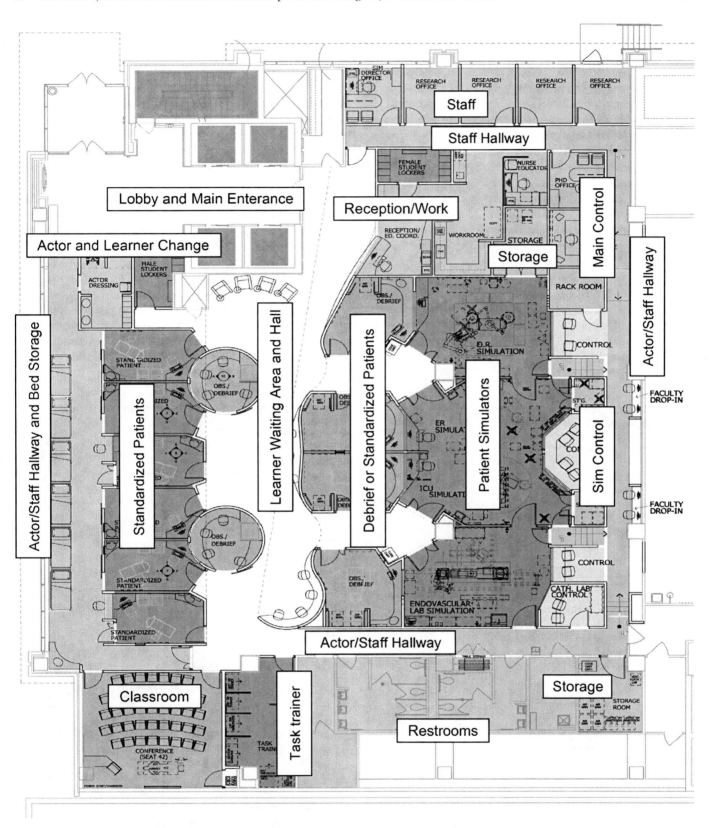

**FIGURE 18.1**  Center floor plan showing public entrance in upper left, public reception area in upper center, private office areas in upper right, public core access down the middle, standardized patient environments on the left, robotic patient simulator environments on the right, class rooms and rest rooms along the bottom, and private actor/staff access to "back stage" around the perimeter.

size. Often, learners are unfamiliar with each other or from differing institutions, and they may be uncomfortable changing in front of each other. All participants would appreciate clothes lockers that can be secured, as it is an unnecessary distraction to look after your purse or wallet during simulation scenarios.

People are curious. Place changing areas adjacent to the reception and waiting areas so that learners aren't traveling throughout the simulation center and possibly disrupting ongoing scenarios or try to sneak a peak, so they can prepare for their session. The pre-event area should be comfortable and give learners several possible attractions so that they do not have to sit and worry about what is going to happen or stare morosely at each other.

Learning space is at a premium. Make you spaces readily adaptable to multiple environments. While it would be nice to have 10–12 standardized patient rooms, most simulation facilities do not have enough throughput to necessitate this much space dedicated to a single-type learning environment. We have placed debrief areas (four areas) for the major simulation space in proximity to the standardized patient rooms (six dedicated), which gives us a total of 10 rooms for standardized patients when needed.

Rooms can look totally different by simply changing key furniture (Figure 18.2). These two images show the adaptability of space to serve different needs. The standardized patient rooms are readily interchangeable from inpatient spaces (complete with working compressed gas lines) to outpatient rooms (equipped with routine otoscopes and ophthalmologic equipment).

The simulation rooms should look like your inpatient or outpatient facilities. Props and medical devices within the room should be exactly the same as is currently in use in your medical facility. (Pennsylvania State University automatically includes extra equipment on the clinical hospital budget for in-service training in their Simulation Lab.) Nothing increases anxiety like knowing what you expect from a device, but you don't know where the switch is, or which way to turn the knob. You detract from the teaching points or educational objectives if the learner is focused on the device, and not on the principles or techniques you are trying to teach. It would be important to include a simulation room familiarization module for learners from outside you institution. When the participants in the scenario are finished, the movement to the debriefing rooms should be logical to determine and easy to perform.

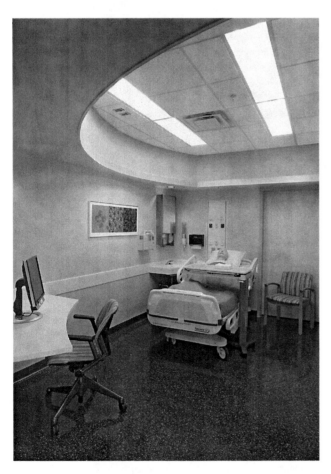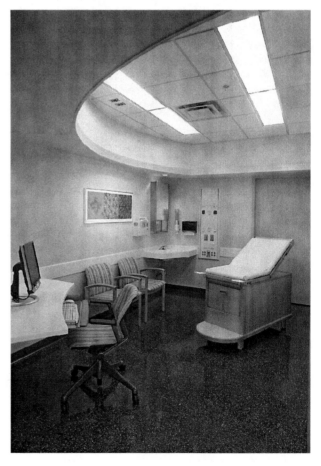

**FIGURE 18.2**   Different clinical teaching needs, different clinical environments, same space.

Simulation center staff should have their offices readily available to the simulation area. This is important in the event that the audio/video equipment, task trainer, high-end patient simulator, etc., malfunction, your staff should be close by and be able to step in and fix the problem rapidly. Waiting for the scenario to start or continue can be frustrating to the learner and educator alike. Time is the most precious commodity and should not be wasted. Easy access by the simulation center staff to their work environment increases their satisfaction and effectiveness and decrease the daily stress and risk of burnout of these key and essential personnel.

## 18.4 Separation of Learners and the Actors/Evaluators

Understanding and controlling people flow within your simulation space is vitally important to the success of your learners and the sanity of your staff (Figure 18.3). Movement control starts from the moment people enter the front door. Participants should go from one space to the next without crossing back through areas with learners who have not gone through the experience yet, as shown by the arrows in the diagram. Thus, similar-purposed simulation rooms should be grouped together. The task trainers should all be in the same area to facilitate both the initial distribution of learners and their subsequent transition from one device to another. Similarly, group the standardized patient rooms together and arrange the connecting hallways so that the learner can go from station to station without becoming distracted or even lost in your center.

If you are going to have some of your simulation room debriefing areas also function as standardized patient rooms, they need to be in the same area as the other standardized patient rooms. Some centers are planning two holding rooms – one for the group going into the standardized patient rooms and another separate room for those who have just come out of the standardized patient rooms. Another standardized patient need is to have quiet places where the participants can write a note about their encounter – might be a paper note, or in the future, more likely, a computer entry note.

Likewise, it is important that learners and the actors/standardized patients do not mix except during their intended encounters in the intended locations. These actors require their own changing areas/restrooms located separate from those for the learners and visitors. After preparing for their simulation, the actors should be able to enter and leave through the back of each room. Seeing a pregnant patient or actor with lacerations or bloody bandages walk through the waiting area decreases the ability of the learners to suspend disbelief and may give the learner a false impression of which direction you want the simulation to go. They may have a difficult or impossible time refocusing on your scenario in the allocated time. Before a live show, stage performers typically don't walk among the audience showing their costumes, spouting their lines, or playing their musical instruments.

Make sure that the hallways and room entrances are large enough to accommodate your maximal number of learners per session. Movement of large equipment can be accommodated through the use of a combination/split hinged door or a sliding door (Figures 18.4a, 18.4b).

Bottlenecks only create more unintended confusion and frustration for the participants. Make sure that the exit from your simulation area is not through the waiting area. Generally, simulation experiences create much excitement among the learners. They will talk about their experiences on the way out, which may spoil the event for those waiting to start their simulations.

## 18.5 Promoting an Educationally Intense Environment

The more the participants feel as if they are in their intended working environment, the quicker they become engaged by the scenario. Stock your rooms with the large volumes of the same clean and fully functional supplies that you use with real patient care. Do not use outdated or dysfunctional equipment. It only adds confusion and frustration during your scenario. However, nonworking equipment is fine, but only so far as those portions not functioning are not required for your teaching needs. For example, a *physically* intact but *functionally* broken X-ray C-Arm is the ideal simulator of an X-ray C-Arm – perfect for use where the actual production of X-rays is not only unnecessary but also undesired. Equipment, supplies, and devices should be in the same place as in their live patient locations so that the learner's mental energies are focused on the scenario and not on where to find the blood pressure cuff.

Elevation of the control rooms by 2 ft moves one-way mirrored glass out of the learner's direct line of sight in the room (Figures 18.5a and 18.5b).

This simple improvement (i.e., higher position of the one-way glass window) can reduce the feeling of being watched by "the man behind the curtain" or the "fish bowl effect." It also provides an improved line of site for the scenario controls augmenting the images received from viewing cameras positioned around the room (Figure 18.6). Since many of these proxy eyes are usually mounted just above standing head height, then why not arrange so that the real eyes of the people sitting in the control room are at about the same elevation. Elevation of the control room can be expensive to make it wheelchair accessible, but has proven to be well worth the additional cost. It provides an unparalleled view of the entire simulation room, especially if you have a movable wall that can be raised between your simulation rooms.

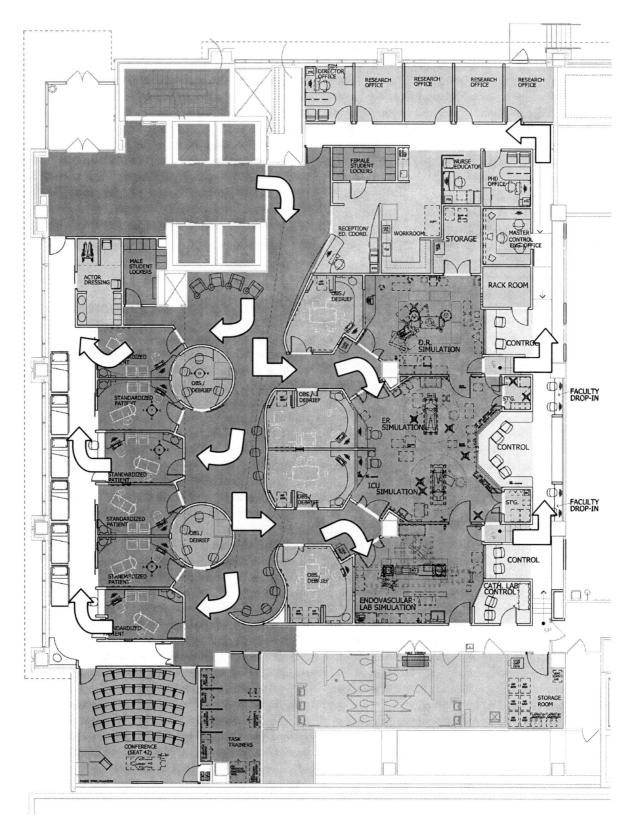

**FIGURE 18.3** Arrows indicate paths for unidirectional flow to reduce participants leaving their simulation experiences from coming in contact with those just entering.

(a)

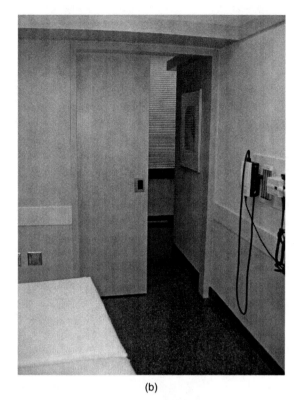

(b)

**FIGURES 18.4a and 4b**   Design doorways and portals to accommodate passage of large objects, like beds. The combination/split hinged door in Figure 18.4a requires a small swing space for people access but opens wide for large objects when needed. The sliding door in Figure 18.4b requires no swing space.

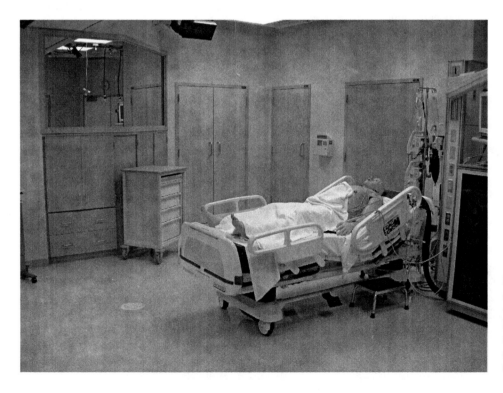

**FIGURE 18.5(a)**   An elevated control room positions the one-way windows above the typical lines of sight of the participants, further minimizing distractions caused by the necessary observational requirements.

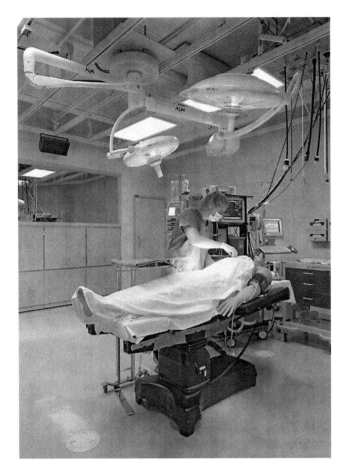

**FIGURE 18.5(b)** With an elevated control room, notice how the reflected image in the one-way window is above the head of the participant, reducing distracting reflections of others in the room.

The area allocated to your simulation center is often less than your imagination. Consider any location in and around your simulation center to be potential teaching space. In real life, codes seldom occur in the cardiac unit and women can deliver anywhere between home and the labor and delivery suite. Place microphones and cameras in the waiting areas/hallways, elevators, office areas, and a loading dock as a location to bring the "patient" to your simulation hospital via ambulance.

One of the best decisions made in the design of our simulation center was to place a movable wall covered in whiteboard (which has sound proofing material in it) that would retract upward (Figure 18.7).

Now, we are able to operate multiple patients at the same time in the same space (Figure 18.8). You can have two patients in an ICU simulation or simulate an emergency department where the second patient is admitted during the course of your scenario. Each mannequin can be controlled from a different control room station and handler. This forces the learners to triage the patients' problems and work with other team members to coordinate the care provided. We only wish this movable wall was present between each of the simulation rooms.

While we tend to focus on the lesson-*doing* within the simulation scenario, the real lesson-*learning* happens in the debrief rooms. These rooms should be spacious, with comfortable chairs and one table large enough for everyone to sit at (Figure 18.9).

Do not relegate planning for this room to an afterthought, as it is where the conversion from short-term experience to long-term learning occurs. Also, do not use this space as a secondary storage room for overflow equipment. This clutter

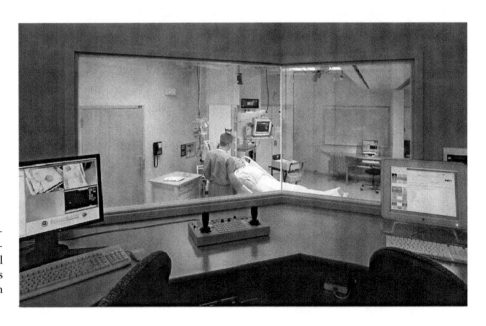

**FIGURE 18.6** An elevated control room provides an excellent overall picture of the events within the simulation scenario.

**FIGURE 18.7**   Full-width, full-height whiteboard and soundproof movable wall – all in one. Note the audio/video system inserts permanent location, date, and time stamps.

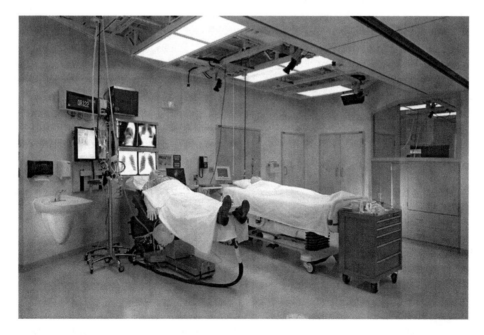

**FIGURE 18.8**   Two patients in a single space support learning to focus in the midst of competing and distracting patients.

**FIGURE 18.9** Room designed primarily for debriefing, but with the preinstalled utilities and infrastructure required for simulated patient care if the need so arises.

not only serves as a distraction after an emotional intense time in the simulation room, but also "teaches" the message that the activities performed in this room have little value. Position a large video-viewing screen in a position relative to the large table so that all can easily see, and place speakers so that all can clearly hear. The controls for this equipment should be so easy to use that time and attention is spent focusing on the content presented in the audio/video reproduction, and not on the audio/video delivery equipment itself. This principle of *simplicity in use* extends to include tools that allow the instructor to mark pertinent frames or sections during recording for easy review afterward.

## 18.6 Information Flow Within Your Center

Eliminate paper whenever possible. It is hard to keep track of, wastes trees, and takes up precious storage space. Consider the use of wireless laptops or Personal Digital Assistants for providing the participants the essential prescenario context information about the impending event, dissemination of information during the scenario (students use Personal Digital Assistants around real patients, so why not develop familiarity with doing so when around simulated ones), and as a resource tool during the debriefing period. If your institution uses an

electronic medical record system, it would be good to use the same system within your scenarios. Our electronic digital devices system can be used in the simulation center to access the patients' past medical care history, review laboratory data, and view radiologic images of the patient.

Include equipment in each simulation room that would be routinely used in your hospital. Telephones should be appropriately placed for paging, ordering stat labs or radiologic exams, obtaining consults, etc. Include view boxes for film X-rays and/or video displays for online radiological images.

We have placed whiteboards throughout the patient simulator, standardized patient, debrief, and control rooms. These are readily accessible on the walls and the backs of the doors. Instructors, participants, and learners are encouraged to jot down thoughts and questions during the scenario so that these can be used for review and teaching during the debriefing sessions (even though your mother never let you write on her walls).

## 18.7 Communication, Tracking and Recording Within Your Center

Monitoring, collection, storage, and reproduction of pertinent actions during the simulation scenarios and debriefing sessions is *paramount* to the simulation center mission. The importance of reviewing what actually happened, as opposed to the recollections of what the learner and educator thought or perceived to have happened, is the foundation to experiential learning. Wall-, ceiling-, or mannequin-mounted microphones and speakers collect and produce the audio portion of events. In chaotic, noisy scenarios, each learner can have an individual headset microphone. Two-way headsets allow selective voice communications between the scenario director, instructors, and helpers in the simulation room and simulator controllers in the control room. Visual collection of the simulations can also include wall-, ceiling-, or even mannequin-mounted cameras (Figure 18.10).

Movable camera mounts and remote pan and camera zoom control that can be directed from within the control room allows visualization of most portions of the simulation action. Consider portable all-in-one camcorders for certain scenarios and unusual locations where the mounted cameras cannot capture the best view.

There is no upper limit to the price of audio/video equipment for sale. Consider building your A/V system in stages matched to both budget and user-competence levels. Even if at first you cannot afford to place cameras, microphones, and speakers in all areas you would like to use, consider installing the necessary data and power cables and brackets during the construction phase (or at the very least, empty conduit). It is much easier and cheaper to add audio/video capability to rooms already prewired and/or preplumbed.

## 18.8 Serving the Learners and Educators

The success of your simulation center depends on the learners' and educators' patronage. Your relationship with them starts well before they enter your door. Simple procedures such as making sure they know where your center is located, where to park, and which door of the building to enter presents a friendlier image. This information can be delivered well before they come. You can provide them with links to your center's web site, which should give them prebrief materials and let them know what to wear and what to expect during their sessions.

Careful scheduling of your orientation, simulation, task trainer, and standardized patient rooms is important. Many folks will not be willing subjects in evaluation simulations required (or mandated) by the hospital or departments. Nothing is more frustrating than standing in line waiting for your turn to enter the next room or a debriefing area.

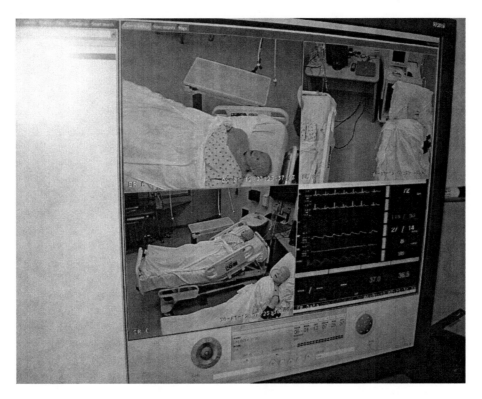

**FIGURE 18.10**  Collected and combined visual information is essential for both controlling the scenario during its execution as well as reviewing the events during debriefing.

With the restrictions placed on residents' work-hours, creative solutions need to be considered for your learners. For instance, place your video camera recorders on motion activation as a way to enhance security in your simulation area or use a key card approach. If your institution has key card access to enter building or rooms, this can be incorporated into your simulation center. This would allow participants to have self-directed learning on the task-training devices after completion of their duties or while they are on call and no activity is occurring. This 24/7 approach optimizes the time spent on call and fill down times with beneficial learning experiences or practice on task trainers.

Maximize the value from everyone's simulation experience. Document learners' experiences and then give them their own copies to review learning points presented during their simulation and debrief experiences. Pittsburgh makes videos of your own sessions available over the Internet – streaming video – password protected. Obtain written anonymous feedback from both the learner and educator prior to their departure from the simulation area. This provides an ongoing evaluation of your facilities, equipment, and instructors. A follow-up letter about their simulation experiences captures the learners' reflections about this often intensely emotional experience that also may lead to improvements in your curriculum or scenario presentation for the future.

## 18.9 Storage Rooms and Storage Space

There is never enough storage room. Repeat. You can never have too much storage space in your simulation center. Ideally, your storage area equals the combined area of the simulation areas, the control areas, and the debriefing areas. Unfortunately, for some pathological reason, in most institutions, *storage* is actively repressed and eliminated where found. However, *utility* is tolerated. Thus, consider listing your space requirements for areas designated for storage and utility with the actual uses reversed. Rarely will any large area designated for *utility* receive the intense scrutiny that those small ones marked for *storage* will attract. Let the storage-phobes exhaust themselves on trying to eliminate a few small closets and let you have large utility areas. You must take into account where you are going to store the examination tables, operating room tables, anesthesia machines, ventilators, dialysis units, CT scanner and gantry, C-arm, hospital beds, etc., used in your standardized patient and simulation rooms. For those centers planning on teaching with part-task trainers, consider space for operating microscopes for microvascular surgery training, endoscopy trainers, endovascular trainers, and a multitude of virtual reality devices that are in the pipeline. All these are quite large and cumbersome. Also, consider your largest props and (orthopedic)

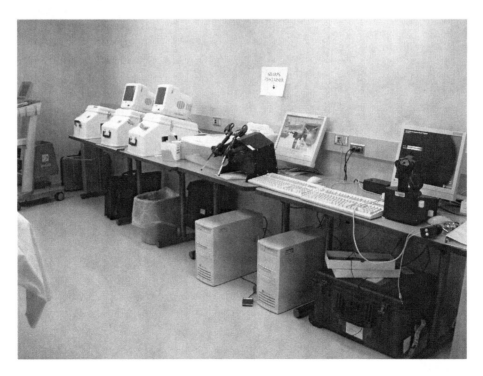

**FIGURE 18.11**   A dedicated task-trainer room requires as much forethought as any other room, particularly power, heat, noise, and maneuver space surrounding each device.

**FIGURE 18.12** Storage in the simulator room below the one-way windows of the control room. The shallow upper cabinets are under the desktop. Elevating the control room allows space for very long drawers suitable for storage of entire and intact mannequins.

hospital beds when designing the doors and hallway corners and the paths these props travel as you move them back and forth between you storage rooms and the areas of their use.

Task trainers take up a large space and have to be interchanged depending on which users are currently using the simulation space. You need to plan a room to be dedicated to task trainers, with a large number of electrical outlets, sufficient electrical power, and adequate ventilation to accommodate the heat load from so many powered devices (Figure 18.11).

Sufficient space and power sources allow the possibility of leaving more task trainers out. By doing this, you decrease the start and calibration times for the equipment most often used. This also facilitates the use of these trainers during nights and weekends.

Mannequins are quite large and hard to move. As we built an elevated control room, a large space was available under the desk and walk areas. We were able to place long drawers (10-feet deep) in the simulation room for storage of these mannequins and other equipment when not in use (Figure 18.12). These readily accessible storage areas drastically reduce setup and take down times between different scenarios. While the simulation areas are being drafted, utilize any potential dead space for storage closets, cupboards, or drawers. Consider reviewing how naval architects find and use every possible storage space.

## 18.10 Last But Not Least

Once you have prioritized the requirements for you simulation space, start designing the layout that will serve you and your primary customers' needs. We had educational representatives from more than 20 disciplines involved in reviewing the plan for our simulation center. We asked each one to mentally walk through several different scenarios and events utilizing the space in our layout to look for problem areas. It is much easier to make accommodations on your layout design than to remodel your space after the construction is complete. Major remodeling can be more expensive than the initial construction costs of your space.

Consider what the future holds for you and your simulation center. Anticipate growth and plan for a logical expansion of your center as the demands on your physical space increase.

## 18.11 Conclusions

This is your space. Design it to fit your needs and the needs of your primary customers. While we have shared our floor plan with you in this chapter, remember that we had space and budget limitations just like you have and it was designed to fit our needs. Visit other simulation centers like we did, and take the best that applies to your needs, thus avoid repeating obvious mistakes. Spend a couple of days watching their centers in action and find out what will work for you. Be sure to watch not just the scenario, but also all the preparatory and behind-the-scenes action. Most important is to understand the culture in which their simulation program exists, compare this culture with yours, and extract the workable ideas where the two overlap.

Be creative. Don't limit yourself to the examples in this chapter. By emphasizing the structure-function relationships of your culture, you can establish meaningful links in the design of your simulation center between your simulated clinical care and your real patient care. The members of the simulation community are truly creative individuals, and we look forward to hearing about your experiences in creating structure-function relationships in your simulation centers.

## 18.12 Example Sequence for Defining Structure-Function Relationships for Simulation Centers

a. Addressing learner anxiety – nonthreatening environment
   i. Greeter/reception areas
   ii. Changing/locker areas and restrooms
   iii. Classroom or orientation space
   iv. Waiting areas
   v. Having the simulation space look familiar; looks like your hospital, clinic, office with equipment where you would find it in your hospital
b. Grouping of spaces serving similar function
c. Adaptable/convertible space (multifunctionality)
   i. Consider alternative learning spaces (hallways, waiting areas, loading docks, elevators, office areas)
   ii. Storage and movement of equipment and furniture
   iii. Debrief rooms that can double as standardized patient rooms
   iv. Movable walls
d. Learner/actor flow through the center
   i. Coordinated flow of learners through the center
   ii. Avoiding bottlenecks
   iii. Flexible seating areas
   iv. Actor prep rooms
   v. Task-trainer grouping

e. Information flow within the center
   i. Eliminate paper (wireless laptops for prebrief, educational resource utilization, notes by the learners, ease of review of the clinical case, and information gathered to date etc.)
   ii. Electronic record
   iii. Laboratory, radiologic, etc.
   iv. Contact between rooms (consult, stat labs, blood ordering)
   v. Utilization of whiteboard space (walls, back of doors)
f. Promoting and educational intense environment
   i. One-way glass in debrief and control areas
   ii. Realism of the environment
g. Consistent communication, tracking, and recording of events
   i. High-end AV equipment pros and cons
   ii. Headset utilization for staff or all participants
   iii. Debrief areas
   iv. Elevated control room
h. Easy and timely access to learners
   i. 24/7 access to center
   ii. Scheduling of events
i. Filling the area space
   i. Donated equipment
   ii. Ancillary props
j. Storage Rooms and storage space
   i. Never enough
   ii. Know what equipment will be used in your space and accommodate (hospital beds, anesthesia machines, OR tables, task trainers, mannequins, props, supplies, and equipment)
   iii. Utilize any dead or unutilized space in your floor plans (if using an elevated control room, use long drawers that fit completely under this space, line the halls or control rooms with storage, next to columns etc.)
   iv. Location of staff/administrative areas
k. Now that you have designed the requirements for you simulation space, start designing the area that will serve you and your learners' needs. Once the design layout is complete, mentally walk through several different scenarios and events, utilizing each space to look for problems spots or areas. It is much easier to change your design than to remodel your space.
l. Also consider what does the future hold for you.
   i. Expansion
   ii. Remodeling
   iii. Anticipating growth
m. Conclusions
   i. This is your space; design it to fit your needs or the needs of your primary customers.
   ii. We have shared our floor plan, but it was designed to fit our needs and the requirements placed on our center.

iii. Visit other simulation centers that will serve a similar purpose you have in mind for your group. Spend a couple of days watching it in action and find out what will and won't work for you.

iv. Be creative and don't limit yourself to just the principles in this chapter. We look forward to your input.

By emphasizing logical structure-function relationships, you can establish meaningful links between scientific principles and clinical care in the design of a simulation center.

By emphasizing logical structure-function relationships important to your simulation space, you can establish meaningful links between scientific principles and clinical care in the design of a simulation center.

# 19

# Evaluating, Prioritizing, and Selecting Simulators

Brian C. Brost
Kay M.B. Thiemann
William F. Dunn

## 19.1 Now that You are Responsible for Equipping a Simulation Center, What Do You Do?

Congratulations! You have obtained permission to develop or expand a clinical simulation center. Now you are faced with the daunting task of filling the rooms with props and purchasing equipment with a finite budget. You have already scoured the hospital warehouses to obtain unused clinical equipment and hospital furniture to increase realism in your simulation spaces. The increasing use of simulation in clinical education has attracted a growing number of companies offering an ever larger number of similar devices at various levels of price and performance. How do you approach this difficult purchasing decision?

Most centers are tasked to serve multiple departments and divisions within the institution, and will be used by a wide variety of allied health staff, nurses, medical students, residents, fellows, and their faculty. With careful planning, you can purchase a core set of patient simulators and task trainers to provide the most effective start to your simulation center. The needs and expectations of this diverse population of users can be hard to satisfy without their active cooperation and collaboration. By thorough evaluation prior to buying your equipment, you have a greater chance of keeping these expensive devices from being relegated to a corner as a coat rack.

The process of evaluating, and ultimately purchasing, simulation equipment can be intimidating. It is hard not to be impressed with the quality of current devices and the incredible graphics of screen-based interactive programs that are currently available. With the announcement of your simulation center being approved or opening, you will be confronted by numerous vendors explaining why you cannot succeed without their company's devices. They will show you all the bells and whistles of their devices. Or, you can go to simulation equipment fairs and not know quite where to start. The desire is to not miss out on a crucial capability; therefore, in your anxious ignorance you want to stock your unit with "everything an educator could desire." By building the world's most expansive and expensive simulation center, why wouldn't everyone want to come to the world's best simulation center? We would suggest that the best way to ensure the effective and regular use of purchased equipment occurs when requests are brought forward from the actual educators who intend to use the equipment, not the owners or the managers of the simulation center, and not the vendors of simulation equipment.

## 19.2 Membership of the Simulation Operations/Equipment Committee

Your center will become either a novelty or a thriving educational environment, depending upon the rigor of your evaluation of the educational devices; the extent that you ensure a curriculum guides your decisions, and how well you prioritize the limited financial resources. Through attention and effort, it is possible to increase the likelihood for developing a successful simulation center.

Identify individuals in your institution who believe in an education process and the incredible potential of improving health care education through simulation. Our entire Operations/Equipment committee consists of people who volunteer their time to ensure the success of this program. Selected members include representatives from accounting, medical device maintenance, engineering, physicians, nursing, administration, purchasing, educational leadership, audiovisual/computers, etc. The Operations/Equipment committee is selected by and responsible to Simulation Center Leaders and is co-chaired by a physician and an administrator. Each member of the committee has an equal voice and vote in committee decisions and expected to bring their area of expertise to meetings.

Encourage your team members to challenge the *status quo*, to be creative, to think outside the box, to seek the highest quality solutions for raising the educational level at your institution. You will be rewarded with a group that will reap great dividends for your simulation center. Our Operations/Equipment committee was awarded the Mayo Clinic Excellence in Teamwork Award for their dedication, problem-solving abilities, creative energy, and innovative approaches to maximizing the financial resources of our budget and learning potential of the center.

We describe the experience of our oversight committee as a framework to evaluate, prioritize, and purchase simulation equipment. The process we have developed involves a three-step procedure, requiring justification from the person or division/department proponent requesting a simulation device, equipment vendors' company description and disclosures of their products, and a final review by an equipment committee member before the entire committee. This committee seeks a match between the stated goals/objectives to be performed with the devices to the projected learning and teaching objectives of the users of your simulation center. Utilization of a scoring system allows the members of the committee to rank and prioritize the purchase of equipment. The scoring system helps to level the playing field for simulation equipment that often is very dissimilar in nature. Similarly, the end-users (departments or individual users) who ultimately use the equipment have varying learning levels and needs.

## 19.3 Request from a Proponent or Department for New Simulation Equipment

When a request for a particular simulation device is forwarded to our committee, the first step in the evaluation is completion of a simple one-page proponent request form for all simulation equipment costing more than US$3000 (Appendix 19A.1). This form has been very helpful in rapidly assessing the possible use of the device within the simulation center. It also forces the proponent to formally state exactly what for, and who, and how many times, the simulation equipment will be used. This proponent request form includes the unit price (or fair market value for accounting purposes if donated), remodeling costs if applicable, the current availability of the same or similar devices within the institution (if this is an upgrade), and the projected annual number of trainees who will use the device.

The proponent should convey a clear view of how the device will be used, how often they will use it in their training sessions, and the number of learners that will be trained on that device each year (you may want to add a section to the form about expected/desired consequences or ability to track learning outcomes on the device). With devices that can be used by multiple disciplines, the ultimate cost per user is significantly decreased. Sometimes, these devices will be required for a submitted or funded educational grant, which should increase the ultimate score the device receives but will not override other factors. Because these devices often require maintenance or service contracts, the Operations/Equipment committee looks out for the hidden costs of all equipment including depreciation, which are not usually covered on donated or grant funded equipment that comes into the simulation center.

With an increasing number of vendors entering the arena of health care simulation, more task trainers and patient simulators are becoming available. When more than one task trainer with similar teaching capabilities and features are available, we have found a "vendor fair" to be a useful event. While many manufacturers initially are apprehensive about a side-by-side evaluation of their products on the same day, we have made it a policy that their device will not be considered for purchase unless it is thoroughly and openly evaluated by our simulation center's learners and educators.

We invite all potential users (from novices to instructor experts) to evaluate the devices in a head-on evaluation (Appendix 19A.2). The learners and experts evaluate each device on user friendliness, learner potential, graded and progressive tasks, realism, etc. These anonymous evaluation results are collated and presented with the proponents' request to the subcommittee where the device is assessed, compared, and prioritized against other products for possible purchase. Vendors have not requested these evaluations, but we would be glad to supply the reviews of their equipment and software, but not of that of their competitors.

## 19.4 Review of the Simulation Device Manufacturer

When a device is being considered for purchase by the center (all simulation device purchases go through this committee), a crucial piece is the Vendor Bid Summary (Appendix 19A.3). This summary provides an overview of what the vendor company intends to deliver, who are the vendor's primary contact people, price/maintenance/warranty information, plus a list of references of other organizations that have purchased and used devices from their company. This form also requires important safety, equipment design, durability, required customer upkeep, utility support, and other pertinent information about the simulation device.

While how the product addresses nationally established safety standards is important and expected, the interface capability with other equipment/computer operating systems used within your simulation center, to say nothing of institutional knowledge, is critical. Having multiple platforms or operating systems *interacting* increases the risk of your entire simulation center shutting down or key components not working. However, having only a single platform or operating system also increases the risk of your entire simulation center shutting down or key components not working from shared single-point failures. As with most decisions, the better decisions take into account the environment, at least as much as the devices. Not all these questions on the vendor bid summary form are applicable to each piece of equipment or device. Unexpectedly, customer service and support is turning out to be one of the most important components of this form. You want to know what you can expect from the company if the device goes down. Who will fix it, and how long will it take before they arrive to do so? Does the device need to be sent back to the manufacturer, or will they assist your engineering or maintenance department in fixing it on site? Will they be giving you a loaner device while the repairs are happening? The greatest blunder is to create an educational service so valuable to your institution that everyone can't live without it, and then fail to assure an uninterrupted supply, because the device won't turn on (and what of the generally unquantifiable cost of wasting the learners and educators time?) What is the company's return policy? These questions are important particularly if your center operates on a tight schedule, or you are using the simulation device for Continuing Medical Education credits, or a high-stakes evaluation/examination.

Knowing this background information may help you negotiate for the price and service level that matches your institution's needs. Bargaining is much more profitable if you know all the facts. The items in our vendor bid summary should not be considered all inclusive to meet the assessment needs of all simulation equipment, but it does give you a starting point. These forms are constantly being updated to reflect the changing experience of the committee as we purchase more and more equipment.

## 19.5 Evaluation and Prioritization of Equipment Requests by the Committee

Once all this information has been collected, it is forwarded to the entire Equipment/Operations committee for review prior to the meeting. An unbiased member of the committee is selected to be the primary reviewer and proponent representative during subsequent committee discussions. This reviewer ensures the accuracy and completeness of the proponent and vendor information and provides the committee with an overview. If necessary, the primary reviewer contacts the proponent and/or vendor to develop the most accurate picture of the simulation device and its proposed use.

Because of the multidisciplinary nature of simulation center, our simulation center leadership thought it was necessary to find an objective means of comparing and prioritizing the purchase of different task trainers and simulation equipment. The committee ranks the equipment or device using the form in Appendix 19A.4. Factors considered important to our simulation center included multidisciplinary or multischool use, potential for use for accreditation, quality and safety initiatives, research and innovation, prior experience by the proponents, a developed curriculum, meeting or exceeding the learners' needs and the educators' desires, and the strength of the vendor. Do not forget to consider installation, compatibility, or remodeling costs when considering each piece of equipment. Last but not least, consider the costs from disrupting the established traditions, climate, and culture. These can be the most expensive of all.

The committee members rank each objective on the review form. Then, a consensus objective ranking is agreed upon by the committee membership. The equipment with the best score receives priority for purchase. For our institution, further review and approval in central committees is needed, prior to obtaining funding for purchase. Our thorough and intensive level of scrutiny and review is generally viewed favorably by the main funding committees at our institution. More importantly, no equipment evaluated through this process has failed to be utilized effectively in our center. The pieces of this evaluation and review process have been used at our institution, but not combined into as complete a process as used by the Simulation Center Equipment Committee.

If you are only purchasing a single simulation device for your department, some of these steps may not be necessary. It is also important to modify this form on the basis of the criteria you select as most important based on your department, center, or institutional strategies and initiatives. We have modified these forms during their use over our early years and anticipate that these living documents will undergo additional changes as new situations are encountered and insightful input obtained.

## 19.6 Alternate Sources of Equipment and Props

Like the stereotyped supply sergeant in the Army, consider alternative means of acquiring equipment for your simulation center. Do not spend money on examination tables, code carts, or scenario props without first visiting your institution's warehouse. Sometimes, departments have already purchased simulation equipment that can be utilized by other disciplines. Talk with other simulation centers to acquire their unused equipment (but only if you have a plan in place to use it and care for it). Generally, outdated or prior renditions of high-end devices do not end up being a cost savings in the long run, any more than buying a used sports car.

We have "found" a lot of useful props from areas in the hospital that were remodeling. A lot of the basic medical devices and equipment were clean, functional, and not clinically outdated; it was just being replaced. An alternate source of equipment includes companies willing to donate equipment (such as ICU monitors, or beds, etc.) for educational use only, or purchase at a significant discount if the entire hospital currently uses their products, or a major purchase is soon to take place. Also, talk to the hospital's equipment committee – convince them to automatically buy an extra monitor, defibrillation device, etc., when buying for the clinical areas. Another alternative is to purchase nonfunctional equipment cheaply if it is just going to be used as scenario prop (C-arm, CT scanner frame and gantry, neonatal assessment beds, etc.).

This being said, even donated equipment is not "free." Storage space, maintenance contracts, upgrades are additional costs, including possible downstream dollars. It is important to determine how each piece of equipment donated or purchased will be used within the simulation center. Each piece of equipment is considered an asset to the institution and depreciation costs will need to be paid by your simulation center unless it has already been depreciated within your institution. Equipment that is not functional to your educational scenarios will clutter your classrooms, clog your hallways, fill your storage space, and otherwise take over like kudzu.

We have made a concerted effort to ensure that the rooms, equipment, and props are the same in the simulation center as are currently used in our inpatient and outpatient settings. This reduces barriers to learning transfer between simulated and real clinical venues. Also, this allows the institution to use the simulation center to reenact sentinel events or model quality and safety initiatives.

Collaboration with biomedical, bioengineering, or medical device maintenance departments to develop staging or new equipment not currently available on the market has been successful at our institution. Examples of this include refurbishing of equipment for staging simulations such a CT scanner or a C-arm (with movable parts and gantry but with reduced weight by removing the radiation producing sources etc.). We are also in the process of making simulation devices that are needed by specific areas that are not available currently on the market. Properly designed, developed, and marketed innovative simulation equipment could serve as a potential revenue stream for your center.

## 19.7 Conclusion

All simulation centers, whether well established or just opening, are faced with the financial pressures of limited funding to purchase, maintain, and update equipment for health care simulation. Too often, simulation equipment is purchased without determining who will be responsible for taking the numerous planning decisions and how simulation-based learning will fit into the numerous preexisting clinical educational programs. By careful evaluation of the proponent proposal, manufacturer or company vendor, and review by an informed and interested Simulation Equipment Committee, you will maximize the centers' financial resources and minimize the likelihood the equipment will gather dust or be used as an expensive simulation museum exhibit.

We encourage utilizing these forms as a starting place to allow an informed review and prioritization for purchasing mannequins and task trainers for your simulation center. You might wish to modify these forms so they address issues important to your simulation center's funding and purchasing needs.

## Acknowledgments

Many thanks to the following for making this chapter possible through the development of this equipment review and implementation process: Jacqueline Arnold RN, Kevin Bennet PhD, Christopher Colby MD, Julie Doherty, Donny Dreyer, Steven Gusa, Lisa Hurley, Cynthia McCabe, Laurence Torsher MD, Jeffrey Ward.

# Choosing Full-function Patient Simulators, Creating and Using the Simulation Suite

David H. Stern

## 20.1 Selecting Simulators, Creating and Using Simulation Environments

In order to design a simulation suite and choose a specific simulator, you must determine how both will be used. For example, a center designed for Crisis Management training will typically require a separate "control room" from which the simulation can be discretely directed, controlled, and recorded; a one-way window to hide observers without distracting the participants; and a nearby private conference room for comfortable and confidential debriefing. A simpler type of simulator will suffice for Crisis Management that is limited to medical resuscitation training, while more complex "model-driven" simulators with gas analysis are best for realistic intra-operative crisis training involving anesthetic gases and complex anesthetic or surgical scenarios. Likewise, the simulation room design must match the intended setting for optimal realism. An operating room will have a much more complex design than a patient room. A suite designed for many different uses must have a flexible configuration and storage space for equipment and furniture that is not part of the current simulated environment.

## 20.2 Types of Simulators

Interactive computer screen–based simulators and part-task dedicated skills simulators serve an important role in clinical education (Table 20.1). However, a maximally immersive and invasive experience is generally best achieved with full-function patient simulators with their life-size mannequins that present a fuller range of clinical signs and symptoms of real patients. Clearly, this comes at increased cost, both in terms of equipment outlay and in operating personnel time and expertise.

**TABLE 20.1**  Two common types of simulator technology

| Type of patient simulator | Definition and examples[a] |
|---|---|
| Interactive computer screen–based | Operates on a personal computer platform with no external mannequin (Anesthesia Simulator Consultant, Gas Man, Rapid Response Sim Networked CRM, Ty Smith's BODY |
| Full-function, life-size | Mannequin driven by computer equipment (see text for examples) |

[a] See reference section.

# 20.3 General Selection Principles

The following list of selection criteria is not exhaustive, but reflects the major considerations in choosing a simulator. It includes many features that we have found are necessary for most simulations, others that are on our "wish list," and a few that are desired but not yet commercially available. Selecting and funding a full-function simulator is very similar to selecting and funding a motor vehicle. First, determine those features and capabilities that support well-defined teaching objectives. Second, compare your requirements with the proven performance of commercial offerings. Then, assess the magnitude and kinds of financial support required to establish and sustain a specific full-function simulator within your full-featured simulation facility.

## 20.3.1 Interface with Clinical Monitors

- Do you need your students to obtain vital signs on real, fully functional clinical monitors, typically identical to those your faculty and students use with real patients (ECG, noninvasive blood pressure, invasive pressures, temperature, oxygen saturation, cardiac output, expired $CO_2$ and volatile gas analysis), or will it be sufficient to obtain this same information from a simulation of a clinical monitor display? In either case, the patient's data is the same, however, your students' experience of the data will differ. The use of real clinical monitor displays can improve the realism of the scenario as well as the transfer of skills from the simulator to actual clinical care. This improved realism has a price: only the most expensive patient simulators offer this capability, and they require a fully functional clinical monitor, itself costing many times that of a generic computer video display.

## 20.3.2 Ventilation

- How sophisticated will you need the mannequin to be with regard to ventilation? For example, do you need to demonstrate changes in airway resistance and compliance

using peak and plateau pressure curves using the ventilator as a diagnostic tool, or the gas exchange response to Positive End Expired Pressure?

- Is the mere presence or absence of exhaled $CO_2$ adequate, or do you need the simulator to present varying exhaled $CO_2$ concentrations that reflect changes in physiologic status? Do you need to be able to produce various condition-specific exhaled $CO_2$ waveforms, such as an upward sloping capnographic trace as typically found with bronchospasm [1]?
- Will you need the simulator's oxygen saturation to automatically detect and respond to different ventilation modes, differences in inspired oxygen concentrations, and differences in student proficiency in mask placement or manual ventilation, or will direct operator control of oxygen saturation changes suffice?
- How much control do you need over the rate of the oxygen saturation change? Very rapid changes as with an obese patient with low FRC, or to emphasize the urgent nature of intubation, i.e., only as long as you can hold your breath? Or do you wish to present very slow changes as after denitrogenation ("preoxygenation")? For teaching purposes, simulation time is often much faster than real time. For example, we demonstrated a case where we wanted to show the effect of a high-heel shoe on a large person occluding the oxygen supply hose – based on a real life incident – in our simulation we had this go on and off several times – we needed the saturation to go up and down quickly, otherwise we would have been there an awfully long time to show just one tiny teaching point.
- Remember that shunt changes are simulator operator activities specifically made to drive changes in the presentation of arterial oxygen saturation and ECG indications of cardiac ischemia. On some simulators, you must change the shunt, while on others you must change the saturation and ECG directly. In any case, with simulated or real patients, students will not directly experience (nor directly perceive) the shunt itself, only the manifestation of its consequences.
- Can the simulator present the types of respiratory problems you need, and in the ways you need them to be?

## 20.3.3 Airway

- Consider compatibility with different types of airway instrumentation: endotracheal tubes of varying diameters and insertion depths, airtight sealing with laryngeal mask airways, jet ventilation, cricothyrotomy, fiberoptic-assisted intubation, double lumen tracheal tubes, advanced techniques such as retrograde wire intubation.
- Can the simulator present the types of mouth, nose, head, and neck problems for airway access that you need? Can

you demonstrate tracheal shift? Jugular venous distension? Jugular venous pulsation?

- Do you need an anatomically realistic upper airway, trachea, and lung model that will allow diagnostic findings via bronchoscopy? Partial-task trainers (Oxford box, New Zealand "Replicant" fiber-optic intubation trainer, among others) are far better for learning eye-and-hand coordination required for bronchoscopy, both for their fidelity and their cost of acquisition, ownership, and use.
- Does the airway anatomy match that of the age/size of the rest of the mannequin?

## 20.3.4 Portability

- Will the simulator be used in a fixed location, moved only occasionally, or routinely taken to other areas within or outside the facility?
- Consider size and associated equipment racks, cabling, hoses, compressor, and/or gas tanks needed to operate the simulator.
- Electric Power requirements: mains line power versus battery, quantity, and quality.

## 20.3.5 Specific Features

- Quality and range of heart and breath sounds, freedom from cross talk and extraneous mechanical noise. Some simulators, such as Harvey, were designed to use special stethoscopes with an electronic signal detector in place of the usual acoustical detector in order to eliminate nonclinical mechanical noises that can easily distract and confuse students. While not being able to use a standard stethoscope may alter realism, many clinicians already use electronically amplified stethoscopes with their real patients. Furthermore, anytime students use low-cost, simulation center-provided stethoscopes instead of their own personal high-quality stethoscope, they are altering clinical realism.
- Regional block anesthesia (spinal, epidural, nerve blocks), with or without realistic tissue penetration feedback (*haptic* technology is just another tool, not a clinical education goal).
- Voice realism (fidelity and apparent location).
- Needle decompression and chest tube insertion sites for treating pneumothorax: are the anatomic locations correct, and are they accompanied by the desired sights and sounds from the body whether performed correctly or not? (An example of illogical features: one patient simulator model had a factory-installed chest tube port on the right side and a needle decompression port on the left. This might only confuse students who, unable to decompress a left pneumothorax with a needle, could place a chest tube only on the right. Fortunately, with minimal

tuning, either the chest tube port could be moved to the needle side of the thorax or a second, spare chest tube port could be added to the needle side.)

- Pericardiocentesis location with desired sights, cardiogenic vibrations, and ECG signal.
- Transthoracic cardiac pacing.
- ECG signal via defibrillator paddles or patches.
- Additional trauma options.
- IV and central catheter sites: does the very expensive mannequin need to have the capability for trainees to practice inserting venous and arterial catheters, or can they be provided with preinstalled access sites? Will a separate part-task simulator handle these invasive skills learning needs without placing the very expensive electrical and pneumatic components within the mannequin at risk of flood damage?
- Cough (sounds and abrupt chest movement).
- Single deep-breath capability (when requested prior to anesthetic induction or listening to the lungs preinduction).
- Arm, leg, and head motions: to reflect a light plane of anesthesia, Glasgow Coma Scale assessment, seizure, fasciculations due to depolarizing neuromuscular blocking drugs, and/or movement with cardiac shock and defibrillation.
- Simulation of valvular heart disease: Murmurs and hemodynamic behavior that reflects the stenotic or regurgitant lesion.
- Cardiopulmonary bypass simulation [2].
- Gastric and airway fluids (vomiting, aspiration, hemorrhage, pulmonary edema). This is usually impractical because the fluids can damage the mannequin's internal parts, but some simulators have incorporated this.
- Mannequin skin appearance: color, temperature, texture, sweating (e.g., "cold and clammy").
- Pulse volumes/strength – bounding pulse, waterhammer pulse, weak and thready pulse.
- Neurological – coma examination: eyes turn in response to head rotation, cold water in the ear causing nystagmus.

## 20.3.6 Costs

- Initial purchase price.
- Annual service contracts.
- Periodic major upgrades, both software and hardware.
- Future additions such as a pediatric or neonatal mannequin, or trauma features. Do the additions require purchasing new "control boxes," monitors, and/or software, or can they use the existing ones?
- Facility to support the specific type of simulator, including gas pipelines, plumbing (sink), electrical supply, lighting, audio/video, associated clinical equipment.
- Operator staffing requirements and training.

- Faculty time – probably the most significant cost in the long run.
- Success: If you become successful, do you have immediate access to sufficient funding to rapidly replace a high-demand simulator, if for any reason it were to cease functioning? Will you purchase a backup simulator?

## 20.4 Examples of Current Devices: Operator-dependent and Model-assisted

Current full-function life-size simulators (such as the Laerdal SimMan, METI Human Patient Simulator (HPS), Leiden Anesthesia Simulator, Sophus Simulator, and PATSIM-1) can be separated into two general categories. One type is wholly operator-dependent: every response to every action by the students is generated by the operator. This mainly requires a dedicated operator for all but the simplest scenarios. The other type is enhanced by mathematical physiologic and pharmacologic models (sometimes described as "model-driven," "model-enhanced," or "model-assisted"), where a software "assistant" usually can reduce the operator's workload by automatically generating fairly accurate physiologic responses to many of the students' actions. These differences affect how the operator uses the simulator, as will be discussed below. Both types are preprogramable, so that during a scenario, a predetermined ("scripted") sequence of physiologic changes can proceed with minimal operator intervention.

As in all forms of simulation, the operator must always retain control of the parameters. Modeling can help the operator by reducing the need to specify each vital sign and parameter and every change of them throughout the scenario. These, however, can usually be overridden when necessary to obtain the desired results. The modeling becomes a tool that gives the operator more time to observe the trainee and deal with other problems.

We will refer to simulators with mathematical model-generated vital signs as "model-assisted," even though the operator can still control the response, and those without this as "operator-dependent." Since the purchase price of the model-assisted type is in the US\$ 150,000–200,000 range while that of the operator-dependent type is in the US\$ 25,000–40,000 range, it is important to understand the advantages and limitations of each type. Very simply, the model-assisted simulator is somewhat like an automobile with an automatic transmission, while the operator-dependent simulator is like one with a manual transmission. The presence or absence of the model, from the *students'* perspective, is almost irrelevant (passengers don't care if the transmission is manual or automatic, just so long as they arrive at their destination on time). On the other hand, the operator will have strong needs for one or the other type of simulator, given the nature of the

"driving" that the scenario calls for. In all cases, the ultimate responsibility for moment-by-moment *presentation* lies in the hands of the operator. The ultimate responsibility for overall scenario *teaching objectives* lies in the hands of the clinical instructor.

With model-assisted simulators, once the patient's starting conditions are created, administered drugs and changes in ventilation, inspired gases, and other parameters should automatically cause appropriate changes in vital signs, expired gas concentrations, and other measurable physiologic variables. Furthermore, it is easy to write "scripted" programs that will automatically cause progression of a disease process or its resolution, depending on administered drugs and ventilation, all without additional data entry during the scenario. In a sense, the operator is less occupied with data entry and has more time to concentrate on the trainee for performance evaluation and teaching.

For routine resuscitation scenarios, operator-dependent simulators have advantages. For the price of a single model-assisted simulator, one can purchase several operator-dependent simulators, making multiple-patient disaster scenarios more affordable. Lacking short umbilical attachment to a large and noisy equipment rack, operator-dependent simulators let you fit more "patients" in a limited floor area. Their more compact size and low weight is also an advantage when portability is needed, such as for simulations done in rooms outside the simulation suite, outdoors, or in remote locations. In many such scenarios, the *content* fidelity of specific vital signs is far more important than the *context* fidelity of their presentation. However, operator-dependent simulators' lack of integration with standard portable monitors that are typically used in the field might be a drawback under some conditions or circumstances.

On the other hand, intraoperative scenarios tend to be more complex, with more diverse causes and interventions, and with higher expected ability levels of the students. Thus, in the midst of producing such scenarios, the many automatic support functions provided by the model software can be a great asset to the operator. There is minimal need for the operator to continuously and rapidly perform pharmacologic and physiologic computations, all the while rapidly entering commands to pilot the patient throughout every nuance of the scenario.

Also, some instructors encourage students to pay more attention to the mean BP (blood pressure) than to the systolic. The physiological model gives a more realistic mean BP, based on the pathology programed into the models. The vast majority of operator-dependent scenarios relies heavily on the systolic BP as the main cue, and may confuse people using the mean BP to try to figure out the underlying stroke volume, peripheral vascular resistance, etc. Particularly, when you "ping the system" with a fluid bolus, for example, the physiologically driven model is more useful.

What is the downside of model-assisted simulators, other than cost and complexity? The models are very good, but far from perfect. In some cases, it may be difficult to duplicate a real clinical scenario because of modeling limitations. For example, in creating a septic or distributive shock scenario, we might desire to present a very low systemic BP and a specific heart rate and cardiac output. This can require extensive trial and error in setting systemic vascular resistance, baroreceptor responses, cardiac chamber contractilities, venous capacitance, arterial elastances, intravascular fluid volume, and other intrinsic parameters, until the desired presentation is achieved. In some cases, it may not be possible to present the intended values within the limits of the models. Tricks can be used to work around problems like this, such as changing the calibration for BP in the clinical monitor's BP module, or change the injectate volume value or catheter calibration constant used by the thermodilution cardiac output module. For example, if the simulator can only produce BP signals that are twice as high as desired for the students' perception, then one could halve the gain on the real clinical monitor's BP inputs. Since the clinical monitor is blind to such manipulations, it will present the desired BP values, and all subsequent computations, like systemic vascular resistance that depend on BP values, will also present as desired. Since all clinicians are blind to the sensor's signals entering their clinical monitors, whether they be from real or simulated patients, no harm is done to the students by such fakery if it produces the desired learning.

---

**Editors' Note:** An approach to overcoming the present limitations of model-assisted simulators is to use everything at the operator's disposal, no matter what the manufacturers had in mind for their products, to make the presentation of the desired scenario as effective as possible for helping the students learn the intended lessons. Our students only experience what their senses detect – our job as *simulation-using clinical educators* is to stimulate their senses so that *their* internal mental constructs grow and develop toward behaviors useful to their patients. Whatever it takes for us to properly stimulate our students is fair game.

---

With operator-dependent simulators, the desired vital signs are simply selected. Sometimes, this is exactly what the instructor needs to be able to do. However, the operator must diligently and promptly enter the changes in vital sign values that should result from administered drugs and maneuvers such as ventilation changes. This is far more labor intensive and subject to individual operator variation and bias. It is entirely possible to inadvertently select physiologically unrealistic numbers, which can be confusing to trainees. Like their more extensive, more expensive brethren, operator-dependent simulators can be preprogramed, thus reducing the workload on the operator. However, as with all preprograming, whenever the student takes a different path than expected, the operator must be immediately ready to take over and "hand fly" the patient.

In order to ensure that the clinical presentation makes sense, it is almost an axiom that instructors operating or guiding the operation of the simulator must be far more clinically advanced than their students.

Real-time gas analysis is both a benefit and a limitation of model-assisted simulators. Real-time analysis of $CO_2$, $O_2$, $N_2$, nitrous oxide, and volatile gas anesthetics makes the simulator responsive to changes in ventilation, and a clinical gas analyzer can then be used intraoperatively to measure expired gases just as it would be in a real patient. This enables scenarios with a degree of realism that is not possible without real-time gas analysis. But because gas composition generation from the simulator, as well as gas analysis by both the simulator and clinical monitor, are real-time processes controlling the mixing and detection of real molecules, lengthy pausing of the simulation for teaching purposes can allow gas concentrations to drift. This may reduce the predictability of the scenario. In fact, perturbations in model-assisted simulators do result in some variability in the generated parameters even if the same scenario is repeated with apparently identical treatment interventions. On the other hand, some operators find that generated vital signs are *too* uniform and have wished that they could *add* some degree of variability to vital signs to simulate real-world measurements. While the rate of time passage in both types of simulators can be made either faster or slower, the values produced by operator-dependent simulators usually show even less variability than those from the model-assisted types.

Specific brands of simulators may have other limitations. For example, the accuracy and variety of ECG waveforms are dependent on how these are digitized and regenerated, regardless of who or what selects the waveform. For example, with some simulators, premature ventricular contractions can be hard to recognize as the heart rate rises, and the relatively fine ventricular fibrillation waveform may be misinterpreted as asystole with baseline noise.

Both types of simulator need regular updating, cleaning, maintenance, expendable parts replacement, and repairs. The staffing and employment requirements to meet these responsibilities are a major expense in treasure for finding, teaching, and retaining them.

### 20.4.1 Lessons Learned

There is no single type of simulator that is ideal for all uses, just as there is no single motor vehicle that would meet everyone's needs. Each type has its own strengths and weaknesses. Your specific application, objectives, and budget will guide the choices. Above all, do not expect perfection. Unlike the history of patient simulation, flight simulation began soon after the first powered flight, most likely because training errors were hard on instructors. Commercial patient simulators have been under continual development since the early 1990s and the most expensive, most elaborate of them are much cheaper and much less accurate than the most expensive flight simulators. Clinical simulators are currently of low reliability compared with other comparably priced devices we use every day. This is especially true of the more complex model-assisted simulators. Technical problems will crop up unexpectedly during tightly scheduled simulations. Do not underestimate the value of a competent, experienced simulator technician who can quickly fix these problems, or devise inventive work-arounds for both sudden failures and inherent deficiencies.

## 20.5  Evolution of a Simulation Suite

In 1994, we purchased the second commercial simulator ever produced by Loral Data Systems (the HPS is now made by Medical Education Technologies Inc, of Sarasota, Florida, U.S.A). At that time, only two commercial companies were producing simulators. We visited the two companies to evaluate their simulators. (Visiting one or more user installations, then inviting their facility/program designers to visit your institution is probably a better approach today.) Our decision to purchase from Loral was influenced by our concern about the highly proprietary nature of a competing product made by CAE-Link, of Binghamton, New York, U.S.A. We were interested in developing software and making our own modifications to hardware and software, and felt that a proprietary product would be less flexible. Loral encouraged (and as we later learned, depended upon) user input, and was willing to share details of the simulator and allow us some access to internal models and programing. The CAE-Link product's new owners, Medsim-Eagle, subsequently discontinued production of patient simulators, leaving their users without support and parts. The fact that many of these machines have continued to be used successfully since then is a testament to the power of users sharing parts and exchanging knowledge.

### 20.5.1  The Closet-sized Room

The $180,000 initial cost of our simulator was shared equally by the anesthesiology department, medical school, and hospital. Originally in a closet-sized room within the operating suite adjacent to one of our cardiac rooms, the simulator was able to be used mainly for one-on-one teaching. To our surprise, we found ourselves acting as a beta test site. With limited production and few units being sold, the Loral HPS simulators were in a continual state of development, and each was somewhat unique. Sessions were frequently interrupted by unstable behavior with repeated hardware malfunctions and software crashes that were partly related to memory usage under the then-used MS-DOS operating system. Today's simulators are far better in every respect, but purchasers should be aware that simulators are of low inherent reliability when compared with aviation equipment, consumer electronics, and

automobiles. This point warrants repeating: assurance that your simulator will be ready at the start of each and every scheduled session is very dependent upon the quality of your people responsible for using and maintaining it.

### 20.5.2  Designing Within a Dedicated Space

In 1998, when our hospital central supply was outsourced, about 1000 sq. ft. of space became available to build a dedicated simulation center. An architect was employed to develop the design, but he had difficulty in understanding how medical simulators are used. The result was that his designs tended to be artistic but impractical. Many different configurations were sketched, the goal being to include a multipurpose room (to serve as OR, ICU, or other patient locations), adjacent control room and debriefing room (each with one-way glass to allow observation of the OR), and a separate education laboratory and storage area (Figure 20.1).

Approximately, $100,000 was budgeted for developing this space into our center ($100/square foot), however, there was some additional expense for installing medical gas piping (oxygen, air, nitrous oxide, and vacuum), air handling equipment, fire sprinklers, and other equipment that was shared with the surrounding offices and storage areas that were not part of the simulation center. Note that it is not absolutely necessary to use nitrous oxide; nitrogen will provide the same $FiO_2$ values to both the clinical and simulator's gas analyzers, and the anesthesia machine will work with $N_2$ in place of $N_2O$. The advantage of real $N_2O$ is that the gas analyzers will read the proper $N_2O$ concentrations. Using $N_2O$ and volatile agents of course requires proper scavenging, which adds to the expense and potential risks in the simulated environment.

If you are designing a new, freestanding center, or have a very large area available, there is flexibility to design the space to fit the needed environments. In many cases, like ours, where there is limited space and outside access from only one side, it

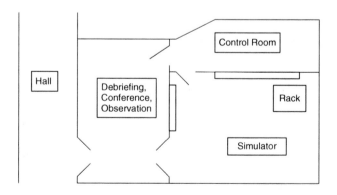

**FIGURE 20.1**   Schematic floor plan for First Simulation Suite, with one-way glass observation windows shown as narrow rectangles. Storage area is to the left of the hallway.

is much more challenging to renovate space and make it work well. Some specifics of the designs that had to accommodate our limitations include:

1. The architect originally planned to elevate the control room floor to give a "top-down view" of the OR. This sounded like an advantage, but in our case the need to move equipment between rooms through a single access path and frequent running back and forth would have been a serious encumbrance and safety risk. Also, floor space would have been lost to the stairs, and the standard one-story ceiling height would have limited the wall space for shelves and cabinets.

2. With an area of only 25 × 38 ft. for three rooms, and only one short wall accessible from the hallway, it was difficult to design three rooms that all had independent access to the hallway. The result is that the rear portion of the debriefing room serves as a pass-through from the hall corridor to the OR and control room. By not creating a separate corridor, we were able to maximize the size of the debriefing room, which can comfortably seat 10 at the table and another 10–20 in the surrounding space when needed. A temporary hallway can still be created by adding a screen or wall to separate the conference area from the pass-through, but we have never found this necessary.

   The architect did promote one design that allowed access from all three rooms to the corridor, but this necessitated making all three rooms odd, angular shapes with nonsquare corners and 5–6 walls each. Our architect resisted rectangular rooms as mundane; he needed artistic freedom. Unfortunately, the furniture and equipment that would occupy those spaces were decidedly rectangular, and much usable space would have been lost. We had to fight for every square inch of space to meet out needs.

3. Sometimes, details are overlooked. For example, originally the ceiling was to match our standard OR ceiling height of 10 ft. However, during construction, an overhead steel catwalk prevented the ceiling tiles from being installed more than 9 ft. above the floor.

Our suggestions for working with architects: First, clearly define your needs well before you start to interview them. Start with the absolute requirements, then you can discuss the less-important aspects where compromise is possible. Second, go over the plans as they are revised. Some things you thought were cast in stone may have changed. Third, make sure you've thought everything out and that everyone is happy before approving the plan. Making changes is very expensive once the plans are done, and an order of magnitude more expensive once construction starts.

Given the space limitations, we elected to create a single multipurpose room of about 400 sq. ft. that could serve as OR, ED, ICU, or any other location. To this end, an ICU headwall was constructed along one wall, and overhead surgical lights and additional gas connections on an adjacent wall were installed to enable setup as an OR as well. The lights are not essential, of course, but have been very useful in our experience. You will have to decide whether this is worth the cost. The ceiling structure must be able to support the weight and torque of these heavy fixtures, and they generally require welded or bolted support beams. Careful camera placement prevents the lights from blocking the camera's view, and we keep the lights pushed up close to the ceiling. The surgical lights do contribute to the realism of an OR, and can become an intrinsic part of some scenarios. In fact, we conducted a Grand Rounds featuring a two-way teleconferencing link to the "OR," so that observers could participate in managing the "case." Many of those present in the conference room were fooled into thinking it was a real OR with a real patient. The two surgical lights can be maneuvered into position to mimic a biplane X-ray setup that limits access to the patient, perfect for off-site cases in the radiology suite or cardiac catheterization laboratory. The anesthesiologist can be asked to adjust the light as a distraction, or these can be aimed in a way that introduces glare and makes it hard for the anesthesiologist to see, as sometimes happens in clinical practice. As a bonus, the bright, focused lights are handy during repairs and maintenance on the simulator. If we had been unable to obtain donations, we would have used an old surgical light removed during OR renovation.

You should carefully plan the electrical layout. In our case, we placed additional switches for the fluorescent lighting in the control room, which is also the location of the subpanel that controls all the equipment outlets in the simulator room. This allows us to simulate power failures from the control room. A separate electric circuit powers the simulator itself so that it remains powered during these simulations. In order to educate students about isolated power systems and line isolation monitors, these were also included in the simulation room.

Donations from manufacturers and suppliers were solicited for the surgical lights, anesthesia equipment, stretcher, blood warmer, and other equipment. We found an old but functional OR table stored in the hospital basement. Gas lines for $O_2$, $N_2O$, and Air proved to be expensive elements, costing tens of thousands of dollars to install. $N_2$ and $CO_2$ are provided by local tanks. Had we not needed the gas piping for a nearby anesthesia technical room that is used for maintenance and repairs, we might have considered using nonmedical grade piping. A cheaper alternative was used in another institution, where standard hardware store grade, 3/8-inch internal diameter, 300 psi–rated air compressor hose delivers "house" compressed air of 50 psi from a "wet" laboratory 75 ft. away. Since most industrial compressors contaminate the air with oil and water, it is necessary to use an efficient filter to remove these before they reach the simulator and anesthesia machine. A similar type hose works well for connecting "house" vacuum lines to clinical suction within the simulation room. However, since our anesthesia machines and gas monitors are donated units that are periodically replaced, and then returned to clinical use by the manufacturer, we decided it is appropriate to use medical grade gases in order to avoid the possibility of contamination with oil or other debris.

I provided the design and installation labor for the analog audio/video system, which initially cost about $10,000. This included three small video cameras with power supplies, two videocassette recorders (SuperVHS format), a digital video mixer, 12-channel audio mixer, wireless intercom system with four headsets (Telex Corp.), two video monitors, and a large-screen presentation monitor. A video format converter takes the RGB-formatted, progressive scan, analog VGA-resolution video from the interface connector on the patient monitor, and converts it into composite-formatted, interlaced scan, NTSC-resolution video (which is the same format used by the camera/recorders). Using the video mixer, we can then combine the clinical monitor display as a picture-in-picture along with the live video for recording and subsequent replay.

As with real estate, location is everything. We located our simulation suite close to the OR's, so that faculty, residents, and nurses could get there and back to work quickly. This facilitated operating simulation sessions during the clinical day. Faculty members who are covering an operating room can still come to the simulation suite and be readily available in the OR when needed. Although the suite is close to the OR's, it is separated physically to prevent simulation items from finding their way into clinical use. Since we have sterile anesthetizing locations outside the OR suite in our hospital, students readily accept the simulation room as a sterile environment. We can make use of scrub clothes, masks, hats, opened gowns and drapes, and other materials that would otherwise be discarded from the nearby OR's, thereby avoiding the expense of ordering these separately. A different approach is used at some centers, such as in Boston, which have their simulators in a building remote from the hospital. This can be an advantage in terms of space availability, design flexibility, and in some cases leasing expense, and avoids visitor traffic near patient care areas. It also means that the simulation center can be on neutral ground, rather than being owned by, or associated with, any one hospital. A small simulation program can be started within the hospital, and then moved to an external location if and when usage is sufficient to warrant building a larger center.

Our suite design has worked well, but is limited to two concurrent simulations, using a METI HPS model-assisted type simulator in the main OR and a Laerdal SimMan operator-dependent type simulator in either the main OR or in an adjacent room. As you would expect with most operator-dependent simulators, the SimMan is far more portable since it lacks the large rack and many associated gas tanks, but the METI HPS has also made many trips to large lecture halls in the medical school (see Example at end for a detailed description). If portability is important to you, you will need to evaluate the proposed equipment for this capability, and match it with your transport abilities.

## 20.5.3  Lessons Learned

If we were redesigning the suite today (Figure 20.2), our first requirement would be more space so that we could operate multiple-patient disaster scenarios and have more flexibility in arranging the room and accommodating more equipment. Each simulation room should have one-way mirrored windows from the control room and perhaps also the observation area (which can be the hallway). Access from the hallway to simulation room and control room should be direct, without passing through other rooms.

Having a multipurpose simulation room has not been a problem for either instructors or students. Only one portion of the room is in use at a time, and the equipment and furniture arrangement makes clear what is "in bounds." We hide a lot of in-place but "out of bounds" gear under standard clinical blue/green sheets. We use two portable folding screens to wall off some areas when needed, such as the corner of the room where the simulator console, simulator support rack, and gas tanks are located.

Medical student scenarios are fairly simple in most cases, with few supplies needed beyond the simulator on a bed. On the other hand, operating room simulations using a team of

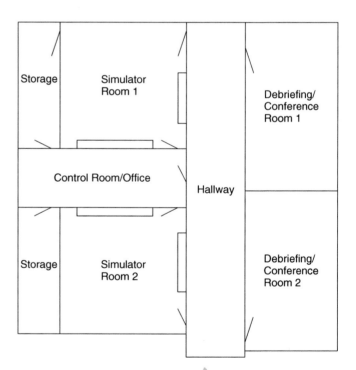

**FIGURE 20.2**  Sketch of an improved simulation suite setup for two Simulation Rooms, with a hallway between the conference rooms and simulator rooms, and one-way observation windows between the simulator rooms and both the hallway and the common control room (as indicated by narrow rectangles). The simulator rooms should be large enough to accommodate several beds or bays for several simulators, for multi-patient environments and disaster scenarios.

anesthesiologists, surgeons (played by residents or actors), and nurses can be very complex, and the supplies and equipment are dictated by the particular case being done. With careful planning, the simulator can be repositioned and set up for the next case during the debriefing of the previous case, without the participants being aware. Your equipment and its configuration are only as good as your choreography of execution.

Many people have asked how we were able to get most of the faculty to teach using the simulator. We had our simulator engineer/coordinator attend a training program and operates the simulator for the faculty, so they didn't really have to learn much about its technical aspects. Each faculty member was asked to provide a clinical scenario in their area of expertise or interest. The engineer then worked out the programing details in coordination with the faculty members, and operated the simulator for the sessions. Each instructor could then concentrate on teaching as if they had a real patient. Without this assistance, only a handful of faculty members would have used the simulator for teaching. This load sharing enables the clinical instructor to act as a "navigator," directing the scenario toward the desired clinical events and goals, while the simulator operator "pilots" the simulator and handles the details as well as any technical problems.

---

> **Editors' Note: The pilot/navigator model for workload sharing** – the clinical instructors are the navigators, they have to know where the scenario starts, where it ends, and the desired path between these two points; the simulator operator is the pilot, who has to know everything about how to fly and maintain their craft; together they must learn to communicate, collaborate, and cooperate to present the desired experiences for their students.

---

## 20.6 A Crisis Resource Management Success Story

We received a grant from our malpractice insurance consortium, which was supplemented by an internal patient safety grant. The latter was created as part of a legal settlement stipulating that a grant was to be created with the goal of improving patient safety.

We conducted a 2-year study involving the training of rapid resuscitation teams (each team to consist of: three internal medicine residents, two nurses, a respiratory therapist, and an anesthesiology resident). The study looked at whether crisis resource management (CRM) skills were improved by the simulator sessions, and the results indicated improvement when newly learned skills were reinforced with additional practice within a few weeks. Improvements in crowd control, leadership, communication, and resource management were soon carried over into practice in the hospital. We received anecdotal

feedback from several individuals about a year after the study started, suggesting that it was already having a beneficial impact on the management of in-hospital resuscitations. However, it is difficult to prove statistically that this has resulted in improved patient survival or outcome. Internal Medicine residents began requesting simulation participation prior to being placed on the rapid resuscitation team. The simulator program is currently a part of the internal medicine resident curriculum, with clinical support by the department and financial support by the hospital.

The current CRM course for Internal Medicine Residents and Nurses serving on the adult resuscitation team consists of a structured mixture of didactic teaching, practice using the full-task simulator, and debriefings. The didactic segment is a slide presentation and discussion based on textbook material [3] and additional information specific to our own program, simulator, and hospital resources. This is followed by an introduction and orientation to using the simulator, including its capabilities (e.g., EKG, BP) and limitations (e.g., lack of skin color change with hypoxemia, limited airway realism). Everyone is given a chance to feel pulses, listen to breath and heart sounds, and become familiar with the mannequin's other features.

The medicine residents then complete two scenarios, working as a group but with different residents serving as the leader. In each case, a nurse who is taking care of the patient is asked to call in a physician when deterioration begins. Shortly after the physician arrives, the patient suffers a respiratory or cardiac arrest, and the rest of the "code" team is called. Each resuscitation is stopped arbitrarily after 15–20 minutes, and the team members meet in an adjacent conference room to debrief under the guidance of a facilitator (see Chapter 69).

During the debriefing, team members review recorded segments of the scenario and discuss their impressions of the resuscitation, particularly their own performance, and where there were problems such as avoidable delays, communication lapses, or errors in judgment or execution. They are shown a "trigger video," in this case a short movie recreating an aviation crew's behavior and interactions preceding a spectacular crash (which was due entirely to human error). The purpose of the movie is to stimulate excitement and discussion about the value of team cooperation and resource management in handling crises, solving problems, and averting disaster. The key points in the airline disaster, such as fixation errors, are quite similar to those we face when taking care of patients in a crisis situation.

The PBS special "Why Planes Crash," originally broadcast in the 1980s, has a 5-minute segment that is useful as a trigger video. It is unfortunately no longer available for purchase from WGBS in Boston, and the rights for commercial use cannot be obtained owing to the complexity of obtaining permission from the many people involved. We, therefore, are able to use this video only for internal, noncommercial courses. However, the PBS transcript and the NTSB report are available, and could be incorporated into course material or even used to produce another video recreating the material. We are considering creating a "best and worst practices" medical video, to use as a trigger video for commercial courses.

The potential impact on patient safety led the hospital's legal department to recommend adding the internal grant to extend our outside grant. In fact, the legal department had previously used the simulator to help their staff learn about tracheal intubation and other aspects of intraoperative patient care while preparing a case, so they were already aware of the value of simulation.

While anesthesiology faculty members initially taught the didactic sessions, supervised the simulations, and facilitated the debriefings, the internal medicine department faculty and chief residents are now able to handle these tasks, with the assistance of our simulator coordinator. The result is that the program has taken on a life of its own, which is rewarding for all involved. Had other departments been willing to pay our salaries, we might have continued teaching these sessions. This was the best solution from our standpoint, allowing us to expand the reach of simulation and see it used by more departments outside our own, while limiting the drain on our faculty resources. However, it was crucial to ensure that those teaching and debriefing were trained and capable. At this point, the respiratory therapy department has made extensive use of simulation, but other departments have not yet become involved except to the extent that surgical residents and nurses already participate in our surgical CRM scenarios.

### 20.6.1  Lessons Learned

Simulation usage remains well below capacity, with most of the fixed costs still falling upon the anesthesiology department. Our goal is to encourage other departments to incorporate simulation into their curriculum. It was initially called the "Anesthesiology Simulator," and our task now is to convince others that, by sharing the cost, they can also benefit from this unique resource and use it to make improvements in education, patient safety, and eventually in credentialing.

## 20.7  Conclusion

Simulator suite design and choice of simulator are driven by the intended use of the simulation center.

We have presented the basic capabilities of different simulator types encompassing a wide variety of needs. Choosing one entails compromise: each simulator has its own set of features and problems, and none is ideal for everything. Choice of simulator really depends on the specific applications.

Likewise, designing a simulation suite is fraught with pitfalls. If you haven't thought out exactly how the suite will be used, money will be wasted in redesign, and there will be frustration in trying to work around design problems that could have been

avoided. Even with careful planning, mistakes will be made, and your needs will probably change. Flexibility is the key.

Two examples of specific applications in our suite are Crisis Resource Management training programs for medical and surgical teams. The medical team course does not require the full capabilities of the main simulator that are needed for our surgical team course (the latter involving anesthesia and surgical residents, operating room technicians and nurses, and medical students). Still, we have the flexibility to modify the workspace for different more or less demanding applications.

Anyone preparing to build a simulation center should visit functioning centers and talk with experienced users; it would be folly to ignore the vast sea of experience that already exists. Though still in its infancy when compared to aviation simulation, clinical simulation is developing rapidly. At most centers, you will find a driving enthusiasm and contagious desire to share experience.

## 20.8  Favorite Problem Solvers

For emesis, we use a mixture of betadine and small fragments of Styrofoam. This keeps indefinitely and has no smell, but it can stain. While it is possible to squirt this mixture into the airway, this is not advisable because of potential damage. Instead, a confederate simply empties a small container of the brown mixture next to the patient's mouth. We have also preempted the container and covered the emesis until needed. To suction the material, simply hide a small reservoir next to the mouth so it appears that the suction is from the mouth.

Avoid leaving standard adhesive tape on mannequins for more than the length of a session – it leaves a sticky mess that is difficult to clean off. As a "minimal standard of care," all tape of any kind should be removed at the end of each workday. Likewise, fluid collection containers should be emptied.

Actual IV fluid bags and drug vials provide realism. We keep the cost down by getting expired drugs such as dantrolene from our pharmacy (powdered drinks can be used, but must be difficult to dissolve). You can refill vials and IV bags with water, which also decreases the risk of spilled salt water on electronic devices. There are two caveats: precautions must be taken to ensure that these will never find their way into accidental clinical use, and actual drugs, even if expired, need to be secured just like those in real clinical use.

Real bandages and dressings with fake blood (home-made or purchased through theatrical supply outlets) are excellent cues in trauma scenarios, and they can be reused. Unlike real blood, the dyes will remain red (as with fresh bleeding) rather than turning brown. However, as the dyes will stain mannequin skin, it's a good idea to place an impervious barrier between the bloody bandage and mannequin skin. Also, housekeeping and trash disposal services will not distinguish between real and fake blood coming from a (fake) clinical facility. Hence, dispose of all such faked waste as if it were real.

When preparing realistic bank blood for transfusion, a red–blue tinted solution of cornstarch can be boiled to create a realistic viscosity that limits transfusion speed. This should not be stored for long periods of time, as it will grow mold and develop an extremely repulsive odor. Real Propofol will likewise become a foul, thick material if left in the drainage container under the bed. Fake Propofol from milk of magnesia avoids this problem, but leaves annoying stains. Food coloring makes pretty colorful urine.

Many annoying *features* in commercial simulators that operators fight against have solutions, and often cheap too. For example, the noise pollution from fans in the nonclincial equipment rack of high-fidelity patient simulators not only distorts the clinical environment you works so hard to establish, but also corrupts audio recordings of voices and clinical audio cues. For example, the entire vendor-supplied multi-blower high speed/high noise fan assembly was replaced by one very quiet desktop fan placed deep inside the rack and aimed upward. To reduce the other nonclinical noises from this device, the lid and walls of the cabinet were lined with short nap industrial carpeting scraps provided by the institutions' facilities personnel. Finally, a typical blue–green clinical sheet was wrapped around the sides of the rack as a visual blind. Now, when folks are talking in the room, their voice sound level overwhelms what little noise sounds coming from the rack and not the other way around, it never exceeds body temperature inside the rack, and the entire device visually disappears from the participants' consciousness.

## 20.9 Example: Use of the Simulator in a Large Lecture Hall

This scenario was used to augment the teaching of respiratory physiology in a way that would literally bring the subject to life. The students are given this summary: The patient is a 35-year-old male with a diagnosis as an asthmatic. He has sustained chest contusions and fractures of three ribs, radius, tibia, and fibula during a motor vehicle crash. He is alert but inebriated with normal vital signs. The plan is general anesthesia for open fixation of his fractures.

The students are then talked through the rapid sequence induction and intubation, after which the simulated patient develops severe bronchospasm, making the peak inspiratory pressure rise precipitously. Medical students are invited to come forward, listen to breath sounds, and assist in diagnosis and treatment, and then recheck the simulated patient after treatment with albuterol. A question and answer format involves the students and keeps them thinking about the physiological basis, but they can also appreciate how it relates to a "real" clinical case. The value of the simulator is that the case is immediate and ongoing, yet easy to control in this setting. This would not be possible with a real human.

It is a challenge using the simulator in this way, i.e., in a large lecture hall for a very large group of medical students. One must have an adequate crew to transport the simulator on a stretcher, accompanied by compressed gas tanks, interface rack, a bulky and heavy anesthesia machine, and associated supplies. The rather long route has to be planned to avoid stairs and ensure adequate clearance to get the equipment into the lecture hall. Time must be allowed for attaching and testing the gas connections and other hoses and cables. Extension cords and outlet strips may be needed to reach power outlets. It is also necessary to carefully plan what will be needed in advance, as there is no way to quickly retrieve any missing supplies or equipment. At the end of the demonstration, help will again be needed to remove the equipment before the next scheduled lecture. Allowance for equipment setup and breakdown will reduce the available lecture time. In spite of the challenges, the demonstrations have been rewarding for both students and faculty.

## References

1. Elser, C. and Murray, W. B. A workshop to illustrate time constants, differential lung-unit ventilation, and basic principles of capnography. *Anesthesiology* 2002. **97**(3A), A-1109 (2006).
2. Miller, A. and Power, J. Using Work Domain Analysis to analyse perfusionists' conceptualisation processes during routine and failure cardiopulmonary bypass scenarios. To appear in *Proceedings of the 50th Annual Meeting of the Human Factors and Ergonomics Society*. 16–20 October, 2006. San Francisco, CA. 2006.
3. Gaba, D. M., Fish, K. J., and Howard, S. K. *Crisis Management in Anesthesiology*. Churchill Livingstone, New York, 1994.

## Access to Information and Specific Products Listed

University of Queensland, CERG – Cognitive Engineering Research Group: http://www.itee.uq.edu.au/~cerg/cergpubs.htm Accessed 6 September 2007 http://www.virtual-anaesthesia-textbook.com/vat/sim.html

Anesoft http://www.anesoft.com/

Gas Man.http://www.gasmanweb.com/

Rapid Response Sim SilverTree Media, Pittsburgh, PA http://www.silvertreemedia.com

Ty Smith's "BODY" http://www.advsim.com/biomedical/what_is_body_simulation.htm Accessed 6 September 2007

# 21

# Survival Guide to Successful Simulation When Located Far Away

Ramiro Pozzo

## 21.1 Location, Location, Location

When a simulation center purchases one or more high-fidelity patient simulators, the manufacturer provides technical support, which may include online help over the phone, spare parts shipment in 24 hours, a repairman visiting the simulation service, etc. This may be true when:

The simulation center is large, prestigious, and commands a large budget;
The client has several simulators from the same manufacturer;
The center is located close to the manufacturer;
The manufacturer believes it has an opportunity of selling more simulators.

In our case, none of the above was true.

## 21.2 Good Fortune/Bad Luck

We have a few Laerdal simulators, a couple of Nasco simulators, and a MedSim-Eagle full-size high-fidelity patient simulator. The latter was purchased in 1997, and we have little hope of replacing it in the near future (maybe in 2012, when the World Congress of Anesthesia will be hosted in Buenos Aires). We are thousands of miles away from any manufacturer, and we are certainly not a famous institution known all over the world. When we purchased the MedSim-Eagle Patient Simulator, we had problems from day one. Here are some of those problems:

A technician was sent from MedSim-Eagle to install the simulator and to give training to the Simulation Operator Professionals (there were two of us).[1] During installation, a part of the electronic circuitry caught fire. A major section of the circuitry had to be replaced. The installer couldn't sit around in Argentina spending money in a hotel, and therefore departed back to the home base. The company shipped the replacement part, but the installer had already gone back home before it had arrived. Thus, the simulator's first repair was done by us, novices, even though the product was still under warranty, and before it had ever been used, and with practically no training for the installers and simulator operators. Since complete training wasn't possible in our country, MedSim-Eagle offered training 2 months later at their factory, at our expense.

The simulator uses a Sun Computer. One day, some of the keyboard's keys stopped working, including one of the keys that correspond to the login password. Sun Computer in Argentina only sells computer parts to registered clients. Since our computer was not directly bought by us, we were not a registered client of Sun Computers, and therefore could not buy a replacement keyboard! Result: simulation courses were interrupted for 2 weeks, while we waited for MedSim-Eagle to send us another keyboard.

The simulator's gas analyzer stopped working while still under warranty. The manufacturer could only respond to this by sending a replacement gas analyzer. They didn't have any in stock. The gas analyzers had to be shipped from Sweden to the U.S.A where they had to be adapted by MedSim-Eagle. Result: simulation was interrupted for a month until the replacement had been shipped, first from Sweden to the U.S.A, and then to Argentina.

Nonurgent technical support was provided by e-mail. Strangely, MedSim-Eagle technicians stopped answering e-mails. When a phone call was made, nobody answered. That was when we found out (not from MedSim-Eagle but from colleagues from other simulation centers) that the patient simulator segment of the company no longer existed! The minimal technical support we had was gone, and spare parts from then until now have had to be "improvised" locally.

When the simulator was purchased, we had a Datex AS3 monitor and the manufacturer sent special cables for the simulator to be used with this monitor.[2] Now, unless we are able to design our own cables, we are forced to keep using this Datex monitor "forever."

## 21.3 Words to the Wise

Obviously, we have had really bad luck. There is no reason for the same or similar problems happening in other simulation centers... or is there? Below are some suggestions and issues to take into consideration when you buy an expensive simulator.

- Take special care in selecting the right simulation professionals. They will play a fundamental role in keeping your simulator functioning. In many ways, their "capabilities and features" are more important than those of the simulation devices (see Chapter 53).
- Carefully review the technical support contract with the manufacturer. Ask them about:
  - Spare parts supply, and who pays for shipping
  - Training in maintenance and troubleshooting
  - Forms of communication, as well as turn-around time
  - Technical visits and who pays
  - Updates – will software updates be free? (Especially, updates correcting "errors" and malfunctions.) Will minor malfunctioning hardware updates be free? Will we get a discount for major hardware upgrades? Will the company trade in the older model for a discount on the newer model?
  - Warranty – specific details are quite important – duration, breakages, replacement parts, etc. Annual cost of maintenance service is relevant.
- Then, ask current customers about the actual follow-through by the specific simulator company.

- Study the possibilities of training your simulation operations professionals.
  - Where will it be?
  - At whose expense?
  - Will it be just user training or will they also get troubleshooting tips?
- Consider essential accessories (computer, clinical monitoring).
  - How specific are they?
  - Can you use any monitor?
  - Can you make a backup of the software?
- Try to foresee and predict any additional expenses in your budget: Spare parts, traveling expenses, etc.
- Purchase spares of all parts expected to wear out quickly, based upon your discussions with other Simulation Centers.

Stay connected to your colleagues: simulation operations professionals around the world have proven to be extremely helpful with contributions to solving all kinds of problems. We have received many interesting suggestions from the Society for Simulation in Healthcare listserv (http://ssih.org).

- Tips for choosing the simulator.
  - Do you know what your needs are? Does the proposed simulator meet them?
  - Don't be an unwitting pioneer. Take advantage of the experience of other Simulation Centers. Ask around and buy a simulator recommended by your peers.
  - Minimize dependency on the manufacturer, necessary consumables, and associated technology.
  - Does the manufacturer supply a hotline?
  - Does the manufacturer supply a specific person allocated to your center, and who knows you and your requirements?
  - Does the manufacturer have a mechanism of recording and responding to your (and other users) inputs, suggestions, and requests?

## 21.4 Conclusion

A full-scale high-fidelity patient simulator is a complicated and expensive teaching tool. Purchasing one is a big decision, as it requires considering the cost of the simulator itself, plus the cost of the simulation space, audio-visual equipment, medical equipment, qualified human resources, etc. Needless to say, the right simulator for your needs is essential for the success of your Simulation Center. But the great amount of technology involved in maintaining everything functioning with as little downtime as possible makes it very important to have the right simulation operations professionals to trust on, all the more important when your Center is not the Simulator manufacturer's "favorite customer."

# Endnotes

1. Our simulation professionals are biomedical engineers. They have an engineering and medical background, which makes it possible for them to interpret the clinician's needs and have the tools to try to meet them. Since technologies in medicine and in simulation centers suffer very dynamic changes, they must be very versatile. Also, they should have an open-minded attitude to get along with professionals of different disciplines, mainly clinicians. They had no simulation experience or training, but had experience and familiarity in working with medical equipment in critical care areas, namely operating room and intensive care.

2. Our high-fidelity simulator has vital signs output that is to be read by many commercially available monitors: (pulse oxygen, invasive blood pressures and temperature). But depending on the monitor you use, you will need monitor-specific and compatible cables to interface between the "patient" and the monitor.

# 22

# Retrofitting Existing Space for Patient Simulation: From Student Lounge to Acute Care Patient Unit

Michael C. Foss

## 22.1 Initial Conditions

Once you get past the discussion about possible benefits of using patient simulation in the health care curriculum, it is time to find a way of making it happen. This chapter relates our experience at the School of Health at Springfield Technical Community College, Springfield, Massachusetts, U.S.A, where we created a virtual hospital. Within the School of Health, an average of 500 students are enrolled in 15 programs, ranging from clinical laboratory technicians to surgical technology and of course, nursing. There is a budget determined by the State for program operation, but over the past 5 years, the budget has been decreased or held "even," leaving no money for new initiatives. The result is that any new program has to be developed and implemented within the existing budget, facilities, equipment, and personnel. Adding to our challenge is an aging facility that used to house the Springfield Armory, manufacturers of the Springfield rifle, used in World War II.

In the 1960s, the armory closed and most of the buildings were turned over to the new community college. The School of Health now resides in a five-story brick-and-steel building of just over 108,000 square feet.

## 22.2 Previous Experience with Simulated Patients

Some years ago, while operating a medical Sonography program at a community college in Florida, I experimented with using dogs as substitutes for real human patients. Working with several veterinary surgeons, we managed to perform ultrasound examinations on dogs, cats, and even an exotic bird. Unfortunately, we found that long hair and slightly different anatomical location of organs in humans combined to discourage using canine animal substitutes for imaging. However, the real killer of the project was the wide range of canine temperament. After one "Oh, he's so friendly" dog tried to bite the transducer in half, we went back to having students scan each other and human volunteers. After joining the College, I became aware of the MedSim Ultrasim sonographic simulator (MedSim LTD. Kfar Sava, Israel and Fort Lauderdale, Florida, U.S.A). This machine looks and is operated like a real sonographic unit. Images from real patients are stored in a memory system that allows you to "pretend" you are scanning a human. Shortly thereafter, I talked the College into getting a METI (Medical Education Technologies, Inc, Sarasota, Florida, U.S.A) adult and pediatric HPS patient simulator and building a "simulation center." The faculty and students in the Respiratory Care Program were the primary users of the new center, with a few nursing instructors coming to visit and beginning to think about how patient simulation might be used in the nursing curriculum.

## 22.3 Deciding to Build

In late 2003, the critical elements began to come together that led to the decision to build a "virtual hospital." The Dean of the Nursing School retired and the college decided to merge nursing and health sciences into the School of Health. That resulted in both health and nursing coming under the one dean (one with previous simulation experience – me). At the same time, the College was awarded a US$ 3.2 million grant, with about one-sixth of that sum going to the School of Health for upgrades of laboratories used to provide hands-on experiences for students. A former student of the nursing program also approached us to host his research project on how to integrate patient simulation into the nursing curriculum. He would provide a Laerdal SimMan patient at no charge to the College for the duration of the project. It gets better. Several years before this nexus, the State decided to remove some asbestos from the fourth and fifth floors of the health building, leading to a loss of over 50,000 square feet space. Unfortunately, the loss appears to be permanent. That meant moving the health programs located on those two floors down to floors 1, 2, and 3. Being the only dean left standing, I was conscripted into the team responsible for planning and designing how the remaining three floors would be used. It became obvious that we needed an environment that would provide the best hands-on (including simulation) experiences possible and bring the various disciplines together. At that time, the Simulation Center was a single-bed affair that tried to replicate the students' future work environment. To do the job right, we needed a virtual hospital. Sometimes, I get to make unilateral decisions, and one was to name the envisioned new facility "SIMS Medical Center™" – a virtual hospital at the College. The idea was conceived, but the delivery was a long way off.

## 22.4 Deciding *What* to Build

The research project involved using a single Laerdal SimMan with a limited number, about five, nursing students. When I first saw our new patient, he was lying on a conference table in a spare room, naked, blemish free, and pale from head to toe. There were cables and tubes of various types running everywhere, some connected to a laptop computer lying next to the patient's head and one tube connected to a gently puffing compressor. When the patient did respond, it was through a limited menu of "voices" within the laptop. This would not do.

The first thought was to add the research patient simulator to the existing Simulation Center, where there was something akin to a hospital environment. Concerns were raised over scheduling and who would be "in charge" of the patient, so we started to look for an alternative. The virtual hospital idea was still developing, and I was hoping that somehow we could create several spaces for patient simulation that encouraged the use by a variety of disciplines. It would not do to retain the practice of designating existing spaces for individual educational programs. The Respiratory Care Program space became the Critical Care Unit, and the existing nursing laboratory eventually became the Basic Care Unit. The Critical Care Unit has about 500 square feet and the renamed nursing lab has about 800 square feet. One step at a time, we were moving toward the goal of a "virtual hospital."

## 22.5 Deciding *Where* to Build

Since we could not place the new research patient in with the existing Critical Care Unit simulated patient, a new space had to be found. We, however, did not have that luxury. Someone was using every space, and none of the room dimensions could be altered. For instance, no walls could be moved, and no walls or plumbing could be added. What we had was what we would get. Across the hall from the old nursing laboratory was a student lounge with the typical vending machines, tables, and some mismatched chairs. It was about 500 square feet, just big enough to fit four beds and two small workstations. It would be a bit cramped, as but the health faculty kept telling people, so were many real hospital units.

Once the students vacated the "student lounge," the room was painted, and temporary office style partitions were installed to separate the "front" from the "back" beds. There was some money available, so we purchased brand-new patient curtains to further define patient areas. Old wall-mounted computer furniture modules from the retired dean's office were moved to the new space providing two small work areas with a telephone, writing, storage, and task lighting. Of course, we could not plug in the task lighting since the one wall plug was already overloaded with the simulator and AV equipment. There was no electricity in the metal partitions, so we used many extension cords. It was only a matter of days before we "ran out" of electricity, and had to be careful about what was on at the same time. During one of our first simulations, one of the electrical outlets overheated and shut off. Fortunately, it happened in between simulations for groups of students and the College electrician performed a heroic change out of outlets in just under 10 minutes. There was no delay for the next group.

## 22.6 Finding and Installing Infrastructure and Utilities

The original plan was to put four beds in the new space, which we named "Acute Care Unit." We had some beds, but only two were deemed worthy for the new unit and besides, we needed to set up the computer, audiovisual equipment, and work area in the "back" of the Acute Care Unit. So, we began with two beds, one of which had our loaned research patient. I found an old camcorder in a closet and hung it over the office partition, pointing toward the patient. It had an acceptable microphone so we could intelligibly record almost all speech and a sufficient amount of activity immediately about the patient bed. We did

find, however, that additional microphones were needed to clearly hear patient and student interchanges. The live video output of the camcorder was connected to a spare video display placed in the control area next to the simulator computer display. With these two displays side by side, settings for the patient software and room activity could be seen at the same time (Figure 22.1). In addition to the camcorder, audio was captured through several more microphones we found and hung from the ceiling grid directly above the bed areas. An audio amplifier controlled the sound levels of these microphones as well as provided a means to hear room sounds through headphones. Both headphones were found in a closet and appeared to have been unused. To allow the operator to speak for/as the patient, another microphone was connected directly to the audio input of each simulator's computer. These patient voice microphones were found in boxes in which other computers had been previously shipped. For some reason, the microphones had been left behind. Operator "as patient" volume was controlled through the patient simulation software provided by Laerdal. Producing and collecting quality sound is a continuing issue. Even today, we continue to experiment with various microphone configurations in an ongoing effort to improve sound capture.

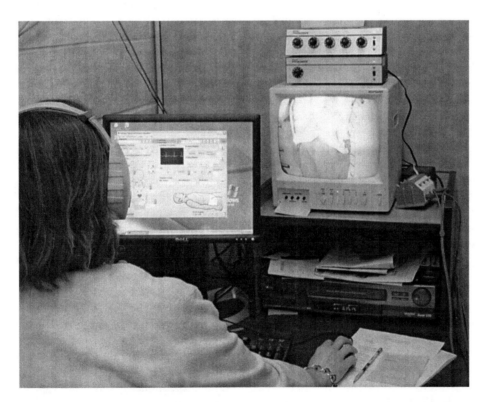

**FIGURE 22.1** A typical setup for controlling a Laerdal SimMan patient simulator. The camera system is a simple "security-" style system with four camera input and built-in video switching. An inexpensive VHS tape deck records the events in the actual patient care area, which is next door to this control room. The operator wears a headset that allows them to hear what is happening in the patient care room and also has a built-in microphone, allowing the operator to "speak" for the patient.

Even after a lot of testing, we were never sure just how long any of the old equipment or jury-rigged connections would last. Even more frightening was the fact that I was the only one who actually knew which cable was supposed to go where. Well, almost every cable. As I look back on that time now, it is very clear that I should have labeled each cable and color-coded each connection point. My only defense is that I was in a huge hurry, doing most of this work after-hours, usually alone, trying to make sure we were up and running for the first batch of students. I also did not want to burden the health faculty with technical issues since they were hard at work developing patient scenarios. We have a part-time laboratory assistant now, and he is wonderful at doing things the right way (including labeling every wire and every connection). He has even managed to "correct" much of the hurried wiring I had done a couple of years ago.

With our "back room" computer and AV setup ready, we held our first "real" patient scenario. Nursing students were oriented to the patient simulators first, finding out what the simulators could and could not do. For instance, some simulators have blood pressure measurements available on one arm while the opposite arm is reserved for "wet" applications like IVs.

Since the partitions separating the "back" from the patient area did not reach the ceiling, unwanted sound became an immediate issue. It took about two scenarios for the students to realize that something was going to change every time a faculty member walked "behind" the partition, or when they heard the keyboard or mouse click. With only a partial partition, students could also clearly hear faculty or staff speaking into the microphone that presented the "patient's" voice. At times, it was easier for the trainees to "cock an ear" toward the partition rather than stay focused on the patient. Some faculty found it difficult to whisper to each other, so discussions about how students were doing could also sometimes be heard. We had created more of a Hollywood sound stage than replicating a hospital setting. This would not do. The control portion of the scenario had to be moved elsewhere. Help was on the way.

In order to begin asbestos removal from the fourth and fifth floors of our building, we had to move all programs, laboratories, classrooms, and offices off those floors. That meant taking space away from programs on the remaining floors. Somehow we got lucky. A large portion of the existing nursing laboratory had to be converted to faculty offices. What was left was just a medium-sized open room of about 800 square feet. There were too many beds for the space and they were scattered about, headboard to headboard, and not looking very hospital-like. If this area were to be useful to the nurses or anyone for that matter, it would have to be modified.

We convinced the College to let us put up real partitions, made of steel studs covered with half-inch plywood, followed by 3/8-inch wallboard. Having the first layer of the new wall with half-inch plywood allowed us to hang any number of items on the wall anywhere we pleased. We asked for a "trough" in

the top of the walls in which to run cables and hoses and a "drop" to allow for cable and hoses from the ceiling down into the wall. Knowing that students could sometimes hear the operator, I asked for the new partitions to run from the floor to ceiling. This is when I found out about HVAC, electrical, and fire sprinkler system codes. There was no way we could build real walls. Each new partition had to be held back from the ceiling by at least 2 feet. It was a little bit better than Acute Care, but still not exactly what we had hoped (Figure 22.2).

Additional electricity supply was a high priority, and it was provided in the form of double duplex outlets installed just above headboard height at each bed station as well as a single outlet located lower on the new wall for powering the electric position/orientation beds. The old nursing laboratory was transformed into two separate four-bed units with workstations, sinks, and storage cabinets. The nursing faculty continued to use task trainers located in the new Basic Care Unit and did not want to give up any of that space to simulated patients. The decision was made to use the Acute Care Unit exclusively for simulated patients. When the time came, we could easily outfit the Basic Care Unit for patient simulation.

Our luck held, and the College agreed to also replace the office style partitions in Acute Care with the new type of walls. The sound issue improved somewhat, but to this day, we get comments at times about "cross talk" over the partitions. The Acute Care area also had an electrical upgrade to handle the power needs of four beds. Two additional beds came from the old nursing laboratory, meaning that all the control equipment had to go elsewhere, or at least that was the excuse to force the construction of a new control room for the Acute Care Unit. A faculty lounge was located directly next door to the Acute Care Unit (the old student lounge). Now, this is dangerous territory. It was one thing for me to take the student lounge for patient simulation, but taking over the faculty lounge for a simulation control room was almost a bit much. Fortunately, the faculty that would be the most inconvenienced were nurses and they appreciated the fact that having a separate control room would greatly enhance the student-patient experience. Using the old office partitions removed from the Acute Care Unit, little control cubicles were set up. All new, extra long cables were installed so that two Laerdal SimMan patients could be operated at the same time from the control room. Several faculty members suggested that a mirrored one-way window be installed so they could watch the students during simulation events. I have used one-way windows before and do not find them as useful as a well-designed camera system. In this particular circumstance and design, the new partial wall would also completely block the view of the two "back" beds (Figures 22.3 and 22.4).

Demand for simulation drove us to add patients to the two "back" beds, but we had already committed a new patient to the Critical Care Unit. Faculty felt that the Laerdal Vital-Sim level patient was appropriate for many scenarios, so we placed two Nursing Anne patients in the "back" of Acute Care. These

**FIGURE 22.2** New wall construction. Metal studding was used to form the skeleton of the wall, then half-inch plywood was attached, followed by 3/8" sheetrock. A trough was created at the top of the partition to hide hoses and wires between the patient and the control area. Hollow stud boxes extending from the wall into the ceiling provided both a pathway for hoses and wires and support for the wall.

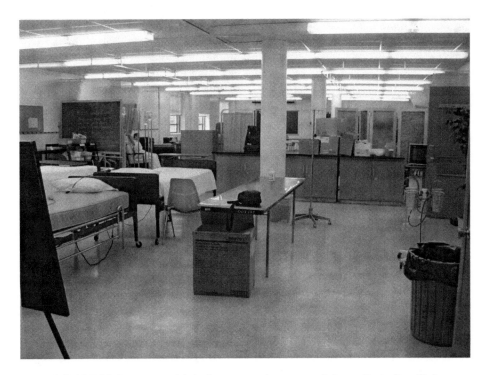

**FIGURE 22.3** Nursing lab before renovation to turn it into a Basic Care Unit.

**FIGURE 22.4** Newer Basic Care Unit with new partition after about half the nursing lab space was taken for faculty offices. A large portion of the old nursing lab had to be used for faculty offices. That forced a complete rearrangement of existing space. Shown here is the new partition allowing for four beds to be used at the same time for either task training or patient simulation.

patients are controlled by a hand-held unit on which the radio frequency may be changed so there is no "cross talk" between two patients in the same area. Two problems quickly emerged. The hand-held control units had to be reset to different frequencies by someone. That was easy. What became the problem was that we did not note on the unit which frequency was in use, so we had faculty grabbing whatever was available. Even if we had labeled the units, distance was the bigger problem. The control room for the "back" beds number 3 and number 4 is some 30 feet away, with one concrete as well as one metal and wallboard wall in between. No way could we reliably get the signal through. Working with the company, custom cables were made for us that we could connect to the hand-held unit in the control room (next to the camera system), and then up into the ceiling and down to each bed. The other end was attached to the control box. To date, they have worked perfectly, and it does not matter which hand-held control is used. Operator provided, interactive voice for these patients is accomplished with wireless microphone systems. To date, even with experimentation with various preamplifiers, we have been unable to "hardwire" Laerdal Vital-Sim patients for speech. We continue to use low-cost NADY (NADY Systems, Inc, http://www.nady.com/) or similar battery-powered wireless microphone systems to speak "for the patient." The

microphone contained within each camera allows the operator to hear what is being said in each bed area.

There was a brief flirtation with remote control–operated "high-end" cameras, but it proved to be more than faculty should have to deal with. In an attempt to capture better sound, we hung decent-quality unidirectional microphones from the ceiling directly above each bed. That meant the faculty had to deal with sound in addition to video and the computer. It did not take long to see them getting very frustrated. They wanted, rightfully, to concentrate on teaching and the student's experiences, leaving the technology issues to someone else. Faculty needed to be faculty, and someone else would have to "operate" the equipment unless the whole operation was simplified. I removed the remote control cameras and all the monitors required for their operation from the patient areas and control room. They were replaced with fixed, wide-angle, black-and-white "security" cameras hardwired to a proprietary video monitor. This specific monitor allows four cameras to be operated with a simple switching system and no special effects generators or video mixers were needed. A proprietary AV cable provided an output for recording.

Not wanting to spend a lot of money on anything not directly related to our student's learning patient care, I decided to use existing analog VHS tape decks. We were almost tripping over them and a local discount store had decent-quality tapes for just a few dollars. Simply operated, low-priced digital systems will be available soon enough, but were not available at this time. The microphones over the beds were also removed along with the audio controls. Faculty tell me they are pleased with the simpler, easier-to-operate controls. Typically, we mounted three of the fixed cameras for each patient area. One at the head and low to see a wide shot of the room, a second directly above the patient to see what the student is doing, and the third camera mounted on the ceiling, but pointing from the foot to the head of the bed. There is almost never a time when one of the cameras does not capture the student-patient interaction. Sadly, the sound problem got worse. When you switch from one camera to the next in the "security-" style system, the sound gets louder or softer depending on the distance of the camera (with microphone) from the patient. As of the writing of this article, I am working on a plan to install ceiling-mounted *boundary*-type microphones and not using the camera-mounted microphones. That will mean adding mixers and amplifiers, but I think I will preset the volume and tone levels and then hide the audio control units from the faculty.

## 22.7 Buying and Using Simulators

Grant money was beginning to be available and we purchased a Laerdal SimMan to keep the research patient company and also ordered two Vital-Sim Nursing Anne simulators. With the adult and pediatric METI patients, the research patient, and the three new Laerdal patients, we had a total of five patients that could be used at the same time. We did not have the money for the extra "background" simulator equipment to operate both METI adult and pediatric mannequins in the Critical Care Unit at the same time. Selected respiratory students had been using the Critical Care Unit for about 1 year and the Respiratory faculty wanted two adults available, not one adult and one pediatric patient. The only solution was to acquire a new adult that would fit within our budget. We eventually added a Laerdal SimMan to the Critical Care Unit.

A new respiratory faculty member interested in patient simulation joined our crew. It did not take her long to almost eliminate the old part-task laboratory exercises and replace them with full-immersion experiences in the Critical Care Unit. As an example, instead of having the students learn how to place a patient on a ventilator in the part-task laboratory, students have to actually "connect" a full-sized, full-function simulated patient to a ventilator and manage the patient. The student has to check the patient identification, determine the status of the patient, and make the appropriate changes to the ventilator as needed. The faculty member can change the status of the patient, hopefully eliciting an appropriate response by the student. Hospital personnel who work with our students

tell us that there has been a definite improvement in student performance in the "real" clinical setting. The College owns two ventilators and leases a new unit for the respiratory students to use. It is wonderful to see those students handling two ventilated patients at the same time. Respiratory faculty make sure that something will go wrong so the students learn to handle a wide variety of "events" (Figure 22.5).

Some time during all the construction, a decision was made to increase the number of first year nursing students from 64 to 100. Respiratory Care faculty had made great advances incorporating patient simulation into their teachings and they were interested in having all first year students replace simple part-task laboratory exercises with more complex, more complete patient simulation experiences. The result was that a total of some 150 first year health students were now expected to have at least some experience with patient simulation. Previously, only a handful of selected students had had experience working with simulated patients in the Critical Care Unit. Assuming we could work out personnel, equipment, and scheduling issues, we were still not at the level of capacity to provide simulation experiences as needed for the increased volume of students. We needed more simulated patients and that meant more space.

To increase our capacity to provide simulated patient experiences, we developed portable control carts that could be wheeled, as needed, into the old nursing laboratory, now called the Basic Care Unit. Most of the time, the Basic Care Unit was used for task trainers. The new portable control units allowed nursing faculty to provide a much more realistic environment for even simple tasks. A good example is administering a drug by injection. Instead of practicing on just an IV injection practice arm, the students must introduce themselves, identify the patient, and demonstrate all the other skills needed in the actual clinical setting. I have seen a student practice injection on a part-task trainer arm perfectly, and then falter when having to perform the same task on a simulated patient. The student could perform (out of context) only the injection component of the larger skill of administering a drug to a patient (in the context of a patient with a disease). Simulated patients can provide the environment where all components must be simultaneously demonstrated. With the portable control carts, we can operate two more Vital-Sim patients and video-record the activities (Figures 22.6 and 22.7). When the simulation is complete, the patients and control carts are moved to a new storage area that used to be a student locker. This storage area is about 200 square feet and is used as a "central supply" for all the health programs. Eventually, we hope to install a computer-based inventory system to keep track of all the supplies we use for laboratories and patient simulation. The inventory system might also provide useful experiences for medical assisting and even accounting students.

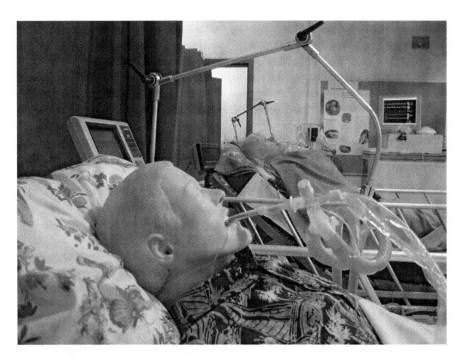

**FIGURE 22.5**   Critical Care Patient Unit. A storage area was emptied and modified to house two beds, a control room, debriefing, and "nurses station." There is still work to be done in this space, but it is currently used weekly by Respiratory Care students. Shown are two patients from different manufacturers on ventilators. The door to the left leads to the room where both patients are controlled.

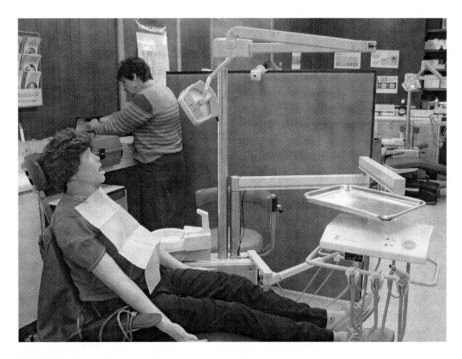

**FIGURE 22.6**   Portable Simulator. A simple portable control unit is used to allow us to provide patient simulation in areas that do not have a dedicated control room. The example shows a portable unit (behind the blue partition) being used in the Dental Assisting clinic. Students had to identify and react to various medical emergencies provided by a simulated patient.

**FIGURE 22.7** Portable Simulation Control Unit. A single camera is used, sometimes attached to the ceiling or even to an IV pole to provide remote viewing of student interaction with the simulated patient.

## 22.8 Success and Growth

Our capacity to provide patient simulation has grown. In addition to the Basic Care Unit and Critical Care Unit, a Trauma room has been added with one patient, and one or two patients can be set up in the Surgery department as needed, again using portable control carts (Figure 22.8).

A second adult Laerdal SimMan patient is now in the two-bed Critical Care Unit. More fixed cameras have been added and both patients can be operated at the same time and also recorded. Since all patient care units are located on the same floor of the building, we can operate six patients in three locations with relative ease. Depending on the type of experience desired, these six simulated patients can accommodate up to 18 students at one time. There is a potential of having 12 simulated patients operational in five separate locations, if we can recruit willing and able volunteer operators. Very recently, eight new computers were networked in an old storage room so that any health student can access screen-based interactive software (e.g., MicroSim, from Laerdal). This little computer laboratory is only about 150 square feet but allows for one group of eight students to work with the software programs, while other students are working with simulated patients. The two groups then switch experiences. Faculty have integrated the use of MicroSim with lecture, laboratory, and now patient simulation exercises.

Considering that we have tried to replicate a fairly sophisticated environment, a hospital, using only bear skins and stone knives, we have not fared too badly. Looking back, there were some real mistakes, but none that we did not either correct or minimize. We have "borrowed" beds, equipment, carts, supplies, audio and visual equipment, and whole modular-furniture "workstations" from other areas of the School of Health. Faculty and staff have made special efforts to "humanize" the patients with pictures of relatives, get-well cards, flowers (introducing the whole "sensitivity issue" with students), and even visits by relatives. A small portion of our grant money has gone to new purchases like crash carts, portable vacuum units, and newer stretchers ("gurneys"). State money was used to erect the new walls and provide upgraded electricity. Most of the grant money to date has gone to acquisition of patient simulators and updating of laboratory equipment (Figures 22.8 and 22.9). Visitors and guests come by about once a week, and we feel very gratified by their comments, but there is much that needs improvement. Some of the improvements relate to patients, but most issues relate to facilities and personnel (e.g., time, funding, promotion and tenure, etc.).

## 22.9 Lessons Learned

Facility issues and very little money constrained our dreams for a spectacular "virtual hospital," but not our resolve. Faculty, staff, and administration, working together, managed to build a very usable and educationally sound environment in which

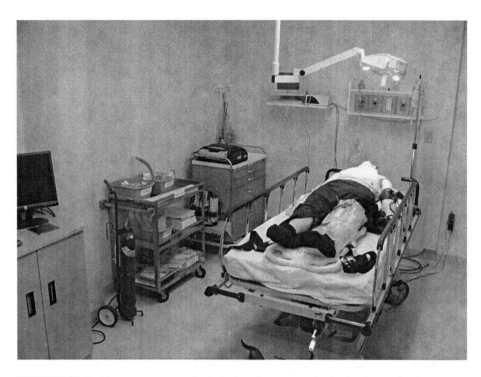

**FIGURE 22.8**   Trauma Room. A single-bed trauma room was built as a result of construc-tion related to asbestos removal on the two top floors of our building. All programs located on those floors had to be moved down to the remaining three floors.

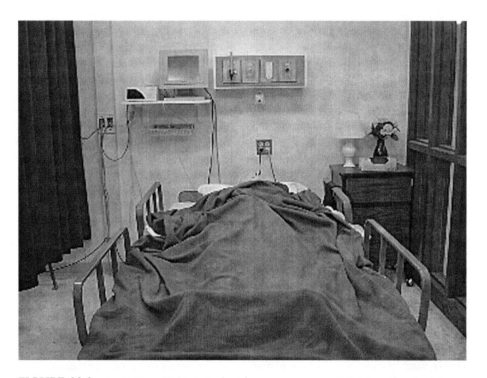

**FIGURE 22.9**   Acute Care Unit. A student lounge was converted into a four-bed Acute Care Unit to provide a dedicated space for patient simulation. This is one of the four patient areas within the Acute Care Unit.

**FIGURE 22.10** Students use the workstation to write progress notes, call for tests, and if needed, call for assistance.

students learn and then demonstrate a large number and variety of competencies. Our center has the capacity to provide realistic patient simulation to a variety of students at the same time. Even though individual clinical educational programs may "house" patients, we encourage interdisciplinary patient care. We have no reason to hang our heads in disappointment.

Should there be a next time, our experience will be used to plan a slightly different environment. The most important factor would be to encourage new construction, complete with the infrastructure found in a typical clinical facility. Instead of using aging compressors, computers, and other equipment "found" in closets, the new facility would have numerous data/voice cables ("drops") and electric power outlets in every space, storage galore, offices, patient simulator preparation and repair rooms, control rooms with sound-deadening surfaces, conference rooms for debriefing, distributed audio/video systems, and patient care units that accurately replicated the students' real patients' clinical care places.

### 22.9.1 There Is No Such Thing as Enough Space

A 500-square-foot room feels like the size of a broom closet in short order. Our Acute Care Unit should have been no smaller than 800–1000 square feet. Even then, the number of patients and students would have to be carefully monitored and controlled. The ideal number for a 500 foot space is two patients, one workstation, and no more than three students per

patient. It is amazing how many supplies and pieces of equipment faculty thinks should be close "at hand." Even though we have seemingly purchased a warehouse of equipment and supplies, a percentage seems to get squirreled away in offices, closets, and other cubbyholes attached to specific clinical educational programs. Eventually, a special storage area had to be built to centralize inventory control. Debriefing, a requirement for each scenario, had to be held "over" the patient until we found empty classrooms nearby. We would prefer not to debrief "over" a real patient in real clinical settings, therefore, we would prefer not to debrief "over" a simulated patient in simulated clinical settings.

---

**Editors' Note:** Some centers reproduce the scenarios as a variation of debriefing methods, especially if evaluation of vital signs is the main educational objective.

---

While we do have the ability to operate four patients at the same time in the Acute Care Unit, we rarely do so because of the noise level. At times, the noise is just part of working in the hospital, while at other times, the noise is a detriment to learning. The ideal is to build a new facility as though you were building a hospital-themed theatrical stage, and then make sure there are plenty of classrooms linked to the simulation areas, computer labs, and small-group meeting spaces.

### 22.9.2  To Build and Operate a Hospital, You Need Personnel

Even though we had challenges with equipment, budget, and space, an equal challenge was getting appropriate personnel. To build and operate a virtual hospital, you need people who know how to teach with patient simulation, people who know how to create simulated clinical environments, and people to manage the operation and growth of the program. A patient simulation center is a stand-alone, separate business within a business. This is true whether the "host" institution is a hospital or a college. It is not good enough to have just clinical faculty. Simulation centers need an organizational structure that includes, at the minimum, an administrator for the business side, a lead educator to ensure quality, technical personnel to make things happen, and simulation-savvy faculty to teach. It wouldn't hurt to have for the entire simulation team a visionary role model, e.g., Walter Elias Disney. Never forget, there is magic in patient simulation.

Our faculty quickly learned how to use patient simulation for teaching Nursing and Respiratory Care students. We were fortunate that one new nursing faculty had previous simulation experience at a local hospital. She mentored three other faculty in the Nursing program, and combined with the two Respiratory Care faculty, we had six faculty familiar, if not comfortable, with simulation. At the time of this writing, eight faculty are working with patient simulation and there are at least two more faculty interested in learning how to integrate simulation into their program curriculum. What we did not have were the "other people" who knew how to replicate clinical facility and then manage clinical faculty as well. Everyone pitched in, and we managed to do a credible job. Again, if there would be another time for us, I would want to have an administrator who would coordinate the whole affair, including hiring of personnel, equipment acquisition, budget, advertising, strategic planning, and quality control. At least two full-time laboratory technicians are needed to operate at the capacity needed for hundreds of students on and off campus, and thousands of patient events. There should be a "coordinator for clinicians" responsible for scheduling all events and making sure that "users" know what they are getting into.

In short, you build a small clinical facility when you embrace patient simulation. Not an easy thing to do, but I think we owe it to the live patients.

## 22.10  Conclusion

With grant money, willing faculty, and a supportive administration, we have managed to create a virtual hospital at the College. We converted space that had previously been used for storage and faculty and student lounges to patient care units that closely replicate the "real" hospital environment. Early in our development, patient simulation was difficult due to lack of experience, borrowed equipment, and inadequate facilities. Still, students were provided enriching patient simulation experiences and encouraged us to provide more such experiences. Hospital personnel began telling us that our students who had had patient simulation experiences appeared to perform better in the clinical setting. Greater demand for simulation experiences meant expanding our capacity. Fortunately, additional faculty members were willing to learn the use of patient simulation and, just as important, we were able to hire a laboratory assistant just for simulation. Right now, the greatest restriction to growth is not money, but lack of organizational infrastructure. For now, I am the Director of our simulation center in addition to my "regular job" of being the Dean. I realize now that when you build a virtual hospital, you need a full-time administrator with a full-time support staff. Perhaps one day that will happen, but for now, students are benefiting from patient simulation experiences. I believe that equates to better patient care, so we will continue to operate with too few personnel knowing that we do make a difference.

## 22.11  Favorite Problem Solvers

Fake medications: www.pocketnurse.com

Cleaning: orange-based cleanser for patient simulator skins

# VI
# Functional Forms at the Institutional Size

*Delay always breeds danger; and to protract a great design is often to ruin it.*

Miguel de Cervantes

Passionate and energetic individuals have reached for simulation as a tool to help solve their clinical education problems, and have done so at many different scales. Clinical simulation is a means to an end. Some ends are as small as the needs of one instructor; others are larger than an entire nation. At every size, there is the common theme of crafting answers to the perennial questions of "who are the students?" and "what are the educational objectives?" The repetition is intentional.

At the most intimate scale is the all-in-one-room facility, with the foreground and background simulation functions all housed within a single (and often small) room. The first simulation facility at many institutions is often at this scale and usually based upon a demand for minimal risk by the space/resource allocation czars. This is a great place from which to initiate a simulation program, because you quickly and cheaply find out the difference between what you thought you wanted and what you actually use, and as usual there is little treasure invested by your institution, you have the supreme luxury of making most of your learning mistakes out of sight. Judith Hwang and Betsy Bencken present the essential components of a simulation facility and how all of them can be arranged to fit within the confines of a single room. Guillaume Alinier describes the ideal and then the actual transformation of available space into a useful simulation environment. Jill Sanko and Amy Agrawal share the steps they took transforming a simple rectangular box of a conference room into an intensive care learning environment.

Most programs that start in all-in-one-room facilities soon grow into a dedicated suite, with real walls separating the various foreground and background functions. Real walls in simulation facilities are there for the same reason we have them anywhere: environmental isolation. In clinical simulation, the single most important isolation provided by real walls is sound isolation: isolation from the discussions of the scenario progression and the students' performance by their instructors in the control area and isolation from the excited discussions by their fellow students in the observation area. Judith Hwang and Betsy Bencken share how they took the lessons learned with their initial all-in-one-room facility and created a suite of dedicated spaces. Guillaume Alinier successfully made a similar transformation.

The longer any one simulation suite is dedicated to a single clinical topic, the more optimized it becomes to support teaching that single topic. This is not in and of itself a bad or good change. It just happens. Likewise, the more successful any single topic simulation program becomes within an institution, the more call there is for other topics to consider incorporating simulation into their teachings. Soon, a variety of different clinical "stages" are proposed, both to optimize furnishings unique to different clinical environments as well as to handle the increase in workload. Judith Hwang and Betsy Bencken have successfully led the expansion at their institution and describe their experiences in what it took to create a variety of different stages within their institution. Claudia Sun and Steve Howard have the good fortune of being contributors to one of the longer-running multistage programs in the world, and share their collective wisdom so acquired. Mike Goodrow presents the design and use of a facility that employs both robotic and human actor simulators located in adjacent, yet functionally dedicated spaces.

# 23

# The One-Room Schoolhouse for Simulation: Adapting to the Learning

Judith C.F. Hwang

Betsy Bencken

## 23.1 Novel Tools in Familiar Environments

With space at a premium at most institutions, having all simulation functions within a single room is definitely the easiest – relatively speaking – to achieve. If the simulator is portable and has a location for overnight storage, then the one room could be a different patient room or area for each use. Using an actual patient location is beneficial because the environment is automatically realistic, though finding an empty patient location may be a challenge. The room could also be just that, a repurposed room that is adapted to mimic a variety of clinical settings.

The all-in-one-room schoolhouse setting is essentially a modern version of bedside teaching, especially when it comes to debriefing. Sometimes, when adverse events have occurred, the learners may perceive debriefing in the same room where the event occurred to be stressful. On the other hand, debriefing in the same room may be helpful in jogging the learners' memories regarding the sequence of events, especially if a video replay of the scenario is not available.

A downside to having functions and personnel in the same room is that there will be extraneous noise during simulation that will absorb and distract some of the students' attention. This detracts from the realism that the instructor is trying to achieve. In addition, if the room is not large, it may be difficult to move about to provide care for their patient.

The primary potential problem with this setup is that if you as the teacher/operator are trying to allow the learners to function independently, the fact that they can see you in the room often makes that reality harder to achieve. Just as the students need to adapt and adjust to make the most of their simulation experiences, so too must their instructors. All too often, instructors are so accustomed to protecting their real patients from potential harm by their students that the instructors fail to allow their students full access to their simulated patients. By not having a dedicated control area in which to observe from it is all too easy for the instructors to "help" by intervening, and thus reducing the greatest added value of clinical simulation: the students owning the results of their actions.

In addition, depending on how the scenario has been programed to execute, students may be aware of when changes are going to occur if the instructor is going back and forth to the simulator controller. A potential solution would be to set up a video camera with a monitor so that you can be in the same room but behind a curtain out of sight, just like the Wizard of Oz. This would tend to limit the ability to change rooms, if only because it would take time to reset up the video equipment. However, if the same room is being repeatedly used in the same way, this would be a moot point.

## 23.2 Clinicians-in-training

If the patient simulator is portable, then it can be taken to any patient care location, with his condition at the start of the scenario already in operation. This means that the patient could be: an adolescent gunshot victim rolling into the emergency room who requires immediate attention from the trauma team; a 60-year-old patient who has undergone a carotid endarterectomy and is now arriving in the postanesthesia care unit where he subsequently experiences respiratory compromise that requires treatment; a pediatric patient transferring from an outside institution to the pediatric care unit who has a cardiopulmonary arrest shortly after arrival; or a 45-year-old patient going to the primary care physician with a 2-day history of palpitations who quickly becomes hemodynamically unstable. If the patient simulator is not portable, the same scenarios can still be used with the patient's starting condition already in place, and the participants asked to report to the patient's bedside. In either case, by using scenarios that require the learners to become quickly engaged in providing care to the patient, the instructor is able to focus their attention on their patient's condition. This helps the instructor and/or operator to "fade" into the background. However, these types of scenarios assume that the participants have some experience providing clinical care; that is, they are at least clinicians-in-training though they can also be clinicians-in-practice.

As mentioned earlier, using the work environment of the learners provides maximum realism in terms of the surroundings and the equipment available. If fostering teamwork is part of the educational objective, already being in a familiar work environment helps to ensure that the multidisciplinary participants will be present versus trying to coordinate the release of the various learners from their clinical responsibilities. The use of an actual patient location decreases the possibility that students will excuse their behavior or inaction on their unfamiliarity with their surroundings. However, using a patient care location – especially in a potentially busy area such as the emergency room – does require coordination with a supervisor in the area to ensure that not only is space available but also that sufficient personnel are present to permit a simulation to be done without interrupting the provision of care to genuine patients.

Video recording sessions when all you have is one room can be a challenge. Setting the video camera up on a tripod may make the participants self-conscious as well as take up space needed for them to move about as they provide care. An alternative solution is to hang the video camera from the ceiling, off a ring from the curtain used to provide patient privacy, or from an intravenous pole. This assumes, of course, that you will also have a system available to replay the video recording during debriefing. Conversely, if you are in a patient care area, the participants could review the video recording at a later time in a room with replay capabilities.

For clinicians-in-training or experienced clinicians, the all-in-one-room schoolroom can also be effectively used to teach or practice specific skills in a clinical context. For example, the instructor may be teaching the students to place central venous catheters with ultrasound guidance and to appropriately treat potential complications of central line placement. Or, the educational objective may be to learn two different ways (such as fiber-optic intubation and retrograde intubation) to approach a patient who has an unexpectedly difficult airway to intubate and whose oxygen saturation is decreasing. In these instances, the deteriorating condition of the simulated patient combined with the learner's focus on performing a particular skill help to distract the learner from the instructor and/or operator's presence in the room. As before, the operator will want to have the patient simulator already functioning at the scenario's baseline status so that the participants can immediately begin to interact with the patient simulator. In this teaching situation, the use of unobtrusive video recording is still important to provide the learners with feedback as to how they managed the patient's clinical condition while performing a given procedure.

When more mundane clinical scenarios are used (e.g., evaluation of a patient in the emergency room and complaining of dizziness) with the clinician-in-training or experienced provider, the instructor and/or operator's presence in the room is more difficult to ignore unless the students participate in simulation sessions so regularly (e.g., weekly) that they simply become habituated to their presence.

## 23.3 Student Clinicians

In contrast, if the participants are student clinicians, then the all-in-one-room schoolhouse setting is ideal, because they often require significantly more guidance as to the care they should provide and the realism of the surroundings while nice to have is less essential. Once again, a variety of scenarios can be produced and presented depending on the educational objective of the instructor and the level of the students (e.g., first semester versus last semester). For example, the learners could be evaluating and treating a patient complaining of chest pain in the emergency room, or be evaluating and treating a medical intensive care unit patient who has become hypotensive, or be inducing anesthesia in a patient in the operating room.

In this situation, the scenario is often started after the students have had the opportunity to familiarize themselves with the patient simulator's capabilities and with a normal patient's vital signs. Particularly for the junior students, the changes in the simulated patient's condition after the scenario is started help them create a list of differential diagnoses. As the students progress, the scenarios can be started prior to their arrival so that the situation more closely mimics a realistic clinical encounter with a patient. It is less critical – and potentially detrimental to the learners – for the instructor to fade into the background, because he/she is often a resource for them. Similarly, the operator of the mannequin does not need to fade into the background. In fact, the knowledge that their decisions are now being carried out by the operator often causes the learners to pay closer attention to their simulated patient as they look for the effects of their orders. The need to videorecord is also less of an issue in these cases because the students are receiving immediate feedback from the patient simulator and their instructor. Over the course of their education, the recurrent use of simulation sessions will help the students become accustomed to the all-in-one-room schoolhouse setting. Hopefully, this means senior students will be able to participate in scenarios that allow them greater independence in providing care without being distracted by the presence of the instructor and/or operator in the same room.

## 23.4 Operations

With everything in one room, having a full-time technical support person could be considered a luxury rather than a necessity. Frequently, at this stage of a simulation center's development, there is neither enough usage of the patient simulator nor enough funds to justify a full-time technical support person. Instead, the faculty often serves as both instructor and operator. In part, this also reflects the faculties' learning curve as they become more comfortable with the patient simulator's functions and capabilities and its integration into the curriculum. It is important, however, to have at least a part-time technical support person who can troubleshoot mechanical problems with the mannequin or issues with the operating software. The exception to this would be if you have faculty available at your institution who are not only clinicians, but also technically savvy, and willing to expend the time to become proficient in the care and feeding of robotic patient simulators.

If you are fortunate enough to have a full-time technical person, care must be taken to minimize the potential distraction caused by the operator's presence. This is particularly true if the instructor and the operator have frequent side conversations regarding the direction of the scenario. Preprograming the scenario and using subtle hand signals are two methods to decrease the verbal communication between the instructor and the operator. A portable dividing wall can also be set up between the operator and the participants. In this case, a video camera with a monitor for the operator to use may be helpful to provide feedback to the operator that the mannequin is responding as expected.

## 23.5 Management

Frequently, when a simulation center is an all-in-one-room schoolhouse setting, the reason is that the simulation center was founded to serve the educational needs of only one group of learners, such as the department of anesthesiology or pediatrics. The benefit of this model is that who owns the simulation center is immediately clear; it is the department who has purchased the mannequins, supplies the space, equipment, faculty, and maintains the center's operations. Scheduling is rarely an issue because it only has to accommodate one group who controls the operations of the simulation center. Similarly, there is less need to acquire additional space for the center because the capacity for simulation sessions is rarely exceeded. Funding is only a problem if the department does not have sufficient funds to continually support the center's operations. The disadvantages of this model include that full utilization of the simulation facilities is unlikely, one department has the entire financial burden of the center, and multidisciplinary scenarios are a challenge to achieve. For an institution, another disadvantage is that multiple all-in-one-room schoolhouse simulation centers may exist throughout the institution with each operating in isolation, perhaps unaware that other similar centers exist. This means that an institution is forgoing opportunities for synergy as well as possibly duplicating expenses.

The Center for Virtual Care in the University of Davis, California Medical Center, Sacramento, CA, differed in two respects from the above described simulation centers. First, its creation was the result of two groups joining together, The Department of Anesthesiology and Pain Medicine and the Division of Nursing. Second, from the beginning, the Center for Virtual Care was intended to be an institutional resource for all disciplines of healthcare. Please see Chapter 11 for more details on the evolution of the Center for Virtual Care. In this case, the all-in-one-room schoolhouse setting reflected the early stage of the Center's development.

If you are in a similar circumstance, once you have a high utilization rate of the simulation center in the all-in-one-room schoolhouse setting, and you are beginning to have more requests than you can handle given the limited capacity, the demand for the simulation sessions becomes the driving force for expansion. Depending on the resources of you and your partner(s) in simulation, you may be able to expand by reassigning space you already have. On the other hand, you may need to approach the institution for space. If this is the case, depending on your institution's resources, you may need to be creative in converting existing space to meet the center's needs for more than one simulation stage each with an accompanying control room, for a debriefing room, and for a utility/storage area. Or, you may be fortunate enough to have an institution that has the capability to raise the funds and has or will create the space on campus to build a new education center focused on the integration of simulation and education.

When a simulation center is being run by more than one discipline, several challenges lie in the issues of who controls the simulation center's facilities, who manages the simulation center's personnel, who controls the scheduling, and who provides funding. If one partner is contributing more financially, that may confer more control over the facilities, the management of personnel, and access to time in the center. If the partners are sharing everything equally, then ideally the operational issues are also jointly managed. While utilization is low, access to the all-in-one-room schoolhouse, can be on a first-come, first-served basis. When utilization is high, however, access to the simulation center potentially becomes a contentious issue. Ideally, the schedule would permit each partner equal access and assigned time in the center. However, if one partner has more sessions than can be accommodated in the allotted time, they may offer compensation to another partner for access to the partner's assigned time. Similarly, if one partner has more simulation sessions than the other(s), they may contribute additional funds to cover the expense of supplies that are used during the simulation sessions, particularly if the supply usage exceeds their share of the supplies budget.

## 23.6 Conclusion

The all-in-one-room schoolhouse for simulation is a modern, high-tech version of bedside teaching that can be used for all levels of students. The major challenge faced by the instructor is overcoming the presence of the mannequin's operator and the noise of the equipment supporting the mannequin's functions to ensure that enough realism is achieved to allow learners to suspend their disbelief during the simulation session. Operationally, however, this type of simulation center is relatively easy to achieve, particularly with a limited budget, since it does not require much space or support personnel. Consequently, the all-in-one-room schoolhouse simulation center is the model most frequently used when the center is established primarily for the use of one department. Over time, the all-in-one-room schoolhouse may become insufficient to meet the students' demand. Then, management issues of obtaining additional space, managing support personnel and scheduling, and securing funding will need to be addressed.

# 24

# All-in-one-room Schoolhouse: Clinical Simulation Stage, Control, Debrief, and Utilities All within a Single Room

Guillaume Alinier

## 24.1 Starting Conditions

The development of simulation centers across the globe has been blessed and impeded by different factors. Some have had the opportunity to start out with vast resources, while others, over a period of time, have had to make their way up the priority ladder of their institution's strategic development plan in order to achieve their aspirations. Some centers purchased the latest technology and devoted space to dedicated simulation, observation, and debriefing rooms. Others made do with a simpler, yet very operational, initial setup where everything happens within the walls of a single room.

One is rarely designing a center from a blank canvas, no matter the size. Even for a new construction, the size and capabilities of your center will be bounded by limitations of finances, neighboring structures and facilities, location, and the needs and imagination of the host institution. Quite often, your center may be based in preexisting facilities, which limit some of your options. The major drawback with an "all-in-one-room facility" is that participants might be able to see and/or hear the patient simulator operator making changes. Students' awareness of their instructors' presence is a great distracter,

and all but eliminates their ability to assume responsibility for their patient. The simulation center will be a great attractant for guests, and their presence is another irritating distraction for the students. Instructors, guests, and fellow students have to remain silent in order not to disrupt their peers taking part in a scenario. Such a setup makes it all too easy for new faculty to prompt participants, again diminishing the learning potential of the simulation experience.

## 24.2 A New Venture

Once upon a time, in 1998, a simulation center was founded at the University of Hertfordshire, in Hatfield, County of Hertfordshire, in the United Kingdom, by a joint initiative between the Department of Nursing and Paramedic Sciences and the Department of Electronic, Communication and Electrical Engineering. This center was an all-in-one-room simulation facility called the Hertfordshire Intensive Care and Emergency Simulation Center. Part of the room (up to $6.0 \times 7.3$ m) was primarily a simulated adult Intensive Care Unit

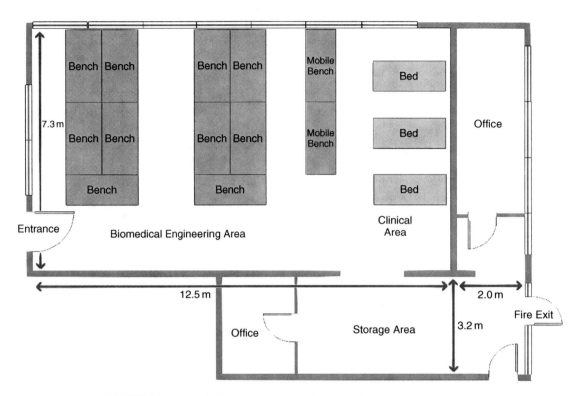

**FIGURE 24.1**   Original layout of the single-room simulation center in 2000.

with low-fidelity patient simulators (Advanced Life Support mannequins) used by nursing students for basic scenarios such as bedside skills and patient monitoring. The other part was set up with benches for about 24 biomedical engineering students to allow them to see clinical equipment in a simulated clinical context and carry out basic maintenance procedures (Figure 24.1). Because of the dual use of the room, the benches and clinical equipment were pushed to one side of the room depending on the group of students taught. This was developed on a limited budget spent mostly on acquiring clinical equipment to recreate the simulated clinical environment. There was no dedicated staff working solely for the center.

## 24.3  A Newer Venture

Although this setup was very appropriate for bedside demonstrations and simple scenarios, it was less than ideal for realistic scenario-based clinical training because of all the visual and audible contamination issues. Toward the end 2000, a non-clinical researcher was appointed to the center[1]. Some minor arrangements were made to improve the setup in early 2001, such as partitioning a corner of the room to configure a control room. This was achieved by using a simple mobile office divider with a transparent upper panel and a desk with a microphone connected to a speaker at the head of the patient

simulator. It adequately prevented participants from being tempted to talk to the operator instead of directly communicating with their simulated patient. Similarly, students were no longer forewarned by seeing the operator changing the parameters of the patient simulator. Additional mobile opaque office dividers were used to separate the engineering section from the clinical area of the laboratory. This engineering area, which became a computer laboratory, was used for briefing and debriefing as well as an observation area during the scenarios (Figure 24.2). The Audio–Video system consisted of a camera on a tripod with audio and video cables strung over the mobile office divider directly connected to a large television screen. In addition, a monitor was displaying the physiological data of the patient simulator.

Given the limited resources, it was the best we could afford and it yet proved very valuable to the students as the results of a study carried out under these circumstances showed [1]. The physical barrier allowed observers not to be seen, but it had no effect on the acoustics aspect, as they could be heard by the scenario participants. Although this limitation was acceptable at the time, it prevented observers, instructors, and operators from freely expressing themselves and commenting about the actions of trainees during the scenario. The main advantage was that the room was multifunctional and flexible, and we had not resorted to do any permanent physical modifications. A detailed image with all the major furniture and equipment is shown in Figure 24.3.

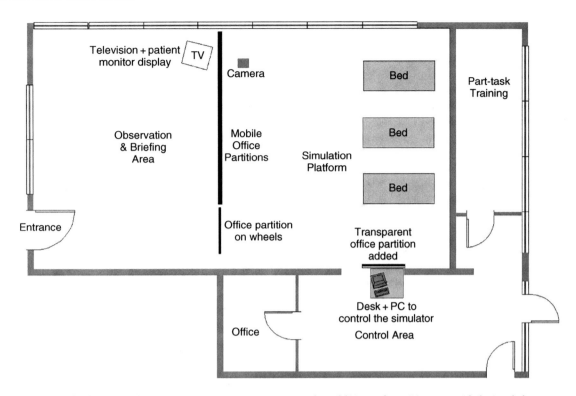

**FIGURE 24.2**   First alterations to our Center in 2001. The addition of partitions provided visual, but not audible, isolation between the students around their patient and the simulator operator/clinical instructor area.

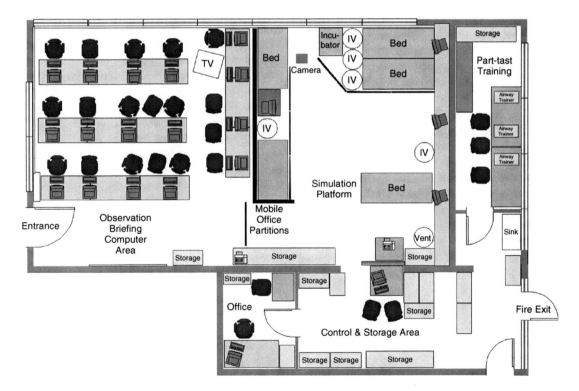

**FIGURE 24.3**   All-in-one Center in 2001 with mobile office partitions between the observation/debriefing room, the simulation stage, and the control room.

The rationale for installing the mobile office dividers was that we did not want to make major physical alterations to the room. The adoption of a physical barrier for the control room was essential to fully use the interactive capability of modern patient simulators by allowing instructors to stay away from the scene of action in the scenario while controlling the physiological parameters and state of consciousness. Not staying near the trainees to prompt them about the state of the patient simulator makes it more realistic. The scenario development relies more on the trainees carrying out proper patient assessment, monitoring, and care. On the trainees' side, not being directly observed "over the shoulder" by an instructor reduces some of the anxiety and allows greater autonomy of action on the trainees' part. Similarly, the adoption of office dividers and a camera link for the other trainees not involved in a scenario enables them to observe their peers tackling a scenario in an unobtrusive manner. It helps scenario participants concentrate on treating their patient more realistically with a reduced impact of peer pressure. The overall approach makes it more realistic than a fully open all-in-one room and forces participants to think more realistically and take, hopefully, appropriate action as they would do with real patients.

## 24.4 Producing Scenarios

The main difference between a multiroom and an all-in-one simulation facility resides in the extent to which the sights and sounds of the observers, clinical instructors, and operators intrude on the participants in the actual simulation stage. A basic physical barrier such as a curtain can make a huge difference for the scenario participants and help them to concentrate only on dealing with the case presented to them. The scenarios we have been producing in the 2001 configuration (Figure 24.2) and in our current center (Figure 24.3) with a separate control and observation room have remained almost identical. They range from patients arriving in the Emergency Department with trauma injuries, chest pain, or other general medical conditions to hypo-volemic postoperative patients. The biggest difference between these two setups has been for the observers who are now able to freely discuss the actions of their peers without disturbing the participants. This has enhanced the learning of the observers and reduced the stress of the participants who were bothered by being directly observed. Also, individuals can come and go from the observation and control rooms without their movements distracting the participants.

## 24.5 Staffing

Opting for an all-in-one simulation facility might have been the only option in terms of physical resources. In accord to the somehow modest investment probably made in setting up such a facility, there is an evident temptation not to appoint a full-time person to direct its day-to-day operations. In contrast, when we consider the larger simulation centers with multiple simulation and observation rooms, it is a lot easier to recognize the importance of having a dedicated team operating its activities. From our experience, the type and number of activities taking place in our initial center significantly increased. There has been a significant improvement in the number of sessions taking place in the center as well as in the quality of the learning experience reported by the students.

The appointment of a researcher dedicated to the center greatly expanded the quantity and quality of the simulation activity. It is a must for the development and practical functionality of any center to have a least one full-time dedicated employee regularly using the technology to remain familiar with the equipment. This permanent and consistently available person is the focal point for the center and becomes the link between for the institution's community of simulation users and the worldwide community of simulation users. This crucial link brings the best practices from around the world into even the smallest of clinical simulation centers.

## 24.6 Conclusion

Just like any hands-on clinical skill, anyone not using simulation technology often enough to sustain their abilities will rapidly forget how to make best use of the patient simulator and the appropriate facilitation approach. I believe it is preferable to appoint someone with a technical background for such a permanent function. In addition, this full-time person should participate in, if not be totally responsible for, the maintenance and development of any aspect of the center. This person can become the point of contact for the center to coordinate its activities and will be able to assist and train other colleagues wanting to make use of the center with their trainees. This is evidence of the fact that this person should not only assume the basic technical responsibilities, but also have a very proactive approach and be involved in the teaching or facilitation of the sessions with the diverse participants' groups.

## Endnote

1. Funded by the British Health Foundation Project Edcomm/Oct98/9d.

## Reference

1. Alinier, G., Hunt, B., Gordon, R., and Harwood, C. Effectiveness of intermediate-fidelity simulation training technology in undergraduate nursing education. *J. Adv. Nurs.* **54**(3), 359–369 (2006).

# The Clinical Simulation Service at NIH: Our Journey

Jill Steiner Sanko
Amy Guillet Agrawal

## 25.1 Characterization of the Initial Program; How We Got Started

The National Institutes of Health Clinical Center, Bethesda, Maryland, U.S.A., is a 250-bed hospital that is unique in that its primary mission is research. It has select training fellowships but no residency programs and no affiliation with a medical or nursing school. A believer in simulation-based education, the chairman of the Critical Care Medicine Department started our simulation program with the purchase of a Laerdal SimMan and audiovisual package in 2003. So far, all of the funding for the program has come from this single department. We started our program with the simple, focused goal of teaching critical care medicine fellows, and planned eventually to expand it to postgraduate nursing education as well.

We feel we have gained a unique perspective on how to make an effective program with a small budget and a small space work. It is our hope that by sharing our experiences we can help others who are embarking on this sometimes trying but worthwhile effort.

## 25.2 Our First Goals

1. Incorporate human patient simulation into an already established critical care fellowship program, with an emphasis on rarely encountered emergency events.
2. Establish a core group of individuals responsible for sustaining the program.

It was not until later, once our core team was established, that we broadened our mission to include enhancement of patient safety via education of nurses and other health care professionals throughout the hospital. For this, we are establishing a larger group of educators from various disciplines whom we will call upon as simulation session instructors. Just as our students start with basic concepts and then progress in their learning, we started with minimal complexity, and only after gaining confidence in the fundamentals did we challenge ourselves by expanding both the range and the scope of our offerings.

## 25.3  A Slow Start

Interested faculty worked on becoming more comfortable with using the simulator and getting more adept at programing and writing scenarios. Since there were no staff members *dedicated* to simulation in these early months, it was slow going. A few scenarios were written and one or two were produced, but the program languished.

Our very first scenarios were Advanced Cardiac Life Support types written by several cardiologists and a clinical nurse specialist for the intensive care unit. A major obstacle was that these staff were struggling to program scenarios with the "flowchart" type approach in the Laerdal SimMan software, with minimal instruction on how to do so. This led to a great deal of frustration as it was time consuming and complex to try to anticipate all trainee moves and plan a response to them in a preprogramed way. Eventually, these volunteer educators gave up and refused to continue to do it this way, thinking they would just have to change vital signs "on the fly." Since simulation was not a major emphasis of their careers, these individuals had little motivation to struggle with the simulator and for several months, it was not used at all.

Two critical events happened in the second year: The Critical Care Medicine Department assigned two interested staff members (authors, AGA and JSS, a Physician and a Certified Nurse Practitioner, respectively) to do simulation at least part time, and a permanent space in the form of a large, single room was acquired. A key component to getting the program moving was professional simulation training at the Harvard Center for Medical Simulation. This training helped to give us a clear philosophy about how to teach using simulation, and our confidence soared. By the third year, the new room was named the Clinical Simulation Classroom, and was finally looking more like a clinical setting. We had regularly scheduled simulation sessions where we trained the critical care fellows, and our program generated some interest among other disciplines.

## 25.4  Starting a Program from the Ground Up

1. Decide on long-term goals for your program, then determine the short-term activities that will move you toward those goals.
2. Seek help from established programs:
   Visit other simulation centers: there is a great deal to learn from established centers; particularly, what is possible for *them* given *their* resources. Visit programs that are bigger, smaller, and about the same size as the one you are starting; much can be gained by seeing how all three function.
   Invite experienced simulation professionals to visit you to learn what is practical for *you* with simulation given *your* resources.
3. Establish a core group of committed individuals to build the program from the beginning, and provide them with formal training.
4. Create a mission statement. It will set the tone for what you are trying to accomplish and guide activities for the program. Mission statements give the program a degree of legitimacy. Seeming established and organized, even when you may not quite be there yet, helps you to seize any early opportunities for help and gifts. Mission statement examples:
   *A. From NIH Clinical Simulation Service*
   1. To establish an interactive clinical simulation service as a resource for training and competency assurance at the NIH Clinical Center.
   2. To develop standardized scenarios for continuing use by the Clinical Center's health care professionals. Primary goals include:
      • Maintenance of technical skills for rarely performed or new procedures
      • Maintenance of mental/critical thinking skills required for emergencies
      • Development of leadership and team cohesion
   *B. From the Harvard Center for Medical Simulation*
      • Lead health care in providing vehicles for clinicians to always practice risky maneuvers on simulated patients before practicing on real patients.
      • Be the organization known for changing the culture of medicine to be more open and honest about mistakes and encouraging of teamwork.

## 25.5  Budgets: Large, Small, and In-between

Simulation programs and budgets come in all shapes and sizes. We started with no formal budget, and continue to teach on a very small budget. After the initial capital outlay of US$ 50,000 for the purchase of a simulator and audiovisual equipment, another US$ 5000 has been spent on formal training and attending simulation conferences. The clinical equipment budget is next to nothing. The main recurring costs are human resources: 50% of a physician's time and 40–50% of a nurse practitioner's time are dedicated to the simulation program. Bids were obtained for upgrades to the room (specifically, the creation of a separate control room with a one-way mirror) and for an upgrade of the audiovisual system, but neither of these items were funded. Once the initial patient simulator warranty ends, several thousand dollars per year for an extended service warranty will need to be factored into our budget.

You can operate a great center on a small budget; it just takes a little creative thinking and a lot of energy. Remember that all simulation is designed to convince your participants to suspend their disbelief. If they can believe that a mannequin is a "real" patient, then surely they can be made to believe that an oxygen flow regulator attached with duct tape to a piece of Styrofoam "wall" really does work. It is almost better to start out with limited resources and be forced to work with what little you have than to have too much money before you are competent enough to use it wisely!

Essentially, there are only three items that you must have to start a program:

1. You need people.
2. You need a simulator.
3. You need space.

The rest you can beg, "borrow," or trade. You do need some start-up funds as well as some money to maintain the program, but simulation can be done well on a shoestring budget.

### 25.5.1 Tips on Doing it Cheaply

1. Research your simulator options: prices are commensurate with features. Your established goals and target audience will define the features in the simulator that will best suit your needs.
2. Find space that does not cost anything to use; no rent payments of any kind, in any form. Our first space was a locker room. Our simulator lived there on a morgue gurney that we "borrowed" for many months. When it came time to produce simulation sessions, we simply wheeled him and all the ancillary equipment (compressor, laptop computer, etc.) into any empty intensive care unit bed space and ran scenarios there. Although this created a lot more setup and take down time on our part, there were some benefits in producing scenarios in the real intensive care unit:
   - It was an actual clinical setting (highest fidelity at the lowest direct cost);
   - We had much of the equipment we needed at our disposal;

- We could easily do "surprise" scenarios; and
- We could easily pull staff and students out of their clinical duties for training.

3. Find enthusiastic educators and offer your services in exchange for their help in producing simulation sessions. Their desire to teach can go a long way in making the scenario successful. As an example, we have worked closely with the "tracheostomy education team." With them, we developed a scenario involving emergency tracheostomy decannulation management (see Appendix 25A.1). They provided reliable actors and co-debriefers for the training of multiple groups of nurses and physicians. Once this was functioning smoothly, it became a showpiece scenario to which we could invite observers who wanted to see what simulation was all about. Having an impressive scenario that is both relevant to the needs of the institution and easy to produce at a moment's notice for nonclinical guests is an ideal way for them to appreciate your value and extend you their support.

## 25.6 Researching Your Simulator Options

Decisions regarding purchases for your lab will be based on your target audience, what teaching goals you have set, as well as on your budget. If most of your customers are going to be students or novice practitioners who need a lot of realistic hands-on procedural training, then acquire lower-cost partial-task trainers. Our target audience was postgraduate nurses and doctors and our focus was critical thinking and leadership skills: we thus determined that higher-cost full-body simulators were best for our needs.

Full-body simulators are more costly, but offer the greatest flexibility. As a general rule, they are not the best way to teach specific physical skills but are good for creating realistic scenarios to test mental, teamwork, and critical decision-making skills under pressure.

To date, there are not very many different makes and types of clinical simulators on the market to choose from. Thus, there is no excuse not to learn everything about the various features, limitations, and possibilities of all of them before any purchase.

## 25.7 Train the Trainer, Education for the Staff

"With thy getting, get understanding"

Malcom Forbes

There are a variety of courses that are offered for simulation program staff. These courses can produce great returns on expenditure for beginners as well as seasoned personnel. They provide instruction in topics such as scenario writing, debriefing, crisis resource management, troubleshooting equipment, setting up a space, and securing funds to maintain a program. They are noteworthy not only for their educational offerings, but also for their networking opportunities. The newness of simulation makes traditional educational resources somewhat limited, so networking provides great opportunity for learning new ideas and problem solving. The comprehensive week-long workshop for clinical simulation course offered at the Harvard Center for Medical Simulation was very stimulating, and as mentioned earlier, gave us the push we needed to get started. Other courses are offered both there and at several other major simulation centers. Attending the annual International Meeting for Simulation in Healthcare is a great way to network as well as a way to learn about new topics in clinical simulation. Lastly, one of the most helpful things is to find willing individuals at other established simulation centers to trade stories and to receive advice. Take the time to elicit the "how we got started," "how we execute sessions," and "how we debrief" stories from multiple simulation centers.

## 25.8  Obtaining Space and Making it Work for You

Obtaining a space and making it your own, although not vital to being able to have a simulation program, makes this much easier. We did not originally have permanent space, yet were able to produce some sessions. There are significant headaches associated with not having a permanent space, but this should not be a limiting factor in starting a simulation program.

Our permanent space enabled us to produce simulation sessions on a regular basis. Our new home was a conference room that was allocated to our department. We seized the opportunity to utilize this prime real estate. This converted conference room now houses all of our equipment, is the stage for all of the simulation sessions, contains a "control area" that has no walls and no one-way window, and is used to debrief the participants. It truly is an all-in-one-room simulation center, measuring 19×30 ft. We have two closets, which hold some of the medium-sized equipment. Curtains and bedsheets conceal very large items like a crib and a ventilator for when they are not in a scenario (Figure 25.1).

To convert a conference room with some clinical equipment in it into a mock clinical environment on a small budget required resourcefulness and creativity. Through our liberal use of duct tape, Velcro, and other materials, the lab functioned, but it was somewhat difficult to get participants to suspend their disbelief in a setting that still looked like a conference room.

Our simulator lives on an old hospital bed. We created a clinical "wall" behind the patient's head out of a piece of heavy-duty Styrofoam to which we attached the oxygen and suction regulators; the whole setup was then taped to the wall behind the head of the bed. Neither the oxygen nor the suction worked, but at least there were places where participants could attach suction and oxygen tubing (Figure 25.2).

The simulator's monitor, compressor, and communication box were placed on a rolling cart behind the head of the bed and were hidden with a sheet. The control computer sat on a borrowed bedside table: with the use of a 50-foot computer cable laid along the perimeter of the room, the simulator operator then could be at least out of the way of the clinical action (Figure 25.3).

Our one camera was set up at the foot of the bed on a tall tripod. The audio/visual tower was then set up next to this. The room did not look exactly like a hospital room, but it was quite functional. After campaigning for months, we finally convinced the maintenance department to install a spare patient head wall (a 4-×8-foot piece of wall that has receptacles for an oxygen flow regulator and suction canisters). With the installation of the head wall, we were able to hide the compressor and communication box (Figures 25.4–25.6). The oxygen and suction regulators can now be fully functional by snaking cables to a portable oxygen tank and a portable suction machine out of view behind the wall, and the real wall itself makes a huge difference in the feel of the room.

Maintenance also installed patient-privacy curtains on tracks in the ceiling that hide the simulator operator as well as the audio/visual tower. The curtains are the same as ones found in the patient rooms throughout the hospital, so not only do they help to hide equipment and the driver, they also make the setting feel like an actual patient room (Figure 25.4). We made decisions regarding placement of items in the room based mainly on utilizing the space in the most effective manner. We placed the simulator in the middle of the room because it allowed for optimal utilization of fixed features in the room: On the left side of the room, there are cabinets and a counter top. We use this for housing all of our scenario props and small equipment. Now, we have a patient curtain that does a better job of concealing the materials. On the far right side of the room are two closets that house bigger equipment, and a filing cabinet. At one point, we toyed with the idea of converting the closet into a control room by having a one-way window put in one of the doors. We have since decided that this is not feasible. Without an actual control room, we could have placed the simulator operator anywhere in the room that would be out of the way of clinical action. With the installation of the head wall however, it made sense to install a patient room curtain from the end of the head wall to the opposite side of the room (see floor plan, Figure 25.1), which would create a control room and still give us access to the closets and the area behind the head wall should we need to make adjustments during scenarios. There is also a door located at the front right of the room that is contained within the curtained area of the control room; this is a convenient way to slip in and

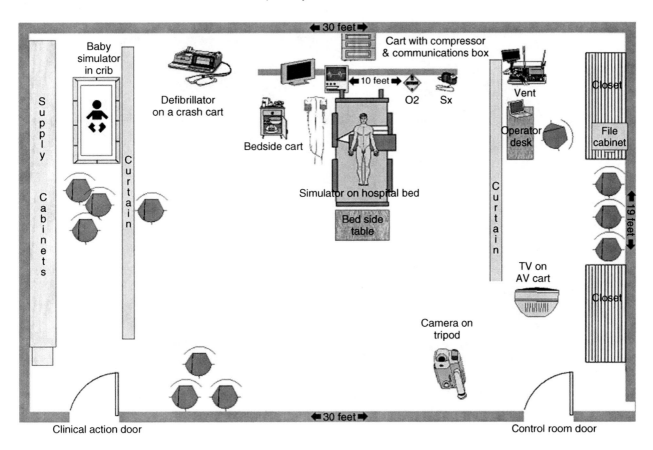

**FIGURE 25.1**  Floor plan of all-in-one-room clinical simulation facility. Simulated clinical area in the center, hidden control area to the upper right, hidden utility area along the very top, debriefing area in lower left, storage in the two sides.

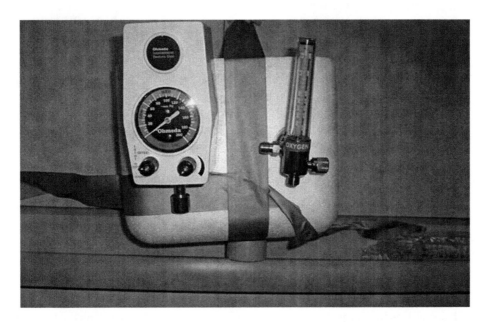

**FIGURE 25.2**  This is a mock-up of what the original head wall looked like.

**FIGURE 25.3** "Control room." The simulator operator sits here to use the laptop computer and can see the clinical action on the television, as collected by the camera on tripod seen through upper part of curtain.

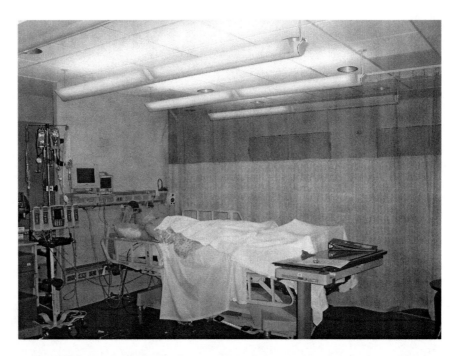

**FIGURE 25.4** Clinical "Action" area with newly installed head wall providing a place for the oxygen and suction regulators and a shelf for the monitor. The head wall now hides the simulator's compressor and signal box. Point of view is from Clinical Action doorway in Figure 25.1.

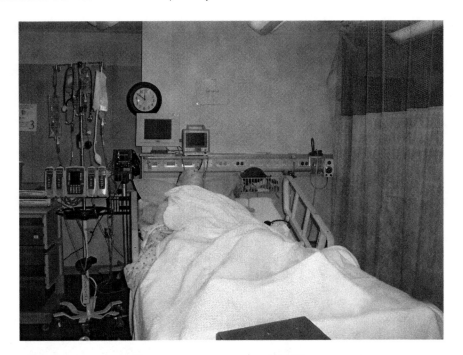

**FIGURE 25.5** This is a photo from the foot of the bed; in this photo you can see the head wall. Also, on this wall we can place a sign that indicates where the scenario is taking place. We have a number of laminated signs for commonly occurring areas (ICU, CT, Medical Surgical Floor), which can be hung with the help of Velcro on the wall and on the back of the sign.

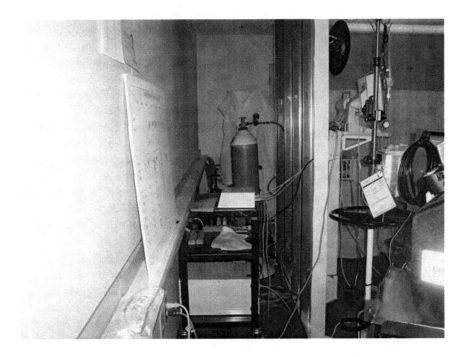

**FIGURE 25.6** "Behind the wall," a 2-foot wide area with the compressor, signal box, and room for an oxygen tank.

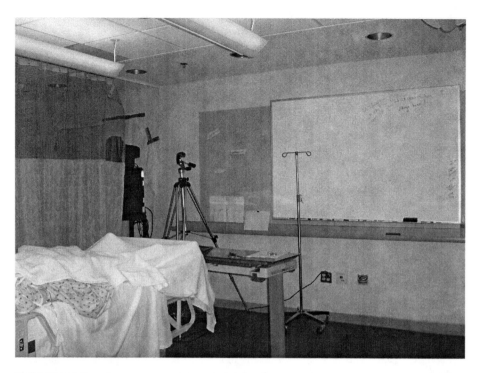

**FIGURE 25.7** This is a view of the room from the Beside Cart, near the head of the bed. Here you can see our dry-erase board, and the location of the camera. Behind the curtain is the "control room."

out of the room without walking through the clinical action. There is a second door on the left side of the room that is utilized for coming in and out of the clinical action during a scenario. The utilization of our simulator and the quality of our simulations increased dramatically once we obtained and outfitted this permanent space.

## 25.9 Obtaining Equipment and Supplies (Begging, Borrowing, and Trading)

A well-equipped space adds to the realism of your lab and provides greater possibilities for scenario development and execution. Our equipment budget was essentially zero; we purchased two items: a large dry-erase board, which was installed on the wall opposite the mannequin's feet, and a 50-ft. computer cable to remove the computer operator from the action area. Nevertheless, we have duplicated much of the equipment in our intensive care unit. Our big "finds" were a code cart, which is stocked exactly like the advanced cardiac life support carts found throughout the hospital, a ventilator, a new vital signs monitor, and an intravenous (IV) pump. We obtained the code cart from the intensive care unit; they used it for the "code blue" training of their nurses. We traded the use of the room and help with various nursing educational projects for

use of the cart. The respiratory department "lent" us a functioning ventilator and a 12-lead electrocardiogram, which we use as a visual prop more than a functional one, as well as disposable respiratory equipment such as endotracheal tubes, etc. In exchange for their donation, we allow them to use the room for both training as well as for some administrative functions. Obtaining equipment has been an ongoing process and we continue to be on the look out for more.

Making friends with people throughout the hospital has served us well. A contact in biomedical engineering has helped us to get the hospital bed, a hospital crib, the vital signs monitor, a defibrillator, and an IV pump. Personnel from the pharmacy gave us an empty medication tray for the top of the code cart. The tray in the code cart houses most of the medication that would be needed in an emergency. We stocked it with empty and expired drug vials that were collected with help of the ICU staff. A recycling bag was placed on the ICU and the nursing staff were asked to place empty and expired vials of medication in it. We relabeled some to simulate rare or unobtainable medications. We order a few other disposables such as syringes, gauze, and tape from the central hospital supply.

Since we stock IV fluids, expired medications, and other medical items that are not suitable for use on patients, we try to prevent these from appearing on actual patient care units. We deal with this problem is three ways. Distance: the simulation classroom is a good distance away from any patient care units, we did not plan on this, but our location works in our favor for this purpose. A Locked door: we keep the classroom locked. This also protects the costly equipment that is housed in the room from theft. Parties wanting access to the room need to sign the departmental key in and out on a log. We instituted the key log recently after a theft of the video camera. Lastly, the medical equipment, medications, and IV fluids are marked with "Not for clinical use" labels.

We also have other contacts in the hospital who collect expired items (sterile equipment, drugs, IV fluids, pacing pads, etc.) for us. We use most items over and over again, very little is thrown out; you never know when you might need miscellaneous "junk." IV fluid bags are refilled; pacing pads are salvaged whenever possible. We also make a deal that if groups come to use the lab and go through a lot of supplies that we cannot easily replace, they help us restock them.

## 25.10 Getting a Program Up and Going

We developed a number of scenarios for critical care fellows and used our department members as trial subjects for many of our early simulations (Table 25.1). Once we felt comfortable, we began offering our services to other departments in an effort to make the program a more institutionally based patient safety initiative. To this end, we invited managers and other key personnel from respiratory therapy, nursing, surgery, anesthesia, and the director of graduate medical education for the hospital. Of those guests who have visited the classroom, about half have returned to utilize our services. The program has not fully blossomed, but we now have a solid foundation from which to work. We have advertised our services; people know we are here, so if they

**TABLE 25.1**  Scenario topics for the clinical simulation service at the NIH

| Category | Title | Main objective/focus | Target audience | Involved staff |
|---|---|---|---|---|
| Leadership | Code leader, DNR* Torsades | Information gathering, establishment of a clear code leader | MD | Simulation service staff |
| Leadership | Code leader, Vfib arrest | Establishment of a code leader, role clarity | MD | Simulation service staff |
| Leadership | Code leader, Trauma SICU* patient | Calls for help, verbalizes global assessment | MD | Simulation service staff |
| Pulmonary | Choking difficult airway | Bagging | MD | Simulation service staff |
| Pulmonary | Mucous plug | Changing ETT | MD, RT, ICU, RN | Simulation service staff and pulmonologist |
| Pulmonary | Pneumothorax | Identifying the problem and calling for help | RN | Simulation service staff |
| Pulmonary | Rapid sequence Intubation | Special topics in RSI | MD | Simulation service and anesthesiologist |
| Pulmonary | Emergent trach decannulation | Replacing trach, securing an airway | MD, RT, RN | Simulation service staff, and trach team |
| Pulmonary | Pediatric respiratory failure | Management pediatric respiratory failure | MD, RN | Simulation service staff, pediatric CNS, pediatrician |
| Pulmonary | Broken ETT* cuff | Troubleshooting vent alarms, treatment for broken ETT | MD, RT | Simulation service staff |
| Cardiac | Wolfe–Parkinson White | Treatments for WPW | MD | Simulation service, cardiologist |
| Cardiac | ACLS*, Code Blue PEA*arrest | Recognition of PEA and its treatment | MD | Simulation service staff |
| Cardiac | ACLS, Code Blue, Vfib arrest | Treatment of Vfib arrest, nursing leadership when MD is acting unsure of next action | RN | Simulation service staff, ICU CNS |
| Cardiac | Transvenous pacing | How to static scenario | MD | Simulation service staff, Cardiologist |

*(Continued)*

**TABLE 25.1**   (Continued)

| Category | Title | Main objective/focus | Target audience | Involved staff |
|---|---|---|---|---|
| Cardiac | Bradycardia, transcutaneous pacing | How to | RN | Simulation service staff, ICU CNS |
| Cardiac | Cardiac medication teaching | Special topics for administering select cardiac drugs | RN | Simulation service staff, Cardiology |
| Other emergency | Anaphylaxis | Treatment for anaphylaxis, emergency cricothyroidotomy | MD | Simulation service staff, +/− allergist |
| Other emergency | Sepsis | Management of the septic patient | RN | Simulation service staff |
| Other emergency | Posttransplant bleeding | Recognition of internal bleeding, preparation for surgery | RN | Simulation service staff and renal transplant CNS* |
| Other emergency | Posttransplant decreased urine output | Recognition of intra-abdominal urine leak, prep. For OR | RN | Simulation service staff and renal transplant CNS* |

DNR = Do Not Resuscitate; SICU = Surgical Intensive Care Unit; ACLS = Advanced Cardiac Life Support; ETT = Endotracheal tube; CNS = Clinical Nurse Specialist.

approach us we are able to work with them. We continue to work with interested parties and establish a larger library of scenarios.

## 25.11 Writing Scenarios/Executing Scenarios

Scenarios are written on the basis of requests by interested educators, or potential learners, or a recognized need (mainly within Critical Care Medicine or Nursing). About two-thirds of the time we write the scenarios ourselves, otherwise we collaborate with educators or content experts outside the simulation staff. A clinical story or "stem" sets the scene, pertinent information may be given to participants at the outset or withheld until they ask the right questions of a clinical actor. We often use a mock patient chart.

Starting with a limited list of major and minor learning objectives sets the path that the scenario will follow (see Appendix 25A.1). This is essential for maintaining focus. We try to keep our scenarios fairly simple, always keeping these learning objectives in mind. Experience has taught us that one of the things that make or break a scenario is the simplicity or complexity of the script. Too much complexity distracts from the point of the objectives, weighs the scenario down, making it difficult to reproduce accurately for multiple learners. We typically write a first draft, discuss it and make changes, and then arrange a dress rehearsal with the help of a naïve "guinea pig" participant who has agreed to critique it. Inevitably, after this

first performance is through, some "debugging" and rewriting for clarification is needed. Sometimes we need a second dress rehearsal. After this, we feel pretty comfortable presenting the new scenario for real participants. Even after the dress rehearsals, the first few sessions with real participants always highlight aspects that need to be modified. We continue to make modifications on the basis of our observations and participant feedback as necessary.

An important consideration when writing scenarios is to establish how much the simulated patient will be allowed to deviate from its script in response to the student's actions, and how the simulator operator will "dance" with the students. The simulator operator can manually control the scenario, meaning the simulator operator will change vital signs and "patient" reactions as the scenario is unfolding, based on what the participant does. The alternative is to *program* the scenario in the computer to make these changes on the basis of a predetermined instruction set. In this approach, as the scenario unfolds, the operator does not intervene but merely records actions that are not detected and recorded by the simulator's own sensors (e.g., "calling for help"). There is also the option of doing a hybrid, having a programed scenario in which the operator will let the program execute but change some parameters as needed.

Really, there is no right or wrong, it is simply what works best for you and what you feel most comfortable doing to achieve your objectives. Each method has advantages and disadvantages. Most of our scenarios are hybrids, with a main vital signs trend programed in, but a simulator operator changing variables "on the fly" in response to participant actions.

### 25.11.1 Fully Manually Controlled Scenarios

Advantages:

> No prior programing required.
> Easier to tailor to skill level of learner especially for basic skill levels.
> Easy to learn.

Disadvantages:

> Harder to reproduce a standardized scenario.
> In-depth knowledge about the subject being taught is needed to accurately change the variables based on learner actions.
> Operators have to "run" the physiological and pharmacokinetic models in their heads (for instance, they have to remember when the epinephrine or muscle relaxant wears off).

### 25.11.2 Preprogramed Scenarios

Advantages:

> Easy to reproduce a scenario.
> A less experienced and less knowledgeable operator can control the simulator.

Disadvantages:

> Much time must be spent learning how to program.
> Participants make unanticipated moves.

### 25.11.3 Hybrids

Advantages:

> Ease of a programed scenario with the ability to intervene when necessary.

Disadvantages:

> Same as for the programed scenario.

### 25.11.4 Lights, Sometimes Camera, Action

Most of our scenarios are active or "live" (scenarios where participants are thrown into a simulated clinical scenario and the action is allowed to unfold as if it were a real crisis). This creates "experiential learning" and puts participants under mild stress, which helps with the retention of the learning objectives. This is better for teaching rapid critical thinking and decision making. We also use the classroom for "static" scenarios (scenarios in which participants and an instructor talk and work through a clinical problem with multiple interruptions for teaching purposes). This is better for teaching a physical technique for the first time to a learner – e.g., insertion of a transvenous pacemaker wire, insertion of a laryngeal mask airway. We call these "static" because often the action is frequently stopped and participants are taught during the session. The term "static" is more succinct than "slowed down time" and it makes intuitive sense. Both methods work well for their intended purposes.

We have found that a video recording is useful when there is a specific physical task that is one of the learning objectives. A single camera can then be set up and trained on the appropriate body part of the mannequin in advance: for example, on the neck, if an emergency surgical airway will be required. For global action or for unanticipated moves on the part of participants, we have found reviewing scenario recordings to be less useful (and often we don't use it) because the time it takes to search for the relevant part of the recorded session leads to awkwardness and delays in the debriefing. Tools do exist that allow "on the fly" marking of interesting portions during recording for easy access during replay. However, they are not free and incur a tax on the most precious resource during any simulation session: the clinical educators' attention to their students' behaviors.

We recently have developed a critical actions checklist and a global scoring tool to evaluate participants' performances (see Appendix 25A.2). Since this scoring is hard to do during a live session, we rely on the camera so that we can review and score the session later. This taping and review method also allows us to have two or more independent scorers.

Debriefing in the same room as the clinical action is perfectly feasible – while initially we felt it would be better to use a different room to have a new "psychological space," it has not been as big a liability as we thought to just step away from the simulator, sit down with participants and talk through their thoughts. We simply silence all clinical noises (alarms, pulse oximeter, etc.), unfold chairs, and invite actors and participants to pull up a chair off to the side of the simulator at the end of each session.

Some centers debrief while replaying the scenario in slow time to indicate the changes in physiological parameters – especially when the status of the patient is the main teaching objective.

### 25.11.5 Lessons Learned

Regardless of how attractive the courses you offer are among students and teachers, the program will go nowhere without a mandate from above (i.e., leaders dictate that simulation will occur). Likewise, it can be difficult to find volunteers to be clinical actors or content experts (co-instructors) for particular scenarios unless their bosses require and reward their participation. Without this mandate from leadership, finding actors and co-instructors when needed can be a point of great frustration.

In the infancy of our program, many thought that we could produce a program with just the two staff we have. We could, but doing so does not produce a well-rounded program. Producing sessions with one person operating the computer and another being an actor does work, but having content experts as part of the debriefing sessions really adds to the learning experience, as well as the discussions. Some of our best scenarios are taught and written with the help of other experts. These scenarios give our program depth. Examples include a cardiologist who teaches during the transvenous cardiac pacing and complex arrhythmia scenarios, an anesthesiologist who presents a rapid sequence intubation scenario, and members of the otolaryngology department who helped develop and teach a tracheostomy scenario. These experts have proven to be excellent resources; and once they've gotten involved, have "gotten into" simulation and are excited about future possibilities. Establishing a core group of people who will teach will pay dividends. They round out a program and broaden its potential.

One should also be cognizant of who the help is in relation to the participants: many learners may become flustered during scenarios or may not be as forthright about areas of ignorance in debriefing if one of their actual supervisors is an actor or debriefer. The consequences are more significant for postgraduate trainees than for undergraduate students, from whom less competency is expected.

Building a program takes time and perseverance. It can involve more people than just the simulation program staff. Plant seeds; make contacts that can add to the depth of your program and that will expand the possibilities. Fertilize often; the "care and feeding" of these ever-important relationships will enable your program to flourish and grow.

## 25.12 Measure the Impact

The trainees who have come through our classroom have predominantly been critical care fellows, nurses, and respiratory therapists. We have kept track of the use of the room and have logged the number of trainees per month and the number of trainee-hours (Figure 25.8). We also administered written evaluations to most of the trainees we had and kept track of their responses (Figure 25.9). We will go on to use this information in building our program.

## 25.13 Conclusion

We feel we have come a long way in our comfort level with simulation, and in the level of sophistication of our offerings. We started with very little, but as the program continues to

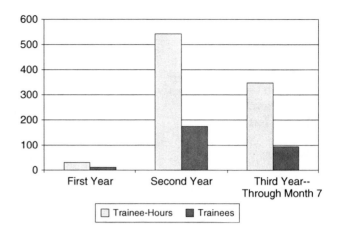

**FIGURE 25.8**   Distribution of participants and hours of simulation provided in our first 2 years of operation.

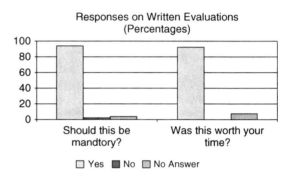

**FIGURE   25.9**   Participant   evaluation   of   their   simulation experiences.

prove itself, we anticipate that more resources will come our way. For the current academic year, we are embarking on several new projects: A competency assessment of the critical care fellows on a core group of scenarios with assessments done at baseline and after 1 year of training and the incorporation of formal Advanced Cardiac Life Support courses into the simulation format. We encourage anyone who is contemplating starting with limited institutional support to jump in and try—provided there is enough money for a simulator, a place to do simulations, and at least one or two administrators who will ensure that trainees attend sessions in the beginning, that is enough to start.

## Acknowledgments

To those who gave freely of their time and expertise to help us get started:

From the Wiser Institute for Simulation Education and Research at the University of Pittsburg Medical

Center: Tom Dongilli, John Schaffer, MD, Paul Rogers, MD.

From Beth Israel Medical Center, New York City: Paul Mayo, MD.

From the National Capitol Area Medical Simulation Center: Jennifer Fellows, BA, EMT and Col Mark Bowyer, MC USAF.

From the Uniformed Services University of Health Sciences Patient Simulation Laboratory: Richard Kyle, MS.

From the Mayo Clinic Multidisciplinary Simulation Center: William Dunn, MD.

From the Harvard Center for Medical Simulation: Dan Raemer, PhD.

# 26

# The Single, Dedicated Clinical Simulation Suite

Judith C.F. Hwang
Betsy Bencken

## 26.1 The Simulation Suite

Having a dedicated clinical simulation suite is a wonderful way to bring a variety of learners into the simulation center and ensure that the focus is on the simulation-based education. In leaving the clinical arena, the participants realize that all of their attention and all of their instructors' attention is now being directed toward their education rather than the delivery of patient care.

The stage can be configured to accommodate the needs of each group, though getting enough clinical environmental cues to create a realistic setting can occasionally be challenging. An adjacent dedicated control room eliminates the distraction of having the teacher/operator in the room during the course of the simulation if the students are to be functioning independently. If the students are working alongside the faculty member, the adjacent dedicated control room allows a second person to operate the scenario so that students are not able to anticipate new changes and developments in the patient's condition when cued by the operator's actions (Figure 26.1).

More importantly, as part of our simulation center, the adjacent dedicated debriefing room serves two functions. First, it allows us to do a debriefing away from the simulation stage, thus removing the stress of being in the same setting where adverse events occurred. Second, the video connection to the dedicated clinical simulation stage enables a second group to watch the scenario unfold and the actions of the group in the scenario. This increases the number of students that can learn at one time. It also permits us to exchange students midway through a given scenario if we so desire without a loss of continuity in the scenario.

Having a separate utility room allows storage of items not required for a given scenario as well as visual separation of the mannequin from its accompanying background-support equipment like compressors, gas tanks, and computer racks.

For students of all types, the biggest advantage of having a single dedicated clinical simulation stage with an adjacent dedicated control and debriefing rooms is the flexibility the instructor has in teaching the students. The instructor can be inside the simulation room guiding the students or the students can be on the simulation stage by themselves, fully responsible for their patient's care.

## 26.2 Instruction in the Simulation Room

The second year medical students come to our Center for Virtual Care for part of their Doctoring 2 course. In addition to procedure stations, the students also have the opportunity to take care of a patient who presents to the emergency room with chest pain. Although they have received didactic teaching on the management of chest pain and hypotension, this is the first time they have a chance to actively practice what they have so far only heard about. An instructor is in the simulation room with the students. If necessary, the instructor helps the

**FIGURE 26.1** Example of the nonclinical devices and clutter best hidden from the students by placing in the separate control room.

students to formulate a differential diagnosis list as well as the accompanying interventions that would be beneficial for each possible diagnosis. The instructor, however, tries to give the students as much independence as possible in managing the patient's care and leaves the actual decision of which action to take up to the students. In the meantime, our full-time support person is in the adjacent control room keeping the scenario progressing, ensuring that the patient simulator presents the lesson-appropriate responses to the treatments the students have performed. When the simulation session is finished, the instructor and the students go to the debriefing room where they review the video recording of the session. At this time, the students can discuss the effectiveness or lack thereof of their interventions and the reasons why an intervention was or was not therapeutic.

## 26.3 Instruction Transitioning Out of the Simulation Room

Clinicians-in-training of all levels and specialties also come to our Center. One group that comes at various levels of training throughout their residency is the anesthesiology residents. During their first month of training, the anesthesiology residents spend time in the Center for Virtual Care learning a variety of basic skills such as mask ventilation and intubation. In groups of four, they also undergo several simulation sessions to learn about induction of anesthesia and emergence from anesthesia. During the first few sessions, at least one faculty member is in the simulation laboratory with the residents, just as they would be in the operating room. Initially, the teacher gives guidance at each step of the induction or emergence sequence with the residents putting into action the instructions. In subsequent sessions, the faculty gives the residents more autonomy during similar scenarios by stepping into the adjacent control room during the session. This helps the residents take fuller responsibility for their learning the steps of induction and emergence and fosters their confidence in their abilities to provide care for their real patients undergoing anesthesia. Other simulation sessions conducted during this first month include patients with hemodynamic instability or respiratory compromise. Usually, the residents are given the opportunity to independently manage the patient's care, because this mimics the situation in the operating room until their attending anesthesiologist arrives.

For the first few sessions when the faculty is providing instructions along the way, video recording is rarely done, because the residents are receiving immediate feedback from the instructor during the scenario. However, as the residents are given more independence during simulation scenarios, video recording is done and shown during a formal debriefing.

This gives the residents the opportunity to observe their actions and review all the consequences of their decisions. The teacher also uses the debriefing to provide feedback about the residents' anesthetic plan and its execution. Following the scenarios with critical events such as hypoxemia or hypotension, the instructor takes the opportunity during debriefing to reinforce the differential diagnoses and accompanying interventions that the residents should always consider in those circumstances. In essence, the didactic lecture on optimum management of critical events is given after the residents have been challenged to do so. This is effective in teaching the residents because they are eager to either fill in the knowledge gaps that have just been revealed to them or validate their thought process and interventions.

Audio communication devices in the simulation room and in the control room (for example, telephones and/or two-way wireless radios) used by either the instructor or the operator provide information to the students that the patient cannot provide, such as results from the laboratory if blood tests were sent or the response of a physician who is on his way to the hospital but is still a 20-min car ride away. At the end of the scenario, the students and their instructor go to the debriefing room where they review the video recording of the session. This gives the students the chance to review their actions and/or inactions, to observe their teamwork and communication skills, and to consider what changes they would implement the next time they encounter a similar situation. The debriefing also helps the students to cope with the emotional impact of dealing with a medical crisis.

## 26.4 Instruction Outside the Simulation Room

Senior nursing students come to our Center for Virtual Care during their last semester to take care of a patient who has a sudden onset of palpitations during his stay in a small community hospital. They have already had several didactic lectures on management of cardiopulmonary problems, including cardiac arrest, and have studied the Advanced Cardiac Life Support algorithms. In this instance, the educational objective is to put the students into a situation they may experience after graduation and provide feedback on their reactions to the situation. Consequently, the students are left alone in the simulation room with the patient simulator once they are given an introduction to the patient simulator's features and capabilities. As the scenario begins, both the instructor and the operator are in the adjacent control room where they can observe the students as they interact with and care for the patient simulator. The scenario has been written with the expectation that the students will take certain actions. Depending on how well the students are meeting the expectations, the scenario may be adjusted "on the fly" to make it either easier or more complicated.

Sometimes, the scenario has a patient with multiple medical problems. In this instance, we will split the students into two groups. The first group will do the initial assessment of the patient and treat one or two of his problems while the second group is in the debriefing room watching the first group on the video display that is linked to the cameras in the simulation room. The scenario is then temporarily paused while the two groups switch places, and the second group now does a secondary assessment and treats the patient's remaining medical issues. This scenario is particularly useful when the student group in the simulation room would otherwise be too large (i.e., more than four or five) for the students to each have the opportunity to interact with their simulated patient.

## 26.5 Operations

Ideally, each simulation session has the support of a technical person to operate the responses of the patient simulator, regardless of where the teacher is planning to be during a simulation session – in the simulation room or in the control room. However, if a technical support person is not available, or the faculty members are comfortable with the software, then they can operate the patient simulator themselves. This works well if the faculty person is planning to be in the control room during the simulation session.

On the other hand, being both the faculty person and the technical support can be challenging if the instructor plans to remain in the simulation room. The use of a wireless remote control (e.g. laptop with simulator control software) in the simulation room is useful for the instructor who then does not have to go back and forth to the control room. The instructor can simultaneously facilitate the students' actions as well as ensure that the patient simulator responds without having to leave the simulation room. The primary downside of this technique is that the participants are more likely to be aware that "something is going to happen" every time the instructor goes to the remote control to change the simulator. For more experienced learners, in particular, this awareness interferes with the suspension of disbelief that is an integral and crucial part of their simulation experience.

Another option for the instructor who plans to operate the simulator while teaching in the simulation room is to have the entire scenario and the possible interventions preprogramed, such that the patient simulator will respond "spontaneously" to a given intervention initiated by the learners. The faculty still needs to be ready to make a change to the program "on the fly" if the students chose to do something not already programed and to troubleshoot if the scenario should fail to progress as expected.

## 26.6 Management

To obtain the space and funds to create a single dedicated clinical simulation suite, with adjacent dedicated control, debriefing, and utility rooms usually entails having institutional support or securing an outside grant. The institutional support can come from the dean of the school, from the chief executive officer of the hospital, or from similar sources. Support outside a department is often needed for the single dedicated clinical simulation suite because the significant space involved in creating such a suite is unlikely to be available within a given department and would be very costly to build from the ground up (assuming the land is available).

Having institutional support, though, does have implications for the management of the simulation center. If an all-in-one-room schoolhouse center has already been established and operated by one department, the department brings their experience to the partnership but that will be counterbalanced by the financial clout of the other party. The founding department may nominally own the simulation suite and even retain control over the facilities and the management of the personnel. However, access to time in the center may now have to extend to other groups who are either identified by the institution's leadership or request access to an "institutional resource." While the simulation center's capacity for classes is not fully utilized, extending time to other learner groups is easily done. When utilization is high, however, access to the simulation center potentially becomes a contentious issue. Decisions will need to be made whether the simulation suite is controlled by the founding department while receiving institutional financial support is an institutional resource solely supported and operated by the institution, or whether each department who

wants to use the simulation suite contributes to its operational costs with either the founding department retaining nominal control or the institution in control of operations.

If the simulation suite is funded by an outside grant, the department receiving the grant initially retains greater control over the facilities, the personnel, and the schedule. However, preparatory plans will need to be made in anticipation of the simulation center's financial needs when the grant expires. Besides approaching their home institution, the department may decide to seek other groups – inside or outside their institution – who are also interested in integrating the use of simulation into their curriculum as partners in the funding, operation, and use of the simulation suite.

## 26.7 Conclusion

The single dedicated clinical simulation suite is extremely adaptable educationally. The stage can accommodate the needs of a variety of students with a change in configuration and clinical environmental cues. The presence of an observation/control room gives the instructor the options of allowing the student to care for the patient independently or being in the simulation lab with the students. The adjacent dedicated debriefing room facilitates discussion after a simulation session as well as enables a second group of students to simultaneously learn during one simulation session. Operationally, however, the single dedicated clinical simulation suite may be constrained by its financial support, which will impact who has control over the facility, the personnel, and the schedule.

# 27

# The Patient Simulation Suite: A Single Dedicated Clinical Simulator Stage Surrounded by Dedicated Control, Observing/Debriefing, Utility, and Office Rooms

Guillaume Alinier

## 27.1 Definition

This chapter defines the Patient Simulator Suite as one room dedicated to the clinical simulation stage surrounded by rooms dedicated for control, observing/debriefing, utility, and office needs. First presented are the purposes, functional components, and prototypical architectural models for a suite. Second are several example configurations of different clinical environments and how the suite plan can be adapted to accommodate different patient simulator–based teaching needs. Third is a description of an actual suite, which illustrates how the suite principles can be honored within a floor plan that looks very different from the basic model.

## 27.2 Purposes, Functional Components, and Prototypical Architectural Models

Effective teaching with any robotic full-body patient simulator will entail the following activities: hands-on teaching and learning with the mannequin, group observation, discussion and debriefing amongst the teachers and learners, control of the patient simulator, office space for the simulation professionals, and utility space for the nonclinical portions of the simulator plus storage of props and equipment not part of the current scenario. Ideally, these activities will take place in dedicated rooms, located adjacent to each other, but separated by walls and ceilings made of sound- and light-blocking materials. Such a concert of mutually supporting and essential simulation activities distributed into adjacent rooms could be called a patient simulation suite. Yes, all these activities can be and are done regularly all within a single room. But with a suite of rooms, effectiveness of simulation-based education is improved, because isolation of sight and sound from the activities in one area prevents cross-contamination into the others. For example, in the control room, instructors and operators can freely discuss both the trajectory of the session and the performance of the students while the scenario unfolds around the simulated patient. In the observing/debriefing room, students can engage in the essential task of learning by watching and discussing their peers currently engaged in a scenario. Subsequent self-evaluation can take place while the clinical

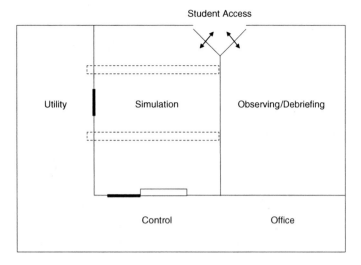

**FIGURE 27.1** An example layout for the five functional areas in a patient simulator suite, their relative positions, and space proportions. Students enter the Simulation and Observing/Debriefing rooms through the two diagonally oriented doorways along the top edge while instructors and operations professionals move freely and unobserved between Control, Office, and Utility. The two dashed rectangles indicate the under floor channels for all the hoses and cables interconnecting the foreground and background components of the patient simulator(s). The open rectangle indicates a one-way window; the two heavy dark lines are doorways for faculty and staff.

simulation room, "the stage," is being reconfigured privately for the next scenario.

As shown in Figure 27.1, solid barriers between the utility, control, and office areas are intentionally absent, as passage between these "backstage" areas occurs frequently during both performances and their preparation. Locating the two student access doors next to each other provides a clear association between the two interdependent teaching environments while reducing the amount of distractions that may occur when traversing from one environment into the other. Using two doors and an intermediate "neutral zone" between these two environments provides both greater sense of psychological separation as well as sound separation than that provided by a single door directly between the two. Placing these doorways on diagonals greatly facilitates access for large objects like beds without having to expend precious floor space on very wide halls or precious funds on special extra-wide doors. Not including a one-way window between the observing/debriefing room and the simulation room eliminates another avenue for easy sound transmission, and minimizes the "fish bowl" effect.

This floor plan and the rationale for it are offered as a starting point for your own planning and design efforts. Every program and every institution have their own set of preferences, requirements, and resources. Typically, the first and foremost limitation on simulation facility floor plans is the fixed boundaries of the existing structure within which your simulation facility will be housed. To date, few institutions have

offered their own simulation community an unlimited budget and a clean sheet of paper to craft a design out of unbounded imaginations. While there are no must-have dimensions for each of the areas in Figure 27.1, the following values are suitable as starting material for consideration:

| | | |
|---|---|---|
| Utility: | $10 \times 30$ ft | $3 \times 10$ m |
| Simulation: | $15 \times 20$ ft | $4.5 \times 6$ m |
| Observing/Debriefing: | $15 \times 20$ ft | $4.5 \times 6$ m |
| Control: | $15 \times 10$ ft | $4.5 \times 3$ m |
| Office: | $15 \times 10$ ft | $4.5 \times 3$ m |

At 300 sq. ft. (approximately $30 \text{ m}^2$), the simulator room can easily accommodate one OR, one ICU, or one labor/delivery patient, and up to two emergency or pre-/postoperative patients. Too much larger, and the simulation room will start to fill with clutter. Too much smaller, and it becomes too crowded for small group teaching. Students do tend to carry a large amount of gear, often in backpacks or rolling suitcases. This "luggage" needs a temporary place to be safely deposited. Thus, set aside a portion of the observing/debriefing room for "garage" space for these objects.

A special mention needs to be made about the utility room. Since few institutions allow for any storage spaces, but they do allow for utilities like communications, heating, and ventilation, and since all robotic patient simulators come with background utility needs and you must have storage space, thus exclusively refer to this area by the name *Utility and Clinical Supplies*. Both for their essential functionality and as further proof that every robotic patient simulator needs a neighboring utility room, include under floor channels (no smaller than $4 \times 8$ inches or $10 \times 20$ cm in cross section) for all the hoses and cables interconnecting the foreground and background components of the patient simulator(s) in the simulation and utility rooms, respectively. Even though the technological trend is for the background components to shrink as ever more of the functionality is packaged within the mannequin itself, there are always additional teaching needs that are only satisfied by adding hoses and cables from "backstage" out to the patient. For example, these channels can be used to pass tubing from the control room to the patient simulator to control urinary flows or Central Venous Pressure, as presented in Chapter 24. These under floor channels remove a dangerous trip hazard, allow "normal" flow around the patient and within the room, and boost the overall fidelity of the simulated environment.

Sound is a significant environmental cue for clinicians and an essential communication channel for clinical educators. In addition, within simulation environment are the nonclinical portions of robotic patient simulators generating loud and nonclinical mechanical noises, instructors in the control room chatting with each other and with the simulator operator, laughter coming from those in the observing/debriefing room, and the general hum from all the cooling fans in all the backstage equipment. The teaching rooms, the simulation rooms, and the observation/debriefing rooms must have sufficient sound proofing to acoustically isolate them from any adjoining rooms or hallways. Without this soundproofing, undesired sounds will contaminate these teaching areas, students will lose focus, lose suspension of disbelief, and your entire effort in simulation-based learning will be greatly diminished. As the utility room will house much of the noise-producing equipment in the suite, the debriefing and office areas are located as far away as possible from these sources of noise.

People flow and separation of distracters are key requirements and warrants revisiting. Student access is confined to the simulation and debriefing rooms, which they enter through doors on the outer edge of the suite. There is no direct, single doorway access between the simulation and debriefing rooms. The purpose is two-fold: (i) better sight and sound isolation and (ii) greater psychological separation. Since their expected behaviors and learning methods within the two rooms are vastly different, when a student is called from the debriefing room to enter the simulator room, the intermediate transition through the outer hallway provides them a neutral zone in which they may prepare themselves for the awaiting adventure. There is no direct, single doorway access between the debriefing room and the office/control areas for all the same reasons that often there is no direct access from the seating areas to the backstage areas in any live theater.

## 27.3 Example Configurations for Different Clinical Environments

Many patient care environments have no preferred handedness and the placement and arrangement of the patient's bed within the room is usually symmetric. For them, the layout as shown in Figure 27.1 could just as easily been mirrored left to right. For example, this suite could be configured with two patient simulators as shown in Figure 27.2. While the floor channels extend much further across the room than needed for this particular configuration, an alternate use for this suite in the future could make their extension essential. The added cost for a few more feet of these channels during *construction* is minimal insurance to future proof your "stage."

However, in a few clinical environments, like the Operating Room (OR), there is a placement preference for some large and essential devices like the anesthesia machine. The anesthesia machine is usually located above the patient's right shoulder except for some operations like neurosurgery. The positional preference for this crucial piece of OR equipment will dictate the placement of all the rest of the areas and rooms in the simulation suite, as shown in Figure 27.3.

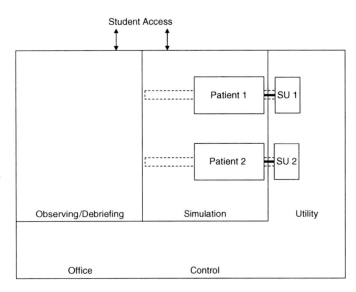

**FIGURE 27.2**  An example layout for two simultaneously functioning patient simulators connected through umbilical hoses and cables via an under floor channel to their respective nonclinical Support Utility (SU) boxes hidden outside the clinical simulation room. Note that this arrangement is a left–right mirror of the plan as shown in Figure 27.1.

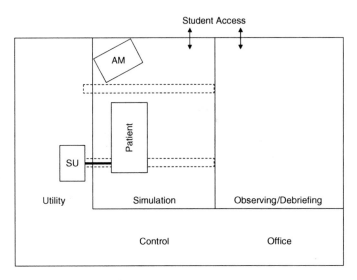

**FIGURE 27.3**  An example layout for an OR environment with the preferred asymmetric location of the Anesthesia Machine (AM) above the right shoulder of the patient. The relative location of these two items, combined with a short umbilical between the mannequin and its Support Utility (SU) box determines where the Utility area needs to be. Placement of the Utility area relative to the simulation stage determines placement of the Observing/Debriefing room.

This basic suite design can be replicated for multiplex environments, as shown in Figure 27.4. Note that the observing/debriefing rooms are placed at the remote corners to isolate them from each other while the central core of control and utility allows for easy flow of people and resources.

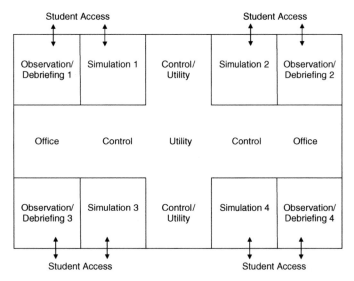

**FIGURE 27.4** An example layout for four Simulation Suites, each with their own dedicated Simulation and Observing/Debriefing rooms, but sharing a common core space for control, utility, and office needs. Simulation Suites 1 and 4 share the same handedness as shown in Figure 27.2, while Simulation Suites 2 and 3 share the same handedness as shown in Figure 27.3.

## 27.4 In the Real World: From an All-in-one Facility to Separate Dedicated Spaces

The previous section described the idealized patient simulator suite. This section describes what happens to that model when it collides with reality. In 1998, the University of Hertfordshire Intensive Care and Emergency Simulation Centre was started with all its functional components in a single room. Six years later, it was relocated on a temporary basis to a different area and a center coordinator was appointed. The setup consisted of a large simulation platform with a very practical raised floor and a small storeroom. We were only allowed to make minor alterations to the room and could utilize part of the adjacent room that was an open-access computer laboratory. We obtained the permission for an audio and visual installation to link the simulation stage with this adjacent classroom to use it for observation and debriefings, as illustrated in Figure 27.5.

The control room was not a real separate area but was positioned in a corner of the simulation room using a smoked glazed office partition along with dark curtains suspended

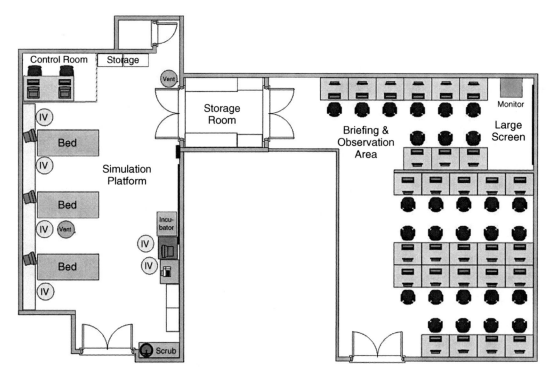

**FIGURE 27.5** Hertfordshire Intensive Care and Emergency Simulation Centre 2004 setup with dedicated observation/debriefing room and simulation stage. Note that while the appearance is vastly different than that offered in the stylized model floor plans, the functional features are very similar.

from the ceiling. As there was no light source in control area, it was significantly darker than on the simulation side of the partition and students could not see the simulator operator. The curtain allowed the operator to hear the trainees working on the simulation platform. The overall new setting and arrangement was considerably better than the previous all-in-one-room facility. It very favorably improved the learning opportunities of the secluded observers, as they were now free to discuss the scenarios in private as they happened on the video displays in front of them.

This setup made the center look more impressive and high-tech to all users and visitors. It rapidly attracted requests for the arrangement of training sessions for our local hospital staff. On many occasions, it proved to be a valuable marketing asset to the University. It was constantly used and presented as a recruiting tool for potential (aspirant) health care students, for the competitive bidding of new medically related programs regulated by the Department of Health, and for the opening of the Bedfordshire and Hertfordshire Postgraduate Medical School in partnership with other local Universities. The only drawback of this setup was that at times, because the observation room was shared and dual purpose, we were sometimes limited in terms of access and could not produce as many sessions as we would have liked to.

## 27.5 Conclusion

The center was funded from a very restricted Faculty budget, small research grants, and a participation in the operating cost by local hospitals using the center, but its activities were strongly supported by the Senior Managers of the Faculty of Health and Human Sciences and of the University. The role of the simulation center coordinator encompassed aspects of operations of our center including administration, time-tabling, technical maintenance and setup, research, web design, marketing, training of faculty, development of scenarios, delivery of short courses, open days, and operating the patient simulator during all the sessions. Although it would have been preferable to have more staff dedicated to the center, this was not an option and being a shared facility would still have been an issue limiting its use. Eventually, faculty users from other University departments and hospitals became such regular users that they also developed satisfactory expertise in facilitating realistic simulation sessions with the coordinator mostly acting as an organizer and patient simulator operator. In this new venue, the demand from internal and external users to access the center increased so much that we obtained University and Higher Education Council funding for the development of a new multidisciplinary simulation facility with multiple simulation stages [1].

## Reference

1. Alinier, G., Hunt, B., and Harwood, C. 2006. *Development of a new multi-professional clinical simulation centre at the University of Hertfordshire*. Abstract presented at the 12th Annual Conference of the Society in Europe for Simulation Applied to Medicine, 28th June – 1st July 2006, Porto, Portugal.

# Multiservice, Single Institution Simulation Center with Multiple Simulation Suites

Judith C.F. Hwang
Betsy Bencken

This chapter is divided into two groups. The first (Sections 28.1–28.7) explores the general concerns of having a simulation center with multiple simulation suites. The second (Sections 28.8 and 28.9) discusses the experience to date at the Center at the University of California, Davis, Medical Center.

## 28.1 Capabilities

Having multiple dedicated clinical simulation stages each with its own dedicated control and debriefing rooms enables us to have multiple educational programs operating independently. Each clinical simulation stage can be configured for a variety of scenarios and schedules. Many of the issues for a simulation center with multiple simulation suites (a clinical simulation stage with adjacent dedicated control, debriefing, and utility rooms) are identical to those for a center with one simulation suite, just on a larger scale. Please see Chapters 26 and 27 for details on the types of scenarios that you can do with this type of clinical simulation configuration when each stage is independently operated.

The simulation stages can also be configured to mimic the transfer of care that a patient would undergo as they move from the doctor's office, admissions or the emergency room to their hospital room, the operating room, the intensive care unit, etc. In this situation, the scenario could be managed with different groups that would move from simulation stage to simulation stage as they provide care in a different environment. For example, a patient who presents in the emergency room with an acute abdomen might be initially evaluated by the emergency room physician who then consults the general surgery service. A general surgery resident would then evaluate the patient in the emergency room, participate in the patient's operation in the operating room, and follow the course of the patient's care as the patient is taken from the operating room to the postanesthesia care unit and then the hospital room.

On the other hand, the patient could be seen in the doctor's office by one group of learners who decide to admit the patient to the hospital. The patient would then be reevaluated in the emergency room by the admitting medical team, a second group of students. However, proper transfer of information and care would need to occur between the doctor's office and the hospital-based team. As the patient's condition worsened, the patient is admitted to the intensive care unit (ICU) where the ICU team, a third group of participants, takes over. Again, good communication to ensure transfer of care goes well is required. And so on.

## 28.2 Operations for Independent Sessions

Operationally, having several clinical simulation stages is a scheduling and logistical challenge. Let's say that your simulation center has six simulation rooms, with each one dedicated to a particular clinical environment, for example: the emergency room, the operating room, the intensive care unit, a telemetry unit, a regular floor bed, and a labor and delivery suite. The benefit of this configuration is that each room can be sized and equipped with the appropriate props and clinical equipment that are needed for the scenarios produced in each room. The use of an adult or pediatric patient simulator is a simple way to change the patient population being treated. With each given simulation room used solely for one type of simulation, it is important to schedule carefully – unless multidisciplinary simulation sessions are planned – to ensure that each class has the appropriate simulation room available for use. For instance, if the pediatric intensive care fellows and residents have an educational session scheduled, the medical intensive care unit fellows and residents session will need to be scheduled at a different time so that both groups can use the intensive care simulation room. However, the surgical intensive care unit fellows and residents may want to have a joint simulation session with the surgical intensive care unit nurses and then both groups would be scheduled simultaneously to use the intensive care simulation room.

Although it may be useful to only have emergency medicine scenarios produced in one room and operating room scenarios produced in another, this decreases the utilization of each simulation room if it cannot be temporarily used for a different specialty's simulation. In addition, it artificially limits the number of students that can be participating in the same scenario simultaneously, because you would ideally have 4–5 students in the simulation and maybe an additional second and/or third group of students watching in the debriefing room. (The actual number of additional student groups will be limited by the size of the debriefing room as well as the educational goals of the scenario.) In contrast, if you used several simulation rooms for the same scenario simultaneously, you would be able to teach a number of student groups at one time. Of course, this assumes that you have sufficient clinical instructors, simulation operators, and additional key props to be able to support each simulation room simultaneously and identically.

If the simulation rooms are not configured to only support one specialty's simulations, the utilization of the rooms by different health care professions is greatly expanded. However,

it will be a logistical challenge to ensure that the rooms are correctly configured with props and clinical equipment for each day's classes. Probably, the best use of the simulation rooms is a combination of dedicated and nondedicated rooms. Let's say that the simulation center's learner groups are dominated by emergency room, intensive care, and operating room user groups. Then, having one room dedicated to each of these areas would be practical because the utilization would be high and the logistics of regularly configuring the appropriate simulation stage would be minimized. If the learner groups are occasionally large in size, having one or two other simulation rooms configured for ancillary use is also beneficial. While not all the necessary props and clinical equipment would be present in the secondary rooms, having some of the equipment will decrease the work needed to transition the simulation stage from a nondesignated or different specialty to the one desired. Meanwhile, the nondedicated rooms are available to all other users of the simulation center and provide the flexibility to provide the appropriate environment for a wide variety of health care professionals.

## 28.3 Operations for Virtual Hospital Sessions

In this case, the simulation center is most likely configured to ensure that the critical parts of a hospital are each represented by at least one simulation room. The challenge for the center lies in scheduling the learners for the sequential sessions that follow one patient's hospital course. Let's say that a patient is being admitted after seeing their primary care doctor for chest pain. Initially, he is sent from the primary care doctor's office to the emergency room via ambulance. From the emergency room he is taken to the cardiac catheterization suite for angioplasty and stent placement before being admitted to the coronary intensive care unit. Two days later, he is transferred to telemetry where he experiences some dysrhythmias. He is eventually discharged 5 days after admission from the hospital floor dedicated to patients with cardiac disease.

## 28.4 Single Specialty Participants

The learners to be taught might all be from one specialty such as family practice or internal medicine. In this case, the simulation sessions might be broken into one scenario at the doctor's office, one in the emergency room, one in the intensive care unit, and one in the telemetry unit. One option is to have all participants arrive simultaneously. They can be divided into groups that take care of the patient at one of the various stages of his hospital course. At the end of the simulation, each group would then go to a debriefing. The benefits of this option are that all the rooms are utilized, the scenarios for each stage of the patient's hospital course can be preprogramed, the simulation rooms do not need to be reconfigured, and all the students have a session to attend. The last point is particularly important if the learners have only one set time to come to the simulation center. If only one simulation session is done, the participants do not actually treat the patient throughout his hospital course (Appendix 28A.1–6, Table 28A.1). If four simulation sessions are held, all the students have the opportunity to treat the patient throughout his stay, but only one group of learners actually does so in the correct sequence (Appendix 28A.1–6, Table 28A.2).

Another option is to divide the participants into groups small enough to use half the simulation rooms at any one time and stagger the groups' arrival. Assuming the total number of learners is 30 and the total number of simulation rooms available is 6, the group could be divided into half with 5 participants in one of three simulation rooms. However, if the total number of students is 45 and the total number of simulation rooms available is 6, the students would have to be divided into three groups of 15 with 5 learners in each simulation session. This works best if a half a day or more has been set aside for the simulation sessions or if multiple session days are planned (Appendix 28A.1–6, Table 28A.3). Using the first example of 30 learners, if a long session is being used, the first 15 students would evaluate the patient in the office and then have a debriefing session. When the second half of the participants comes, they too would evaluate the patient in the office while the first half would now reevaluate the patient in the emergency room. At the end of the simulations, both groups would have a debriefing session. This pattern would continue until all four scenarios have been completed by each half. If multiple session days are planned, the schedule would have only the first half of the group show up for the first day. During the second through fourth session days, both halves of the group would be involved in a scenario though the second half would be one scenario behind the first half. The fifth and last session day would be attended only by the second half of the group. The benefits of this approach is that at least half of the center is used at any one point in time, all the students have the opportunity to manage the patient's hospital course in the correct time sequence, and rather than simply having the learners participate in a preset scenario, the scenario can be tailored to reflect their decisions in the previous session(s). The disadvantages to this option are that several of the simulation rooms will need to be reconfigured to appropriately support the clinical scenarios produced, the schedule needs to be carefully managed and followed by the groups and faculty, an extra simulation session is needed to account for the staggering of groups, and, if group-specific decisions are to have a long-term impact on a given patient's course, there must be a way to record and track the specific changes for each group.

## 28.5 Multiple Specialty Participants

If the learners are to be from multiple specialties, the simulation sessions for the same patient might be broken into the following scenarios: doctor's office, emergency room, cardiac catheterization suite, intensive care unit, telemetry, and hospital floor. The primary care team might be responsible for the patient's care when he is in the doctor's office, on telemetry, and on the hospital floor. While the patient is in the other areas, other teams are responsible for his care, but they must communicate their decisions and treatment plans to the primary care team. Since the participants are coming from different subspecialties, one of the early decisions needs to be whether this series of simulations is to be done as a one-time course for everyone or if it is to be repeated every X months.

If the patient's entire hospital experience is to be presented as a one-time event for all disciplines, then the simplest approach is to schedule the different groups to come to the simulation center at the approximate time of their part in the patient's care (Appendix 28A.1–6, Table 28A.4). In the meantime, they might be attending didactic or small group sessions focused on their specialties' educational objectives. The advantages are that all the simulation stages will be utilized, the students will provide care in the correct time course of the patient's hospitalization, the importance of good communication between teams can be emphasized, and, if desired, the learners may have to cope with the effects of the previous team's decisions and interventions. The disadvantage is that the simulation rooms will need to be reconfigured to mimic the correct clinical setting since, at most, only one room will be already existing in the correct configuration. Thus, careful scheduling will be necessary to ensure that the right groups are present at the correct times, and group-specific decisions need to be tracked so that the necessary changes are made to the baseline scenario.

As before, another method would be to divide the participants into groups small enough to use half the simulation rooms at any one time and stagger the groups' arrival (Appendix 28A.1–6, Table 28A.5). However, with multiple groups involved, scheduling this can be further complicated if the various learner groups vary significantly in size. Likewise, even if you have several days for everyone to go through the simulation sessions, dividing each group of participants into groups of 4–6 and then starting one group can be a scheduling challenge with an increase in the number of simulation sessions needed to accommodate everyone (Appendix 28A.1–6, Table 28A.6). This can be made more complex if there are didactic sessions and/or small group sessions that also need to be scheduled for each specialty. The one benefit of this last option is that the center's simulation stages can often be used as they are designed. Otherwise, the advantages and disadvantages of these two options are similar to the one previously listed for single specialty participants who are similarly divided.

On the other hand, if the series of simulations is to be repeated over the course of a year, then scheduling is somewhat easier. Often, a group from each subspecialty can be identified as being most appropriate for the scenario by virtue of the rotation they are on. For example, the emergency medicine residents primarily assigned to the chest pain section of the emergency room or the internal medicine residents on a coronary intensive care unit rotation. By identifying such groups in each subspecialty, the number of participants is automatically limited. However, their availability to come every X months is predetermined and scheduled, and they benefit from a simulation session that reinforces their current clinical responsibilities. A challenge with this option is that although the teams may be coming in the correct time course of the patient's hospitalization, the actual sessions may not be contiguous. Thus, there needs to be a way to ensure that the teams still have to communicate with each other as the patient's care is passed along. This may be done with video-recorded oral reports and/or with written progress notes. Feedback from the receiving team to the team giving report should be done with a written evaluation.

## 28.6 Technical and Nontechnical Support Personnel

The technical support needed for a center with multiple dedicated simulation stages depends on the faculty using the center and the funds available to the center. In the ideal world with an unlimited budget, there would be a technician for each simulation stage. Realistically, though, it is more likely that there might be one technician for every two to three rooms, and some of these rooms containing more than one simulator. The ability to operate the numerous patient simulators then depends on the extensive use of preprogramed scenarios and the availability of some faculty who are comfortable operating the simulators.

In addition to technical support, the center may also need a nontechnical support person who is available to help ensure the simulation rooms are properly configured for each simulation session. The number of nontechnical support personnel needed depends on the number of classes held each day during a given week and the complexity of the setups. Fewer personnel will be needed if only 1–3 classes are held each day versus 6–8 different classes each day. Conversely, if the configurations for the three classes are fairly complex (e.g., involves multiple procedure stations in addition to the simulation room), then more personnel would be needed than if the configurations for the three classes consist of simply ensuring the simulation rooms have the correct environmental props, patient simulator, and clinical equipment.

## 28.7 Management

Ideally, the multiple simulation suite center would have started as a collaborative effort between the different users (both actual and potential) in the planning stages such that many of the issues regarding who owns the facility, manages the personnel, does the scheduling, contributes to the budget, etc., will already have been discussed and decided upon. While changes may need to be made after the center becomes operational, the preceding discussions will help determine what changes are made and how they are implemented.

## 28.8 Our Experience

Although we started with two groups, the Department of Anesthesiology and the Division of Nursing, our vision at the Center for Virtual Care (the Center) at the University of California, Davis, Medical Center has always been to be a multiservice simulation center for the entire University of California, Davis, Health System. By gradually introducing different groups to the benefits of simulation, we have been able to expand our user base to include paramedics, respiratory therapy, pharmacists, nursing and medical students, and residents in pediatric intensive care, pediatric emergency medicine, adult emergency medicine, family practice, and internal medicine. Recently, fellows in pulmonary critical care and pain medicine have also integrated the use of simulation in their training. Groups we are just beginning to incorporate include general surgery and obstetrics and gynecology. For more details on the integration of simulation into these users' curriculum, please see Chapters 7–15. To date, most of the services have done simulation sessions in isolation. That is, the participants have practiced solely with others in their own specialty. Our goal is to begin to produce multidisciplinary sessions that will allow us to focus on teamwork and communication skills.

At the same time the simulation user base has expanded, the Center has also grown. We began, in 2002, as an all-in-one-room schoolhouse operation that a year later grew into a single simulation suite with a dedicated clinical simulation stage and adjacent control, debriefing, and utility rooms. At that time, we also assimilated a cardiac catheterization simulation suite and the hospital's robotic surgery program. The new location of the Center reflected the support provided by the University of California, Davis, Medical Center's Chief Executive Officer who approved the assignment of precious hospital space. Meanwhile, the Dean of the University of California, Davis, School of Medicine also provided the Center with educational funds used toward the purchase of additional simulation equipment. The management of the simulation facilities, operations, schedule, and personnel, however, continued under the auspices of the Department of Anesthesiology and

Pain Medicine, which provided the salaries and the staff for the Center's administrative support, technical support, and faculty.

In the third year of operations, the Center's day-to-day operations remained under the control of the Department of Anesthesiology and Pain Medicine, in part because the majority of the staff came from that department and in part because that was the department experienced in managing the Center. However, its chief administrative officer began reporting not only to the Chair of the Department of Anesthesiology and Pain Medicine but also to the University of California, Davis, School of Medicine's Associate Dean for Outreach, Continuing Medical Education, and Graduate Medical Education. The latter was a reflection of the gradually growing impact of simulation on medical education within the University of California, Davis.

As we enter our fourth year of operations, we are increasing our facilities to include a new surgical simulation center as well as an additional simulation suite. These additions allow us to not only explore the application of simulation to numerous surgical subspecialties, but also increase our capacity for simulation sessions to accommodate the growing number of regularly scheduled sessions for the multiple departments. Although the Department of Anesthesiology and Pain Medicine continues to provide the salaries and staff for the Center, this is not a viable way to maintain the operations of the Center in the future, especially as the Center begins to fulfill its vision of being a multidisciplinary center. The Center needs to establish ongoing funding that can be used to cover the expenses of purchasing and maintaining the simulation equipment and the salaries of the faculty who teach in the Center. The Center needs to be supported by all its constituent users, whether that is through a combination of renewed financial support from the Medical Center and the School of Medicine or is through funds provided by the various clinical departments that use the Center or through some other mechanism; the road we will take has yet to be decided. In the meantime, we are negotiating contracts with a variety of interested parties outside the University of California, Davis, that are interested in the training and support we can provide

them as they explore the incorporation of simulation into their curriculum or start their own simulation centers.

While the details of funding still need to be determined, the plan is for the Center to evolve into a virtual hospital (with multiple simulation suites) that will provide the optimal environment for all clinicians to learn and practice their skills prior to treating actual patients. The financial support structure of the Center will ultimately have an impact on who controls the facility, manages the personnel, and oversees the schedule. However, the magnitude of the change may be muted because the focus of concern will be ensuring that the overall operations of the Center continue to strive toward fulfilling its vision rather than managing the daily operations and, to date, the current management personnel have always done just that. The Center and its equipment are available to all disciplines upon request on a first-come-first-serve basis though over time many specialties have established regularly scheduled simulation sessions. With additional purchases and growth of the facility, the Center should be able to accommodate multiple overlapping requests for the simulation suites without conflict. The personnel may report to a different management tree but their responsibilities will remain similar to their current job descriptions.

## 28.9 Conclusion

The challenges of being an institutional resource for multiple disciplines lie in the issues of who controls the simulation center facilities, who manages the simulation center's personnel, who controls the scheduling, and who provides funding. Each institution will develop its own solutions that take into account the local politics of operating a simulation center, the capacity and utilization of the simulation center, and the financial support given to the simulation center. These issues may need to be periodically revisited as the simulation center grows in response to growing demand for its value added to clinical education.

# 29

# Operations and Management at the VA Palo Alto/Stanford Simulation Center

Claudia Sun
Steve K. Howard

The Veterans Affairs Palo Alto/Stanford Simulation Center, Palo Alto California, U.S.A., is a large, multiservice Simulation Center whose purpose is to provide simulation expertise to multiple clinical disciplines and to assist in the development and teaching of educational programs. The primary service the Center provides is to develop, organize, and produce simulation courses for participants (medical students, residents, staff, nurses, and allied health care professionals) in varied clinical specialties. Research continues to be an important component of our overall program.

## 29.1 Students and their Courses

The courses that we offer at the Center range from courses for medical students to entire ICU teams. A brief overview of our courses follows (Figure 29.1).

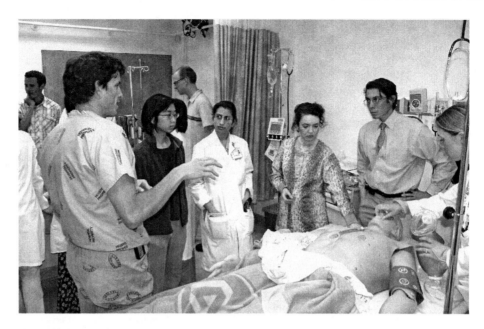

**FIGURE 29.1** Dr. Lighthall teaches a medical student course at the Center. Note the use of support bars to easily reposition cameras to adapt to view actions in different locations within the room.

### 29.1.1 Simulator Training for Acute Resuscitation Skills (STARS)

This course is designed to teach medical students on an ICU rotation how to manage dynamic situations and to identify treatment priorities in the critically ill patient. The students are given a drug recognition exercise to reinforce textbook learning of pharmacology followed by two clinical scenarios and debriefings (see Example in Section 29.13).

### 29.1.2 Anesthesia Medical Student Clerkship

This course exposes medical students on an anesthesia clerkship rotation to the anesthetic care of surgical patients in the operating room. The students are introduced to basic anesthetic concepts followed by a drug recognition exercise during the first simulation session. In the second session, the students are given two simulated surgical patients to anesthetize that include the development of minor complications.

### 29.1.3 Introduction to the Management of Ill Patients (IMIP)

This course uses two to three short scenarios to introduce preclinical medical students to the dynamic clinical management of a variety of patients. These simulations allow the students to practice the interleaving of medical diagnosis and treatment as a dynamic process that is different from what students experience in clinic or ward-based medicine.

### 29.1.4 Stanford Course on Active Resuscitation, Evaluation, and Decision Making (SCARED)

This course was developed for interns who showed an interest in additional simulator training in managing emergency situations and cardiac arrests. It consists of two to three emergency scenarios followed by debriefing sessions.

### 29.1.5 Sort-of-SCARED (SOS)

The miniversion of our SCARED course, this is a short 1-hour course for interns to review Advanced Cardiac Life Support and the management of cardiac arrests.

### 29.1.6 Anesthesia Crisis Resource Management (ACRM)

Stanford anesthesia residents are required to take one full day ACRM course during each year of their residency. The course stresses CRM principles – behavioral techniques used to improve the management of acute, dynamic situations (teamwork, communication, resource management, etc.). Each level includes precourse reading, didactic sessions, and scenarios followed by debriefing sessions. Each resident in the course rotates through one of four parts: the primary anesthesiologist or the "hot seat," the first responder, scrub technician, and observer [1, 2].

### 29.1.7 Emergency Medicine Crisis Resource Management (EMCRM)

The CRM principles can also be applied to emergency medicine. This course adapts the general structure of the ACRM course using scenarios specific to emergency medicine [3].

### 29.1.8 Improving the Management of Patient Emergency Situations (IMPES)

This multidisciplinary, team training course applies the CRM principles to managing emergency situations in the intensive care unit. The course includes a 1-hour didactic session that introduces these principles and provides a group exercise as a forum on how to apply the principles to a patient emergency situation. During the simulator session, the participants are divided into two ICU teams of medical students, interns, residents, fellows, nurses, pharmacists, and a respiratory therapist. Each team participates in one scenario followed by a debriefing session [4].

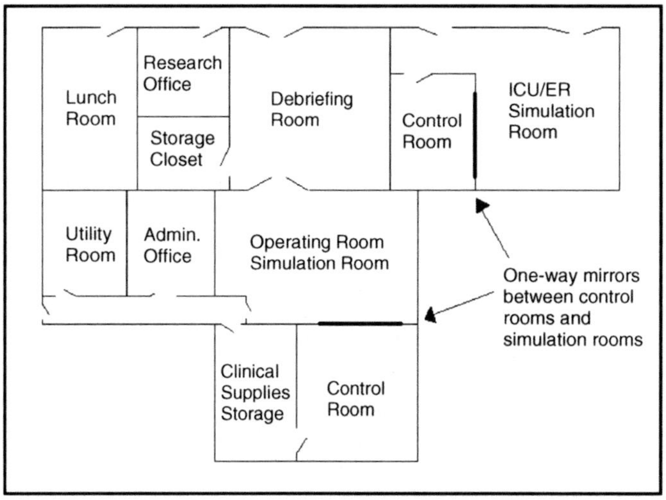

**FIGURE 29.2** Layout of the Stanford/VA Palo Alto Simulation Center. The Clinical Supplies Storage room works well as a "sound barrier" between the Operating Simulation Room and its Control Room.

## 29.2 Preparation and Logistics

Many training programs have limited time and resources (human and monetary) for educational activities. Preparation is necessary to assure that courses execute efficiently under these and other limitations. There can be dozens of tasks to complete up to a year in advance for each course offered by the simulation center. These tasks will range from scheduling sessions and participants to restocking supplies for a scenario to configure the audiovisual system to record the session.

Checklists organized by "due dates" can be very useful in order to guarantee that all the necessary tasks are completed in time for each session. Many of the courses start with the same basic checklist of tasks. The checklist is expanded and customized according to the particulars of each course (see Example in Section 29.13).

The logistics of operating a session smoothly depend on many variables: the sizes and arrangement of spaces of the Simulation Center and its equipment, the audiovisual setup and its capabilities, and the schedule the center maintains to accommodate clinical educators, students, researchers, and visitors.

## 29.3 Layout and Equipment of the Simulation Center

There are three teachings spaces: the simulation rooms, their control rooms, and the debriefing room; and three support spaces for storage, administrative, and research purposes (Figure 29.2).

## 29.4 Simulation Rooms and Equipment

Our simulation rooms are equipped to recreate fully functional and realistic clinical environments such as operating rooms, delivery rooms, ICU, or ER bays. The equipment in these rooms reflect the equipment that would be available in the real clinical environments such as gas supply, overhead surgical lights, anesthesia machines, code carts, ventilators, etc. Besides improving the fidelity, teaching with the same make and model equipment (e.g., defibrillators) used by our participants in real clinical environments improves transference of lessons learned and will likely enhance its proper use in real clinical situations as well. For this reason, some instructors opt to bring the actual equipment that is used in their hospital area. An inexpensive way to obtain some equipment is to salvage excess equipment from the clinical center or to acquire from manufacturers as donations.

The number of mannequins in the rooms also reflects the real clinical environments. For example, we have multiple simulators in our simulation room designated for our ER and ICU courses, whereas our simulated OR has just one simulator (Figures 29.3 and 29.4). The type of simulator used in each room is chosen on the basis of their functions and instructor preference. In general, for our more basic courses, we use lower-fidelity simulators, whereas we use the higher-fidelity simulators for advanced courses.

Another important part of simulating a realistic clinical environment is to provide all the clinical supplies that may be needed by the participants. This includes medications, IV tubing, oxygen tanks, central lines, surgical tools, etc. The

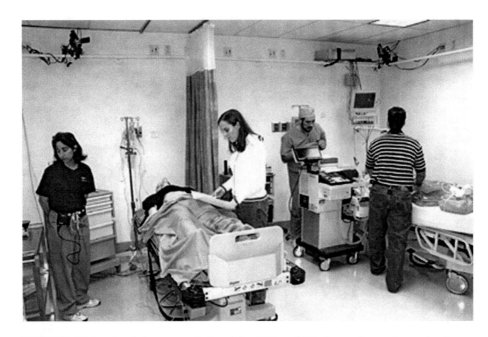

**FIGURE 29.3**   Simulation room designed for ICU and ER classes. A curtain can be drawn between simulators to divide the room.

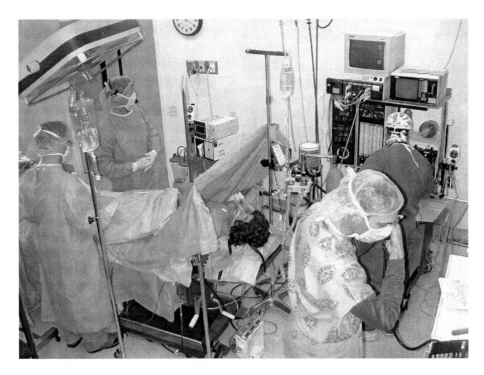

**FIGURE 29.4**   Simulated operating room. Anesthesia machine is completely functional and anesthesia cart is stocked with real drugs and equipment to increase realism.

anesthesia and code carts are fully stocked as they would be in the actual clinical environment. Using the real equipment and drugs allows the simulation to progress in "real" time by making the participants open packaging, draw up drugs, configure equipment for procedures, etc., as they would in a real situation. This may not seem like an important part of simulation, but the amount of time actually expended on each task often leads to interesting debriefing points. For example, if a team determines that they would like to place a central venous line, the participants have to "go through the motions" and take the requisite amount of time and attention to perform this type of procedure. The amount of resources and attention redirected from their patient that is required for these types of procedures is often underestimated, especially when compared to what new information it brings to the scenario.

Whenever possible, our local Pharmacy Service provides us with outdated medications that are then used to supply the Center. We employ safeguards to prevent these drugs from reentering the clinical environment. In cases of controlled substances or drugs that are otherwise required, we label drug ampules/vials with the desired drug information. The Center Manager takes an inventory of supplies before each session and restocks supplies as needed. We have a storage room to warehouse extra supplies. This room is next to the simulator room and allows easy access for confederates to obtain supplies when they are needed.

It is essential that the Center's clinical instructors familiarize themselves with the supplies, equipment, simulators, and rooms available to the course ahead of time. This is especially true for new instructors or ones that participate infrequently. The effectiveness of simulation courses often reflects the degree to which instructors are familiar with all the aspects of the simulation center. Instructors help the Center Manager to decide what equipment and supplies will be needed on the basis of the learning objectives and the types of scenarios that will be produced during each course. Instructors and confederates have also been very successful at obtaining old equipment and supplies from their work units.

Utilizing nurses, respiratory therapists, and pharmacists as confederates in scenarios is an effective way to ensure that the equipment will be operated by people who are familiar with the machines. Although we are interested in how participants manipulate the machines and monitors in their environment, the confederates can be used as a resource to help manage any particular scenario. For example, an anesthesia technician could be used to help set up an infusion pump but they would not select, mix, nor administer any particular dose of drug. The confederates are familiar with the course and its scenarios, the equipment and simulators, as well as the setup of the simulation room. They assist the students in performing procedures on the simulators, operating machines, and getting the supplies they need. Their assistance facilitates movement through the scenario without awkward and unrealistic pauses due to an unfamiliar environment and equipment. Confederates can be trained to help keep the rooms in order, keep the center updated on the supplies that need to be restocked or repaired, and help protect the simulator from overzealous students.

## 29.5 Control Rooms

Our control rooms are situated directly adjacent to the simulation room with a one-way window in between to allow the students to be observed and to minimize intrusiveness. As seen in Figure 29.2, each simulation room has its own control room. The simulators and their audiovisual systems are managed from their respective control rooms (Figure 29.5).

The control room also functions as an information center. In a real clinical environment, an operator might be called to page a consultant or to call a code. In our courses, this is replicated by having a phone in the control room that the participant can call for help. This allows the instructor to manage how the scenario progresses by controlling actions such as when help is activated or when patient laboratory values are delivered.

The instructors activate help (if requested by the participants) by calling our lunchroom, which we use as a staging area for participants. During scenarios, the lunchroom functions as a "soundproof booth" where participants can be sequestered until they are needed in a scenario. This keeps them unaware of the scenario while other participants are either involved in the scenario as a scrub tech or in the debriefing room watching the scenario develop via television monitors. The participants sequestered in the lunchroom are activated by the instructor in the control room.

In our control rooms, we usually staff a minimum of three positions: a simulator operator, a debriefing instructor, and a telephone operator. The simulator operator is responsible for controlling the simulator and works closely with the debriefing instructor to decide how the scenario progresses. While watching the scenario in real time, the debriefing instructor annotates the video recording of the scenario to be used during the debriefing. The telephone operator answers the phone when participants or confederates call the control room, and they also have the task of providing supplies, clinical laboratory results, or any other assistance the participant or confederate may need. The job of answering the telephone can be done by the debriefing instructor but this takes attention away from watching the scenario. In addition to these positions, it is also useful to have extra people in the control room who go into the simulation room in different roles such as anesthesia technicians, respiratory therapists, pharmacists, relatives, etc. These types of courses are personnel intensive, which may be a significant limitation at some centers.

**FIGURE 29.5**   Control room for the ICU/ER simulation room. The views on the television monitors can be configured to accommodate both simulators and their monitors.

## 29.6 Debriefing Room

We believe that the cornerstone of immersive and simulation-based learning is the immediate and thoughtful debriefing of clinical scenarios that students have just participated in [5]. A debriefing room that facilitates this should be readily accessible to the simulation room and control room. Our debriefing room is large enough to seat everyone comfortably so that all students and instructors can see each other and all visual aides (e.g., television monitor, whiteboard). The visual aides that are used in each debriefing session are determined by the instructors and their particular teaching style. A dry-erase board or chalkboard can be used for didactic sessions, overhead projectors or computers used for slides, and video players used for video recordings. Instructors are also able to play back video recordings of the scenario from the debriefing room (Figure 29.6).

## 29.7 Audiovisual System

The audiovisual needs of each course vary depending on the mode of debriefing and teaching objectives. Most often, the simulation sessions are recorded for replay and review during the debriefing session immediately after each scenario. This allows participants to observe their actions and to evaluate themselves. Multiple camera angles and different views can be recorded and manipulated from the control room. We are able to adjust the camera angles and to switch between the different views while recording. The different angles and views are optimized according to the goal of the video recording, whether it is to capture the overall team and scenario landscape, to capture important communication and actions, or to capture interesting equipment usage. There are several microphones strategically placed throughout the simulation room and on the participants (via wireless microphones) so that the audio is clearly captured for future playback. We also use wireless headsets to allow communication between the control room and the confederate, nurse, and surgeon; this allows important communication between the control room and the simulated environment while keeping the participants out of the loop.

### 29.7.1 Observation

Our audiovisual system allows participants not actively involved in the scenario to watch their colleagues from the debriefing room. This allows them to evaluate performance without having to actually perform tasks and provides a unique view on the management of the clinical event. They are expected to participate in the debriefing along with the other

**FIGURE 29.6** Anesthesia residents debrief after a clinical scenario using video clips of their performance.

participants who were engaged in the scenario. Our single debriefing room is multifunctional, allowing participants to watch live and replayed video from either simulation room. Our debriefing room is also equipped with a camera and microphone, which allows visitors, instructors, or other staff to watch a debriefing session from the main control room. Our approach has been to outfit most of the Center's rooms with audiovisual capture and reproduction capabilities for potential future use. For example, a camera in our lunchroom allows us to use the room as a consultation area to role-play breaking bad news after catastrophic medical problems.

### 29.7.2 Annotation

Our new digital recording and annotation system uses a software program (Studiocode) that allows the instructor in the control room to annotate the video recording while it is being captured and saved. This allows the instructor to mark specific points in the video recording that they would like to show during the debriefing session. The instructor can open the video file and show the segments that they want to use in the debriefing session. A central hard drive, which is accessible from both control rooms and the debriefing room, stores the annotated video recording for this purpose. Although not perfect, this all-digital random access system saves time and effort over our previous time-consuming linear access method of forward/rewind of 8 mm and VHS tapes.

### 29.7.3 Managing Video Recordings

Management of video records has also gotten much easier and convenient with digital recording capabilities. Storage of 8 mm and VHS tapes took up space, and had to be well organized in order to readily find specific video segments of interest. With our new digital recording system, video records are stored as files on a hard disk. This takes up much less room, and with search features and file organization, it is very easy to locate specific video segments. An organized collection of the videos is extremely valuable to researchers when searching for specific scenarios. We ask participants to complete video consent forms after each course so that the videos can be used for research or other purposes.

## 29.8 Scheduling Courses

Managing the schedule of a busy simulation center can be a delicate balancing act. Experience has taught us to schedule courses as far in advance as possible for all of the individuals involved. The scheduler (in our case, the Center Manager) must balance between the time limitations of the students, clinical instructors, and confederates since they often have clinical duties and a limited number of hours for education and training.

## An Example: Scheduling Anesthesia Crisis Resource Management (ACRM) Courses

For ACRM courses, the process begins with selecting dates for the courses and adding them to the master schedule of the Center. Different constraints are discussed at this meeting and the dates are chosen on the basis of these limitations. Course directors, Training/Education Coordinators, Center Manager, and the Center Director are required to facilitate this process. Other individuals involved in trainee scheduling and education (e.g., the Chief Residents) participate in this meeting to discuss special aspects of scheduling as it pertains to them. After dates are chosen, the participants are assigned to specific dates by the Chief Residents according to the participants' rotations. Residents are scheduled during a normal working day and we attempt to have them participate when they are rested and not postcall. If participants are unable to attend on their assigned date, a specific plan of substitution is followed to prevent having less than a full course for any given simulation activity. The Center Manager sends an e-mail reminder 2 weeks before the session that includes date, time, directions, and precourse reading to participants. Attendance is confirmed with the participants a week ahead of time.

## Anesthesia Crisis Resource Management Precourse Checklist

2 Weeks
  • Send out invitations
1 Week
  • Confirm attendants and nurse
1–2 Days
  • Fill out scenario schedule (e-mail template to the Center's Clinical Director)
  • Paperwork:
    – Video consent
    – Research consent
    – Confidentiality
    – Evaluation
  • Other Paperwork:
    – Familiarization
    – Nurse timesheet
  • Configure American Society of Anesthesia ACRM DVD
  • Check/replace/charge batteries for wireless headsets and microphones
  • Prepare charts
  • Check drawers, restock OR and code cart
  • Clean up OR
  • Prepare blood bags
  • Restock pharmacy
Day of Course
  • Put charged batteries in wireless headsets and microphones

  • Configure simulator for familiarization scenario
  • Configure Audio-Video, Studiocode

In addition to scheduling participants, course instructors and nurse confederates must be scheduled for the anesthesia resident course. Course instructors work with their administrators ahead of time to ensure that they will not have clinical duties the day of the course. Nurse confederates are scheduled by the Center manager and their attendance confirmed a week ahead of time. If they are unable to attend their assigned dates, the nurses usually find a substitute for themselves within their group.

### 29.8.1 Number of Participants and Length Of Course

We feel that it is important for every participant in our simulation courses to be an active participant in the "hot seat." This can be difficult in large courses, which is our main rationale for dividing large numbers into many small group courses. Our ACRM courses have 3–5 participants per session. This small number allows for a course of reasonable length (about 8 hours), allows every participant to rotate through each role (Table 29.1), and optimizes debriefing dynamics so that all members can participate.

For courses other than ACRM, the Center Manager works with Course Directors and Training/Education Coordinators to schedule dates for the courses. Participants, instructors, and confederates are scheduled and confirmed by the Training/Education Coordinators without the involvement of the Center. The number of participants, instructors, and confederates in other courses depends on the type of course and frequency of courses. The courses range from just a few medical students per session to an entire ICU team for the ICU team training course.

Along with video consent forms, participants are also asked to complete confidentiality forms, research consent forms when applicable, and evaluations. Their evaluations of our sessions are important to the continual development and quality management of simulation courses. The confidentiality forms are important for two reasons. One reason is to ensure that

**TABLE 29.1**  Participant roles for a four-person ACRM course

| Role | Participant location |
|---|---|
| "Hot seat" or primary participant | Simulated operating room |
| First responder | "Soundproof booth" until activated |
| Scrub technician | Surgical assistant – facing the monitors |
| Observer | Debriefing room watching scenario on monitors |

participants do not divulge information on the scenarios to other participants. The secrecy of scenarios is important so that participants cannot anticipate events and are forced to deal with the management of the scenario as it develops, much as they would in real life. Scenarios are difficult and time consuming to develop, so we ask the participants to maintain confidentiality. The second reason we have participants sign confidentiality forms is to protect the participants from judgments and opinions of their performance in the simulation session. By signing the confidentiality form, both the instructors and the participants are "pledged" to not discuss each other's performance in the scenario outside of the Simulation Center.

## 29.9  Scheduling Visitors

We have many visitors throughout the year, and they are scheduled in a number of ways. Since we are located at the Veterans Affairs Hospital in Palo Alto, many of our visitors are associated with the Veterans groups. These visitors are scheduled through the public affairs office, our administrator, and the Center Manager. Many of the visitors are not clinicians so their visits are not specifically scheduled to coincide with a course. If there is a course at the time of their visit, we allow them to watch from the control room, otherwise, we give them a tour of our center and demonstration of the simulators.

Other visitors come from other clinical institutions and are specifically interested in simulation. They may already be familiar with simulation and interested in working with simulators. For these visitors, we try to schedule their visit while there is a course. The Center Manager works directly with these visitors to schedule their visit. They are sent a master schedule of courses so they are able to visit during the course in which they are most interested. Course directors and instructors are notified ahead of time about the visitors.

## 29.10  Scheduling Research

The instructor group at the Simulation Center remains involved in simulation research. We have a human use Institutional Review Board approved protocol for our teaching courses that allows us to graft research projects onto the throughput work of producing courses. The library of recorded performance in the simulator contains over 15 years of courses and has been and continues to be an invaluable source of research. Many simulation groups have cataloged the performance of practitioners exposed to varied clinical events. Simulation allows the presentation of the same event to multiple teams, which allows evaluation and common fault pathways to be determined.

Research questions that are difficult to pursue in real clinical environments can be proposed and achieved with simulation techniques. Our group has been successful in completing simulation studies such as evaluating performance assessment metrics [6], human factors design of medical equipment [7], effects of clinician fatigue [8], and study of cognitive aid use [9].

## 29.11  Conclusion

Our work at the VA Palo Alto/Stanford Simulation Center is expanding at a rapid pace. We are continually developing courses for the many different phases of medical education and training including medical students, residents, fellows, and allied health professionals, both within the walls of the Center as well as the provision of *in situ* training.

Our simulation efforts have been graciously supported by the VA Palo Alto Health Care System, the Anesthesia Service at the VA Palo Alto, the National Center for Patient Safety, and Stanford's Department of Anesthesia. Without this support, the success of our laboratory and our continued efforts would be impossible.

## 29.12  Example: Drug Recognition Exercise

Students are given the following baseline vital signs on a monitor:

    HR 80
    BP 120/80
    RR 10
    Sat. 97%

They are then instructed to inject the simulator with a syringe labeled "Drug A." Changes to the vital signs are made to the simulator according to the following chart. The exercise is repeated for Drugs B–F.

|   | Drug | HR | BP | Sat. | RR | Time frame |
|---|------|----|----|------|----|-----------|
| A | Epinephrine | 120 | 190/110 | – | – | 5–10 sec |
| B | Atropine | 135 | 130/90 | – | – | 5–10 sec |
| C | Neo | 60 | 160/115 | – | – | 10 sec |
| D | Esmolol | 40 | 90/60 | – | – | 10 sec |
| E | Fentanyl | 60 | – | 89% | 2 | 10 sec |
| F | Nitroprusside | 110 | 80/20 | – | – | 10 sec |

# References

1. Gaba, D. M., Fish, K. J., and Howard, S. K. *Crisis Management in Anesthesiology*. Churchill Livingstone, New York, 1994.

2. Gaba, D. M., Howard, S. K., Fish, K. J., et al. Simulation-based training in anesthesia crisis resource management: a decade of experience, *Simulation and Gaming* 32, 175–193 (2001).

3. Reznek, M., Smith-Coggins, R., Howard, S., et al. Emergency medicine crisis resource management (EMCRM), pilot study of a simulation-based crisis management course for emergency medicine, *Acad. Emerg. Med.* 10, 386–389 (2003).

4. Lighthall, G. K., Barr, J., Howard, S. K., et al. Use of a fully simulated intensive care unit environment for critical event management training for internal medicine residents, *Crit. Care Med.* 31, 2437–2443 (2003).

5. Dismukes, R. K., Gaba, D. M., Howard, S. K., So many roads: facilitated debriefing in healthcare, *Simulation in Healthcare* 1, 23–25 (2006).

6. Gaba, D. M., Howard, S. K., Flanagan, B., et al. Assessment of clinical performance during simulated crises using both technical and behavioral ratings, *Anesthesiology* 89, 8–18 (1998).

7. Sowb, Y. A., Howard, S. K., Raemer, D. B., et al. Clinicians' recognition of the Ohmeda Modulus II Plus and Ohmeda Excel 210 SE anesthesia machine system mode and function, *Simulation in Healthcare* 1, 26–31 (2006).

8. Howard, S. K., Gaba, D. M., Smith, B. E., et al. Simulation study of rested versus sleep-deprived anesthesiologists, *Anesthesiology* 98, 1345–1355 (2003). (discussion 5A)

9. Harrison, T. K., Manser, T., Howard, S. K., and Gaba, D. M. Use of cognitive aids in a simulated anesthetic crisis. *Anesth. Analg.* 2006 Sep; 103(3): 551–6.

# Health Care Simulation with Patient Simulators and Standardized Patients

Michael S. Goodrow

## 30.1 The Community

The University of Louisville School of Medicine, Louisville, Kentucky, U.S.A., has two nationally recognized simulation programs – a standardized patient program and a mannequin-based simulator program. Both programs support not only the educational goals of the medical school, but also several other schools within the University. This includes the nursing, dental, and public health schools at the Health Science Campus, as well as the social work and engineering schools at the main campus. These two programs also support education programs for staff at the hospitals at the Health Science Campus, as well as first responder training for various local, state, and national agencies.

In this chapter, I describe the organizational structures of these two programs, the facilities used, and how the programs operate. I will also discuss the ramifications of having the two programs as separate entities within the medical school. I will also include some of the "lessons learned" from the first eight years that these two programs have been in existence. I will begin with a brief history of the two programs.

## 30.2 Beginnings

Implementation of clinical simulation at the School of Medicine began in 1999. Several factors contributed to this activity:

- The Liaison Committee for Medical Education had just completed its regular accreditation review of the medical school. One of the Committee's recommendations was to consider alternate approaches in assessing students beyond the traditional written-exam format.

- Two companies were selling computer-controlled patient simulators with full-sized mannequins as an alternative to traditional teaching approaches. Although marketed primarily toward anesthesia applications, these mannequins had sufficient features and were versatile enough to be used for a variety of applications.
- The use of Standardized Patients was becoming more widely accepted in medical education. The course director for the Clinical Practice Science course began using Standardized Patients; however, since there was not a Standardized Patient program at the University, he contracted with another school in the state to provide them.

In 2000, two decisions were made at the medical school that would ultimately lead to the development of our simulation programs. One, the medical school decided to make use of the emerging commercial robotic patient simulators, committing to purchase patient simulators and to build a simulation center to house them. A clinical director was designated from the Department of Anesthesiology and two simulators were purchased. Two, the medical school decided to create its own Standardized Patient program. Gina Wesley, PhD, was hired to be the director of the Standardized Patient program – her first task was to create and build this Standardized Patient program. Management of this program was placed within the Office of Medical Education, which is part of the Office of the Dean.

These two initiatives were strongly supported by the Dean of the medical school. He had recently arrived at Louisville, and he was open to new approaches in medical education. He provided the crucial initial support that enabled these two programs to get off the ground: funds for hiring essential personnel funds for purchasing necessary equipment and supplies, as well as encouragement for faculty to incorporate the use of standardized patients and patient simulators in undergraduate and graduate medical education. Also, he also encouraged both programs to look beyond the medical school, to find applications in other schools within the University, and other institutions outside of the University.

## 30.3 Patient Simulator Program

In 2000, the medical school purchased two Human Patient Simulator (HPS) systems, manufactured by Medical Education Technologies, Inc. (METI, Sarasota, Florida, U.S.A.), with two "C" model adult mannequins and one pediatric mannequin. These were selected after a formal evaluation/bidding process. At the time of this purchase, there were only two companies offering computer-controlled, full-function, full-sized robotic

patient simulators for clinical training, and both submitted bids. Although the two competing systems were similar in functionality and price, the HPS product was selected over the competitor primarily because METI offered a formal warranty program with their products, while the competitor did not. These two HPS systems were originally located in an unused operating room at University Hospital located about one block from the medical school. They were used for some training of anesthesiology residents, and one was temporarily moved into a lecture hall of the medical school for teaching one pharmacology lecture.

In the spring of 2001, renovation commenced on a 2500 sq. ft. space on the third floor of the Instructional Building at the medical school. This was funded in part through a generous donation by an alumnus, Dr. John Paris. Dr. Paris was very interested in innovative approaches to teaching medical students, and had a long history of giving back to the community. In July 2001, the University hired a Patient Simulation Specialist. His responsibility was to work with the Clinical Director to properly program the simulators for the faculty to use. He was also responsible for maintaining and servicing the simulators. In August 2001, the medical school purchased two additional HPS systems with the "D" adult mannequins, bringing the total to four simulators with five mannequins. The John M. and Dorothy S. Paris Simulation Center officially opened in September of that year. The design of this facility (the Center) is described in Section 30.6.

Use of the Center quickly grew during the first 2 years of operation. A significant contribution to the rapid growth was the incorporation of the simulators as a formal part of the medical physiology course for the first year medical students [1, 2]. The success of these sessions has encouraged other faculty to use the simulators in their courses. The center currently averages about 1200 instruction hours and 4500 student encounters each year. Over half of these are for undergraduate medical students, with the rest evenly divided between residents, nursing students, hospital staff, and outside users. With this increased level of activity at the Center, the staffing level was increased. In addition to the part-time Clinical Director, there are two full-time staff people, the Director of Operations, and the Patient Simulation Specialist. The Director of Operations works with faculty to enable them to use the simulators during training sessions. The Simulation Specialist maintains and operates the simulators.

## 30.4 Standardized Patient Program

Dr. Wesley arrived at the University of Louisville in July of 2000. She quickly began building the Standardized Patient (SP) program. She began recruiting and training people to be SPs, while also working with course directors to incorporate the use of SPs into their curricula. The first user was the medical students' Clinical Practice Science course. Within 2 years, SPs were fully integrated into this course, as well as several other courses. However, the SP program did not have its own dedicated space at the medical school. The student encounters with the SPs were held in various shared places, particularly a set of small rooms intended as study rooms for the preclinical medical students. In late 2001, renovation began on a 1500 sq. ft. space on the third floor of the Instructional Building at the medical school, next to the Center. This dedicated Standardized Patient Clinic opened in March 2002. The design of this facility (the Clinic) is described in Section 30.7.

The Standardized Patient program has grown considerably since its inception. Currently, each medical student has over 100 SP encounters during their 4 years of medical school. The majority of these occur during the first 2 years, particularly within the Introduction to Clinical Medicine course. Standardized Patients are also used in many of the clerkship courses, including most of the required clerkships. The nursing, dental, and public health schools at the Health Sciences Campus make significant use of SPs to train their students as does the school of social work at the main campus.

The Standardized Patient program also provides a significant amount of training outside of the University. This is primarily focused on first responder training, with emphasis in bioterrorism training. The SPs are particularly effective for these types of applications. Since they are trained in both presentation and evaluation, they make the scenarios more realistic and can provide meaningful feedback. The SPs can also be "made up" with moulage to make their injuries look realistic for the scenarios.

With the growth of the SP program's use, additional staffing was required. In addition to the director, the SP program gained an assistant director, a moulage specialist, and a program assistant. The assistant director is responsible for hiring and training SPs, selecting them for specific training sessions, and monitoring their activities. The moulage specialist works with various types of makeup and prosthetics to make the SP's appearance consistent with their condition. The program assistant ensures that paperwork, particularly the payroll for the expensive SPs, is properly accounted for. With over 60 SPs ranging from age 8 to 72, and with over 6000 SP/student encounters each year, our SP program is one of the more active programs of its kind.

# 30.5 Facilities

The Patient Simulation Center and the Standardized Patient Clinic are adjacently located in the same building and directly above the preclinical lecture halls and teaching laboratories used by the first and second year medical students. Their presence in the Instructional Building puts them within a block of the nursing, dental, and public health schools. It is also within two blocks of University Hospital and three other hospitals.

# 30.6 Patient Simulator Center

The Center is 2500 sq. ft., adjacent to the Clinic and down the hall from the gross anatomy laboratory. The primary feature of the center is the four simulation "suites." Each suite consists of a classroom, a control room, and a simulation laboratory with its own patient simulator. The Center was established in an existing building space, built between an existing hallway and the exterior wall of the building. This produced some limitations and challenges in the floor plan design (Figure 30.1) [3].

## 30.6.1 Simulation Laboratories

Each simulation laboratory "stage" is about 18 ft. × 18 ft. One wall of each laboratory contains a cabinet and undercounter drawers for storage of equipment and supplies. There is also an X-ray view box on this wall. Each laboratory has two dry-erase whiteboards, a large 8 ft. unit, and a smaller 3 ft. unit. As indicated on the floor plan in Figure 30.1, there is a "cut out" in one corner of each laboratory. This corner is for the control room and this diagonal wall has a large one-way mirrored glass that enables the operator to observe the laboratory without being observed by the learners. Each laboratory has one entrance, from either the lobby or a classroom. The two outermost laboratories also have entrances from the control rooms, but students seldom use them.

The centerpiece of each laboratory is an HPS system with an adult mannequin. Two of them have the "C" mannequin, and two have the "D" mannequin. We also have a pediatric mannequin (METI PediaSim) that can be used in place of the adult mannequins. Each of the four HPS systems also has the METI Trauma Disaster Casualty Kit (TDCK) unit that enables them to lacrimate, salivate, and bleed. There is additional equipment in the laboratories including anesthesia ventilators, clinical vital signs monitors, "crash" carts with defibrillators, instrument trays, carts, bedside tables, and IV poles. All of these items are completely functional.

In addition to the simulators in each laboratory, we have other simulators such as three portable METI Emergency Care Simulator (ECS) units. These are used for training at remote locations, and occasionally for overflow capacity at our center. We also have several part-task trainers, including arterial and vein arms, airway mannequins, CPR mannequins, and a labor/delivery simulator. As we identify new training requirements, we will expect to procure additional part-task trainers.

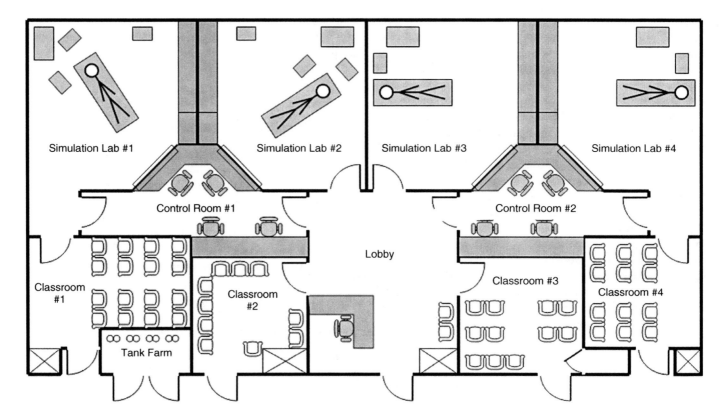

**FIGURE 30.1** Floor plan of the Paris Simulation Center. This is a 2500 sq. ft. facility, with four simulation labs, classrooms, and simulators. Each lab is about 18 ft. × 18 ft. The sizes of the classrooms vary. The two control rooms support two labs each. The tank farm is accessible from the main hallway, to allow deliveries without interfering with simulation sessions.

While some simulation centers have topic-specific simulation suites, at our center, each of our four are equipped and configured for hosting a number of different clinical educational purposes. However, we have made a modest attempt to differentiate the areas. Two laboratories are configured more like operating or trauma rooms, while the other two laboratories have utility "wallboards" like those typically found in hospital patient rooms. These wallboards have functional oxygen, air and suction hookups, as well as electrical outlets. To increase the realism, the patient vital signs monitors are mounted on the wallboards. This enables these two rooms to be easily and realistically used as patient rooms.

### 30.6.2 Classrooms

There are four classrooms; each can comfortably seat 8–16 people, depending on the specific room. The unique configurations of each of these rooms are a consequence of non-adjustable boundaries common when renovating an existing space. These classrooms are used for pre- and postbriefings of simulation sessions. They are equipped with video displays for replaying the recordings of the simulation sessions.

### 30.6.3 Control Rooms

The simulation center has two control rooms. Each control room overlooks two simulation laboratories through one-way mirrored windows. A mirrored window enables the operator to see the entire room, receiving a "feel" of the room. In particular, it allows the operator to see things that are not necessarily in the field of view of the cameras.

While this intentional design makes it possible to produce two simulations in the two laboratories at the same time, this can be somewhat challenging for one operator to manage. We only share one operator between two rooms when we are presenting the same scenario in both rooms. This way, the operator only has to attend to a small set of particular cues common to both rooms and respond appropriately. If different scenarios are being produced in the two rooms, then we use one operator dedicated for each. The control rooms contain the computers that control the simulators, as well as the controls for the A/V system described below (Figure 30.2).

There are advantages and disadvantages to this design. One major advantage is that this layout easily allows faculty to leave the clinical "stage" and observe the students from the control

**FIGURE 30.2** The control room contains the computer that controls the mannequin, as well as the audiovisual system controls. Each control room looks into two simulation labs.

room. With the ability to see two rooms, one faculty member can instruct two separate groups. Typically, an instructor does not remain in the control room – they move between the two stages. The operator can advise the instructor as to the status of the other group during the transition between groups.

### 30.6.4 Audiovisual System

It could be argued that the audiovisual (A/V) system is the most important piece of a patient simulation center. Without an effective A/V system, all you have is some really cool toys. With a good A/V system, you have a real teaching system. The A/V system provides two important functions for the simulation center. First and most obvious, it is the mechanism to provide feedback to the learners. This is what makes it an educational experience, when the students can examine their actions and learn from them. The second function of the A/V system is to enable control of the simulation activity. The A/V system is how the operator knows what is happening, and how to respond to actions. A more detailed description of the functions of an A/V system is provided in Chapters 73 and 74.

Each of our simulation laboratories has two cameras, a wall-mounted camera directly above the mirrored glass, and a ceiling-mounted camera over the mannequin. These cameras can be panned, tilted, and zoomed from the control room.

Each classroom also has two cameras, one near the large display facing the students and the other in the opposite corner from the first camera facing the large video display. The classroom cameras are used to collect actions in the classrooms, for example, during live videoconferencing.

Signals from all cameras are connected to an A/V mixer in the control room. The mixer enables the operator to select which sights and sounds are to be recorded. In our system, the signals are first mixed and then recorded. This makes playback quicker, but does have the risk of losing data. If something significant happens within the field of view of a camera, but is not selected at the mixer, then it is lost. (Systems that first record then mix avoid this problem. This issue is discussed in more detail in Chapter 73.) We frequently use the mixer to produce picture-in-picture images, particularly with the vital signs monitor image inserted in the corner of a camera view of a laboratory. The mixer output is connected to displays in the classrooms as well as the control room, the latter so that the operator can monitor what is being recorded as it is happens, a "what you see is what you get" recording system. Representative views from two different cameras are shown in Figures 30.3a and 30.3b.

The patient vital sign monitors' video output signal is split into two at its source: one connects to a typical clinical display for the students use in the laboratory; the other connects to the A/V rack in the control room. Within the A/V rack, the

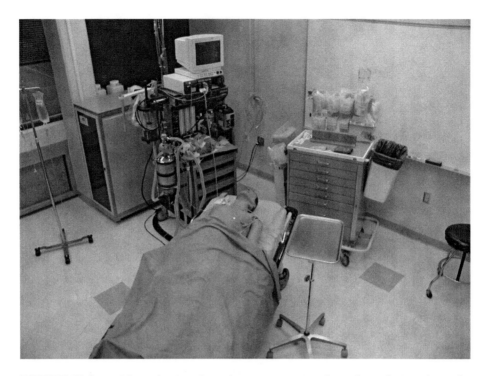

**FIGURE 30.3a**   Wide-angle view from the camera mounted on the wall, just above the one-way mirrored glass. Note the location of the large simulator support rack out of the way behind the anesthesia machine in the upper left, and the presence of a large whiteboard for posting teaching points in the upper right.

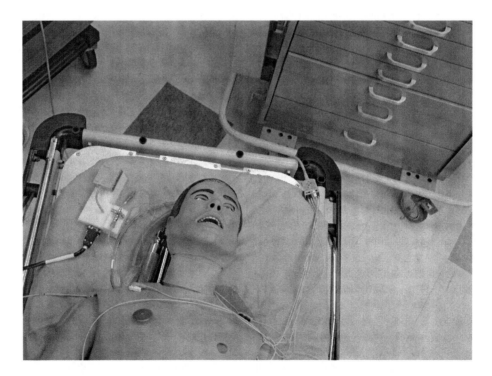

**FIGURE 30.3b**   Close-up view from the ceiling-mounted camera in the center of the lab. Note that the status of the patients' eyelids is captured as well as any student's IV drug administration and actions to or around the head and upper torso.

signal is split three more times. One connects to a dedicated display in the associated classroom. Another connects with a dedicated display in the control room so the operator and clinical instructors always have an identical view of the clinical vitals signs as available to their students. The third connects with the mixer.

The laboratories and the classrooms each have one ceiling-mounted microphone. Their signals pass through a set of volume controls, and then into the mixer. The volume controls allow us to select which microphones are recorded. This enables us to record the conversations during debriefing sessions as well as the simulation sessions.

The A/V output from the mixer goes to a professional-quality VHS recording deck. From the VHS deck, the A/V signal goes to a router. The router is used to send the A/V signal to the displays and speakers in the classrooms. The router enables us to "broadcast" from one laboratory into several classrooms at the same time. The video router also has a Polycom videoconference system attached to it. This enables us to use our classrooms for videoconferencing over the Internet. Since the Polycom is attached to the router that receives signals from the mixer, we can actually videoconference from any classroom or simulation laboratory. This does require an A/V monitor in the laboratory, so that the people in the laboratory can see and hear the observers.

One of the outputs from the router goes to the general A/V distribution system throughout the Health Sciences Campus. This enables us to "broadcast" to sites within the campus, including all of the lecture halls. We have used this capability to broadcast live images from the patient vital sign monitors into the lecture halls, to introduce the medical students to these monitors before their first simulation session.

### 30.6.5 Additional Spaces

In addition to the simulation suites, the Center contains two additional spaces: a lobby and a tank farm. The main entrance to the center is the lobby area indicated on the floor plan. This area has proved essential for traffic control, particularly when multiple events are scheduled at the same time. The lobby provides a space where the participants orient and organize before entering their particular laboratory or classroom. The other unexpected advantage to having a lobby is for tours, primarily as a place to organize and start the tour. It has also enabled us to host tours even when teaching sessions are occurring in the center – we can talk to the visitors, and then show them the activity as seen from the control rooms or through the displays in the classrooms.

There is also a tank farm within the footprint of the center (Figure 30.4). This space contains the compressed gas tanks

**FIGURE 30.4**  The interior of the tank farm. In addition to the eight tanks always connected to the four regulators, there are several spare tanks located in this area.

for the simulators and clinical devices. Access to the tank farm is from the main hallway in front of the center. This enables us to have tanks exchanged without interrupting the activity within the center. The tank farm has large "H" cylinders of compressed gas to supply the simulation center. Each cylinder holds 6000–7000 L of compressed gas when at a full pressure of 2000 PSI, and is 9 in. in diameter, 55 in. tall, and 110 pounds empty. These tanks supply the oxygen, nitrogen, and carbon dioxide required by the simulators and clinical devices, alike. For safety reasons, the nitrous oxide used for anesthesia scenarios is simulated with nitrogen.

The compressed gas tanks are attached to regulators that serve two functions. First, they reduce the pressure from the tank to a constant 50 PSI. Second, they have valves that automatically switch from the selected tank to the spare tank. The second tank, therefore, serves as a backup to the first, preventing a loss of gas pressure. When the first tank is empty, that tank is replaced, and a valve on the regulator is turned to make the second tank as the primary, and the fresh tank is now the backup. Using this method reduces the likelihood of ever unintentionally running out of compressed gas. The tank farm is large enough to store several extra compressed gas tanks. For our operations, we aim to have three extra oxygen and two extra nitrogen tanks at all times. The nitrogen tanks that simulate the infrequently used nitrous oxide serve as additional spare nitrogen tanks required by the HPS simulators.

The simulators also require dry, compressed air to function. During periods of high use our facility consumes over 100 L/min. This would empty one "H" cylinder in 1 hour. Obviously, a continuous compressed air source is necessary. Thus, we use the compressed air system that was already installed for the building. The advantage of this system is that it provides a continuous supply, and it is managed by physical plant personnel. There are two drawbacks, however. First, the building's compressed air contains significant amounts of moisture. Therefore, the compressed air for the simulation center is filtered and dried in the tank farm room before its use. Second, the compressed air pressure is about 42 PSI. Our simulators specify 50 PSI, so we had to adjust our simulators to function with the lower pressure.

# 30.7 Standardized Patient Clinic

The Clinic is 1500 sq. ft., adjacent to the Center, and down the hall from the gross anatomy laboratory. The primary feature of the clinic is the eight patient examination rooms (Figure 30.5). Like the Center, the Clinic was installed into an existing building space, and thus its design was constrained by fixed boundaries. It too was constructed between an existing hallway and an exterior wall, between the simulation center and a large conference room.

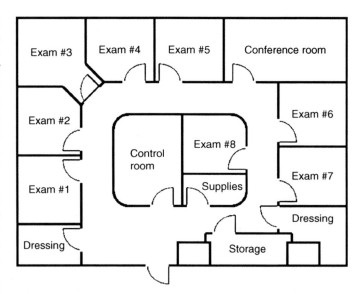

**FIGURE 30.5**  Floor plan of the Standardized Patient Clinic. This is a 1500 sq. ft. facility, with eight exam rooms, a conference room, and a centralized control room.

## 30.7.1  Examination Rooms

Each exam room is about 8 ft. × 12 ft., and is designed to resemble a typical patient examination room (Figure 30.6). Each has one entrance door. They have a fully functional set of typical examination room equipment, including patient table, otoscope, pan-optic opthalmascope, and blood pressure cuff. They also have cabinets and a functional sink. Each exam room also has a camera and microphone connected to a monitor/recorder system in a common control room. There is also a speaker, part of the public address system for the clinic.

## 30.7.2  Control Room

The control room is centrally located in the SP clinic (Figure 30.7). It contains display monitors presenting sights and sounds from each of the SP exam rooms, enabling faculty to remotely observe the student/SP encounter. These monitors have built-in VHS recording capability, so the SP encounter can be recorded. The control room also contains the controls for the public address system. The PA system is used to control movement within the clinic; students are told when to begin and end their sessions using the PA system.

## 30.7.3  Conference Room

There is a conference room in the SP clinic. It is primarily used for training the SPs. It is also used for debriefing students, and to provide a "break room" for SPs during long training sessions.

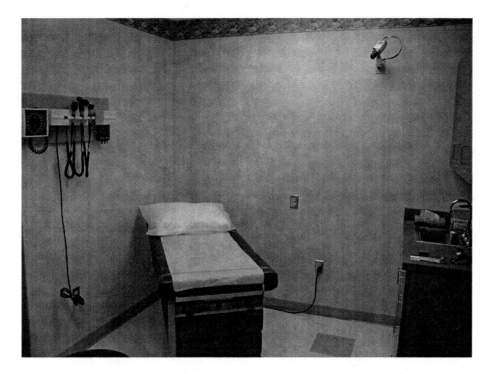

**FIGURE 30.6** View into one of the exam rooms. There is a camera mounted on the back wall of the exam room, attached to the audiovisual system.

**FIGURE 30.7** View of the Standardized Patient Clinic control room. The monitors enable the faculty to observe the SP–student interactions in real time.

## 30.8 Functional Issues

The two simulation programs support a variety of users within the University, as well as many outside institutions. As such, it is considered a large, multi-institution simulation program. There are several issues related to the operational requirements necessary to provide simulation in clinical education institutions, irrespective of the size of the simulation program or the institution. I will present these in the context of clinical simulation as a whole, considering both SPs and mannequin-based patient simulators. Our solutions might not fit your specific needs, but the issues themselves are universal to clinical simulation programs. There are three major issues:

- First, the operational model – how a simulation program is operated and managed. The operational model defines the scope of the program, the customer base for the program, and the growth potential of the program.
- Second, the organizational position. This is driven by the operational model, and affects the funding model for the simulation program.
- Third, personnel – what roles need to be filled, and what are the characteristics of people to fill those roles.

### 30.8.1  Operational Models

The operational model for a simulation program describes how that program is operated and managed. It also defines the possible funding approaches, and the scope of the program. Let me describe different operational models by using libraries as an example.

Many clinical departments have their own libraries. These are collections of journals, reference books, and other resources all located in a single room. The room, although it might be called "the department library," is actually a conference room that has a large number of books and journals in bookcases or shelves along a wall or two. There is not a dedicated, professional "librarian" *per se*; the department's secretaries or assistants put new journals in the "stacks," residents and faculty members check journals out when they need them and return them to the proper place when they are done. If someone does not know where to find something, they will "poke around" until they find it or ask someone else for to help them. The "library" does not have its own budget; it is supported within the department; if a new bookcase is needed, the department purchases one.

This model works great within a clinical department. It must, since so many departments locate and operate their libraries like this. However, this is not how the library for the university operates. The university's library has a dedicated space, usually its own building. The library staff handles the physical tasks within the library such as putting books back in the stacks. Professional librarians are available to answer

questions and direct people to the resources they need. Policies and procedures are developed to enable the library to operate effectively for all of its customers. The library has defined resources (budget, space, personnel), and these must be managed to support the needs of the university.

Simulation programs are similar to libraries. Many schools have mannequin-based simulators that are owned and operated by specific clinical departments. In most cases, these departments do not have dedicated staff to support the simulators – interested faculty members use them when they want, and others don't. When something breaks, then the department finds a way to repair or replace the simulator. There are some small SP programs that are also managed this way, with the SPs managed within a clinical department, separate from the primary function of the department.

Although this model works well for many institutions, it does limit the potential of these simulation programs. Much like the university's library, by creating a separate simulation program, the program can have the staff, funding, and capability to support multiple groups within the institution, and other agencies outside of the institution. This is the model our University has pursued for its clinical simulation programs.

### 30.8.2  Organizational Position

The Center and the Clinic both operate out of the Office of Medical Education, which is part of the Office of the Dean of the School of Medicine at the University. Both the Center Director of Operations and the Clinic Program Director report directly to the Associate Dean for Medical Education, who is in charge of the Office of Medical Education. Both programs receive their funding from the Dean's office. The Office of Medical Education also provides administrative support for the two programs. The Office of Medical Education is not a department, and it does not have any assigned faculty members. This limits some of the possible autonomy for the Office and these two programs. On the other hand, being part of the Office of the Dean provides a level of financial security and stature that departments do not typically have.

At this time, we do not charge fees to clinical members of the University for training sessions. This applies not only to the medical school students and residents, but also to the nursing, dental, public health and graduate students located at our campus, as well as students from our other campuses. However, both programs do charge fees for training sessions for groups outside of the University. This includes training staff from local hospitals, first responder training, and bioterrorism drills.

The Center and the Clinic are separate entities within the Office of Medical Education. They have separate management structures and are independently funded. This separation is a result of the history that created these two programs. Although there is significant amount of cooperation and joint training sessions between the two programs, and new programs at other schools are creating joint programs, we expect that the separation between these two programs will remain for the foreseeable future. There are advantages and disadvantages to this arrangement.

The two main advantages in keeping the two programs separate are the nature of the expenses and the type of work the staff performs. The major expense in the SP Program is the salary expense, particularly for the SPs. This expense is dependent on the amount of activity with the SPs, which can vary significantly year to year. In contrast, after staff salaries, the biggest expenses for the Center are warranty costs and new equipment, which do not vary significantly year to year. So, keeping the two programs apart makes it easier to track and manage these expenses.

The staff at the Center must be very comfortable working with equipment and technology. Although they also work with students and faculty, their primary focus is working on and being comfortable with machines. The SP Program staff works almost exclusively with people, particularly the SPs. So, they need to be more "extroverted" and "people oriented" by nature than the simulator staff.

There are two major disadvantages to keeping the two programs separate. First, if the two programs were combined, it would be easier to schedule certain activities, particularly those activities that use both facilities. At our institution, the two programs expend the effort required to cooperate. Second, we lose "cross training" opportunities. It seems reasonable to expect that if the two programs were combined, then some of the staff could be cross-trained to work in both. However, as mentioned before, the personalities necessary for the two areas are different and the effectiveness of the cross training would likely be very limited. We have attempted to do some cross training, but have found it difficult to maintain sufficient levels of competency.

### 30.8.3  Personnel

Using a university library model, the simulation programs have specific personnel-performance requirements related to the specific simulation programs, and specific positions to meet those requirements. These positions are similar to those at other schools that have clinical simulation programs.

### 30.8.4  Personnel for Robotic Simulation

When the Center opened, there was one full-time staff person, a Patient Simulation Specialist. As the use of the simulators increased, this person became the Director of Operations, and a Patient Simulation Specialist was hired. These two positions are described below.

The **Director of Operations** is responsible for the day-to-day operations of the center. This includes, but is not limited to:

- Working with faculty to identify applications within their curriculum
- Teach faculty how to incorporate simulation into their teaching sessions
- Program simulators to present scenarios designed by clinical faculty
- Operate simulators during teaching sessions
- Resolve scheduling conflicts
- Manage the budget for the simulation center
- Promote the center, internally and externally.

The **Simulation Specialist** maintains and operates the simulators. This includes making repairs as necessary and setting up the simulation laboratories for sessions. The specialist also schedules the simulation sessions, maintains records of sessions, and performs bookkeeping records of financial transactions.

There is a significant amount of overlap between the duties of the operations director and the specialist. To summarize the differences between the positions, the specialist makes sure that the simulator is ready when the session starts, and the director makes sure the simulator does what the faculty expects it to do. Both of these are full-time positions within the university, both needs first and foremost an interest in education. For these two positions, it is difficult to find new people that already perform this type of work. Therefore, it is necessary to do some in-house training. To facilitate this, and given the nature of the work environment, there are two types of people that I would suggest considering:

- Technically oriented people who are willing and interested in learning clinical topics
- Clinically oriented people who are not afraid of technology.

One question that I ask when interviewing for new technical staff is "How comfortable do you feel about holding a screwdriver in one hand, a telephone in the other hand, and taking apart a $200,000 box full of electrical/mechanical/pneumatic components while being directed over the phone?" A short list of possible candidates would include computer engineers or technicians, anesthesia or respiratory technicians, emergency medical technicians, and paramedics.

The simulation center also has a **Clinical Director**. This position is currently a 20% position. The primary job of the Clinical Director is to provide oversight, advice, and guidance in the operation of the center. This includes, but is not limited to:

- Working with clinical faculty when designing scenarios
- Explaining clinical issues to the simulation staff

- Assisting in the selection of equipment, supplies, etc., for acquisition (and disposal)
- Meeting with prospective customers, donors, etc.
- Teaching faculty how to teach using simulation
- Developing long-term plans for the center.

The Clinical Director provides a valuable "reality check" on clinical issues for the nonclinical staff at the center. Having a Clinical Director also provides some reassurance to potential new users that the educational experience will be relevant, appropriate, and as realistic as possible. Thus, the Clinical Director needs three characteristics: must be interested in teaching as a profession, must be dissatisfied with traditional clinical education, and must be very enthusiastic about using technology to improve teaching methods. Since this person must be a "technophile," they tend to come from clinical areas with significant technology, such as anesthesia or surgery. The Clinical Director must have training status at least equivalent to that of the other faculty using the simulator in training. This is a primarily an issue of respect – respect to the faculty and respect from the students. So, for a medical school, the Clinical Director must be an MD. At a nursing school, the Clinical Director must be at least a Master's-trained RN.

### 30.8.5  Personnel for Standardized Patients

The Standardized Patient Program has three full-time personnel: the Director, the Assistant Director, and a Program Assistant. The **Director** is responsible for the overall operations of the SP Program. The Director also actively works to secure funding through grants and contracts for external users of the SPs. The **Assistant Director** is responsible for daily activities of the program. This includes:

- Designing and implementing training sessions for SPs
- Supervise the SPs
- Assisting faculty with case development
- Scheduling activities in the clinic
- Promoting the program at regional and national conferences and meetings
- Prepare statistical reports on SP activities for course directors.

The **Program Assistant** supports the activities of the other two by providing typical office functions, including reception and the extensive communications always required when scheduling and organizing large numbers of dissimilar people (instructors, students, and SPs). As with the simulation center, the growth of the SP Program required the increase in full-time personnel. When the program first started, the Director did the scheduling and training activities. As the program grew, particularly with increasing demands by external customers, it became necessary to separate the coordination/training activities into a different position.

## 30.9  Lessons Learned

The two clinical simulation programs at the University have both been operating since 2000. During this time, we have learned many lessons regarding how to organize and operate simulation programs. Also during that time, new techniques, technologies, organizations, and resources have become available as well.

### 30.9.1  Engineering

#### Simulator laboratory size

I personally believe that our 300 sq. ft. "stages" are a good size for the patient simulators. If anything, this size might be a little larger than necessary. They can comfortably accommodate 10–12 people, as well as the mannequin, rack, anesthesia ventilator with monitor, and several other equipment carts, without becoming too crowded. So, when there are half that number of people, they have plenty of room to move around, and when there are extra people, students can't move too far away to "hide." For a typical simulation learning session, we generally don't have more than about six learners at a time that are actively involved in the session. Six is about the limit for the number of tasks in an acute clinical emergency. For example, in a "code" situation, there are people responsible for (i) leadership, (ii) airway, (iii) circulation, (iv) defibrillation, (v) drugs, and (vi) recording. So, if there are significantly more participants than that in the room, they become simple observers, and their involvement and interest level drops quickly.

For someone designing a new center, I would suggest clinical "stages" in the 250–300 sq. ft. range. The smaller rooms keep the students engaged, and they don't have anywhere to go to "hide." It also discourages faculty from attempting to put too many students in any one session, maintaining small group sizes, and reduces the likelihood of using areas of this room to store unused devices and props.

#### Mirrored glass

The mirrored glass (also known as one-way window) between the control rooms and the laboratories has been a tremendous boon to the operation of the simulators. First and foremost, it provides a glance at what is going on in the whole laboratory. It is possible to see how many students are off having private conversations, how many are paying attention, etc. It lets us see areas that the cameras don't necessarily see, and it lets us see without moving the camera, which is particularly important when recording actions.

There are two different reactions that students and faculty have to the mirrored glass. Many students seem to forget that there is someone behind the mirror. They act like no one is watching them, and that they are completely on their own. Some students even use the mirror as a mirror, to check their appearance! On the other hand, some students and most of the instructors are constantly aware of the mirror, and their first reaction is to look at it when something does not go as expected. At least one simulation center has found an approach to improve the use of one-way mirrored glass. The simulation center at the Mayo clinic has elevated their control rooms by several feet, which raises the mirrored glass as well (see Chapter 18). This should reduce the trainees' awareness of the mirror and their sense of being "watched," producing more realistic action.

## Oxygen shutoff

When the center was built, we had valves installed in the gas supply pipes that allow us to shut off the oxygen supply to the laboratory. There is also another valve that lets us substitute nitrogen instead of oxygen. There is one set of these valves for each laboratory. This arrangement has been very effective for simulating construction mistakes, such as cutting a supply line or cross connecting lines. It is simply amazing how unprepared residents are for a total loss of oxygen on the anesthesia ventilator!

When this crossed piping was installed, we had to convince the contractors that the blueprint was correct, and that the oxygen and nitrogen were supposed to be connected. This was completely contrary to their experience, and probably against code. We had to assure them that this was not a real clinical space, and that we needed this function for training purposes.

On the other hand, we did make one mistake when this was installed. Because of the way that the METI HPS is mechanized, it creates the appropriate exhaled gas mixture for each breath. When the oxygen is shut off, the HPS does not have oxygen to create a physiologically correct gas mixture. This problem is worse when the oxygen is replaced with nitrogen. The better approach would have been to designate separate oxygen lines for the simulator and the ventilators, and connect the shutoff valves just to the ventilator oxygen.

While installing shutoffs, it might also be useful to include shutoffs for other resources in the clinical stage area. This could include (but is not limited to) electricity, compressed air, and suction. This would greatly expand the ability to simulate adverse events. Again, these should be configured so that turning off anything used by the students does not stop the simulator from functioning.

## Compressed gas "tank farm"

You need to consider how you will monitor the compressed gas supply and how the tanks will be delivered. You need the tanks to be close enough that you can readily check them during a session and change tanks if necessary. However, you also need to consider how they will be delivered. Putting the tank farm in a closet at the far back corner of the center might look like a good idea at first, especially if it reduces the cost of pipes. However, when the delivery person wheels several wet, dirty, dripping tanks back to that corner during a major simulation session, that corner might not seem like such a good location anymore.

Anesthesia ventilators require a constant flow of oxygen, even when they are turned completely down. Some even when they are turned off. The HPS rack leaks all three gasses, oxygen, nitrogen, and carbon dioxide, even turned off. Because of this, you should shut off the tanks every evening and turn them back on in the morning. When we were leaving our oxygen tanks on overnight, we found that we lost 10–20% of an "H" cylinder each night.

## Economics of compressed gas

Three points about the economics of compressed gas:

- With the METI simulators, it is very expensive to use tanks to supply the required amount of compressed air. Use your building's compressed air supply or install a compressor (more about compressed air is discussed below).
- It is cheaper, in the long run, to buy the tanks instead of renting them. (By the way, tank rental is called demurrage.) We use the large "H" size cylinders. In our case, the breakeven point was about 3 years. So, spend a year learning how many tanks you need, and then buy the whole set.
- When placing a refill order, look at delivery cost, refill cost, and the upcoming schedule before placing the order. We've had situations where, knowing that we would be using many oxygen tanks in very short order, we would replace a partially full tank so as not to disrupt the next delivery. Or, it might be cheaper to order an extra tank and pay the rental for a few days.

## Compressed air

The compressed air in our building is only at 42 PSI, which is below the pressure recommended by the simulator manufacturer. It is possible to adjust the HPS rack to function at this pressure. The ECS functions fine on this pressure. If you install your own compressor, there are two major things to consider – capacity and noise. The required capacity (pressure and flow rate) is determined by needs of the types and numbers of simulators. Compressors can be large, heavy, hot, and very noisy. Incorrect mounting can distribute instead of dampen their sound and vibration pollution. I know of a school where they firmly mounted their heavy-duty compressor to one of the structural beams in the building. Whenever the compressor was turned on, everyone in the building knew it!

## Gas outlets

We did two things wrong in our compressed gas outlets: numbers and locations. First, we did not put enough outlets in the laboratories. The HPS uses air, oxygen, nitrogen, and carbon dioxide. The anesthesia machine uses compressed air, oxygen, and nitrous oxide (nitrogen). We have two oxygen outlets in each laboratory, but only one for compressed air. So, we had to purchase splitters for each air line to deliver air to both the HPS rack and the anesthesia machine.

The larger mistake was the location of the gas outlets. Ours are mounted in the ceiling over 4 ft. from the wall. This location is practically in the middle of the room! For the simulation center, mounting these outlets closer to the walls, or even on the walls, would have been more convenient. The outlets for the HPS rack would actually be better very low on the wall, between 1 ft. and 4 ft. off the floor.

## Lobby

The lobby has proved essential to enabling us to move people around, particularly during sessions when we have 40 or more students in the center. The lobby also gives people unfamiliar with the simulation center a logical place to start when they arrive. This includes tour groups, delivery people, and new faculty members.

## Doors and passageways

About a year after we opened our center, we acquired two standard hospital beds. We have put them in two of the laboratories, to make them better resemble a patient room. Unfortunately, most of the doors in the simulation center are too narrow for a hospital bed to pass through. To move them in or out of the rooms, we have to lower them all the way down and turn them on their side. This is very inconvenient, to say the least. So, all of the doors for simulation laboratories and storage areas should be at least 42 in. wide, plus all passageways and doors configured to allow easy movement of a full-size clinical bed.

## Lights

The lights in our laboratories do a great job of illuminating the laboratory. There are just two problems with them. First, they are too bright. We have eight banks of lights, each with four 48 in. fluorescent bulbs, as shown in Figure 30.8. When these are turned on, it is too bright in the laboratory to see the patient monitor. The lights do have a dimmer switch, which we seldom set above the halfway point. Second, the light banks are arranged in a square around the center of the room. This arrangement prevents us from doing anything with the ceiling. In particular, we can not mount any tracks for curtains in the laboratories without having the lights moved first. My suggestion, to alleviate both of these issues, would be to use recessed floodlights, with zone controls to provide

**FIGURE 30.8**  The banks of lights in one of the simulation labs. There are eight sets of lights, each with four florescent bulbs. The square arrangement around the room limits the ability to alter the configuration of the ceiling, or to mount curtain tracks on to the ceiling.

better control of the illumination levels. Also, direct most of the illumination on the patient's location within the room.

The light controls for the laboratories are in the control rooms. This has been a mixed blessing. On the plus side, we can set the illumination to a level that enables the monitor to be seen effectively, and to still prevent them from seeing into the control room through the mirrored glass. On the other hand, few other people know where the control is, so they wander around the room looking for a switch that is not there.

## Storage

Storage is always an issue for any simulation center. There are many things that you may want to store for occasional use. These include gurneys, stretchers, ventilators, and equipment carts. You may also have simulation supplies, part-task trainers, and other things that only are used a few times each year. Also, you might want to store some shipping supplies, such as the large box necessary to ship an expensive mannequin.

At our medical school, we are politically not able to have areas that are "just storage." We did not designate any significant amount of closet space at our center. A large walk-in closet would have given us space to store extra gurneys, ventilators, and carts. (Of course, when we built our center, we never imagined that we would have simulators that are primarily used for outside activities.) If I had the opportunity to do this over, I would cut 2 ft. off of each laboratory, and put an 8-ft. wide storage space between laboratories 2 and 3, just opposite the lobby. I know of one school that got around their prohibition against storage spaces by calling a large closet "backstage space," which was apparently acceptable. Other possible designations might be "utility, repair, support, or supplies."

The other storage problem at our center is the walls – they were not built to support overhead cabinets or bookshelves. When I asked, the principle architect told me that our drawers and cabinets would be plenty of storage space. All of that space was full within 12 months of that conversation. The simulation center at Mayo used an interesting approach. Since their control rooms are elevated they built long drawers that go under the control room floor (see Chapter 18).

## Standardized patient clinic design

As indicated earlier, our Clinic is a very busy facility. During the 4 years it has been open, it has served us well. During this time, we have learned a number of things that would make the clinic much more effective. The top three considerations when designing a new SP clinic:

- First, the SPs need access to a lavatory that is separate from the students. At our facility, the only lavatories on the floor are the common restrooms/lockers for the students. The SP/student learning encounter is often a very emotional experience for both the student and the SP.

Subsequent "social" encounters in the lavatory immediately after such an encounter can be awkward, at best.

- Second, the SPs need their own "space" within the facility. This space functions similar to a faculty lounge. It would provide a place to wait before and during sessions, decompress after sessions, and to generally "be themselves." Like a faculty lounge, it needs to off limits to the general public, particularly the students. Ideally, it should contain small lockers for secure storage of personal effects, comfortable chairs, bulletin board, refrigerator, microwave oven, and a sink. Access to the SP lavatories should be through this room as well.

- Third, provide separate hallways and access for the SPs to eliminate any contact between the SPs and the students. Several schools have incorporated this into the design of their facility, and it is very effective in maintaining realism and professionalism. However, providing separate access requires careful design, and can significantly increase the space requirements for the facility.

## 30.9.2 Operations

### Large-scale teaching sessions

Having a large simulation center makes it possible to provide simulation-based teaching sessions for a large number of students over a relatively short period of time. Each medical school class at the University of Louisville has 150–160 students, and the nursing school admits 100–110 new students each semester. We have 10 sessions each year involving an entire medical school class, and four similar sessions for the nursing school. Executing these sessions requires close management of the logistics of moving students through the center. These sessions have been described in detail elsewhere [1, 2]; I will provide a brief overview of one of the physiology sessions to illustrate how this is accomplished.

We have designed four 1-hour simulation-based training sessions that are part of the first year medical physiology course. All of the students attend these four sessions. To accomplish this, we divide the class into 16 groups of 9 or 10 students. Four of the groups arrive and enter the four simulation labs. Each group spends 1 hour with two faculty members, working through a structured simulation exercise. At the end of the hour, those four groups leave, and the next four groups enter the simulation center. Repeating this process four times, the entire medical physiology class experiences the simulation session over a 4-hour period. This is done four times during the semester.

We preprogram the simulators for these cases. There is very little clinical variation between the groups. This is done to provide a consistent educational experience for the students. It also makes it easier on the faculty, as they know what to expect for each session. We recruit clinical faculty for these sessions. They guide the students through the clinical scenario. We provide them with the scenario, as well as questions (with answers) for them to ask the students. The clinical cases require 15–20 minutes to complete. The remaining time in the hour is used for discussion. This provides the students an experience to integrate the basic science knowledge with the clinical case.

### Documenting cases

During our 5 years of operation at the Center, over 200 clinical cases have been developed by faculty members for use with the simulators. We produce these cases by preprograming them into the simulator before the simulation session. By preprograming the cases, the faculty members know what to expect during the case, which enables them to focus on teaching, not attending a simulator program.

Since the cases are programed for each simulation, we have a record of each case. We have been documenting these cases, and keeping a copy in a set of three-ring binders. Each case contains at least three pieces of documentation – the clinical case description, the simulator program, and a summary page for the case. Some cases contain other information, such as reference documents, example patient records, or explanatory notes on the case.

The clinical case description is the document that the faculty member writes to describe the case. This is the starting point for developing the simulation session. It contains the information that will be provided to the students, as well as the status of the patient at different times during the case. If appropriate, it also contains notes describing what adverse events occur if particular actions do not occur. The simulator program contains the preprogramed changes that will occur in the simulator. For METI simulators, these are known as the "scenario" files.

We have designed a summary sheet that we use for each case. This sheet essentially has two parts. The top half contains an overview of the case and the major steps that occur. It also lists the original designer and the intended audience/learners. This part of the summary is used by faculty to decide if this is a case that fits their students and learning objectives; after reading this part, they can decide whether to read through the actual case description. The bottom half of the summary sheet contains logistical information for running the case. It lists the hardware, monitors, drugs, and other materials necessary to run the case. This part of the summary is used by the operators when setting up the simulation session.

Preprograming and documenting the cases enables us to provide consistent learning experiences for the medical and nursing schools. The patient simulators have been integrated into several clerkships. Using preprogramed cases, the students in each clerkship rotation experience similar learning experiences. An example scenario with all three documents is provided in the Appendices.

The documentation also enables faculty to build on the efforts of others. We have several cases where a faculty member used a previously developed case as the starting point to develop their own case. We have four different versions of "septic shock with pupura." Each version is targeted at a different learner audience. This was possible because we documented the case the first time it was used. Other faculty saw it, and adapted it for their students. In this manner, they did not have to completely "reinvent the wheel" – they only had to modify it.

### Preprograming scenarios

Computer-controlled patient simulators are very versatile, sophisticated devices. They provide tremendous training opportunities. When used by creative and imaginative instructors, the possibilities are almost endless. Changes to the patient simulators can be preprogramming before the training session, or the changes can be performed "on the fly" during the session. We preprogram almost all of our scenarios before the sessions begin. Preprograming the scenarios requires a significant amount of "up-front" work before the sessions. However, during the sessions, the operator only has to follow along through the scenario and monitor the simulator, to be ready to adapt and respond when the unexpected happens.

As previously discussed, we have four simulation suites, and we often produce simultaneous simulation sessions. Having the scenarios preprogramed enables us to have consistent experiences for the students. Preprograming the scenarios also means that the scenarios are available for use later. Several of our clerkships are using the simulators as part of their training activities. We use the same cases throughout the academic year. Having the cases preprogramed makes these sessions consistent throughout the year.

### 30.9.3 Management

### Recruiting faculty as simulation users and instructors

As any "service organization" that supports a medical school and wants to grow, we must encourage faculty members to incorporate simulation into their training activities. This is an ongoing effort for us. We do not require that faculty members undergo a formal training session before teaching with simulation. We recognize that some centers do have such a requirement, but we feel that this does not encourage people to use the resource. Instead, we focus on making the use of the simulation as "painless" as possible.

We begin by providing an overview of the functionality of the simulators. This usually occurs during the "recruitment" phase, when we are working with the faculty to induce them to agree to use the simulators. We also do this for the whole-class sessions when we recruit many clinicians for a large number of students in a single activity, such as presenting clinical representations of basic physiological principles.

As part of the overview for new clinical educators, we present some of the existing cases that we have previously developed and used. These examples provide them a framework to experience what is possible and what their colleagues have done with simulation. For some educators, this is as far as they wish to go with simulation. They teach in the simulation laboratory using existing cases that have been developed by others. There are a few of these "level C" educators.

Most of our clinical educators step up to the next level when they realize that the existing scenarios do not exactly meet their needs, and they are willing to expend the effort to make the changes. At that point, they work with us to develop new cases, and the simulator operators then program these scenarios into the simulators. We currently have over 180 cases and internally developed scenarios. We have published a group of these scenarios [4], and we are working on assembling other groups to publish. These "level B" educators make up the majority of the clinical faculty using the simulators.

Several educators have progressed to "level A" users. At this level, they not only create their own scenarios, but they know the functionality of the simulator and the input parameters well enough to actually program the scenario themselves. At this level, these educators are very comfortable with the actions of the simulator, and will even adjust the simulator parameters during the case if necessary. These three levels are summarized below:

> Level A – Knows what parameters are in simulator, and the capabilities and limits of those parameters;
> Level B – Understands how to write scenarios appropriate for the simulator;
> Level C – Understands and uses scenarios written by others.

**Managing events**

The Center has four simulation suites and seven full-sized patient simulators. Its size and configuration was determined by our requirement to accommodate simultaneous sessions and multiple users. The primary "trick" to successfully managing a busy schedule in a multistage simulation facility is in managing people movement. We use the lobby and classrooms as staging areas while making the transition between back-to-back sessions, just like in amusement parks. Also, all scenarios are preprogramed in the simulator before any session starts. This preparation greatly increases the reliability and predictability of our productions. Although the simulator operator can make changes to the simulator "on the fly," we do not start any sessions to execute like that. Obviously, not every possible action can be anticipated and preprogramed. When unexpected events happen, then the operator will adjust the simulator. This is usually done in conjunction with the faculty member teaching that session. For identical scenarios,

preprograming also enables one operator to simultaneously control simulators in two adjacent rooms. This amplifies the productivity of a scarce resource: the competent simulator operator.

The function of the Center is to support the medical school and the other organizations within the university. Since the Center is part of the medical school, courses for the medical students receive top priority for scheduling events, followed by the residency and fellowship programs. Scheduling medical school sessions even preempts scheduling potential paying customers.

There some issues that arise when scheduling simulation sessions within a medical school's schedule. The clinical students' schedules are based on an academic year, from July to June. Their training sessions can be reliably scheduled before the beginning of the academic year. Unfortunately, the preclinical courses at our institution are usually not established before the start of the academic year in July. Some of the sessions for the spring semester are not scheduled until December. This would not be a particular problem, but these sessions tend to involve the entire class, about 160 students. To accommodate the class, these sessions usually require the entire simulation center for an entire morning or afternoon. Obviously, nothing else can be scheduled during that time.

The months of December and May are very slow for us, with the holiday break and spring graduation. June and July are very busy. This is filled primarily by Advanced Cardiac Life Support (ACLS) classes and resident orientation. We have ACLS classes for the medical students and for incoming residents in June, and we have orientation sessions for several clerkships at the beginning of July.

## 30.10 Conclusion

The University of Louisville School of Medicine has successfully incorporated simulation in the medical school curriculum, using both mannequin-based simulators and standardized patients. Although based in the medical school, these programs also support educational activities within the nursing, dental, and public health schools at the Health Science Campus. These programs also support education programs for various hospitals in the region, as well as first responder training for local, state, and national agencies.

Both of these programs have grown quickly since starting in 2000. This has been accomplished through strong institutional support from the central administration and aggressive promotion of the programs, both within the university and to external agencies. Both programs continue to grow and evolve to meet new challenges and requirements. We have learned many things during our first eight years, and we have made some mistakes along the way. However, we have incorporated these lessons along the way, which has resulted in both programs being stronger for it.

# References

1. Anderson, G. and Goodrow, M. Simulation in physiology, In: Loyd G., Lake, C., and Greenberg, R. (eds.), *Practical Health Care Simulations*. Elsevier, Philadelphia, PA, 2004, pp. 117–141.

2. Goodrow, M. S., Rosen, K. R., and Wood, J. A., Using cardiovascular and pulmonary simulation to teach Undergraduate Medical Students: Cases from two schools. *SCVA* 9(4), 275–289 (2005).

3. Olympio, M. Space considerations in health care simulation. In: Loyd, G., Lake, C., and Greenberg, R. (eds.), *Practical Health Care Simulations*. Elsevier, Philadelphia, PA, 2004, pp. 49–74.

4. Goodrow, M., Whatley, V., Godambe, S., and Paul, R. Simulation cases for pediatric emergency medicine (PEM) fellows. *MedEdPORTAL* (2006). Available from http://www.aamc.org/mededportal, ID=193.

# 31

# Educational Needs Dictating Learning Space: Factors Considered in the Identification and Planning of Appropriate Space for a Simulation Learning Complex

Eileen R. Wiley
W. Bosseau Murray

## 31.1 Introduction

Simulation is an example of a "disruptive technology" – it will totally change the way we do things – therefore, it needs totally new and different teaching and learning methods, which in turn need purpose-built innovative buildings.

Why do we need to change, and what are the advantages of learning with hands-on simulation? Many other chapters in this book go into this in depth. Here, we describe the educational need, then we describe how we "converted and transcribed" the educational needs into floor space and bricks and mortar (i.e., a building).

Many years ago, medical education consisted of the apprenticeship model of a single trainee going with a single physician to do house calls. The teaching model of multiple trainees with a teaching physician required patients to be "collected" in a single building, and hospitals were born.

We are now in the midst of another drastic change in medical education with an ever-shrinking exposure by our students to sufficient numbers and kinds of patients to fully assure that they are as qualified as they should upon graduation. The fixed number of years in any given course of study, combined with the reduction to an 80-hour work week, reduces the raw amount of time available for their learning. The rapid advances in high performance perioperative care has had the paradoxical effect of reducing patient stay in hospital, thus limiting the amount of learning available to the next generation of clinicians. Thus, to even sustain current quality and quantity of clinical graduates, let alone improve one or both, more efficient and more effective means for using student time is required by their clinical educators. Clinical simulation, acting as the "disruptive technology," can be such a means for changing the way we do things to meet our current and near future obligations.

To improve safety, and as an ethical imperative, much of clinical education will now be moved to a time period before patient contact, i.e., health care workers will learn, practice, and become "competent" in a simulation environment *prior* to attempting the procedure or intervention on a real live patient. Simulation will be a *preparation*, not a replacement, for learning with (and practicing on) patients.

Simulation can be the stepping stones in the gaps between textbook learning, hands-on skills acquisition, and professional attitude development. Live patients' safety is placed first by placing live patients last in the students' curriculum. Students will prepare better when using a well-designed curriculum that makes the best use of the latest in simulation tools and techniques. This major change in thinking requires a specially designed and dedicated physical space in which this new kind of learning can occur.

The magnitude of this "major change" is difficult to describe to others who are new to simulation. Perhaps, one could consider the massive impact the invention of printing had on learning (and compare this to learning with simulation): rather than almost all facts being communicated by personal communication and learning from an elite few experts, everyone could learn on their own using the printed books, and then gain practical hands-on experience in the presence of the master clinician. A similar incredibly large magnitude of change is occurring with learning by simulation: trainees can practice on their own prior to encountering the experience with the master teacher. Time can now be spent with the master teacher learning the most advanced skills and attitudes. Students take ever more responsibility for their own maturation as they convert themselves from the passive child learner into the active adult learner. The master teacher no longer courts burn-out from the drudgery of repetitively teaching basic psychomotor skills over and over again. The administrator no longer pays top clinical teacher salary rates for time expended during all the necessary and numerous low level practice sessions. Advanced skills that previously could only be taught by random happenstance can now be taught at an appropriate date, time, place, and suitable challenge for the level of the trainee. Furthermore, all trainees can be assured of similar exposure to similar experiences.

This chapter describes how one group, Pennsylvania State University College of Medicine and Milton S Hershey Medical Center[1] made this change in thinking, and developed Phase I of the Learning Complex. The emphasis in this chapter will be the "why" of the decisions to help other groups apply the lessons learned during this experience to their local circumstances.

## 31.2 The Need for More Space: Which Educational Needs are the Most Pressing?

In the previous section, we described in general terms the requirements for new and different teaching modalities, which require a different type of space. Now, we describe this educational need as experienced at our institution.

In 2002, the Academic Team[2] decided to support innovative education as a system-wide initiative, i.e., for both the College of Medicine and the Medical Center. They developed the concept of a comprehensive, overarching Learning Complex. This complex would consist of separate ("noncontiguous") units that could be located at different locations and have diverse organizational structures, but they would all be developed to cooperate with each other in support of the common teaching mission while using various task-specific forms of simulation. On the basis of a campus-wide survey of educational needs for teaching using simulation, a decision was also made to start three simulation programs[3] with the most urgent needs for space (see Appendix 31A.1). Each of these teaching efforts was hosted in areas and spaces that were overcrowded and not conducive for optimal learning.

It was realized that this would only be a start, and hence the project was designated as a Phase I effort. Three educational needs ("programs") were included in Phase I of the Learning Center.

1. The educational needs for a high-fidelity environment using full-human simulators and part-task trainers. This need is presently fulfilled by the "Clinical Simulation Laboratory," a joint effort to support education by the departments of Anesthesia, Nursing and Surgery;
2. Testing using standardized patients for medical student evaluation under the rubric of Objective Structured

Clinical examination (also called Objective Skills Clinical Evaluation and both abbreviated as OSCE); and

3. Clinical skills training, which includes nursing skills and licensing courses like ACLS, BLS, PALS, and ATLS[4].

All of these areas required more space and became the focus of an integrated approach to education.

The Learning Complex is envisaged to be based on expanding and amplifying an existing strength in simulation-based clinical learning as well as the synergy of colocating space for medical students, residents, and nurse training and testing. Providing an area with focused concentration on these activities will provide opportunities for better training of health professionals, resulting in improved patient care due to better decision-making proficiency and clinical competency. Other expected results include a reduction in errors and improved team work.

## 31.3 Existing Space

It might be instructive to the reader to follow the path of how we went about selecting a new site for the learning spaces. We will first show the existing space and indicate its deficiencies as well as its value to the organization to be used for a different purpose. Then, the lessons learned will be much clearer to the reader.

The college of medicine and the medical center are adjacent and meet at the library, which is in the center, see Figure 31.1.

At present, teaching with full-human simulators and part-task trainers occurs in a 2000 sq. ft. facility on the second floor of the Biomedical Research Building (BMR), see Figure 31.2. This is conveniently located close to:

- The medical student learning areas including their Problem Based Learning (PBL) rooms.
- The clinical areas and the operating rooms via a bridge on the second floor from the BMR building to the medical center.

In the early 1990s, it was realized that the educational needs for residents going into anesthesiology would include an area where the residents and medical students could practice hands-on techniques prior to performing these maneuvers on patients in the clinical environment. The design of the Biomedical Research Building (BMR), a 250,000 sq. ft. facility physically connected to existing college and hospital buildings, provided an opportunity to design and build such a learning facility.

The BMR building housed academic offices and research laboratories ("wet labs" with scavenging facilities of gases and special disposal for contaminated liquids) on floors 2–7 for basic and clinical science departments. The first floor contains Medical Education administrative offices that provide student services, student support spaces, and seminar rooms and classrooms. This first floor space represented the first new

educational space since the institution's inception in 1967 and supported the change to problem-based curriculum.

With the opening of the BMR building in 1993, on the second floor, the Department of Anesthesia provided 1200 sq. ft. for teaching using simulation. This was adjacent to their 1000 sq. ft. lecture room. This was called the Simulation Development and Cognitive Science Laboratory and was supported by the Department of Anesthesiology.

In 1995, the Departments of Nursing and Surgery joined the simulation teaching effort and the Simulation Lab was then supported by the Departments of Anesthesiology, Nursing, and Surgery. As the number of users and uses of the Simulation Lab increased, and the number of simulation trainers increased, the space became limiting, and teaching was compromised in the overcrowded space. Similar to most universities, "storage" is not allowed on the campus, but has to occur off-site in rented space. Given the frequent use of the part-task simulators, it was not feasible to store equipment off-site. In 1999, the Simulation lab expanded into an adjacent 1000 sq. ft. room and this area was also developed as a realistic clinical learning environment with one-way glass and video cameras.

As learning by hands-on practice using simulation has greatly expanded, another expansion is needed. Additional space is required so that we can provide multiple simultaneous sessions using more types of simulators that would allow for multiple large groups to use the facility simultaneously as well as provide for more specific discipline-focused simulators. Additionally, the current simulation lab is located in a section of the Biomedical Research building that is designed and designated to house wet labs (as mentioned, this is very expensive space to recreate in another location, and the institution decided rather to move the simulation facility). The institution desperately requires additional wet lab space. Construction of a new research building on campus, while in the master plan, currently is not within the 5-year planning horizon. A decision has been made to convert this existing simulation area to wet labs as soon as the simulation laboratory is moved.

Another educational need that was to be addressed was learning and teaching using actors ("standardized patients"). This OSCE testing is done in an off-campus location in five examination rooms. This presents scheduling and logistical issues. Owing to the lack of space (only five rooms), only half the class of around 140 medical students can be accommodated on a given day. This presents scheduling issues as well as the need to have multiple OSCE stations to minimize students discussing cases. As OSCE testing increases both in frequency during the year and with additional classes using this evaluation method, the space deficiency will become even more pronounced. The use of OSCE and standardized patients is expected to increase massively as ACGME competency requirement training and testing become standard in residency training and evaluation. It will be very difficult to integrate OSCE testing and evaluation into the curriculum if the only available space is both undersized *and* at a remote location.

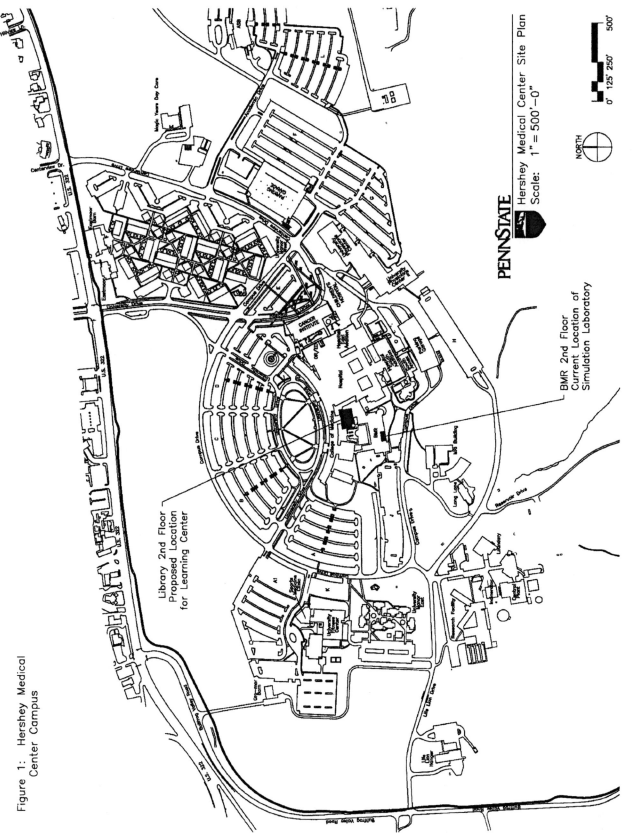

Figure 1:  Hershey Medical
Center Campus

FIGURE 31.1   The overall crescent-shaped design with the college of medicine on the West (left) side and the medical center on the East (right-hand side).

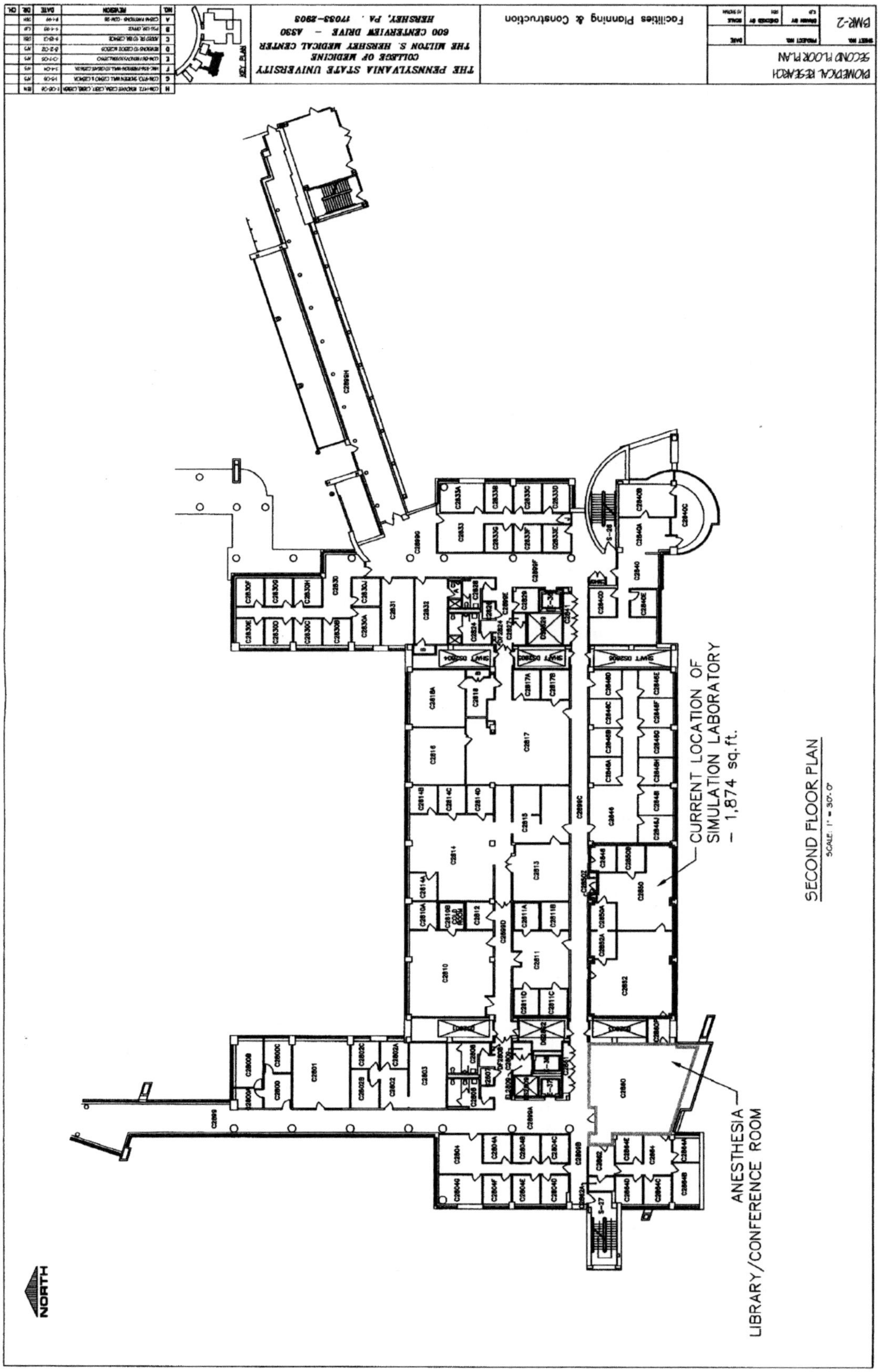

FIGURE 31.2 The location of the present Simulation Lab within the school of medicine structure.

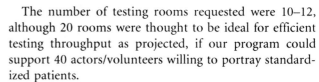

The Clinical Skills training (e.g., ACLS) typically uses low-fidelity simulators such as ResusciAnne. The instructor schedules and uses existing rooms in the hospital, and some stations end up in the corridor or in a room distant from the main lecture room. The trainees are nurses, physicians, respiratory technicians, etc., basically everyone who needs these skills in their work or teaching duties. This massive task is run by the Nursing Department's Nurse Educators with help from other departments. A list of courses and requirements are in Appendix 31A.2.

All of these deficiencies and the desire to provide an excellent educational product point to the need for additional space. A subgroup consisting of members of the Physical Resources Team, the Academic Team, and "Facilities"[5] was formed to develop solutions to support these learning systems in the future.

## 31.4 Converting Educational Needs to Space Requirements

The first step in understanding space requirements was to describe the educational activities that needed "better" space. We analyzed and described not only the existing way of teaching, but also how we would like to teach, given ideal circumstances and ideal space. We produced a narrative of the flow ("traffic patterns") of trainees, instructors, actors, visitors, personnel moving simulation equipment, etc. We then developed a space program or list of rooms required for these functions. Part of this analysis included gathering information on the frequency of various activities. The list of space requirements for each educational function is included in Appendices 31A.1 and 31A.2. The frequency of activities is also summarized in Appendices 31A.1 and 31A.2, with more details given in Appendix 31A.4. The space program includes size and quantity as well as special attributes of rooms, including required adjacencies.

1. Teaching with high-fidelity full-human simulators required an estimated space of 7321 sq. ft. (for purposes of convenience, we called this space "The Simulation Lab"). It was envisaged to consist of four high-fidelity clinical environments with their associated debriefing rooms, control rooms behind one-way glass, utility, locker and office rooms, etc. See Chapters 23–30 for descriptions of similar facilities. The educational needs were translated into space requirements to support the learning, see Appendix 31A.3.

2. Training and testing with human actor–based learning ("standardized patients") was estimated to require about 5900 sq. ft., as outlined in Table 31.1 and illustrated in Figure 31.3. For purposes of convenience these were called "OSCE" rooms – Objective Structured Clinical Examination, where examination could mean examination of a patient as well as an "exam" of the students, formative as well as high-stakes evaluative examinations.

The number of testing rooms requested were 10–12, although 20 rooms were thought to be ideal for efficient testing throughput as projected, if our program could support 40 actors/volunteers willing to portray standardized patients.

To save space and money installing and maintaining real, functional sinks in every OSCE room, most of these rooms would have antiseptic dispensers (Purell type hand sterilizing/hand washing dispensers). To accommodate the learning of correct "scrubbing" techniques, some rooms would have washbasins ("real scrub sinks"). The rooms would also include an examination table and at least three chairs (patient, student, and faculty observer, possibly a family member or two), a computer station for computerized records and digital imaging, fold-down writing surface, and a mural on the wall with painted cabinets. When these rooms would be used for clinical skills, the rooms will be occupied by 4–6 people standing and observing the exam table with a patient or a simulator.

For the OSCE testing space, "traffic flow" is very important. There are three types of users using and moving through the OSCE space.

a. The standardized patients or actors arrive and require a space to store their coats and personal belongings and to prepare their makeup for the role they will play. Ideally, there are twice as many actors as interview rooms. Playing a patient can be exhausting, and it is beneficial if the actors are in the interview rooms for every other student only. The actors also need a meeting room to be briefed, to write or type evaluations, and to gather to rest (take a break) when they are not in an interview room. Typically, there are refreshments available in this room.

b. The faculty who will be evaluating the students. They also need a place to meet, rest, and write evaluations (paper or computer entry). There are three methods used for observation of the students during their evaluation.

   – In some centers, there is one-way glass and a space for faculty to observe the students from directly outside the room.

   – In some centers and occasions, the Faculty are actually in the room with the trainee and the standardized patient.

   – The plans at our center are based on video monitoring in each interview room with viewing monitors in a single large control room with multiple monitors (one for each room). Note that more than one person might wish/need to watch the interaction: for interrater reliability studies, for training new faculty. The control room should be close to the faculty room.

c. Students arrive and require space to store their coats, laptop computers, PDAs, and/or book bags. They are registered and wait as a group in a preshow/briefing

**TABLE 31.1** The table indicates the OSCE educational activities and the space needed to support the activities. Note that the lecture hall requirement was not deemed to be possible given space constraints, and existing lecture halls would have to be located quite close by to support optimal needs

| | | | Additional Room/Space Requirements for OSCE Program | |
|---|---|---|---|---|
| Room description | Quantity | Square foot – each room | Total square foot | Comments |
| Meeting room | 2 | | 0 | Up to 80 people will use lecture room. Meet before and after simulation. ACLS is often 80 people with 10–15 breakout stations |
| **OSCE testing requirements** | | | | |
| Interview rooms | 9 | 120 | 1,080 | One-way glass or video monitoring. Used for standardized patients. |
| Interview/PBL room | 1 | 300 | 300 | Larger room needed to support some of the clinical skills programs. |
| PBL room | 2 | 260 | 520 | |
| Registration area | 1 | 222 | 222 | 10 students at a time, need place for coats and book bags. |
| Pretesting waiting area for students | 1 | 250 | 250 | Needs to accommodate up to 20 students. |
| Posttesting area for students | 1 | 250 | 250 | Needs to accommodate up to 10 students, with area for writing. |
| Changing and makeup room for standardized patients | 2 | 250 | 500 | Need one each for male/female. Rest room, sinks with mirrors, and changing area with lockers. For 10 rooms, there are 20 standardized patients (actors), allowing them to take a break during testing (every other test). |
| Meeting room for standardized patients | 1 | 225 | 225 | Needs to accommodate up to 20 standardized patients – 10 standardized patients alternate sessions. |
| Meeting room for faculty | 1 | 225 | 225 | Needs to accommodate up to 20 instructors/faculty – 10 faculty alternate sessions, or trainees. |
| Storage/clinical supplies | 1 | 150 | 150 | Need space to keep supplies, props, makeup, equipment, etc., needed for the sessions. |
| Video control room | 1 | 150 | 150 | Need a central room for the controller to see all OSCE rooms, and coordinate recordings. |
| TOTAL FOR OSCE | | | 3,872 | |
| TOTAL WITH DEDICATED, CONTROLLED CIRCULATION | | | 5,957 | |

area. They will then individually proceed to an interview room. After completing the scenario and the post-OSCE questionnaire/written exam/feedback station/write up tasks, the students will go to the postshow/debriefing room as a group. It is important to provide separation between students who have completed the OSCE and those who have not completed each scenario. Figure 31.3 shows the flow of students in the space with arrows.

3. The space required for Clinical skills was estimated at about 1000 sq. ft.[6] This includes five rooms at least 120 sq. ft. that resemble inpatient rooms and/or outpatient examination

rooms, as well as five rooms at 90 sq. ft. Storage[7] is very important as there is a lot of material and equipment used for the different courses. Clinical Skills space needs to support teaching, practicing, and assessment for many different and varied groups including nurses, high school students, student nurses, medical students, and physicians, and includes continuing education and recertification.

4. The total space requirement for these combined education programs was estimated to be between 13,000 and 19,000 sq. ft. depending on whether minimal requirements or ideal requirements were used for the calculations.

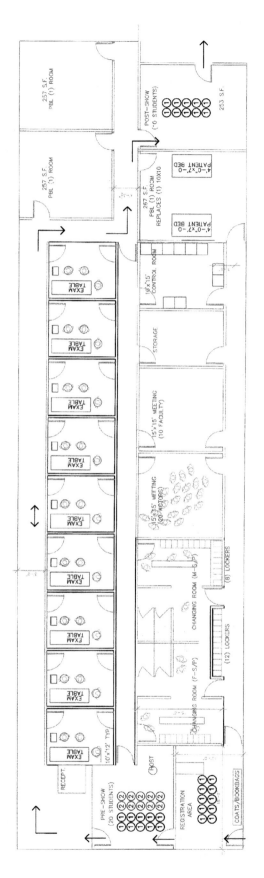

**FIGURE 31.3**   The teachers' requirements for human actor sessions translated into physical space.

## 31.5 Ensuring Educational Needs will be Met by Dual Use of Rooms, i.e., Development of Options

Each of the three groups listed their ideal space requirements. When added together, even the minimal sum total of about 13,000 sq. ft. exceeded the realistic expectations of the amount of space that could be made available. As a result, there was rather quick recognition that some sharing of space would be necessary. Each group provided schedules and use data as well as absolute minimum space requirements (see Table 31.1, Appendices 31A.2 and 31A.3). This allowed re-evaluating the space requirements and development of rooms that could be of multiuse. By enlarging the typically tiny OSCE rooms to allow sufficient space for a standard hospital bed, these rooms can also support the Nursing skills requirements. When the rooms are not being used for OSCE or nursing skills, they can be used for part-task skills teaching purposes ("Sim Lab"), especially since many of the Nursing skills courses require the use of simulators.

The educational needs could be fulfilled by dual use of space, and this also fulfilled the overall vision of the Learning Complex as outlined by the Academic Team, i.e., a vision of joint educational projects by multidisciplinary groups, see Figures 31.4 and 31.5.

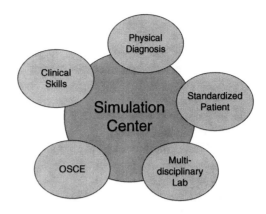

**FIGURE 31.4** The figure shows the users groups and their educational needs for learning using simulation.

## 31.6 Educational Requirements Dictating Options for Location of Learning Space

*Location* has a huge effect upon Return of Investment (ROI): the educational efforts must occur close to where students and instructor spend most of the rest of their time, yet not so close to real patient care so as to contaminate the clinical

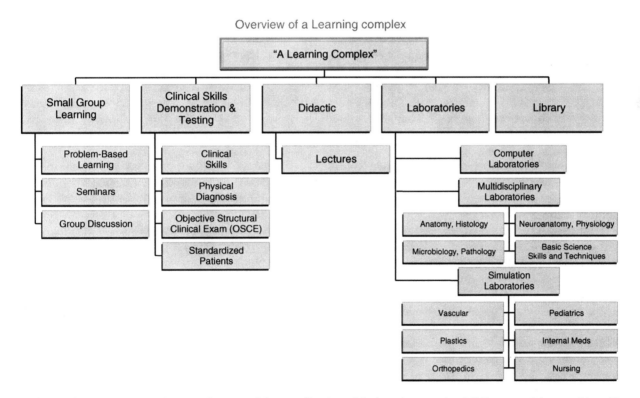

**FIGURE 31.5** The figure shows a schematic of the overall vision of the learning complex. While many of these entities will function under separate organizational structures, it is proposed that all work together to improve the learning environment.

environment with sights, sounds, smells, and inert, possibly dysfunctional simulation items.

We will describe our process of selecting a specific space. This will give the reader an idea ("modus operandi") on how to approach these very sensitive issues in their home environments. Very few decisions engender such passion as the reassignment of somebody's "space." This is often seen as a very strong vote of "no confidence".

Selecting more than one location also helps to negate the feeling of a specific individual or a specific department being targeted for "downgrading."

Therefore, the subgroup took the decision to evaluate three locations for the learning center. They included the first floor of the Biomedical Research building, the first floor of the teaching wing, and the second floor of the library stack area. The options were reviewed by Academic Team, and FROG[8].

**Option 1:** First Floor of the Biomedical Research Building (see Figure 31.6). As mentioned previously, the first floor of the Biomedical Research Building contains seminar or PBL rooms and two large classrooms, totaling 10,993 sq. ft. The space is in a prominent campus location and is easy to access. This option was not chosen because it is already used for educational purposes and replacement space is not available. It would be more expensive to renovate replacement space for classrooms and seminar rooms, and to renovate this area for a learning center. Also, the space is broken up by corridors, presenting challenges to meet the required program flow.

**Option 2:** First Floor of the teaching wing (see Figure 31.7). The first floor of the teaching wing contains academic offices for the Department of Humanities on one side of the corridor and the Department of Neural and Behavioral Sciences on the other side of the corridor. The total square feet of area for the Humanities offices is 2367, and for Neural and Behavioral Sciences it is 2,833 sq. ft. While Neural and Behavioral Sciences has faculty engaged in wet bench research, Humanities does not require wet labs for their office-based research. Therefore, initial thoughts were that Humanities could possibly move their offices. It would be more challenging to move Neural and Behavioral Sciences offices.

The first floor space of these departments, on "main street," is a prominent campus location and is easy to access. This option was not chosen because there was not sufficient space to meet the program needs of the Learning Center and replacement space would be needed for the department offices within the main building. Adjacent to Humanities offices is the College Bookstore (1906 sq. ft.). This area was considered as a potential option for additional space under this option. Whether the bookstore needed to be on campus or could function as a virtual service reserving conference room space when needed was evaluated. It was decided that it was desirable to maintain a traditional bookstore. In addition, the combined square feet (4276) is not sufficient to support the space requirements for all three programs, i.e., the simulation lab, the OSCE testing, and clinical skills area.

**Option 3:** Second floor of the Library stack area (see Figure 31.8). As mentioned before, the Medical Center consists of a long crescent-shaped building that connects to other facilities behind it. The western side is the College of Medicine and contains rooms that support the education of medical and graduate students and offices and research labs mostly for basic science departments. The eastern side is the clinical side. It is here where clinical science departments have their offices and labs and the hospital are located. The Medical Center Library is physically located in the middle of the crescent building, integrating the clinical and basic sciences. The library is located on three floors and is named after our founding dean, George T. Harrell (see Appendix 31A.5 for graphics of the layout of the existing three floors of the library).

1. The main entrance to the library is located on the first floor. This area of the library also contains librarian offices, reference books, journal stacks, and open study areas for students.
2. The second floor contains stacks, a computer training room, computer stations, and study carrels. The study carrels and computers are available to students 24/7 by accessing a door on the second floor bridge that connects the Biomedical Research Building to the hospital. The doors between the library stack area and the study carrels are locked when the library is closed, to protect the collections. On the second floor outside the library stack area are two former offices that since the original study have been converted to PBL/seminar rooms for medical education. These rooms have recently become available to students to schedule for group study at the times they are not used for classes.
3. Since the original study, the ground floor stack area has been converted to offices. The remaining space on the ground floor of the library contains videoconferencing rooms that are managed by the media services group. The media services group is envisaged to play a large role in the information technology (IT) needs of all facets of the hands-on learning (see Chapters 73 and 74 for more details on information infrastructure requirements).

In close proximity to the library are, on floors 3–7 above the library, four lecture rooms. There is also a large meeting room on floor 3. The lecture rooms are on two floors on one side of the building wing. For example, Lecture Room A is on floors 3 and 4, Lecture Room B is on floors 4 and 5, etc. While the library is physically close to the lecture rooms and there is a stairwell connecting floors 2 and 3, there is no direct access at present to the lecture rooms from the library due to library security needs. The connecting stair is locked.

Space and book access in the library has been underutilized for many years. As mentioned, recently the ground floor stack area was converted to offices (including offices for personnel working in the simulation facility) as well as an additional PBL

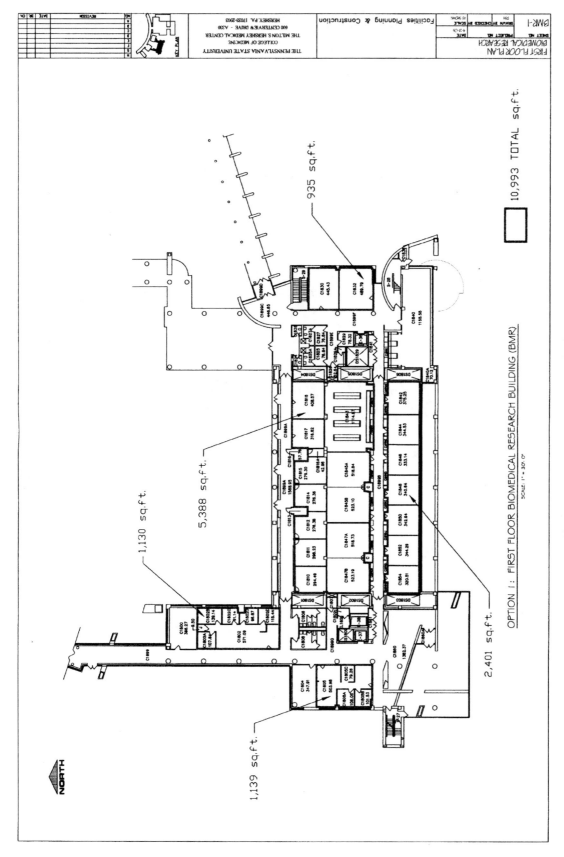

**FIGURE 31.6**   The first floor of the Biomedical Research Building, which was considered as Option 1.

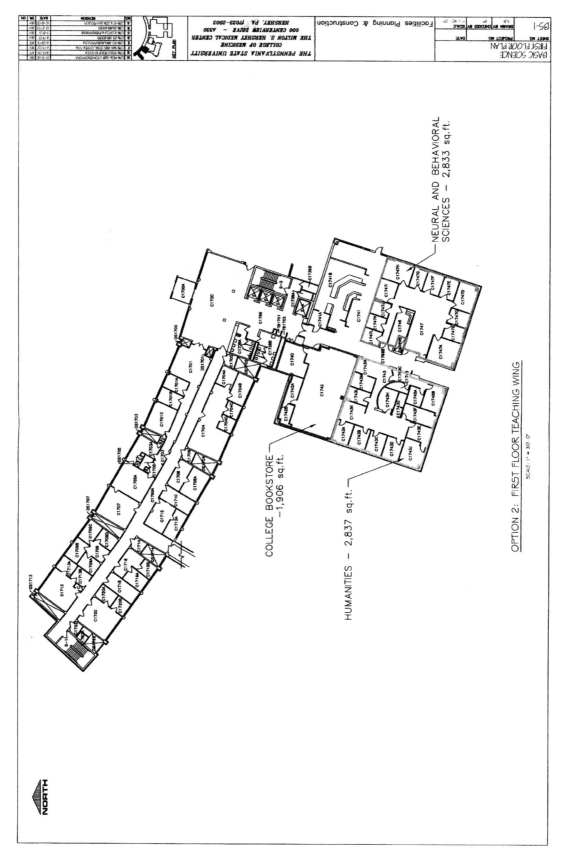

FIGURE 31.7   The first floor of the teaching wing, which was considered as Option 2.

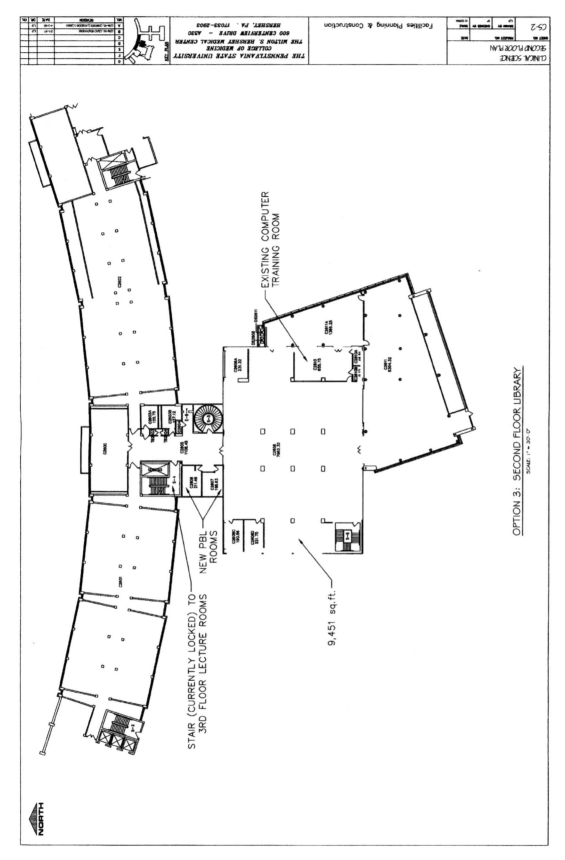

OPTION 3: SECOND FLOOR LIBRARY
SCALE: 1" = 30'-0"

**FIGURE 31.8** The second floor of the Library wing, which was considered as Option 3, became the recommended location.

room to accommodate increases in the medical class size. The first floor of the library (see Appendix 31A.5) has sufficient space to add shelving as well as study tables. The card catalog, which has not been used for years, was located on the second floor but has since been removed.

The area identified in Figure 31.8 shows the 9347 sq. ft. determined as fulfilling the educational needs for Phase I of the program. It was determined that compact shelving, off-site stacks, or provision of online texts would be used for the journals on the second floor of the library.

This option was the selected option for further development. Two PBL rooms are located in the area outside the stacks. Stairs are present that lead to a large lecture room and a large meeting room on the third floor. With these elements (large lecture rooms) so close by and available to the Learning Complex, they could be excluded from the requirements of the Learning Complex. Therefore, the planning group thought it was feasible to fit the program within the second floor library space.

It is feasible to install a card reader and camera at the stairwell so that the connecting stair will provide access to the third floor lecture rooms and still maintain security of the library holdings. The library already is a student-driven space that supports learning activities at the medical center. The use of the second floor of the library for a learning center is consistent with the goal of changing the library into more of a community space for collaboration, contemplation, and socialization. Librarians are used to managing vast numbers of objects and classifying the data about these objects. There are many partial-task simulation devices and software simulation programs (thousands to tens of thousands respectively). Librarians are well suited to manage this massive amount of information.

## 31.7  Refining Options

Utilizing the space programs and requirements, Facilities staff developed several test-fits, see Appendix 31A.6, which shows the development of various combinations of high-fidelity simulation requirements and clinical teaching requirements. These drawings also indicate which requirements were left out (not included) in each of the six drawings.

One test-fit showed all the Simulation space requirements being met and the other showed all the OSCE and Clinical Skills training space being met (see Appendix 31A.6). The Facility staff included on each option a list of the rooms that were not provided. Facilities used this method to clearly illustrate to each group that it was not possible to provide every room for each group within the identified space boundaries. If only one group was emphasized over the other groups, then those other groups would not have a functional program. This was a team approach and there needed to be some flexibility by each group to develop a space plan that supported all program needs.

At this time, there was pressure from others to complete the study, pressure from library staff who did not think it was feasible to consolidate the stacks, and the realization that we had to make this particular space work or the opportunity for more space for these programs could be lost. There certainly was much discussion that this was not enough space. The group agreed that this would be a successful Phase I enhancement and that additional requirements and space would be addressed in a future phase. Historically Facilities provided the increases in medical education space at the Medical Center. The first significant new education space since the institution's inception in 1967 was 26 years later in 1993 when the Biomedical Research Building was opened with new seminar rooms and class rooms (and the Simulation Laboratory). The Simulation Laboratory increased space by 100% in 1999. The Learning Center would result in another significant increase of education space much sooner than the prior increase in space.

The test-fits (see Appendix 31A.6) were reviewed at several subgroup meetings. The planning phase is extremely important and the group was dedicated to spend the time needed. To appease those that wanted a quick resolution, many updates on progress were provided to others. This included review of options for replacement space for the journal stacks. Three options were reviewed. They were:

1. Consolidation of the stacks in compact shelving on the first floor of the library,
2. Off-site stacks, and
3. Providing the journals online.

Costs, advantages, and disadvantages of each option were reviewed. The use of off-site stacks was considered but thought to be impractical by the library staff. The library staff did not think faculty would come off-site, even if the off-site area was staffed and did not think the library staff would be able to meet the journal requests if they had to deliver the journals from off-site to faculty on campus. This was because the requestors want the information right away and would not wait for a daily or weekly run for the journal. The potential for increasing information that is available online was considered but the annual cost of providing electronic access to replicate all the books currently owned and in the second floor stack area is cost prohibitive.

The option that was selected is to provide compact shelving on the first floor of the library to accommodate the second floor journals. While the library staff understand the importance of the learning center and that they are partners in this effort, they are concerned that study spaces for students will be compromised due to the amount of space required for the compact shelving. It has been agreed that the relocation of the second floor journal stacks in compact shelving will require an analysis of library space that may result in other physical changes. A library planning expert will need to be part of the commissioned architect design team. It is intended that some of the librarians will participate in the compact shelving location as well as be a member of the learning center design team.

The discussion surrounding the test-fits that occurred at the Learning Center subgroup meetings was fascinating. Once the stakeholders realized that there was not enough space to include every room that was desired, the creativity began to flow. What evolved was a consensus plan that eliminated several spaces and developed many areas into multiple-use and multiple-user areas.

Several areas were identified that could be close to the Learning Center but not adjacent or within the center. These included offices for the Simulation Lab medical director and operations director offices, which will be located on the ground floor of the library (in the stack area that would be converted to an office suite). The conference/lecture rooms on the third floor will be used for the briefing/debriefing space to support simulation programs and OSCE.

The OSCE interview rooms were increased in size from 10 × 12 ft. to 15 × 15 ft. so that they can be used for clinical skills and OSCE testing. In addition, these rooms will have a curtained area[9] that will contain simulators that are not provided with their own space. The simulators will be out of sight when the rooms are used for testing and may be used for some of the training courses. Instead of a dedicated preshow room, it was decided that the reception room or the room containing Harvey, the cardiac simulator, or the project development room can be used to support this activity when needed. Two rooms in the lobby area outside the learning center have been identified for PBL rooms and can also be used for OSCE testing and training when not scheduled for PBL classes. The locker rooms/bathrooms for the OSCE actors were eliminated since there are restrooms in the area outside the stacks. The final test-fit developed by the team is included as Figure 31.9.

Please note that these **"test fits"** are **"space planning drawings"** that only illustrate the size of rooms and their relationship to each other. Once the project is approved to proceed to design, an architect design team will be commissioned. That team will actually design the space and conduct a regulatory code review to include making sure exit distances are sufficient, hallways are sized adequately, and accessibility requirements are met. It is difficult for the novice building designer to recognize that **a test-fit is not a design** but just a space plan. For example, there are structural columns within the space that may be less intrusive when an architect designs the space. Also, new electrical, telecommunication, and mechanical systems will be needed to support this renovation. Space for these areas is not provided within the program space. The design team, including the engineers, will need to resolve this issue.

Furthermore, the design team should include professional theater designers. Medical School and clinical space designers have been proven to be all but oblivious to the theatrical nature of simulation events, and thus don't provide instructors with spaces that support this theatrical form of education. The same goes for requirements of lighting, sound, video, people flow, video and sound recording with immediate replay capabilities, long-term storage of huge video files, the need for security of data from internal personnel[10], etc.

## 31.8 Cost and Schedule

A project management firm was hired to provide a cost estimate and schedule. We anticipate that it is likely to take 2 years from the time approval is provided to select a design team to the time the learning center would move in to their new space. The project renovation cost (bricks and mortar) is estimated at $6 million. This does not include furniture or equipment.

## 31.9 Mechanism of Funding

There are proposals to fund the renovation through philanthropy. It was anticipated that the test design would be approved to begin architectural design in January 2006. Architect design was expected to require 12 months, and building another 12 months. It was emphasized many times by experienced facilities personnel that adequate time should be available for the design phase, which would include visits to other successful (and not so successful – to learn from mistakes) centers.

As of September 2007, approval had not been provided. This is an important Medical Center initiative and the chief financial officer of the College is reviewing other possible funding sources. This initiative supports all our primary missions, education, research, patient care, and community service. Once the simulation laboratory moves, its current location will be converted to wet bench research labs. It is one of the few remaining areas in the facility that can be easily converted to wet labs at a reasonable cost.

Another mechanism of funding is through the patient safety office. If you have a Patient Safety program/office/guru, be sure to consider including them in this design and funding process; their title and function, PS (Patient Safety), will lead the charge forward for many PS (Patient Simulation) programs across the globe, as they climb out of individual personality–born infancy.

## 31.10 Lessons Learned

It is important to build trust on your planning team. For our team, all three groups wanted to make sure that their program needs were met, and providing the opportunity to describe the ideal space became a tremendous asset. All of the groups were willing to be flexible and were not interested in empire building or grabbing all that they could get for their own program.

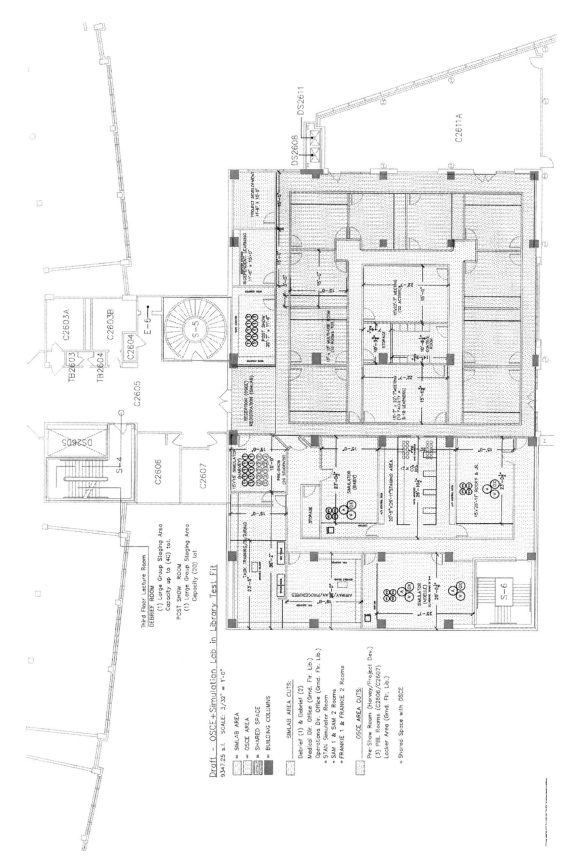

FIGURE 31.9 The final test-fit that includes as many as possible of the educational requirements. Areas not included are also indicated.

There was a focus on the combined effort and providing the best programs for all the constituents.

Do not rush the planning process. There will be external pressures to finish. There are many layers of complexity and it is important to provide the participants with time to think creatively about different ways to provide the program and to get input from other people. We had many meetings. We hit obstacles, we recessed, and we met again and together provided creative solutions.

It is critical to involve all of the stakeholders. Make it clear that everyone will not get everything that they want but they will get what the program requires. Once the amount of space was defined, all of the groups had to cooperate to fit the programs in that amount of space.

While stakeholders are important, it is also essential to have a nonstakeholder participate and act as process facilitator. For our group, the Facilities representative played that role. Her focus was on making sure program needs were articulated and met and on building consensus. Facilities knowledge of the space planning process and experience in understanding what will and will not fit in space was essential. To identify the space needs, stakeholders described what they do in the space. Next, they were asked to list the rooms that were necessary, who uses the rooms, and what needs to be in the rooms. With that information, the sizes of rooms and number of each room were determined, resulting in the total department sq. ft. needed. An adjustment factor is applied to that space to include corridors, mechanical spaces, etc., and resulting in the total (gross) sq. ft. required. Understanding the number of hours the rooms are used helped to identify those rooms that could serve multiple purposes. By providing a picture, this process of simulating with a test-fit helps the stakeholders confirm the flow within and between spaces. When there is not enough space for all your wants, the data collected and the test-fit support the decision-making process as to what rooms are included in the final program.

Selecting space close to other educational spaces like library resources, computer training, problem-based learning rooms and the lecture rooms is beneficial. Then, those resources, as well as their supporting utilities like restrooms, do not need to be duplicated in both areas.

If you are lucky and have planned well, your program will be successful and you will need more space in the future. By acknowledging and providing a voice that this space planning is not the end, and identifying this as Phase I, provided more freedom for flexibility and reduced anxiety.

# Endnotes

1. Located in Hershey Pennsylvania, U.S.A., our College of Medicine has a 4-year MD program with around 150 medical students per year, as well as an active Graduate Medical Research program. The Medical Center has around 500 beds with almost 20,000 operations per year, including many trauma patients from our own system (Level I trauma designation), as well as many referrals from surrounding smaller medical centers.

2. Background to the team structure – the PennState College of Medicine and Hershey Medical Center has eight Teams: three resource, three mission, and two integrative. The Academic Team is one of the mission Teams with the overarching aim of improving the educational milieu in our organization. The Physical Resources Team includes several personnel from "Facilities," and is a resource team charged with all aspects of space and building.

3. The learning objectives, course names, and student levels in these three items are described in Appendix 31A.1. The actual numbers of students taught and fraction of each graduating class of the entire clinical educational endeavor at HMC is included in Appendix 31A.4.

4. See Appendix 31A.1 for a full list of abbreviations.

5. "Facilities" describes personnel who have been involved in prior space planning building design and construction. They have the experience of building design, talking to clinicians and researchers, and translating their needs into a space program that can be represented in physical space via computer assisted design/drawings (CAD). This gives an idea to the novice users of what their ideas would look like, once translated into bricks and mortar. Often, it is not anything as they envisaged, and they have to go back to the drawing board. Experienced "Facilities" staff play a very important role in any venture such as this.

6. The educational requirements also included the need for two large lecture rooms. As neither space nor funding would be available, the group voiced this need as a requirement for the Clinical skills space to be close to existing lecture rooms.

7. Storage is often not allowed "on site," and many centers label such space for clinical supplies, props, simulators, etc.

8. FROG is the Facilities Recapture Oversight Group: this group is charged with ensuring that all lab space is used for optimal lab purposes. For instance, the very expensive "Wet Lab" space should not be used for offices. Areas such as the present Simulation Lab, which is not making full use of *all* the services of a wet lab, is also to be moved to "cheaper space."

9. We are considering the use of a lockable system, similar to a roll-up garage door that could totally protect the simulation equipment.

10. Note that typically, only the simulation teaching personnel are able to access training videos. For instance, neither the Chair nor the residency director has access to teaching videos. These are kept solely for teaching and research purposes.

# VII

# Functional Forms at the State and Nation Size

*We do not know, in most cases, how far social failure and success are due to heredity, and how far
to environment. But environment is the easier of the two to improve.*

J. B. S. Haldane

The *value* of clinical simulation that attracts scarce resources is the same at all program sizes:
after simulation, your first time for real is not your first time. However, the *management* of
those resources is unique at different sizes due to differences in established cultures of command,
control, communications, and expectations. Several programs have been successful in harnessing the
economies of scale and created simulation programs that span across institutional borders. Like those
programs that were initiated in nothing more lavish than a single room, these interinstitutional programs
were the result of hard work by passionate individuals. Note that in every one of these chapters, the
authors start by defining the unmet challenges faced by their clinical educational community, and then

they describe a plan of action, including the various participants and their inter-relationship that is necessary for success. Only after these non-visible but nonetheless vital foundations are established do they start the tangible work of designing and building schools and selecting and buying teaching tools.

Neil Coker describes how a number of different types of institutions within the large state of Texas combined their complementary resources to create an efficient program. Sharon Denning *et al.* extends this theme with their program successfully combining the needs and resources of public and private clinical institutions despite differences in their respective cultures. Stefan Mönk *et al.* present a program that matched a common and efficient solution to the common needs of numerous institutions within a nation-state. Michael Seropian and Bonnie Driggers do the same for a state within a nation. Fabian Purcell and Denis French illustrate how an institution as large, diverse, and traditional as military clinical care can adapt and adopt clinical simulation to improve the clinical care for their nation's warriors. Ronnie Glavin examines successful integration of clinical simulation into his nation's requirements for clinical competency. Amitai Ziv *et al.* share their experiences of the same for their nation.

# Designing and Developing a Multi-institutional, Multidisciplinary Regional Clinical Simulation Center

Neil Coker

## 32.1 Benefits and Challenges

Clinical simulation is a powerful tool for teaching health care professionals. Placing learners in situations that mimic actual clinical events enhances educational efficiency and effectiveness by:

- Stimulating a desire to learn,
- Enhancing depth of information processing,
- Evoking emotional states similar to those occurring in reality,
- Expanding awareness of and attention to environmental cues (mindfulness),
- Encouraging active efforts to create meaning from experience, and
- Promoting group interaction, mutual feedback, and shared problem solving.

During effective clinical simulations, learners "suspend disbelief" and interact with the simulator and the environment as if they were real. A good clinical simulation resembles a good theatrical production. Participants become fully immersed in the experience because costumes, makeup, scenery, props, lighting, and other stagecraft elements support their illusion of reality. Therefore, a well-designed clinical simulation facility should not simply provide access to mannequins that reliably reproduce patient presentations. It should also duplicate the key sights, sounds, smells, and other sensations associated with the clinical environment.

Simulation centers that reproduce patient care environments with high fidelity are expensive to build, equip, and operate. Additionally, anticipated use by a single institution is frequently too low to justify these costs. The experiences of Temple College (Temple, Texas, U.S.A.) in developing a center that meets the simulation needs of several academic institutions and hospitals in Central Texas demonstrates one strategy for overcoming these difficulties.

## 32.2 Our College, City, and State

The College is a comprehensive community college with 4000 students in Temple, Texas, a city of 59,000, located 120 miles south of the Dallas-Fort Worth Metroplex on Interstate

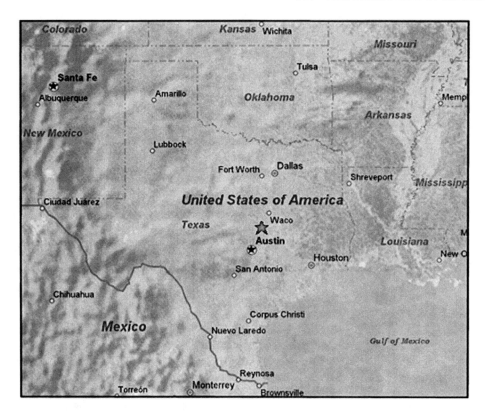

**FIGURE 32.1**   College location and major communities in the surrounding region.

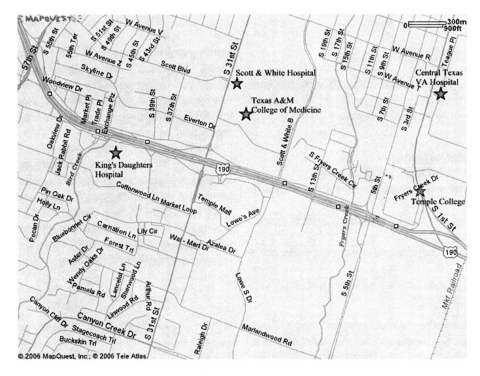

**FIGURE 32.2**   College location and clinical institutions within the city.

Highway 35 (Figure 32.1). A regional center for health care, Temple is the home of Scott & White Memorial Hospital, Central Texas Veterans Health Care System, King's Daughters Hospital, and Texas A&M University System Health Science Center College of Medicine's clinical campus (Figure 32.2). Health care contributes US$ 750 million annually to Temple's economy and supplies over 15,000 jobs.

In the summer of 2001, the College offered associate degrees or certificates in seven health care disciplines – associate degree nursing, dental hygiene, emergency medical services, licensed vocational nursing, medical laboratory technology, respiratory care, and surgical technology. Although these programs had developed reputations for academic excellence, they were scattered across three campuses. Several were housed in temporary facilities with inadequate space and outdated equipment. The community had expressed strong expectations that the College would continue to educate skilled health professionals to meet an aging population's needs, accelerate recruitment and graduation of professionals to meet existing and projected area health care workforce shortages, and support rapidly developing community initiatives in medical research and biotechnology.

## 32.3 Joint Decision on a Joint Venture

After exploring several alternatives for meeting these expectations, the College's administration decided that the best solution was to build a new Health Sciences Center on a site adjacent to the existing Nursing Education Center. This would provide adequate instructional space and updated equipment for all programs and promote greater collaboration and resource sharing among the health professions programs. The College's Board of Trustees accepted this plan and agreed to call a bond election to raise US$ 7.23 million for construction the Health Sciences Center and renovation of nursing and mathematics/biomedical sciences facilities.

In preparation for the bond election, the College's administration engaged the services of an architectural firm to draw preliminary plans for the new Health Sciences Center. To provide input for the architects, health sciences faculty visited several institutions that recently had built new health sciences facilities. On one of these visits, two department chairs saw a series of simulation laboratories designed to resemble an emergency department, an intensive care unit, and an operating room. Although the fidelity of the individual laboratories was very high, the hallways connecting them looked like those typically found in a college classroom building. During their return trip from this visit, the two chairs began to discuss how interconnecting these spaces would have created a much more effective instructional environment. When these observations were presented to the rest of the health sciences and nursing faculty, they catalyzed further discussions that resulted in

plans to develop facilities and resources in the College's new facilities that would support multidisciplinary, integrated laboratory experiences with expanded use of clinical simulation. Subsequently, integrated instruction and clinical simulation became the two central themes for the College's bond election campaign.

As the campaign progressed, administrators and faculty began to give presentations on the proposed project throughout the community. After hearing one of these presentations, administrators from Scott & White Memorial Hospital approached the College with an offer to assist in developing a state-of-the-art Clinical Simulation Center in the new Health Sciences Center.

The Scott & White Memorial Hospital had been considering building a clinical simulation facility to support education of its resident physicians and third and fourth year medical students from Texas A&M University System Health Sciences Center's College of Medicine. Initial analysis had indicated that utilization by only its own trainees would be too low to justify the projected expense. Partnering with the College to develop a shared facility offered a cost-effective alternative. Initial discussions between the College and the Hospital quickly expanded to include representatives from the College of Medicine.

The College, the Hospital, and the College of Medicine approved a formal agreement for joint use of the Health Sciences Center, and collaborative planning began, which included staff from each institution. The planning team produced the Clinical Simulation Center's functional design, developed and approved scheduling procedures, and prepared the foundation for a permanent clinical simulation advisory council. The College, the Hospital, and the College of Medicine agreed that the College's activities in the Clinical Simulation Center would have highest priority. After the College's activities were accommodated, uses by the Hospital and the Medical School would be scheduled. Since the College's use of the Simulation Center primarily would take place during the morning hours, Monday through Thursday, the Hospital and Medical School would schedule their use of the facility during the afternoon and on Fridays. Hospital and Medical School activities also could be scheduled during the College's Christmas and Spring breaks. Other organizations would be allowed to purchase time in the Clinical Simulation Center once all College, Hospital, and Medical School activities were accommodated. This cooperation among the College, the Hospital, and the College of Medicine was well received by the voters of the Temple Junior College District who passed the proposed bond issue by a 3 to 1 margin on November 3, 2001.

As planning progressed, the partnership expanded again to include Laerdal Medical Corporation, a leading manufacturer of patient simulators with facilities 35 miles from Temple at Gatesville, Texas. After learning of the plan for a multi-institution, multidisciplinary, state-of-the-art Clinical Simulation Center at the College, Laerdal committed to providing ongoing support in return for the opportunity to use the

Center as a reference facility for prospective customers and receive feedback on its products from the College, the Hospital, and the College of Medicine faculty and students.

## 32.4 Simulation Center Design

The Health Sciences Center opened in January 2004. The ground floor of the 30,000 square foot facility is divided into three areas – a dental clinic, a commons area with three general-use classrooms and a student lounge area, and the Clinical Simulation Center. The second floor houses faculty

offices, a faculty workroom, a records storage area, a faculty lounge, and a large conference room (Figure 32.3). Separating instructional areas from offices by placing them on different floors significantly reduces noise and other distractions for faculty who need uninterrupted work time. However, the separation of office and instructional space initially was a problem for simulation center staff that often was unable to be reached by telephone during much of the day. This problem was resolved by the College's acquiring a cellular phone that is carried during work hours by one of the Center's staff.

Located at the west end of the Health Sciences Center's ground floor, the dental clinic supports the dental hygiene department's curriculum. The facility includes a reception/

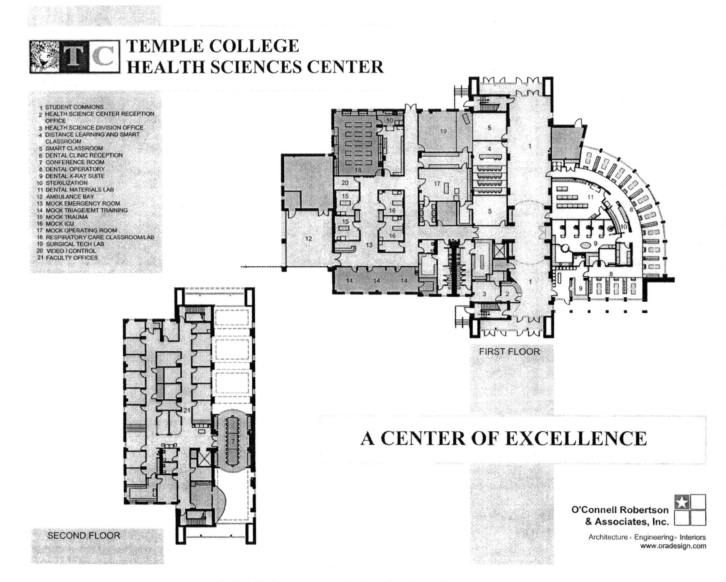

**FIGURE 32.3**   Floor plan of the College's Health Science Center.

waiting area, a radiography suite with five bays, an instrument cleaning/sterilization area, a dental materials laboratory, and 12 dental operatories. One operatory is equipped with a ceiling-mounted dome camera and microphone connected to a small conference room where instructors can observe and record a student's management of patients. This operatory is equipped to support operation of a Laerdal SimMan, which can be used to teach dental hygiene students appropriate responses to dental office emergencies.

The center portion of the Health Sciences Center's ground floor contains a large atrium area that serves as a student lounge. Adjacent to the lounge are three classrooms shared by the Health Sciences Center's users. Each of these rooms is equipped as a "smart" classroom, with a computer, VHS tape player, DVD player, AM/FM stereo receiver, desk top gantry-mounted camera allowing projection of images of 3-dimensional opaque objects, and ceiling-mounted video projector. The largest of the general-use classrooms has tiered seating with a power outlet and network connection installed at each student seat (Figure 32.4). Another of the general-use classrooms is equipped for interactive videoconferencing.

At the east end of the Health Sciences Center's ground floor is the Clinical Simulation Center, a state-of-the-art "mini-hospital" designed for multidisciplinary health care professions education using patient simulation technology. The Clinical Simulation Center includes approximately 9800 square feet

with an ambulance bay, a receiving area with nurses' station, two emergency major treatment rooms, two intensive care rooms, an operating room with adjoining scrub room, a simulation control room, and part-task training laboratories for emergency medical services, respiratory care, and surgical technology. Nine Laerdal SimMan simulators, Two Laerdal SimBaby simulators, and a Medical Education Technologies Incorporated Human Patient Simulator comprise the core of the Clinical Simulation Center's capabilities. Simulators are moved between the Health Sciences Center, the Nursing Education Center, and a simulation laboratory at the College's off-campus center in Taylor, 40 miles south of Temple, as dictated by user needs. Clinical Simulation Center staff provides training and technical support to simulator users at all three locations. However, program faculty operate the simulators during most activities.

The Clinical Simulation Center is accessed through an emergency department (ED) entrance from an ambulance bay that houses two fully equipped ambulances (Figure 32.5). A small box on the outside of the building near the ambulance bay entrance contains ports for compressed air and carbon dioxide that allow emergency medical services faculty to operate simulators outside for extended periods of time by using the building's compressed gas supply. A second box containing compressed gas ports for simulators is located near an exit on the opposite side of the building.

**FIGURE 32.4** General-purpose "smart" classroom with full audio/video teaching facilities and network access for all students. Note the overhead video projector and speakers, and the 3D object podium camera.

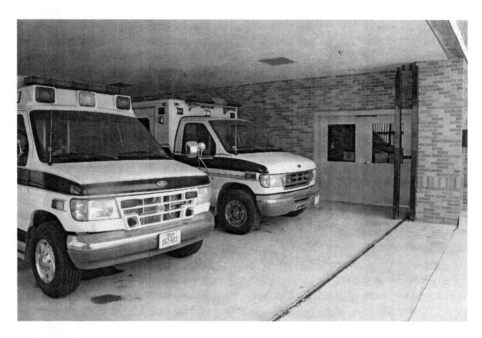

**FIGURE 32.5** Ambulances outside entrance to Clinical Simulation Center's Emergency Department. Hidden from view are the ports for compressed air and carbon dioxide to operate simulators while using the building's compressed gas supplies.

Immediately inside the ED entrance is a receiving area and nurses' station (Figure 32.6). On a hallway beyond the nurses' station are four 15 foot × 15 foot simulation rooms (Figure 32.7). On one side of the hallway, two rooms simulate ED major treatment rooms (Figure 32.8). On the opposite side of the hallway two rooms simulate intensive care unit (ICU) rooms (Figure 32.9). However, the only significant differences between these areas are the headwall configuration and the presence of a ceiling-mounted procedure lamp in the ED rooms. Built-in cabinets have been minimized with equipment and supplies stored either in carts or in baskets and on clips placed on wall rails. Larger pieces of equipment such as extra beds, stretchers, and wheelchairs are placed against the walls in the hallway, enhancing the learner's illusion of being in a hospital. By rearranging treatment tables, beds, and carts, all four rooms can be configured to look like either an ED or an ICU room.

The simulation room headwalls include fully functional ports for vacuum and medical-grade oxygen and air. Because all gases are medical-grade, the facility can provide surge capacity for area hospitals during disasters. The Clinical Simulation Center's capabilities for use to deliver real care also have resulted in its being designated as one of the Bell County Health District's emergency public health facilities for dispensing vaccine on antibiotics in the event of a bioterrorism incident requiring release of the Strategic National Stockpile.

A second set of gas ports near floor level provides oxygen, carbon dioxide, nitrogen, and compressed air for operating patient simulators (Figure 32.10). This feature eliminates

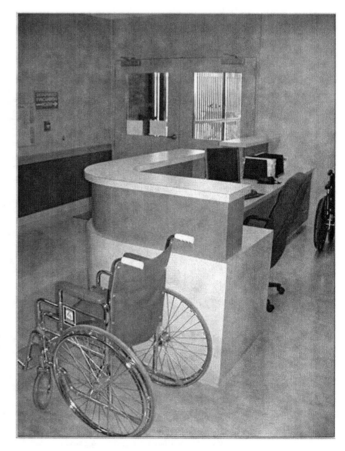

**Figure 32.6** Entrance from Ambulance bay into Clinical Simulation Center's Emergency Department.

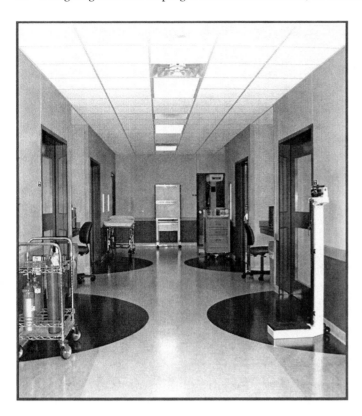

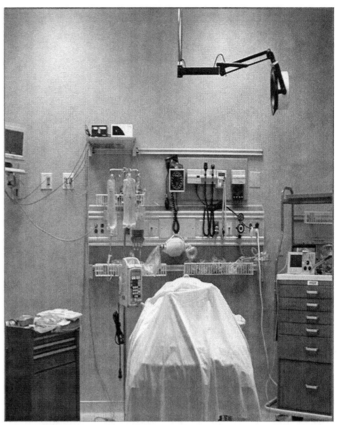

**FIGURE 32.7**   Hallway and entrances to two ICU and two Emergency Department simulation environments.

**FIGURE 32.8**   One of two simulated Emergency Department major treatment rooms.

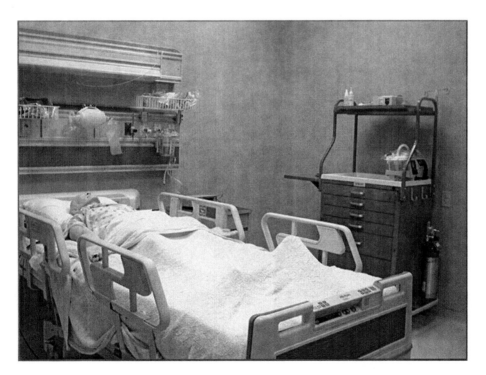

**FIGURE 32.9**   One of two simulated ICU rooms.

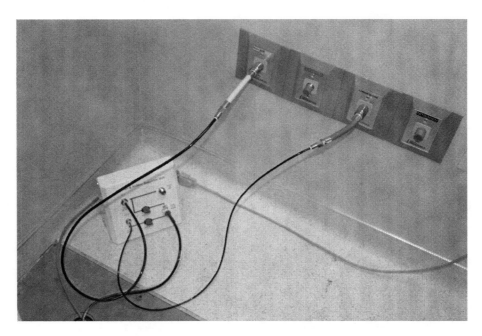

**FIGURE 32.10**   Gas ports for patient simulators.

distractions that would result from locating compressors or gas cylinders needed to drive simulators in the rooms. Compressed gas zone valves in the hallway add to the visual realism of the environment as well as allow the oxygen, vacuum, and air to be interrupted during simulations.

Data and audio ports in each room permit patient simulators to be operated from a central control room (Figure 32.11).

To eliminate trip hazards and distractions, all simulator gas lines and data cables are installed in a covered chase in the floor from their points of origin at the wall to a point directly under the treatment table or bed (Figure 32.12).

Each simulation room is equipped with two cameras to allow operators in the control room to observe and record activities while they control the simulators. Use of cameras

**FIGURE 32.11**   Simulation Center control room.

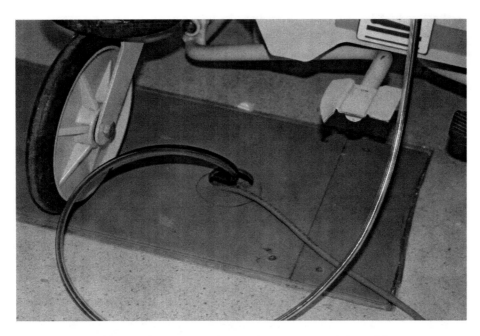

**FIGURE 32.12** Chase for simulator gas line and data cable.

eliminates the need for one-way glass windows, which detract from the realism of the environment and are relatively ineffective because the operators spend much of their time looking at the learner's backs. Additionally, use of one-way glass in a facility using a mock hospital design philosophy would have required multiple control rooms, consuming large amounts of floor space. A fixed camera above the head of the bed or treatment table allows unobstructed observation and recording of airway management procedures. A pan-tilt-zoom dome camera allows activities elsewhere in the room to be observed and recorded. Cables connect the cameras to control room computers that record digital video. The cameras are mounted on ceiling panels with approximately 20 feet of slack in the cables, allowing them to be relocated easily. An additional cable with approximately 20 feet of slack is installed above each room's ceiling and is connected to the control room to support future installation of additional cameras. Audio is recorded using an omnidirectional boundary microphone mounted in a ceiling panel above the head of the bed or treatment table. The microphones connect to the control room computers though an audio mixer. The audio mixer allows use of several microphones in a given room, such as wireless microphones worn by participants as needed to improve the quality of voice recordings and provides uniform sound levels for both live listening and recording.

All audio and video is recorded in digital format by the Laerdal AVS®, a software application that also maintains a log of simulation events that are either sensed by the mannequin or entered by the operator. The digital format allows video and audio from the simulation rooms to be accessed using the College campus intranet. Under instructor control, a group of learners in a classroom elsewhere in the Health Sciences Center can watch another group participate in a simulation in real time. Or, a simulation can be recorded and then played back during a debriefing session so learners can review and discuss their performance. Video can be archived by transferring it to offline storage media.

Speakers in the simulators allow operators to be the voice of the patient and interact with learners during simulations. A telephone in each simulation room permits participants to receive calls from actors playing referring physicians, consultants, concerned family members, laboratory staff, operating room control desk personnel, and other individuals with whom they would interact during management of actual patients. Two-way radios with ear pieces allow operators to speak privately with one person in the simulation room, allowing information not available from the mannequin (such as skin color or temperature) to be communicated without the distraction of a voice coming through an overhead speaker. Operators also have the ability to simulate a power failure by activating a switch that interrupts all power in the hallways and simulation rooms for 8 seconds (with the exception of one outlet that stays on to power the simulator) followed by return of power to only the red emergency power outlets.

A fully equipped operating room (OR) is located a short distance from the ED/ICU area (Figure 32.13). In the OR, an anesthesia gas column provides medical-grade oxygen, compressed air, vacuum, nitrogen, and carbon dioxide. For safety purposes, nitrous oxide is not available from the anesthesia gas column. As needed, anesthesia faculty bring appropriate

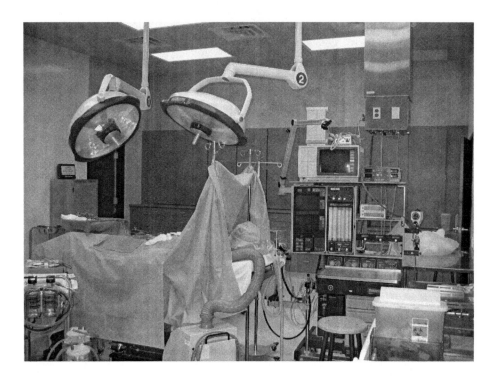

**FIGURE 32.13**   Simulated OR showing many of the standard devices and their positions expected in a real OR.

volatile anesthetic agents with them. Dual ports for each gas on the column allow for both an anesthesia machine and either a Laerdal SimMan or METI HPS to be operated using the building compressed gas supply. Data ports near the head of the operating table allow a Laerdal SimMan to be controlled either from a laptop computer connected to a set of ports just inside the OR's door or from the Clinical Simulation Center control room. The METI HPS can be controlled either from inside the OR or from the control room using a dedicated network connection.

Two ceiling-mounted pan-tilt-zoom dome cameras and a ceiling-mounted omnidirectional boundary microphone allow activities in the operating room to be observed and recorded. The operating room connects through storage areas on either side into classrooms, so groups can move between the simulation area and a debriefing area without having to go into the simulator center's hallways. One OR wall is a floor-to-ceiling whiteboard that provides ample space for instructors to write notes or draw. This feature was specifically requested by the College's surgical technology department to help overcome instructor inability to use facial expression as a communication tool during OR simulations. A scrub room with six sinks is located adjacent to OR. A ceiling-mounted camera over one of the sinks allows an instructor in the control room to observe and record a student's surgical scrub technique (Figure 32.14).

The success of the Clinical Simulation Center as a multi-institutional, multidisciplinary resource for simulation

education has continued to provide opportunities to Temple College. During construction of the Health Sciences Center, a local foundation challenged the College to furnish and equip the facility in return for a substantial gift that would fund renovation of the Nursing Education Center to the same capabilities as the new Health Sciences Center. The College and community responded to this challenge by raising US$ 1.4 million in gifts and donations in a 6-month period.

The renovated Nursing Education Center, which was dedicated in September 2005, contains a simulation laboratory that complements the Clinical Simulation Center in the Health Sciences Center. The nursing simulation laboratory includes a nurses' station, a partial-task training laboratory with six beds arranged in an "open ward" configuration, four hospital rooms for use of patient simulators, and a home-care simulation room (Figure 32.15).

The hospital rooms simulate facilities that might be found on a general medical/surgical unit, complementing the intensive/emergency care spaces in the Health Sciences Center Clinical Simulation Center. Learners from all disciplines move between the Nursing Education Center simulation laboratory and the Health Sciences Center Clinical Simulation Center depending on the type of space needed by faculty for simulation experiences.

Data ports near the headwalls allow Laerdal SimMan patient simulators to be operated in these rooms from control stations in alcoves just outside the rooms' doors.

**FIGURE 32.14**   OR scrub room with ceiling-mounted camera for observing students' hand-washing performances.

Ceiling-mounted fixed cameras and pan-tilt-zoom dome cameras allow operators to observe and record activities in the room. Additionally, a one-way glass window at each control station allows the operator to look directly into the room. One-way glass was installed in the nursing simulation laboratory because it was anticipated that a larger portion of the faculty in this area would be infrequent users of simulation who would be less comfortable attempting to observe students using only an image on a computer screen. A ceiling-mounted omnidirectional boundary microphone allows recording of audio from each simulation room.

Headwalls in the simulation rooms are equipped with appropriately labeled ports for oxygen, vacuum, and air. However, the oxygen ports only simulate oxygen flow using compressed air. The more expensive oxygen and medical-grade compressed air were not installed because there was no use planned in this laboratory for equipment that required these more expensive gases. A second set of ports for compressed air and carbon dioxide located at floor level below the headwalls

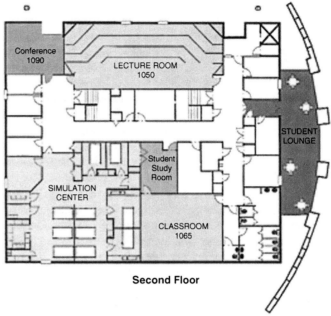

**FIGURE 32.15**   Nursing Education Center simulation laboratory.

provides gases for the simulators, eliminating the noise and obstruction to movement created by having a compressor in the room.

## 32.5 Success Leads to Growth

In March 2006, Texas A&M University Health Sciences Center announced a plan to expand the curriculum at its Temple campus to a 4-year medical curriculum. This created a need for rapid expansion of the Health Sciences Center Clinical Simulation Center to accommodate an anticipated increase in medical school enrollment from 150 third and fourth year students to 400 students at all 4 years of the curriculum. There was an additional need for a clinic to support instruction of medical students using standardized patients.

Once again, the College's response was to propose sharing of space on its campus with the expanding medical school. The College had recently constructed a 30,000 square foot metal building, the Pavilion, adjacent to the Health Sciences Center using several abandoned tennis courts as the foundation. Approximately, half of the space in the Pavilion had been dedicated to construction of a large room for community activities that would seat 350–400 persons in a theater-style arrangement and three general-use classrooms that would each seat 50–75 students. The remainder of the Pavilion was unallocated except for a tentative plan to relocate the College's fitness center from the adjacent Gymnasium building into part of the unfinished space.

The Health Sciences Division Director and the Clinical Simulation Center Director proposed a plan to convert approximately 2600 square feet of unfinished space in the Pavilion into a space that could be shared by the College's health sciences and nursing divisions, the College's physical education department, and the College of Medicine's standardized patient program.

Approximately, 1700 square feet would be finished as a standardized patient clinic that would contain 10 exam rooms and a control room for monitoring and recording patient interviews. This area would be connected to a 900 square foot room that would serve as a task laboratory for the College's emergency medical services department. The Emergency Medical Services department's old task laboratory in the Health Sciences Center would then be retrofitted to create 650 square feet of simulation space and 100 square feet of additional storage for the simulation center without having to expand the Health Sciences Center. The remaining space in the Pavilion, approximately 6000 square feet, would house the College physical education department's fitness center with an understanding that the standardized clinic area and EMS task laboratory space also could be scheduled by the physical education and athletic departments for classes and training activities when it was not being used by health sciences, nursing, or medical students.

The development of the Clinical Simulation Center as a partnership between a community college, a state-supported medical school, a private teaching hospital, and a private manufacturer of medical education equipment has demonstrated the value of collaboration as a strategy for expanding local and regional capabilities to provide initial and continuing education in the health professions that meets industry needs and accreditation and quality standards.

# 32.6  Lessons Learned

As the Temple College Clinical Center completes the first three years of operations, the following lessons learned are particularly significant.

## 32.6.1  Staffing

- The Center's Director should be selected early and involved in the project from its beginning.
- The best Director has a clinical education background, is committed to the use of simulation as a teaching tool, is comfortable with (but not necessarily and expert in) computer and audio/visual technology, and has training/experience in managing the logistics and politics of a multi-user facility.
- The Director's position should be at a place in the organizational structure where it can be insulated from user pressure and ensured access to a level of management within all consortium member organizations to deal with problems efficiently and effectively.

- The Director's clinical and teaching responsibilities as well as "additional duties as assigned" should be limited but not eliminated.
- Since utilization is likely to increase rapidly, management of simulation consortium members should have a plan in place for adding additional staff quickly enough to avoid Director burnout.

## 32.6.2  Planning

- Incorporate the ideas from as many potential users as possible in planning the facility. It is easier to build in special features and opportunities for future expansion than it is to retrofit.
- Encourage a wide range of potential users from the health care and public safety sectors.
- Consider alternative uses, such as surge capacity during disasters or public health emergencies.
- Keep simulation spaces as generic as possible, so they can be quickly reconfigured by moving beds and carts.
- Avoid a "have it your own way" approach to design. Users need to have maximum input into facility capabilities, but they need to be prepared to live with the consensus of the groups that will be using the facility.
- Decide on priorities for use by consortium members early, and document the results of these decisions in a formal memorandum of understanding.
- Determine facility use charges by organizations that are not part of the simulation consortium as soon as possible, but be open to "creative financing" arrangements such as bartering use of the center for equipment donations.

## 32.6.3  Information Technology and Audio/Visual Infrastructure

- Select architects, engineers, and simulator vendors with experience in theatrical facility design and construction.
- Put the architects, engineers, and simulator vendors in communication with one another early in the planning process.
- Choose the contractors who do the cabling for the simulators and A/V equipment very carefully, check their work against specifications regularly as the project progresses, and include a provision in their contract that they will not be paid until all systems perform to user satisfaction.
- Have someone who understands the requirements and the planned uses of the simulators and A/V equipment meet directly with the subcontractor personnel who will be doing the installation. The work is most likely to be done correctly the first time if the personnel installing the microphones, cameras, and cabling understand how it is going to be used.

- Buy control room computers from well known, reputable dealers.
- Overspecify as much computer RAM, processor speed, and storage capacity as the project budget will allow.
- Insist that the simulator vendor certify that your computers, A/V equipment, and associated cabling will accommodate any software upgrades or simulator system architecture changes for the next 5 years.
- Ensure that the simulation center's information technology and A/V support staff are comfortable with the selected hardware and meet with the simulation vendor's staff to discuss user support.

### 32.6.4 Logistics and Supplies

- Determine supply needs for the simulation areas as early as possible.
- Determine whether supplies will be used frequently enough to justify buying them out of a common supply budget or so specialized they should be provided just by groups that use them.
- Identify sources of expired supplies for use in simulations. Identify "leftovers" from used trays and kits, such as syringes, unused needles, and extension sets, that can be scavenged for use in future simulations.
- Develop a strategic planning process involving all simulation center users that avoids duplication of equipment requests and gifts and that ensures items purchased with simulation center funds benefit the maximum number of users.

### 32.6.5 Scheduling

- Establish a single point of contact at the simulation center for scheduling use.
- To the extent possible, have all requests from other organizations pass through a single point of contact at that organization.
- Maintain a master calendar that is visible online. Allowing users to see when time is available before they make their requests makes the scheduling process more efficient.
- Handle scheduling requests through e-mail, not telephone to ensure that there is an easily accessible record of requests for use.
- Track utilization of the center from the beginning. Keeping the data from the start of operations is easier than reconstructing it after months of operation.
- Establish and enforce deadlines for scheduling activities and delivering scenarios to the center's staff. Trying to accommodate last-minute requests leads to poor-quality simulations and exhausted center staff.

- Block out time on the schedule for cleaning, restocking, preventive maintenance, and administrative activities. Left uncontrolled, simulator use will rapidly expand to fill all available time, resulting in deteriorating reliability and quality.

### 32.6.6 Storage

- Adequate storage is critical to preventing wear and tear on simulation center equipment and staff.
- Determine the amount of storage space you think you will need, then double it. Having between 20 and 30% of the simulation center's total square footage dedicated for storage is not an unreasonable goal.
- Think of the simulation rooms as auxiliary storage space for larger equipment and small equipment/supply items that can be kept in carts, cabinets, or wall baskets, but do so only when and where such items will not detract or distort the intended learning.

### 32.6.7 Miscellaneous

- Do not go operational until everything has been tested thoroughly and is working reliably.
- Do not let full-function patient simulators be used for partial-task training.
- Similar items all purchased at about the same time will all start breaking at about the same time.
- On any given day in a large, active simulation center, at least one major piece of equipment probably will not work.
- Try to accumulate an "extra" one of everything.
- Always be thinking about alternatives for accommodating current simulation center users in the event of an equipment failure.
- Never let users (faculty or participants) near the simulators without a briefing on equipment capabilities, limitations, and safety issues, regardless of how many times they have been in the center. Consider video recording this briefing so everyone hears the same information every time. Fully expect to regularly update this message as new, unanticipated "misadventures" occur.
- Be prepared to accept a phase of sporadic and less than rigorous use of simulation as a teaching tool followed by sudden jumps in volume of utilization and user sophistication.
- Establish a policy of "open access" to scenarios by all users of the center. Frequently, a simulation developed by one department can serve the needs of another with minor adjustments.

- Store and password code scenarios on control room computers by department/course, not instructor. Any instructor from a department should be able to access any of that department's materials; otherwise, a faculty member's absence can turn into a disaster.
- Familiarize as many users with operating and programing the simulators as quickly as possible. Too much other work is required to keep operating on a day-to-day basis for the center's staff to be tied up sitting at computers in the control room.
- Seventy percent of computer and A/V equipment problems can be solved by restarting the system. Twenty percent more can be corrected by double-checking the connections, valves, switches, and settings.

# 33

# Partners in Simulation: Public Academic–Private Health Care Collaboration

Sharon M. Denning

Constance M. Jewett Johnson

Dan Johnson

Marilyn Loen

Carl Patow

Cathleen K. Brannen

## 33.1 Purpose and Players

In 2003, the HealthPartners Simulation Center for Patient Safety at Metropolitan State University, St. Paul, Minnesota, U.S.A, was established through a unique partnership between HealthPartners, one of the three largest care delivery systems in Minnesota, and Metropolitan State University, a mid-sized urban university that is part of the Minnesota State Colleges and Universities system. Our goal was to create a simulation center that would be a community clinical resource based upon an effective partnered approach between commercial and academic institutions.

The benefit of collaboration to build a shared simulation center, on the face of it, seems obvious: the resources of one partner complementing the challenges of another, pooled skills and experience, an enlarged network of connections, and the inherent appeal of partnership to communities and foundations. It makes good sense. But while the water may seem inviting, hidden boulders are to be found beneath the surface, and sudden white water with an occasional waterfall must be anticipated and navigated. These are the inevitable results of meshing differing organizational cultures and in the coming together of the health care industry and academia. This chapter describes an effective partnered approach to building a simulation center as a shared clinical community resource. It describes critical dimensions needing attention and lessons learned in relation to each dimension.

When this center was first contemplated, simulation centers were developing across the country with the intention to use them to improve patient safety. With national attention on patient safety and prevention of avoidable deaths, simulation had a message that was welcomed by the public – let students practice on a "patient" to whom they could not do harm before practicing on real live patients. Communication and team work, as well as clinical skill development, were being recognized as equally important to patient safety and essential to effective patient care. Simulation could provide the medium for teaching these kinds of skills.

Along with recognition of an opportunity to engage in this exciting work came a sense of urgency, a need to respond before the opportunity disappeared. The luxury of a long planning horizon did not seem to exist. It was at this point that the differences in approach to planning and decision making became apparent. Health care delivery (industry) tends to be responsive to the financial market and change occurs somewhat more readily with a greater degree of freedom. Academic organizations lean toward stability, longer planning horizons, and greater degrees of tradition. The middle ground that emerged was a modified rapid cycle planning process, a spiraling approach that allowed the planning group to move forward in a timely fashion based on their best combined understanding of a reasonable first step. It enabled evolution of the project based on the evaluation of actions taken and with corrective action in response to key points learned. The "middle ground" both protected us from major errors and proved to be a reasonable methodology for the two very different cultures.

This development process revealed seven critical dimensions essential to any simulation based endeavor: nature of the partners' relationship; space; funding; equipment; staffing; supportive infrastructure; and focus.

## 33.2 The Partnered Approach

Location and mission are key factors in the partnership. The University is centered in a large metropolitan area with several hospitals in the vicinity, including Regions Hospital, a component of HealthPartners.

HealthPartners is the largest consumer-governed, nonprofit health care organization in the nation. HealthPartners is focused on improving the health of its members, its patients, and the community. It has physicians, dentists, and employees in more than 70 locations in the Twin Cities of Minneapolis and St. Paul, St. Cloud, Duluth, and Western Wisconsin. It is made up of two hospitals, more than 50 primary, specialty, and dental clinics, a health and dental plan (insurance), research, and educational foundations. The health plan serves more than 640,000 insured lives. HealthPartners has one of the largest networks in Minnesota with more than 27,000 providers to choose from. As an enterprise, HealthPartners has an annual operating budget of US$ 2.3 billion with 10,000 employees. Its owned clinics care for more than 400,000 open charts and the larger of the two hospitals is a level one trauma center with 435 beds and 21,000 annual admissions.

The HealthPartners Institute for Medical Education (the Institute), the clinical learning arm of this health system, is among Minnesota's four largest providers of medical education, training more than 500 resident physicians each year. The Institute oversees residency programs at Regions Hospital as well as provides continuing medical education programs for physicians, nurses, and other health care personnel.

The State of Minnesota has two public postsecondary systems:

- The University of Minnesota is a major research institution with four campuses providing undergraduate, graduate, and professional education. The main campus and central administration are also in Minneapolis and St. Paul.

- The Minnesota State Colleges and Universities system has seven 4-year universities and 25 2-year colleges located throughout Minnesota, providing occupational, general, baccalaureate, and graduate education. Metropolitan State University (the University) is one of the 4-year universities, a nonresidential campus located in St. Paul, the second largest city in Minnesota. Together with its neighbor across the river, Minneapolis, the area's population is approximately 3 million.

The Institute was the initiator of the conversation about a simulation center and is the principle link to the University. This collaborative effort enjoyed critical buy-in and support from a range of strategic points across the HealthPartners enterprise.

The mission of the University is congruent with the philosophy and mission of HealthPartners; both institutions advocate for being deeply committed to the people they serve, with respect that embraces and values diversity. The mission is grounded in community partnership. With special emphasis on accommodating the needs of adult learners, the University offers classes evenings and weekends in over 40 major topics and seven graduate programs. The largest numbers of degrees are granted in business, accounting, psychology, nursing, criminal justice, and law enforcement. The University's medium size (nearly 10,000 students enrolled) enables it to respond quickly to opportunities, and the University was amenable to engaging in the development of an innovative project for the School of Nursing.

Partnering organizations face two challenges: defining their shared work and aligning their interests so as to take advantage of the unique strengths and capacities of each. It requires some modification of the rules to allow for new possibilities (Figure 33.1).

Organizers recognized a need to develop a formal structure to carry out the anticipated vision and plan. The two

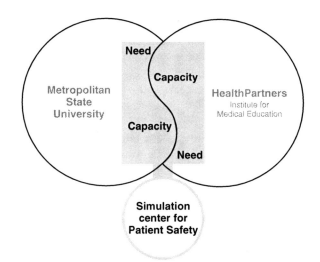

**FIGURE 33.1** Schematic of relationships between the three partners.

organizations, through a representative Steering Committee (three key leaders from each organization), took up these tasks by immediately making available capacity and resources from each side to provide what neither organization could secure alone. The Steering Group became an ideal vehicle for taking up the shared work ahead.

Aligning two organizations around strengths and weaknesses or assets and deficits enables the development of those highly desirable synergies possible in an effective collaboration. Academic institutions are often found to be cash poor and unable to fund the purchase of equipment initially. This is especially true, as is the case with the University, when the academic partner offers a single health care program and is not part of a larger complex of health professions majors and/or a medical school. The School of Nursing was the only health professions program on the campus. The curriculum was designed as a Bachelors' of Science in Nursing completion program and an advanced practice graduate program. These programs do not require basic skills labs and, therefore, there was no campus precedence for addressing equipment needs in the campus operating budget. The very fact that there is a limited demand for nursing student access to simulation, leaving more room for the HealthPartners staff and residents, makes this a good match. The academic side often feels that the hospitals and health care providers have large financial resources, oftentimes a significant misconception. In this instance, HealthPartners experienced an unusual opportunity to direct grant funds to the cost of the purchase of initial equipment required by the center. Once the partnership had formed, the ability to jointly attract funding, grants, gifts, and foundation funding was significantly enhanced.

One of the first projects in establishing the partnership was to formalize the relationship between the two organizations. A Collaborative Agreement between the University and the Institute was developed by the Steering Committee and refined by legal departments from both organizations. The Agreement outlined the financial obligations; governance structure; operational, capital funding, and budgets; use of the Center; ownership of intellectual property; terms of agreement; dispute resolution; and other miscellaneous provisions. The Agreement established a framework to guide the development of the project and prevented possible misunderstanding that might have occurred without a written document. The Agreement satisfied the requirement of the Institute's parent organization, HealthPartners, for legal contracts when entering into such business relationships. It also satisfied the University's obligation in meeting the contracting requirements of its owner, the Minnesota State College and University System.

*Lesson Learned*

*A formal agreement is essential when applying for grant funding and dealing with other external parties, and it should be executed as soon as possible before developing a simulation center in partnership with others.*

## 33.3 Space

The identified simulation center space was 3550 square feet on the University campus that had served as a community clinic for the nurse practitioner program. At the time options for a center were being explored, the space was underutilized, and the university was planning to reassign it for other uses. This space was dedicated for use by the Center by the University. Because it remained unclear as to how simulation space could be best utilized, minimal investment was made in renovation. At that time, few were deeply experienced in simulation, and space-related needs were not precisely understood. On the basis of the information available at the time, the best decisions were made, reevaluated along the way with changes made as needed. Within the space, it was agreed that no walls would be moved. The space was cleaned up, painted, and carpeted. In a single afternoon, representatives from both partners walked through and envisioned what could be done with the space as it was. An operating room, a critical care room, four exam rooms, a partial-task trainer skills lab, a classroom, a reception desk, a special procedures room (endoscopic exam), a storage room for medical gases, an office, and general storage areas were defined (Figure 33.2).

*Lessons Learned*

- *If you do not know exactly what is needed, defer investment in bricks and mortar. What you think is important today may turn out to not be so important tomorrow.*
- *You can never have enough storage. The space needed for storage will significantly exceed what is predicted.*

## 33.4 Funding

As is often the case with health care or education-related ventures, funding of anything is a challenge. In this instance, three fortuitous events occurred simultaneously to allow the Center to access the necessary resources to begin operation: the care delivery partner had the unusual opportunity to directly grant funds to the purchase of the initial equipment; the academic partner had the available space; and the University agreed to place the Center's needs high on the priority list for the University's Foundation, the fund-raising division of the University.

Ongoing cost of operation will be addressed within the business plan which, as currently conceived, anticipates revenue from three sources:

- **Fees:** Participant fees and contracted projects with other health care organizations and academic institutions in the region.

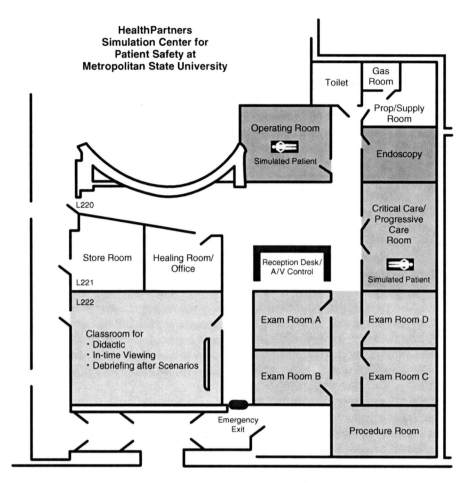

**FIGURE 33.2**   Simulation Center floor plan.

- **Grants:** The Foundation has been successful in helping to obtain local grants to support the work of the Center. There has been strong interest through the community in the possibilities it affords. In the first year of operation, a significant grant was obtained through the Minnesota Job Skills Partnership, which helps Minnesota businesses and schools train a competitive workforce. Grants are awarded to educational institutions that partner with businesses to develop new-job training or retraining for existing employees.
- **Partner contributions:** Any expenses not covered by grant funding or fees are equally divided by the two partners.

At this point, only a small percentage of the budget is based on fees; grants and partner support provided the initial funding and continue to be an essential component of funding.

*Lessons Learned*

- *Sometimes your best opportunities are the result of serendipitous events that can not be replicated. Watch for them and act when they present themselves.*

- *Initial excitement about the simulation center may provide start-up funds. These will not last indefinitely, especially as other organizations integrate simulation into their health professional education and compete for funding. To sustain the center, a business plan is essential.*

## 33.5 Equipment

In the approach to the purchase of equipment, a few guiding principles were identified and were frequently reviewed and evaluated. The first principle was to buy large costly items first when the most resources were amassed. It was believed that in the future it would be easier to purchase less costly items. The large items could be unreachable if resources were initially dispersed over too wide an area. A high-end high-fidelity patient simulator was purchased first, and the critical care space was equipped, including a cardiac monitor as a first priority. The audio-visual infrastructure was established, which allows those in the control area to view several aspects of the scenario (Figure 33.3). It also provides extension of the

**FIGURE 33.3** Clinical instructor observes session at centralized audio/video control and recording console.

visual awareness to those in the observation/debriefing area. An intravenous access simulator and an endoscopic exam simulator (colonoscopy, flexible sigmoidoscopy, and bronchoscopy) were purchased. This equipment created a powerful visual array that provided an introduction of the concept of simulation to visitors. It was enough for the critical stakeholders to "get it," to actually see the potential in this new way of learning, and people became excited. This potential was recognized by funders, increasing the likelihood of support for the Center.

A caveat here is that equipment should not be purchased too early. In many cases, a simulation center or project begins when money is earmarked to purchase equipment. The use of the equipment, however, does not begin until a plan is developed and personnel are identified and trained in the use of the mannequins. The warranties for the higher-fidelity mannequins are very expensive, and many months of the warranty can be wasted if the mannequin is not being actively used.

The second principle guiding the acquisition of equipment was, "How people train is how people will practice." One goal of simulation is to create an environment that is close enough to the real thing to produce a suspension of disbelief. In order to simulate the clinical patient care environment, many pieces of equipment are needed in addition to human patient simulators. Pumps for intravenous solutions, cardiac monitors, defibrillators, and blood glucose meters are among the many

tools that add to the realism of the scenario. Initially, it was believed that we would be able to obtain equipment through donations from hospitals or vendors. It was soon apparent that this would not be an acceptable solution to the equipment problem. To have an optimal learning experience, participants need to be using the same equipment in the simulated setting as they use in the actual patient care setting. Unfortunately, donated equipment is often outdated and very dissimilar from the equipment in current use. Depending on the level of the learner and the objectives of the session, the equipment may not even need to be in working order. For example, the glucometer will not work with the artificial blood used for the mannequin, so using a nonfunctional meter and "simulating" the test may be sufficient.

*Lessons Learned*

- *Do not rely only on equipment donated by vendors and/or hospitals.*
- *The clock starts ticking on the warranty as soon as you receive the equipment; do not purchase expensive items too early.*
- *Choose to make one big strategic purchase over purchasing many smaller useful items. In the future, the bigger blocks of funds may be infrequently available.*

## 33.6 Staffing Model

Buying the equipment is actually the easiest part of creating a simulation center. The real challenge is developing educational programs that will be utilized and that can generate revenue. Curriculum and training events must be built that can be integrated into traditional learning activities. Such utilization rests on identifying a staffing model that can support it. This was one of the more significant challenges.

There are five personnel skill sets that were deemed essential for operating a simulation center: project manager, educational designer, advanced clinician, technology support, and administrative support.

Initially, a senior member of the Institute was assigned the role of project manager for the start-up of the center. He was partnered with the director of the school of nursing. Between them they had contact with the critical individuals in both organizations and had a significant degree of autonomy for getting the center established.

The early staffing model assumed that if the center provided space, equipment, and staff presence, each of the "customers" would identify one of their own instructors to learn the simulation equipment and would integrate that into the learning activities involving their learners. Ten key clinical educators were provided a 2-day intensive course to learn how to operate the human patient simulator, and then a user group was formed to provide ongoing support to this group. After a few weeks, the ten had dwindled down to three and then to one. It soon became apparent that the technology was a barrier in and of itself. It required a significant period of time not only to learn the equipment and the design of scenarios, but also to incorporate this new approach into current instructional activities. More importantly, it required moving to a new paradigm for teaching – a student-centered teaching style as opposed to the more traditional teacher-centric approaches that take place at the bedside in most settings today. It was discovered that the primary instructors were clinicians who also serve as preceptors and clinical instructors, weaving teaching into a full load of patient care activities, and the expectation that they would be able to master the simulators was unreasonable. It became clear that this clinical faculty had to be provided with a partner, someone who could control the simulation equipment to respond as necessary given the objectives of the learning event. This combination became the functional unit for much of what the Center provided. In a few cases, emergency medicine physicians mastered the use of the simulator and were able to work independently in teaching emergency medicine residents.

A nurse educator with a master's degree was then hired and trained on the use of the high-fidelity patient simulator and on construction of scenarios. The role was a difficult one, being required to do it all, and at times was a poor use of a highly skilled clinician.

The third iteration of the staffing model was based on a sense of the need to expand the staff to adequately respond to the emerging demand. There was some obvious financial risk to pursuing a direction that calls for staffing up to the level of anticipated demand. The roles include a director, a PhD nurse educator, a half-time physician, and a program assistant who could provide simulation assistance as well as carry out clerical duties.

Current challenges include:

- Creating backup and relief coverage to the key roles without incurring excessive cost in doing so.
- Securing a half-time faculty member to champion integration of these tools into the nursing curriculum.
- Maintaining ongoing professional development for the staff.

*Lessons Learned*

- *A key to moving quickly is having one or more individuals dedicated to driving change. They should be linked to high-level support of both organizations. Assigning center development to someone with heavy operational responsibilities will make for slow going.*
- *As demands for building curriculum and sponsoring education activities increase, the number and types of personnel may vary. Be prepared for changes in staffing models.*
- *The decisions about selection and purchasing of simulation equipment and development of staff capacity to utilize the equipment effectively need to be a simultaneous process.*

## 33.7 Supportive Infrastructure

A range of infrastructure components is essential to organize work and maintain records. Prior to beginning the actual work of simulation, it may be difficult to predict these needs. The Minnesota Job Skills Partnership grant mentioned earlier was the motivation behind establishing one of these components, a user database. The grant reports required an accurate record of participants, with various data points included. The implementation of a database helped to meet the reporting needs of the grant and also established a good method of record-keeping.

An easily accessible, user-friendly schedule was also determined to be a necessity. The initial scheduling process used for the Center was labor intensive. After significant exploration, an online scheduling program was selected that could be adapted to meet scheduling needs. The scheduling program is accessible at any time and from anywhere via the Internet, and it is relatively user-friendly. It allows the user to check availability of space and services and, after being provided with access, the ability to register for sessions. An ideal situation for seamless data collection and reporting would be to have the scheduler integrated with the database.

Although a web site for the center was established early on, it provided limited information and was not interactive. As the staff grew, time was allotted to develop a more informative and dynamic web site. The web site is used as a reference for visitors and participants for directions to the center, contact information, viewing access to the schedule, and general information about the center.

Many of the issues that arise are difficult to foresee in the initial stages of development. A process for the development of clear, simple, customer-centered policies is essential. Some equipment pieces, for example, are portable so individuals or departments may request to borrow equipment for use at the hospital or other training site. Policies should be developed so that there is a clear understanding of where equipment can be used and by whom. Another example is policy related to intellectual property rights. With the partnership between industry and academia, which partner owns intellectual property developed in the Center? Or is the Center a separate identity with its own policies? What about the staff and/or faculty developing the materials? What about customers who work with Center staff to develop programs? A policy on ownership of intellectual property should be developed prior to dissemination of curricula and other materials.

The Steering Committee and staff are currently involved in a significant strategic planning process, which will result in a business plan, pricing structure, goals, and strategies for the future.

*Lessons Learned*

- *Try to predict what your data collection and other system needs will be and proactively establish an infrastructure to meet those needs.*
- *Policy development should not precede the learning that comes from experience nor should it be so delayed that the critical guidance it affords is unavailable when needed.*

## 33.8 Focus

Initially, the focal point of the Center was patient safety. By providing a safe environment in which providers could practice high-risk or low-volume critical events, the Center can make a contribution toward increased patient safety in a variety of settings.

Because simulation provides endless opportunities, a wide variety of requests for programs was encountered. With limited staff resources, it is not possible to produce quality curricula on a wide range of topics and for a wide range of audiences. It is essential to determine what the focus of the Center will be and limit that focus to what can be done well. In a medical school setting, the focus of simulation will clearly be on medical students and/or residents. In a community college setting, the focus may be on entry-level associate degree nurses, practical nurses, nursing assistants, and/or paramedics. The equipment and curriculum needed for these two different student types is clearly different.

Because the Center was designed to serve both academic and industry needs, with potential for a variety of customers in the community, the focus was initially less clear. As the Center evolved, however, the primary participants have been experienced providers. The high-fidelity patient simulators allow development of challenging scenarios that can be made more or less complex, depending on the objectives of the course.

Initially, it was difficult to attract customers because simulation was a new concept and the value was not well understood. Once providers had visited the Center and could see the applications of simulation, interest in using the center increased significantly. By the end of the first 2 years, the Center's schedule was busy with a variety of courses. At present, the Center is being used at close to half the potential capacity; well over 2000 participants have attended offerings at the Center. Offerings included crisis resource management, rapid response team training, orientation to higher levels of care, anesthesia induction for nurse anesthesia students, air medical transport, and learning of procedures. Participants represent a variety of disciplines, composed of approximately 80% practicing registered nurses, 15% medical students and residents, 11% miscellaneous other disciplines (respiratory therapists, pharmacists, physicians assistants, paramedics, nurse practitioner students, etc.), 8% practicing physicians, and 3% nurse anesthesia students.

Use by the School of Nursing has been limited because the baccalaureate program for non-RN students was only recently initiated; the largest program offered by the School of Nursing is for returning RNs with associate degrees seeking the Bachelor's of Science in Nursing or Master's of Science in Nursing. The potential for expansion in using the Center is high; in the future, there could be as many as 100–200 students requiring simulation experiences each semester.

As a result of its partnership in the Center, the University is viewed as a leader within its parent statewide university system and has served as a resource for other state colleges and universities interested in incorporating simulation into nursing or other health care programs. The University hosted a conference in 2005 on simulation to inform faculty of its benefits and encourage its use in curriculum.

*Lessons Learned*

- *Lack of familiarity with the technology and approach to learning may cause utilization to lag in the beginning. Inviting potential users to experience simulation will, in time, lead to a somewhat exponential rise in utilization. Owing to limited capacity, no center can serve all desires of all interested parties. As soon as possible, define the target population and focus for educational activities before filling the schedule.*

- *The purpose of the center lies at the intersect of the center's capacity and the needs of the community of interest. This will take into account faculty, facility, the partners, funding, and equipment. Visitors will have a multitude of ideas about how to use the center, many of which may not be aligned with the center's vision.*

## 33.9 Example Scenario

Groups of participants come to the Center from many different hospitals in the regional and state area for a variety of purposes. Although it often involves traveling and disrupting the normal work schedule of the staff, employers believe that the simulation experience is worth the effort. One group that has participated in simulated learning is an air transport unit.

Orientation of experienced critical care nurses and paramedics to a new role as helicopter transport staff presents unique challenges. Leaders need to be confident that new staff can manage a wide variety of situations, but opportunity to experience such situations with real patients under supervision is not always feasible. Simulation is an excellent vehicle for creating realistic scenarios without risk to patients. The following is an example of how simulation can be used for this purpose.

An air transport crew, consisting of a registered nurse and a paramedic, receives word that they are to transport a pediatric near-drowning victim from a small rural hospital to a trauma center. Upon arrival at the hospital's emergency department (the Center's emergency department/critical care area), they receive a brief report on the patient, a 7-year-old female who has fallen off the dock while her family was visiting their lake cabin. She may have been in the water up to 15 minutes. She is unresponsive, cool, wet, and pale, and her pupils are dilated (Figure 33.4). Cardiopulmonary resuscitation is in progress, with ventilation by bag-valve-mask. Intravenous access has not yet been established. The crew works with the hospital emergency department staff to intubate, administer intraosseous medications such as atropine, and establish intravenous access. A pulse is established and she begins to respond. A high-fidelity patient simulator is used in this situation to produce responses

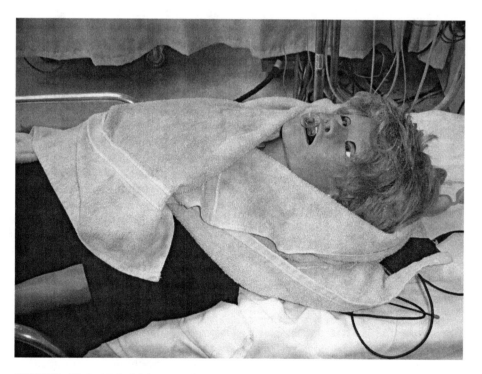

**FIGURE 33.4** High-fidelity simulator configured as a cold water–immersion patient: unresponsive, cool, wet, and pale, and dilated pupils.

to treatment that mirror human responses. The simulator is operated via a laptop computer outside the clinical room, so the simulation "pilot/navigator" is not visible to the participants. In other scenarios in which the simulated patient is conscious and able to speak, the person at the remote console speaks for the patient, again increasing realism.

During this scenario, the learners seek to increase proficiency in managing the stabilization and transport of a pediatric near-drowning patient. Critical elements expected for the care of this patient include recognition of the need for sedation and ventilation with positive end expiratory pressure. Goals for this scenario include demonstration of effective team process and decision making; the team needs to determine whether to transport quickly or stabilize the patient at the hospital prior to transport. The helicopter team is expected to demonstrate tact and diplomacy in working with the local ambulance and/or hospital staff; customer service is a significant aspect of maintaining good relationships and assuring future referrals.

Staff or actors often serve as confederates, playing a variety of roles to enhance the scenario and the learning. Social clinical problems are sometimes introduced into the scenario, as the staff needs to learn how to diplomatically manage difficult situations. By playing the role of the emergency department physician, the manager of the flight staff can interject various challenges, such as pressuring them to transport quickly, questioning their skills, and being reluctant to surrender management of the patient.

Another confederate is used to play the role of the patient's mother, who is distraught and demanding to ride along in the helicopter. While actively managing the patient and preparing for transport, the staff needs to assess the situation and determine whether or not the mother can safely ride along.

Creating a realistic situation helps the participants to suspend disbelief, and the experience becomes more meaningful. Mannequins are dressed in clothes and wigs to fit the situation, and other props are used to increase realism. In this scenario, the child is wet, wrapped in a wet beach towel, and smells of Coppertone sunscreen. She is wearing summer attire, including sandals. The flight crew is supplied with an actual flight bag, so they are working with equipment as they would in a real transport.

Once the patient is stabilized, the crew prepares for transport. Although an actual helicopter simulator is not available, the crew is directed to two chairs that are located in the same proximity to the patient that they would be in the helicopter and they are reminded that their space is limited (Figure 33.5). They then have to manage the patient as they would during transport.

During each scenario, other participants are observing from the classroom (Figure 33.6). Scenarios are video recorded and reviewed during the debriefing. The debriefing time allows the manager to include teaching and guidance on managing transport situations.

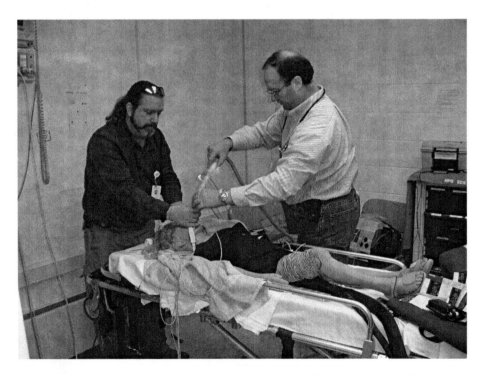

**FIGURE 33.5**  Patient care during air transport (simulated). Note that only those clinical resources normally used during air transport are provided for the participants.

**FIGURE 33.6**   Participants in the observation/debriefing room study a live scenario from the simulation room.

This approach to preparing flight crews for their new role has had very positive results. The staff members report increased confidence in their ability to function effectively, and the manager gains valuable insight on the performance of the staff. Feedback from participants is generally very positive (see Chapter 62).

## 33.10 Conclusion

Since its inception in 2003, our simulation center continues to experience increasing demand for its programs and its expertise in curriculum development. The strength of the partnership has been its collaborative approach, alignment of interests, and willingness to adapt to relatively rapid changes. By consciously defining the center as a place for educating clinicians who are providers of advanced levels of care, with the primary emphasis being on critical care, emergency medicine, and inpatient nursing, the center has been able to retain its focus and not become distracted in producing disparate programs.

An academic–industry collaboration can be a successful model for establishing a simulation center supporting the educational needs of both entities. Awareness of operational, educational, and community requirements and continual monitoring of the progress of the center's programs, personnel, and policies are required.

# 34

# A National Simulation Program: Germany

Stefan Mönk
Jochen Vollmer
Wolfgang Heinrichs

## 34.1 Early Adopters

The first German university anesthesia departments pursued the concept of simulation with their own training centers as early as in 1997. Eight centers (Aachen, Berlin, Erlangen, Hamburg, Heidelberg, Mainz, Tübingen, Würzburg) were founded soon after that and began to educate their own universities staff as well as outside participants and undergraduate medical students. The level of activity differed notably from institution to institution, but the existing core group managed to produce a constant albeit not huge flow of training events.

As a means of defining quality criteria, in 2002, the early adopters produced a list of requirements for simulator trainings in anesthesia that was submitted to and accepted by the executive board of the German Society for Anesthesia and Intensive Care Medicine (the Society, Deutsche Gesellschaft für Anästhesiologie und Intensivmedizin, DGAI) [1]. After board approval, this paper was published in the Society's journal. Later, however, the Society failed to establish a method to control these requirements, and the publication had no further consequences. At that point, simulation did not move forcefully beyond anesthesia, nor did it create huge impact within the specialty.

In 2003, the German law was changed and now required more hands-on training, small-group education, and problem-oriented learning for undergraduate medical students [2]. Through the same law, anesthesia, for the first time, became a compulsory subject for medical students, augmenting the educational role of the specialty. Traditionally, emergency medicine is practiced and taught in many German clinical and academic institutions through anesthesia. Prof. Jürgen Schüttler, anesthesia chair of Erlangen University, saw an opportunity to improve education through simulation for a large number of trainees. He decided to take a bold step and convinced his colleagues at the Society executive board to allocate money. A commission was inaugurated with the goal of nationwide introduction of simulator-based education for undergraduate medical students in emergency medicine. This goal was to be achieved by providing a simulator for each medical school and by supporting its use.

## 34.2 Planning Phase

Germany has 35 medical schools. To reach every medical student, each school would have to be equipped with at least one simulator. It became clear very soon that this was a major undertaking and that financial management, planning, and organization would be crucial for the success of the project.

One of the first steps the simulation commission underwent was the selection of a suitable simulation device. The budget clearly aimed for a price of less than 50,000 € per unit, which immediately ruled out the Human Patient Simulator (HPS, METI, Sarasota, Florida, U.S.A). The HPS was considered the most useful simulator for the training of anesthesia since it provides a mechanical lung model as well as gas exchange. It was abandoned however, and the selection of a mechanically less complicated model could be accepted especially since the main topic to be taught was emergency medicine rather than anesthesia. Two bidders made a tender, and it was up to the commission to make a choice between SimMan (Laerdal, Stavanger, Norway) and the Emergency Care Simulator (ECS, METI). After some initial and fundamental discussion, two users of the devices were invited to present the products. Both were anesthesiologists and members of the commission. Space and time on a commission meeting was allocated for each presentation and a subsequent question and answer round and a final discussion. The discussion focused on two major topics: Hardware and software. Concerning the hardware, it was decided that both simulators did have mannequins and other hardware that differed in shape, feel, and look but offered similar functionality. A decision based on hardware was not possible.

An extensive and vivid discussion within the commission followed about the software. Both devices stood for a different approach. Laerdal prefers a direct control of vital parameters and symptoms, while METI offers a model that allows control of the signs and symptoms based on the underlying mathematical models of physiology. The discussion evolved not around the question whether the physiological models were of higher quality but around the motivation to use them. Some preferred a device that seemed very easy to use, while others preferred the physiological models as their application would lead to more realistic results. Paradoxically, both sides claimed ease of use: "Nonmodelists" explained that no knowledge of physiology was required in SimMan, while "modelists" reasoned that the automatisms within the ECS lead to less burden on the facilitator. Both fractions, however, agreed that training on physiology and pharmacology would only make sense with models of physiology. After the discussion, the commission expressed its preference for a physiology-based simulator and the executive board of the Society decided to obtain ECS systems. They were to be handed out to the participating institutions but would remain in Society ownership.

By making this decision and forwarding the order of about 40 units to the manufacturer, the Society initiated a process in which several active partners were involved:

- The Society itself as buyer from METI and provider to the universities and as overall and principal overseer of the program

- METI as the manufacturer and provider of the systems
- The Mainz simulation center as on-site coordinator and responsible entity of the installation process and provider of train-the-trainer events
- The simulation centers of Berlin, Erlangen, Heidelberg, Tübingen as providers of additional training elements
- One of the authors (SM) was named program coordinator by the Society and METI. The coauthors and other members of the Mainz simulation center supported his activities.

It quickly became clear that this would be a major undertaking. We aimed to install a large number of simulation devices, to train a large number of mostly new users of these systems who in parallel would have to establish their own centers. We were also aware that the Society gave the simulators to each medical school that did not refuse the offer. While this was exactly the purpose of the project, it meant that in some instances we might meet new users who had not been waiting for a simulator and in some cases may never have heard of simulation in medicine before.

The Society was interested in the results of the process but left the details very liberally to the conducting entities, which certainly helped the implementation because we could concentrate on the creation of good results. Short reports were given to Prof. Schüttler when milestones were reached or problems occurred in order to make him and the Society executive board aware of the state of the project.

The implementation phase of the program was a combination of several – partly parallel – processes with the shared goal of granting the Society program a successful start:

1. Definition and planning process
2. Installation process
3. Training process

The latter two were seen as the implementation phase. This was to be followed by a later application phase during which the use of the simulators was to be supported.

## 34.3 Definition and Planning Process

This phase started immediately after the decision to buy the simulators was made. A visit to the manufacturer in September 2003 by two of the authors (SM, JV) helped to finalize the technical specifications of the product: METI recognized the German anesthesia community as a large and important group and took their requests very seriously. Some changes to the ECS were granted by the manufacturer for the project and later introduced into the standard product. METI would also increase its customer-support capability in Germany and provide two additional ECSs in the country to guarantee immediate support if a device would fail in a critical moment. During this phase, the logistics of the installation were planned and the duration and the several subprocesses were defined. March 2004 was decided as the delivery date of the last patient simulator.

From the very beginning, it was agreed that the transfer of knowledge and content about simulation would be crucial because most of the centers were expected to have little or no prior knowledge about simulation or about the specific product. Each center was to receive a short immediate technical training upon installation. Later, during on-site and during regional meetings, additional contents were to be conveyed. The regional meetings were also a platform to generate a German user community by making sure that the people involved came to know each other. An additional method was the creation of a "buddy system" by which each of the existing "old" centers would act as coach and first place to ask questions for the new centers.

It was decided that the patient simulators would be delivered and installed and users would be trained in parallel to guarantee a fast process. Thus, different recipients could be in different phase of the process: while some centers were awaiting delivery installation and technical training, others would already be equipped and await technical training, while others already would receive application trainings or participate in regional user meetings.

## 34.4 Installation Phase

Once the installation phase commenced, a steady stream of patient simulators were shipped to Germany (Figure 34.1). Since there were no large storage facilities and also arrival of the devices were awaited by the users, the whole process of installation and training had to work continuously and without interruption. A bottleneck would have seriously disturbed the project. The latter more so, because the different steps of the implementation could only be executed in the right order. It did not make any sense to train a group that did not have a simulator yet.

Each patient simulator was delivered to a central facility in Mainz where it was tested and, if necessary, minor repairs were conducted. This was also an opportunity to equip the devices with some material that had to come from local sources in order to follow German specifications. This additional equipment included German gas connections with pressure regulators and also batteries to provide power for the patient simulators. We learned that some effort was necessary to schedule installation visits with each of the participating university hospitals. The installation visit was, in most cases, the first contact of the new users with a simulator and simulation. We wanted to make sure that this would be a positive

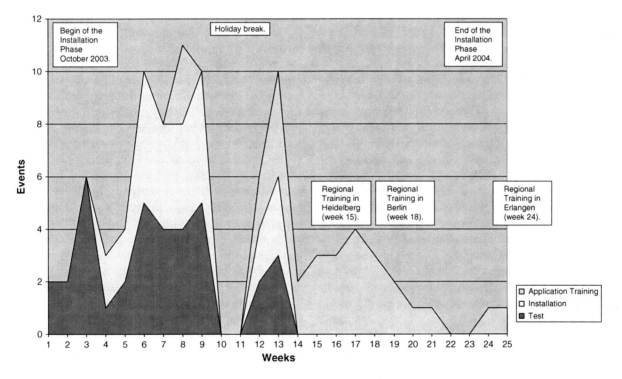

**FIGURE 34.1** Timeline of the installation phase: Each week saw up to 10 patient simulator tests or site visits for installation or training. While testing and installation happened almost concurrently during the initial weeks, scheduling of trainings took longer and lasted until April 2004. Immediately after the installations were completed, the series of regional meetings was started.

experience. We also wanted to put the groups in a position where they felt confident and motivated enough to start using the system and understand that it was very easy to use. Therefore, we asked each group to allocate staff to the installation that should consist of those clinical educators who would be expected to use the system later. In some cases, medical technicians were sent by their hospitals to participate in these meetings.

The technician who delivered and installed the system would meet with the local group at the future training center. This installer was a METI-trained technician with detailed knowledge about the system. The organizers wanted to make sure that the new users (i) learn as much as possible about their new device through presentations and answers to questions and (ii) understand that there is an infrastructure behind the product that could be accessed if necessary. In most cases, the hospital was able to provide rooms for this purpose, mostly unused operating theaters or intensive care unit rooms or other unused patient facilities. The ongoing decrease in in-hospital patient numbers and extensive construction effort after the German reunification were helpful coincidences. In some instances, previously existing training rooms were now becoming home of the simulators. Only in isolated incidents the technician could not install the system in a location that could be considered adequate for training. One group that lacked educational facilities had to configure the simulator in an induction room where it had to be put to the side during clinical use. From a clinical as well as from an educational perspective, this was considered an inappropriate but temporary solution. Such situations never led to delay or nonconductance of the installation.

The technician installed the simulator together with the trainees to ensure that the familiarization with the device began immediately. The purpose was to prevent members from forming as their first impression that their patient simulator system was too technical and an excessively complicated device. This misconception could have created a hurdle and hindered the trouble-free use of the system. Installation included unpacking and setup with respect to the local needs, and their training commenced. The technician explained the different components, their use, and the power on and power off procedure. The reminder of the day was dedicated to a formal and preliminary application training. During this training, a description of the software was given and the installed patient and scenario files were introduced. The technician outlined possible and simple applications and recommended the playful use of the new device.

The installation was followed by a break for the trainees that could last several weeks. It was difficult to find another workday during which the future simulation facilitators would be available again for a train-the-trainer event. Our initial hope was that the groups would use their patient simulator playfully in the meantime. We were disappointed in most cases. It seemed to be very difficult to find time next to clinical work that allowed the use of the simulator. And in some sites, especially when no previous simulation interest existed, the installation failed to create enough enthusiasm to accept the new device and perform the tasks required in its use.

## 34.5 Application Training

The goal of the application training was twofold:

1. **To give the group the technical capability to use their patient simulator in the intended way.** The members should be able to use their patient simulator in a productive way, i.e., in a training situation. They also should be able to generate trainings with the device, i.e., define learning goals, and create scenarios and patients. In general, the participants should be able to make use of the new tool for educational purposes.
2. **To generate interest and motivation.** From the beginning, it was strongly felt that sites with knowledge *about* but no interest *in* simulation would not become active simulation centers on their own. Only an enthusiastic group could overcome the unavoidable hurdles. We tried to give the new group a feeling for the opportunities that simulation had to offer. This was considered to be just as important as the technical training aspects.

The application training would last one full day, and was intended for the whole group of one hospital. Although this was the second time during the project that centers were expected to let staff work outside of the operating theater with the goal to improve education through simulation, we did not encounter severe scheduling problems. It sometimes took several weeks until an appropriate date was found, but since we tried to plan well ahead this never posed a serious problem.

The training covered the HPS software of the ECS system, its user interface, and the programing technique. It also covered extensively the underlying physiology. It was important that the trainees would develop a good understanding of the physiological concepts and how to influence the models. In most cases, this was a fascinating exercise because the groups had a good understanding of physiology and felt great fun by manipulating variables and conducting little experiments.

Toward the end of the training, the users began to practice the programing and – if time allowed – conducted a short scenario to see the success of their previous work and to understand that role of the facilitator. We also wanted to generate a feeling for the need of feedback and debriefing sessions. This was prerequisite to make sure that the simulator would not be judged as just another task trainer but rather a device that opened new opportunities and that it was possible and important to implement new methods for education. Prior to the debriefing training, very little had been known about facilitation and other education methods typical of simulation. Thus, the new users were asked urgently to keep or – when necessary at least begin – working with the new system.

Last but not least, the participants were informed that after this training, they should feel as new members of a community: The existing simulation centers, their knowledge and experience, as well as their willingness to help. The users were also introduced to the various meetings and societies that provide an infrastructure for the further development of simulation (German Society for Anesthesia and Intensive Care Medicine and the Society in Europe for Simulation Applied to Medicine, SESAM).

## 34.6 Regional Meetings

The third and last elements of the training and installation process were regional user meetings (Figures 34.2 and 34.3). These trainings had several purposes:

### 34.6.1 Training

The regional meetings were intended as a continuation of the training process that commenced during the previous visits. They were an opportunity to pose questions that had arisen during the initial use of the system.

Also, additional programing practice could be gained by taking part in facilitated programing sessions with an experienced trainer available to help solve problems.

It was especially important to also resume work on topics that before had only been mentioned briefly: Human

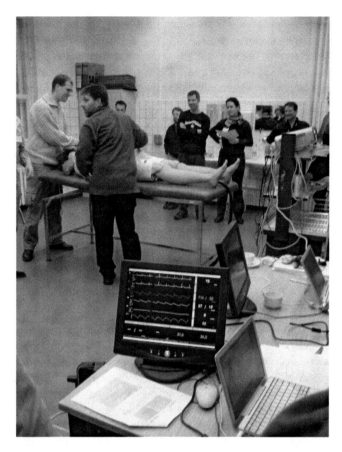

**FIGURE 34.2** The regional meeting at the Charité university hospital in Berlin. Scenarios were developed during the event. Depicted is the dry-run of a simulation scenario that is tested on participants who did not participate in its creation. Some participants act as observers for the later debriefing.

**FIGURE 34.3** Regional meeting at the Charité university hospital in Berlin. A group of participants is debriefed while others observe the debriefing process.

factors aspects and Crisis Resource Management. Therefore, we offered simulation sessions and debriefing sessions that gave everyone the opportunity to be a facilitator, participant, and observer. Heidelberg and Tübingen shared their specific knowledge about human factors aspects and learning methods.

### 34.6.2 Experience

We knew that this would be the last opportunity during the introduction and installation phase for the users to gain experiences in a safe and friendly "protected" environment. After the regional meetings, the official implementation phase would be over and the productive use of the new patient simulator systems was expected. It was a unique opportunity to prepare, conduct, and watch simulator sessions and debriefings, to exchange thoughts, and to fully concentrate on simulation.

### 34.6.3 Community Shaping

Experience does not have to be gained alone but can be shared. Learning from others was considered to be very important. The new centers began to talk about what they and others had achieved so far. They also found an opportunity to talk to representatives of the "old" and experienced German simulation centers. To avoid the impression that all knowledge and wisdom in simulation came from Mainz, it was assured that members of other groups had an active role in these sessions. Especially, the nontechnical topics were introduced by these presenters. Sharing of scenarios was encouraged and facilitated and the value of the German Simulator e-mail list (DESI – "Deutsche Simulanten") was repeatedly reinforced. These meetings were also an opportunity to meet representatives of the coaching centers to begin or continue contact.

This function of group forming was considered another important aspect of the regional meetings. The message was: Nobody is alone, help is available, facilitators help each other, most encounters are open and friendly, there are solutions, problems are often similar, and there are concepts and experiences available. The positive mood certainly helped to shape a community of simulator users out of an isolated and disseminated number of anesthesiologists who had the new task of changing their teachings by using a new and possibly complicated educational tool that sometimes, was even unwanted.

The regional meetings took part in different cities to allow for short travel times and to share responsibilities: Each meeting was hosted by another simulation center. Each participating group was asked to send up to three members to the meetings and it was best if these would attend in one group. Heidelberg, Berlin, and Erlangen were selected as meetings sites as these could be reached with acceptable effort and were experienced training centers. The latter provided access to rooms and equipment so that installations could be compared, and the local facilitators were an additional resource of knowledge and experience. Each meeting took 2 days, which

allowed an evening of social gathering and informal contacts. Arrival on the evening before the meeting was encouraged for the same reason. Some special relationships originated here (Table 34.1).

Similar to the previous trainings, a mixture of theoretical and practical sessions was offered. This time, however, practical aspects became more important. Also, sessions on nontechnical skills were provided by centers other than Mainz. In summary, the changing venues and different presenters and the buddy system helped to make clear that the *society* project was a combined effort with contributions from many sites rather than an isolated effort of the group in Mainz. A very active role in these meetings and other parts of the whole process was provided through the centers in Aachen, Berlin, Heidelberg, Erlangen, and Tübingen.

One thread in the regional meetings was breakout sessions. In small groups, the new facilitators would gather to work on a course element or training scenario. Groups were formed around common needs or interests. For example, if in a regional meeting, two centers expressed the need for a case of neurological trauma, they would form a group and cooperate on this. Each group was encouraged to formulate detailed training needs and learning goals for their future trainees. Around these ideas, a scenario was drafted in each group. In a later session, the scenario was programed and tested. After some additional background information on human factors and debriefing techniques, the groups became trainees in other groups' scenarios and vice versa. Each facilitating group debriefed the training group while others observed the process. In a later step, the case was debriefed with the help of the observers. This process was called meta-debriefing. This guaranteed the roles of trainee, facilitator, observer, debriefing person, and debriefed participant for as many participants as possible.

The scenarios that were developed during the regional meetings were collected and posted for downloads at the Mainz simulation center web site (www.Simulationszentrum-Mainz.de).

## 34.7 Application Phase

After the regional meetings, the implementation phase of the program ended. In follow-up analyses, the participating parties agreed that the following goals were met:

- A patient simulator was installed in each participating university.
- Each site had a core group of trained facilitators who were willing and able to use the patient simulator in the intended way.

TABLE 34.1 Schedule of the 2-day regional meetings. Intensive lecture work is interrupted by many breaks. Breaks were also considered important for the group-forming process. Lectures are highlighted with a dark background, workshops are highlighted with a light background. Note the increase in practical work toward the end of the meeting (day 2)

| Begin | End | Dur. | Title | Topic | Contents | Goal |
|---|---|---|---|---|---|---|
| 9:00 | 9:45 | 45 | Welcome, Introduction | | Who has which experience by now?? | |
| 9:45 | 10:15 | 30 | V57 Inside Emergency Care Simulator | Hardware | A look under the hood of ECS | Basic technical Know how |
| 10:15 | 10:45 | 30 | V58 ECS Simple Trouble Shooting | Hardware | Simple Trouble Shooting | Solve little Problems without Support |
| 10:45 | 10:55 | 10 | break | | | |
| 10:55 | 11:25 | 30 | V59 OSX Details | Software | Users, File Management | Safe Multi-user Operation |
| 11:25 | 11:35 | 10 | break | | | |
| 11:35 | 12:05 | 30 | V44 Principles for Scenarios | Didactics | Learning Goals, Target Audiences | Training Know how |
| 12:05 | 12:15 | 10 | break | | | |
| 12:15 | 13:15 | 60 | V61 Contents of Simulator Trainings | Didactics | Possible Applications of ECS | Know the Curriculum |
| 13:15 | 14:15 | 60 | break | | | |
| 14:15 | 15:15 | 60 | V61 Contents of Simulator Trainings | Didactics | Possible Applications of ECS | Know the Curriculum |
| 15:15 | 15:30 | 15 | break | | | |
| 15:30 | 16:00 | 30 | Definition of a Patient | Software | Practice before Work in small Groups | Be able to write Program |
| 16:00 | 16:30 | 30 | Definition of a Scenario | Software | Practice before Work in small Groups | Be able to write Program |
| 16:30 | 16:45 | 15 | break | | | |
| 16:45 | 17:15 | 30 | Definition of a Patient: Work in small Groups | Software | Practical work in small Groups | Write Program |
| 17:15 | 17:45 | 30 | Definition of a Scenario: Work in small Groups | Software | Practical work in small Groups | Write Program |
| 17:45 | 18:00 | 15 | End of day, Remarks | | | |
| 18:00 | 18:00 | 0 | | | | |
| 18:00 | 18:15 | 15 | | | | |
| 18:15 | 9:15 | 900 | | | | |
| 9:15 | 9:45 | 30 | Welcome, Summary of Day 1 | Didactics | Lecturing, Media, Video-Debriefing | Use new Tool |
| 9:45 | 10:15 | 30 | V48 Basics of Didactics | Software | | Have Material for Training |
| 10:15 | 11:15 | 60 | V47 Patient Programming | Software | | Have Material for Training |
| 11:15 | 12:15 | 60 | V47 Scenario Programming | Didactics | | Understand: soft skills are a Topic for Simulation |
| 12:15 | 12:45 | 30 | V49 Basics in Human Factors | | | |
| 12:45 | 13:45 | 60 | break | | | |
| 13:45 | 14:30 | 45 | Run cases | Software | | Gain Experience |
| 14:30 | 15:00 | 30 | Run Debriefings | Didactics | | Gain Experience |
| 15:00 | 15:15 | 15 | Analysis of case and debriefing | | | Gain Experience |
| 15:15 | 15:30 | 15 | break | | | |
| 15:30 | 16:00 | 30 | Run cases | Software | | Gain Experience |
| 16:00 | 16:30 | 30 | Run Debriefings | Didactics | | Gain Experience |
| 16:30 | 16:45 | 15 | Analysis of case and debriefing | | | Gain Experience |
| 16:45 | 17:15 | 30 | Questions, Feedback, Final Remarks | | | |

- Almost all the centers had found adequate – in some cases impressive – facilities for simulation in their respective institutions.
- With different levels of intensity, each university hospital had commenced undergraduate medical student training with the patient simulator. In some cases, this became part of the regular education process and in other cases, this was voluntary for the students.
- A group had been formed that had the opportunity to bring simulation to new shores and create new opportunities.
- Ironically, the most active groups in the implementation phase – Erlangen, Heidelberg, and Mainz – did not benefit materially from the project: Being "old," experienced, and relatively well equipped centers, they postponed their own installation until the other centers had received their patient simulators. By that time, the project money of the Society was spent and the Society informed the heads of these centers that they would not be eligible for a Society-sponsored patient simulator.

# 34.8 After the Revolution: The Current State of the Project (Spring 2006)

After the enthusiastic and powerful start, the project continued in a style that seems to be adequate for the use of a well-accepted tool:

- Usage did not decrease.
- Maintenance of the systems is guaranteed through the contract between METI and the individual centers.
- Simulation in Germany has become a fact and a normality, certainly in the medical student education in anesthesia and intensive care medicine.
- The Society commission for simulation has been integrated with the commission on education, forming the new commission on simulation and education. This can be interpreted as a sign of normality: Simulation is no longer such an exotic topic that it requires a dedicated commission. Also, education was the goal of the Society's program and it only seems logical to follow on this path now.
- The Society's project has resulted in several follow-up installations, e.g., some universities procured additional ECSs (without Society support) to further improve their training. Some universities embraced simulation even more and installed Human Patient Simulators (HPS, METI), the more sophisticated version.
- The fact that some teaching hospitals procured ECSs to participate in the Society's project is an impressive example of the spread of simulation.

- Some institutions became strong advocates of simulation and established very active centers with ambitious programs.
- Staff, rooms, and other resources are distributed unevenly. While some centers are very active and have good resources to rely on, others are still struggling. In situations with very limited resources, a minor change (e.g., a staff member leaving) can lead to problems as some groups are still very small.

## 34.8.1 Ownership Aspects

The contract contained the following rules:

- The Society buys the simulators and keeps ownership.
- The universities receive simulators and can keep them if the obey to the following rules: (i) They sign a maintenance contract with the manufacturer. (ii) They pay for the maintenance of the systems. (iii) The simulators are used for education of undergraduate medical students. (iv) The universities are not allowed to use the simulators in a way that generates a profit as the Society is not allowed to generate profit.

If recipients failed to follow these rules, the Society is entitled to remove the patient simulator. So far, this has not happened.

## 34.8.2 The Willingness to Accept a Present

As one can easily imagine, the whole project involved so many individuals, institutions, contracts, and individual tasks that conflicts and problems were unavoidable. Of particular interest, however, may be that the patient simulators of the Society, which actually were a present, were not always greeted with enthusiasm by the head of departments. Although this was overcome in the end, it is worth while citing a few of the initial remarks. These remarks probably do not exist in writing and were conveyed during the project phase orally only. Therefore, they are listed only in a narrative below. At the time of the project, however, their reluctance was very real and had to be dealt with. Most of the comments came from heads of departments and not from the staff who would be the users. Examples: "We are not willing to use simulators." "We are not willing to use simulators with physiological models." "We are not willing to install a Society device in our department."

While such comments may have to do with personalities, personal relationships, power games, the very hierarchical German academic medicine system, they could not be ignored. It is possible that simulation is resented by some exactly because it is one way to teach differently and to grant more access to training. Simulation is disruptive because its use naturally leads to nontechnical skills, to questions about decision making, communication, resource management, to accept mistakes as facts to deal with, to accept that everyone is fallible.

### 34.8.3 Achievements

- With some everyday struggle, but without harsh fights, simulation is applied every single day for the German undergraduate medical education in all universities and in many academic hospitals. Today, each medical student in Germany receives simulation education, has a hands-on experience in a safe environment, as we all wished it would be.

- Simulation meetings are held in Germany several times per year and medical conferences have more sessions that are related to simulation and education. The number of German members in international simulation societies (SESAM: Simulation in Europe Applied to Medicine and SSH: Society for Simulation in Healthcare) has grown and now probably ranks second or third after the U.S. and the U.K.

- A group has been formed. Many contacts have grown between the centers; we really have a simulation community. Many opportunities are used to hold informal meetings.

- The centers tend to grow. Not everywhere, but in general, centers have more experienced staff, they give more hours of education with the simulator, they have better rooms, they are better equipped and better accepted in their institutions.

- Typical topics of simulation education, e.g., nontechnical skills have been adopted in German undergraduate education.

## 34.9 Failures

We certainly believe that achievements outweigh the failures, especially because we have a community that is strong enough to improve. But we have also failed in several ways:

- The Society's project has not kept up the quick pace it originally started with. The Society's commission meets only once per year, small working subgroups are not supported. This is too slow to keep up momentum.

- After about a year into the project, data was collected to make a snapshot record, but the data has yet to be published. A data collection was undertaken to make snapshot after about a year into the project, but the data has yet to be published.

- We do not have, nor have we even started, a formal and multicenter evaluation. This would be very valuable as a nation's progress could be measured.

- We have not founded a formal or informal German simulation society. No group of individuals has stood up to do this.

- We have yet expand beyond anesthesia and emergency medicine into new clinical domains. Nurse education and other specialties should become part of the effort.

## 34.10 Conclusion

The revolution is over. This is good and bad. Good, because the changes that were brought are now normal everyday reality of German undergraduate medical education, bad because some of the initial enthusiasm and momentum was lost.

This may change health care. We all hope that future clinicians and their patients will profit from the improved education. Personally, the authors hope that this new training tool, one that allows mistakes-based learning, will also be helpful in shaping a new kind of physicians. Such physicians may grow up not with visions of infallible semigods. Instead, they may have a more modest and patient-centered approach. It would be so beautiful if this project were a catalyst that changes the shape of medicine toward less hierarchical structures and less belief in authorities, with new approaches to deal with errors and with better nontechnical skills. Then, it will have been a true revolution.

We are proud to have been part of this. We know that we achieved a goal through the cooperation of institutions (the Society, the German academic anesthesia departments, and METI). We also know that without the special support and the "can do" attitude of individuals, we would have gotten nowhere. Thank you Dr. Georg Breuer, Dr. Benjamin Gillman, Dr. Christoph Grube, Chip Nichols, Dr. Torsten Schröder, Ronny Schürer, Prof. Jürgen Schüttler, and all the others we forgot.

## References

1. Schüttler, J. Anforderungskatalog zur Durchführung von Simulatortraining-Kursen in der Anästhesie. *Anästhesiologie & Intensivmedizin.* 43, 828–830, 2002.
2. Beckers, S., Bickenbach, J., Fries, M., Killerstreiter, B., Kuhlen, R., Rossaint, R. Die Anästhesiologie als scheinpflichtiges Fach in der neuen Approbationsordnung. *Anästhesiologie & Intensivmedizin.* 45, 33–41, 2004.

# 35

# Statewide and Large-scale Simulation Implementation: The Work of Many

Michael Seropian
Bonnie Driggers

## 35.1 Scale and Goals

Creating a simulation facility in a single institution is not a simple feat. The political, logistical, and fiscal issues often mire a project down. Now, magnify that to not just one institution, but an entire state. The need for simulation-based education is widespread. However, it is the need for infrastructure and operational expertise that far outstrips the supply of money or manufactured simulation devices. Our Statewide approach is not magical, but is just a means to an end. It is a method that will be applicable to some states but not others. It is a model that can be used locally or nationally – the concepts are remarkably similar, just the scope differs. However, the larger the scale, the greater the consequences and implications.

It is commonly accepted that a functional team will outperform a single individual. The notion of a statewide implementation can take various forms, which may include a centralized governing structure. This is not absolute, and in fact may not be desirable. Models could include a body that acts as a resource to develop the infrastructure skill set required for maximal access to simulation in the state. There are a number of core elements that are critical to implementing clinical simulation at any scale:

- Interested parties getting together.
- A common consensus and the need to develop a common goal.
- A more formal group is formed to act on the need and to develop a mechanism to meet the need and goal(s).
- Lobbying for funding for seed money.
- Site visits provide a method to:
  - Deliver the core message
  - Provide education about simulation and its implementation
  - Assess a locality's readiness for simulation
- Success measured by overall success rather than individual success.
- Collaborative individuals with good communication and listening skills.
- Skill sets consistent with the defined objectives.
- A vision and mission.
- A good understanding of the political mind-set and landscape.

## 35.2 A Nongoverning Model and Structure: The Statewide Oregon Simulation Alliance

The notion of statewide simulation in Oregon, U.S.A., was born out of need and politics. The effort was initiated through the benign collision of multiple agendas of multiple groups:

- The Governor's office
- Oregon Health and Science University (OSHU, the state's medical school)
- Oregon Consortium for Nursing Education
- Department of Public Health
- Community College Health Action Plan (CCHAP)
- Association of Oregon Hospitals and Health Systems
- Area Health Education Centers
- Public/Private Universities

The model that was to be developed was not *de novo* but drew on the experience of key people in the group. These individuals, however, could not have succeeded alone without the skills and talent sets of others at the table. Common need, collaborative personalities, and complementary expertise saw the formation and success of the Oregon Simulation Alliance.

## 35.3 The Alliance

The Oregon Simulation Alliance (the Alliance) was formed in November of 2003 with the goal of acting as a point of access for Oregon institutions and collaboratives stated above. It also functioned as a delivery mechanism for needs assessment, training, seed funding, and a framework for sharing. The group defined success as success in simulation irrespective of location, specialty, or sector. The definition of successful simulation is not absolute but rather refers to the ability of the user to achieve their desired objectives effectively and efficiently.

The Alliance recognized that there were many different ways to approach the need but decided on the development of a statewide network as the most efficient and effective model. The purpose was to create a statewide network of simulation users that would have access to expertise to not only help with start-up but with sustainability. Their programs would not need to recreate the wheel and would have access to best practice and resources not otherwise available. The level of sophistication of each program would not be limited by local talent but by the ability of the statewide system to effectively deliver a high level of modeling and expertise. Theoretically, the pool of expertise should reach a critical mass to effectively sustain the effort over time.

The need to develop this system was born not only out of need but passion for simulation. The model that was developed did not rely on a *geographic* regional concept. That is, the state was not split up into areas and possible sites determined solely by geography. The model, instead, allowed groups to define their own regions of *interest*. This type of model acknowledged the political, geographic, and budgetary diversity between areas. Because the model that was developed was all about access and not about control, coalitions maintained their autonomy while gaining from becoming part of a larger network. Local institutions were strongly encouraged to collaborate (as coalitions) with either natural partners, such as partner schools or hospitals, or those that they felt would be a good fit. The facilities that would be created would be self-governed and responsible to their respective political and funding stakeholders, and constituents. The exception to this was if localities received monies from the Alliance, then certain obligations to share and collaborate were imposed.

Some economists would argue that fewer centers with a concentration of resources would be a sounder approach. Although this may be true, it assumes the same overall utilization independent of the total number of centers. It does not take note of commuting as well as the geopolitical and historical factors that may, in fact, preclude effective utilization. There are several examples worldwide of this very circumstance. There are examples of multiple centers existing in the same hospital that function completely independently of each other, not out of need but out of the inability to share a common vision. At this early stage, without any other example to follow, it seemed wise to develop the local talent first and then allow larger regional centers to develop in future years. Those centers would not only be based on need and relationships, but would have the benefit of substantial experience and expertise.

## 35.4 Site Assessments

The Alliance contracted with Bonnie Driggers and Michael Seropian (through OHSU) to do a comprehensive statewide need assessment that combined local education along with the assessment using a proprietary readiness tool. The assessment looked at a variety of factors, and over 100 data points, which were considered important to successful local implementation:

- Local level executive buy-in
- Faculty development
- Access to expertise
- A training plan
- The local vision
- Space commensurate with the vision
- Funding above and beyond the monies from the Alliance (both for start-up and maintenance)
- Business planning.

Nineteen separate sites were visited. A final report was published in February 2005 that outlined the major needs of the state and the dominant positive and negative features that were relevant to a successful implementation of simulation at the statewide level.

## 35.5 Funding Distribution

Simultaneous to the assessments, the Alliance had secured US$ 1,050,000 of seed funding from a variety of sources. Half of this money was set aside for equipment with no one institution receiving more than US$ 50,000. Preference was given to applicants that represented collaboratives, but single institutions were also considered. The monies were distributed through a Request for Proposals process that specifically required applicants to address elements that were thought to be essential for the successful implementation of simulation.

Fifteen institutions were granted funding through this process over a period of 7 months. Equipment purchases were made by the institutions themselves. The funding was granted entirely to local coalitions throughout the state. The makeup of the coalitions was diverse and often included multiple disciplines and sectors. Although difficult to quantify if these coalitions had been successful, it represented more than 75% of the state's health care workers and students. The monies were specifically intended for equipment alone. The burden of the cost of implementation was borne by the coalitions themselves. This was purposeful, as the Alliance believed that simulation was not something that could be sustained through grant funding alone in most cases. Coalitions, therefore, had to demonstrate fiscal buy-in from their respective executives. The actual dollar amount of this buy-in was not the most relevant issue. The most relevant issue was that the dollar amount matched the initial vision of the specific coalition. Once a community had the opportunity to successfully implement simulation with a coalition structure, the Alliance was confident that this structure would spur further investment. To date, it is estimated that over 7–8 million dollars have been spent by institutions for supporting the use of simulation. This is compared to the 1 million dollars in seed money the Alliance secured. By any standard, this would be considered a successful use of seed funding.

## 35.6 Funding with Purpose

The types of institutions involved ranged from community colleges, health systems, nursing schools, to the only medical school in Oregon. For the purpose of illustration, the makeup of the aggregate coalition groups initially included at least nine health systems (representing many hospitals), ten community colleges, and multiple schools of nursing. The majority of the coalitions started with full-size mannequin-based simulation technology. One coalition (surgical) purchased a laparoscopic virtual reality task trainer. The Alliance encouraged, when appropriate, coalitions to invest in mannequin-based simulation. The intent was to expose them to equipment that permitted a level of simulation fidelity they likely had never seen before. With this experience, other simulation modalities could be used with a high level of understanding.

The Alliance did not act as a direct buyer or decision maker for any purchases. The Alliance did however pursue pricing plans with manufacturers for the Alliance membership. These plans offered substantive discounts that further helped the process. The Alliance, although at the time not a formal legal entity, had successfully created a statewide dialog that allowed the simulation customers (the need) to mature much more quickly than would have otherwise occurred.

The other half of the seed money was set aside for training and faculty development – teaching educators and faculty from hospitals and schools respectively, about simulation methodology and infrastructure. The Alliance contracted with Oregon Health and Science University to provide statewide training. The training was three-tiered with tier-1 representing foundations training (2 days), tier-2 represented hands-on training at established simulation centers, and tier-3 that was specific for individuals who would train others in the future. Tier-3 training would not only include hands-on training at an established simulation facility, but a more in-depth study of simulation principles, and on-location facilitation (going to their site). The training plan that was developed was based on sound education principles that recognized the importance of experience in the development process. The plan gave institutions an alternative to the standard 2–5-day course. In this respect, the Alliance acknowledged that simulation specialist training was a 1–2-year process per individual to develop the necessary skill level. Respect for the education of simulation specialists greatly enhanced the inherent level of expertise in the state and formed a significant talent pool of specialists. Long-term sustainability of the project will be dependent upon a pool of simulation specialists capable of training others. Monies for training were not restricted to in-state training alone. Specialized groups with expertise above and beyond that which can be found in the state have provided sessions of a more specialized nature (e.g., debriefing). In this way, the state is able to adopt "best practices," and not become victim to a self-limiting inbred mentality.

Further funding of US$ 300,000 has been secured to continue to target training of individuals to: deliver simulation education, evaluate the outcomes of the Alliance, create a website, and to develop personnel capable of training other trainers to a high level of sophistication. The monies for the education process have done several things: (i) highlighted the importance of education and training; (ii) demonstrated a long-term commitment and plan on the part of the Alliance; (iii) shared the fiscal burden of training.

## 35.7 The Glue

All the simulation facilities/programs in Oregon (21 as of this writing) are self-governed but enjoy a healthy relationship with other facilities in the state (Figure 35.1).

The Alliance fully understands that competitive pressures will periodically reveal themselves. This in itself is not a bad thing as long as each institution has had the opportunity of access. The Alliance does not delude itself to believe that these forces will not become an issue in the future. However, there are methods that can be used to help mitigate these issues. Continuous reevaluation and visioning is critical. Much like a company, the Alliance must understand its customer (the health care institutions in Oregon) and update its strategic plan regularly. A Web-based portal dedicated to sharing best practice, information, resources, and scenarios is on line (www.oregonsim.org). Similarly, the Alliance hosts a simulation summit each year in the fall. This summit is dedicated to and limited to Oregon and southwest Washington institutions alone, in an effort to support collaboration and networking. This allows the numbers (100–150)

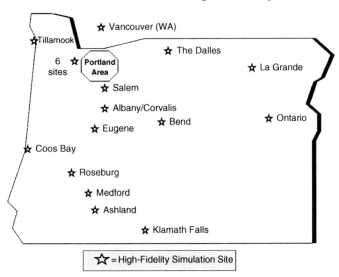

**Simulation Facilities and Programs in Oregon**

★ = High-Fidelity Simulation Site

**FIGURE 35.1** Distribution of simulation sites within the Alliance, including one site in the neighboring state of Washington. The concentration of sites in the western half of the region is a reflection of population density.

to remain at a level that is consistent with interactive and educationally balanced sessions.

## 35.8 Conclusion

Although this description has been rather condensed, several themes were at the center of its success. The Alliance did not focus on how many people it could expose to simulation (i.e., the participant), but rather focused on the people delivering the education experience. This was intentional and acknowledged the fact that expertise in producing simulation-based clinical education is often one of the more significant bottlenecks. Also, by making this venture about the local entities and not about the Alliance, thus, these new simulation communities were empowered. The Alliance in many respects acted as a contractor and consultant. It shared the success but did not own it. It did not govern, but offered a state-level buying power, and a set of guiding standards that were reasonable. In many respects, this worked, because the state did not already have several well-established programs that would be more likely to act as natural competitors rather than collaborators. The end result was to have independent, highly functional, and sustainable simulation programs, in large and small settings across the state. This process requires years to mature, and needs to be evaluated and reevaluated regularly. It is the work of many.

There was a time when many believed simulation belonged in academic centers. Oregon has demonstrated that the need exists well beyond academia and capacity needs to extent beyond academic borders. The Oregon experience is an example of how large-scale simulation networks can be created. The fundamental principles can be applied to other venues and scales. The foundation of much of these efforts relies on relationships, openness, a willingness to share and give, as well as a passion for the process. Variations of this model will appear throughout the globe and will further define best practice in approach.

## Acknowledgments

We would like to acknowledge funding for the statewide project from the Northwest Health Foundation, and the Oregon Workforce Investment Board.

# 36

# Implementing Military Health Simulation Operations: The Australian Defence Force

Fabian E. Purcell
Denis B. French

## 36.1 Strategic Implementation

### 36.1.1 Background – Program Initiation

All conventional military combat forces incorporate some form of simulation into their policy, doctrine, infrastructure, training, and curricula. The Health Services of the Australian Defence Force (ADF) formally and systematically began this process in 2001. At that time, five independently administered "intermediate-fidelity" mannequin systems were already in operation in the ADF. These independent systems were instigated by "early adopter" personnel at Defence health establishments, catering to the local education and training needs of their Commander and the enhancement of individual course curricula.

The role of simulation was identified as part of a broader process exploiting opportunities to identify and incorporate technologies to enhance the provision of health care during the conduct of military operations (Joint Project 2060). This 13–14-year project, divided into three phases including health simulation, has the wider focus of addressing contemporary advances in clinical treatment, surface and air evacuation, disease and injury prevention, health

information management, and deployment flexibility [1]. Public information available for the JP 2060 Project can be found at http://www.defence.gov.au/dmo/lsd/jp2060/jp2060.cfm.

The overall intention of the project is to enhance military-deployable health capability utilizing the structure of the ADF-defined fundamental inputs to capability, which are categorized as collective training, command and management, facilities, major systems, organisation, personnel, supplies, and support [2]. The practical outcomes expected of Defence's health services of this strategic implementation are:

1. maximizing the number of personnel who were fit in all respects for operational duty,
2. minimizing preventable injury,
3. ensuring the prompt and effective treatment of casualties and facilitating their return to duty,
4. enhancing the morale of deployed personnel, and
5. minimizing the incidence of postdeployment and long-term psychological illness.

Rather than the traditional view of utilizing simulation for training in acute illness, the benefits of simulation were considered in both the acute and preventative health sense.

The simulation modalities considered in this project were planned to complement, not replace planning activities at strategic level where other tools, models, and applications apply.

The enhancement of individual/team competencies improves deployed health support operationally, which in turn logically supports the wider current Australian Defence strategic objectives [3]. In the military sense, it also conforms to the military paradigm of all activity at the three levels of war, strategic, operational and tactical, in turn supporting each other and the overarching strategic goal.

The incorporation of simulation training from the top-down allows the overall net value of simulation on skills, knowledge morale, and ultimately practical real-world competency to be maximized for the Department of Defence.

### 36.1.2  Sequence of Events

**Phase One** (completed) was a Project Definition study that delineated the scope of the project, determined the required capabilities, and considered the options on acquisition and management processes to maximize deployed health capability. This allowed the developed management plan to maintain currency and adapt concurrently with the wider revolution in military affairs.

**Phase Two** is the ongoing Acquisition phase, which has been marked by the early purchase of intermediate-fidelity mannequin systems because of their perceived immediate education and training benefit. Later components of this period concentrated on acquisition of capabilities other than patient simulation systems.

**Phase Three** will be a period of review, reengineering, and quality improvement to achieve the Project's aims. This will allow the simulation process to reevaluate and improve its capability to enhance deployed health support in increasingly diverse operational tasks. JP 2060 has progressed under the financial auspices of the Defence Materiel Organisation (DMO). The current (forward projected) estimated budgetary bands for each remaining phase (Australian dollars) are:

1. Phase Two (remaining) $50–75M
2. Phase Three $250–350M [4]

Placed in a comparative fiscal context, the projected budgetary allocation available to the entire Department of Defence in 2004–05 was $19,541M [5], whilst the cost of ongoing military operations (actual deployed ADF personnel) in the same period was 388.9M [6].

## 36.2  Operational Implementation

All systems within the ADF have "life cycles" and effective life cycle management is required to develop and maintain capability. The acquisition life cycle of simulation infrastructure began with the identification of current or prospective capability gaps and opportunities for improvement of capability.

The responsibilities for managing the individual phases of each life cycle are shared sequentially within the ADF. The Capability Development Group administers the Needs and Requirements Phases of the life cycle [7]; Defence Materiel Organisation is responsible for the Acquisition Phase, whilst individual Units, as end-user managers of the mannequins, are accountable during the "Through Life Support and Development" phase. This final period is important because it is where the operational effect is achieved [8].

A typical life cycle for any Defence system will be about 20 years [9]. It is postulated that costs for Through Life Support and Development will be 2–4 times the cost as the Acquisition Phase or 15% per annum [10]. This assessment is based on the financial history of military Electronic Warfare Systems that are inherently more complex than military health simulation, but it emphasizes the need for extensive and persistent funding beyond the initial capital expenditure.

### 36.2.1  Value Added by Simulation

The initial concept paper for Defence simulation operations, defined in a JP2060 Phase One concept paper, defined a group of related functional systems that contribute to Defence-deployed health capability [11]. These were identified as:

1. Command, control, communication, and information
2. Health support services
3. Medical evacuation
4. Preventative health
5. Treatment

Notwithstanding simulation's potential role in preventative health issues, day-to-day simulation operations were placed under the Treatment system.

### 36.2.2  Simulation Topics and Students

Simulation can obviously be used for training all levels of health providers, but with the strategic emphasis on improving capability to deployed health support, there is a deliberate focus on deployed or deployable personnel. This includes training in basic competencies, ongoing skills training, equipment familiarization, and predeployment familiarization in field protocols and procedures. The initial training accent has been on full-time permanent personnel involved in level one to level three of casualty care resulting from combat operations:

> **Level One.** This level in the hierarchy includes the location and removal from danger of casualties and provision of immediate first aid.

**Level Two.** This level involves the collection, sorting, treatment and evacuation of casualties, and provision of resuscitative procedures where appropriate. The focus at this level is on sustaining care, resuscitation, stabilization, and evacuation.

**Level Three.** At this level, the first formal surgery, including initial wound surgery, is performed and hospitalization is provided for medium- and high-intensity nursing of the wounded, sick, and injured. Level Three medical facilities are able to prepare for evacuation those patients who require care beyond the scope and management of the facility.

These are the sailors, soldiers and aircrew, and specific health-support formations who are posted to frontline, high-readiness units. They are often the most inexperienced but practically the most frequently deployed. Some current ADF courses in place that train these people include:

Basic Medical Assistant Course
Advanced Medical Assistant Course
Field Nursing Course
ADF Peri-operative Nursing Course
Airfield Ambulance Operations Course

The current annual training of personnel using simulation operations is estimated as:

Basic Medical Assistant Course
  Army, 90
  Navy, 18
Advanced Medical Assistant Course
  Army, 60
  Navy, 20
  Air Force, 32
Field Nursing Course
  Army, 30

Nearly all specialist doctors (i.e., physicians, anesthesiologists, surgeons) are reservists who can (and do) access centers with high-fidelity mannequins in their civilian practice. Higher level operations, such as ADF operations in East Timor in 1999, usually involve the deployment of medical specialists to Level Three/Four health facilities. These individuals have little opportunity to practice with other deployed team members. To overcome this training shortfall, the ADF is considering generating policy and doctrinal papers to consider formal training links between each military simulation center and a geographically convenient civilian center.

### 36.2.3 Simulator Evaluation

To determine which simulation product(s) was most appropriate, the available "Commercial off the Shelf" systems were matched to a list of mannequin criteria stratified desirable, important, and essential. A list reflecting the author's bias-demonstrating features that could be considered is in the Appendix 36A.1.

The overall acquisition of an entire simulation system considered other features deemed vital to simulation education such as audio-visual equipment and debriefing facilities. Simulation systems are usually integrated into existing well-established health infrastructure programs, personnel, personalities, and cultures facilities at Defence establishments. Base commanders have overall responsibility for the day-to-day delivery of clinical care, health administration, and the occupational health and safety of personnel in the simulation facilities. To facilitate smooth integration, the deleterious impact on training, doctrine, logistic supply, personnel numbers, and operations of the host establishment must be minimal. Common features required for mannequin operation included operational realism, reliability, robustness, flexibility, ease of use, and ongoing maintenance and servicing were also considered.

This leads to the question of level of mannequin fidelity required. Levels one and two first aid military personnel work in hostile, austere, and noisy environments with minimal advanced resuscitation equipment. And often in great personal danger! Unlike the conditions of levels four and five military health environments, the need for a complex underlying mathematical physiology model and volumetric lung function is questionable for training military first aid response. Of far more importance is the impact of the environment, personal equipment, and emotional readiness of the individuals involved. In this sense, the construct of the environment is the more important trigger to realism, as mannequin requirements for the frontline battlespace, although not perfect, have largely been met by industry. For land-based forces particularly, constructing these environments will become more taxing as battlespace environments increasingly vary.

The implementation of Defence health simulation operations also encompasses the integrated use of both part-task and screen-based modalities in concert with mannequin-based scenarios. Part-task trainers were already extensively utilized and the initial acquisition focused on mannequin systems because of their perceived greater ability to enhance deployed health support. The operational introduction of these mannequins considered the following features:

1. The minimal manpower or establishment changes required to support the introduction,
2. Training of operators by the commercial supplier,
3. Clarification of personnel liability within each site of operation,

4. The development of training scenarios and curricula that match the intended outcomes,
5. The mannequin's ability to operate as a stand-alone system that will not directly affect current computer hardware/software already in place,
6. The need for commercial service/support contracts to the simulation system, including the development of an equipment pool providing redundancy for the whole of Defence simulation operations.

### 36.2.4  Curriculum Integration

Simulation and supporting infrastructure must be linked to curricula development. Indeed, the acquisition of mannequins and other infrastructure without thought to curricula integration and disruption has been a recurring error of other simulation-acquisition programs. This is significant because there are unique features of the military first aid training and operational environments that will influence the subject content presented. These features include:

- They are a training cohort whose time is highly managed (to a degree, they are a captive audience). Practically, this allows for more repetitive simulation exposure, hopefully making permanent the knowledge and behaviors required in the battlespace;
- They often train in the same teams with definitive hierarchies, especially for unit-based military organizations;
- Highly standardized equipment, which is utilized in a protocol driven manner; and
- The expectation to work *repeatedly* in personally hazardous environments under possibly extreme duress. This is the outstanding feature of the military environment. It enforces a requirement for strong mental models of duties and responsibilities, and promotes the emphasis on precompiled responses. In short, there must be carefully controlled duress, including the influences of injury or death to team personnel.

Ultimately, each nation will have differing influences and emphases on curricula development appropriate to their tactical, operational, and strategic requirements.

Ultimately, 16 identical commercial off-the-shelf systems were purchased by the DMO on behalf of the Defence Health Services. The number acquired and placement was partially influenced by the Australian feature of a very large geographic area and dispersed personnel elements. Simulators were supplied to both formal training establishments and high-readiness units beyond the five already in operation (Figure 36.1). Upgrade of existing mannequin systems was also included in the contract and delivery was subsequently proceeded.

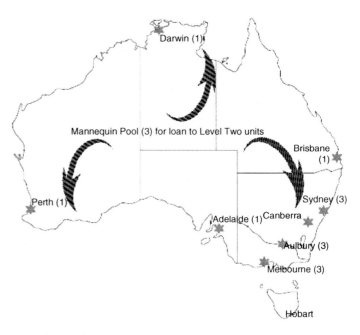

**FIGURE 36.1**   National Distribution of Patient Simulators.

## 36.3  Tactical Implementation – Lessons Learned

Subsequent use of Defence patient simulator systems throughout Phases One and Two of JP2060 has provided anecdotal evidence of improved learning outcomes. Military simulation like all other areas of health care training suffers from the lack of a universal, systematic, codified, and valid assessment tool for assessing improved learning outcomes [12].

Over the span of several years, recurrent challenges to providing ongoing high-quality simulation services have been identified. These are:

1. Maintaining a cadre of trained simulator operators is difficult given the military posting cycle environment. This ongoing dynamic movement of personnel means those initially trained by the commercial provider under acquisition contractual obligations, soon no longer work in simulation centres. Establishing civilian trainers and/or implementing recurrent train-the-trainer obligations may provide effective solutions.
2. Debriefing has been recognized as a separate skill, which also must be included in ongoing simulator instructor training. Such skill is similarly degraded by frequent personnel relocations.
3. Providing all the diverse battlespace environments that military response options encompass can be very difficult. This may be overcome by the construction of a specific multipurpose site able to provide varied conditions.

4. Adequate audio/visual infrastructure and computer network bandwidth is required for on-site debriefing purposes as well as enabling remote instruction and debriefing. The components of this system should be specifically codified and separately funded in all acquisition programs, to emphasize their value and criticality to the educational impact of simulation. Anecdotal experience has indicated that, if not identified separately, audio-visual capital acquisition costs are undervalued, inadequately funded, and subsequently inadequate for the needs of scenario playback and review.

5. Mobility of mannequins allows for simulation to accompany units into the field for training. A clear stressor from postdeployment reports is the experience of having to undergo familiarization with equipment after arrival in the Area of Operations. The role of simulation preventing this occurrence is made easier due to the high level of medical equipment standardization between services and units in the ADF.

6. Utilizing screen-based simulation systems in support of mannequin operations can impact on the computer protocols of host establishments. This has made implementation of network-based screen-based learning complex and time consuming.

7. Highly recurrent use of simulation services especially during extended "live in" courses requires a high number of established and robust scenarios. Developing these scenarios imposes a high curricula development load on staff. *This is the greatest administrative burden and most time consuming task for trainers.*

Despite these challenges, simulation services are in high demand placing pressure on the mannequin use, testing its reliability, and strength of construction. The following issues have been identified:

- Yearly servicing is required; however, only minor replacement of parts is usually necessary.
- Good training in use and train-the-trainer programs prevent user damage to the mannequin.
- Most damage is from incorrect procedure rather than normal wear and tear.
- Buying service contracts at the time of the initial purchase can reduce initial servicing costs.

Some desirable features that remain outstanding for Defence clinical simulation operations include the incorporation of research methodologies into improving simulation training. The well-established human ethics processes and committees of the ADF, combined with a stable cohort of available students contrasted with the lack of published work, reflects the opportunities for research in clinical education have not being fully exploited.

Despite being administratively "owned" by their single service host establishments/bases, greater formalized collaboration needs to occur between the centers. Military simulation user groups need to be established (even internationally) to share the global impetus for creating outstanding military health care simulation services. A Military Chapter of the Society for Simulation in Healthcare is one possibility.

## 36.4 Conclusion

The strength of the multiphased (JP2060) approach ensures that difficulties with implementation and changing strategic and operation directives can be incorporated to optimize the application of simulation services.

The keys to success for the acquisition and ongoing implementation phases for this project are:

- The acquisition of health care simulation fit with higher strategic goals.
- The project was recognized to be a long-term process exploiting the opportunities of, and evolving with, the ongoing adaptations in Defence planning on the long term.
- The place and goals of simulation in the operational level of management were realistic, and focused on enhancing capability.
- The product chosen met with Defence needs and training shortfalls.

In reality, providing current, relevant, and effective simulation services is a role that transcends the life of project mentality. It is a role without end; especially for those cursed to live in "interesting times." And therein lies the challenge.

## References

1. http://www.defence.gov.au/dmo/lsd/jp2060/jp2060.cfm [Accessed 7 Sept 2007]
2. Department of Defence. *Defence Capability Development Manual*. Canberra, Defence Publishing Service, Commonwealth of Australia, 2005.
3. Department of Defence *Defence 2000: Our Future Defence Force*. (Defence White Paper 2000) Commonwealth of Australia, 2000, p. 120.
4. Defence Capability Plan 2004–2014 (Public version) Department of Defence, Published by the Industry Division of the Defence materiel Organisation, Defence publishing Service DPS: NOV016/03.

5. Defence Annual Review. Department of Defence, Defence Publishing Service. http://www.defence.gov.au/budget/04-05/dar/downloads/0405_DAR_02_ch1.pdf, 2004–2005, p. 11.

6. Defence Annual Review. Department of Defence, Defence Publishing Service. http://www.defence.gov.au/budget/04-05/dar/downloads/0405_DAR_02_ch1.pdf, **2004–2005, p. 32.**

7. Department of Defence. *Defence Procurement Review 2003.* Canberra: Defence Publishing Service, Commonwealth of Australia, 2003, p. v.

8. Department of Defence. *Defence Capability Development Manual.* Canberra: Defence Publishing Service, Commonwealth of Australia, 2005, p. 4.

9. Heilbronn, D. *The DSTO View of the Transition from Self Protection to Information Superiority* [online]. Adelaide, AOC Convention. Available from: http://www.oldcrows.org.au/download/convention_2000/heilbronn.pdf [Accessed 7 Sept 2007], 2000.

10. Ibid.

11. Department of Defence. *JP 2060 ADF Deployable Health Capability Simulation Health Operating System Concept 2001.* Commonwealth of Australia.

12. Byrne, A., and Greaves, J. Assessment instruments used during anaesthetic simulation: review of published studies. British Journal Anaesthesia. 86(3), 445–450 (2001 Mar).

# Disclaimer and approval

The views/ideas expressed in this article are those of the authors and do not reflect the official policy or position of the Royal Australian Navy, the Australian Defence Force, or the Australian Government.

DHSD DHCD 2001/31171

CMDR Purcell:

Fabian,
You requested clearance of a (chapter) manuscript Implementing Military Simulation Operations – The ADF. HDHS (AVM Austin) has provided clearance for this to be published; DGSHPP (CDRE Walker) has asked me to convey that clearance to you ASAP because of a publishing deadline (hence the e-mail).

Regards,
John Hatfield
WGCDR
SO1HHPR
DHCD, DHSD
CP2-7-042 (Campbell Park Offices)
ph: 02-62663818, fax:02-62663784
e-mail: John.Hatfield@defence.gov.au
on behalf of DHCD, GPCAPT Robinson.

# 37

# A National Simulation Center Influences Teaching at a National Level: Scotland

Ronnie J. Glavin

## 37.1 Outcomes Measures for Whom?

This is a story about how a national center was able to influence teaching at a national level. Feedback is a frequently mentioned term in this book, and mostly it is used to refer to individual participants on courses watching recordings of themselves under the guidance of a facilitator. However, in this chapter, feedback will be used in a different setting, i.e., feedback to directors of training programs: the feedback will refer to performance of trainees in simulation, which indicates weaknesses in their training programs.

Our tale begins in 1999. Our national center in Stirling, Scotland – The Scottish Clinical Simulation Centre – had been in operation for 1 year. As part of an evaluation exercise, we invited all first year anesthesia trainees or residents in Scotland to attend a simulator-based course on anesthetic emergencies. The intended outcomes of this course were to prepare them for out-of-hours, on-call duties. At that time, the practice in the UK was for all first year trainees to spend their initial 3 months working side by side with a more experienced anesthetist on both elective and emergency cases, both during daytime and out-of-hours. During the third month, a decision was taken as to whether the trainee could participate alone in out-of-hours duties. In the U.K., this means that although a more senior anesthetist would be available for help, the trainee would be expected to manage simple ASA 1 or 2 patients undergoing minor surgical procedures without a more senior anesthetist being physically present in the hospital. A more detailed description of UK training in Anesthesia is listed in Appendix 37A.1. The intentions of the simulation-based course were to prepare these trainees for the type of emergency event that could arise during the perioperative management of such cases.

## 37.2 Designing the Course Based upon a Needs Assessment

Our center was staffed in 1999 by two educational codirectors (both anesthetists) and a technician (who had come from a clinical background)[1].

We were short of time and so could not conduct a proper needs assessment exercise. We asked some of our consultant anesthetist colleagues and some of the trainees working with us in our real operating rooms for a list of clinical problems that had given cause for concern during the first year of anesthetic training. We decided to explore the background of the course attendees by asking all participants to complete a questionnaire on arrival at the center on the day of the course. This asked about the level of teaching, both formal and informal, that they had received in their own hospital departments on a range of possible emergency events. We sent a similar questionnaire to the directors of training (College Tutors) in the major hospitals in Scotland and asked which of the emergency events had been covered as part of the formal departmental teaching program for first year trainees (Table 37.1).

## 37.3 Producing the Course

Each course had five participants: anesthetic trainees with around three months experience. We had devised five scenarios, which we had tested on our faculty. Each participant played the part of the anesthetist during one of the scenarios. One of the educational codirectors operated the mannequin from the control room, while another faculty member played the part of surgeon. Our technician played the part of the anesthetic assistant. This role is essential in UK practice. No anesthetist should begin a case without an adequately trained assistant present. Other operating room personnel were played by the other course participants. After each scenario, we reviewed performance with the aid of a video-recording replay. We were quite surprised by the substandard performances of the trainees (Table 37.2).

This probably reflected that our expectations were of a level of performance more suited to a trainee with more experience. Changes in working practices resulting in both less time spent in training and more supervision by senior anesthetists may have given the trainees less opportunity to respond to changing patient parameters, thus reducing their responsiblity to make decisions related to these changes.

We collated the information from the questionnaires and common events from the trainees' performances during the scenarios (Table 37.3). Our students are not experiencing many of the common and well-characterized emergency conditions that are potentially life threatening and are the primary responsibility of anesthesiologists. Also of interest was a

**TABLE 37.1**   List of emergency conditions in questionnaire

Laryngospasm
Regurgitation and aspiration
Failed intubation
Increased airway pressure
Desaturation
Failure to breathe
Unexpected blood loss
Hypotension
Ventricular tachycardia
Fast atrial fibrillation
Third-degree heart block
Perioperative arrest
Failed spinal
High spinal
Local anesthetic toxicity
Anaphylaxis
Malignant hyperpyrexia

**TABLE 37.2**   Examples of substandard performance by trainees

**Preop**
Lack of structured approach to taking history from patient.

Very little exploration of the impact of the surgical condition on the patient such as checking for hypovolaemia, signs of sepsis, etc.

Very inconsistent approach to assessment of the airway.

**Intraop**
Fixation on one cause for change in physiological variable, e.g., tachycardia developing during a case was usually interpreted as the patient being "light," and so the anesthetic vaporizer would be adjusted to give a higher inspired concentration.

Very little in the way of a systematic approach to dealing with other conditions, e.g., if oxygen saturation dipped a little, there was no systematic review of possible causes.

Failure to anticipate possible changes in surgical plan and to prepare for them were also seen. For example, very little discussion with the surgeon about how confident he/she was with the surgical diagnosis.

**TABLE 37.3** Responses to questionnaire from first year Trainees (Trainee Responses; n = 21)

| Emergency conditions | Experienced, % | Been taught, % | *Neither, %* |
|---|---|---|---|
| Laryngospasm | 52 | 57 | *0–43* |
| Regurgitation & aspiration | 5 | 33 | *62–67* |
| Failed intubation | 29 | 62 | *9–38* |
| Increased airway pressure | 14 | 24 | *62–76* |
| Desaturation | 48 | 38 | *14–52* |
| Failure to breathe | 38 | 38 | *24–62* |
| Unexpected blood loss | 14 | 19 | *67–81* |
| Hypotension | 43 | 52 | *5–48* |
| Ventricular tachycardia | 29 | 43 | *28–57* |
| Fast atrial fibrillation | 29 | 38 | *33–62* |
| Third-degree heart block | 24 | 33 | *43–67* |
| Perioperative arrest | 5 | 19 | *76–81* |
| Failed spinal | 5 | 5 | *90–95* |
| High spinal | 5 | 10 | *95–90* |
| Local anesthetic toxicity | 5 | 5 | *90–95* |
| Anaphylaxis | 10 | 38 | *52–62* |
| Malignant hyperpyrexia | 5 | 38 | *57–62* |

Experience = reporting they actually encountered such a condition in their practice
Taught = reporting they had didactic teaching on this condition
*Neither = an after-the-fact computation based upon the reported results from the Experience and Taught columns. The smaller values assume zero overlap, that is, each individual received either Experience or Taught but not both; the larger values assume total overlap of Experience and Taught.*

**TABLE 37.4** Comparison of responses to questionnaire from first year Trainees (Trainee Responses; n = 21) and from College Tutors[a] (Tutor Responses; n = 15) for Condition Recognition. Recognize Condition = were they confident that they would recognize this condition developing

| Emergency conditions | Recognize Condition | |
|---|---|---|
| | Trainee responses % | Tutor responses % |
| Laryngospasm | 90 | 100 |
| Regurgitation & aspiration | 76 | 100 |
| Failed intubation | 100 | 100 |
| Increased airway pressure | 71 | 100 |
| Desaturation | 95 | 100 |
| Failure to breathe | 90 | 100 |
| Unexpected blood loss | 81 | 100 |
| Hypotension | 100 | 100 |
| Ventricular tachycardia | 100 | 100 |
| Fast atrial fibrillation | 100 | 100 |
| Third-degree heart block | 90 | 93 |
| Perioperative arrest | 90 | 100 |
| Failed spinal | 67 | 100 |
| High spinal | 52 | 87 |
| Local anesthetic toxicity | 62 | 93 |
| Anaphylaxis | 81 | 93 |
| Malignant hyperpyrexia | 48 | 80 |

[a]The College Tutor is a consultant anesthetist who has responsibility for ensuring that training is being carried out to the standard required by the national body – The Royal College of Anaesthetists.

discrepancy between the trainees' responses to the questionnaire and their trainers' responses (Tables 37.4 and 37.5). We asked the trainees to elaborate upon this at subsequent visits and there were several reasons offered – didactic sessions that should have taken place were cancelled. Trainees were unable to attend some sessions due to clinical placements (some training departments provided anesthetic services for two separate hospital sites). On some occasions, the anesthetist to whom teaching had been delegated may not have been fully briefed or may have misinterpreted the requested instruction. For example, "increased airway pressure" may have resulted on a session on "asthma" and so ignored changes in compliance and did not offer a structured approach to diagnosis or management.

In retrospect, this was an important for us because it did bring about a minor paradigm shift. As a former College Tutor, I had made the mistake of assuming that the very structured course I ran was typical and representative of the wider world. There was much more variation out there than we suspected (Tables 37.3–37.5).

**TABLE 37.5** Comparison of responses to questionnaire from first year Trainees (Trainee Responses; n = 21) and from College Tutors[a] (Tutor Responses; n = 15) for Treatment Initiation. Initiate Treatment = were they confident that they would be able to begin the appropriate treatment

|                              | Initiate Treatment      |                       |
| ---------------------------- | ----------------------- | --------------------- |
| Emergency conditions         | Trainee responses, %    | Tutor responses, %    |
| Laryngospasm                 | 81                      | 100                   |
| Regurgitation & aspiration   | 52                      | 93                    |
| Failed intubation            | 71                      | 100                   |
| Increased airway pressure    | 48                      | 93                    |
| Desaturation                 | 76                      | 93                    |
| Failure to breathe           | 81                      | 100                   |
| Unexpected blood loss        | 67                      | 100                   |
| Hypotension                  | 81                      | 93                    |
| Ventricular tachycardia      | 71                      | 93                    |
| Fast atrial fibrillation     | 71                      | 80                    |
| Third-degree heart block     | 52                      | 67                    |
| Perioperative arrest         | 81                      | 93                    |
| Failed spinal                | 14                      | 80                    |
| High spinal                  | 29                      | 80                    |
| Local anesthetic toxicity    | 38                      | 87                    |
| Anaphylaxis                  | 67                      | 93                    |
| Malignant hyperpyrexia       | 38                      | 87                    |

[a]The College Tutor is a consultant anesthetist who has responsibility for ensuring that training is being carried out to the standard required by the national body – The Royal College of Anaesthetists.

## 37.4 First Phase of Feedback

Broad inferences should not be made from one performance of one individual in a simulation-based course. Likewise, broad inferences should not be made about that individual's training. However, when 30 trainees from different institutions perform in a consistent manner, then one becomes more confident about the strengths and weaknesses, not of the trainees, but of the training scheme itself. Fortunately, a meeting of the directors of the training programs (College Tutors) is held annually and we were invited to present our findings (Appendix 37A.2). We did so, and waited with interest on our next group of first year anesthetic trainees. Would the points that we raised at this meeting bring about a change in practice in the teaching and preparation of the next cohort of novice learners? We

hoped that the trainees would perform a more structured approach to the management of patients during the scenarios.

## 37.5 Next Set of Information

The following year, the next group of trainees duly arrived and we repeated the questionnaire information. We decided that we also needed to acquire more data from the performance of the trainees, which we obtained from reviewing the recorded scenario performances. Certain areas had improved. For example, in the original cohort, there was consistently poor performance in assessing the airway of the patient – trainees were either not doing an airway assessment or were not carrying it out in a structured manner. In contrast, the second cohort was much more methodical and much more likely to assess this component, and the trainees were much more willing to impart information to the patient about the process of anesthesia and what to expect in the postoperative period. Already we were beginning to see some impact from the feedback that had been given to the annual meeting of the directors of training programs.

However, there were still some areas where poor performance featured. Trainees had not improved in their willingness to explore the impact of the surgical condition on the patient. Many trainees seemed to lack a structured approach to dealing with abnormal physiological variables such as a fall in saturation or tachycardia. Many trainees seemed to have no plan B to fall back onto if plan A was unsuccessful. The literature on feedback encourages those giving it to be specific [1]. Using phrases such as "that was very good" or "that was very bad" are not as well received as giving more specific feedback – "that was not an easy intubation, as you had correctly predicted, and your choice of equipment and approach were appropriate. Your technique was very smooth, with the endotracheal tube being in the correct place at first attempt and with no trauma to the prominent upper incisors."

If we were to convey this approach to the training directors, we needed a more detailed breakdown of the performances of the trainees. Once more, we were invited to comment on our findings at the annual meeting of the training directors. We were able to report that we had seen some progress, especially in the approach to airway assessment and communication of the perioperative course of events but that we wanted to conduct further analysis on the performances by spending more time scrutinizing the video recordings of the management of the scenarios.

## 37.6 There can be too Many Ways to Skin a Cat

In 2001, a new cohort of first year trainees attended our course. We asked ourselves some questions, "Did they know what they were supposed to do, but couldn't translate that into action?"

Or, "Did they now even know what to do?" "Did they have a set of protocols or guidelines that they could follow?" We decided to give the participants on the course a written test to determine how well they could recall protocols or management strategies. The trainees were given a written question after each scenario but before debriefing had taken place. For example, after a scenario involving a failed intubation, the written question was "You are inducing anesthesia in an ASA 1 patient undergoing emergency appendicectomy. On insertion of the laryngoscope, your best view is Cormack and Lehane grade 3. Describe how you would proceed with the management of this patient."

The information from this phase was most helpful. We found that the problem was that the trainees did not have plans for the majority of the five clinical scenarios on the course. During debriefing, we discussed with each group of participants why there was a lack of a plan. A common theme emerged – there was not a lack of teaching; indeed there had been too much *teaching* but not enough *learning*. There was no consistent message being given in how to manage these emergencies. Different senior anesthetists were giving different and sometimes contradictory messages to the trainees. For example, some trainees were being taught to perform rapid sequence induction by using propofol and rocuronium by one consultant, only to be told to use thiopentone and suxamethonium by another consultant. Each was very vocal in condemnation of the approach of the other, according to the trainees. Now, we had something on which we could act.

## 37.7 Designing Teaching Material

In 2002, we asked each of the training directors if they could provide copies of protocols that were given to their first year trainees in the management of these emergencies featured in the questionnaire in Table 37.1. We received no copies of protocols. In the absence of such documentation, we, at the simulation center, decided that we needed to develop a set of guidelines. We also appreciated that there was no point in confining training in this area to the simulation center – the courses at the center had to be consistent with the teaching being provided at the hospitals in which the trainees worked [2, 3]. We invited anesthetists involved in the teaching of first year trainees, including training directors and other people connected with the Royal College of Anaesthetists in Scotland. Around 20 people attended, and they were divided into groups and each group was given a set of draft protocols that we had developed at the simulation center. We did not want to impose these draft protocols upon learners and teachers because we appreciated that the group had to develop a sense of ownership of the material. The net result of this was a collection of protocols. Two of the group volunteered to edit the collection of protocols, which was in turn sent to the other members for

comment. The final booklet was distributed to the directors of training to distribute to the trainees. The booklet allowed a common approach to the management of these emergencies that could be taught at the level of the hospital department as well as at the simulation center (Appendix 37A.3). Indeed, we proposed that the teaching be carried out at hospital department level prior to the trainees attending the emergencies course at the simulation center.

## 37.8 Studying the Impact of Teaching Material

The next phase was to review the impact of this material on the subsequent performance of the next cohort of trainees to attend the course at the simulation center. A more consistent approach to several areas, such as rapid sequence induction, the management of failed intubation, the management of the unexpected hypotension in the anesthetized patient has been noticed, but the recordings of the performances have not yet been fully analyzed. The feedback process to the training directors and others with administrative responsibility to the training program continues. There are some questions that we would like to answer – Why do some hospital departments succeed in teaching their trainees the protocols? What techniques and teaching methods do they use? Can these be transferred to the other departments? Our work is not finished.

## 37.9 Conclusion

Simulation centers are engaged in a dialog with those who have educational responsibility for potential participants. "Why should our learners attend your center? What will they gain from it? What will we gain from it?" The educational advantages for the individual learners are discussed elsewhere in the book, but this chapter has shown that simulation can also help with the quality control of the teaching program. We chose possible emergencies that could arise during the perioperative care of patients. These events are not common, so it would be difficult to collect meaningful outcome data from real events. By producing a course that allows practically all of the trainees for one area to attend, and by observing the performance of those trainees, one can provide feed back information to the organizers of the teaching program. I have given a brief outline of the steps that we undertook. This process is described as "evaluation" in educational circles but will more familiar to clinical staff as the audit cycle.

In the audit, cycle one determines the desired end point, compares where the target group are in relation to that end point, introduces an intervention that will help move the target group nearer to the desired end point, and then repeats the comparison. It is an iterative process. I have described our experiences in those terms. Some other important points have also emerged; the process is not quick. It has taken us almost 6 years to reach where we are at present, and our journey is not complete. It cannot be done alone. Changes to teaching practice will only occur if those responsible for organizing teaching and those responsible for delivering teaching in the workplace are committed to the project and are willing to change their practice. This is similar to clinical audit – if there is no willingness to change practice, there is little point in beginning an audit project. Finally, this type of analysis of performance is very time consuming. Analyzing recorded performances of 60 or so scenarios not only takes a long time, it also requires perseverance on the part of the person doing the analysis. We chose a set of performances that could be measured against a set of protocols, i.e., the protocols that we devised. This is still labor intensive but less so than analysis of such tasks as team working or decision making.

Simulation-based training can offer useful information to those responsible for designing and running training programs by reviewing the performance of a population of trainees.

# 37.10 About Our Center

I have described the above so that others working in their own centers may benefit from our experiences. There will be questions of a practical nature, such as funding, etc., that require to be addressed. When we began, our center had received funding for a 3-year project whose aim was to determine the contribution of our center to teaching of anesthetists. We were able to invite the first cohort without any fees having to be charged. However, it was important to win over the directors of training so that we could obtain their support. Newsletters, open nights, invitations to speak at departmental meetings (grand rounds) were sent to enlist support. Although we had a full-time staff, we realized that we needed people to help produce courses, and we were fortunate to receive such support. We relied on goodwill. The arrangements for conditions of employment in the UK have changed since the project began, but all consultant anesthetists have some flexible time in the working week to allow them to fulfill the nonclinical components of their job. These include teaching, so we did not need to reimburse either our colleagues or their departments for their time. Nevertheless, there are many demands on the time of our consultant colleagues and we could not take their involvement for granted. The first few months after our center opened were spent recruiting and training our faculty of consultant anesthetists so that they could help us produce courses at our center.

Although our first cohort of trainees were not charged a fee for the course, this was not sustainable because part of the overall simulator project was to explore ways of obtaining funding to offset the costs of the center. When our center opened, we agreed that we would only be able to attract trainees or consultants to courses if they felt there was a worthwhile product. We decided to charge very low fees for the first year to encourage people to attend and so build up the reputation of the center and the courses. I have chosen to comment on the course for first year anesthetic trainees because it enjoys a very good reputation and we have no difficulty recruiting participants to the course[2]. First year trainees are encouraged by their second year peers and their trainers to attend. The course would not enjoy such a reputation if we had not been able to attract that first cohort. More than 95% of the first year trainees in Anesthesia in Scotland attend so we can be confident that what we observe in the performance of those trainees reflects strengths and weaknesses of the training program.

# Endnotes

1. Our technician had a background as an Operating Room technician in a New Zealand hospital, setting up invasive monitoring etc., and then an Operating Department Assistant in Scotland assisting anesthetists setting up equipment for endotracheal intubation etc.
2. Every trainee has an entitlement to 20 days per year available for personal study. The activity must meet with the approval of the local program director. The trainee also has a budget for study, around £500 sterling, which is managed on their behalf by the local training organization. The decision of which of the approved educational activities to attend lies with the trainee. Trainees can pay for courses from personal funds if they so choose. Indeed, to maximize the use of the 20 days, they will have to access personal funds because the course fees will not be met by the budget. At our center, we charge £75 for one trainee attending a 1-day course each year. Attendance at an Advanced Life Support course will cost over £300 sterling for personal study.

# References

1. Kurtz, S., Silverman, J., and Draper, J. *Teaching and Learning Communication Skills in Medicine.* 2nd Ed. Radcliffe Publishing Ltd., Abingdon, UK, 2005.
2. Olympio, M. A., Whelan, R., Ford, R. P. A., and Saunders, I. C. M. Failure of simulation training to change residents' management of oesophageal intubation, *Br. J. Anaesth.* 91 312–318 (2003).
3. Glavin, R. J. and Greaves, D. G. The problem of coordinating simulator-based instruction with experience in the real workplace, *Br J. Anaesth.* 2003; 91, 309–311 (2003).

38

# Clinical Simulation on a National Level: Israel

Amitai Ziv

David Erez

Haim Berkenstadt

## 38.1 The Nation and its Health Care System

Israeli health care is a traditional westernized system employing more than 27,000 physicians and 45,000 nurses, and serves a population of some 6 million citizens, in a geographical area the size of the USA state of New Jersey. Until recently, the Israeli health care system barely recognized or acknowledged the advantages of simulation-based clinical education. The extent of clinical simulation was limited to the utilization of partial-task skills trainers and basic life-support mannequins in clinical schools, hospitals, and in emergency medical services resuscitation courses.

In a geographically small country, such as Israel, a centrally located, multidisciplinary, and multimodality simulation center was viewed as a cost-effective means of providing simulation-based clinical education on a national level, enabling better matching of simulation modalities to educational training goals, and facilitating cross-transference of the learning experience and knowledge from one health care profession to another.

## 38.2 The Israel Center for Clinical Simulation

The Israel Center for Medical Simulation opened in late 2001 and was designed to teach important clinical educational concepts and act as a vehicle for change in training and assessment culture within the entire Israeli health care system. The 1600 m² facility occupies two floors of the National Education Center for Health Professionals building (Figure 38.1) at the Sheba Medical Center, which is ideally situated in the center of the country. The center was designed as a virtual hospital with multiple clinical in-hospital care environments including emergency rooms, operating rooms, intensive care units, and outpatient facilities. It also provides the capability to create various prehospital care environments.

**FIGURE 38.1**   The Israel Center for Clinical Simulation building.

## 38.3 Cultural Change

The greatest advantage of simulation-based clinical education is its ability to decrease the reliance on vulnerable live patients for training. This awareness is promoting a gradual change in the culture of clinical education within the Israeli clinical health care system. Empirical observations suggest that when used appropriately, simulation-based clinical education has the potential to improve the communication skills of health care professionals, decrease the numbers of clinical errors, and enhance patient safety.

## 38.4 Educational Approach

The center uses an "error-driven" educational approach, which recognizes that errors provide an opportunity to create unique learning experiences. Hence, the center designs and produces experiential educational scenarios for health care professionals ranging from the mere challenging to the extreme "nightmare,"

yet in a safe environment. Trainees commit errors and learn from their mistakes in a constructive and formative atmosphere. Training at the center provides an opportunity for both current and would be health care professionals to experience a variety of clinical scenarios/vignettes and procedures, including typical and atypical diseases, critical incidents, near misses, and crises, while protecting "real" patients from the elevated risks and dangers associated with traditional learning while caring for patients in academic care facilities.

## 38.5 Simulation Instructors

The center has a dedicated multidisciplinary staff with educational and clinical expertise in a wide spectrum of health professions – amongst others, physicians, nurses, clinical psychologists, social workers, occupational therapists, paramedics, and medics. All staff were recruited on the basis of their former educational and/or training positions and experiences. Most of the staff have been trained as educators/

instructors and maintain clinical practice in their specific care disciplines.

Training of instructors is one of the major steps in the process of developing any simulation-based teaching. Since the center's inception, more than 800 physicians, nurses, paramedics, and other health care professionals participated in instructor courses and rater programs, which are prerequisites for teaching or rating activities at the center. These instructor courses, rater programs, and "train-the-trainer" courses aim to produce dedicated clinical educationalists who have their own individual educational styles while sharing a common core curriculum. It is difficult to measure the added benefits health care professionals gained through the exposure to concepts and values such as clinical error identification, patient safety, trainee-centered debriefing, and performance assessment. However, there is no doubt that such professionals are a key element in the future success of positive changes to the culture of clinical education in Israel as well as any other clinical simulation training environment.

## 38.6 Simulation Resources

A unique aspect of the simulation center is the routine use of over 50 types of simulation models and modalities. These range in a continuum starting from low-tech task trainers, ranging from simple noncomputerized skill trainers and part-task mannequins, to sophisticated multimedia screen-based simulation software, high-tech computerized interactive patient simulators, and realistic procedural/surgical virtual reality simulators. In addition, the center employs more than 100 simulated patients (actors), known as Standardized Patients (SP) ranging in age from 17 to 80 plus. About half of the programs at the center make use of Standardized Patients, either for solely communication skills courses or as a component of multimodality courses, which combines both high-tech and low-tech simulators in different scenarios. Standardized Patients are used to promote a more ethically sensitive approach to communication with patients in attitude-challenging scenarios such as informed consent and delivering bad news.

An extensive network of digital audio-visual equipment, including, more than 30 cameras (average of two cameras per room) and microphones, one-way window observation booths, and customized debriefing software collect, record, and replay participants' activities. This is an effective and unobtrusive setup for debriefing and constructive feedback to trainees as well as an ideal environment for performance testing. These debriefing sessions are an essential component of the center's "error-driven" educational approach, and facilitating the debriefings are regarded as the most difficult role for instructors.

## 38.7 Bringing the Center to Life Through Collaboration

A key element in the center's operational strategy was the close collaboration established with Israeli health authorities as well as national clinical and academic societies. For example, in a partnership between the center and the Israeli Surgical Association, national-level surgical programs including basic surgical skills, basic and advanced laparoscopic skills, and communication skills were developed for surgeons. The center's personnel contributed their expertise in simulation-based clinical education exercises, while the surgical association determined the surgical skills, techniques, management, and treatment protocols, as well as providing outlines for the various communication skills scenarios. Once the surgical association defined this content, the center's experts in program development, surgical simulation, and communication skills training wrote and tested scenario "scripts," defined debriefing points, and tailored the curriculum to meet the defined program content and learning objectives. In addition, the center organized the logistical aspects of the courses, including staffing (instructors, simulator operators, and support staff), preparation of simulators and additional equipment, as well as Web-based learning materials, which are completed by trainees prior to attending the programs. The surgical association was responsible for recruiting program participants from surgical departments around the country. Together, this effort presented the trainees with deliberate and planned curricula that reinforced quality and safety in care. Similar programs were designed for other clinical associations and organizations such as the medical and nursing schools, the Ministry of Health, the Israeli Emergency Clinical Services, and different branches of the Israeli Defense Force Clinical Corps.

## 38.8 Performance Assessment/ Accreditation

An important benefit of simulation is the reproducible, standardized, and objective setting it provides for both formative and summative assessment. A number of programs became mandatory. For example, in 2004, Israel's Ministry of Health and the Medical Schools Deans' Association began requiring all Interns (450 annually) to participate in a 5-day simulation-based course in their first week of internship. The course focuses on enhancing the clinical and communications skill competencies necessary to handle crisis scenarios commonly encountered during internship (i.e., crises encountered and/or created by the inexperienced during their numerous first-time exposures to real clinical events). The major benefit observed is that this type of training allows immediate feedback about patient care decisions, and appears to lead to better performance and communication skills, in at least the short term. Other programs evolved into summative assessments and board exams. The Israeli Anesthesiology Board Examination Committee incorporated a simulation-based hands-on section into their Final Board Examination for Anesthesiology and the Israeli Emergency Clinical Services Governing Body incorporated a simulation-based component into the final examination for paramedic registration [1, 2].

The center is also using simulation-based assessment methods to help Israeli medical schools select candidates with better humanistic qualities and personal and interpersonal characteristics. As part of the admission process to medical school, candidates are assessed while undergoing a series of brief encounters with standardized patients who present the candidates with ethical and humanistic challenges and dilemmas. The expectation is that such simple yet relevant experiences will afford both the medical schools and the applicants an opportunity to determine the likelihood of "unsuitability" ("not a good fit") before either expends scarce resources.

## 38.9 Readiness to Meet Emergencies

Terrorism is now a global reality, and health care systems must prepare to meet this threat. With this in mind, the center developed simulation-based programs in collaboration with Israel's civilian and military clinical authorities to improve emergency preparedness for conventional and nonconventional threats. These programs emphasize both the clinical and technical skills required to meet the special challenges of managing terror-related casualties. We had over 7000 military clinical team members in simulated battlefield environments, over 700 civilian paramedics, in prehospital settings, and over 1000 physicians and nurses, from 20 public hospitals, in in-hospital ER environments, participating in such programs. The programs included training for the management of casualties from both nonconventional (chemical, biological, and nuclear) and conventional threats, and in particular, all programs focused of crucial team management strategies (Figure 38.2).

Large-scale simulations of mass casualty chemical exposure allowed decision makers to observe the limitations of traditional clinical protocols in meeting this challenge, and this opportunity allowed them to make the appropriate modifications to management protocols. The training of prehospital military teams in trauma management also provided an opportunity to learn the limitations of conventional Advanced Trauma Life Support (ATLS) training, and this also led to a change in the method of training these teams, i.e., more practical and hands-on training (compared to previous mainly lecture-based training).

**FIGURE 38.2** Paramedic, in full-body chemical protection suit treating a wounded patient that had also been exposed to a chemical agent. This scenario was set in a food store, hence the presence of food on the floor as both a location cue and a slip hazard.

## 38.10 Money, Money, Money...

The establishment and maintenance of a high-tech simulation center like ours is expensive, and requires large initial setup costs and ongoing financial stability. The center was established as a nonprofit educational organization with the initial funding secured from a large private donation, which purchased the clinical equipment and a small number of simulators and mannequins. In addition, the Sheba Medical Center provided the building and paid the staffing costs.

The center successfully secured continuing financial stability through a financial model based on long-term "fee for service" contracts with national institutions, such as the Ministry of Health, hospitals, medical schools, Health Maintenance Organizations, and private bodies. These fees for service contracts cover only the actual costs of course development, training, and operational costs. Furthermore, the center secured other lines of financial support for sectors that typically do not have money for training, such as nursing, by creating partnerships with the pharmaceutical and clinical device industry, philanthropic foundations, academic and research grants, and soliciting private donations.

## 38.11 Is It All Worth It?

Within our first three and half years, the center has trained over 23,000 individuals from all of Israel representing most clinical professions and health care sectors in the country: 20% of practicing physicians, 80% of medical school candidates, 100% of clinical interns prior to beginning their clinical internships, 100% of graduating paramedics, and 100% of military physicians. Such a high penetration produces a high demand. The exposure of numerous military and civilian clinical teams to simulation-based clinical education and training is influencing a cultural change in the approach to clinical hands-on training, performance assessment, patient safety, quality of care, and as well as improving communication skills of the country's health care providers.

## 38.12 Conclusion

Although the use of simulation-based clinical education has passed its "infancy," it is still in the early stages of development and the center is trying to lead this development within the Israeli health care system. Internationally, national-level clinical simulation centers are a rarity and may only be viable in very limited circumstances. However, where the circumstances allow the establishment of national (state or regional) level centers, they have the opportunity to quickly introduce better standards and innovative approaches to patient safety and health care training through creating a cultural change in the delivery of an entire nation's clinical education.

## References

1. Berkenstadt, H., Ziv, A., Gafni, N., and Sidi, A. Incorporating simulation-based objective Structured Clinical Examination (OSCE) into the Israeli National Board Examination in Anesthesiology. *Anesth Analg.* 102(3), 853–8 (2006).
2. Berkenstadt, H., Ziv, A., Gafni, N., and Sidi, A. The Validation Process of Incorporating Simulation-Based Accreditation into the Anesthesiology Israeli National Board Exams. *IMAJ*, 10(8): 728–733, 2006.

# VIII

# The Big Picture: Sum of Many Smaller Views

*Better be wise by the misfortunes of others than by your own*

Aesop

The big picture is necessary but not sufficient for your students to gain value from exposure to your passion for simulation-based clinical education. The following chapters provide usable solutions to a number of the many smaller challenges in helping your students change their knowledge, skills, and attitudes. Yet, even in these up close points of view, you'll find echoes of the big-picture themes: the right level of complexity in your simulations matches the level of competency of your students; use of the smallest, simplest, and cheapest task trainer to clearly support your most important educational goals.

Selecting equipment for simulation programs is always a resource-limited challenge. We all start out with a finite amount of time and money and understanding and equipment, which can be exchanged for more understanding and additional or different equipment. We all want to optimize this exchange in our favor. Despite the wide range of different starting and ending points, there are a few perennial principles to consider:

- Exposure to obsolete (i.e., no longer in clinical use) equipment is false training;
- Exposure to the most recent, most complex equipment might delay learning the underlying basic concepts (automation is desired during a product's *use* because it hides so much of the details, but this same masking handicaps learning);
- Ideally, before learners begin encounters with their patients, they should have sufficient opportunities to master the exact same environment and equipment that they will use for real.

Leonard Pott shows how the initial equivalent appearance of actor patients and real patients can be directed toward improving attitudes for treating the cares and worries of the least familiar member of any clinical care event: your patient. Remember, patients don't care what you, the clinician, know until they know that you care. Students think the same about us, their instructors. Craig Balbalian shares his successes and challenges in training advanced providers to perform advanced skills in austere environments. No matter if you don't intend similar teachings in such simulated environments, he offers excellent examples of how to think about how much and what kinds of realism are useful, safe, and reproducible. Judith Hwang and Betsy Bencken show how the simple things in life are still often the best – how basic, cheap, easy-to-use one-task skill trainer simulators should be a regular contributor to any simulation-based skills learning program. Larry Cobb extends this message with examples of how you can easily make and use your own simulators that are identical to any current hand-held clinical device your students are expected to master.

As a general principle, your editors have intentionally selected and arranged the content of this book to be independent of specific clinical disciplines, to only include them as reference points where they reinforce specific teachings about simulation. We've made an exception for Pediatrics, because of the relatively large emotional component when dealing with pediatric patients. Now that having parents being present during resuscitation is becoming the norm, and is encouraged, imagine how your students will prepare for this kind of event. Simulation, with its ability to present "scheduled disasters" is an ideal mechanism for formal clinical education to deal with emotional issues – "today, we instructors are going to shine the light on X and you, the students, are going to learn to deal with X," where X might be our own emotions, our emotions when dealing with parents, grandparents, and siblings, etc. Passive forms of clinical education are ineffective in helping our students learn such important lessons and this kind of training using real patients is never ideal – there are too many extraneous aspects to learn safely and efficiently.

Edmundo Cortez provides an overview of pediatric education and how simulation has supported the acquisition of understanding. Chris Chin focuses upon how a key strength of simulation, the ability of presenting scheduled disasters, is employed to care for the most fragile of patients. Paul Severin *et al.* share their rich experience in pediatric simulation, and include a collection of scenarios for your consideration. Carol Vandrey describes how a simulation facility originally configured for Operating Room environments can easily be adapted for Intensive Care Units and critical care skills experiences. Mike Foss shows how simulation can provide positive first impression exposure to a very common event, albeit one fraught with potential catastrophe: patient transport.

# 39

# The Invisible Standardized Patient

Leonard Pott

## 39.1 Definition

Standardized patients are used extensively in both training and assessment. As the name implies, the patients' presentations are standardized, which allows each learner to be exposed to the same patient scenario. This is particularly important for assessment because each candidate will be exposed to the same clinical case, improving assessment reliability. No matter how good the actor, an unknowable amount of realism is always lost because the candidate *knows* that it is an actor. What is more, the candidate also knows that he is being examined. By integrating the actor completely within the clinical environment, and not informing the candidates when they are actually being examined, we can make the Standardized Patient invisible, thus revealing the students actual behaviors!

This technique is an innovative modification of the familiar Objective Structured Clinical Examination (OSCE) using a standardized patient. Many of the issues will be the same. This article will describe the use of the Invisible Standardized Patient in assessing the preanesthesia visit by anesthesiology residents,

as an example of the general technique. A brief description of the technique used by our Department of Anesthesiology is:

- Residents are informed that they will be examined on their preop visit at some stage during their training.
- On the selected day, the resident is sent by the Chief Supervising Anesthesiologist to see a "patient" requiring surgery.
- The patient is a standardized patient, who is familiar with the clinical and social scenario script.
- The location where the resident will see the patient has been identified and the computer system, nursing staff, and other relevant medical personnel have been informed of the impending Invisible Standardized Patient evaluation session.
- The resident sees the patient, discusses the anesthetic plan with the patient and answers any questions, obtains anesthetic consent, and writes or dictates a medical report.

- The Invisible Standardized Patient, using a checklist (see Example 2), scores the encounter immediately after the resident has left the room.
- The medical report is also scored, using a checklist (see Example 3), by faculty involved with this project.
- A faculty member gives feedback in very general terms at a later time.
- Once a resident has seen the patient and completed his documentation, the next resident is sent by the Chief Supervising Anesthesiologist to see the same Invisible Standardized Patient. Obviously, none of the residents know that the patient has been seen before.

## 39.2 Motivation

The Accreditation Council for Graduate Medical Education (ACGME) has requested that residents be evaluated in terms of core competencies. Patient care and medical knowledge are relatively easy to assess but interpersonal and communication skills, professionalism and systems-based practice have proved more challenging. Part of the problem is that when the candidate is informed that he is being examined, his behavior tends to change. If he is assessed as part of his daily clinical practice with real patients, then the assessment environment is not standardized, thereby decreasing intersubject reliability. It would be very useful to have a standardized health care encounter that can be reliably assessed while not altering routine practice. By making the Standardized Patient "invisible" to the candidate, these goals can be met.

There is an additional benefit to using Invisible Standardized Patients. Residents are informed at the start of their residency that Invisible Standardized Patients will assess them at some stage during their training. They are not told when this will occur, or how frequently. In addition, they are not given immediate feedback, so if the assessment executes as planned, they never know when they have been assessed! If, however, concerns are raised about the resident's communication skills and professionalism, then early identification is possible with feedback, support, and remediation. Because the resident does not know if the next patient he sees constitutes an assessment or not, he will be highly motivated to treat *every* patient as though he is being assessed. We believe that this will encourage the development and persistent use of appropriate behaviors.

## 39.3 Method

### 39.3.1 Identify a Patient Encounter Environment

An extremely important patient encounter is the preoperative anesthesia assessment. During a relatively brief period the anesthesiologist needs to assess the patient, decide on an appropriate anesthetic technique and obtain anesthesia consent. The patient frequently needs reassurance, and it is essential to develop a good rapport with the patient. ACGME core competencies such as Professionalism, Interpersonal and Communication Skills, and Systems-based Practice are essential in the preoperative evaluation.

The identification of a suitable encounter is not always easy because of the tightly integrated systems in which we work. By placing an assessment patient into the real clinical environment, many other components of the system may be affected. Since an important component is to keep the true status of this patient a secret, not many people can be informed of the simulation, yet many individuals may be affected. Encounters must be chosen that will not place a significant workload on others, or expose nonparticipating patients and staff to harm. For example, in a busy Emergency Department, the Invisible Standardized Patient must not deprive real patients of triage resources (bed space, nursing care, etc.).

### 39.3.2 Determine Whether a Standardized Patient can be Integrated into This Encounter

Not all encounters allow for Invisible Standardized Patients. Performing painful or invasive procedures exclude the use of actors, so inserting a central venous catheter is not a suitable encounter. Some clinical conditions can only be simulated with great difficulty. Examples would include significant trauma, abnormal EKG in an ischemic patient, as well as pediatric cases. However, the preoperative assessment of adults is considered suitable by our Department.

Logistic issues include:

- The physical location. Most hospitals have very limited bed space, and do not make provision for purely educational bed space. It is therefore difficult to persuade administrators to provide a regular ward bed. Result: our Invisible Standardized Patients are usually located in the Emergency Department. What is important is that the location chosen must be clinically compatible with the scenario script.
- The medical record. Either the paper record or the hospital's computerized record-keeping system must include the Invisible Standardized Patient scenario. It is essential that the Invisible Standardized Patient appear to the students identical to a real patient.
- Interdisciplinary consultations. Clinical practice entails a team approach. It is therefore essential to inform other team members of the assessment event. For example, the preoperative patient assessment scenario required that the surgeon be informed that a patient was admitted under his name, in case the anesthesiology resident directed questions to the surgical team.

- Schedule an appropriate time for the encounter. For anesthesiology, this point is not a relevant concern, but for some scenarios, the assessment needs to be scheduled during office hours.
- Make absolutely certain that there is an effective mechanism to clear the record of any evidence of a prior workup by another resident. This is particularly important if multiple residents will sequentially see the same patient.
- There must be a clear plan to terminate the session, which usually includes multiple resident evaluations. This involves informing the Information Technology department, the Head Nurse, and the Chief Supervising Anesthesiologist. The Standardized Patient completes the evaluations, is thanked, and goes home. All clinical records of the preop visits are removed from the computer system and the room.

Imagination and creativity are required to ensure that all systems integrate seamlessly. A trial production is essential, preferably with candidate at the same level of training as those who you plan to evaluate. There is no point is using an attending as a candidate (who may very seldom see a patient in the ward) to test a simulation that will be used to evaluate a resident. Testing will frequently identify unexpected problems (see below).

### 39.3.3 Chose a Scoring System

It is essential to choose an appropriate scoring system. The choice of subjective/objective, checklist or not, and nominal/ordinal scale depends on the circumstances. We chose to use an objective checklist similar to that which would be used for an OSCE. We decided that no one would fail the assessment but that the faculty would address those areas identified as "needing improvement." At this stage, we believe that this technique is optimally suited to formative assessment, rather than summative assessment. Once reliability and especially validity have been shown, then perhaps this technique may be suitable for high-stakes assessment. However, there are many other alternatives, and the issue of scoring simulations is beyond the scope of this article.

### 39.3.4 Train the Actor

It is important that the actor be provided with a standardized script. Essential features are:

- History of presenting complaint
- Other significant medical history
- Physical signs remain consistent for each interview

- Same affect and demeanor during each encounter
- Ask the same scripted questions

The actor needs to be instructed how to use the checklist for evaluation purposes. We felt it appropriate to allow for an additional subjective comment and feedback by the actor. We do not use any form of surreptitious recording device such as video or voice recording because the installation costs involved were too high. Instead, we request the standardized patient to record a limited number of specified behaviors, which they are trained to do. It is not essential to use the same Standardized Patient on each occasion. What is important is that each encounter be very similar to allow for intersubject comparison.

### 39.3.5 Feedback

It is well known that adequate, timely feedback is an important part of education. Ideally, candidates ought to be provided with immediate feedback after an assessment. We elected not to provide early feedback because that would alert the candidate to who the actor was. It can be expected that the resident would then inform his colleagues, thereby exposing the OSCE. We felt that delayed feedback would still be appropriate because the assumption is that the assessment accurately reflects daily practice. Feedback is given in very general terms, and includes both positive and negative comments. The crucial issue of what is and how to provide appropriate feedback is beyond the scope of this article.

## 39.4 Problems

In our experience, the biggest single problem has been arranging for a fictitious patient to be included in the hospital computer system. For very obvious reasons, the Information Technology department is not keen to contaminate their data with fictitious patients. This has implications for hospital performance data, privacy rules and regulations, patient billing, and medical records. It is only with the excellent cooperation of our Information Technology department that the Invisible Standardized Patient program is possible. Our hospital computer system is provided by a commercial vendor and modified for local circumstances. The major vendors do not provide for assessment patients in their systems, and to date, there has been no incentive for them to do so. If, however, enough institutions request this facility, then the vendors will certainly build it, making the Information Technology integration far simpler. Our Information Technology department developed a protocol to block the transmission of patient data to billing and other departments (lab, radiology, etc.) and to manually remove the assessment patient data.

Setting up, and especially producing the simulation proved very educational for the computer system designers. We are all aware that we practice our profession as part of a system (ACGME Systems-based Practice) but only when we tried to simulate a patient in the system did we realize just how many components the system had. Examples of items that we did not take into account were:

- The Housekeeping personnel in the Emergency Department (where the patient was located) noted that the patient had been there for a particularly long time and started asking questions. While this is a highly commendable part of our system, it proved to be frustrating to the simulation.
- The charge nurse in the OR noted that there was a patient scheduled for an emergency operation that she had not been informed about. Again, highly observant of her and commendable, but difficult for the scenario.
- On an occasion, one resident who was asked to see the patient had been in the Emergency Department earlier, and actually recalled seeing a colleague speaking to the same patient. He then concluded that he did not need to see the patient since there had already been a preoperative visit.

A normal preoperative anesthetic visit takes about 20–30 minutes, although the duration may be significantly longer if the patient has a complex medical condition or asks very many questions. Theoretically, therefore, we should be able to assess 2–3 residents per hour but this has not been possible in practice. Because of the requirement to remove all evidence of the previous assessment, we can assess no more than one resident per hour, at best. The duration of our assessment sessions depends on the Standardized Patient availability. The Standardized Patients do not find the sessions arduous because they can read between assessments.

## 39.5 Results

We have been very encouraged by the results of the assessments on our residents. The Standardized Patients, who also work with other departments, have frequently commented on how friendly, approachable, and reassuring the residents are. We have, however, had one adverse assessment to date. In one case, the resident documented a normal airway assessment but the Invisible Standardized Patient stated that the resident had made no attempt to examine the airway at all. (The Invisible Standardized Patients are specifically instructed to assess the airway examination because this is a vital component

of the preoperative assessment.) This incident was brought to the attention of the Residency Program Director and the Departmental Chairman, who took appropriate disciplinary action.

## 39.6 Conclusion

Invisible Standardized Patients are difficult to arrange and require meticulous attention to detail in order to execute effectively. However, the resultant realistic assessment of resident physician performance is well worth the effort. This technique truly provides an accurate objective assessment of daily practice with very high validity, and is probably suitable for many clinical disciplines. In addition, it seems reasonable to assume that knowing that this form of assessment exists will improve practice habits.

## 39.7 Examples

### 1. Simulated Patient Case Scenario

Location: Emergency Department
50-year-old, male.
Operation: Laparoscopic cholecystectomy
Allergies: No known drug allergies
Last meal: 3 hours ago, is nauseated, not vomiting
Medications: None
Habits: Nonsmoker, social drinker, no drugs
Remainder of history: Essentially normal
Has beard.
Medical record:

- Complete Blood Count: shows leucocytosis
- Surgical report: requires decompression
- Background history; as per patient (nonsignificant)

Relevant features:

- full stomach
- semiurgent procedure
- beard; potentially difficult mask ventilation

Inform:

1. Emergency Department: Dr A
2. Surgeon: Dr B
3. Emergency Department Head nurse: Ms C
4. Information Technology Department: Mr D
5. Chief Supervising Anesthesiologist: Dr E
6. Admissions: Ms F

## 2. Patient Checklist (completed by patient immediately after resident visit)

Date:_____

Resident name:_____

|  | Yes | No |
|---|---|---|
| Did resident introduce him/herself? | | |
| Did resident confirm patient identity? | | |
| Was eye contact appropriate? | | |
| Was body language appropriate? | | |
| Was easily understandable language used? | | |
| Was affect friendly/approachable? | | |
| Did the resident appear rushed? | | |
| Were questions invited and answered? | | |
| Did the resident ask about medications? | | |
| Did the resident ask about the last meal? | | |
| Did the resident ask about allergies? | | |
| Did the resident examine the mouth/airway? | | |
| Were different anesthetic techniques discussed? | | |
| Was the consent form reviewed? | | |
| How did the resident make the patient feel? | | |

## 3. Anesthesia Record Assessment Checklist (used to assess the resident's medical record)

Date:_____

Resident name:_____

|  | Yes | No |
|---|---|---|
| Patient's name | | |
| Allergies? | | |
| Medications? | | |
| Last meal? | | |
| Airway and beard issues? | | |
| Type of anesthetic? | | |
| Consent? | | |

# 40

# Prehospital and Tactical Simulation: More than Just a Mannequin

Craig Balbalian

## 40.1 Tasks, Conditions, and Standards

This book has a tremendous amount of information regarding simulators and simulations. The majority of simulations tend to focus on either the patient or the prevailing conditions of the current patient location. This is entirely proper when first introducing providers to concepts and actions, allowing them to focus on the acquisition of new skill sets. Since the goal of training should be to prepare the participants for real-life events, additional elements must then be added to increase the complexity of their environment. Call it positive habit transfer or psychomotor habituation. We want the participants to avoid, as much as possible, their first experience occurring in real life. An Army sergeant once told me that "Experience is something you gain right after you need it." Life may not come with a reset button but simulators do, allowing the participant to perfect behaviors for a given event.

The primary focus of this chapter is on training advanced providers to perform advanced skills in austere environments. This educational goal places great technical demands on training resources and great logistical demands upon the instructors. Simulations for prehospital and tactical medical care offer both tremendous challenges as well as opportunities for educators, since they can cover a wide variety of skill levels, equipment, patients, and conditions. Prehospital and tactical medical care is a widely variable term that encompasses a lot of

territory. Police, fire, rescue, Emergency Medical Service, and military all operate in some aspect of this field. The common thread through all of them is that they operate outside of a controlled clinical environment. Since the goal of simulation is to allow the participant to gain experience in a controlled situation, the contradiction is immediately obvious: How do we, in a controlled situation, simulate an uncontrolled one? This chapter includes some concepts to keep in mind when developing simulators and scenarios for use outside of a hospital environment.

The fundamentals of identifying your participants and their training needs (phrased as educational objectives) have been addressed in Chapters 12–15. These educational fundamentals drive your scenario development, which in turn provide the basis for your simulations and simulators. Training for Prehospital and Tactical Providers tends to be organized along the lines of Task, Condition, and Standard. A Task is the goal to be accomplished: "stop the bleeding, extract patient from vehicle." A Condition is the environment and resources at hand: "patient is trapped inside wrecked car, car is upside down, it is night, only illumination is what you bring." A Standard is the level of proficiency desired: "patient extracted and bleeding rate reduced enough so that the patient survives to next phase of care." These Tasks, Conditions, and Standards are usually very simplistic at first and increase in complexity as the skill level of the students increases. For example, the separate tasks of checking a pulse, measuring blood pressure, and counting respirations eventually merge with other skills into a combination task called "patient assessment." Scenarios, and therefore simulators, should have varying levels of complexity as well. The high dollar, high tech, model assisted, electricity requiring, compressed gas–consuming robotic mannequin patient simulators are great for pharmacological training in a critical care unit setting, but aren't going to fare well in the rain and mud of a wilderness rescue scenario. Compromise obviously must be made, but only a thorough evaluation of training needs and resources is going to provide you with the information you must have to make a good selection decision.

## 40.2 Equipment

Once training needs have been identified, then scenarios can be developed to present the conditions under which the standards must be performed. These scenarios may require the use of several different kinds of simulators to complete the overall simulation. Areas of your attention when formulating a scenario to support training a prehospital task (or any clinical training task for that matter) should encompass equipment, the operating environment, and the patient.

Equipment is a broad topic encompassing outfitting both the students as well as the patients. Since by definition all providers will perform tomorrow just as they learned today, formal prehospital provider education should include training with the equipment that they expect to wear as well as any specialized equipment that their patients will have. Students should be utilizing their basic issue uniform, bunker gear, flight suits, battle dress, whatever the normal working clothing is for their tasks. In addition, they should include typical Personal Protective Equipment as well as any specialized protective ensembles such as hazardous material nuclear, biological, chemical ballistic protection (body armor, helmet, etc.), structural firefighting gear, and so on. Use of simulated or training equipment depends on cost, wear and tear, and protection/safety. The ideal training equipment will provide the student with all the experience necessary for them to acquire the desired lessons, and no distractions leading them to undesired lessons, and no risks to the life and limb of anyone. Since your simulation will include no *real* hazardous material or ballistic threats, the costs of brand new, fully mission-capable gear is often not required. The purpose is for the students to become very familiar operating with the bulk, weight, and restrictions of their standard gear. Needless to say, if there are realistically dangerous elements in your simulation, then real safety gear is a must.

Equipment goes far beyond personal clothing though. It includes items used to locate, treat, and transport the patient. Oxygen tanks, aid bags of various descriptions, litters or gurneys, etc., the list can go on and on. Focus on the equipment that you want them to gain competency with and that they can expect to have. Mission-focused equipment not essential

to patient care may be needed as well, such as a specific types of wrecked vehicles common with their intended patients. While the temptation is there to have every cool gadget seen in trade magazines (and this includes patient simulators as well), use just enough realism necessary for the immediate learning. For example, a firefighter/paramedic would have their own Self-Contained Breathing Apparatus while searching the burning building for the patients, but would not need manual ventilation devices nor a simulated patient that could be ventilated until the patient was removed from the simulated fire. If search and rescue techniques are the focus of the iteration, then using the Self-Contained Breathing Apparatus while dragging a human-sized and weighted mannequin from the building is realistic and valuable training. But, manual ventilation devices and patients that can present realistic ventilation challenges become vital once the rescue from the fire is complete and treatment for smoke inhalation begins. How about any equipment the patient had been wearing? Are you simulating an injured firefighter, perhaps unconscious who will be wearing a full set of bunker gear? Patient gear can be an even bigger issue in military settings where weapons, ammunition, contaminates, radios, maps, and other sensitive items must be secured and transported from the scene to prevent them falling into enemy hands. Again, the specific tasks you are training will determine your scenario, which in turn determine the specific items necessary for your students to gain familiarity with performing the tasks.

## 40.3 Environment

We have considered provider equipment and clothing, mission equipment, and patient equipment. Next consideration is environment or setting. Again, we must clearly determine the task to learn. A common task simulator is the intravenous practice arm. The device allows the student to become familiar with the anatomy and basic mechanics of starting a peripheral IV, and is first introduced in a calm, clean, well-lit classroom setting. Repeat until muscle memory makes the procedure almost automatic. There is minimal risk of bloodborne pathogens and all of the other concerns associated with live patients are greatly diminished if not eliminated outright. Ideal circumstances. Is real life ever like that? Hardly. But it is a great first step for beginning providers. We now have a good foundation to build upon. We can then alter the goals and environment slightly to provide ever-greater challenges. Place the same simulator in the back of an ambulance. The lighting isn't as good, and the space is somewhat cramped, and the provider has to position himself just so . . . Mastered that? Great! Now have some other student who needs ambulance drivers training operate the vehicle (hopefully with a driving instructor present) while your beginning provider practices IV starts in a moving ambulance. Good to go? Now you can get really inventive. Have the two of them carry the human-weight mannequin down the stairs, across the parking lot, load it into the vehicle, and then try to start the IV in a moving vehicle while still out of breath, huffing and puffing. Or, while the provider wears a biohazard suit because of simulated infectious disease. Or . . . well, you get the idea. Break the complex real-life challenge into smaller, digestible subtasks that your students can clearly experience success in mastering.

The setting or environment can provide invaluable learning. Lighting is a huge factor. Humans are by nature visual. Functioning in dim light, under complete blackout or light discipline, in smoke or some other obscurant all takes practice. Most times this can be handled safely indoors with lighting adjustments, so something as simple as a dimmer switch on the overhead lights can make a big difference. Tactical training for law enforcement and the military can occur indoors but quite often is outside, at night. Smoke houses (structures with nontoxic smoke–generating equipment) can be used as well, especially for firefighters. No budget? How about blindfolds or safety goggles coated with opaque spray paint or a permanent marker? If the purpose of your training is to teach visually impaired operations, then any number of techniques can be used. On the other end of the budget spectrum would be use of real night-vision devices under blackout/low-light conditions. The other end of the illumination spectrum involves bright sunlight on either sand or snow and, other than preventive measures for sunburn and snow blindness, is rarely of enough concern to warrant duplication. Night and day are easily simulated indoors, but what about weather conditions? Temperature extremes are pretty much beyond the capability of normal indoor climate control systems. If, for whatever reason, you have to train in cold weather conditions, you might consider some sort of arrangement with a cold storage warehouse or refrigerated/frozen food processor. Hot conditions are probably best left until the summertime since any indoor locations already hot enough (like a heating plant) are most likely also too dangerous for novices. Precipitation is pretty much limited to rain and then only if your simulation area is equipped with floor drains and is capable of dealing with high humidity. Outdoors (desired weather permitting) is preferable. Work with the fire department for your water supply. Perhaps, they need to do some hazardous materials decontamination work that involves setting up water streams or fog systems and you can piggyback and/or add to their training.

## 40.4 Real Learning, Real Risks

The next factor to consider is noise. Indoor clinical settings are typically quiet in addition to being clean and well lit. This allows auscultation to be used to gain information about patients. Breath sounds, heart sounds, and bowel sounds, along with percussion, are all part of the repertoire of Prehospital and Tactical Providers. While students are usually able to assess each other for normal sounds, abnormal sounds usually require some sort of prerecorded media playback or scarce and very expensive live Standardized Patients. Several of the newer high-fidelity patient simulators produce some sounds in appropriate body areas to provide training experience (yet, today, most of them have pathetic sound sources and awful sound contamination from all the washing machine–like parts inside the mannequin). Unfortunately, these same mannequins, which tend to be rather expensive, are generally too fragile to use in any scenario involving movement or an austere environment. A clinical setting in the back of an ambulance is about the environmental limit for these devices. But Prehospital and Tactical Providers also operate in areas with high ambient noise levels. Crowds, concerts, sporting events, traffic, and transport (either road, boat, or helicopter) are all good examples. The military can add combat and tracked vehicle ambulances and patient evaluation aircraft to the above list. Fortunately, a good sound system can fulfill these ambient noise requirements. Need to go cheap? Try five or six portable stereo "boom boxes" in the same room, tuned to different radio stations with the volume raised. Just watch the ambient noise level. If you start getting too loud, you'll need to ensure that everyone is wearing hearing protection (which they should on a loud scene anyway). Just think: a good sound system, dim lights, a strobe light or two, and a smoke machine . . . Dude! It's an Emergency Medical System standby at a rock concert! Now all you need is stage divers, mosh pit, and crowd surfers; add liberal quantities of ethanol and *you* are there.

Seriously though, while sound and lighting (or lack thereof) goes a long way toward setting the stage for your simulator, there are still some other aspects of environment that you need to consider. Among these are the specific scenes or tactical situations for which you want to train. Vehicle entrapment after a crash, hazardous materials, combat, structure fire, and explosion are but a few of the potential scenarios that you may be tasked to support with simulations. Let's look at postcrash vehicle entrapment as an example. If the training requirement is to actually tear open a car, then the best approach is to take your class to an automotive junkyard, pull out the power tools, and start cutting up cars. If the focus is on patient assessment, extrication, and medical care, then an old car body that has been rendered safe (fluids drained, motor and drive train removed, fuel tank removed, etc.) and moved into an open bay or garage is ideal. Appropriate lighting and sounds will, just like a theatrical production, set the stage for you. Quite often, the goal here is to reproduce the physical demands of patient extraction and care without real hazards. These environments will typically demand more assets and facilities but often can be used concurrently to support other training. You could, for instance, have your firefighter paramedics practice fire suppression at a burning building as part of a company or battalion training event and then have them work a patient scenario with a clinical simulator that you have prepositioned in the relatively climate-controlled setting of an ambulance compartment. The physically draining experience of firefighting, search, and rescue can be done with a human-weight mannequin with no electronics that is carried out to the ambulance. The hands-on patient care and assessment can be done with a computer based, model driven, high fidelity, full-human simulator that can't handle the smoke, heat, water, and physical abuse that comes with being pulled out of a burning building. Needless to say, a scenario that is this involved and complex is not for beginner instructors. How about aggressive people on the scene? You will most likely use an actor to simulate the unfriendly person but how is he dressed? What is he carrying? Common sense would dictate that simulated weapons be used in a clinical scenario: plastic wiffle ball bats instead of real bats or clubs and the proverbial rubber knife instead of cold steel.

Combat, because of the emphasis on weapons fire and maneuver, tend to have minimal patient care involved. Care under weapons fire is very limited in scope with a much greater emphasis placed on movement techniques to clear the wounded from the immediate danger zone. Only then can emergency care begin. This need to move is similar to that of the firefighting scenario, so a similar approach should be taken in that providers should be made to move the patient prior to starting advanced care. Failure to do so may result in providers getting killed in real life. How do you drive the point home? A low key, low-tech approach would be the instructor walking up, pointing, and saying, "You, you, and you. Lie down, you're dead." This doesn't really add the memorable adrenaline rush that drives the lesson into the student's psyche but it is cheap, easy, and can be done on a spur of the moment. You want cheap, easy, and somewhat dramatic? Walk up, pop a balloon to simulate gunfire, booby trap, Improvised Explosive Device, or a secondary explosive device, and make the same announcement. You'll get the startle reflex with the advantage of popping cheap balloons that pose relatively few safety hazards. Another approach is to use remote controls or motion sensors to turn on sound-producing devices (remember that sound system we discussed earlier?) instead of lights. Go to the local electronics/hobby store and let your inner geek out.

But let's face it, most people still like things that go BANG and no amount of safe noisemakers can satisfy the cravings of the adrenalin junkies that self-select for prehospital and tactical care providers. If you've got the clearance, access, and facilities to do so, pyrotechnics really drive the point home. Nothing like flash bangs, blanks, or grenade simulators point out that they should have moved before starting patient care. But pyrotechnics demand a *very* high level of safety monitoring that may detract from the overall setting. Jury-rigged, field expedient, or improvised pyrotechnic devices are *not* a good idea. Even something as simple as Alka-Seltzer™ in a plastic drink bottle with a little water can be unpredictable and therefore dangerous. "Watch this" or "Want to see something cool?" are dangerous words when fire and explosives are

involved. Make sure you *do a risk assessment* before using any kind of pyrotechnics.

Yet another subject under the "Things that go BANG" category is firearms and their kin. Firearms are a whole separate issue by themselves. Outside of a tactical training scenario, only nonfiring replicas should be used and they must be readily identifiable as such. Replicas such as the Redman™ system, where the weapon is made of bright red plastic, prevent possible confusion with real weapons. Brightly colored water pistols accomplish the same thing with the added benefit of your confederates playing "the bad guys" being able to squirt careless providers. The key component isn't necessarily going to be the prop so much as the well-briefed actor wielding it and the well-written and practiced scenario or script that puts the well-defined learning objective first. Real weapons are more in the purview of tactical operations and should be treated in a whole different light.

Projectile weapons of any variety are an issue in and of themselves. The safety requirements of running any type of live fire scenario are far outside the scope of this chapter, but I will touch on paintball, simulated munitions or simunitions, and blanks. Paintball and simunitions both involve hollow projectiles filled with a marking compound that are propelled either by compressed gas (paintball) or a charge of gunpowder (simunitions). Paintball requires the purchase of a weapon usable only for training as well as the projectiles and propellant. In contrast, an actual service weapon may be used for simunitions. This eliminates the cost of a separate weapon system but comes with an increased ammunition expense. Both require safety equipment that would not normally be worn during day-to-day operations. This safety equipment can actually detract from the realism of the scenario. Blanks do not have a projectile *per se*, but still have the potential to push particulate matter out of the muzzle at injury producing high velocities. Simunitions and blanks should require the same interlocking and overlapping safety requirements as a live fire range. Why? Because they are real weapons capable of firing real ammunition, the effects of which would be catastrophic for your simulation program! Again, you need to consider your training requirements. Unless you are actually part of a police Special Weapons and Tactics team or in the military, then the use of projectile weapons is probably not worth the cost, aggravation, and increased risk.

# 40.5 Patient Selection

Having considered participant equipment and environmental aspects, we finally come to the patient. Patient simulators run the gamut from basic to highly technical and complex. One end of the spectrum is the section of tree trunk that approximates a patient's weight and size. This is appropriate for use when your students are learning how to negotiate a litter obstacle course where simulator durability takes precedence over appearance. The other end is the simulator system involving a high-fidelity computerized mannequin with all the vital signs, simulated body fluids, and the ability to respond, within the programed scenario, to medical and pharmacological interventions. A quick Internet search for "patient simulators" will reveal a bewildering array of companies and devices. Costs range from free for the tree trunk to hundreds of thousand dollars for the highest-fidelity models and associated clinical equipment and devices. Live patients with simulated injuries (moulage) are a part of this continuum as well. How do you decide what to use? The type and level of training, the number of devices needed, the ever-present budgetary constraints, all need to be taken into consideration. Portability, durability, and flexibility are the major factors that I look at once I've figured out the who, the what, and the how much.

Portability is a consideration on several levels. One consideration of portability is that unlike a fixed facility like a formal classroom where the students come to the course, the "traveling road show" goes from location to location bringing the course to the students, with all the latter's attendant/consequential/accompanying problems (see Chapters 60–62, 81). If any travel is involved (even just across the hall), then modularity is a big concern, allowing easy and rapid breakdown and setup in a variety of settings. High-fidelity patient simulators can require gas cylinders, compressors, power sources, and possibly audio-video and Information Technology support. Basic Cardio-Pulmonary Resuscitation (head and thorax) mannequins can be packed with as many as 10 or 12 to a man-portable case. Obviously, the mode of transportation may be a consideration as well. Gas cylinders, for example, are considered hazardous material and usually can't be transported by airplane without a lot of paperwork. Another level of portability involves the simulator's use during the scenario. Are you expecting your students to move the simulator as part of the training iteration? If evacuation/movement more than a few feet is part of the scenario, then you can't use something tethered to gas lines, power cords, and data cables. Conversely, the lightweight Basic Cardio-Pulmonary Resuscitation mannequin won't be realistic for transport training either, so a human-weight mannequin with all appendages and replaceable trauma modules may be required. You could use another student or faculty member instead, but then predictability and durability becomes issues.

High-fidelity simulators generally require gentle handling and do not fare well in settings outside of a nice, clean, clinical environment. Live subjects, with or without moulage, are considerably more adaptable, but even they don't take very well to being dropped. Full-size articulated mannequins can be found that are very durable and capable of being used in a variety of settings such as vehicle extrication, removal from a burning building, etc. These offer flexibility as well. Add bulky clothing and no-longer-serviceable body armor to increase weight. You can add the IV arm for fluid resuscitation, the thorax simulator for needle chest decompression, and the airway mannequin for advanced airway techniques. If you need to go cheap, have the students practice starting IVs on tangerines or oranges (not navel oranges – the skin is too thick) and use a section of pork ribs taped to an old inner tube filled with "fix-a-flat" tire repair sealant for a chest decompression simulator.

## 40.6 Conclusion

Any clinical simulation should have clearly defined educational goals with well thought out tasks, conditions, and standards. All clinical simulation consists of trade-offs between fidelity, suitability, usability, cost, and acceptance. Simulation is ideal for training prehospital and tactical providers, so that they gain most of their learning experiences before their real patients provide it. The low costs required for finding or creating austere environments for prehospital and tactical provider training make simulation an easy option for the instructors. The nonclassroom-like conditions of temperature, humidity, light, and sound in those austere environments challenges the creativity and inventiveness of the instructors. Be imaginative and experiment, play around and try new things. Between real life and your fertile imagination, not to mention discussions with your peers, you'll be amazed at the resources and variations that you discover.

## 40.7 Teaching Resource

BreakAway Games Ltd. offers *Incident Commander*, a training simulation "game" free to municipal emergency departments, public safety agencies, and incident response training programs: http://www.incidentcommander.net

# Value Added by Partial-task Trainers and Simulation

Judith C.F. Hwang
Betsy Bencken

## 41.1 Value Added

Partial-task trainers are wonderful tools for teaching particular skills and allowing a novice learner to progress toward mastery of those psychomotor skills. They also allow the more advanced learner to maintain and fine-tune their skills. The partial-task trainers also eliminate wear and tear on mid- to high-fidelity mannequins that could be used to teach the same skills but at a much higher cost when the labor of replacing the arm on the mannequin is included (Table 41.1). We have used partial-task trainers for teaching airway management, intravenous (IV) catheter placement, central line placement, and pelvic examinations. We also have purchased a partial-task trainer to facilitate learning of lumbar punctures, as well as spinal and epidural anesthesia.

Once manual dexterity skills have been acquired, many of these partial-task trainers can be used in conjunction with a live person who acts as the patient. This hybrid pairing encourages and ensures that the student not only masters the technical skill, but also learns and practices appropriate communication with the patient. For example, the pelvic simulator helps the student to become comfortable with the steps of the pelvic examination and with the different instruments that are used during an examination (i.e., speculum, culture swab). The student also learns how to ergonomically position himself in regard to the patient and the tools used during the examination. By having a person give a voice to the patient, the student now has to introduce herself to the patient – check to make sure that the patient has no questions about the upcoming examination and has voided – and explain what she is doing as she proceeds with the exam.

**TABLE 41.1**  Relative costs of trainers

| Item | Cost, US$ |
| --- | --- |
| 4-Vein venipuncture arm pad | 250 |
| Injection training arm | 500 |
| Arterial and venous patient training arm | 675 |
| Patient simulator IV arm | 40–725 |
| Central lines injection trainer (no ultrasound capability) | 670 |
| Central venous access head neck & upper torso model (with ultrasound capability) | 1500–2300 |
| Pelvic trainer (without software to evaluate exam) | 3500 |
| Pelvic trainer (with software to evaluate exam) | 20,000 |
| Airway trainer | 535–1800 |
| Mid-fidelity airway trainer | 10,500 |
| Patient simulator head | 6100 |
| Lumbar puncture trainer | 700–1350 |
| Spinal and epidural injection trainer | 840–1100 |

## 41.2 Intravenous Access

Partial-task trainers do not need to be perfectly realistic, though they should enable the student to actually perform (almost) every step of the procedure. Sometimes, multiple sequential partial-task trainers are needed to teach all the steps. For instance, IV boards that have multiple veins of varying sizes and at varying depths do not permit students to practice tying and untying a tourniquet (Figure 41.1). However, students not only can easily learn tourniquet tying upon each other, but should do so. The risk is negligible, while at the same time it is the most effective way for them to learn both how much tension is required to occlude flow and how annoying

such tension can be. Furthermore, the IV boards do provide the learner with the opportunity to cannulate a variety of veins that differ in size and lie at different depths below the skin, an experience that is not necessarily available on typical IV arms and hands. Unlike the pseudo-noninvasive use of the tourniquet, the mechanics of venous access should first be practiced on synthetic arms before practiced on live arms (Figure 41.2).

The IV arms and hands and the IV boards allow the student to become comfortable wearing gloves, "prepping" the area, maintaining sterile technique, inserting different-sized catheters, attaching the intravenous extension, flushing the catheter, securing the catheter and tubing in place, and applying a protective dressing.

Having several venous access trainers available increases the modalities that can be taught and practiced simultaneously. For example, some central line placement task trainers are adaptable for the use of ultrasound guidance, whereas others are not. Having both types available encourages the student to learn how to place a central line using both anatomic landmarks as well as ultrasound guidance.

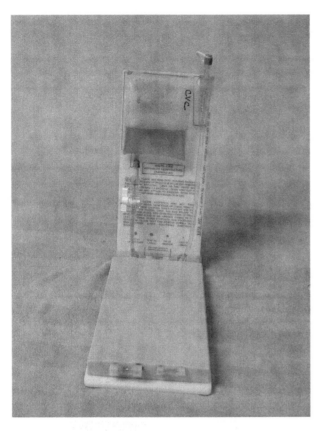

**FIGURE 41.1**   IV access arm board with differing sizes and depths of veins.

## 41.3 Airway Access

Having a variety of partial-task trainers for the same task is invaluable as each trainer has its own advantages and disadvantages. By providing a variety of trainers, the student gains the opportunity to experience the differences, much as performing the skill on multiple patients would. For example, some airway trainers are good for mask ventilation and intubation (Figure 41.3), while others are good for cricothyroidotomies and tracheostomies (Figure 41.4). On the other hand, other airway

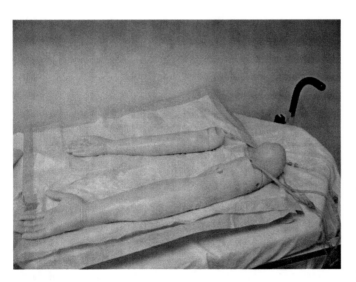

**FIGURE 41.2**   Pediatric and adult intravenous trainer arms.

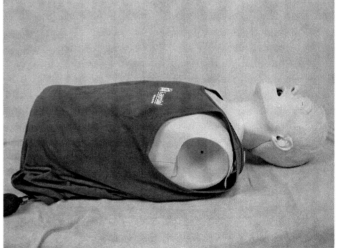

**FIGURE 41.3**   Airway trainer that facilitates learning mask ventilation and intubation skills.

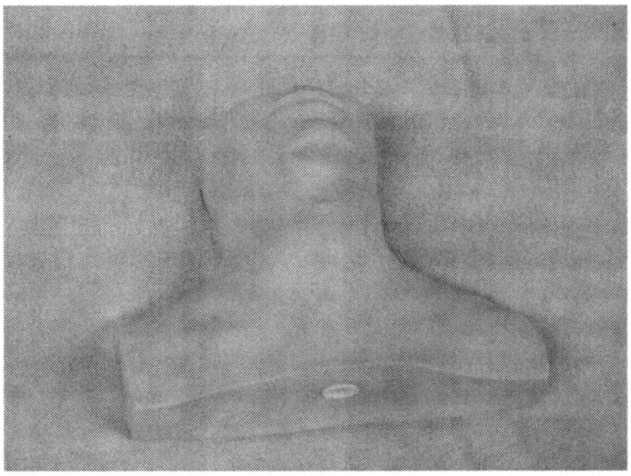

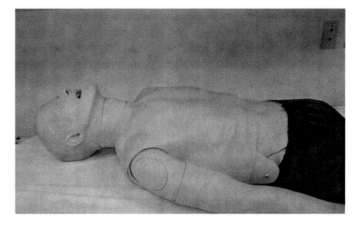

**FIGURE 41.4**  Cricothyroidotomy and tracheostomy trainer.

**FIGURE 41.6**  Airway trainer with anatomically correct bronchial tree for fiberoptic bronchoscopy.

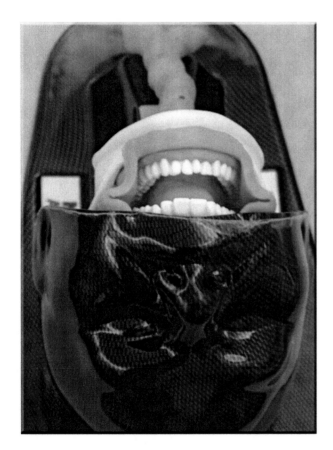

## 41.4 Pelvic Examination

The pelvic simulator is a valuable task trainer for introducing the process of performing a pelvic examination in an objective manner (Figure 41.7). The student learns not only to use the equipment (e.g., speculum), but also to correlate their findings on palpation with the underlying anatomical structures. The learner can also be introduced to both normal and abnormal

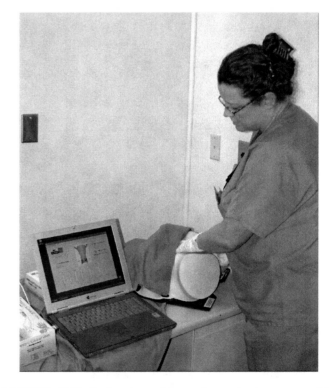

**FIGURE 41.5**  Airway trainer with movable lower jaw, allowing for better appreciation of crucial anatomy hidden in full-function patient simulator mannequins, and anatomically correct oropharyngeal airway.

trainers are better designed for use with intubation, placement of laryngeal mask airways, and fiberoptic intubations (Figures 41.5 and 41.6). Having several different airway trainers available increases the modalities of airway management that can be taught and practiced simultaneously.

**FIGURE 41.7**  Performing a bimanual examination on a pelvic simulator that provides real-time feedback on palpation location, force, order and duration.

physical findings (see Chapter 68). We are currently using a pelvic simulator to introduce first year medical students to the pelvic examination, and many are returning for a refresher course prior to beginning their third year clerkship in obstetrics and gynecology.

## 41.5 Lumbar Puncture

The spinal anesthetic (lumbar puncture) partial-task trainer will enable a variety of disciplines including emergency medicine, pediatrics, anesthesiology, internal medicine, neurology, and physical medicine and rehabilitation to learn and practice lumbar punctures (Figure 41.8). While it is not perfect, it does provide a start to the basic mechanics of how to set up for the procedure, landmarks to look for, and steps to take to perform a lumbar puncture.

## 41.6 Conclusion

Task trainers are very useful for novice learners and as well as advanced learners who want to maintain their skills. When looking to purchase partial-task trainers, look at the expensive equipment when it offers an additional function such as ultrasound capability, but do not be afraid to invest in the low-end equipment that will allow students to learn the basic manipulation steps of a procedure or a skill. Students can successfully master the skills needed while functioning in a simulated environment without expensive high-fidelity simulators. Finally, having more than one brand of a partial-task trainer provides different "patient" experiences for the students.

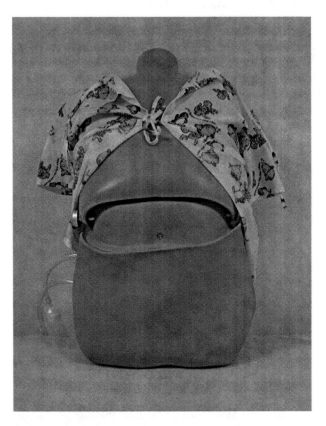

**FIGURE 41.8**   Lumbar puncture trainer.

# 42

# Implementing Partial-task Trainers in Simulation

Larry A. Cobb

## 42.1 Right Tools for the Job

In this chapter, we'll look at implementing partial-task trainers in clinical simulation exercises. Numerically, partial task trainers are the most commonly used type of clinical simulator. The reasons are obvious: lower cost, smaller size, and lower risk. Purchase and use costs not only place limits on the numbers of devices that can be acquired, but also limits on the numbers of institutions that can afford them. Simulators, of every kind, take up space during use, take up space when not in use, and the term "Storage" is all but outlawed in most clinical educational institutions. Since funds and space are the two most contested resources in clinical education institutions, and since simulation is still considered by many to be an "unproven" disruption, those who advocate diversion of scarce resources for simulation are risk takers. For these very same reasons, partial-task trainers are an excellent starting point for anyone new to simulation. For example, let us consider the use of an airway mannequin (often, a simple "Intubation Head" or a head and chest) for teaching, practicing, and evaluating airway management skills. Once these skills are mastered on the partial-task trainer, they can be transferred to more extensive care and teamwork scenarios with the full-body mannequins of high-fidelity patient simulators.

## 42.2 Plan for Success

Successful simulation programs start with planning. Poor or no planning results in excessive work, failed and nonrewarding teaching/learning experiences. Out of organization comes success, hence the importance of the logistics of simulation planning. The following skills laboratory example illustrates a planning and execution process with the key questions and example answers. First, it is essential to develop a brief list of detailed context and content (learning/skill objectives) to be distributed prior to/or in each station. One version of this list is a reminder for the instructor/proctors' emphasis during their teachings and another version is a guide for the learners' self-assessments of their progress. These lists define the size and scope of the exercise. The following questions (and example answers) illustrate typical list components:

- Who is the intended audience? (Paramedics and Nurses)
- How many learners/attendees are planned/scheduled? (15 learners)
- How many skills labs are you conducting? (three skills labs)
- Establish learner groups/lab (# learners ÷ # of skills labs: 15 ÷ 3 = 5 learners/group).

- How many hours have been allocated for this simulation session? (3 hours)
- Establish the duration of each lab session (3 hours ÷ 3 sessions = 1 hr/session).
- Thus, each group/team of five learners will rotate amongst the three lab sessions on an hourly basis (actually schedule each session for 55 minutes to allow 5 minutes for moving from skill station to skill station and bathroom visits).

In this example, we will use the following three different skills labs:

> Station #1: Basic-Life Saving airway skills: Bag-Valve-Mask, Nasal Pharyngeal Airway, Oral Pharyngeal Airway
> Station #2: Oral Endotracheal Intubation skills
> Station #3: Rescue Airway skills: Laryngeal Mask Airway, Multilumen Airway.

We list the three different skills labs, and combine them with answers from the above questions and create a who/what/where/when matrix (Table 42.1).

**TABLE 42.1**   A who/what/where/when matrix of skills, people, and places

| Time | Room #1 Basic Life-Saving Airway Skills, Proctor: | Room #2 Oral Endotracheal Skills, Proctor: | Room #3 Rescue Airway Skills, Proctor: |
|---|---|---|---|
| 13:00–13:55 | Red team | Blue team | Green team |
| 14:00–14:55 | Green team | Red team | Blue team |
| 15:00–15:55 | Blue team | Green team | Red team |

## 42.3 Successfully Execute the Plan

For each station, make a doorsign stating the skill station set up in that room. Ideally, these signs are printed on stiff paper and then laminated so they can be used repeatedly (out of organization comes repeatable success). Print both sides of the sign, one side could be "BLS Airway Skills – Practice Station" and the other could be "BLS Airway Skills – Testing Station." Use neutral designations for student group identifications like colors because numbers, and letters too, have ranking significance, and some participants and instructors may not want to be "ranked" as 2nd or 3rd place. Whereas, there is no ranking in color: blue team, red team, but do stay away from gold or silver or bronze colors. When configuring your rooms (for training, followed later by testing), where feasible, keep similar skills in the same room (simply turn your laminated doorsign over); this greatly reduces turnover/setup time, and reduces confusion for returning instructors and reduced loss and damage when moving simulators.

Make a "tools of the trade" list for each station listing mannequins, equipment/supplies, etc. Include the quantity/sizes needed, as you may have multiple setups at each station. You might even want to make a sketch of the room setup to assist you or others in the future. Remember to save the list in your computer for frequent updates and improvements (out of organization comes success). Today's planning will make the next course even smoother, for everyone.

The configuration of each skills station is dependent upon the learning objectives of the audience. For example, we can place the intubation head upon a table/cart or a gurney/bed for hospital-based skills. Likewise, we can place the intubation head on the floor, under a table, beside the bed, by the toilet, in the tub/shower, in a tunnel (a common fiberboard tube with length and diameter about that of an adult) for prehospital/EMS skills, or any combination thereof for an enriched experience. Obviously, simulating a realistic setting will enrich the experience; however, the experience level of the learner also dictates and/or influences the setting. Too "realistic" a setting may intimidate or confuse the inexperienced, "entry" level learner. Every distracter should be intentional and support a well-defined teaching objective. After the learner acquires the psychomotor skills, adding realistic scenario environment settings will challenge them to expand their competence.

Let us look at establishing a setting for the learner who will be introduced to Basic Life Support and Advanced Life Support Airway skills for the first time. The body parts and mannequins are placed on tables large enough to hold them and associated equipment required for the station. Don't use too large a table that the learner must stretch over and have to extend to reach the "tools of the trade." We use 18″ wide × 60″ long × 29″ high folding tables, arranged in a "T" configuration. Setting up two "T" configurations will facilitate one instructor assisting two students simultaneously. You might choose to "define" the space with towels, pads, or drapes. This provides an area upon which to place/arrange the equipment, laryngoscopes, endotracheal tubes, stylets, syringes, etc. This setting of boundaries also helps keep the individual stations inventories separate. When starting the individual lab session, define the goals, review the objectives, demonstrate the skill, then proceed to facilitate the learners' experience and then guide their self-assessments. To facilitate ease of setup of our skills labs, we've established equipment "pre-packs." For example, our Adult Airway teaching prepacks include:

- Plastic "shoe box" or "tote box" storage container labeled "Adult Airway"
- Laryngoscope handle
- Laryngoscope blades: #2, #3, #4 Macintosh, #2, #3, #4 Miller
- # 6.5, 7.0, 7.5, 8.0, & 8.5 endotracheal tubes
- Stylets for various sized endotrachel tubes

- 10cc syringes for inflating endotracheal tube cuffs
- #4 Laryngeal Mask Airway and 30cc syringe
- Home-made, resin-cast ETCO$_2$ capnometer prop
- Colormetric capnometer
- Tonsil suction tip
- French suction catheter
- Bulb and Syringe endotracheal tube placement confirmation devices.

## 42.4 Make What You can't Buy

In our airway management labs, we stress familiarity with choosing and using alternate airway adjuncts and tube confirmation as much as the actual intubation process. To facilitate these skills in our simulation settings, we have made or modified a number of tools to facilitate learning opportunities. When teaching or evaluating a skill and the learner does not perform the task as expected, the question arises: does a learner not perform the task due to lack of knowledge, or due to the lack of the tool and/or resource required to demonstrate existing knowledge? An example: real capnometers are fairly expensive, thus not always accessible for training purposes. In our Airway prepacks, we include a "home made" resin capnometer that connects between the endotracheal tube and the bag-valve-mask. This capnometer has a nonfunctional colormetric or numerical display. Upon successful intubation, the learner is expected to attach the inline capnograph, and assess the numerical reading. With successive ventilations (based upon

FIGURE 42.1b "Home-made" resin glucometer and pulse oximeter capnometer from castings of real devices.

scenario ventilation and perfusion parameters), the capnometry is expected to be continually reassessed. We've also made pulse oximetry and glucometers, wall-mounted gas flow meters etc., in this manner, as shown in Figure 42.1a and 42.1b. This inexpensive way to make "training perfect" replicas of expensive "real" devices assures that when our students' have their first hands-on experience with functional devices, their eyes and hands will already be very familiar with what to expect. As real clinical devices are upgraded, our teaching program can easily and inexpensively stay current. Given the proper training tools, the students can demonstrate to their instructors that they are performing as desired, as well as gaining familiarity with the act. Lowering the cost to provide clinical items increases the frequency of their use, and familiarity through use of these training aids has greatly increased the learners' compliance in assessing these clinical parameters.

## 42.5 Adapt Tools to Your Teachings

The Laryngeal Mask Airway is one of the broadly accepted Advanced Life Support Rescue Airways (in addition to primary Basic Life Support/Intermediate Life Support Air Way adjunct). Some learners have difficulty conceptualizing the proper placement of the Laryngeal Mask Airway. When one presents the concept of a "rescue" airway device, the proverbial "picture is worth a thousand words." To facilitate their understanding, we've modified three Airway Structure models, as shown in Figure 42.2. We cut a "window" on the one side of the airway at the level of the glottis and larynx, demonstrating the proper placement of an airway. In one of the models, a standard endotracheal tube is placed, in another a Laryngeal Mask Airway, and the third contains a Multilumen Airway.

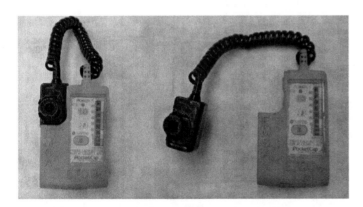

FIGURE 42.1a "Home-made" resin capnometer from casting of real device. Actual "real" capnometers sell for hundreds of dollars, are all too easily "misplaced," and don't take well to being dropped. We found that we can mold a master (using one real capnometer) in Room Temperature Vulcanizing Rubber for approximately $11.00. Then, from this mold, we cast multiple Synthetic Resin duplicates for as little as $5.00–$6.00 each.

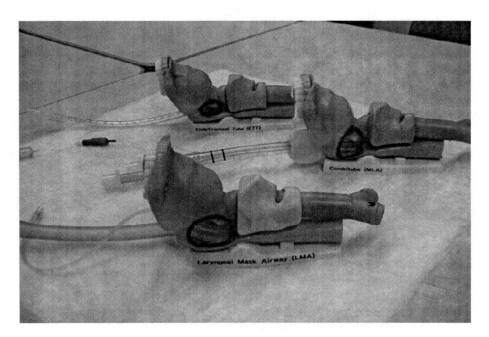

**FIGURE 42.2**  Three airway structure models modified to allow easy visualization of the correct placement of three different ventilation access devices. The partial-body nature of these simple devices accentuate the essential anatomic features pertinent to airway insertion. The nonphysiological modifications made to them improve their teaching utility.

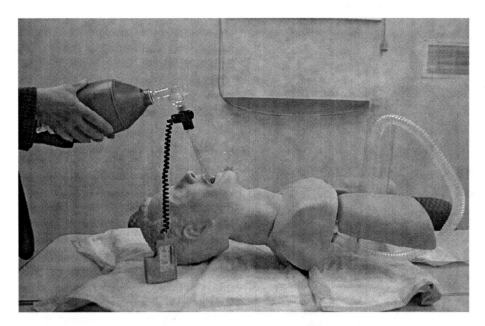

**FIGURE 42.3**  Demonstration of airway insertion placement with an airway task trainer modified to indicate undesired esophageal ventilation. The esophagus was replaced with common ventilation circuit tubing and the stomach replaced with a column of water. The depth of the water above the open end of the tube creates the equivalent gastric opening pressure as well as an audible and visual indicator of undesirable stomach ventilation.

## 42.6 Make Tangible the Ethereal

Finally, despite illustrating their placement, some learners "can't believe something that simple can be effective!" Again a picture, or better yet, a demonstration really is an eye opener. Using one of our airway training heads, we've substituted common respiratory circuit tubing for the esophagus and stomach, as shown in Figure 42.3. This tube extends into the bottom of a transparent cylinder containing water, and the depth of water equates exactly to esophageal opening pressure. We use these modified mannequins throughout bag-valve-mask training, endotracheal intubation, and Laryngeal Mask Airway insertion. With the audience "doubting" the efficacy of the simplistic Laryngeal Mask Airway, upon insertion by a novice, we exaggeratedly manipulate the airway while hand-ventilating the Laryngeal Mask Airway. No bubbling in the column indicates both correct insertion placement and proper amount of bag compression. The proof is in the direct experience while doing.

## 42.7 Conclusion

The organization, planning, and imaginative application of partial-task mannequins in your simulation program can provide your learners with outstanding educational opportunities. Given their low cost, small size, and low risk, partial-task trainers are an excellent route to start, as well as to expand any clinical simulation program. As described, they lend themselves to easy modification to improve your teaching needs. When real, fully functional clinical devices are beyond your budget, you can easily and inexpensively make exact replicas suitable for initial hands-on familiarity. Thus, when your students do have their first experience with fully functional devices, it won't be their first-time handling them. This allows them faster progression to overall competency, the goal of any formal educational program.

# 43

# The Role of Patient Simulators in Pediatric Education

Edmundo P. Cortez

## 43.1 History of Patient Simulators in Clinical Education

In the 1950s, Dr Peter Safar, whom many consider one of the pioneers of modern emergency medicine, developed a technique for mouth-to-mouth resuscitation in combination with chest compressions that could be utilized by emergency personnel as well as by the lay person. He had hoped that this technique would be easily taught if a life-sized mannequin was utilized [1]. Dollmaker Asmund Laerdal took up Dr Safar's challenge and hence "Resusci Anne"™ (Laerdal Medical) was born in 1960. Since then, various "Resusci Anne" (also commonly referred as "Resuscitation Annie" or "CPR Annie") have been used for Basic Life Support (BLS), Advanced Cardiac Life Support (ACLS), and Pediatric Advanced Life Support (PALS), courses in the teaching and evaluation of trainees in the techniques for appropriate resuscitation.

As many clinical personnel can recall from their own experience with "Resusci Anne," use of this life-size mannequin does not easily or always translate to a "real-life" resuscitation experience that one may encounter. Limitations of mannequins like "Resusci Anne" include lack of "real-life" physiologic responses for the trainee and dependence of real-time verbal feedback from the instructor. Simulation can be referred as "techniques in the artificial replication of sufficient elements of real-world domains to achieve a stated goal (e.g., training and testing the work capacity of individuals or teams)" [2]. More recently, patient safety has come to the forefront. As technology has developed and educational models to train clinical personnel have evolved, application of mannequin-based simulators since the 1960s has become the standard.

The pioneering use of mannequin-based patient simulators in anesthesia resident training provided a safe environment for demonstration of adverse operating room events, improved scores of faculty reviews of trainee skills, identified judgment errors, and evaluated decision-making skills of trainees [3, 4]. This is due to the ever-changing dynamics of patient physiology in the operating room, the high likelihood of patient morbidity and mortality due to human error, and the team cohesion (physicians, nurses, technicians) required for appropriate patient care.

Through the experience of anesthesiology with patient simulators, the concept of Crisis Resource Management (CRM) was developed in order to refine the team/group management concept that is necessary in the coordinated care of an acute patient emergency [5]. CRM principles have applied to other fields of healthcare including, but not limited to, critical care, emergency and trauma care, neonatal resuscitation, and emergency response services. Integration of CRM and patient simulators is becoming common in pediatric personnel training and the care of pediatric patients.

## 43.2 Clinical Skills Needed to Provide Care for the Pediatric Patient – Are They Adequate?

The field of pediatric care encompasses a wide range of patient age groups – from the premature newborn infant to the adult with a chronic congenital disease – and a wide scope of pathophysiologic processes – from the healthy well-child to a child in extremis. Pediatric residency training programs vary from each other in the population demographics they serve and the census of patients and disease processes encountered. This has posed a problem for medical educators in pediatrics. How does a pediatric trainee acquire experience, if during their training it is not possible to encounter all types of patients and diseases? This is especially true in patient crises that a pediatric trainee should, but may not encounter.

Pediatric cardiopulmonary (CPR) or trauma resuscitations are quite rare, whether at a small community-based hospital or at a major tertiary care pediatric teaching institution. A review of over 80,000 admissions to a large pediatric emergency department showed that patients requiring acute resuscitation are rare, 0.23% [6]. Therefore, a resident trainee's experience in acute resuscitation outside of their critical care rotations is usually very limited.

PALS and Neonatal Resuscitation Program (NRP) have done a very adequate job at formally preparing the pediatric provider in the acute management of a patient crisis. Informal training occurs quite frequently during pediatric residency training through the use of bedside teaching and mock arrest scenarios. PALS and NRP certification can give the provider confidence in their skills and abilities prior to a pediatric cardiopulmonary arrest event [7].

However, does this confidence correlate with good performance results when it comes time to perform in a mock or actual pediatric cardiopulmonary arrest? The answer would seem to be no. Pediatric residents have shown difficulty in performing even basic, let alone advanced, airway techniques in mock codes [7, 8]. Delay in identification of the need and administration of oxygen has also been shown in mock code scenarios [9]. Deficiencies in advance life support skills were not only limited to airway skills, but also included cardiac compression, vascular access, and electro-cardioversion skills. Given their very limited reserves, early recognition of respiratory versus cardiac arrest and appropriate provision of therapy markedly improves survival in pediatric patients [10].

The adult population represents the vast majority of patients who present for care in an emergency department, outside of a specialized children's center. Practitioners in such settings who deal with both adult and pediatric patients rely on their residency training experience when treating pediatric patients. Although nonpediatric clinicians may have been formally trained in advanced skills during PALS or NRP, survival and patient outcome relate to the ability to apply forgettable skills, such as endotracheal intubation [11]. Also, it is not uncommon for pediatric patients first be seen at outside referral institutions and are then transported to a second, pediatric-specific institution for appropriate care and personnel for further stabilization [12]. Thus, pediatric needs, whether personnel or equipment, might not be met by first care providers.

## 43.3 Application of Patient Simulators in Pediatric Training

Objective measures of competency and frequent exposure of pediatric care givers–in-training to the uncommon cardiopulmonary arrest are becoming the standard. Patient simulator use in anesthesia and emergency medicine training has been well implemented and evaluated. The role patient simulators are taking in teaching and evaluating skills in pediatrics are becoming more common. The concept of simulator-based CRM probably applies even more so in the field of pediatrics given the fact that pediatric resuscitations are rare, team members are likely unfamiliar with each other, and poor management of a crisis can yield catastrophic patient results.

In this new millennium, public and governmental bodies are scrutinizing how graduate medical education has an impact upon patient care and errors. Various medical bodies such as the Accreditation Council for Graduate Medical Education (ACGME) and the American Board of Medical Specialties (ABMS) have taken the lead by shifting medical education to a learner-centered and competency-based system driven by desired training outcomes [13].

Since 2006, the ACGME has required residency program graduates to have demonstrated competency in six areas: patient care, medical knowledge, practice-based learning and improvement, interpersonal and communication skills, professionalism, and systems-based practice [13]. Evaluation of areas of competency now calls for direct observation and assessment of trainees performing tasks of "real-world" practice [14]. Patient simulators have gained credibility as a teaching and evaluation tool of pediatric resident competencies in "real-world" scenarios [15, 16].

The American Board of Pediatrics (ABP) and the ACGME lists a number of patient care issues or disease processes that a resident must encounter during their training in order to be eligible for board certification [17]. Because experience and patient volume differs between training programs, it can be difficult for a resident to encounter patients with certain disease processes. At our institution, even though it is a major tertiary pediatric care center, trauma resuscitation is very infrequent. Our training program's monthly mock resuscitation codes for residents have simulated pediatric trauma victims. This mock session fulfills resident requirements for patient encounter and training in patient care.

The American Academy of Pediatrics (AAP), Committee on Pediatric Emergency Medicine, and the American College of Emergency Physicians (ACEP) have provided guidelines for care of children in the emergency department (ED) [18]. Physician coordinators of the ED must ensure adequate skill and knowledge of ED physicians. They must also facilitate pediatric emergency education for ED care providers and out-of-hospital providers associated with the ED. In turn, the ED physicians must have the necessary skill, knowledge, and training to provide emergency care in all pediatric age ranges. The Pediatric Chemical Agent Simulation Workshop that we have held in our institution to train adult emergency department care providers throughout the Chicago region has resulted in significant positive feedback on the training through the use of the pediatric patient simulator[1]. Many community-based emergency departments who had participated have requested further experience and training with the patient simulators for other ED personnel.

There is an increased use of standardized patients in the formal evaluation of clinical skills during medical school and residency training. A standardized patient is a live person who has, or a trained person who simulates signs and symptoms of a particular disease process. From a pediatric standpoint, standardized patients as young as 5 years old, if properly trained, can be utilized. Currently, the United States Licensing Examination (USMLE) Part II and the Federal Licensing Examination (FLEX) for foreign medical school graduates utilize standardized patients as part of their examinations. Objective Structured Clinical Examinations (OSCE) using standardized patients have been shown to identify deficiencies in resident skills that otherwise could not be identified by other evaluation tools (i.e., in-training examinations) [19]. Given the fact that patient simulators can simulate numerous physiologic, pharmacologic, and anatomical processes like that of a real patient, consideration for the use of patient simulators in such high-stakes examinations is likely.

## 43.4 Conclusion

From 1994 to 2005, over 356 simulation centers were established just in the United States [20]. At the end of 2005, the top two vendors of robotic patient simulators had combined cumulative, worldwide sales of over 4000 devices. Various institutions from medical schools to emergency medical technician training centers have made purchases of simulators for enhancement of their training programs. Also during this period of time, objective measurements of clinical competency in medical training are being required. OSCEs and standardized patients are commonly being used. Patient simulators are becoming easily available and accessible to train care givers. Infant simulators can bridge the gap of patients when the use of standardized patients is not feasible. In our institution's pediatric residency program, we have used the pediatric patient simulators to augment the training and experience that our residents have received (see Chapter 45). Future plans for our program include collecting data regarding the usefulness of patient simulators in identifying deficient clinical skills and teaching the importance of CRM during pediatric codes.

## Endnote

1. *Preparing Hospitals For The Littlest Victims Of Terrorism* http://www.medicalnewstoday.com/medicalnews.php?newsid=37414&nfid=crss

## References

1. Safar, P. and McMahon, M. Mouth-to-airway emergency artificial respiration. *Journal of the American Medical. Association.* **166**(12), 1459–1460 (1958).
2. Gaba, D. M. A brief history of mannequin-based simulation and application, Dunn, W. F., (ed.), *Simulators in Critical Care and Beyond.* Society of Critical Care Medicine Des Plaines, IL., 2004, pp. 7–14.
3. Abrahamson, Denson, J. S., Wolf, R. M. Effectiveness of a simulator in training anesthesiology residents. *Journal of Medical Education* 3, 80 (1969).
4. DeAnda, A. and Gaba, D. Unplanned incidents during comprehensive anesthesia simulation. *Anesthesia and Analgesia.* **71**, 77–82 (1990).
5. Murray, W. B. and Foster, P. A. Crisis resource management among strangers: principles of organizing a multidisciplinary group for crisis resource management. *Journal of Clinical Anesthesia.* **12**, 633–638 (2000).
6. Schoenfeld, P. S. and Baker, M. D. Management of cardiopulmonary and trauma resuscitation in the pediatric emergency department. *Pediatrics* **4**, 726–729 (1993).
7. Nadel, F. M., Lavelle, J. M., Fein, J. A., et al. Assessing pediatric senior residents' training in resuscitation: Fund of knowledge, technical skills, and perception of confidence. *Pediatric Emergency Care* **16**(2), 73–76 (2000).
8. White, J. R., Shugerman, R., Brownlee, C., and Quan, L. Performance of advanced resuscitation skills by pediatric housestaff. *Archives in Pediatric and Adolescent Medicine.* **152**, 1232–1235 (1998).
9. Palmisano, J. M., Akingbola, O. A., Moler, F. W., and Custer, J. R. Simulated pediatric cardiopuomonary resuscitation: Initial events and response times of a hospital arrest team. *Respiratory Care* **39**(7), 725–729 (1994).
10. American Heart Association. *Pediatric Advanced Life Support Provider Manual* (2002).
11. Gausche, M., Lewis, R. J., and Stratton, S. J. Effects of out-of-hospital pediatric endotracheal intubation on survival and neurologic outcome: A controlled clinical trial. *Journal of the American Medical Association* **283**, 783–790 (2000).
12. Seidel, J. S., Hornbein, M., and Yoshiyama, K., et al. Emergency medical services and the pediatric patient: Are needs being met? *Pediatrics* **73**, 769–772 (1984).
13. Accreditation Council for Graduate Medical Education. *Outcomes Project 2001.* Available online at http://www.acgme.org/Outcome/.

14. Sectish, T. C., Zalneraitis, E. L., Carraccio, C., and Behrman, R. E. The state of pediatrics residency training: A period of transformation of graduate medical education. *Pediatrics* **114,** 832–841 (2004).

15. Halamek, L. P., Kaegi, D. M., Gaba, D. M., et al. Time for a new paradigm in pediatric medical education: Teaching neonatal resuscitation in a simulated delivery room environment. *Pediatrics* **106,** 45 (2000).

16. Weinstock, P. H., Kappus, L. J., Kleinman, M. E., et al. Toward a new paradigm in hospital-based pediatric education: The development of an onsite simulator program. *Pediatric Critical Care Medicine.* **6**(6), 635–641 (2005).

17. *Preparing Hospitals For The Littlest Victims Of Terrorism.* Available at http://www.acgme.org/acWebsite/downloads/RRC_prog Req/320pediatrics07012007.pdf

18. American Academy of Pediatrics: Care of Children in the Emergency Department: Guidelines for preparedness. American Academy of Pediatrics, Committee on Pediartric Emergency Medicine and American College of Emergency Physicians, Pediatric Committee. *PEDIATRICS* **107**(4), 777–781 (April 2001).

19. Joorabchi, B. Objective structured clinical examination in a pediatric residency program. *American Journal of Diseases in Children* **145,** 757–762 (1991).

20. McIndoe, A. and Jones, A. Simulation centers established world wide 1994–2005. Available online at http://www.bris.ac.UK/Depts/BMSC.

# 44

# Simulation Training for Pediatric Emergencies

Chris Chin

## 44.1 Big Challenges Learning to Care for Small Patients

The first National Confidential Enquiry into Perioperative Death in the United Kingdom in 1989 recommended that surgeons and anesthetists should not undertake pediatric cases if such patients are only occasionally part of their practice and those who do focus their care on children should keep up to date and be competent in the management of children [1]. A follow-up report in 1999 has shown a consequent decrease from 84% to 42% in the proportion of anesthetists who anesthetize infants of less than 6 months [2]. Another report in 1997 (Paediatric Intensive Care – A Framework for the Future) from the National Coordinating Group on Paediatric Intensive Care to the Chief Executive of the United Kingdom's National Health Service Executive has led to similar changes, with increased centralization of pediatric intensive care work in specialist centers [3]. In addition, changing to a 7-year anesthesia training program (in 1999) from a previous average of 8 years and the introduction of the European Working Time Directive reducing the working week to 56 hours (with a further reduction to 48 hours by 2007) has resulted in less exposure of anesthesia trainees to the management of pediatric emergencies [4]. It is likely that these reforms may have led to some "deskilling" of medical and nursing staff in managing the small child and infant.

Guidelines from the Royal College of Anaesthetists for pediatric training have also listed some professional qualities that should be developed, including the ability to communicate clearly, the ability to work as part of a team, priority setting, and the ability to lead others and delegate appropriately [5]. Simulation training provides an ideal opportunity to teach and reinforce these crisis resource management qualities, as well as refresh management protocols and technical skills.

The recent introduction into the simulation market of a 9-month old toddler simulator has provided an additional tool for training in the management of Pediatric Emergencies (Figure 44.1).

At Barts and the London Medical Simulator Center, I have developed a 1-day course, focused on emergency room and

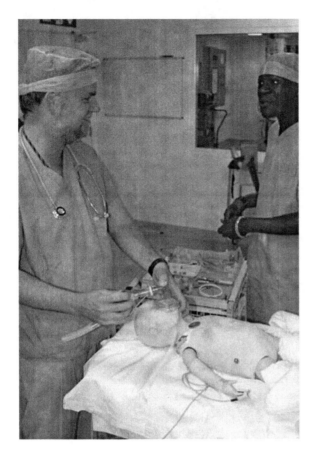

**FIGURE 44.1**   Nine-month-old toddler mannequin.

anesthetic pediatric emergencies, and aimed specifically at non-pediatric specialist medical staff and trainees who would be expected to manage the initial care of a critically ill toddler brought into their nonspecialist hospital. Scenarios selected are representative of actual cases suggested by anesthesia trainees, emergency room consultants, and pediatric anesthetists.

## 44.2 Simulation Course Goals

- To challenge the candidates with realistic pediatric emergency scenarios without risk to patients.
- To refresh existing, and develop new, pediatric management skills (Appendix 44A.1).
- To familiarize candidates with simple and effective management plans for the critically ill child.
- To develop crisis resource management skills and improve appreciation of teamwork's contributions to improving patient care.

### 44.2.1 Course Location and Staffing

Our Simulation Center consists of a simulator room with a separate control room. We also have a separate observation/debriefing room. As many of the scenarios involve the management of the critically ill child presenting to hospital for the first time, we have opted to simulate an Emergency Room environment for many of the scenarios (Figure 44.2).

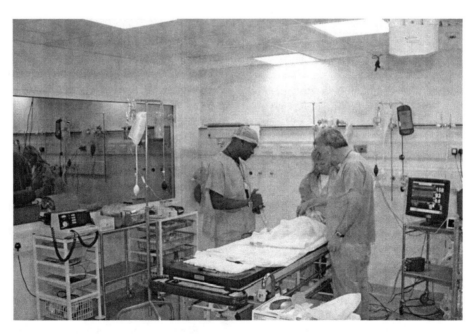

**FIGURE 44.2**   Simulated Emergency Room with pediatric patient.

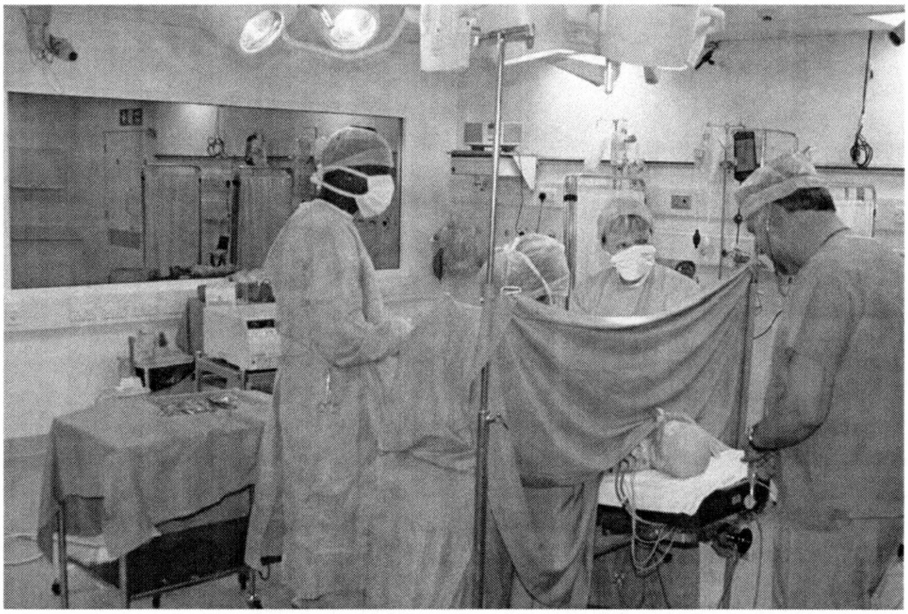

**FIGURE 44.3**  Simulated Operating Room with pediatric patient. Note that this is the same simulation stage in Fig. 2, but configured for a different clinical teaching environment.

We also include some scenarios that involve anesthetizing a child, and such scenarios take place in an operating room (Figure 44.3).

As we only have one simulation room, logistical planning requires that Emergency Room scenarios take place in sequence before undertaking the Operating Room scenarios, as this minimizes the movement of equipment required for creating the two "different" environments.

### 44.2.2  Course Design

The day begins with a short introduction talk as to the course goals and the timetable for the day (Table 44.1, Appendix 44A.2 – Course timetable). During the introduction, we emphasize that this is a training day and no assessment is undertaken. Introduction is followed by a familiarization session during which the candidates are introduced into the simulator room and allowed to handle the baby mannequin

**TABLE 44.1**  Planned sequence of events

- Introduction
- Familiarization session
- Simulated scenarios
- Debrief: discussion of management guidelines and nontechnical skills (following each scenario)
- Final wrap-up

and all the available equipment. The familiarization session is supervised by any one of the course instructors, and in addition to explaining features of the mannequin, we also allow all candidates to intubate the mannequin. Familiarization generally takes between 30 and 45 minutes, and is complete once the candidates have confirmed that they are comfortable with the environment and equipment. We have not found it necessary to *exactly* duplicate the equipment and environment of the trainees' own hospitals – they do learn very adequately with different equipment, if the familiarization period is sufficient. Feedback from trainees has not suggested that this has any effect on the educational experience.

### 44.2.3  Scenarios

We allow only a short delay following familiarization before starting the first Emergency Room scenario (5–10 minutes). Scenario configuration is undertaken by any clinical staff not involved in debriefing and, apart from the first scenario, all scenario preparations (in the simulator room) are undertaken whilst the debriefing sessions are carried out (in our separate observation/debriefing room). As feedback from other simulator courses has shown that candidates like more time spent with the simulator, we allow two candidates to be involved in managing the scenario. All other candidates observe from the observation/debriefing through a large one-way mirrored window. Audio-visual linkage allows display of the patient monitoring and reproduction of all sounds for the observers.

**TABLE 44.2**   Scenarios presented

- Septicemia
- Preparation for retrieval
- Status epilepticus
- Cardiac arrest
- Anaphylaxis
- Burns
- Airway obstruction
- Trauma
- Laryngospasm

Our Pediatric Scenarios are listed in Table 44.2 (see Appendices 44A.3 and 44A.4 for examples of scenario recipes).

Whilst we will take a maximum of nine candidates on a course, if we have fewer than nine candidates, we are usually able to cover all topics. For example, the "septicemia" scenario can be merged with the "preparing for retrieval" scenario. Likewise, a "cardiac arrest" scenario can include all cardiac arrest rhythms (e.g., asystole converts to ventricular fibrillation before converting to sinus rhythm) or the burns scenario can also include an airway burn/inhalational injury (to then incorporate management of the airway obstruction).

Each scenario is followed by a debriefing session during which management guidelines are presented and discussed with the group. This usually takes the format of allowing the candidate to describe the scenario and the management stratagems that they used. The discussion is then opened to the group before a final summary accompanied by a 2–3 minute slide presentation of a recommended management algorithm is made. The recommended management algorithms are regularly reviewed and updated with multidisciplinary input from all trainers and using a literature search in an attempt to produce a "best practice" guideline.

Digital video recording takes place during scenarios and playback of segments may be used during the debriefings that take place after each scenario. Signed consent is obtained from candidates at the start of every course day for these recordings to be made for the purposes of debriefing (and research) only. These recordings are fully "off the record," no copies are made, and they are not accessible outside of the center. Generally, if no research is being carried out, we delete all recordings from the hard drive at the end of each week.

Approximately 20 minutes is allowed for debriefing. As described in the introduction talk, the majority of time is allocated to discussion of technical skills and management plans. However, where appropriate and if time allows, some aspects of nontechnical skills such as team working, leadership, communication skills, and anticipation and preparation can also be discussed.

### 44.2.4   Faculty

Our permanent paid clinical staff includes a Center Manager, two or three Simulator Fellows (usually anesthesia trainees

on a 6–12 month secondment) and three part-time Operating Department Practitioners (anesthetic trained technical assistants). In addition, we recruit a mix of unpaid volunteer Pediatric Anesthetists, Pediatric Emergency Room Physicians, and Pediatricians to help produce the course. This is invaluable in both improving the realism during the scenarios and providing a multidisciplinary approach to the teaching.

### 44.2.5   Candidates

We believe that this course is suitable not only for anesthesia staff and trainees, but also for emergency room physicians, surgeons, and pediatric trainees. In short, anyone who may be required to occasionally provide initial care for the management of the critically ill child. We strongly believe that there is increased value when each person on the course has a role in the management of at least one of the scenarios, and thus will limit the maximum number to nine candidates per course. The cost of the course is based on the daily cost of the simulation center but does not cover faculty time. Candidates generally pay for themselves, although I have been able to negotiate funding from our own hospital postgraduate center to pay for trainees from the hospital. We currently advertise the course directly to hospitals in the region (flyers sent direct to the departments of anesthesia and word-of-mouth) but will soon be advertising the course nationally (in the advertising section of medical and anesthesia journals).

### 44.2.6   Course Impact

Since it's inception, we have produced the course every 3 months. As with all our courses, outcome assessment on the impact of this course is undertaken by asking the candidates to fill in simple pre- and postcourse questionnaires (see Appendix 44A.5 for our current questionnaires). A trauma scenario, not originally included in the course, was added as a result of suggestions made on the questionnaire. The initial results of these questionnaires (presented as a poster at the Annual Meeting of the Society of Simulators Applied to Medicine in June 2005) have suggested that this course improves confidence in the management of the critically ill child for the events covered during the course (Table 44.3).

## 44.3   Conclusion

The Pediatric Emergencies Anesthesia Training in the Simulator course has been well received. Initial feedback suggests that confidence in the management of the pediatric emergencies covered by the course is improved. The course can easily include any specialty where management of pediatric emergencies is required and it is also possible to use this course for multidisciplinary training. Evaluation of this course is ongoing, and the questionnaire is regularly reviewed and modified as necessary.

**TABLE 44.3** Some initial results from responses to the pre- and postcourse questionnaires

| | Precourse average | Postcourse average |
|---|---|---|
| **How confident are you in your own ability to manage the following:** (1 = not at all confident, 10 = very confident) | | |
| Pediatric cardiac arrest | 6.0 | 8.7 |
| Critically ill child | 5.8 | 7.9 |
| Difficult pediatric airway | 5.5 | 8.3 |
| Pediatric burns | 4.8 | 8.2 |
| Fitting child | 5.8 | 8.4 |
| Anaphylaxis | 6.3 | 8.8 |
| Will today's course assist you in your daily practice? | | 8.5 |

# References

1. The report of the National Confidential Enquiry into Perioperative Deaths (1987). NCEPOD, London 1989. (http://www.ncepod.org.uk/sum89.htm)
2. Extremes of Age. The Report of the National Confidential Enquiry intoPerioperative Deaths (1997–98). NCEPOD, London 1999. (http://www.ncepod.org.uk/1999.htm)
3. National Coordinating Group on Paediatric Intensive Care. *Paediatric intensive care: a framework for the future.* London: Department of Health, 1997.
4. Directive 2000/34/EC of the European Parliament and Council. *Official Journal of the European Community* L195, 41–5 (2000).
5. CCST in Anaesthesia I: General Principles. A manual for trainees and trainers (Second Edition April 2003).

# Information Resources

The following web sites have been useful in formulating up-to-date algorithms for pediatric emergencies:

1. http://www.erc.edu

   Web site for the European Resuscitation Council. Contains the latest European guidelines on Resuscitation, as well as a full overview of the ERC educational tools such as manuals, posters, and slides.

2. http://www.health.state.ny.us/nysdoh/ems/nysemsc.htm

   Web site for the New York State Department of Medicine Emergency Medical Services for Children. Contains guidance on pediatric assessment and treatment, and provides medical algorithms and charts for BLS and ALS providers.

3. http://www.cps.ca/english/publications/EmergPaediatrics. htm

   This is the web site for the position papers on emergency pediatrics produced by the Canadian Paediatric Society (CPS).

4. http://www.bestbets.org

   Best Evidence Topics web site developed in the Emergency Department of Manchester Royal Infirmary, UK, to provide rapid evidence-based answers to real-life clinical questions. Whilst not dedicated to pediatrics, it does contain some useful information on a range of problems, and the growing database of topics is easy to browse.

5. http://www.resus.org.uk/SiteIndx.htm

   Web site providing education and reference materials on effective methods of resuscitation. Includes adult and pediatric material.

# 45

# Considerations of Pediatric Simulation

Paul N. Severin
Edmundo P. Cortez
Christopher A. McNeal
Jane E. Kramer

## 45.1 Differences of Approach to Pediatric Simulation

As we all agree, "Children are not little adults [1]." In the clinical arena, this becomes quite apparent when a pediatric patient arrives in acute crisis. An entire emergency room or even hospital may be hurled into crisis mode if they are not prepared for an acutely ill pediatric patient. Along the same lines, even if the adolescent patient is close to adult size and age, there is still great urgency to transfer care immediately to a pediatric specialist. In either case, the responses of many practitioners are related to unfamiliarity of pediatric developmental differences versus adults. On the basis of current theory and technology, simulation offers an avenue to educate individuals [2–7], especially regarding differences in pediatric anatomy and physiology, behavioral development, psychological aspects, and therapeutic management. Although the broad review of pediatric differences is beyond the scope of this chapter, some pertinent aspects merit discussion.

Respiratory differences in the pediatric patient are essential to stabilization of the acutely ill child (Scenario 1, in Appendix 45A.2). The most common etiology for cardiorespiratory arrest is respiratory pathology. Most of the airway resistance in children occurs in the upper airway. Nasal obstruction due to congenital (choanal atresia) or acquired (adenoidal hypertrophy) lesions can lead to severe respiratory embarrassment due to infants being obligate nose breathers. Their relatively large tongue and small mouth can lead to airway obstruction quickly, especially when neuromuscular tone is abnormal such as during sedation or encephalopathy. In infants, physiologic (i.e., copious secretions) and pathologic (i.e., vomitus, blood, foreign body) factors will exaggerate this obstruction. Securing the airway in such events can be quite challenging. Typically, the glottis is located more anterior and cephalad. Appropriate visualization during laryngoscopy can be further hampered by the prominent occiput, which causes neck flexion and, therefore, reduced alignment of visual axes. The omega-shaped epiglottis in young infants and children is quite susceptible to inflammation and swelling. As in epiglottitis, the glottis becomes strangulated in a circumferential manner, leading to dangerous supra-glottic obstruction. Children also have a natural tendency to laryngospasm and bronchospasm. Finally, due to weaker cartilage in infants, dynamic airway collapse can occur especially in states of increased resistance and high expiratory flow (e.g., bronchiolitis or asthma). Along with altered pulmonary compensation and compliance, a child may rapidly progress to respiratory failure and possibly arrest [8, 9].

Cardiovascular differences are critical in children (Scenarios 2, 3, and 4, in Appendix 45A.2). Typical physiological responses tend to allow compensation with a seemingly normal

**TABLE 45.1** Pediatric modified Glasgow coma scale

Eye opening
    Spontaneous  4
    To speech  3
    To pain  2
    None  1

Best verbal response
    Oriented (coos, babbles)  5
    Confused (irritable)  4
    Inappropriate words (cries to pain)  3
    Nonspecific sounds (moans to pain)  2
    None  1

Best motor response
Obeys (normal spontaneous movements)  6
Localizes (withdraws to touch)  5
Withdraws to pain  4
Abnormal flexion  3
Abnormal extension  2
None  1

homeostasis [10]. With tachycardia and elevated systemic vascular resistance, children can maintain a normal blood pressure despite poor cardiac output and perfusion (compensated shock). Since children have less blood and volume reserve, they progress to this state rather quickly. The unaccustomed clinician may be lulled into complacency since the blood pressure is "normal." All the while, the child's organs are being poorly perfused. Once these compensatory mechanisms are exhausted, the child rapidly progresses to hypotension and therefore, decompensated shock. If not reversed expeditiously, this may lead to irreversible shock (ischemia), multiorgan dysfunction, and death of the child [11].

Alterations in neurological status should be carefully evaluated (Scenario 5, in Appendix 45A.2). The modified Pediatric Glasgow Coma Scale (Table 45.1) is useful in this regard. Using the AVPU scale (**A**lert/Responsive to **V**erbal Commands/Responsive to **P**ain/**U**nresponsive) can also be helpful for ease of mental status determination [11]. It is important to remember that hypothermia is a mortal enemy of children. Owing to a larger body surface area to mass ratio, hypothermia can occur rapidly despite a temperate environment [12]. Apnea, bradycardia, coagulation abnormalities, and a host of other derangements often are associated with hypothermia in the acutely ill or injured child. Appropriate thermoregulation should be addressed with the remainder of initial assessment and resuscitation. Table 45.2 summarizes how we use infant and child simulators to highlight anatomical and physiological developmental differences seen in pediatric patients [13].

One of the biggest challenges in pediatrics is the lack of the patient's ability to verbally convey complaints. We are often faced with a caregiver's subjective assessment of the problem. Although it can be revealing and informative, this may not be available in an acute crisis situation. It will take the astute clinician to determine, for example, if an inconsolably crying

infant is in pain from a corneal abrasion or something more life threatening such as meningitis. Often, children will reflect the emotional state of their caregivers, parent, or otherwise. The psychological impact of an event will vary greatly with the child's development and experience. Children tend to have a greater vulnerability to Post-Traumatic Stress Disorder especially with disaster events [12, 14, 15]. And as any pediatric clinician can tell you, children tend to also have greater levels of anxiety especially while undergoing procedures [12]. We integrate these factors into each scenario. For example, our infant simulations begin with normal babbling and cooing and then can be rapidly changed to agitated crying and screeching, stridor, coughing, or agonal breathing. As our participants have learned, progression to absence of verbalization is a very ominous sign.

"First, do no harm" comes to mind when formulating a therapeutic plan for every patient, especially a child. Relative to population distribution, there are fewer specialized care areas for children than adults. Many adult practitioners rely on their introductory medical school or adult-focused residency training. Also, pediatric expertise and resuscitation education like Pediatric Advanced Life Support (PALS) varies amongst institutions. The predominant solution is to expeditiously transfer the child to a tertiary pediatric institution. Even though many hospitals deal with children in the emergency department, pediatric hospital "preparedness" cries out for simulation [16]. The lack of transfer protocols, unfamiliarity of pediatric medication dosages, and inappropriate use of pediatric equipment reduce access to effective and efficient care of the pediatric patient [12]. Multiple areas in the country are working to improve their pediatric emergency treatment by pediatric faculty and facility recognition including standardization of interfacility transfer [17].

The use of various scenarios and a pediatric simulator (child or infant) can be used as a tool to improve the care of children. With the recent improvements in this technology, many medical schools and residency programs are considering pediatric simulation as a vital tool to enhance and broaden education. In this protected environment, nonpediatric clinicians can apply learned material, observe specific responses, and review the appropriate standards of care. This also may address requirements of hospital and residency credentialing programs [18, 19]. Furthermore, it can be used to form a collegial interaction (team building) amongst various care teams including physicians, nurses, pharmacists, and other allied health professionals [20–24]. Some professional societies offer specific courses or breakout sessions to supplement core material. Finally, the use of actors, portraying parents and relatives, may facilitate the practice of delivering dire information. The American Heart Association [11] has identified the importance of appropriately "breaking bad news" to families by incorporating education into their Pediatric Advanced Life Support curriculum and modules.

**TABLE 45.2**  Summary of pediatric developmental differences reproduced by infant simulators

| Organ systems | Clinical features |
|---|---|
| General | Represents a 3- to 6-month old infant<br>Skin elasticity and turgor |
| Airway | Direct laryngoscopy (oral, nasal intubation)<br>Use of standard devices (endotracheal tube, laryngeal mask airway)<br>Laryngospasm<br>Bronchial occlusion |
| Respiratory/pulmonary | Respiratory rate and tidal volume<br>Chest rise<br>Chest and lung compliance<br>Oxygen consumption<br>Carbon dioxide production (end-tidal $CO_2$)<br>Pulse oximetry and oxyhemoglobin saturation correlation (pH, temperature, intrapulmonary shunt fraction)<br>Cardiopulmonary interactions<br>Abnormal respiratory patterns (e.g., paradoxical breathing)<br>Asymmetric lung ventilation (e.g., pneumothorax, bronchial intubation)<br>Absence of breath sounds, chest excursion, and $CO_2$ with abdominal distension (esophageal intubation)<br>Dose-dependent response to pulmonary medications<br>Spontaneous and mechanical ventilation modes |
| Cardiovascular | Heart rate<br>Heart sounds<br>Brachial and femoral pulses<br>Cardiac monitoring<br>Blood pressure measurement (invasive and noninvasive)<br>Vascular access (intraosseus, intravenous)<br>Chest compression measurement<br>Electrical therapy (cardioversion, defibrillation)<br>Dysrhythmias (e.g., asystole, bradycardia, tachycardia, and ventricular fibrillation)<br>Hemodynamic responses to dysrhythmias<br>Independence of brachial and femoral pulses (e.g., coarctation of aorta)<br>Hypovolemia and hypervolemia<br>Right and/or left heart failure<br>Dose-dependent response to cardiac medications |
| Metabolism | Temperature measurement<br>Blood gases (arterial and venous)<br>Alveolar concentration of oxygen and carbon dioxide<br>Metabolic acidosis and alkalosis<br>Bowel sounds<br>Gastric distension |
| Genitourinary | Male and female external genitalia<br>Urinary flow |
| Neurological | Responses to sympathetic and parasympathetic tone<br>Pupil sizes<br>Blinking<br>Anterior fontanel (soft or bulging)<br>Preconfigured phonation (e.g., cooing, crying) |

And one final argument in favor of pediatric simulation: safety. Our simulation laboratory has been founded on the following: "*Patient safety without patient risk$^{SM}$*." It is estimated that approximately 1% of all hospitalized children experience an adverse event [25]. In fact, one analysis found that about 11% of "medically complex children" experienced an error [26]. It is believed that child-specific factors, such as physical characteristics and development [27], affect the delivery of health care in such a manner that contributes to the occurrence of patient safety problems [28]. On the basis of the Chicago Pediatric Patient Safety Consortium, hospitals should implement strategies to improve pediatric clinical communication, acuity assessment, adherence of a care chain of responsibility, and timely notification of attending physicians [29–31]. Training methods suggested include the use of scenarios and simulation exercises. By advancing the use of pediatric simulations, hopefully, we can improve pediatric patient care and their safety.

Our pediatric simulation laboratory floor plan is shown in Appendix 45A.1. The first image is our current configuration, the second is a Proposed Operating Room Area, and the third is a Proposed Intensive/Acute Care Area.

## 45.2 Scenarios

Until the advent of pediatric simulators, the instructors were faced with predominantly using adolescent/young adults. Now, with commercial child and infant simulators, the possibilities are greatly expanded. As technology improves, one can envision a term or preterm neonatal simulator to enhance realism. In any case, objectives are essential to the scenarios. The course director and facilitators must agree upon a core set of objectives to stay on track and complete the task of the session. We have found that more than four or five objectives can be cumbersome. Furthermore, other teaching points arise during the session, which adds to the experience. Also, objectives can aid in the assessment of the course. Using a pre-and a posttest should be considered especially if didactic material is presented or there are ongoing competency requirements. On the basis of our objectives and participants, our cases will vary. The format is usually modeled after mock codes texts (Table 45.3) [32], PALS instructor scenarios [11], or real-life cases. The scenarios usually last between 10 and 15 minutes. If the group is struggling, the facilitators will assist by providing well-placed cues. Once again, the goal is to enhance learning of the student without hampering the experience. Finally, we never allow the scenario patient to expire. This can be counterproductive to the morale of any group. The only case we have performed outside of these rules was a Sudden Infant Death Syndrome case. However, in this case, death is the end point of the disease. The debriefing session follows with the use of video/audio replay. Depending upon the objectives and expectations of the participants, we attempt to limit the amount of lecture material. We will use the debriefing session as our method for relaying curriculum. However, in some circumstances, participants would like to have reference material. A hard copy of slides, web-based material, and customary texts may also supplement learning [33]. Certainly, it should supplement but not detract from the primary method of simulation-based education.

**TABLE 45.3**   Mock code scenario template

Title
Author
Objectives
Brief presenting history
Initial vital signs
Pertinent requested information
Weight
Initial physical exam (including general appearance, ABCs)
Further history given on request
Expected interventions and their complications
Repeated vital signs
Progression
Laboratory tests
Options
Additional potential complications
Disposition
Discussion of objectives
Additional comments
References

**Legend:** The mock code template is used to organize each scenario and provide a general outline for faculty members and facilitators. Each scenario is modified to reflect current resuscitation and therapeutic guidelines as found in literature (Modified with permission by the copyright owner: Ref. 32).

## 45.3 Debriefing Methods

The process of debriefing remains a cornerstone of simulation (see Chapter 69). Without earnest review and constructive criticism, the participant is faced with their own subjective interpretation of their actions and the actions of others. As mentioned earlier, objectives are essential to all aspects of the simulation experience. Teaching and learning objects assist in the overall education and benefit that is derived from the scenario. Furthermore, objectives provide a guideline to the facilitators during the debriefing process. On the basis of the objectives, debriefing usually focuses on one of the three following aspects of performance: proper use of procedures and techniques of care delivery; crisis resource management and group dynamics; or a combination of both. In pediatrics, both are equally important no matter the objective of the session. Debriefing must be viewed as an opportunity to educate and enlighten the participants. Our method integrates many general aspects of debriefing during pediatric sessions, whether participants are in training or postgraduate practitioners: safe and educational environment; verbal feedback and constructive criticism; reflective learning; promotion of communication and debriefing among participants; sponsor culture change regarding medical errors; and positive modifications to enhance performance [34]. The importance of a safe and educational environment cannot be overemphasized. An entire simulator session can become eroded quickly if someone is chided for an action. It is absolutely vital to assure a nonthreatening session without sacrificing the participant or their education. Providing verbal feedback promotes attention to detail, allows the identification of areas of improvement, and fosters a collegial student/teacher relationship of trust and respect. By fostering interaction during debriefing, participants become actively involved in their education. It is always challenging to address the extremes of students (i.e., overzealous versus withdrawn). One must balance the group by promoting everyone's right to participate and be heard. As humans, it is difficult to place ourselves under scrutiny and inspection. However, as leaders in medical education, health professionals must strive to improve patient care and safety [35, 36]. Clinical communication is becoming increasingly important to address medical errors. In fact, the most common management issues identified are communication and liaison problems [37]. Performance will be greatly enhanced by correction of behaviors in a constructive manner in a safe environment. This is commonly translated to mind-set adjustment when approaching common or complex pediatric patients in the clinical areas [38].

It is important to remember that all simulation sessions, pediatric or adult, are for the participant and not the facilitator. Failure to keep the participant first will be disastrous. It is quite useful to enlist facilitators recognized for their capabilities as educators in the practice of clinical pediatrics. Being well versed in the management of the acutely ill or traumatized child, Pediatric Advanced Life Support, and academic pediatric medical education form the core of our team. Also, the facilitators must have experience in the debriefing process during their clinical practice. We offer and encourage debriefing sessions after any pediatric resuscitation or death. As in simulation, it does promote solidarity amongst the group, whether a single discipline or multidisciplinary [34]. We apply caution in our pediatric debriefing sessions to stay on track but allow for flexibility. Tangential topics may induce distraction and rapid loss of focus or interest. However, it is accepted by the facilitator group not to squander a potential teaching point or pearl. Finally, we strive to "practice what we preach," especially crisis resource management. As colleagues, we respect each other's opinions and actually debrief ourselves after each session to examine how we can improve the experience for our participants. The allowance of self-inspection has improved the quality and effectives of our courses.

## 45.4 Lessons Learned

### 45.4.1 Pediatric Resident Mock Codes

The amount and type of pediatric codes vary amongst all hospitals. Most occur in the emergency department or acute care treatment areas (e.g., pediatric intensive care units, PICU). The vast majority are related to a respiratory pathology. However, if the institution cares for many congenital heart disease patients, then a cardiac cause for an arrest is more frequent. In either case, due to sheer numbers of trainees, it is possible that a resident-in-training may never participate in or lead a pediatric resuscitation. Furthermore, they may not become familiar with pathologic cardiac rhythms such as ventricular fibrillation. Initially, this challenge was addressed by providing PALS training to our pediatric interns and recertification during their senior year. With the changes in PALS educational format in 2002, the emphasis on practical application was well received but the execution suffered from lack realism. Adult and pediatric mannequins were used but the emergency resuscitation environment was deficient. Even though there were the stressors to perform (i.e., oral and written examinations), there was no sense of urgency by participants. They were not in a clinical environment. Therefore, we began to provide monthly pediatric mock codes on the units. Each pediatric resident participated during their general and PICU rotations. It was held in an unoccupied patient room with familiar hospital equipment, including pediatric personnel and a resuscitation cart. An in-room debriefing session followed the code. Once again, the "urgency" aspect was lacking despite being in the hospital. Furthermore, we could not review everything because we did not see everything. Finally, the proof of a positive or negative maneuver was not recorded. We began searching for another medium to mimic a real pediatric code.

In 2003, our simulation laboratory acquired a child simulator. As we became familiar with its use, we began holding our monthly pediatric mock codes in the simulation lab. As a mandatory experience, there would be two teams with four or five residents per team. To replicate a "code" environment, pediatric seniors, juniors, and interns performed the scenario together. The cases were unknown to each participant. The students did not know the scenario objectives and were blinded to the actual case. However, individuals knew the objective of performing a pediatric resuscitation with the application of PALS skills and algorithms [11]. After a brief orientation to the child simulator capabilities, laboratory environment, and equipment, the first team would perform the resuscitation. The participants are allowed to use any resource available including, but not limited to, a Pediatric Broselow Tape and PALS Algorithm Cards. While performing, the other team would monitor via video/audio links in real time. Once completed, debriefing follows the code with replay of the scenario. Once debriefing is complete, team roles are reversed and the second pediatric mock code begins. Of course, the second case was not the same as the first. We wanted each group to perform a different scenario to limit bias in management and knowledge of the "right answer." Certainly, we could do the same scenario and adjust parameters, e.g., patient goes into pulseless electrical activity versus pulseless ventricular tachycardia. The rest of the process is repeated including classroom debriefing session.

We typically have three to four facilitators for each session. Each faculty member is a PALS Instructor of whom some are also PALS Course Directors or Regional Educators. Furthermore, the Pediatric Residency Director and Chief Resident participate. Our focus on education includes the appropriate application of PALS guidelines and use of algorithms mirrored during the course. One faculty member is in the clinical action room performing as scenario presenter and circulating nurse. Another faculty member, typically a PALS Course Director/Regional Educator, modifies clinical parameters in the control room. Our simulator technician provides rapid adjustment of physiologic variables. And the final faculty member is in the classroom during the scenario. Their task is to identify key PALS points and facilitate discussion of the clinical problem. We communicate through headsets to be consistent with our material presentation and areas of focus. This format has been well received. It has also stimulated the beginning of simulation sessions for visiting pediatric residency programs. Pediatric residency directors from both programs are pleased with our methods to date. However, as with any algorithm-driven process, recommendations do change. PALS material is no different. We strive to teach the most current recommendations [39]. To achieve such excellence, we continue to enlist faculty members not only Instructors in PALS, but also Course Directors (Table 45.4).

## 45.4.2 Chemical Agent Workshop

In 2002, Rush University Medical Center received grant funding through the Chicago Department of Public Health and Health Resources Services Administration to advance Bioterrorism and Disaster Preparedness efforts. Along with only one other hospital in the City of Chicago, Rush received the highest grant designation as a Center of Excellence. Throughout the year, it quickly became apparent of the many vulnerable populations present in Chicago and the lack of specific plans. Of those populations, pediatrics was considered a priority. Since 25% of the population consists of those in the pediatric age range, pediatric specific requirements were added to the 2003 grant Request for Proposal. Initially, a Pediatric Subcommittee was formed and didactic lectures were created for citywide dissemination. However, it did not promote psychomotor integration of educational materials. Practice and hands-on training were desired by the Chicago Department of Public Health. However, it was uncertain of the appropriate educational format and how to advance it. Therefore, on the basis of the outcry for such an educational experience, pediatric representatives from Rush created the Pediatric Chemical Agent Simulation Workshop.

Medical researchers have used baby primate models to illustrate the various responses to chemical agents. Under those

**TABLE 45.4**   Pediatric resident mock codes

| | |
|---|---|
| Learners/participants | Pediatric residents in training including interns, junior residents, and senior residents |
| Objectives | 1. Perform initial assessment and evaluation during resuscitation.<br>2. Provide treatment for life-threatening issues in the acutely ill child.<br>3. Utilize Pediatric Advanced Life Support algorithms and techniques during resuscitation.<br>4. Identify key physiologic and pathophysiologic processes in the pediatric patient.<br>5. Apply appropriate communication skills and techniques of crisis resource management.<br>Other objectives were based upon specific scenarios (see above). |
| Key events and conditions | Conditions included pediatric emergency department, pediatric intensive care unit, and general pediatric units. Location was the simulation stage. |
| Roles | Residents were poised as responders to a pediatric resuscitation. The circulating assistant aided in scenario description, answers to questions, and finding equipment. |
| Functional equipment | All equipment was available to perform all aspects of Pediatric Advanced Life Support. Antidotes were available for use (autoinjectors and medication vials). |
| Nonfunctional props | Simulation stage configuration dependent upon each scenario (e.g., patient room). |
| Functional features of simulator | The baby and child simulator were appropriate for size of scenario patient including airway and vascular access. |
| Script of scenario | See above. |
| Functional floor plan | See accompanying disk. |
| Location of session | All pediatric residents perform their duties on the simulator campus or within a short walking distance. As part of residency educational objectives, the participants were relieved of their duties for the session. Faculty members are academic attending physicians. |

conditions, the animals responded in a similar fashion to human infants (e.g., size constraints, copious airway secretions, airway manipulation resistance). Owing to limited grant resources and our experience with pediatric simulation, we proposed the purchase of a baby simulator for chemical agent training involving infants and children. Our target audience was composed of emergency medicine physicians and nurses from more than 30 area hospitals. We created three scenarios to illustrate the core knowledge required to diagnose and manage various pediatric chemical exposures. The scenarios included nerve agents (sarin), pulmonary irritants (chlorine), and cyanide. However, we adapted the scenarios to highlight the unique anatomic and physiologic vulnerabilities of children (Scenarios 6, 7, and 8, in Appendix 45A.2). The course began with a summary of specific pediatric vulnerabilities, their clinical response to each agent, and review of essential management. Each group consisted of no more that five individuals and had at least one physician. A case was presented to the participants in the clinical action room but the type of agent was not revealed. The group then proceeded to resuscitate the infant. During the scenario, the other participants were observing events of the clinical action room via real-time video/audio. Each scenario lasted no longer than 10–15 minutes. Upon completion of the first scenario, the entire class reconvened in the debriefing room. Prior to reviewing the scenario, an algorithm developed to approach pediatric chemical agent exposures was reviewed [40]. Debriefing followed soon after. This was repeated for the second and third cases minus the review of the algorithm. Educational material (slide presentation, nerve agent antidotes including use of autoinjectors, and copy of the algorithm) was supplied to the participant for classroom review. However, they were allowed to use any resource in the clinical action room including, but not limited to, the American Heart Association Handbook of Emergency Cardiovascular Care [39] and Pediatric Advances Life Support algorithms [11].

During each workshop, we provide three faculty members well versed in pediatric resuscitation, critical care, and emergency medicine. Furthermore, each faculty member had experience in pediatric disaster preparedness and chemical agent pharmacotoxicology. As a group, we thoroughly reviewed the content of the course and made modifications as needed. One faculty member was in the clinical action room performing as scenario presenter and circulating medical attendant. Another faculty member modified the clinical parameters in the control room. Our simulator technician provided rapid adjustment of physiologic variables. And the final faculty member was in the classroom during the scenario identifying key educational points and facilitating discussion. We communicated through wireless headsets to be consistent with our material presentation and areas of focus. During the debriefing sessions, we were able to review all aspects through video/audio replay (Table 45.5). The response to the sessions was overwhelmingly positive. Satisfaction surveys revealed that multiple participants wanted more cases, longer sessions,

**TABLE 45.5** Chemical agent workshop

| | |
|---|---|
| Learners/participants | Practicing clinicians including emergency department physicians, clinical nurse specialists, staff nurses, nurse practitioners, interfacility transport personnel, and emergency medical technicians. |
| Objectives | All were dependent on the chemical agent exposure and their implications in a pediatric patient (see below). |
| Key events and conditions | Location was a receiving emergency department during the daytime hours. Location was simulation stage. |
| Roles | Clinicians were poised as the pediatric resuscitation team. The circulating assistant aided in scenario description, answers to questions, and finding equipment. |
| Functional equipment | All equipment was available to perform all aspects of Pediatric Advanced Life Support. Antidotes were available for use (autoinjectors and medication vials). |
| Nonfunctional props | Simulation stage configured as emergency resuscitation area. |
| Functional features of simulator | The baby simulator was appropriate for size of scenario patient including airway and vascular access. |
| Script of scenario | See scenarios in Appendix 45A.2. |
| Functional floor plan | See disk with book. |
| Location of session | Participants were from area hospitals. Most were permitted time off from daily duties. Parking at simulator facility was required. Faculty members are academic attending physicians. |

and further opportunities for practice. Also, many were able to perform procedures not customary to their clinical practice (e.g., intraosseus placement). On the basis of the results of our course, Chicago Department of Public Health has asked that we move forward in designing a Pediatric Blast/Thermal Injury Workshop and participate in a multi-institutional pediatric hospital simulator drill.

### 45.4.3 Advanced Life Support (Pediatric and Neonatal)

As illustrated thus far, the use of pediatric simulation definitely has a role in Advanced Life Support, including PALS and Neonatal Resuscitation Program. Taking into mind the current methods of teaching, child and infant simulators are

**TABLE 45.6**   Advanced life support (Neonatal)

| Learners's objectives | The Neonatal Resuscitation Program has been designed by the American Academy of Pediatrics to teach health care providers an evidence-based approach to resuscitation of the newborn. Mostly, participants are nurses and physicians. |
|---|---|
| | **Lesson 1**<br>Overview and principles of resuscitation of a Neonate<br>Changes in physiology that occur when a baby is born<br>Flow diagram showing all steps to follow during resuscitation<br>Risk factors that can help predict which babies will require resuscitation<br>Equipment and personnel needed to resuscitate a newborn |
| | **Lesson 2**<br>Initial steps in resuscitation<br>Decide if a newborn needs to be resuscitated<br>Open the airway and provide initial steps of resuscitation<br>Resuscitate newborns when meconium is present<br>Provide free-flow $O_2$ when needed |
| | **Lesson 3**<br>Use of a resuscitation bag and mask<br>When to give bag and mask ventilation<br>Differences between flow inflating and self-inflating bags<br>Operation of each type of bag<br>Correct placement of masks on the newborn's face<br>Testing and troubleshooting each type of bag<br>Evaluating the success of bag and mask ventilation |
| | **Lesson 4**<br>Chest compressions<br>When to begin chest compressions<br>How to administer chest compressions<br>How to coordinate chest compressions with positive pressure ventilation<br>When to stop chest compressions |
| | **Lesson 5**<br>Endotracheal intubation<br>How to prepare the equipment needed for resuscitation<br>Using the laryngoscope to insert an endotracheal tube<br>Determining if the ET tube is in the trachea<br>Using an endotracheal tube to suction meconium from the trachea<br>Using the ET tube to administer positive pressure ventilation |
| | **Lesson 6**<br>Medications<br>How to administer epinephrine through an ET TUBE and Umbilical vein<br>When and how to administer fluids IV to expand blood volume<br>When and how to administer other resuscitation medications |
| | **Lesson 7**<br>Special considerations<br>Special situations that may complicate resuscitation and cause ongoing problems<br>Ethical considerations<br>How the principles of this program can be applied to babies who require resuscitation beyond the immediate newborn period |
| | **Lesson 8**<br>Megacode<br>Integration of all skills in a simulated resuscitation situation. |

**TABLE 45.6**  (Continued)

| | |
|---|---|
| Key events and conditions | Delivery room after a complicated delivery. |
| Roles | Individuals were resuscitating the newly born infant after a complicated deliver (vaginal or cesarean). Faculty members are certified instructors of the Neonatal Resuscitation Program. |
| Functional equipment | All equipment was available to perform all aspects of neonatal resuscitation except umbilical vein catheterization. |
| Nonfunctional props | Simulation stage configured as delivery room. |
| Functional features of simulator | The baby simulator was appropriate for size for a larger infant but not a term or premature newborn. However, anatomical and physiologic parameters were simulated for the newborn regardless of age and size (e.g., airway intubation and chest compressions). |
| Script of scenario | The scenario was a megacode that integrated all skills of Neonatal Resuscitation Program (see above objectives). |
| Functional floor plan | See disk with book. |
| Location of session | All participants were from the simulator medical center. As part of their educational objectives, the participants were relieved of their duties for the session. |

excellent teaching and testing tools. For PALS, many of the content including the modules are case based. Therefore, there is no need to create scenarios *ab initio*. The PALS instructor has a multitude of cases that can be adapted for the office, emergency department, general wards, or the PICU. The material is well written and has logical flow. Furthermore, the simulation can more accurately depict pathophysiologic changes and the rapidly declining course of an acutely ill child. For Neonatal Resuscitation Program (Table 45.6), the course reviews all pertinent algorithms in neonatal resuscitation such as bag/mask use, chest compressions, endotracheal intubation, and medication usage. Once didactics and practice were completed, a megacode is performed to integrate all skills learned by the participant. Once again, the sense of urgency is accentuated and the participants feel totally immersed in a real-life neonatal resuscitation.

# 45.5 Conclusion

Recent advent of pediatric simulators has allowed for many new educational and training opportunities for pediatric clinicians. On the basis of experience, it would be simple to presume that it increases knowledge, improves safe care, and determines competency. However, evidence should be sought to objectively evaluate the added value in using pediatric simulation [41]. The next step is to advocate for such research and foster it.

# 45.6 Favorite Problem Solvers

*Deep Survival* [42]: The book is a quick read on various individuals and how they survived despite their lack of training or died despite having training. It familiarizes the reader with behavioral concepts one can use in stressful situations to calmly think through things. It is perfect for those in highly stressful, rapidly changing medical environments. Plenty of references are also contained.

*Effective Phrases for Performance Appraisals* 10th edition [43]: Helps you to gain verbal/written phrases to describe performance and articulate an excellent evaluation.

*Thoughts on Leadership* [44]: The book includes various inspiring phrases from well-known leaders throughout history, including presidents, authors, and other key figures throughout the world. For example, one inspiring quote was from CS Lewis who wrote that "the only people who achieve much are those who want knowledge so badly that they seek it while the conditions are still unfavorable. Favorable conditions never come." Ideally, this would characterize every clinician before, during, and after training.

*Monsters Inc.* [45]: There is a simulator used in the movie to "train" monsters how to be frightening. The scenes contain many relevant aspects of simulation including stage setup, control room, facilitator, equipment function, how things can go really good or really bad, and even how the experienced "practitioner" can succumb to the realism of the scenario.

# References

1. Fiedor, M. L. Pediatric simulation: A valuable tool for pediatric medical education, *Crit. Care Med.* **32**(2), S72–74 (2004).

2. Blum, R. H., Raemer, D. B., Carroll, J. S., et al. A method for measuring effectiveness of simulation-based team training for improving communication skills, *Anesth. Analg.* **100**, 1375–1380 (2005).

3. Cravero, J. P. and Blike, G. T. Review of pediatric sedation, *Anesth. Analg.* **99**, 1355–1364 (2004).

4. Dalley, P., Robinson, B., Weller, J., and Caldwell, C. The use of high-fidelity human patient simulation and the introduction of new anesthesia delivery systems, *Anesth. Analg.* **99**, 1737–1741 (2004).

5. Fiedor, M., Kochanek, P., Watson, R. S., and Angus, D. Preliminary evaluation of SimBaby, a novel pediatric simulator, *Crit. Care Med.* **32**(12), A62 (2004).

6. Weinstock, P. H., Kappus, L. J., Kleinman, M., et al. Toward a new paradigm in hospital-based pediatric education: the development of an onsite simulator program, *Pediatr. Crit Care Med.* **6**(6), 635–641 (2005).

7. Schwid, H. A., Rooke, G. A., Michalowski, P., and Ross, B. K. Screen-based anesthesia simulation with debriefing improves performance in a mannequin-based anesthesia simulator, *Teach. Learn. Med.* **13**, 92–96 (2001).

8. Perkin, R., van Stralen, D., Mellick, L. B., and Rothrock, S. G. Managing pediatric airway emergencies: anatomic considerations, alternative airway and ventilation techniques, and current treatment options, *Pediatr. Emerg. Med. Reports* **1**(1), 1–12 (1996).

9. Dalton, H. J. Airway structures and functions. In: Slonim, A. D. and Pollack, M. M. (eds.), *Pediatric Critical Care Medicine*, Lippincott Williams and Wilkins, Philadelphia, PA, 2006, pp. 243–244.

10. Goodwin, J. A., van Meurs, W. L., Sa Couto, C. D., et al. A model for educational simulation of infant cardiovascular physiology, *Anesth. Analg.* **99**, 1655–64 (2004).

11. American Heart Association. *Pediatric Advanced Life Support Provider Manual*, 2002.

12. American Academy of Pediatrics. The youngest victims: disaster preparedness to meet children's needs. Children, Terrorism, and Disasters Toolkit. 2002. Available at:www.aap.org/terrorism

13. Medical Education Technologies, Incorporated™. Baby-SIM® User Guide. 2005.

14. American Academy of Pediatrics. How pediatricians can respond to the psychosocial implications of disasters (RE9813), *Pediatrics* **103**(2), 521–523 (1999).

15. Brown, E. J. Clinical characteristics and efficacious treatment of posttraumatic stress disorder in children and adolescents, *Pediatr. Annals* **34**(2) 138–146 (2005).

16. American Academy of Pediatrics. Care of children in the emergency department: guidelines for preparedness (RE069904), *Pediatrics* **107**(4), 777–781 (2001).

17. Illinois Emergency Medical Services for Children. http://www.luhs.org/depts/emsc/index.htm (Accessed Aug 29, 2007).

18. Epstein, R. M. and Hundert, E. M. Defining and assessing professional competence, *JAMA* **287**, 226–235 (2002).

19. Renuka, M., Hudson, V. L., and Fisher, L. G. Assessment of a mock code curriculum to teach pediatric residents resuscitation skills. *Crit. Care Med.* **32**(12), A62 (2004).

20. Barret, J., Gifford, C., Morey, J., et al. Enhancing patient safety through teamwork training, *J. Healthcare Risk Manag.* **21**(4), 57–65 (2001).

21. McCartney, P. Human patient simulators in maternal-child nursing. *Maternal Child Nursing*, May/June 2005, p. 215.

22. Murray, W. B. and Foster, P. A. Crisis resource management amongst strangers: Principles of organizing a multidisciplinary group for crisis resource management, *J. Clin. Anesth.* **12**, 633–638 (2000).

23. Murray, W. B. Simulators in critical care education: educational aspects and building scenarios. In: Dunn, W. F. (ed.), *Simulators in Critical Care and Beyond*, Society of Critical Care Medicine, Des Plaines, IL, 2004, pp. 29–32.

24. Risser, D. T., Rice, M. M., Salisbury, M. L., et al. The potential for improved teamwork to reduce medical errors in the emergency department, *Annals Emerg. Med.* **34**, 373–383 (1999).

25. Woods, D. M., Thomas, E. J., Altman, S., et al. Adverse events and preventable adverse events in children, *Pedaitrics.* **151**, 155–160 (2005).

26. Slonim, A. D., LeFleur, B. J., Ahmed, W., and Joseph, J. G. Hospital reported medical errors in children, *Pediatrics* **111**, 617–621 (2003).

27. Woods, D. M., Bhatia, M. M., Ogata, E., et al. Child-specific risk factors and patient safety, *J. Patient Safety.* **1**, 17–22 (2005).

28. Woods, D. M., Holl, J. L., Mohr, J. J., et al. Anatomy of a patient sadtey event: A taxonomy for pediatric patient safety, *Qual. Safety in Healthcare.* **14**, 422–427 (2005).

29. Lingard, L., Espin, S., Whyte S., et al. Communication errors in the operating room: an observational classification of recurrent types and effects, *Qual. Safety in Health Care* **13**, 330–334 (2004).

30. Beckmann, U., Gillies, D. M., Berenholtz, S. M., et al. Incidents relating to the intra-hospital transfer of critically ill patients, *Intens. Care Med.* **30**, 1579–1585 (2004).

31. Soglin, D. and Severin, P. Chicago Pediatric Patient Safety Consortium (http://www.chicagopatientsafety.org/); 2006 (personal communication)

32. Roback, M. G., Teach, S. J., First, L. R., and Fleisher, G. R. (eds.), *Handbook of Pediatric Mock Codes*, Mosby Yearbook, St. Louis, Missouri, 1998, pp. 211–213.

33. Schwid, H. A., Rooke, G. A., Ross, B. K., and Sivaraian, M. Use of a computerized advanced cardiac life support simulator improves retention of advanced cardiac life support guidelines better than a textbook, *Crit. Care Med.* 27, 821–824 (1999).

34. Mort, T. C. and Donahue, S. P. Debriefing: The basics. In: Dunn, F. (ed.), *Simulators in Critical Care and Beyond*, Society of Critical Care Medicine. Des Plaines, IL, 2004, pp. 76–83.

35. Blike, G., Cravero, J., and Nelson, E. Same patients, same critical events-different systems of care, different outcomes: Description of a human factors approach aimed at improving the efficacy and safety of sedation/analgesia care, *Quality Manag. Health Care* 10(1), 17–36 (2001).

36. Blike, G. T., Christoffersen, K., Cravero, J. P., et al. A method for measuring system safety and latent errors associated with pediatric procedural sedation, *Anesth. Analg.* 101, 48–58 (2005).

37. Sutclife, K. M., Lewton, E., and Rosenthal, M. M. Communication failures: an insidious contributor to medical mishaps, *Acad. Med.* 79, 186–194 (2004).

38. Dannefer, E. F. and Henson, L. C. Refocusing the role of simulation in medical education: training reflective practitioners. In: Dunn, W. F. (ed.), *Simulators in Critical Care and Beyond*, Society of Critical Care Medicine, Des Plaines, IL, 2004, pp. 25–28.

39. American Heart Association. *Handbook of Emergency Cardiovascular Care for Healthcare Providers*, 2006.

40. Henretig, F. M., Cieslak, T. J., Madsen, J. M., et al. The emergency department awareness and response to incidents of biological and chemical terrorism. In: Fleisher, G. F., Ludwig, S., and Henretig, F. M. (eds.), *Textbook of Pediatric Emergency Medicine*, 5th ed. Lippincott Williams and Wilkins, Philadelphia, PA, 2006, p. 159.

41. Eppich, W. J., Adler, M. D., and McGaghie, W. C. Emergency and critical care pediatrics: use of medical simulation for training in acute pediatric emergencies, *Curr. Opin. Pediatr.* 18, 266–271 (2006).

42. Gonzales, L. *Deep survival*, W.W. Norton and company, New York, 2003.

43. Neal, J. E. *Effective pharases for performance appraisals: A Guide to successful Evaluations.* 10th ed. Neal publication, Inc., Perrysburg, Ohio. 2004.

44. Forbes. *Thoughts on Leadership: Thoughts and reflections from History's Great Thinkers*, Triumph Books, Chicago, IL, 1995.

45. Doctor, P., Silverman, D., Unkrich, L., et al. Monsters, Inc. [DVD]. Disney/pixer, 2001.

46. Aaron, C. K. Organophosphates and carbamates. In: Ford, M. D., Delaney, K. A., Ling, L. J., and Erickson, T. (eds.), *Clinical Toxicology*. W.B. Saunders, Philadelphia, PA, 2001, pp. 819–828.

47. Galloway, R. J. and Smallridge, R. C. Nerve agents and anticholinesterase insecticides. In: Chernow, B., Brater, D. C., Holaday, J. W., et al. (eds.), *The Pharmacologic Approach to the Critically Ill Patient*, 3rd ed. Williams and Wilkins, Baltimore, MD, 1994, pp. 935–956.

48. Henretig, F. M. and McKee, M. Preparedness for Acts of Nuclear, Biological, and Chemical Terrorism. In: Gausche-Hill, M., Fuchs, S., and Yamamoto, L. (eds.), *Advanced Pediatric Life Support: The Pediatric Emergency Medicine Resource*, 4th ed. Jones and Barlett Publishers, Sudbary, Massachussets, 2004, pp. 568–591.

49. Rotenberg, J. S. and Newmark, J. Nerve agent attacks on children: diagnosis and management, *Pediatrics* 112(3), 648–658 (2003).

50. Rotenberg, J. S., Burklow, T. R., and Selanikio, J. S. Weapons of mass destruction: the decontamination of children, *Pediatr. Annals* 32(4), 260–267 (2003).

51. Burklow, T. R., Yu, C. E., and Madsen, J. M. Industrial chemicals: Terrorist weapons of opportunity, *Pediatric Annals.* 32(4), 230–234 (2003).

52. Ford, M. D., Delaney, K. A., Ling, L. J., and Erickson, T. (eds.), *Clinical Toxicology*, W. B. Saunders, Philadelphia, PA,. 2001.

53. Rotenberg, J. S. Cyanide as a weapon of terror, *Pediatr. Annals* 32(4), 236–240 (2003).

# 46

# Critical Care Simulation: A Nursing Perspective

Carol I. Vandrey

## 46.1 Increase Learning, Decrease Risks

The shortage of proficient, experienced, critical care nurses, proliferation of new medical technologies, decreasing length of hospital stays, and escalating patient acuities put patients at risk for iatrogenic injury. The Joint Commission on Accreditation of Healthcare Organizations places patient safety at the top of its priority list. The challenge for health care educators is to foster the development of psychomotor, critical thinking, decision-making skills and teamwork in nurses and their clinical colleagues who are learning to care for patients in a real clinical environment, without putting either patients or learners at risk.

## 46.2 Historical Clinical Education

Historically, clinical education relied on didactic instruction, case studies, and skills laboratories to prepare clinical practitioners. The traditional classroom environment was relatively passive and emphasized the presentation and memorization of facts. Case studies encouraged the use of critical thinking to problem-solve paper-and-pencil scenarios. Skills laboratories isolated individual components of a task and improved psychomotor proficiency. All of these methods of instruction have benefits, but also limitations. Passive learning disengages the learner, case studies don't exercise clinical responsibility, and skills laboratories lack the distractions of an actual patient's physical, emotional, and psychological needs.

## 46.3 Learning Theory

Learning theory postulates that the acquisition of complex skills involves repetition of behaviors and thoughts within an appropriately defined environment. The more effectively an educational method incorporates these learning principles, the more effectively a student will learn. Educators favor interactive learning experiences that encourage students to apply newly acquired facts and skills to patient scenarios, while protecting both the learner and the patient from harm. Computer-controlled patient simulators improve response times to complex, dynamic patient crises and enhance the critical care nurses' confidence in their ability to work with unstable, critically ill patients.

The challenge to the educator includes determining the objectives to be learned during the simulation experience and imbedding them in a realistic scenario. The early full-function, high-fidelity human simulators were useful for, among other things, emergency airway management and allowed for airway stabilization, bag-valve-mask ventilation, and intubation. Furthermore, current simulators can be modified after purchase to present additional learning experiences. These include recent airway adjuncts such as supra-glottic airway devices (e.g., the Laryngeal Mask Airway), sterile insertion of a real flow-directed pulmonary artery catheter followed by real cardiac output fluid injection.

Future challenges to both commercial vendors of high-fidelity patient simulators and educators alike include enhancements for arterial blood draws, and 12-lead ECG generation/recording realistic types and ranges of body motions.

## 46.4 Preparation for Instruction

The instructor has several important tasks to complete before allowing learners into the simulation laboratory. The first responsibility is to determine what is to be learned. Overplanning and underplanning are common pitfalls. Students are easily overwhelmed during their first simulator experience, and a common expectation is that they will do everything they studied in the classroom. The other extreme is to just put them in the simulator room by themselves, present a scenario, and see what happens. This is experiential learning run amok. An experienced, confident clinician may enjoy this approach; but as Lyons and Milton noted, a student with a low degree of self-efficacy tends to be more apprehensive and often lacks the ability to recognize and respond to complex events. This student may freeze and simply not learn clinical care from the experience. A more reasoned approach is to determine two or three learning objectives, tell the students ahead of time what they are, prepare a scenario that incorporates those objectives, and then take the time to allow the students to demonstrate their improved proficiency to both you and themselves.

For example, the students are instructed to review the components of cardiac output, the ports of a pulmonary artery catheter, and the pressure waveforms generated by the chambers of the heart. The students arrive at the patient simulator and care for a patient who develops cardiogenic shock and needs a pulmonary artery catheter. The learners configure the equipment, insert the catheter, interpret the pressure waveforms, and discuss the values obtained. Courses of action based on the patient's clinical presentation are carried out and the students immediately experience the outcomes of the actions they take.

The students will need an orientation to the game rules of simulation, which includes both what the simulator can and cannot do, how to request help from the instructor or simulator operator while maintaining the integrity of the scenario, and assurances that their actions cannot harm either the simulator or themselves. During the initial lesson, the instructor needs to be in the room to oversee execution of

psychomotor skills and immediately correct errors. For example, if the learner forgets to flush the ports of the pulmonary artery catheter prior to insertion, an air embolus would result. Or, if the operator inflates the balloon for more than three respiratory cycles, pulmonary infarction can occur. Muscle memory of bad habits is much harder to overcome than learning to do a skill properly from the outset. A useful approach is to demonstrate the correct technique for a skill, observe and make on-the-spot corrections during the practice stage, and then step out of the room and watch through a one-way window and/or closed circuit audio/video while the skill is being practiced.

A typical visit to a simulation laboratory might focus on emergency airway management with assessment of breathing difficulties, correct head positioning, bag-valve-mask ventilation with 100% oxygen, intubation and evaluation using pulse oximetry, auscultation, exhaled $CO_2$ detectors, and arterial blood gas results. This simulation consists of seven separate psychomotor skills and one purely cognitive skill. When incorporated into a scenario where the patient progresses from spontaneous breathing to labored breathing to apnea with the accompanying cardiac dysrhythmias that coincide with hypoxia, the novice practitioner has ample opportunity to do some critical thinking, apply principles from didactic instruction, and enhance psychomotor skills. Many of the best scenarios are ones composed from events the instructors have observed in actual clinical practice. Not only are they are simply recreated in the safe atmosphere of the patient simulation laboratory with no risk to the patient or the learner, but also their real-life basis helps the learners accept the validity of the scenario.

## 46.5 Practical Application Examples

The defibrillator challenge (avoiding shocks to trainees and instructors) can be met by internally redirecting the energy-producing portion of the device from the paddles to the internal discharge power dissipator, leaving intact the appearance and operational features. The learner, seeing the rhythm on the defibrillator screen, can charge the device and then shock the simulated patient without the electrical risk of a shock injury to themselves and the other participants. Presently available mannequins do not jerk[1] in response to the energy as a human would, but all other cues and teaching points required for gaining competence in operating this life-saving device are present.

Learning pulmonary artery catheter insertion can be enhanced by placing a jugular vein access device into the neck to facilitate the insertion movements of the catheter. One learner advances the catheter while teammates announce the location of the catheter tip by watching the pressure waveform as it appears on the monitor. Passage of the catheter with an inflated balloon along with fluid injection for cardiac output determination is possible by connecting the distal end of the venous access device to a large bore tube that terminates in a hidden waste fluid collection container.

Even if arterial blood sampling and 12-lead ECG generation are beyond the mannequin's factory-installed capabilities, the learner can use sterile techniques while configuring the hemodynamic monitoring equipment for the arterial line and the pulmonary artery catheter at the bedside as well as attach five leads for ECG monitoring. The instructor provides arterial blood gas, other blood and clinical laboratory results, and preprinted 12-lead ECGs, as requested by the learners during the scenarios.

## 46.6 Evaluation of Performance

Evaluation of the learning experience is essential for both the students and the instructors. Students in the laboratory are asked to describe their thoughts and actions as they work, in order to cue the simulator's operator and to give the instructor insight into the student's thought processes. Video recording the students during the scenarios while the skill is being practiced is extremely helpful. While watching their actions on the video replay, they can self-critique as well as receive suggestions for performance improvement. After watching herself on video replay, one student said, "I feel more comfortable and confident with the steps I took. It was a good experience to see the outcome of each intervention." Equally, video recording the instructors during scenario development, as well as during a session is extremely helpful. They can self-critique as well as receive suggestions for ways to improve their performance as instructors.

Student response to training with the patient simulator can be measured with a Likert scale. Students consistently give the highest ratings, indicating a high level of satisfaction with the learning experience and frequently request more time with simulation scenarios. A typical student's comment was "It helped to practice concepts in lifelike situations because nothing goes like the textbook says it does." Another student said, "We don't have to worry about killing our patients, but still received the benefit of actually doing all of the interventions. We could see what the real-life consequences of our actions would be. Videos are helpful because they show good and bad decisions and allow us to consider other options or choices that could be made."

## 46.7 Conclusion

Patient simulation continues to evolve as a viable teaching strategy. Simulators are becoming more lifelike with each generation, and will continually reduce many of the risks from the real patient on-the-job training model that typified clinical education in the past. Simulation addresses both clinician competency and patient safety in a risk-free environment for patient and learner alike.

## Endnote

1. Some Simulation laboratories have used an instructor or actor to provide the jerking motion, but only if the defibrillator had previously been rendered harmless.

# 47

# Transporting a Patient: Interdisciplinary Simulation Exercises

Michael C. Foss

## 47.1 Who We Are

The School of Health at the state-supported, public Springfield Technical Community College in Springfield, Massachusetts, U.S.A, made a commitment in 2003 to integrate patient simulation into the health curriculum. The School of Health houses 16 health-related programs, including nursing, respiratory care, surgical technology, medical imaging, and medical assisting. Typical enrollment is about 500–600 students, with graduates numbering a total of over 6000. With a small grant and modification of existing facilities, the School created a very usable "virtual hospital," which we have named SIMS Medical Center™. The term "SIMS" has no meaning, but is rather a reference to the first patient simulator used, whom many students called "Mr. Sims." A business decision was made to trademark the name since we are planning to market SIMS Medical Center™ for simulation training and continuing education. Over the past 3 years, the use of patient simulation has grown, prompting the school to convert old laboratories and even a student lounge into patient simulation space. Today, our center consists of a two-bed Critical Care Unit, four-bed Acute Care Unit, eight-bed Basic Care Unit, Trauma Care, a Surgical Suite, a fully energized Radiology room with fully operation table and chest rack, and several multipurpose classrooms (Figure 47.1). There is also a Health Clinic, Respiratory Lab, Nuclear Medicine, Clinical Lab Science and Sonography room with an ultrasound (Ultrasim, Medsim, Ft Lauderdale, Florida, U.S.A) simulator. Medical Assistant faculty have recently become interested in patient simulation, and their current space will most likely be renamed and become a

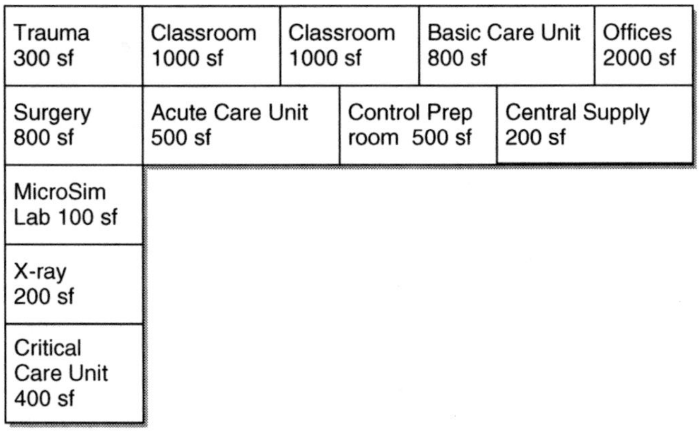

| Trauma 300 sf | Classroom 1000 sf | Classroom 1000 sf | Basic Care Unit 800 sf | Offices 2000 sf |
|---|---|---|---|---|
| Surgery 800 sf | Acute Care Unit 500 sf | | Control Prep room 500 sf | Central Supply 200 sf |
| MicroSim Lab 100 sf | | | | |
| X-ray 200 sf | | | | |
| Critical Care Unit 400 sf | | | | |

**FIGURE 47.1**   Conceptual diagram is intended to give you an idea of the layout of facilities. Numbers indicate room area in square feet. Not shown are the Health Clinic, Respiratory Lab, Nuclear Medicine, Clinical Lab Science and Sonography, each located in their own spaces.

part of our center. It has been the willingness of talented health faculty to change their teaching methods that has driven the development of our program.

## 47.2  Total Application Scenario

The purpose of creating our center was to provide a safe and effective learning environment in which students would prepare themselves to provide quality patient care. It has also become a place to train other faculty how to integrate patient simulation into their health curriculum, with the hope that patient care will be enhanced. It was decided that to reach both goals, we would have to replicate, to the best of our ability, a fair portion of the actual clinical experience expected of our students. For years, we had been using part-task trainers such as injection pads, injection arms, and strap-on wounds, but with mixed results. Even though these task trainers are very useful for introducing a student to part of an entire procedure, not all faculty feel that the skills learned is easily transferred to actual clinical practice. Instead of using additional task trainers, we decided to invest in full-size full-function patient simulators and associated full-function environments. What we did not want to do was turn our new facility into one big teaching laboratory where students worked on "disembodied simulated human parts." Instead, we wanted to give a "real feel" to the total patient experience. For example, expecting students in a part-task laboratory do everything they should do if they were changing a postop bandage, they would replace the bandage and dispose of the old in the proper receptacle while using sterile technique. Clinical care with real patients involves appropriate application of many layers of skills and behaviors, not just one stand-alone skill. Thus, within a full-function environment with full-function simulators, the student would change

the bandage, but not before introducing themselves, getting positive identification of the patient, explaining what was to be done, getting permission from the patient, etc. Students still learn about bandaging in lecture and part-task trainers are still used, but these are steps before the students progress to full-function simulation. In the latter, the emphasis is now placed on having the students perform every skill and behavior expected for care of real patients. The idea is to have the students perform all required skills appropriately and in the order needed before they ever encounter real patients. Of course, it does not always work that smoothly. There are times when scheduling issues prevent timely use of full-function patient simulation. Still, the aim is to provide each student a "total application scenario" for each topic.

We have had success using the full-function simulation with particular groups of students, but we recognized that when nurses and respiratory therapists care for real patients, they do not do so in isolation from each other. When we reexamined how we were conducting our full-function simulations, we saw that interdisciplinary team building was not being addressed. Nurses practiced completely alone, usually in the Acute Care environment, while Respiratory practiced almost exclusively in the Trauma Care and Critical Care environments. Truly full-function scenarios could not be performed until we found a way to bring various disciplines together around a single patient. This sounds obvious, but presimulation attitudes about how the disparate groups of students should learn their craft are well established and difficult to change. Skill evaluation protocols to determine student skill proficiency are still developed and exercised in isolation by the individual clinical fields. Many of our different health programs have been located in one area for 30 years and felt that they "owned" the space and everything within. However, during some construction, we had the opportunity to rename new spaces. To complete the change, over one weekend I removed all "program-oriented" signs and replaced them with "care-oriented" signs, such as Surgery, Trauma Care, Acute Care, and Critical Care. When someone asks who "owns" a space, I reply, "The Dean owns everything." That would be me. Fortunately, the health faculty agreed to share all spaces.

## 47.3  Creating a Team-building Opportunity: The Transportation Scenario

Sometimes, a "nonphysician, non-nurse" is needed to bring clinical people together. In our case, as the Dean of the School of Health, I was fortunate enough to act as a go-between in helping three specific disciplines come together for an interdisciplinary event. Prior to the development of our transportation scenario, faculty of Radiography, Respiratory Care, and

Nursing, for the most part, all kept to "their" respective simulated clinical environments. Despite this self-imposed isolation, there was some crossover. Nursing students did visit the Critical Care environment when in use by Respiratory faculty to see how simulated patients were put on ventilators and Radiography students had visited the Critical Care environment to learn how to handle patients while performing portable X-rays. Still, there was no single event that brought all these different team members together for the good of one patient. The goal was to bring an interdisciplinary team together to care for a single simulated patient, much like happens every day with real patients.

## 47.4 Setting the Stage

Finding the one "place" to bring several disciplines together was not as easy as first thought. The Acute Care environment seemed the most likely place, since you would expect nurses and respiratory therapists to at least sometimes cofunction within the same "real" space in hospitals (Figure 47.2). Respiratory Care scenarios need oxygen or at the very least,

compressed gases via typical hospital gas outlets to simulate some treatments. Even though we had the gas outlets, we had not yet connected them to the compressor and vacuum systems. Thus, there was little that respiratory students could do in Acute Care. Without a real or simulated portable X-ray unit, it was difficult to attract radiography students to see a patient in Acute Care. Nurses did visit patients in the Critical Care environment, but rarely when respiratory students were present (Figure 47.3).

I remembered that in the past, the Radiography faculty asked if we could provide a patient simulator for his students to practice moving a patient onto the X-ray table. Our Laerdal MegaCode Kelly was configured with a fractured leg. Just the leg, without the body, had been X-rayed before, and showed a beautiful fracture (Figure 47.4). In fact, we had digitized the image and used it on our prototype patient information system (Appendix 47A.1). Eventually, it became clear that the best location to bring nursing, respiratory, and radiography together might be Trauma Care. The only problem was that we did not have a simulated Trauma Care environment.

We took advantage of some construction in progress and acquired the use of a small room, about 300 sq. ft., which proved to be about the right size for a one-bed Trauma Care

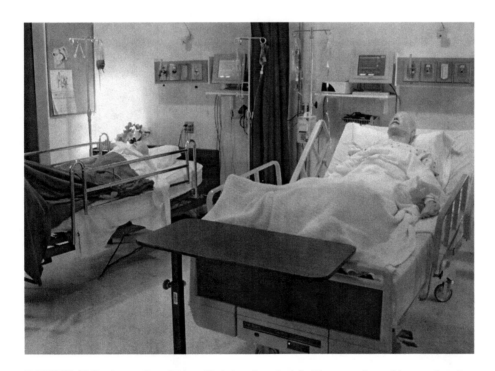

**FIGURE 47.2** Acute Care Patient Unit is a four-bed facility, complete with a workstation for the nursing students as well as crash cart, drug storage, linen and other typical patient unit equipment. All of this is contained in about 500 sq. ft. The partition is specially built with metal studs, covered by 1/2-inch plywood, followed by 3/8-inch sheetrock. This construction allows us to install wires and hoses anywhere needed. The plywood provides a solid surface for mounting equipment to the wall.

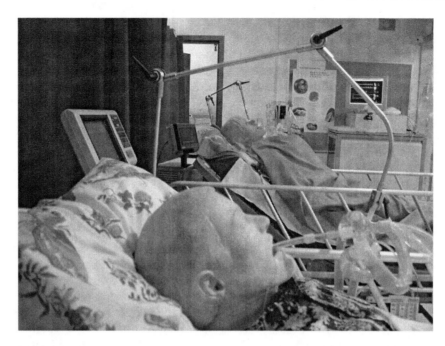

**FIGURE 47.3** The Critical Acute Care Patient Unit is a two-bed facility mainly used by the Respiratory program faculty and students. Shown are two patients connected to ventilator units. Respiratory students use the Critical Care Unit on a weekly basis.

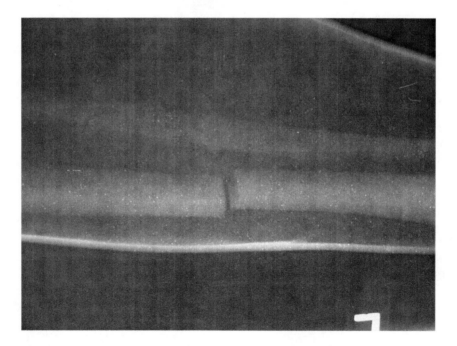

**FIGURE 47.4** Radiography taken by our Radiography students. Shown is the left lower limb of a Laerdal MegaCode Kelly patient simulator. The fractured leg is one option during purchase. We discovered that it could be X-rayed and have used it in our transportation scenario. Recently, we had a portable X-ray unit donated to our facility, and will develop a new scenario where Radiography students take the portable unit to surgery to X-ray the fractured leg.

environment. With some grant money, we purchased most of the usual equipment found in a trauma bay/area, e.g., a crash cart, gas outlets, and typical supplies. A small surgical light was mounted on the ceiling. I must add a word about the surgical light. We currently provide patient simulation experiences in the new trauma care room for our respiratory and nursing students. A local, large medical facility has inquired about using our facility to provide patient simulation experiences for their physicians. I felt it best, while we still had grant money, to install the surgical light. You never know when it might be needed in the future. Since we have no fewer than three wall- and ceiling-mounted cameras in this room, there has been no problem with the surgical light blocking our view. Of course, that may change when emergency physicians begin using our facility in the future. Since the room was new construction, there were no issues with asbestos above the ceiling, and there was sufficient electric power available for the new overhead light. Creating proper support for installing the light was also not an issue.

The Department Chair of our respiratory program was involved in the development of the Trauma Care environment and kept asking for gas outlets like the specific type we had in the Acute Care environment (Figure 47.1). Since his budget was not terribly large, he was hoping that I could use grant money to provide the desired gas outlets. What he did not realize is that we could not afford the fancy built-in headwall units found in a real facility. We also did not, and probably never will, have the industrial-sized vacuum, medical air, and oxygen supplies required for numerous gas outlets all in use simultaneously. What we could afford was a small air compressor that we found in the old respiratory lab. Free is good. For a few dollars, we ran industrial-grade compressor hoses from wall-mounted gas outlet boxes back to a closet in the adjacent control room. This is the same closet that contains two air compressors used to operate two Laerdal SimMan patients in the Acute Care environment. None of these oxygen outlets in the Acute Care environment actually provide 100% oxygen, but the students do not know the difference since our Laerdal patients do not automatically react to real differences in inspired oxygen content. The gas outlet boxes had to be built from scratch. Our local hardware store sells standard, white laminate shelving material. Four shelves can be used to create a decent enough "box" in which to mount brand-new (i.e., not free) real gas outlets. We had enough room to install ports for oxygen, vacuum, medical air, and a mount for the suction canister. A "face plate" was made from some leftover thin cabinet material (Figure 47.5). For mounting gas outlets in the new Trauma Care environment, it was back to the hardware store for more shelving material.

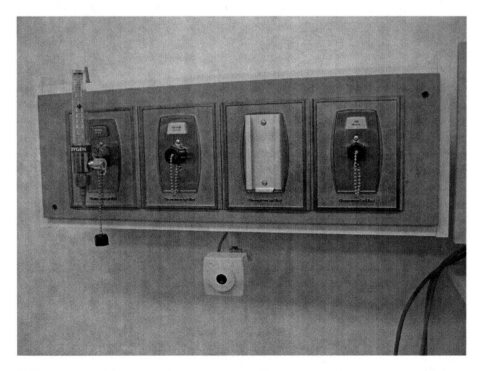

**FIGURE 47.5** "Home-made" gas outlet box used in Basic, Acute Care, and Trauma Patient Units. The material for the box is standard shelving material and the outlets were ordered through a local medical equipment supply company. At the time of this writing, the oxygen and medical air outlets produce room air provided by a compressor in the closet of the control room next door. Plans have been made to install a vacuum line for suction. Note the small camera mounted underneath.

Immobile black-and-white "security" cameras were mounted for video collection and a separate control room was created next door. Many people have asked why I did not spend the money on a large one-way window between the control room and patient room. Two reasons come to mind. The cost was just beyond reason, and I have yet to see a window view that was not at least partially blocked by someone working with the patient. Sufficient numbers of wall- and ceiling-mounted cameras overcome this "blockage" problem.

Now we were getting brave. The room we used as the Control room for our new Trauma Care environment was intended to be an office for a physician in our new Campus Health Clinic. He now shares the office with a complete simulation control console (Figure 47.6). Not knowing which type of patient simulator would be used in the new Trauma Care environment, we decided to install the hoses and cables needed to operate both the Laerdal Vital-Sim or SimMan patients. You will see later how we decided the actual patient to use for the first interdisciplinary experience.

Since we were also getting ready for an open house, we placed a Laerdal Vital-Sim MegaCode Kelly in the room. He was easy to move and came with some interesting inserts representing various injuries. The only drawback to him was that unlike his SimMan brother, he did not come with a patient monitor. We borrowed the patient monitor and link box from one of our SimMan patients. As there was no shelf to put the monitor on, I bought a standard, prefinished shelf and brackets, then installed it near our gas outlet box. It took a bit of coordination, but we were able to display correct patient vital signs on the SimMan monitor almost immediately after changing the same vital signs of the patient through the hard-wired Vital-Sim remote control. It looked "real," and later, visitors thought MegaCode was actually attached to the patient monitor. Of course, that meant the SimMan patient did not have a link box nor monitor and was kept "off line" longer than anticipated. It was worth it (Figure 47.7).

An almost-new wheeled stretcher ("gurney") was purchased to give a more realistic feel to the room and to give students

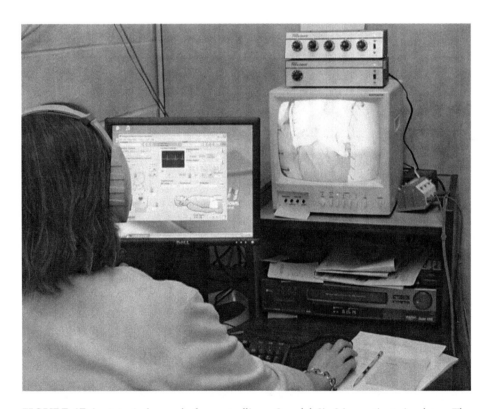

**FIGURE 47.6**  A typical console for controlling a Laerdal SimMan patient simulator. The camera system is a simple "security" style system with four camera input and built-in video switching. An inexpensive VHS cassette tape deck records the events in the actual patient care area, which is next door to this control room. The operators wear a headset that allows them to hear what is happening in the patient care room and also has a built-in microphone, allowing the operator to "speak" for the patient.

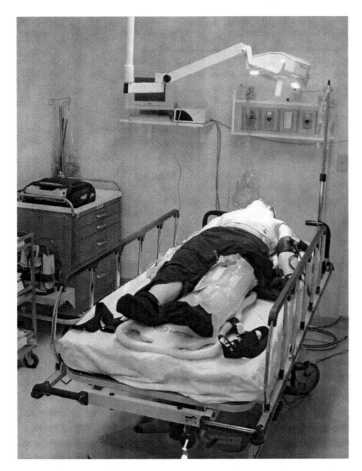

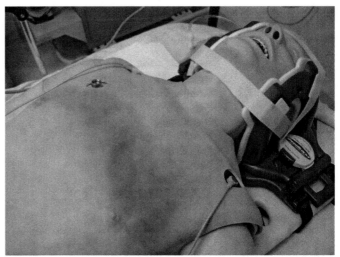

**FIGURE 47.8** Transportation patient prepared for simulation scenario. Common powder-based cosmetics were used to create the bruising from the patient hitting the steering wheel of his car.

**FIGURE 47.7** Our Trauma Care Room for patient simulation. A double system is used for this Laerdal MegaCode Kelly patient simulator. In addition to the usual controls for MegaCode Kelly, we also use the SimMan software and patient monitor to display changes in the patient's vital signs. The SimMan software does not control MegaCode Kelly.

the chance to become familiar with stretchers that are used in hospitals. An ancient gurney would not do. We learned early on that students who had been to local clinical affiliates had difficulty using our much older equipment during patient simulation. We now had something that looked akin to a real Trauma Care Room.

## 47.5 Getting People on Board

The open house for the School of Health gave us an opportunity to show off our new Trauma Care Room to a variety of guests. Current and potential students came by and talked with our patient. We used an inexpensive wireless microphone with its receiver connected directly to the voice input of the Laerdal MegaCode Kelly. Wall- and ceiling-mounted cameras provided visual feedback to the control room of anyone talking

to our patient. More importantly, many of our own faculty came by to see the new Trauma Care Room and told us they were impressed with the space, equipment, and even with the patient who looked pretty rough due to a car accident. Powder-based over-the-counter makeup was used to create realistic looking bruising on the patients face and chest (Figure 47.8). Our patient, named Willard Randolph Scott, was on a backboard, with neck brace, fractured left leg, and on the monitor. The ease of transportation and how the leg could actually be X-rayed was emphasized. It took about another week or so, but faculty began to form ideas on how to use the new patient and room. There was strong interest from the Respiratory Care and Radiography faculty almost immediately, and it did not take long for the Director of Nursing to join in. The willingness of these three groups to work together led to establishing the required performance competencies, which in turn led to the scenario development.

## 47.6 Developing the Scenario

Scenarios contain a collection of performance competencies that are to be either learned or demonstrated during some specified time. The more performance competencies involved, the longer the scenario and probably the more complex. The number and level of performance competencies is determined by the overall educational goal. In this instance, the goal of the scenario was to safely transport a trauma patient to the Radiology "Department" and back to the Trauma Care room. It seems very simple, but there is a fairly long list of performance competencies for each of the different clinical teams.

**TABLE 47.1**  Performance competencies for various clinical disciplines: 1 is lowest proficiency required, 4 is highest proficiency required

| Competency | Nursing | Radiography | Respiratory |
|---|---|---|---|
| Communication with patient | 4 | 4 | 4 |
| Communication with team | 4 | 4 | 4 |
| Patient ID | 4 | 4 | 4 |
| History taking | 4 | 3 | 4 |
| Basic vital signs | 4 | 1 | 3 |
| ECG change | 4 | NA | 3 |
| Record data | 4 | NA | NA |
| Act on data change | 4 | NA | NA |
| Act on patient Change | 4 | 4 | 4 |
| IV management | 4 | NA | NA |
| Intubation | NA | NA | 4 |
| Ventilation | 1 | NA | 4 |
| Team work: Coordinated effort | 4 | 4 | 4 |
| Body mechanics | 4 | 4 | 4 |
| Environmental safety: Tube, wire management, safe transport | 4 | 4 | 4 |
| Personal safety | 4 | 4 | 4 |
| Film quality/data | NA | 4 | NA |

Table 47.1 shows particular performance competencies and the clinical disciplines in which that skill would need to be demonstrated. A relative "level" of competence scale is included for teaching purposes, but is not intended to be terribly scientific. It is provided only as a guide about the possible level that must be obtained for "passing." A "1" under, say, "Nursing" and corresponding to "communication" indicates the lowest level of proficiency required. A "4" indicates the highest level of proficiency that is required for "passing." "NA" indicates a competency Not Applicable for this particular patient scenario. What is interesting is how many of the competencies are rated as "high" across disciplines. "Communication," "patient identification," and "act on patient change" are all rated as "4" for all disciplines. Faculty members from each discipline were to be present during the transportation scenario performance and provide feedback to their students.

## 47.7 Selection and Presentation of the Patient

Patient selection was based on the goal and related performance competencies. What level and type of performance competencies determined the type of patient needed for this particular scenario. As an example, we wanted the students to talk to the patient, but only have the patient respond with groans, a cough or two, and a pretty good scream while being moved onto the X-ray table. This is simple to achieve by using the built-in voice responses provided by the simulator manufacturer. If we had wanted a more sophisticated care

provider – patient interaction, we would have outfitted our patient with a battery powered, wireless, two-way voice communication system that would allow for real-time conversation. The operator would have the microphone in one hand and the remote control unit in the other. Sometimes, standardized patients or other faculty "play" or "act" the voice. This leaves the simulator operator free to control the simulator and video, etc. During patient transport, a very realistic patient–student interaction can be achieved by having the operator/speaker trailing close enough to the gurney to see what is happening yet far enough away so the students hear sounds only from their patient.

The goal was to safely transport a trauma patient, so it made no sense to use a "high-end" patient tied to nonportable components via short umbilicals. This transported patient needed to be completely self-contained, completely portable. Since we had just prepared one of our Laerdal MegaCode Kelly patients for the open house, he was almost ready to go "as is." The computer box and hand-held control can be battery powered, enabling transport of the patient without "wires." We wanted the focus to be on taking care of their injured patient, not accommodating the limitations of the patient simulator.

Our first production of the scenario was a bit rough for both students and faculty, but proved to be a valuable learning experience for everyone involved. The rehearsal consisted the faculty and me walking through the scenario, trying to make note of any difficulties that might arise. We waited until Radiography and Nursing students were present, and then found volunteers. Three nursing students formed a transport team. They performed all the required competencies expected of them with some prompting. It could be expected that some

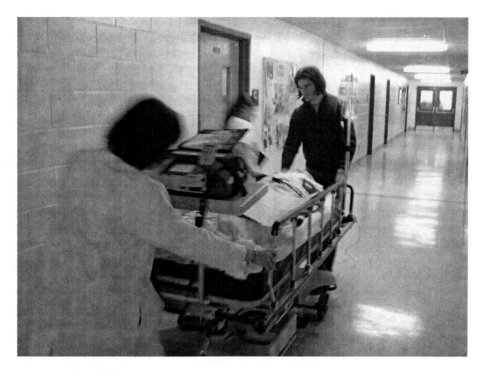

**FIGURE 47.9** Two nursing and one Respiratory Care student returning a patient to the Trauma Care Room. These students are coming from X-ray, where Radiography students actually X-rayed the chest, neck, and left lower leg of the patient.

prompting would be needed since they had never been in charge of a trauma patient, let alone the transport of one. The patient was unresponsive, and was kept stable throughout the entire time. He was not on a backboard, but did have a neck brace. No oxygen, no ECG, no blood, nor gore. Emphasis was on getting that patient to Radiology and back to the Trauma Care Room (Figure 47.9). We learned a lot.

I believe that all three nursing students had been in on a patient simulation at least once before and were somewhat familiar with the type of patient simulator used for this particular scenario. They did have trouble measuring the blood pressure, so I increased the volume of the pulse to help them out. Everyone, from students to faculty was pretty nervous. The Radiography students had never even seen a patient simulator before their patient Willard showed upon their doorstep. Only three Radiography students were selected by the Radiography faculty to perform the X-ray examinations. The rest of the class of about 20 students crowded about the control area trying to see what was happening through the small leaded glass window. Unfortunately, many of the Radiography students had some difficulty seeing what was happening in the X-ray room. In the future, it would be useful to mount wall and ceiling cameras and microphones in the X-ray room to allow projection of video and sound in the classroom. There was a bit of trouble getting the patient over to the X-ray table, and eventually, the entire gurney pad was used to move the patient. The nursing students did a good job of communicating with their patient (even though he was unresponsive). It appeared that everyone was taking responsibility for the safety of the patient, using proper body mechanics, coordinating movements, and talking with each other.

It was loud, took way too long, and almost "freaked" out a couple of students. On debriefing, every single student involved stated that they learned a lot and wanted more. The Radiography faculty and Director of Nursing were impressed with the results and said it was a good idea for an interdisciplinary scenario. Encouraged, we went back to the drawing board.

## 47.8 Presentation of the Patient: Version 2

Even though this first try at an interdisciplinary scenario went well for the patient, it left us wondering what could be improved. The single most important factor is the need for more prescenario instruction for the students. I found out, only after the scenario, that none of the Nursing students had ever transported a patient before. None of the Radiography students had to handle a patient from the emergency department. Unknowingly, I had created a pretty stressful situation for all the students. The good news is that the students did a remarkable job of taking care of our patient in spite of this being a first-time experience for them. Better instruction to the students on what is expected of them and more careful selection of students would have made for a less stressful learning environment. I think for the first-time transport scenario, we will keep the patient stable, with less monitoring needs, and focus more on the safe transport piece. For these same students' next experience, we can add another skill layer or two and again more for the third time and so forth, until it is as complex a scenario as they will be expected to be fully responsible for with their real patients.

Respiratory Care faculty were absent in the trial run, but they were really keen on having students participate the next time around. With input from Nursing, Respiratory Care and Radiography faculty, a new, improved scenario was constructed. This time, two Nursing students and one Respiratory student made up the transport team. Even though the two Nursing students had not performed patient transport, they were "comfortable" with monitoring a patient from the Emergency Department. The Respiratory student was working part-time in a local hospital and had performed airway management during real patient transport several times. Our simulated patient was intubated and hand-ventilated with 100% oxygen and connected to an ECG monitor through a Lifepack unit on the stretcher. There was an IV to manage. He was wearing street clothes and on a backboard with a neck brace. The transport team was told to observe and record vital signs and move their patient to Radiology quickly and return to the Trauma Care Room as soon as possible. The patient was to be "cleared" for neck brace removal by an X-ray of the neck since CT was unavailable. Furthermore, a chest film and X-ray of left lower extremity (for possible fracture) were to be performed. The left lower leg was in a pneumocast (an inflatable limb brace) placed by paramedics at the scene of the accident.

The Respiratory student wanted to stay with the patient during the X-ray exposure and, while that might happen in a "real" hospital with a lead apron provided, we could not allow it here at our center. Our lead-lined X-ray room is fully energized and is used by the Radiography program to teach students basic X-ray procedures using phantoms and a fully operational dark room. Students actually make exposures, develop the films, and then have them critiqued by Radiography faculty. Our patient's left lower extremity contains a radiographically correct fracture. Thus, the Respiratory student kept hand-ventilating the patient until the very last second before the exposures were made. After about the third exposure, the flow of movement between all parties became almost a well-choreographed dance. All students communicated with their own clinical colleagues as well as with the larger team formed by the three disciplines. Nursing students noted changes made to the patient's vital signs and documented them properly. Radiography students quickly realized that the stretcher would have to be turned around in order to align the proper angles for their films.

(This was a problem for the trial run.) The time it took for the X-ray room to be cleared, exposure made, and the resumption of hand-ventilating the patient fell from some 45 seconds in the previous trial production all the way down to 15 seconds. Everyone was becoming more efficient, helping each other, and working for the benefit of the patient. The last execution was very encouraging.

Recently, a local television news station came to our center searching for a story. What they saw on a video recording were two Nursing and two Respiratory Care students taking care of a badly injured patient in our Trauma Care room. The scenario was a slight upgrade to what we had done before "off camera." This time, the patient was brought to our Trauma Care Room by paramedics, represented by one of the nursing students who also happened to be a paramedic. Our Director of Health Services, a Physician Assistant (and Emergency Medical Technician) played the Emergency Department doctor. The Director of Clinical Education volunteered to be the patient's wife. It was a thing of beauty as the nurses, respiratory care, and ED doctor worked together to save the patient.

## 47.9 Effective Team-based Learning

Every student participating in these transportation scenarios has said that it was a good experience, they learned something, and would like more experiences like that. Hospitals use teams for most real patient care, and it would seem logical to do the same in simulated patient care. In the higher education arena, health programs tend to teach in isolation, with few opportunities to expose their students to interdisciplinary environments. I believe that patient simulation can change that approach to some degree. It is possible to provide at least some learning experiences that involve more than one discipline. The transportation scenario described here is one possibility, but it cannot create all of the "day-to-day" interaction seen in the hospital environment. Many of our patient scenarios are short lived, lasting for about 20–30 minutes, not enough time to provide a realistic experience of the extensive interactions between multiple clinical disciplines that real patients are exposed to.

A local health system is now talking to us about providing "minishifts" for use by their doctors. Instead of short scenarios, we are being asked to consider providing patient scenarios lasting hours at a time. Our goal of having the most effective team-based learning environment for students will be greatly enhanced by these "minishifts." They will be of sufficient duration and complexity to have Respiratory Care students give treatments, Medical Assisting students perform EKGs, and maybe even have Radiography students take portable X-rays right in the Acute Care or Critical Care environments.

# 47.10 Conclusion

Patient simulation is the only method of which I am aware, that will allow us to build a true interdisciplinary environment where all members of the health care team can learn and practice safely together without involving "real" patients. If there is even just a chance that creating interdisciplinary learning environments will result in better patient care, then I feel that we have to take the chance. That is what we are here for: saving lives and improving the quality of life.

# Make Your Own

*Always listen to experts. They'll tell you what can't be done and why. Then do it.*

Robert Heinlein

We all use tools made by others, and often find ourselves applying them to tasks not envisioned by their makers, or modifying them to make them more useful for our own needs. The acme or peak characteristic of the simulation professional is achieved when one can instantly transform anything into whatever is needed to support the teaching objectives. Some cultures call such individuals "Wizards." For most of us, this creative activity usually extends to selecting furnishings for our simulation environment to appear in clinical configurations appropriate for our teachings. For a few of us, this includes creating from scratch an entire simulator when what one needs is not commercially possible or affordable.

Ramiro Pozzo and Alfredo Guillermo Pacheco created an entire Emergency Care simulation environment *and* an Emergency Care Simulator to use in it. Just their environmental design and its fabrication would be a story in itself. Far more inspiring is their self-made robotic patient simulator that includes a dynamic mannequin, clinical monitor presentation, and controlling software. Most extraordinary, they accomplished this not embedded within a Silicon Valley neighborhood, and not expending a National Insitutes of Health–sized budget. In the absence of resources from others, they applied their own resourcefulness to create what they needed.

Ty Smith is a clinical resource in and unto himself. He not only understands the essences of anatomy, physiology, and pharmacology, he also describes their internal actions and interactions in the most powerful of all languages: mathematics. In addition, he is highly skilled in algorithms and computation. He has linked his mastery of these challenging domains to single handedly create one of the planet's more sophisticated clinical models. If ever the community of clinical educators wished to adopt and adapt one core engine for describing the physical interactions within and between human organs, we would be hard pressed to find a better tool than Ty Smith's creation.

48

# Development and Implementation of a Low-budget Simulation Center for Clinical Emergencies (Ambulance in a Box)

Ramiro Pozzo
Alfredo Guillermo Pacheco

## 48.1 Purpose and Means to its Fulfillment

At our hospital in Argentina, called the "Red SIES Sistema Integrado de Emergencias Sanitarias: Emergency Health System of the Province of Buenos Aires," the main use of our center called the "Simulation in Stabilization and Transport of Critical Patients," is to train emergency clinicians, nurses, and technicians in the management of critical events during the different phases of prehospital and interhospital critical patient transport. Our emphasis is on critically ill patients who unexpectedly develop serious conditions that should be rapidly diagnosed and treated given the limited resources during transport.

The director of our emergency system, Dr. Daniel Farias, an anesthesiologist, worked for years in an anesthesia patient simulation center, and decided to create a similar program for critical care transport. There were two main challenges to this: our budget was very limited and we were not able to find any commercially available simulators suitable for our very specific objectives and at prices we could afford. We considered the following simulators in our assessment (contact information URLs in Section 48.8):

The Simulation Group's VIRGIL
Nasco and Laerdal's Mr. Hurt Head Trauma Simulator
Laerdal's Crash Kelly
Laerdal's Ultimate Hurt
Laerdal's SimMan
Laerdal's ALS Simulator
Laerdal's ALS Skillmaster 4000
METI's Human Patient Simulator

Our solution to this dilemma was to create our own "trauma patient simulator" and build the simulated transport clinical care environment. This approach required us to:

1. Create a computer program to simulate physiologic data and generate the presentation of the patient's vital signs;
2. Construct the simulator body by joining a commercially available intubation head (Nasco) to an inert mannequin;
3. Adapt the back end of an unused ambulance for the teaching environment.

## 48.2 Inventing a Trauma Patient Vital Signs Simulator

Real clinical monitors are designed to detect, analyze, and present signals only from real sensors attached to real signal generators (live patients) that are under the "control" of real physiology and pharmacology. Since we had none of this, we invented all of it. We chose to use the clinical instructor's knowledge for both the control and the presentation decision-making functions. A computer program, written in

Pascal using the Delphi programming tool, was created to provide the operator with various clinical values and signal shapes in accordance with required real vital sign information values and display formats. For simplicity in programing and ease of use, we chose to exclude any models within the program. All vital sign alterations are done manually through a standard PC keyboard. The clinical signals we chose to include are those typically used for real trauma and shock patients during ambulance transport (Figures 48.1 and 48.2). This image can be observed on monitors in the control room, the ambulance, the shock room, and the debriefing room, and can present the following information:

EKG waveform shapes
Heart rate in beats per minute
SaO₂ waveform shapes and pulse oximetry saturation in %
Arterial pressure values (systolic, diastolic, and mean) in mm Hg
Temperature in degrees centigrade

EKG frequency and amplitude can be changed and different arrhythmias can be simulated (ventricular tachycardia, ventricular fibrillation, asystole, cardiac tamponade, etc.). With a

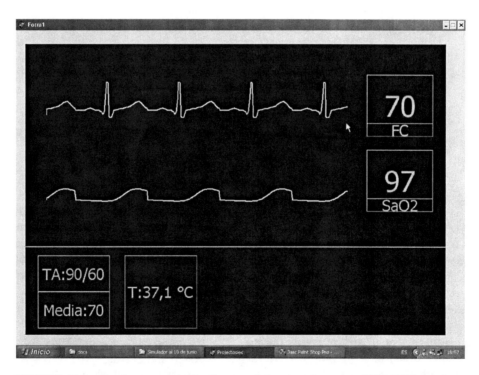

**FIGURE 48.1**   Simulator screen with all sensors "placed on the patient." The EKG waveform indicates a normal sinus rhythm; a heart rate of 70 bpm, an oxygen saturation of 97%, a systemic blood pressure 90/60 mm Hg, with a mean arterial pressure of 70 mm Hg, temperature 37.1°C. FC is abbreviation of Frecuencia Cardíaca, Spanish for heart rate; TA is an abbreviation for Tensión Arterial, Spanish for blood pressure; Media is Spanish for mean, in this case, mean blood pressure. SaO₂ is oxygen saturation.

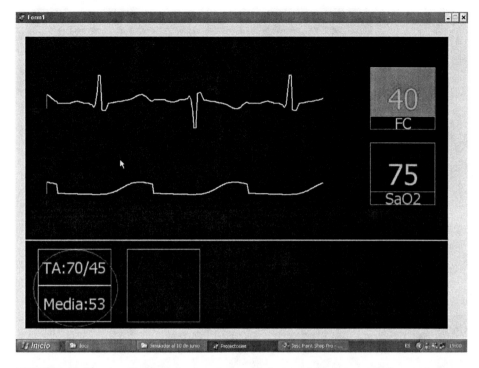

**FIGURE 48.2** The EKG waveform indicates an abnormal rhythm; a heart rate of 40 bpm, an oxygen saturation of 75%, a systemic blood pressure 70/45 mm Hg, with a mean arterial pressure of 53 mm Hg. Same units and abbreviations as in Figure 48.1.

specific Hot Key, values and curves can be turned on or off, or be changed (listing of Hot Key functions in Section 48.9). When a sensor is placed on the patient, the operator clicks on the screen in the appropriate place and the corresponding signal shows up. When selected again, the signal disappears. This allows for the operator to:

- Set the patient's physiological values.
- Simulate disconnection or malfunctioning of a sensor.
- Simulate nonregistration of a sensor because of physiologic problems, for example, no pulse oxygen signal when the blood pressure is too low.
- Simulate a measurement that is in the process of being measured – for instance, cuff blood pressure. A Hot Key

is used by the operator and shows the measuring sign (midiendo is Spanish for measuring) until the operator inputs the desired pressure values.

### 48.2.1 EKG

The electrocardiography pattern and the corresponding heart rate value can be selected and presented by the instructor:

- Heart rate changes
- Extra systoles (Figure 48.3)
- Ventricular fibrillation (Figure 48.4)
- Cardiac tamponade (Figure 48.5)
- Asystole (Figure 48.6)

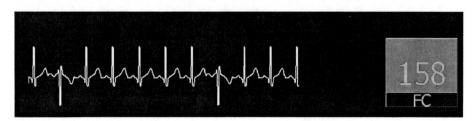

**FIGURE 48.3** Sinus rhythm with extra systoles.

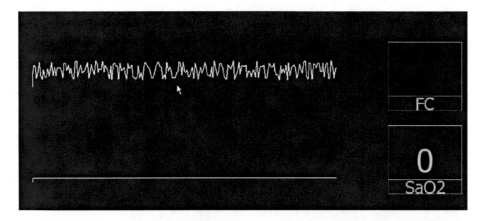

**FIGURE 48.4**   Ventricular fibrillation with no associated heart rate and an oxygen saturation value indicating zero perfusion.

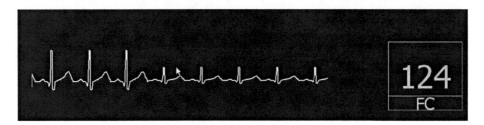

**FIGURE 48.5**   Pericardial tamponade.

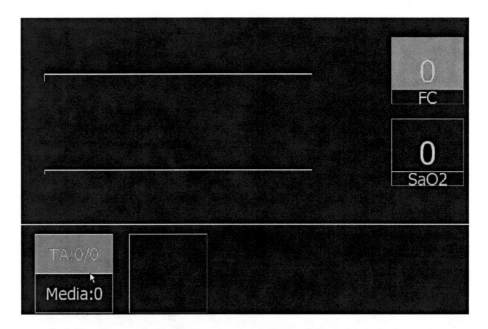

**FIGURE 48.6**   Asystole with an associated heart rate of zero and an oxygen saturation value indicating zero perfusion.

## 48.2.2 Pulse Oximetry

You click with the mouse on the screen once you observe the student has placed the sensor on the patient's finger and the pulse oxymetry trace and associated saturation value appear (Figure 48.1). If there is asystole or ventricular fibrillation, or when systolic pressure falls below 60 mmHg, the curve automatically disappears.

## 48.2.3 Noninvasive Arterial Pressure

Both systolic and diastolic pressures can be modified independently. When this happens, the simulator calculates the average pressure and updates it on the screen. A Hot Key presents the measuring sign (midiendo is Spanish for measuring) until the operator inputs the desired pressure values (Figure 48.1). First, "measuring" arterial pressure appears, followed by operator-provided systolic and diastolic and computed mean noninvasive blood pressures.

## 48.2.4 Body Temperature

The operator can select and display a value of a temperature sensor (Figure 48.1).

## 48.2.5 Alarms

When any parameter is below or above an operator-determined value, a sound and visual alarm is activated:

High or low arterial pressure
High or low heart rate
Low oxygen saturation

# 48.3 Our Simulation Center

Our Simulator Center has six basic physical areas on two floors (Figures 48.7 and 48.8):

1. Ambulance Simulator
2. Urban Rescue and Stabilization Area
3. Stabilization and Life Support Area, "Shock Room"
4. Debriefing Room
5. Skills training and procedures with task-training mannequins, heads, etc.
6. Control Room

## 48.3.1 Interconnectivity

Communications for sound, visuals, and data between these areas are shown in Figures 48.9, 48.10, and 48.11, respectively.

## 48.3.2 Ambulance Simulator

This "simulation room" is composed of the rear box of an old ambulance, normally called assistance compartment. Its shape, dimensions (3 m × 2.5 m), logo, light system, and equipment reproduce those of a regular UTIM (Spanish for Mobile

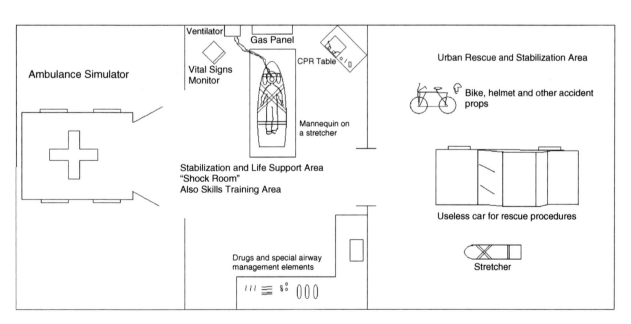

**FIGURE 48.7** Lower level with Urban Rescue Area, Shock Room, and Ambulance Simulator.

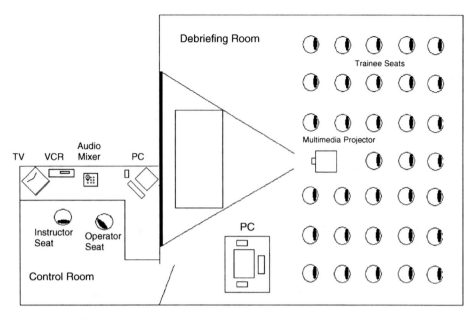

**FIGURE 48.8**  Upper level with Simulator Control and Debriefing rooms.

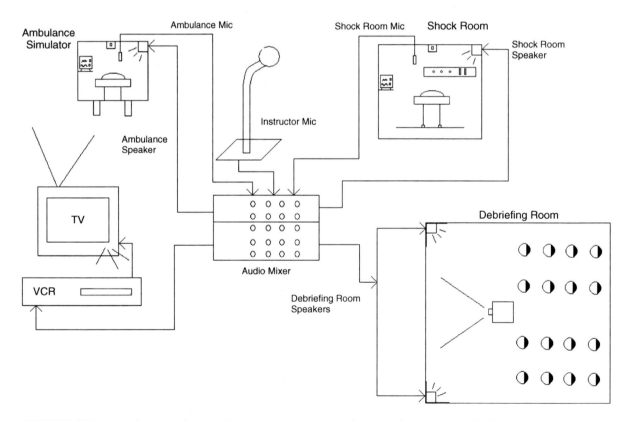

**FIGURE 48.9**  Sound system diagram showing microphones in the control and various clinical areas connected to a central audio mixer, which in turn sends these sounds to speakers in the hands-on and observation rooms as well as a recorder.

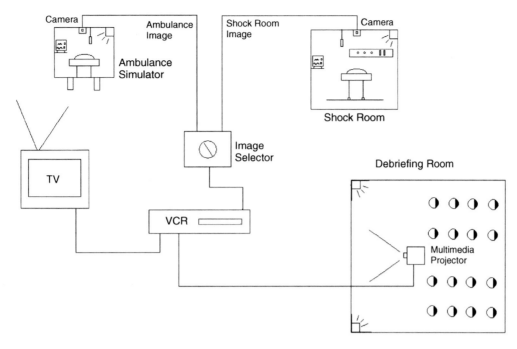

**FIGURE 48.10** Video system diagram, showing two cameras connected through a selector box to a recorder, which in turn sends these images to a local display and a projector in the debriefing room.

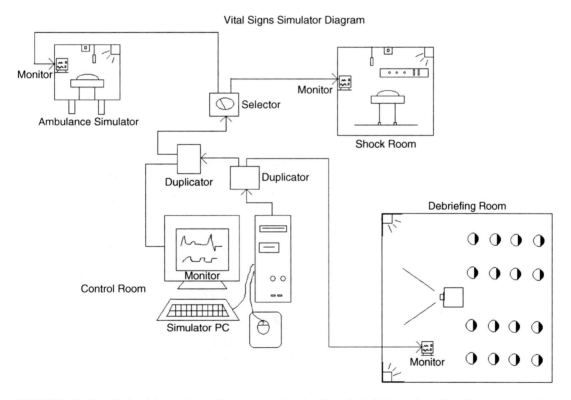

**FIGURE 48.11** Vital signs simulator diagram showing the clinical vital signs selected by the operator in the control room and sent to displays in the hands-on and observation rooms.

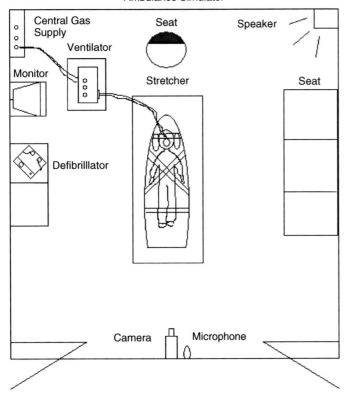

**FIGURE 48.12** Ambulance simulator diagram showing the centrally located patient surrounded with clinical care devices typical in kind and location in our real ambulances.

Critical Care Unit) used for transporting critical patients (Figure 48.12):

- Central oxygen panel: connections for transport ventilator, aspiration, or oxygen therapy
- Defibrillator
- Vital signs simulator
- Transport ventilator
- Six liter transport oxygen tube
- Fire extinguisher
- Drug box
- Airway box
- Other consumables (gauze, syringes, needles, serum solutions, antiseptics, etc.)
- Biosafety kit

# 48.4 Ambulance Sights, Sounds, and Vibrations

Inside the ambulance there is a fixed camera pointing at the general working area as well as portable camera that can collect through the windows for zooming in for close-ups of specific maneuvers. An environmental microphone fixed in the middle area of the roof captures sound. The instructor's voice is heard from a speaker placed in an upper corner of the ambulance box. The ambulance box rests on a mechanical system (industrial fan) that turns on when the ambulance is "on the road," adding realistic motor noise and vibrations. A key teaching objective for using patient simulation within the confines of a real ambulance is the minimal space and awkward postures within which patient care takes place. A real Ambulance box makes an excellent environment for such learning (Figures 48.13a–d).

**FIGURE 48.13a**   Outside view of Ambulance Simulator.

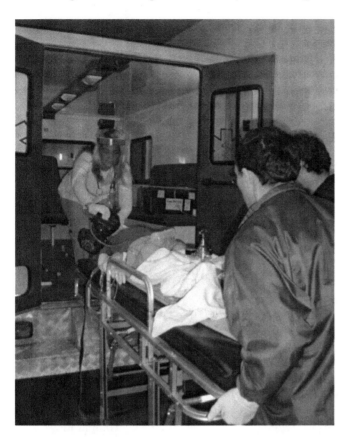

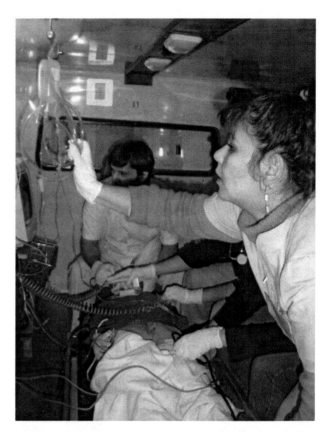

**FIGURE 48.13b** Loading patient into Ambulance Simulator.

**FIGURE 48.13d** Defibrillation inside Ambulance Simulator. When "Clear" is called, how far away can you move?

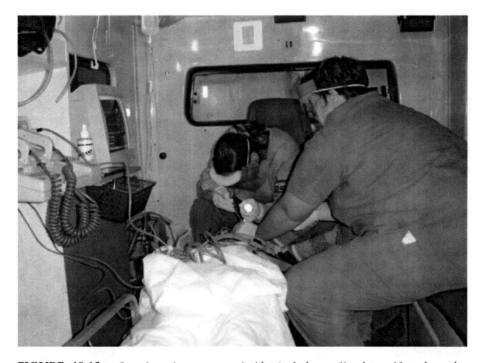

**FIGURE 48.13c** Securing airway access inside Ambulance Simulator. Note how the position of the student with respect to the patient within the Ambulance exacerbates an already challenging task.

## 48.5 Scenarios and Spaces

Each of our teaching spaces were designed and equipped for supporting critical teaching objectives.

### 48.5.1  Urban Rescue and Stabilization Area

This is a simulated outdoor space that recreates a public area where different types of vehicles can be placed in accordance with the workshop's teaching objective – such as cars, motorcycles, bicycles, etc. Here, the scenario is developed and victims (mannequins and actors) are placed all over, inside, and outside the vehicles.

### 48.5.2  Stabilization and Life Support Area "Shock Room"

A 5 × 6 m surface that replicates a shock room or stabilization room of a hospital emergency service. At the head of the patient bed is a defibrillator and a central gas panel connected to an oxygen supply (Figure 48.14). Connected to the gas panel is a pressure-controlled ventilator with a patient circuit, a flow meter for connecting to an oxygen mask, or a mask for nebulization and inhalation therapy. At the other side is a Trauma Patient Vital Signs Simulator. There is a table with a full complement of airway management elements and equipment, such as oxygen masks, nebulization masks, intubation equipment, and ampules that simulate drugs (Figure 48.15a). Another table contains equipment for special airway access and maintenance (laryngeal mask airway, Combitube, cuffed oropharyngeal airway) and for special intubation (Fastrack) and rescue procedures (crycothyroid puncture, crycothyroidotomy), tracheostomy tubes, and other shock room props (Figure 48.15b).

### 48.5.3  Debriefing Room

Many sights and sounds collected in the simulated clinical care areas are captured and replayed in the Debriefing room, including the patient's vital signs (Figure 48.16).

### 48.5.4  Control Room

Just like in the hidden control rooms of most Simulation Centers, the operator and instructor work together to operate the simulator, control the audio-visual system, record and capture images, interact with the trainee, etc. (Figures 48.17a and 48.17b). The main problem with our particular control room is that it is upstairs, while simulations are downstairs, which means, a stressful 30-second sprint downstairs when there is a technical problem in the middle of a simulation session. It also meant a lot of cabling complications when building the simulation center.

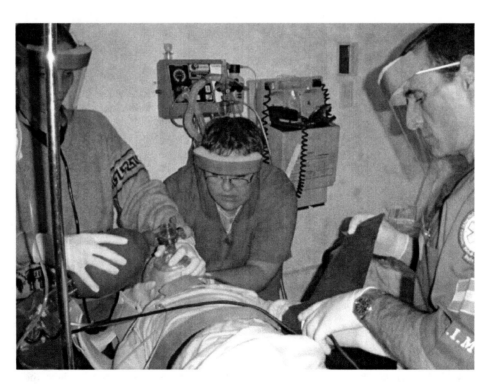

**FIGURE 48.14**  Shock Room with typical emergency care devices for ventilation/oxygenation and defibrillation.

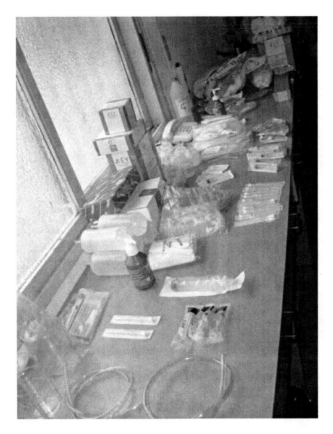

**FIGURE 48.15a**  Shock Room common airway management
elements and equipment.

**FIGURE 48.15b**  Shock Room special airway access and maintenance devices.

**FIGURE 48.16**   The scenario as shown in the Shock Room (Figure 48.14) is presented in the Debriefing Room.

**FIGURE 48.17a**   Control Room with video recorder, display operator microphone, and audio mixer.

**FIGURE 48.17b**   Control Room with operator controlling the patient's clinical vital signs.

### 48.5.5 Mannequin

The patient consists of a commercially available intubation head that we combined with a body (Figures 48.18a and 48.18b). The body is that of a polyurethane mannequin used to exhibit clothes for sale. This makes it easy to work with without worrying about it being deformed (Figures 48.18c and 48.18d). Modifications have been done on its thorax, arms, abdomen, and genitals.

This partial-task trainer doll, as he is a trauma victim, is usually presented with immobilization elements (cervical collar, immobilization vest, lateral immobilizers, etc.) (Figure 48.19). At the thorax, carving has been done to place tubes in the neck to simulate trachea and esophagus. The cardiac area has been hollowed out to enable placement of a 100-ml plastic bag that is connected to the surface through a tube under the xiphoid appendix to simulate pericardiocentesis. The bag acts as the pericardiac sac, and can be filled with fake blood though a tube hidden under the left shoulder. Also, in the thoracic cage, there are two urine collector bags acting as lungs. They assume the shape of the anterior surface of the thorax from the clavicles to the diaphragms: they are fixed with adhesive tape proportionally, so that the insufflation of the thorax behaves like a normal breath. In each hemithorax, an 11-mm tube was placed to enable simulated placement of a 9- or 10-mm intrathoracic drain tube. In each hemithorax, at the intermammarian line in the axilla gap, these tubes were exteriorized, simulating the trajectory where drain tubes are placed. These tubes are inside the thorax, and both come out of the mannequin, behind the shoulders. Each one is connected to a three-way stopcock, so that through it fake blood, air, or oxygen can be sent, simulating hemothorax pathologies, pneumothorax, or both when an underwater drain (Bulleau jar) is connected to the system.

On both arms, rubber tubing was placed to simulate veins, where peripheral IV lines are placed for the administration of parenteral solutions. This tubing also goes through the back of the shoulders with a three-way stopcock to regulate the "blood volume:" either letting it go to the circulatory system or diverting it outside to simulate renal failure or a severe hemorrhage. Through this tubing on both arms, other tubes are connected, which are in the posterior and dorsal part of the trunk that at the level of the groin converge in a receptacle at the hypogastria region simulating a bladder, through which a tube going to the exterior can be connected to a Foley catheter no.14, and finally to a urine collector bag.

At the abdominal region, the epigastrium has been carved out to place a receptacle simulating the stomach in which fake blood, secretions, or vomit is placed. These fluids can be evacuated through the nasogastric catheter. Also, on the abdominal region, a receptacle has been placed at the level of the left iliac fossa, exteriorized through tubing with an orifice on the abdomen's skin, 2 cm under the belly button, through which a large (e.g., subclavian K-10) catheter can be placed, to simulate the procedure of peritoneal washout or diagnostic peritoneal lavage.

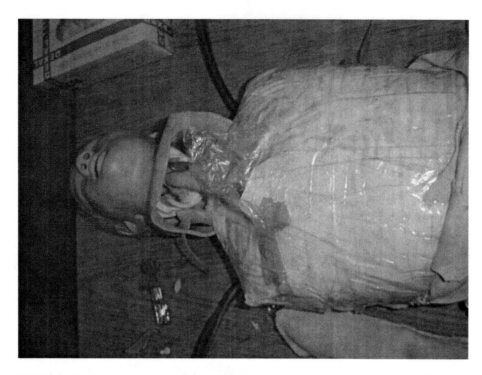

**FIGURE 48.18a**  Our mannequin is a hybrid, composed of an airway trainer head grafted onto the body of a store-clothing mannequin.

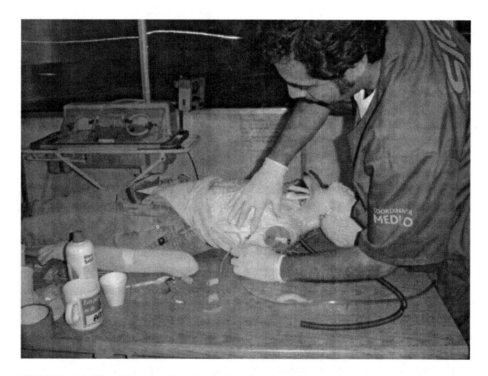

**FIGURE 48.18b**  Transforming the body of a typical store-clothing mannequin into a clinical simulator suitable for emergency transport and care.

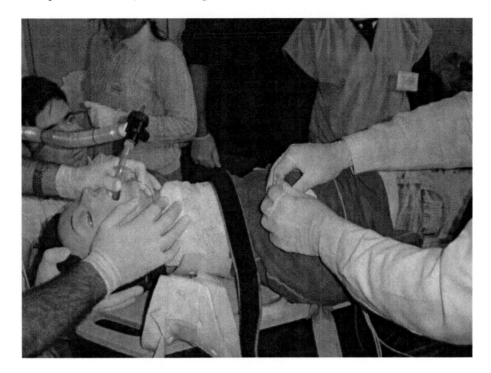

**FIGURE 48.18c**  The chest material allows the student to clearly indicate how well they know needle decompression placement.

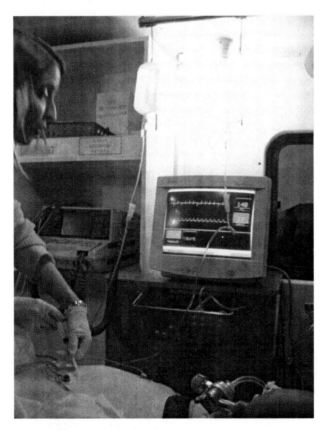

**FIGURE 48.18d**  The indicated physiological responses indicate the effectiveness of treating Pericardial Tamponade.

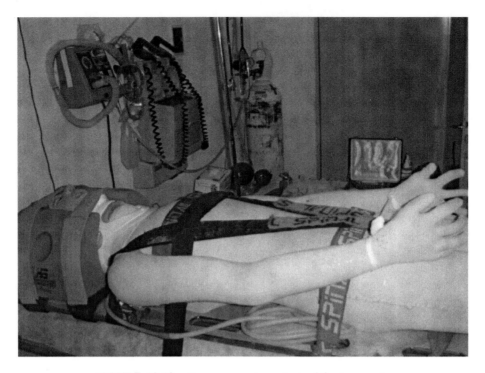

**FIGURE 48.19** Our mannequin, restrained for transport.

## 48.6 Scenario Execution

At least two instructors and an operator who controls the software are required to produce the scenarios. One of the instructors is at the control room next to the operator. He is usually the head instructor and takes decisions on how the case develops. With some practice, this instructor can control the simulator software and control the case at the same time, although he is more likely to make mistakes with this high workload.

The other instructors are at the scene of action (Ambulance or Shock Room). They participate as guides for the trainees or as part of the scenarios doing role-playing. Camera images, the vital signs monitor, and sound from the scenario are seen and heard in the debriefing room by up to 50 people and are recorded.

## 48.7 Conclusion

The immediate success obtained using this simulator center for training our emergency health care staff from all over the country has encouraged us to start working on a project composed of two stages:

Development of a mobile version of the simulator: Using a real ambulance, the adapted mannequin, a computer with the vital signs software, and a few audio-visual props, we will be able to take our simulation cases practically anywhere in the country.

Development of other similar centers: Other emergency centers in the more populated cities of Argentina (Córdoba, Tucumán, Mar del Plata) have expressed their interest in instituting our idea and adapting it to their needs and requirements.

There are many aspects of the software, the ambulance, emergency room, and the body that can be improved to be more realistic and to add teaching variables, and we are also working on many of these aspects:

Software: new arrythmias, different alarm sounds, capnography simulation

Ambulance and Shock Room: better image and sound, better ambience and surroundings

We have not yet performed methodological or scientifically designed research on the academic and clinical outcome results of this simulation technique as a teaching tool. However, comments from instructors who have used it in several workshops, responses from the trainees, and our own observations suggest that it is sufficiently close to reality to be used as a teaching tool for several important learning objectives.

## 48.8 Simulator Vendor Contact Information

The Simulation Group's VIRGIL:

http://www.medicalsim.org/ virgil.htm

Nasco and Laerdal's Mr. Hurt Head Trauma Simulator:

http://www.enasco.com/healthcare/ProductDetail.do?sku=LG02152U

http://www.laerdal.com/document.asp?subnodeid=7423393

Laerdal's Crash Kelly:

Laerdal's Ultimate Hurt: http://www.laerdal.com/document.asp?subnodeid=7423385

Laerdal's SimMan:

http://www.laerdal.com/document.asp?subnodeid=7320252

Laerdal's ALS Simulator:

http://www.laerdal.com/document.asp?subnodeid=16358619

Laerdal's ALS Skillmaster 4000:

http://www.laerdal.com/document.asp?subnodeid=7320302

METI's Human Patient Simulator:

http://www.meti.com/Product_HPS.html

## 48.9 "Hot Key" Keyboard Commands for our Trauma Patient Vital Signs Simulator

Note the manageable number of instructions to control the device:

A: raise systolic pressure
S: lower systolic pressure
Z: raise diastolic pressure
X: lower diastolic pressure
F3: measure arterial pressure (simulates an automatic noninvasive arterial pressure monitor (Figure 48.2))
N: raise temperature
M: lower temperature
K: raise $SaO_2$
L: lower $SaO_2$
O: raise heart rate
P: lower heart rate
F1: sinus rythm with extrasystoles (Figure 48.3)
F2: sinus rythm
F5: venticular fibrillation (Figure 48.4)
F7: tamponade (Figure 48.6)
F6: end of tamponade
F9: asystole (Figure 48.5)

# 49

# Physiologic Modeling for Simulators: Get Real

N. Ty Smith

## 49.1 Models: Of You, With You, In You

The purpose of this chapter, as it is of all the chapters in this book, is to help you buy and use simulation products more intelligently. "Buy" and "use" are different concepts. You can buy the best simulation products in the world, but if you or your students use them in an unstructured fashion, they can impede more than they help. The chapter also brings out questions that you have been dying to ask, such as "Why a physiologic model?" "Do all simulators need a model?" or "Is there anything to a physiologic model besides physiology?" (Hint: There is.) We'll spend considerable time to remind you why a good model is so important for good simulation. Finally, this essay arrived in the editors' mail engendering a slight tinge of cynicism. It left the editors' desk with some additional healthy cynicism and usually helpful constructive criticism.

The following glossary is optional reading. You may refer to it when you feel you don't understand a term or when you want to learn more about a term. This is a new subject for many of you, and some of the terms that I use may seem as if they had come from an exotic foreign language.

## 49.2 Glossary

**Buyer and User.** The buyer is the customer, the one who raises the cash or allocates the funds for simulation, while the user is the consumer, the one who has to make a silk purse out of the pigs' ears that the customer bought and foisted upon the users. In other words, the user is the one who implements and structures the simulation center. Rarely are buyer and user the same. Examples of customers include university presidents, deans, departmental chairs, alumni associations, and parents. Examples of consumers comprise mainly students and their teachers.

**Student.** A student is anyone in the trenches, the one who learns, in our case, with simulation. Age and experience do not enter into the definition. Students include a wide variety of health care workers such as nurses and prehospital paramedics, not just "medical students."

**Instructor.** The instructor is the one who stands over the trenches, making sure that the simulation is being used for the optimal good of the student.

**Model.** When I say, "model," I mean "mathematical model." Thus, I shall ignore these models: animal, *in vitro*, chemical, structural (Harvey's observations on the circulation), qualitative (Starling's Law of the Heart), and mental. The last models are those models used by instructors to try to compensate for the lack of an adequate physiologic model in their simulator, and possibly gaps in their own *internal* model of physiology.

**"Our model."** "Our model" is the BODY Simulation model. More detail can be found at our web site[1] [1].

**Agent.** An agent is anything that is transported by the physiologic model: $O_2$, hemoglobin, isoflurane, propofol, cyanide, histamine, endotoxins, etc.

**Mathematical model.** The following is like reading the fine print. A mathematical model consists of elements each describing in mathematical terms the relation between two or more quantities. If all the descriptions are correct, the model will simulate the behavior of real-life processes (systems, in physiologic terms). Especially when many different subsystems are involved, the models are useful in helping understand all the complex interactions among these subsystems, for example, cardiovascular and respiratory interactions. A useful characteristic of a mathematical model is that its predictions are quantitative. Bassingthwaighte [2] points out that these models are therefore refutable, thereby facilitating the entire process of science. What's good for science is good for simulation.[2]

**Transport model.** Transport models are accurately named. Essentially, they transport something from one place to another. That "something," in a physiologic model, we call an "agent." In the case of the circulation, transport models, of course, carry agents around and around. Transport modeling is incredibly powerful. One has to think optimistically and creatively when developing transport models. Assume, until proven otherwise, that a transport model can carry anything, anywhere, and by any route.

**Multiple model.** Most good complex models are modular. Each of these modules can be a model, and the entire model is called a "multiple model." The physiologic systems themselves are submodels, for example, cardiovascular, respiratory and liver systems. Put another way, a model with two or more modules is called a "multiple model." Essentially, anything that can be transported (in our case an agent) can be a submodel in a multiple model. This includes anything from $O_2$ to a bacterial species to a specific atom. Thus, our model has about 115 modules, or submodels. Ideally, each model can be sufficient unto itself, and can therefore be tested before it is incorporated into the main model.

**Whole-body model.** We arbitrarily define a whole-body model as one that includes the circulatory and respiratory systems, plus at least two other major systems. There is no whole-body model that includes all the major systems, much less any one system in detail. The good whole-body models have not been used extensively in education, while most simulators have used *ad hoc* models assembled to meet the perceived needs. Although our topic is physiologic modeling, to be useful for clinical simulation, a model should also incorporate pharmacology, toxicology, and physics. Through physiologic whole-body models, one can advance to these topics, as well as scientific concepts and clinical situations.

**Hooks.** Hooks are connections within a model or with the outside world. Hooks can be uni- or bidirectional.

**Part-task trainer.** A part-task trainer simulates part of the total system. It often consists of a computer and one or many pieces of hardware. It is a self-contained system. Examples in flight simulation could be the hydraulic system or the flight management system. We are trying to be broader than just anesthesia. Examples include an ECG simulator used to mimic ECG abnormalities during ACLS training, a lung model, an intravenous catheter insertion simulator (physical or virtual reality), a ventilator, a right-heart catheterization system, or a "body part" that approximates human anatomy. A body part could be an intubation head or a bladder catheter ("Foley") insertion trainer. By definition, a part-task trainer can include any part of the whole, but in medicine, the trainers are usually broken down into small parts – often body parts.

**Parameter.** A user-adjustable quantity that governs some aspect of a device's performance. I recently wrote an encyclopedia article on Physiological Systems Modeling [1] in which was described a model as a device, partly because the encyclopedia's topic was medical devices.

# 49.3 Modeling Versus Simulation

The difference between modeling and simulation is important. Modeling attempts to identify the mechanisms responsible for experimental or clinical observations, while with simulation, anything that reproduces experimental, clinical, or educational data is deemed acceptable. For the purposes of this chapter, when you "run a model," you are performing a simulation. Thus, "simulation" is not necessarily an overworked word, although it does extend far beyond for clinical or flight simulation. Most models are not "simulators," however. That word (simulator) should be reserved for those models that allow a nonmathematically oriented user, teacher, or student to interact with the model and perform in a simulation.

In a previous publication [1], one reviewer made the following request: "Tighten the contrast between modeling and simulation; maybe point out that 'simulare' originally meant 'to do as if,' 'to cheat'." As the reviewer suggested, the backgrounds and connotations of "simulate" and "model" are strikingly different. Reading the etymology of "simulate" and "simulation" is a chastening process, emphasizing as it does the unsavory past of the terms. The Latin "simulare" can indeed mean "to do as if," "to cheat," to "feign," or "to counterfeit," as the reviewer suggested. Hence, a set of its meanings, from the Oxford English Dictionary, includes "To assume falsely the appearance or signs of (anything); to feign, pretend, counterfeit, imitate; to profess or suggest (anything) falsely", and "The action or practice of simulating, with intent to deceive; false pretence, deceitful profession." In contrast, "model" has a more virtuous past.

The Middle French "modelle," on the other hand, implies "perfect example worthy of imitation." Thus, one definition, again from the Oxford English Dictionary, is "Something which accurately resembles or represents something else, especially on a small scale; a person or thing that is the likeness of another. Frequently, 'in the (very) model of.'" Another definition is "An idealized description or conception of a particular system, situation, or process, often in mathematical terms, that is put forward as a basis for theoretical or empirical understanding, or for calculations, predictions, etc.; a conceptual or mental representation of something [see Glossary: 'model']." Our final definition is "A person or thing eminently worthy of imitation; a perfect exemplar of some excellence. Also: a representative specimen of quality."

If this comparison between simulation and modeling seems cruel, outlandish, or irrelevant, consider the following excerpts from the Society for Simulation in Healthcare listserv. The problem posed was this. "Has anyone succeeded in creating a good septic shock scenario on the BLANK simulator? We've been adjusting a variety of parameters and have been unable to set the cardiac output above 10 L/min while keeping the arterial BP below 100 mmHg systolic. It seems impossible to set the SVR [systemic vascular resistance] as low as 400 dyne seconds/cm$^5$. Any tricks?" Note the word "tricks." And note the "Wizard of Oz" and other deceptive approaches to the problem described below in the next six paragraphs, all direct quotes, albeit slightly edited to protect the innocent.

> One thing you might try is tricking the monitor, since what you really want is a high cardiac output reading. Some combination of probe temperature and internal settings might do the trick....
>
> However, what we do, for most cases, is not let them measure the cardiac output. Instead, we tell them the value, and let them do the calculations....
>
> First rule of simulation: do what ever it takes (out of sight) to generate the story you want your students to experience.
>
> If you are using the BLANK (simulator) to generate real vital sign data for a clinical monitor (specifically HR, CO and BP), from which the values are

collected for the numerical analysis to generate SVR and stroke volumes, and your students will never know what the internal, behind the curtain BLANK values are.... then perhaps you might consider altering the 1:1 ratio of BP "out" from the BLANK to BP "in" on your clinical monitor (i.e. change the calibration on your clinical monitor BP input).

> For example, make your clinical monitor BP read/display 50 mm Hg for a 100 mm Hg BLANK input value. Now the BLANK simulation program can present a patient that is healthy enough to generate the CO values at the HR values you want, and which the clinical monitor will collect and display 1:1, but the clinical monitor will compute SVR based upon BP values half as much as the BLANK is producing.
>
> Try this, and see if this helps you produce the septic shock scenario you desire.

With a good model, all this deception is not necessary, and the time spent on the deception could be better spent in other more productive areas. With both endotoxic shock and anaphylactic shock, our model can produce SVR values as low as 350 dynes/sec $\times$ cm$^{-5}$, mean arterial pressure of about 50 mm Hg, and cardiac output over 10 L/min, i.e., a true hyperdynamic state. Infusing fluid during endotoxic shock can accentuate the hyperdynamic state.

The remarkable difference between the two terms simulation and modeling can be transmuted to our basic philosophy: a simulation is no better than the model that drives it. Put another way, the model is arguably the most important part of the simulation. It helps prevent the simulation from being something that "feigns."

Clinical simulators need a model to drive them. Although practically every simulator that you buy should have a model somewhere, I'm not sure how many actually do. You should therefore ask questions of the sales person, or better yet, the current users, about any model that might be involved with the simulation, just as you would kick the plastic in the mannequin, try out a few canned scenarios, or try to craft a scenario – by yourself, without somebody looking over your shoulder. Don't be satisfied with glib, vague representations of how great the model is, and that includes ours. Ask specific questions about the model. Try out a difficult scenario or two and see if the results appear "real." I'll discuss in this chapter a few examples of where to look for realism; for example, the shapes of the arterial and central venous pressure wave forms, "septic shock," CPR, what's happening to brain oxygenation, quantitatively, during hypoxic or ischemic conditions. When does brain damage begin during CPR?

Every simulator that you buy does *use* a model, but most simulators rely upon the consumer to provide most, if not all, of the modeling. While the cost of a model provided by a vendor is very obvious, the cost of the model provided by the minds of the instructors using clinical simulators might not be as obvious. For example, one full-body patient simulator vendor assumes that the user will provide all the modeling, during both scenario design and presentation. Their device can provide a very effective learning experience when guided by an operator with far greater knowledge and wisdom than the students. Another full-body patient simulator vendor offers an internal model that can be used to assist the operator. Today, the fact that either type of commercial product is in any way useful in clinical teaching (as opposed to clinical entertaining) is the overwhelming amount of modeling brought to the lessons within the minds of the instructors: a mental model. Again, see Glossary: "Model."

All whole-body simulators need a whole-body physiologic model. Many part-task trainers should also incorporate "modified" whole-body models. For example, a simulator of the catheterization of the heart could simply model the pertinent vessels and chambers of the heart, but the realism would be enhanced if the entire cardiovascular system were modeled, because of the complex interactions among the components of the cardiovascular system, as well as its control. Catheter-induced arrhythmias, for example, affect the entire cardiovascular system. A part-task trainer of a ventilator should incorporate a whole-body physiologic model, so that the student could learn how the changes in ventilator parameters, for example airway pressure, affect the patient's respiratory and cardiovascular systems. These changes can be observed on the patient monitor. Another example of the need for a model in a part-task trainer is in the intubation head. Medics, in practicing on a "head," can take up to 2 1/2 minutes to successfully intubate the head. This is clearly not acceptable – even less acceptable when hemorrhage and fear accompany the apnea [3]. Although anyone practicing intubation should be made to understand all the important techniques that pertain to intubation, the bottom line is time. They must understand when it is time to quit a given attempt and regroup.

Surgical simulators are not exempt from this rule. If a surgeon cuts a blood vessel in a laparoscopic simulator, the vessel should bleed and the hemodynamic changes should be evident to the student, especially if an artery is nicked and there is an artery-to-floor shunt, producing a precipitous fall in blood pressure. Early in the training, the instructor may turn off the model and concentrate on the mechanics. The model should, however, be used eventually to complete the student's education. Without a model, many part-task trainers provide blind exercises without any physiological and/or clinical context.

Although most full-body clinical simulators need a physiologic model, physiologic models do not necessarily need a full-body clinical simulator. Screen-based simulators are eminently suitable for many purposes, and are much less expensive. With a screen-based simulator, one can spend more time learning the rich amount of information that the model provides: basic physiology, for example.

## 49.4 Why Does a Simulator Need a Good Physiologic Model?

What you see in simulation is what the model gives you. If the physiology isn't realistic, the simulation isn't realistic. Those who make simulators cannot, yet, do much about the trade-offs between high-cost, low-durability tissue-like plastics and low-cost, high-reliability unnaturally stiff plastics that they use in their mannequins, but they can about an inadequate mathematical model. The size, detail, and good planning of a model will contribute to the correctness and realism of the physiology, in turn, enhancing the realism of the simulation. What are some of the details that should be included? A beating heart, breathing lungs, and lots of compartments, such as brain, heart, kidney, and liver, are essential. Temperature modeling is extremely important – and extremely difficult. Body temperature influences most physiologic and pharmacologic phenomena. Specialized models, such as obstetrical/delivery models, require exquisite detail and creativity.

## 49.5 Modeling as an Hypothesis

A model is an hypothesis, albeit an hypothesis that is presented mathematically. The hypothesis for clinical models and simulation, however, is "Is it realistic or is it not realistic?" The goal for simulation realism *vis-a-vis* the model should be to pass the Turing test: If the instructor or student cannot tell the difference between the output of the model in a simulation and that from a real patient, under a variety of circumstances, it passes the Turing test, and it is "real." Thus, a clinical simulator should have an accurate physiologic model as its engine. If the arterial pressure waveform from the model does not look real, for example does not have a notch and/or the proper shape, the simulation presentation fails the Turing test in that area.

Similarly, if the model cannot achieve very low vascular resistances during "septic shock," it does not pass the Turing test for septic shock. These failed-Turing-test models, unfortunately, can teach inaccurate information to the student. Other Turing-test questions to ask of a model include the following. Do all patients respond in the appropriate way to a given type of shock? Does age make a difference in this response? Can you create a wide variety of realistic patients, including the aged?

## 49.6 Modeling as a Compromise

Even the best model is a compromise, as is a simulator, for that matter. Not every little detail can be included in the model, especially if it is expected to execute real time, or faster, on a machine that we all can afford. Modeling compromise is an art, and those people constructing models for clinical simulation should have detailed knowledge of physiology, pharmacology, toxicology, and clinical realism to assure that any compromise does not have an adverse effect on the simulation realism. Anything that affects realism affects instructors' teaching objectives and students' learning objectives. It is often very tempting, for computational, logistical, and fiscal reasons, to reduce a model (remove some of its components) or to simplify it (emasculate its assumptions). This can be very hazardous and should be discouraged with models that drive simulators. The overall cost of the model is probably very low compared with the cost of the mannequin, equipment, space, personnel, etc.; and the gain in realism with a good model is much greater than the time and effort necessary for the modelers to achieve model-driven realism.

## 49.7 Modeling as a Tool for Simulation: How Does a Physiological Model Work in Simulation?

Primarily, the model must execute reliably and at least in real time. It needs to act as an "engine" for the simulator. Analogous to a car engine, a physiologic engine is useless if it cannot connect with the simulator both ways, in and out. The real-time connections in a multiple model are called "hooks," (inputs and outputs) and they go back and forth between the model and the simulator. The transmission in a car is an example of a hook. Inputs are of various types. If a student injects a drug, the model needs to know which drug has been injected, the drug's concentration, and the rate(s) of injection. The model rapidly chews up all the data and spits out a continual stream of data about what the body is doing to that drug (its kinetics) and what the drug is doing to the body (its dynamics). Some of the information is shown to the student by the patient monitor – which is connected to the model via outgoing hooks. The student need not see all the generated data on the simulator's control monitor – just those on the patient monitor. But the student should be able to access the model's richly detailed data, which can help explain, in physiologic and pharmacologic terms, the clinical course that has occurred. A similar set of processes of data chewing and spitting out takes place with bleeding, apnea, and toxins, for example. The engine needs to drive something, and hooks can supply meaningful outputs – blood pressure and respiratory rate for example, to drive the patient monitor. The data can also be displayed in educational graphs, plots, etc. In addition to information display, the model outputs could drive many features in a mannequin: the lungs, pupil size, vocal output, airway resistance, etc.

What makes a good model? It should be accurate: in the case of a clinical model, it should be physiologically, pharmacologically, toxicologically, and physically accurate. It should

also have as much detail as possible. A model with a liver and a kidney is much more useful than one without them. I prefer a model that describes each patient or agent in great detail. Our model has up to 100 *user-settable parameters* that together describe each patient and up to 100 *previously defined, but user-settable* parameters that describe each agent. (Default values are provided with each patient and agent.) A four-chambered model of the heart is better for generating realistic arrhythmias and the hemodynamic consequences of those arrhythmias, especially atrial arrhythmias.

The model should be flexible and adaptable. One can implement such models in several different ways. For example, one can construct models so that it is relatively easy to add new features, such as additional agents (nerve agents, for example) and new compartments (skin or atria). This is possible mainly through the transport and multiple model features. Transport models offer considerable flexibility. The slogan "We can transport anything, anywhere" is not far fetched. If a multiple model has been skillfully implemented, it is simply a matter of the programer's adding a new agent, as a submodel, using the same template as the one used to build the other agents. The newly constructed submodel can be hooked to the rest of the model. Yes, hooks can be strictly internal, too. The finest minds in the world could *never* perform the roles of clinical instructor and simulator operator while simultaneously executing over a hundred interconnected pharm/physio models reliably and repeatedly – from day to day, week to week, year to year, and city to city. Especially, if we want to go "high stakes exams," we absolutely have to have some consistency over time and place.

A model should be prescient. Prescience is implemented, in part, by "giving them more than they want." "Them" includes the ones who define the specifications for the simulator, as well as the user, the instructor, and the student. If the developer builds the model correctly, unpleasant surprises should rarely occur when the user develops or encounters new needs or situations. Shock is a good example. The model should be able to handle and differentiate between shock caused by blood loss, dehydration, heat, sepsis, anaphylaxis, tachycardia, bradycardia, certain arrhythmias, or heart failure. These forms of shock are different. They have different physiology, different prognostic implications, and different requirements for therapy.

The following paragraphs describe a few of the many ways a detailed model helps make a simulation more realistic.

### 49.7.1 A Liver and Kidney

The liver and kidney help eliminate many foreign agents – drugs, poisons, and toxins. These removal functions help implement both drug and physiologic interactions. For example, an agent that decreases hepatic and/or renal blood flow will impair the disposition of any agent that is metabolized by the liver and/or excreted by the kidney, including the agent itself, if applicable. This impairment of agent disposition leads to a greater retention of the agent, with a resulting higher peak action and a more prolonged action than otherwise. Similarly, shock indirectly inhibits the metabolism and excretion of many agents. Any trauma that compromises the function of either the liver or the kidneys will have altered the effects of drugs and toxins.

## 49.7.2  A Portal Circulation

Modeling a portal circulation is important when a drug is administered by mouth in a simulation. This circulation allows "first-pass" metabolism, during which the drug is absorbed from the GI tract and passes via the portal vein into the liver where many agents are metabolized. The result of first-pass metabolism is that a lesser portion of most agents reaches the patient's circulation, for several reasons. Drugs can be given orally during ACLS simulations and as preanesthetic medications. Examples of the former include aspirin and captopril; the latter include many agents in pediatric patients.

## 49.7.3  Intramuscular or Subcutaneous Injections

How can a model help demonstrate the administration of agents by routes other than intravenous? A good model will realistically modulate the release of an agent from the muscle or subcutaneous tissue by taking mathematically into account different factors, many of them related to the ability of the body to shut down the circulation to nonvital organs and tissue, including muscle. During hemorrhage, these factors include muscle vascular constriction, which in turn depends on the changes in blood volume caused by the blood loss, fluid replacement, transfer into the third space, etc. The effectors of the changes in muscle vascular tone include the baroceptors. The vigor (slope) of the baroceptors' responses will vary according to the age of the patient and the blood concentrations of certain agents, inhaled anesthetics for example. Other agents, such as dopamine or norepinephrine, affect muscle bed circulation directly and considerably, and the effects vary with the age of the patient.

Thus, intramuscular injections require a detailed and accurate model. Let's examine why this is so, and why it is important. We know that it is potentially dangerous to administer IM morphine, for example, to a patient in traumatic hemorrhagic shock. Repeatedly administering morphine because the drug is having no effect on the pain can lead to a dangerous situation when fluid is administered, and good circulation is restored. All the morphine that has been stored in the muscle is suddenly released. The converse phenomenon can occur with a vasodilated muscle bed, and I was fooled on that one in the simulator that incorporates our model. I was using IM epinephrine while testing out our new anaphylactic shock scenario. Essentially, I was unintentionally killing

the patient, or causing serious problems, with the dose that I thought was correct. I explored several avenues to try to solve the problem. Perhaps, the epinephrine agent parameters had been changed during software revisions. No, they hadn't. I tried IV epinephrine, and it worked realistically and safely at the recommended concentration and infusion rate. Our model plus simulation allows one to dig into what is going on – in this case, after IM injection of epinephrine during anaphylactic shock. I noticed that the blood epinephrine level was shooting up, along with the heart rate and blood pressure, and ventricular arrhythmias began to appear. Next, I administered IM epinephrine to a normal patient, who showed the anticipated relatively gradual increase in blood concentration, and the considerably smaller changes in heart rate, blood pressure, as well as no arrhythmias. When all else fails, go back to reliable sources, which in this case emphasized two points. (i) The IM dose of epinephrine for anaphylaxis is less than that used in most other indications for the agent, certainly less than with CPR. (ii) Real patients have died following IM epinephrine injection for anaphylactic shock. Our effect indeed was the opposite of the phenomenon seen with traumatic hemorrhagic shock: more, not less, epinephrine was being released from the muscle, since epinephrine was going out of the muscle at a faster rate than normal, because of the *profound vasodilation during anaphylactic shock*. When I used the IM dose that most sources recommend for anaphylactic shock (0.3 mg), the results were realistic. This was a sobering lesson but isn't the first time that something like this has happened to me. It was a reminder that sometimes you just have to trust a good model – unless you have the facts to back up your distrust, not just a feeling or a hunch, or "clinical experience." I find that difficult to do, i.e., to trust the model when it seems to go against my clinical or scientific experience. Time and time again, the model has turned out to be correct.

---

Editors Note: In this day and age of widespread concern of nerve agent toxins, a basic and widely taught therapy is self- or buddy-administered 2-PAM Chloride and epinephrine or atropine into the thigh musculature with autoinjectors. We use inert autoinjector devices to teach the psychomotor *skill* in the *handling* of these devices, but what is the fate of these drugs once injected into an under-perfused thigh muscle? What of the fate of those casualties that actually need those drugs in their heart and lung tissue, not thigh tissue? This real-word clinical training/treatment failure would not occur if the training system included an accurate transport model as described here by the author.

---

## 49.7.4  Cardiopulmonary Resuscitation (CPR)

Shortly before we implemented CPR in our model, the American Heart Association released its 2005 guidelines on CPR, including BLS and ACLS. Several significant changes in the American Heart Association guidelines are applicable for this chapter [4].

(1) The ratio of chest compression to ventilation increased from 15:2 to 30:2.
(2) After giving two rescue breaths, rescuers should not stop to check for signs of circulation before starting compressions.
(3) Instead of applying defibrillator pads and shocking up to three times before beginning CPR, rescuers should give one shock and then do 2 minutes of CPR, beginning with chest compressions, before trying the defibrillator again.
(4) Shock less frequently and resume CPR as soon as possible after a shock.
(5) Chest compressions should be at the rate of 100 per minute. The reasons given for the changes are simple. "When chest compressions are interrupted, blood flow stops. Every time chest compressions begin again, the first few compressions are not as effective as the later compressions. The more interruptions in chest compressions, the less is the victim's chance of survival from cardiac arrest" [4]. To oversimplify it, some flow is better than none.

Our model confirms that the rate of decrease in mean arterial pressure after stopping CPR is much more rapid than the buildup. The gradual buildup in arterial pressure, including mean arterial blood pressure, is easily seen in the simulation's "patient" monitor, as well as the plots. In our model, the decrease in mean pressure after stopping CPR is much faster than its restoration upon reinitiation of CPR. When one stops CPR, one can also observe more rapid increases in brain and heart damage (see **Quantitative brain and heart damage** section below).

Isn't the $O_2$ reservoir in the lungs important, however? Doesn't it need replenishing with frequent breaths? Doesn't the need for whole-body perfusion need to be balanced against the need to provide more $O_2$ (to be available to) be extracted from the lungs to supply the rest of the body? Because we can examine in our model the concentration of $O_2$ in the lung – with "breathing" pie charts, plots, and numeric values, I was able to observe that, during the 32 chest compressions, the $O_2$ was disappearing very slowly from the lung's FRC. This is because the total flow (i.e., pulmonary blood flow) is very low. If there is no perfusion (during ventilation, patient examination, defibrillation shock), no $O_2$ is being extracted from the lungs. Two good breaths, about every 20 seconds, or six a minute, supply all the $O_2$ that is needed, with room air, in our model. Again, our model supports the rationale of the new AHA guidelines.

It is also educational to examine the difference between giving two good breaths compared with two inadequate breaths, and also compared with increasing the inspired $O_2$ concentration in small increments during CPR. Exercises like these can help the student understand and implement the finer points of CPR. How useful is a higher $O_2$ concentration during CPR, for example?

### 49.7.5 Quantitative Brain and Heart Damage

Patients do die during simulation. The final common pathway to realistic death is insufficient $O_2$ in the heart and brain, resulting in an $O_2$ deficit. This deficit can occur for many different reasons, each of which is taken into account by our model. Brain damage in our model is essentially a time-severity process, as in real life. Our model computes two variables, one that reflects the $O_2$ deficit in the heart and one that reflects the deficit in brain. We can continually compute these variables that compare the $O_2$ supply with the $O_2$ demand, because our model computes regional $\dot{V}O_2$ values ($O_2$ demand), as well as regional blood flows, $PaO_2$, the concentration of Hb, $SpO_2$, plasma $O_2$ concentration, and the configuration of the Hb dissociation curve (all related to supply). When demand exceeds supply, our model starts calculating the deficit. The value of the deficit determines the extent of the organ damage: mild, moderate, severe, or irreversible, for example. With the last, the patient will either remain comatose, or die.

Still another example of the usefulness of a detailed model is the physiology and pharmacology of human aging. Geriatric medicine and anesthesia are becoming increasingly important, and simulation must accurately reflect the physiology of aging. The anatomic, physiologic, and pharmacologic changes that occur as we age are widespread, general, complex, and depressing (apologies for the double pun). To create an elderly patient in our model, we change, directly or indirectly, 50 patient parameters, each as a function of age [5]. The parameters range from tissue or organ masses to lung FRC to baroceptor sensitivity. Imposing stress on an older person, such as anesthesia or bleeding, demonstrates that while the resting, unstressed elderly person may appear normal on the monitor, the appearances are deceptive. The elderly do not respond to stress as well as a younger person since their reserves are proportionately less, and a stress that might show little disturbance in a healthy young person can severely disturb, or even kill, an older person. For example, in our model, bleeding a recumbent 30-year-old healthy, awake young male 1 L in 10 minutes will barely perturb the cardiovascular system, at least on the patient monitor, while the same hemorrhage can kill the healthy (no disease) elderly patient in his 80s or 90s [5]. In other words, the margin of safety is less in the elderly.

# 49.8 What Do You Want Out of a Model – Whether You Know it or Not?

### 49.8.1 Learning as an Example

I have a prejudice. I believe that learning and understanding are enhanced if one understands the mechanisms behind the event(s). Why does blood pressure usually decrease with inhaled anesthetic agents? How does cyanide kill a victim? How do hemorrhagic and cardiogenic shock differ? Understanding the mechanisms not only helps retention, it allows the students to approach other problems that they have not dealt with before, with a greater chance of success.

### 49.8.2 Do we Stop at Physiologic Models?

I have described physiologic models as if they were sufficient in and of themselves. But, like people, no physiologic model is an island. I shall briefly discuss pharmacology, toxicology, and physics and their relationship to a physiologic model. In our model, we use detailed physiology to help make the pharmacology more real. We also use "pharmacology" to aid the physiology. For example, in our model, the physiologic reaction to pain is produced by releasing epinephrine and norepinephrine. Part of the realism with this maneuver is that when the pain trigger has subsided, the two agents still circulate until they are taken up or metabolized. Thus, the patient's reaction to pain doesn't simply shut off when the pain source disappears.

Similarly, models of pharmacology become better models when accurate physiology and anatomy are incorporated into the model. The drug kinetics of obese patients is different from that of muscular athletes, and that of elderly patients is different from that of young patients. Many compartments are needed to implement these distinctions. I have already described the usefulness of liver and kidney compartments to help implement the realism of pharmacology.

Toxicology is often simply an extension of pharmacology. After all, if you administer enough of any agent, it becomes toxic. If you give a very small amount of some poisons, they can become therapeutic, or at least useful (curare!). Anesthesiology, as you know, is the clinical practice of intentionally administering poisons for a beneficial purpose. If a good physiologic/pharmacologic model is already configured, it becomes much easier to implement a toxicologic model, since most of the principles are similar. (A major exception is cyanide.) This relative ease of implementing toxicology is important because of the increasing emphasis on terror agents and mass terror casualties.

Physiologic models contain considerable physics, especially with the cardiovascular and respiratory systems. In addition, models for anesthesia, intensive care, the emergency room, and the trauma unit, for example, require very detailed physics to implement an anesthesia machine and/or a ventilator. Many years ago, we modeled an Engstrom ventilator and superimposed the plots from the model over the plots from the real ventilator, and they matched almost exactly (unpublished data).

As a more clinical example, the physics (impedance) of the ventilator must "match" that of the "patient." The good physics in our model helped our clinicians understand the following case, which took place at our institute. The case involved an expiratory valve that was stuck in the closed position, in the anesthesia circuit, a very dangerous circumstance, since airway pressure increases extremely rapidly and can rupture the lungs.

In addition, the increased airway pressure severely depresses the cardiovascular system via decreased venous return. Briefly, the patient experienced a difficult endotracheal intubation during induction of anesthesia. Mask ventilation was also difficult, due partly to the high inspiratory pressures required and partly to a poor mask fit, with a resulting considerable leak between the mask and the face. The first and subsequent attempts at intubation were greeted by a stiff bag and a lack of a $CO_2$ waveform on the capnograph. Each time, the diagnosis was an esophageal intubation, and the tube was quickly removed. After each intubation, mask ventilation became progressively more difficult. Throughout this episode, the patient became intermittently hypotensive. About 30 minutes after induction, a perceptive technician passed by and pointed out that the expiratory valve was stuck. The problem was quickly resolved, and the patient successfully intubated. We were able to simulate the respiratory and cardiovascular changes that occurred in this case, partly because in our model the cardiovascular system is connected to the intrapleural pressure, so that an increase in airway pressure impedes venous return. We were also able to show that a leaking mask probably saved the patient's life, by relieving the extreme buildup in airway pressure, and subsequent hypotension that occurred with each intubation [6].

By now, it should be obvious that good physiologic modeling is difficult and tricky. Why? Although there are many similarities between flight and clinical simulation, the analogy fails here. Boeing, for example, makes a plane's models available to those companies doing simulations of the plane. God made the model for the human body, but, unlike the model for the Boeing aircraft, the human model has not yet been divulged. This means that we all have to start from scratch.

## 49.9 Conclusion: Simulation without a Model

What is simulation without a model like? Besides being less accurate and realistic, the simulation could be more expensive. If no model is available in the simulator, expensive and extensive human intervention becomes necessary to teach and train – scripting, twisting dials, making adjustments, introducing personal backgrounds and biases, changing the monitoring variables seen by the students, etc. There would be more variability with the same patient and with the same scenario, because of the timing and the vagaries of human adjustments. This would make teaching, learning, and certainly testing less structured and reproducible. It would increase the temptation to "fake it." Without a model, we would wind up where we started at the beginning of this chapter: "simulation" as cheating, faking, counterfeiting. That's not the way to go.

## 49.10 Information Resources

The following three resources are among the standout resources. As with "a student of simulation," age of the publication is not a consideration.

Cummins, R., ACLS provider manual. 2004: American Heart Association. 315 pp.

This manual taught me a lot about ACLS and what was needed to model it. It also gave me a healthy respect for first responders. They make my job (anesthesia) look easy.

Bonica, J., Principles and practice of obstetric analgesia and anesthesia. Fundamental considerations. Vol. 1. 1967, Philadelphia: F. A. Davis Company. 837 pp.

Modelers have to think differently – strangely different, the editors of this book would say – and John's classic book helps steer thinking in the right direction on a very difficult modeling challenge, one that includes obstetrical, delivery, and neonatal-resuscitation modeling.

Evers, A., M. Maze, and Ed, Anesthetic pharmacology. Physiologic principles and clinical practice. 2004, Philadelphia: Churchill Livingstone. 1,000 pp.

One can collect all sorts of wisdom on the anesthetic trade by flipping through this wonderful textbook. It's a pity that there is no CD version of the book, so that the reader could easily bring ostensibly disparate material together into a cohesive study unit.

## Endnotes

1. Starko K. and N.T. Smith, Body simulation web site <http://www.advsim.com/biomedical/body_simulation.htm>. 2006, as well as in the published, peer-reviewed literature.

2. Everyone, through the ages, has experienced force, mass, and acceleration. No one understood their interrelationships until Newton developed the simple mathematical model F = MA. Because of Newton's model, people could not only understand, they could also predict and develop. More importantly, they could refute, and the model was finally refuted – by Einstein.

## References

1. Smith, N. T. and Starko, K. 'Physiological systems modeling.' In: Webster, J. G. (ed.), *Wiley Encyclopedia of Medical Devices & Instrumentation.* John Wiley & Sons, Inc.: Hoboken, NJ. 5, 299–321, 2005.

2. Bassingthwaighte, J. B. A view of the physiome. <http://physiome.org/files/Petrodvoret.1997/abstracts/jbb.html> 1997.

3. Smith, N. T. and Starko, K. Worst-case scenario: Battlefield injury/can't intubate. In: *Medicine Meets Virtual Reality.* Long Beach, CA, IOP Press, p. 65, 2006.

4. Anonymous, Highlights of the 2005 American Heart Association guidelines for cardiopulmonary resuscitation and emergency cardiovascular care. *Currents in Emergency Cardiovascular Care.* 16(4), pp. 1–28, 2006.

5. Smith, N. T. and Starko, K. The physiology and pharmacology of growing old, as shown in body simulation. In: Medicine Meets Virtual Reality *The Magical Next Becomes the Medical Now.* pp. 488–491, 2005.

6. Smith, N. T. and Starko, K. Anesthesia circuit, In: Atlee, J. (ed.). *Complications in Anesthesiology.* W. B. Saunders, Co.: Philadelphia. pp. 543–548, 1998.

# Buy from Others

*Progress isn't made by early risers. It's made by lazy men trying to find easier ways to do something.*

Robert Heinlein

We all use tools made by others, and both tool users and maker/sellers share a common theme for our mutual interests in the tools themselves: to solve a problem. However, unlike the educators' problem of helping others profit by the latter accepting the gift of learning from the former, the maker/sellers' problem is to help themselves profit by others purchasing their wares. This fundamental difference in their goals, along with profound differences in their environments, shifts the simulator maker/sellers to such a different perspective that all too often educators can't imagine that there would be any valuable learning from them. Yet, like the warp and woof fibers in woven fabric, different directions and different shapes interlinking, but not intermelding, create together utility that no amount of mere thread could ever provide. Where educators often expend years helping a few students within a single institution, maker/sellers often spend moments helping create many customers within numerous institutions. This broad, yet thin exposure to many different teaching cultures provides the maker/sellers with an exposure to clinical simulation that few educators can ever hope to attain. In addition, the knowledge and wisdom available from a business outlook is well worth considering by educators.

The following two chapters are from early risers who have responded to the call of folks like us who would rather buy easy ways to do our jobs instead of make our own. William Lewandowski provides a commercial view encompassing the big-picture thinking and questioning that all successful simulation program leaders and managers must address. Thomas Doyle *et al.* present a similar commercial view from a more detail-oriented perspective. The financial success that their two respective firms have enjoyed is one key benchmark of the value of their offerings here.

# 50

# Success with Clinical Simulation = Assessment + Planning + Implementation

William E. Lewandowski

Eye of newt, and toe of frog, Wool of bat, and tongue of dog, Adder's fork, and blind-worm's sting. Lizard's leg, and howlet's wing, For a charm of powerful trouble, like a hell-broth boil and bubble.

From Macbeth (IV, i, 14-15)

## 50.1 Successfully Select, Implement, and Integrate Simulation into Your Clinical Teachings

If you are reading this chapter, you are probably considering purchasing a clinical simulator. You consider yourself a clinical educator firmly grounded in science, and you have probably done your due diligence in conducting a literature search regarding how other institutions have used clinical simulation. While you may have found a number of studies that speak to their conclusions regarding the training effectiveness of different simulations under various conditions, what you probably have not found is much information on choosing the right simulator, or successfully implementing that simulator into your training program. In fact, what little you had been able to glean from your research, in this regard, may have sounded a bit like the witches brew from Macbeth, instead of the "evidence-based medicine" type approach you hoped to uncover.

My goal in this chapter is to provide you with the tools to accomplish your mission – to select the right simulator and to successfully implement and integrate it within your organization or institution. The tools I am suggesting to you are drawn from professional writings in the field of instructional and performance technology, tempered by my own experience in observing what works and what does not work. The first of these tools is "Mega Planning," which is based on the work of Roger Kaufman [1], and addresses how organizations plan and how planning can be implemented to ensure organizational acceptance and success. The next tool is Human Performance Technology, which is an "an engineering approach to attaining desired accomplishments from human performers by determining gaps in performance and designing cost-effective and efficient interventions [2]." Human Performance Technology is solidly based on systems theory, and is in use in industry and government around the world. In this chapter, I will be using the Human Performance Technology model described by Van Tiem, Mosely and Dessinger [3]. To this, I have added aspects from the work of Jack Phillips [4] on his (Return on Investment) "ROI process," which helps you to identify not just the financial aspects of Return on Investment, but also those other critical benchmarks to help you to assess the relative success of your program. I propose that if you use these tools to guide you in your selection, planning, implementation, and evaluation of your use of clinical simulation, then you will be on a prudent course of action leading toward success.

## 50.2 So, You Think You Need a Clinical Simulator?

To date, I do not know of anyone who has purchased a clinical simulator for personal use. Perhaps this statement may sound a bit facetious, but the point is that clinical simulators are generally purchased for use by institutions or organizations to meet their real or perceived organizational needs. If a simulator is purchased and used to meet a real need, as long as it is purchased and utilized as part of a well thought out plan, the program stands a good chance of success. If, however, it is purchased without a well-defined intention to have a direct, positive impact on an institutional goal or objective, it will soon fade into obscurity, to be stumbled upon one day in the forgotten recesses of a storage closet.

The first step away from consignment to a storage closet is to understand the stated goals and objectives of your institution or organization, and then to identify how a clinical simulator can materially assist in meeting those goals and objectives. A model that you might find useful in visualizing the hierarchy of goals and objectives in your organization or institution is the Organizational Elements Model developed by Roger

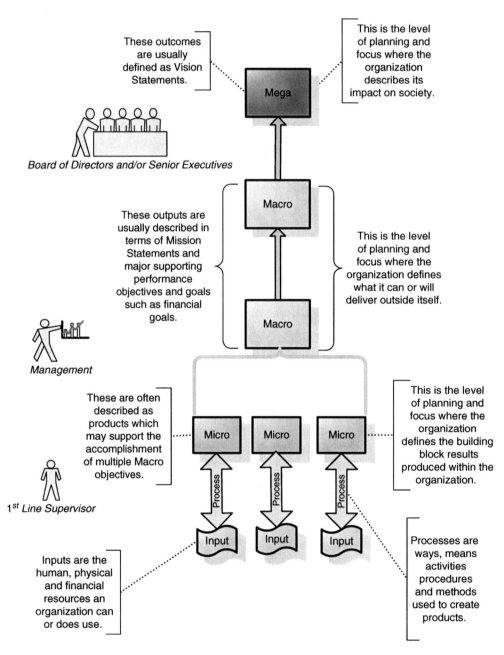

**FIGURE 50.1**  The Organizational Elements Model.

Kaufman, which defines and links what any organization uses, does, produces, and delivers [5]. The Organizational Elements Model provides a framework for looking at an organization and understanding how it determines what is important. In Figure 50.1 is a graphic representation of the Organizational Elements Model adopted from Kaufman's work.

The important point is to identify the relationships between what you are doing (or want to do) and the formal hierarchy of your institution's or organization's goals and objectives. The more connections you can draw between you and your project to supporting organizational strategic (Mega) goals and tactical (Macro) objectives, the more important your project becomes (or is perceived to be) to the overall success of your institution or organization.

Goals and objectives define future end states. There is always a gap between where the institution or organization wants to be at some future date and where it is at today. Usually, this gap is related to a desire to improve performance as correlated to current costs. For example, let us say that a health care institution decides that its error rate in a particular clinical procedure is unacceptable. The institution is not only paying the base costs associated with performing the procedure, it is also paying any direct financial penalties (lawsuits, wasted labor/materials, etc.) as well as indirect financial penalties (decreased patient satisfaction, inferior reputation, decrease in employee moral, etc.) associated with the errors. This is the current cost of doing business. The performance gap between the current state of affairs and the desired end state expressed in the goal or objective represents both a financial metric and a performance metric against which any proposed intervention must be judged.

So, what does this all mean?

- Regardless of why *you* think you need to have a clinical simulator, you must be able to tie its acquisition and use back to an institutional and/or organizational goal or objective. *Even though we usually think of clinical simulators in terms of how they can be used to improve individual performance, their performance does takes place in an institutional or organizational context.*
- *You must be able to show how simulation relates to improving the performance of the institution or organization.* You must be able to demonstrate how acquiring and using a clinical simulator will bridge the gap between the current state of affairs (at the current cost of doing business) and the desired end state as expressed in an institutional or organizational goal or objective. *Once you have identified the institutional or organizational goal to which the use of simulation relates, you must be able to demonstrate how the use of simulation will have a positive impact in reducing the implicit performance gap in the stated goals and objectives.*

Up to this point, we have been looking at the impact of simulation from the perspective of meeting formal institutional and organizational goals and objectives, as a critical mechanism for obtaining buy-in from executives and senior management. There are, however, critical supporting and implementation objectives that are key to the decision process of choosing and acquiring the appropriate simulator, and then successfully implementing its use.

## 50.3 Selecting the Right Simulator and Simulation Vendor

Jack Phillips of the "ROI Institute" has identified six categories of implementation objectives for you to consider when developing an implementation plan for performance interventions such as the use of simulation. These categories, along with a brief explanation of each, are presented in Table 50.1.

Implementation objectives perform a number of critical roles in selection and acquisition of a clinical simulator:

- They define the functional parameters or selection criteria when choosing a clinical simulator.
- They define the benchmark or evaluation criteria for evaluating the relative success of the simulation program.
- They form the key objectives when creating the implementation plan for your simulation program.

To help you formulate your implementation objectives and to assist you in evaluating the various simulator vendors and products, a list of questions [6] and comments based on the Phillips' categories are presented in Table 50.2. These are questions I suggest you should ask yourself, your institution, and any potential simulation vendors as you assess and plan for your simulation needs.

Defining the implementation objectives can be a daunting task when planning for simulation, because going through this process forces you to examine thoroughly all aspects of the "why?" in selecting a particular simulator. Once you have defined success (in terms of the implementation objectives), and determined the mechanism for proving success (in the form of evaluation requirements), you have then created clearly defined parameters on which to base your decision regarding which simulator will be used in your program and your detailed plan to implement that program.

Creating evaluation requirements, at this point, will help guide you in creating a complete implementation plan. If you decide early on how you are going to collect information to assess your program's relative success, you will find it much easier to create an implementation plan that is focused on success, and which has the collection mechanisms in place to provide you the feedback required to make adjustments during implementation. The other benefit to this approach is that by having established evaluation requirements, you then

**TABLE 50.1**   Categories of implementation objectives based on Phillips' "ROI process"

| Objective(s) | Comments |
| --- | --- |
| Reaction and satisfaction | This defines the level of reaction and satisfaction you want to achieve with both the target audience and major stakeholders. An example of this in a patient safety setting is wanting to achieve a 95% positive acceptance rate by patients, staff, and management with the intervention. |
| Learning | This defines specific changes in skills, knowledge, and attitudes in the target audience. Learning objectives are particularly applicable (but not exclusively) for training interventions. |
| Implementation | This defines the level of success for the intervention, often in terms of utilization and sustainment over time. An example would be setting an objective of achieving a 99% utilization of the intervention, by all staff members, within 6 months of implementation and sustaining that rate for a period of at least 3 years. |
| Impact | Also referred to as the Business Impact, this generally comes from the Performance Gap (see Section 50.2) and defines the specific impact or change expected from the intervention. |
| ROI (Return on investment) | This defines the actual cost versus benefits expected from the intervention. It is generally defined as ROI(%) = Net Benefits/Net Costs × 100. |
| Intangible benefits | This defines those effects, which are not addressed by the other objectives above. Some examples are: decreasing employee complaints, increase in innovation, and increase in teamwork. It is not unusual for potential intangible benefits not being identified prior to implementation, but surfacing during postimplementation evaluation. |

**TABLE 50.2**   Questions and comments regarding implementation objectives

| Objective(s) | Questions | Comments |
| --- | --- | --- |
| Reaction and satisfaction | <ul><li>How important is user and stakeholder acceptance to you?</li><li>Aside from the simulated procedure, how difficult is it to operate the simulation?</li><li>Can the vendor provide independent documentation on face validity?</li><li>Is the vendor open to suggestions on product content and new applications?</li><li>Will the simulation be considered relevant by your users and stakeholders (language, culture, local procedures, etc.)?</li><li>How difficult (time & money) is it to make the simulation relevant to your situation?</li></ul> | *User and stakeholder acceptance is critical to the long-term success of a simulation project.* |
| Learning | <ul><li>What are the performance objectives (task, conditions, & standards) the learners must master?</li><li>What specific aspects of the target procedure does the device simulate?</li><li>How realistic is the simulator?</li><li>Has the simulation content been validated by recognized subject matter experts?</li><li>Does the simulation provide performance feedback?</li><li>Has the simulation been proven to show transfer of learning?</li><li>Does the simulation have validated metrics?</li><li>Has the simulation been proven to distinguish between master and nonmaster performers?</li><li>Has the simulation been proven to have predictive validity?</li><li>Does the simulation have a user database that records performance?</li><li>What are the vendor's future plans for product additions? Upgrades?</li><li>How focused is the vendor on your area of clinical education?</li></ul> | *Obtaining design documentation related to learning objectives is important.*<br><br>*Simulation should only be used as part of a systematically designed and validated training and testing program.* |

**TABLE 50.2** (Continued)

| Objective(s) | Questions | Comments |
|---|---|---|
| Implementation | • Do you have implementation objectives or parameters (when, where, how many, how much)?<br>• Can the vendor demonstrate quality control in the production process (ISO 9000 or similar process)?<br>• Does the simulation equipment meet local government standards (FCC, UL, CE, etc.)?<br>• Will the simulation work in your local conditions (electrical, humidity, temperature, space requirements)?<br>• What is the failure rate of the simulation equipment?<br>• Do you have the infrastructure to control and operate the simulation equipment?<br>• What type of warranty and/or service support does the vendor offer? Is it convenient and timely for your location(s)?<br>• Can the vendor provide proof of business stability and long-term sustainability?<br>• Does the vendor offer suitable payment models? | *Always ask for, and check, references – especially those of users in situations similar to your own.* |
| Impact | • Are there specific gaps, in either individual or organizational performance, you need to address (complication rates, learner time to proficiency, etc.)?<br>• Can the vendor provide you examples of how others have used the vendor's simulation equipment to bridge similar performance gaps? | *It is important to demonstrate to decision makers that simulation will help to bridge the gap between current conditions and desired organizational outcomes.* |
| ROI (Return on investment) | • Do you have specific, financial ROI targets?<br>• Can the vendor provide independently verified ROI case studies of situations similar to yours?<br>• Does the vendor have an independently verified ROI model? | *When considering Net Benefits, always consider positive financial Impacts. When considering Net Costs, consider all Implementation costs.* |
| Intangible benefits | • Can the vendor provide independently verified examples of Intangible Benefits in situations similar to yours? | *These are benefits not specifically identified in previous objectives and are difficult to put in to financial terms.*<br><br>*These are often not identified until after implementation of a performance intervention such as simulation.*<br><br>*Examples might include: an increase in new enrollment, an increase in innovation, or an in increase in teamwork.* |

have the detailed data necessary to ask simulation vendors about embedding collection mechanisms into the simulator itself. Some make and models of clinical simulators already have some sort of system in place to collect some data on user performance and simulator usage. However, if you have defined your specific requirements prior to implementation, it will make it much easier for the simulation vendor, or some other commercial entity, to accommodate you.

## 50.4 Purchasing and Fielding a Clinical Simulator

### 50.4.1 Creating a Plan

Every clinical simulation vendor would like to believe that their device will meet the majority of your needs regarding a given clinical procedure, but the vendors know that each institution and organization, while sharing a common subset of needs, also have a subset of needs and circumstances that are unique to each. The clinical simulator vendors also know that because of the unique set of needs and circumstances of each of their customers, that *how* a simulator is implemented is just as important as *what* simulator is implemented.

Perhaps you have noticed that much of what we have discussed in this chapter has been more focused on asking questions than providing answers. That is because one of the main purposes of planning is to come up with reasonable questions to guide you toward the right answer. The other main purpose is to translate that answer into action. In a perfect world, the perfect plan would answer all questions, address all unforeseen contingencies, and guarantee success before the plan was implemented. In the real world, planning is an iterative process, just as learning is, where despite our best efforts, we do not always start out knowing all the questions to ask, or what the right answer will look like. The clinical simulator vendors want you to ask reasonable questions, because they will only be successful if your implementation of their systems is successful.

Up to this point, we have addressed the following questions:

- What are the goals and objectives of your institution or organization?
- Given those goals and objectives relevant to your position in the institution or organization, what are the gaps in performance between the current state affairs (and the related cost of doing business) and the desired end state reflected in the goals and procedures?
- What are your implementation objectives regarding clinical simulation?
- What questions should you ask yourself, your institution, and the simulation vendor to help you to assess the appropriate course of action(s) regarding clinical simulation?
- What are your evaluation requirements to determine the relative success in meeting your implementation objectives?

I would now like to offer some additional questions. These are intended to assist you in creating the implementation plan itself – addressing acquisition of the simulator through evaluation of the program. It is a relatively short list of questions drawn from various sources and which I have modified on the basis of my experience. For those of you who have served any time in a military service from almost any country, you will recognize the key elements as being from the time-proven methodology of operations planning:

- Who?
  *(Who is the plan affecting? Who is performing what tasks? Who is in charge? Who is providing the funding?)*

- What?
  *(What will occur? What is the sequence of events? What controls will be used? What will the plan cost? What are the operational requirements of the simulators?)*
- When?
  *(When is the plan effective? When do the events occur? When does the plan stop?)*
- Where?
  *(Where will the events of the plan occur?)*
- Why?
  *(Why is the plan being implemented and what are the intended results?)*
- How?
  *(How will the plan be implemented? How will it be supported and resourced? How will the equipment be maintained? How will success be evaluated?)*

If you use these questions as a guide to develop a plan that is focused on achieving your implementation objectives, you will find that you have covered those items critical for your program to achieve success.

## 50.4.2  Executing the Plan

As you execute your implementation plan, there are key follow-through items that you should focus on to ensure the success of your program:

- Monitor execution. *When driving, there is no substitute for keeping one's eyes on the road, and the same is true when executing an implementation plan for clinical simulation – especially in the early phases of executing the plan.*
- Establish channels of communication with users, the simulation vendor and primary stakeholders. *You will find that the vast majority of minor problems and issues that degenerate into major problems and issues do so because of a breakdown in communications. Ensure that all key personnel in the implementation process know who each other are and how to contact each other to ask questions and report issues. All the participants must be made to feel comfortable with asking questions, reporting issues, and with using the channels of communication you establish. If not, the channels of communication will be useless.*
- Identify and fix problems. *You cannot anticipate all future implementation problems, but you should assume that they will occur. Encourage everyone involved in your implementation program to report problems and issues as soon as possible. Create a mechanism that disseminates information regarding problems and their resolutions. Clearly establish and maintain an attitude where reporters of problems are rewarded and not punished.*
- Evaluate success. *Earlier in this chapter, we discussed the role and importance of establishing evaluation requirements*

*for your program. It is important that you evaluate whether or not your plan is achieving your implementation objectives. Evaluation should never be an afterthought, but an integral part of your program. Just as clinical students are always being evaluated throughout their educational progress, so too should evaluation occur during the entire progress of your clinical simulation program.*

- Provide feedback mechanisms. *This item is related to the "Establish channels of communication" bullet, because feedback is transmitted through those channels. Performance feedback is the lifeblood of any systems approach, because it is the mechanism by which a system stays on course to achieving its goal. Feedback channels should be multidirectional, flowing to and from users, stakeholders, and simulation vendors; however, the key to ensuring that these feedback channels will function is creating the environment where all participants feel that their feedback is welcomed and valued as being vital to the success of the project.*

The final part of any systems approach to planning is revision. Once a plan is launched, you need to be prepared to make revisions based on the input you receive from various sources and your own observations. Document the changes you make to the plan and ensure that you disseminate them to all concerned.

In summary, executing your implementation plan for a clinical simulation program should be accomplished in a deliberate manner using the systems-based approach described in this chapter. There is an old saying that a well-executed, poorly designed plan is better than a poorly executed, well-designed plan. A well-executed plan can overcome much, as long as it is adoptive in its reaction to problems and issues, and everyone knows his or her role in the plan, and any subsequent changes to that role.

## 50.5 Conclusion

In this chapter, I have described a systematic approach to assessing, planning, and implementing clinical simulation within your institution or organization, based on the time-proven work of leaders in the field of Human Performance Technology and Return on Investment analysis, tempered by my own experience in clinical simulation.

Even though you usually consider the use of clinical simulation within the context of training individuals, since that training takes place in an institutional or organizational context, it is important to ensure that the acquisition and use of clinical simulation is shown to be directly supportive of institutional or organizational goals and objectives. Using the Organizational Elements Model in Figure 50.1 is one way to show these relationships. Another is by demonstrating how the use of clinical simulation can be instrumental in reducing the performance gap between stated goals and objectives, and current conditions with their accompanying cost of doing business.

There are six categories of implementation objectives to consider when creating a plan to develop and implement any type of performance-improvement intervention (including training). They are Reaction and Satisfaction, Learning, Implementation, Impact, Return on Investment, and Intangible Benefits. In Table 50.1, I described these in more detail and suggested that you use them to guide you in selecting the appropriate simulator and in developing the implementation plan. In Table 50.2, I provided a list of suggested questions to assist you in this regard.

When developing the implementation program, focus on meeting the implementation objectives and use the questions: "Who?", "What?", "When?", "Where?", "Why?", and "How?" to guide you in addressing the important details. Strongly consider establishing evaluation requirements early in the process and concurrently developing your evaluation plan with the implementation plan.

Finally, focus on monitoring execution, establishing channels of communication, identifying and fixing problems, evaluating success, and providing feedback mechanisms. Be prepared to revise the plan as the situation dictates. A strong follow-through is critical to the success of executing a plan and will help you to overcome areas that you might have missed in planning.

## Acknowledgment

The author would like to thank Mentice AB, Gothenburg, Sweden, for its support and contribution to this chapter.

William E. Lewandowski Consulting, Daytona Beach, Florida, USA. (bill@welewandowski.com)

## References

1. Kaufman, R. *Mega Planning: Practical Tools for Organizational Success.* Sage Publication Inc., Thousand Oaks, CA, 2000, pp. xi–xiii, 34, 111–132.

2. Harless, J. *Performance technology skills in business: Implications for preparation.* Performance Improvement Quarterly. 8(4), 75–88, 1995.

3. Van Tiem, D. M., Mosely, J. L., and Dessinger, J. L. *Fundamentals of Performance Technology: A Guide to Improving People, Process and Performance*, 2nd ed. International Society for Performance Improvement. Silver Spring, MD 2004, pp. 2, 6–20, 48.

4. Phillips, J. *The Consultant's Scorecard: Tracking results and bottom-line impact of consulting projects.* McGraw-Hill, New York, 2000, pp. 20–22, 37, 204, 220.

5. Kaufman, R. *Mega Planning: Practical Tools for Organizational Success.* Sage Publication Inc., Thousand Oaks, CA, 2000, pp. xii, 34.

6. Lewandowski, W. A Return on Investment (ROI) Model to Measure and Evaluate Medical Simulation Using a Systematic, Results-Based Approach. *Medicine Meets Virtual Reality 13* Conference, Long Beach, CA, 2006.

# 51

# Successful Simulation Center Operations: An Industry Perspective

Thomas J. Doyle
Ronald G. Carovano
John Anton

## 51.1 The Industry

The Health Care Simulation Education Industry (defined here as the commercial firms who provide technology, services, and educational solutions to users of healthcare simulation) is an often-underutilized resource as organizations investigate, install, and integrate patient simulation into their educational programs. Most firms in the industry have hundreds or thousands of customers, and thus can serve as a beneficial resource to other organizations as they create their Patient Simulation Centers. Additionally, industry firms know the best practices that exist across their user base, the way to access these organizations, and the key individuals within them. As organizations embark upon the journey of patient simulation, they should ask industry representatives for this contact information. Many companies also have internal expertise that can assist organizations as they design their facilities and integrate the technology into their curriculum and programs. Some companies also offer turnkey solutions (for example, educational content or consulting) to assist the faculty in more readily integrating the use of patient simulation into their curriculum. Numerous educational conferences sponsored by industry and clinical societies now have as their primary focus the use of simulation in health care. Smaller, user-organized meetings also offer learning opportunities. Any organization pursuing the use of patient simulation should avail themselves of all of these resources.

This chapter presents information from representatives of a leading health care education technology firm that organizations should consider as they contemplate clinical simulation.[1] First, considerations for facilities preparation will be discussed. This will then be followed with strategies to engender and promote successful faculty, leadership, and staff buy-in and engagement. Finally, suggestions for the actual implementation and first year of operating a Patient Simulation Center will be presented.

## 51.2 Facilities Preparation: Key Questions

When it comes to preparing facilities, it can be helpful to ask a series of fundamental questions about the simulation center (Table 51.1). Answering each question will provide insight into requirements and offer guidance as facilities are prepared for both the short term and long term.

**TABLE 51.1** Fundamental questions

- What is the mission of the simulation center?
- Who are the learners?
- What is to be taught?
- What is to be measured?
- What are the clinical environments to be created?
- What clinical equipment is needed?
- What is the budget?
- What space is needed and/or available?
- What simulation technologies will be used?

The following paragraphs discuss each question, offering a top-down approach that starts with the broadest issues and ends with the most detailed ones.

### 51.2.1  What is the Mission of the Simulation Center?

Conceptually, this is the most important question to answer. Knowing the overall reason for embarking on integrating simulation into the enterprise defines the direction and tone for everything else that follows. If an organization does not know why it is creating a simulation center, then the organization will surely be lost as it tries to address practical issues. Perhaps, too often, the decision to establish a simulation center is driven by the availability of funding, competition with a neighboring institution, or some other vague notion that simulation is "the right thing to do." In the 1990s, innovators and early adopters surely were pioneers who invested in simulation because they had foresight and an entrepreneurial spirit. Today, the health care simulation community has advanced far beyond that exploratory phase, to where the majority of users are ready to benefit from mainstream simulation applied to their enterprise. Whatever the level of individual experience or community maturity, all users require a clear mission direction and scope.

As an example, one mission in nursing education is clear. In the United States, there is a nationwide shortage of nurses, and the demand for nurses is expected to increase over the next decade [1]. Schools of Nursing are responsible for producing nursing graduates, but they are limited by the availability of clinical placement sites to meet patient contact requirements. Many state Boards of Nursing are accrediting or increasing the percentage of time that simulation can be substituted for

contact with actual patients. Thus, a nursing school simulation center may express their mission as:

> Increase matriculation of nursing graduates by integrating a clinical simulation program into the curriculum thus enabling more learners to gain educationally accepted patient care experiences in both the simulated and live patient environments.

Another example comes from health care provider organizations. Quality and patient safety are of paramount interest clinically and economically (errors are expensive, some even more so than the education required to prevent them). Today, many hospitals have an excellent grasp of their quality issues, but identifying those issues is only half of the battle. What hospitals *do* to educate providers on these issues and to change behaviors is another critical component. Indeed, adult learning theory and practice have shown that *doing* leads to effective learning. Thus, a hospital-based simulation center may express their mission as:

> Enhance patient safety by incorporating simulation into ongoing professional education and link it to the quality management program.

Of course, many simulation centers will have multiple facets to their core mission. Hospitals, for example, may include continuing professional development as part of their mission as a way to retain providers and increase professional satisfaction. Academic institutions may include research in addition to the educational missions. Caution, however, should be exercised to avoid trying to have the simulation center "be all things to all people." This is particularly true at the outset. User experience has shown that once the doors to the simulation center open, many will come knocking. Having a clear mission provides the necessary top-level guidance to stay focused.

### 51.2.2  Who are the Learners?

If education is part of the core mission, then identifying the learners comes next. Again, several questions come to mind:

(1) Is the focus on students and trainees, health care providers in practice, or both?
(2) Is the focus on a particular provider category, such as physician, nurse, paramedic, or combat medic?
(3) Is there interest in a particular discipline or clinical venue, such as surgery, anesthesia, emergency medicine, or critical care?
(4) Is there interest in individual providers, clinical teams, or "teams of teams" in the chain of patient safety?

Going back to the example of the nursing school, the simulation center would primarily cater to the needs of learners in undergraduate nursing. Contrastingly, with the example of the hospital, the simulation center would primarily cater to the needs of adult learners at the graduate level. Although seemingly self-evident, the abundance of opportunities can distract from the design and resourcing of the simulation center in meeting the core mission.

### 51.2.3  What is to be Taught?

Careful consideration must be given to identifying how learning domains (cognitive, psychomotor, and affective) are addressed and how the design of the simulation center supports this. Newcomers to simulation often overly emphasize the psychomotor domain (e.g., technical skills), neglecting opportunities to address cognitive (e.g., knowledge) and affective skills (e.g., professional attitudes, situational awareness). Additionally, different simulation technologies are more appropriate than others, depending upon the learning domains of interest [2].

### 51.2.4  What is to be Measured?

Simulation experiences are conducted to address many different learning objectives. For example, the objective may be to convey and reinforce specific knowledge about the characteristics and usage of a new medication. Alternatively, the objective may be to offer the opportunity to learn and practice a new procedure. Or, the objective may be to offer the opportunity for clinical teams to hone their performance [3]. The success at achieving the educational goals can be assessed by creating learning objectives that are definable, observable, and measurable. Thus, the design of the simulation center must facilitate observation and measurement. Many users have instrumented their simulation centers with distributed digital audio-visual systems, real-time electronic checklists, and learning management systems integrated with patient simulation technologies and clinical environments.

### 51.2.5  What are the Clinical Environments to be Created?

An understanding of who is being taught, what is being taught, and what is being measured leads to consideration of the simulation environment. It is self-evident that the environment accommodates the answers to the three preceding questions. However, the fidelity of the environment also must be considered [2]. Environment fidelity concerns the extent to which the simulation replicates motion cues, visual cues, and other sensory information like a real clinical environment. For example, if the educational objectives are to maximize the initial learning of teamwork skills, then it is advantageous to minimize undesirable distractions. Alternatively, if the primary goal

is to maximize the transfer of trained behaviors into clinical practice, then it is advantageous to introduce intentional distractions [2]. The ultimate aim is to achieve a desired and appropriate level of psychological fidelity, which concerns the "degree to which the trainee perceived the simulation to be a believable surrogate for the trained task" [2]. We want learners to "suspend disbelief" and interact much as they would in the real world. *"Sufficient" fidelity in the environment aids in effective simulation-based learning, whereas too much or too little fidelity can hinder learning and assessment.*

Simulation environments vary among users, reflecting how users' needs and resources vary. Examples include an operating room, an emergency department trauma care room, an intensive care unit, a patient home setting, or an emergency transport vehicle. Some users have elected to create multiple clinical environments, such as the Riverside Methodist Hospital Center for Medical Education & Innovation [4] (Columbus, Ohio, U.S.A.), which has an emergency department, cardiac catheterization suite, operating room, intensive care unit, and patient care room. Another example is the Mecklenburg EMS Agency [5] (Charlotte, North Carolina, U.S.A.), which has replicas of a patient home setting, prehospital accident/disaster scene, ambulance, and emergency department. Most centers (at least initially) implement a multipurpose design that can be quickly reconfigured to represent a variety of patient care settings. An example is Penn State University [6] (Hershey, Pennsylvania, U.S.A), which uses several different painted backdrops in one room to create the illusion of various settings: indoors and outside.

### 51.2.6  What Clinical Equipment is Needed?

In addition to the simulation equipment, clinical equipment is required to outfit the simulation environment. Table 51.2 lists examples of such equipment for an Operating Room:

TABLE 51.2  Example clinical equipment list

- Patient bed
- Patient monitor
- Ventilator
- Anesthesia delivery system
- Suction device with vacuum
- Compressed air and oxygen
- IV pole
- Infusion pump
- Supply cart
- Defibrillator
- Resuscitation equipment
- Disposable clinical supplies (e.g., IV fluids, syringes, venous and bladder catheterization, airway management)
- Surgical/procedures equipment (e.g., scalpels, forceps, electrocautery)

Furthermore, the selected equipment must be congruent with:

- The learning objectives of the simulation
- What is typically found in the real clinical environment being replicated
- What would typically be used clinically by the learners
- The technical capabilities supported by the simulation technology.

Many users have found that a walk through a real clinical environment can be beneficial in identifying essential clinical equipment required to support the teaching objectives in the simulated clinical environment.

### 51.2.7  What is the Budget?

Integrating simulation into the enterprise is not a one-time event. Readdressing the questions previously cited provides guidance as requirements, funds, and abilities change over time. Although not always the case, funding is usually the rate-limiting factor, particularly in the early stages of a developing simulation center. Of course, the key is to balance short-term spending constraints while keeping an eye on long-term learning objectives. Limited funding is not always a bad thing, however, since this has the positive impact on putting emphasis on the most important activities and rewards development of essential skills in adaptability and creativity.

Newcomers to simulation often have a concern about start-up costs for equipment acquisition and facilities preparation. Of equal or greater concern, however, are the recurring costs of sustaining operations for the years ahead. These include items such as costs for personnel, consumables, and equipment maintenance, service contracts, replacement of equipment, and other recurring operational expenses. Costs for initial acquisition and sustaining operation vary depending upon the size and scope of the simulation program. Fortunately, firms within the simulation industry can assist programs with determining these costs, including consultation with industry and other users.

### 51.2.8  What Space is Needed and/or Available?

After budget, space is probably the second most important resource to evaluate. Ideally, the space should reflect the form, fit, and function of the clinical environment to be replicated. It should accommodate the patient simulation equipment, clinical equipment, learners, and instructors. More space offers the opportunity to create additional clinical environments, as well as ancillary spaces, such as those listed in Table 51.3.

Some users do not have a dedicated simulation performance environment, such as those whose mission it is to bring

**TABLE 51.3**   Ancillary simulation facility spaces

- Simulation control rooms
- Observation areas
- Facilities for postsimulation debriefing
- Audio-visual and other back-end systems
- Storage for supplies and equipment (perhaps the most overlooked need!)
- Office space
- Reception and staging areas
- Project development space

simulation to users in geographically disparate locations. In these cases, the only dedicated space is for equipment storage.

Again, firms within the simulation industry can offer guidance regarding technical considerations such as essential infrastructure requirements, physical space, equipment dimensions, supporting power and gas supplies, and educational considerations such as how to make best use of available space.

Industry offers assistance that is even more valuable when *new space* is being constructed or renovated. Architectural design is a very costly process, particularly when redesign is required to meet an unanticipated requirement. The bottom line is that many users will have few opportunities to design new space, while industry works with a collection of users doing just this on a daily basis. Additionally, industry insight into what the future holds (e.g., coming technologies, services, and educational solutions) can help users maximize their investment for years to come. *It is never too early for users to start talking with industry about future plans.*

### 51.2.9  What Simulation Technologies will be Used?

Although it is often the simulators themselves that capture the initial attention of the user, final selection of simulation technologies should address all of the preceding questions, and users should think of simulations technologies in a broad sense. Simply, simulations can be categorized as shown in Table 51.4 [2].

Another way to analyze these categories is to compare the equipment, environment, and psychological fidelity (Table 51.5) [2].

Many contemporary simulation centers incorporate all of these categories of simulation to achieve their learning objectives.

Once all the user requirements have been identified, then technology selection comes down to identifying commercial products and services that best meet the user needs and provide the greatest return for the investment. In some cases, industry may offer a number of alternatives. In others, industry may offer only one viable solution, or even none at all.

**TABLE 51.4** Categories of simulations

- **Case studies and role-plays.** Very basic forms of simulation that users present fictional examples of cases to reinforce trained material. Most require nothing more than paper and pencil and are low cost. They primarily promote competencies in knowledge and attitudes.
- **Part-task trainers.** These can take many forms, from intubation mannequins to endoscopy trainers to standardized patients, and are designed to focus on a specific clinical task. They primarily promote competencies in knowledge and skills. Most device-based methods are of low to moderate cost. Most people-based methods are of moderate to high cost.
- **Full-mission simulations.** These are designed to simulate the entirety of a complex task, such as anesthetizing a patient or delivering a baby. They primarily promote competencies in knowledge, skills, and attitudes. Most are of moderate to high cost.

**TABLE 51.5** Simulation fidelity comparison

| Category | Equipment fidelity | Environment fidelity | Psychological fidelity |
|---|---|---|---|
| Case studies/ role-plays | Low | Low | Low to medium |
| Part-tasktrainers | Medium | Low | Medium |
| Full-mission simulators | High | Medium to high | High |

# 51.3 Faculty, Leadership, and Staff Preparation

## 51.3.1 Obtaining Buy-in and Engagement of Faculty, Leadership, and the Community: What are the Fundamental Questions to Address and Why is this Necessary?

One of the essential concerns any organization must focus on is the involvement and buy-in of faculty in the adoption of patient simulation technology to facilitate learning. This is imperative as the utilization of patient simulation affects faculty workload and instructional design. Faculty should be engaged early in the decision-making process and organizational plans for integration of this technology. Leadership should consider including those key faculty who tend to be "techno savvy" and innovators. It is also important to include representatives from all areas of the curriculum or health care programs. One sure way to fail is to not include the faculty in the process early. Care must also be taken to ensure that faculty does not perceive the use of patient simulation "as one

more thing to do." The leadership should consider rewarding participation of the faculty in the process by recognition as a scholarly activity for promotion and tenure. Organizations that do this will ensure their success. If faculty involvement is seen as an activity worthy of consideration when they are presented for promotion and/or tenure, they will be more willing to buy-in and become engaged in this learning innovation.

Consideration also needs to be given to the impact on faculty workload. Since faculty are responsible for the curriculum, they must be given adequate time for the planning and instructional design of simulations. Many successful organizations have done this with "release time." Other organizations have obtained small grants to fund faculty time for development of the various simulations and integration work. This can take the form of supplemental summer contracts or "overload" during any given semester. If faculty is not given adequate time, the organization will not experience the immediate adoption and use of their simulation investment.

Organizational leadership support for the adoption of patient simulation as a learning strategy is important due to the ongoing financial commitment needed. The costs of operating a successful patient simulation program go beyond the initial capital outlay. There are ongoing program support costs such as clinical supplies, medical gases, warranty programs, software updates, and personnel. Leadership needs to clearly understand the continuing financial commitment that is required by the organization. In many organizations, the simulation program is identified as a cost center, or in some cases, a profit center. (The Bristol Medical Simulation Center, Bristol, United Kingdom, Berkeley Equipamentos Medicos, Rio De Janeiro, Brazil, and Dies Group, Rome, Italy, are notable examples of the latter.)

Additionally, as the leadership team interfaces with the community, it needs to be able to articulate what patient simulation is and how their organization is utilizing it. The leadership team needs to clearly appreciate their simulation program's capabilities and limitations, understand how the technology works, and how it is being used to facilitate learning. This can be accomplished by the leadership attending demonstrations of the equipment, having ongoing discussions with the faculty as to how the technology is being used or could be used, and actually participating in simulation sessions with their learners. Leadership can also interface with others at professional meetings to share and learn from each other. Industry contacts can also facilitate discussions between prospective or new users, with those who have already adopted simulation technology. An organization's leadership should ask industry for a list of current user sites and contacts at those organizations. This will help facilitate dialog, answer many questions from a like perspective, and assist with buy-in and engagement of an organization's leadership.

Support from the community to obtain and use simulation can have many rewards. This may take the form of actual financial support, in kind support (e.g., architectural, facility construction), or donations of clinical equipment. Thus, it is important for the community to know of an organization's plan to adopt and integrate patient simulation into their learning activities. An organization's leadership and faculty must begin the dialog early with the community. This is especially important with the health care community. Health care facilities need to understand how the technology will be utilized and the benefits of using it. If enrollment capacity will be increased, then it is imperative to communicate this to the community. If the community understands that by adopting patient simulation as a strategy to increase enrollment, they are more likely to support the organization's efforts and this may take the form of financial donations.

Many "naming" opportunities exist within a patient simulation center. The actual "patient room" may be named for those who fund the equipment contained within it. Other opportunities exist for naming support includes space such as conference rooms used for prebriefing or debriefing. The ultimate show of support from a donor would be to name the entire patient simulation center after that individual, family, or corporation. A successful strategy to accomplish this is to have actual plans and renderings when meeting with potential donors. A clearly defined list of equipment that is needed is helpful to facilitate their support. Involvement of an organization's Foundation or Development Department is important to consider in the early stages of planning. An organization may also wish to involve industry in executing demonstrations of the various patient simulation technologies to potential donors and others in the community.

Once the actual patient simulation center is constructed, open houses for both the community outside and the "community" within an organization are an important means by which to garner continued support. When people can actually see, touch, and experience simulation technology firsthand, they are more likely to show their support through donations of additional funds. Organizations also should consider employing the use of demonstrations by faculty with actual learners. This is a very successful strategy in obtaining community support.

### 51.3.2  Establishing an Executive Committee for Simulation Integration: Why is this Necessary, What are the Responsibilities, and What are the Reporting Relationships?

From an industry perspective, the most successful patient simulation programs have established some configuration of an "Executive Committee for Simulation Integration." While this takes many different forms, the basic structure involves faculty representation from the various courses or programs within an organization. This committee should commence upon the decision to investigate and adopt simulation technology.

Furthermore, this committee should be given decision-making authority over center operations. This committee designs the organizational "road map" for how patient simulation will be integrated into the curriculum or programs and makes recommendations to the organization's leadership. It also addresses logistical concerns such as the scheduling and coordination of the center's activities.

The composition of the committee should reflect the organization's structure and culture. At a minimum it should include key faculty representatives from each course or program. While some faculty may argue that they do not imagine simulation being used in their course or program, at least initially, these individuals should be included from the beginning. In many cases, this ultimately helps with garnering their support and buy-in. It also serves for dissemination of what is being planned and has actually occurred across the curriculum or programs. Involvement of ancillary staff such as key personnel from Information Systems and Administration is at the organization's discretion. Successful models exist that include these individuals and successful models exist that do not. If an organization will have full-time or part-time center leadership or technical support, these individuals should be brought on early in the process to ensure that their input and assistance occurs in a timely fashion. In many cases, if the center will have a Director or faculty member who has been given responsibility for the initiative, then this person will serve as the chair of the committee. The committee may also be cochaired by the center Director and a member of the faculty leadership. If the Director does not have faculty rank, this is important to ensure that curriculum integration can occur in a seamless fashion.

A variation on the management arrangement is to have the Simulation Center Director function as a nonvoting guide *for* the Committee instead of the boss *of* the Committee. An informative position may reduce the vested interests of a supporting program from subordinating the goals and resources of the host institution. However, as the Committee will require administrative resources and effort to function, the Simulation team may offer themselves to fulfill these needs, including a facilitator for the Committee's meetings.

The importance of such a formal committee structure cannot be overemphasized. The committee addresses almost every aspect of the Patient Simulation Center's operations, including how the technology will be used to facilitate learning. This group is given the important charges for its mission as outlined in Table 51.6.

Upon its creation it is important to address reporting relationships of this committee to those existing within an organization. In most successful models, the "Executive Committee for Simulation Integration" is given full binding decision-making authority. In an academic setting, this must be agreed upon by the faculty as a whole in a school or program. It must also have the necessary administrative support for their decisions. It is not reasonable to establish such a committee

**TABLE 51.6** Responsibilities of Executive Committee for Simulation Integration

Facilities and preparation
- What technology will the organization adopt?
- What optional equipment and educational content will be included in the acquisition?
- How is the physical plant configured to replicate the actual clinical environment in which patient care occurs?
- Who will interface with those responsible for facility design and construction?
- What clinical equipment and supplies are needed to utilize the patient simulators?
- Who will be trained initially to operate the simulator(s) and how will this be accomplished?
- What will be the role of industry?

Operations and management
- What hours will the simulation center operate?
- How will scheduling of the simulator(s) be handled?
- Who will be responsible for ordering, stocking, and maintaining clinical supplies and equipment?
- Who will be responsible for setting up the simulation equipment and environment for actual use?
- Who will be responsible for cleanup and dismantling following actual use?
- How will routine maintenance and cleaning of the equipment and environment be handled?
- Who will be primarily responsible for interfacing with industry representatives for ongoing care and maintenance, updating of software, etc?
- How will the initial and ongoing budgets for operating the patient simulation center be developed?
- How will an annual report summarizing the center's operation be completed and disseminated?

Curriculum integration
- How will the patient simulators be used to simulate clinical experiences?
- What course(s) or program(s) will the patient simulators be used in?
- Who will train the trainers in those elements unique to simulation-based education?
- What specific simulated clinical experiences will be done in phase I?
- What specific simulated clinical experiences will be done in phases II, III, etc., of implementation?
- How will simulation sessions be developed and designed?
- How will the various simulations be programed or scripted with the simulation technology?

Evaluation and research
- How will evaluation of learner and faculty experiences with simulation be done?
- How will evaluation data be compiled, analyzed, reacted to, and reported?
- What research will be done in the center and how will this occur?
- How will application for research funding occur?
- If the simulators will be utilized for evaluation, how will confidentiality be maintained?

without this authority, as it is counterproductive for these individuals to then have to seek approvals of their decisions. Clear guidelines should be established that outline what the committee may decide and what they may not. A good example of this is budgetary input and authority. While the committee may recommend certain expenditures, ultimately, approval may be required by the administration. In summary, the responsibilities and boundaries must be clearly identified upon creating this important committee.

## 51.3.3 Simulation Center Staffing: Important Considerations in Determining Staffing

The size and scope of the simulation program is the primary driving force used to determine actual personnel requirements. Programs range in size from one simulator used by one faculty member to large, full-scale, replicated health care environments with multiple simulators. In these large centers, usually multiple full-time equivalent staffing models are employed. The staff mix includes technical, clinical, and administrative support roles. Most are funded by the organization, some include those on "soft money" (where grant funding is constantly being sought to continue the employment of these individuals).

In moderate to large-scale facilities with multiple simulators, an organization should consider the use of full-time technicians. Most clinicians and academics do not wish to be concerned with the care and support for the technology. Individuals with a background in information systems, electronics technology, or biomedical engineering are usually employed in this role. Their responsibilities range from installation, ongoing care, and maintenance, to the repair of the equipment. Some companies offer a technical support course that assists these individuals in the execution of their role. In many organizations, this technical support role may also be responsible for the actual programing or scripting of scenarios into the simulator software. When this model is employed, this software interaction is performed under the guidance and direction of clinical educators.

Of importance to consider is the need for a Director or Coordinator of the center. In most successful centers, there is a key individual who fulfills this role, be it in a full- or part-time capacity. The importance of this role is that one individual serves as the "Champion" who ultimately becomes the organization's authority on the use of patient simulation. This individual can build support for the use simulation both within the organization and out in the community. Additionally, this individual becomes the organization's spokesperson on patient simulation that can interface seamlessly with media, donors, and others from the community. Furthermore, this individual serves as the administrative gateway to the technology that facilitates its utilization across the curriculum or programs. As previously discussed, this individual may also serve as the chairperson or cochair of the Executive Committee for Simulation Integration, as well as principal investigator for grants and research.

An organization will also need to address the requirements for administrative assistance in operating a simulation center. Initially, many centers begin without this human resource. This is mainly driven by financial constraints. Many of the duties that such a role fills are delegated to the Director or Coordinator in the beginning of operations. As a center grows, the importance of this role becomes very apparent. With increasing use, scheduling resources, especially time in the center becomes a logistical concern that requires close attention. This individual may also be given responsibility for ordering supplies, maintaining inventories, and compiling the various reports that are required by organizations. Additionally, this person may serve as the "traffic cop" to ensure the smooth logistical flow of the various groups that will use the center on a daily basis.

In summary, there is no one ideal staffing structure that is recommended. Many centers operate successfully with an individual who is identified as the Coordinator or Director by either title or responsibilities on either a full- or part-time basis. Other centers have very structured organizations that include a Director, technicians, and administrative assistants. Others employ clinicians who facilitate learning with the simulation, while others utilize their full-time faculty members to do this. Many academic centers, especially at the Community College level, begin with giving a key faculty member release time to coordinate the center's activities. As utilization increases, most centers employ full-time staff. It is recommended that the fundamental questions discussed earlier in this chapter be asked to determine actual staffing needs. It is also recommended to ask industry for contact information from their user base to explore various models employed by others to determine a center's actual staffing needs.

## 51.4 Considerations for Implementation and Initial Operations

### 51.4.1 Preopening Preparation

Preopening preparation encompasses all events leading up to the grand opening of the center. Activities include facility preparation, curriculum integration planning, initial training of faculty and staff, and design of statistical data gathering and reporting process.

As discussed previously, facility preparation involves answering several fundamental questions. With these questions addressed, a structured plan and timeline need to be developed in preparation of the center opening. This plan should include the major milestones outlined in Table 51.7.

To plan for curriculum integration, the previously suggested Executive Committee for Simulation Integration should be in place to facilitate this. As this group is composed of representatives from the faculty, specific questions of how simulation will be incorporated into the curricula may be addressed. These include those identified in Table 51.8.

For an example of how introduction of simulation into a clinical program triggered the first-time development of nationwide curricula, please see Chapter 37.

**TABLE 51.7** Major milestones to consider in planning center grand opening

- Assigning and hiring of personnel
- Consultation with industry
- Consultation and site visits with established simulation centers
- Experienced simulation center designers and operators visit your institution, advise on your needs and plans
- Design and construction of simulation center
- Procurement and installation of equipment (simulators, clinical equipment, furniture, etc.)

**TABLE 51.8** Curriculum Integration Planning Considerations

- Who will be taught, and how many are they?
- Identification of what will be taught (e.g. skills acquisition, building critical thinking skills, team resource management, etc.)
- Development of a phased plan for integration into each course or program
  - Who will develop the learning content?
  - Who will facilitate the simulated clinical experiences?
  - How will instructors become prepared to facilitate learning with simulation?
- Establishment of an institutional process for simulation-based education development and implementation which is:
  - Independent of discipline or individual
  - Documented in a uniform format

Adequate consideration needs to be given related to the initial training of those who will use the technology. This initial training should include both faculty and technical support individuals. In addition to simulation product training (typically provided by industry), there are numerous courses that focus on the fundamentals of simulation-based education. These programs cover topics ranging among the underlying principles, debriefing techniques, instructional design, training program design, design for test, evaluation, and research along with other contemporary best practices. Most include experiential learning activities for course participants. Additionally, there are courses designed for preparing facilitators for specific content such as crisis resource management. Centers pioneering these courses include the Stanford University-Veterans Administration Palo Alto Simulation Center (Palo Alto, California, U.S.A) and the Harvard-MIT Center for Medical Simulation (Cambridge, Massachusetts, U.S.A).

Preopening preparation planning needs to also occur around statistical data gathering and evaluation. From the beginning it is important to gather the center's productivity measures. These include types of courses offered, frequency of the various courses, numbers of learners per course, and total learners trained. Consideration must also be given to the design of evaluation tools as previously discussed in addition to the collection, analysis, and reaction to this data. All of this is done to ultimately demonstrate the measures of a center's activity and productivity, as identified in Table 51.9.

Additionally, a digital camera should be used to capture images of the center in operation. Users should secure participant permission and release, so that these photos may be used legally in center reports, marketing, and professional presentations.

## 51.4.2 Center Grand Opening

The opening of a simulation center is an important milestone in any organization, deserving of special recognition and celebration. This offers a single time point and focus to seize upon the energy generated by the opening of the center. Planning

**TABLE 51.9** Productivity and Activity Measures for a Simulation Center

Is the mission of conducting simulation based learning programs accomplished?
- Is the simulation program effective (e.g. are learning objectives being met)?
- Is the program growing and advancing as demonstrated by increased activity, research, publication, and participation within the host institution?
- Is the program growing and advancing as demonstrated by increased participation in the larger simulation community?

should occur to accommodate several distinct groups: institutional staff and learners, administrators, funders, other VIPs, and the community. Depending upon the scope of the center, this may be one combined celebration or multiple events. Of primary importance is to ensure the recognition of the funders and leadership who have made the center possible in addition to those who did the actual work of making the center a reality. Inviting the staff and learners offers the opportunity to share in the excitement and pride brought to their organization, even for those not directly involved in creating the center. Additionally, such grand opening activities can also serve to open the door to future funding sources. These may include Foundations, the institution's Board of Trustees, local industry, health care leaders affiliated with the organization, and individual benefactors.

# 51.5 Considerations for the First Year of Operation and Beyond

After the excitement of building and opening the center has passed, the hard work of day-to-day operations begins. Organizations should anticipate the emotional let down following the grand opening. This further demonstrates why the planning previously described is of vital importance. With the adequate planning that has occurred, integration of simulation into learning promptly begins. This is now the time for executing the plan, not developing it. It is important to anticipate that changes will need to be made to the plan to adjust for unanticipated realities; however, this is normal for any start-up operation. It should also be recognized that organizations will be most adaptable early in their operations, and thus, it is important to frequently assess performance and address issues in a timely manner. For example, has adequate instructional time been allocated for each course? Has adequate time been scheduled for care and maintenance of the equipment? Have adequate supplies been procured? Is gas consumption more or less than expected? These are straightforward questions. The bigger questions for consideration include: Are the training and learning objectives being met? Is the training and learning effective? Is there a positive reaction to simulation learning from all involved?

Ongoing analysis and financial planning needs to occur in relation to operational expenditures and future capital investment for program expansion. It is important to adequately analyze expenditures to identify the cost of providing simulation learning. This should include personnel costs, depreciation, consumable expenditures (gases, clinical supplies, etc.), warranties, and maintenance. This financial analysis married with the utilization statistics can easily provide insight into the hourly and annual cost of providing this type of learning. This also positions the organization to establish a fee structure for use by external entities, research budgeting, and cost transfer within the organization. However, experience has shown that the most active and successful centers are those that are open and freely accessible to the entire organization as an institutional asset similar to a library.

Finally, the simulation center should create and disseminate an annual report. This report would summarize overall activities and attainment of identified goals, provide detailed statistics and performance metrics, summarize operational expenditures, and identify the following year's plan and budget.

Development of the simulation enterprise is an ongoing activity. Growing centers will always have the opportunity to pursue new pathways, and are continuously seeking funding sources to expand their program. This may include additional simulation equipment, facility expansion, additional personnel, and ancillary equipment for the simulation environment.

## 51.6 Conclusion

Launching a successful simulation center and integrating it into an organization can appear to be a daunting challenge. Fortunately, new users have the benefit of learning from their peers' collective experience over the past decade. Industry offers an invaluable resource as the nexus of all simulation users. This resource should be utilized early and often in planning, launching, and sustaining a simulation center. This chapter has summarized over a decade of best practices, lessons-learned, and resources available for new users.

## Endnote

1. The company provides patient simulators for clinical clients. Medical Education Technologies, Inc., corporate headquarters is located in Sarasota, Florida, mailing address is: 6000 Fruitville Road, Sarasota, FL 34232, U.S.A., Phone: 941-377-5562; www.meti.com.

## References

1. Projected Supply, Demand, and Shortages of Registered Nurses: 2000–2020. U.S. Department of Health and Human Services, Health Resources and Services Administration, Bureau of Health Professional National Center for Health Workforce Analysis. July 2002.
2. Beaubien, J. M. and Baker, D. P. The use of simulation for training teamwork skills in health care: how low can you go? *Qual Saf Health Care.* 13 Suppl 1, i5–6, Oct (2004).
3. Devita, M. A., Schaefer, J., Lutz, J., Wang, H. and Dongilli, T. Improving Medical Emergency Team (MET) performance using a novel curriculum and a computerized human patient simulator. *Qual Saf Health Care.* 14(5), 326–31, Oct (2005).
4. OhioHealth Riverside Methodist Hospital. Center for Medical Education & Innovation http://www.ohiohealth.com/medicaleducation/riverside/cmei.htm. Accessed 29 Aug 2007.
5. Mecklenburg EMS Agency. <http://www.charmeck.org/Departments/Medic/home.htm>. Accessed 29 Aug 2007.
6. Penn State College of Medicine. Simulation Development and Cognitive Science Lab http://www.hmc.psu.edu/simulation/. Accessed 29 Aug 2007.

# Funding, Funding is What Makes Simulation Go On

*No one is ready for a thing until he believes he can acquire it.*

Napoleon Hill

It is amazing what we will buy with other peoples' money. Unless we just happen to be independently wealthy and have a penchant for owning and running a simulation program, like it were some kind of sports club, then all our funds for all our simulation efforts come from someone else. Usually, these other people dilute their investment in your simulation with their interests in supporting other forms of clinical education. Thus, there always seems to be a gap between the magnitude of financial resources we have been given and with what we can imagine we could spend. The size of this gap can be reduced in two ways. Guillaume Alinier describes creative thinking and energetic actions for working within academics' tight financial constraints. John Gillespie describes creative thinking and energetic actions to attract outside financing to your program.

# 52

# Prosperous Simulation Under an Institution's Threadbare Financial Blanket

Guillaume Alinier

## 52.1 Follow the Money

Most simulation centers based within an academic institution will be a not-for-profit entity and rely on educational or charitable funding for establishment and daily operations. Operating a center is an expensive business due to the cost of the clinical equipment, the simulation technology, the staff and faculty time and expertise, and the important demand on faculty time to facilitate sessions because of the relatively low student to staff ratio required. Breaking even in financial terms is a challenge that many of us are constantly struggling with.

## 52.2 Networking, Negotiation, and Collaboration

One of the key principles to minimize cost is good information, and most of that can be obtained through effective communication. In all fields and professions, good networking skills can prove to be a rewarding activity. Whether you are in need of a piece of information or require help, having established links with a range of people in your field or other related fields is a useful asset. In my case, I have greatly benefited from

advice from colleagues, either simply by discussing with them at conferences or via interest group list servers like the one operated for the Society for Simulation in Healthcare[1]. For a little investment, directly visiting other simulation colleagues in their work environment is an even better way to understand the context in which they operate and how they manage and organize the activities of their center in general. It is very likely that somewhere someone else will have faced and solved the problems you may have now. This may include how to obtain educational funding to produce projects, how to reduce resources consumed in specific scenarios, and inexpensively prepare the appropriate props so it can be realistic enough to not confuse the trainees.

You will quickly learn that negotiating and "selling" skills are also vital. You need to sell the activity of your center very favorably, to different people and for different reasons. Successful industry is ruled by business people who often know and believe that they have to give to obtain something in return. This is where you should try to take reasonable advantage of your position as an educator of future health care professionals, who will potentially become customers of clinical products and services. Companies are much aware of this fact, and hence "painting a nice image" of your center and brushing up on your negotiation skills should enable you to obtain some expensive equipment at a reduced price or maybe acquire some demonstration or loaner equipment, which is then regularly updated as new products come out on the market. This enables companies to demonstrate their equipment to trainees at an early stage. As they will certainly become accustomed to using this equipment, they may be more inclined to advise the purchase of similar equipment when working in clinical practice. When negotiating for equipment, emphasize the number of trainees that access your center, their geographic origin, the disciplines represented, the access by qualified professionals for continuing professional development, and any other external visitors that may come to have a look at your center.

You should also think of raising the profile of your center within your community by putting out press releases whenever something significant happens. It is most likely that some of the equipment will at least appear in the background and will be noticed by some expert eyes. Remember to think carefully about what you are doing and receive the approval from anyone involved, as you do not want to compromise future potential collaborations with other companies or important visitors. In our case, we have had journal press releases, radio and television interviews on a number of occasions regarding the visits of Health and Defense Ministers from several countries, Members of UK and EU Parliament, and other celebrities interested in our activities in relation the political, social, and educational context. Other publicized events have included the addition of new patient simulators in the center, the start of new collaborations with local hospitals, the launch of new courses relying heavily on the use of simulation training, and, more recently, the fund-raising campaign for the construction of our new multiprofessional simulation center. The presence of the equipment in your center will create publicity for the companies and that is exactly what they will be looking for. On a collaborative note, as part of the deal, some companies will certainly be expecting feedback about their products from the users (the simulation participants and you) in order to improve their usability and functionality.

Other collaborative aspects you might want to explore could include offering access to your center to local institutions that could also benefit from simulation training (i.e., hospitals, clinics, and emergency services). By putting in place a charge for using your center, you might be able to alleviate some of the financial burden. Local clinical facilities and specialty societies have ongoing needs for training their employees and constituents. Potential *paying* trainees that experience your offerings during an open house will certainly encourage their host institutions to adopt your center as a regular training venue, where they will now spend a significant proportion of their educational or training budget.

Establishing links with the outside community can help you improve the image of your institution with your local community through job fairs, educational events, and open days. Although the magic of patient simulators has probably been transformed for those of us who use them everyday, the general public always find them intriguing and impressive. Presenting simulation-based presentations with groups of applicants is still a powerful recruiting tool. Your admissions department should compensate you in some way to help them fulfill their obligation to the institution.

A different strategy, which does not preclude you from adopting the other ones, involves carrying out projects in partnership with other centers. This could include codeveloping scenarios and courses and producing them together. It can help you improve the research activity of your center, its reputation, and your experience. Such activity can be very rewarding and help you to become a strong contender when it comes to applying for educational or research funding. These examples will hopefully encourage you to extensively explore this venue, which is about establishing external links to open up new opportunities to increase the activity and success of your center.

## 52.3 Using Existing Resources

Cost avoidance is an important issue, and making use of internal resources can be very helpful in that way [1]. You may be able to use the expertise of other departments within your institution by, for example, by providing research project ideas for upper-level students. In our case, we have recently used business school trainees to do a small market research project concerning the interest of doctors in using our center. This was a mutually useful project, for the business school, the students, and our center. The other benefits to the department concerned is that it reduces the number of project titles they have to create and potentially their supervision input as you might be able to share in the supervisory responsibility.

Increased activity means extra work. On a human resource aspect, I would recommend the consideration of offering placement opportunities to college or university students depending on the type of projects available in your simulation center. It is a mutually beneficial arrangement that will give work experience to a young mind and at the same time probably relieve some work pressure from you, or at least enable you to do these simple things that you have meant to do for so long, but have not had the time to do. This works very well for technical as well as secretarial support tasks. High school students might be satisfied to assist in simple tasks like tidying up your center, the scenarios and their props, and doing equipment inventories, while university students might be more interested in web site design, production of publicity leaflets, statistical analysis, or even small electronic or mechanical engineering projects. On the financial side, it is often a very advantageous for the center as you may not have to remunerate the trainee depending on your institution's regulations for obtaining credits and/or approved training experience. One of the determining factors will certainly be the duration of the work placement, which can be from a few weeks to several months. Having benefited myself from similar experiences while at school and during my University studies, I found it an extremely useful experience and I have since kept good links with these companies.

## 52.4 Keeping Things Simple

Since you are not operating for profit, you do not really need to impress an audience of customers with very fancy equipment and high-tech audio-visual facilities. Your participants will primarily be looking for a valuable learning experience; hence your setup can remain simple enough to be functional and reliable while producing results of appropriate quality [2]. It will be less likely to cause you any problem and will be easier to use and troubleshoot should anything go wrong. This will also probably reduce the risk of burglary in your center. Similarly, if you try to raise funding for your center and receive a panel of visitors, displaying the latest audio-visual gadgets and luxurious chairs and furnishing could put them off and be an embarrassing experience for you.

## 52.5 Building a Good Profile

In order to obtain external funding for the operation of your center, you will certainly need to prove the value of the center and the educational offerings, and you cannot do that unless you are already operating and have some supporting evidence and a track record of excellence in simulation training. It is important to ask all trainees to complete a standard evaluation form, especially asking for free text comments that can be used in your motivations. Once again, networking and collaboration might be the keys to the problem. Doing simple collaborative research with another center would allow you to progressively build up your profile in the field of clinical simulation training, and hence your track record of good practice. As mentioned above, good publicity from the press is also an important aspect of building your profile and it will indirectly help you with the financial aspects of your center.

## 52.6 Staffing

All programs that make use of your simulation facility should directly contribute in some way. Human nature being what it is, we do not value what is free nearly as much as what is earned. Another significant cost-saving option, although not always optimal, is to use staff resources from other departments when they are actually putting their students through your

center. We have always operated this way in our center, and the staff most regularly involved (3–4 times per month) have developed some real simulation facilitation expertise that really makes the difference in comparison to other staff who attend on a less regular basis (3–4 times per year). This is very favorable to the smooth production of the simulation sessions and this is noticeable to students who attend several sessions produced by different faculty. We are now encouraging the involvement of only those faculty who are committed to regularly assist producing simulation training sessions so that they can keep up their simulation facilitation and debriefing skills. Faculty can also have some very useful contacts in the industry and help you in your negotiations to obtain some equipment.

## 52.7 Minimizing Costs, Maximizing Returns

Regularly view every source of cost, and assess if the amount you are spending and the way you are paying for it is producing the value you expect in return.

## 52.8 Conclusion

Not receiving a sizable amount of regular funding can slow the development of your center, but it will also encourage you to develop solutions and make use of your contacts. Adherence to a few basic principles as outlined in this section can make the available funds stretch further, as well as generate even more funds and support.

## Endnote

1. www.ssih.org

## References

1. Alinier, G., Hunt, W. B., Gordon, R., and Harwood, C. Enabling students to practise nursing skills using simulation technology: Could your institution do it? In: Beldean, L., Zeitler, U., Rogozea, L. *Nursing Today*. Alma Mater, Sibiu, RO. 2005. pp. 161–164. ISBN 973-632-224-6.
2. Alinier, G. Investing in sophisticated medical simulation training equipment: Is it really worth it? *Journal Medical Brasovian*. 4(1), 4–8 (2005).

# 53

# Creative Procurement for Your Simulation Program

John Gillespie

## 53.1 Key Components to Simulation

The following lists the essentials required for simulation. They all need to be acquired, maintained, and replaced as and when exhausted. They all can be bought with money. Not all of them *have* to be bought with money. Most of them are already available but are used or misused or underused. You just need to develop a strong trading network to find, convey, and convert them to your purposes.

Time
Energy
Enthusiasm
Attention

Acceptance
Approval
Creativity
Teaching objectives
Students
Instructors
Simulation operations professionals
Simulators (human or plastic)
Space
Clinical devices, equipment, and furniture
Clinical consumables
Clinical clothing
Audio/video devices, equipment, and recording media
Computing and communicating devices and equipment
Office equipment and furniture

## 53.2 Reality of Simulation Procurement

In the next few paragraphs, we will address the reality of simulation procurement. Discussed in this topic will be what your center needs and how to be a good steward of your available funds. We will look at some of the do's and don'ts of establishing and equipping your center.

## 53.3 How are You Going to Use Simulation?

What are your learning objectives? How does your curriculum drive the simulations and how are you going to make simulation effective? This question and its three-part answer are the basis to both *the what* and *the how* you procure for your simulation program. Your initial acquisitions should be the internal customers of the institution that have asked to use simulation, the faculty whose curriculum you are enhancing, and the students who will be participating in the scenarios. Your follow-on acquisitions are the devices and equipment required to satisfy these first three.

As the person charged with making simulation happen, whatever your official title is, you will have to become a sales person in order to acquire and/or procure. Time is the most precious asset that any of us have. Use simulation to sell time in exchange for what your program needs. You will find that you are not only selling simulation to administration, but you will be selling your thoughts and ideas to the faculty as well. In some sense, they are a harder sell than the administration is. It is very much a "what can I do for you" approach. Frequently, faculty feel that they are giving up lecture time to support simulation. If seeing a picture is worth hearing a thousand words, then just how much more valuable is actually participating in the action? Your goal is to show that together, you can enhance their current curriculum by using the schedulable, predictable, and transportable clinical reality that is inherent in simulation. For example, if they are spending a week on Cardiac Reflexes via lecture and a live animal laboratory, you show where they can improve their student's comprehension in less time with a simulation designed to complement the weaker aspects of these two historical ways of teaching. Perhaps, you can deliver this experience via live, bidirectional audio/video links into the lecture hall so that the inelastic teaching schedule of 180 students is not disrupted.

Actually, I feel that all of these components are intertwined. They each rely on the other and are equally important to having a successful program. Your most important customers are the school that has asked to use simulation, the faculty who you are working with to enhance their curriculum, and the students who will be participating in the scenarios.

## 53.4 Suspension of Disbelief

How proficient would you have been at your driver's license test if all that you did to prepare was sit on your couch with nothing more than a book in your hands? Picture yourself on the couch, book in hand, looking at imaginary mirrors trying to find imaginary traffic encounters using imaginary pedals, with imaginary distractions and imaginary sights, sounds, vibrations, and smells. I would have failed miserably at the time of the in-car test without an interactive in-car learning program that filled the gap between reading the book and being fully responsible for the safe operation of a car. Simulation captures our imagination while we work with our bodies. Safety, time, imagination and money limit how you fill this gap.

How real should you make it, and how do you know when it is real enough? The setting should *appear* real enough so that your students never guess incorrectly about where they are. If your scenario is in a patient room in the hospital, it should look like a patient room in a hospital: place your patient in a hospital bed with the overbed table, night stand, and guest chair. You ideally would have a head wall unit for gases, the plastic tub and sputum cup, ice pitcher, straw, water glass, and if your patient is well liked, plastic flowers and get-well cards.

Is your patient on any life support or clinical equipment? Perhaps a ventilator or an IV pump? Perhaps an NIBP monitor? Do you have a chart for your students to read? These are all important for both their hands-on teaching purposes and for linking with their imaginations. When your students are ready for the added distraction, you might want to pipe in environmental sounds such as hospital pages and other noises that you would hear in this setting. Should there be any smells? I'm not talking about just the nasty smells, but consider the cleaning solutions that are used as well.

If your patient is in the chair at the local Dentist's office, then put him in a dental chair complete with the bright light in his eyes and the funny little napkin on a string. If he is at home, configure your room as a living room complete with the hazards that we all have in our houses (minus the territorial dog: safety issue). Is this an ER? Use a gurney instead of bed. If your setting is an OR, then use an OR table and an Anesthesia Machine. However, when your OR becomes the living room, make the effort to remove the Anesthesia Machine from the room, except for those very rare scenarios when you *want* to recreate the living room of someone who just happens to keep their Anesthesia Machine in their living room.

It has been shown that the olfactory gland is the strongest trigger of a distant memory. As a result, I feel that smells play an important role in simulation. However, this is also an area where it is easy to become unsafe, too real, and simply offensive, so care should be taken to balance against the desire for your learners to remember the patient with gangrene in the lower right leg. He is in septic shock and the surgeon needs to perform a below-the-knee amputation. What smells can we expect when the knife is laid to the skin? There should be a rotting flesh smell much like what you smell in a slaughter house. Nasty. As a result, we will use Evergreen (very strong pine-scented air freshener) to offset the stench. Now, throw into the mix a faint odor of isoflurane and you now have the right smells. If your area is adjacent to offices, you might not be very popular. However, I promise that the first time one of your learners walks into a room with sepsis, they will know it just by the smell!

If you are in a prehospital setting that calls for a bomb simulation such as an ADLS course (Advanced Disaster Life Support), you might want ask your local fireworks supplier to donate and demonstrate the detonation of some training munitions. This will provide realistic smells, sounds, and the smoke of an explosion. Again, utmost caution is required here but the result will be *"When can we do this again?"* These are both extreme examples of what you can do. There are many other smells that are equally as good at triggering memory. Again, you are only limited by your imagination, money, and safety.

The point is simple. Students know that simulation is a sensory experience. They will assume that everything they sense is intentional. Thus, you control what touches as many of their senses as you can. It will go a *long way* toward the learners walking away from simulation saying, "Wow! That was amazing!" If they and their instructor have that feeling, do you think that they will (i) remember it? and (ii) want to do it again?

## 53.5 Environmental Safety Issues

What type of environmental safety issues could simulation possibly have? Too much isoflurane in the room air, and that could be a big enough environmental issue to get your center on the national news the hard way! What do you think would be the result if your center was near the Dean's Office and you had the sepsis cocktail too strong? Well-faked bloody bandages look just as real as do real bloody bandages when they appear on the evening news as they are discovered with municipal waste. Use real medical waste disposal for all faked items that if real, they would be disposed as medical waste.

Just as you never want nonfunctioning simulates to intermingle with real patient care equipment, you don't want real patient care hazards contaminating simulation. Even if your simulation center will never be used for real patient care, most, if not all, of your support equipment has been used for real patient care at some time. While we all know that hospital staff does an amazing job of keeping things clean and safe, we also know that hospitals are notoriously dirty places that attract and concentrate an amazing assortment of toxins and infectious agents. Do you think that you might want to give that newly acquired piece of used equipment that you just got from surplus a very good and thorough cleaning to eliminate any possible biohazards before you start teaching with it without wearing gloves, before your next group of guests tours your center?

Another important concern for environment hazards is the equipment itself. Let's take a moment to look at the defibrillator. You and I should know how to safely use this potentially lethal device. But, as guides or facilitators to simulations, are we the ones that are going to use it? No, the whole purpose of simulation is to allow learning in a safe environment so that when the learner performs for the first time in a real setting, they will already know what to do, how to do it, and how to behave. As a result, there will be times when you have to stop or pause a simulation to teach how to safely do something or use a particular piece of equipment such as a defibrillator so that you are not forced to perform Advanced Cardiac Life Support for real. Whenever possible, vasectomize defibrillators by cutting the power output to the paddles and rerouting it to the internal discharge component. They can still sense and be charged, but they shoot blanks. No instructor should ever treat *any* defibrillator with anything other than the full respect any potentially lethal device deserves, especially in front of students. No student ever has to know that the simulation center's defibrillators shoot blanks, otherwise they may not develop the proper respect for real loaded weapons.

## 53.6 When Does the Scenario Begin?

Now, you have your room configured with all of the appropriate equipment in place. Your environment is tooled to the right setting and your learning objectives are defined. You are ready to start when your learners arrive.

When do you start your scenario? Do you wait until your learners are all in the room next to the patient and then you say something like, "OK, let's start the scenario?" If you do this, then you just wasted all of the time that it took you to set the stage. The scenario starts when the learners' imaginations enter the room. This is an integral part of setting the stage as well. Again, the thinking behind this is to make it feel real. If you tell the students that you are starting the scenario now, then you will be causing the suspension of disbelief to be lost.

Another example would be that if you walked into a real patient's room on rounds and found the patient not breathing, blinking, and no pulses, you would call a code. If the learners find this in the simulation center, they can have mixed messages as to what the learning objectives are. Likewise, just how alive a patient do you want for your students' last impression as they leave the lesson?

## 53.7 Timing

This leads us to timing. There are several different ways to handle the timing of any given scenario. As a general rule, the teaching objectives, time allotment and type of learning event will dictate the timing of a scenario. One of the best scenarios that I have ever seen was a scenario that was "allowed" to proceed in real time. In other words, there was no time constraint dictating how long we could go. It was a malignant hyperthermia case in the postop care area for clinically experienced students. When the nurses keeping the simulated patient alive asked for things like an arterial blood gas, we paged an Anesthesia Technician to come for the "sample" that we provided and asked them to perform the tests and bring us the results. The scenario progressed while the test was being performed. We did the same thing when they asked for an Anesthesiologist to intubate and then again when they needed a ventilator from respiratory therapy, which happened to be three floors up. The whole time, these nurses worked furiously to keep the patient alive. It was an absolutely amazing event to watch.

Our objective in this scenario was to task the staff in the Postanesthesia Care Unit with the challenges they would be dealing with in real life if they were to experience the fairly rare case of Malignant Hyperthermia. Several questions arose. Where are the drugs housed that are used to treat this? How are those drugs mixed? How many people do I need to mix the drugs? Where can I find sufficient ice? Where do I find large bags to place the ice in? How long will it take to receive an arterial blood gas analysis or a have a ventilator delivered? What are the clinical symptoms of Malignant Hyperthermia? What can I expect from my interdisciplinary coworkers during this crisis? On all fronts, these were met by presenting this scenario in real time. Had we "accelerated" the case, we would have not learned so many crisis management components inherent within this scenario.

## 53.8 Curriculum-driven Buy-in

As I mentioned earlier, you are now a sales person. There will be many times when you have to show your faculty how to use simulation in their curriculum. The easiest, but also the most rare, is to have a faculty member approach you to ask how you can help them show a certain event in simulation. This is easy because you know how your simulator works and what the capabilities are. So you'll be able to say, "Yes, I can do that", or "No, I can't show that just yet."

The harder sell is when the faculty has been told that simulation will be incorporated into the curriculum. Usually, the faculty will then fear the unknown world of simulation. It can be a very adversarial-based fear. They usually have no idea what is expected of them. In our shop, we approach this empathetically and reinforce that all they need to do is teach. We will produce the simulation for them. Together, we will make the clinical instructors look brilliant in front of their students because unlike the unexpected from real patients, we deliberately plan and rehearse the scenarios before the students ever join in.

The third hard sell is when we approach faculty who don't want anything to do with simulation. This is the equivalent of a cold call. If the faculty don't recognize anything broken in their curriculum, then why would they need to change and adopt simulation if they are not being told to do it?

All of these are "normal" human responses. As a result, you eventually will improve at detecting and responding professionally to these feelings. In summary, the teaching objectives in your curriculum and learning objectives of your students are the two most critical components to simulation.

## 53.9 Learning must be Objective Driven

When you are designing and refining your scenarios, one of the most important things to think about is the learning objective. This is one of the most basic items that your customer is in need of. When a faculty member is talking with me about developing a scenario, I ask them for all of the teaching points and/or learning objectives. This allows me to design the scenario to meet their needs. By doing this, you ensure that the faculty will leave the simulation exercise feeling satisfied, and they will want to do this again. One of your goals is to have the faculty tell the other faculty what a wonderful, positive experience that this was. This will ensure growth in your Simulation Center.

## 53.10 What are You Going to Need?

### 53.10.1 Space

The most limiting factor to what you can create with your center is the amount of available space. With the many objects used in a diverse collection of scenarios comes the need for "off-stage" storage. I am frequently "flipping" and rearranging our laboratory to support all of our activity. As the setting for our scenarios change, so does our lab. I simply do this by exchanging the simulators in the room, as well as the support equipment. The doors have to be wide enough, the objects have to be, or are made to be, portable enough, and the storage space has to be large enough. As the latter never is, I am able to "share" some spaces of my neighbors, only so long as I am sensitive to their needs. For example, whenever possible, I procure for them items that they need. As a result, I "borrow" space from the Learning Resource Center. I have two rooms that they have graciously loaned. As a result, I am very sensitive to their needs. If they need a piece of equipment such as hospital beds or blood pressure monitors, they will ask me to procure it for them. We have worked out a very nice arrangement. Recently, my department chair and I went on a driving tour of other centers in the region to learn how they are building/operating their centers. Overwhelmingly, lack of space was the number one answer when asked what they would change in their center.

### 53.10.2 Creativity

Let's assume that you actually politicked your way into lots of space. Let's also assume that you have tons of creativity and a very willing faculty. Your creativity is one of the components that will be driving your needs for support items. The way that I look at the need for support equipment is by looking at the setting when creating the learning objectives.

### 53.10.3 Environmental Fidelity

The clinical settings will dictate your inventory, and how you use it. Recently, while supporting Dental Continuing Education Unit courses, I left our anesthesia machine in the lab. I should have taken it out of the laboratory before they arrived, but I thought that it would be a good way to deliver $O_2$. During the "come in and familiarize yourself with the simulator" phase of the class, one of the Dentists asked if he could "use" the anesthesia machine. I asked him if he would normally use a full anesthesia machine in his practice? His response was, "no." So, starting out, I had lost the suspension of disbelief with the Dentists. The setting was wrong. What I should have done was look to see what type of $O_2$ delivery system they are currently using in a Dental setting. Now, I know that I don't have to have on hand a complete set of air-driven drills, but I do need to keep the inappropriate devices out of the room.

### 53.10.4 Real Items, Devices, and Machines

When you configure your laboratory for the first time, you should not dedicate a tremendous amount of time and energy to procuring supporting devices and machines. Examples of this are: hospital room furnishings, ventilators, dialysis machines, IV pumps, etc. If you have an anesthesia program that you will be supporting right away, then one possible exception to this is an anesthesia machine. One other exception is a bed for your mannequin to present on. (However, in pre-hospital settings, the floor might be the best place to present.) You will find that "gathering" equipment will be an ongoing process that will evolve from and with your learning objectives over time.

With the anesthesia machine, there are two schools of thought. One is that if the students can learn to operate an old machine that is all manual, then they certainly will be able to operate an automated machine. This is particularly true when the automation fails. The other thought is that the students should learn on what they will use with real patients.

The reason that I bring this up is that if you need a machine right up front, and have the money, you could buy a new one for around \$40,000. If however, you need that \$40,000

elsewhere, you certainly can use an old machine. In our shop, even with just one lab, we have a new machine and an old one. We will be able to teach the use of both. Prioritize your equipment needs and create a wish list to work from. Keep it current because you never know when gifts of funds will arrive.

As a footnote, some centers serve several hospitals, and accept trainees from outside their systems. It is impossible to keep a "full set" of, for example, all types of anesthesia machines in a center. However, the trainees still gain the full value from the sessions, even though it is not identical to their home base. Furthermore, residents and trainees need to be willing to learn new equipment, because when they have completed their training and leave, they will most certainly encounter unfamiliar equipment.

## 53.11 Prioritize Your Fund Types and Sources

There are several funding sources out there. The important question is, "Where is your money coming from and what do they (the donors) think is a suitable use for it?" This will have a big impact on how your center will look and operate. For example, you intend to produce OR scenarios, so you order at least one of everything that is in the latest "OR catalog." Some funding sources might applaud that their name is over the door to this extravagant display of their generosity, while another might nail your hide over the same door as a warning to others about your fiscal irresponsibility. The list that I work from when looking for money is: Institutional funds, government grants, nongovernmental grants (hospitals, medical companies), endowments as well as international, national, and local companies.

The way that I look at hospital grants is a fee for service type of annual grant. I do not want to replace in-house education for any hospital, but I want to enhance it through simulation. As a result, the hospital still uses their in-house educators. As a part of that arrangement, I try to negotiate into the grant the agreement that I can borrow hospital-owned equipment, as needed, subject to availability, for the purpose of teaching the hospital staff. In the past, I have used this to borrow items such as fiber optic intubation equipment. With this, economy of motion and maintaining sterile fields can easily be taught. Learning the lack of ability to "position" your patient during a difficult intubation is another area of benefit to doing this in simulation. Learning to use fiber optics take time and practice to perfect the use and "feel" of the equipment. It is very much a learned art. As a result of the ability to perform this procedure over and over, there is benefit to doing this in simulation. The drawback is that the airways in some high-fidelity simulators are not anatomically correct past the vocal cords.

In addition to the equipment, I also ask for the supplies that will be needed to support their simulation. Once this type of mutual relationship is developed, it is easy to find the hospital educators for each department and ask them to hold expired or damaged supplies for the Simulation Center. This is a great way to acquire items like central lines for example, and will help lower operating costs.

### 53.11.1   Coordinate With the Corporate/ Foundation and Development Personnel at Your Institution

One thing that you don't want to do is solicit large companies or local endowments without first checking with your university Development personnel. The thinking behind this is that they may already be in contact with these specific donors, working on a different project for your institution. If that is the case, you are now in "competition" with your own university for the same funds. If the "endower" or corporation elects not to offer to either of the "great programs" that are asking for them, then who do you think will be in trouble?

The other issue is that the Development office does this for a living. They are very good at it and have better information and sources than you or I do. The Development office loves to work on a project like this and should be considered an asset to you.

### 53.11.2   The Fine Art of Begging: SELL, SELL, SELL

Again, the one thing that you can and will have to do to create a successful center is sell. You will be selling your services to everyone that you can think of. One of the first places that you need to sell is your administration. Dr. Michael Seropian from Oregon found that the higher the buy-in person was in the campus administration, the better the chances are that the center will be successful [1]. While on the surface, this sounds logical, actually it goes deeper than that. If you have buy-in from the top, you are more likely to acquire the things that you need. You will have more faculty willingness to use your center and, as a result, you will have more manpower to run your center. One great risk to a successful center is that if there is only one person that is a champion of simulation, and if that person chooses to leave after the center is started, then there is a possibility that the simulator will become an expensive dust catcher. Therefore, if you have buy-in from the top and have more people driving the project, your chances of your center surviving the departure of your champion are much higher. Another benefit of buy-in from the top is in endowments and corporate donations. If they see that the university is backing your endeavor, then they are more likely to help you create your vision. If you are the only person pushing the project, then you could be a shaky investment for a corporation.

A component of the selling end is Education. You will find that you devote a large amount of time educating the people to whom you are selling your idea or product. While they are, in their own right, very bright people, they may not have any idea what simulation is all about. Or worse, have an inappropriate conception of what this is all about. To them, this might be just an expensive toy. You might have to actually show them what you have in mind or how this can help them.

Another key component of begging is bending to the needs of the person that you are begging from. If you are trying to convince a company to give you a high-dollar item and the only way that they will give it to you is if you are willing to let them use your center as a showcase, then by all means, accept. (However, first gain approval from your institution's legal department.) You want to do this even if you had never entertained the thought of this before now. If you bend, not only do you receive the item(s) that you are after, but also the vendor will be bringing in other companies, hopefully with deep pockets, to visit your center. This is a great opportunity for you to make new contacts.

### 53.11.3   Show Your Need

One of the key components to begging is showing that you have a need. Just saying so won't do the trick. You will have to demonstrate that you indeed have a use for their product. If you are begging for buy-in from the university for your center, then you will want to show them several points: who, where, what, and how the center will be used. You will have to have a business plan! They will be looking at the business plan for your goals, mission, vision, objectives, and return on investment. This is not all-inclusive. Your administration will want more. The business plan will have to be very detailed and an intricate part of selling the need for a simulation center to your administration.

If you are begging for money (through your Development office), you will need to show that there is a need for them to support your center. How will their money be used? What will it enable you to do? What will be the return on investment?

There are several ways to measure the return on investment. The most important, but also the most difficult to prove, is the educational value to our critical customer, the student. This is difficult to do without a long-term study in the form of a survey of past and present student bodies. An example question could be, "Did simulation in the curriculum play a role in your decision to attend here?" Because we don't have years to do a study before opening a center, we chose to approach this from an accounting perspective. Can we generate *any* income from this venture?

The answer is yes. Through billing for center time and services to other departments and schools for use of the center and via Continuing Education Unit Courses, income is available. The largest pitfall to this is the need to wait until the following fiscal year for the other schools and departments to plan into their budget for simulation. You often will find this with organizations outside of your institution as well.

As a result, you have to be very careful with the funds that you have. This is particularly important for the first year of operation as you may not have any income. Soliciting some free advertising via television news programs and local radio talk shows might be helpful in getting customers outside of your organization that you might not have thought of.

As always, safety will be a huge component of your sales technique. Patient safety is a key phrase to use while talking with potential customers. Be creative. Simulation is new enough that the person that you are talking with may not have any idea on how they can use your services.

## 53.11.4 Recognize Your Competition

There are several types of competition. Some are simulation centers that are in your area or region. Others might be other nonprofit organizations that have a medical component to them. Organizations that procure medical equipment for third-world countries is but one example. As with anything, knowledge is power. It is always a good idea to know your competition. If you know who your competition is and what their needs are, you can exploit that by finding a way around them to meet your needs. Now, I'm not talking about anything underhanded or inappropriate here. What I mean is that if you know what your competition is doing in the market place, then you might want to incorporate some of their ideas into your program. Because of confidentiality, I have to be vague here, but, as an example of this, there is a university in Texas that partnered with an organization that provides equipment to third-world countries. On the surface, this sounds as thought they both are chasing the same equipment. However, the nonprofit is funded by a *huge* oil company. As a result, the university followed that trail to the oil company's deep pockets. Because of the financial and political backing, this can lead to you doing a better job of selling your center to all of your contacts.

## 53.11.5 If Your Competition is Formidable, Partner With Them

So, you did the digging and you learned what you needed to know about your competition and found out that there is no way that you can compete with them. What are you going to do now? You still have your needs and they still are not being met. If you quit, your needs still won't be met.

Partner with your competition! So look at how your potential adversary can benefit from partnering with you. They have to receive something out of the deal as well. For example, several neighboring institutions in Boston combined resources and built simulation programs and facilities that service a number of the mutual needs. If they had each made a go of it on their own, they would invariably end up competing for the same funds not just in start-up, but in operating budgets and status as well.

## 53.11.6 Only Make Promises That You can Keep

This is another one of those areas that on the surface seems obvious but is worthy of mentioning. If you make a promise, such as a year's worth of simulation time in exchange for a ton of equipment, you might find that you can't keep your end of the bargain. For whatever reason, you have left your "investor" with less than promised on their return on investment. How would you feel if you were the investor? I doubt that you would be willing to even talk to them again. You certainly would not want to have any further dealings with them. My only other advice is to be careful what you offer. Make sure that your superiors are all onboard before you start bargaining.

## 53.11.7 Resources Available to You

One institution that I visited recently was literally renting out their facility to corporations as a meeting/education facility during off-hours and when they had downtime. I thought that was pretty innovative. To date, they had brought in over $200,000 in rental income through this method. In this case, the university was renting the Standardized Patient component of their center to companies and organizations that need small rooms that have audio/visual capabilities. These were used to teach and evaluate sales techniques and communication skills.

You can offer to enhance hospital education as a means of bartering for equipment. As another means of generating funds, you can look at offering simulation-based Continuing Education Unit courses in your facility. There is a large market for this that appears to be untapped for the most part. You, of course, will need to coordinate and have the cooperation of your faculty to support something like this.

If you are a state or nonprofit school, you have a huge array of tools available to you to both generate income and equip your center. One of the best ways to acquire equipment is via surplus from other local hospitals. I have "made friends" with the property managers for all of the area hospitals. As a result, I will go periodically to their warehouses and look at what is in their surplus inventory for anything that I might be able to use. I have procured over 35 hospital beds, 2 OR tables, 1 NICU incubator, 2 dialysis machines, 3 ventilators, 2 head wall units, 15 stretchers, 1 anesthesia machine, 4 complete sets of patient room furnishings. A new, never-used gynecological examination table was sent to "surplus" (discarded, unusable for clinical purposes) because the padded cover was slashed with a box cutter when the shipping crate was opened. Perfect for plastic patients. I'm still on the look out for more. All of these items are either directly or indirectly tied to scenarios.

Another source is the local Emergency Medical System. They too will have old equipment that they would be willing to donate. I find them particularly helpful with items such as ambulance stretchers, backboards, and C-collars. You can also offer to enhance or augment their education as well. This is a nice trade-off to acquire the items that you need.

If you are a health sciences campus, you also can often find the equipment that you need from the other schools that are on your campus. We are helping the School of Dentistry with a Continuing Education course. As a result, they donated a dental chair complete with the bright light, to help us set the stage for the course that we are supporting for them. As I mentioned earlier, we only have one lab, hence I have put the chair on wheels to move it in and out of the laboratory as needed.

We are looking into setting up a home health room to use the simulators with our School of Allied Health. To support the course, we are asking for living room furniture, a stove, and a refrigerator. All of this will also be used with Emergency Medicine in their EMT and Paramedic courses.

Used Medical Equipment companies are also a source. They are very good at knowing who is getting rid of old equipment. They are in the business of buying that equipment and reselling it at a profit, so they may hold back on telling who actually has the items, but they often will tell you what they have access to that they don't really want to buy. In other words, they usually buy in lots. As a result, to acquire the item that they want, they may have to buy a whole truckload of stuff that they can't resell. By developing a contact, they will contact you to inform you that they have examination tables that need some refurbishment that they will give to you if you want them. *Take them!* They usually deliver for free. If they need to be recovered, ask a local upholstery shop to donate the job in exchange for a tax deduction.

One of the most valuable items that a new or used equipment company can give you is information. I have found two local hospitals that are getting all-new anesthesia machines. They even disclosed who the vendor is. I also found out about a Veterans' Administration Hospital that is refurbishing a whole wing. As a result, I know when surplus items are hitting the "market" and who is selling them and who the buyers are. With this information, you can go directly to the institution that is replacing their property and to the company providing the new property to try to acquire the retired equipment. You would be amazed at what people will give you if you just ask.

As a state-supported school, you are in an enviable position due to the fact that you have a resource that you can draw on that very few people are aware of. This resource is government surplus. On both the state and federal level, the government has web sites that list the inventory that they have on hand in surplus. As a state institution, you are entitled to this surplus for a fraction of the normal cost. You literally can receive a fiber optic intubation bronchoscope tower for $25.00 + shipping! There are several locations where these items are housed, so you obviously want the closest location to you. There are a few hoops to jump through, but nothing that your property management folks shouldn't be able to help you with.

Now, you must be asking yourself, "What can I do to help make people want to give things to me?" Occasionally, you will have to dangle a carrot to acquire the items that you are asking for. If the surplus items are being obtained from a nonprofit institution, then there is very little that you have to offer in terms of carrots. You can barter your Center's time or your Center's services. It is my experience that this will not often be successful. One of the reasons for this is that the personnel in the surplus areas are not in the business of education. If you are talking to a company that is for profit, then you can offer a tax deduction, much like when we donate personal items to charities.

One day, I needed some silicone caulk for a project that we were building in our lab. I went to the local auto parts store, and they just gave me the item that I needed once they heard what we were using it for. You have to be excited and convey that excitement to incite people to be willing to give you things. You want the donor to feel a part of your project. You might also have to offer a liability release. Basically, what you are saying is that it will never be used on real people. When you do finally dispose of the item, you will physically watch it be destroyed by being put into a crusher.

One arrangement that I recently made was to "store" hospital beds for the hospital next door. They had been in the warehouse for 3 years, and the hospital wanted them out of the expensive storage space, but didn't want to lose ownership. They wanted to be able to recover them in case of a local disaster. We were very happy to be a part of that arrangement. All of my clinical monitors are also surge capacity units from the hospital and are currently "in active-use storage" in my center. The hospital's maintenance and update contract still covers them.

One of the final ways for you to find equipment is literally to dumpster dive. You have to have a bit of enthusiasm for your project to do this, but the rewards can be worth it. The example that I'm thinking of is this: You have your center with X amount of dollars to spend. You have to choose between a simulator and support equipment. There is simply not enough money to purchase both. Obviously, you'll buy the simulator. Now you need a bit of support equipment. The hospital warehouse is just down the street and you go on the day that it's open. You really don't find anything that you can afford, because they sell their equipment instead of donating to your good cause. But, you see a dumpster outside with some really high-quality wheels attached to something sticking out of the top. What are you going to do? This looks like it might be something really good like that charting cart that you want. Up the side you go and, sure enough, it's a pristine cart that is perfect for charting. It's already in the "trash" so it's off of the hospitals books. Free, in other words. There you have it, your first dumpster dive. For all of your efforts, you have a great cart for your center, for free. Every now and then, this is something that works as well. I will put in this caveat, even as excited as I am about my actions, I don't do this often . . . but I have done this.

# 53.12 Rearranging Your Environments

## 53.12.1 Flexibility

One of the key goals of your center is to be flexible. By doing this, you will be able to expand your customer base while meeting the needs of your customers and your center. By setting up your center with flexibility in mind, you will, in the long run, make life a lot easier for yourself.

Ideally, I feel that the same rooms should be used for Standardized Patient Exams and small group break out rooms. In addition, they should serve Simulation in several different settings. The settings that I am considering are:

OR
ER
ICU
Patient room
Dental office
Home living room

Here are two ways to accomplish this type of flexibility: mobility and stage-craft. Mobility is one step toward that goal. You can do this by having most, if not all, of your equipment is on wheels. While it might be hard to fasten wheels to a couch, in the long run, you'll be happy that you did.

Another innovative way to insure flexibility is by changing the look of the room. Following the lead from Pennsylvania State University's Simulation Center, Wiser Institute in Pittsburgh is changing the "look of a room" by hanging wall-sized fabric that is actually a life size, high quality, picture of an OR wall. Wiser also can take the same room and make it a street scene by using a floor to ceiling drape that has the same high-quality picture of a street. (Essentially, they are using painted and printed drapes as used in any live theater.) This way, they don't have to take up valuable floor space with actual OR cabinetry. If they need to make this room look like a street for a prehospital setting, then they use the backdrop that has a life-size picture of a busy Pittsburgh intersection on it. Both examples actually look like they have depth. They are of very high quality. I was suitably impressed with the setup. As I said earlier, it is very innovative, effective, and efficient.

## 53.12.2 Size Does Matter

One of the largest obstacles to resetting your setting is doorway size. Imagine having a hospital bed that is 45" wide and a doorway that is 36" wide. Now, you need to reset your room – alone. How are you going to do this? When you were planning your center, hopefully, someone thought of this. I know of one center that was designed and built to be a simulation center, but furniture and equipment movement was overlooked.

Later, I worked there. It took two men and two furniture dollies to roll the hospital bed onto its side and through the door. We had to remove the mattress, head board, and foot board as well. We had four rooms to do this in. As a result, we didn't do this often, and later, we dedicated two rooms as ICU/Hospital room settings. The other two became OR settings.

## 53.12.3 Packing the Laboratory

Now that you have your doorways wide enough and have accumulated all of this equipment, where are you going to put it? All of us look for the easiest way to configure the devices so that we don't have to take valuable time to "reset" our rooms. It's a normal thought process. As a result, we tend to "pack our labs."

I would caution you against "packing" it all into your laboratory. If you do, you will not have adequate room to perform your scenarios with ease and comfort. Even if you have a large three- or four-station laboratory, I don't think that it would be a good idea to do this. If you do, you can't isolate the "stuff" from the area where you will be doing your scenarios. Not only does it place the wrong equipment into your scenario, but it does not present the professional environment that we all expect from someone that we are doing business with. If you must have more equipment than you have off-stage storage space for, then at least cover the items that should be off stage with typical clinical sheets. This applies as well to any and all support equipment that is attached to patient simulator mannequins by an umbilical that is too short to locate "off stage."

Recently, on a tour of Simulation Centers, there were three that we visited that were the large multistation type. Of those, two had extra equipment stored in the teaching areas. Honestly, it looked like the simulation was taking place in a cavernous storage closet. It did not convey professionalism. The third center managed to hide stored equipment behind drapes similar to those used to separate patient cubicles. This also has the advantage of being able to make a large room smaller as needed. It actually was reasonably successful, although not ideal. We have to remember that, like it or not, we are in a customer service industry. As a result, we have to operate accordingly.

# 53.13 Conclusion

We have learned that simulation is not only a beneficial means of teaching, but it is also very doable. Through enthusiasm and creativity, you can accomplish the goals set forth by the curriculum. This can be a very daunting task, if you let it be. Remember, when you are setting up your center, walk before you run. You can not support every request that comes through the door. Once word gets out that you are up and functioning, you will have more requests than you can possibly deal with. Be very careful to choose what you can do well with the people, time, funding, equipment, and experience that you have on hand. Quality supersedes quantity.

# Reference

1. Michael A. Seropian, Kimberly Brown, Jesika Samuelson Gavilanes, and Bonnie Driggers. An Approach to Simulation Program Development. *Journal of Nursing Education* 43 (4) (April 2004).

# Hybrid Vigor: The Simulation Professional

*Men are wise in proportion, not to their experience, but to their capacity for experience.*

James Boswell

The Simulation Professional is a new cast member in clinical education, playing the familiar part of the supporting character. All successful simulation programs have at least one Simulation Professional; no matter what official job title may be assigned to them. The raison d'etre of the Simulation Professional is to make the clinical instructors look brilliant in front of their clinical students during classes in simulated clinical environments. This simple-to-state purpose is as liberating as it is limiting. Anything is possible and worth trying at least once, just so long as the clinical teaching and learning objectives come first, last, and always. On stage, the Simulation Professional portray their persona through robotic plastic. Off stage, they provide all the skills and perform all the tasks required so that the show will go on. To do it all requires a great deal of the scarcest resource of all: time. To do it well requires full-time commitment by individuals dedicated to mastery of this role. Their essential skills and tasks complement those of clinical educators. Individuals formally educated and employed in clinical domains can transform themselves into the full-time role of Simulation Professional, but in doing so, must forgo all clinical care. No matter their past, successful Simulation Professionals are happy being *in* the group, but not *of* the group of practicing clinicians. No matter their place in their institution's organizational structure, the successful Simulation Professional will work well with a wide range of "higher ups" and will promote and champion issues on the basis of demonstrated competency, values, and needs and not by positional authority. In a note to the editors, Kristina Stillsmoking offered these words of wisdom:

> When deciding on the person to fill the position of the director/manager/coordinator of a simulation center, it is not necessarily a great idea to put a physician in that position. I realize that some centers are part of a medical school, but these physicians have their hands full with patients, clinics, residents, research, and anything else that can be heaped on them, i.e., a simulation center. Our staff has found that promises of a "live" presence to guide us is not always possible on any given day. If they do come to supervise, it's usually for signatures on various administrative details, talk about the same things they mentioned the week before, and then of course the beeper goes off, and so do they. Thus, issues of staffing,

equipment, and special projects move rather slowly. For me, a type A personality, it gets very frustrating, as I know that things could move more efficiently. I believe that a board of directors (or whatever you would call it) could work together to decide the direction of the center, pick workable projects/research, and provide medical/nursing support. The director/manager/coordinator should be focused on the daily operations where they make staffing, equipment, and budgeting decisions without the physician's signature and buy off. The clinical director should not be burdened with these operational pieces.

Guillaume Alinier *et al.* share their personal wisdom about the Simulation Professional, from the inside based upon creating this role for themselves, and from the outside based upon inheriting a successful clinical simulation program sustained by the persistent efforts of a dedicated Simulation Professional.

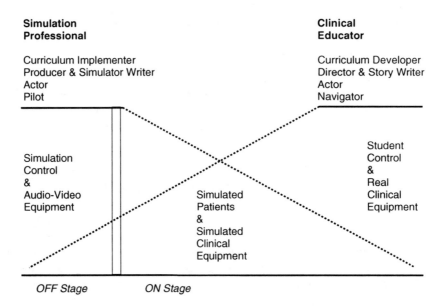

**FIGURE XII**  Areas of responsibility and performance for the Simulation Professional and the Clinical Educator. The Simulation Professional's areas of responsibilities and performance fully encompass all those spaces, issues and items on left side of the figure, and then decrease progressively across to the right. The Clinical Educator's areas of responsibilities and performance are the mirror image, fully encompass all those spaces, issues and items on right side of the figure, and then decrease progressively across to the left. Mostly, there is overlapping with shared responsibilities, shared ideas. This overlap of complementary contributions produces greater returns than from either alone. The individuals fulfilling these two essential but non-identical roles must alloy themselves into a high performance team, so that the students never experience any fissures or gaps in their simulation-based clinical sessions.

# 54

# The Simulation Professional: Gets Things Done and Attracts Opportunities

Guillaume Alinier
Ramiro Pozzo
Cynthia H. Shields

## 54.1 Tasks, Backgrounds, Characteristics, and Compensation

This chapter identifies the multiple tasks performed by simulation professionals, what backgrounds and characteristics are needed and best suited to expeditiously manage and execute these tasks, and what kind of compensation is required to attract and retain such professionals. The key contribution required from the simulation professional is the ability to get things done and attract opportunities. The simulation professional is not a support technician, but is an educational professional who integrates content and concepts from many different disciplines and communicates with and translates for those in many others. The simulation professional is both a day-to-day manager of a high visibility, high value asset, and a long-range leader to drive the program forward.

## 54.2 Desirable Qualities in the Simulation Professional

**Educator** – must have a passion for teaching.
**Creative** – see what can be, rather than what is.
**Autonomous** – be able to make appropriate decisions regarding the simulator and its surrounding environment.
**Flexible** – must be able to work with a large number of different clinical, academic, and administrative personalities.

**Gadgetier** – enjoys getting their hands dirty – be able to tinker, modify, augment, fabricate, and repair simulation and clinical equipment.

**Innovative** – seeks new ways to teach same content.

**Energetic** – willing to take on new teaching projects and challenges.

**Inquisitive** – express a genuine interest.

**Dedicated** – committed to make things happen thanks to a genuine interest.

**Academic** – can take part and eventually lead research projects.

**Physiology** – background to allow them to communicate with clinicians.

**Engineering** – background to allow them to understand the simulator and to communicate with simulator manufacturers.

## 54.3  Tasks Performed

Most of these tasks must be performed in every simulation center, and usually the simulation professional(s) will take most, if not all, responsibility for them. These tasks will vary between large and small centers and also depending on the clinical specialty. Infrequent tasks that require highly specialized capabilities, like computer network configuration and audio/video system design, may require temporarily including outside higher-level expertise than is essential for dedicated simulation professionals.

### 54.3.1  Educator

- Technology: pros and cons in clinical care and patient safety
- Eventual training of clinicians, other simulation operations professions
- Understand clinical teaching objectives
- Know student types and numbers
- Limitations of simulator(s)
- Story telling
- Stage craft
- Art of misdirection
- Communicating, cooperating, and collaborating with clinical instructors
- Style of educational approach appropriate for students' levels of maturity
- Performance assessment of learning *and* teaching

### 54.3.2  Master of Simulator Devices

- Evaluation
- Selection
- Installation

- Operation
- Maintenance
- Troubleshooting
- Modifications
- Scenario programing
- Incorporation of accessory technology

### 54.3.3  Designer of Simulation Center

- Identify users: instructors and their students, their numbers, their needs, and their schedules.
- Define, design, and configure the various simulation areas, their sizes, and respective locations and access ways.
- Define, design, and install audio-visual systems, sound control, and lighting.
- Define, design, and develop clinical and educational equipment and props.
- Define, design, and install computers, printing, phones, faxes, and internet communications capabilities.

### 54.3.4  Researcher

- Ethics
- Human use requirements
- Design and methods of clinical investigation
- Validity and reliability
- Performance assessment of investigative inquiry
- Development of new uses for existing simulators and simulations
- Development of new simulators and simulations

## 54.4  Formal Education

Until programs specific to the profession of simulation operations provide sufficient graduates to meet the demand, such individuals will have to be sought from those who chose to earn degrees in other topics such as biological, clinical, or physical science and engineering disciplines. Since the knowledge, skills, and abilities required for simulation operations far exceeds any current program of formal study, the proven ability *to learn* a complex subject is far more important than already knowing the specific topics of any one formal educational program.

### 54.4.1  Physiology

The simulation operators must understand the clinical instructors' requests, both in their content and context. They must also understand clinical language and "jargon," as well as interpret and translate the requests into specific simulator commands to generate the presentation as desired by the clinical instructors.

### 54.4.2 Engineering

While any one simulation professional will never be *the* ultimate expert across all domains, they must know enough to identify problems, propose solutions, and transmit them to the respective higher-level experts. Conversant if not competent in electricity and electronics, mechanics, physics, pneumatics, sound and video, computers, networks, construction, etc.

### 54.4.3 Communication Skills

Must communicate fluently with the instructors so they can understand the simulator's capabilities and limitations with respect to the teaching objectives they (the instructors) want to present; must communicate fluently with trainees, visitors, VIPs, etc., to complement the instructors.

### 54.4.4 Languages

English is a must, no matter where they live, as it enables them to interact with the international simulation community, and the simulator companies' help desks.

### 54.4.5 Teaching Skills

Need to be able to explain, if not also to teach, about the how and why of simulation and simulators, their supporting equipment, simulation facility requirements, etc., to both your institution's members and their outside guests.

### 54.4.6 Biomedical Equipment Knowledge

- Clinical applications
- Safety
- Limitations
- Knowledge of other brands/purchase assessment
- Common equipment-related errors

### 54.4.7 Audio/Video

Fully understand use and maintenance of installed audio/video system, collection, replay, editing, etc. Basic knowledge of technological options (Analog and Digital) and their functional and economic advantages and disadvantages.

### 54.4.8 Computer Skills

Maintain computerized schedule, collate data, manage high-level e-mails and listservs, advise on content for center's web site, data entry (Excel, database, etc.), simulator controlling and programing, computer troubleshooting (with and without simulator company Help Desk on the telephone).

In short, you might be looking for a biomedical engineer or any other engineer with a clinical background. Likewise, it may

be a health care professional familiar with technology. No matter how capable individual simulation operations professionals are, they will require the diverse services of the institution's technology and service support departments.

## 54.5 Occupational Experience

The ideal simulation operations professional to consider for hiring would preferably be one with experience in other simulation centers, but today that reduces the number of applicants to almost zero. So, other desired proxy experiences might be:

### 54.5.1 Clinical

A big plus would be experience with clinical educators, particularly so if with the same or similar clinical specialty as using simulation.

### 54.5.2 Audio and/or Video

Live production and distribution, and familiar with editing;

> Worked for company that sells or services clinical and biomedical products that will be used in the simulation scenarios;
> Troubleshooting and repairing electronic or mechanical devices;
> Installation and construction trades, either commercial or residential;
> Teaching: Ability to transmit knowledge in a manner that encourage deep learning processes;

### 54.5.3 Computers

- Programing, service, networking, help desk, etc.;
- Theatrical production, particularly the technical aspects of live theater.

## 54.6 Other Special Features

### 54.6.1 Interpersonal Skills

Good interpersonal skills have an impact on all aspects of the day-to-day operations of a simulation center: from retaining and retraining staff and faculty, to negotiating support from colleagues in other departments within the institution or other centers, through to public relations matters to generate publicity and interest in the simulation center. Working in a simulation center requires good communication and listening skills at all levels, regardless of the title and roles of the simulation professional and whom they deal with. It is important for the ideal simulation professional to understand the underlying psychological concepts of human communication to appreciate its potentially favorable or negative effects. Effective communication is especially important when dealing with trainees because creating a positive learning atmosphere helps them achieve maximum benefit from their clinical simulation experience. Poor interpersonal skills will create conflicts and animosity with trainees and colleagues that will impede the learning experience as well as the development of a simulation center. Note that the title "trainee" applies equally well to everyone new to clinical simulation, independent of their official title as clinical instructor or clinical student.

### 54.6.2 Resourceful and Quick Thinker

Local knowledge is another important asset, and can help the simulation operations professional to solve problems in a simple fashion and with minimum effort. Knowing who to contact for services and where to find spare or substitute equipment is essential. Always striving to improve and not just bow down to fate is the difference between having an inert simulation facility and a vital simulation program.

### 54.6.3 Commitment and Dedication

Going the extra mile to get things right, prepared, or repaired on time is essential. This is theater; "the show must go on." People's love for their job often makes them forget their contracted hours and makes them perform beyond the call of duty. Such attitude can make a big difference to the working atmosphere as well as the operations of a simulation center. This quality can be selected for, but then must be nourished by the rest of the team.

### 54.6.4 Creative Jack or Jill of All Trades

Patient simulators are thankfully more complacent than real patients regarding the treatment they receive. They allow for the development of tricks that might not be physically possible or allowed in real clinical environments or with real human beings[1]. A creative mind can create new tricks to enhance a simulation platform and advance the students' learning experience. Problem-solving skills are very useful in troubleshooting the patient simulator, the audio-visual (A/V) system, and/or the medical equipment. However, just be careful not to do anything that could void manufacturers' warranties.

For example, one center needed the information about every IV drug administered to their patient simulator but did not want the internal pharmacological model from acting upon this information. Along with the pharmacological model, the particular simulator manufacturer provided an editing feature and an optical drug recognition device. A simulation professional changed one crucial characteristic variable of each drug, thus rendering them inert as far as the pharmacological model was concerned, but still fully identifiable by the optical recognition and drug administration record-keeping systems. In addition, the single valve handle on/off indicator on the optical recognition device was opposite in appearance and function to that of all such valves used by the institutions' clinicians. This handle was replaced with one regularly used. Thus, the simulation professional applied communication, engineering (software and hardware), and scrounging to fulfill a vital educational requirement.

As a professional in a clinical or academic environment, a Jack or Jill of all trades not only excels in technical abilities, but also in different academic and management aspects. Having some research experience is an asset in designing projects and analyzing findings. On the management side, a simulation operations professional will be required to manage resources: equipment, teaching facilities, and faculty staff. If your center does not benefit from secretarial or administrative support, juggling all these tasks and duties can be a challenge. The duties of a simulation operations professional might even include web site maintenance and marketing, in order to recruit participants and generate income for the center. To carry out all of the above effectively, one should excel in the use of modern technology and especially in Information Technology skills, as it will make some of the tasks more manageable. Proficient use of e-mail, Internet, and all other modern communications and information access tools is fundamental.

## 54.7 Compensation for the Simulation Professional

### 54.7.1 Bread on the Family Table

Given that the simulation professional is first and foremost an educator with knowledge, skills, and abilities that complements the clinically trained educator, then the simulation professional should be financially compensated the same as their nonclinician educator colleagues. Some clinically trained individuals have chosen to shed their clinical duties to become full-time simulation professionals, and in so doing, are no longer compensated at the higher levels that patient care commands. Note that many simulation programs are started from the bottom up by clinicians that are willing to become part-time simulation professionals in order to create and operate their simulation programs. For every moment that these clinicians are fulfilling the role of the simulation professional, their employer is paying full-patient care prices for the services of a part-time "amateur" simulation professional (amateur, as in those clinicians who self-select to become educators rarely have the additional inclination, background, or desire sought in the fully engaged clinical simulation professional).

### 54.7.2 Satisfaction

Seeing participants improving a range of skills over time, receiving thank you letters or e-mails in which they express their gratitude after a session for the efforts put in by the facilitators to run a good sessions brings real satisfaction. Often, participants will at least verbally express how much they enjoyed and learnt from the scenarios and debriefings before they leave the simulation center. After a smooth session run with a team of facilitators, the members always close the session with a nice inner feeling. Participants' positive feedback enhances that sense of satisfaction and makes you feel it was worth making the efforts to develop the scenarios and related props or "physical artefacts/prompts" required to make the scenario more believable.

### 54.7.3 Career Prospects

Because simulation in health care education is still growing, with many centers across the world still at a developmental stage, there are real opportunities for people with the right attributes to join this field, acquire some valuable experience, and climb the ladder of operations and management of a simulation center. For people starting at a fairly low position within a center, the potential career prospects are encouraging. There is a real global community in this field that will support you and provide helpful advice.

### 54.7.4 Job Security

Simulation professionals are still a rare breed, so being one presents an advantage in all aspects of employment and employer selection. As every nation needs to train health care professional and simulation becomes more and more recognized as the way forward, many centers still need to be established and staffed throughout the world. They all require the diverse range of skills of experienced simulation professionals to operate effectively. Given that English is the international language of clinical academia, and that the principles of both clinical education and clinical simulation are universal, the potential job market is truly worldwide.

### 54.7.5 Opportunities for Collaboration and Research

Thanks to the existing and growing community of professionals supporting simulation training, there is a multitude of ongoing research projects taking place, some of which are open for collaboration. Because the penetration and application of clinical simulation in clinical education is still in its infancy, there are as many research opportunities as can be imagined.

## 54.8 Conclusion

This chapter shows that there are great incentives to become a simulation professional, as well as for institutions to invest in them for the benefits they bring to a simulation center. It often makes the difference between being fully functional with having a healthy range of courses regularly produced and having the equipment not used appropriately or the center not existing at all. The simulation professional is the training wizard able to develop solutions, prepare the scene, produce the operations of a center and courses, execute research, and be the local expert with all the equipment used. This list of tasks is broad and requires a wide range of skills and personal attributes on the part of the simulation professional. They will include creativity, resourcefulness, quick thinking, and interpersonal skills. There is currently no formal educational curriculum possibly preparing such individuals, hence their training often has to be sharpened on the job following a qualification in a scientific discipline or having had clinical experience.

## Endnote

1. Use of modified clinical equipment such as a disabled defibrillator, performance of invasive procedures such as putting in a chest drain, making patient simulator alterations to produce unilateral chest rise on spontaneous breathing, . . . etc.

# Good Answers Start from Good Questions

*The deepest sin against the human mind is to believe things without evidence.*

Thomas H. Huxley

The best question to ask of clinical simulation is "what is it good for?" The best answer to this question is "so that your first time for real is not your first time." The next best question to ask of clinical simulation is "what is the best way to do clinical simulation?" The best answer to this question is "there is no such thing as *best*, just *better*, and we are striving to determine how to determine what *better* really is." Usually, this produces the third question "are you sure you know what you are taking about?" This is the rub. Despite the obvious fact that each and every one of us has learned a great deal since our births, and that education is all about sharing our learning with others, very little is known about *how* we learn. However, what is known about any activity that does require conscious thought (unlike digestion, thankfully) is that better performance only comes through deliberate practice. Formal education lives in the gap between this sober "fact" that all improvement takes practice and the "fact" that there is a finite amount of resources (least of all time) available for practice.

Those of us already accepting the value added by simulation, often have a difficult time understanding the reluctance of those who have not. Those of us who are facile in one form of simulation often have a difficult time accepting that we have something to learn from our colleagues that use a different form. Some proponents of simulation like to claim that investing some of these finite resources in their favorite form of clinical simulation will result in greater improvement than not doing so. We can all be persuaded by one form of argument or another, and in some cases, we prefer arguments based upon data. Data comes from research. Better data comes from better research. Research is a formal activity worthy of study and practice like any other nonautonomic activity, can be taught, and on occasion, learned.

Guillaume Alinier, now a Simulation Professional, at one time successfully mastered the arcane world of graduate level physics. He is very familiar with various qualities of data and the challenges in producing better data from a domain as ethereal as education. As students, William McGaghie *et al.* traveled paths that some might say are more difficult to master than physics, since their domains of study lack such convenient attributes as commonly accepted, independently verifiable standards of measurement. These two offer ground truths from their learning experiences about simulation research to help you in your efforts for producing better data leading to better teaching.

# Pitfalls to Avoid in Designing and Executing Research with Clinical Simulation

Guillaume Alinier

## 55.1 Why Investigate?

Simulation offers a great potential not only as an educational tool, but also for diverse research purposes. It can be used to investigate a multitude of aspects such as obtaining feedback regarding the ergonomics of clinical equipment used during simulated clinical cases, to investigate teamwork dynamics, as well as to look at response times to specific incidents or physiological changes. It presents several advantages over using the real clinical environment and real patients.

## 55.2 Advantages

The main advantage of doing research with simulation is that it takes place in a safer, nonthreatening, and more controllable environment than the real clinical arena. The risk of potentially harming a patient has been removed. We can decide on the clinical condition of the simulated patient, the number of participants involved in a case, and their level of experience. The response of the patient can be controlled (actor or patient simulator), and we can decide what equipment should be used, or be at the disposition of the participants. All these aspects that might have been confounding variables in real clinical environments can fairly easily be set as constants in a simulation center study. In terms of ethical considerations, there is normally no patient consent issue when we are using robotic patient simulators.

The only participants in need of consent in such a simulation research project should simply be the scenario participants. Because we are not asking them to perform operations or procedures on each other and those performing the study should be trained to properly debrief the participants, the participants should not suffer any physical or psychological trauma. It is expected that ethical clearance should be fairly easily obtained when doing educational research with simulation.

## 55.3 Participants' Briefing and Data Collection

Although simulation participants (as individuals) inevitably present significant variations in terms of individual skills, knowledge, and experience (which might be what we want to test in a study), the faculty instructor and facilitating team is potentially a very significant and ill-defined variable, unless an effort is made to minimize differences and standardize interactions. Ideally, the entire study team should be trained to the same standard and follow the script of the scenario. Because participants' actions cannot always be predicted, the instructors sometimes have to operate outside of their script. Under ideal circumstances, such variations should be discussed during the planning of the research, and responses prepared for the inevitable variations. The interaction of faculty with participants before the end of data collection could jeopardize a study. For example, the instructors might provide hints related to the study being carried out, or bias the view of participants by their comments (positive or negative comments to individuals, but not to the whole sample studied) during briefing, debriefing, or coffee breaks. One should ensure that there is homogeneity in the information transmitted to all participants in order to prevent any issues of reliability regarding the research carried out. If your study requires participants to complete a pre- and postexperience questionnaire, you should ensure that the same briefing is given at the same time to the participants and for all the sessions over which the study is carried out. If a faculty starts explaining about the concepts of simulation training just before the participants fill in a preexperience questionnaire relating to that aspect, it is likely to skew the findings.

## 55.4 Significant and Contributing Factors: Are We on the Same Wavelength?

Your research should be carefully designed to take into account the participants involved, the resources, and the faculty. These aspects are even more important when carrying out multicenter studies. When reporting the findings of your research, do not be afraid to state the obvious, as what might be second nature to you when facilitating simulation training might not be expected at all from someone else reading your findings. For example, I would strongly encourage clearly defining the way your training is facilitated, especially regarding the degree of interaction and type of help, if any, faculty provide to participants during the simulation session. Not everyone is as familiar as you might be with simulation, and that very word "simulation" could be misinterpreted for a more traditional psychomotor skills teaching approach. In my view, there are not only several degrees of fidelity of simulation technology, but also several degrees of fidelity of producing simulation training sessions. It is fairly common for people to refer to simulation training simply because it takes place in a simulated clinical environment and makes use of a patient simulator, yet the session could simply be about patient monitoring, with a relatively large group of trainees gathered around a bed. I often find myself having to stress a particular point when talking about simulation to other health care educators. I have to refer to it as "realistic scenario-based training," whereby the faculty/lecturer/trainer/instructor is simply a facilitator leaving the students in charge of the scenario and therapy, and also not questioning or prompting the participants' actions. I clearly explain that the "simulation" we are talking about allows students to make mistakes and take wrong decisions. It is a student-centered approach where the facilitator "starts" the scenario by providing some basic information, stands back until the end of the scenario, and then helps the participants analyze how they could improve in multiple aspects.

## 55.5 Comparing Like With Like

When designing research studies making use of simulation, one should carefully consider the validity and robustness of the design. One of the fundamental challenges faced by most researchers in the field is to prove the validity and value of simulation-based learning in relation to more traditional teaching methods. Some study designs rely on an experimental group learning to respond to a particular incident in a simulation medium, while a control group would learn about the proper course of action to the same incident in a classroom, hence in a more theoretical manner. After a period of time, the study continues by then presenting a similar scenario in a simulation environment to both groups and comparing and measuring their performance. Evidently, the group who has had the greater practical experience of responding to a particular incident will outperform the other group.

## 55.6 Conclusion

Experience with simulation-based education has an effect upon how well trainees perform during subsequent simulations. When a simulation-based test is used, it will always favor the group with the most simulation-based training due to their familiarization with the concept of simulation, the simulation environment, and the patient simulator itself. A less biased or more appropriate measure would consist on evaluating the performance of candidates from both groups to a real clinical (patient-related) event. Such a measure is however impossible, as it cannot be planned for and could put patients' life in jeopardy. An objective way of comparing trainees' acquisition of skills and knowledge should make use of a different tool or assessment method to any of the ones employed in the training period of the research study. Having recognized this potential bias of using simulation as part of the evaluation process to compare the performance of two study groups, we tried to evaluate the effectiveness of simulation-based training of undergraduate nursing students by subjecting the students to a pre- and postexperience Objective Structured Clinical Examination practical test. We found that students who had taken part in the simulation sessions made more significant improvements to the practical test than their peers who had not had such practical training [1].

## Reference

1. Alinier, G., Hunt, B., Gordon, R., and Harwood, C. Effectiveness of intermediate-fidelity simulation training technology in undergraduate nursing education. *Journal of Advanced Nursing* 54(3), 359–369 (2006).

# 56

# Fundamentals of Educational Research Using Clinical Simulation

William C. McGahie
Carla M. Pugh
Diane Bronstein Wayne

## 56.1 Components of Simulation-based Research

This chapter has four sections. **Section one** begins with a short review about the background and history of using simulation technology to educate health science professionals, especially physicians. Historical evidence about the utility and value of these simulation-based educational technologies is presented as available. Special attention is given to a recent systematic literature review spanning 34 years on the features and use of high-fidelity clinical simulations that lead to effective learning. **Section two** presents our views about features of a sound educational research project in the health professions, including specific examples. These features are important for individual research studies and for more ambitious cumulative research programs. **Section three** presents a concrete illustration of a cumulative, simulation-based, clinical education research *program* now underway at the Northwestern

University Feinberg School of Medicine and the Northwestern Memorial Hospital in Chicago. This is a stepwise series of education and research studies aimed at helping internal medicine residents acquire and maintain procedural skills in advanced cardiac life support (ACLS) and other clinical procedures. The promise and potential pitfalls of such a research program are also addressed. **Section four** gives a set of practical suggestions about doing simulation-based research in health professions education drawn from our experience. Here, we aim to convey tacit, "insider" knowledge about the clinical education research enterprise that may help other investigators to be more efficient and effective.

The purpose of this chapter is to present a framework and a set of tools for current and future health science educational researchers who plan to conduct studies about new and innovative educational technologies. Of course, the ultimate goal of this work is to figure out the best ways to use health science simulation to produce superb clinicians who can deliver safe and effective patient care. That's why biomedical engineers use rigorous evaluation methods [1] to create and improve simulators and simulation scenarios that mimic *in vivo* clinical conditions. That's also why educational researchers design and conduct studies to evaluate if clinical simulation technology yields measurable and meaningful educational results. Educational researchers also try to determine the conditions where good outcomes are most likely and will be long lasting. Thus, the "bottom line" of this chapter is about simulation technology evaluation. We ask, "How can we study the features and conditions that help clinical simulation yield maximum educational benefits?"

Readers may consider this a brief educational research primer, certainly *not* an approximation to a textbook on educational research methods. A list of useful textbooks and other scholarly resources is found in Appendix 56A.1.

## 56.2 Background and History

The use of simulation technology for the education of health care professionals has ancient roots and modern expressions. Cadaveric and fabricated anatomical simulators were used for medical training in ancient Alexandria and Baghdad and in Renaissance Padua [2]. Since the 18th century, medical education has not only relied on cadavers and animal models but also on inert human patient simulators to enable doctors in training to practice basic clinical skills. Rosen, for example, states "Crude obstetric mannequins were described as early as the 17th century" [3]. The advent of digital computer technology, robotics, full-body mannequins, and virtual reality environments are the latest examples of simulation technology tools used to educate health care professionals.

### 56.2.1  Eight Unanswered Questions on Simulation in Clinical Education

The availability of educational tools, however sophisticated, does not automatically guarantee their effective use in programs aimed at skill, knowledge, and conative acquisition among learners. Everyone agrees that the goal of medical and health professions education is to produce superb clinicians. Health science educators also endorse the idea that simulation technology has an important role to play in clinical education [4]. However, research is needed to tell faculty curriculum planners and educational policy makers how to use clinical simulation technology to maximum advantage. There are many unanswered questions. Here are eight.

1. How much practice at cardiac auscultation is enough to insure learner response competence in real-life situations?
2. Does a public safety officer's performance of endotracheal intubation in a controlled, laboratory environment generalize to real EMT field conditions?

3. Are reading and video observation of thoracentesis sufficient preparation to enable medicine residents to perform the procedure safely on real patients?
4. What is the best blend of simulator practice and *in vivo* experience to insure an obstetrics and gynecology resident's competence at managing a breech delivery?
5. What is the nature and extent of *faculty training* about the use of simulation technology that is needed to ensure effective use of simulation in health sciences curricula?
6. What is the minimum number of team training sessions needed for teamwork skill acquisition and use in patient care settings?
7. What are the patient safety payoffs, in terms of improved patient outcomes, from investments in simulation-based health professions education?
8. What are the health care cost reduction benefits from investment in simulation-based health professions education?

A recently reported systematic research review, conducted under the auspices of the Best Evidence Medical Education (BEME) collaboration, evaluated 670 published research reports on the features and use of high-fidelity medical simulations that lead to effective learning [5]. The review covered a 34-year period, from 1969 to 2003. The original aim of this systematic review was to perform a quantitative, meta-analysis of the available evidence. However, heterogeneity of research designs, small sample sizes, inconsistent journal reporting conventions, and different outcome measurement methods precluded a quantitative research synthesis. Instead, the authors resorted to a qualitative, narrative synthesis of the evidence as the best available alternative.

### 56.2.2  Ten Features of Simulation Leading to Effective Learning

The best evidence from this summary of 34 years of medical education research shows that there are 10 features and uses of high-fidelity medical simulations that lead to effective learning. These are the *primary* findings of the BEME report. In decreasing order of reported frequency, the features and uses are:

1. Feedback is provided during learning experiences.
2. Learners engage in repetitive practice.
3. Simulation is integrated into an overall curriculum.
4. Learners practice tasks with increasing levels of difficulty.
5. Simulation is adaptable to multiple learning strategies.
6. Clinical variation is built into simulation experiences.
7. Simulated events occur in a controlled environment.
8. Individualized learning is an option.
9. Outcomes or benchmarks are clearly defined and measured.
10. The simulation is a valid representation of clinical practice.

These findings should serve as a guide in defining critical areas for research in evaluating simulators as educational technologies. The results of learning outcome studies will continue to be the mainstay in understanding the potential benefits of simulations for training and personnel evaluation.

Secondary outcomes of the BEME report include calls for improved rigor in simulation-based medical education research *and* improved journal reporting conventions (e.g., presenting descriptive statistics before statistical results). These calls are not new. For example, a Joint Task Force of the journal *Academic Medicine* and the Group on Educational Affairs-Research in Medical Education (GEA-RIME) Committee of the Association of American Medical Colleges (AAMC) published a report in 2001 titled "Review Criteria for Research Manuscripts" [6]. This document provides detailed technical suggestions about how to improve medical education research and its sequelae, scholarly publications. Stephen J. Lurie, former Senior Editor of *JAMA* echoed this position in a 2003 essay titled, "Raising the passing grade for studies of medical education" [7]. Lurie documents many flaws in medical education research and calls for common metrics, increased standardization of educational interventions, better operational definitions of variables and, at bottom, more quantitative rigor. Reports like these encourage health science educational researchers, and journal editors who evaluate research manuscripts, to boost the rigor and scholarly quality of simulation-based research publications.

## 56.3 Sound Educational Research

Educational research in the health professions varies in quality and dependability just like research in any other field of scholarly inquiry. The research and practice communities give higher regard to research reports that are based on important questions, addressed by rigorous designs, using measures that yield reliable data, and contain appropriate data analyses. Written presentations that weigh the evidence carefully and report study outcomes in a meaningful intellectual context are highly valued. Flaws in reasoning, research design, or written presentation weaken a research project and distort its results. Careful planning, study implementation, study management, and study reporting are essential to fulfill the goal of reaching useful and trustworthy research outcomes.

Our collective experience in health sciences educational research, both simulation-based and using other instructional technologies, has taught us there is a set of at least 10 key features that embody sound projects. This is tacit knowledge for seasoned investigators with many educational research publications. Novice investigators will benefit from giving these features close attention.

The 10 features of a sound educational research project are presented in Table 56.1. The table lists each research project feature, provides a short definition, and presents a concrete example from the simulation-based educational research literature in the health sciences. Each example contains at least

**TABLE 56.1** Features of a sound educational research project

| Feature | Definition | Example(s) |
|---|---|---|
| 1. Theory-based, hypothesis driven, grounded in an intellectual context | Research embedded in a model or theoretical framework that gives coherence to individual studies and an action plan for a cumulative research program. Use of a model or theoretical framework encourages expression of research hypotheses that focus and direct research efforts. | Wayne *et al.* studied ACLS skill acquisition among internal medicine residents [8]. The investigators used simulation technology in the context of the *mastery learning model* [9–11] including *deliberate practice* [12]. All residents acquired 100% of expected skills (*mastery learning*) with little outcome variation. The duration of *deliberate practice* needed to reach mastery standards varied among the residents. |
| 2. Focused research question | An educational research study needs to address a focused research question that: (i) states conceptual and statistical hypotheses; (ii) expresses research conditions; (iii) describes inclusion and exclusion criteria; and (iv) addresses its limitations. | A *post hoc* study derived from a larger systematic review of nearly 800 studies on the features and uses of high-fidelity medical simulations that lead to effective learning addressed the question, "Is there an association between the duration of deliberate practice and standardized learning outcomes in health professions education? [13]." The answer is yes. The investigation addressed inclusion and exclusion criteria for the constituent studies and acknowledges its limitations. |

*(Continued)*

**TABLE 56.1**  (Continued)

| Feature | Definition | Example(s) |
|---|---|---|
| 3. Structured research design | This is research whose question is addressed by a structured experimental (e.g., randomized trial) or quasi-experimental (e.g., cohort study, case-control study) design as a clear research plan. Designs vary in strength and integrity toward the research goal of generalized causal inference about variables being studied. | Wayne et al. conducted a randomized clinical trial with a wait-list control group to evaluate the effectiveness of simulation technology to promote ACLS skill acquisition among internal medicine residents. An 8-hour educational program produced a 38% ACLS skill improvement compared to baseline performance that was replicated at crossover. No learners were denied the educational program (placebo group) [14]. |
| 4. Sufficient sample size (statistical power, generalizability, etc.) | A research study needs to involve enough participants to allow for an honest outcome evaluation. Small studies lack power and generalizability. Study size should be determined early, at the design phase. | A series of three large studies of cardiac and pulmonary auscultation skills (total $n > 1900$) involving postgraduate trainees in internal medicine, family medicine, cardiology fellows, and a subset of fourth year medical students done by Mangione, Nieman et al. shows that the doctors correctly recognize approximately 20% of the basic cardiac auscultatory findings; they know about 50% of pulmonary findings [15–17]. Residents and fellows were no better than medical students at this key clinical skill. Simulation technology was used for trainee evaluation. |
| 5. Solid measures → reliable data → valid inferences | Good research depends on high-quality measurement methods. Measurement methods and procedures must be developed and refined carefully (or selected and used critically) to yield highly reliable data that promote valid inferences about variables being studied. | Issenberg et al. systematically developed computer-based pre- and posttest measures of clinical skills in bedside cardiology that are equivalent in content and difficulty [18]. The measures have been used in subsequent research to gauge the effects of simulation-based educational interventions [19]. |
| 6, Simple, efficient, data analysis | Research data analysis should use the simplest possible "minimally sufficient analysis." Wilkinson argues, "Although complex designs and state-of-the-art methods are sometimes necessary to address research questions effectively, simpler classical approaches often can provide elegant and sufficient answers to important questions. If the assumptions and strength of a simpler method are reasonable for your data and research problem, use it [20]." | Issenberg et al. analyzed data from a quasi-experiment on the effects of a cardiology course on medical students' and residents' acquisition of auscultatory skills. A $3 \times 2$ (group, occasion) split-plot analysis of variance (ANOVA) was simple and straightforward. Results show unequivocally that simulation-based training is much better than customary ward work for auscultatory skill acquisition [19]. |
| 7. Clear data presentation (tables and figures) | Parsimonious data presentation as tables and figures is a hallmark of effective scientific and professional writing. Huth [21] is an excellent source of advice about how to prepare tables and figures. | Data displays in reports by Issenberg et al. [19] and Wayne et al. [8, 14] leave no doubt that simulation-based medical education programs yield excellent results in controlled studies. Graphic and tabular data presentations may be more convincing than statistical analyses of outcome data. |
| 8. Good writing: organized and crisp | Good scientific and professional writing is organized carefully, crisp in presentation, simple, and declarative. Sources of advice about scientific and professional writing include Huth, Kazdin [22], and Strunk and White [23]. | Murray et al. report a study whose purpose was to "develop simulation exercises and associated scoring methods and determine whether these scenarios could be used to evaluate acute anesthesia care skills [24]." This journal article is a model for its focus, simple structure, tight writing, and (relatively) short reference list. |

**TABLE 56.1** (Continued)

| Feature | Definition | Example(s) |
| --- | --- | --- |
| 9. Professional presentation | The final product of a research project – a manuscript ready for submission to a peer-reviewed scholarly journal – needs to be comprehensive, coherent, and stylish. A truly professional scientific manuscript will read well and look good. | One of the best published reports on the use of simulations (standardized patients) in health professions education was written by van der Vleuten and Swanson [25]. This journal article is a model for other authors of literature reviews to emulate. |
| 10. Tight project management | Research project success depends on controlled management of its many features, e.g., participant recruitment, allocation of participants to study conditions; implementation of research interventions; measuring variables before, during, and after experimental manipulations; establishing data files; data analysis; report writing. | Schwid *et al.* conducted a study involving 99 anesthesia residents at *10 institutions* [26]. The residents were evaluated on their management of four simulated scenarios: (i) esophageal intubation; (ii) anaphylaxis; (iii) bronchospasm; (iv) myocardial ischemia. Internal and external evaluators graded residents' performances using two scoring systems. The study yielded reliable and valid data that separated resident groups by years of experience. |

one citation to a published journal article that illustrates the feature. Interested readers can study each feature in depth and learn how other contemporary scholars have approached and solved potential research problems.

## 56.4 Cumulative Research Program: The "Hothouse Effect"

Using the principles given in Table 10.1, a cumulative teaching and research program in clinical simulation was implemented at the Northwestern University Feinberg School of Medicine and Northwestern Memorial Hospital in Chicago, Illinois, U.S.A. This program educates and evaluates internal medicine residents in the second year of postgraduate training in ACLS skills. Designing, initiating, and maintaining this program illustrates many of the features of building a successful research *program* from the ground up.

The primary goal of this program is to address several basic education and research questions in the field of clinical simulation. We are interested in the acquisition, maintenance, and assessment of *procedural skills* among resident physicians. We chose to study the performance of residents in ACLS because it is a required procedure for internal medicine board certification.

A team was formed in July 2003 to launch the education and research program. The team includes the internal medicine residency program director and associate director, the directors of education and technical support in the Northwestern Memorial Hospital Patient Safety Simulation Center, a statistician, and an educator with expertise in medical education research. This group connects clinical program directors,

Simulation Center faculty and staff, and faculty with expertise in research design, measurement, and data analysis. Team composition assured full access to the Simulation Center and active participation of internal medicine residents as trainees and research participants. Involvement of residency training program faculty was critical to design a simulation-based educational program that would complement, not replace, the residents' clinical curriculum. Active involvement by Simulation Center faculty and staff ensured development of algorithms that were feasible and practical in the simulated environment. Collaboration with research faculty allowed us to study the protocols we developed rigorously. Issues of intellectual property and authorship credit (i.e., inclusion, order of authors) were discussed openly and agreed-upon by the team before the first project started. The team leader was identified (DBW). A common goal of advancing education through research in clinical simulation was endorsed by all team members and won support from their clinical and academic departments.

We envisioned a series of sequential education and research projects from the beginning of the team collaboration. The projects would be consecutive and cumulative, boosting skill acquisition among the residents, advancing knowledge about simulation-based medical education programs based on empirical data, and growing the research team's experience and morale. Team experience and morale are important yet rarely considered outcomes of successful collaboration. They are markers for Barton Kunstler's notion of the "Hothouse Effect [27]" as high-energy teams catalyze and accelerate their productivity.

A graphic representation of the simulation-based ACLS education and research projects both finished and underway in our cumulative research program is shown in Figure 56.1. Some of the projects (**A–D**) are parts of a stepwise, linear progression.

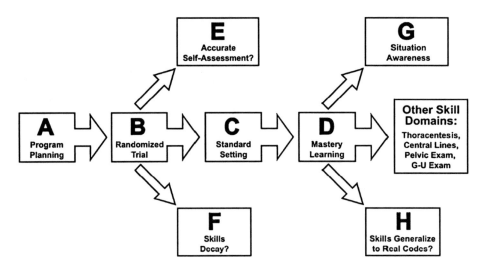

**FIGURE 56.1**  Progression of a cumulative, simulation-based, educational research program at Northwestern University.

Other projects (**E–H**) are "spin-offs" from parent studies. The "spin-off" studies were not anticipated during early program planning yet became research opportunities as the program matured and gained momentum.

### 56.4.1  Formation and Planning

The first step of this education and research program (**A**) was its early formation and planning. The program was formed due to the perception of internal medicine faculty (and residents) that despite each resident's successful completion of an American Heart Association (AHA)–sponsored ACLS provider course, resident responses to in-hospital "codes" were uneven and not standardized. This perception is reinforced by resuscitation research conducted at another Chicago academic medical center [28]. Consequently, the residency program director (DBW) assembled the education and research team and designed an overall program plan including goals, a timetable, equipment and space needs, funding requirements, resident scheduling expectations, and a flowchart of the total program. The flowchart was the best program image *at that time*. It has been changed regularly due to the teaching team's maturation and experience. The hospital administration quickly recognized potential improvements in patient safety outcomes from the program and supplied necessary funding.

### 56.4.2  Developing Outcome Measures

The research program began by developing outcome measures. The team reasoned that it should focus on end-points and then address teaching. They started by identifying six ACLS case scenarios that are (i) commonly encountered by residents in actual cardiac arrest situations and (ii) whose *best clinical*

*responses* are described in the ACLS *Provider Manual* published by the AHA [29]. Adherence to these best responses is the "gold standard" for resident performance. The six scenarios are: asystole, ventricular fibrillation, supraventricular tachycardia, ventricular tachycardia, symptomatic bradycardia, and pulseless electrical activity.

### 56.4.3  Creating Algorithms and Checklists

A checklist was developed for each of the six conditions from the ACLS algorithms using rigorous step-by-step procedures [30]. Within the checklists, each patient assessment, clinical examination, medication, or other action was listed in the order recommended by the AHA and was given equal weight. None of the checklist items was weighted differentially. A dichotomous scoring scale ranging from 0 (*not done/done incorrectly*) to 1 (*done correctly*) was imposed for each item. Checklists were reviewed for completeness and accuracy by ACLS providers and instructors.

### 56.4.4  Initiating a Randomized Clinical Trial

A *randomized clinical trial (RCT) with a wait-list control (WLC) group* was initiated within the internal medicine residency program (**B**) after the program purpose and team composition were determined [14]. The RCT was approved by the Northwestern University Institutional Review Board (IRB). All second year residents were tested individually at baseline in six ACLS skill protocols during July 2003. The residents were then randomized into two groups that received either (i) four, 2-hour ACLS teaching sessions using a medical simulator or (ii) clinical experience alone – the wait-list control (WLC) condition. All residents were tested a second time after a 3-month interval. Residents who received the simulation-based teaching sessions demonstrated a 38% higher adherence to published algorithms for management of ACLS scenarios compared to the WLC group. A crossover was done subsequently, and after a third test session, the 38% skill improvement was replicated. The study design ensured that all residents received the simulation-based teaching intervention because there was no placebo condition. Residents rated the program highly for educational quality and relevance to their training [14].

As shown in Figure 56.1 several other studies resulted from conducting the randomized trial in an orderly flow. After demonstrating that the simulation-based ACLS training program was successful and practical within the residency program (B), we then set minimum passing standards for each of the six simulated ACLS scenarios using an expert panel of judges (C) [31]. This allowed the introduction of a *mastery learning model* [9–11], including *deliberate practice* [12], into the residency training program (D). *This model aims to produce high and uniform learning outcomes while the amount of training time needed to reach the goals can vary.* All mastery learning goals in ACLS were reached by 100% of a new group of residents who also reported that the simulation-based training had very high value as a supplement to clinical education [8].

"Spin-off" studies from the linear progression shown in Figure 56.1 is derived from curiosity spawned by research experience. Two separate projects derived from the initial randomized trial (B) were shaped by discussions in research team meetings that are held regularly. One of these projects studied the ability of the medical residents to self-assess their ACLS skills (E). The evidence shows clearly that the residents cannot self-assess their skills accurately [32]. A second project sought to determine the degree of residents' retention of ACLS skills acquired in the randomized trial over a 14-month period (F). Skill retention is remarkably robust in our sample of residents with little decay even without refresher training [33].

The education and research team continues to see and respond to opportunities as new ideas arise from experience and "Hothouse" enthusiasm. Examples include two projects that are derivatives of the mastery learning study (D) in Figure 56.1. One work in progress is an attempt to measure and evaluate "situation awareness" (SA [34]) in the context of simulated ACLS performance (G). Endsley reports [34], ". . . simply put, SA is knowing what is going on around you" (p. 5). Formally, SA is ". . . the perception of the elements in the environment within a volume of time and space, the comprehension of their meaning and the projection of their status in the near future [35]" (p. 97). A second study, now completed, sought to find out the extent to which ACLS skills acquired in the controlled, Simulation Center laboratory environment generalize to real "code" events on the hospital floors (H). This research involved a retrospective, case-control study of resident team responses to in-hospital cardiac arrests. Code teams led by simulator-trained residents showed significantly

greater adherence to AHA resuscitation guidelines than teams led by residents with clinical training alone [36].

### 56.4.5 Consequences of Success

The success and energy of this cumulative research program has attracted attention and received much interest from other research groups and from hospital administration. Thus, with strong professional and financial support locally, we plan to extend this work to other skill domains including such procedural skills as thoracentesis and central line placement and to physical examination skills including the female pelvic exam [37] and the male genito-urinary (G–U) exam.

To summarize, our research team has been in place since 2003. It has broad membership including clinical and educational leaders, researchers with statistical and methodological expertise, and key personnel from the Patient Safety Simulation Center. With focused expertise and a shared goal, this team has used energy and input from its members, feedback from learners, and an ability to react nimbly to new ideas to build a successful research initiative in clinical simulation.

# 56.5 Practical Suggestions

This section offers practical advice about conducting simulation-based educational research in health professions education settings. These suggestions address frequently encountered problems in research projects; yet, the list is not exhaustive. More detailed advice is available elsewhere [38]. Here we offer six tips: (i) know the literature, (ii) write a detailed research protocol, (iii) obtain methodological consultation early, (iv) discuss authorship openly and early, (v) create a research team, and (vi) make research a routine activity.

### 56.5.1 Know the Literature

Investigators must have a detailed knowledge of the clinical and educational literature before deciding on a research question. Scholarship depends on knowing if a research question has been addressed previously, including the context and methods of past research. Awareness of the research literature *outside* one's discipline is a sign of scholarly excellence. Consultation with a reference librarian is the best way to search the literature to gain knowledge about its scope and depth.

### 56.5.2 Write a Detailed Research Protocol

We urge all educational researchers, whether novice or experienced, to write a detailed research protocol *before* launching a study. The discipline of writing a detailed protocol or proposal before starting a project forces an investigator or research team to address many potential problems that may otherwise go unnoticed. The protocol should include at a minimum:

1. Clear statement of the research question or problem;
2. Identification of the research design including acknowledgment of its strengths and weaknesses;
3. Description of research participants, i.e., inclusion and exclusion criteria, numbers needed, gaining access, etc.;
4. Description of the educational treatment or intervention, i.e., setting, intensity, duration, integrity, etc.;
5. Specification of measurements, i.e., outcome and process, data reliability, validity of inferences, data collection procedures;
6. Data analysis procedures;
7. Likely research outcomes and their implications, often in the form of "dummy" tables without data;
8. The project timetable.

Writing a detailed research protocol before starting a study has several other practical advantages beyond clarifying an investigator's intent. The protocol can be used to prepare documents and (informed) consent forms needed to obtain human use Institute Review Board approval of the project. Parts of the protocol can also be used verbatim in research reports, abstracts, and manuscripts when the study results are being prepared for publication in peer-reviewed journals.

Concrete advice about how to prepare a research protocol is available in textbooks [39] and in a key journal article [40]. This is an essential first step in the conduct of any research project.

## 56.5.3  Obtain Methodological and Statistical Consultation Early

The most important time to seek and obtain technical consultation about research methods and statistical analysis of data is in the project planning phase. This is especially the case for new investigators. Getting early advice about research design options, measurement methods, sample size and statistical power, effect sizes, and many other issues will prevent many research headaches down the road. A poorly conceived and executed study simply cannot be salvaged after the data are collected.

## 56.5.4  Discuss Authorship Openly and Early

Determining authorship credit can become a serious professional problem if not discussed openly before a research project begins. The project leader must exercise leadership in this endeavor. The first author of a research manuscript is not necessarily the one who does most of the writing. The first author may be an individual who conceptualizes the project and project ideas, writes the research protocol, oversees data collection and analysis, and edits the final version of a manuscript.

Conditions for authorship are clearly outlined by most journals. Below is a summary adapted from the *JAMA* Authorship Responsibility, Financial Disclosure, Copyright Transfer, and Acknowledgment Form [41].

### Authorship Responsibility, Criteria, and Contributions

Each author should meet all criteria given below (A, B, C, and D).

A. The author certifies that the manuscript represents valid work and has not been previously published.
B. All authors have given final approval of the submitted manuscript.
C. All authors have participated sufficiently in the work to take public responsibility for either part or all of the content.
D. To qualify for authorship, each person must check at least one box for each of the three categories of contributions listed below:
   i. Category 1
      • Conception and design
      • Acquisition of data
      • Analysis and interpretation of data
   ii. Category 2
      • Drafting of the manuscript
      • Critical revision of the manuscript for important intellectual content
   iii. Category 3
      • Statistical analysis
      • Obtaining funding
      • Administrative, technical, or material support
      • Supervision
      • No additional contributions
      • Or are disclosed in an attachment

Consultants, statisticians, and data collection assistants do not necessarily warrant authorship credit unless they contribute significantly to the research design, data analysis, and manuscript preparation.

## 56.5.5  Create a Research Team

A productive research team should be focused and efficient, operating like a "well-oiled machine." Diverse expertise and leadership are both important but so are teamwork, morale, hard work, and recognition of individual limits. Research productivity will most likely happen when team members "place their egos on the shelf." Dysfunctional research teams are frequently those where personalities take priority over productivity.

As mentioned earlier, management scientist Barton Kunstler describes a phenomenon termed the "Hothouse Effect" [27].

This is an achieved balance of teamwork and individual genius that can maximize research output and quality. Other benefits of the "Hothouse Effect" include improved productivity, morale boosts, and increased team motivation and dedication. All of these factors contribute substantially to high-quality research and research programs. The overall goal of a research team should be to sustain a high level of innovation and creativity. In addition, achievement of the "Hothouse Effect" requires a team leader who has superb interpersonal skills and the ability to communicate effectively.

### 56.5.6 Make Research a Routine Activity

Research in any setting of health professions education need not be considered an *extraordinary* activity. With thoughtful planning and courteous execution, educational research addressing important questions can become a routine endeavor in most clinical, laboratory, and classroom settings. This requires a spirit of collaboration, candid discussion of research protocols and ideas, constantly seeking ways to improve educational practices and outcomes, and sharing credit as appropriate. This requires that individual investigators and members of research teams be aware of the organizational culture where they operate and to respect cultural priorities. Settings where everyone benefits from evidence-based educational improvements are the most likely locations where research becomes routine.

# References

1. Verschuren, P. and Hartog, R. Evaluation in design-oriented research. *Qual. Quant.* 39, 733–762 (2005).
2. Puschmann, T. *A History of Medical Education.* Hafner Publishing Co., New York, 1966 (originally published 1891).
3. Rosen, K. R. A history of medical simulation. In: Loyd, G. E., Lake, C. L., (eds.), *Practical Health Care Simulations.* Elsevier, Philadelphia, 2004, 3–26.
4. Issenberg, S. B., McGaghie, W. C., Hart, I. R., et al. Simulation technology for health care professional skills training and assessment, *JAMA.* 282, 861–866 (1999).
5. Issenberg, S. B., McGaghie, W. C., Petrusa, E. R., et al. Features and uses of high-fidelity medical simulations that lead to effective learning, a BEME systematic review. *Med. Teach.* 27, 10–28 (2005).
6. Joint Task Force of *Academic Medicine* and the GEA-RIME Committee. Review criteria for research manuscripts, *Acad. Med.* 76, 897–978 (2001).
7. Lurie, S. J. Raising the passing grade for studies of medical education. *JAMA* 290, 1210–1212 (2003).
8. Wayne, D. B., Butter, J., Siddall, V. J., et al. Mastery learning of advanced cardiac life support skills by internal

medicine residents using simulation technology and deliberate practice. *J. Gen. Intern. Med.* 21, 251–256 (2006).
9. Block, J. H. (ed.), Mastery Learning: Theory and Practice. Holt, Rinehart and Winston, New York, 1971.
10. Bloom, B. S. Time and learning. *Am. Psychol.* 29, 682–688 (1974).
11. McGaghie, W. C., Miller, G. E., Sajid, A., and Telder, T. V. Competency-Based Curriculum Development in Medical Education. Public Health Paper No. 68. Geneva, World Health Organization, Switzerland, 1978.
12. Ericsson, K. A. Deliberate practice and the acquisition and maintenance of expert performance in medicine and related domains. *Acad. Med.* 79 (10, Suppl.), S70–S81 (2004).
13. McGaghie, W. C., Issenberg, S. B., Petrusa, E. R., and Scalese, R. J. Effect of practice on standardized learning outcomes in simulation-based medical education. *Med. Educ.* 40, 792–797 (2006).
14. Wayne, D. B., Butter, J., Siddall, V. J., et al. Simulation-based training of internal medicine residents in advanced cardiac life support protocols: a randomized trial. *Teach. Learn. Med.* 17, 210–216 (2005).
15. Mangione, S., Nieman, L. Z., Gracely, E., and Kaye, D. The teaching and practice of cardiac auscultation during internal medicine and cardiology training: a nationwide survey. *Ann. Intern. Med.* 119, 47–54 (1993).
16. Mangione, S. and Nieman, L. Z. Cardiac auscultatory skills of internal medicine and family practice trainees: a comparison of diagnostic proficiency. *JAMA* 278, 717–722 (1997).
17. Mangione, S. and Nieman, L. Z. Pulmonary auscultatory skills during training in internal medicine and family practice. *Am. J. Resp. Crit. Care Med.* 159, 1–6 (1999).
18. Issenberg, S. B., McGaghie, W. C., Brown, D. D., et al. Development of multimedia computer-based measures of clinical skills in bedside cardiology. In: Melnick, D. E., (ed.), *The Eighth International Ottawa Conference on Medical Education and Assessment Proceedings. Evolving Assessment: Protecting the Human Dimension.* National Board of Medical Examiners, Philadelphia, 2000, 821–829.
19. Issenberg, S. B., McGaghie, W. C., Gordon, D. L., et al. Effectiveness of a cardiology review course for internal medicine residents using simulation technology and deliberate practice. *Teach. Learn. Med.* 14, 223–228 (2002).
20. Wilkinson, L. [and the Task Force on Statistical Inference] Statistical methods in psychology journals: guidelines and explanations. *Am. Psychol.* 54, 594–604 (1999).
21. Huth, E. J. *Writing and Publishing in Medicine*, 3rd ed. Baltimore: Williams & Wilkins, 1999.
22. Kazdin, A. E. Preparing and evaluating research reports. *Psychol. Assmnt.* 7, 228–237 (2006).
23. Strunk, W., White EB. *The Elements of Style.* New York: Penguin, 2005.

24. Murray, D. J., Boulet, J. R., Kras, J. F., et al. Acute care skills in anesthesia practice: a simulation-based resident performance assessment. *Anesthesiology* 101, 1084–1095 (2004).

25. van der Vleuten, C. P. M. and Swanson, D. B. Assessment of clinical skills with standardized patients: state of the art. *Teach. Learn. Med.* 2, 58–76 (1990).

26. Schwid, H. A., Rooke, A., Carline, J., et al. Evaluation of anesthesia residents using mannequin-based simulation: a multiinstitutional study. *Anesthesiology* 97, 1434–1444 (2002).

27. Kunstler, B. *The Hothouse Effect*. American Management Association, New York, 2004.

28. Abella, B. S., Alvarado, J. P., Mykelbust, H., et al. Quality of cardiopulmonary resuscitation during in-hospital cardiac arrest. *JAMA* 293, 305–310 (2005).

29. Cummins, R. O. (ed.), *ACLS Provider Manual*. American Heart Association, Dallas, TX, 2001.

30. Stufflebeam, D. L. Guidelines for developing evaluation checklists: the checklists development checklist (CDC). Western Michigan University, Evaluation Center, July 2000. Available at: http://www.wmich.edu/evalctr/checklists Accessed 31 Aug 2007.

31. Wayne, D. B., Fudala, M. J., Butter, J, et al. Comparison of two standard setting methods for advanced cardiac life support training. *Acad. Med.* 80 (Suppl.), S63–S66 (2005).

32. Wayne, D. B., Butter, J., Siddall, V. J., et al. Graduating internal medicine residents' self-assessment and performance of advanced cardiac life support skills. *Med. Teach.* 28, 365–369 (2006).

33. Wayne, D. B., Siddall, V. J., Butter, J., et al. A longitudinal study of internal medicine residents' retention of advanced cardiac life support (ACLS) skills. *Acad. Med.* 81 (Suppl. 10), S9–S12 (2006).

34. Endsley, M. R. and Garland, D. J. (eds.), *Situation Awareness: Analysis and Measurement*. Lawrence Erlbaum Associates, Mahwah, NJ, 2000.

35. Endsley, M. R. Design and evaluation for situation awareness enhancement. In: Proceedings of the Human Factors Society 32nd Annual Meeting, Vol. 1, Human Factors Society, Santa Monica, CA, 1988. pp. 97–101.

36. Wayne, D. B, Didwania, A, Feinglass, J, et al. Simulation-based education improves quality of care during cardiac arrest team responses at an academic teaching hospital: a case-control study. *Chest* 131, in press (2007).

37. Pugh, C. M. and Youngblood, P. Development and validation of assessment measures for a newly developed physical examination simulator. *J. Am. Med. Inform. Assoc.* 9, 448–460 (2002).

38. McGaghie, W. C. Conducting a research study. In: McGaghie, W. C., and Frey, J. J. (eds.), Handbook for the Academic Physician, Springer-Verlag, New York, 1986, 217–233.

39. Fraenkel, J. R. and Wallen, N. E. How to Design and Evaluate Research in Education, 4th ed. McGraw-Hill, Boston, 2000.

40. Gordon, M. J. Research workbook: a guide for initial planning of clinical, social, and behavioral research. *J. Fam. Prac.* 7, 145–160 (1978).

41. Authorship Responsibility, Financial Disclosure, Copyright Transfer, and Acknowledgement. *JAMA* 293, 1796 (2005).

# XIV

# Simulation Scenario: Telling the Story – Discussing the Story

*The only films that matter are the movies that begin when the lights come back on.*

Alfonso Cuarón

It is only fitting that O'Henry, the master of the short story, was the son of a clinician. Today, we make liberal use of the short story form in sharing clinical accounts when we write and produce clinical simulation scenarios. Like his short stories, we must very quickly convey time, setting, place, characters, plot, and purpose to our audience, that is, our students, so that they can effectively engage themselves into our "stories." Likewise, we must limit ourselves to include only one or two key memorable points so that our students successfully ingest, integrate, and retain desired lessons. Within our brief scenarios, the plot is often structured like an obstacle course combined with a treasure hunt. The obstacles are essential to exercise deliberate action and reward improved performance in preparation for the real world of boundless impediments. The treasure is essential for exercising and rewarding attitudes.

Experiencing a scenario is like eating a meal. Discussing one's experience is like digesting it, where the essence of the activity is internalized. All scenario creation should start at the conclusion of the debriefing: how do you want your students to be changed by their simulation experience? From these answers come the key points of your desired discussion. From these key points comes the basis for your simulated clinical story. And like all stories, the learning points can be made with comedy, or with tragedy; usually a bit of both. And like all of us, no one really likes to be corrected, especially in such a naked and public form as a simulation experience before one's peers. Thus, all the cleverness of the world put into a scenario execution will be corrupted and flushed away by a hurtful discussion.

Imagine what you would take away from an experience when the authority figure in the room starts their assessment with "Now here is what you did wrong: . . . "

The scenario *is* the embodiment of the teaching objectives, thus everything put into the scenario must support the teaching objectives; everything else must be left out. For example, a highly complex and convoluted history that has nothing to do with the scenario should be simplified (unless the objective is complexity, or the excess information is deliberately intended as a distracter). For example, unless dose per weight is a crucial teaching objective, all adult patients weigh 70 kg so that trainees can use "the usual" dose. If the trainees pull out a calculator to determine the dose for a patient of 56.76 kg, you have lost their train of thought, and they are most likely missing critical learning objectives.

Judith Hwang and Betsy Bencken describe their successful creation and discussion methods, and introduce many of us to the "sugar with the medicine" approach called Plus-Delta. Kristina Stillsmoking offers an excellent checklist scaffolding suitable for adoption into a wide range of different simulation forms. Peter Dieckmann and Marcus Rall built, refined, and now share their extensive scenario-building tools.

# 57

# Scenario Design and Execution

Judith C.F. Hwang
Betsy Bencken

## 57.1 Our Approach

Before creating any scenario, we identify the intended student population. That is, are they paramedic, nursing, or medical students? Are they residents, and if so, in what year of training? Next, we articulate our educational goals and objectives, as these will help guide the process of scenario creation. Both the educational goals and objectives must take into consideration the students' discipline and level of training. Goals are the overarching educational purpose of the simulation session, while objectives are the specific issues we would like the student to learn from the simulation session. Another factor we determine early is how the simulation sessions will be scheduled. That is, will the students be coming in small groups for 1-hour sessions spread out over weeks, with each small group coming only once or twice? Or will the students be coming as one large group for a half or all-day-long session? If it is spread out, only one or two scenarios may be needed, whereas for the all-on-one-day session, 4–6 scenarios may be needed. Once we have defined the issues above, we brainstorm to come up with scenarios that would best highlight our educational goals and objectives, given the constraints of the class schedule.

## 57.2 A Favorite Scenario: The Postanesthesia Care Unit

### 57.2.1 Educational Goals and Objectives

Our *goal* was to teach the anesthesiology residents about managing postoperative complications in the postanesthesia care unit (PACU). We wanted to foster teamwork and good communication between:

> the operating room anesthesia resident and the PACU anesthesia resident,
> the PACU resident and the PACU nurses,
> the patient and the PACU anesthesia resident and nurses.

One of the specific *objectives* was to teach the proper identification and management of a patient with hypoxia secondary to residual neuromuscular blockade. We knew we would have one or two residents at a time for about an hour to an hour and a half, and that the residents would change on a biweekly basis. The residents are on the PACU rotation during their postgraduate years 2 and 4. This scenario is equally suitable for both years of residents as well as with student registered nurse anesthetists in place of anesthesia residents, as our goals of fostering teamwork and good communication remain unchanged.

## 57.2.2 Key Clinical Characteristics

We identified factors that could contribute to the patient's condition, and that the resident would have to consider in their differential diagnosis regarding the etiology of the hypoxia. These included: (i) moderate obesity that would cause decreased functional residual capacity, as well as possibly hypoventilation in the supine position; (ii) hypothermia that could potentially slow down the metabolism of the neuromuscular agent, as well as cause hypoxia if the patient were shivering; and (iii) incisional abdominal pain that could cause the patient to hypoventilate. We wanted to emphasize the maximum dose of reversal agents for neuromuscular blockade that should be given and the consequences of exceeding the maximum dose. Finally, we wanted the resident to learn the optimal management for the patient who requires reintubation.

Compiling all this information together, and knowing we wanted to go from a recently extubated patient to a patient that would need to be reintubated, we began to design the sequential changes of the scenario. For each change in the scenario, we defined the signs and symptoms that the mannequin would demonstrate or exhibit ("give voice to"), the actions that the PACU nurses and the resident might take, and the consequences of those actions (see complete PACU scenario and prop list in Appendix 57A.1).

## 57.2.3 Key Teamwork Challenges

To be able to capture the handoff between the OR anesthesia provider and the PACU anesthesia resident, the scenario starts as the patient is being wheeled into the PACU. The OR anesthesia provider is given a preoperative evaluation and an anesthesia record with all the pertinent information regarding the patient's intraoperative course, including the fact that the patient is moderately obese, underwent a rapid sequence induction, was easily intubated, and has received the maximum doses of 5 mg of neostigmine and 1 mg of glycopyrrolate (example records in Appendix 57A.1). After the PACU nurses follow their usual admission process of obtaining vital signs and briefly evaluating the patient, the OR anesthesia provider gives report to the PACU anesthesia resident and nurses, and then leaves to begin their next case. If any information is not transmitted completely or clearly by the OR anesthesia provider, the PACU team has an anesthetic record to refer to. The PACU anesthesia resident then begins the evaluation of the patient.

## 57.2.4 Patient/Treatment Action/Reaction

The interaction between the resident and the nurses is observed as the patient complains of feeling cold, having pain, and being short of breath. At the same time, the mannequin shivers, his oxygen saturation begins to decrease, his respiratory rate increases, and his respiratory efforts show only minimal to moderate chest excursions. The resident is expected to verify the patient's temperature, consider and order a chemical or external measure to treat the patient's low temperature, as well as order pain medication, place the patient in a reverse Trendelenberg position (head-up) to improve diaphragmatic excursion, and finally, to increase the patient's inspired fraction of oxygen to treat the hypoxemia. The resident may even decide to assist the patient's ventilation. The nurses are expected to communicate and document their observations of the patient's condition to the resident, and administer the ordered therapeutic interventions. In addition, the nurses should request an increase in the patient's inspired fraction of oxygen, if not ordered by the resident. The nurses may opt to set up an ambu bag and mask and/or obtain the PACU's emergency airway cart in anticipation of the potential need to assist the patient's respirations, and/or reintubate the patient. The anesthesia resident and/or the PACU nurses should also be communicating with the patient to assess his pain level with the administration of the medication and to further evaluate his complaint of shortness of breath.

The patient's pain and shivering will resolve with the application of appropriate therapeutic measures. His respirations will initially slightly deepen and slow in response to the pain medications. However, he will continue to complain of feeling weak and short of breath. He will gradually become more tachycardic and tachypneic. The resident should examine the patient, who will have shallow breaths that are clear to auscultation. If a nerve stimulator ("train-of-four") is checked, the train-of-four ratio will demonstrate fade. The PACU resident should confirm the dose of neostigmine and glycopyrrolate given to the patient in the OR. The PACU resident and nurses should not give any additional neuromuscular blockade reversal agents. If additional neostigmine and glycopyrrolate are given, the patient will get significantly weaker and more tachycardic, tachypneic, and hypoxemic. If an arterial blood gas is ordered, the results will be returned after a brief delay, and indicate a relatively low $PaO_2$ (given the patient's supplemental oxygen), and a $PaCO_2$ of 55 mm Hg. During this time, the resident should be communicating with the PACU nurses his thoughts regarding the differential diagnosis of the patient's hypoxemia. Similarly, the PACU nurses should add any suggestions they might have regarding the patient's condition and care. The team should decide that the patient needs to be reintubated.

The PACU resident should inform the patient of the decision to reintubate and explain the reason why. If the PACU resident does not inform the patient of the decision to reintubate the patient and explain why, the patient should inquire what is going on and initially refuse to be intubated. Once the patient consents to proceeding, the PACU resident prepares for intubation by requesting and setting up the necessary equipment and medications. The nurses may assist the resident with explaining to the patient what is occurring, obtaining the requested materials, by providing cricoid pressure, calling for respiratory therapy and a ventilator, and possibly by assisting with the patient's ventilation. After preoxygenation, intubation can be done with any induction agent, minimal if any muscle relaxant, and cricoid pressure. If the resident chooses to use succinylcholine, the prolonged action of the relaxant due to the presence of neostigmine will need to be discussed in the debriefing. Appropriate ventilator settings will need to be chosen by the PACU resident. The PACU resident should also continue to actively warm the patient to minimize any contribution of hypothermia to the prolonged neuromuscular blockade.

With the different stages of the scenario decided, the high-fidelity simulator can then be programed to mimic the desired changes. In general, we tend to control the progression of the scenario rather than program a transition phase. That is, we preprogram the conditions, and then, when the students' behaviors warrant that their patient change from the current presentation of vital signs, we command the simulator to do so. This enables us to tailor the scenario to the actions of the various students.

### 57.2.5 Debriefing: Plus-Delta Method

After the clinical simulation scenario is finished, we proceed to a debriefing of the events that occurred. At this time, we discuss the actions of the students and the reasoning behind their actions, ensure that the educational objectives we had are explicitly elicited, and examine the interactions between the participants. To maximize learning but minimize the time needed during debriefing, we use the *Plus-Delta Method*. We have the participants tell us things that they did well (*Pluses*) during their care of the patient and then the actions they would change (*Delta*) in the future. This way, if we do not have any additional time, we will at least have had a chance to positively reinforce correct actions taken and potentially improve management decisions when the students next encounter a similar case situation. If we do have more time, we use it to explore in greater depth the issues raised by the students list of pluses and changes.

## 57.3 Operations

### 57.3.1 The Paper Work

Before we begin any clinical scenario, we have the participants sign two forms. One is a consent to be videotaped, and the other is a confidentiality agreement. The former provides consent should we use the recording for activities other than the debriefing at the end of the scenario, but is primarily intended to allow us to record for debriefing purposes. The latter is to minimize discussion of the scenario itself outside of the Center for Virtual Care so that we can use the same scenario without too many of the details leaking out. However, the confidentiality agreement also is to minimize discussion of the students' actions and behavior outside of the debriefing session by the teachers as well as the students.

### 57.3.2 The Simulators

Although we use primarily simulators from Medical Education Technology Inc., Sarasota Florida (METI), any patient simulator will do. The advantage of the METI emergency care simulator is that its seizure mode is a good way to simulate shivering. However, we often use the METI human patient simulator (which does not have a "seizure" mode), in which case the patient complains about feeling cold and shivering rather than actually shivering. It works just as well. We have the monitoring equipment mimic the PACU setup so that the nurses feel as if they are in their usual environment. The pulse oximeter is configured so that the connection between monitor and METI rack needs to be connected to give the sense of applying the pulse oximetry monitor rather than just having the oxygen saturation "magically" appear on the monitor at the start of the scenario. We have a thermometer available for the nurses to use to give a sense of realism and because hypothermia is an issue of concern during the scenario. Not using the thermometer and just providing a temperature (either visually or verbally) does not seem to work as well, because the participants either do not see the temperature displayed on the monitor or they do not ask for a temperature. When the PACU nurses take the temperature, they are informed what the temperature of the simulator is at that stage. Ideally, we would have a replica of the PACU emergency airway cart for this scenario. However, we have opted to just provide the same equipment without the cart.

### 57.3.3 Audio/Video

There are two cameras set up in the simulation suite; both are installed in the ceiling but are at right angles to each other to provide different views – one is from the side of the patient and one is from the foot of the bed. Typically, we record using the camera that views the patient from the side (Figure 57.1).

We also record the vital signs on the monitor and overlay their image on the view of the room. In addition to the cameras, simulation operators and clinical instructors have direct view of the simulation room via a one-way window (Figure 57.2).

Audio collection is done via a small microphone near the side-view camera (Figure 57.1). The combined audio/video recording is then replayed in our conference room during the debriefing (Figure 57.3). Usually, the instructor will show the start of the scenario to get the participants used to seeing themselves on video and then focus on a few clips that highlight

**FIGURE 57.1** Simulation room with the side-view camera and microphone mounted on ceiling to minimize distraction to and blockage by participants.

the desired teaching points. Although currently we do not have this capability, it would be useful to be able to mark the recording to assist in locating the desired clips during debriefing.

## 57.4 Conclusion

When beginning to design a scenario, it is crucial to take into account several factors. First, identify who the student participants are, because their educational base and needs will influence the creation of the next factor, educational goals and objectives. Second, clearly delineate educational goals and objectives for the simulation sessions as these will impact the content of the scenario. Third, obtain the schedule for the simulation sessions, because the schedule will determine if a given scenario will be self-contained or be part of a series of interrelated scenarios. Given these constraints, the scenario should be designed such that each element (i.e., the patient, his history, the presenting problem, physical findings, ensuing complications, etc.) reinforces the educational objectives.

In the execution of a scenario, key components to consider include the physical setting, the equipment (i.e., monitors, medications, medial instruments, etc.) available, and the student's perspective. The setting must be realistic enough to enable the student to suspend reality. The equipment provided should be at least similar to, if not identical to, what the student

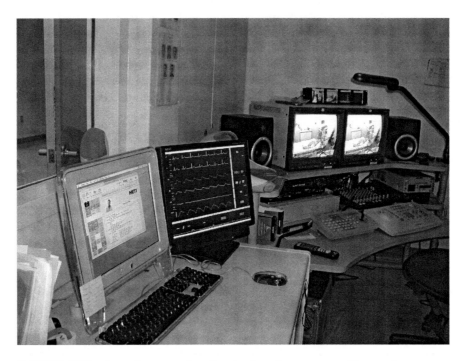

**FIGURE 57.2** Control room and audio-visual equipment setup. The two live camera views are always visible on the small video displays in the upper right, while the one view selected for recording is visible in the larger displays below them.

**FIGURE 57.3** Audio-visual setup in the conference/debriefing room. The single channel audio/video record system in the control room complements the single-channel audio/video replay system in this conference/debriefing room.

would encounter in their clinical settings. While not all equipment has to be immediately in view for an advanced learner who should be knowledgeable enough to look or request a "missing" item, beginners may benefit from having most equipment in view to give them a hint of what can or should be done.

Finally, use the feedback that you obtain from the students themselves; this information is invaluable in refining both the scenario itself as well as how you execute it. This can be explicitly obtained after each simulation session or indirectly obtained by reviewing the students' actions during the session. For example, if several groups of students consistently forget to obtain the patient's temperature, putting a thermometer into the setting may "cue" them. The effectiveness of the thermometer as a visual cue is then determined by the actions of subsequent groups.

# 58

# Simulation Scenario Building

Kristina Lee Stillsmoking

## 58.1 The Scenario – Why Bother

Scenarios are scripts, stories, or algorithms created for instructing the participants, including the simulator (human or robotic), on how to interact with the students. When I first came to work at our center, scenario creation was perceived as a difficult task. Being new to simulation and not knowing any better, I wanted to find a way to make this process more user-friendly. As I began to learn more, I realized that this goal was neither too difficult nor real easy. The key was putting in the time and effort to learn how to use the simulator software and its hidden nuances. From my lessons learned, I describe here a basic framework of questions for crafting a scenario and provide a template for the answers.

I was tasked to facilitate and teach a 4-hour block of an 8-hour facilitator's course we present monthly. This course trains new facilitator's in the basic overview of the four high-fidelity simulators we use for Advanced Cardiac Life Support (ACLS), Pediatric Advanced Life Support (PALS), resident, nursing, and medic training. They also learn the basic steps of creating, building, running, evaluating a scenario. We also include the basic proponents and importance of debriefing.

I must admit that it was scary teaching something I just barely understood myself. Luckily, I was sent to a great week-long workshop, and that gave me a new perspective. Thus, I immersed myself in learning the software and found ways to make the building process easier for me. So began my journey into the depths of simulation software and scenario design. I started by looking at what others had already created, printed their information, and categorized this by areas or types of clinical patient presentations. One great simulation user at a recent workshop was asked how he came to be so proficient in this process; he replied "I slept with the software on a lot." What he was saying was that he spent hours and hours learning how the software worked and how to make it work for him.

As I progressed, I tried to identify ways to make it easier for my students to catch on and internalize the information so they could create their own scenarios. At this point, I didn't have a clue why or how the scripting functions built in our simulators worked. You might be asking yourself why I didn't receive formal or even on-the-job training about my new teaching tool. Like many departments in any setting, the immediate short-term need for my body and work ethic and ability was greater than the long-term value of me learning my surroundings and simulators. As a word to the wise from a long-time educator, invest the time with your new employees for a solid orientation and training period. It will provide a well-trained, versatile employee and good customer service. Likewise, as a long-time employee, seek supervisors that value well-trained help. We expect our students to require time to absorb and integrate newly learned information, then why not so our instructor colleagues?

The hardest part for me was thinking up the physiological responses and student actions to a scenario. It is one thing to intrinsically know clinical information, but harder to put it on paper in a meaningful scenario. Thus, my first patient creation felt very artificial and inept. I finally asked a seasoned simulation person in the center, and she showed me how to build and present a realistically flowing patient scenario. Yes, I did look at the embedded "Help," but I found there were key pieces missing that I needed. Would it not be helpful to have a course where you, as a first-time user, were mentored in creating scenarios and cases for 2–3 days? More ideas for my new course began to evolve.

I began searching the Internet for more information while corresponding with a fellow simulation educator. Gradually, the picture was becoming clearer. The more facilitator courses I taught and the more time spent with the software, the more my comfort level increased. As I mentioned earlier, it takes time to think out, plan, create, and evaluate a solid scenario. Each phase within a scenario should have its own title with physiological aspects within in it. These are built and located with the "did" or "did not" perform a process in mind. With all this in mind, let us take a look at this whole process of scenario creation. I think of it in three steps: the goal, the clinical story, and the integration at the end. Let's take a look at them:

## 58.2 "The Goal is . . . ?"

What is the purpose/goal/objective/outcome you want for your student to learn? Is it teaching a new task, an incorporation of a skill into a process, or is it a competency practice/test?

The birthplace of these questions may have been from a needs assessment, faculty or student suggestion, or a need to comply with a new practice or regulation. Below, I have listed some of the many components to "The Beginning," with an explanation of each.

### 58.2.1 Acquiring New Skills

Individuals needs to learn how to perform specific skills such as listen and assess heart and lung sounds, start an IV, turn on and use a defibrillator, etc.

### 58.2.2 Practice Established Skills

Individuals have been trained on a skill but because of the "high risk/low use" of that skill they need to practice regularly to remain competent. Thus, we hope to reduce the risk of a patient error as defined in the Joint Commission on Accreditation of Healthcare Organizations guidelines for National Patient Safety Goals and increase patient safety. For instance, the use of a defibrillator in a low-risk environment such as a chronic pain clinic, or an allergy specialist's office.

### 58.2.3 Testing for Competency

Individuals must follow correctly a specific pattern of responses according to predetermined criteria. This might be used in sedation, ACLS, PALS, patient history and physical assessment, or other clinical student testing.

### 58.2.4 Teamwork Development

A group of individuals using simulation to improve communication (role clarity) and to foster integration of their clinical problem-solving skills. This allows for a neutral playing field for everyone, no matter who they are, i.e., physician, nurse, respiratory technician, pharmacist, or other players. Often, this is referred to as Crisis Resource Management.

Write down your exact objectives for the training session.

a. **Identify** two or three objectives, and keep the scenario to 5–10 minutes in length.
b. **Remember** that only observable behaviors and actions can be determined by you, the instructor. For example: student will demonstrate their understanding of the Airway – Breathing – Circulation steps of a response to a crisis by actually performing said steps; student will state the need to call for assistance within 1 min.

## 58.3 Matching Simulation Capabilities to Learning Objectives

How can simulation be used in a training session to meet the objectives (purpose/goal/outcome) as desired?

1. Will the use of simulation enhance or detract from the above?
   a. Will the students take the use of the simulator seriously enough to perform "close to" (identically) as they would with genuine patients?
   b. Will the students talk over the mannequin instead of engaging with it?
2. What types of clinical simulators are needed to meet the objectives?
   a. Does the mannequin need to exhibit physiological responses?
   b. Can you use a static simulator to accomplish the objective?
   c. Will a part-task trainer provide the best learning experience (e.g., central IV access)?
   d. In the absence of any invasive procedure, will live actors or standardized patients be the best teachers?
   e. Would a hybrid simulator, combining the human element of a live actor wearing appliances for safely performed invasive procedures be best?

## 58.4 Patient, the Location, and the Environment

Where do you want the simulated patient to be situated? (See Section 58.9, Examples 1 and 2.)

1. Scene or setting where the simulated patient will be found by the student.
   a. CT scanner
   b. Intensive Care Unit
   c. Emergency Room
   d. Ambulance
   e. Hallway
   f. Medical/Surgical
   g. Administrative area
   h. Exam room
   i. Obstetrics
   j. Behavioral Health
   k. Recovery Room
   l. Outpatient Surgery
   m. Field (medic) – many possible variations
2. Where will you find your simulated patient in those types of areas?
   a. Floor
   b. Bed
   c. Steps
   d. Car, truck, airplane
   e. Gurney/litter/trolley
   f. Chair
   g. Wheelchair
   h. OR table
   i. Construction or Military equipment of various types

What physiological responses do you want your simulated patient to manifest to the student?

1. Determine those vital signs that support your teaching objectives. Remember to define both their steady-state values and their rates of changes between these steady states as you take your simulated patient through the scenario.
   a. Breathing rate
   b. $SpO_2$
   c. Blood pressure
   d. Cardiac rhythm/rate
   e. Heart/lung sounds
   f. Vocal sounds
   g. Temperature
   h. Consciousness
2. Describe the medical history and patient's physical findings according to scenario you have created. You can make this as elaborate as you think is necessary to help the student handle the scenario correctly. If you decided to give clinical laboratory results, electrocardiogram (ECG),

or X-ray information, use copies that match the scenario and place them in a medical record folder identical to those used with genuine patients.

What props, equipment, or actors will be needed to create the "realism" and stress you need to have a solid scenario?

1. Are actors needed to enhance the realism of the scenario? You can enlist the aid of other staff or find friends/family who wish to volunteer. You will have to educate them in what their role and response will be in the scenario. Typically, family roles are played as distracting while clinical roles are played as competent but noninitiating.
   a. Distressed husband/wife/child
   b. Emergency Room physician
   c. Nurse/Technologist
   d. Respiratory
   e. Other types of responders
   f. Administrators
   g. Press/Media/Reporters/Camera crews
2. What are the nonclinical props these actors or settings need to enhance the "realism" of the scenario?
   a. Clothing – white laboratory coat, street clothes, weather/local/occupation appropriate
   b. Wigs, jewelry
   c. Identification badges
   d. Room signs match scenario setting
   e. Dummy phones, real phones, cell phones
   f. Crash cart, etc.
   g. Voice in the ceiling – institutional paging, scenario guidance
   h. Video recording
      i. Be sure you have placed a sufficient number of cameras and microphones so that key actions and sounds are always in view of at least one camera and captured intelligibly (audio).
      ii. All audio/video equipment is functioning properly.
      iii. Have signed photographic and video releases for each person who is in the scenario in case video recording is to be shown outside the debriefing session.
3. What are the clinical props needed to enhance the "realism" of the scenario?
   a. Invasive-IV line, Chest tube, etc.
   b. Noninvasive nasal oxygen cannula, blood pressure cuff, etc.
   c. Other considerations
      i. vomit
      ii. bleeding extremity
      iii. IV arm ready with blood system charged
      iv. amputation
      v. body wounds
      vi. various moulage applications
      vii. bladder filled with yellow water for Foley insertion
      viii. medical alert bracelet

4. What supplies will the student need to complete the scenario?
   a. Have available-for-use supplies that the student is expected to use to successfully perform within the scenario, e.g., stethoscope, gloves, Foley catheter, chest tube, IV catheter, fluid bag and tubing, surgical setups, etc.
   b. These can either be purchased outright from vendors or you may have an agreement with the local hospital to supply you with outdated items. Be cautious here as you don't want out-of-use supplies and obsolete equipment. If you are training with obsolete equipment, the training isn't as valuable or may be perceived as unreal.

Look back at your objectives to decide and list what behaviors you expect your student to complete and how they will affect the simulated patient (vitals signs, etc.). Then, decide and list the expected "fail-to-respond" behaviors and how the absence of responses will affect the simulated patient. Together, these will be used in creating your scenario script.

Transfer the scenario plan into the simulated patient software and test if what happens is what was desired. You can save your scenario information, props, actors, and other vital information when you finished building. You can print it and add to your resource book of scenarios. Once the simulator program is written and saved for one scenario, it becomes easier to build others from its components. The initial configuration time may seem daunting but it is well worth it in the long run.

## 58.5 Ending the Scenario

The end of the scenario will be based on answering one of the following questions:

a. Do you want the student to fumble horribly through to the scenario's end? It is more than just a good idea to ensure that the trainees actually know the steps to solve the problem (e.g., have the appropriate skill set), and the purpose of the simulation is "to put it all together" in a simulated clinical environment.
b. Do you want the student to fumble so horribly that they kill the patient? Allowing the student to kill a patient brings with it a new set of potential problems if death were not an intentional phase of the scenario. If death happens because of something the student did or did not do, you would hope that the student experiences physiological and psychological stress. Questions of competence and confidence will occur. Most, if not all, other lessons learned during the scenario may well be lost by the emotionally overwhelming event of death. Thus, only

allow death when the instructors are ready to immediately address the students' role in the patient's demise or if it is part of the scenario's objectives. *Do not* deviate from nondeath objectives to "punish" poor performance – the goal is learning.
c. Do you stop in the middle, pause the scenario, remediate, and then go on?
d. Do you stop the student when the fumbling and ineffectiveness is too great, but the scenario isn't completed? Is your goal to complete a schedule or foster learning?
e. Always keep in mind that simulation training is supposed to be a safe environment where errors made can be non-threatening and the process repeated and practiced with guidance until it is done correctly.

## 58.6 Scenario Map

The simulation production team needs to share a common, clear script that easily conveys the key scenario events and their sequence. A scenario map in the form of a graphical time line is an excellent representation of the scenario, and is suitable for both initial scenario design and actual execution (Figure 58.1). The key variables and their values are traced across the duration of the scenario. Color coding the different variable traces and their magnitude scale values distinguishes them from each other, and using the same colors as displayed on the clinical monitor greatly helps maintaining visual/conceptual links. The horizontal time scale can be very elastic: wide during large changes, narrow during little or no changes.

These maps can be easily shared, stored, posted, revised, and reproduced. During scenario execution, posting the map near the simulator controller provides a common visual reference of the entire scenario for all members of the control room team. Also, the map may be shared during the debriefing to provide everyone with the big-picture.

## 58.7 Debriefing

Debriefing is where the true learning takes place. Simulation is a means to come to the debriefing. I was confused about this topic until I talked with my supervisor. I initially thought that all debriefing had to be structured and in a conference room with a video replay. The confusion came when I saw facilitators with more experience than I talking over the simulated patient with no video recording and no conference room time. Was I to take them to task as not doing a good job of debriefing as I knew it? They explained to me that debriefing, as I learned it, was to teach teamwork and learning together. The debriefing over the simulated patient was more of a teaching process

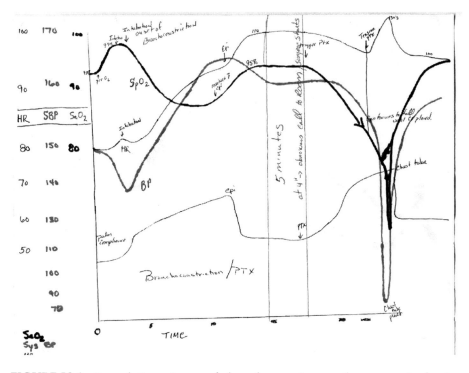

**FIGURE 58.1**  Example Scenario map of a bronchospasm/pneumothorax scenario, showing the changes in oxygen saturation (SaO₂), heart rate (HR), systolic blood pressure (SBP), and pulmonary compliance (Pulm Compliance). Also shown are the key interventions expected by the participants, such as preoxygenation, intubation (which triggers the bronchospasm), administration of epinephrine (Epi), triggering the pneumothorax, and inserting the chest tube. Note that the time scale is elastic; to account for both participant variability and the 5-minute period in the middle where very little change occurs. The crude, hand-drawn nature of this example is its strength: anyone with paper and colored markers can create a scenario map in a form that is as useful as it is durable (this image is a scan of a map drawn over 6 years ago, and both the scenario and this page are still used regularly).

where a student was remediated immediately due to the nature and time constraints of the training, i.e., PALS, ACLS, etc. We now incorporate a quick simulation *with debriefing around the simulator* in our Facilitator's course to demonstrate the process and its meaning. It's exciting to watch them participate and learn. Now that I have a more comfortable and complete understanding of debriefing of simulation training, I personally have separated debriefing into two types, formal and informal:

a. **Formal debriefing** takes place after a scenario session with full audio and video recording. The students move out of the simulated environment and into a conference room with the necessary audio and video replay equipment. A facilitator (an instructor trained in facilitating group discussions) begins the debriefing by showing a clip of the session and begins to draw out comments in a nonthreatening process. Using "I" to describe the facilitator's observations on a problem area that might be

directed toward a specific student, the facilitator elicits a response. For example, "I am wondering what information led you to respond outside the normal guidelines of ACLS." This response may be followed by another student commenting, and hopefully, the comments facilitate the learning process. Maybe, the objective of the training was to foster role clarity, leadership skills, communication, problem solving, and team building, i.e., you have the opportunity to discuss Crisis Resource Management principles.

b. **Informal debriefing** takes place over a simulated patient after the session had been paused or ceased. This may be a facilitator's teaching style. With many of the classes we provide (PALS, ACLS, etc.), time constraints and staffing issues make it difficult to spend extra hours debriefing in the formal way. I believe that there is a greater impact on the student's learning a skill if the debriefing happens during the session and with some questions and humor injected at that time.

## 58.8 Conclusion

It is important to realize that there are multiple ways to create a scenario, and multiple ways to use the simulated experience as a "hook" to engage the trainees. Not all debriefing needs to occur in a sit-down darkened room with a video replay. Much can be learnt from replaying the scenario at the simulator, to point out items that might not have been noticed with the first go-around.

The views/ideas expressed in this article are those of the author and do not reflect the official policy or position of the US Department of the Army, US Department of Defence, or the United States Government.

## 58.9 Examples

### 1. Framework for a Simulation Session

Date Scheduled:

Type of Session:

Department Point of Contact:          Phone Number:

Date of Training Session:          Start/End Time:

Number of Students:

| Learning objective/skill | Type of simulator | Room | Supplies needed | Props | Actors | Comments |
|---|---|---|---|---|---|---|
| | | | | | | |
| | | | | | | |
| | | | | | | |
| | | | | | | |
| | | | | | | |

### 2. Framework for Planning a Scenario

Target Audience:

Objectives: Two or three objectives you want the student to learn

Props: Supplies, equipment, and simulator setup needed

| Hx: Patient's name, age, presenting vital signs and other medical history, allergies, medications, etc. Cause of current problem, how long has it occurred, etc. Family members present. | | | | |
|---|---|---|---|---|
| Phase 0: The first view of the patient. Includes vital signs and handlers. If no handlers used, event actions will trigger to next appropriate frame. | ABC Handler: Names that correspond a student action and subsequent simulator physiological response. Will include responses and the lack of correct response. | Miscellaneous Handler (names) | Medications Handler (names) | Actors: Who's taking care of the patient, family member. |
| Links to next phase Student responses | | | | |
| Phase 1 Physiological response due to student action | | | | |
| Phase 2 | | | | |
| Phase 3 | | | | |

# 59

# Designing a Scenario as a Simulated Clinical Experience: The TuPASS Scenario Script

Peter Dieckmann
Marcus Rall

## 59.1 Value in the Scenario Script

Producing realistic patient simulations entails enacting a clinical case, i.e., a typical problem or illustrative case from different health care environments: operating rooms, intensive care units, prehospital, etc. All these events have in common that they are characterized by a complex interplay of humans, technology, and the surrounding organization [1, 2]. Consequently, incidents are not explainable using a single cause, like a lack of knowledge, bad design of equipment, or difficult working conditions, but are due to a dysfunctional interplay of the different factors involved, as can be seen in other industries [3–16].

The complexity of the medical field is a salient characteristic that influences each case and that should be preserved within simulation scenarios to obtain "ecologically valid" [7] scenarios for crisis resource management (CRM) [2, 4, 8–12] training. The preservation of the complex contexts of clinical cases within simulation scenarios is also in line with theories on situated learning. These theories assume that knowledge and skills are heavily bound to the context in which they are constructed and reconstructed [13, 14].

Participants' experience of the simulation scenarios is influenced by many other aspects besides the technical simulation such as the fidelity of the mannequin and the overall plausibility of the simulated case. Among these latter aspects are role-play of the simulator team and how competent participants felt acting in the scenario, i.e., whether they have the feeling that they could perform up to their usual level of proficiency [15–17]. If participants are not fully aware how to approach the simulator or do not know where to find the material, they might become angry, or might feel lost. Consequently, all these influences should be taken into account in designing and producing simulation scenarios [A, B].

Given these principles, it is clear that it is not enough for realistic patient simulation of emergency cases and teaching CRM to focus only upon realistic anatomical and physiological simulation. It is also important to include social aspects in the scenarios to simulate interactions that are typical for medical practice, and these should further be adjusted to the goals of simulation and the proficiency of simulation participants.

We developed the TuPASS Scenario Script to help with designing and producing scenarios for simulation-based CRM courses with video-assisted debriefing. The script can be adapted to other scenarios as well, e.g., focusing on basic skills. The concept is based on our ideas about a simulation-based program (the simulator setting) as being constructed of different and related course phases (modules) [15, 18].

## 59.2 The TuPASS Scenario Script

The TuPASS Scenario Script is a regular part of our Instructor and Facilitation Training (InFacT) – see Chapter 67. It can be obtained from our web-page for free: www.tupass.de/downloads/scenario.html, and is also included in Appendix 59A.1. The scenario script consists of seven pages: the title page, the summary page, and the five development pages. The summary page covers the same points as the development pages (Figure 59.1). The only difference is the amount of space provided. The development part provides more space for the development and detailed references. The summary page enables a quick overview and reference during a course. In the following, we will explain the components of our Scenario Script.

The title page allows for naming the scenario. It is important to choose a name for the scenario that does not give away the "trick" of the scenario right away. If the scenario is called "Anaphylaxis" or "Vfib," all participants seeing the script will know what the next case is about. We do recommend assigning a unique patient name for every case, especially if a limited number of scenarios is used repeatedly. This allows for "secretly" referring to scenarios and their underlying stories within the simulator team, without giving away too much information. "We do the 'Mister Miller' scenario next" could thus signal to the nurse to prepare the mannequin in a special way, the simulation controller to start the appropriate programed scenario in the computer, and so on. Thus, the scenario name can be a very short and effective basis for a team briefing before the scenario. Certainly, this requires that all members of the simulator team understand this coding. We also included the contact information of the scenario designers for questions that might arise concerning producing the scenario by others.

The second page, with an overview of the scenario, is the core of our scenario script (Figure 59.2). This page should, as all remaining pages, be kept secret from participants in order not to spoil their learning experience by giving away too much information about the scenario. On the top right corner, it is possible to write the *Scenario Name* again. This might make it easier to locate the script in a folder.

The first row, "*Major Problem*," allows for providing keywords about the scenario. This information is helpful, if the scenario name doesn't ring the bell, or the reader wants to recheck what kind of case "Antonia Smith" is representing. There are two columns in this row. The left column is labeled "Medicine" (e.g., "Hypothermia," "Hypoglycemia") and the right column with "CRM" (e.g., "Fixation error," "Workload distribution," etc.). (A)CRM in its original version meant Anesthesia Crisis Resource Management and was adapted from the aviation industry by the group around David Gaba and Steven Howard in Stanford, U.S.A [19]. As crisis resource management is relevant for many domains, we tend to skip the "A" and use CRM only.

The second row contains again the two columns "Medicine" and "CRM," this time with respect to the intended teaching goals for the scenario. The more specific and precise these teaching goals are, the better the scenario can be. However, we are aware that sometimes the prospectively envisioned teaching goals may be set aside to reach some spontaneously arising goals from the dynamic realistic scenario in which participants might do something unexpected. In this case, it might be beneficial to deviate from the preplanned goals in order to teach even more interesting points. This strategy is not about "punishing" good teams by making the scenario harder and finally unsolvable if they perform well. It is about saving and, in fact, improving the scenario as a learning event.

It is important to state the scenario goals in a way so that it becomes clear whether they are reached or not. There is some literature that can help to define proper goals [20]. Ideally, goals should be formulated positively (what should be achieved), stated clearly, be structured (e.g., into superimposed or subordinated goals), ordered according to priorities, checked for discrepancies (e.g., working fast versus working precisely), and flexible (e.g., when participants do something unexpected). For example, a superimposed clinical goal could be to teach participants how to apply current resuscitation guidelines. A subordinated goal would be to perform the Airway Breathing Circulation check, minimize the times without chest compressions, and to use the chosen algorithm consistently. On the CRM side, the superimposed goals might be to manage the team work and cognitive aspects. The subordinated goals would be, for example, to discuss the optimal way of distributing the workload during resuscitation, to use clear communication about which guidelines are used, and to reevaluate whether they are put into practice effectively.

The "*Narrative Description*" contains all the information you would pass on to a simulator colleague. It can be written in prose or keywords. Prose is more work, but might make it easier to be on the same page about the details of scenario (and both the devil and Mr. Murphy love details and especially they love *the lack of details* even more). This description should contain information about the patient, his or her condition, diseases, and so on, but also information about the time and place in which the scenario takes place. The better the simulator team "knows the patient," the better can be following case briefings, scenarios, and debriefings.

The "*Staffing*" is divided into two groups: the simulator team and the course participants (trainees of all levels of competence). A member of the simulator team is usually needed to control the simulator, to direct the scenario, and to conduct

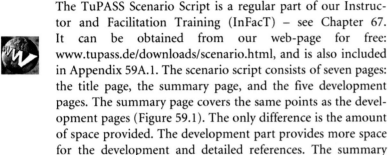

Developed by P. Dieckmann & M. Rall, TuPASS Germany;
Contact: peter.dieckmann@med.uni-tuebingen.de or marcus.rall@med.uni-tuebingen.de

Scenario Name:_____

# TuPASS
# Scenario Script

Keep confidential from participants in order not to spoil their learning experience

Scenario designer:_____

Contact info in case of questions:

Phone: _____

Email: _____

**FIGURE 59.1**  Title page of the TuPASS Scenario Script for Crisis Resource Management–oriented simulation training with video-assisted debriefing. Note that the clinical case/the main problem should not be mentioned as the scenario name (on the front page) in order to prevent unintended reading by participants.

| Quick reference: Scenario name | | |
|---|---|---|
| Major problem | Medicine | CRM |
| Learning goal | Medicine | CRM |
| Narrative Description | | |
| Staffing | Simulator team | Participants |
| Case Briefing | All participants | "Hot-Seats" only |
| Simulator Set up Manikin preparation | | |
| Room Set up | | |
| Simulator operation | | |
| Scenario life savers | | |
| Props needed | | |

**FIGURE 59.2** Summary page of the TuPASS Scenario Script for Crisis Resource Management–oriented simulation training with video-assisted debriefing. "Medicine" is defined as the medical/clinical aspects. "CRM" describes the aspects related to Crisis Resource Management. "Hot-seats" refers to the active participants in the scenario. "Scenario life savers" describe ideas and methods to save the scenario if participants do something unforeseen. See explanations in the text.

the debriefing. However, for presenting realistic scenarios, it is very valuable to work with role-players who take part as "surgeons," "surgical assistance," "relatives," "dispatchers," etc. Both, participants and simulator team members, can be selected to take over such roles. In addition, the roles can be "scripted," meaning that persons enacting the role may receive secret instructions for the scenario, for example to commit a "stupid error." By doing so, it is possible to systematically trigger interesting points for the debriefing without having to wait for the error to happen and without the need to potentially embarrass a simulation participant. While using this method, it is of utmost importance to clarify and explain the scripting of the error immediately after the scripted person steps out of the role (usually after the scenario, in the beginning of the debriefing). Failing to let this person out of the role very clearly might mean that no participant (or simulator team member) is willing to engage in scripted roles in the future. The need to protect one's self-image is most probably stronger than the desire to help with the course or make new insights (a general principle that should be considered throughout all simulation experiences). Clearly, distinguishing between the person and the role that she or he enacted is essential for this protection.

Asking own team members to enact roles allows for more "control" over how the roles are filled – this would be essential whenever standardized scenarios would be needed, for example in assessment. Simulator team members can also better be scripted to do "stupid" things during the scenario because they can be trained to differentiate between roles and their own person – a competence that is not trivial. Working with participants allows for asking them to assume roles during the scenario, normally not their own. Using this "cross training" [21] systematically allows for eye-opening experiences. For example, an anesthesiologist in the role of a nurse can experience how it feels to receive five orders at a time.

*"Case Briefing"* is the information participants receive before they actually start the scenario. It contains information about the patient, his or her history, underlying diseases, the "here and now" of the case, and so on. Despite, time wise, being a short part in the simulation experience, nevertheless the case briefing is very important [15, 18]. The case briefing is different from the narrative description mentioned above, because it does not contain the "trick" of the scenario or its underlying problems, but only information about the patient and the circumstances. Especially when working with scripted roles, it is reasonable sometimes to provide different information to the "hot-seat" participants, and the participants watching the scenario. Some "hot-seat" participants might receive more, less, or different information to enhance the effectiveness of their learning from the scenario.

The case briefing is similar in many ways to clinical "handovers," where the patient and his or her history are also introduced. However, compared to clinical takeovers, case briefings serve additional functions because they must help the participants to understand the "here and now" of the scenario.

Health care professionals in real clinical environments simply know where they are. In contrast, the simulator rooms of a simulation center are often is used for cases in different surroundings, like the operating room, intensive care units, etc. The same *simulator* room "becomes" different *simulated* rooms. Participants should know about these changes, so they form a picture for themselves of what available resources can be expected.

Table 59.1 contains questions that are used in theater acting to establish a "scene," and that can help with the case briefing and establishing a scenario. All simulator team members and all participants should be able to answer these questions for a given scenario. If they can, they are likely to have understood the scenario that is simulated well enough to act competently within it.

The *"Simulator setup, Mannequin preparation"* provides room for describing special preparations, like neck restraints, bandages, wigs, etc.

*"Room setup"* describes how the simulator room should be prepared and what resources should be available to participants. It might be easier for participants to recognize that a room is supposed to represent a different room than in the last scenario, when it looks different, at least slightly. Especially for scenarios that take place in nonclinical (e.g., a home) or "less-clinical" environments (e.g., a typical ward), it might be necessary to reduce the medical equipment that is available, like respirators or infusion pumps. Those could be rolled out or covered using drapes.

The *"Simulator operation"* describes the major steps in controlling the simulator. It should contain the initial vital signs for the simulated patient, the major steps of their development, and the desired state at the end of the scenario. It can be written in prose or keywords. Code or scripts for controlling the simulator are also possible but might not be understood by new instructors in the team.

*"Scenario life savers"* are all elements that can be used to save the scenario once it has started, and if the participants do something unexpected during the scenario: It is about expecting the unexpected or "being prepared for the unexpected," and that participants will do unexpected things *can* be expected

**TABLE 59.1** Questions that can help to provide enough information to establish a coherent scenario during case briefings

- **Who** is acting?
- **What** is being done?
- **Where** does the situation take place?
- **What** resources are available?
- **When** does the situation take place?
- **Why** did this situation evolve?
- **Which motives** do people follow?
- **What** do people want to obtain?

(greetings again by Mr. Murphy!). Scenario life savers can be used to save the scenario as a reasonable experience to learn from. In general, there are two distinct ways to save a scenario that is unfolding against its plan: one can try to direct it back toward its original goal, or one could leave the old scenario that was planned and flexibly create a new scenario on the fly (which requires considerable skills to keep the scenario plausible and consistent). Both ways can be used by making the scenario harder, easier, or changing it.

Bringing participants back on track can be done from within the scenario or from outside of the scenario. "Within" life savers are, for example, sending help into the scenario, calling into the simulated clinical area in the role of a consultant, or let the patient provide helpful information. Outside strategies would be to use the freeze technique and stop the scenario, helpful hints by an instructor but without stopping the scenario, or asking helpful questions as the instructor, etc.

Creating new scenarios on the fly would mean to flexibly adjust the scenario and possibly its original goals. If, for example, one superimposed goal of the training course is to teach the CRM principles, and it was planned to discuss anticipation and planning in this scenario but a very "nice" fixation error shows up, it might be good to engage in dynamic decision making [22] and switch priorities, provided there are still other scenarios left during which one can deliver what was postponed.

In the life saver section, it is possible to continuously collect strategies for both approaches. The best approach would be to be prepared for the unexpected beforehand, and to think about ways in which participants could deviate from the planned scenario. It would be possible to use procedures following the Failure Modes and Effects Analysis [23] approach for important scenarios. This process breaks down processes into steps, analyzes what could go wrong, and what should be done in this case. The scenario script can also be used to collect all the deviations that could be expected, but also the unexpected that happened anyway. The script can become a living document by collecting new scenario deviations and good solutions to bring it back on track.

In addition, generic "scenario life savers" could be collected that can be used in any scenario. For example, it often happens that the simulator team does not recognize immediately what the participants said to the patient. In order to avoid that participants treat an unplanned unconscious patient, a generic "scenario life saver" would be to let the patient say: "Oh, I just did not listen to you. Daydreaming, you know. What did you say?"

"*Props needed*" in the last section of the script can be noted what kind of additional material is needed to present the scenario, like neck restraints, bandages, defibrillators, etc. As it could be possible in several sections of the script that additional material was requested for producing the scenario, this section provides a summarized overview. It can serve as a checklist for preparing the scenario this way.

Providing a tool does not help much, if users do not know how best to use it. For this reason, we describe some ways on how to use the Scenario Script.

## 59.3 Designing Scenarios

The scenario script can be used during the design of scenarios as well as during preparing and producing them. For designing scenarios, it is crucial to start from defining the goals that should be reached with the scenario and work through the template relating all design decisions to these goals. The procedure to actually design scenarios in such a goal-oriented way is not covered in the script itself, as it allows for collecting the result of such a design process. However, considerable work was done to define principles of scenario design in medicine [24–28], the aviation industry [29–33], naval simulators [34], nuclear power plant simulation [35], and in general [36]. On a general level, Hoffman made very valuable suggestions to design scenarios (Table 59.2, adapted from [35]).

During the design process, a single person or a team can work through the script and discuss the single sections of the script. It proved to be beneficial to use an iterative rapid prototyping approach during the design. Whenever a first version of a scenario is prepared, the script can be used to discuss the result with colleagues, or to try the scenario. The scenario can then iteratively be improved on the basis of the experience gained by discussing and testing it.

An interesting idea to develop more scenarios is to ask clinical students or new simulator team members to design scenarios. The clinical students can learn a lot about the clinical issues in the scenario, the new team members a lot about using the simulator.

## 59.4 Preparing and Producing Scenarios

The scenario summary page provides an overview and guides producing the scenario. Using the *Case briefing* section, it is possible to explain the upcoming scenario to participants in terms of the case itself and the "here and now." Before the actual start of the scenario, it can be used as a checklist for the simulator, the simulator room, etc., using a "time out" procedure. During the scenario execution, the *Simulator operation* section becomes especially relevant. It might be helpful to print out more information material about the scenario, like programed trends, model descriptions, or action trees for more detailed reference. In case participants do something unexpected, the simulator team can refer to the *Scenario life saver* section (and hope that the current deviation had been encountered previously and/or thought through).

**TABLE 59.2** Criteria, characteristics, and general goals for designing scenarios (adapted from [35])

| Criteria | Characteristic | General scenario goal | Medical examples |
|---|---|---|---|
| Stressful | High psychological pressure<br>Information overflow<br>Time pressure | Learning of a strategy to deal with the situation | Acute, unexpected bleeding in a young (pregnant) person or a child/baby with too few personnel at the beginning. |
| Complicated | Scientific complex relationships<br>Deep understanding of processes is hard to reach or not yet available | Compensation for difficulties in understanding by collecting experiences with the processes<br>Standard Operating Procedures (SOP) | A patient with preexisting coronary disease is getting hypotensive during the anesthetic and develops ST-depression – might go on to overt heart failure and major rhythm disturbances. |
| Critical | Mistakes would have severe consequences | SOP<br>Double checking | Takeover of a patient who has been intubated into the esophagus by the previous colleague, but is breathing spontaneously.<br>Airway management SOP/Algorithm |
| Rare | Problem usually not experienced in actual work practice | Refreshing knowledge<br>Updating of procedures | Thyroid storm after or during surgery.<br>Acute Prophyria in ICU/PACU.<br>Intoxications |

It could be helpful to use some form of graphical representation of the scenario life savers, similar to action trees. These would graphically represent specific decision points (node) in a scenario and anticipated actions participants could take (branches); for example, in a difficult airway scenario, a node could be the detection of the difficult airway and branches could be (i) another attempt with the laryngoscope, (ii) selecting a different airway tool, (iii) preparing an invasive airway, etc. Another node (life saver) would be needed if participants do not detect the difficult airway at all. Then, there would be different branches to deal with this problem without just letting the simulated patient die. It is helpful to make the script a living document, especially by keeping the life saver section up to date. Every new "deviation," as well as ways how to deal with it, should be recorded – this will increase the script's value with its continued use. For this, it is very valuable or even required to do a team debriefing after training courses so that the experiences can be recorded and thus saved.

## 59.5 Experiences With the Script and Points to Consider During Scenario Design

As stated above, we used the script during our Instructor and Facilitation Training (InFacT) courses (see Chapter 67), besides for training courses we run. Within each course, a "scenario design workshop" is conducted during which 3–5 scenarios are developed. There are some typical issues that arise during these workshops that make the designed scenarios difficult to use in a goal-oriented way. We describe some of these typical problems, on the basis of the experience with approximately 40 developed scenarios in 13 InFacT courses.

### 59.5.1 The "too much" Scenarios

There are too many aspects integrated into one scenario. The patient might experience heavy bleeding, severe anaphylactic shock, and in addition, there might be a power failure, all in a single case. Even if such a combination might happen in a real clinical event, most likely, participants will be overwhelmed during the scenario, and the instructors overwhelmed during the debriefing. Considering that the major problems of a scenario should be covered in the debriefing, lasting mostly less than an hour, it is hardly possible to include more than one or two of these into any one scenario.

### 59.5.2 The "too fast" Scenarios

Often, the physiological development during the scenario is constructed to change very rapidly and often much more rapidly than it would be physiologically reasonable. Participants readily note such an "on-off" character of physiological changes as implausible and unrealistic [15]. In addition, it seems that for most participants, time is passing faster (in any case) during the scenario, therefore, they feel an increased time pressure as well [15–17]. For this reason, it might even be valuable to slow scenarios down so that they are *experienced in real time.*

### 59.5.3  The "find the detail" Scenarios

We encountered scenarios that were designed in such a way that participants had to discover a tiny, minute detail as the key to solve the scenario. For example, participants were required in one prehospital scenario to discover a strange-smelling liquid in an almost empty glass on a living room table in the patient's home and infer an intoxication from this detail. It is most likely that no participant ever would find such a detail, especially when considering that very often it is not easy for participants to judge what is a planned part of the scenario, what could be a mistake in its design or execution, or what they might have overlooked. Problems in scenarios should be designed bold enough so that participants can find them.

### 59.5.4  The "props and whistles" Scenarios

Sometimes, scenarios are full of props and whistles, like a power failure, disconnections, broken suctions, external resources that are *never* available immediately (e.g., the CAT scan), clumsy or choleric role-players, and so on. If there is too many such gimmicks within one scenario, or over several scenarios as the simulator course proceeds, participants might not see the simulation as relevant any more.

## 59.6  Conclusion

Designing simulation scenarios as learning or research events requires taking all the aspects into account that influence the participants [37]. This is far more than the physical, functional, psychological fidelity. The scenario should have a believable story, take place in a reasonable environment, and be plausible in a medical sense. Its different parts like role-players, clinical devices, etc., should function together consistently and oriented to the goal of the scenario. The TuPASS Scenario script can help to design and produce scenarios in such a way.

## Acknowledgments

We thank our instructor colleagues Silke Reddersen and Jörg Zieger as well as the participants in our InFacT courses for valuable discussions of this tool that again helped to improve it.

## Endnote

We have put much energy in designing and refining the Scenario Script, and are happy to share it within this book. In the beginning of the chapter, we have provided the address where to obtain the script. We ask you to provide us with feedback when you use the script (peter.dieckmann@med.uni-tuebingen.de; marcus.rall@med.uni-tuebingen.de).

## References

1. Gaba, D. M., Fish, K. J., and Howard, S. K. *Crisis Management in Anesthesiology*. Churchill Livingstone, Philadelphia, 1994.
2. Rall, M. and Gaba, D. M. Human Performance and Patient Safety. In: Miller, R. D. (ed.), *Miller's Anaesthesia*. Elsevier Churchill Livingston, Philadelphia, 2005, pp. 3021–3072.
3. Cooper, J. B., Newborner, R. S., Long, C. D., and Philip, J. H. Preventable Anesthesia Mishaps: A Study of Human Factors. *Anesthesiology* **49**, 399–406 (1978).
4. Rall, M. and Dieckmann, P. Safety Culture and Crisis Resource Management in Airway Management. General Principles to Enhance Patient Safety in Critical Airway Situations. *Best Practice & Research Clinical Anaesthesiology* **19**(4), 539–557 (2005).
5. Perrow, C. *Normal Accidents. Living with High-Risk Technologies*. Basic Books, New York, 1984.
6. Columbia Accident Investigation Board. *Columbia Accident Investigation Board Report Vol 1.* (online: http://www.nasa.gov/columbia/home/CAIB_Vol1.html), 2003.
7. Bronfenbrenner, U. *The Ecology of Human Development: experiments by nature and design*. Harvard University Press, Cambridge, 1979.
8. Gaba, D. M., Howard, S. K, Fish, K. J., et al. Simulation based training in anesthesia crisis resource management (ACRM): A decade of experience. *Simulation & Gaming*. **32**(2), 175–193 (2001).
9. Rall, M. and Dieckmann, P. Crisis Management to Improve Patient Safety. [Online: http://www. euroanesthesia.org/education/rc2005vienna/17RC1.pdf]. *ESA Refresher Course*, 2005, 107–112.
10. Rall, M. and Dieckmann, P. (2005). Simulation and patient safety. The use of simulation to enhance patient safety on a system level. *Current Anaesthesia and Critical Care*, 16, 273–281.
11. Rall, M. and Gaba, D. M. Patient simulators. In: Miller, R. D. (ed.), *Anaesthesia* Elsevier, New York, 2005, pp. 3073–3103.
12. Gaba, D. M. and DeAnda, A. A comprehensive anesthesia simulation environment: re-creating the operating room

for research and training. *Anesthesiology.* **69**(3), 387–394 (1988).

13. Johnson, E. *Situating Simulators. The integration of simulations in medical practice.* Arkiv, Lund, 2004.

14. Searle, J. R. *The construction of social reality.* Free Press, New York, 1995.

15. Dieckmann, P. *"Ein bisschen wirkliche Echtheit simulieren": Über Simulatorsettings in der Anästhesiologie. ["Simulating a little bit of true reality": about simulator settings in anaesthesiology.] (online: http://docserver.bis. uni-oldenburg.de/publikationen/dissertation/2005/diebis05/ diebis05.html).* Universität, Dissertation, Oldenburg, 2005.

16. Dieckmann, P., Manser, T., Schaedle, B., and Rall, M. How do anesthesiologists experience a simulator setting in comparison with clinical settings? – Results from an interview study (Abstract Santander-02-12). *European Journal of Anaesthesiology* **20**, 846 (2003).

17. Dieckmann, P., Manser, T., Wehner, T., and Rall, M. (2007). Reality and Fiction Cues in Medical Patient Simulation. An Interview study with Anesthesiologists. *Journal of Cognitive Engineering and Decision Making*, **1**(2), 148–168.

18. Dieckmann, P., Manser, T., Wehner, T., et al. Effective Simulator Settings: More than Magic of Technology (Abstract A35). (online: http://www.anestech.org/media/ Publications/IMMS_2003/sta3.html). *Anesthesia and Analgesia* **97**(S1–S20), S 11 (2003).

19. Howard, S. K., Gaba, D., Fish, K. J., et al. Anesthesia Crisis Resource Management Training: Teaching Anesthesiologists to Handle Critical Incidents. *Aviation, Space & Environmental Medicine* **63**(9), 763–770 (1992).

20. St. Pierre, M., Hofinger, G., and Buerschaper, C. *Notfallmanagement – Human Factors in der Akutmedizin. [Emergency management – Human Factors in Acute Care Medicine]* Springer, Berlin, 2005.

21. Salas, E. and Cannon-Bowers, J. A. Methods, Tools, and Strategies for Team Training. In: Quinones, M. A. and Ehrenstein, A. (eds.), *Training for a rapidly changing workplace.* APA, Washington, 1997, pp. 249–279.

22. Gaba, D. M. Dynamic Decision Making in Anesthesiology: Cognitive Models and Training Approaches. In: Evans, D. A. and Patel, V. L. (eds.), *Advanced Models of Cognition for Medical Training and Practice.* Springer, Berlin, 1992, pp. 123–147.

23. An introduction to FMEA. Using failure mode and effects analysis to meet JCAHO's proactive risk assessment requirement. Failure Modes and Effect Analysis. *Health Devices* **31**(6), 223–226 (2002).

24. Kneebone, R. L., Kidd, J., Nestel, D., et al. Blurring the boundaries: scenario-based simulation in a clinical setting. *Med Educ.* **39**(6), 580–587 (2005).

25. Larsson, J. E., Hayes-Roth, B., Gaba, D. M., and Smith, B. E. Evaluation of a medical diagnosis system using simulator test scenarios. *Artif Intell Med.* **11**(2), 119–40 (1997).

26. Murray, B. W. Simulators in Critical Care Education: Educational Aspects & Building Scenarios. In: Dunn, W. F. (ed.), *Simulators in Critical Care and Beyond.* Society of Critical Care Medicine, Des Plaines, 2004, pp. 29–32.

27. Murray, B. W. and Henson, L. C. Workshop on educational Aspects. Educational Objectives and Building Scenarios. In: Henson, L. C. and Lee, A. C. (eds.), *Simulators in Anesthesiology Education.* Plenum Press, New York, 1998, pp. 57–64.

28. Vollmer, J., Mönk, S., Uthmann, T., and Heinrichs, W. Which Scenarios should be included in Simulator Trainings? Empirical decision trees predict crisis during anaeshtesia (Abstract). *European Journal of Anaesthesiology* accepted.

29. Cannon-Bowers, J. A., Burns, J. J., Salas, E., and Pruitt, J. S. Advanced Technology in Scenario-Based Training. In: Cannon-Bowers, J. A., Salas, E., (eds.). *Decision Making Under Stress. Implications for Individual and Team Training.* American Psychological Association, Washington, 1998, pp. 365–374.

30. Gregorich, S. E. What Makes a Good LOFT Scenario? Issues in Advancing Current Knowledge of Scenario Design. In: Jensen, R. S. (eds.) *Proceedings of the sixth international symposium on aviation psychology.* Vol II. Ohio State University, Columbus, 1991, pp. 981–985.

31. Komich, N. CRM Scenario Development: The Next Generation. In: Jensen, R. S. (ed.), *Proceedings of the sixth international symposium on aviation psychology. Vol I.* Ohio State University, Columbus, 1991, pp. 53–59.

32. Prince, C., Oser. R., Salas, E., and Woodruff, W. Increasing Hits and Reducing Misses in CRM/LOS Scenarios: Guidlines for Simulator Scenario Development. *The International Journal of Aviation Psychology* **3**(1), 69–82 (1993).

33. Küpper, K. In welchem Maße läßt sich Gefahr simulieren? Eine arbeitspsychologische Untersuchung zur Ausbildung von Piloten (unveröffentlichte Diplomarbeit) *[To which extend can danger be simulated? A work psychological investigation about the education of pilots (unpublished diploma thesis)].* Freie Universität, Berlin, 1999, p. 126.

34. Mehl, K. and Schuette, M. Simulators – A perspective on what to train and what to analyse regarding human reliability. In: Scheller, G. I., Kaffka, P. (eds.) *Safety and Reliability.* A. A. Balkema, Rotterdam, Brookfiel, 1999, pp. 675–680.

35. Hoffmann, E. Lernpsychologische Aspekte und deren Umsetzung in der Simulatorschulung *[Learn psychological aspects and their integration in simulator education].* In: Schweizerische Vereinigung für Atomenergie (SVA) – Kommission für Ausbildungsfragen, (ed.), *SVA-Vertiefungskurs: Simulatoren für die Ausbildung des Kernkraftwerkspersonals.* SVA, Bern, 1995, pp. 1.2-1–1.2-14.

36. Rogalski, J. From Real Situations to Training Situations: Conversation of Functionality. In: Hoc, J-M., Cacciabue, P. C., and Hollnagel, E. (eds.), *Expertise and technology: cognition & human-computer cooperation.* Lawrence Erlbaum, Hillsdale, 1995, pp. 125–139.

37. Dieckmann, P. and Rall, M. (in Press). Simulators in Anaesthetic Training to Enhance Patient Safety. In: J. N. Cashman and R. M. Grounds (eds.), *Recent Advances in Anaesthesia & Intensive Care 24.* Cambridge University Press, Cambridge.

A. Dieckmann, P., Gaba, D., and Rall, M. (2007). Deepening the theoretical foundations of patient simulation as social practice. *Simulation in Health Care*, 2(3), 183–193.

B. Rudolph, J. W., Simon, R., and Raemer, D. (2007). Which reality matters? questions on the path to high engagement in health care simulations. *Simulation in Health Care*, 2(3), 161–163.

# Location, Location, Location

*No one goes there nowadays, it's too crowded.*

Yogi Berra

All successful simulations, no matter what the subject matter or the students or the simulators involved, shares a common trait: bringing together those students and simulators to experience a deliberate event designed and presented to foster a lesson learned about that subject matter. While this concept may seem obvious, the choice(s) and consequences of your decisions of where you stage your simulations may not be. Until simulation is as essential to clinical education as plumbing, location will be a primary determinant of the rate of adoption and adaptation by others in your institution. Like any business where the customers have choices, including the choice of not partaking at all, ease of access is crucial. No coffee outlet will ever succeed if located on the evening rush side of the road. Ideally, your location will be right next to where all the students and instructors already spend all their time (ease of access), yet nowhere near real patients (patient safety). Such a space, if it exists at all, is expensive and almost never left fallow awaiting you and your simulation program proposal. Therefore, you must be creative both in trading your resources for others' space, as well as evaluating the location characteristics essential to your simulation program's success, and do all this while placing the learning needs of your students first. Derek Leblanc shares his challenges and successes in staging simulations within clinical environments normally used by real patients. Christopher Gallagher *et al.* offer a light-hearted view of the most challenging venue of all: your paying audiences' turf. Marcus Rall *et al.* make use of the axiom "train the way you work" by offering their methods on how to "train where you work."

551

# 60

# Situated Simulation: Taking Simulation to the Clinicians

Derek J. LeBlanc

## 60.1 Defining "Situated Simulation"

The concept of "situated simulation" is seemingly simple: take a human patient simulator into a real care unit, place it in an empty bed, and let the clinicians have at it. However, as we all know, one of the most challenging components of any educational initiative is often bringing it from concept to reality. Conducting simulations in active clinical care areas can, in effect, offer a level of environmental fidelity that one can only dream of recreating in the simulation laboratory. It also comes with its own unique set of challenges. In the following sections, I will address some of these challenges, as well as some of the benefits of bringing simulations to clinicians. I will also discuss some of the theoretical and academic support for utilizing situated approaches, as well as provide some examples of recent and ongoing situated simulation initiatives.

## 60.2 Theoretical Underpinnings of Situated Simulation

For clinicians, the assimilation of clinical knowledge, technical skills, and professional attitudes occurs through a complex blend of didactic and experiential learning modalities.

Brown, Collins, & Duguid [1] write:

> Activity, concept, and culture are interdependent. No one can be totally understood without the other two. Learning must involve all three. Teaching methods often try to impart abstracted concepts as fixed, well-defined, independent entities that can be explored in prototypical examples and textbook exercises. But such exemplification cannot provide the important insights into either the culture of the authentic activities of members or that culture that learners need (p. 33).

Referring to the breadth of academic literature on simulation-based clinical education, Kneebone [2] writes, "In this literature there is a heavy emphasis on individual simulations and the technology that underpins them, but a coherent theoretical structure for their use is seldom presented" (p. 549). While this may be true of much of the literature on simulation-based clinical education, Bradley and Postlethwaite [3] have provided a comprehensive view of simulation-based clinical education from various theoretical perspectives on learning and clinical education. They posit that comprehensive approaches to simulation-based clinical education include behaviorist perspectives on the use of feedback, which is a central component of simulation-based learning (p. 1). They also suggest that simulation-based clinical education benefits from the application of constructivist theories, where experienced clinicians are able to assimilate new information with currently held schemas or "... cognitive structures..." (p. 1), and accommodate experiences that challenge those schemas (p. 1). Daley [4] concurs, "... the interrelationships of knowledge and professional practice appeared to exist in a highly complex, interrelated system that reflected concepts of various forms of constructivist learning" (p. 51).

Bradley & Postlethwaite [3] also discuss Lave & Wenger's [5] theory of situated learning and legitimate peripheral participation (p. 2) stating, "... it might be valuable to regard simulator experience as legitimate peripheral participation in the 'real world' of medical provision for human patients" (p. 2). Within this discussion, the authors pose the question:

> ... since peripheral participation might be empowering (as the student gradually moves to more and more central forms of participation), or disempowering (as the student is constrained to remain on the edge), how can an empowering transition from the simulator to the 'real world' best be managed? (p. 3).

Kneebone *et al.* [6] respond by stating, "Our proposal is to break down the artificial divide between simulation and patient care by providing distributed learning resources alongside the clinical workplace, leading to an explicit contextualization of learning" (p. 1100). While Bradley & Postlethwaite [3] limit their review of simulation-based clinical education to its theoretical underpinnings, Kneebone *et al.* [6] take a more pragmatic approach, arguing that situating simulation at the point of care is crucial to adding the clinical context necessary for clinicians to bridge the gap between learning and practice.

## 60.3 Situated Approaches to Simulation – Pearls and Pitfalls

There are numerous advantages of conducting simulations in clinical areas, beyond the level of environmental fidelity it affords. In the following sections, I will describe some of our recent experiences with situated simulation that highlight some of these benefits, as well as illustrate the need for increased diligence while operating in active clinical areas.

## 60.4 Simulation in the Intensive Care Unit and the Emergency Department

The opportunity to participate in heterogenous team-based learning is most often cited by clinicians and clinical learners as the most beneficial aspect of situated simulation. Most often, we deliver our laboratory-based programs to homogeneous groups – be they paramedics, respiratory therapists, physician residents, specialty nursing students, or practicing physicians, inevitably some of those learners and/or their instructors will be tasked to play the roles of other clinicians. Though this is usually not a problem, it does reduce somewhat the value to the participants. This is not the case when simulations are situated in active clinical environments, where the actual clinicians that would be expected at the bedside are present and engaged in the simulation.

Nowhere is this more evidently illustrated than in our Critical Care programing. Postsession feedback from our critical care participants supports this loose hypothesis. Typically, the critical care residents come with a facilitator to the simulation laboratory with focused clinical learning objectives. On occasion, we will take a simulator into the unit, and produce the same types of scenarios there, including all of the on-duty staff in the session. Comments such as, "I found overall experience to be much more realistic than similar sessions I've attended in the [simulation] lab," and "... staff participation in the unit really made a difference." Participant feedback also suggests that there is a degree of comfort derived from participating within the familiar environments of the clinical practice area. "Being on duty, I was more prepared and felt comfortable participating in the mock resuscitations;" "It helped that I knew where to find everything." Not having to role-play allows for others not participating in the scenario to observe more intently, improving the quality of the dialog during the debriefing phase.

As well received as these sessions in the unit have been, we still recognize that there are some learning objectives that are better suited for laboratory-based delivery. An example of this in the Intensive Care Unit context is Crisis Resource Management training. Though multidisciplinary sessions may be preferable for this type of training, it can be inappropriate to produce scenarios where medical errors or equipment malfunctions are intentionally introduced with actual patients and their family members in the next bed. Simulating oxygen pipeline failures, triggering multiple alarms, or role-play involving chaotic yelling or "harsh" language is better left in the safe confines of the simulation laboratory.

Unlike the Intensive Care Unit, conducting Crisis Resource Management training in the physically divorced yet more expansive confines of a resuscitation room in an Emergency Department allows for more isolation of distracting noise from the ongoing clinical care in the rest of the department. Consistent with comments received from other clinicians, the Emergency Department clinicians from our adult and pediatric centers have expressed higher levels of comfort performing in their clinical environments as compared to their experiences in the laboratory. Staff of both the adult and pediatric facilities have also used these experiential opportunities to orient new staff to the higher-acuity resuscitation environments, and familiarize all staff with new equipment and procedures.

One of the biggest challenges of conducting simulations in these settings concerns the debriefing of scenarios. Taking simulation out of the laboratory requires portability of all equipment, including video recording and playback capabilities if required/desired. Though there are many portable solutions on the market, simple recording with a portable video camera and playback through an audio/video connection to a digital projector works well. Strategic camera placement, or timely panning of the camera toward the patient monitor at crucial moments can make up for the lack of vital-sign monitor recording that one may be accustomed to in the laboratory setting. Often, facilitators will simply opt not to use video in favor of bedside scenario debriefings, or utilize strategic pauses at crucial moments in the scenario to illustrate key learning points.

## 60.5 Examples

### 60.5.1 District Trauma Center Simulation

An observed spin-off benefit of situated team-based learning is the opportunity to uncover and address institutional and systemic issues in the care continuum. Such is the case with the Nova Scotia Trauma Program's mobile trauma education sessions. Supported by our simulation program, the Trauma Program staff travel to each of the district trauma centers in our province to deliver half-day, case-based, multidisciplinary education sessions on the management and transfer of care of trauma patients. On the surface, the objectives for these sessions appear to focus on the clinical management of these patients, but inevitably, the focus eventually shifts toward issues such as timeliness of transfer to tertiary care, accessibility of air transport or trauma services, or local issues of staffing inequities or ground ambulance support. Such opportunities to address local issues, deal with concerns, provide linkages, and dispel misconceptions are often as valuable as the opportunities to address clinical care objectives (Figure 60.1).

The case-based approach taken with these sessions is another important element. Recreating actual cases from the particular institution helps the clinicians to engage early in the simulation, as many of them may actually have been involved in the management of the actual patient. Pausing the case at key moments to emphasize clinical or systemic learning objectives detracts somewhat from the overall fidelity of the scenario, yet it allows those involved, and those observing, to share information on the thinking behind the clinical decisions that were made at the time of the actual event, and to discuss other potential care options available. Vozenilek *et al.* [7] used this

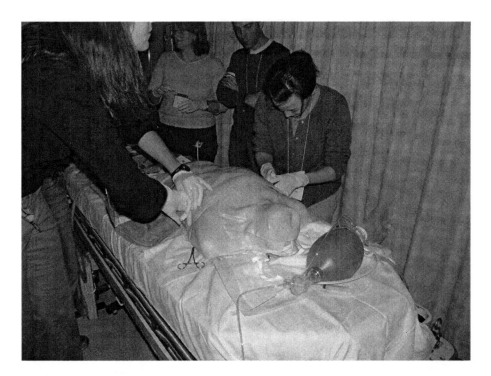

**FIGURE 60.1**  Partial-task Simulation at a District Trauma Center. Dedicated partial-task training devices can greatly reduce overall program costs and allow for parallel learning with large groups.

type of approach for a Morbidity and Mortality conference, and reported positive participant feedback. This finding is consistent with the feedback that we have received from clinicians involved in these types of sessions.

Because we typically travel only with our medium-fidelity simulators, we request to conduct the simulation-based components of these half-day sessions in an actual resuscitation room to improve the overall fidelity of the simulations. This typically is not a problem at most district trauma centers, as patient volumes are somewhat less demanding. This having been said, the requirement always exists for the room to remain ready to receive an actual patient, should the need arise. This requires some forethought to ensure that another bed is readily available, that essential patient care equipment is not contaminated during simulations, and that training supplies such as expired drugs and consumables be clearly identified to avoid mix-up with clinical supplies (identifying simulation supplies with bright red adhesive tape works very well).

## 60.5.2 Paramedic MegaCode Recertification

Kneebone [2] wrote, "There seems a danger, indeed, that task-based simulation may become a self-referential universe, divorced from the wider context of actual clinical practice" (p. 549). This phenomenon is very evident when paramedics come to the simulation laboratory. The unique features of their operating environments can be very difficult to recreate with any level of fidelity in the simulation laboratory, unless purposefully built for such purposes. Context is easily lost when street-level paramedics who are used to performing resuscitations in patient's homes, on the sidewalk, or in the back of an ambulance are asked to perform in the relatively quiet and controlled confines of the simulation laboratory – despite our best efforts to "stage" something more chaotic or realistic (Figure 60.2).

Paramedics in Nova Scotia are required to perform a simulated "megacode" resuscitation in the presence of their physician medical director annually in order to maintain their license to practice. In the past few years, our simulation program has been conducting a number of these sessions by traveling out to paramedic base stations and mobile standby locations in an off-duty ambulance, with the medical director onboard. As with the feedback received from other clinicians, a majority of the paramedics participating have indicated that they are more comfortable performing in this environment, and that they appreciate the opportunity to fulfill this component of their relicensure requirements during on-duty time. Additionally, many of the paramedics have expressed an appreciation for the opportunity to have their medical director out in the field, gaining a deeper appreciation for the unique challenges that their day-to-day practice environment poses.

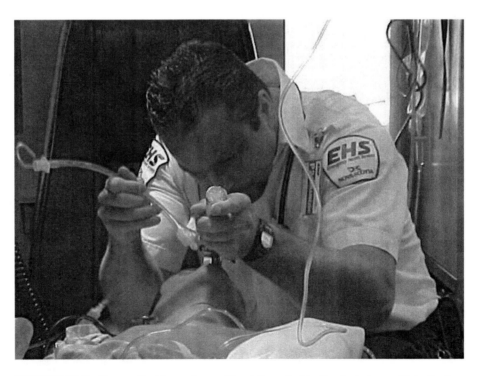

**FIGURE 60.2**  Paramedic Airway Access Simulations inside the Ambulance. Note the low position of the patient relative to the clinician – just one of the many unique challenges faced by clinicians operating in prehospital environs.

## 60.6 Further Considerations

For all of its benefits, situated simulation is not without its own set of challenges that require ample consideration before you decide to take your simulation show on the road.

### 60.6.1 Cost/Benefit

Before all else, a major consideration in determining if your simulation program should offer to deliver content off-site is the cost-benefit analysis. Depending on your program's business model, this may be a no-brainer. From a clinical department's perspective, bringing educational programing to on-duty clinicians represents tangible cost saving. Yet some of those same savings are offset by the added costs incurred by the simulation program. Beyond the obvious costs such as vehicle rental or owned vehicle utilization and other travel expenses, more subtle costs should also be considered, such as additional preparation time, insurance for transportation of capital assets, and higher rates of depreciation on those assets.

### 60.6.2 Preparing to Hit the Road

While we all realize that time to prepare for each encounter with clinical learners is essential, the amount of time and energy needed to prepare to conduct simulation sessions outside of the laboratory is exponentially greater. For every hour that you spend preparing for clinicians coming to your simulation laboratory, you can count on spending at least twice as long preparing to go on the road. Simulation programs with dedicated resources designed for mobile content delivery only have to consider these types of things once, but if your program utilizes the same resources for its laboratory-based and mobile applications as our program does, something as simple as a power adaptor could be the difference between a successful session and an absolute disaster – believe me! Simple checklists certainly help to ensure that everything you meant to take arrives at your destination.

### 60.6.3 Packaging and Transport Decisions

Now that you have decided to take your simulation show on the road, you have to decide how you are going to reach your intended destination. Moving within an institution is pretty straightforward, though one must be prepared for the funny looks, double takes, and often repeated comments one receives when rolling a pale looking, plastic patient heaped with equipment through the corridors. Planning to move outside your hospital or academic institution is another matter altogether.

For ease of mobility, our program only utilizes our medium-fidelity simulator for our mobile course offerings, electing to leave our high-fidelity simulator in the laboratory. Though our medium-fidelity simulator comes equipped with packing containers, the time required to disassemble, package, unpack, and reassemble, not to mention the wear and tear on the simulator itself, limits the perceived utility. Instead, we often elect to leave the simulator assembled and utilize an ambulance stretcher for transport, which consequently fits nicely into the back of a minivan or moderately sized SUV with the rear seats stowed. Suffice it to say that anything you can do to reduce your packaging and setup time, not to mention the likelihood of omitting key pieces of equipment, is beneficial.

## 60.7 Conclusion

Being a multidisciplinary simulation program, we have tried in the past to "be all things to all people," with varying degrees of success. Though we strive to achieve the highest level of fidelity possible for each of our laboratory-based programs, the burden of recreating an Operating Room environment for the anesthesia residents one day, a hospital ward for the respiratory therapists the next, and a living room for the paramedics the day after that can be daunting. By occasionally taking simulation outside of the laboratory, we have been able to bridge the context gap to a degree and add elements of realism that would be otherwise unachievable.

## References

1. Brown, J. S., Collins, A., and Duguid, P. Situated cognition and the culture of learning. *Educational Researcher* **18**, 32–42 (1989).
2. Kneebone, R. Evaluating clinical simulations for learning procedural skills: A theory-base approach. *Academic Medicine* **80**, 549–552 (2005).
3. Bradley, P. and Postlethwaite, K. Simulation in clinical learning. *Medical Education* **37**, (Suppl. 1), 1–5 (2003).
4. Daley, B. J. Learning and professional practice: A study of four professions. *Adult Education Quarterly* **52**, 39–54 (2001).
5. Lave, J. and Wenger, E. *Situated learning: Legitimate peripheral participation.* Cambridge University Press, New York, 1991.
6. Kneebone, R. L., Scott, W., Darzi, A., and Horrocks, M. Simulation and clinical practice: Strengthening the relationship. *Medical Education* **38**, 1095–1102 (2004).
7. Vozenilek, J., Wang, E., Kharasch, M., Anderson, B., and Kalaria, A. Simulation-based morbidity and mortality conference: New technologies augmenting traditional case-based presentations. *Academic Emergency Medicine* **13**, 48–53 (2006).

# 61

# On the Road with the Simulator

Christopher J. Gallagher
Riva R. Akerman
Daniel Castillo
Christina M. Matadial
Ilya Shekhter

## 61.1 Road Trip

We, of the Jackson Memorial Hospital Patient Safety Center at the University of Miami in Miami Florida, U.S.A., did a road trip. Granted, unlike Jack Kerouac's Beat Generation classic *On the Road*, we only drove from downtown Miami to Miami Beach (10 minutes, even with traffic). But it was still unforgettable.

What follows is a little how-to and lessons-learned when you take your simulator Dog and Pony Show on the road. Rather than slather you with bland generalities, we'll tell you exactly what we did for a specific trip so you can "feel the heat" like we did.

## 61.2 Planning

Before you embark on your own simulator road trip, you have to decide what the *thing* is, i.e., what are your goals and objectives.

*Why are you doing this simulation?*

- Train emergency medical technicians how to handle intubating in a Lear jet converted into a flying ambulance?
- Impress high school students with the "ooh aah" of medicine?
- Certify a few office worker in Basic Life Support, and use of the Automatic External Defibrillator?

*Who is your audience?*

- A group large enough to fill up all the seats in the American Airlines Arena? Or, a couple of people sitting around in a circle, close enough so you can smell the coffee on their breath?
- Highly skilled and experienced technicians, who need to adjust to the cramped quarters found in the Lear jet/ambulance? Or, are these beginning EMTs who will have to learn the basics first, and then make the jump to the Lear jet/ambulance scene?
- Honors high school students with high motivation to enter and conquer the Ivy League? Or, is this a remedial class where students have academic trouble and need a jolt of "real world relevance" to get back to their studies?
- Businees executives (MBAs and CPAs) with high educational backgrounds? Or, is this a group of hands-on factory workers with more calloused hands than college degrees?

*What is your audience expecting to take away from the simulation?*

- Skill in handling one specific airway problem in one specific setting?
- An impression of a clinical emergency?
- The confidence to push "hard and fast" on someone's chest when death stalks a coworker?
- The knowledge and training to jump into action when coming across a lifeless, purple human being sprawled

across the factory floor. Tear the shirt off, put those pads on, press those buttons 1, 2, 3, and pop the life right back into someone whose time has just plain not come yet, not if I have anything to do about it. Then afterward, the adrenaline shakes hit you, and you can't believe you actually did it, and you watch them ride off in the ambulance and still you can't believe you actually did it. But you did. Those guys with the mannequin simulator thing showed you how to do it and, if you didn't do it just like they said you should, when Joe was down and looked to all the world to be dead. But you did it.

# 61.3 Our Experience

The American Academy of Allergy, Asthma, and Immunology contacted us about 6 months before their annual meeting. Right from the start, the key question we posed was, "What do you want from this?" Forget which simulator, who will provide staffing, how many, how long, and all those details. What is it you want to take away from this? Why are you wanting us to show up at your meeting?

"Every year, in allergists' offices across the country, something like five or so people die from anaphylactic reactions," the allergist said. "We need to do a drill on reacting to an anaphylactic reaction in an office setting. We need to drill how to keep someone alive long enough for the emergency medical technicians to show up."

"Keep them alive until the cavalry arrive?"

"That's it."

"Got it."

We had the *thing*. We had identified the *objective*.

## 61.3.1 The Scenario

Now that we have the *thing*, we have to weave "a story to fit the need." That is, we had to invent the software (the thinking) that we would program into the hardware (the equipment). As in the **Planning** section, I'll first demonstrate how you might create a scenario to fit "other conceivable road trips" before describing what we did on our "allergy road trip."

To review, the other road trips were:

1. Training EMTs to intubate aboard a Lear jet.
2. Showing high school students a clinical emergency.
3. Training office workers how to do BLS and use the AED.

**The Lear jet:** A 35-year-old man with idiopathic cardiomyopathy requires air transport from his small town to a university center to receive a heart transplant. He develops progressive shortness of breath while en route, and his saturation drops to 81%, even on a 100% rebreather and oxygen blowing at gale force across his face. He needs an endotracheal tube. And there

you are, doubled over in a low-ceilinged Lear jet, barely able to stand up, let alone get into a "normal intubating position" like you have in an operating room. The scenario emphasizes the unique problems specific to the EMT's needs.

**The high school:** "Oh, my chest hurts!" the mannequin says through speakers. You tell the students in the auditorium, "You are at a basketball game and one of the coaches grabs his chest. What's your best move right now?" You anticipate the students will say, "Dial 911." "OK, you called 911 and the EMTs took the coach to the hospital. Now YOU are the doctor in the emergency room. What will you do with this man? What do doctors do when faced with a heart attack?" The scenario is broad enough to get the students thinking, and with the help of the mannequin's speaker adds a touch of drama.

**An office:** After a little introduction on how BLS works, with emphasis on the forceful and rapid chest compressions you need to do, you simply say, "This man is not responding at all and there is no sign of life. Show me you know how to do chest compressions correctly." After they demonstrate that, you pull out the Automatic External Defibrillator, demonstrate its use, and say, "Now, each of you, hook this up, go through the three steps of use, and prove to me that you know how to apply and use this." The scenario focuses on drilling the basics. Fast and hard on chest compressions. Fast and decisive when it comes to applying the defibrillator.

**Allergist's office:** A patient in an allergy office gets anaphylaxis. Hmmm. How could we make this a little "jazzier"? We managed. (Details in the thrilling conclusion of the chapter.)

## 61.3.2 Staffing

We need thinking, instructing, sentient creatures to make the simulation scene come to life. So we plucked the best and brightest from our attending staff.

We produced the simulations on a Saturday and Sunday. Each day, we had three attendings. These attendings were familiar with simulation scenarios, the computer that ran the simulators, "faking it" when glitches happened so that the "audience" never perceive the glitches (this is *the* attribute most essential to any simulator instructor), and debriefing people after the scenario. Each of these instructors had worked with these simulators before and had worked with residents or medical students in a simulator setting. Experience counts.

Why three attendings? We had two simulation rooms. Wait a minute, do the math. Wait, that's one extra attending! Exactly! You always have a spare tire in your *car*, so bring an extra attending to your *simulation*. Breaks, lunch, someone to help troubleshoot, someone to help act out an extra part. Always a good idea to have a cushion in your "instructor pool."

Who else? We brought along nine residents each day as well. Why all the extra people? Two reasons:

1. Extra characters to act in the simulation rooms.
2. To staff the intubation and airway stations.

To augment our two simulation rooms, we also created a station dedicated to intubation and airway management. Most of our target audience consisted of allergists used to working in an office environment. Airway emergencies were rare but catastrophic, so the allergists needed practice with this skill.

We brought along six intubation mannequins, endotracheal tubes, laryngoscopes, laryngeal mask airways, ambubags, and masks. The "airway residents" stayed out by the mannequins, demonstrating and guiding the allergists through the different techniques of supporting ventilation.

Just as we looked for *experience* in our attending staff, we looked for *enthusiasm* in our resident staff. Putting a cattle call out for volunteers, we snagged the best and the brightest residents. These were residents who had already demonstrated a willingness to learn, to teach their fellow residents, to help out medical students. Then, to make sure their enthusiasm stayed at a fever pitch, we tossed some money at them. We didn't tell them we that we were paying the attendings *as well*, and that we were paying the attendings *more*, but such is the way of the world.

### 61.3.3 Equipment

We decided to bring along our two Laerdal adult SimMan mannequins. Laerdal simulators are not prohibitively expensive, not too clunky, and have relatively few connections. Programing "on the fly" is a simple, straightforward process.

For the airway stations, we made sure we had the same thing at each station:

- mannequin
- Macintosh #3 laryngoscope (the easiest laryngoscope for a novice visualizing the adult SimMan airway)
- 7.0 endotracheal tube
- intubating stylet
- ambubag and mask
- #3 and #4 laryngeal mask airway

### 61.3.4 Schlepping

Here's a take-home lesson from the "boy, will we do it differently next time" school of thought: So, you rent a little van, throw the stuff in the back, drive to the place, and your set, right?

Wrong!

If there is one thing worth taking out of this chapter, it is this – *Get special wheeled boxes that are hard on the outside padded on the inside to move your mannequins around. This investment is well worth the price*[1].

Naïve to the gills, we found ourselves struggling like fools to load the two mannequins on to an "ambulance Gurney with easy fold-up legs." So then, we're limping down to the truck, the expensive Laerdals looking like the two-man luge team at the Olympics, ready to fall over at any minute. Once

we get to the truck, we discover that the "ambulance Gurney with the easy fold-up legs" is anything but "easy to fold up." Luckily, no one chopped their fingers off, but it was a close-run thing. Eventually, we manhandled the beast into the truck, our lumbar discs popping like champagne corks on New Year's Eve.

Oy, never again.

Especially, if you're thinking of *shipping* your mannequins, you will definitely need specialized crates. Don't just show up at Mail Boxes R Us, tell them to put the mannequin in a big box with lots of Styrofoam peanuts, and hope everything arrives intact!

Is it worth getting insurance? Yes indeed. Say that truck was involved in an accident or your crates were lost. That's no time to start thinking about insurance!

Plan your move in every detail!

### 61.3.5 Setting Up

The biggest part of setting up was to *take nothing for granted*.

"Where are the electric outlets?" "Oh, you need those?"

"Who brought the extension cords? The only outlets are all the way over there!"

Make sure you have electric outlets, and lots of them.

"This laryngoscope doesn't light up, do we have extra batteries and replacement bulbs?" "Oh, yeah."

Make sure you have extras of everything, and that you think through what will happen with wear, tear, and use.

"This endotracheal tube's cuff is broken."

"Someone walked off with the laryngeal mask airway."

Have extras, extras, extras.

One thing we really blew was not having informational sheets around to answer the deadly FAQs – frequently asked questions.

"How do I get in touch with you?"
- Have plenty of business cards.

"Where do I buy these laryngeal mask airways?"
- They asked us that a million times and we always came up with a limp, "Google it, you'll be sure and find them somewhere."

"Do you have a press release that we can use?"
- We neglected to have a little blurb describing what we were all about. That's the kind of thing the press will ask for and you should have available.

So, given all the missteps and booboos, our final setup looked like:

In the simulation room, we had a Laerdal mannequin, attached computers and pneumatics, resuscitative equipment, a little office furniture to "give the feel" of an office, an IV pole, IV with attached tubing, and syringes of "pretend" epinephrine and atropine.

To add to the fun and give it a "Miami Beach feel," we spruced up things a bit. (If you can't have fun with this, I mean, come on.)

- Loud T-shirt and Miami Beach shorts on the mannequin, plus a floppy hat, sunglasses, and water wings.
- Copies of Soap Opera Digest and Us magazine on a coffee table.
- On a little round table, an open can of Toffee peanuts. What? Peanuts in an allergist's office? Peanuts? The allergic equivalent of a "Weapon of Mass Destruction?" More on this little touch later.

### 61.3.6  OK, So What Happened on the Big Day?

"Come on in doctors." (We had people come in three or four at a time.)

Upon arrival, the allergists saw a typical Miami Beach tourist, complete with beach garb and glasses, lying on an examining bed. In the corner, an instructor sat at another table and worked the simulator laptop. A screen on that same table faced the doctors. On the screen, an EKG, oxygen saturation, and blood pressure appeared. The pulse oximeter beeped. At first, all vital signs were normal.

Lying near the patient's head were airway instruments. Hanging on the IV pole, a bag of normal saline. Two syringes of epinephrine, one had a concentration of 10 mcg/cc, another 100 mcg/cc, were nearby; also available was one syringe of atropine, in a concentration of 1 mg/cc.

*This was much more equipment than would be set up in a typical allergist's office, and the patient already hooked up to all the monitors was also a stretch ("not too realistic"). But we wanted "a lot to happen fast" and felt it better to have things already "ready to leap into action" rather than spend half their time "getting stuff." This was a meeting with 7000 people in attendance, and something like 200 people ended up going through our simulation rooms, so we had to move fast. Sacrificing a little "reality" and taking a little "artistic license," we were able to "make the point."*

"I have good news and bad news for you, doctors. The good news is, you have a brand-new nurse who just came to us from the Intensive Care Unit at a tertiary facility."

We pointed to the resident in the room.

"This nurse can smell trouble, so went ahead and laid out everything you might need in case this patient deteriorates and goes south in a hurry."

"The bad news is, you also have a new receptionist, and she wants to make a good impression on the patients. So she put out a bunch of peanuts in the waiting room." We pointed to the can of Toffee peanuts. "And this patient just had some."

"I don't feel so good," the mannequin spoke through its speakers.

"What do you want to do?," we would say, then step back.

In a few seconds, the saturation dropped, blood pressure dropped, and the patient stopped talking. Heart rate went from 80 to 150 beats per minute and the pulses became weak.

*If you want the simulation to be completely realistic, you take your time and slowly evolve things. But if you need to "get a move on," then GO FOR IT! We lived in the GO FOR IT camp.*

At this point, we stepped back and looked for the allergists to do three crucial things:

- call 911 (as they would do in "real life")
- secure the airway (if they couldn't intubate, then place a laryngeal mask airway or mask ventilate)
- give epinephrine (allergists are used to giving epinephrine intramuscular, but we had this patient have a complete cardiovascular collapse, and intramuscular is not such a desirable method of drug administration. You need to give epinephrine intravenous, something outside the allergists' comfort zone. They were blessed with an experienced, ICU-trained nurse who would instantly provide them with IV access if they merely said the word.)

*Now the fun began. The following are the "twists" we threw in to complicate the scenario, rattle them, and generally shift into "chaos" mode.*

"Here, here's the laryngoscope!"

They open the laryngoscope, no light.

"This doesn't work!"

"Oh my Gosh, I forgot to check it! I think these batteries predate ether!"

*Lesson learned – You need to practice doing codes in your office. Part of that practice includes looking over the equipment and making sure all of it works.*

"Who's in charge here? What should I do? Are you in charge?"

*Playing the part of an agitated nurse, we hammer home the idea of Crisis Resource Management. In an emergency, some ONE has to take charge, you don't just mill around and "rule by committee."*

"Which epinephrine do I give, 10 mcgs per cc or 100, which should I give?"

*When you're used to giving stuff IM, you can often forget the doses needed for IV use.*

"This endotracheal tube is in the stomach, what are we going to do now?"

*Allergists are not airway experts, and few could intubate well. They needed to learn "easier" ways, such as the laryngeal mask airway and mask ventilation. They would often "fixate" on intubating, and forget the mantra "you don't have to intubate, but you do have to ventilate."*

"Hey, I've got two crates of medical supplies out on the loading dock, where do I put this?"

*Extra characters throwing in distractions and irrelevancies always add to the general chaotic mix. Once I came in the room as a "delivery boy" and fainted right on the allergists' feet.*

"Did you call 911? You? Ohmigosh, what's the number to 911?"

*Once the juice got turned up and everyone was shouting and floundering, people would forget the fundamental principle, Call for help.*

### 61.3.7 Evolution

Flexibility and backup plans are the keys to any "simulation foray." Computers can freeze up (ours did, we rebooted, lost our preplanned scenarios, and had to enter everything *a la carte*), participants can sniff at our scenario and walk out (happened, we just kept rolling), and wild unexpected things can come up (one guy leaned against the wall between the simulation rooms, the wall caved in, and the guy and the wall came crashing in like the Walls of Jericho – we all laughed).

But no matter what happens, the show goes on!

Another important thing is the "evolution of the lesson." It was apparent after just a few scenarios that the allergists "best bet" for airway management was the laryngeal mask airway. So we spent a lot of time talking about and demonstrating this useful tool. We'd have them go out to the airway station and spend a lot of time getting the placement of this down on the intubating dummies.

### 61.3.8 Response

Serving up simulations is like serving up ice cream. Everybody loves you. And so it was with us. The allergists believed that simulations helped them out a great deal. Almost all of them were making plans to increase their "code drills" and lay in a stock of laryngeal mask airways. After the training, they all felt more comfortable with the "disaster waiting (lying) in the wings," an anaphylactic reaction in their office.

Will this make a difference? Will someone live, who would otherwise have died, because of our little simulation effort?

We have no way of knowing, but we sure hope so. Perhaps some future e-mail or phone call will contain the magic words, "Thanks to you, I saved someone."

Let's hope so.

## 61.4 Conclusion

We took home some great lessons:

– Next time, have special crates to move your stuff.
– Extra people, extra equipment, always good ideas.
– Simplify, simplify, glitches happen and you might be left with just your wits. But that should be enough.
– Juice it up, make it fun and exciting.
– Roll with the punches. Let your lesson evolve as you see what your students need.

We were pretty proud of ourselves, as we were on the road, and managed to survive, learn, and thrive with every problem.

## Endnote

1. For example, see www.justcases.com.

# 62

# Mobile *"In Situ"* Simulation Crisis Resource Management Training

Marcus Rall
Eric Stricker
Silke Reddersen
Jörg Zieger
Peter Dieckmann

## 62.1 The Concept of Mobile *In situ* Simulation Training

Mobile *in situ* (term coined by David Gaba, personal communication) simulator training is a way to bring the training to participants' workplaces as opposed to bring participants into simulator centers. Thus, the simulator training comes closer to the clinical arena. The idea is to provide training exactly in the location where people normally work ("training where you work" concept), using most of their original equipment used in routine work (*in situ* training). This setting also allows for training of complete teams where team members work together on a regular basis (the "training together–working together" concept). We focus our simulator training on Crisis Resource Management (CRM), a term defined by David Gaba as: "The ability to translate the knowledge of what needs to be done into effective team activity in the complex and ill-structured real world of the OR (patient care)." Crisis Resource Management is most relevant to patient safety and amongst the most beneficial uses of simulation as it uniquely addresses attitudes, as well as knowledge and nontechnical skills of the participants.

Questioning, changing, and even reinforcing attitudes needs insight, tact, and encouragement. Adults cannot be requested to change, but will only change if they think it to be beneficial to *their* interests. In order to use these benefits of the CRM-type courses [1, 2], the live video transmission of the action (we use a quad split screen with three perspectives of the team in action plus the vital-signs monitor along with a mixed sound channel from an ambient and several clip microphones) to observing participants followed by video-assisted debriefings are essential. Live observation amplifies the potential for learning to a larger number than can be accommodated within any one event. Live observation by those expected to perform the scenario soon afterward enhances their self-reflective thinking and offers brand-new insights similar to that of video debriefing with those who just completed the scenario. Debriefing is important in a simulator course to ensure self-reflected learning (see Chapter 67). A typical course format in these mobile *in situ* simulations is mainly that of the CRM course described in Table 62.1.

Our group did courses in many environments (see following example section). For all *in situ* training sessions, the SimMan (Laerdal Medical) simulator was used. For the purpose of

**TABLE 62.1**   Key elements of prototypical realistic Crisis Resource Management (CRM) simulation course

(1)  Small groups (4–10).

(2)  High instructor-trainee ratio (1:2 to 1:4).

(3)  Duration approx. one full day.

(4)  Realistic clinical environment in simulation area.

(5)  All tasks have to be performed practically (e.g., opening ampoules, diluting drugs, preparing infusion devices, etc.).

(6)  The simulator is controlled from outside the simulation room (control room simulation).

(7)  Sessions are videotaped for debriefing. Whenever possible, audio/video of the sessions is transmitted live to the remaining nonactive participants of the training group in the debriefing area (stimulating self-reflection).

(8)  After each session, a video-assisted debriefing takes place (time ratio scenario-debriefing 1:1 to 1:2).

(9)  The CRM scenarios are carefully designed to address certain critical elements and learning objectives.

(10) The atmosphere is blame free and error friendly – analysis with "why" questions.

(11) The focus is on CRM, not on clinical/technical issues.

(12) The format focuses on facilitation and self-reflection, not instruction by the instructors (mentors) [1, 3–5].

our trainings, this simulator proved to be robust and durable, but especially the nonmodel-based control of the simulator allows us to simulate and modify the reactions of the simulated

"patient" exactly according to our didactical goals and not to any given physiological model. We believe that "realistic" models are not always beneficial to achieve the best learning effect. Slowing down a process or adjusting the reactions in order to allow trainees time to rethink and treat the patient might work better for some educational goals. Another aspect is that this simulator's controlling interface is very easy to learn for new or assistant Instructors.

*In situ* mobile simulation can be done wherever clinical treatment is provided. It is especially suitable for environments that offer difficult working conditions, for example, in terms of space constrictions and restrictions and/or noise, etc. (e.g., an ambulance, a catheterization lab, a dentist's chair or even an ICU ambulance, or a small aircraft); see the following Examples section. In order to provide beneficial training, the salient characteristics of these surroundings need to be understood by performing a thorough needs analysis. Furthermore, it is important to explicitly agree on and define the goals and circumstances of the mobile simulation course with the participants and their institutions [6].

## 62.2  Examples

- In cooperation with the German Air Rescue command, we provided a simulator-based course within medical-evacuation helicopters (Figures 62.1–62.5) and fixed wing

**FIGURE 62.1**   External simulator control of mannequin inside helicopter.

**FIGURE 62.2** The patient simulator within a real medical evacuation helicopter. Note the lack of space around and above the patient compared to typical terrestrial ambulance, as well as the nonpermanent mounting of the cameras and microphone.

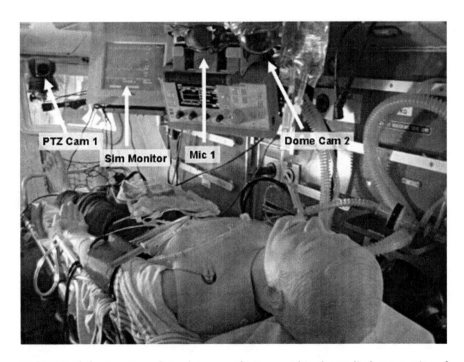

**FIGURE 62.3** Location of simulator-specific items within the medical care portion of a real medical evacuation helicopter.

**FIGURE 62.4**    Close-up view of a remotely controlled camera mounted within a real medical evacuation helicopter.

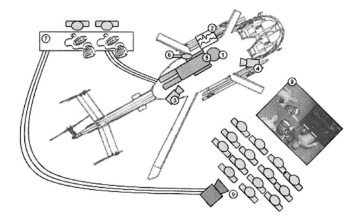

**FIGURE 62.5**    Schematic of patient simulator within a real medical evacuation helicopter, and the supporting control and observation/debriefing areas: 1 – mannequin; 2 – simulator vital sign display; 3, 4, 5 – cameras; 6, 7 – microphones; 8 – projection screen; 9 – video projector. Other than the helicopter, all other items are required for most *in situ* simulation training sessions.

ambulance aircrafts (Figures 62.6–62.10). In both environments, the constricted space is one of the challenges – both as a working and as a training environment. The other salient characteristic of providing clinical treatment is the challenging environmental factors, especially very high amplitude sounds. With the unavoidable vehicle noise, both clinical treatment (e.g., auscultation) and

resource management [7, 8] (e.g., communication) are hampered.

• Dentists do not experience emergencies very often. For this reason, it is very beneficial to use the advantages of simulation's ability to "schedule disasters" to provide experiences with managing emergency cases in their typically independent clinical environments (Figures 62.11–62.13).

**FIGURE 62.6**    Outside view of jet medical transport with simulator control area on the right and the observation/debriefing area on the left.

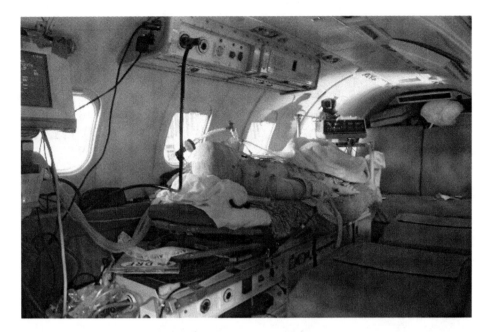

**FIGURE 62.7** Mannequin within jet medical transport, showing slightly more room to maneuver than within the helicopter.

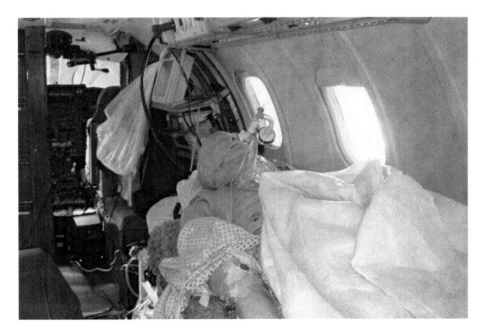

**FIGURE 62.8** Mannequin within jet medical transport, reverse view of that in Figure 62.7.

**FIGURE 62.9** Simulation control area just outside the jet medial transport. Note the extensive amount of cables that must all be fully accounted for and fully functional, yet easily reconfigured for rapid deployment in austere environments.

**FIGURE 62.10** Observation/debriefing area next to jet medical transport: all the functionality of any typical simulation observation/debriefing area, plus the added requirement for rapid deployment in austere environments.

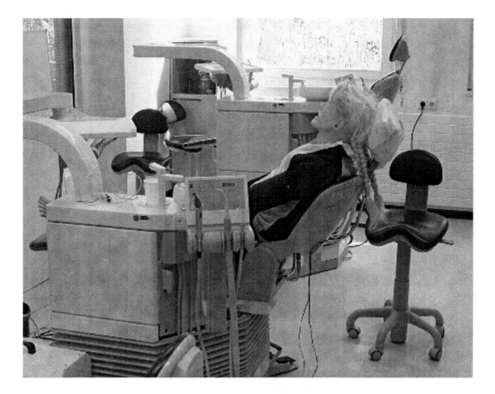

**FIGURE 62.11** Typical dentist's oral care environment.

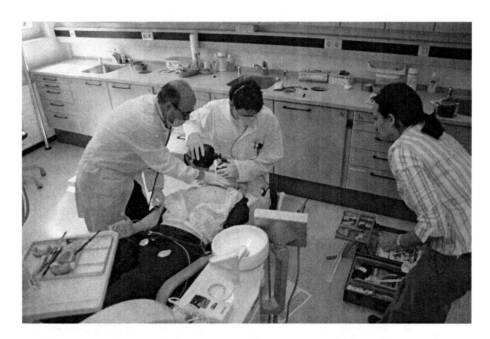

**FIGURE 62.12** Patient receiving prompt resuscitation in dental chair.

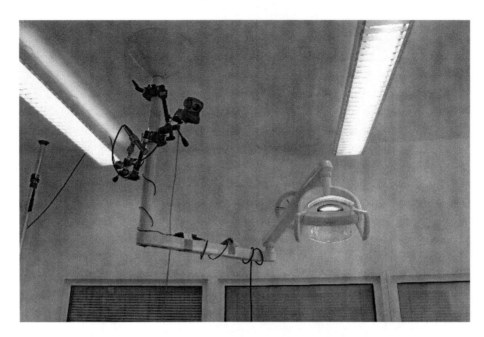

**FIGURE 62.13** Typical dental overhead lamps make good platforms to mount cameras and microphones.

- Training of the Italian White Cross team (the White Cross is the major prehospital emergency care provider in South Tirol (Alto Adige, Italy)) inside ambulances (Figures 62.14–62.17) and in prehospital settings in an apartment (Figures 62.18–62.19).

- Training of the emergency response team inside a cardiac catheterization lab (Figures 62.20–62.22).
- Training of anesthesia teams inside their operating room (Figure 62.23).

**FIGURE 62.14** Outside view of ambulance with simulator control area on the foreground, and out of the way of the clinical areas around the back of the vehicle.

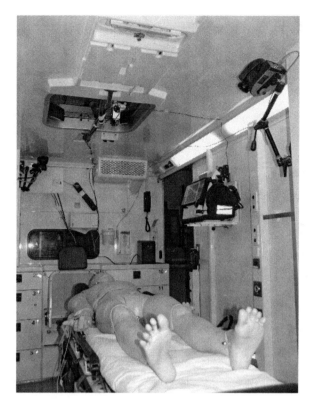

**FIGURE 62.15** Mannequin within ambulance, showing full-height room above the patient.

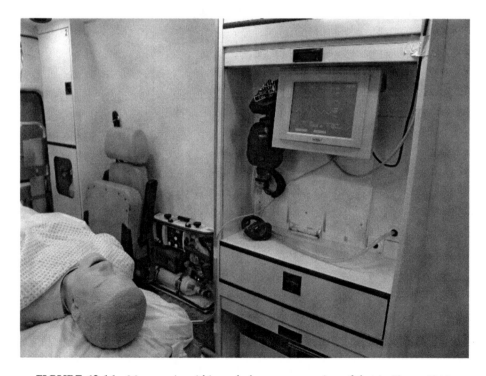

**FIGURE 62.16** Mannequin within ambulance, reverse view of that in Figure 62.12.

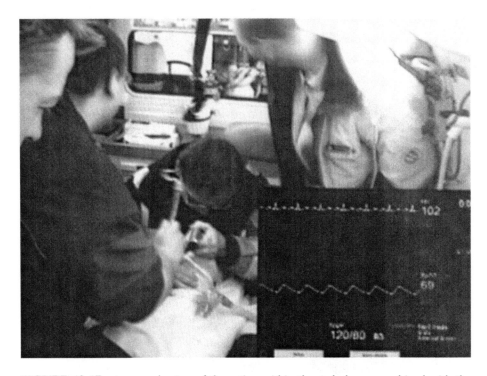

**FIGURE 62.17** A camera's view of the action within the ambulance, combined with the patient's vital signs image.

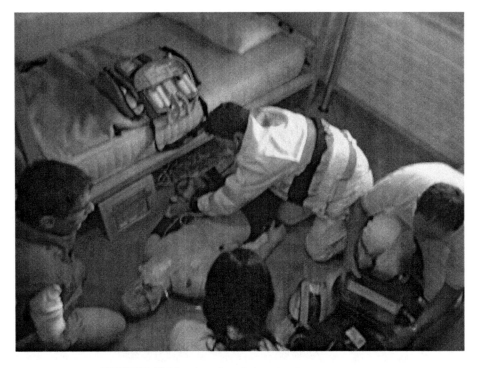

**FIGURE 62.18** A prehospital setting in an apartment.

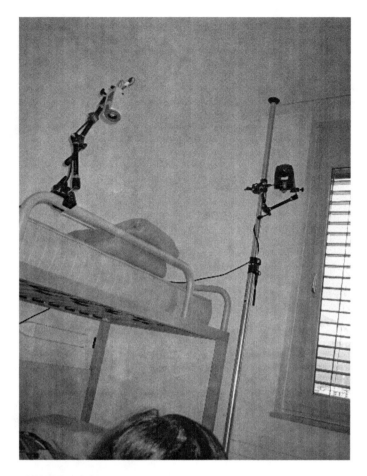

**FIGURE 62.19**  Cameras and microphone mounted above the action in an apartment.

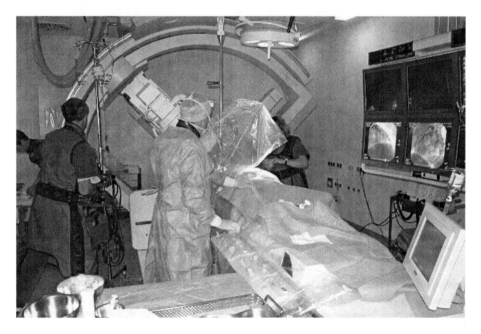

**FIGURE 62.20**  Catheterization laboratory.

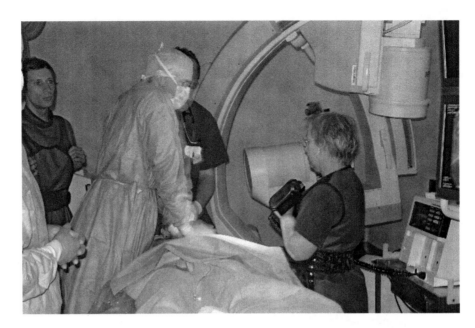

**FIGURE 62.21**   CPR on bed in Catheterization Laboratory.

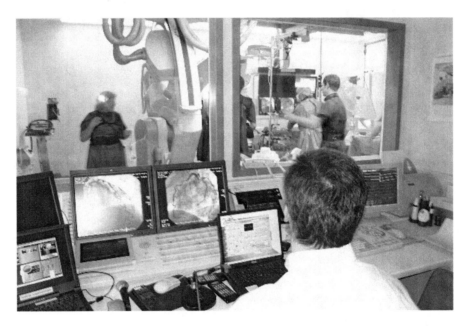

**FIGURE 62.22**   Simulation control area for Catheterization Laboratory.

# 62.3  What is Needed

## 62.3.1  The Technical Side

Setting up a mobile training area requires careful design of the setup of the mannequin, cameras, and microphones. We work with 2–3 cameras in order to assure at least one camera view is not blocked and all critical actions are collected for live display and subsequent replay. Especially for the first time when training in an unknown environment, it is necessary to experiment until suitable positions for cameras and microphones are found (see for example Figure 62.3). For example, it might happen that participants in the scenario bump into your fixed microphone and produce lots of disturbing noise that you haven't foreseen. We recommend to always check sound and

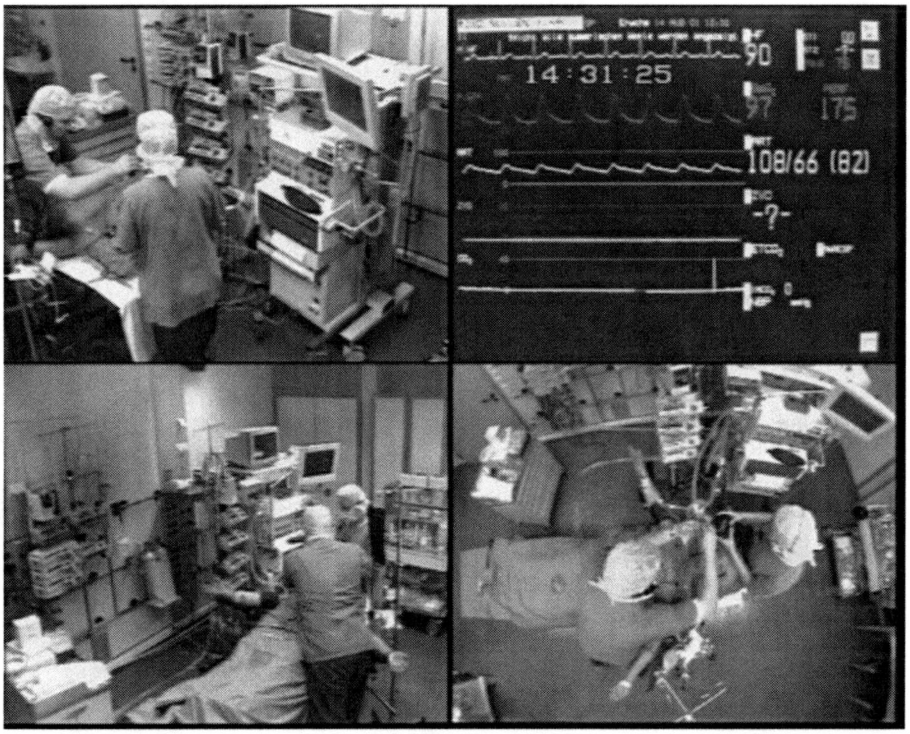

**FIGURE 62.23** Operating Room showing three camera views and the converted clinical monitor image combined into a single video signal by using a quad splitter.

video signals with a full team in the action environment in order to adjust perspectives, zoom range, and positions of cameras as well as sound levels. We use three cameras and a so-called "quad split" (an electronic device that allows for combining four video channels into one, i.e., a two-by-two form of "picture within a picture") to combine the three different camera perspectives as well as the vital signs that are captured from the vital-signs monitor (Figure 62.23). In order to include the vital-signs monitor in this newly created four-in-one video signal, it is necessary first to convert the computer format patient monitor video signal into a video signal format identical to that of the cameras. The great advantage of this four-into-one approach is that all four channels are forever time-locked together, and can be recorded and replayed as one homogeneous and time-locked signal. The great disadvantage of this four-into-one approach is that each of the four signals has only one-fourth the resolution, hence some critical fine details in the four original signals will be lost. So long as details crucial to evaluation are preserved, this is an excellent approach to reduce recording and reproduction devices by three-fourths.

The basic setup of all mobile trainings consists of the mannequin (1) and its patient monitor (2) (see diagram in Figure 62.5). Most simulators need a control computer that should best be placed outside the simulation area. It is important that all nonclinical sights and sounds be as minimally intrusive as possible in order so that the training experience is as valid as possible. Especially, noisy air compressors needed to power the simulation mannequins should be placed so that the noise does not disturb the training. From our experience, good camera positions are at the feet of the patient (3), the head (4), and the so-called "god's view" camera above the simulator mannequin (5).

In general, the sound quality is even more important than pictures. We work with a wired microphone (6) and wireless microphones (7) to collect the spoken words of the participants and the background, environmental sounds. The highest possible audio quality is of utmost importance for good debriefings; money should never be saved with cheap microphones and transmitters. The advantage of a good audio and video collection and transmission system is that from the control area you have an idea of not only what the active participants are doing, but also what can be seen and heard by the nonactive participants in the observing/debriefing area. We use a computer-based system for audio/video recording, time marker insertion, and replay. To execute training sessions without a hitch, we isolate these heavy computational tasks in a computer separate from the one that controls the simulator. In addition to preventing overloading a single computer, the computer devoted to video (located in the control area) also sends the live sights and sounds to the display in the observing/debriefing area.

Audio and video signals are transmitted live to the control area and the observing/debriefing area [6, 7]. In the control area, the incoming audio/video data are mixed and recorded. The three video streams from the different cameras and the video stream coming from the patient vital sign VGA scan converter are compressed into a single quad split video stream. During the session, this combined video stream can be seen live in the observing/debriefing room. The recorded video can be shown afterward during the debriefing. We use a digital video system that allows the time-shifted insertion of markers. When we hit the "mark" button, the mark is set 8 seconds earlier in order to be able to show what we saw when we hit the button minus delay in thinking and reaction time – this is *very* useful when you want to show focused short video clips.

### 62.3.2  The Logistical Side

We use 19-inch transport cases for all the audio/video equipment and the technical equipment (Figure 62.24). With a selection of dedicated cable setups, it is very easy to connect all

**FIGURE 62.25** Adaptable crew and portable patients make for good *in situ* simulation training.

**FIGURE 62.24** Audio/video devices mounted in protective yet portable enclosures.

components rapidly. These cases allow for easy transportation of the entire simulator and supporting equipment and are sufficiently robust to protect the expensive gear. Fortunately, the mannequins themselves are easily carried and placed where desired (Figure 62.25).

## 62.4  Advantages of *In situ* Training

### 62.4.1  For the Trainees

*In situ* simulation can be very relevant for the participants and their organization as it is possible to tailor the training specifically to their needs.

- *In situ* training can help in forming very influential learning episodes that might actually succeed in triggering changes in organizational practice to enhance patient safety. If your colleagues all have experienced the same training (*in situ* training offers the possibilities to teach a critical mass of the employees of a department in a short period), it is much more likely that you are able to apply in the real world what you learned in the simulator course.

- Considering travel expenses and time lost to disruption, it might be cheaper to bring a simulator team (three persons) and the simulator to the organization than it is to bring in a complete team or even many members of a big department to a simulator center.
- Problems already experienced during daily routine or critical incidents can be reenacted, thus analyzed and even solved.
- Problems in the local system of care and the processes may show up during the simulation of emergencies, thus the "*in situ*" training can also serve as a very sharp "hot" test for the real clinical environment.
- Problems that nobody thought of so far can be discovered during analysis of scenarios and can be corrected before they can have negative effects [9].
- Participants are more familiar with their own real clinical environment as compared to center-based courses. Participants know where to find their gear and know how to use it. They know the ventilator, how to get help, whom to call for help, etc.
- During debriefing, the real team is able to reflect on their actual clinical practice. From strengths and weaknesses that are discovered during this process, the people who are involved in these processes can learn directly. They could even decide to change and improve their practice.
- Extending the use of *in situ* simulation might spare each and every different department within one institution from having to buy their own simulator, additional equipment, building facilities, and expensively building up a simulator team that is able to run high-quality courses.

### 62.4.2 For the Simulator Instructor Team

- It is very interesting, while challenging, to see how others work and how their performance regarding patient safety can be improved.
- The scope of the simulation training is widened to include the local system and its organizational factors, as these will always come up during local *in situ* trainings. This makes it more interesting for the simulation team, but also needs a deeper understanding for the systems safety approach on the side of the Instructors (you move from a skill trainer in former mega-code trainings to a team trainer in classical CRM courses to a departmental structures and procedures adviser!).
- It is a very creative work to always adjust and adapt to new locations and domains.
- Lessons learned at one institution can be transferred to the next training facility.

## 62.5 Disadvantages and Problems of *In situ* Training

### 62.5.1 Problems for the Trainees/Local Facility

- Many problems that are rooted in the team's history can hinder effective training. For example, conflicts within the team might make it impossible to openly discuss aspects that can be improved in the organization's clinical practice – especially when the aspects are hidden in the group dynamics.
- Working with participants in their own environment might mean that it is harder for them to concentrate on the course and for the organization to really free them up. Often, they will have to carry their pagers (just in case . . . ). It might take some negotiations ahead of the course to ensure that the participants can actually participate in the course and are not called to other duties during the course or its breaks. Departing and reentering participants also destroy the atmosphere for those who are staying.
- The rooms for the *in situ* trainings need to be available, which often means to block them for other clinical productive use. An operating room in which the simulator course takes place cannot be used to earn money. Often, the local training rooms are restricted in size and functionality, which can affect the training negatively (e.g., long distance between scenario room and debriefing room; small or not available dedicated control area).
- If the training group from one institution is a large fraction of a care team, then this loss of personnel might result in conflicts with patient care.

### 62.5.2 Problems for the Instructor Team

- In mobile simulation endeavors, it is the simulator team who needs to rapidly familiarize themselves with an unknown environment, culture, and resources.
- Most clinical teams quickly develop a specific language ("jargon") that is not easily understood by other persons. The instructors need to be sensitive for this "slang" especially during debriefing. While a sentence like "Hey, we should call the second floor" might be perfectly clear for the local clinical team ("this means calling the resuscitation team"), the simulator instructor team might not have a clue about the "hidden" meaning within this message.
- During mobile debriefing, it is even more important than in center-based training to ask participants what kind of reality they constructed out of the scenario and not to rely on observations only.
- Facilitating the reflection about what can be learned from the scenario gains even more weight during mobile simulations. Working "mobile" means less control of

the training infrastructure for the simulator team in terms of availability of training areas (helicopter, airplane, ambulance!), clinical equipment, environmental conditions (cold hangars), or catering.

- During mobile simulation, a member of the simulator team cannot work with a team member in the role of a clinician within the scenario. Thus, the team has less direct control about the scenario (checking actions, helping participants).
- It is essential that everybody in the team is cross trained on all the technical equipment to avoid canceling a course due to one missing person.
- Sometimes persons who supervise the participants (the head of the department) might show up during the course, influencing the group dynamics. They should be gently but firmly asked to leave as they might unwillingly moderate ("hinder") the training atmosphere by negating the blame-free atmosphere.
- As the challenges for instructors are even larger during mobile simulations, we strongly recommend the education of new instructors with an instructor course (see Chapters 65, 66, and 67).

Examples of Centers who run several International Instructor Courses:

- Stanford, California, U.S.A.
- Harvard, Boston, U.S.A.
- DIMS, Copenhagen, Denmark
- TuPASS, Tuebingen, Germany (see Chapter 59 about this concept).

## 62.6 Mobile *In situ* Simulation Training Pearls

### 62.6.1 Organizational Pearls

- Have a single contact person at the training site that is responsible and able to reliably organize the necessary things. ("Your" training will be rated badly if there is nothing to eat or no coffee in the morning.)
- Get detailed information about the logistics (room size/shape/location, light, heating, power, noise, restrooms, parking, unloading) of the training site.
- Inform yourself thoroughly about special conditions in the field of training (your instructors must know the "secrets" of flight medicine if they train air rescue teams). Also know the complete set of available equipment (e.g., what kind of difficult airway equipment is stored in the training site). Gaps in this knowledge always lead to big problems during debriefings.
- Perform a thorough needs analysis including as many professions and disciplines of the local team as possible in order to understand the salient characteristics and safety relevant particularities of the environment.
- Have a good replacement insurance policy for all your equipment (simulation and especially audio/video) including theft and damage by trainees and instructors. On-site liability insurance is recommended as your team might damage equipment or other structures in the remote site. Last but not least, if a real patient is harmed after your training session, will your training be blamed?

### 62.6.2 Technical Pearls

- A good idea is an additional loudspeaker with some special effects. For example, for simulator-based training in a helicopter or a jet airplane, the noise and vibration are salient characteristics and need to be included in the training for this reason. During training, influences on communication patterns and diagnostic steps (auscultation!) can be addressed. We replay previously recorded engine noises with heavy speakers placed on the bottom of the aircraft (12).
- It is recommended to use water proofing to protect all power and signal connectors from water damage. This could be helpful if the simulator gets wet. For example, if you do trauma training and the simulator is placed on a vacuum mattress that tends to form a nice swimming pool if fluid volume is administered forcefully and a vein is leaking ... we lost a mainboard inside the body with that.
- The best way to convert the patient monitor into a video stream is to use a device called a Scan Converter or Media Converter. (The most expensive is not necessarily the best for vital-sign signals! Check the quality of the ECG and $SpO_2$ curves before you buy anything.)
- The original patient monitor that comes with the simulator may be inappropriate for use in small, cramped spaces (see Figure 62.2). However, its vital-sign information can be reproduced on another display, one that is both better sized for small spaces and cheaper to replace if/when damaged from accidental impact (but do consider if you want to forego the touch screen functionality). Ideally, the vital signs would be displayed on a device that appears and functions identically to the real vital-signs monitor as used by the participants.
- As the instructors are not in the "simulation room," it is necessary to have an additional loudspeaker in there to convey their voice from a microphone in the control room [9]. This loudspeaker, located far away from the patient voice speaker, has been termed "the voice of god" by Gaba and Howard (not associated to any religious topic of course) and used in many centers around the world. This allows instructors to provide verbal commands (e.g., report on findings like skin color) without

their voices seeming to come from the patient (which is absolutely inappropriate in our view).

- Always have enough cables (always have one spare of each kind used and all at least 50% longer than you think!), connectors, gender changers, tape, etc., with you (Figure 62.9). A small tool kit with a soldering iron is vital!
- A very good sound system for recording and replay is more important than a good video system.

### 62.6.3 Personnel Pearls

- Our team must be ready to travel and carry and be flexible to adapt to changing and unexpected conditions (Figure 62.25).
- Instructors for mobile training sessions must even be more aware about group dynamics (e.g., group structure, norms, the culture in the group, hidden rules, coalitions, status, etc.) as they are dealing with existing teams (the instructors are the "aliens").
- Especially in mobile *in situ* training, all members of the actual team should be included (nurses, doctors, paramedics, technicians, etc.), if the maximum training effect is to be obtained. Otherwise, the training might produce resistance from some groups.

# 62.7 Potential Value of *In situ* Simulation Training

From an ergonomic point of view, the mobile simulation equipment can be used as a mobile test-bed for layouts and processes in newly built or recently reconstructed facilities. Likewise, the mobile simulation equipment can be used as a mobile test-bed for newly created clinical teams and refresher training for exiting clinical teams.

A further variant of mobile simulation is to conduct mobile instructor courses. This concept has the advantage that the instructors-to-be can work within their own, new simulation environment. They can practice the use of their technical equipment and simulated environment under the supervision and guidance of the instructor trainer team. Their decision makers can see the needs for necessary additions in technique (e.g., audio-visual) and other equipments (clinical, props, etc.) and then perhaps are more likely to finance these needs. The authors have produced successful mobile instructor courses (Netherlands, Italy). Of course, the challenge in doing so is tremendous, because the success of the Instructor course heavily depends on a functional Infrastructure for the course as such (organizational topics like coffee breaks, lunch, air conditioning, space, etc.) as well as the simulations to train the new instructors (simulator and props and audio-visual system). We

bring our own full equipment if we are in doubt about the new center's capabilities. In this case, you know what you have and there are fewer "surprises" (In any case, we recommend you bring a soldering iron with you . . . ).

# 62.8 Conclusion

Mobile "*in situ*" simulator-based training is a very promising concept that is feasible for many different acute care settings. Bringing simulation to the users can help to spread simulation more quickly and make it possible for learning in locations not easily reproduced in simulation centers. There are good reasons to assume such active educational effects will improve patient safety. These are very well worth the big efforts necessary to successfully manage these interesting endeavors.

# References

1. Rall, M. and Gaba, D. M. Patient simulators. In: Miller, R. D. (ed.), *Anaesthesia*, Elsevier, New York, 2005, pp. 3073–3103.
2. Dunn, W. F. (ed.), Simulators in Critical Care and Beyond. Society of Critical Care Medicine, Des Plaines, 2004.
3. Rall, M. and Dieckmann, P. Simulation and Patient Safety. The use of simulation to enhance patient safety on a system level. *Current Anaesthesia and Critical Care* 16, 273–281 (2005).
4. Howard, S. K., Gaba, D., Fish, K. J. et al. Anesthesia Crisis Resource Management Training: Teaching Anesthesiologists to Handle Critical Incidents. *Aviation, Space and Environmental Medicine.* 63(9), 763–770 (2001).
5. Gaba, D. M., Howard, S. K., Fish, K. J. et al. Simulation based training in anesthesia crisis resource management (ACRM): A decade of experience. *Simulation & Gaming.* 32(2), 175–193 (2001).
6. Dieckmann, P., Gaba, D., and Rall, M. (2007). Deepening the theoretical foundations of patient simulation as social practice. *Simulation in Health Care*, 2(3), 183–193.
7. Rall, M. and Dieckmann, P. Crisis Resource Management (CRM) to Improve Patient Safety. [Online: http://www.euroanesthesia.org/education/rc2005vienna/17 RC1.pdf]. ESA Refresher Course 2005, pp. 107–112.
8. Rall, M. and Gaba, D. M. Human Performance and Patient Safety. In: Miller RD, (ed.), Miller's Anaesthesia. Elsevier Churchill Livingston, Philadelphia, 2005, pp. 3021–3072.
9. Dieckmann, P., Wehner, T., Rall, M. and Manser, T. Prospektive Simulation: Ein Konzept zur methodischen Ergänzung von medizinischen Simulatorsettings [Prospective Simulation – A concept for the methodological complementation of medical simulator settings]. Zeitschrift für Arbeitswissenschaft 59(2), 172–180 (2005).

# Move the
# Learning,
# Not the Learners

*"The people I distrust most are those who want to improve our lives but have only one course of action."*

Frank Herbert

Hands-on simulation experiences require bringing the students and the simulators together. While a few of your students are learning by doing with their hands, what are your other students learning while they wait their turn? A key advantage of simulation is the ability to "schedule disasters." Most simulation facilities include some form of audio/video capture and presentation capabilities. Most simulation facilities have Internet if not also Intranet access. Most simulation programs function within institutions that already have some form of two-way live audio/video teleconferencing. All of our students started out being located some distance from our institutions. Suppose you could simultaneously convey to most, if not all your students, some of the rich learning that comes from interacting with just the sights and sounds of a clinical event? Suppose your students could participate in such experiences without actually having to relocate to where you happened to have created your simulation facility? Could such amplification of your simulation resources improve your clinical educational offerings? Would such distant learning change whom your institution considers suitable for inclusion as students?

Andreas Meier shows how learning the essentials of making surgical decisions can be improved by linking students with simulation via the Internet. Dag von Lubitz *et al.* reinforce this concept with their examples of simulation-presented content and globe spanning tele-education.

# 63

# Creation of a Combined Surgical Curriculum Using the Internet and Patient Simulation

Andreas H. Meier

## 63.1 Purpose and Goals

The rapid advances in surgical technology combined with substantial changes in health care delivery present a significant challenge for surgical education. Therefore, it is necessary for surgical educators to consider nontraditional approaches to the surgical curriculum, for instance, the combination of prior acquisition of Internet-based cognitive information with the hands-on application of this theoretical knowledge using simulation. Our experience with patient simulators are now sophisticated enough to allow the creation of realistic scenarios which can include pre-, intra-, and postoperative complications. Furthermore, the rapid development of the Internet provides us with ubiquitous access to up-to-date information, databases, images, and videos. This makes the online delivery of curricular content very attractive, especially as it can be designed to be interactive. The overall goal of our project is the integration of a web-delivered and hands-on patient simulator–based curriculum into our surgical residency program. At this initial stage, the main focus is directed toward clinical judgment and decision making, not pure technical skills training. The author chose the incoming first year surgical residents as the initial target group for a pilot study with the following objectives:

1. Provide a reassuring, less stressful learning environment for young surgeons prior to assuming clinical duties.
2. Evaluate the subjective impact of the curriculum on the individual participants.
3. Show that the implementation of such a curriculum is feasible.

## 63.2 Planning the Project

### 63.2.1 The Status Before Implementation

In the past, surgical interns arriving at the Hershey Medical Center, Penn State University, Hershey, Pennsylvania, U.S.A., were basically thrown into their on-call duties with little preparation, a practice still widely prevalent across the globe. This represents a fairly stressful experience for the young physician and also cannot be considered ideal as far as patient safety is concerned. To alleviate this problem, our department had previously implemented a curriculum prior to starting on-call duties, "The First Three Days in Surgery" [1]. This program includes a variety of orientation sessions, an ATLS and ACLS course, and mock nursing calls for the incoming house staff. To improve this program further, the author thought about ways to provide a curriculum to the incoming house staff prior to arriving, specifically exposing them to potentially critical ward emergencies before these young surgeons would take responsibility for the lives of real patients. Using the web as a tool for distance education appeared to be a logical solution, as was the integration of full-patient simulation into the First Three Days in Surgery program to provide challenging scenarios in a safe environment.

### 63.2.2 Project Goals

Our immediate goal for the expanded pilot project was to further improve this curriculum by:

1. The implementation of a "preemptive" web-based curriculum for new surgical interns prior to their arrival at our institution,
2. The exposure of the incoming interns to common postoperative complications during the First Three Days in Surgery curriculum in a safe and reassuring environment by using simulated scenarios.

The driving force of the project was a junior surgical faculty member with an interest in education and expertise in advanced educational technology using the Web (AHM).

### 63.2.3 Creating the Infrastructure

The Simulation Development and Cognitive Science Laboratory, founded in 1993, and supported by the Departments of Anesthesiology, Nursing and Surgery, provides access to a large variety of simulators, including multiple high-fidelity patient simulators. The cultural and material infrastructure for the simulation component of the curriculum was therefore already in place. The author developed an educational web site (http://surgery.psu.edu) and integrated the online curriculum (Table 63.1). Open source software was used wherever possible to minimize expenses.

## 63.3 Implementation and Description of the Project

### 63.3.1 Online Curriculum

Shortly after match day[1], the new surgical interns receive an e-mail with information on how to access the online curriculum. They also are asked to fill out an online questionnaire to assess their self-efficacy [2] prior to working through the curriculum (Figure 63.1). The curriculum focuses on postoperative complications and contains several content areas, which includes an introduction and six crisis events (Figure 63.2):

- General tips for taking call (tap into resources, such as nurses, ask for help at appropriate times, etc.)
- Myocardial infarction
- Postoperative hemorrhage
- Pulmonary edema
- Cardiac arrhythmias
- Septic shock
- Pulmonary embolism

**TABLE 63.1**  Hardware and software used for the web-based curriculum, including URL for download and description of function

| Hardware/Software | URL | Function |
| --- | --- | --- |
| Xserve dual-processor G4 with Mac OS-X Server 10.4 | http://www.apple.com | Computer hardware to run website |
| Apache web server | http://httpd.apache.org | Web server software (Open source) |
| PHP Programing environment | http://www.php.net | Web development software (Open source) |
| MySQL database | http://dev.mysql.com | Database to support dynamic web development (Open source) |
| Postnuke | http://www.postnuke.com | Content Management System (Open Source) |
| ContentExpress | http://www.xexpress.org | Postnuke module to publish content (Open source) |
| FormExpress | http://sourceforge.net/projects/pn-formexpress | Postnuke modules to create web forms (for questionnaires, Open source) |

**PGY I Preparation Study**

<u>Self Assessment of Confidence</u>

Pre Orientation

**Enter your personal information please!**

| First Name:* | |
|---|---|
| Last Name:* | |
| Specialty* | ⏷ |

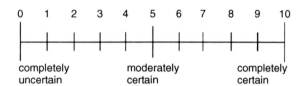

completely
uncertain

moderately
certain

completely
certain

Judge your level of confidence regarding the tasks below using the scale! When done, press the submit button.

We hope that we can use this information throughout the year to improve our educational program, which should be in your interest!

*Andreas H Meier, MD*

**Here are the statements**

I can provide the initial management for a patient with an acute myocardial infarction.*   [ 5 ▾ ]

I can provide the initial management for a patient with postoperative bleeding.*   [ 5 ▾ ]

I can provide the initial management for a patient with postoperative pulmonary edema.*   [ 5 ▾ ]

I can provide the initial management for a patient with postoperative cardiac arrhythmias.*   [ 5 ▾ ]

I can provide the initial management for a patient with sepsis.*   [ 5 ▾ ]

I can provide the initial management for a patient with pulmonary embolism.*   [ 5 ▾ ]

I can place peripheral iv access successfully on an adult patient in need of fluid rescussitation.*   [ 5 ▾ ]

I can place a central line successfuly on a post-operative patient in need of invasive monitoring.*   [ 5 ▾ ]

I can successfuly intubate a patient in respiratory distress.*   [ 5 ▾ ]

I can successfuly place a chest tube in a trauma patient.*   [ 5 ▾ ]

[ Submit Questionnaire ]

**FIGURE 63.1** Online form to assess self-efficacy. The form is based on a Likert scale. The curriculum and simulation sessions only covered the patient management portion of the questionnaire, whereas the skills-related statements were covered by an ATLS and ACLS course, which occurred during the First Three Days of Surgery as well. The asterisks mark fields that are required to be filled out.

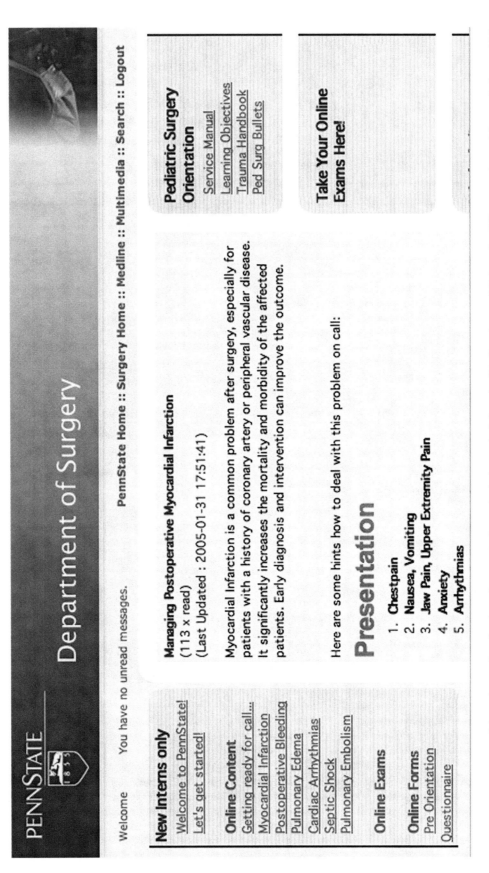

**FIGURE 63.2**   Excerpt of the web-interface for the online curriculum. The links on the left under "Online Content" link to the different sections of the curriculum, which are presented in the middle. This example shows the beginning of the section on myocardial infarction.

## 63.3.2 Patient Simulator Sessions

During the First Three Days of Surgery curriculum, the 15–20 new surgical residents are distributed into three groups of 6–8. Six patient simulator scenarios had been created, each covering one of the postoperative complications described in the online curriculum. A staff member of the simulation laboratory and a surgical faculty member present the simulations. Simulation scripts (see Appendix 63A.1) are available for the teaching faculty and have the following structure:

1. Description of the clinical scenario (given to the resident when the simulation begins)
2. Information the resident can obtain during the scenario (given to the resident depending on questions asked and actions taken)
3. Expected actions by the resident (action is considered "standard of care" and will likely lead to an optimal outcome)
4. Acceptable actions by the resident (action may, or may not, influence outcome in a positive way, but is not harmful)
5. Unacceptable actions by the resident (action is potentially harmful or negligent)
6. Information and details for the instructors on how to operate the simulator during the scenario.

Each of the groups rotates through each of the six different scenarios. For each scenario, one of the interns is the primary physician responsible for the patient, whereas the rest of the group is available to provide assistance as requested. The rest of the group is in the same room as the primary student to observe and learn as much as possible. Clinical nuances about the subsequent patients are changed to prevent boredom and mindless repetition of the previous student's replies. The scenarios start with a simulated phone call from a nurse describing the problem, and the resident is then asked to manage the patient initially over the phone and then at the bedside. Each scenario ends with a group discussion and feedback from the surgical faculty member. Feedback includes aspects of cognitive performance, but also touches upon communication and overall management, including calling for help at the appropriate time (i.e., Crisis Resource Management – CRM – principles). Asking for help and appropriate utilization of resources (e.g., nurses) as a positive feature had already been emphasized during the web curriculum. After about 10 days on service, the residents are asked to fill out the same self-efficacy questionnaire.

## 63.4 Current Status of the Project

The curriculum and simulation sessions have become an integral part of the First Three Days of Surgery program. Since its inception 3 years ago, all new interns have participated. The feedback from the interns has been overwhelmingly positive (Figure 63.3). The online curriculum and the scenario scripts are updated annually to ensure that new or changed treatment modalities are reflected appropriately. At this time, the project is maintained with only one faculty member as the instructor. The project and its results were presented at the Association of Surgical Education meeting in Houston in 2004 and published in the American Journal of Surgery in 2005 [3]. Even though other institutions have shown interest in the concept of this program, preemptive curricular activities prior to assuming clinical duty does not appear to be a widely used practice in surgical education yet.

## 63.5 Where are We Headed?

The rapid acceptance and positive feedback the project received from its participants should encourage efforts to broaden the application of prior online content delivery and hands-on simulations throughout the surgical curriculum. At the current state of robotic patient simulation capability, two content areas appear logical next steps: acute trauma management and intensive care. These two areas represent aspects where overall patient management, clinical judgment, communication skills, and even systems-based practice can be taught and assessed. To be ultimately integrated into the standard surgical curriculum, however, it will require rethinking of traditional surgical teaching concepts and broader acceptance of simulation technology within the surgical faculty.

## 63.6 Lessons Learned

### 63.6.1 Accomplished

The introduction of a preemptive curriculum, and the opportunity for the residents to practice common on-call problems in a safe, reassuring environment is a big step forward for the educational mission of our department. The high marks the project received from its participants is reassuring, and most likely speaks for a rising awareness and acceptance by surgical trainees that simulation in surgical education is helpful and beneficial. I also think that the improvement of self-efficacy scores after the intervention is directly related to the curriculum and simulator sessions (Figure 63.3).

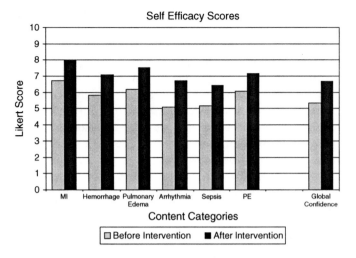

**FIGURE 63.3** Self-efficacy scores of participants based on online questionnaire pre- and postintervention [3]. The details of the questionnaire are depicted in Figure 63.1.

### 63.6.2  Future Directions

This pilot project was intended to show feasibility, which has been accomplished. Further work needs to be done in the following areas:

1. Expand the curriculum to cover the whole academic year.
2. Analyze the impact of the simulator session and the online curriculum independently.
3. Encourage more faculty participation within the curricular development process.
4. Include higher-level residents into the curriculum.
5. Broaden the scope of content areas.
6. Use scenarios not only for initial teaching, but also for subsequent assessment and remediation.
7. Develop metrics that objectively assess the trainee's clinical judgment and decision-making capabilities.
8. Develop metrics that objectively assess the instructors' educational judgment and decision-making capabilities.
9. Align curriculum with concepts of cognitive theory with the goal to expedite development of expertise in the trainees, and the trainers.

## 63.7  Conclusion

The project shows the feasibility and potential benefits of simulation and other novel educational technologies in surgical education. The biggest challenge to implement this concept more broadly is to convince faculty of its advantages and to

motivate them to participate. Learning to accept that more of surgical teaching has to occur outside of the traditional (expensive) clinical environment is not an easy process, yet necessary due to the increasing constraints of our health care environment. It appears that colleagues in other specialties, especially anesthesia, have already realized the paradigm shift toward some separation between service requirement and educational activities for residents and are leading the effort in clinical simulation. The very recent activities of the American College of Surgeons establishing and accrediting centers of excellence in surgical education (including the PennState Simulation Facility) are encouraging. They will almost certainly promote similar efforts within the surgical community in front of a national audience, and provide the fuel necessary to raise simulation in surgical education to the level it deserves.

## Acknowledgments

The project received internal financial support through a surgical feasibility grant and would have not been possible without that funding.

Mentoring by the previous director of the Cognitive Science and Simulation Laboratory (W. Bosseau Murray) provided crucial support during the planning phase of the project.

Special thanks to Jody Henry for her inspiring support through all the simulation sessions.

## Endnote

1. Match day is a fixed day in March every year when all USA 4th year medical students find out at which institution they will start their post-graduate training by July 1st of the same year.

## References

1. Marshall, R. L., Gorman, P. J., Verne, D., et al. Practical training for postgraduate year 1 surgery residents, *Am. J. Surg.* **179**(3), 194–196 (2000).
2. Bandura, A. and DH, S. Cultivating competence, self-efficacy, and intrinsic interest through proximal self-motivation, *J. Pers. Soc. Psych.* **41**, 586–598 (1981).
3. Meier, A. H., Henry, J., Marine, R., and Murray, W. B. Implementation of a web- and simulation-based curriculum to ease the transition from medical school to surgical internship, *Am. J. Surg.* **190**(1), 137–140 (2005).

# 64

# Distributed Simulation-based Clinical Training: Going Beyond the Obvious

Dag K.J.E von Lubitz
Howard Levine
Frédéric Patricelli
Simon Richir

## 64.1 Issues in Clinical Education and Training

The past 50 years witnessed probably the most explosive rate of growth of clinical knowledge in recorded history [1, 2]. New discoveries in pharmacology, molecular biology, neuroscience, or reproductive biology to mention just a few disciplines, were accompanied by similarly dynamic development in clinical practice, new diagnostic instrumentation, new surgical techniques, and treatment modalities. Improved understanding of the process of disease combined with a revitalized interest in preventive medicine resulted in new, often international, treatment guidelines while the emergence of new threats imposed further changes on the practice of health care. Not surprisingly, the issue of life-long clinical learning rose to an unprecedented prominence and intensifying exploration of the means to improve the efficiency of clinical education and postgraduate training [3–5]. According to the Entrez-PubMed literature database, over 100,000 papers on clinical education have been published to date. Many of these papers describe highly innovative methods that frequently involve utilization of very sophisticated technologies. Predictably, most advanced training techniques are employed by, and for the realtively few, at the major clinical training centers [6–8]. On the other hand, access to continuing clinical education remains to be problematic for the far more numerous practitioners in rural and remote regions of the globe [9–12]. Isolation, inconsistent quality of training programs, and variation in the allocation of training resources are among the principal issues determining quality of care for most of the world, and need to be addressed to produce measurable changes [13].

The emergence of new clinical threats such as bioterrorism [14] poses additional challenges in educating large numbers of prehospital and emergency room personnel necessary to ensure the maximum level of readiness [14–17]. The fact that clinicians, such as general practitioners or dentists (http://dentistry.uic.edu/research/demrt/), who typically do not participate in Emergency Medical Services (EMS) operations may have an active role in interventions associated or following an act of bioterrorism [18] is a clear indicator that the required training must account for the existing differences in the baseline knowledge. Moreover, despite the very large number of the national and international trainees, the required training programs must be based on reproducable curricula that are based on best practices, and with the training itself conducted at a consistently high-quality level if it is to fulfill the mission of developing and sustaining adequate preparedness against emerging threats of bioterrorism or disasters associated with mass casualties [19–23]. Finally, training must be repeated frequently enough to maintain a consistent operational competency for infrequent events. In summary, even a perfunctory review of the existing literature will clearly indicate the persistent and pervasive need for continuous education and training of both pre- and in-hospital personnel [24–26], and the significant role of federal, state, local, and nongovernmental agencies in developing robust tools and systems required to provide such training both on the continuous and "just-in-time" basis [27–29].

Many technologies have been adapted to or specifically designed for simulation, telecommunciations, and education. The glossary of less than universal terms is included in Section 64.8 Abbreviations and Definitions.

## 64.2 Challenges in Training Prehospital Personnel

Rapid growth of clinical knowledge, its ever faster translation into changes in practice, increasing specialization [30–32], and multidisciplinary approach to the treatment of disease [33–37] require a very wide range of specialized education at all levels of clinical competence. Problems ranging from communication and definition of professional identities and roles within the clinical management team [38–41], through procedure difficulties [42–44], missed or wrong diagnoses [45–48], to errors in clinical command of EMS operations or inadequacy of prehospital provider training [49–51] have been described. The majority of adverse incidents appear to be related to inadequate knowledge and/or insufficient training seen as much at the prehospital (e.g., [49–51]) as at the in-hospital environments [44, 46, 52, 53]. One of the predominating issues that await effective and efficient solutions is that of inadequate training of nonspecialist health care workers, particularly in the rural areas, in adult and pediatric emergency and trauma medicine [52–62] and in surgery [63–67].

Clinical and procedural errors resulting from training deficiencies at the first responder/paramedic level are common [68, 69]. The adverse effects of less than optimal teamwork caused by poor team training [70], substandard mastery of essential (even basic) resuscitation diagnostic skills and procedures [71–75], and unreliability in the delivery of such

common services as Advanced Cardiac Life Support [76] have been well documented. Even more problematic are the substantial inadequacies in pediatric resuscitation skills resulting from inadequate training and infrequent exposure to pediatric emergencies [77–80]. The intense need for continuous training based on standardized "best practices" protocols is emphasized by the recently reported lack of reliability in performance rating during Emergency Medical Technician licensing examinations [81] – a failure that allows personnel with potentially substandard knowledge and skills, but spuriously rated as "fully qualified," to participate in field operations. Inadequate entry-level preparation combined with the demonstrated skills decay [79, 80, 82] provide a further argument supporting the need for the development of initial high-level clinical competence followed by its vigorous maintenance through the process of life-long learning among all health care professionals [83–92].

The financial costs that accompany each incident of faulty patient management constitute another significant motive to provide adequate training. It is estimated that the average price of a resuscitation attempt at a British hospital reaches approximately £1.000–£2.000. In the U.S.A., the expenditure for the resuscitation following out-of-hospital cardiopulmonary arrest may reach $6000–$10000 [93–95]. Clearly, any error (be it a the pre- or in-hospital level) that results in serious aggravation of the presenting complaint will automatically increase the final cost of care by making it more complex and more resource-demanding [96]. Thus, while the argument of "I am certified" may still be heard from many who attain a Pass on their infrequent, paper-pencil pass/fail BLS-ACLS-ATLS-PALS recertification examination, the wealth of existing data on errors committed by "certified" personnel indicates that the presence of certification may offer false security in one's own clinical prowess. The real cost of such unchallenged (and therefore potentially unwarranted) certitude may, indeed, be quite extreme both in terms of the outcome and in the related expenditure.

## 64.3 Existing Training Methods

Despite all the advances in didactic technology and methods in other domains, health care education continues to be conducted in a manner that has not significantly changed during the recorded history of medicine [1]. In the broadest terms, training of health care professionals is conducted either as a totally passive assimilation of the existing body of knowledge such as books [97, 98] or lectures [99, 100], or through an active hands-on approach. The latter may involve either the combination of passive and active methods or the hands-on methods alone [101–103]. Significantly, while it is often claimed that electronic dissemination of medical knowledge provides "interactivity," many existing platforms represent nothing but technologically advanced forms of traditional, essentially passive, training based on old didactic principles [104–106]. The advantage of the effective use of information technology in the didactic electronic-training (E-training) packages rests with the ease of access to the appropriate sources of information, rapid cross-referencing of information and the supporting data, and the ability to organize information derived from various resources into easily cataloged logical units that assist in the subsequent assimilation and solidification/retention of the acquired knowledge.

Rapid growth of Internet connectivity in the technologically advanced countries is associated with the most important attribute of electronic health care knowledge dissemination – rapid erosion of distance as the main obstacle in accessing postgraduate professional education among health care personnel in rural and remote regions of the world [107]. The existing Internet/Web-based clinical training and/or consultation programs cover a wide range of topics [108–113], satisfy almost every need for specialized knowledge, and, with the increasing sophistication of the existing models, may involve a large variety of approaches spanning from e-mail exchange [114–116] to videoconferencing and multimedia offerings [115–120]. The main disadvantage of didactic distance learning is its essentially static nature that may not reflect the dynamism of clinical care, particularly in the context of invasive specialties like emergency/trauma and military medicine, or surgery [121]. The existence of inaccurate, obsolete, or incomplete content of many clinical information sites provides another major problem that continues to affect the overall quality of Internet-based learning resources [107, 122].

Hands-on training based on the trainee-patient contact, while highly effective in the development of the necessary clinical skills [123, 124], is also associated with a number of challenges [125–129] and risks [130]. In the past 80 years [131], the aviation community, where both training and operations are associated with many characteristics similar to the high-paced tempo of clinical specialties such as emergency medicine, trauma medicine, or perioperative care, has used simulators both to increase the efficiency of training and to minimize the dangers associated with training [132–135]. Although the history of simulation in medicine is significantly shorter, its importance in education and training, and the level of sophistication is increasing very rapidly [136–140].

Immersive and semi-immersive virtual reality (VR) simulation [141–143] represents technologically the most expensive level of clinical simulation, and is particularly suitable for training surgical techniques [143]. High-fidelity Patient Simulators (HFPS) and single procedure simulators provide the less technically complex yet highly sophisticated and more accessible training tools that serve as highly complex "clinical flight simulators" in a wide variety of training tasks. The use of both HFPS units and single procedure training devices in hands-on training eliminates the risk factors associated with training on live patients, permits improvement of diagnostic skills by

allowing practical understanding of the involved steps, hones clinical team training [144–147], and helps in punishment-free learning from unintentional and intentional errors [148, 149].

In addition to the training personnel and physical training facilities, the complex nature of almost all clinical simulation devices requires dedicated technical support staff. Hence, despite decreasing acquisition price (including VR environments) the purchasing and operating costs are high [150–152], placing simulation-based training beyond the means of many smaller organizations [107].

# 64.4 Access to Remote Training: Equipment and Expertise

Although progressive reduction of simulator costs provides a highly encouraging step toward wide dissemination of simulation-based training, access to the simulation centers poses probably the most significant obstacle. The vast majority of simulation centers are associated with institutions of higher clinical education (medical schools, colleges of allied health professions, major teaching hospitals). While readily available to students and the personnel of the school (and frequently also the locally affiliated hospitals and emergency services), access to personnel outside the immediate geographical vicinity remains often difficult, if not impossible. The problem of access is virtually unsolvable for those working in rural and remote regions and, needless to say, for essentially all clinical personnel in the countries where simulation-based education remains (and will remain for a long time) an unattainable luxury. However, the fact that simulators are endowed with sufficient degree of interoperability to allow external manipulation and control by remote devices that, in turn, use similar operating software (Microsoft or Apple), combines with the revolution in telecommunications and visualization technologies, making remote access to simulation-based clinical training available to essentially anyone with an Internet connection. Paradoxically, adoption of remote access philosophy may prove to be critical in providing entry to the most advanced form of clinical training in the countries where currently even its most rudimentry elements are faltering due to insufficient resources [107]. In many of the less-developed countries, the adequate infrastructure required for remote simulation training is already in place, due to the needs of the expanding foreign business operations for both wired and wireless telecommunications [153].

The concept of distance-based simulation training is based on the remote access to the central simulation facility and its HFPS from any number of distant training sites located practically anywhere in the world. Training can be conducted under the guidance of an expert teacher who may be physically present at the HFPS facility, at the location of the trainees, or at a place that is completely removed from any of the other sites.

The principal constraint of such training [154, 155] is the difficulty in coordinating successful operations of more than a single HFPS unit. Yet, the need for orchestrated, simultaneous operation of several HFPS (using either as a simulator LAN or WAN – local or wide area network) is essential when training involves mass casualties, or when the machines used as the constituents of the training federation (i.e., the assembly of heterogenous simulation devices and environments that operates as a single, albeit physically dispersed unit, Figure 64.1, Ref. [156]) are built by more than one manufacturer. The Medical Application Software Provider (MedASP) model offers a solution to the problem of large-scale simulation-based training. The MedASP model combines the principles of Distributed Interactive Simulation (DIS) and of the Application Service Provider (ASP); Figure 64.1 and Refs. [156–159].

Practical implementation of the MedASP concept allows execution of highly realistic training in often complex and stress-filled environments of field emergency and trauma medicine. It also obviates the need for the trainees to leave their physical locations that may be separated by thousands of miles from the simulation facility [154, 155, 159, 160]. The additional advantage of the model is its preeminent suitability for real-time dissemination of world-class training expertise in the form of practical "patient demonstrations" rather than the commonly encountered didactic format of a lecture (Ref. [155] and Figure 64.2).

Consequently, learner audiences in rural and remote regions or in the economically disadvantaged countries may readily benefit from both advanced training technology and expertise until now available only to the most privileged students [107].

MedASP operations allow preparation and delivery of highly customized training services virtually immediately, whenever they are needed, and delivered wherever the need exists (on-demand operations). Experiments have shown [155, 158–160] that this type of access may be of special significance for Emergency Medical Services and military health care providers, particularly in cases of the acute requirement for "just-in-time" training of large numbers of personnel stationed/ deploying to geographically dispersed sites [121] as seen during bioterrorist threats (Figure 64.3), military conflicts, or large scale rescue operations [160].

When several dispersed High-fidelity Human Patient Simulators are controlled from a central training facility or accessed by the remote learners, the only means to ensure uniformity of training is to assure full compatibility of both physical and operational characteristics of the simulation devices., i.e., the range and complexity of simulated conditions, similarity of outputs, etc. This aspect becomes critical when the training center has either the overriding remote control of all distributed HFPS units, or when the center serves as the "expertise headquarters" [121, 156, 158–160] during simultaneous, real-time training of large numbers of dispersed learners, e.g., in

multisimulator training of international medical intervention teams (e.g., just-in-time preparation for mass casualties caused by acts of terrorism, natural disasters, etc., [18, 20]).

While the two principal High-fidelity Human Patient Simulator systems that are commercially available today use mannequins with comparable anatomical features and functions, and generate a very similar profile of training-relevant physiology and pathology, the conceptual basis of their software/hardware interaction with the operator differs significantly. As a result of these differences, combining both systems into a unified, remotely accessible training environment poses practical difficulties. The simplest approach to the remote control of multiple simulators are voice commands. Moreover, from the clinical point of view, when such commands are given by the remote trainee to the personnel at the simulator host site (the central training facility), a certain degree of reality is attained since voice is used to direct real-life actions of a clinical team. However, voice control of several remote units by the control center is unpractical since it either requires the presence of a technician operating each remote simulator unit or requires the trainees to both control simulator behavior *and* deal with the clinical situation presented by the machine, which entirely disposes of the "suspension of disbelief" aspect of simulation. The use of separate computers located at the control center where one machine is slaved to the control of only one simulator provides the easiest automated solution to the operations of simulator federations where individual units are manufactured by different companies. Yet, in a fast-paced environment of a multipatient scenario (Figure 64.3) such control becomes confusing and cumbersome. Moreover, in scenarios where the trainees are required to assume machine control through Web-based interfaces, confusion will most likely ensue (particularly if remotely executed by the trainees with little or no background in computer operations) since, in trying to overcome the need for machine control, the trainees' attention will rapidly shift from the main subject (clinical training) to the frantic attempts at mastering the unfamiliar interface. Clearly, the essential attribute of simulation – situational realism – will deteriorate and decrease effectiveness of training.

Simulator-bridging software that automatically translates commands given by the control interface of one system into the commands that are understood by the simulator of a different and otherwise incompatible brand is the most effective solution to multisimulator environments. It is also technically the most complex since, in the absence of commercially available products, the software bridge must be developed as a private venture at the user's facility. Thus, in accordance with MedASP concept, and both from the technical and the fiscal point of view, the most suitable placement of the software bridge is at the central (hub) control facility at which all signal processing takes place [150]. MedASP-based activity simplifies signal traffic and, by providing more effective processing, eliminates the annoying time lags that may render distance-based simulation training exceedingly unrealistic. MedASP participation assures

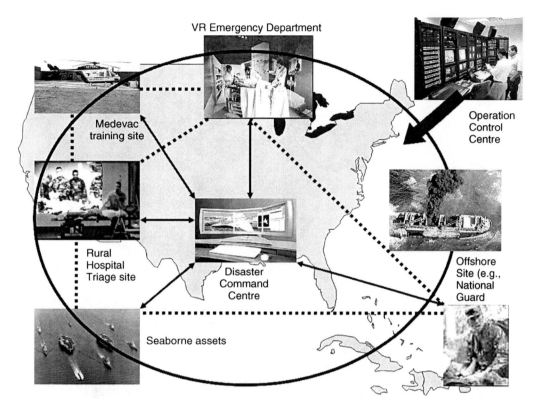

**FIGURE 64.1**  Federation of widely distributed High Fidelity Human Patient Simulators (HFPS), virtual reality (VR), clinical management, and control and command ($C^2$) sites (surrounded by an oval) for training Emergency Medical System response to major disasters (in this case, a burning ship with a cargo of explosives). The Operational Control Center has the overall technical control of all federated elements (solid black arrow) but does not have direct involvement in the conduct of the exercise. Apart from purely technical support of the activities within the federation (voice/video communication, monitoring and maintenance of communications stability, data recording, technical support, etc.), the Operational Control Center acts as a "background control" of simulator scenarios (introduction of unpredictable clinical complications, etc.), determination of "virtual resources" (e.g., availability of transport, weather, state of the roads, etc.), introduction of unpredictable factors (e.g., sudden "weather changes" that hamper aerial transport). The Disaster Command Center, on the other hand, performs in the manner identical to real-life events and it coordinates all acivities performed by the subordinate members of the federation. The Disaster Command Center has direct contact with all field units of the federation (solid arrows) and manages their activities, allocates resources depending on need (or availability), and makes oparational decisions (e.g., Medevac casualties or not, transfer from one treatment facility to another, shift assets.) The distributed members of the federation although devoted primarily to direct interaction with the casualties (triage, stabilization, initial treatment, requests for transfer to higher echelon treatment facilities, etc.) can, in similarity to real-life events, communicate with each other (stippled lines.) Despite wide geographical dispersion of individual training sites, the entire exercise can be performed as if all participating units were colocated due to the bridging functions provided by the Operational Command Center. This type of distributed training represents the highest level of complexity and requires very substantial materiel and human resources to function. Although such resources are presently available only to the armed forces of few nations, complex synthetic environments allow *realistic* training of participants who must respond in a coordinated manner to a wide variety of challenges. Synthetic federated environments allow training in real time, with minimal or no preparation, and with utmost flexibility necessary for training in management of a seemingly uncontrollable crisis scenarios involving large numbers of victims, a wide variety of participating organizations (medical personnel, law enforcement, military, relief organizations, etc.). The additional advantage of synthetic environments is their ideal suitability for training of Observation, Orientation, Determination, Action (OODA) Loop-based thinking. Reproduced by permission of MedSMART, Inc. Copyright© to all figures and photographs except where noted: MedSMART, Inc.

**Medical Application Software Provider**

- Access
- Live expertise
- Optimization
- Technology
- Cost reduction

**FIGURE 64.2** The Medical Application Software Provider (MedASP) concept is based on concentrating pertinent knowledge, information, data, technology, and expertise within a dedicated operational facility that can be remotely accessed, and which can provide the required services remotely. MedASP centers can either exist as separate entities or be attached to major sites of clinical expertise (e.g., universities) and allow their clients access to all training resources at a fraction of the cost required for construction of own facilities (the remote sites require a stable Internet connection and adequate visualization platforms rather than specialized physical sites housing simulation units and together with the associated staff). Providing adequate infrastructure is in place, organizations with fewer resources (on the left side of the composite picture) obtain training by using the central expertise and simulation facilities of the MedASP (center). Clients with partially developed simulation facilities (upper left and lower right pictures) can access simulators and training experts of MedASP remotely from their location. Clients with fully developed facilities (upper right and lower left) or combine into larger federations with other users, thereby creating synthetic high patient volume environments (e.g., Emergency Department operations) that are ideal for safe training and education of personnel staffing such facilities. All groups can access training by MedASP experts disseminated from the central facility that operates its own simulators and also assumes functional control of the machines at the client sites. Experimental model of MedASP has been successfully tested during a series of transatlantic exercises (see Ref. [160]). Reproduced by permission of MedSMART, Inc. Copyright© to all figures and photographs except where noted: MedSMART, Inc.

**TRANSATLANTIC BIOTERRORISM**
**READINESS EXERCISE**

- 22 year old man with fever and malaise for the past 24 hrs
- Rash appeared during the past 24 hrs
- Temp 39.4°C, Blood pressure 106/78, Cardiac rate 116 bpm
- Respiration 18/min
- AP Chest X-Ray lungs clear
- Scattered maculae and vesicles present with predominantly thoracic distribution

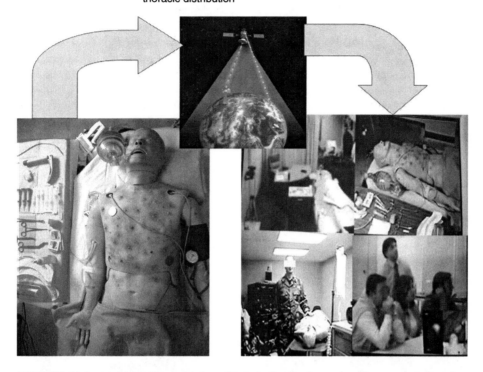

**FIGURE 64.3**  Multisimulator training of U.S., Italian, and French military and civilian clinical personnel in the management of bioterrorism. In this real time, free-play exercise (i.e., the exercise with minimum amount of predetermined content, where the course of action determined by the trainees' response to random, unpredictable events that introduce unforeseen complications that may demand implementation of additional, nonclinical skills such as team leadership, decision making, collaborative problem solving, etc.), the action involved several HFPS units manufactured by two different companies and placed at three widely separated U.S. locations. The trainees and experts were located on both sides of Atlantic and operated at three layers of clinical command and control (two in the U.S. and one in Italy), while the technical control center (in U.S.A.) maintained overall background control of the exercise through the random introduction of "confounders," maintenance of transatlantic simulator synchronization, operation of remote voice/video systems, telecommunications support, etc. Despite great publicity surrounding bioterrorism and substantial resources spent on the development of preparedness, the exercise revealed surprising deficiencies in readiness to manage bioterrorism-related events. Only distributed synthetic environments allow conduct of similar free-play exercises aimed at the training of large numbers of personnel that is performed on a regular basis and at a minimal cost and minimal inconveninence to participants. Reproduced by permission of MedSMART, Inc. Copyright© to all figures and photographs except where noted: MedSMART, Inc.

that only the meaningful commands are passed within the simulator federation, and that the exchange occurs at the maximum speed allowed by the available bandwidth.

### 64.4.1 Concept of Telepresence and Visualization Devices

In its most restrictive definition, "telepresence" is defined as "the experience of being fully present at a live real world location remote from one's own physical location" [161], where, under the most ideal circumstances, the observer would be able to be stimulated by a variety of sensory inputs and execute social interactions as if physically present at the remote site [162, 163]. While many aspects of telemedicine, particularly teleconsultation, are rooted in the loosely defined sensation of telepresence [164–168], the most advanced clinical approaches utilize the strict definition of *the operator experiencing real presence at the remote location* [163–167]. Hence, the most essential aspect of simulation-based distance training is visualization of the simulating devices (Figures 64.4 and 36.5).

The simplest and most readily available visualization forms are video displays projecting two-dimensional images (2-D) ranging in cost and complexity from the simple analog CRT ("television") devices to digital flat panel displays offering very high image quality (Figure 64.5).

However, in order to derive the true benefits of the latest 2-D display technology, both image quality and the field of vision must be identical at all sites, and the cameras must collect all the essential visual information required for both remote control of the sim and accurate assessment of the student's behaviors. Even if endowed with pan/tilt/zoom capacity, a single camera allows only a limited view of the entire action field, and multiple camera systems need to be used. In order to ensure maximum flexibility and view uniformity, all cameras must be under remote control from the central control station, and the control devices must allow simultaneous operation at both on-site and remote cameras (Figure 64.6). This is particularly important in the federated simulator environments in which individual machines may be separated by large geographical distances.

**FIGURE 64.4** Global Telepresence Room at the ultramodern College of Health Professions at Central Michigan University during preparations of a Human Patient Simulator training session. The curving wall display consists of several, individually controlled plasma displays that permit virtually unlimited imagery manipulation, interaction, and support with other media. The technology visualization and telecommunications technologies of the Global Telepresence Room's unprecedented flexibility in simultaneous, real-time training of on-site and globally distributed audiences. The room and its associated support facilities are ideal for MedASP operations. Reproduced by permission of MedSMART, Inc. Copyright© to all figures and photographs except where noted: MedSMART, Inc.

**FIGURE 64.5** Two-dimensional displays. The simple VGA screen of a laptop (top left) provides arguably the cheapest and most portable approach, and it also provides a direct interface for the control of the remote simulator. The wall-mounted LCD monitor (top right) provides a much higher picture quality although the relatively small size of the monitor precludes its simultaneous viewing by a large number of trainees. Moreover, the display is passive, and it is supported by "point and touch" devices, it will not allow remote simulator control. A large, multipanel flat screen display at one of the classrooms at the College of Health Professions at Central Michigan University (bottom left) offers full imagery manipulation capabilities (such as multiangle viewing, seletive magnification, presence of explanatory text and diagrams, etc.), either from the central computer at the lecturer's desk or via computers plugged into the viewer-accessible ports. Note, however, that in cases of crisis events like the depicted aircraft fire (in this case, a simulated one), and when "just-in-time" refresher course/information/training is needed, the laptop display serves as well as the most sophisticated unit at a prime training center. Reproduced by permission of MedSMART, Inc. Copyright© to all figures and photographs except where noted: MedSMART, Inc.

Effective, multicamera control systems are commercially available, and their use offers the most practical solution, even if (at times) it may be quite expensive and technologically complex. The most critical aspect of camera operation is the calibration of the individual units' field of view, particularly in the context of multisite activities when centrally located simulators are operated from a large number of remote learner

sites. Under such conditions, the view of the simulator at each remote site must be identical to that displayed at all other sites.

When only restricted bandwidth is available [153], limitations of typical videoconferencing systems allows a single camera unit to feed optimum imagery to not more than 3–4 sites. Addition of further camera sets fragments the available bandwidth and results in a substantial deterioration of image quality

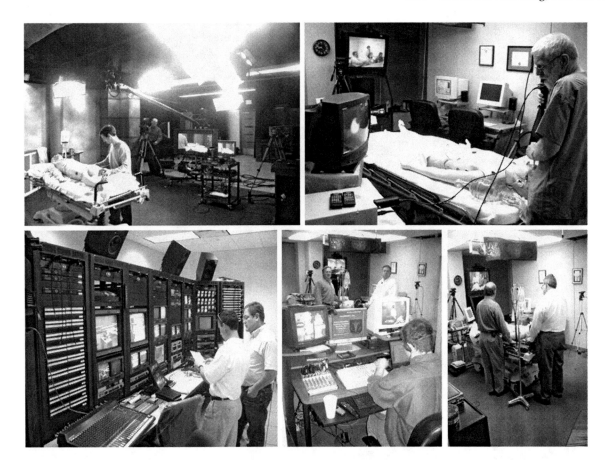

**FIGURE 64.6** Engineers at the advanced, TV-like production studio at the College of Health Professions at Central Michigan University (top left) prepare for a training session on trauma resuscitation. A camera mounted on a remotely controlled boom is suspended directly above the simulator. Note the complexity of the lighting stage that eliminates shadows and provides uniform illumination optimized to the capabilities of the camera lenses, multiple cameras on tripods, and soundproofing on the walls. In the top right picture, Dr. Ludwig, a visiting German anesthesiologist places a microcamera of a fiberscope into the trachea of an HFPS at MedSMART, Inc. facilities in Ann Arbor, U.S.A., during remote training in fiberoptic intubation for Italian students at L'Aquila School of Medicine. The studio at Central Michigan University is almost fully automated and can be remotely controlled from a sophisticated monitoring room (bottom left) located elsewhere at the University. The monitoring room has the capacity of direct, simultaneous output to multiple remote locations all over the world. The control station at MedSMART in Ann Arbor, MI (bottom center) is equally sophisticated although smaller. Its reduced input mixing capability allows control and mixing of inputs from not more than eight separate sources (video, text, still pictures, etc.). Nonetheless, during cardiac resuscitation training shown in the picture to the right, the simulator and two training experts interacting with it are viewed by as many as five simultaneously operating cameras whose input is mixed and delivered to the remote transatlantic trainees in real time. The control station shown in the lower center picture can relinquish camera control to the remote locations, allowing distant students full control over the viewing angles and zoom capability. Reproduced by permission of MedSMART, Inc. Copyright© to all figures and photographs except where noted: MedSMART, Inc.

delivered to the remote sites. Therefore, if more sites are to be served, additional camera units will need to be employed, and in order to assure identical field coverage, the cameras will need to be mounted as close to each other as possible. Mounting the units in the angle of view-corrected racks may provide a simple but rather cumbersome solution. The use of either wavelet compression software such as WINVICOS [140] or integration

with comprehensive medical communication systems may be also contemplated [168], and would provide a much more optimal and robust approach. It must be remembered, though, that the need to have compression executed at the transmission site limits its use to the more technologically advanced remote sites. The same is true of integration into complex communication systems that will require ready access to

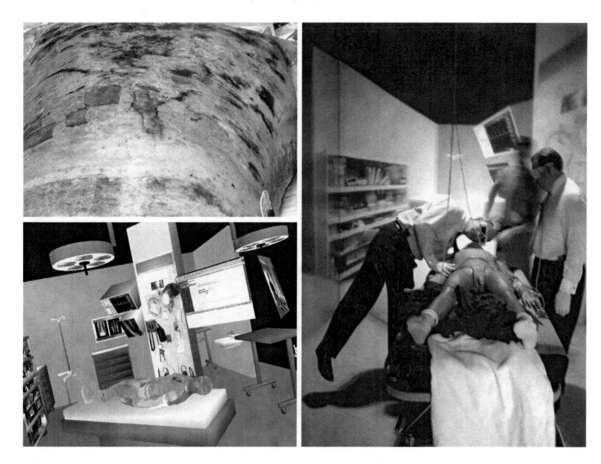

**FIGURE 64.7**  Treatment of burns (upper left) is difficult and requires special skills. Immersive VR permits synthetic but static recreation of the environment in which a badly burned patient is treated (lower left). However, the combination of HFPS and VR (portal) such as developed by the Medical Readiness Trainer and the Virtual Reality Laboratory at the University of Michigan permits interactive practical skill development (right). Reproduced by permission of MedSMART, Inc. Copyright© to all figures and photographs except where noted: MedSMART, Inc.

telecommunications expertise at the remote site and frequent monitoring of platform performance within the training network. Naturally, in such environments (rare at present) where the available bandwidth greatly exceeds uncompressed video flow rate rquirements, substandard bandwidth allocation to each camera unit ceases to be an overriding issue.

While 2-D displays are the cheapest and often most practicable means of visualization, 3-D systems such as virtual reality "CAVE" (Cave Automatic Virtual Environment), head-mounted displays (HMDs), or planar 3-D viewing devices (e.g., ImmersaDesk©) offer the greatest spatial realism and the "surround" or immersive effect. However, despite the emergence of very low-cost VR systems, e.g., "gloves & goggles" and PC stations [169–171], the operational complexity and cost of physical facilities associated with advanced immersive VR operations may be prohibitive for users without the required financial and technical support (see [172]). The most important practical drawback of the immersive VR-based telepresence is, however, the fact that

VR environments need to *be created* (Figure 64.7), while in true telepresence, the viewer is telepresent in the space that is real and already *exists*. Also, most of VR systems need special viewing devices (shutter glasses, head-mounted displays). There is no doubt, however, that immersive VR offers a unique capacity for training complex skills [173, 174] and preparation for executing complex procedures in the environments where execution of on-site training is highly impractical or dangerous (wilderness, ships, battle field environment, etc., Figure 64.8). One of the best U.S. federated medical VR training systems has been developed by the University of New Mexico [175].

A compromise solution between virtual reality–based telepresence and true telepresence has also been proposed. In these environments, a "portal" concept is used as the means of bridging the gap between passive and active awareness of one's surroundings [176]. The "portal" concept provides high-fidelity static VR reconstruction of physical space as a background to the dynamic activity of the remote "collaborator" [177, 178].

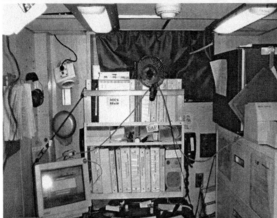
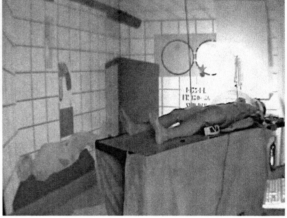

**FIGURE 64.8**  Tight, cramped spaces of warships present special training problems. The cramped sick bay of a small frigate (bottom left) can be recreated as a VR portal (bottom right) to allow medical personnel to acclimatize to work in a ship at sea. Additional realism can be provided by visual cues of the roll and pitch of the vessel, sounds of a ship at sea, and even occasional whiff of burnt fuel, hot machinery, etc., leading to equally realistic seasickness and disorientation. Reproduced by permission of MedSMART, Inc. Copyright© to all figures and photographs except where noted: MedSMART, Inc.

The "portal" concept has been developed to its ultimate level at the University of Michigan where a combination of immersive VR, human patient simulation, and Internet-based resources allowed human operators (emergency room physicians, nurses, paramedics) to execute training activities within an almost entirely synthetic environment of an emergency room patient bay (Figures 64.7, 64.8, and Refs. [121, 179, 180]).

As a variation of the portal discussed above, augmented VR devices may, in combination with advanced wireless telecommunications, become highly portable, cheap, and flexible clinical training and mentoring devices. Using the "blue table" concept (Figure 64.9) as the background for their work, a group of French and U.S. scientists is currently exploring augmented VR in education, training, and telementoring of field emergency medical personnel, First Responders, and nonexpert physicians acting in emergency scenarios.

Autostereoscopic devices [181–183] present a completely different class of viewing platforms that can present "real life" 3-D images whose viewing does not require additional devices such as goggles or display helmets. In the context of medical training, telementoring, or operational telemedicine, autostereoscopic systems offer a very significant advantage over other 3-D viewing platforms: in addition to offering the best optical approximation of the real object [182], they support the physiological foundations of spatial vision. There are, essentially, no differences in viewing objects rendered in an autostereoscopic display from viewing them under natural conditions [184]. The source of images in the autostereoscopic systems can be computer files, stereo film or video recording, or live transmission from stereo cameras. The pseudo-hologram images where the light source appears to come from the model or subject itself can be viewed as a floating object suspended in "free space" without special glasses (Figure 64.10).

Freedom from special viewing devices (e.g., 3-D glasses, head-mounted displays) is particularly important in clinical applications, where prolonged viewing of objects in

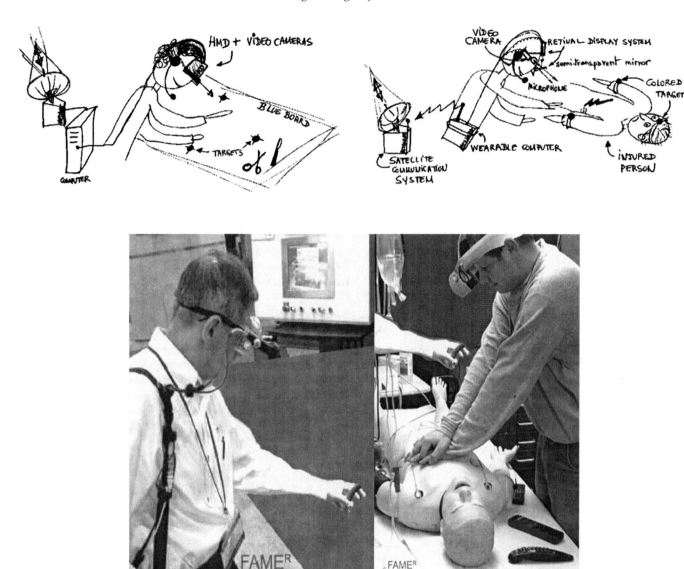

**FIGURE 64.9**  The original sketch (2001) by Prof. Richir of ISTIA Innovation at the University of Angers in France showing the principle of augmented VR system for training and mentoring of field emergency clinical personnel developed by ISTIA and MedSMART, Inc., Ann Arbor, U.S.A. The "BlueBoard"© technology-based system allows the expert (upper drawing) see the object as viewed by a paramedic in the field who follows the advice using voice commands and the retinal display system. The two lower photographs show the practical operation of the system, with the instructor on the left and the clinical trainee wearing a headset on the right side. Note the hand of the instructor appearing in the right picture. The trainee sees it as a semitransparent, "floating in the air" overlay on the simulator and follows its movement. Essentially, the instructor "guides the hand" of the trainee despite thousands of kilometers that may separate them physically. Reproduced by permission of MedSMART, Inc. Copyright© to all figures and photographs except where noted: MedSMART, Inc.

preparation for, e.g., surgery may result in significant fatigue of the user exposed either to these or immersive VR. From the practical point of view, the emerging autostereoscopic systems are independent of format, brand, or software issues, and all sources that provide optical stereoscopic output that can be converted into common video output format (e.g., NTSC, composite video, S-video, RGB, VGA, etc.) can be viewed. Since the optical/electronic sources include stereoscopic cameras, VCRs, computers, DVDs, etc., the flexibility is substantial, and the range of uses covers the entire spectrum of Web-based material from basic science to real time clinical training, consultation, and, in probably not too distant future – telepresence-based surgery and telesurgery (Figure 64.11 and Refs. [174, 185]).

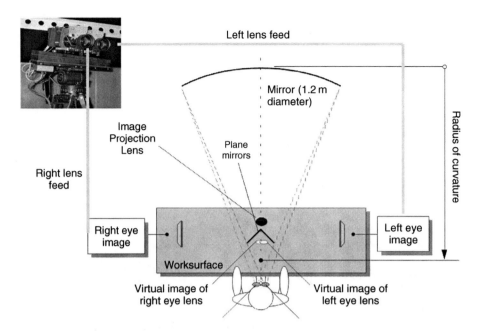

**FIGURE 64.10**   Diagram of an autostereoscopic system developed by Ethereal Technologies, Inc., Ann Arbor, MI. Left and right image is fed (stippled lines) from the stereoscopic camera (Thompson, upper left inset) to two computer screens on the left and right side of the desk, transmitted through a prism and projected on a Mylar mirror. The reflected 3-D image appears suspended in front of the viewer. This is one of several methods of offering depth perception of images produced on flat surfaces that are offered commercially by several companies. Modified from a diagram by Ethereal Technologies, Inc., Ann Arbor, MI., U.S.A. Reproduced by permission of MedSMART, Inc. Copyright© to all figures and photographs except where noted: MedSMART, Inc.

## 64.4.2  Telecommunication Infrastructure

Depending on the level of complexity, the per-channel bandwidth needs of remote simulation range from relatively modest (128 Kbs for HFPS systems) to very substantial (>10 Mbs for VR systems). In the former case, many of the European and U.S. Internet training requirements can be fully satisfied due to the exponential development and popularization of broadband access to the Internet even within relatively remote regions [153]. However, the development of federated and fully integrated VR training systems requires high priced and sophisticated networks with the bandwidth capacity of Internet 2 (Figure 64.12 and see, e.g., Ref. [175]). Nonetheless, the pioneering effort of OP 2000 Group at the Charity Hospital in Teams in Berlin [140, 186, 187] demonstrated that long-distance transmission of complex collaborative VR clinical training environments is feasible not only in purely technical terms but also economically.

It must be remembered, however, that only very few training scenarios (e.g., combat medicine) will require a dynamically changing VR environment. In most cases (car crash

site, explosion aftermath, etc.), the environmental envelope will be static or almost static. In such cases, the "bandwidth-hungry" VR component will be significantly reduced and the VR setting can be generated by the central controlling facility and distributed throughout the network (Figure 64.13). Access from the periphery to the central facility and vice versa can be obtained either by using point-to-point connectivity, with each remote site having its own IP address and an allocated fast Internet connection, dedicated fixed lines (ISDN, ADSL) or wireless ones (GPRS, EDGE, UMTS, Wi-Fi, WiMax), or through a Web-based portal hosted at the central training facility.

The Internet-based access without Quality of Service (QoS), although the simplest one, may become unreliable during extended (more than 1 hour) continuous transmission due to frequent connection interruptions and slow-downs, or up- and download loss of transmission speeds that are particularly annoying during long- or very long-distance operations (e.g., transcontinental or global). Access through a Web portal necessitates its creation [189] – a matter of technical complexity that is best accomplished by the technical personnel

**FIGURE 64.11** The upper row shows a desk-mounted autostereoscopic system using a laser optical element as the image-generating device (Intrepid Inc., Southfield, MI. U.S.A.) and the image of a trachea model it produces. The large Mylar mirror shown in the lower left picture serves as the reflecting surface in the autostereoscopic unit by Ethereal Technologies, Inc. (operating diagram is shown in Figure 64.10). The 3-D image of human cranium seen in the lower right picture is the reflection of the stereoscopic image projected on the mirror and then reflected back at the viewer who sees the cranium "suspended in the air" in front of the mirror (silver surface behind the cranium). Routine distributed resuscitation training using Ethereal Technologies devices has been performed by the Medical Readiness Trainer Team of the University of Michigan at distances of up to 1500 km (see Refs. [107, 159]). Reproduced by permission of MedSMART, Inc. Copyright© to all figures and photographs except where noted: MedSMART, Inc. Lower row figures courtesy of Dr. Robert Andrews, Ethereal Technologies, Inc.

at the central simulation facility. With the portal located at the training facility servers, and a significant part of the operational software necessary for the efficient training (simulator control/translation software; remote camera control software, training scenario programs, etc.) accessible through such a portal, multisite activities become greatly facilitated [189]. The peripheral sites are provided with a simple, intuitively understood simulator interface tool displayed at the remote computer monitor. Control of the simulator is performed either via point-and-click mouse operation or, at a more sophisticated level, by touching appropriate controls on the touch-sensitive screen.

In summary, the central site acts as a broad-concept MedASP (Figure 64.13) that, in addition to the simulation-centered software, provides other training elements, e.g., access to more traditional didactic tools, archives of

previous simulation-based courses, testing materials, etc. In such a configuration, prior experience indicates that even ISDN transmission speeds of 128 Kbs in both directions (upload/download) are adequate to fulfill all the required tasks without any deterioration in the quality of image/voice/data elements although inclusion of higher-end visualization systems (e.g., VR portals) will demand higher (Mbs) bandwidth.

During Internet-based operations, one of the critical factors is the dependence of the overall quality of sound and imagery on the available bandwidth and latency (delay) of the Internet connection. A minimum of 128 Kbs sustained transfer rate is required for real-time videoconferencing. If the round-trip latency exceeds 200 milliseconds, then the conference participants will notice a delay in the conversation, similar to the delay encountered over a satellite telephone call. In practical terms, sustained latencies below 100 milliseconds

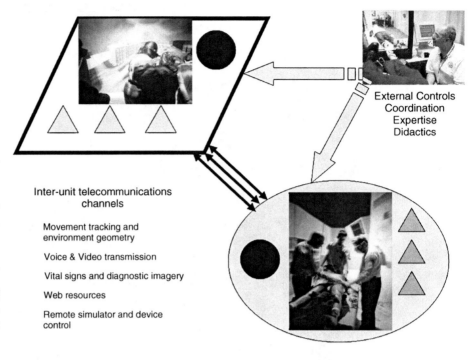

**FIGURE 64.12** While federated VR environments (particularly VR portals) offer major advantages in training in near-real environments, the complexities of distributed federations and their bandwidth requirements are acceptable only for the organizations with substantial technical and fiscal resources. Presently, although a few academic institutions are engaged in predominantly experimental activities, only the armed forces use such federations routinely as a part of their training programs (examples: tank team driving, tank team fighting). Symbols: Rhomboid – the controlling VR environment; oval – peripheral federated VR site; triangles – trainees, circle – training expert. Reproduced by permission of MedSMART, Inc. Copyright© to all figures and photographs except where noted: MedSMART, Inc.

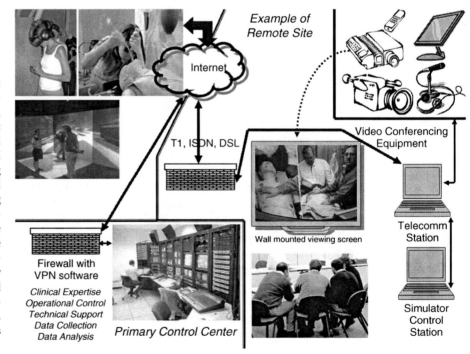

**FIGURE 64.13** Complex federation consisting of classrooms with limited levels of interactivity (right side of the figure), VR environments (portal VR), dynamic peripheral sites, and remotely controlled peripheral sites. Federation is controlled by the Primary Control Center providing subject expertise, multimedia educational support, technical support, and exercising overall program control. The Control Center may switch from structured to "free play" training mode without alerting the participants. It is the latter form of training (i.e., "free play") that is particularly suitable for the development of operational readiness (see Refs. [151, 188]). Reproduced by permission of MedSMART, Inc. Copyright© to all figures and photographs except where noted: MedSMART, Inc.

round trip, and sustained bandwidth greater than 300 Kbs allow continuous and contiguous, uninterrupted transoceanic training sessions lasting up to 4 hours each. This assures that the remote trainees have a very high-quality learning experience based on a fully interactive simulation technology [190].

# 64.5 The Goal for Simulation-based Distance Training

Contrary to live aviation training, where altitude means safety, and where given enough altitude the pilot can safely practice recovery from what would otherwise be catastrophic manevers at low altitudes, live patients cannot be brought up to a "safe altitude" for training purposes. Much can be learned from didactic programs, observation, and peripheral participation in the actions of a clinical management team. Nonetheless, the dilemma of "treat or teach" is real, and the teacher's choice is always obvious – treat. Morally and ethically indisputable, the choice results in insufficient experience by the student to gain ideal mastery of critical clinical conditions. As the result, the novice "clinical pilot," while in posession of the essential knowledge to convert a potential disaster into a safe "landing," often fails under the stress of the real-life circumstances, especially when treating and acting without expert supervision. This is due to being deficient in the ability to translate this knowledge into a chain of appropriate responses needed to avert the adverse outcome.

In similarity to aviation and maritime operations, a broad introduction of simulation-based training is probably the only means of safely equipping clinical personnel fresh from their colleges and schools of medicine with sufficient skills to stabilize any "clinical crisis" long enough for the experts to take over the management. However, in order to be useful, skills (whether manual or cognitive) need to be routinely practiced to be maintained. Hence, exposure to appropriately challenging practice must be sufficiently frequent if its educational impact is to translate into sustained proficiency. In reality, distance from simulation centers (or even nationwide absence of simulators), often with a significant cost associated with travel and participation, or even disruptions of professional and private life caused by their absence for training purposes, pose a severe obstacle in adequately frequent training. Problems associated with simulation-supported training are particularly severe for many of the prehospital personnel (particularly paramedics and firefighters) whose salaries are lower and who frequently perform emergency duties as volunteers, while being full-time employed with other jobs. Distance simulation-based training (Figure 64.14) fills the gap by providing frequent and easy access to the simulators, to superior training expertise (training sessions can be conducted under the supervision of world's foremost specialists), and, when properly conducted, provides the distant trainees to a telepresence experience that

can not only expose their deficiencies, shatter the "I already know it" attitude, but also help by building and strengthening their expertise and cognitive skills.

## 64.5.1 Preparedness and Readiness

Historical examples of military defeats indicate that the critical thinking of the commander under pressure, his ability to extract relevant information from multiple sources, and formulate flexible responses based on such information, as well as his ability to predict the course of action a few *tempi* ahead of the current state of affairs, provide the essential operational advantage that is a prerequisite to victory [e.g., 191]. While some commanders have the intuitive grasp of the principles involved in successful field operations, the majority need training in leadership and crisis resource management. Goal oriented "Free-play" exercises provide the best instrument for the development of the required skills.

There are many obvious differences between military and clinical operations. Among the principal ones is the fact that management of a clinical crisis is, by its very nature, *reactive*. Contrary to the military action that can preempt the perceived or emerging danger, the clinical response is always the consequence to the event that has already taken place: In military terms, it is the response to an ambush. The sudden emergence of the threat (i.e., the crisis) and the dynamic nature of its evolution need to be countered with the flexibility and fluidity based on the appropriate assessment of the principal danger(s) posed by the nature of the emergency. This is followed by the allocation of the available resources to contain the immediate threat(s), subsequent execution of all appropriate and relevant containment actions based on threat evolution, and suitable deployment of all resources necessary to reduce threat expansion. The outcomes of the immediate actions must be then reevaluated, and the next steps, i.e., reiterating assessment, disposition, counteraction, etc., must be taken.

In similarity to the military, intensive cognitive training is essential for all clinical personnel in preparation for dealing with a suddenly arising crisis, and the development and maintenance of the instinctive mastery of operational skills need to be continuously practiced. It is these skills, after all, that are the fundamental aspect of readiness, i.e., instantaneous ability to respond to sudden, unexpected threats [151]. To assure an adequate level of such a response, training must not be based on predetermined patterns and prescribed scenarios known to all participants prior to the training session, but on the development of situational awareness and assessment abilities, as well as on the decision-making and command capability of the trainees, i.e., their ability to think and act clearly and rapidly in the environment saturated with stressors.

Training must emphasize the ability to perform continuous assessment of the crisis evolution within the changing environmental envelope of the event. It must also put emphasis on the flexibility necessary by participants for the development and

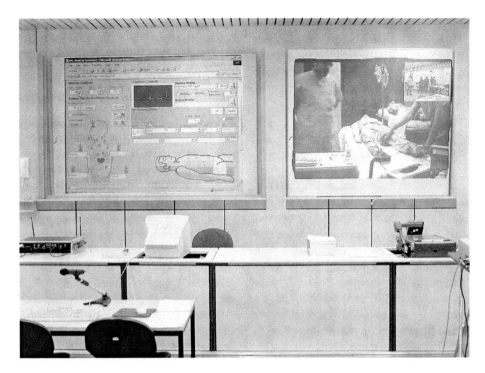

**FIGURE 64.14**  Remote training site in L'Aquila, Italy. Although the site, located at the training facility of Telecom Italia, has abundance of bandwidth, there are no simulators. Senior students from the School of Medicine at L'Aquila University train on human patient simulators in Ann Arbor by accesing the machines from their Italian site. The machines are remotely controlled by the trainees through the standard simulator interface on the left panel of the viewing screen. Using the interface, the student can administer the full range of clinical interventions the simulator supports. However, control of the physiological parametres of the simulator is in the hands of the central site in Ann Arbor, allowing the training expert to introduce unpredictable events (e.g., ruptured aneurysm, adverse reaction to a drug, etc.). The students have full view of the simulator (right panel) and can control viewing cameras (pan, tilt, and zoom) allowing them to see the details of the procedures performed by the personnel in Ann Arbor, detect mistakes in their execution, direct corrective measures, etc. The inset in the upper right corner of the viewing panel shows the view of the Italian site – a new team is preparing to start the exercise under the direction of a Spanish physician visiting the central facility in Ann Arbor, U.S.A. Although distance training provided to the facilities that have no access to simulators precludes practicing manual procedures, this form of training leads, nontheless, to a significant improvement in diagnostic and clinical management skills, drug familiarization, management of clinical crises, and in clinical communication skills. Reproduced by permission of MedSMART, Inc. Copyright© to all figures and photographs except where noted: MedSMART, Inc.

implementation of actions commensurate with the ongoing evolution of the adverse event. Essentially, and in similarity to military and aviation training, in order to be truly effective, clinical simulation-based training (at least at the command level) must be based on the principles of the "free-play" exercises, where little is known at the onset of the exercise, and much information must be collected and analyzed, decisions taken and implemented while the exercise evolves.

For the most part, clinical simulation-based training develops and challenges *preparedness*, defined as *the availability of all resources, both human and physical, necessary for the management of, or the consequences of, a specific type of clinical crisis* [188]. Training is often conducted in a carefully preplanned setting, where surprises and operational stressors are either eliminated or significantly reduced. Yet, major procedural and operational errors are common and frequently repeated, and their persistent existence may indicate that, while the overall level of preparedness may even increase, the level of *readiness* defined as *the instantaneous ability to respond to a suddenly arising major crisis that is based on the locally available/unprepositioned and unmobilized countermeasure resources* is either unchanged or, possibly, decreased due to the inherent flaws built into the current philosophy of training [188]. Moreover, the common error of using preparedness as a synonym for readiness is among the major sources of slowly developing complacency in the attitudes to training [192], although simulation-based studies have clearly demonstrated that highly trained (i.e., *prepared*) personnel, exposed to a sudden crisis whose nature falls outside the scope of prior preparation (i.e., *readiness*), commit grave errors of judgment and procedure [193–195]. It thus appears that the operational doctrine behind current clinical simulation training [196] largely fails to address the realities of crisis development. Unpredictability, confusion, inadequacy of immediately available resources, involvement of unplanned, and seemingly irrelevant factors that continuously change the development of the temporal patterns are all inherent constituents of each crisis, as illustrated by clinical events during and following Hurricane Katrina. Furthermore, the impact of sheer physical and mental stress of the participants [194–200] also promote stress. In other words, while the level of preparedness may be at maximum, the level of readiness may be far from optimal.

### 64.5.2 Critical Thinking and the *Observation, Orientation, Determination, Action* Loop

The theory of interaction with complex environments and the response to their rapidly changing (dynamic) and unpredictable nature has been developed by Boyd [201]. Boyd's Loop – the Observation, Orientation, Determination, Action (OODA) Loop, provides the essential framework for knowledge-based, multidimensional critical thinking and rapid decision making. The Loop's operational consequence (action) is executed in real time, and determines the operational tempo and direction of the entire interaction process [202, 203]. It is this type of behavior that constitutes the foundation of response readiness and successful interaction with unpredictable and dynamically changing environments.

The concept of the Loop [201–203] is based on a cycle of four critical and interrelated elements: **Observation → Orientation → Determination → Action** that revolves in time and space. Observation and Orientation segments of the Loop are the stages at which the plurality of implicit and explicit inputs determines the sequential and subsequent Determination and Action steps. The outcome of the Action stage affects, in turn, the character of the next initiation point (Observation) in the forward progression of the sequence (Figure 64.15). The Orientation stage specifies the characteristics and the nature of the environment necessary to determine the "center thrust" of the planned action (Determination stage), i.e., the site of the most significant (decisive) activity. The subsequent Determination stage defines the nature and characteristics of the action(s) to be taken, and at the Action stage, the planned activity is fully implemented. The Loop does not progress in a linear fashion along the time axis. Instead, the individual stages are often nearly or completely simultaneous: Action does not interrupt Observation, nor does Decision halt Orientation. Since the operational environment is spherical, the domain of the loop is also multidimensional and embraces all constituents of that environment.

Time is only one of those constituents and the ability to shorten the time needed to execute each stage of the Loop reduces the time needed for the full completion of a cycle. The loop becomes "tighter" (i.e., faster or shorter). One stays

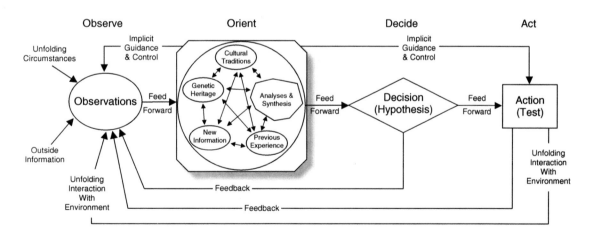

**FIGURE 64.15** Note the complexity of the involved processes that goes far beyond the "Observe-Orient-Decide-Act" cycle typically depicted in the literature. The Loop accounts for both tacit and explicit factors and also for the frequently highly complex, dynamic, and often unpredictable interaction of people, processes, and technology. From "The Essence of Winning and Losing," J. R. Boyd, Jan 1996. http://www.d-n-i.net Reproduced with the permission of the Estate of COL. John Boyd, USAF (2005). Reproduced by permission of MedSMART, Inc. Copyright© to all figures and photographs except where noted: MedSMART, Inc.

"inside the loop" of the unfolding events, and by increasing the speed of one's own Loop revolutions, *the future tempo* and *the course* of action are "driven" (shaped) in the desired direction. Actions become simpler to decide and perform, their specificity and target orientation improve, and one ceases to remain a mere passive subject of pressures generated by the imposed environment. Proper implementation of the Loop principles allows attainment and sustainment of *operational mastery of the unpredictable environment* that constitutes the essential ingredient of a successful clinical crisis management.

Every environment (a patient in a clinical crisis scenario, a disaster scene in the context of multiple casualties, etc.) is characterized by the information it contains. Typically, the environment holds operational advantage – although one must interact with it, very little may be known about the characteristics of that environment. Hence, the initial actions are exploratory and aim at the extraction of meaningful information from that environment (initial assessment). The cumulative advantage of broad-based probing of the environment (i.e., looking beyond the obvious and determining relations among seemingly irrelevant facts) rests in the elevated perceptiveness and sensitivity to *all* inputs that the operational environment provides. Once the state of information superiority is reached (i.e., we have extracted all information from the environment, determined the existing interrelationships and dependencies, and can anticipate with relative certainty what the unavailable data may represent), all responses become more flexible. They can now be based on the analysis of *all* relevant factors rather than those that are either the *most obvious* or the most *traditionally accepted* as relevant: one knows more, and the accumulating knowledge has increasingly greater precision. One attains the state of *information superiority* (Figure 64.16).

Attaining information superiority depends on the speed of OODA Loop revolutions, and constitutes one of the crucial challenges in constantly changing environments that are full of multiple situational inputs and potential dangers which are, in turn, endowed with their own temporal and physical instability (operational chaos or "fog of war" – [204, 205]). In practical terms of medicine, the state of information superiority allows correct assessment and stabilization of the critical patient and a much faster and precise establishment of preliminary diagnosis (Orientation stage). The latter then serves as the basis for the subsequent set of intermediate range actions. The initial treatment decisions (Determination stage) are then made and executed (Action). The results of the Action stage will have a direct bearing upon the next revolution of the Loop. However, information superiority allows continuously increasing predictability of all plausible events that are increasingly more distant from the present. The ensuing forward temporal shift of action leads to a lasting *initiative control* that not only brings order to the originally chaotic environment, but also allows the conduct of all future operations strictly on one's own terms.

Contrary to the exposure provided by hands-on simulator experience, where tactile inputs play a significant role,

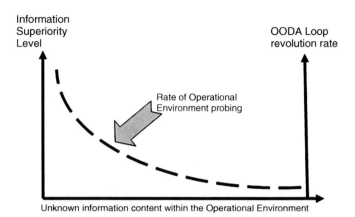

**FIGURE 64.16** Dependence of the state of information superiority on the speed of OODA Loop cycles, which determines the information extraction (probing) rate of the operational environment. However, extraction of information alone is not enough. It has to be compared against the preexisting tacit and explicit knowledge and then transformed into pertinent knowledge, i.e., the knowledge necessary for the solution of the crisis at hand [203, 204]. Networkcentric operations described in this chapter are essential for such transformation in the context of national/global range health care activities, which also involve appropriate training of the relevant personnel. Reproduced by permission of MedSMART, Inc. Copyright© to all figures and photographs except where noted: MedSMART, Inc.

distance simulation-based training aims predominantly at the development of cognitive and leadership skills, the ability to respond to sudden crises, and the ability to assume complete command and control of an environment characterized by rapidly escalating complexity and/or chaos. As already mentioned in Section 64.2, many of the often fatal errors occurr not because of missing skills but due to the lack of knowledge, inadequate individual and team knowledge management skills, inadequate communication within the team and with the essential support personnel external to the team itself (field unit ↔ dispatcher ↔ ambulance crew ↔ Emergency Department ↔ other hospital services, etc.) Distributed simulation environments have the flexibility to permit both direct interaction with the simulator and the verbal/visual contact not only with a very large group of relevant but clinically uninvolved personnel (e.g., dispatch crews, ambulance drivers, rescue pilots, etc.) that can actively participate in the training session, but also with the personnel operating simulators at the simulation center. Since the distant trainee is provided with the full range of diagnostic signs provided by the simulator (or, if need be, the required laboratory results or imaging studies stored at the electronic library of the center), the trainee exercises conceptual solutions based on the ability to extract information, correlate it with the preexisting knowledge base, and implements the correct sequence of actions. The trainer oversees management of the patient and the patient's status changes only in response to actions initiated by the remote student. Verbal and visual contact with the ancillary personnel and the staff of the simulation center allow development of communication, resource mangement, and leadership skills. The student *manages* the case and is forced to provide guidance to the entire personnel involved in the exercise based on knowledge, awareness of best clinical practices pertaining to the crisis at hand, control confusing and often conflicting demands imposed by the environment, availability or resources, personnel friction, etc.

Telepresence is the constituting element of distance-based training. When implemented properly, telepresence provides the trainee with the convincing sense of "being there," the emphasis shifts from the expert trainer leading the trainee to the trainee focusing on the entire spectrum of the available knowledge on the solution of a crisis which, unless addressed appropriately, will deteriorate into an irreversible chaos. Training can be accessed at a very moderate cost, whenever it is needed, at intervals that are most appropriate to the professional needs of the trainees, and reduces constraints imposed by the availability of individual financial resources and/or time for travel to the distant training site. Simulation-based training that exploits full advantages of telepresence (some of which are still at the very early stages of experimentation, i.e., haptic and olfactory telepresence) may eventually fit into another, larger concept, of health care–related activities: Networkcentric Health Care Operations.

# 64.6 Networkcentric Health Care Operations

At the moment, simulation-based clinical training is conducted along the principles of platform-centricity – it centers on the available simulation platform, and despite involvement of modern technology, employs the most traditional, archaic forms of clinical education [202]. Training activities are performed within nonfederated, physically isolated centers, using specific and nonfederated platforms (simulating devices), delivered in an uncoordinated manner, and based on independently developed curricula. While providing sterling service to the immediate users (and a potential revenue stream to the providers), the effect of simulation-based training at level of the national, international, or worldwide scale of preparedness and readiness is essentially nil.

Despite the unprecedented educational and training potential of clinical simulation, platform-centricity eliminates almost all advantages that simulation has in the development of a national/international health care force with uniformly high quality of education based on uniform standards of best clinical practices, and whose training is maintained using uniformly accepted standards. There is no doubt that international personnel trained to such standards would be capable

of delivering high-quality clinical services anywhere and whenever required.

The *networkcentric approach* stands in direct opposition to platform-centricity [201, 205–207]. While the networked use of Information, Computer and Communications Technologies is by no means new, the critical conceptual, architectural, and operational elements of their applications in turbulent and unpredictable environments has been provided and implemented by the U.S. Department of Defence in form of the "doctrine of networkcentric warfare" [208].

The doctrine of "networkcentric health care" is defined [202] as *"unhindered networking operations within and among all domains that govern all activities conducted in healthcare space and are based on free, multidirectional flow and exchange of information, without regard to the involved platforms or platform-systems, and utilizing all available means of $IC^2T$ [Information, Computer and Communications Technologies], including legacy platforms, to facilitate such operations."* In practical terms, the doctrine calls for the development of interconnected information grids that, together, constitute a powerful and well-structured network, promoting and facilitating information sharing among all participants existing within the operational continuum (or "space," see Refs. [208–210]). Consequent to the improved information sharing is the enhancement of information quality and flow. Cumulatively, improved and intensified information management escalates the level of situational awareness that, in turn, serves as the foundation for efficient, real-time collaboration among the involved entities, their self-synchronization, and operational sustainability. Network-centricity has been embraced by the European Union whose rapid transition toward the "networked society" is based on interoperability of different systems, mutually agreed operational standards, and the development of high speed information grids linking the entire continent [211]. Networkcentric health care operations, i.e., education, training, delivery, and administration are a part of the broader European concept of the networked society and may, eventually, contribute to the reduction of health care costs.

Although European health care philosophy and policies differ significantly from those in the U.S.A., many of the problems concerning the quality of the workforce, competence across regions, nondisruptive access to meaningful continuous education and training, remain the same. Simulation-based distance clinical education and training are ideally suitable for networkcentric health care operations. The concept is rooted in Information, Computer and Communications Technologies-facilitated telepresence, interoperability of individual simulation platforms, federation of platforms into larger entities, and the ultimate operation of international, globally accessible multiplatform "mega-systems" (systems of systems, see [212, 213]). Most importantly, it is this form of education and training that will most readily satisfy the demands of multinational entities (e.g., the European Union, Pan American Health Organization) for wide distribution and operational adherence to the uniform standards of knowledge and professional proficiency within the entire health care force whose mobility is, at least within the European Union, essentially unrestricted.

## 64.7 Conclusion

The present world of clinical simulation, exciting as it may be, continues to remain the "toy of the rich." A predominant number of countries of the world has no access to human patient simulators, virtual reality training devices, or massive and complex simulation federations. Yet, it is precisely in these countries that the burden of treatable disease is the greatest, and the needs for training very large numbers of health care providers the most intense. Distance simulation-based medical education and clinical training may address the issue of access to the latest technology, the leading experts, and the most advanced treatment techniques by large numbers of those for whom such things are today beyond their dreams. Rapid expansion of telecommunication networks that follows the needs of global economy results not only in the enhanced variety of means to access the Internet, but also in reduction of prices that, until recently, were beyond the reach of the poor. The concept of networkcentric health care operations that begins to gain momentum in Europe, and soon will spread to the rest of the world as probably the most effective means of reducing costs, assuring quality access to health care, and promote the quality of its delivery encompasses E-education as one of its principal components. With today's major centers of health care operating as the principal centers for simulation-based education and training, the same organizations will transform in the future into the nodes of the vast health care network that will continue to provide education and training but with a rapidly growing emphasis on bridging the distance by means of technology.

What we have described in this chapter is a start, a pioneering effort to demonstrate that what many considered impossible or ineffective can be done, that it is indeed not only possible but also preeminently useful. At the moment, practically all simulation centers concentrate on their small, almost myopic, vision of global reality. The sophisticated, and often very expensive, facilities are accessible only to the very (almost chosen) few, and even within a comparatively small distance from the physical location of such facilities, clinical simulation is a grand theoretical concept rather than a practical reality. Yet, it is not the major centers that need simulation. Within confines of their walls, any and every form of clinical expertise is available at any time for virtually any purpose. Simulation-based training, refresher courses, or competency evaluation will have their greatest impact "out there." There where experts are scarce or simply never show up, where clinical crises that at major centers are dealt with cool aplomb because of amassed know-how become nerve-racking incidents because such know-how is missing. There, where routine training is impossible because of the lack of resources.

Distance simulation-based clinical training can change all that. It can close the distance not only in physical, but also intellectual terms. It can help to eliminate the condescending attitude of well-heeled big brothers and sisters to the "provincial hicks" out there who are derided in the literature because of errors they make, and because of their persisting inadequacy of knowledge and skills. Technology can change a lot, and change it rapidly. But in order for this to happen, a change of mind must take place first, and it is at here that essentially all simulation centers fail miserably by forgetting that the greatest advantage of simulation is the fact that today it can be projected outward, and reach those who can not afford own facilities, own experts, and own technology. It is therefore the duty of the existing simulation centers to think in terms broader than the classical clinical didactics blandly reproduced on plastic cadavers filled with electronic gadgetry. It is at these centers where the realization that the world became smaller must attain its most acute form: Ebola, HIV-AIDS, SARS, avian influenza, and a potential host of other diseases have not emerged from the leisurely suburbia of Europe or U.S. but from regions where health care is poor, training often subliminal, and clinical knowledge outmoded in comparison with the affluent Western world. It may be therefore a highly cynical but, nonetheless, appropriate suggestion that by providing the best there where it does not exist but is most needed may be simply in the interest of saving our known skins. Distance-based simulation may serve as a great tool in such rechanneling of energies. Surely, bold thought and bold action are needed, and the world of medicine, famous for its bold men and women, has also a large number of its less adventurous citizens. Yet, it has been also said that "bold was the man who first ate an oyster." It is the time for the simulation community to stop thinking of cakes and reach for the oyster of the "outer world." Actually, the taste may turn out to be extraordinarily rewarding.

## 64.8 Abbreviations and Definitions

**ADSL:** Asymmetric Digital Subscriber Line is a form of DSL, a data communications technology that permits faster data transmission over copper telephone lines than provided by a conventional modem.

**Autostereoscopy:** Method of displaying three-dimensional images that can be viewed without the use of head-mounted displays or glasses. Autostereoscopy allows perception of depth perception of images produced on flat surfaces (screens).

**BlueBoard:** BlueBoard display technology created by IBM, designed to let people in physically separate locations to collaborate on the same project irrespectively of the distance that separates the participants.

**CAVE:** Acronym of Cave Automatic Virtual Environment. An immersive virtual reality environment in which image projectors located outside the CAVE direct images at four, five, or six of the walls of a room-sized cube. A lifelike visual display is created. Stereoscopic shutter glasses convey the created 3-D image whose relation to the viewer is controlled by his physical movements inside the CAVE. The name Cave in the long version of the acronym has uncertain origins, but may refer either to the cave-like appearance of interior of the visualization space or to the philosopher's contemplation cave in Plato's "Republic."

**CRT:** Cathode Ray Tube

**DIS:** Distributed Interactive Simulation

**EMS:** Emergency Medical Services

**Federation:** In the context of simulation activities, a union of several partially self-governing entities (simulators, CAVES, etc.) united by a central ("federal") control facility. In a simulation federation, the individual components (sites) have self-governing status independent of the control facility (i.e., the activities *within* the federated entity are not subject to the central operational control, although the central facility may impose condition states upon the members of the federation (e.g., introduce unpredictable elements into the simulation environment, change scenario parameters, etc., in order to force the participants to respond to the unexpected).

**HFPS:** High Fidelity Human Patient Simulator. Historically, full-body simulators were known as Human Patient Simulators. However, the name could easily lead to the confusion, with humans simulating diseases in order to develop clinical skills in medical students and also known as Human Patient simulators (or "Standardized Patients"). Hence, "High Fidelity" was added to the name as the means of clear identification of the artificial simulation device.

**HMD:** Head Mounted Display. An HMD has either one or two small displays with magnifying lenses incorporated into a helmet, glasses or visor. With two displays, stereoscopic images can be produced by displaying an offset image to each eye. Eye strain is prevented by the lenses that induce illusion of the images are produced at a much larger distance than in reality. Head-mounted displays may also be coupled with head-movement tracking devices to allow the user to "look around" inside the virtual reality environment using the natural head movements.

**ImmersaDesk:** A drafting-table format virtual device developed in 1994 at the Electronic Visualization Laboratory (EVL) of the University of Illinois at Chicago. Using stereo glasses and sonic head and hand tracking, this projection-based system offers a type of virtual reality that is semi-immersive

**ISDN:** Integrated Services Digital Network. A type of circuit-switched telephone network system that allows digital transmission of voice and data over ordinary telephone copper wires. ISDN-based transmissions have better quality and higher speeds than available with analog systems.

**EDGE:** Enhanced Data rates for GSM Evolution. A digital mobile phone technology that acts as an input to General Packet Radio Service (GPRS) networks. EDGE (also known as EGPRS) can function on any GPRS deployed network (providing necessary upgrades are made).

**GPRS:** General Packet Radio Service (GPRS): a mobile data service available to users of GSM mobile phones. It is often described as "2.5G" (technology between the second (2G) and third (3G) generations of mobile telephony. GPRS provides moderate speed data transfer.

**MedASP:** Medical Application Software Provider. Organization that provides medical knowledge management resources including clinical training expertise, remote access to the electronic-training devices, and technical training support. MedASP organizations serve as portal nodes in the networkcentric (see below) system of operations.

**Network-centricity:** Unhindered networking operations within and among all domains that govern (health care) all activities conducted in the domain of health care space, and are based on free, multidirectional flow and exchange of information, without regard to the involved platforms or platform-systems, and utilizing all available means of $IC^2T$, including legacy platforms, to facilitate such operations. Network-centricity prevents redundancy of resources and promotes reduction of operational costs.

**Platform-centricity:** Operations concentrating on the functionality of one specific system (hardware or software) that cannot be integrated into a federation of disparate systems, even if such systems address the same issue (e.g., a wide variety of incompatible and nonintegratable electronic patient records).

**Portal (Web):** A content management system web site. Typically, it is password protected, permitting the site administrators to edit or update content (text, images and other resources) whenever needed. This allows easy updating of the web site content without the need to learn programing code.

**UMTS:** Universal Mobile Telecommunications System. One of the third-generation (3G) mobile phone technologies. UMTS is sometimes marketed as 3GSM, emphasizing the combination of the 3G nature of the technology and the GSM standard which it was designed to succeed.

**VGA:** Video Graphics Array. An analog video system introduced by IBM and serving as the basis of PC computer graphics. The term VGA is often used to refer to a resolution of $640 \times 480$, regardless of the hardware that produces the picture. VGA was officially superceded by IBM's XGA. However, extensions to VGA made by clone manufacturers are now collectively known as "Super VGA."

**VR:** Virtual Reality. Technology that allows interaction with computer-simulated environments. Most VR environments offer visual experiences displayed either on a computer screen or through special stereoscopic displays (see CAVE, HMD, and ImmersaDesk above). Some simulations provide auditory information through speakers or headphones. In the most advanced tactile information (force feedback), olfactory stimuli are also generated by means of devices coupled to the primary VR system.

**Wi–Fi:** Wi–Fi was intended to be used for mobile computing devices, such as laptops, in LANs (Local Area Networks). It is now often for other applications, such as Internet access, gaming, and basic connectivity of consumer electronics (television sets, DVD players). A Wi-Fi-capable computer, telephone, or personal digital assistant (PDA) can connect to the Internet when in proximity of an access point ("hot spot"). Wi-Fi allows connectivity in peer-to-peer mode that permits devices to connect directly with each other.

**Wi-Max:** Worldwide Interoperability for Microwave Access, defined to promote conformance and interoperability of the standard IEEE 802.16, also known as WirelessMAN. The purpose of Wi-MAX is to employ standards-based technology that will allow delivery of last mile wireless broadband access as an alternative to cable and DSL.

# Acknowledgments

The chapter summarizes our work since 1996. Much of it was groundbreaking, performed for the first time, and at times against staunch opposition. Obviously, such work could not be done without the participation of many other scientists, and the fact that only three names appear on the front page indicates simply that we provide the summary; we have directly involved, but are not the ones who "did it all."

We owe gratitude to Medical Education Technologies, Inc., Sarasota Florida, U.S.A., for placing at the very early stages of our work two of their simulators at our disposal. The machines allowed us to perform the initial studies of the concept at the University of Michigan Department of Emergency Medicine, whose chairman Dr. William Barsan was indisputably the first who recognized the value of the concept, and offered

unflinching support – often in the face of significant opposition, and despite frequently buccaneering style of operations conducted by the rest of us – the "Young Turks." None of the initial work could be done without the selfless assistance of many of the Department's physicians, Drs. James Freer, David Treloar, and William Wilkerson, who found time in their extremely busy schedules to come and "play." The play was also supported by Dr. Jose-Marie Griffiths, then the CIO of the University of Michigan, and now one of the deans at University of North Carolina in Chapel Hill, and a member of the Presidential Scientific Advisory Board.

Subsequently, nothing would be done if not the trust and help that we received from Laerdal, Inc., and in particular its chairman, Tore Laerdal. The generosity that we have received from him and his company (Rick Ritt and David Johnson in particular) goes far, far beyond the "call of duty." Many of our operations would have simply never happened if not their selfless, often last moment, assistance with equipment, technical help, and even personnel.

Similarly, Dr. Marvis Lary, the dean of the College of Health Professions at the Central Michigan University offered us not only the financial support of the College but also put at our disposal her splendid facilities as the command center during our operations with the Air National Guard in Alpena, MI., U.S.A., and during the work performed with Telecom Italia in L'Aquila, and for which we owe her our gratitude.

We were fortunate in obtaining generous support of Citrix Inc., Ft. Lauderdale, FL, U.S.A. The assistance of their "ambassador," Dr. Traver Gruen-Kennedy opened many subsequent doors. Incredible as it may seem, the support was obtained without meticulously formatted, voluminous applications, but merely on the basis of a seven slide-long Power Point presentation of the idea that, at that time, was not even fully crystallized. It is heartwarming to know that in the world of current bureaucratic nit-picking of applications for research funding, the true renaissance sponsors of bold ideas are still around. Unfortunately, there are very few of them. We have been extraordinarily lucky to find one of them.

Our special thanks go to the countless members of the U.S. and NATO Armed Forces who participated in our activities, but especially to CAPT. Brown and Montgomery of the U.S. Navy, and CAPT. French of the U.S. Coast Guard for making many of the pioneering efforts possible. The hospitality of CDR. Sikorsky, USCG, his wardroom, and his crew aboard USCGC Forward will never be forgotten. It was in their vessel that the world's first distance simulation-based training at sea was performed in mid-Atlantic.

Benjamin Carrsaco, the President of Digital Realm, Inc., Ann Arbor, MI, U.S.A., put unhesitatingly both telecommunication and fiscal resources of his company at our disposal, and his intellect and unsurpassed knowledge of everything concerning computer and telecommunications technologies helped us to make the impossible into actually a fairly easy endeavor. So did Madeline Leigh with her unmatchable organizational skills necessary to steer the often stormy administrative. While she "fought the elements," we played knowing that Madeline would always made sure the game would succeed.

Drs. Peter Beier, Richard Levine, Timm Ludwig, Tim Pletcher, Caleb Poirier, Simon Richir, Eric Wolf (alphabetically, but all superb friends and professional colleagues in any order) were all indispensable contributors to the effort. Their names can be found on numerous publications that emerged as the result of our work, and these are probably the best acknowledgment of their significant part in the arena of distance simulation-based training.

The list of all those to whom we owe our gratitude is, unsurprisingly, a very long one. It always is when activities take place on several continents, and with different groups of people. There are too many of them, scientists, engineers, paramedics, administrators, soldiers and sailors, physicians and nurses, to mention all of their names here. We remember them all, and will never forget their work and help that we received from them.

In the end, albeit it may seem facetious, we would also like thank our opponents and critics. They helped us to look at things differently, they helped us to think harder and to devise new solutions, and ultimately – they helped us to forget frustrations and "keep on playing."

# References

1. Porter, R. The Greatest Benefit to Mankind: A Medical History of Humanity, Norton & Co., 1997, pp. 3–831.
2. Kegley, J. A. Genetics decision-making: A template for problems with informed consent, *Med. Law* 21, 459–471 (2002).
3. Colliver, J. A. Educational theory and medical education practice: A cautionary note for medical school faculty, *Acad. Med.* 77, 1217–1220 (2002).
4. Hans, K. Global standards in medical education for better health care, *Med. Educ.* 36, 1116 (2002).
5. Billingham, K., Howe, A., and Walters, C., In our own image – a multidisciplinary qualitative analysis of medical education, *J Interprof Care* 16, 379–389 (2002).
6. Leitch, R. A., Magee, H., and Moses, G. R. Simulation and the future of military medicine, *Mil. Med.* 167, 350–354 (2002).
7. Harter, P., Krummel, T., and Reznek, M. Virtual reality and simulation: Training the future emergency physician, *Acad. Emerg. Med.* 9, 78–87 (2002).
8. Letterie, G. S. How virtual reality may enhance training in obstetrics and gynecology, *AM. J. Obstet. Gynecol.* 187, S37–S40 (2002).
9. Hoyal, F. M. "Swallowing the medicine": Determining the present and desired modes for delivery of continuing medical education to rural doctors, *Aust. J. Rural Health* 7, 212–215 (1999).
10. Davis, P. and McCraken, P. Restructuring rural continuing medical education through videoconferencing, *J. Telemed. Telecare* 8, 108–109 (2002).
11. Delaney, G., Khadra, M. H., Lim, S. E., et al. Challenges to rural medical education: A student perspective, *Aust. J. Rural Health* 10, 168–172 (2002).
12. Booth, B. and Lawrance, R. Quality assurance and continuing education needs of rural and remote general practitioners: How are they changing?, *Aust. J. Rural Health* 9, 256–274 (2001).
13. Rourke, J. T. and Rourke, L. L. Rural family medicine training in Canada, *Can. Fam. Physician* 41, 993–1000 (1995).
14. Darling, R. G., Huebner, K. D., Noah, D. L., and Waeckerle, J. F. The history and threat of biological warfare and terrorism, *Emerg. Med. Clin. North Am.* 20, 255–271 (2002).
15. Brocato, C. E. and Miller, G. T. The next agent of terror? Understanding smallpox and its implications for prehospital crews, *J. Emerg. Med. Serv.* 27, 52–55 (2002).
16. Willaims, B. Bioterrorism: Are we prepared?, *Tenn. Med.* 94, 413–417 (2001).
17. Cunha, B. A. Anthrax, tularemia, plague, ebola or smallpox as agents of bioterrorism: Recognition in the emergency room, *Clin. Microbiol. Infect.* 8, 489–503 (2002).
18. Fahlgren, T. L. and Drenkard, K. N. Healthcare system disaster preparedness, Part 2: Nursing executive role in leadership, *J. Nurs. Adm.* 32, 531–537 (2002).
19. Guay, A. H. Dentistry's response to bioterrorism: A report of consensus workshop, *J. Am. Dent. Assoc.* 133, 1181–1187 (2002).
20. Rubinshtein, R., Robenshtok, E., Eisenkraft A., et al. Training Israeli medical personnel to treat casualties of nuclear, biologic, and chemical warfare, *Isr. Med. Assoc.* 4, 545–548 (2002).
21. George, G., Ramsay, K., Rochester, M., et al. Facilities for chemical decontamination in accident and emergency departments in the United Kingdom, *Emerg. Med. J.* 19, 453–457 (2002).
22. Baker, D. J. Management of respiratory failure in toxic disasters, *Resuscitation* 42, 125–131 (1999).
23. Peleg, K., Reuveni, H., and Stein, M. Earthquake disasters-lessons to be learned, *Isr. Med. Assoc. J.* 4, 361–365 (2002).
24. Cowley. R. A., Myers, R. A., and Gretes, A. J. EMS response to mass casualties, *Emerg. Med. Clin. North Am.* 2, 687–692 (1984).
25. Doyle, C. J. Mass casualty incident. Integration with prehospital care, *Emerg. Med. Clin. North Am.* 8, 163–175 (1990).
26. Watterson, A. E. and Thomas, H. F. Acute pesticide poisoning in the U.K. and information and training needs

of general practitioners recording a conundrum, *Public Health* **106**, 473–480 (1992).

27. Lalich, R. A. The role of state government, local government, and nongovernmental organizations in medical innovative readiness training, *Mil. Med.* **167**, 367–369 (2002).

28. Koplan, J. CDC's strategic plan for bioterrorism preparedness and response, *Public Health Rep.* **116**(2), 9–16 (2001).

29. Levi, L. Michaelson, M., Admi, H. Bregman, D., and Bar-Nahor, R. National strategy for mass casualty situations and its effects on the hospital, *Prehospital Disaster Med.* **17**(1), 12–16 (2002).

30. Donini-Lenhoff, F. G. and Hedrick, H. L. Growth of specialization in graduate medical education, *JAMA* **284**, 1284–1289 (2000).

31. Schroeder, S. A. Primary care at a crossroads, *Acad. Med.* **77**, 767–773 (2002).

32. Gulesen, O. Specialization of doctors, general practice and the training system, *Can. Sociol. Demogr. Med.* **41**, 386–396 (2001).

33. Buckingham, C. D. and Adams, A. Classifying clinical decision making: A unifying approach, *J Adv. Nurs.* **32**, 981–990 (2000).

34. Connor, M., Ponte, P. R., and Conway, J. Multidisciplinary approaches to reducing error and risk in a patient care setting, *Crit. Care Nurs. Clin. North Am.* **14**, 359–367 (2002).

35. Scheen, A. J., Rorive, M., Letiexhe, M., et al. Multidisciplinary management of the obese patient: Example from the Obesity Center at the University of Liege, *Rev. Med. Liege* **56**, 474–479 (2001).

36. Hazard, R. G. The multidisciplinary approach to occupational low back pain and disability, *J Am. Acad. Orthop. Surg.* **2**, 157–163 (1994).

37. Schriefer, J., Engelhard, J., DiCesare, L., et al. Merging clinical pathway programs as part of overall health systems mergers: A ten-step guide. Spectrum Health, *Jt. Comm. J Qual. Improv.* **26**, 29–38 (2000).

38. Burd, A., Cheung, K. W., Ho, W. S., et al. Before the paradigmshift: Concepts and communication between doctors and nurses in a burns team, *Burns* **28**, 691–695 (2002).

39. Sherwood, G., Thomas, E., Bennett, D. S., and Lewis, P. A teamwork model to promote patient safety in critical care, *Crit. Care Nurs. Clin. North Am.* **14**, 333–340 (2002).

40. Lingard, L., Reznick, R., DeVito, I., and Espin, S. Forming professional identities on the health care team: Discursive constructions of the "other" in the operating room, *Med. Educ.* **36**, 728–734 (2002).

41. Cooper, J. B., Newbower, R. S., Long, C. D., and McPeek, B. Preventable anesthesia mishaps: A study of human factors, *Qual. Saf. Health Care* **11**, 277–282 (2002).

42. Ruppert, M., Reith, M. W., Widmann, J. H., et al. Checking for breathing: Evaluation of the diagnostic capability of emergency medical services personnel, physicians, medical students, and medical laypersons, *Ann. Emerg. Med.* **34**, 720–729 (1999).

43. Lefrancois, D. P. and Dufour, D. G. Use of the esophageal tracheal combitube by basic emergency medical technicians, *Resuscitation* **52**, 77–83 (2002).

44. Bair, A. E., Filbin, M. R., Kulkarni, R. G, and Walls, R. M. The failed intubation attempt in the emergency department: Analysis of prevalence, rescue techniques, and personnel, *J. Emerg. Med.* **23**, 131–140 (2002).

45. De Lorenzo, R. A. Prehospital misidentification of tachydysrhythmias: A report of five cases, *J. Emerg. Med.* **11**, 431–436 (1993).

46. Trzeciak, S., Erickson, T., Bunney, E. B., and Sloan, E. P. Variation in patient management based on ECG interpretation by emergency medicine and internal medicine residents, *Am. J Emerg. Med.* **20**, 188–195 (2002).

47. Herlitz, J. Hansson, E. Ringvall, E., et al. Predicting a life-threatening disease and death among ambulance-transported patients with chest pain or other symptoms raising suspicion of an acute coronary syndrome, *Am. J. Emerg. Med.* **20**, 588–594 (2002).

48. Linn, S., Knoller, N., Giligan, C. G., and Dreifus, U. The sky is a limit: Errors in prehospital diagnosis by flight physicians, *Am. J. Emerg. Med.* **15**, 316–320 (1997).

49. Holliman, C. J., Wuerz, R. C., and Meador, S. A. Medical command errors in an urban advanced life support system, *Ann. Emerg. Med.* **21**, 347–350 (1992).

50. Chiara, O., Scott, J. D., Cimbanassi, S., et al. Trauma deaths in an Italian urban area: An audit of pre-hospital and in-hospital trauma care, *Injury* **33**, 553–562 (2002).

51. Cupera, J., Mannova, J., Rihova, H., et al. Quality of prehospital management of patients with burn injuries-a retrospective study, *Acta. Chir. Plast.* **44**, 59–62 (2002).

52. Cone, K. J. and Murray, R. Characteristics, insights, decision making, and preparation of ED triage nurses, *J. Emerg. Nurs.* **28**, 401–406 (2002).

53. Hodgetts, T. J., Kenward, G., Vlackonikolis, I., et al. Incidence, location and reasons for avoidable in-hospital cardiacarrest in a district general hospital, *Resuscitation* **54**, 115–123 (2002).

54. Tye, J. B., Hartford, C. E., and Wallace, R. B. Survey of continuing needs for nonemergency physicians in emergency medicine, *JACEP* **7**, 16–19 (1978).

55. Luiz, T., Hees, K., and Ellinger, K. Prehospital management of emergency patients after previous treatment by general practitioners – a prospective study, *Anasthesiol. Intensivmed. Notfallmed. Schmerzther.* **32**, 726–733 (1997).

56. Tollhurst, H., McMillan, J., McInerney, P., and Bernasconi, J. The emergency medicine training needs of

rural general practitioners, *Aust. J. Rural Health* 7, 90–96 (1999).

57. Somers, G. T., Maxfield, N., and Drinkwater, E. J. General practitioner preparedness to respond to a medical disaster. Part I: Skills and equipment, *Aust. Fam. Physician* 28, S3–S9 (1999).

58. Johnston, C. L., Coulthard, M. G., Schluter, P. J., and Dick, M. L., Medical Emergencies in general practice in south–east Queensland: Prevalence and practice preparedness, *Med. J. Aust.* 175, 99–103 (2001).

59. Dick, M. L., Schluter, P., Johnston, C., and Coulthard, M. GP's perceived competence and comfort in managing medical emergencies in southeast Queensland, *Aust. Fam. Physician* 31, 870–875 (2002).

60. Dick, M. L., Johnston, C., and Schluter, P. Managing emergencies in general practice. How can we do even better?, *Aust. Fam. Physician* 31, 789–790 (2002).

61. Simon, H. K., Steele, D. W., Lewander, W. J., and Linakis, J. G. Are pediatric emergency medicine training programs meeting their goals and objectives? A self-assessment of individuals completing fellowship training in 1993, *Pediatr. Emerg. Care* 10, 208–212 (1994).

62. Simon, H. K. and Sullivan, F. Confidence in performance of pediatric emergency procedures by community emergency practitioners, *Pediatr. Emerg. Care* 12, 336–340 (1996).

63. Mil'kov, B. O., Shamrei, G. P., Stashchuk, V. F., et al. Training of the general physician in the problems of emergency surgical care, *Sov. Zdravookhr.* 7, 46–48 (1988).

64. Kelly, L. Surgical skills for family physicians. Do family physicians make the cut?, *Can. Fam. Physician* 44, 476–477 (1998).

65. Reid, S. J., Chabikuli, N., Jaques, P. H., and Fehrsen, G. S. The procedural skills of rural hospital doctors, *S Afr. Med. J.* 89, 769–774 (1999).

66. Sohier, N., Frejacques, L., and Gagnayre, R. Design and implementation of a training programme for general practitioners in emergency surgery and obstetrics in precarious situations in Ethiopia, *Ann. R. Coll. Surg. Engl.* 81, 367–375 (1999).

67. Girgis, A., Sanson-Fisher, R. W., and Walsh, R. A. Preventive and other interactional skills of general practitioners, surgeons, and physicians: Perceived competence and endorsement of postgraduate training, *Prev. Med.* 32, 73–81 (2001).

68. Linder, G., Murphy, P., and Streger, M. R. You did what? Clinical errors in EMS, *Emerg. Med. Serv.* 30, 69–71 (2001).

69. Cayten, C. G., Herrmann, N., Cole, L. W., and Walsh, S. Assessing the validity of EMS data, *JACEP* 7, 390–396 (1978).

70. Williams, K. A. Rose, W. D., and Simon, R. Teamwork in emergency medical services, *Air Med. J.* 18, 149–153 (1999).

71. Eberle, B., Dick, W. F., Schneider, T., et al. Checking the carotid pulse check: Diagnostic accuracy of first responders in patients with and without a pulse, *Resuscitation* 33, 107–116 (1996).

72. Coontz, D. A. and Gratton, M. Endotracheal rules of engagement. How to reduce the incidence of unrecognized esophageal intubations, *J. Emerg. Med. Serv.* 27, 44–50, 52–54, 56–59 (2002).

73. Bradley, J. S., Billows, G. L., Olinger, M. L., et al. Prehospital oral endotracheal intubation by rural basic emergency medical technicians, *Ann. Emerg. Med.* 32, 26–32 (1998).

74. Hubble, M. W., Paschal, K. R., and Sanders, T. A. Medication calculation skills of practicing paramedics, *Prehosp. Emerg. Care* 4, 253–260 (2000).

75. Liberman, M., Lavoie, A., Mulder, D., and Sampalis, J. Cardiopulmonary resuscitation: Errors made by pre-hospital emergency medical personnel, *Resuscitation* 42, 47–55 (1999).

76. Peacock, J. B., Blackwell, V. H, and Wainscott, M. Medical reliability of advanced prehospital cardiac life support, *Ann. Emerg. Med.* 14, 407–409 (1985).

77. Seidel, J. S., Henderson, D. P., Ward, P., et al. Pediatric prehospital care in urban and rural areas, *Pediatrics* 88, 681–690 (1991).

78. Seidel, J. S. Emergency medical services and the pediatric patient: Are the needs being met? II. Training and equipping emergency medical services providers for pediatric emergencies, *Pediatrics* 78, 808–812 (1986).

79. Su, E. Schmidt, T. A. Mann, N. C., and Zechnich, A. D. A randomized controlled trial to asses decay ion acquired knowledge among paramedics completing a pediatric resuscitation course, *Acad. Emerg. Med.* 7, 779–786 (2000).

80. West, H. Basic infant life support: Retention of knowledge and skill, *Paediatr. Nurs.* 12, 34–37 (2000).

81. Snyder, W. and Smit, S. Evaluating the evaluators: Interrater reliability on EMT licensing examination, *Prehosp. Emerg. Care* 2, 37–46 (1998).

82. Zautcke, J. L., Lee, R. W., and Ethington, N. A. Paramedic skill decay, *J Emerg. Med.* 5, 505–512 (1987).

83. Graber, M., Gordon, R., and Franklin, N. Reducing diagnostic errors in medicine: What's the goal?, *Acad. Med.* 77, 981–992 (2002).

84. Elkin, P. L. and Gorman, P. N. Continuing medical education and patient safety: An agenda for lifelong learning, *J Am. Med. Inform. Assoc.* 9, S128–132 (2002).

85. Sultz, H. A., Sawner, K. A., and Sherwin, F. S. Determining and maintaining competence: An obligation of allied health education, *J Allied Health* 13, 272–279 (1984).

86. Taylor, H. A. and Kiser, W. R. Reported comfort with obstetrical emergencies before and after participation in the advanced life support in obstetrics course, *Fam. Med.* 30, 103–107 (1998).

87. Hall, W. L. and Nowels, D. Colorado family practice graduates' preparation for an practice of emergency medicine, *J Am. Board Fam. Pract.* **13**, 246–250 (2000).

88. Wise, A. L., Hays, R. B., Adkins, P. B., et al. Training for rural general practice, *Med. J Aust.* **161**, 314–318 (1994).

89. Forti, E. M., Martin, K. E., Jones, R. L., and Herman, J. M. An assessment of practice support and continuing medical education needs of rural Pennsylvania family physicians, *J Rural Health* **12**, 432–437 (1996).

90. Orient, J. M., Lindsay, D., and Whitney, P. J. Educating primary physicians in emergency surgical procedures, *South Med. J.* **75**, 852–854 (1982).

91. Kanz, K. G., Sturm, J. A., and Mutschler, W. Algorithm for prehospital blunt trauma management, *Unfallchirurg.* **105**, 1007–1014 (2002).

92. el-Tobgy, E. and Rupp, T. Anaphylaxis. Vicious chain reaction, *J Emerg. Med. Serv.* **27**, 84–88, 90–93 (2002).

93. Gage, H., Kenward, G., Hodgetts, T. J., Castle, N., Ineson, N., Shaikh, L., Health system costs of in-hospital cardiac arrest, *Resuscitation* **54**, 139–146 (2002).

94. Ronco, R., King, W., Donley, D. K., and Tilden, S. J. Outcome and cost at a children's hospital following resuscitation for out-of-hospital cardiopulmonary arrest, *Arch. Pediatr. Adolesc. Med.* **149**, 210–214 (1995).

95. Rosemurgy, A. S., Norris, P. A., Olson, S. M., et al. Prehospital traumatic cardiac arrest: The cost of futility, *J. Trauma* **35**, 473–474 (1993).

96. Zack, J. E., Garrison, T., Trovillion, E., et al. Effect of an education aimed at reducing the occurrence of ventilator–associated pneumonia, *Crit. Care Med.* **30**, 2407–2412 (2002).

97. Lancaster, T., Hart, R., and Gardner, S. Literature and medicine: Evaluating a special study module using the nominal group technique, *Med. Educ.* **36**, 1071–1076 (2002).

98. Fritsche, L., Greenhalgh, T., Falck-Ytter, Y., et al. Do short courses in evidence based medicine improve knowledge and skills? Validation of Berlin questionnaire and before and after study of courses in evidence based medicine, *BMJ.* **325**, 1338–1341 (2002).

99. Murphy, A. W., Bury, G., and Dowling, E. J. Teaching immediate cardiac care to general practitioners: A faculty-based approach, *Med. Educ.* **29**, 154–158 (1995).

100. Rourke, J. T. Rural advanced life support update course, *J. Emerg. Med.* **12**, 107–111 (1994).

101. Loutfi, A., McLean, A. P., and Pickering. J. Training general practitioners in surgical and obstetrical emergencies in Ethiopia, *Trop. Doct.* **25**(1), 22–26 (1995).

102. Sanson-Fisher, R. W., Rolfe, I. E., Jones, P., et al. Trialling a new way to learn clinical skills: Systematic clinical appraisal and learning, *Med. Educ.* **36**, 1028–1034 (2002).

103. Moseley, T. H., Cantrell, M. J., and Deloney, L. A. Clinical skills center attending: An innovative senior medical school elective, *Acad. Med.* **77**, 1176 (2002).

104. Gold, J. P., Verrier, E. A., Olinger, G. N., and Orringer, M. B. Development of a CD-ROM Internet Hybrid: A new thoracic surgery curriculum, *Ann. Thorac. Surg.* **74**, 1741–1746 (2002).

105. De Leo, G., Krishna, S., Balas, E. A., et al. WEB-WAP Based Telecare, *Proc. AMIA Symp.* 202–204 (2002).

106. Dornan, T., Carroll, C., and Parboosingh, J. An electronic learning portfolio for elective continuing professional development, *Med. Educ.* **36**, 767–769 (2002).

107. Von Lubitz, D. K. J. E., Levine, H., and Wolf, E. The Goose, the Gander, or the Strasbourg Paté for all: Medical education, world, and the internet. In: Chin, W., Patricelli, F., and Milutinovic, V. (eds.), *Electronic Business and Education: Recent Advances in Internet Infrastructures.* Kluwer Academic Publisher, Boston, 2002, pp. 189–210.

108. Greengold, N. L. A Web-based program for implementing evidence-based patient safety recommendations, *Jt. Comm. Qual. Improv.* **28**, 340–348 (2002).

109. Tichon Jennifer, G. Problem-based learning: A case study in providing e-health education using the Interent, *J. Telemed. Telecare* **8**, 66–68 (2002).

110. Mann, T. and Colven, R. A picture is worth more than a thousand words: Enhancement of a pre-exam telephone consultation in dermatology with digital images, *Acad. Med.* **77**, 742–743 (2002).

111. Casebeer, L. Allison, J., and Spettell, C. M. Designing tailored web-based instruction to improve practicing physicians' chlamydial screening rates, *Acad. Med.* **77**, 929 (2002).

112. Poyner, A., Wood, A., and Herzberg, J. Distance learning project-information skills training: Supporting flexible trainees in psychiatry, *Health Info. Libr. J.* **19**, 84–89 (2002).

113. Fieschi, M., Soula, G., Roch, G. J., et al. Experimenting with new paradigms for medical education and the emergence of a distance learning degree using the Internet: Teaching evidence-based medicine, *Med. Inform. Internet Med.* **27**, 1–11 (2002).

114. Deodhar, J. Telemedicine by email – experience in neonatal care at a primary care facility in rural India, *J. Telemed. Telecare* **8**, 20–21 (2002).

115. Pastuszak, J. and Rodowicz, M. O. Internal e-mail: An avenue of educational opportunity, *J Contin. Educ. Nurs.* **33**, 164–177 (2002).

116. Marshall, J. N., Stewart, M., and Ostbye, T. Small-group CME using e-mail discussions. Can it work?, *Can. Fam. Physician* **47**, 557–563 (2001).

117. Haythornthwaite, S. Videoconferencing training for those working with at-risk young people in rural areas of Western Australia, *J. Telemed. Telecare* **8**, 29–33 (2002).

118. Davis, P. and McCraken, P. Restructuring rural continuing medical education through videoconferencing, *J. Telemed. Telecare* **8**, 108–109 (2002).

119. Allen, M., Sargeant, J., McDougall, E., and O'Brien, B. Evaluation of videoconferencing grand rounds, *J. Telemed. Telecare* **8**, 210–216 (2002).

120. Allen, M. Sargeant, J., McDougall, E., and Proctor-Simms, M. Videoconferencing for continuing medical education: From pilot project to sustained programme, *J. Telemed. Telecare* **8**, 131–137 (2002).

121. Pletcher, T., Beijer, P., Levine, H., et al. Immersive virtual reality platform for medical training: A "killer application." In: Williams, J., et al. (eds.), *Medicine Meets Virtual Reality.* IOS Press, Amesterdam, 2000, pp. 207–213.

122. Lamminen, H., Niiranen, S., Niemi, K., et al. Health related services on the Interent, *Med. Inform. Internet Med.* **27**, 13–20 (2002).

123. Fay, V., Feldt, K. S., Greenberg, S. A., et al. Providing optimal hands-on experience. A guide for clinical preceptors, *Adv. Nurs. Pract.* **9**, 71–74, 110 (2001).

124. Hicks, G. D. An Appeal for more "hands-on" surgical training and experience, *Plast. Reconstr. Surg.* **107**, 1612–1613 (2001).

125. Gonzalez, Y. M. and Mohl, N. D. Care of patients with temporomandibular disorders: An educational challenge, *J. Orofac. Pain* **16**, 200–206 (2002).

126. Robinson, G. Do general practitioners' risk-taking propensities and learning styles influence their continuing medical education preferences?, *Med. Tech.* **24**, 71–78 (2002).

127. Girdler, N. M. Competency in sedation, *Br. Dent. J.*, **191**, 119 (2001).

128. Mandavia, D. P., Argona, J., Chan, L., et al. Ultrasound training for emergency physicians – a prospective study, *Acad. Emerg. Med.* **7**, 1008–1014 (2000).

129. Haponik, E. F., Russell, G. B., Beamis, J. F. Jr., et al. Bronchoscopy training: Current fellows' experiences and some concerns for the future, *Chest.* **118**, 625–630 (2000).

130. Friedrich, M. J. Practice makes perfect: Risk-free medical training with patient simulators, *JAMA* **288**, 2811–2812 (2002).

131. http://www.flightofthephoenix.org/link_trainer.htm

132. Knowles, W. B. Aerospace simulation and human performance research, *Hum. Factors* **9**, 149–159 (1967).

133. Krebs, W. K., McCarley, J. S., and Bryant, E. V. Effects of mission rehearsal simulation on air-to-ground target acquisition, *Hum. Factors* **41**, 553–558 (1999).

134. Ricard, G. L. Acquisition of control skill with delayed and compensated displays, *Hum. Factors* **37**, 652–658 (1995).

135. Brannick, M. T., Prince, A., Prince, C., and Salas, E. The measurement of team process, *Hum. Factors* **37**, 641–651 (1995).

136. Nagoshi, M. H. Role of standardized patients in medical education, *Hawaii Med. J.* **60**, 323–324 (2001).

137. Vardi, A., Levin, I., Berkenstadt, H., et al. Simulation-based training of medical teams to manage chemical warefare casualties, *Isr. Med. Assoc. J.* **4**, 540–544 (2002).

138. Greenberg, R., Loyd, G., and Wesley, G. Integrated simulation expereinces to enhance clinical education, *Med. Educ.* **36**, 1109–1110 (2002).

139. Bond, W. F. and Spillane, L. The use of simulation for emergency medicine resident assessment, *Acad. Emerg. Med.* **9**, 1295–1299 (2002).

140. Shapiro, M. and Morchi, R. High-fidelity Medical Simulation and Teamwork Training to Enhance Medical Student Performance in Cardiac Resuscitation, *Acad. Emerg. Med.* **9**, 1055–1056 (2002).

141. Graschew, G. Roelofs, T. A., Rakowsky, S., and Schlag, P. M. Intereactive telemedical applications in OP 2000 via satellite, *Biomed. Tech.* (Berl) **47**, 330–333 (2002).

142. Patterson, P. E. Development of a learning module using a virtual environment to demonstrate EMG and telerobotic control principles, *Biomed. Sci. Instrum.* **38**, 313–316 (2002).

143. Agazio, J. B., Pavlides, C. C., Lasome, C. E., et al. Evaluation of a virtual reality simulator in sustainment training, *Mil. Med.* **167**, 893–897 (2002).

144. Seymour, N. E., Gallagher, A. G., Roman, S. A., et al. Virtual reality training improves operating room performance: Results of a randomized, doubleblinded study, *Ann. Surg.* **236**, 463–464 (2002).

145. Pittini, R., Oepkes, D., Macrury, K., et al. Teaching invasive perinatal assessment of a high fidelity simulator-based curriculum, *Ultrasound Obstet. Gynecol.* **19**, 478–483 (2002).

146. Weller, J. M., Bloch, M., Young, S., et al. Evaluation of high fidelity patient simulator in assessment of performance of anesthetists, *Br. J., Anaesth.* **909**, 43–47 (2003).

147. Kanter, R. K., Fordyce, W. E., and Tompikns, J. M. Evaluation of resuscitation proficiency in simulation: The impact of simultaneous cognitive task, *Pediatr. Emerg. Care* **6**, 260–262 (1990).

148. Small, S. D., Wuertz, R. C., Simon, R., et al. Demonstration of high fidelity simulation team training for emergency medicine, *Acad. Emerg. Med.* **6**, 213–223 (1999).

149. Mackenzie, C. F., Jefferies, N. J., Hunter, W. A., et al. Comparison of self-reporting of deficiencies in airway management with video analysis of actual performance. LOTAS group. Level one Trauma Anesthesia Simulation, *Hum. Factors* **38**, 623–635 (1996).

150. Garden, A., Robinson, B., Weller, J., et al. Education to address medical error-a role for high fidelity patient simulation, *N Z. Med. J.* **115**, 133–134 (2002).

151. Lary, M. J., Pletcher, T., von Lubitz, D. K. J. E. Distance-based mass medical readiness training for prehospital providers, *SSGRR Internet Conference* L'Aquila, Italy, 2003.

152. Morgan, P. J. and Cleave-Hogg, D. M., Cost and resource implications of undergraduate simulator-based education, *Can. J. Anaesth.* **48**, 827–828 (2001).

153. Schaefer, J. J. III and Grenvik, A. Simulation-based training at the University of Pittsburgh, *Ann. Acad. Med. Singapore* **30**, 274–280 (2001).

154. von Lubitz, D. K. J. E., Wickramasinghe, N., and Yanovski, G., Networkcentric healthcare operations: The telecommunications structure, *Int. J. Networking Virtu. Operations* **1**, 60–85 (2006).

155. Treloar, D, K. P. Beier, J. Freer, H. et al. On site and distance education of emergency medicine personnel with a human patient simulator, *Mil. Med.* **166**, 1003–1006 (2001).

156. von Lubitz, D. K. J. E., Carrasco, B., Gabbrielli, F., et al. Transatlantic medical education: Preliminary data on distance-based high fidelity human patient simulation training. In: Westwood, J. (ed.), *Medicine Meets Virtual Reality*. IOS Press, Amsterdam, 2003, pp. 379–385.

157. Proctor, M. D. and Creech, G. S. Object-oriented modeling of patients in a medical federation, *IEEE Trans. Inf. Technol. Biomed.* **5**, 244–247 (2001).

158. von Lubitz, D. K. J. E., Montgomery, J., and Russell, W. Just in time training: Emergency medicine training aboard a ship. *Navy Med.* **3–4**, 24–28 (2000).

159. von Lubitz, D. K. J. E. EMERGENCY! medicine and modern education technology, In: *Proceedings Intl Conf. on Internet, Business, Science, and Medicine*, L'Aquila, August 2002, CD-ROM, SSGRR (L'Aquila), ISBN 88-85280-63-3.

160. von Lubitz, D. K. J. E., Carrasco, B., Fausone, C. A., et al. Bioterrorism: Development of a large-scale medical readiness using multipoint distance-based simulation training. In: Wood, J. D., Haluck, R. S., Hoffman H. M., et al. (eds.), *Building a Better You-The Next Tools for Medical Education, Diagnosis, and Care, Medicine Meets Virtual Reality* 12 (Studies in Technology and Informatics, Vol. 98), IOS Press, Amsterdam, 2004, pp. 221–227.

161. Transparent Telepresence Research Group, http://telepresence.dmem.strath.ac.uk/telepresence.htm

162. Mantovani, G. and Riva, G. "Real" presence: How different ontologies generate different criteria for presence, telepresence, and virtual presence, *Pres. Teleop. Virtual Environ.* **5**, 540–550 (1999).

163. Kim, T. and Biocca, F. Telepresence via television: Two dimensions of telepresence may have different connections to memory and persuasion, *JCMC* **3**, (1997) http://www.ascusc.org/jcmc/vol3/issue2/kim.html

164. McKay, M. Cooperative multi-agent robotics. http://www.inel.gov/capabilities/robotics/Emerging/multi.html,1999.

165. Butera, R., Locatelli, C., and Gandini, C. Telematics equipment for poison control surveillance. Its applications in the health management of relevant chemical incidents, *G. Ital. Med. Lav. Ergon.* **19**, 42–49 (1997).

166. Lemley, J. R. and Curtiss, J. A. Technology for managing risk during international inspections, *Trans. A. Nuc. Soc.* **72**(Suppl.1), 27 (1995).

167. Potter, C. S., Carragher, B., Chu, H., et al. A testbed for automated acquisition from TEM, *Communications* **14**, 1263–1279 (1998).

168. Holcomb, G. Coherent medical informatics, http://www.som.tulane.edu/oit/documents/CMI.PDF, 2002.

169. Brenner, D. W. Development of cost-effective virtual reality tools for engineering education. Educ. Symp., Fall Meeting of the Materials Res. Soc., Boston, Dec. 1998, http://www.mse.ncsu.edu/CompMatSci/overheads/MRS98/

170. Vangelova, L. Virtual reality turns inside out. Government Exec. Magazine October 1, http://www.govexec.com/tech/ARTICLES/1096INFO.HTM, 2000.

171. McLellan, H. Virtual reality: Visualization in three dimensions. In: Bradan, R. A. (ed.), *Art, Science and Visual Literacy: Selected Readings from the 24th Ann. Conf. of the Intl.* Visual Literacy Assoc., Pittsburgh PA, Sept. 30-Oct 4, 1992, VA Polytech. State University Press, Blacksburg, VA, 1993, pp. 142–150 (1993).

172. Cliburn, D. Virtual reality for small colleges, *J. Comp. Sci. in Coll.* **4**(April) (2004) available at: http://delivery.acm.org/10.1145/1060000/1050235/p28-cliburn.pdf?key1=1050235&key=0641003411&coll=Portal& dl=ACM&CFID=71893857&CFTOKEN=9076533

173. Waterworth, J. A. Virtual reality in medicine: A survey of the state of the art. http://www.informatik.umu.se/~jwworth/medpage.html, 1999.

174. Szekely, G. and Satava, R. M. Virtual reality in medicine, *BMJ* **319** (1999) accessible at: http://bmj.bmjjournals.com/cgi/content/full/319/7220/1305

175. Alverson, D. C., Saiki, S. M. Jr., Jacobs, J., et al. Distributed interactive virtual environments for collaborative experiential learning and training independent of distance over Internet2. *Stud. Health Technol. Inform.* **98**, 7–12 (2004).

176. Giachino, L. Activity sensing through portholes images: A bridge between passive awareness and active awareness. http://www.dgp.toronto.edu/tp/papers/9308.html, 1993.

177. Chen, W. W., Towles, H., Nyland, L., et al. Toward a compelling sensation of telepresence: Demonstrating a portal to a distant (static) office. *Abstr. In Proc. 11th Ann. IEEE Visualization Conf. (Vis 2000)* **155** (1993).

178. Gibbs, S. J., Arapis, C., and Breitender, C. J. TELEPORT – towards immersive telepresence, *Multimedia Systems* **7**, 214–221 (1999).

179. Day, P. N., Ferguson, G., O'Brian Holt, P., et al. Wearable augmented virtual reality for enhancing information delivery in high precision defense assembly: An engineering case study, *Virtu. Real.* **3**, 177–184 (2005).

180. Harasaki, S. and Saito, H. Virtual overlay onto uncalibrated images for augmented reality, *Syst. Comp. Jpn.* **5**, 47–55 (2003).

181. Starks, M. Stereoscopic imaging technology. 3DTV Corp., http://www.3dmagic.com/, under Support/Articles heading, 1996.

182. Halle, M. Autostereoscopic displays and computer graphics, *Comp. Graph.* **31**, 58–62 (1997).

183. Hattori, T. Sea Phone 3D display. http://home.att.net/~SeaPhone/3display.htm

184. Perlin, K., Paxia, S., and Kollin, K. An autostereoscopic display, In: *Proceedings of 27th International Conference for Computer Graphics and Interactive Technologies.* 2000, pp. 319–326.

185. Cosman, P. H., Cregan, P. C., Martin, C. J., and Cartmill, J. A. Virtual realtiy simulators: Current status in acquisition and assessment of surgical skills, *ANZ J. Surg.* **72**, 30–34 (2002).

186. Graschew, G., Rakowsky, S., Balanou, P., and Schalg, P. M. Interactive telemedicine in the operating theatre of the future, *J. Telemed. Telecare* **6**(Suppl. 2), S20–S24 (2000).

187. Graschew, G., Roelofs, T. A., Rakowsky, S., and Schalg, P. M. Interactive telemedical applications in OP 2000 via satellite, *Biomed. Tech.* **47**(Suppl. 1 Pt 1), 330–333 (2002).

188. von Lubitz, D. K. J. E. Medical readiness for operations other than war: Boyd's OODA loop and training using advanced distributed simulation technology, *Intl. J. Risk Assess. Manag.* (in press), (abbreviated version can be accessed at the Defense and National Intelligence Network, http://www.d-n-i.net/fcs/pdf/von_lubitz_1rp_ooda.pdf

189. von Lubitz, D. K. J. E. and Wickramasnghe, N. Network-centric healthcare: Outline of entry portal concept, *Intl. J. Electr. BusinessMngmt.* **4**(1), 16–28, 2006.

190. von Lubitz, D. K. J. E. and Levine, H. Distributed, multiplatform high fidelity Human Patient Simulation environment: A global-range, simulation based medical learning and training network, *Int. J. Healthcare Technol. Mgmt.* **5**, 16–33 (2005).

191. Baldwin, H. W. *Battles Lost and Won.* Koneck & Konecky, New York, 1966.

192. von Lubitz, D. K. J. E. and Lary, M. J. Bioterrorism, medical readiness, and distributed simulation training of first responders. In: Johnson, J., Cwiek, M., and Ludlow, G. (eds.), *Community and Response to Terrorism (Vol. II): A Community of Organizations: Responsibilities and Collaboration Among All Sectors of Society.* Praeger (Greenwood Publ. Group), Westport, CT, 2005 pp. 267–312.

193. Smith, M. S. Medical errors: Lessons from aviation, *Med. Econ.* **81**, 60–65 (2004).

194. Pizzi, L., Goldfarb, N. I., and Nash, D. B. Crew Resource Management and its applications in Medicine, http://www/ahrq.gov/clinic/ptsafety/chap44.htm

195. *The 9/11 Commission Report, Authorized Edition*, W.W Norton & Co., New York, 2004.

196. Dunn, W. (ed.), Simulators in Critical Care and Beyond, *Soc. Critical Care Medicine (Des Plaines (Il))*, 2004, pp. 1–130.

197. Baldwin, D. C., Jr. and Daugherty, S. R. Sleep deprivation and fatigue in residency training: Results of a national survey of first- and second-year residents, *Sleep* **15**, 217–223 (2004).

198. Papp, K. K., Stoller, E. P., Sage, P., et al. The effects of sleep loss and fatigue on resident-physicians: A multi-institutional, mixed-method study, *Acad. Med.* **79**, 394–406 (2004).

199. Bailey, B. P., Konstan, J. A., and Carlis, J. V. The effects of interruptions on task performance, annoyance, and anxiety in the user interface, http://interruptions.net/literature/Bailey-Interact01.pdf

200. Gawande, A. A., Zinner, M. J., Studdert, D. M., and Brennan, T. A. Analysis of errors reported by surgeons at three hospitals, *Surgery* **133**, 614–622 (2003).

201. Boyd, J. R. COL USAF, In: *"Patterns of Conflict,"* unpubl *Briefing (accessible as "Essence of Winning and Losing," http://www.d-n-i.net).* 1987.

202. von Lubitz, D. K. J. E. and Wickramasinghe, N. Healthcare and technology: The doctrine of networkcentric healthcare, *Int. J. Electron. Healthcare* **2**(4), 322–344 (2006).

203. von Lubitz, D. K. J. E. and Wickramasinghe, N. Network-centric healthcare: outline of the entry portal concept, *Intl J. Electron. Business* **4**(1), 16–28 (2006).

204. von Lubitz, D. K. J. E. and Wickramasinghe, N. Dynamic leadership in unstable and unpredictable environments, *Intl. J. Innov. Learn.* **4**, 339–349 (2005e).

205. Alberts, D. S., Garstka, J. J., and Stein, F. P. *Network Centric Wardare: Developing and Leveraging Information Superiority.* CCRP Publication Series (Dept. of Defense), Washington, DC, 2000, pp. 1–284, (available at http://www.dodccrp.org/publications/pdf/Alberts_NCW.pdf

206. Alberts, D. S., and Heyes, R. E. *Power to the Edge: Command and Control in the Information Edge.* Center for Advanced Concepts and Technology, Washington, DC, 2003, pp. 1–256.

207. Alberts, D. S. Network centric warfare: Current status and the way ahead, *J. Def. Sci.* **3**, 117–119 (2006).

208. Stein, P. Observations on the emergence of network centric warfare, http://www.dodccrp.org/research/ncw/stein/_observations/steincw.htm, 1998.

209. Ibrügger, L. NATO: The Revolution in Military Affairs, Special Report of the NATO Parliamentary Assembly, http://www.naa.be/publications/comrep/1998/ar299stc-e.html, 1998.

210. Newby, G. B. Cognitive space and information space, *J. Am. Soc. Info. Sci. Technol.* **12**, 1026–1048 (2001).

211. Rifkin, J. *The European Dream: How Europe's Vision of the Future is Quietly Eclipsing the American Dream*, Tarcher/Penguin Publisher, New York, 2004, pp. 1–434.

212. Catudal, J. T. Artificial Intelligence Support for the United States Armed Forces' "System of Systems" Concept, U.S. Army War College, accessible at http://www.carlisle.army.mil/usacsl/divisions/std/branches/keg/98TermIII/aipr1.htm, 1998.

213. Fukimoto, T. The Evolution of Manufacturing Systems at Toyota, Oxford University Press, New York, 1999, pp. 3–349.

# XVII

# We Teach in the Style that We Learn

*"My education was dismal. I went to a series of schools for mentally disturbed teachers."*

Woody Allen

Generations will have to pass before most clinical instructors will have received most of their clinical education with simulation so pervasive that it will no longer seem remarkable. Until that time, we who want to harness the power of simulation-based teachings must change our teaching behaviors the hard way: after we have established "successful" teaching habits that didn't recognize the value nor employed the concepts and tools of simulation, we now have to change and embrace this new simulation technology. Furthermore, many of us will have to change our teaching behaviors numerous times as we discover for our selves better and worse ways of doing so. Firsthand learning experiences are often the most powerful, and we readily employ simulation in ways of providing firsthand yet safe clinical experiences of deliberate content for our clinical students. The power of firsthand learning also applies to educators in their role as students learning how to employ simulation. The value of making use of deliberate content provided by the more experienced applies equally. Fortunately, for all of us, a few early adopters have made a concerted effort in studying how to facilitate this essential change in ourselves as teachers so we can become better in making the desired changes in our clinical students. If you don't know how to use the tools, why have them?

Jochen Vollmer *et al.* describe their formal course specifically designed to help established clinical educators to become simulation facilitators. Roger Chow and Viren Naik share their fundamental tasks and their purposes in an easy to adopt step-by-step fashion. Peter Dieckmann and Marcus Rall also offer a formal course specifically designed to help established clinical educators to become simulation facilitators.

# Staff Education for Simulation: Train-the-Trainer Concepts

Jochen Vollmer
Stefan Mönk
Wolfgang Heinrichs

## 65.1 Clinical Teachers to Simulation Facilitators

Clinical Educators interested in simulation are often good and experienced clinical *teachers* as well as highly motivated for teaching in simulation environments. But in the first years of Mainz Simulation Center, Mainz, Germany, we had to learn that just being willing teachers and even working at a university department with many traditional teaching responsibilities did not make our staff excellent simulation *facilitators*. We found out that there were additional skills needed to become a good simulation facilitator, and that a common educational basis and constant training was of essence to present classes at the same high teaching level, independently of the teacher.

This led us to the development of a set of Train-the-Trainers (TTT) classes that took care of the individual needs of the simulation center's staff members to learn all the additional skills to work at the center. This curriculum was also offered to new colleagues joining the group: developed and presented as a curriculum of all the simulator-specific skills and the experiences that they needed to become qualified teaching and facilitating members of the simulation center. In this chapter, we describe the need for Train-the-Trainer classes, discuss the key components of such classes, and illustrate this with concrete examples from our Center [1].

## 65.2 Functions and Staff in Simulation Centers

Clinical simulation training is still a new and developing field. Simulation-based clinical education involves skills in many disciplines such as medical, psychological, technical, and organizational. None of the currently available curricula prepares a professional with all the necessary skills to competently perform the tasks needed in a simulation center (see Topic XII and Chapter 54). The organization of centers and their economic background may be different, but all have the same needs in terms of providing education and training with the use of simulators. We will focus on these educational tasks in Chapter 67, and not mention other important branches such as administration or marketing. To analyze what mix of skills is best suitable for the work in a simulation center, we need to have a look at the tasks that arise in a simulation center. The main functions in a simulation center can be categorized as: Education, Operation, Technical, and Organization.

## 65.2.1  Education and the Educators

Teaching using simulation and simulators is the main objective of a simulation center and so, most of the staff of a center is involved in providing training or education with a simulator (Figure 65.1). In contrast to traditional teaching methods, simulation training typically involves many different teaching styles such as lectures, interactive teaching, workshops, or monitored self-experiences. The pure training of technical skills, of algorithms, procedures, or clinical knowledge is enhanced with nontechnical skill training[1]. For the training of clinical skills, clinical knowledge and expertise is indispensable. For the education of certain clinical skills, a trained person in that area of interest is needed. Suturing can best be explained and demonstrated by a surgeon – or other individuals (e.g., surgical assistants) with expertise in this skill. Intubation, difficult airway management, or respiratory therapy typically are domains where anesthetists are specialists in.

Training of nontechnical skills is more challenging as the instructor needs to not only have an expertise in communication skills and/or debriefing, but also needs to be knowledgeable in the clinical branch and application in which the training is embedded. Psychologists may have experience in observation or providing feedback, but their education and careers do not necessarily provide sufficient understanding of clinical problems or clinical workflow. Educators also need to understand the possibilities, limitations, and the basic concepts of operation of the equipment used in the training and how the training element taught relates with the overall concept of the class or curriculum.

## 65.2.2  Operation and the Operators

Simulation operation means the control of the simulator and other equipment during the classes. This can be the physiologic control of a full-scale simulator plus the management of the audio-video system. The person that operates the equipment needs to understand the how-to-do-it of the equipment, and also to be able to manipulate and control the equipment during a teaching session. Often, this requires both some technical skills as well as some clinical knowledge to be able to convincingly control the equipment (Figure 65.2). Additionally, training sessions require some preparation of the clinical equipment – e.g., the monitor or ventilator needs to be configured for the next clinical case – and after the end of a training session some cleanup is always required.

Depending on the working philosophy of the center and the type of training, which often has an impact on the complexity of the operational part of the simulation, the control may be performed by the clinical instructor of the session "on-stage"

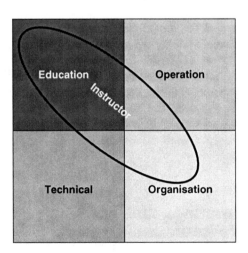

**FIGURE 65.1**  Instructors need to focus on educational topics, but also need to understand technical, operational, and organizational conditions.

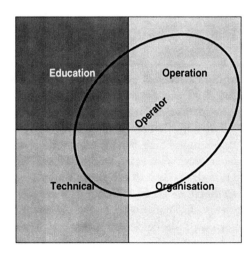

**FIGURE 65.2**  Operators need to have a broad understanding of educational goals and technical and organizational conditions, however, Operators are mainly involved in controlling the simulation.

using a remote control for the simulator or by the instructor or a second person that manages the simulator operations "off stage." During a live performance, the operator directs the course of a simulation session. The operator of the simulator typically observes and controls the action at the simulator from a neutral point of view. As the operator has the overview of the performance of the group, and the equipment as well as knowing the technical constraints of the equipment and the learning goals of the session, the operator can influence the course of a training session, as necessary.

### 65.2.3  Technical Support and the Technicians

The technical contribution of operating a simulation center is easily underestimated. Different types of highly technical devices such as audio-video equipment, partial-task skill trainers, full-scale simulators, medical devices, and computer systems form a communicating, interacting, complex, and heterogeneous technical environment. All this equipment needs care and maintenance, and if something goes wrong, somebody needs to have the overview and understanding of functions, connections, and basic performance (Figure 65.3).

It is good to have a person with a real technical background to take care of these problems, but often a person with skills and interest in technical tasks is sufficient. The major manufacturers of simulation equipment provide extensive technical customer support. This ranges from e-mail contact and telephone hotlines to on-site visits for maintenance and repair of the systems. But, classes in the local simulation center are often taught outside of common office hours, on weekends, at remote locations and it might be difficult to contact customer

support directly once the problem is recognized – which often happens right before or during a class. Your own on-site technical person can fix problems, can help the customer support to troubleshoot the system remotely, and perform preventive maintenance checks to guarantee a minimum downtime of the systems.

Besides deeper technical knowledge into the systems, the technical person needs to understand the basic operational part (how to operate the systems) and to have an idea of the educational application of the system to be able to prepare the training systems as well as the medical equipment and other technology for facilitating the class for the training purpose and to judge how relevant and time critical an issue is.

### 65.2.4  Organization and the Managers

The organization of a simulation center ranges from the definition of a class, didactic and logistical preparation of the course elements, programing of simulator cases, design of evaluation tools, the concrete planning of a class to be taught, the number of instructors, operators, and technicians needed as well as, possibly, the concrete planning determining which instructors will teach each element of the class. This requires didactic skills for the information basis of a class, administrative skills for the planning of a course, and clinical and physiologic knowledge for the programing of the cases (Figure 65.4).

The organizer of a class needs to understand what is feasible with the given equipment and the constraints of the course, and also needs to make sure that at the time of the training, all resources needed such as training materials for the participants, as well as technical or clinical equipment and educators, are available.

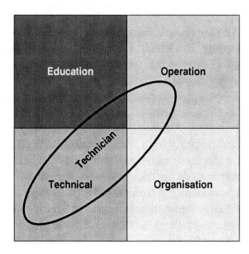

**FIGURE 65.3**  Technicians must be domain experts in, and managers of, clinical and simulator specific equipment. They also have to consider educational, operational, and organizational aspects of the center in their work.

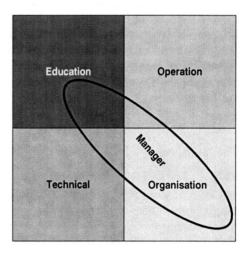

**FIGURE 65.4**  The Manager of a center administrates and coordinates the activities at the center.

## 65.3 Who Does the Work?

In many simulation centers, a mix of fully employed staff and part-time employees can be found. Often, the functions of operation and organization are provided by the full-time staff, and the educational staff and instructors come on an on-demand basis. In most of the centers, the majority of staff members have a clinical background. The mix of the clinical disciplines involved often reflects the range of classes taught at the center – and vice versa. Full-scale-simulation is an area that was first developed for and with anesthesiologists, so they still are a prominent group. But specialists in emergency medicine and intensive care as well as pediatricians, gynecologists, internal medicine specialists, or surgeons also often form part of the team. In many centers, nurses play a major role in teaching as well as in administration and operation of the simulators. To a lesser extent, nonclinical professionals such as psychologists and engineers or technicians contribute. All successful teams are interdisciplinary. All of the professional specialities have their value in a simulation center, and the staff members stimulate each other through their different professional backgrounds.

None of the professions mentioned solely fulfills all the requirements for each of the functions a simulation center requires. Facilitating classes in a simulation center demands from each of the aforementioned professionals a new and enlarged set of personal skills for a maximum exploitation of the educational possibilities that simulation offers.

The above-mentioned functions often cannot be fully separated from each other. A good educator needs to understand the capabilities and the limitations of the equipment to facilitate a good class even though an additional person operating the system supports him. An operator of the simulator has to be also knowledgeable in the educational topics as well as in some of the technical concepts[2]. A good technician also needs to have the applications of the equipment in mind. These additional skills form the link between the different professional specialties and functions in the simulation center. Establishing these links contributes greatly to the success of the center as every member of the team works to reach the same goal within the means of his specialty.

## 65.4 Why Train-the-Trainer is Necessary – A Fundamental Decision

In the latter years of the previous century, educational "explorers" founded clinical simulation programs across the globe. During this early phase, all of these pioneers sustained themselves with their own enthusiasm about this new tool: simulation. Everything needed to be invented and created from scratch: from the ideal design and configuration of a simulation suite, to course designs and teaching styles, from observation to providing feedback, from physiologic modeling to real-time simulator operation. Clinical specialists discovered similarities with airline pilots and became interested in the psychologists' work on crew teamwork (also known as crew or cockpit resource management – CRM). Engineers studied human physiology, and psychologists learned about clinical practice. Everybody involved invested their individual strengths and mixed them with newly adopted skills. Almost every day, new, brilliant, and sometimes very individual ideas were born and shared with the growing simulation world for consideration. Now, with thousands of simulators in use worldwide, this initial phase of wonder and doubt is over. Even though there is still a lot of room for development, simulation has already become an accepted educational method. For example, if a new group joins the world of simulation, they do not need to start from scratch – they can visit "reference centers," discuss with specialists, visit scientific conferences and even hire companies to help them design, build, and start using their simulation facility. The same applies for the preparation of the staff of a simulation center. While in the early years, each member of the simulation community needed to discover and research which skills were needed to adopt or to enhance to make the most out of the possibilities that simulation education offers, now certain skills have been established as necessary, and accepted as helpful for producing a simulation program and using it efficiently for education.

Summarizing the experiences of what skills are needed to simulate successfully became of importance when new members joined our group. On a local scale, we needed to summarize and teach the lessons-learned about how to operate a simulation program when new members joined our group. For the preparation of these new members, we have condensed the essentials of clinical simulation into a curriculum called Train-the-Trainer classes. Before going into the details of this course, we want to address our rationale for establishing a standardized Train-the-Trainer curriculum.

By defining a formal educational program for the trainers of our center, we wanted to guarantee rapid learning for each of the new trainers. Instructors are always a limited resource, and we consider well and homogeneously educated trainers to be the basis to maintain a high level of training quality for each class, independently of who the actual facilitator is. Requesting each person new to our program to take our Train-the-Trainer course emphasizes our seriousness and dedication to its principles. Our desire for a formal approach was also motivated through the seriousness of the impact that simulation will have in shaping generations of future clinical professionals and their treatment of patients. Just as we use simulation to supplant the traditional clinical teaching paradigm of "see one, do one, teach one," we avoid using "see one, do one, teach one" in teaching how to use simulation.

## 65.5 Which Skills are Needed?

For each type of class taught, a specific set of skills is needed. We have developed a method to keep track of all required skills, expertise, and mandatory requirements needed for each of the classes (Table 65.1). We found that some of the skills were common requirements for several classes and that the profiles of skills for some of the classes were similar.

By teaching these commonly required skills first, we could prepare the facilitators efficiently. This became the basis for the our Train-the-Trainer curriculum and so now, by participating in the Train-the-Trainer courses, a new member of the team can quickly obtain a broadly applicable set of tools for the work in simulation. In addition to the list of requirements for each of the classes, we keep track of the level of the qualification for each of the members of our staff, to be able to plan our courses efficiently. Controlling of all the qualifications for all our instructors helps to decide which staff members are available to facilitate which kind of class. We can simply match the profile of requirements for a class with the profiles of the staff and immediately understand who is a possible facilitator for which part of the course. Furthermore, this method gives the

**TABLE 65.1** Example of instructor qualification checklist. In the first column of the table, all the educational, operational/technical, and course-specific skills are listed that have been identified as a requirement for one of the classes. In the other columns, we can mark for each instructor if he possesses one of the listed skills

| | Task | Instructors and Qualifications | | | | | |
|---|---|---|---|---|---|---|---|
| | | NN | NN | NN | NN | NN | NN |
| Educational Topics | Has taken part in a simulator training | X | X | X | X | X | X |
| | Has taken part in TTT Human Factors seminar | X | X | X | X | X | X |
| | Has read relevant literature on CRM | X | X | X | X | X | X |
| | Has taken part in TTT Debriefing seminar | X | X | X | X | X | X |
| | Has accompanied a trainer in several scenarios and debriefings | X | X | X | X | X | X |
| | Has conducted debriefings together with an experienced trainer | X | X | X | X | X | X |
| | Has worked as trainer accompanied by a tutor until both are satisfied | X | X | X | X | X | X |
| | Has taken part in TTT Presentation Techniques | X | X | X | X | X | X |
| Operational/ Technical | Has taken part in TTT Course development | X | X | X | X | | |
| | Has taken part in TTT Quickstart Simulator | X | X | X | X | X | X |
| | Has operated the simulators and feels technically secure | X | | X | | | X |
| | Has developed a scenario | X | | X | | | |
| | Has directed a class | X | | | | | |
| | Has taken part in TTT Technical Troubleshooting HPS | | X | X | | | X |
| | Has taken part in TTT Technical Troubleshooting ECS | | | X | | | X |
| Course specific clinical expertise | Has experience as 'surgeon' | X | X | X | X | X | X |
| | Has theoretical background on TIVA | X | | X | | X | |
| | Has theoretical background on VIMA | X | | X | | | |
| | Has theoretical background on special respiratory therapy | X | | X | X | | X |
| | Knows the animal model | X | | X | | | X |
| | Is a bronchoscopy practician | | X | | X | X | |
| | Knows about intensive care for organ transplantation | X | X | X | | | |
| | Is a specialist on respiratory equipment | X | | X | | X | X |
| | Knows Trendcare equipment | | X | | X | | X |
| | ... | | | | | | |

Abbreviations: TTT, Train-the-Trainer; CRM, Crisis Resource Management; TIVA, Total Intravenous Anesthesia; VIMA, Volatile Induction and Maintenance of Anesthesia.

**TABLE 65.2**  Example of Anesthesia Crisis Resource Management course requirements

| Anaesthesia Crisis Resource Management Course Requirements Profile |
|---|
| **Task** |
| Has taken part in a simulator training |
| Has taken part in TTT Human Factors seminar |
| Has read relevant literature on CRM |
| Has taken part in TTT Debriefing seminar |
| Has accompanied a trainer in several scenarios and debriefings |
| Has conducted debriefings together with an experienced trainer |
| Has worked as trainer accompanied by a tutor until both are satisfied |
| Has taken part in TTT Presentation Techniques |
| Has taken part in TTT Course development |
| Has taken part in TTT Quickstart Simulator |
| Has operated the simulators and feels technically secure |
| Has developed a scenario |
| Has directed a class |
| Has taken part in TTT Technical Troubleshooting HPS |
| Has taken part in TTT Technical Troubleshooting ECS |
| Has experience as, surgeon' |
| Has theoretical background on TIVA |
| Has theoretical background on VIMA |
| Has theoretical background on special respiratory therapy |
| Knows the animal model |
| Is a bronchoscopy practician |
| Knows about intensive care for organ transplantation |
| Is a specialist on respiratory equipment |
| Knows Trendcare equipment |
| ... |

potential for a step-by-step qualification of the new member of the simulation group as he might already be ready to facilitate certain simulation sessions after successfully participating in a part of the Train-the-Trainer curriculum.

Many of the required qualifications are covered through the seven Train-the-Trainer classes. However, we also list in the Instructor Qualification Checklist, additional theoretical knowledge of Crisis Resource Management (CRM) or clinical topics such as the theoretical background of certain classes (e.g., TIVA: Total Intravenous Anesthesia, VIMA: Volatile Induction and Maintenance of Anesthesia). Additionally, practical experience such as supervised debriefings, development of scenarios, or direction of class form part of the list. On the other hand, to teach one specific class requires only a certain subset of all the skills listed in the instructor checklist. So, each of the courses define a mask on this list of all the possibly required skills that blinds out the skills not needed for the course. Table 65.2 shows a mask of the required skills for an Anesthesia Crisis Resource Management facilitator.

To identify which of the instructors working for the center are actually now qualified to teach this class, we can superimpose the Anesthesia Crisis Resource Management mask to the Instructor Qualification Checklist, and easily identify who fulfills the requirements for this class, becomes eligible as instructor, and who does not fulfill the criteria (Table 65.3).

**TABLE 65.3**  Identification process of qualified instructors for a class. Combining the information in Table 65.1 with Table 65.2 produces this table that readily illustrates the various topics each instructor has completed, which are the necessary prerequisites to teach

**Instructors and Qualifications**

| Task | NN | NN | NN | NN | NN | NN |
|---|---|---|---|---|---|---|
| *Educational Topics* | | | | | | |
| Has taken part in a simulator training | X | X | X | X | X | X |
| Has taken part in TTT Human Factors seminar | X | X | X | X | X | X |
| Has read relevant literature on CRM | X | X | X | X | X | X |
| Has taken part in TTT Debriefing seminar | X | X | X | X | X | X |
| Has accompanied a trainer in several scenarios and debriefings | X | X | X | X | X | X |
| Has conducted debriefings together with an experienced trainer | X | X | X | X | X | X |
| Has worked as trainer accompanied by a tutor until both are satisfied | X | X | X | X | X | X |
| Has taken part in TTT Presentation Techniques | X | X | X | X | X | X |
| *Operational/Technical* | | | | | | |
| Has taken part in TTT Course development | X | X | X | X | | |
| Has taken part in TTT Quickstart Simulator | X | X | X | X | X | X |
| Has operated the simulators and feels technically secure | X | | X | | | X |
| Has developed a scenario | X | | X | | | |
| Has directed a class | X | | | | | |
| Has taken part in TTT Technical Troubleshooting HPS | | X | X | | | X |
| Has taken part in TTT Technical Troubleshooting ECS | | X | | | | X |
| *Course specific clinical expertise* | | | | | | |
| Has experience as, surgeon' | X | X | X | X | X | X |
| Has theoretical background on TIVA | X | | X | | X | |
| Has theoretical background on VIMA | X | | X | | | |
| Has theoretical background on special respiratory therapy | X | | X | X | | X |
| Knows the animal model | X | | X | | | X |
| Is a bronchoscopy practician | | X | | X | X | |
| Knows about intensive care for organ transplantation | X | X | X | | | |
| Is a specialist on respiratory equipment | X | | X | | X | X |
| Knows Trendcare equipment | | X | | X | | X |
| … | | | | | | |

**+**

**Anaesthesia Crisis Resource Management Course Requirements Profile**

| Task |
|---|
| *Educational Topics* |
| Has taken part in a simulator training |
| Has taken part in TTT Human Factors seminar |
| Has read relevant literature on CRM |
| Has taken part in TTT Debriefing seminar |
| Has accompanied a trainer in several scenarios and debriefings |
| Has conducted debriefings together with an experienced trainer |
| Has worked as trainer accompanied by a tutor until both are satisfied |
| Has taken part in TTT Presentation Techniques |
| *Operational/Technical* |
| Has taken part in TTT Course development |
| Has taken part in TTT Quickstart Simulator |
| Has operated the simulators and feels technically secure |
| Has developed a scenario |
| Has directed a class |
| Has taken part in TTT Technical Troubleshooting HPS |
| Has taken part in TTT Technical Troubleshooting ECS |
| *Course specific clinical expertise* |
| Has experience as, surgeon' |
| Has theoretical background on TIVA |
| Has theoretical background on VIMA |
| Has theoretical background on special respiratory therapy |
| Knows the animal model |
| Is a bronchoscopy practician |
| Knows about intensive care for organ transplantation |
| Is a specialist on respiratory equipment |
| Knows Trendcare equipment |
| … |

**=**

Instructor A eligible for main instructor

Instructor B not eligible as instructor, as course specific clinical expertise in missing

Instructor C eligible for second instructor

**Anaesthesia Crisis Resource Management Course Requirements Profile — Instructors and Qualifications**

| Task | NN | NN | NN | NN | | |
|---|---|---|---|---|---|---|
| *Educational Topics* | | | | | | |
| Has taken part in a simulator training | X | X | X | X | | |
| Has taken part in TTT Human Factors seminar | X | X | X | X | | |
| Has read relevant literature on CRM | X | X | X | X | X | X |
| Has taken part in TTT Debriefing seminar | X | X | X | X | X | X |
| Has accompanied a trainer in several scenarios and debriefings | X | X | X | X | X | X |
| Has conducted debriefings together with an experienced trainer | X | X | X | X | X | X |
| Has worked as trainer accompanied by a tutor until both are satisfied | X | X | X | X | X | X |
| Has taken part in TTT Presentation Techniques | X | X | X | X | X | X |
| *Operational/Technical* | | | | | | |
| Has taken part in TTT Course development | X | X | X | X | | |
| Has taken part in TTT Quickstart Simulator | X | X | X | X | X | X |
| Has operated the simulators and feels technically secure | X | | X | | | X |
| Has developed a scenario | X | | X | | | |
| Has directed a class | X | | | | | |
| Has taken part in TTT Technical Troubleshooting HPS | | X | X | | | X |
| Has taken part in TTT Technical Troubleshooting ECS | | X | | | | X |
| *Course specific clinical expertise* | | | | | | |
| Has experience as, surgeon' | X | X | X | X | X | X |
| Has theoretical background on TIVA | X | | X | | X | |
| Has theoretical background on VIMA | X | | X | | | |
| Has theoretical background on special respiratory therapy | X | | X | X | | X |
| Knows the animal model | X | | X | | | X |
| Is a bronchoscopy practician | | X | | X | X | |
| Knows about intensive care for organ transplantation | X | X | X | | | |
| Is a specialist on respiratory equipment | X | | X | | X | X |
| Knows Trendcare equipment | | X | | X | | X |
| … | | | | | | |

# 65.6 Formal Requirements for Educational Staff

As the importance of simulation grew, the need for a well-defined concept of what simulation encompasses became more necessary. In 2002, the German Society for Anaesthesia and Intensive Care's (DGAI [2]) commission on simulation education published a set of guidelines for the "Standards and Requirements for a Qualified Anesthesia Simulator Training" [3], which was developed by a joint convention of all German simulation centers. These standards and requirements were presented to the Society in Europe for Simulation Applied to Medicine (SESAM [4]), which approved and adopted these as a guideline for clinical simulation in Europe, but still are to be published. This formed the first formal definition of an international standard related to simulation in medicine.

Besides defining the training content and minimal requirements of technical equipment, the guidelines list the minimally required qualification of a clinical instructor of a simulation course, and by this, it defines the content of a career curriculum for simulation instructors.

- Expert in the clinical field that is to be discussed
- Knowledge about adult education/psychology
- Knowledge about nontechnical skills issues, human factors, crisis resource management
- Knowledge about the functional principles, the possibilities, and limitations of the systems used

According to the German Society for Anaesthesia and Intensive Care/Society in Europe for Simulation Applied to Medicine (DGAI/SESAM) guidelines, the clinical instructors should have the following minimum qualifications:

- Clinical competence for the scenario
- They teach generally accepted standards and factual knowledge
- They have been participants at Crisis Resource Management courses
- They have been familiarized with their simulation center
- They participated at an introductory course for instructors; courses which are being offered by several simulation centers. Topics should cover:

  - Debriefing techniques
  - Development of scenarios
  - Psychology of training
  - Communication
  - Team interaction
  - Anaesthesia crisis resource management (ACRM)[3]
  - Error evolution (human error)

# 65.7 Topics for Staff Training Courses

Skills that simulation center staff need to master can be grouped in the following ways.

## 65.7.1 Course Development

Classes that have been identified as worthwhile being taught need to be planned and designed. Designing a class typically starts with clearly identifying a training audience and learning goals. This requires knowledge in didactics, how to create a training curriculum, how to combine lectures to transfer information with hands-on experiences to challenge the learners, when to repeat, and where to give new input. Decisions have to be made if it makes sense to use the simulator as the educational tool in a certain course element, and where possibly a different method is more effective. Concepts on how the resources in the center can be applied most efficiently should be respected in this process. Course development skills form a part of pedagogy.

Scenario development plays an important role in the course development part. Clinical cases have to be chosen, tailored for a certain time frame, and have to be then condensed into a feasible and realistic scenario for the class and students it is designed for. Often, it takes some effort and a few repetitions until all the parameters and possibilities to control the scenario have been explored well enough and programed into a repeatable and functional course element.

## 65.7.2 Human Factors/Crisis Resource Management/Root Cause Analysis/ Human Error/Decision Making/ Workload Management/Leadership

Training of nontechnical skills is probably one of the most interesting aspects of simulation training, and is one of the areas that is rarely taught outside of simulation. The key concepts in this field come from psychology and were initially developed for aviation simulation training. Some of these concepts have been translated into the clinical context [5]. The knowledge of these concepts does not form part of our formal clinical curriculum of the instructor, but needs to be obtained for successful simulation.

## 65.7.3 Debriefing

Debriefing is an essential part of a simulation training event, and requires human psychological skills from the clinical educator. Again, it represents a method seldom used in clinical practice and education outside of simulation. Many aspects outside of the pure clinical background play a role in good debriefing: focused observation of a trainee or a training situation, the clear understanding of a group dynamics, the possibility of managing a discussion, providing feedback, and respecting cultural and personal attitudes, but all in the context of the underlying clinical case (see Chapter 69).

### 65.7.4  Operation of the Simulators

This is clearly of essence for the operational part of the simulation center. Somebody needs to operate the equipment. In many centers and training environments, the same person instructing the class also starts the simulator and loads and presents the scenario. However, the operation of the equipment must not distract the focus from the educational target. Controlling the simulator in a virtuoso (expert) manner can be a challenge. Thus, during complex scenarios, dedicated simulator operators take responsibility for operating the simulated patient, allowing all the instructor's attention to focus upon the clinical and social behaviors of the students.

### 65.7.5  Deeper Insight into Technical Aspects: Modeling Details, Hardware Details, Programing Techniques

"Know your environment" is one of the key points in Crisis Resource Management. This applies in the same way to the staff of a simulation center. Understanding the potential and limitations of the equipment used is a key to convincingly utilize the simulators. In the case of full-scale simulators, a clear understanding of the underlying physiologic models is of essence, as well as understanding the hardware interfaces. Clinical educators are typically not trained programers, but certain logical programing concepts can be applied to the simulators and help to create reliable and controllable scenarios.

### 65.7.6  Presentation Techniques

Simulation offers teaching possibilities that are often different from the traditional teaching methods. This is partly due to smaller groups and a more interactive style of the course. Besides the performance-observation/debriefing type of class, classes are often taught including bedside-teaching scenarios. Depending on the size of the group and the level of the performance, this teaching becomes more like a moderated discussion among the trainees. Possibly, an additional specialist assists the instructor and alternate techniques like "double-moderation" need to be applied[4]. Successful utilization of presentation material, as well as aspects of body language and teaching style can be trained.

### 65.7.7  Technical Troubleshooting

The members of the staff need to have some insight into the technical modeling of the equipment used. This is meant partly to troubleshoot the equipment in the event of a technical problem, and partly to understand the possibilities and limitations of the systems. Often, small problems can have a major impact on the systems and in the worst case lead to cancellation of a class. The basic knowledge of technical troubleshooting of the devices used can make the group more independent from a support infrastructure, and often allows support to be more efficient. Preventive maintenance and the anticipation of problems help minimize downtime of the equipment (just as it does with our own health).

### 65.7.8  Clinical Topics

Last but not least, credibility of a clinical instructor depends on profound knowledge and practice in the clinical field of interest. Clinical topics often can only be taught by clinical professionals as only these have a background of own experiences in the clinical field to be taught.

Thus, the instructors need to have the most profound knowledge on the clinical topics they are teaching. This means that the most current literature, guidelines, and scientific discussions need to have an impact on the training. This also applies to patient safety issues, fatigue, and political questions related to clinical care.

## 65.8  Detailed Description of the Train-the-Trainer Curriculum

The above-mentioned ideas were incorporated in a family of Train-the-Trainer classes: our curriculum currently consists of seven elements, which are basically independent of each other, but add to a logical sequence. In the case of our Simulation Center, all the simulation equipment–oriented classes were implemented with METI HPS and METI ECS systems, (METI, Medical Education Technologies, Inc., Sarasota Florida, U.S.A.) but the concepts are device independent.

Titles of our Train-the-Trainer classes:

1. "Human Factors"
2. "Quick Start: METI Simulators"
3. "Presentation Techniques"
4. "Course Development"
5. "Debriefing"
6. "Help Needed: METI ECS/METI HPS"

The training initially has been designed for the education of our Center staff, but the sessions have become accessible to external interested persons. The seminars are held regularly with groups of participants up to 20 persons and last 2 days each. The courses are taught in German, or in English for an International group of participants. After each of the courses, the participant is rewarded with a certificate of completion of the course.

## 1. "Human Factors"

In this course, the basic concepts of Human Factors are defined, explained, and practiced in workshops. The participants of this class are trained to become facilitators for human factors elements of a simulation class. Through the course, the participants receive, discuss, and adopt materials and applications ready to use in their own teaching context. Key topics of this course are Decision Making, Errors, Situational Awareness, and Observation. Examples of real clinical case studies have been chosen to underline and apply the theoretical aspects. The target group for this workshop includes clinicians teaching in a simulation center. For a detailed course schedule, see Table 65.4: Course schedule: "Human Factors."

## 2. "Quick Start: METI Simulators"

This course explains the operational part of a simulation class on the example of METI HPS and ECS simulators: all operational aspects of the systems and the physiology are explained and practiced in real simulation class-like workshops. The course prepares the participants to produce simulation courses from the perspective of the operator, and to manage the equipment unassisted. The course is designed for educators, operators, and technicians. Details on the course schedule can be found in Table 65.5: Course Schedule: "Quickstart METI Simulators."

## 3. "Presentation Techniques"

The main topics of the course on Presentation Techniques are related to the different teaching styles in a simulation environment. The participants are introduced to techniques for traditional lecturing as well as experience-oriented workshops, or the seminar-style of discussion rounds. The different styles are compared on the basis of their efficiency of information absorption and level of self-experience of the participant, and identified for which type of course element which presentation style fits best. The participants experience several practice runs on presentations in different contexts, applying different presentation styles, and receive feedback through video debriefing techniques. Introduction to presentation tools, aspects of body-language, visualization of information, and tips and tricks complete the course. A detailed schedule can be found in Table 65.6: Course Schedule "Presentation Techniques."

## 4. "Course Development"

The main focus of this class is to introduce the organizational aspects in establishing a simulation class and the didactic tools for designing simulator cases. The theoretical parts are combined with practices in scenario writing for METI HPS/ECS systems. The content of this class is mainly of interest for the operators and the educators of a simulation center. The participants of the course have the possibility to apply what they learnt to a concrete example from their teaching needs. This class starts with the definition of the learning goals for the course and each course element, and structuring the course based on didactic principles. It also includes the definition and implementation of individual scenarios. The participants of this class receive course planning tools and scenario development schemes as means for future unassisted developments. Details on the schedule can be found in Table 65.7: Course Schedule "Course Development."

## 5. "Debriefing"

The theory of debriefing and practical examples based on real simulation class experiences have been collected for this course. Central topics are: providing feedback, conduct of debriefing sessions, facilitation of self-awareness in the participants, conscious observation, and purposeful reproduction of observations. Each participant experiences debriefings from the point of view of the debriefer and the person debriefed. Practical hints on how destructive circumstances during debriefing can be avoided, as well as communication strategies are discussed. This class is designed for educators actively debriefing trainees inside and outside of the simulation context. For more details, please compare to Table 65.8: Course Schedule "Debriefing."

## 6. "Help Needed: METI ECS/METI HPS"

These two classes are designed for the immediate troubleshooting the METI HPS and ECS systems. The courses explain the basic technical functions and interactions of the system and give practical hints for troubleshooting. The operators and technicians participating in these classes practice the concepts for troubleshooting and smaller repairs in workshops during the course. Concrete Schedules of these classes can be found in Table 65.9: Course Schedule "Help needed METI ECS" and Table 65.10: Course Schedule "Help needed METI HPS."

# 65.9 Course Schedules

**TABLE 65.4**  Course schedule for Train-the-Trainers #1: Human factors

**Course Outline: TTT1 Human Factors**

**Day 1**

| Begin | End | Duration | Type | Title |
|-------|-----|----------|------|-------|
| 12:00 | 12:30 | 30 | | Welcome, Come together |
| 12:30 | 13:30 | 60 | Lecture | Definition Human Factors |
| 13:30 | 14:00 | 30 | | break |
| 14:00 | 15:00 | 60 | Workshop | Saving "Geha 17"! |
| 15:00 | 16:00 | 60 | Workshop | Debriefing |
| 16:00 | 16:30 | 30 | | Break |
| 16:30 | 17:00 | 30 | Lecture | Errors |
| 17:00 | 17:15 | 15 | | Break |
| 17:15 | 17:45 | 30 | Workshop | Titanic |
| 17:45 | 18:15 | 30 | | Summary Day 1 |

**Day 2**

| Begin | End | Duration | Type | Title |
|-------|-----|----------|------|-------|
| 09:00 | 09:30 | 30 | | Welcome, Questions? |
| 09:30 | 10:00 | 30 | Lecture | Error Case Studies |
| 10:00 | 10:35 | 35 | Practice | "Peanut-Problem" |
| 10:35 | 11:20 | 45 | Practice | Debriefing |
| 11:20 | 11:35 | 15 | | Break |
| 11:35 | 12:05 | 30 | Lecture | Decision Making |
| 12:05 | 12:50 | 45 | Workshop | Blackbox |
| 12:50 | 13:50 | 60 | | Break |
| 13:50 | 14:05 | 15 | Workshop | Basketball |
| 14:05 | 14:35 | 30 | Lecture | In our midst |
| 14:35 | 14:50 | 15 | | break |
| 14:50 | 15:25 | 35 | Lecture | Violations |
| 15:25 | 15:55 | 30 | Workshop | What to do? |
| 15:55 | 16:25 | 30 | | Summary Day 2 |

Simulationszentrum Mainz
Wernher von Braun Straße 9
55129 Mainz
Tel. +49-6131-97 13 546
e-mail: info@aqai.de
www.simulationszentrum-mainz.de

**TABLE 65.5**   Course schedule for Train-the-Trainers #2: Quickstart METI simulators

## Course Outline: TTT2 Quickstart METI Simulators

### Day 1

| Begin | End | Duration | Type | Topic |
|-------|-----|----------|----------|-------|
| 12:00 | 12:30 | 30 |  | Welcome – Come together |
| 12:30 | 13:00 | 30 | Lecture | Introduction into HPS 6 |
| 13:00 | 13:15 | 15 |  | break |
| 13:15 | 13:35 | 20 | Lecture | Introduction into Mac OS |
| 13:35 | 14:20 | 45 | Workshop | Features of HPS Application |
| 14:20 | 14:35 | 15 |  | break |
| 14:35 | 15:35 | 60 | Workshop | Power on HPS |
| 15:35 | 16:20 | 45 | Lecture | Cardiovascular Parameters |
| 16:20 | 16:35 | 15 |  | break |
| 16:35 | 17:20 | 45 | Lecture | Playing Scenarios |
| 17:20 | 17:35 | 15 |  | break |
| 17:35 | 18:20 | 45 | Lecture | Respiratory Physiology |
| 18:20 | 18:30 | 10 |  | Questions, End of Day 1 |

### Day 2

| Begin | End | Duration | Type | Topic |
|-------|-----|----------|----------|-------|
| 9:00 | 9:30 | 30 |  | Welcome – Come together |
| 9:30 | 10:30 | 60 | Lecture | Writing Scenarios |
| 10:30 | 10:45 | 15 |  | break |
| 10:45 | 11:30 | 45 | Workshop | Operating the Simulator |
| 11:30 | 12:00 | 30 | Lecture | Multiple Scenarios |
| 12:00 | 12:45 | 45 |  | break |
| 12:45 | 13:30 | 45 | Workshop | Quickstart ECS |
| 13:30 | 15:15 | 105 | Practice | Writing scenarios |
| 15:15 | 15:30 | 15 |  | break |
| 15:30 | 16:15 | 45 | Practice | Playing scenarios |
| 16:15 | 16:30 | 15 |  | Questions, End of Day 2 |

Simulationszentrum Mainz
Wernher von Braun Straße 9
55129 Mainz
Tel. +49-6131-97 13 546
e-mail: info@aqai.de
www.simulationszentrum-mainz.de

**TABLE 65.6**   Course schedule for Train-the-Trainers #3: Presentation techniques

## Course Outline: TTT3 Presentation Techniques

### Day 1

| Begin | End | Duration | Type | Title |
|-------|-----|----------|------|-------|
| 12:00 | 12:45 | 45 | | Welcome, Introduction |
| 12:45 | 13:15 | 30 | Lecture | Introduction |
| 13:15 | 14:15 | 60 | Lecture | Presentation styles |
| 14:15 | 14:30 | 15 | | Break |
| 14:30 | 14:55 | 25 | Lecture | Fundamentals of Feedback |
| 14:55 | 15:55 | 60 | Practice | Presenting |
| 15:55 | 16:10 | 15 | | Break |
| 16:10 | 16:30 | 20 | Lecture | Structuring the presentation |
| 16:30 | 17:15 | 45 | Workshop | Tools |
| 17:15 | 17:25 | 10 | | Break |
| 17:25 | 18:10 | 45 | Lecture | Moderation techniques |
| 18:10 | 18:55 | 45 | Practice | Planning Moderation |
| 18:55 | 19:00 | 5 | | Summary |

### Day 2

| Begin | End | Duration | Type | Title |
|-------|-----|----------|------|-------|
| 9:00 | 9:20 | 20 | | Welcome, open questions |
| 9:20 | 10:20 | 60 | Lecture | Do and Don't |
| 10:20 | 10:35 | 15 | | Break |
| 10:35 | 11:05 | 30 | Workshop | Body language |
| 11:05 | 11:15 | 10 | | Break |
| 11:15 | 12:15 | 60 | Practice | Individual Presentations |
| 12:15 | 13:00 | 45 | | Break |
| 13:00 | 14:00 | 60 | Workshop | Debriefing of individual presentations |
| 14:00 | 14:15 | 15 | | Break |
| 14:15 | 15:15 | 60 | Practice | Double moderation |
| 15:15 | 15:45 | 30 | Lecture | Visualization of Information, Materials |
| 15:45 | 16:15 | 30 | | Questions and Answers |
| 16:15 | 16:30 | 15 | | Summary and Goodbye |

Simulationszentrum Mainz
Wernher von Braun Straße 9
55129 Mainz
Tel. +49-6131-97 13 546
e-mail: info@aqai.de
www.simulationszentrum-mainz.de

**TABLE 65.7**   Course schedule for Train-the-Trainers #4: Course development

**Course Outline: TTT4 Course Development**

**Day 1**

| Begin | End | Duration | Type | Topic |
|---|---|---|---|---|
| 12:00 | 12:45 | 45 | | Welcome, Introduction |
| 12:45 | 13:15 | 30 | Lecture | Introduction |
| 13:15 | 14:15 | 60 | Workshop | Collection of ideas |
| 14:15 | 14:35 | 20 | | Break |
| 14:35 | 16:05 | 90 | Workshop | Definition of course elements and schedule |
| 16:05 | 16:25 | 20 | | Break |
| 16:25 | 17:55 | 90 | Workshop | Definition of scenarios |
| 17:55 | 18:15 | 20 | | Summary |

**Day 2**

| Begin | End | Duration | Type | Topic |
|---|---|---|---|---|
| 9:00 | 9:30 | 30 | | Welcome, open questions |
| 9:30 | 10:00 | 30 | Lecture | Housekeeping |
| 10:00 | 11:00 | 60 | Workshop | Definition/programming |
| 11:00 | 11:20 | 20 | | Break |
| 11:20 | 12:20 | 60 | Workshop | Programming |
| 12:20 | 13:20 | 60 | | Break |
| 13:20 | 14:20 | 60 | Workshop | Programming |
| 14:20 | 14:40 | 20 | | Break |
| 14:40 | 15:10 | 30 | Workshop | Plans for debriefing |
| 15:10 | 15:40 | 30 | Workshop | Finetuning |
| 15:40 | 16:10 | 30 | | Summary, Good bye |

Simulationszentrum Mainz
Wernher von Braun Straße 9
55129 Mainz
Tel. +49-6131-97 13 546
e-mail: info@aqai.de
www.simulationszentrum-mainz.de

**TABLE 65.8**   Course schedule for Train-the-Trainers #5: Debriefing

## Course Outline: TTT5 Debriefing

### Day 1

| Begin | End | Duration | Type | Topic |
|-------|-----|----------|------|-------|
| 12:00 | 12:45 | 45 | | Welcome, Introduction |
| 12:45 | 13:15 | 30 | Lecture | Introduction |
| 13:15 | 14:15 | 60 | Lecture | Debriefing Functions and Techniques |
| 14:15 | 14:30 | 15 | | Break |
| 14:30 | 14:55 | 25 | Lecture | Providing Feedback |
| 14:55 | 15:55 | 60 | Practice | Case, Debriefing, Debriefing |
| 15:55 | 16:10 | 15 | | Break |
| 16:10 | 16:55 | 45 | Workshop | Facilitating Self-awareness |
| 16:55 | 17:40 | 45 | Lecture | Communication Conduct |
| 17:40 | 18:00 | 20 | | Summary Day 1 |

### Day 2

| Begin | End | Duration | Type | Topic |
|-------|-----|----------|------|-------|
| 9:00 | 9:20 | 20 | | Welcome, open questions |
| 9:20 | 10:20 | 60 | Lecture | Conscious Observation |
| 10:20 | 10:35 | 15 | | Break |
| 10:35 | 12:05 | 90 | Practice | Case, Debriefing, Debriefing |
| 12:05 | 12:50 | 45 | | Break |
| 12:50 | 13:35 | 45 | Workshop | Special Situations |
| 13:35 | 13:45 | 10 | | Break |
| 13:45 | 15:15 | 90 | Practice | Case, Debriefing, Debriefing |
| 15:15 | 15:30 | 15 | | Break |
| 15:30 | 16:00 | 30 | Workshop | Do and Don't |
| 16:00 | 16:15 | 15 | | Summary Day 2 |

Simulationszentrum Mainz
Wernher von Braun Straße 9
55129 Mainz
Tel. +49-6131-97 13 546
e-mail: info@aqai.de
www.simulationszentrum-mainz.de

**TABLE 65.9**   Course schedule for Train-the-Trainers #6: Help needed METI ECS

## Course Outline: TTT6 ECS Troubleshooting

### Day 1

| Begin | End | Duration | Type | Title |
|---|---|---|---|---|
| 12:00 | 12:30 | 30 | | Welcome/Introduction Day1 |
| 12:30 | 13:00 | 30 | Hands-On | Setup and PowerOn |
| 13:00 | 14:00 | 60 | Hands-On | Technical Components |
| 14:00 | 14:30 | 30 | Lecture | Problems/Tools |
| 14:30 | 14:45 | 15 | | break |
| 14:45 | 15:15 | 30 | Lecture | Network Layout/Computer Tools |
| 15:15 | 15:30 | 15 | Workshop | Troubleshooting Methodology |
| 15:30 | 15:45 | 15 | | break |
| 15:45 | 16:45 | 60 | Lecture | Pneumatic System |
| 16:45 | 17:00 | 15 | | break |
| 17:00 | 18:00 | 60 | Hands-On | Troubleshooting the Network |
| 18:00 | 18:30 | 30 | | Questions Day 1 |

### Day 2

| Begin | End | Duration | Type | Title |
|---|---|---|---|---|
| 9:00 | 9:15 | 15 | | Come together Day 2 |
| 9:15 | 10:00 | 45 | Hands-On | Trauma Disaster Casualty Kit Setup and Operation |
| 10:00 | 10:45 | 45 | Hands-On | Trauma Disaster Casualty Kit Network/Maintenance/ Technical Components |
| 10:45 | 11:00 | 15 | | Break |
| 11:00 | 11:30 | 30 | Lecture | Electrical System |
| 11:30 | 12:30 | 60 | Hands-On | Calibration of pneumatical sytem |
| 12:30 | 13:30 | 60 | | Break |
| 13:30 | 14:15 | 45 | Hands-On | Calibration of Defibrillation/Pacing |
| 14:15 | 15:15 | 60 | Hands-On | Practice Troubleshooting |
| 15:15 | 15:30 | 15 | | Break |
| 15:30 | 16:15 | 45 | Workshop | Maintenance |
| 16:15 | 16:30 | 15 | | Questions End of Day 2 |

Simulationszentrum Mainz
Wernher von Braun Straße 9
55129 Mainz
Tel. +49-6131-97 13 546
e-mail: info@aqai.de
www.simulationszentrum-mainz.de

**TABLE 65.10**  Course schedule for Train-the-Trainers #7: Help needed METI HPS

**Course Outline: TTT7 Help needed HPS**

**Day 1**

| Begin | End | Duration | Type | Title |
|-------|-----|----------|------|-------|
| 12:00 | 12:30 | 30 | | Welcome/Introduction Day 1 |
| 12:30 | 13:30 | 60 | Hands-on | Setup and PowerOn |
| 13:30 | 14:30 | 60 | Hands-on | Functionalities and Components |
| 14:30 | 14:45 | 15 | | break |
| 14:45 | 15:15 | 30 | Lecture | Introduction HPS5 |
| 15:15 | 15:45 | 30 | Workshop | Methodology of Troubleshooting |
| 15:45 | 16:15 | 30 | Lecture | Problems and Tools |
| 16:15 | 16:30 | 15 | | break |
| 16:30 | 17:00 | 30 | Lecture | Technical modelling of Lung |
| 17:00 | 18:00 | 60 | Workshop | Tools for the lung |
| 18:00 | 18:30 | 30 | | Closing Day 1 |

**Day 2**

| Begin | End | Duration | Type | Title |
|-------|-----|----------|------|-------|
| 9:00 | 9:15 | 15 | | Welcome Day 2 |
| 9:15 | 10:15 | 60 | Lecture | Tools for Calibrations |
| 10:15 | 11:00 | 45 | Hands-on | Calibrations 1 |
| 11:00 | 11:15 | 15 | | break |
| 11:15 | 12:15 | 60 | Hands-on | Troubleshooting the Lung |
| 12:15 | 13:15 | 60 | | break |
| 13:15 | 14:15 | 60 | Hands-on | Calibrations 2 |
| 14:15 | 14:30 | 15 | | break |
| 14:30 | 15:00 | 30 | Lecture | HPS Computer/Network Tools |
| 15:00 | 15:30 | 30 | Hands-on | Troubleshooting the Network |
| 15:30 | 16:00 | 30 | Workshop | Maintenance/Quicktest |
| 16:00 | 16:30 | 30 | | Closing remarks |

Simulationszentrum Mainz
Wernher von Braun Straße 9
55129 Mainz
Tel. +49-6131-97 13 546
e-mail: info@aqai.de
www.simulationszentrum-mainz.de

# 65.10 Discussion

Our view on Train-the-Trainer concepts has been influenced by three major components: The first is a cooperation with Lufthansa Flight Training experts [6], who apply similar train-the-trainer concepts to aviation simulation facilitators. The second component includes experiences through training of the participants of the German Anesthesia Society's simulation initiative [7], and the third is experiences in facilitating Train-the-Trainer seminars in more than 22 countries for more than 750 participants partly in the context of a cooperation with METI [8]. We are aware that other groups also have thought about Train-the-Trainer concepts and are successfully training trainers for simulation. Among these, a starting program with credits in a master's course level at the University of Hertfordshire, U.K. [9], or a certification process for simulation instructors at the Center for Medical Simulation, Cambridge, U.S.A. [10], but among all these activities, no common basis has been established yet, nor have any standards been agreed-upon by the simulation community.

# 65.11 Conclusion

The need for training the trainers in the world of clinical simulation should encourage all of us toward the goal of developing a universal standard for simulation center staff qualifications. This will help Simulation to become a universally utilized educational tool with comparable levels of training standards. The idea of specifying the minimal requirements of capabilities of staff for certain positions within a simulation center – as done in Mainz with the "staff-qualification-checklist" or in the Center for Medical Simulation with "guidelines for instructor certifications" – will eventually lead to degrees and internationally accepted "simulation facilitators."

The above described topics for a Train-the-Trainer curriculum just point out the basic simulation-specific skills, and they can easily be widened and deepened into each of the branches of interest and the specific training needs. Despite the trend of clinical simulation to become more and more interdisciplinary between the different branches of clinical practitioners, the presented Train-the-Trainer concept does not yet take other simulation equipment, such as partial-task trainers or surgical trainers into account. A stronger focus on Human Factor aspects including role-plays and active reviews or including techniques for course and participant evaluation would be desirable to add to the program.

Currently, no examination of the success of a class or a formal recertification process has been established, and an implementation of a second level of Train-the-Trainer training for repeating and intensifying the introduced topics is missing.

The appropriate platform for the definition of Train-the-Trainer standards and minimal educational requirements for certain positions at a simulation center would surely be one of the international scientific organizations such as Society for Simulation in Healthcare (SSH) [11], Society in Europe for Simulation Applied to Medicine (SESAM), or other societies such as European Society of Anaesthesiology (ESA) [12], the patient safety foundation (NPSF, APSF) [13], or the educational societies of clinical specialties.

Establishing standards will, beside leveraging quality of simulation trainings, encourage the international exchange of ideas and expertise and eventually create an international job market for staff members with simulation-specific skills and wisdom.

# Endnotes

1. Nontechnical skills (NTS) are the cognitive and social skills required in any operational task involving decision making, management, communication, workload management, leadership, and team work.

2. We call this the navigator/pilot model: the navigator (clinical instructor) has to know were the scenario starts, ends, and its desired path between the two but does not need to be an expert in making the craft do his bidding, whereas the pilot (simulator operator) needs to know everything about the craft, and how well its capabilities as well as their own matches the navigator's plan. Together they succeed as a team only through respect, communication, collaboration, and cooperation.

3. While these guidelines specifically refer to anesthesia crisis resource management – ACRM, many simulation centers are now offering generic, multidisciplinary and interdisciplinary crisis resource management – CRM – training sessions.

4. Double moderation: Two facilitators moderate a session together. In the mentioned case, both of the facilitators moderate the session from the view point of their professional specialization (e.g., a psychologist and a clinician), and in an alternating manner they address the group of trainees. The coordination of who is actively moderating and who stands in the background and the consistency of the content of the moderation is very important and can be practiced. A well-performed double moderation gives the trainees a very interesting session as each of the moderators will be identified with a certain role or position in the session and as the focus of the group alternates between the two moderators, the session will be more lively for the listener.

# References

1. Simulation Center Mainz, Simulationszentrum Mainz, Germany, http://www.simulationszentrum-mainz.de/.
2. DGAI, Deutsche Gesellschaft für Anästhesie und Intensivmedizin, http://www.dgai.de/.
3. DGAI, Ad-hoc-Kommission Simulatoren and Schüttler, J. Anforderungskatalog zur Durchführung von Simulatortraining-Kursen in der Anästhesie (Beschluss der DGAI 02/2002), *Anästhesiologie und Intensivmedizin* 43, 828–830 (2002).
4. SESAM, Society in Europe for Simulation Applied to Medicine, http://www.sesam.ws/.
5. Gaba, D. M., Fish, K. J., and Howard, S. K. *Crisis Management in Anesthesiology*, Churchill Livingston, New York, 1994.
6. LFT, Lufthansa Flight Training, http://www.lft.de/lft/deu/data/index_start.php?flashan=1.
7. Moenk, S. Development of the German Simulation Program. 4th Annual International Meeting on Medical Simulation. Society for Technology in Anesthesia. 2004, Albuquerque/Santa Fe, New Mexico.
8. METI, Medical Education Technologies Inc., http://www.meti.com.
9. Concepts of simulation training in healthcare education as part of Master in Health and Medical Education, University of Hertfordshire, Contact: Guillaume Alinier, G.Alinier@herts.ac.uk; http://perseus.herts.ac.uk/.
10. Center for Medical Simulation, Cambridge, MA, U.S.A., http://www.harvardmedsim.org/cms/.
11. SSH, Society for Simulation in Healthcare, formerly SMS, Society of Medical Simulation, http://ssih.org/index.html.
12. ESA, European Society of Anaesthesiology, http://www.euroanesthesia.org/index.php.
13. NPSF, National Patient Safety Foundation, http://www.npsf.org/, APSF, Anaesthesia Patient Safety Foundation, http://www.apsf.org/.

66

# Experiential Training for New Simulation Coordinators

Roger E. Chow
Viren N. Naik

## 66.1 Balancing the Art with the Science

Our Patient Simulation Center is located at St. Michael's Hospital in downtown Toronto, Canada. This inner-city teaching hospital is fully affiliated with the University of Toronto, and is one of the busiest trauma and emergency centers in Canada. The Simulation program services all disciplines and is under the Department of Clinical Education.

Our simulation experience has been gained since our beginning in 1995 from interactions with a supportive and sharing simulation community locally and internationally.

The challenge in learning to become a simulation professional is similar to the struggle of learning any clinical specialty – how to balance the art with the science? The *science* is easy to define as the technical operation of a patient simulator, peripheral equipment, audio-visual system, etc. The *art* is more difficult, and it isn't always intuitive. Initially, the technical aspects seem most difficult, but in our experience, it's the art of simulation that takes far more time to achieve competence (see Topic XII and Chapter 54).

With the sheer volume of information to pass onto new coordinators, it's overwhelming for us and for them. They have to learn how to handle the technical operations, become familiar with the educators and their groups, with the different simulation scenarios each group uses, and with how each group uses simulation differently. On top of that, there's the art of transforming the technical aspect of simulation into "clinical reality" for learners. Gone are our naïve days when we would tell them everything we knew about simulation all at once, have them participate in any simulations happening at the time, and wait for them to fly on their own. With deliberate organization of the information, we can simplify the process, present a more focused orientation, and give them a foundation to build their learning on.

## 66.2 The Training

In the "hands-on" spirit of Patient Simulation, we want the new coordinators to have training that's experiential. First, they need to receive an orientation to simulation and then produce their own simulation *session*. We define a simulation *scenario* as a simulated "clinical event" that involves a patient simulator operated by a coordinator. In contrast, a *session* encompasses the scenario, the designing, planning, and ultimately, execution of the simulation.

## 66.3 The Orientation

New coordinators will become familiar with the technical capabilities of the patient simulators, the software interface, and the simulation sessions currently happening (examples to gain familiarity with). We'll sit and debrief about the versatility and limitations of simulation through the live examples and viewing previously recorded scenarios of different learner groups. We are a teaching hospital affiliated with a University, and therefore, service a broad range of learners at different levels from students to hospital staff. These live and prerecorded scenarios highlight different ways we customize simulation for each learner group. Most groups will manage a scenario without interruption, followed by a debriefing (usually senior-level learners and staff), while other groups will pause the scenario for didactic-like instruction during the simulation (usually the junior levels).

We start with an orientation first, but this process continues in the form of an apprenticeship for an extended period of time (6–12 months depending on the individual). While building their own simulation session, they will continue to watch and participate in actual scenarios and witness the logistics of some current sessions. In designing a session, the new coordinator will write a scenario, program it, develop a needs list, figure out the logistics of production, rehearse it, and troubleshoot throughout. When the scenario is ready to test, they "get behind the wheel" in the control room, and drive the same scenario several times while operating the software. An actor will play the part of learner and present different responses each time, providing the new coordinators with practice, and illustrating how the same scenario is dynamically unique each time. In addition to producing the scenario from the control room, the new coordinator will be placed into the scenario to participate as the learner. This important step allows them to personally experience how the learner feels in the simulated environment, and improves their ability to relate to the person on the receiving end. This being simulation, we like to debrief following each of these practice scenarios.

# 66.4 The Anatomy of a Simulation Session

We present the new coordinators with an organized framework to guide them in their development of a session:

- The logistics
- The scenario
- The software
- The theater setup and a props/to do list
- Rehearsing
- Presentation of the session

## 66.4.1 Logistics of Delivery

When designing a new session for a learner group, these are the typical questions we consider first:

- How many learners?
- How much time is available?

- What are the learning objectives?
- Is there research involved?

### Hints

- Determine if patient simulation is indeed the best way to deliver the learning objectives.
- Learning objectives and whether research is involved can determine whether individual or group participation is required for the simulation.
- The number of learners and the time available will dictate how the session is organized: for example, 1 learner doing 3 scenarios in an hour, a group of 5 doing 3 scenarios in 2 hours, a group of 40 rotating 10 learners through every 15 minutes for didactic-like simulation, etc.
- We service many different disciplines in the hospital; sometimes educators have to organize larger groups with significant time limitations, and therefore cannot take full advantage of what simulation can offer. We respect these circumstances and strive to provide the most effective simulations for everyone.
- Availability of time is a common limitation; we can suggest using simulation not as something extra to be added onto a full curriculum, but that maybe some aspects in the curriculum can be more effectively learned through simulation, i.e., replace existing material to become fully integrated into the syllabus.
- Ultimately, we find ways to deliver a session trying to capitalize on the advantages of experiential learning by maximizing learner participation.

## 66.4.2 Developing the Scenario

*Always* identify the following factors before you *ever* write a scenario:

- Who are the learners?
- What are the learning objectives?
- What is their education level?

### Hints

- Avoid scenarios that are too complicated; try developing simple ones that will best achieve the objectives. Remember, it's about the learners and how your creativity with simulation can help deliver an experiential education for them to take away a lesson *learned*.
- Coordinators are always trying to reproduce the technical aspects in order to achieve a clinical accuracy in their presentation of scenarios. They may go to great lengths to outfit the simulation theater with equipment and props, dress up the mannequin with clothes, prepare and apply wounds, etc. However, there is a need to strike

a balance with the learners, taking into account their level of education, as it has a direct bearing on how the simulation is presented. Learners in their early stages are easily overwhelmed with highly realistic simulations. It is also a waste of resources to make scenarios too sophisticated for learners who are not yet ready for that level of experiential education. Be aware of how simulation can affect them, not only on an educational level, but consider psychological effects as well.

- The best scenarios are based on real events, and modified with the key learning objectives extracted and transformed for simulation. Use your past clinical experiences or those of an affiliated clinician.

## 66.4.3  Simulation Software

For most of our simulations, we have been using a Laerdal SimMan, and less frequently our Eagle MedSim.

- Preprogramed scenarios
- Manual, operator controlled scenarios
- The combination of both

### Hints

- Programing a scenario is time consuming; requiring intensive troubleshooting while creating and constant readjusting for timing changes, smooth transition of patient monitor vital signs, unforeseen events, etc.
- For research purposes, preprograming becomes essential for reproducibility.
- With preprogrammed scenarios containing reasonable decision/action branches, you have the option of using simulator operators without extensive clinical backgrounds.
- Manually controlled scenarios work well for early-stage learners where you can slow down or hold patient deterioration. You can maintain control of the scenario, do bedside teaching, and not lose sight of the learning objectives.
- Manually controlled scenarios provide added versatility of quick improvisations when you need to respond outside of the game plan, like increasing the complexity of the scenario.
- Scenarios can be preprogramed to include components of manual, operator modified maneuvers, and sometimes this combination is the most effective way. It's a good way to become familiar with the software interface too.
- Gaining experience with the patient simulators and different user groups will tell you the most effective way to deliver the scenarios; totally preprogramed, totally manual, or the combination of both.

## 66.4.4  The Theater Setup and a Props/To Do List

The simulation theater will be decorated and configured on the basis of where the scenario takes place, and the List will help minimize oversights (missing items) during presentation of the scenarios.

- Equipment and props needed?
- How much realism?

### Hints

- To develop a list, ask yourself what the mannequin needs before the case starts (i.e., cuts on the right arm, bruises on the left ribs, backboard, cervical collar, etc.). Now add to the list by visualizing the execution of the scenario in your mind, and anticipate equipment the learner may ask for (i.e., intravenous infusions, intubation equipment, ventilator, warming blanket, units of blood, etc.).
- Involvement in multiple scenarios for many different learner groups makes it difficult to remember specific scenarios as discreet entities. Including a list with the electronic and hard copy of each scenario will simplify your job. During a quick turnover between scenarios, the list will remind you of the props and equipment required.
- The list will assist you in the short term, and in the future when new coordinators come on to take over producing the scenarios.
- You may have to search for your own props: for example, X-rays with common clinical conditions can be downloaded and printed from the Internet. You may have to customize or make your own props: for example, reusable wounds like cuts and bruises. Please see Chapters 10 and 72 for techniques for "making wounds" (moulage).
- Theatrical supply stores have many resources and expertise to help with the props.
- Friends: a simulation center needs them. Developing relationships with different disciplines in the hospital and community is invaluable to the acquisition of props and equipment. "Dominic" the Respiratory Technician is an indispensible resource, "John" in pharmacy can supply expired medication vials/bottles, "Jane" in biomedical engineering can provide you with an electrocautery machine that has been disabled, "Sue" in the intensive care unit can provide disposables in exchange for simulation services, "Bob" in private industry may donate equipment that would otherwise be too expensive to purchase, etc.
- Level of realism/fidelity: we believe the closer to reality the better, on all levels of presentation simple or sophisticated. Having said that, there is a need to balance the use of equipment, props, and actors. Have enough to make it sufficiently realistic for suspension of disbelief. Too much is unnecessary, and eventually the cost will outweigh the

benefits. There is no formula for just how much realism is the right amount for any given simulation, we learn and adapt as we go. Reactions and feedback from your learners provide good benchmarks.

### 66.4.5 Rehearsing the Scenario

- Talk and walk through the scenario. This "technical rehearsal" will provide you the opportunity for coarse-tuning.
- Troubleshoot the scenario in real time with actual interventions while employing a scenario-naïve confederate. This "dress rehearsal" will provide you the opportunity for fine-tuning.
- Troubleshoot the simulation session.

**Hints**

- Like a rough draft in need of editing, a rehearsal reveals what works and more importantly, what does not work.
- For the scenario, you can judge the realism by looking for the clinical accuracy and plausibility of the event. Ask yourself questions: Are any components of the scenario misleading? Are the intentional distractions working? Are the simulator's preprogrammed transitions working well? Are changes in the dynamic patient monitor vitals signs realistic? What are my improvisations for "unpredictable" actions?
- For the session, look at the logistics of delivery. Have we recreated a realistic environment? Do we have all the props needed? Are they functioning as *required* for the teaching objectives? How is the timing, the cues, the script, the placement and accessibility of props, etc.?

### 66.4.6 Presentation of the Session

All that hard work and preparation, now it's time to deliver the complete package.

- Before starting, always present ground rules (guidelines) to the learners on how to approach (behavior) the simulation scenarios.
- Along with the ground rules, provide an orientation to the simulation environment, specifically the boundaries of what is and is not "in play".

**Hints**

- Power up early, and do your system checks before you go to get coffee. Always remember, *time* is our most precious resource – use yours wisely so you don't waste others' time.
- Providing ground rules is essential; the learner needs to have a similar approach and expectations to simulation, as the coordinator and clinician instructor. All the work and preparation of a session may be compromised if the learner is unclear about the ground rules.

- Keep the ground rules short and sweet; otherwise you risk confusing the learner. We want them to treat the patient and the scenario like its real, and to behave as natural as possible. We demonstrate this desired behavior during our orientation. Your patient simulator is always referred to as *their patient*, never *the simulator*, and is always as alive and as animated as their patient should be when they are in *their patient's* room.
- An unfamiliar environment is not usually a desired part of the scenario, it can be a distraction or used as an excuse by learners. We always try to have a coordinator in the environment acting in a clinical role (i.e., a non-initiating clinician who just happens to know where everything is) to eliminate unfamiliarity felt by the learner.
- Having coordinators role-play provides many benefits: number one is protecting the mannequin from damage, usually from over enthusiastic learners. Their acting can enhance the realism, they can cause distractions, they can instantly troubleshoot technical problems, and try to solve them without interrupting the scenario, they provide a communication link to the control room in person, by wireless radios or by telephone, and they can watch for the safety of the learners (live shocks from a fully functional defibrillator, fluid spills on the floor, etc.).
- Driving the scenario in the control room is a powerful place to be, stay focused on the objectives and the learners. When you have people participating in simulations, you're asking them to suspend disbelief, and you're telling them that this is a safe place to make mistakes. When they're willing they become vulnerable and place their trust in you, in return we honor that.
- Coordinators play a significant role in the control room and in the simulation theater. Success depends on how well they present the scenario and how well a clinician instructor debriefs the learner.

Once the new coordinators have been orientated to simulation and the production of a session, they'll continue with their apprenticeship gaining familiarity with the operations and developing their skills in the science and art of patient simulation.

## 66.5 Conclusion

Learning to become a competent simulation professional as challenging as learning any clinical specialty – both require a balance of art with science. Traditional clinical programs explicitly acquaint students with science and technology, but not so much with art and theater. Hence, mastering the technical operations of patient simulators, clinical equipment, audio-visual system, etc., is easier than the less familiar but no less essential artistic creativity. Like board and card games, the rules of play are easily grasped, the nuances of play takes far more time to achieve competence, let alone expertise.

# 67

# Becoming a Simulation Instructor and Learning to Facilitate: The Instructor and Facilitation Training (InFacT) Course

Peter Dieckmann
Marcus Rall

## Statement on Conflict of Interest

Most of the InFacT courses we conducted were held and financed in cooperation with Laerdal Medical or SimuLearn. Nevertheless, the InFacT course is designed to be simulator independent, and the course does not incorporate simulator-specific topics. It could as well be used for users of other makes and models of clinical simulators.

## 67.1 Why Instructor Courses are so Important

Realistic simulation training is an intense experience and a powerful agent for change [1, 2]. Participants dive into the scenario and take it for real – if not always, at least very often. What happens to them during the simulation should affect their self-image – and will do so positively, especially if the scenarios are well matched to the participants' level of competence and maturity, and if they are purposefully designed and presented to appropriately engage in them. In particular, Crisis Resource Management (CRM) scenarios are the most socially complex and, therefore, often the most stressful [3–6]. Every powerful agent has many effects, both desired and undesired. Users of simulation as an agent should be aware of its power. This holds true especially for instructors using simulation for basic education, advanced training, as well as other purposes like research [7–9] and assessment [10, 11]. Being a simulation instructor is a complex task that changes dynamically over the short duration of any simulation-based course. Conducting good simulation education and training means more than buying a high-fidelity simulator and writing scripts and producing realistic scenarios. "The key is the program, not the hardware" [12] and "[i]t is not how much you have but how you use it" [13]. As simulation instructors are the persons whom the participants look to for instruction, guidance, and approval, their competence in using simulation is amongst the most important "elements" needed in any simulation-based learning, and that is why instructor courses are so important.

We designed the Instructor and Facilitation Training (InFacT) course to provide experienced health care professionals with a jump start for the life-long learning process of becoming a simulation instructor. The course is aimed at instructors-to-be in simulation-based courses oriented to teach CRM with video-assisted debriefing (see Chapter 69). In our view, both the CRM-orientation and the video-assisted debriefing are amongst the most valuable and unique benefits of simulation-based training. In this chapter, we will present the InFacT course concept that addresses the different roles of simulation instructors in educational-simulation settings.

# 67.2 Tasks of Simulation Instructors During Simulation-based CRM Courses

For understanding the tasks of simulation instructors and the related requirements, it is necessary to understand what the instructors do. We conducted an extensive needs analysis in this regard. We interviewed simulation instructors in six different domains (anesthesiology, aviation, naval, nuclear power plants, truck driving, and computer screen-based simulations) in Germany about their goals and ways they used simulation [14]. We interviewed 18 simulation instructors from 11 anesthesia simulation centers in Germany and Switzerland about their goals, success factors, and difficulties in designing and producing simulation courses [15]. We also visited many centers in order to observe instructors in their different roles [16]. We conducted an interview study with simulation

scenario participants about how they experienced the scenario [17]. Finally, we have our own experiences from 6 (PD) and 10 (MR) years of simulation training, both being instructors and researchers working with simulation.

A simulation course consists of different modules or different phases (e.g., setting introduction, simulator familiarization, scenarios, debriefings) [15, 18]. The different modules are interrelated, and problems arising in one module can affect other modules. For example, it might be difficult to conduct an open, nonconfrontational debriefing if the instructor did not succeed in creating an open atmosphere in the beginning of the course and again at the beginning of the debriefing [29, 30]. The different phases require different roles that the instructors must fulfill. At the beginning of the course, there is mostly instruction and explaining, for example, on what the course is about, how to use the simulator, and possibly new theoretical input (e.g., clinical concepts). In the middle of the course, the instructors are responsible for creating relevant learning experiences during scenarios. During debriefings and toward the end, instructors are responsible for helping participants to learn as much as possible from the simulation experience – the instructors become facilitators. This mix of roles is reflected in the name of our course – the Instructor and Facilitation Training.

# 67.3 Aim and Objectives

The main aim of the InFacT course is to enable health care professionals to start providing simulation-based courses with video-assisted debriefing, addressing the principles of CRM in acute care medicine for members of their own clinical professions. This means that nurses would mainly provide courses for other nurses at least in the beginning of being an instructor, likewise for paramedics or physicians. We emphasize that the InFacT course graduates are not "ready for everything" – simulation instructors. Becoming a skilled instructor is a life-long learning process, similar to other train-the-trainers concepts, as well as in all clinical disciplines.

The InFacT objectives state that participants shall:

- understand how it feels to act within a simulation scenario and be debriefed on their experience,
- acquire the knowledge and skills necessary to conduct and time manage a simulation course with its different phases,
- reflect on their changing roles within a simulation course,
- and understand basic concepts of patient safety and CRM in relation to simulation-based courses.

A special emphasis is placed on conducting debriefings.

## 67.4 Basic InFacT Outline and Course Concept

Figure 67.1 shows the basic format of the InFacT course. The course concept comprises three blocks in a two-by-two-plus-one format. The first block of 2 days focuses on the theoretical basis as well as the experience of CRM-oriented scenarios and video-assisted debriefings. After the first block, there is a break of about 4 weeks to allow new concepts and ideas to sink in. The second block of 2 days focuses on practice and application of what was learned during the first block. The third block is optional but recommended and lasts 1 day for a course visit and coaching in the home town simulation center of the InFacT participants. Coaching is also done independently of InFacT. Up to 16 participants can be in one InFacT group led by two InFacT instructors. We will provide more details on the single blocks in the remaining text.

## 67.5 Contents

We designed the contents of the course to fit the tasks of instructors and the related needs to perform them. We present the concepts here according to the chronological order of a prototypical InFacT course.

### 67.5.1 Theory of Human Errors, Patient Safety, and CRM

Amongst the most unique possibilities simulation offers is to sensitize participants to systemic issues of patient safety and crisis resource management and to provide them with some opportunities to practice related skills. For this reason, we included basic related issues into the InFacT course [5, 6, 19–23].

### 67.5.2 Debriefing

Debriefing is one of the most important parts of any simulation learning experience [24, 25]. It allows for integrating the experience of difficult moments during the scenarios. For this reason, it differentiates simulation-based courses from clinical duties and augments the reflection, as there is usually much experience during clinical duties but hardly any time to reflect on these experiences. The InFacT course provides a specific debriefing structure to be practiced during the course debriefings (see Chapter 69). Alternative structures for debriefings are also presented as well as the need to balance CRM aspects and clinical-technical issues. Particulars of video-assisted debriefing and annotation systems are discussed. There is some input on difficulties during debriefings, while the focus is on "regular" debriefings in order not to overwhelm participants. Another focus includes ways of delving into an analysis of what were the causes of what happened during the scenario, in contrast to staying with a superficial description of what happened only. The main tools to do so are open questions and – probably the most difficult technique to accomplish – providing time to answer.

### 67.5.3 Importance and Methods of Simulator Familiarization

Only if participants understand the possibilities, limitations, and ways to use the simulator can they feel competent while using it. For this reason, we thoroughly familiarize InFacT participants with the simulator, and explain what to take care

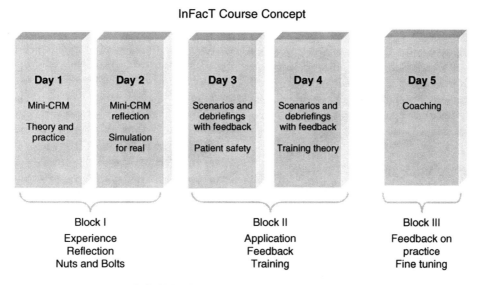

InFacT Course Concept

**FIGURE 67.1** Basic InFacT course format.

of when they do the same for their participants. During this familiarization, the simulator's features, capabilities, and limitations are reviewed (e.g., pulses, auscultation, etc.) as well as the simulated surrounding (e.g., carts, communication devices, interaction with role-players, etc.).

### 67.5.4  Simulation Setting

We propose that the different phases of a simulation-based training course are interrelated within the "simulation setting" [15, 18]. We present the different phases and explain how they are interlinked and influence each other. Especially during a scenario, there are different reality levels involved that need to be understood by both instructors and participants, in order to best use simulations, engage into the "fiction contract" [26], and perform the "willing suspension of disbelief" [27].

### 67.5.5  Designing and Producing Scenarios

Using the TuPASS scenario script (see Chapter 59), participants do have time to develop their own scenario in the break between Block I and II. In the beginning of the second course block, scenarios are discussed, and five of the scenarios that were created by the participants are selected and produced during the remaining course. Participants are given tips in using the script and also for designing further scenarios.

### 67.5.6  Instructor's Role in the Different Phases of a Simulation Course

Being a simulation instructor requires the instructor to reflect on the dynamic role changes over the presentation of a course. InFacT participants are helped to do so after experiencing simulation scenarios and debriefings and also in a special workshop during which the different tasks mentioned above are discussed in their relation to the changing instructor role. We discuss the differences between instruction and facilitation. In order to understand the potential and pitfalls of simulation-based training, it is necessary to experience taking part in simulation-based and CRM-oriented simulation scenarios with video debriefing. Only this direct experience and change of perspective can open an emotional and cognitive understanding of how participants feel during simulation scenarios, which are very complex social interactions.

### 67.5.7  Blended Learning, Part-task Trainers and Computer Screen-based Simulators

The basic idea of blended learning is to tailor the educational endeavor to target groups or even individuals. It is about the right learning, at the right time, in the right manner, for a specific person. We present this idea as well as other methods that might be used to enhance simulation-based training by selecting different tools also to provide basic knowledge (e.g., lectures) or procedural skills (e.g., part-task trainers or computer screen-based training scenarios).

## 67.6  Methods

The InFacT course uses a mixture of different methods to help participants working with the content of the course; all contributing to the goal of becoming a simulation instructor and learning to facilitate. Basic methodological principles concern the mixture of theory and practice, practice and reflections on this experience, and the gradual role shift from being a *participant* in a simulation course toward being an *instructor* in a simulation course. We summarize these basics by referring to the InFacT course as "double featured." InFacT participants have both: possibilities to experience how it feels to be part of a simulation course as well as to reflect on implications for the instructor role. Participants gain hands-on experience in both roles. This is accomplished in a way that a team of two InFacT participants takes on the role of instructors and presents a scenario for two to four other InFacT participants who take on the roles of simulation course participants. The remaining InFacT group watches the scenario and the debriefing and later provides feedback.

Lectures help participants to construct new, or to augment existing, knowledge about patient safety, CRM, debriefing, the simulation setting, designing scenarios, and using video-assisted debriefing.

Workshops are provided to discuss potentials and pitfalls of debriefing, the role of simulation instructors, and observing and discussing CRM aspects in simulation scenarios.

Practice is provided for executing scenarios and debriefings. In the beginning of the course, participants "only" do the debriefing, while, in the second block, they produce their own scenario from a control room and debrief it later – gradually shifting into the role of simulation instructors. During the InFacT course, each participant will have the opportunity to practice the debriefing at least once and to take part in several scenarios.

Feedback is one of the essential methods during the InFacT course. After a team of two InFacT participants performed a debriefing (and possibly a scenario), they provide themselves with feedback on their own debriefing, receive feedback from the "hot-seat" participants whom they debriefed, the remaining InFacT course group, as well as from us as the InFacT tutors. Feedback rules are enforced so that the feedback remains descriptive, precise, constructive, and friendly. Feedback content is not discussed outside the debriefing.

The Break between the first and second block is a methodological break in order to avoid overwhelming InFacT participants with too much information. Overload happens to some degree in any case, as becoming a simulation instructor is a complex task, and 4 days for providing a jump start is actually a very a short time.

## 67.7 Course Schedule

The InFacT course schedule is tailored to individual projects. The sequence can be shifted as needed, but in the end, the course will cover all topics mentioned above. The overall working time in the course excluding any breaks is about 26 hours over the 4 days. Table 67.1 provides a prototypical course schedule.

**TABLE 67.1**  Typical InFacT course schedule

| Duration in hrs. | InFacT Block 1–Day 1 |
|---|---|
| 0.75 | Course introduction |
| 1.00 | Simulator familiarization |
| 1.00 | Patient safety and crisis resource management |
| 1.25 | Simulation scenario with Debriefing by InFacT tutors |
| 1.50 | Simulation scenario with debriefing by InFacT participants and feedback |
| 1.00 | Debriefing workshop |
| | **InFacT Block 1–Day 2** |
| 0.25 | Open questions |
| 3.75–6.25 | 3–5 Simulation scenarios with debriefing by InFacT participants and feedback |
| 1.00 | Simulation setting |
| 0.50 | Scenario design and TuPASS scenario script (see Chapter 59) |
| | **InFacT Block 2– Day 1** |
| 0.25–0.50 | Open questions |
| 1.50 | Scenario design workshop |
| 0.50 | Debriefing refresher |
| 3.00–4.00 | 3–4 Simulation scenarios with debriefing by InFacT participants and feedback |
| 1.00 | Debriefing workshop |
| | **InFacT Block 2–Day 2** |
| 0.25 | Open questions |
| 0.50 | Lecture blended learning and microsimulation |
| 3.00–4.00 | 3–4 Simulation scenarios with debriefing by InFacT participants and feedback |
| 1.50–2.00 | Microsimulator practice |
| 1.00 | Course summary/discussion |
| 0.25 | Evaluation forms |

## 67.8 Experiences with Producing the InFacT Course

We started to conduct InFacT courses in April 2004. By October 2006, we had conducted 13 courses in Italy, Germany, and The Netherlands, with approximately 240 participants in all, with an average of 18 participants per course. We evaluated the course using questionnaires and by asking for oral feedback. The questionnaire focused on the five areas: "Clarity of instructor role," "Overview of designing simulation settings," "Ability to produce a simulation-based course," "Debriefing skills," and "CRM skills." Both methods (questionnaires and oral feedback) indicated a high value as perceived by the participants [28]. At the Italian simulation center, SimuLearn in Bologna, we trained all of their simulation instructors, who subsequently produced simulation-based clinical courses for approximately 10,000 physicians in 2005. In this center, a needs analysis is underway for accessing the needs for the second-level instructor course, Ad-InFacT$^2$.

## Further References

A poster with a short description of the InFacT course is provided on the website of this book (www.books.elsevier.com/9780123725318/companions) and can be downloaded from our web site (www.tupass.de/downloads/InFacT.html).

## Acknowledgments

Our colleagues Eric Stricker, Silke Reddersen, Joerg Zieger, Christian Karcher, Patrizia Hirsch, Thomas Armbruster, Bettina Nohe (TuPASS), former colleagues Albrecht Wengert (TuPASS) and cooperation partners Gerson Conrad (German Air Rescue – DRF) ran InFacT courses with us. A special thanks to the staff of SimuLearn, Bologna, Italy, for hosting eight InFacT courses very professionally. Another special thanks to Laerdal Medical, Germany for the cooperative work in providing InFacT courses in our simulation center TuPASS in Tuebingen, Germany.

## References

1. Rall, M. and Dieckmann, P. Simulation and Patient Safety. The use of simulation to enhance patient safety on a system level. *Current Anaesthesia and Critical Care* 16, 273–281 (2005).

2. Rall, M. and Gaba, D. M. Patient simulators. In: Miller, R. D. (ed.), *Anaesthesia*. Elsevier, New York, 2005, pp. 3073–3103.

3.  Gaba, D. M., Fish, K. J., and Howard, S. K. *Crisis Management in Anesthesiology*. Churchill Livingstone, Philadelphia, 1994.

4.  Gaba, D. M., Howard, S. K., Fish, K. J., et al. Simulation based training in anesthesia crisis resource management (ACRM): A decade of experience. *Simulation & Gaming* 32(2), 175–193 (2001).

5.  Rall, M. and Dieckmann, P. Crisis Management to Improve Patient Safety. [Online: http://www.euroanesthesia.org/education/rc2005vienna/17RC1.pdf]. ESA Refresher Course 2005, pp. 107–112.

6.  Rall, M. and Gaba, D. M. Human Performance and Patient Safety. In: Miller, R. D. (ed.), *Miller's Anaesthesia*. Elsevier Churchill Livingston, Philadelphia, 2005, pp. 3021–3072.

7.  Dieckmann, P., Reddersen, S., Wehner, T., and Rall, M. Prospective memory failures as an unexplored threat to patient safety: results from a pilot study using patient simulators to investigate the missed execution of intentions. *Ergonomics* 49(5–6), 526–543 (2006).

8.  Howard, S. K., Gaba, D. M., Smith, B. E., et al. Simulation study of rested versus sleep-deprived anesthesiologists. *Anesthesiology* 98(6), 1345–1355 (2003).

9.  Manser, T., Dieckmann, P., Rall, M., and Wehner, T. Comparison of anaesthetists' activity patterns in the operating room and during simulation. *Ergonomics* 50(2), 246–260 (2007).

10. Weller, J. M., Bloch, M., Young, S., et al. Evaluation of High Fidelity Patient Simulator in Assessment of Performance of Anaesthetists. *British Journal of Anaesthesia* 90(1), 43–47 (2003).

11. Forrest, F., Taylor, M. A., Postlethwaite, K., and Aspinall, R. Use of a High-Fidelity Simulator to Develop Testing of the Technical Performance of Novice Anaesthetists. *British Journal of Anaesthesia* 88(3), 228–344 (2002).

12. Caro, P. W. Aircraft Simulators and Pilot Training. *Human Factors* 15(6), 502–509 (1973).

13. Salas, E., Bowers, C. A., and Rhodenizer, L. It is Not How Much You Have but How You Use It: Toward a Rational Use of Simulation to Support Aviation Training. *International Journal of Aviation Psychology* 8(3), 197–208 (1998).

14. Dieckmann, P. Simulatortraining: Eine Bestandsaufnahme in verschiedenen Anwendungsfeldern [Simulator training: a survey in different domains]. MMI-Interaktiv [Online: http://www.mmi-interaktiv.de/ausgaben/11_00/dieckmann.pdf] 2000, p. 4.

15. Dieckmann, P. "Ein bisschen wirkliche Echtheit simulieren": Über Simulatorsettings in der Anästhesiologie ["Simulating a little bit of true reality": On simulator settings in anaesthesiology]. [Online: http://docserver.bis.uni-oldenburg.de/publikationen/dissertation/2005/diebis05/diebis05.html]. Universität, Dissertation, Oldenburg, 2005.

16. Dieckmann, P. and Manser, T. Praxis des Simulatoreinsatzes in der Anästhesiologie: Begründung einer Bestandsaufnahme und erste Ergebnisse [Practice of simulator use in anaesthesiology: Reasons for a survey and first results]. In: Manser, T. (ed.), *Komplexes Handeln in der Anästhesie*. Pabst, Lengerich, 2003, pp. 46–75.

17. Dieckmann, P., Manser, T., Wehner, T., and Rall, M. Reality and Fiction Cues in Medical Patient Simulation. An Interview study with Anesthesiologists. *Journal of Cognitive Engineering and Decision Making* 1(2), 148–168 (2007).

18. Dieckmann, P, Manser, T, Wehner, T, et al. Effective Simulator Settings: More than Magic of Technology (Abstract A35). [Online: http://www.anestech.org/media/Publications/IMMS_2003/sta3.html]. *Anesthesia and Analgesia*. 97(S1–S20), S 11 (2003).

19. Rall, M. and Dieckmann, P. Safety Culture and Crisis Resource Management in Airway Management. General Principles to Enhance Patient Safety in Critical Airway Situations. *Best Practice & Research Clinical Anaesthesiology* 19(4), 539–557 (2005).

20. Kohn, L. To err is human: an interview with the Institute of Medicine's Linda Kohn. *Joint Commission Journal Quality Improvement* 26(4), 227–234 (2000).

21. Committee on Quality of Health Care in America. Crossing the quality chiasm – A new health system for the 21st century. National Academy Press, Washingtion D.C, 2001.

22. Bogner, M. S. (ed.). *Human Error in Medicine*. Erlbaum, Hillsdale, 1994.

23. Cooper, J. B., Newborner, R. S., Long, C. D., and Philip, J. H. Preventable Anesthesia Mishaps: A Study of Human Factors. *Anesthesiology* 49, 399–406 (1978).

24. Rall, M., Manser, T., and Howard, S. K. Key Elements of debriefing for simulator training. *European Journal of Anaesthesiology* 17(8), 516 (2000).

25. Dismukes, K. R. and Smith, G. M. (eds.), *Facilitation and debriefing in aviation training and operations*. Ashgate, Aldershot, 2000.

26. Eco, U. *Six walks in the fictional woods*. Harvard University Press, Cambridge, 1995.

27. Coleridge, S. T. Lyrical Ballads. In: Engell. J. and Bate, J. (eds.). *The collected Works of Samual Taylor Coleridge*. Princeton University Press, Princeton, 1801, 5–18.

28. Dieckmann, P. and Rall, M. Becoming a Simulator Instructor and Learning to Facilitate: The Instructor and Facilitation Training – InFacT. *Simulation in Health Care* 1(2), 103 (2006).

29. Dieckmann, P., Gaba, D., and Rall, M. (2007). Deepening the theoritical foundations of patient simulation as social practice. *Simulation in Health Care* 2(3), 183–193.

30. Rudolph, J. W., Simon, R., and Raemer, D. (2007). Which reality matters? Questions on the path to high engagement in health care simulations. *Simulation in Health Care* 2(3), 161–163.

# XVIII

# Assessment: Why, What, and How

*"Nothing focuses the mind like knowing that you are to be executed in the morning"*

The purpose of formal education is to help students improve their behaviors in a designated activity through the deliberate guidance of their instructors. The flow is from the instructor (and all other learning materials) into the students, takes place over extended periods of time (digestion and integration of novelty is a slow process), and is intentionally designed to provide numerous opportunities for refinement through repetition and discovery of alternate paths. In contrast, formal testing helps instructors distill their students' behaviors into a single metric that is both easily communicated and easily ranked. The flow is from the students to the instructor, takes place over a brief period of time, and is intentionally designed to provide few or no subsequent opportunities for refinement through repetition or discovery of alternate paths. And yet, by definition, every experience must include some form of evaluation by the participants in order for education to occur. Furthermore, every formal testing event can prompt study beforehand and reflection afterward in both students and instructors.

What are the differences between formally tested simulation sessions and ones that are not? Mainly, the magnitude of the consequences and the relative ease of recovery from errors. What are the differences for the instructors, for the students, for the simulation production team? Again, mainly the consequences, and the relative ease of recovery from errors. What are the differences for the simulation facility's design and resources? Once again, mainly the consequences, and the relative ease of recovery from errors. It is the elevated consequences and reduced ease of recovery from errors that make formally graded simulation sessions more like real life, more challenging to produce, more challenging to experience. Thus, a formally testing simulation session can be seen as a welcome learning "stepping-stone" to help our students convey themselves across that gap between the recoverable world of the clinical student and the more irreversible world of the clinician.

Ideally, progression of all clinical students toward and through graduation would be based upon demonstrated competencies, "proven" proficiency both as an isolated individual and as a care team member. Today's teaching is like the automated car wash, where every vehicle, no matter the degree of filthiness, is pulled through the washing machinery at the same rate and is exposed to the same treatment. At the exit, time inside is the only metric; determination of overall cleanliness as the criteria for "successful completion" is implied but never measured overall. Furthermore, each vehicle is intentionally treated in isolation from all those that come before and follow after. This model for removing grime from vehicles is very popular, as many of the human-hands-on keep-working-until-seen-as-clean car wash businesses have either shut down or converted to the automatic-unevaluated type. If learning were dependent only upon exposure without guided reflection, and all of us learn the same amount with the same exposure, and if clinical care were a solo activity, then perhaps this automatic car wash model would be acceptable for organizing clinical education. But this is not the case. Learning does require examination. Learning rates and methods are diverse. Patients (and their survivors) do judge the care they receive. Patients do seek professional clinical care after their own, solo attempts at treatment fail. All clinical care is a team effort, even when the "team" consists of the patient and just one clinician. A crucial requirement to do away with the popular but questionable automatic car wash model of clinical education is the development and regular use of overall evaluation. However, if this were easy, we would already be doing it.

Debriefing is to simulation as digesting is to ingestion. This repetition is intentional. Debriefing integrates the activity. Debriefing is an active student-centered event. Listening to the instructor point out desired and undesired behaviors is not a useful debrief. The *students* should have been sufficiently prepared before the simulation so as to *lead* their discussion about their behaviors and how they fit with essential teaching objectives *assigned* for the event. All learning objectives within the simulation event should be clearly detectable via the replay apparatus. The students are now observing and watching from a higher level to evaluate higher-level functioning. By leading the debriefing, they are taking responsibility and owning the consequences of their actions. Such higher-level functioning is the essence of any professional. What better time to reinforce this intangible but essential behavior than within each and every simulation session. During debriefings, each and every time the instructor speaks, they are taking away learning opportunities from their students. In the ideal debriefing, the instructor is very attentive but remains mute, until giving the final words of guidance and encouragement. Unprepared students, pushing themselves through an unanticipated simulation event, only to be talked at about their failures by instructors is not a useful application of anyone's time and attention, least of all yours.

Carla Pugh provides a big-picture view of those differences between formally tested simulation sessions and ones that are not, and illustrates those differences with three common clinical domains, each employing a different kind of simulator. Peter Dieckmann *et al.* address those first crucial steps in any evaluation: collect and present data on "what did happen" during the simulation session. This is not a trivial task for the simulation production team to execute at all, let alone do well. Before simulation, most clinical instructors never had to question how they collect and review the performance data of others' behaviors, let alone formulate a functional answer on how to do so. Guillaume Alinier and Lance Acree focus this issue of evaluation upon us, the simulation-using educators, and state why we all need to formally assess of our program's performance and describe how we might go about doing so. They start by addressing the question: "why bother to collect feedback?" They describe the design features of a useful questionnaire. They conclude with an example of simulation with surgical ophthalmology residents. Paul Williamson *et al.* offer their formal framework for creating formal evaluations. Their tool and its application are useful for evaluating both our students' performances as learners and our performances as teachers.

# 68

# Simulation and High-stakes Testing

Carla M. Pugh

## 68.1 What is High-stakes Testing?

High-stakes testing is the process of using test scores for the purpose of making decisions that have major consequences for the individual being tested. The examination scores are often used by schools, training programs, or companies where the number of candidates exceeds the number of positions. In addition, the scores may be used to obtain a license or privilege for professional practice where proof of competency is necessary.

In many cases, high-stakes examinations are administered at multiple stages during one's progress within a profession. Each time, the results have major consequences for the examinees' future. Thus, legitimate high-stakes testing must have maximal accuracy and minimal misclassification. To achieve this, the tests are often limited to domains that are well defined, and given in a format that can produce highly reliable scores. Commonly, this translates into examinations that test "knowledge" domains using multiple-choice questions (MCQs).

## 68.2 What is the History of High-stakes Testing in Health Professions?

High-stakes testing is part of the licensing and/or credentialing process for many of the health professions [1]. A well-known example of high-stakes testing in medicine is the United States Medical Licensing Examination (USMLE), sponsored by the Federation of State Medical Boards and the National Board of Medical Examiners (NBME). High-stakes state licensing examinations grew out of much needed reforms in medical education and regulation in the late 19th century [2].

The NBME was established in 1915 to elevate the standards of qualification for the practice of medicine and to provide a means for recognition of qualified persons to practice anywhere in the US without the need for further examination by other licensing boards. The NBME delivered its first high-stakes examination in 1916. The examination required a full week, and covered a range of topics. It included written and oral examinations, preparation of laboratory specimens, and a suturing examination with strict criteria for successful anastomosis of a severed dog intestine. Of the 32 who applied, only 16 were found to have satisfactory credentials to take the test, only 10 showed for the examination; five passed, and five failed. While that early successful pass was a measure of distinction, passing the USMLE is now required by all U.S. jurisdictions as one of two prerequisites for licensure (the other being graduation from an accredited medical school).

In addition to driving what many trainees study and what many medical schools consider for their curricula, the examination establishes a standard within medicine about professional values. USMLE is a three-step examination comprising multiple-choice questions, computer-based case simulations, and a newly implemented (June 2004) clinical skills examination based on standardized patients. **STEP 1** of the USMLE assesses student's understanding and ability to apply important concepts of the **sciences, basic** to the practice of medicine, with special emphasis on principles and mechanisms underlying health, disease, and modes of therapy. **STEP 2** focuses on the principles of **clinical science** that are deemed important for the practice of medicine **under supervision** in postgraduate training. The purpose of **STEP 3** is to determine if a physician possesses and can apply the medical knowledge and understanding of clinical science considered essential for the **unsupervised practice** of medicine, with emphasis on patient management in ambulatory care environments. Table 68.1 shows when the examinations are usually taken and the general topics that are covered. Students eligible to take the examination include: medical students or graduates of an accredited U.S. or Canadian medical school, or non-U.S. or Canadian enrolled students or graduates meeting Educational Commission for Foreign Medical Graduates (ECFMG) requirements. Candidates seeking certification can take Step 1 and Step 2 in any order, but must pass both in order to take Step 3. Currently, there is no limitation on the number of times a test may be taken. However, the

**TABLE 68.1**   Description of USMLE STEP Examinations

| USMLE step | When exam Taken | Topics covered |
|---|---|---|
| **Step I** | 2nd year in medical school | **Basic science:**<br>Hematopoietic and Lymphoreticular Systems \| Central and Peripheral Nervous Systems \| Skin and Related Connective Tissue \| Musculoskeletal System \| Respiratory System \| Cardiovascular System \| Gastrointestinal System \| Renal/Urinary System \| Reproductive System \| Endocrine System |
| **Step II** | 4th year in medical school | **Clinical science:**<br>Infectious and Parasitic Diseases \| Neoplasms \| Immunologic Disorders \| Diseases of the Blood and Blood-forming Organs \| Mental Disorders \| Diseases of the Nervous System and Special Senses \| Cardiovascular Disorders \| Diseases of the Respiratory System \| Nutritional and Digestive Disorders \| Gynecologic Disorders \| Renal, Urinary, and Male Reproductive Systems \| Disorders of Pregnancy, Childbirth, and the Puerperium \| Disorders of the Skin and Subcutaneous Tissues \| Diseases of the Musculoskeletal System and Connective Tissue \| Endocrine and Metabolic Disorders \| Congenital Anomalies \| Conditions Originating in the Perinatal Period \| Symptoms, Signs, and Ill-defined Conditions \| Injury and Poisoning |
| **Step III** | 1st Postgraduate year | **Application of medical knowledge:**<br>Diseases/Disorders of: Central Nervous System \| Eye \| Ear/Nose/Mouth/Throat \| Respiratory System \| Circulatory System \| Digestive System \| Musculoskeletal System \| Skin/Subcutaneous Tissue \|f Endocrine/Nutrition/Metabolism \| Kidneys and Urinary Tract \| Male Reproductive System \| Female Reproductive System \| Blood/Blood-forming Organs. Behavioral/Emotional Disorders, Pregnancy/Childbirth, Neonate/Child, Infectious/Parasitic Diseases, Injuries/Wounds/Toxic Effects/ Burns, Health Maintenance, Ill-defined Symptom Complex |

NBME requires all tests be passed within a 7-year period. Since 1999, the USMLE has been administered as a computer-based examination. After registering with USMLE, the examinee may schedule a test date at a local testing center. For updated information, please visit the USMLE web site (http://www.usmle.org).

## 68.3 The Pros and Cons of High-stakes Testing

A definite outcome of high-stakes testing is the performance standards that are defined. Without agreed-upon standards, it is difficult for individuals and institutions of learning to gauge their performance. The results from high-stakes tests provide feedback on what is being done well and where improvement is needed, both at the individual and program level. Moreover, the results have some predictive validity for performance on other knowledge tests [3–4]. High-stakes testing drives individual learning, enables self-assessment, and can provide a focus for medical school curricula. Over time, such tests allow for the capture of large performance databases, which can be used for educational research and programatic feedback.

In addition to setting standards, high-stakes examinations may define values within a profession. In 2004, the NBME added a clinical skills examination using standardized patients to the three-part examination. In the 2004 Annual Report, the NBME stated that their objective is to "protect (and enhance) the health of the public through state-of-the-art assessment" of all the requisite competencies of health professionals. The addition of the clinical skills examination is a step toward increasing accountability in the medical profession by setting higher standards for assessing competency. While their decision to do this was not met without controversy (largely due to examinee costs nearing US$1000 per candidate), the reality is that it shows the medical profession is, once again, moving beyond assessing medical knowledge to assessing skills and behaviors, which will ultimately provide a better measure of clinical performance.

A major limitation of high-stakes testing is the high accuracy, minimal misclassification requirement that often limits the examinations to multiple-choice "knowledge" assessments. The problem with limiting the examinations to "knowledge" domains is that other domains such as skill, attitude, and behavior are ignored. According to Bloom's Taxonomy (Figure 68.1), "knowledge" is a lower-order thinking skill, while "application of knowledge," and "synthesis" or "evaluation" of information are examples of higher-order thinking skills [5–6]. This is why some testing organizations have a high-stakes oral examination component in addition to the written test[1]. Moreover, as part of the hiring process, many programs include face-to-face interviews when selecting

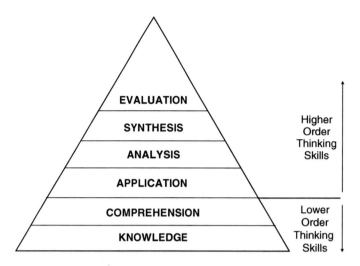

**FIGURE 68.1**  Blooms Taxonomy.

candidates in addition to using performance scores generated from high-stakes examinations [7].

The results of multiple-choice "knowledge" examinations have limited evidence for practice implications. And, because multiple-choice examinations often ignore skills and behaviors, it may send a message that these skills are less important. As a result, students may be less motivated to learn or focus on skills that will not be tested, and training programs may not provide a curriculum that emphasizes skills and behaviors that will not be tested. Overall, current high-stakes examinations are not able to provide a broad measure of different types of actual proficiency or "performance" and continue to have limitations in the ability to measure practical, applied knowledge.

## 68.4 What is it About High-fidelity Simulation that Makes it Attractive for High-stakes Testing?

For the purpose of this chapter, we use the term "high fidelity" as it relates to high psychological fidelity or the extent to which the simulation requires the user to perform or cognitively progress through a real world task [8]. Using this definition, a written test question or case scenario where the examinee chooses the single best answer would be considered a low-fidelity simulation. Likewise, a mannequin-based trainer used to practice intubation may be considered a "moderate-fidelity" simulation. A mannequin-based or robotic trainer that responds to intravenous medications, talks, has a pulse, and generates appropriate respiratory and physiologic responses if the student does not secure the airway in time would be considered a "high-fidelity" simulation. In essence, hands-on trainers and robots that are situated in realistic clinical environments and require the user to respond or react to clinical findings are in the realm of "high fidelity."

High-fidelity simulation has many features that make it attractive for use in high-stakes testing. The following is a list of some of the attractive features:

- *Simulation is more closely related to actual clinical practice.* Thus, real world clinical tasks can be performed with physical simulation, robotic patient simulators and/or standardized patients, in a more realistic fashion than written examination questions.
- The recent profusion of new forms of clinical simulation has increased the awareness of health professionals about the potential uses of simulation. *Familiarity will make it easier to convince test sponsors and examinees that additional assessment formats are needed and feasible.*
- High-fidelity simulation will help in making arguments about the validity of performance scores. Computer-generated *scores from simulators are less subjective, and some behaviors and actions are automatically detected and documented.*
- *Simulation-based examinations can complement the results achieved on high-stakes knowledge examinations by assessing psychomotor skills.* The addition of this information should enhance the quality of decisions made on the basis of high-stakes testing outcomes.
- Preparing for simulation-based examinations, including "mock" examinations, *may provide examinees with more relevant feedback about their ability and needs for improvement.* If feedback is sufficiently detailed and provided in a constructive and formative manner, it can guide individual learning based on strengths and weaknesses and allow for deliberate study and review [9].
- Detailed, constructive feedback may also *guide mentor oversight, and provide detailed programatic feedback to educators.* By aggregating individual feedback, educators can decide how to redesign both the simulation and other aspects of training to achieve optimal educational outcomes with available resources.
- *Simulators can depict multiple variations of normal and abnormal conditions.* This provides opportunities for changing the test content through a constant supply of "unexposed items," which keeps costs lower and decreases the chances of examinees informing one another about the examination.
- *Students will be motivated to practice specific tasks* as they will recognize that performance on the simulator will be required for the high-stakes examination. To the degree that this is the same task that must be done in real life, educational ends are met perfectly with clinical skills needs.
- *The same simulators that are used for high-stakes testing can also be used for day-to-day instruction,* and for lower-stakes developmental and formative assessments. In fact, the seamless integration of instruction and low-stakes assessment should be a logical end-point of simulations used for high-stakes purposes. Self-operated forms of simulation, made widely available at all hours, can allow students to develop mastery through self-motivation and self-study, while still providing educators with the information they need to monitor performance improvement and provide active guidance to trainees.

## 68.5  What are the Potential Difficulties in Using Simulation for High-stakes Testing?

Simulations also pose some challenges for high-stakes applications. The following list reviews some of those challenges:

- Simulators by definition are physical or virtual depictions of reality. The depiction might include visual and auditory stimuli, as well as realistic feel or haptic feedback. Because the human brain can be exquisitely sensitive to minor variances in the portrayal of reality, *developers must make deliberate design decisions to ensure that they model fidelity exactly where it is needed to represent sufficiently the task and task challenges.* For instance, at the hardware end of the spectrum, haptic devices typically function with a feedback rate of 1000 Hz (a limitation of affordable present day hardware). This leads to a slight delay between visual and tactile stimuli. Trainees quickly learn to move a bit slower than in real life to avoid confusing their senses. This may lead to negative training transfer or a lack of focus on the task at hand. While at the software end, most physiopharmacologic models associated with robotic patient simulators all too easily fail the Turing test.
- *When a simulator is developed, it must still be turned into a simulation.* The difference here is that the simulator is the vehicle through which the simulation is depicted. The simulator may be a robotic patient simulator, or a computer-generated representation on a video monitor interacted and interfaced through input devices like a joystick. The simulation is the full depiction of the scenario the examinee must confront and the potential tasks that the examinee must execute.
- *Developing realistic simulations requires some knowledge of computer programing and extensive input of content experts* who happen to know how to parse their expertise into digestible components suitable for the given level of student.

- In assessing content validity of simulations, it is important to note that abstract tasks such as placing small doughnuts on a peg or putting animated green balls in a box will raise more questions than simulations of real tasks such as performing a dissection of an animated gallbladder and cystic duct. While unproven, *simulations of real tasks will appear to have much greater relevance in assessing fine motor skills than simulations of abstract tasks.* Although, simulations of abstract tasks may possess a high amount of reliability and predictive validity, the potential for questions about content validity will continue to be high.

- The linkage between testing, test preparation, and real-life skills places high demand on the validity of simulation-based training and assessment. *There must be some assurance that behaviors gained during testing and test preparation are in fact representative of behaviors desired in real life.*

- One of the major challenges is that *simulations require more testing time with more expensive people per unit of measurement than other performance assessments.* For example, 10 multiple-choice questions, on average, may take 10–15 minutes for the examinee to complete. Once complete, the 10 items are easily scored as right or wrong in less than 1 minute. A single low-cost nonclinical person can administer both phases. However, a simulated task such as performing a code may also take 10–15 minutes, but producing the scenario and then scoring the user's performance requires, at a minimum, an expensive clinical rater to review the 10–15-minute performance recording to generate a score. In addition, because inter-rater reliability is important, it may take two (or more) independent raters to score one examinee's recording. This brings the time for scoring to 30 minutes. In comparing the two methods, the multiple-choice question examination will take approximately 16 minutes of low-cost time. The simulated task will have a total of 45 minutes of high-cost time. Differences in total test time multiply when you consider that a class of students needs to be tested. The multiple-choice examination can be administered to over 30 examinees in the same 15 minutes. Moreover, one rater may grade all of the students as inter-rater reliability is not an issue. In fact, a scoring machine can be used if students mark their answers on machine-readable forms or enter their answers on direct-input devices. With simulations as the number of students increase, total scoring time (and likely test time) increases. Overall, scores derived from simulations typically will have lower reliability per unit of test time than lower-fidelity formats due to inter-rater reliability factors and rating times. This difference may limit the practicality of introducing some forms of simulations for large-scale high-stakes testing.

- *Simulations also present particular challenges in scoring.* In a high-stakes examination wherein someone passes or fails, hundreds or even thousands of data points must be reduced ultimately to one side of the dividing line. Is their performance good enough or not? To score a simulation, complex information within one testing unit, for example, a single simulated scenario as part of a 12-scenario Objective Standardized Clinical Examination, must be evaluated. With multiple-choice questions, each testing unit or test question typically has a value of 1 if correct or 0 if incorrect. The number of correct answers are aggregated and converted into a measure using psychometric techniques that can correct for the relative difficulty of the items. An aggregate score can be produced and the pass-fail line defined based on industry-accepted methods for established performance standards. With more traditional testing, techniques for proceeding from test questions to pass/fail classification are well established. In contrast, techniques for reducing the complex information in a simulation are not well developed. One simulation might provide hundreds of data points. For example, if a simulated performance is video recorded, every frame of video, at 24–30 frames per second, *could* be a point of extraction for performance measures. Moreover, one frame may have several possible information points from all the visual data in the image. To add another layer of potential performance measures, there is also the possibility of extracting information from audio recordings during the simulation. Putting this all together and determining the quality of this information, parsing the examinee's performance deficits from the audio-video collection and reproduction deficits, and then converting it into a score or individual performance is not straightforward.

- To assure that test results are more on content mastery and less on test-taking ability, test takers should be well versed in testing rules. Providing clear instructions on how to answer a question is the mainstay of multiple-choice examinations. Samples of properly answered questions with an emphasis on how to indicate your answer are often provided as part of the instructions for multiple-choice examinations. Ideally, simulation-based assessments should provide the same opportunity. To this end, *examinees facing high-stakes simulation-based examinations should have access to comparable pre-examination familiarization experiences with the simulation equipment.*

- For all of the above reasons and more, *simulations can be very time consuming and expensive to develop.* At a minimum, there needs to be a relevant, reliable, and reproducible set of simulator devices that are then turned into a simulation. Then, reliable and valid scoring systems must be developed.

## 68.6 Examples

### 1. Standardized Patients

A specific example where high-fidelity simulation could enhance high-stakes assessment is in the USMLE STEP 2 Clinical Skills Examination, an 8-hour examination composed of 12 Standardized Patients (SPs). Standardized Patients are people who portray health care challenges and clinical diseases for the purpose of training and assessment. Examinees rotate through examining rooms taking histories, performing physical examinations, and writing patient notes. Scoring occurs through checklists completed by the Standardized Patients based on communication and information sharing, professional manner and rapport, clarity of spoken language(s); and through the patient notes written by examinees and scored by highly trained physician raters.

Validity of the USMLE might be enhanced if formats were found wherein additional tasks could be evaluated, and a broader range of pathologic states could be covered. Typically, Standardized Patients are healthy people without overt clinical pathology. As a result, Standardized Patients are helpful in being able to assess whether one knows how to perform an examination. However, being able to perform an examination does not equate to being able to detect abnormalities or make a diagnosis. Robotic and hybrid simulations offer the possibility of introducing abnormal clinical findings, allowing examinees to perform intrusive or intimate physical examinations, and conduct routine procedures as would be required in clinical practice. Currently, the examination largely consists of history taking, interpretation of laboratory results and conducting noninvasive, and mostly nonintimate examinations such as pulmonary or cardiac auscultation and ophthalmic examinations. There is no opportunity for examinees to suture wounds, or insert catheters or tubes on the examination. In addition, rectal, pelvic, and breast examinations are not part of the USMLE Step 2 Clinical Skills examination, as it is not feasible to do this with Standardized Patients given the examination format, the number of students that need to be tested, and the need to standardize the interactions. In essence, it is not practical to expect an Standardized Patient to allow 100 examinees to perform a rectal examination in one day.

### 2. Pelvic Examination Simulator

Currently, a pelvic examination simulator [10] is being evaluated by the NBME for potential use on the USMLE STEP 2 Clinical Skills examination. The simulator is a partial mannequin, umbilicus to upper thigh, made in the likeness of an adult human female (Figures 68.2a–d). The simulated organs have sensors embedded in them, which allows the capture of hands-on performance data. During a simulated examination, data may be collected on where the examinee is touching and with how much pressure, as well as the number of times and length of time (duration) a given anatomical location was palpated.

The process for evaluating this simulator as a potential high-stakes assessment tool takes into account that the NBME has a preexisting panel of subject matter experts whose goal is to write questions for the multiple-choice examination. As the subject matter experts have thought about the competencies relating to pelvic examinations for a number of years, it would be a natural fit for them to assess the validity of the simulator. Their preliminary evaluations have outlined how the simulator might be used and established a framework for scoring performance. Owing to the early stages of this evaluation, details on simulator use and scoring cannot be disclosed. While the results of the preliminary evaluation are positive, many steps need to be taken to develop the scenarios that will be simulated, ensure validity and reliability, and finalize the pass/fail scoring. As with the written examinations, the resulting simulator scores must maximize accuracy and minimize misclassification.

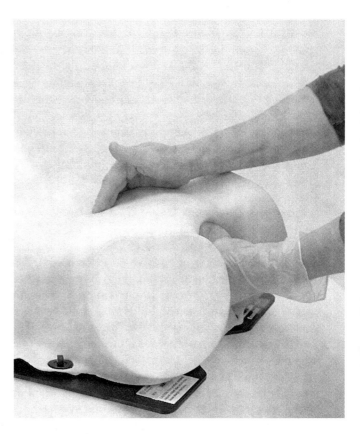

**FIGURE 68.2a** The pelvic examination simulator: mannequin portion for students' procedures.

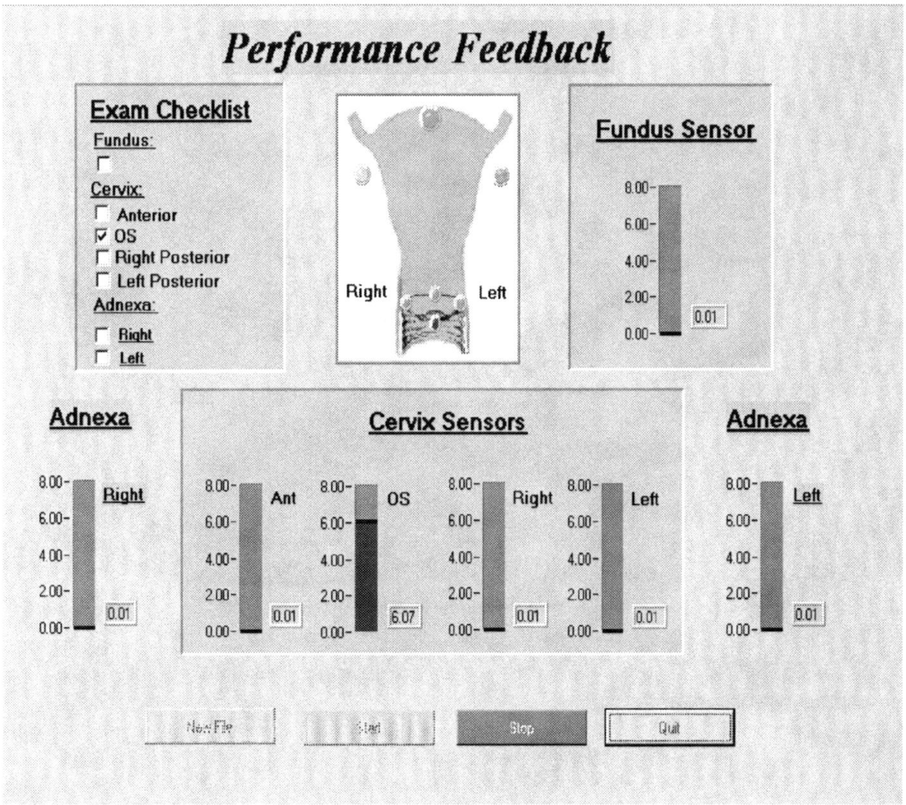

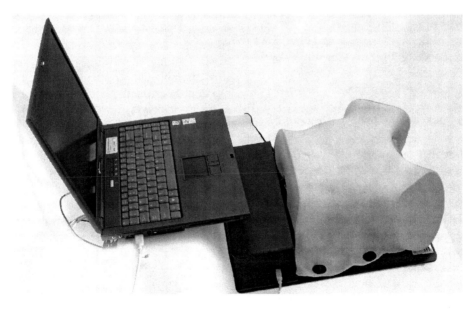

**FIGURE 68.2b**   The pelvic examination simulator: Operator/Instructor interface showing location of touch/force sensors and their readings. The blue dot in the anatomical diagram indicates a region of palpation, and correlates with the blue bar indicating amount of applied force.

**FIGURE 68.2c**   The pelvic examination simulator: entire portable package.

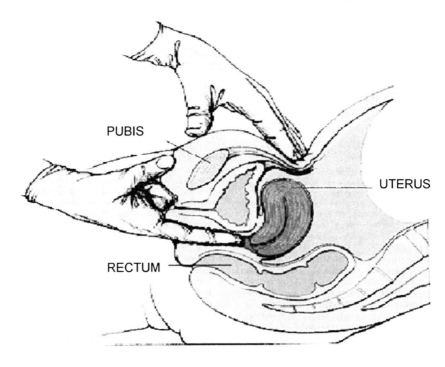

**FIGURE 68.2d**  Anatomy of the pelvic examination recreated by the pelvic examination simulator.

## 3. Carotid Stenting Certification

Another example where high-fidelity simulation could enhance high-stakes assessment is in Carotid Stenting. Although this is not a licensing examination, obtaining or not obtaining certification has major consequences on allowing a physician to perform these procedures and obtain employment in a hospital needing these types of services.

When carotid stenting was reviewed as a new medical procedure by the U.S. Food and Drug Administration (FDA), it was approved (August 2004) on the basis that some form of training and certification be developed. It is not uncommon for the FDA to require that applicants (equipment manufacturers) outline specific plans for medical device training. In the FDA acceptance letter to Guidant Corporation, they state "... to ensure the safe and effective use of the [carotid stenting] device, the device is further restricted.... insofar as the labeling specify the requirements that apply to the training of practitioners who may use the device." In addition, the letter underscores postapproval studies that must be done; requiring the inclusion of patients that have been treated by physicians from low-, moderate-, and high-volume centers. The intention here is to understand the effects of physician case volume on patient outcomes.

The potential for simulation to address the FDA's training requirements and affect patient outcomes in low-volume centers (by augmenting low case volumes or number of real-life procedures performed with successfully performing simulated procedures) is great. The FDA was advised by an independent advisory committee to mandate proficiency-based training, including a tiered curriculum. The tiered curriculum would include: (i) Online didactic cognitive training and assessment, (ii) metric-based simulator training to proficiency prior to working on patients, and (iii) onsite proctoring.

Currently, the simulator is being used in a trial, which will assess the efficacy of simulation in training physicians to place a carotid stent. In this trial, physician trainees with no prior carotid stenting experience will be trained to an objectively established level of proficiency on a virtual reality simulator prior to performing stenting in a patient. Patient outcomes for the virtual reality–trained physicians will be compared with a case-matched group of patients who had their carotid stent inserted by experienced operators [11]. The same issues of scoring, content validity, and reliability will be issues for this simulator as well. Figures 68.3a–e portray two examples of the virtual reality carotid stenting simulators and depict the concept of carotid stenting.

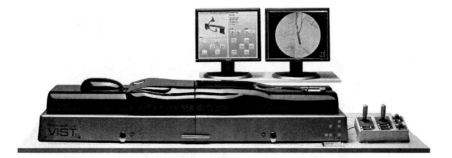

**FIGURE 68.3a**   Example of a brand of Carotid Stenting Simulator.

**FIGURE 68.3b**   Example of another brand of Carotid Stenting Simulator.

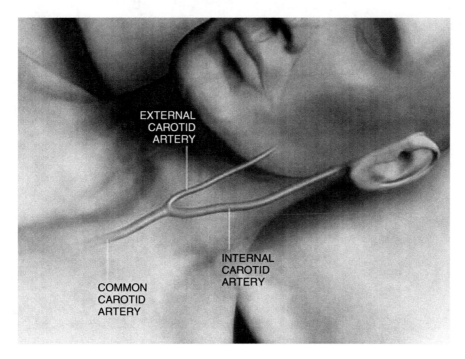

**FIGURE 68.3c**   Anatomy of carotid artery.

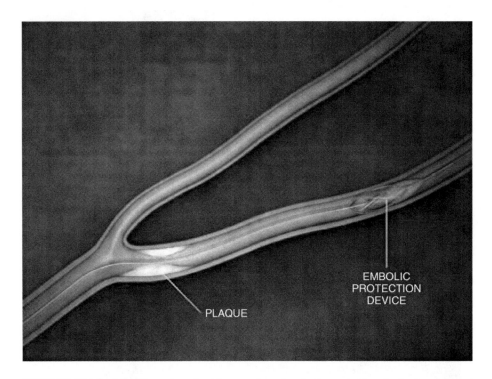

**FIGURE 68.3d**   Initial insertion of guide cannula and embolic protection device into internal carotid artery.

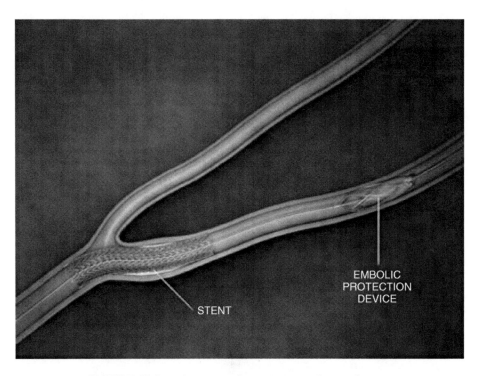

**FIGURE 68.3e**   Placement of stent in internal carotid artery.

## 68.7 Industry's Role

While industry's role may seem obvious from the development perspective, there are a few pearls that might prove helpful. For example, the technology and materials used to make realistic tissues has been available for decades. However, ease of manufacture, durability, and shelf life are inversely proportional to their fidelity. Hence, the tissue materials in many of the current simulated body parts are selected and bought based primarily on the financial side of market forces, not the education and learning side.

For those companies whose goal it is to have their simulators used on high-stakes examinations, it is imperative to understand the pros and cons, as outlined in this chapter. In addition, an understanding of the history of high-stakes testing and the needs of test developers is critical for success. Before a high-stakes examination using simulation on a national level can be developed, there needs to be sufficient numbers and types of simulations that represent the essentials of current clinical practices. For example, while the laparoscopic approach is the future fate of many surgical procedures, open procedures are still commonly performed. Today, there is a paucity of open procedure simulators to fill this void. In addition, solely focusing on the technical and psychomotor aspects of an operation will continue to limit the use of simulators for assessment and training. Most of the major errors made in the operating room are not as a result of a trainee or surgeon not knowing *how* to use an instrument or perform a task but rather not knowing *when* to perform a task and *why*. It is one thing to know how to cut, but knowing where and where not to cut given the clinical presentation, including pathology and anatomical anomalies, is extremely important in preventing major surgical errors.

The most positive outcome of the simulation industry is that it has forced the health care professions to discuss training and assessment issues that previously have been ignored or left for the "hidden curriculum" as something "you will learn in time." Well, the time for learning is now, and innovative methods such as cognitive task analysis and other research and data collection methods meant to capture expert decision making must be incorporated into the development of simulators to ensure that training and assessment goals are well defined and made explicit.

## 68.8 Conclusion

In summary, we have reviewed a definition of high-stakes testing, discussed, via specific example, the history of high-stakes testing in the U.S. and provided the pros and cons of high-stakes testing. We have also discussed what makes simulation attractive for high-stakes testing and the potential difficulties. Lastly, we have provided two specific examples

(licensing and certification) where two currently developed simulators have high applicability and are being evaluated for use in high-stakes testing at a national level.

We did not discuss the use of simulators at local programs, where there are many examples of simulators being used for moderate- to high-stakes testing where the results have important consequences. For example, some programs require that trainees achieve a certain level of performance in the simulation laboratory before proceeding to the operating room or clinical environment to perform procedures or care for a live person. These decisions, made on a local level, require a pioneer (usually an individual faculty member who is passionate about teaching) to turn the simulator into a simulation, define pass/fail scoring, and develop a training program for the purpose of remediation.

The topic of *remediation* is also a very important one. If high-stakes testing has the potential to point out deficiencies in individuals and training programs, a pathway for remediation should also be determined. This will help to maintain the morale of training programs and facilitate learning.

Given that most technology endeavors always have a steep production/cost curve, it is to be expected that subsequent simulators and developments will become less expensive and less time consuming to develop and integrate in the assessment and training process. Overall, the future remains promising and very exciting as the nature of training and assessment will forever be changed by simulation. Moreover, sole use of multiple-choice, "knowledge" examinations for making the major decisions that affect so many lives will be balanced by examinations that assess skill, behavior, and attitude.

## Acknowledgment

I would like to thank Drs. Stephen Clyman and Robert Galbraith of the NBME for their assistance and guidance in writing this chapter.

## Endnote

1. For example, the American Board of Surgery (ABS) is an independent, nonprofit organization founded in 1937 for the purpose of certifying surgeons who have met a defined standard of education, training, and knowledge. Surgeons certified by the ABS have completed a minimum of 5 years of surgical training and successfully completed a written and oral examination process. The ABS currently certifies surgeons in the following fields: general surgery, pediatric surgery, vascular surgery, surgical critical care, and surgery of the hand. The ABS is one of the 24 member boards of the American Board of Medical Specialties.

# References

1. David B. Swanson, Geoffrey R. Norman, and Robert L. Linn. Performance-Based Assessment: Lessons from the Health Professions. *Educational Researcher* **24**(5) 5–11+35 (Jun–Jul, 1995).

2. Hubbard, J. P. and Levit, E. J. The National Board of Medical Examiners: The First Seventy Years – A Continuing Commitment to Excellence. *National Board of Medical Examiners*. Philadelphia, 1985.

3. Julian, E. R. Validity of the Medical College Admission Test for Predicting Medical School Performance. *Acad Med.* **80**(10), 910–917 (Oct 2005).

4. Salvatori, P. Reliability and validity of admissions tools used to select students for the health professions. *Adv. Health Sci. Educ. Theory Pract.* **6**(2), pp. 159–75 (2001).

5. Bloom, B., Englehart, M., Furst, E., Hill, W., and Krathwohl, D. *Taxonomy of educational objectives: The classification of educational goals. Handbook I: Cognitive domain.* Toronto, Longmans, Green, New York, 1956.

6. Anderson, L. W. and Krathwohl, D. R. (eds.). *A Taxonomy for Learning, Teaching, and Assessing: A Revision of Bloom's Taxonomy of Educational Objectives.* Longman, New York, 2001.

7. Albanese, M. A., Snow, M. H., Skochelak, S. E., Huggett, K. N. and Farrell, P. M. Assessing personal qualities in medical school admissions. *Acad. Med.* **78**(3), 313–321 (Mar 2003).

8. Salas, E., Wilson, K. A., Burke, C. S. and Priest, H. A. Using simulation-based training to improve patient safety: what does it take? *J. Qual. Pat. Safety.* **31**, 363–71 (2005).

9. Ericsson, K. A. Deliberate practice and the acquisition and maintenance of expert performance in medicine and related domains. *Acad. Med.* **79**, S70–81 (2004).

10. Pugh, C. M. Medical Examination Teaching System – U.S. Patent Number 6,428,323.

11. Gallagher, A. G. and Cates, C. U. Approval of virtual reality training for carotid stenting: what this means for procedural-based medicine. *JAMA.* **292**, 3024–3026 (2004).

# 69

# Video-assisted Debriefing in Simulation-based Training of Crisis Resource Management

Peter Dieckmann
Silke Reddersen
Jörg Zieger
Marcus Rall

## 69.1 Debriefing: What is it Good for?

Debriefing is considered "the heart and soul of simulator training" [1–5]. The time and institutionalized framework to reflect on an intense experience during the simulation scenario is what most clearly differentiates realistic simulator trainings with video-assisted debriefings from other forms of training as well as the clinical practice, where, often, the time for conducting debriefings is not valued enough to merit inclusion.

As debriefing is a complex task, a structured approach can help facilitate the processes that take place during the debriefing. There are many ways to accomplish a beneficial educational and reflective experience during debriefing [6–12, 13, 14]. We will present our way to conduct debriefings in Crisis Resource Management (CRM)–oriented simulator courses [15–19] with video-assisted debriefings [20]. We conduct these courses in our simulation center as well as outside our center, using a mobile simulator. This latter training in the clinical environment is the so-called "in situ" training (termed "in situ" by D. Gaba, Stanford), e.g., in the heart catheterization laboratory, the ambulance car, inside the helicopter, the ambulance jet, etc. During these team training courses, we mostly focus on "common" emergency scenarios, less on major catastrophes. We emphasize that CRM is good, not only for dealing with a crisis, but also to prevent it.

## 69.2 Debriefing Aims and Objectives

In general, debriefings should serve two major functions: make the active participants in the "hot-seat" deal with the emotional strain that the scenarios put on them and to enable participants to learn from the scenario via self-reflection. As a rule of thumb, our debriefings last a little longer than twice the time of the scenario. So, a scenario of 15 minutes would be followed by a debriefing of approximately 40 minutes. One might say that debriefing is to performing within a simulation scenario as digesting is to eating: without digestion, nothing internally changes, and digestion does take much more time than does eating (RR Kyle and WB Murray, personal communication).

Looking closer at the learning, our overall aim is to sensitize health care professionals to human factors or CRM. Even if they know the relevance of these issues for patient care from their daily work, most of our participants do not have, as of yet, any theoretical background that they can use to reflect on these experiences at a more abstract, "big-picture" level.

For an "early exposure," we also integrate CRM aspects in our courses for medical students. They are treating emergency cases that are very challenging for them, but offer the opportunity not only to discuss clinical issues, but also CRM issues. Typically, the medical students do not have sufficient clinical knowledge to solve the scenario completely on their own and are asked to focus on stabilizing the patient's vital signs. However, it is astonishing that during the scenario sometimes they are not even able to use what they already know. For this reason, an early exposure to CRM is very appropriate for this target group.

As a teaching objective during the debriefing, we discuss with participants what they could change in order to improve the safety for "their" patient and what they already did well in this regard. Our experiences show that most of the problems with clinical issues originate more from suboptimal CRM application, rather than from a lack of clinical knowledge or skills. For this reason, we keep the teaching objectives for a given debriefing flexible and pick up the most impressive CRM points in a scenario (e.g., different mental models that show up in the debriefing, communication problems, fixation errors, limitations in situation awareness) and discuss them with participants in relation to the diagnostic and therapeutic steps they took.

Concerning the emotional side of debriefing, we offer "hot seat" participants the first words in order for them to have an "emotional washout" during which they can talk about how they felt during the scenario. This serves three purposes. Firstly, it allows participants to really leave the scenario behind. By talking about how they felt during the scenario, they start talking *about* the scenario and not within it any more, thus preparing for the analysis of their experience. Secondly, it allows participants to be reintegrated into the training group. As they have acted during the scenario in a prominent position – being intensively watched and video recorded – the first friendly and supportive words for and by them in the debriefing should open the way back into the group. The reintegration should further be supported during the debriefing itself. Thirdly, a very open-ended question in the beginning of the debriefing about how participants felt during the scenario, and how realistic it has been for them, helps to discover which "reality" participants constructed from the scenario (and their version is the only one that counts in the debriefing). The way participants understood the scenario, and the reality they constructed on the basis of this understanding, might be different from what the instructors intended during designing and presenting the scenario. It also helps to discover any problems participants might have had with the scenario itself, for example, having been implausible to them. If this happens, the debriefer should react and adjust the debriefing accordingly. If, for example, the participants did not recognize a specific medical problem (e.g., unequal breath sounds), then it does make little sense to discuss why they did not react to this. In such a case, it would probably be more beneficial to discuss the reasons why participants did not recognize this issue – which might often be in fact a simulator artifact. Further, it is necessary to realign with the participants asking for their agreement that the scenario and its debriefing still is a valuable learning opportunity despite the unequal breath sounds not being recognized. Note that we do not imply that scenarios are necessarily worthless if participants think them to be not realistic. *Unrealistic* scenarios can often help in reaching valuable learning points, but they require a different treatment during the debriefing.

## 69.3 Instructor Versus Facilitator

As we are running courses for adults and adults-in-training, we apply basic adult learning principles in all our courses, from medical students to senior consultants. We try to stimulate self-reflected learning in participants, and help with this process by providing a methodological framework for doing so. This means that we especially try to reduce the amount of instruction and maximize the use of facilitation. In the following, we differentiate both approaches.

**A facilitator** defines the structure and some overall topics for the discussion and helps participants to stay on target, being a kind of moderator. The facilitator is not responsible for the contents with which participants fill this structure. For example, the facilitator might ask participants to discuss the workload distribution during the scenario but would, as a facilitator, accept all inputs that are related to this CRM principle. A facilitator could also move the discussion away from the specific scenario (e.g., when participants did perform very poorly) and ask participants to discuss CRM aspects more in general. The facilitator could ask an open-ended question to the group such as: "What does a leader in general do?" and then would wait for the group to discuss these issues, and only help if participants lose the focus on this topic. In this way, the facilitator steers the discussion, without directly providing content input. In the case that a participant proposes something that the facilitator thinks is (absolutely) wrong, the facilitator can certainly offer a different view. This means sometimes to leave the role as a facilitator and to become an instructor.

**An instructor** provides both, the structure and the content and during the debriefing, for example, offers a mini input on ways to distribute the workload. We try to focus on facilitation, but add issues that we think are most important during the debriefing in an instructor mode, if participants do not cover points that we think are essential. We switch between the role of facilitators and instructors if we think it is appropriate. The "mixture" depends on target groups, educational ideas, and to a certain extent also on the teaching style of the instructor. Some simulator instructors emphasize the "instructor mode" when working with medical students.

## 69.4 Proposed Structure

Our proposed structure for CRM-oriented debriefings is depicted in Table 69.1. In the following section, we explain this structure and how it can be used. There are four phases in

**TABLE 69.1** Proposed structure for CRM-oriented and video-assisted debriefing

| Phase | Goal | Methods/Questions |
| --- | --- | --- |
| Structuring | Provide overview of debriefing method and time frame | Explain debriefing as a method, its goals, how it works, as well as roles involved |
| De-roling | Transition of participants into the analysis mode Reintegration of participants into training group Discovery of potential problems with the scenario (e.g., in terms of realism) | "How did you feel during the scenario?" |
| Description and Analysis | Provide feedback Analyse strengths and weaknesses Focus the discussion | "What happened?" "What was the main problem?" "What was good? Why was it good?" "What was not good? Why was it not good?" "What would you do differently now? How would you do it differently?" |
| | Integrate the remaining group | "What is your perspective on the scenario?" |
| | Watch selected video clips | Show key scenes (e.g., intubation, takeovers, good examples or bad examples, etc.) Ask participants what they want to see |
| Integration | Facilitate the learning Active, participatory ending of debriefing | "Do you have any final statements?" "How would you summarize the major points from this scenario?" "What is your take-home message?" |

the structure: Structuring, De-roling, Description and Analysis, and Integration. In the following, we will explain each phase in more detail.

During the **first phase**, the **structuring**, the facilitators briefly explain what will be done during the debriefing and what the time frame is for the current debriefing. This phase shortens with each succeeding scenario and in some cases can be omitted later on, if both instructors and participants know and use it. However, it is important to reestablish this structure from time to time.

After this very short opening of the debriefing, during the **second phase**, the **de-roling**, hot-seat participants have the chance for their first word. The first question focuses specifically on feelings during the scenario, concerning, for example, the workload or the experience of the scenario. This question serves as an ice breaker. It is important to ask this question to all participants who were active in the simulation scenario (especially also persons who came in later, e.g., when they were called for help during the scenario). A tricky task for the facilitators is to find a trade-off between the possibility for a first word for participants and a common tendency of participants to mention all the difficulties during the scenario within their first statement. It might often be necessary to stop participants in order to let the other "hot-seat" participants speak their minds. In this phase, all hidden instructions to participants should be uncovered that might have been used for role-players during the scenario (especially if a participant was asked to

play a negative role during the scenario – see Chapter 59 – The TuPASS Scenario Script). The de-roling phase enables all participants to really step out of their roles and leave especially negative characters behind them. It is further important, particularly with medical students, to address their emotional concerns when they think that they did not act well during the scenario because they do not have an established frame of reference from actual clinical work about a "normal" case and their "normal" performance.

In the **third phase**, the **description and analysis**, the facilitator should give participants most of the time for talking by asking questions and letting participants answer. If there is an answer that is totally wrong, the facilitator should first rephrase the question (maybe the instructor did not phrase the question correctly, or the question was too obtuse), and only then switch into an instructor mode in order to explain why an answer was wrong (if this is still deemed necessary). Especially when working with experienced health care professionals, it will be necessary on occasion to state a different view, but accept differences in procedures in order to avoid long, fruitless struggles about who is right.

First, a description of the major problem and topic of the scenario should be facilitated, and later, an analysis of what was good and what could have been improved and in which way. A short description of the major clinical points of the scenario can help to cover these issues and to focus on CRM aspects. However, it is important to also emphasize the close relation of clinical treatment and CRM. They are not two separate concepts, but are interrelated quite closely. Almost always, clinical treatment is made better or worse because of better or worse CRM skills – facilitators should try to highlight this relationship.

Beginning with the "hot-seat" participants, one could work through the scenario (mostly) from the beginning, in a chronological order, discussing important points during the scenario. From the description ("what" happened), the discussion should move on to the analysis for both negative and positive aspects ("Why did this happen?", "Why were you good?", "How can you be good next time again?", "How can you change what you want to change?"). As a general principle, the discussion should start with the "hot-seat" participants and integrate the remaining group where possible. Especially, questions about what could be changed or improved next time should be directed to the "hot-seat" participants first, before the group is involved. As in most cases participants know exactly what could be improved after the scenario, it might feel awkward to be told by others what they already know for themselves. "New" aspects can be added later by the remaining group or the instructors. It is important and beneficial to really analyze positive examples, and why they were good. With such an analysis, it can become clear whether positive examples happened "accidentally only," or were the result of conscious decisions. In this way, the learning from positive examples can be emphasized (see Plus-Delta in Chapter 57).

For **integrating** the remaining group, the facilitator could ask them for their view and comments on the topics discussed. During the debriefing, selected video clips should be shown to participants. The clips should be selected to show "new" aspects, which are mostly CRM-related issues and often aspects that participants did not notice during the scenario. With this approach, it easily happens that participants do mention additional aspects that the instructors did not see. However, this opportunity can only be used if the facilitator provides room for such initiatives from the course group. Clinical topics can easily be covered by discussion alone. The video should neither be used to prove that the instructor was right and the participant was wrong, nor to rewatch the complete case (the time this takes can be spent more effectively by focusing on key points).

Usually, it is possible to discuss the case in a chronological order, going deeper into the analysis of two to three selected points. It may be difficult to be content with so little "action" as a facilitator, but less is more, and if participants learn one or two points by their own insight, it is worth the effort rather than talking superficially on too many points.

At the **active ending** of the debriefing, participants should have a formal last word to say, any comments or concerns, and can then be asked to summarize their major learning points (or "take-home-messages") in order to provide an active ending.

## 69.5 Ways to Use the Structure

The proposed structure is used during our own simulator courses as well as in our Instructor and Facilitation Training Courses (see Chapter 67). It is important to make this structure obvious to participants: in the beginning of a simulator course, the so-called "setting introduction" [21] as well as in the beginning of individual debriefings. If participants know what they are facing and when they will have the opportunity to make their points, it seems to be easier for them to wait – and actually use the given structure. Thus, applying the CRM principle "anticipate and plan" to the conduction of debriefings can model its use as well as help everybody involved to be on the same page.

## 69.6 Additional Tools During the Observation Phase

In order to use the proposed or any other structure during the debriefing, instructors/facilitators need material from the scenario that they observed. To reduce the burden of observation and note taking, we are working with preformatted paper and pencil tools. Depending on the goals and target group, we use different structures. The different tools that we use are provided on website for this book and can also be downloaded from our web page at www.tupass.de/downloads/debriefing/structures.html.

## 69.7 Problems in Using the Structure

One difficulty – especially when addressing CRM-related topics – may be the danger of a superficial character of discussions by the facilitator. It happens frequently that CRM aspects are discussed in descriptive manner (e.g., "you communicated well") without sufficient detail (e.g., what exactly was good?). Especially reasons underlying good CRM behavior (e.g., "Why did you communicate well?" – "Because we do not know each other and tried to be very explicit what made our communication very clear") are omitted and the debriefing stays with descriptions only. If the facilitator keeps asking "why" questions, it might be easier to delve into the analysis of what happened. A "Five Why Approach," as known in root cause analysis, might be helpful for "why" over and over again to analyze the events (cf. http://www.npsa.nhs.uk/health/resources/root_cause_analysis).

When trying to integrate the remaining group, it happens very easily that other participants start to "pick on" the "hot-seat" participants. When this happens, we try to overcome these negative comments and tendencies by asking for a different tone or a clear message of exactly what can be learned from the critique. We further try to be very explicit with the questions we ask the audience. "What do you think of what the 'hot-seat' participants did?" might trigger more critique and discussion than "What are the tasks of a leader if we take the specific segment of the scenario as a starting point?"

Even as a facilitator, it is necessary to intervene rather frequently to bring the group back to the topics that were agreed upon to discuss (i.e., CRM aspects). The discussion very easily goes back to clinical issues [22]. Being a facilitator thus means not to act in a laissez-faire style – but the facilitator does not have to know or pretend to know everything.

## 69.8 Conclusion

Using structures for debriefings makes this complex task easier for the instructors/facilitators as well as for the participants. If the structure is explained to participants, they know what will happen during the debriefing and might thus feel more comfortable and willing to stick with what the instructor/facilitator proposes to do. However, it is also important to be open to the often very dynamic processes during the debriefing. Instructors/facilitators should thus also be willing and able to put aside a structure – if only temporarily – to enable them to make very valuable learning points to participants.

## 69.9 Words of Wisdom

In the debriefing:

- There is only one reality, that of the students in the hot seat.
- Ask about what happened and why or why not, before you judge anything.
- Let the participants talk. You only stimulate and facilitate.
- Ask open ended questions. Keep them open. And wait! Wait!
- Be the last one to answer your own questions.
- There are always other opinions than yours.

- You might be wrong, too.
- Trainees might know more about a specific subject than you do, let them talk.
- You are not a god, just an instructor/facilitator.
- You can just help them learn, you have no further power, so help them learn and no more.
- "You must. . ." starts resistance, resistance hinders learning, learning is all.
- Redirect questions from the trainees to the hot seater, the other participants or the observing audience.
- Simulation is about learning, that is, improvement, and not performance. So do not talk about performance, even if it is good.
- What is not necessary to reach a learning goal, is not necessary.
- Learning in simulation is not only about learning clinical and/or technical aspects. Experiencing stress, personal limitations and human factors in action may be a big learning success and goal.
- You might not know what they learn.
- Learning might go on and continue after the course.
- Never trick your participants, always tell the truth.
- Simulation is no adventure park ride – use special effects with care.
- A good clinician is not a good instructor – per se.
- A good clinician instructor is not a good clinical simulation instructor – per se.
- Years of clinical experience are not a requirement to become a good CRM-instructor – it is about facilitation of learning by others, not instructing at them.
- Sound quality is key – video quality is not important – spend hundreds for video and thousands for audio.
- Do not simulate simulation – they really have to practically do all the little things (pulling up syringes, infusions etc.). Otherwise you are at "mental simulation" which can be done without a simulator in the classroom.
- Before you start a scenario, perform a formal stop-procedure: Check video, state of the simulator, the team, the participants. Only after this quick final check give a "go".
- Apply CRM principles with your simulation production team before attempting to impose them on your clinical students.

# Example Checklists

## *Simulation Training*

### Event-List

| Date: | Participants: | Case: |
|---|---|---|
| Proj. Nr.: | Tape/File: | |
| Start Time: | End Time: | |

| Time | M | Action/Event | Comments |
|---|---|---|---|
| | | | |
| | | | |
| | | | |
| | | | |
| | | | |
| | | | |
| | | | |
| | | | |
| | | | |
| | | | |
| | | | |
| | | | |
| | | | |
| | | | |
| | | | |
| | | | |
| | | | |
| | | | |
| | | | |
| | | | |
| | | | |
| | | | |
| | | | |
| | | | |
| | | | |
| | | | |
| | | | |

For CRM-Principles see: Rall, M., & Gaba, D. M. (2005). Human Performance and Patient Safety. In R. D. Miller (Ed.), *Miller's Anaesthesia* (pp. 3021–3072). Philadelphia: Elsevier Churchill Livingston.

Center for Patient Safety and Simulation – TuPASS, Tuebingen, Germany
(http://www.tupass.de/downloads/TuPASS_Event_List.pdf)
marcus.rall@med.uni-tuebingen.de

## Open ACRM Principle Orientation
Scenario- /Debriefing Protocol             Date._____

| Course | Scenario | Debriefer | Trainees |
|--------|----------|-----------|----------|
|        |          |           |          |

Make short notices on observations during scenario for debriefing. Mark points to be covered in debriefing, use time stamps if available to find spots on the video

| Principle | Keypoints | | Stamp |
|-----------|-----------|----------|-------|
|           | Positive  | Negative |       |
|           |           |          |       |
|           |           |          |       |
|           |           |          |       |
|           |           |          |       |
|           |           |          |       |
|           |           |          |       |
|           |           |          |       |

For CRM-Principles see: Rall, M., & Gaba, D. M. (2005). Human Performance and Patient Safety. In R. D. Miller (Ed.), *Miller's Anaesthesia* (pp. 3021–3072). Philadelphia: Elsevier Churchill Livingston.

Center for Patient Safety and Simulation – TuPASS, Tuebingen, Germany
(http://www.tupass.de/downloads/TuPASS_Good_Bad_CRM.pdf)
peter.dieckmann@med.uni-tuebingen.de

## CRM-Orientation Focussing on Key scenes in scenario
Scenario- /Debriefing Protocol                          Date._____

| Course | Scenario | Debriefer | Trainees |
|--------|----------|-----------|----------|
|        |          |           |          |

Make short notices on observations during scenario for debriefing. Mark points to be covered in debriefing; prepare with key scenes from scenario in advance.

| Principle | Key scene 1: | Key scene 2: | Key scene 3: |
|-----------|--------------|--------------|--------------|
|           |              |              |              |
|           |              |              |              |
|           |              |              |              |
|           |              |              |              |
|           |              |              |              |
|           |              |              |              |
|           |              |              |              |
|           |              |              |              |
|           |              |              |              |
|           |              |              |              |
|           |              |              |              |

For CRM-Principles see: Rall, M., & Gaba, D. M. (2005). Human Performance and Patient Safety. In R. D. Miller (Ed.), *Miller's Anaesthesia* (pp. 3021–3072). Philadelphia: Elsevier Churchill Livingston.

Center for Patient Safety and Simulation – TuPASS, Tuebingen, Germany
(http://www.tupass.de/downloads/TuPASS_Key_Scenes_CRM.pdf)
peter.dieckmann@med.uni-tuebingen.de

## Chronological Order
Scenario- /Debriefing Protocol                                    Date._____

| Course | Scenario | Debriefer | Trainees |
|--------|----------|-----------|----------|
|        |          |           |          |

Make short notices on observations during scenario for debriefing. Mark points to be covered in debriefing. Adapt timeline to scenario.

|                            | Positive | Negative | Stamp |
|----------------------------|----------|----------|-------|
| Pre Anesthesia             |          |          |       |
| Introduction to Patient    |          |          |       |
| Anamnese                   |          |          |       |
| Preperation for Anesthesia |          |          |       |
| Induction                  |          |          |       |
| Intubation                 |          |          |       |

Center for Patient Safety and Simulation – TuPASS, Tuebingen, Germany
(http://www.tupass.de/downloads/TuPASS_Scenario_Chronology.pdf)
peter.dieckmann@med.uni-tuebingen.de

# References

1. Rall, M. and Gaba, D. M. Patient simulators. In: Miller, R. D. (ed.) *Anaesthesia*, Elsevier, New York, 2005, pp. 3073–3103.

2. Rall, M., Manser, T., and Howard, S. K. Key Elements of debriefing for simulator training. *European Journal of Anaesthesiology* 17(8), 516 (2000) (Abstract).

3. Dismukes, K. R. and Smith, G. M. (eds.), Facilitation and debriefing in aviation training and operations. Ashgate, Aldershot, 2000.

4. Garden, A., Robinson, B., Weller, J., et al. Education to address medical error – a role for high fidelity patient simulation. *New Zealand Medical Journal* 115(1150), 133–134 (2002).

5. Schaedle, B., Dieckmann, P., Wengert, A., et al. The role of debriefing in simulator training courses for medical students (Abstract Santander-02-20). *European Journal of Anaesthesiology* 20, 850 (2003).

6. McDonnell, L. K., Jobe, K. K., and Dismukes, R. K. Facilitating LOS Debriefings: A Training Manual – Part 1: An Introduction to Facilitation. Moffett Field: Ames Research Center, 1997.

7. McDonnell, L. K., Jobe, K. K., and Dismukes, R. K. Facilitating LOS Debriefings: A Training Manual – Part 2: Getting Started. Moffett Field: Ames Research Center, 1997.

8. McDonnell, L. K., Jobe, K. K., and Dismukes, R. K. Facilitating LOS Debriefings: A Training Manual – Part 3: Facilitation Techniques. Moffett Field: Ames Research Center, 1997.

9. McDonnell, L. K., Jobe, K. K., and Dismukes, R. K. Facilitating LOS Debriefings: A Training Manual – Part 4: The C-A-L Model in Action. Moffett Field: Ames Research Center, 1997.

10. Mort, T. C., and Donahue, S. P. Debriefing: The Basics. In: Dunn, W. F. (ed.), *Simulators in Critical Care and Beyond*. Society of Critical Care Medicine, Des Plaines, 2004, pp. 76–83.

11. Peters, V. A. M. and Vissers, G. A. N. A simple classification model for debriefing simulation games. *Simulation and Gaming* 35(1), 70–84 (2004).

12. Schwid, H., Rooke, G., Michalowski, P., and Ross, B. Screen-based anesthesia simulation with debriefing improves performance in a mannequin-based anesthesia simulator. *Teaching and Learning in Medicine* 13(2), 92–6 (2001).

13. Dismukes, R. K., Gaba, D. M., and Howard, S. K. Facilitated Debriefing in Health Care. *Simulation in Health Care* 1(1), 23–25 (2006).

14. Rudolph, J., Simon, R., Dufresne, R. L., and Raemer, D. B. There's No Such Thing as "Nonjudgmental" Debriefing: A Theory and Method for Debriefing with Good Judgment. *Simulation in Health Care* 1(1), 49–55 (2006).

15. Gaba, D. M. Dynamic Decision Making in Anesthesiology: Cognitive Models and Training Approaches. In: Evans, D. A., and Patel, V. L. (eds.), *Advanced Models of Cognition for Medical Training and Practice*. Springer, Berlin, 1992, pp. 123–147.

16. Gaba, D. M., Fish, K. J., and Howard, S. K. *Crisis Management in Anesthesiology*. Churchill Livingstone, Philadelphia, 1994.

17. Rall, M., and Dieckmann, P. Crisis Management to Improve Patient Safety. ESA Refresher Course 2005, pp. 107–112. [Online: http://www.euroanesthesia.org/education/rc2005vienna/17RC1.pdf]

18. Rall, M., and Dieckmann, P. Safety Culture and Crisis Resource Management in Airway Management. General Principles to Enhance Patient Safety in Critical Airway Situations. *Best Practice & Research Clinical Anaesthesiology* 19(4), 539–557 (2005).

19. Rall, M., and Gaba, D. M. Human Performance and Patient Safety. In: Miller, R. D. (ed.), *Miller's Anaesthesia*. Elsevier Churchill Livingston, Philadelphia, 2005, pp. 3021–3072.

20. Howard, S. K., Gaba, D. M., Fish, K. J. et al. Anesthesia crisis resource management training: teaching anesthesiologists to handle critical incidents. *Aviation Space Environment Medicine* 63(9), 763–770 (1992).

21. Dieckmann, P., Manser, T., Wehner, T., et al. Effective Simulator Settings: More than Magic of Technology (Abstract A35). *Anesthesia and Analgesia* 97(S1–S20): S 11. (2003) [Online: http://www.anestech.org/media/Publications/IMMS_2003/sta3.html]

22. Debriefing in Anaesthesia Crisis Resource Management Training. Abstracts of the Annual Meeting of the Society on Europe for Simulation Applied to Medicine (SESAM), Stockholm, Sweden, June 2004. [Online: www.sesam2004.se/complete_abstracts.pdf]

70

# Questionnaire Design and Use: How to Craft Tools to Determine How Well Your Simulation Program Objectives are Being Met

Guillaume Alinier

## 70.1 Why Should We be Collecting Feedback?

Just as all formal education includes student testing in some form to create data for feedback on their learning or collecting feedback data on the performance of the *teachers*, subsequent analysis and reflection is just as essential. The results of this effort can be used to not only help you to improve your teaching approach, but also to demonstrate your relative worth within your academic community.

However, like all tools, feedback is only valuable if it is well designed and provides you with valid information to the questions you ask. Remember that you cannot wholly rely on the feedback you receive through any one assessment method or single application of a method. At times, discussing and probing candidates on particular points will help you to obtain additional and important insights. I (GA) have found aspects of the way some faculty facilitate scenarios that participants did not particularly enjoy, yet they did not report it in writing on the feedback forms! They might think the tutor, "plant," or actor is prompting them too much, or that the scenario is turning into bedside teaching or a question and answer

case study, hence distracting them from providing care to the patient. Ultimately, I have been able to address the problem by getting participants to write their concerns on the feedback forms. Note that anonymity might be a relative concept depending on the way the feedback is processed and given that every simulation participation is well documented, and the low student-to-teacher ratio in most clinical education programs assures that such unique identifiers such as discipline, gender, and hand writing are recognizable and identifiable entities. In the case of our center, faculty do not see the feedback after every single session but only periodically, which prevents them from being able to recognize who the originator of a particular comment could be.

## 70.2 Questionnaire Design

When designing your questionnaire, you have to consider the type of information you will be collecting and the form in which it is to be presented. The information collected can be objective or most probably subjective, and quantitative or/and qualitative. Your questionnaire might be handed out at the beginning of the session with all the other paperwork, with some portions to be completed at the start and others to be completed at the end of the session. The simulation session might have been an intensive training experience but you do want to collect some information to evaluate the impact of your training and determine if anything can be improved. If you want constructive criticism, do not annoy your simulation participants with a questionnaire that is too long. A typical goal is to keep it to a single page in length in order not to discourage the respondents. It should be as short as you can make it, easy to fill in and as concise as possible without using a specialist vocabulary. You should only ask questions for which you will actually make use of the answers. Do not ask questions you do not want to learn the answers to. For example, do not collect information about their experience of performing a particular procedure if you are not going to use that piece of information for a particular research project. For the design of the questionnaire, try to create a visual structure that is pleasing to the eyes, unambiguous, and rapid to fill in. By the end of your session, participants will want to chat with their peers about their exciting simulation experience, or they might have to rush back to work or home. There are many good textbooks and publications about questionnaire design, but something you might want to consider is to design something you can process with an optical mark reader. It is a machine that scans pages and then produces a summary of the results. It is very fast at analyzing results and provides basic statistical results. Any written ("text") feedback would need to be entered separately using a word processor and be used as qualitative data.

Elements of your feedback forms might be useful for research purposes, but once again, make sure your questions are well phrased and unambiguous. The goal is to eliminate the chance that the question will mean different things to different people. In some cases, it is easier to think of what you would like to find out and how you would report it. This can then more easily be turned into the exact question you are looking for to be included in your questionnaire. Similarly, such approach can help you in your broader research design approach. When collecting information regarding the frequency of an occurrence, it is more valuable to quantity choices objectively (i.e., 2–3 times, 3–8 times, more than 8 times…) rather than offering subjective choices open to interpretation (i.e., rarely, sometimes, very often…). There is nothing more annoying than collecting a vast amount of information to notice in the end that the question was not understood by participants. This happened to a colleague who tried to evaluate students' attitudes toward clinical skills training versus simulation-based learning. When asked how many simulation sessions they had attended, over a quarter of the respondents provided an exaggerated answer as they confused some of the sessions they have had in the clinical skills laboratories as being simulation-based learning. This highlight

the facts that you should provide basic definitions about any specialist terminology you may employ in your questions. It can prevent some of the information collected being invalid or inadequate and possibly completely failing your research.

The basic questions often found at the start of a questionnaire help the study participants to focus onto the questionnaire without having to think too much. They often include demographic questions (age, gender) but can very appropriately explore their educational background, qualifications, professional disciplines, previous exposure to simulation-based training, and number of years of experience. Then, you can introduce more profound questions that require the participant's personal opinion. Often, because of these initial personal questions, you will have to seek ethical approval to gain formal authorization to subject the session participants to your questionnaire. An ethics committee will scrutinize your proposal (reasons and aims) and the questionnaire itself to evaluate its potential psychological impact on respondents. If you are intending to use the information collected for research purposes, it is also advisable to have initially obtained ethical approval. Your proposal will have been examined by the ethics review panel, making the research more rigorous and easier to present and defend your work. Research papers for peer-reviewed journals often require authors to include a heading about ethics.

When designing your questionnaire, you should carefully think of the way your questions or statements are phrased. This is especially important if you are using rating or agreement scales. How do your positive and negative questions balance? The participants might read questions too rapidly and circle a whole column at the same rating without noticing that an odd question was phrased differently and required, for example, a "strongly disagree" to refute your negative statement. It is a good idea to keep your Likert scale the same throughout the questionnaire, for instance, keep low or poor scores left, and high or good scores on the right. For example, define your Likert scale with: 1 = strongly disagree and 5 = strongly agree.

Another aspect to think of when designing your questionnaires is to leave a special space for the administrator or investigator input to mark simple but potentially significant things. This can include the date, the name of the faculty present or leading the session, whether it was a multiprofessional session, if the team was composed of participants who knew each other, or if there was any special incident that occurred during the session that could have had an impact on the participants' feedback. Such information is difficult to retrieve retrospectively and can be very rapidly added after a session. It might come in very useful at a later stage if, for example, you are trying to assess whether your implementation of particular changes (based on the feedback forms) caused significant improvements in the feedback from the next series of simulation sessions.

Once you think you have finished preparing your questionnaire, it should be evaluated by being piloted using a sample group of participants. They might be able to highlight weaknesses such as omissions, repetitions, or poorly phrased questions. This phase is very important as it should still be the time to make changes to the questionnaire so that you end up with the most suitable tool. Any changes to its design or formulation implemented at a later stage might affect the feedback collected and may invalidate previously collected information.

- Pilot or pre-test your questionnaire to find out if any question could be misunderstood, have been repeated, or omitted.
- Make use of the information you collect through your feedback forms and questionnaires to improve your sessions.

## 70.3 Summary

The take-home messages of this chapter are the following:

- It is good educational practice to evaluate your session by collecting feedback.
- The design of your questionnaire should be appealing and easy to fill in.
- Try to keep it to one page in length even if it means having two columns.
- Limit yourself to only ask questions you will actually use in your research.
- Find out whether you require ethical approval from your institution before subjecting participants to your questionnaire. It is worth going through the process if you eventually want to publish you data.
- Always offer some blank space and invite for open feedback and comments.
- Allow some space for administrator input and comments on the forms.
- Ask yourself what you want to find out to help you formulate the questionnaire.

## 70.4 Conclusion

Designing a valid and useful questionnaire is not as straightforward as it looks. Well formulated, it can provide valuable information, however, when poorly developed it can be a waste of time for the respondents as well as the researcher inputting and analyzing the feedback. Real thoughts need to be employed in the design process as it will be a determinant factor in the value of the results collected through your research tool, i.e., the questionnaire or feedback form. Make use of the feedback you collect at least to inform your teaching practice. If you obtain particularly good or interesting results and feedback from your teaching approach, share it with your colleagues through publication in journals and conferences. Similarly, when you find out that something does not work, it is as valuable to share this kind of findings with others so that it is not attempted again. Collecting feedback is also very valuable internally, for the center, to evaluate self-worth, but it can also be used when writing grant applications. Quotes of supportive comments or interesting research findings from the data collected can be reported as evidence of good practice for example.

# 71

# Planning and Assessing Clinical Simulation using Task Analysis: A Suggested Approach and Framework for Trainers, Researchers, and Developers

Paul Williamson
Harry Owen
Valerie Follows

## 71.1 The Curriculum Procedures Frames Simulation Framework

This Chapter introduces the Curriculum Procedures Frames Simulation (CPFS) framework for use in the development, analysis, and assessment of clinical simulation. This framework is based on Task Analysis – a commonly used set of analysis tools common in training design – and systems analysis methods that are essential to business systems analysis and used by consulting experts. By the end of this Chapter, you should have an understanding and some tools to apply these four steps of the framework to your needs.

(C) Curriculum. Be able to look at the requirements of your students and divide them into procedural elements that can be taught.

(P) Procedures. Understand the five Ps of Planning, Preprocedure, Procedure, Postprocedure, and Professional development and the range of tools available to organize them into effective teaching units.

(F) Frames. Be able to identify and use the interrelationships between the six elements of each training "Frame" and use your resources to best effect.

(S) Simulation. Put simulation technologies, techniques, and outcomes clearly into context with a method to objectively assess their Characteristics and Capabilities.

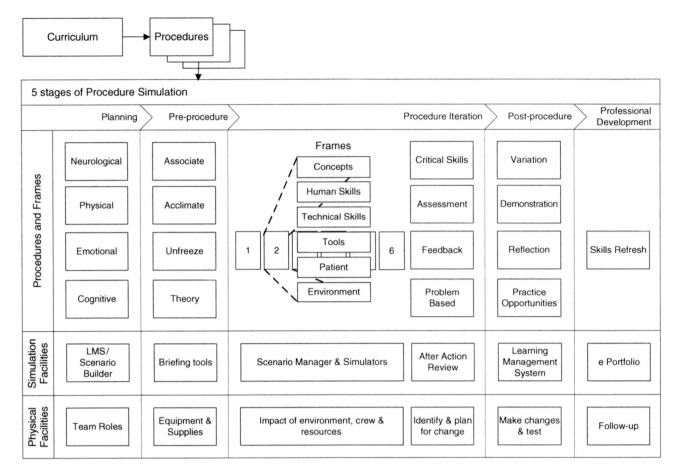

**FIGURE 71.1**  Curriculum Procedures Frames Simulation framework.

These steps are shown graphically in Figure 71.1 and are explained later in a step-by-step fashion – so don't panic if it looks a bit complex to begin with.

Likely, your clinical simulation program will have one or more of the following roles requiring your clinical educators and simulation professionals to fulfill:

- A Curriculum or a subject matter expert who defines the knowledge, skills, and behaviors required for competent professionals.
- A Planner who guides the creation of training elements that meet the needs of students and curriculum.
- A Facilitator who must coordinate and deliver the training sessions.
- A Simulation Professional or technician/expert who must find, develop, and provide the simulation technologies that meet the aims of the teaching plan.

This Chapter is structured around a workflow to help you perform your best in these roles.

### 71.1.1  Unfreeze!

The term "unfreeze" is commonly used in management consulting to refer to activities engineered to break a person's bonds to the "old" way of doing things and putting them in a receptive state for making a change. Failure to "unfreeze" a person or organization can result in the changes not "sticking" or even being resisted from the outset. Here, a successful unfreeze requires us to recognize a number of the pressures being experienced in clinical simulation training and therefore the need for methods of allowing training and technology designers to better cope with them.

- The successful use of clinical simulation methods and devices depends on your ability to match the right resources to the needs of the clinical students of any level. This is not as easy as it sounds.

- The number and complexity of simulation devices are rapidly increasing, with similarly numerous opportunities to spend money on them. Your resources, however, are likely not increasing at the same rate.

To answer this challenge, the Curriculum Procedures Frames Simulation (CPFS) framework is a tool that can be used as follows:

- (C) Curriculum. Do what is most important first. Use a proven tool to inventory the critical tasks for a group of students, define them in manageable procedures, and prioritize those to attack first.
- (P) Procedures. Analyze the procedures to be taught and define them in stages that can then be planned, developed, and reviewed.
- (F) Frames. Objectively resolve the various interrelated elements of a student's actions and be able to plan and influence them to achieve the outcomes you need.
- (S) Simulation. Be able to understand how the simulation technologies you are using, planning to use, or are developing will be used in your training procedures, spot the gaps, and avoid wasting money, time, and reputations.

## 71.2 About Task Analysis and Framework

Task Analysis is a method that you can use to identify the most basic components of your most essential activities. Used in conjunction with a framework, Task Analysis is a powerful tool to achieve effective results faster. In this context, a framework is a collection of tasks, functions, or activities in groupings and processes that represent real-life systems and procedures. By manipulating how you address the elements of the framework, you can determine how your actions and resources will be used. Frameworks can be developed or adopted for any sort of activity; they're certainly a tool that enables professional consulting firms to make money by allowing their consultants to quickly analyze complex situations and determine a course of action to recommend. As opposed to a Model, a framework does not necessarily have direct causal links between its elements. It cannot be easily computerized or simulated. It does, however, provide an algorithm that allows

a complex or "fuzzy" concept to be analyzed, understood, and communicated.

## 71.3 C is for Curriculum

> Planner – you will need to lead this process. Once you have identified what you want to teach, circulate the worksheet with your Facilitators to construct a comprehensive list of Procedures involved in your chosen curriculum.

### 71.3.1 Selecting your Tasks

The professional area in which you choose to work has its own domain of essential skills and knowledge required for a practitioner to be regarded as competent. In many cases, the body in charge of professional standards will publish its list of essential competencies and the standards required to qualify. Clinical simulation is a developing trend however, and the skills and knowledge of your particular area of interest may not have been published yet. So, whether you have the published guide or not, you can choose an intervention or competency that you want to focus on, according to your priorities.

Here, for example, we have chosen an aspect of the Trauma curriculum, cervical spine (C-spine) stabilization after an actual or suspected injury. This is an essential skill for many health professionals and requires mastery of concepts as well as motor skills. In Figure 71.2, we show cervical spine stabilization set in context to 3rd year curriculum and broken down into four key procedures.

The following section, Complete the Task Inventory, helps define the components of the curriculum using Task Analysis methods [1] and offers a framework to prioritize elements and choose how to teach them. The result is a catalog of elements, prioritized and classified into learning procedures, which can then be developed into learning modules. This is represented in the Curriculum Procedures Frames Simulation framework as a breakdown of Curriculum elements, which are then selected to be developed further into Procedures (Figure 71.2).

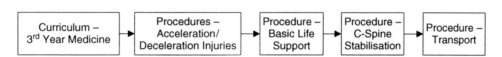

**FIGURE 71.2** Curriculum stream with selected Procedure.

**Worksheet 1**  Example of the task selection

| TASK SELECTION CRITERIA WORKSHEET Situation: Body system: Trauma of nervous system **Basic Life Support training essential **Crisis Resource Management training desirable | CRITICALITY | UNIVERSALITY & FREQUENCY | STANDARDIZATION | FEASIBILITY | DIFFICULTY | TOTAL | PRIORITY | SEQUENCE |
|---|---|---|---|---|---|---|---|---|
| **WEIGHTING** | 40 | 10 | 10 | 10 | 30 | 100 | — | |
| C-spine stabilization for symptomatic spinal cord injury | 40 | 1 | 7 | 9 | 25 | 83 | High | 2 |
| C-spine stabilization for risk of spinal cord injury | 35 | 3 | 7 | 9 | 25 | 81 | High | 1 |

## 71.3.2  Complete the Task Inventory

The following process is demonstrated in Worksheet 1. This is modified to suit clinical training from Jonassen [1]. Determine, on the basis of the total score and priority, the method of instruction. Typically, high-scoring tasks will need Full Mission Simulation to teach effectively.

1. **Specify a referent situation.** Conduct your analysis of a particular set of skills within the context of a particular situation and location. That is, state the situations in which this task may be used.
2. **Complete the skill inventory.** A complete task inventory is needed because the ultimate goal here is to select and rank tasks to decide which will be developed for certain types of formal instruction.
3. **Assess the criticality of the skill or task.** Consider the severity of the impact of unlearned skills on the mission of the organization – in this case, our Patient may never walk again if you damage their spine! Rank the task in terms of its criticality to that mission 0–40 (0 = unnecessary, 40 = invaluable, absolutely critical).
4. **Assess the universality and frequency of the task to be performed.** How often will the practitioner encounter a need for this task? Consider:
   - Percent or number of contexts within the organization where task is performed.
   - Percent of similar job positions requiring performance of task.

   - Percent of employees, students, or other personnel who will be given the task over time.
   And then rank the universality of each task 0–10 (0 = never performed by anyone, 10 = performed by everyone).
5. **Assess the required standardization of skill performance.** Consider the consequences of task or skills not being standardized for:
   - mission or goals of the organization
   - ability to operate in normal manner
   - safety and well-being of personnel and patients
   Specify minimum standards/conditions necessary for task performance and rank the standardization of each task 0–10 (0=idiosyncratic, 10=standardized methods).
6. **Assess feasibility for learning and performing the task or skill.** Consider:
   - Is adequate instructional support available for each learning task?
   - Will task be used in its application context?
   - What is the attitude of the learners toward learning the task or skill?
   - Will follow-up training be necessary?
   Now rank the feasibility of each task 0–10 (0=not able to be learned, 10=learning and performance supported and feasible).

7. **Assess difficulty in learning tasks.** Consider:
   - Amount of time average student would need to learn the entire task?
   - What is degree of danger to patient, personnel, equipment, or environment if task/element were taught entirely by on-the-job learning?
   - Likelihood that the learners will retain skills or knowledge?
   - Probability that the skill can be learned on the job?
   - Do learners possess the aptitude, prerequisite skills, maturity, and motivation to learn?
     Rank the difficulty of each task 0–30 (0=easy, 30=extremely difficult to learn).
8. **Total the scores for the five criteria in the worksheet for each task.** On the basis of these scores, prioritize these tasks. The highest scores indicate the highest need for investment in training.
9. **Group or sequence the learning elements into Procedures.** Determine the order and grouping of the delivery of each element. Many will be able to be grouped together to be taught at the same time. The determination of the teaching method helps group Procedures together as shown in Worksheets 2 and 3.

---

Facilitator, you will need to help complete this list and help the Planner prioritize what can be taught with the resources at hand. It will take all players a few iterations together to get it right. Hence the "Required by Date" field – plan ahead!

---

Table 71.1 describes a number of common training delivery methods ordered by their level of interactivity. An abbreviation is suggested for use in Worksheet 2.

**Worksheet 2** Example assessment of elements and appropriate delivery methods

| PROCEDURE TYPE | DOC | CBT | DID | PST | SP | FMS |
|---|---|---|---|---|---|---|
| Risk factors for spinal cord injury (Scene assessment) | ✓ | ✓ | ✓ | | | |
| Knows signs & symptoms of spinal cord injury | ✓ | ✓ | ✓ | | | |
| Can prioritize treatment – BLS may need to be undertaken ahead of motion restriction of spine | | ✓ | | | | ✓ |
| Knowledge of spinal anatomy and changes with age and disease | ✓ | | ✓ | | | |

**Worksheet 3** A completed Simulation Curriculum planning worksheet

| **CURRICULUM** | Trauma curriculum | **STUDENTS** | Medicine, 3rd Year |
|---|---|---|---|
| **Planner** | H. Owen | **SESSIONS** | 3 × 2 hr sessions |
| **Facilitators** | V. Follows, A. Vnuk | | |
| **Technicians** | R. Forbes, H. Owen | | |
| **To be Ready by:** | 1 July, 2007 | | |

**TABLE 71.1**  Common training delivery methods

| Score | Code | Method | General use |
|---|---|---|---|
| Low | DOC | Documents | Low interactivity. (Covers the range from most printed and electronically distributed material, including most web pages, video, and presentation files.) |
| | CBT | Computer-based Training | Some interactivity and testing to identify critical learning elements |
| Med | DID | Didactic | Interactive and flexible using tutorial, AV, and other technologies |
| | PST | Part-skills Training | Focused skills development including part-skills simulators |
| | SP | Standardized Patient | Highly interactive but limited in range of application |
| High | FMS | Full Mission Simulation | Use of skills under stress and highlights areas of complexity and risk |

| PROCEDURE TYPE | DOC | CBT | DID | PST | SP | FMS | Session |
|---|:---:|:---:|:---:|:---:|:---:|:---:|:---:|
| Risk factors for spinal cord injury (Scene assessment) | ✓ | ✓ | ✓ | | | | 1 |
| Knows signs & symptoms of spinal cord injury | ✓ | ✓ | ✓ | | | | 1 |
| Can prioritize treatment – BLS may need to be undertaken ahead of motion restriction of spine | | ✓ | | | | ✓ | 1,3 |
| Knowledge of spinal anatomy and changes with age and disease | ✓ | | ✓ | | | | 1 |
| Knowledge of pathophysiology of spinal cord injury – mechanism of injury/injury kinematics | ✓ | | ✓ | ✓ | | | 1,2 |
| Decision points for treating as spinal cord injury | ✓ | ✓ | ✓ | | ✓ | ✓ | 1,2,3 |
| Knowledge of motion restriction techniques, rationale, and limitations | ✓ | | ✓ | | ✓ | | 1,2 |
| Can safely and effectively use motion restriction techniques available locally | | | | | | ✓ | 3 |
| Patient questioning and explaining skills | | | | | ✓ | | 2 |
| Can use leadership/teamwork skills in this setting, incl. listening skills and conflict resolution | | | | | | ✓ | 3 |
| Knows how and when to undertake screening physical exam; for other injuries, undertake brief neuro exam | ✓ | | ✓ | ✓ | | ✓ | 1,2,3 |
| Can interpret and act on findings of screening physical and neuro exam | | | | | | ✓ | 3 |
| Can restrain C-spine manually in neutral position | | | | | ✓ | ✓ | 2,3 |
| Can recognize contraindications to neutral position | | | | | | ✓ | 3 |
| Can apply C-collar safely and effectively | | | | | ✓ | ✓ | 2,3 |
| Knows methods of transferring spinal cord injury victim to rigid spine support device | ✓ | | ✓ | | | | 1 |
| Can safely transfer to rigid spine support device using most appropriate method: (a) Log roll, (b) Flat lift, (c) "clam shell" scoop, and (d) extrication device – vacuum splint or Kendrick | | | | | ✓* | ✓ | 2,3 |
| Communication with other health professionals and other emergency services | | | | | | ✓ | 3 |
| Can restrain head to toe for transport to care facility | | | | | ✓ | ✓ | 2,3 |
| Can modify care for children (<7) | | | | | | ✓ | 3 |
| Knows complications of spinal cord injury that can worsen outcome (secondary spinal cord injury), especially (a) inadequate ventilation & (b) spinal shock | ✓ | | ✓ | | | | 1,2,3 |
| Looks for/monitors and treats when appropriate: (a) inadequate ventilation & (b) spinal shock | | | | | | ✓ | 3 |
| Knowledge of controversies in management of spinal cord injury | ✓ | | ✓ | | | | 1,2,3 |
| Knows primary health care measures to reduce incidence of spinal cord injury | ✓ | | ✓ | | | | 1 |

| TASK SELECTION CRITERIA WORKSHEET Situation: Body system: Trauma of nervous system **Basic Life Support training essential **Crisis Resource Management training desirable | CRITICALITY | UNIVERSALITY & FREQUENCY | STANDARDIZATION | FEASIBILITY | DIFFICULTY | TOTAL | PRIORITY | SEQUENCE |
|---|---|---|---|---|---|---|---|---|
| **WEIGHTING** | 40 | 10 | 10 | 10 | 30 | 100 | — | |
| C-spine stabilization for symptomatic spinal cord injury | 40 | 1 | 7 | 9 | 25 | 83 | High | 2 |
| C-spine stabilization for risk of spinal cord injury | 35 | 3 | 7 | 9 | 25 | 81 | High | 1 |

# 71.4 P is for Procedures

Planner – you will need to be sure that the worksheet is being translated into specific plans for each training session. Get to understand the "Placemats" described here – they are an easy and repeatable tool for building training Procedures.

Facilitator – you will need to lead this process. Once you have identified what you want to teach, construct your teaching sessions with this structure in mind. You'll also need to collaborate with a Simulation expert to effectively script SP or Full Mission Simulation.

Simulation Professional – your simulation methods and technologies need to fit into the needs of the Facilitators'. Work with them ahead of time to create the "Placemats" for each session. That will ensure that they get full value from your technologies and you have time to prepare.

## 71.4.1 Structuring Procedures

A Procedure is a single training event based on Documentation, Didactic presentation, Part-skills Training sessions, or Full Mission Simulation, for example. Each of these procedures requires a measure of preparation and resources to be effective. We won't try to reflect an entire body of expertise in this area. This section simply explores a way of representing the scope of what is involved in training procedures and offers a framework for you to cover all that you need to cover.

## 71.4.2 Five Ps of Procedures

For ease of recollection, we define the concept of Procedures in five stages. These are described in Figure 71.3: framework showing the Procedure stages. In the diagram, the five Ps represent the time-based stages of a training procedure. Beginning with the psychological and logistic considerations of Planning and Pre-procedure, we execute the event itself and then review it in the Post-procedure stage. Later, we consider Professional Development.

Each of these stages has a series of conceptual layers or resources relevant to them, and the activities involved can be fully considered using these layers. Horizontally, the layers loosely relate to each other, in increasing "level" with the most critical contributors to competence being at the top.

### Planning

Planning is the crucial stage where all of the elements are brought together by the training professional to create a successful training event. In this stage, they consider a range of inputs that are classified here as Cognitive, Emotional, Physical, and Neurological.

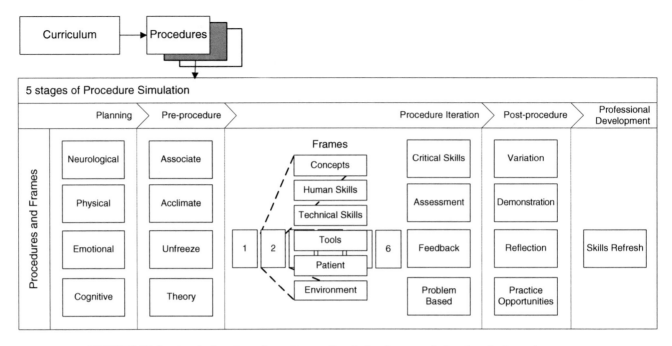

**FIGURE 71.3**  Curriculum Procedures Frames Simulation framework showing the Procedure stages.

1. **Cognitive.** The most obvious of the inputs is the theory and logical progression of the training procedure, covering the content of what is to be taught. This is handled at the intellectual level and covers the concepts and goals of the session.
2. **Emotional.** Experienced trainers understand that students come to the sessions with a range of emotional states, some conducive to training and others disruptive. These states can be reset, adjusted, and manipulated to achieve desired outcomes with careful planning.
3. **Physical.** The move from intellectual to physical is a significant one and consideration must be made of the logistical and spatial requirements of the session. Where people are required to start acting, using devices, and moving patients bodily, the choices in room configuration, equipment and team dynamics become very important.
4. **Neurological.** In order for the right competencies to be learned, the students must experience the right cognitive, emotional, and physical stimuli at the right point in order to maximize understanding and recall of the points you are trying to make. Associating a negative response to your actual desired outcome can give the student an unconscious disincentive to practice the right skills when the time comes in real life.

A simple yet effective tool for this is Worksheet 4, normally printed laminated with plastic – hence the name "Placemat." This Training Session Map is a proven, reusable tool that has been effective in hundreds of Full Mission and Partial-Task simulations.

**Preparation**

The planning stage can, and should be, executed well in advance of the training procedure. Closer to the event, the session itself can be resolved into stages that are useful for easing students into the session itself.

1. **Theory.** The first step in the actual procedure sets up the Problem, the Setting, and the necessary science that is involved. This gets the students grounded in what they are to be doing for the next couple of hours.
2. **Unfreeze.** Considering the emotional states and preconceived ideas of the students, an "unfreeze" may be required to clear away their biases and resistive states and get them receptive for learning. This may be a simple discussion or video presentation showing what is wrong with the status quo. More powerfully, in the example of our Dental Emergency, we prefer to run each student group through a crisis event that shows them conclusively that their current procedures cannot work [8]. This must be done carefully to avoid an overly defensive response.
3. **Acclimate.** The unfrozen student is more receptive and also more vulnerable than previously, having had some of their defences stripped away. To ease them into a simulation, we spend some time introducing them to the equipment and people involved, to get them used to working with the unfamiliar environment. Once the strangeness has been removed as a distraction, we can proceed with the learning exercise itself.
4. **Associate.** As mentioned previously in "Neurological," it is important to get those "Ah-Ha!" moments supported

by the right emotional and physical stimuli. That is one of the reasons that there are only three outcomes in the planning sheets shown in Worksheet 4. Care must be taken to make sure that the students will get the right messages and not be diverted by less useful conclusions.

## Procedure

The training procedure itself is a fairly complex construction that requires expertise in the content and real-life applications, to be taught properly. The components of this procedure itself are divided into "*Frames*" using task Analysis. These are discrete chunks of time that demonstrate a certain element of the problem. These are described in more detail in Section 71.5 "F is for Frames."

The Task Analysis used to resolve the details can employ a different method every time depending on the nature of the procedure being taught. Notwithstanding the wide range of methodology options, the sequence is often not linear. The trainer will often address critical concepts and events relating to the problem *before tracing the series of steps that lead to them.* As shown below, this can take the form of complex, or reiterative process models.

Simulation technologies possess Characteristics that support the effectiveness of training procedures. These are described in more detail in Section 71.6 "S is for Simulation."

## Post-procedure

The Post-procedure Review or After Action Review session (also called debriefing) is regarded as the most important stage of the training procedure. Experienced trainers in the aviation, defence, and medical industries hold that 70–80 % of the learning and association takes place in this session (Worksheet 5).

A great debriefing checklist is provided by the mnemonic GREAT shown in Table 71.2 [2].

**TABLE 71.2**   Debriefing checklist

| | |
|---|---|
| G | Guidelines: the facilitator must have the most recent best-evidence guidelines and local policy guidelines for managing the situation being simulated. |
| R | Recommendations: if comprehensive guidelines are not available, then recommendations such as those contained in published reviews should be used. |
| E | Events: participants are given time to reflect on the simulation and identify what were the important events. |
| A | Analysis: were signs identified promptly, how did treatment given compare with guidelines and recommendations, were resources used effectively, etc.? |
| T | Transfer of knowledge to clinical practice; what has been learned to improve patient care? What are the key "take-home" messages? |

The "placemats" come into play again here. On the reverse of Worksheet 4, we print Worksheet 5 (laminated, remember) and use this to deliver our GREAT feedback in an quick and timely manner.

Other proven processes include manufacturing practices like the Toyota Production System that are being employed with great effect in clinical institutions around the world [3]. Practices that involve instant postevent review, and empower teams to make changes to prevent such problems, have been credited with 90 % reductions in the specific errors targeted, especially hard-to-simulate teamwork errors. Analysis and understanding of "Kaizen" and "Total Quality' procedures already mastered in business can have huge benefits to clinical staff and trainers alike.

In the Curriculum Procedures Frames Simulation model, a series of Capabilities are set aside that support trainers in the Post-procedure stage of training. These are described in more detail in Section 71.6 "S is for Simulation."

## 71.5 F is for Frames

Planner – OK – now we're getting technical. You probably should know how these things get put together, and how to determine if they actually result in learned competence or not. For the day-to-day workflow, however, you'll probably let the Facilitators and Simulation Professionals worry about how to get this done.

Facilitator – these are your tools of trade, should you wish to use them. Simple application of these concepts will help you lead and observe students in your sessions. With mastery, you'll be able to use them to deliver more powerful and efficient learning experiences with better results.

Simulation Professional – these concepts can help you "layer" your technologies, methods, and innovative concepts to deliver truly effective learning experiences. Being aware of how all the elements can interact to reinforce learning can yield startling results. You should be a master of these concepts and how to use them.

A Procedure is more than just a series of tasks, linear or nonlinear. Better to consider them as a series of "Frames" – like in a movie. Each Frame consists of:

- Environment – the "set," the room, the props, and surrounding gear
- Actors – the patient people you have to deal with
- Tools – your tools of the trade
- Processes – the scripts that the different players are working to
- Human skills – the "soft" skills that you need your student to demonstrate

**Worksheet 4**   Simulation training session map

### TRAINING SESSION MAP

DATE   . . . . . . . . . . . . . . . . . . . . . . . . . . . . . . . . . . . . . . .

SETTING   . . . . . . . . . . . . . . . . . . . . . . . . . . . . . . . . . . . .

AT COMPLETION OF THIS SESSION, THE PARTICIPANT SHOULD BE CAPABLE OF:

1. Prioritizing elements of emergency patient care to provide best outcome

. . . . . . . . . . . . . . . . . . . . . . . . . . . . . . . . . . . . . . . . . . . . . . . . . . . . . . . . . . . . . . . . .

2. Initiating motion restriction for SCI as soon as appropriate

. . . . . . . . . . . . . . . . . . . . . . . . . . . . . . . . . . . . . . . . . . . . . . . . . . . . . . . . . . . . . . . . .

3. Using leadership/teamwork skills to ensure effective patient care being delivered

| SPECIFIC LEARNING OUTCOME | TEACHING PLAN | ACTIVITY PLAN | DURATION OF SESSION | PRESENTER | RESOURCES REQUIRED |
|---|---|---|---|---|---|
| 1. Need to prioritize treatment, even delaying motion restriction | Victim may have SCI but needs to breathe now. Here jaw thrust = breathing | Fall from ladder. Victim on side, not breathing. Student first on scene – SimMan not breathing. | 3–5 mins | | Ladder, paint can, backdrop. Dressed SimMan |
| 2. Distinct decision points for treating as Spinal Cord Injury | Must initiate motion restriction now and look for other injury. Call EMS | SimMan starts moaning. Help arrives but they want to move Simulator into "coma" position. | | | Helper |
| 3. Good leadership/teamwork skills are needed for optimal patient care | Must be assertive and explain plan and why. | Helpers question plan but will follow instructions for log roll onto stretcher. | 7–15 mins  Total time: 15–20 mins | | Aids for motion restriction arrive with another helper. |

MAP DEVELOPED BY VAL FOLLOWS, FLINDERS UNIVERSITY CLINICAL SKILLS TRAINING CENTER [7]

**Worksheet 5** Simulation training map filled in and used for feedback

POSTACTIVITY FEEDBACK/DEBRIEFING

DATE .........

NAME OF PARTICIPANT/GROUP.........

SESSION TYPE .........

LEARNING OUTCOMES

1. Prioritizing elements of emergency patient care to provide best outcome

----------

2. Initiating motion restriction for Suspected Cervical Injury (SCI) as soon as appropriate

----------

3. Using leadership/teamwork skills to ensure effective patient care being delivered

| NOTES ON PERFORMANCE | LOG TIME | ACHIEVED | NOT ACHIEVED | PLANNED FEEDBACK |
|---|---|---|---|---|
| 1. Student overly concentrates on making scene safe (moves ladder, paint can, etc.) and on presumed SCI | | | 1<br>2 not really achieved | Review guidelines for immediate trauma care. No breathing will lead to severe injury whether or not SCI is present. Emergency care is dynamic and priorities can change. |
| 2. Excellent communication, organization and teamwork skills displayed | 4'40" | 3 achieved | | Good demonstration of this when Tim went to move SimMan. Use as model demonstration to ensure that in group, they know how to function at this level. |
| 3. Had trouble with adjustable C-collar | 7'05' | | | Remind group that formal teaching shows them examples of equipment and they need to seek out additional practice opportunities for comprehensive training. |

MAP DEVELOPED BY VAL FOLLOWS, FLINDERS UNIVERSITY CLINICAL SKILLS TRAINING CENTER

- Technical skills – the actual skills you are trying to teach/test/demonstrate
- Concepts – the concepts and understanding behind everything happening in the Frame

These should be considered as layers, as shown in Figure 71.4, to be determined and manipulated to obtain the ideal outcome. Most simulation technologies and methods will address some of these in different ways. Knowing where they fit in helps you assess their effectiveness.

For example, in our Emergency C-spine situation, we can articulate the various interacting layers as shown in Table 71.3.

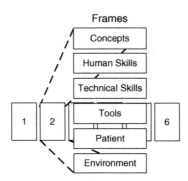

**Figure 71.4**   Layers in Frames.

**TABLE 71.3**   Layers in a particular frame

| Frame | 1. Patient is lying down, immobile, and unresponsive |
|---|---|
| Concepts | What are the essential Basic Life Support steps? Will the student recognize the signs of C-spine injury and act accordingly? |
| Technical skills | Successfully provide motion restriction. |
| Human skills | Be able to manage the patient and a team to coordinate BLS and stabilization in what is a delicate and stressful procedure. |
| Processes | Don't get distracted from the BLS guidelines while ensuring that the stabilization procedures are followed carefully. |
| Tools | The C-collar, other motion restriction procedures and tools. |
| Actors | Patient (mannequin) goes limp and utters a moan, or if a SP starts having a convulsion ("fitting"). |
| Environment | People need to be moved from floor, beds, cars, ladders, etc. There are many standard BLS dangers and specific challenges for stabilization that can be delivered with clever prop usage – but don't overdo it. |

### 71.5.1  Concepts

The underlying concepts of a clinical training procedure provide the starting points for the training design. These include

the procedural and scientific elements of the problem and provide the clinical inputs and outputs that the trainee's skills must handle and produce.

### 71.5.2  Technical Skills

The technical skills are the primary focus of most training sessions and students. The ability to isolate these skills is the greatest challenge of the clinical trainer. Proven methods for teaching these critical skills involve varying the degree of difficulty and conditions of the procedure. This forces the student to focus on the basics of the procedure rather than "learning the simulation."

### 71.5.3  Human Skills

"Analyses of lawsuits show that there are a lot of highly skilled doctors who get sued a lot and doctors who make lots of mistakes and never get sued. At the same time, the overwhelming number of people who suffer an injury due to the negligence of a doctor never file a malpractice suit at all. In other words, patients don't file lawsuits because they've been harmed by shoddy medical care. Patients file lawsuits because they've been harmed by shoddy medical care and *something else* happens. What is that something else? It's how they were treated, on a personal level, by their doctor." [4]

### 71.5.4  Processes

Processes are generally recognized as an integral part of clinical training. Poor team processes or the lack of "Crisis Resource Management" is the source of the majority of errors as described by the often-cited "To Err is Human" [5] and other reports. The solutions to these are often hard to simulate and separate from the day-to-day rush of patient care.

Process mapping methods and critical disciplines such as Kaizen or the Toyota Production System can have a huge impact here as well. For example, a major predictor in Dental emergencies is the patient's history. Unfortunately, this history is often recorded in the patient's file and never referred to again, until it is too late. One approach to address this is the following simple process:

1. Patient makes a booking and arrives at the practice.
2. The Receptionist checks the patient's records and affixes a color-coded paper clip to the patient's file before the Dentist receives it. Green means no risk factors, red means heart or circulation conditions, blue means breathing, and yellow means allergies.
3. If there is no clip, the Dentist must go out to collect it before acting on or treating the patient. If he/she ignores the clip, then he or she is breaking written procedures and is culpable – a great incentive to following the process.

4. If there is an emergency, there is a bright red, blue, or yellow card hanging conspicuously with the response procedure printed clearly on it to be followed. This helps in the event that the adrenalin response causes indecision.

5. The receptionist has a similar card, and with a shout of "Emergency – Red" from the dentist can immediately make the emergency call using the script on the card and help the dental technician find the oxygen tank and mask, which is also indicated on the card.

Not only do such processes help in emergencies, they force discipline and provide comfort and confidence for the trainees in "peacetime," that when called, they will know what to do.

### 71.5.5 Tools

The tools available to the trainee include the devices of the trade.

### 71.5.6 Actors

The actors in a frame are primarily the trainee and the patient, with influence from the team, if a larger simulation is being conducted. In a training scenario, the patient is often the main application of simulation. This can be in the form of a device in part-task training, a mannequin in high- or low-fidelity simulation, or a standardized patient in interpersonal skills training.

### 71.5.7 Environment

As with Processes above, the working environment exerts a powerful influence on the trainees and how they act in the future. This is why it is so important to match the training to the actual day-to-day environment of the practitioner (Figure 71.5). This is part of the neurological association mentioned previously. Any influence upon performance, whether it be from the environment, personalities, or intended training devices, becomes part of the curriculum by virtue of their influence on behaviors.

Replication of the rooms, street scapes, noises, and dangers of the real-life workplace are essential, but only when the students are ready for their inclusion. Also useful are the disciplines that can be employed to address these dangers and turn the environment into a reinforcement of good skills.

Again the Japanese have pioneered *disciplines* employed with great success in business. The "Kaizen" principle of 5S is a *discipline* that transforms cluttered, inefficient workplaces into clean, tidy, and organized powerhouses.

Figure 71.5 maps the environmental factors to the five stages of Procedures.

1. Planning for the roles the team and facilities must play,
2. Preparing the equipment and supplies,
3. Demonstrating the impact of environment on the procedure,
4. Reviewing the impacts and making plans to change,
5. Implementing the changes in the Post-procedure stages, and later
6. Following up to make sure the changes have endured and were effective.

## 71.6 S is for Simulation

Planner – This is going to make your job easier. Assuming you have resource restrictions to reach your training goals, it would help to be able to put these expensive simulation technologies in context, understand how they might work together, and what it might be worth to you.

Facilitator – you are required to fit these devices into your training experiences. The framework here should help you work out what does what, and how they might be used together. Clarity is power, and you can use that power in your teaching if you are clear about your tools.

Simulation Professional – many is the time that a much anticipated new simulation technology acquisition has turned out to be a failure, because you didn't understand exactly what you needed it to do. You can avoid this, and make it easier to justify your future plans financially, by being able to translate your technologies into learning outcomes. While this section "S is for Simulation" is high level, it should provide a reusable tool that you can refine to better measure and value your simulations.

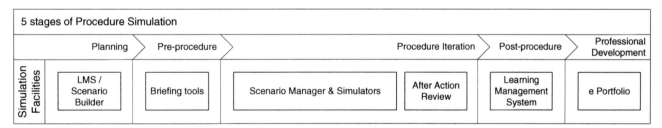

**Figure 71.5** Environmental factors and changes.

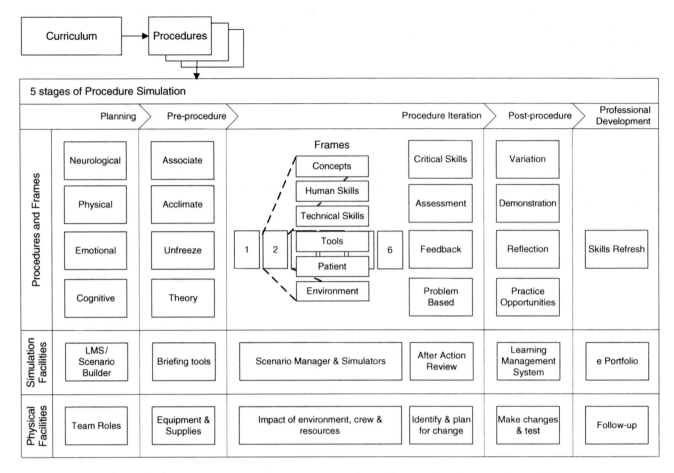

**Figure 71.6**   Curriculum Procedures Frames Simulation framework with supporting Facilities.

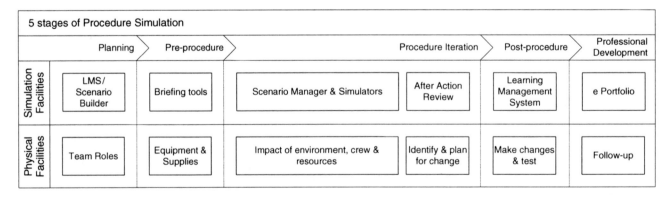

**Figure 71.7**   Simulation facility functional areas.

Simulation is a special Procedure that is becoming increasingly accepted as an effective method for teaching hard-to-learn and high-risk competencies. In the Curriculum Procedures Frames Simulation framework, simulation technologies are integral to the procedures, frames, and resources that deliver them.

The horizontal layers of the framework shown in Figure 71.6 are ordered with Curriculum resolved into Procedures. Each specific Procedure is resolved into the five stages, all starting with "P" and forming the five columns of the diagram. In the core, the Frames are shown, and to the right are the Characteristics and Capabilities of Simulation technologies.

The two facility layers of Environment (Physical Facilities) and Simulation are highlighted in Figure 71.7. The Environment layer has been described previously, but the Simulation layer needs discussion and linking to the Characteristics and Capabilities that support your simulations. Read on.

The hugely confusing range of clinical simulation technologies can be mapped in a way that is helpful when trying to put them in context and classify their functions. The first mapping is to where they lie in the continuum of the 5Ps of Procedures.

1. **Learning Management Systems (LMS).** With much of the focus on high-tech and partial-task patient simulators, it is easy to miss the role of Learning Management Systems (LMS). These systems essentially function as a database of Curriculum requirements and the Procedures that comprise them.
2. **Scenario Builders** or Content creators are a wide range of applications that can develop rich computer-based content for interactive use (or printed and used as hard copy), all the way to dedicated scenario creators for specific simulation devices. We can download new procedures for use on SimMan™ full-body and Harvey™ partial-task trainer. We can also exchange content packages between Learning Management Systems by use of Shareable Content Object Reference Model (SCORM) compliant tools. SCORM is a set of standards and tools allowing the exchange and sharing of training content between various systems. This reduces the cost and duplication of effort in developing training content. At present, the current version of SCORM does not support interactive simulation, but this is being addressed.
3. **Briefing Tools** are those that help in the preparation of trainees for the training session. These can range from our laminated "placemats" to DVD and videos of previous sessions. More complex tools incorporate Virtual Reality or 3-D anatomical models and animations to demonstrate the clinical concepts involved in a problem.
4. **Scenario Manager/Simulation Professional.** The training scenario itself often involves a Simulation Professional and in some cases – for complex "high-fidelity" simulations – a scenario manager. Such systems are often inte-

grated with audio-video equipment and provide results that can be reviewed in the After Action Review.
5. **After Action Review**, the Post-procedure stage, or debriefing is the essential stage of simulation training. Most technologies here use simple video recordings with a variety of ways to flag and return to specific places in a recording that can be used to quickly move to any point in the training session to point out a useful lesson.
6. **Learning Management Systems** (again!). All of these results can sometimes be output to and stored for later recovery in our Learning Management System. These systems might provide an enduring record for a student's results to be reviewed over the passage of several years – very useful for a supervising professor to spot trends in performance or glean indicators of positive or negative attitudes. Some links may exist to University Executive Information Systems (human resources, records, etc.).
7. **Portfolio.** A practitioner's professional record is the subject of current discussion about having the capability to track a student's portfolio of competencies. This is gaining acceptance internationally and will likely become an important consideration in the measurement, tracking, and reporting of simulation outcomes in the future.

Having put the technologies in context, we now must determine how useful they are for our purpose. A useful tool is to use the Validation Methodology for Medical Simulation Training Committee Educational Framework [6] adapted below:

1. Assess the needs of your simulation against the VMAS framework.
2. Consider how the different elements of your simulation facilities and training practices contribute to these elements.

The Validation Methodology for Medical Simulation Training framework was proposed as guide to incorporating simulation effectively in clinical training. We have adapted it to use it as a guide to the functionalities that our simulations should incorporate.

In Figure 71.8, you will see the Validation Methodology for Medical Simulation Training elements listed under Simulation Characteristics and Capabilities. Of these, Curriculum Integration is actually the starting point of the Curriculum Procedures Frames Simulation framework, which is covered earlier in Section 71.3 "C is for Curriculum."

The reason for the Characteristics/Capabilities split is:

- **Characteristics** tend to be delivered by direct configurations or technical features of the simulation technology. Problem-based characteristics are the realism of shape

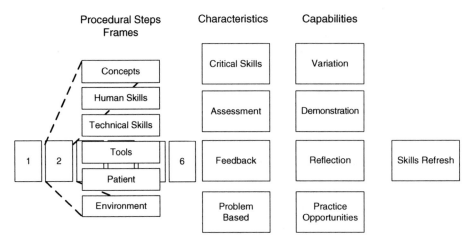

**Figure 71.8**   Simulation context, characteristics, and capabilities.

and feel, where Feedback is delivered by realistic clinical and physical response to a student's actions. With Assessment, sensors can indicate success or failure and so guide a student's acquisition of Critical Skills.

- **Capabilities** are generally one step removed from the physical simulation, and are delivered by the Scenario Controller or After Action Review (debriefing) functions. Practice Opportunities are a function of repeated iterations of a range of simulations, as is Variation. Demonstration and Reflection map the student's actions against critical criteria to both demonstrate and better understand why things went better or worse.

These elements are useful in providing a classification of most technological functions. Table 71.4 explores each one in more detail, and is a useful checklist to see if you've forgotten something in your simulation design.

**TABLE 71.4**   Clinical simulation worksheet – Simulation requirements checklist using the Validation Methodology for Medical Simulation framework

| |
|---|
| Application: C-spine stabilization using standard mannequin |
| ADAPTED FROM THE "Validation Methodology for Medical Simulation Training" FRAMEWORK BY KAY HOWELL & GERALD HIGGINS (2004) |
| **Curriculum Integration.** Clinical training expertise is essential to create new and modify existing curriculum to take full advantage of simulation technology.<br>  Integration to Learning Management Systems for student performance against curriculum standards<br>  Curriculum should be built with the simulation in mind, rather than trying to fit simulation into existing curriculum<br>  Integration with Learning Management Systems for curriculum management |
| **Skills Refresh.** The simulation should be suitable for skilled practitioners to use for skills maintenance and Continuous Professional Development.<br>  Accessibility for professionals<br>  Tracking of required standards against curriculum elements<br>  Reporting of compliance/noncompliance against standards<br>  Privacy and data protection<br>  Adjustment for skill differences between trainee and out-of-practice "expert" practitioners |
| **Practice Opportunities.** The students must be able to practice to get enough experience to fully absorb the concepts and skills being delivered.<br>  The learning environment should include many opportunities for practicing performance.<br>  Learners should practice simple to complex versions of a whole task,<br>  with instructional methods that promote just-in-time information presentation to support the recurrent aspects of the whole task.<br>  While at the same time, instructional methods that promote elaboration should be used to support the nonrecurrent aspects of the task. |

**Reflection.** The students should be able to "explore" the simulation in flexible ways to gain full understanding of its concepts and the consequences of their actions.

Research shows that training learners to self-explain examples consistently improves learning outcomes.

Learners' thinking should be made visible through discussions, text, or tests,

and feedback must be provided.

Learning is promoted when the instruction demonstrates what is to be learned, rather than just telling about what is to be learned.

**Demonstration.** The simulation must allow students to demonstrate their understanding and skill levels to an observing instructor and/or another student.

Demonstrations should be consistent with the learning goal.

Should include varied examples concepts,

Demonstration for procedures,

Visualizations/simulations for procedures,

And modeling for behavior.

**Varied and contrasting examples.** The simulation must explore the critical concepts through a range of examples, forcing students to grasp the underlying principles rather than just learning the simulation.

Examples that look different on the surface but that illustrate the same guidelines.

Help the learner understand which features are relevant or irrelevant to a particular concept.

A sequence of problems

Demonstration and application are integrated as a whole rather than as distinct parts.

**Critical Skills.** The simulation must provide a focus on the critical skills to be transferred and a clear conceptual framework supporting why. Different levels of skill must be met with different levels of difficulty.

Identify expert-novice skill differences

Define skills within hierarchy of constituent skills and problems

Highlight the Key steps

Incorporate skills hierarchy for demonstration, practice, and assessment

Identify general skills required

**Assessment.** The achievement or failure to demonstrate competence must be clearly measured and reported against the clinical criteria.

Use Task Analysis techniques for evaluation process

Define key steps in performance of the clinical procedures

Define the underlying skills that enable trainees to perform those procedures

Formulate appropriate test scenarios

Related performance measures

Verify the simulation, which provides a fair assessment

Verify the simulation provides effective training

Scenarios to provide complete coverage of critical skills and areas of interest

Provide feedback to the Learning Management Systems to enable future improvements to simulation

**Feedback.** Immediate, accurate feedback is required for the students to learn effectively from their actions.

Allow revision of thinking during task

Provide continuous feedback

Provide unintrusive feedback as part of instruction

**Problem-centred learning.** The simulation needs to immerse students in a scenario that realistically explores the concept to be learned. It needs to be recognized as authentic – a student needs to feel that it is a skill that might be needed tomorrow.

Use Cases to define Learning Issues

Use Cognitive Task Analysis to derive problems and solutions

Use simulation

Incorporate Coaching

Provide Maps for learners to review their problem solving

Allow comparison with "expert" maps

## 71.7 Conclusion

The Curriculum Procedures Frames Simulation framework was developed as a tool to allow the authors to link their respective specialities and provide a common method for organizing the new developments in simulation technologies. Of greatest importance was the need to link the high-level Curriculum needs to a process for planning and evaluating technologies. So much time and resource can be wasted if this process is not clearly understood.

### 71.7.1 Re-freeze!

In change management methodologies, the term "re-freeze" is used to refer to the process of locking in useful changes to your day-to-day operations. You have "unfrozen" your methods, learned how to change them, and are now prepared to make the changes permanent by re-freezing them in their new shapes.

Try applying the four steps to some of your current simulation processes:

(C) Curriculum. Be able to look at the requirements of your students and divide them into procedural elements that can be taught.

(P) Procedures. Understand the five Ps of Planning, Pre-procedure, Procedure, Post-procedure, and Professional development and the range of tools available to organize them into effective teaching units.

(F) Frames. Be able to identify and use the interrelationships between the six elements of each training "Frame" and use your resources to best effect.

(S) Simulation. Put simulation technologies, techniques, and outcomes clearly into context with a method to objectively assess their Characteristics and Capabilities.

**Share your results!**

Early drafts of the framework have been successfully tested with applications in mental health as well as driver training in the transport industry; so we hope that the overall tool is robust and flexible enough for your applications. Further discussion, examples, and resources are welcome, and can be shared online at http://www.paulwilliamson.net/cpfs.

## References

The following resource is offered as a useful breakdown of simulator characteristics.

Rosenthal, E. and Owen, H. An Assessment of Small Simulators Used to Teach Basic Airway Management. *Anaesthesia and Intensive Care* 32(1) February (2004).

1. Jonassen, D. H., Tessmer, M., and Hannum, W. H. *Task Analysis Methods for Instructional Design*. LEA Publishers, London, 1999, pp. 18–21.
2. Owen, H. and Follows, V. *GREAT simulation debriefing*. Medical Education, Blackwell Publishing, 2006, p. 1.
3. Spear, S. *Fixing Healthcare from the Inside*. Harvard Business Review, September 2005, pp. 78–91 Available from Harvard Business Review On-Point Product Number:1738.
4. Gladwell, M. *Blink*: *The Power of Thinking Without Thinking*, Penguin, 2005, ISBN 0-7139–9844-X.
5. Kohn, L. T, Corrigan, J. M., and Donaldson. M. S. (eds.), Committee on Quality of Health Care in America, Institute of Medicine – To Err Is Human: Building a Safer Health System. ISBN: 0-309-51563-7, 312 pages (2000). This PDF is available from the Academies Press at: http://www.nap.edu/catalog/9728.html.
6. Howell, K. and Higgins. VMAS – A Proposed Educational Framework. Federation of American Scientists. 2004 Whitepaper – The VMAS Educational Framework was developed as part of the work of the Validation Methodology for Medical Simulation Training (VMAS) Committee for the Telemedicine Advanced Technology Research Center (TATRC) of the US Army Medical and Materiel Command.
7. Follows, V. Training Session Map. Flinders University Clinical Simulation Unit in collaboration with Flinders Medical Center, OTS Educational Consultant, 2006. Instructors' tools and templates developed for internal use.
8. Owen, H. Emergencies in Dental Practice. Being Cool, Calm and Prepared. Flinders University Clinical Skills Training Unit, Flinders University, 2006. Course instruction notes.

# XIX

# Tricks of the Trade

*"In real life, unlike in Shakespeare, the sweetness of the rose depends upon the name it bears. Things are not only what they are. They are, in very important respects, what they seem to be."*

Hubert H. Humphrey

If it were not for our capacity to place importance on what things seem to be, to use, and be used by symbols (e.g., letters and numbers), we would still be functioning at the unicellular level. Simulation works because we are willing to be persuaded to modify our behaviors, based upon our acceptance of *what seems*, in place of *what is*. The better simulation producers recognize this premise, accept it, embrace it, and use it to the fullest of their abilities. However, wielding such magic-like power requires responsibility: responsibility that we don't become seduced by our abilities to manipulate the senses of others, or overly enamored with our acquisition of tools and tricks we employ for doing so.

We expose our clinical students to simulation to help them improve their clinical behaviors, not to improve their clinical simulation behaviors. This is a crucial difference, one that we should always remain mindful of, one that defines boundaries for our imagination, one that provides a guide for our goals. For example, Shakespeare's plays have been recast in a wide assortment of historical times, with costumes and settings coincident with the different eras. His words have been translated in many different languages. In each case, these modifications were made to help the audience accept the story. Yet, left unmodified was the essence that we know of as Shakespeare: the stories themselves. Our task as producers, directors, cast, and crew of clinical theatrics is to use whatever will help our students learn the perennial essence of our story – our lessons. Without a clear story agreed to by all participants, all the fancy stagecraft in the world will come to naught. Each of you must create your own story; it is the inner skeleton that anchors all inner motives and outer appearances. Once you have your story, the sky is the limit to how you choose to present it, and you should freely consider employing the stagecraft invented and perfected by others. When done well, both the stagecraft used during productions and the tools used to collect and review your productions will be invisible to your students, and all but invisible to your instructors while your story will be very, very obvious to all.

Guillaume Alinier offers numerous examples of how to convert what *are* common items, into what *seems* to be common clinical presentations. Mike Goodrow *et al.* provide an extensive overview of how and where audio/video supports clinical simulation, and include details sufficient to caution everyone that the audio/video domain is about as complex and diffuse as any clinical one. Guillaume Alinier returns with pragmatic advice on how to work with audio/video vendors and their catalogs to create a functional audio/video system that supports *your* production goals within *your* budgets of money, time, and competency.

# 72

# Professional Stage Craft: How to Create Simulated Clinical Environments Out of Smoke and Mirrors

Guillaume Alinier

## 72.1 Introduction and Disclaimer

This chapter presents a number of general-purpose ideas, tips, and tricks that can be adapted easily for use in your simulation scenarios to make them more realistic and engaging. They are all nonspecific to any particular make or model of robotic patient simulator, and many can even be applied to human actors and standardized patients. They include basic audio and video systems, control room configuration, central line insertion, venous pressure monitoring, fluid output, hemorrhaging, makeup for simulation of injuries and bruising, and lastly, safe fluid suctioning from within wounds. Each suggestion has been built, tested, adjusted, used, refined, and are offered with no reservations about their safety other than where noted.

Please note that the author of this chapter, the editors, and the publisher do not accept liability for any damage or injury you might cause to your self, your students, colleagues or friends, your patient simulators, other equipment, your facilities, or reputations as a result of the installation and application of any of the described. Be aware that some might also affect the manufacturer warranty of your patient simulator or other warranty-covered items.

## 72.2 Fundamentals of Fluids

Real patients generate messy and smelly fluids. The more and the messier, the worse off the patients usually are. Students and teachers alike would benefit from their simulated patients' ability to reproduce these crucial signs of pathology associated with such fluids. Simulator operators can easily make their mannequins produce very accurate renditions of real human fluids, but most likely only once. Subsequent cleanup and repair from such realism can be very costly in lost time and ruined equipment.

The ideal approach is to create a circuit, which is as closed as possible between the source of fluids and the collection unit, but at the same time presents clinically accurate signs, smells, and sounds of the fluid's type and amount in ways that your students will grasp the meaning of. Although electronic pressure transducers are more frequently used and water column manometers are slowly disappearing, it is always beneficial to expose and introduce trainees to the fundamentals of instrumentation as it can help them understand underlying principles and units of physiological measurements (Figure 72.1) [1].

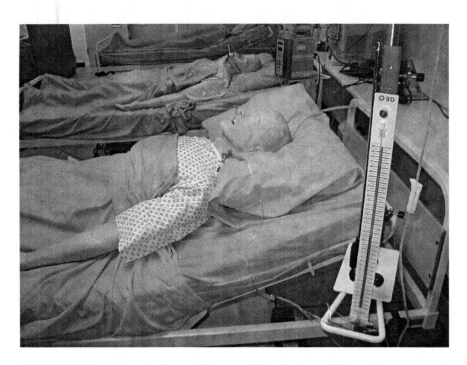

**FIGURE 72.1**   Patient simulator with a central line for Central Venous Pressure monitoring [1]. The equipment required to make this adaptation is identical to what would be used on a real patient from the manometer to the central line.

However, some additional tubing and a syringe are required to control the water level on the manometer from the control room (Figure 72.2).

The internal diameter of this long additional tubing should approximately coincide with the outer diameter of the CVP catheter so that it can be partly fed inside the former and glued with an appropriate adhesive, e.g., LockTite #201. Make sure *all* the lumens of the catheter are within the tube. This portion of the catheter should be hidden under the patient's neck skin while the remainder can protrude as it would on a real patient. The rest of the tubing should go under the mannequin's chest skin and exit through a discreet location, and then installed

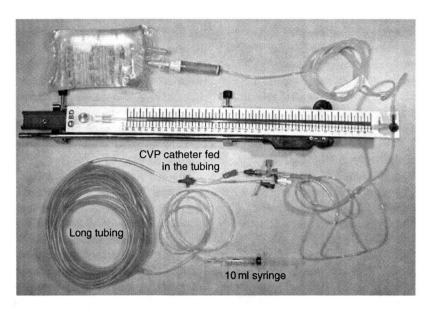

**FIGURE 72.2**   Central Venous Pressure monitoring equipment required for the patient simulator adaptation [1].

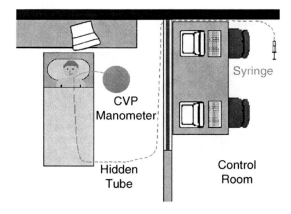

**FIGURE 72.3** Setup of the Central Venous Pressure adaptation in the simulation center. Tubing between the patient and the control room installed below a raised floor [1].

out of the way of foot traffic to reach the control room. In the control room, connect a syringe to this tubing with which to control remotely the CVP reading on the manometer at the bedside (Figure 72.3).

You should make sure that your circuit does not leak, even under a maximal pressure, as fluids could result in damage to the electronics of your patient simulator or become a slip hazard. Fill the circuit with clear liquid such as filtered water that has a fungicide added. A small amount of common laboratory alcohol prevents the growth of ugly, dark contamination, while it is compatible with most tubing and does not form bubbles like detergents do.

For greater accuracy and easier control of the water level in the manometer, it is preferable to use a small syringe (e.g., 10 ml) and stiff wall (poly-propylene) tubing. If you notice undesired variations of water level in the CVP manometer, you simply need to put a clamp near the syringe to stop the fluctuations when required. Alternatively, instead of a syringe to change the height of the column of water, use an almost full IV bag. To change the pressure presented by the patient, just change the height of the bag in the control room. The operator should have a view (direct or via video camera) of the side of the manometer scale to appropriately set the water level during the scenario, or another identical manometer located in the control area. The use of colored markings on the side of the manometer scale makes it easier to see and to set the water level. It is preferable to immobilize the manometer at the bedside so that trainees are not inclined to move it or turn it away from the sight of the operator.

This adaptation is easy to install and remove from the patient simulator, as once in place, you can simply purge the tubing, disconnect the manometer line going to the catheter, and store the latter tubing inside the chest of the patient simulator. Overall, this CVP line setup is inexpensive, very realistic, and can be adapted to any patient simulator with a removable chest or neck skin. It can really add to the realism of a scenario.

## 72.3 Hemorrhage

You might want to be able to make your patients bleed in a controlled manner to produce more realistic trauma, intraoperative and postoperative scenarios. For trauma scenarios, you might have opted to purchase optional wound modules that allow for the direct connection of fluid reservoirs to the wound sites. Alternatively, you can create your own wounds, as explained later in this chapter, with some integrated invisible (or at least thin and discrete) tubing to enable the bleeding to occur where and when required. This is also valid for the simulation of an intraoperative or even postpartum hemorrhage. In all cases, you need to carefully consider where the fluid will flow and collect, especially when your trainees do not appropriately manage the bleeding, in order not to damage your patient simulator or create a slip hazard. The tubing should have an open end at the wound site(s) while the other end should be connected to a bag of simulated blood situated in your control room. This connecting tubing should be hidden under the floor or in a floor gutter in order not to confuse your trainees and prevent any trips and falls.

For slow hemorrhage rates, raise the source bag above the wound sites on the patient and gravity will provide the motive force. When you want the hemorrhage to cease, lower the bag below the wound sites. For faster flows, place the source bag within an IV pressure bag, and inflate as necessary to produce the desired hemorrhage rate (Figure 72.4).

For a postoperative bleeding scenario, you do not require any artistic skills. Instead of having your tubing connecting to a wound site, you can simply connect it directly to a wound drain. If it is a vacuum drain, this should be pierced at the top to allow the air to escape for free filling with the simulated blood. In addition, to make it more realistic, you need to thread a wound catheter from the bandaged operation site and tape it to the drain inlet where your tubing enters. This simulated and bandaged wound does not need to be connected for real as no fluid will actually flow through it. It is simply used to make trainees believe that the blood is actually coming from the operation site. In this configuration, you have two tubes going to the vacuum drain. Similarly as with the trauma tubing for the simulated blood, it should be hidden from the trainees. This should be quite easy as the tubing does not need to go to the patient simulator, but simply from the control room to under the bed where it can directly connect to the drain, which is usually attached on the bed frame.

In the case of a postoperative hemorrhage for a patient with a wound drain, the risks of potentially damaging your patient simulator are minimal. Overall, the setup is derived from the same philosophy as shown in Figure 72.3, with the syringe being replaced by a bag of simulated blood inside a pressure bag as shown in Figure 72.4. The proposed wound drain arrangement is shown in Figure 72.5. It is preferable for the drain to be positioned in such a way that the operator can observe the level of blood in the drain to appropriately control the apparent hemorrhage using the pressure bag.

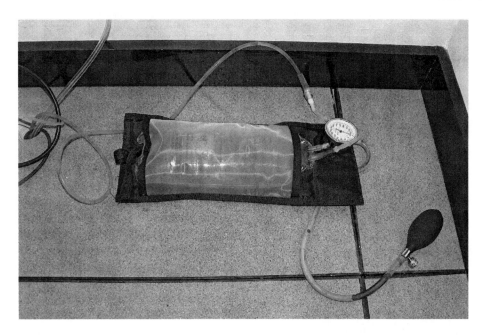

**FIGURE 72.4**   Pressure bag in the control room to remotely control the hemorrhage rate [2].

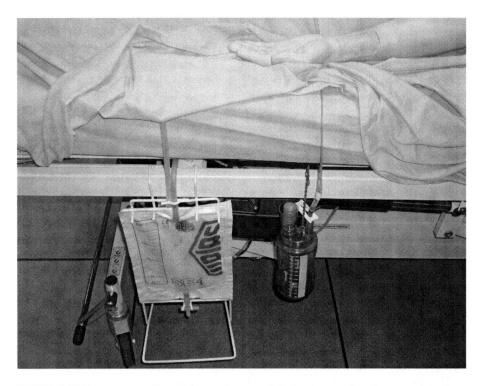

**FIGURE 72.5**   Urinary collection bag and a wound drain connected to their respective bags of colored fluids and attached to the side of the patient's bed.

For perioperative hemorrhage, the sights and sounds of blood suctioned from the surgical site can be created easily and safely for the mannequin. Hide a container of red fluid under the patient. Install a tube from this container to a site where you want the confederate playing the "surgeon" to suction the fluid. Make sure the lumen of this tube is large enough to accept the distal end of your suction probe. Now, whenever you want participants to become aware of a hemorrhage, a standard and familiar surgical suction system will easily provide all the unmistakable sights and sounds of bloody fluid being suctioned.

## 72.4 Urinary Output

Hospitalized patients often have urinary catheters ("Foley catheters") placed in their bladders, and it should be realistically rendered during scenarios presented with patient simulators where urine output is an important teaching parameter such as monitoring fluid output on peri- and post-operative patients. This simple adaptation will enable you to produce an easily controllable urinary output/collection system.

The minimal equipment required includes a urinary catheter, a bag of colored fluid, a basic peristaltic volumetric infusion pump, some tubing, a few connectors, and a urinary

collection bag. As with the adaptation for hemorrhage, it can be very safe and not put the electronics of your patient simulator at risk.

The volumetric infusion pump should be installed in your control room with a bag of simulated urine of the desired color. This tube can be directly connected to the urinary collection bag. The urinary collection bag can be attached to the side of the patient's bed. To make it more realistic, the urinary catheter should be in place and appear to be going to the urinary collection bag (Figure 72.5).

This simple setup allows the operator to precisely control the urinary output using the volumetric infusion pump from the control console during a scenario. This can be adapted to any appropriate patient simulator and even human actors and standardized patients to enhance the realism of a scenario. Alternatively, the infusion pump can be eliminated, and gravity or a hand-powered syringe used to propel the fluid from the reservoir in the control area to the patient's collector bag (Figure 72.6). If several different urine color choices

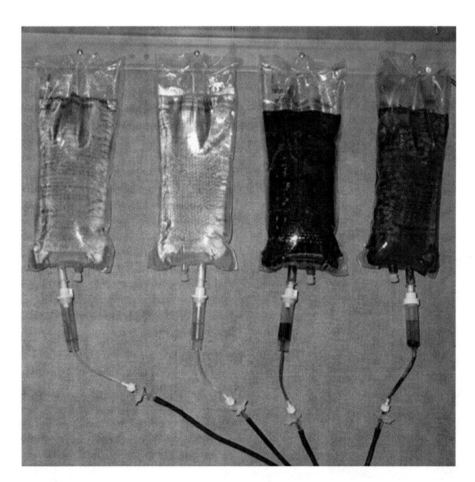

**FIGURE 72.6** Four different colors of "urine" (water with food coloring), each delivered by gravity feed, each flow rate independently controllable, with the resultant color and volume in the clinical collection container determined by the selection(s) made in the control room.

are desired, create one reservoir for each color, and install for each an on/off stopcock and tubing from its reservoir in the control room to the entrance of the collection container. Once all the tubing are prefilled with their respective colored waters, any volume of any given color that leaves its reservoir will immediately appear in the urine collection container.

## 72.5 Fluid Administration

Most patient simulators allow for placement of intravenous catheters for the injection of drugs and fluids. To that effect, they have an intravenous (IV) access arms with rubber tubes under their rubber skins. For scenarios in which trainees are expected to perform appropriate fluid management, it is quite important for them to actually spend the time and their attention while performing the procedures required for gaining IV access in and delivering fluids to their patient. Although the following adaptation will not allow them to have a reliable fluid return when trying to withdraw blood samples, it will however enable them to safely deliver any amount of fluid into the patient simulator.

The simple adaptation involves blocking one end of the under-skin tubing, while extending the other end so that any fluid flowing through it can be collected into a bucket or other type of container hidden below the patient simulator (Figure 72.7).

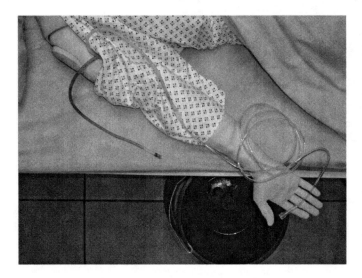

**FIGURE 72.7**   Intravenous fluid administration for your patient simulator (in this case, the bucket would be hidden under the patient's bed) [3].

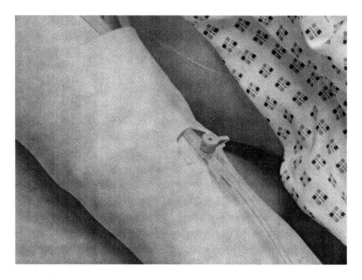

**FIGURE 72.8**   No-leak intravenous access port directly connected via tubing to a waste fluid collection container (e.g., bucket) hidden under the patient.

Unfortunately, the tactile feedback learning that occurs during a cannulation of a rubber blood vessel under rubber skin is the antithesis to that which we want our students to learn and experience while placing a real IV catheter. Thus, many simulation centers do not attach too much importance to their students performing the actual cannulation on the patient simulator. However, your students can still perform all the other essential IV access thoughts and actions (e.g., deciding to gain IV access, collecting all the correct access equipment, opening the packaging and connecting all the access equipment parts). When they are at the instant of actually piercing the skin, reveal to them a no-leak fluid connector permanently installed in the skin (e.g., female Luer port), connected to the waste fluid connection container via tubing hidden within the body (Figure 72.8). This bypass technique enables savings on the maintenance cost of regularly replacing the arm. A limitation, due to the fact that the liquid is not pressurized with the "vein," is that there is no venous return if a blood sample is to be collected.

## 72.6 Makeup for Injuries and Bruising

One of the constraints with patient simulators is that they do not clean themselves autonomously. They are fairly easy to maintain on a regular basis, but more effort is required when it comes to removing makeup used to simulate bruising, for example. A very practical and economical method of making easily removable and reusable bruises and wounds was developed by Chow and colleagues [4]. The following description is based on their work. The material required to produce

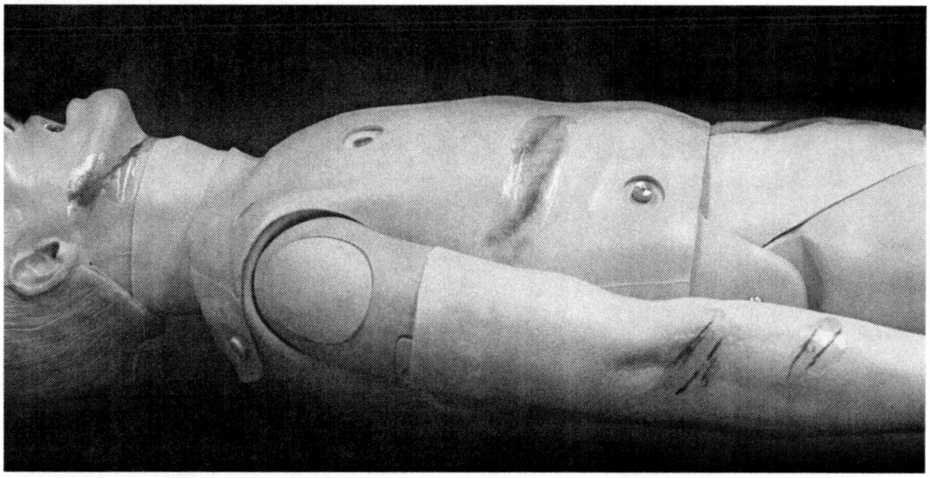

**FIGURE 72.9** Examples of easily removable wounds created by Chow *et al.* [4].

these home-made patches include transparent/clear paint for application on glass, normally used for making "sun catchers," acrylic paint of appropriate colors (red, blue, purple, black, and white), small paint brushes, and letter-size pieces of paper.

To make wounds, bruises, or scratches, you simply apply some of the clear paint on a piece of paper and spread it out to obtain the patch of the desired shape and size. Within 24 hours, this will dry into a transparent base onto which you will be able to paint the bruises or wounds with the required colors. You can create any type of superficial wounds or skin condition. These can then be peeled off the paper and placed onto the patient simulator skin.

Since these wounds, bruises, or scratches are painted onto transparent surfaces, the color of the skin from the patient simulator can be seen through them. This creates a more natural blend and makes the traumas appear more realistic (Figure 72.9). Another big advantage they have is that they are easily removed leaving the mannequin's skin unsoiled as these patches can be peeled off and reused. These visual cues can really help making scenarios more lifelike, and help trainees treat the patient simulator more like a real patient.

## 72.7 Suctioning Vomit or Blood

Owing to the high cost and difficulty of cleanup, it is not advisable to allow any fluids to be poured into the airway of any patient simulators. At the very best, this would create a horrendous cleaning problem to remove such liquids and associated particulate from the mouth, airways, and lungs. At worst, even clear liquids such as pure water could cause irreversible damage to the mechanical lungs and pneumatic equipment. When real patients produce vomit or other substances that need to be rapidly cleared to prevent any airway obstruction, the desired action would be to promptly lower or turn the patient's head to the side and to suction the airway. However, with patient simulators (or live "patients"), even if vomiting noises are produced, no fluids can be withdrawn from the airway. Once again Chow and colleagues [4] have had an innovative idea that allows fluids to *seem* to be suctioned from the airway, as described in the following paragraph.

The material required includes a suction unit with its tubing and a Yankauer suction tip, a 3-way stopcock, some small bore tubing, some IV tubing with a clamp, a bag of fluids for the simulated vomit or blood, and a pressure bag. The idea is to create a closed circuit between the bag of fluids and the suction unit, thus preventing any fluids going into the airway of the patient (Figure 72.10). The innovative trick of this circuit is that a tiny incision needs to be made about 115 cm (4 feet) away from the Yankauer suction tip. Through this incision, you insert about 100 cm of lubricated small bore tubing until it reaches just a few cm (an inch or so) from the end of the Yankauer's tip. Just outside the incision point, the small bore tubing should be fitted with a simple on/off stopcock that is itself connected to the IV tubing attached to the source bag of fluids inside the pressure bag.

In use, an actor should operate this system as the stopcock needs to be opened after the suction has been activated. The pressurized bag of simulated vomit or blood then acts as a self-feeding supply that will flow through the small bore tubing to very near the tip of the Yankauer and then be aspirated to the

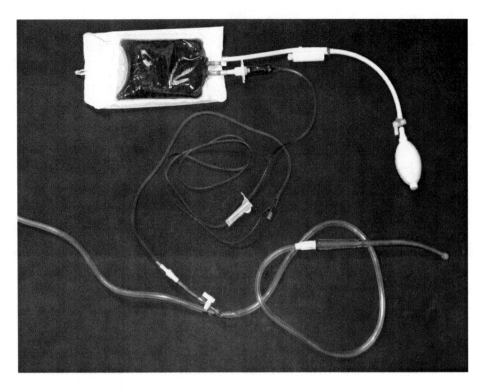

**FIGURE 72.10**   Circuit for blood or vomit suctioning of a patient simulator [4].

suction unit before it flows into the airway. The stopcock then needs to be closed before stopping the suctioning. Although it might be possible to extend the circuit to enable the operator to operate the stopcock from the control room, it is safer to use an actor to perform the actual suction as they can ensure that no fluid actually flows into the airway. This setup very realistically portrays the impression that the blood or vomit is being aspirated from the patient simulator's airway. Another advantage is that fluids can appear to be suctioned from wherever the confederate using the suction probe chooses to employ it.

## 72.8  Clinical Gas Ports

Many simulation laboratories are installed in spaces that were never designed or configured for patient care. Existing electric power, lights, and ventilation suitable for most office needs are readily adaptable for most simulation needs. However, many simulation environments and simulators require gasses, and ideally the simulation experience includes the sight, touch, and sound of clinical gas ports on a wall or ceiling. A certain amount of creativity is required to link these clinical gas ports to the sources of compressed gasses and suction, whether from a local tank farm, compressor, and vacuum pump, or a building/institution-wide utility system. Access to the back side or inside of the walls provides a way to hide the pipes and hoses behind those gas ports (see Chapter 25, Figure 6; Chapter 22, Figure 2). Lacking such interior access requires another application of smoke and mirrors.

In this example, tanks provide $O_2$, $N_2$, $CO_2$, while a central service provides compressed air and vacuum. The simulator requires $O_2$, $N_2$, $CO_2$, and air, all of which should not be easily disconnected by students or instructors. The clinical devices require $O_2$, $N_2$ (in place of $N_2O$), compressed air, scavenging, wound and airway suction, all of which should be easily accessed through familiar clinical connectors and ports. In addition, for teaching purposes, the composition of the gas delivered through the oxygen ports should be easily changed, but without leaving any obvious cues.

To meet all these requirements, a shallow box was designed to mount the visible gas ports, hide the hoses, and support the $O_2$ composition valves (Figure 72.11). It was intentionally painted with the same paint as used on the wall, so as to minimize its visual intrusion into the room.

The box was securely attached to the wall by several long bolts into a wall stud, allowing easy installation and removal (Figure 72.12), yet firmly holding the ports in place during hose connection/disconnection.

Three valves, mounted low in the front of the box, allow easy changes in the oxygen composition delivered through the

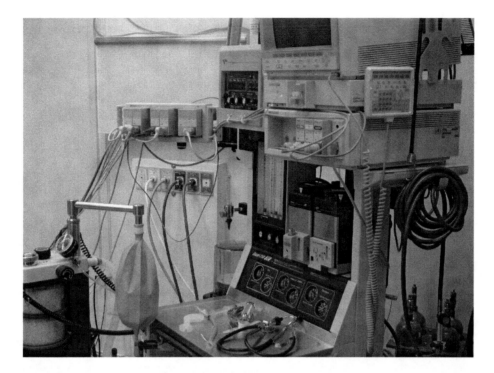

**FIGURE 72.11** The visual presentation of fully functional clinical gas ports. Their height and room position were intentionally devised to allow easy detection, identification, and use, yet remain out of the usual traffic flow pattern. The nonclinical features and design requirements are hidden from the students' line of sight.

**FIGURE 72.12** Typical clinical gas ports on an atypical but sturdy mounting.

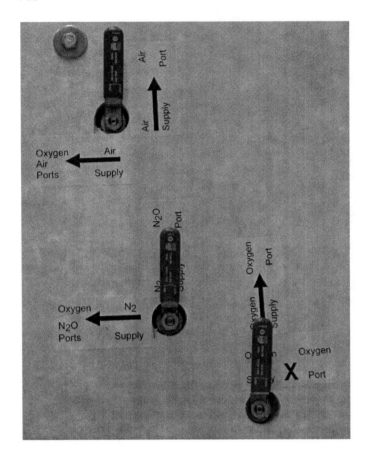

**FIGURE 72.13** Valves for changing the composition of the gas provided by the $O_2$ ports. Note the additional bolt securing the box to the wall, located on the box near where forces will be applied when moving the handles.

two $O_2$ ports: 100%, 21% (air), and 0% ($N_2$). These valves were oriented so that when all three handles are vertical, the $O_2$ ports deliver 100% $O_2$ (Figure 72.13). When the lower right handle is turned 90° clockwise and the upper left handle 90° counterclockwise, the $O_2$ ports deliver 21% $O_2$, or with the middle handle 90° counterclockwise, the $O_2$ ports deliver 0% $O_2$. In any case, the Air port always provides air, and the $N_2O$ port always provides $N_2$.

As none of the simulation-specific hoses or cables could be installed within or behind the wall, they are laid along the baseboard of the wall (Figure 72.14). The lower sides of the box are open to accommodate this requirement. Hoses and cables can be hidden from view using inexpensive, easily installed cable raceways, for example: http://cableorganizer.com/surface-raceways/ Note that simply Velcro-mounting a white plastic tray over the valves visually hides the very existence of these valves from the students. This covering also provides a modest protection from feet and carts. As clinical students are so often taught to consider the floor nonsterile, they naturally avoid anything placed on or near the floor. This phobic attitude of theirs is just more smoke that simulation professionals can freely make use of.

**FIGURE 72.14** Cutouts on the lower sides allows for access and pass-through of hoses and cables that cannot be installed inside walls. On the far right of the image, hidden under a typical blue clinical sheet, is the tank farm. Out of view on the left are the simulator support rack and the incoming hoses for compressed air and vacuum. The $O_2$ gas composition changing valves are hidden behind the white plastic tray affixed near the bottom of the box.

Flexible hoses allow easy installation, while metal pipes and valves provide strength for items receiving repeated manipulations (Figure 72.15). Note that since the maximum pressure difference driving suction can never exceed 1 atmosphere, vacuum lines should be larger than that required for compressed gasses typically provided at 3 atmospheres or more (in this example: 3/8″ versus 1/4″ internal diameter hoses, respectively).

## 72.9 Arterial Blood Gas Analyzer Simulator

Another very interesting idea has been the recent independent development of a full arterial blood gas (ABG) analysis simulator with a touch screen interface developed by Neal Jones from the Cheshire and Merseyside Simulation Center, Liverpool, UK [5]. As illustrated in Figure 72.16, it enables the clinical

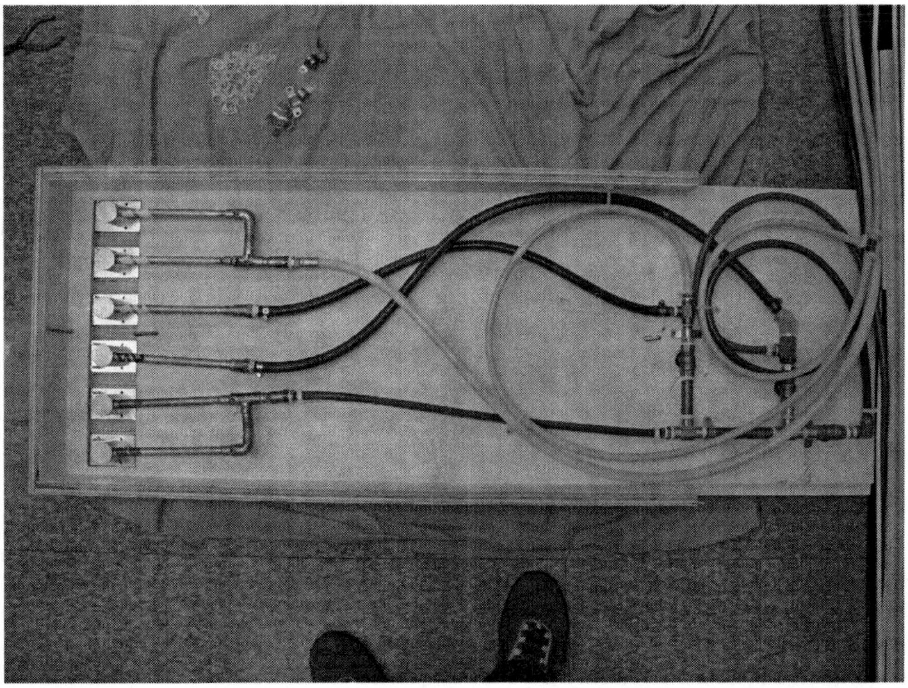

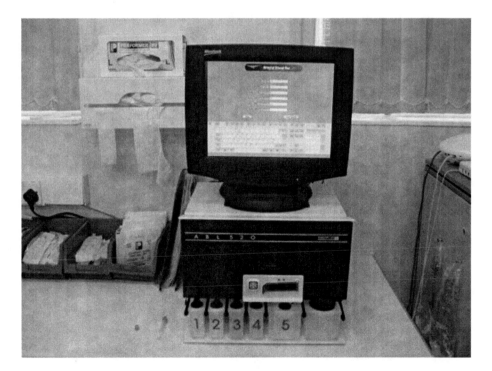

**FIGURE 72.15** Inside the box, with the backsides of the clinical gas ports on the left, pipes and valves for changing the $O_2$ composition on the right, and the flexible hoses connecting them together. The entire assembly rotates down from the wall for easy hose attachment and future modifications.

**FIGURE 72.16** An adapted blood gas analyzer in the simulation room engages scenario participants to perform the task for real.

team to perform a "real" blood gas analysis within a scenario, hence enhancing the realism. This simulator could put an end to the small piece of paper often handed out to the scenario participants by a plant or faculty after a request for the test has been made. This simulator is quite revolutionary and ecological as it can be made of mostly recycled components. The actual ABG case is almost empty as it primarily contains some tubing connected to the syringe port and an IV bag to collect the sample to be "analyzed."

The system developed allows for real-time blood gas analysis of simulated blood that is consistent with the patient simulator physiological state. It can force the person doing the test to enter patient information. A while after the scanning button has been pressed, the machine simulates the processing before displaying the appropriate results. This is achieved by relaying the desired data to the touch screen monitor from a networked computer in the control room, where the data is entered by the patient simulator operator. More information about the software developed and hardware adaptation can be freely obtained by e-mailing Neal (neal@patientsimulation.co.uk) or visiting the web site [6]. Another good news is that it all can be achieved for less than $300 if you can reuse an old ABG machine.

I would strongly encourage the simulation community to share their ingenious ideas through the medium provided by the Patient Simulation web site as well as the annual simulation conferences (www.ssih.org and www.sesam.ws). One may notice that most of the above-mentioned tricks as well as many more are freely shared on the patient simulation web site thanks to the goodwill of the creative problem solvers.

# References

1. Alinier, G., and Harwood, C. A touch of added realism: Preparation of your patient simulator for CVP monitoring. *Anaesthesia and Analgesia*, 2005, 101(6S), S2.
2. Harwood, C. and Alinier, G. A touch of added realism: Preparation of your patient simulator for internal haemorrhage. Poster presented at the 10th Annual Scientific Meeting of Society in Europe for Simulation Applied to Medicine (SESAM). Stockholm, Sweden, pp. 17–19 June 2004.
3. Alinier, G. and Harwood, C. A touch of added realism: Preparation of your patient simulator for urinary catheterisation. Poster presented at the 10th Annual Scientific Meeting of Society in Europe for Simulation Applied to Medicine (SESAM). Stockholm, Sweden, pp. 17–19 June 2004.
4. Chow, R. E., Naik, V. N., Savoldelli, G. L., and LeBlanc, V. R. Simulation props: Enhancing simulation on a budget. Poster presented orally at the 6th International Meeting on Medical Simulation (IMMS). San Diego, CA, USA, pp. 15–17 January 2006.
5. Jones, N. 2006. How to build your own arterial blood gas simulator. Workshop presented at the Annual Meeting of the National Association of Medical Simulators, London, UK, pp. 14–15 September 2006.
6. Jones N, 2006. Tips and tricks on patient simulation website: http://www.patientsimulation.co.uk accessed on 7/09/07.

# 73

# Professional Audio/Video for Clinical Simulation

Michael S. Goodrow
Michael Seropian
Judith C.F. Hwang
Betsy Bencken

## 73.1 Good Answers come from Good Questions

### 73.1.1 Functionality Before Complexity

This phrase should be the guiding principal in designing audio-visual systems for use in simulation-related activities. When designing and using the audio-visual system for a simulation center or facility, there are six questions that need to be addressed about the audio-visual needs for simulations being performed:

- What do I need to know about the simulation activity?
- When do I need to know it?
- How can I get that information?
- Who else, and where else, needs this information?
- What is the skill level of the end user (personnel)?
- What is my budget?

The answers to these questions will help to define the complexity of the audio-visual (A/V) system installed in your center. Note that these questions do not ask what type of simulation you are doing. This information is inherently imbedded in the answers to these questions. Also, A/V signals flow through a chain of devices, and like any chain, the lowest quality component defines the highest possible level of results. Thus, rigorously answer these six key questions, and only then select equipment of matching quality. We present the use of these questions by starting from the most general, pass through the very specific, and conclude with the broad picture. As the myriad concepts and details of audio and video can all too easily become overwhelming, we offer both user-centered and function-centered viewpoints of the same topic. Some repetition is good. Finally, helpful hints on how to address these six questions appear at the end of major sections. The chapter concludes with some "lessons-learned" about designing, building, and using an A/V system for simulation.

## 73.2 Matching the Tools to the Teachings: Three Examples

Here are three examples to demonstrate how the answers to these questions affect the design of the system. The A/V requirements for these three cases are dramatically different.

**Case A:** You have a standardized patient center with six examination rooms. You want to be able to tell if the standardized patient is still in the midst of a student encounter, which can be determined by visually observing the room.

In this first case, the observer only needs to be able to see who is in the room in real time. There is no requirement for storing the images. Also, there is not a requirement for high-quality images. Thus, the "A/V system" could be a small one-way mirrored window, allowing observation from another room or hallway. Or, it could be a set of low cost, black and white, immobile closed-circuit cameras that feed six small low-cost black and white TV monitors, or a single low-cost black and white TV monitor with either a simple, inexpensive device to select which of the six rooms is being observed, or a simple inexpensive device to combine all six signals and display them simultaneously. Note the lack of any sound required to fully meet the user's needs.

**Case B:** You have a full-body patient simulation room set up to resemble an intensive care room. A patient is in the bed ventilated using a new, portable ventilator. The current research project is examining ventilator settings, and the responsiveness of the ICU team to changes in respiratory parameters. The response times and appropriateness of actions and settings will be judged by faculty members after the activity is completed.

In this second case, there are several additional implied requirements. The video images must be recorded, so that the faculty members can review them after the event. At least three video images need to be recorded – the clinical monitor, the setting on the ventilator, and the actions of the ICU team. These three images must somehow be presented to the reviewers in a consistent, time-synchronized manner. The quality must be high enough for the reviewer to distinguish changes in the monitor and the ventilator. The video recording also needs a "clock," so that the reviewers can easily determine how long various actions take. And lastly, this recorded video must be in a format the reviewers can easily use.

This system will require at least two cameras in the room, and the simulator controller must be able to "see through the cameras" during the exercise. The field of view (direction and zoom) of these cameras will probably have to be remotely controllable to keep particular regions of interest within their frames. To capture the activity in the entire room, that camera must have a wide-angle (or short) lens, and might suffice with a fixed wall mount and manually set focus and zoom. To capture the portable ventilator settings, that camera must have a narrow-angle (or long) lens, auto or remote focus, as well as remote pan, tilt, and zoom. To record the clinical monitor, a video feed must be pulled from the monitor signal source

(a camera pointed at the monitor can also work, but not as well, as participants might stand in the way and obscure the camera's view of the monitor). If the participants are able to move the clinical monitor, then that camera also needs remote pan, tilt, and zoom.

The three video images must be combined and recorded. The images can be combined during the event in real time, and then the combined image stored. Alternately, the three video streams can be recorded individually, and then combined after the event. If the latter, then some mechanism must be used to synchronize the three recordings. Furthermore, time should be clear to the reviewers – this can be a physical clock in one of the camera's views, or a digital time created by the video recording system and superimposed over the camera's recorded video data.

After the combined video is produced, it might have to be converted to another format. For example, the reviewers might want only VHS tapes or DVD disks. The final recordings may have to be copied to other media and formats appropriate for long-term storage. If there is a requirement for placing timing markers at specific points, for researchers and debriefers to quickly go back and forth to specific events of interest, then a digital format might be the most appropriate.

**Case C:** You have a task-training room that has multiple surgical laparoscopic task trainers. You want to observe who is in the room, what they are doing, and be able to not only record them but also the image they are seeing. In addition, you want to be able to send those live images to a remote location for others to observe.

Recall everything you thought about in Case B. We can now build on those concepts in Case C where we see a mixture of complex requirements: (i) surveillance; (ii) real-time viewing for the faculty/educator; (iii) real-time broadcasting; (iv) recording of audio and video; (v) playback; (vi) combining incompatible signals. As we walk through the different requirements, you will see that the design of this type of system becomes inherently more complex.

The surveillance will require sufficient cameras to pan/tilt/zoom to each and every participant in the room. The cameras should be of sufficient quality to produce a video signal that contains essential details following broadcast/distributing and/or recording and replay. Your needs define just what those essential details are. Generally, broadcast quality is directly related to the camera quality, and if digital, the extent of compression(s) that are most likely to occur in the camera prior to output, and/or in the broadcast/distribution method, and/or in the recording/replay methods. Each camera (yesterday: analog, interlace scan, narrow aspect ratio, low resolution; tomorrow: digital, progressive scan, wide aspect ratio, high resolution) and computer video/clinical monitor feed (yesterday: analog such as VGA; tomorrow: digital such as DVI) is routed to a video switcher (also called a matrix) that has X number of inputs and Y number of outputs. Each and every input can be assigned to one or multiple outputs.

You can begin to see that with the complexity of this system comes many more choices for selecting what is to be seen, by whom it will be seen, and where it is seen. Some solutions will capture and compress the data stream in real time. These compressed images are then broadcast, usually with the sacrifice of quality. The main advantage of this method relates to the wiring and simplification of design. Sending compressed signals encoded in Internet Protocol (IP) format over network cable to a number of distributed sites is often easier and cheaper than sending unprocessed video signals over shielded, UL-listed A/V cable. The user must decide what quality will meet the goals for the activity.

Recall that all A/V broadcast/distribution systems are chains, and the highest delivered resolution is no better than the lowest resolution component along that chain. In the world of analog signals, all manipulations, even sending data over cables between devices, degrade and distort the signals. Thus, generally, the sooner that your signals are converted from an analog format into digital, the less undesirable alterations are created.

Recording video and audio can be accomplished in a myriad of ways. It can be a simple VCR or a complex system that uses real-time compression and saves digital A/V files to a hard disk-based recorder or data server. In Case B above, the three video signals are likely to be of two types. The first type is that produced by the two cameras viewing the participant and the ventilator. These might be in the form of the common, lowest cost, lowest quality composite analog television-type format. The other image is generated by the computer that is being used by the simulator controller, or by the clinical monitor. This computer-type image is of higher resolution, and in a different format than from any analog television-type camera. Therefore, these two types of images will be recorded differently, unless converted to a common format. You now have to decide what to do with these three images. If you want to combine these images, you would need specialized equipment such as a multiplexer or video mixer.

Next, you need to consider the quality of the recording. Common recording methods discard some to most of the data collected by, and sent from, the video source. The analog VHS became the dominant standard over analog Beta for home use, because the former was specifically configured to record 2 hours on one tape at the expense of only recording about half the data collected by the TV (about 480 lines of horizontal resolution received, but only about half that recorded/replayed). Likewise, almost all digital video recording systems make extensive use of the compression capabilities inherent in computational manipulation of any digital data. The greater the compression, the poorer the quality of the recording. Digital systems use different recording formats (e.g., AVI, MPEG, MOV). These formats are commonly called CODECs (COmpression/DECompression). Finding a format that is widely used across different platforms is essential in the decision-making process. Assuming you are using digital video technology, solutions such as the Video Toaster 5 by NewTek® provide

high-resolution/low-compression-ratio algorithms that use huge amounts of bandwidth and disk space. Other solutions that use MPEG4 compression consume much less storage space and transmission bandwidth, but store and reproduce a lower resolution image. Ultimately, the solution here is a compromise between quality, processing power, and available archiving space. This is a trade-off decision you have to make, and then revisit periodically, as your needs may change.

Just when you thought you were done, you have to decide where, how many, how large, and how to send your different captured video signals to a remote (on- and/or off-campus) location(s). In each instance, your needs and preferences are the deciding factor. Each of the following example solutions has its own advantages. Video projectors are versatile and will give the largest image. Recall however that poor-quality images that are made larger look progressively worse as they increase in size. The digital image that looked great in a $16 \times 9$ inch window on a new computer may look terrible when seen on your 42-inch plasma display or the institution's 42-foot auditorium projector screen. So, compression and signal quality rears its head in the selection and use of your display equipment – consider it the weakest link. As for sending live video to off-campus locations, several well-established protocols exist. This will require a call to your information technology department to find out what teleconferencing system your institution uses.

A small, readily available and well-defined object such as a syringe is an ideal evaluation-and-test object for your entire video system chain. If the presented image at the final point of use, either live or replayed from a recording, is clearly identified as the same syringe in front of the camera, then most likely, all other visual details are being captured, processed, and presented with sufficient resolution for typical simulation needs. Note that the fine markings and numbers on syringes are great items to challenge vendor's claims for their A/V systems.

In each and every example, the system must fit the answers to the questions above. Some people will argue that the skill level of the end user is the most important factor to consider. If a system is so complex – either because of its overwhelming sophistication or because of its patched together unreliability – that no one can use it, then it is time, energy, and money ill spent.

## 73.3 Simulation-specific A/V issues

With this as an introduction to some of the issues surrounding the A/V system, we will now describe some of the particulars of A/V related to simulation. We will start with descriptions of the various players' needs in an A/V system, then the general uses of an A/V system, followed by the functions of the various components in an A/V system. Then, we will discuss the control, communication, recording, and feedback requirements for the A/V system.

The A/V equipment you purchase is frequently dictated by finances and all too often by what your local vendor suggests.

Recording a simulated event in order to facilitate debriefing can be done with very basic A/V equipment, e.g., a single, strategically mounted video camera plus VCR, if the budget is very limited, or if the budget permits, this can be done with more sophisticated equipment, e.g., multiple mounted high-definition (HD) digital cameras, an integrated ability to mark events of interest, an editing suite, and digital HDD/DVD playback. There are many other information resources available that can provide more details about A/V systems. Chapter 74 lists several. We recommend that you look into some of them after reading this chapter.

# 73.4 Match the Tools to the Users

There are several key players common to most simulation activities. These players have specific requirements for an A/V system. These requirements are also specific to certain types of simulation within which they contribute.

## 73.4.1 Simulator Controller

The **Simulator Controller** of mannequin-based simulation – The controller needs to be able to discern what all participants are doing and saying in order to interact with them during the simulation. For example, if the simulator has the plumbing to enable the participants to administer intravenous drugs, the controller needs to know which drug, its concentration and the dose administered. One brand and model of patient simulator does use an optical technology to determine syringe content and concentration in combination with a volume-administered detector to report the drug, total delivered volume and dosage, and duration of injection. However, this system requires that the students be trained in a syringe-handling technique that is rarely required for their use with real syringes and real patients. In contrast, having the students state the drug or fluid they administer is not necessarily a bad thing during their learning. However, this too would seldom be done in daily practice.

An imbedded radio Frequency Identification (RFID) system like that being introduced for tracking each and every drug vial might eliminate any such unreal, simulation-specific syringe-handling requirements on the part of the students. It might also eliminate their verbalizing drug administration during examinations if such behavior is not expected of them with their real patients.

Today, the most clinically faithful, universally applicable, and most reliable way to gather any drug or fluid administered is through direct video and/or audio capture using typical simulation A/V systems or occult communication with confederates and actors within the scenario using two-way radios.

The controller also needs to be able to have one-way and two-way communication with participants, actors, and other personnel involved in the scenario. This communication may be broadcast to everyone, or it might be to specific individuals. Intelligible sounds from the simulation room must be broadcast within the control room. This can be accomplished within the control room either by speakers or headsets. Speakers obviate the need to be tied to a headset, but pose the potential problem of feedback through the microphones for the patient's voice and the Public Address/Paging system.

## 73.4.2 Clinical Instructor

The **Clinical Instructor** (or **Director** of the Session) – All sights and sounds that the operator needs are equally vital for the instructor. In addition, the instructor will have unique communications needs that depend heavily on the specific kind and format of simulation. If instructors are in the room with the patient and participants, they might have to signal the controller when something is happening. The instructor might have to signal other actors in the simulation room as well. These signals may need to be covert, to avoid alerting the learners. A telephone is another excellent way to communicate either way between the control room and the simulation room. The instructor can also utilize concealed two-way voice communication systems similar to that used by law enforcement.

## 73.4.3 Learners

The **Learners** – The learners need to communicate with each other while providing intelligible communication to the controller. The controller's actions are often influenced by the actions and conversations of the learners. Audio equipment that ensures good sound pickup is paramount. The learners may also need to communicate directly with the controller, such as when the students are confused and uncomfortable with the simulation itself (this is infrequent but no less vital). They may need reassurance that particular events actually "happened." For example, if the learners are administering a drug, they might be requested to state the drug and its dosage to their fellow students and those in the control room. The controller might then respond with an affirmation that the dose of medication was given. This communication may also have to be recorded.

## 73.4.4 Observers

The **Observers** – It is reasonable to expect that you will have observers during simulation activities. These include potential instructors learning more about simulation, people assessing the instructors, and other learners observing to participate in the debriefing. This last group is probably the most demanding from an A/V standpoint. These learners need to be able to see what the participants are doing and how they are doing it. Lack of clarity and directional sound are the two most common complaints. Depending on the type of course, you might also want the observers to be able to communicate with the participants.

### 73.4.5 Actors

The **Actors** – Some simulation activities include Standardized Patients or other actors as part of the simulation activity. For some scenarios, it might be necessary for the controller to communicate with these actors during the simulation, similar to the need to speak covertly with the bedside instructor. Solutions include two-way radios and headsets, as well as clever use of the telephone system.

### 73.4.6 Simulator

The **Simulator** – The simulator should be considered another participant in the learning process. As such, someone, such as the controller will "talk" for the simulator. Some systems require the operator to actually do all the talking into a microphone connected to a speaker in or near the mannequin's head, while other systems have predefined phrases that the mannequin's speaker will produce. Either way, a solution to the patient's voice problem raises another interesting problem: how to record and capture the sounds originated in the control room and produced by the simulator's speaker. There are two primary methods: (i) directly patch the audio signal that is going to this speaker into a record-in port in your A/V system; or (ii) capture the sound via the microphones already around the mannequin in the room. Each of these solutions is viable, but feedback can become an issue when a lavaliere microphone is placed too close to the mannequin's speaker.

The second "communication" with the mannequin is less obvious. Unless the mannequin is set up to run on "autopilot" for the entire case, the operator will have to make the simulator do things during the case. This requires the ability to control the simulator, in real time. The controller needs to know how to initiate the changes (speak the language of the simulator) and how to recognize that the desired change has occurred. This might seem a bit esoteric, but this is the essence of training a new simulation operator.

When developing the A/V requirements for your simulation center, it might be helpful to develop a functional requirements matrix. An example of such a matrix is shown in Table 73.1.

Make similar lists for each simulation facility that you visit during your information-gathering visits to other simulation centers. Selection of specific A/V devices should occur only after you have completed this matrix.

## 73.5 Using the A/V system

The A/V system is only as good as it can be controlled. All functional A/V systems have some form of communication and control network. The most important components in this network are the participants and their required communication needs that were described above. Within this network, there are certain links that must be available and fully functioning. These links are shown in Figure 73.1.

Three different approaches can be used to achieve this communication and control. **First**, a one-way mirrored window

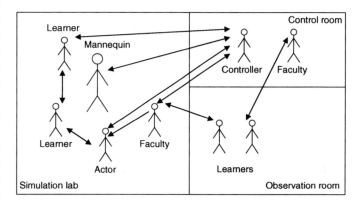

**FIGURE 73.1**  Lines of communication for a simulation center. This center contains three rooms: a simulation room, a control room, and an observation room (conference room). The arrows indicate which way the communication needs to occur. While this schematic shows the three rooms sharing common walls, this is neither typical nor necessarily desirable due to sound contamination.

**TABLE 73.1**  Example of a portion of an A/V functional requirements matrix

| Players | Sound captured from | Sound generated or broadcast | Images captured | Images generated or broadcast to |
|---|---|---|---|---|
| Simulator controller Clinical instructor Learners Observer #1 Actor #1 Simulator (plastic or people) | Learners, Instructors, Actors, Monitors, Patient voice | Patient voice, Captured simulation room sound broadcast in control room | Learners, Monitors, Simulator Console | Control room monitors, Recording device |

may be placed in the wall between the control room and the simulation room. Although this is not "technically" a piece of A/V equipment, it does provide visual information to the people in the control room. The controller can watch what the faculty and students are doing, and respond appropriately. The glass also enables simple, discrete communication from faculty to operator. For example, an instructor gives a simple hand gesture, such as thumbs up, to indicate that the learner has completed a specific task and it is time to start the next phase of the scenario. The other advantage of having a mirrored window is that the people in the control room can get a "sense of the room." This "sense of the room" is invaluable. Other than out and out breakage, these mirrored windows never fail, and everyone already knows how to use them.

**Second,** remote-controlled cameras are used in the simulation room. Frequently, cameras are mounted just above head height on the walls to provide good views of the room. The cameras are the eyes of the operator and the other learners. The capability of auto focus and the ability of remotely controlled pan, tilt, and zoom is therefore indispensable. Zooming in on particular actions, such as setting the desire respiratory rate and tidal volume on a portable ventilator, can provide valuable information. Cameras are sometimes obscured by students, particularly near the head of their patient. This can be obviated by having multiple cameras strategically placed throughout the room. Don't forget that your room may have varying configurations for different clinical environments that mandate different camera positions. In rooms that have multiple clinical orientations (i.e., use different "headwalls"), numerous cameras with easy repositioning may be necessary.

Having a camera mounted on the ceiling directly above the mannequin is useful. This view is typically unobstructed by participants, and provides a good view of their activity. Ultimately, camera positions will be determined by the intended use of the room, how many configurations are anticipated, and the objectives of having the cameras in the first place.

**Third,** an intercom system with speakers separate from that of the patient's voice in the A/V setup can provide several functions. First, the operator and instructor in the control room can use it to speak to the students without confusing the students about the identity of the person speaking. This enables the operator to provide guidance, confirm administered drugs, etc. The intercom system also serves to provide environmental fidelity, for instance when a code is called. The intercom or overhead paging system is also helpful in providing direction for larger sessions that involve mass task-training stations. The overhead paging system should not be confused with the ability to communicate via wireless radio headsets or earphones with actors and faculty. These systems are often discrete and serve a different function. There are some intercom and paging systems that provide both functions, but it is best to conceptually consider them as discrete entities that may be bundled together.

## 73.6 Matching the Tools to the Functions

There are several reasons for having an A/V system. The most obvious ones, particularly for educational applications, are to provide feedback, assessment, and remote access. Removing instructors and simulation operators from the students' immediate surroundings is a powerful step toward the students' assuming total responsibility for their patients. But once out of direct sight and sound range of their students, they are also out of direct sight and sound range of controlling and directing the simulator and simulation, respectively. Cameras and microphones, displays, and speakers extend eyes and ears into the midst of the simulation action. Then, during debriefing, being able to "roll the tape" does a wonderful job of providing immediate and memorable feedback for the participants. In addition, a student's formative or summative evaluation may require that a record of the simulation session be retained. Lastly, there are the "other" uses. Being able to record the activities in the simulation center can provide a valuable resource for research purposes as well as for promoting the center's accomplishments. The A/V system could also be part of the security system for the center. If the cameras record 24/7, then a visual log will be available should something happen in the center (other than outright theft of the recorder itself). Your simulation setting, complexity, and purpose all determine type, quality, quantity, location, and deployment of audio and video support devices.

### 73.6.1 Remote Access: Simulator Control/Simulation Direction

The A/V system can be a powerful tool to control and direct simulation activities. This might be as simple as confirming that a session is completed, as shown in Case A above. Or, it might be essential to the proper execution of the case, as in Case B, where the operator needs to see the actions of the ICU team. This might also require multiple simultaneous views to follow all the action. There have been many different design approaches, but core functional concepts prevail:

- The quality (i.e., individual price) of cameras and microphones should be no greater than that sufficient to capture the details in those sights and sounds essential for meeting your current teaching requirements. Tomorrow's audio and video devices will be more capable and available at a lower price than what you can buy today. Purchase today just the functionality that you know that you will use. It has been observed in many centers that bought very high-end systems, that there is a tendency to underutilize their equipment because of the inherent complexity of the user interface. Nothing is gained by paying for features that remain dormant, or worse, distract and disrupt.

- The number of cameras and microphones should be slightly greater than that sufficient to capture those sights and sounds essential for meeting your current teaching requirements, as unexpected distracters and obstacles *will* occur. Redundant cameras and microphones will assure that key actions will be captured when the unexpected happens, and the unexpected always happens. Also, spares assure that the show will go on when an essential device fails.
- The individual video images presented in the control room must be large enough and have sufficient resolution to allow the operator to discern what is transpiring rather than having to guess. The images' sizes and placements should minimize the need for the operator to move significantly closer to the display to see essential details.
- Control room displays must be placed in an ergonomically sound way. Having your operators and instructors crane their necks to see the display(s) would be counterintuitive. Out of sight is out of mind.
- Everyone in the control room needs to see the exact same image that is presented on the clinical vital-sign monitor that the participants see within the simulation. Most vital sign monitors have a video output port for a second,

repeater display. Most of these video output ports use the very common analog computer-format known as VGA and its corresponding HD-15 connector. A high-quality video extension cord (with compatible connectors and pin-to-pin wiring configuration) and a computer-type display will reproduce the vital-sign image. A distribution amplifier is commonly used in circumstances where there is a single VGA output but several are needed to feed other displays or recording devices. The distribution amplifier creates both the boosted signal and the extra output port(s) required for your additional display(s).

- Video signals from several sources can be synchronized and combined using equipment such as a video multiplexer (Figure 73.2). The image sources may be cameras and/or a mixed processed signal (camera view with monitor superimposed, Figure 73.3). This approach will assure that subsequent recording and replay will always contain all the combined visual elements as well as reduce the number of displays needed. As many debriefing and classrooms have only a single video projector, a precombined signal is a great way to simultaneously present a number of synchronized views. In addition, precombining also reduces the number of recording devices. The concern is that if you combine too many images

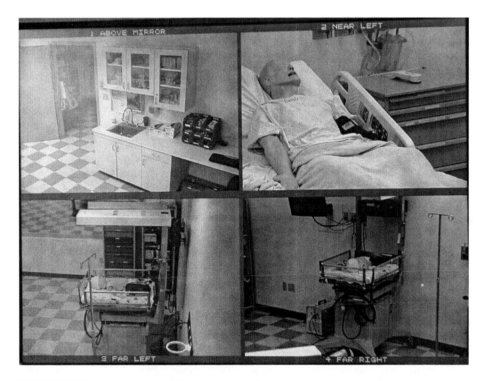

**FIGURE 73.2** Quad view of four camera angles. Despite the displayed resolution of any one view containing less than a fourth of its original resolution, starting with high-quality images such as in this example assures that condensed presentation will retain resolution sufficient for its intended uses. Each image source is labeled on the screen.

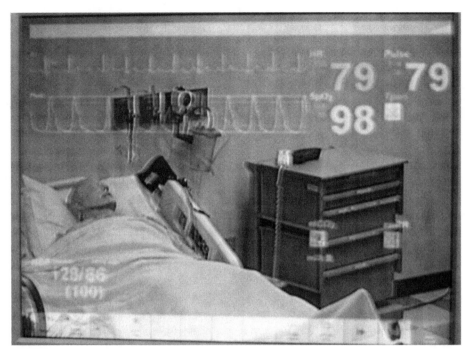

**FIGURE 73.3**   Mixed image output – note the vital signs from the clinical monitor superimposed on a single camera image. The displayed resolution retains most of the resolution of the two original signals.

prior to display or recording, then the individual images become too small for your intended use, thus breaking the visibility rule stated above. It is our experience that four good-quality images multiplexed on a 21-inch flat panel display will present a satisfactory image in the control room (Figure 73.2).

- In addition, the A/V operator needs to be able to see what is being broadcast to other participants (if this is applicable) in local and remote observation rooms. So, the broadcast signal (often called program out) needs to be displayed somewhere that the A/V operator can easily see it. All too often, someone forgets to turn on the broadcast mode or forgets to mix signals when a superimposed combination is required.

## 73.6.2  Debriefing: Recording and Playback

As simple as it sounds, being able to perform playback for assessment and feedback has several critical implications in the design of the A/V system. Typically, most or all of the same sights and sounds required for live observation are recorded for subsequent replay. There are many methods and devices that provide recording and playback. These two functions are usually performed by the same, or identical, or at least very similar devices sharing a common storage format (e.g., DVD recorder/players and DVD players).

You will need to address three questions about recording – what to record, how to record it, and how long to keep it. Obviously, you want to capture what the learners are doing. This is most often the visuals already collected by the cameras during their clinical scenario. Note that the clinical scenario may take place outside of the simulation laboratory, such as in real clinical spaces (known as *in situ* simulation), in elevators, in vehicles, outside, etc. (for example, see Chapter 62). Such environments require A/V systems similar to those designed and used for live events or news gathering instead of A/V systems suitable for a studio or simulation laboratory. No matter the venue, multiple cameras viewing different perspectives ensure collection of all the activity within the room. The more cameras positioned around the action, the less likely that *all* of them will be blocked by bodies or equipment at any one moment. You also want to record what is being said as well as any other sound sources like alarms and ringing telephones. Usually, audio is collected by microphones placed about or on the patient(s) and may even be worn by participants.

In some instances, "to capture what the learners are doing" includes their behaviors during the debriefing, as indicated in Figure 73.1. Thus, all the A/V requirements for the simulated clinical environments as described above apply equally to the debriefing environments.

There are simple and inexpensive devices that combine several images (Figure 73.2). This type of solution is often used in commercial security settings when maximizing the number of images per recorder and per display is desired. In this example, two cameras are providing different views of the same infant while two other cameras are each viewing separate areas. Which views are combined and in what pattern of presentation are completely determined by your needs. However, this method does reduce visual resolution after combination and requires a trained eye to scan all the images. The tendency is to focus on one image and scan to others when more information is desired. Some people find this solution very distracting; others find it very useful. In any event, all combined signals are permanently synchronized together.

There are several options available to combine two or more images together for projection and recording. One option is to combine the images during the session using a video multiplexer and then record the output of that unit. However, the video multiplexer solution provides more ease of use and less signal loss when images are combined. The advantage to a video multiplexer is that it is very straightforward in use and in what it produces. However, it cannot produce effects like superimpose or picture-in-picture – these features come mainly with video mixing equipment (Figure 73.3). At the "pro-sumer" level are products like the MxPro DV by Focus Enhancements (formerly Videonics). More sophisticated professional systems made by manufacturers such as Extron, Panasonic, Sony (to name a few) exist that have the advantage of higher quality and the ability to be controlled via touch panel interfaces using RS-232 connections. (The RS-232 is a standard through which high-end A/V products can be controlled remotely without touching the unit or being close to it.) Many video recorders are large, noisy, and produce heat, all of which are best kept far away from the simulation control console.

After the cameras, what other video signals should be recorded? If the simulation activity includes clinical monitors, then recording the image from the clinical monitor will be very helpful for debriefing. The ability to simultaneously record and broadcast the participants' actions together with the mannequin's vital signs is extraordinarily helpful because the patient's perspective is gained. The patient's vital signs can be separately recorded and replayed in conjunction with the camera recordings of the students' activities, or combined with one or more camera views and then recorded.

In either case, the signal that comes from the clinical monitor to the control room is usually a VGA signal that can either be used in its native form (depending on your display and recording equipment), or more likely, needs to be converted into a common television-type signal via a device called a scan converter before it is broadcast and recorded. This item can range in price from US$ 50 to US$ 2000. We find that we get good results at the US$ 250 price point. You might want to record the physiologic data that is presented on the display of the computer controlling the simulator, as it usually contains information not always presented on the clinical monitor in the simulation laboratory (for example, beat-by-beat blood pressure). Since this computer-generated image is often identical in format to the clinical monitor–generated image, you can apply the same devices and methods for both.

A note about extending and splitting signals: all audio and video signals, whether they be analog or digital, television-type or computer-type can travel only finite distances in cables before corruption exceeds acceptable limits. The exact cable length is very dependent upon the kind of signal, the quality of source, receiver, and cable, and the strength and proximity of interference. A rule of thumb is that whenever you want to greatly exceed the distance as typically used, expect the need for some form of amplification. For example, most computer and clinical monitor video signals are designed to travel through a 6-foot or shorter cable to the display. Sending this same signal to a second display in the control room over an extension cable four times this length, even with a very high-quality extension cable, may exceed the source and receiver capabilities and require a booster amplifier in between them. Furthermore, for many of the same reasons, most clinical monitor and computer video signals cannot simply be split without creating significant artifact and ghosting. The use of a distribution amplifier is required. This inexpensive device can "replicate" a signal so that several others can be sent to a variety of locations for display, recording, teleconferencing, etc.

The clinical monitor signal and camera images can be synchronously combined with the help of a video mixing system. If a video mixer is used, you then have the option of recording via split screen (where one side of the screen shows the participants and the other side the patient's vital signs), picture-in-picture (where the participants or the vital signs are located in one corner of the entire screen), or with the vital signs superimposed over the participants (Figure 73.3). Some centers prefer to use the superimposed method because it allows the viewers to shift their attention between the participants' actions and the patient's condition effortlessly. Some people find this solution very distracting.

As attractive as mixing signals sounds, be it by mixer or multiplexer, the major disadvantage is the potential for data loss. If the mixed recording fails and the individual source streams are not independently recorded, then all sources that were used for mixing are lost! An alternate approach is to split the original signals before they go to the mixer, which will allow the user to record native and mixed audio and video signals simultaneously. This avoids the problem of data loss and provides the security that all information will likely be recorded. However, different challenges arise from using this method:

- This requires significant amounts of storage to achieve. If the data is stored in a digital format, several terabytes may be necessary to record a week's worth of simulation.
- This requires that the information be suitable for mixing after the event. One major issue is that the video be

synchronized correctly, so that the images from two or more sources are matched correctly. Also, the sound must be appropriately mixed with both time synchronization and volume uniformity as well. This is particularly problematic if the video is from two different rooms (such as the simulation laboratory and a classroom). There are technologies that allow sources to synchronize. This can be complex and cumbersome. Automated systems can be purchased to do this, thus making user input minimal after complex installation and extensive (i.e., costly) debugging.

- The information from the different sources needs to be easily accessible. Debriefing sessions can involve simply rewinding a tape and hitting play. This is a very simple approach to retrieving the video, but requires multiple tape sets. The video data management system must also allow for rapid replay. In this respect, digitally recorded sources are superior to tape in that they allow instant and random access, as well as easy index marking of important events.

Splitting the original signals before they go into the mixer and recording both the individual and the mixed signals speaks to the issue of having redundancy built into your system. In domains where this information is vital, such as research and high-stakes testing, this becomes particularly important. The disadvantage is storage space, complexity, and cost. Also, how long does one keep all the video feed recordings? Which ones do we keep and which ones do we delete? Protection of privacy and persistence of recording media fidelity are not trivial problems to solve and keep solved.

There are computer-based packages that will provide sophisticated video data transmission, storage, and replay solutions. There are also A/V vendors that will build a solution for your particular application.

Recording and Playback can use various formats of storage media, each with its benefits and limitations: VCR, DVD, DVR, and Video Server.

- **VCR**, Video Cassette Recorder – A simple and familiar stand-alone unit that commonly uses VHS tapes. After rewind, the VCR tape is immediately ready to be carried to the adjacent room where the debriefing is to be performed. Visual resolution for standard units is limited to half that of low-quality cameras but provides a very low price, very easy-to-use video storage/replay method readily available today. This modality does not allow for instantaneous or random access playback, making playback and bookmarking cumbersome. Challenges such as finding the last recorded session, or a specific point in a recorded session, can be time consuming. Furthermore, if a single tape is used for the recording of multiple sessions and the recordings are to be retained, there is always the danger of inadvertently "wiping out"

valuable data by rerecording over a previously recorded session.

- **DVD**, Digital Video Disk – A simple and familiar stand-alone (or computer-based) unit that plays back content from an optical medium (a plastic DVD disk). The technology has been around for greater than a decade, and has large penetration into the market place. It allows for instant or random access, is more robust than tape, and stores more recorded time in a smaller package at higher resolution. Recording to DVD requires specialized software or a stand-alone DVD player/burner. This is a good medium for archiving. The data can be stored either as computer data files or using standard DVD movie protocol using VOB files. The latter allows DVDs to be played on any DVD player, while the former requires a computer and video file–specific software player. Playback is easy and familiar. If the DVD has to be physically carried to another room for immediate replay during debriefing after the scenario, then first the DVD has to be "ended and locked," or "finalized" so that another machine can access the DVD. Time and people required for this finalization process must be provided. For a given group of participants, this might mean that multiple consecutive sessions will each have to be recorded on a new DVD disk. If instead of the disk itself the data signals are sent from the DVD recording device to the debriefing room, then some form of remote control of the device must be available in the debriefing room. Rewritable DVDs are reusable like VCR tapes, but unlike VCR tapes, these disks usually require time-consuming "refreshing" (or reformatting) prior to rerecording.

- **DVR**, Digital Video Recorder – These units often combine a record/player DVD with a record/player hard disk drive (HDD). Typically, information is temporarily stored on the hard disk and can be transferred to a DVD for archive or sharing. The DVR is a powerful and fast method of playback. It is simple and intuitive. Choose units that have the capability of allowing video files to be renamed to reflect the session. This method of playback makes for easy instant and random access, as well as bookmarking. The major A/V manufacturers offer a range of DVRs with various amounts of hard disk drive storage time and video input format options. The Windows-based Media Center PC and TiVo are based on the DVR concept. The HDDs in DVRs are usually not designed for easy removal like DVD disks. Thus, all HDD-based replays require that the data *and* the remote control signals for the DVR be conveyed via cables or wireless transmission to the debriefing and review rooms. Any DVD disk created in a DVR is subject to the same time and effort for finalization, and rewriteable disks are subject to the same time and effort to "refresh" prior to reuse.

- **Computer** or server based – This is arguably the most powerful and potentially most complex playback modality. If the software used to playback is intuitive and familiar, then playback and instant/random access are as easy as a DVR. The advantage of these units is the ability to use files of multiple formats. For those who are computer-phobic, this medium may be problematic without an excellent, intuitive-for-the-user and easy-to-use user interface backed up by very competent and very convenient support personnel. There are vendors emerging that are providing such a product. The downside, of course, has to do with ease of use, likelihood of system crashes, and disk failure. The first two are really the most significant. A system that is well maintained and not loaded with extra, unrelated software is generally very stable and reliable. If every signal is simultaneously mixed and then recorded to save storage space, then the system must be robust enough to mix during the recording. If everything is recorded and then mixed during replay to minimize unanticipated data loss, then the playback system must be able to maintain synchronization to mix during playback. Or, you may want the option to do it either way.

## 73.6.3 Presentation

Your choice of video presentation is dependent on the way the information will be used, the size of the room, your budget, the lighting in the room, and the numbers and types of signal sources. Recall, your A/V system is a chain. An optimal system is one with a common level of usable quality and features shared by every device and interconnection, with an emphasis on the display devices. When selecting loudspeakers for a home-theater system, the best way to know which ones are best for you is to listen to them. When selecting a display, watching it in your typical setting and lighting *with your signals* will be the single most important determining factor. Do not let specifications in brochures, nor inappropriate test signals in showrooms, trump what your eyes see when you transmit your signals in your environment. For example, don't select a display based upon its ability to show flashy, monochromatic animated fish colors if you are going to present real human flesh tones.

A flat screen mounted on a wall, or video projector and roll-up screen, allows both economy of space and quality video viewing. If your debriefing room is relatively small, yet you need one or more displays greater than 30 inches, standard CRT-based unit (traditional TV set) will have a relatively large footprint relative to the room size. Projectors and flat screens allow you to have a large screen without the CRT depth issue. All this being said, if your budget is an issue, the standard TV is almost always a viable option.

## 73.6.4 Transmission for Remote Presentation: Live and Replay

Another consideration is whether or not you plan to transmit live or prerecorded simulation sessions within the simulation center (e.g., to another room where other students or reviewers are watching), within the medical center or affiliated university, or to outside sites as part of distance learning. As outlined in Case C, there are many considerations that go into determining the method of transmission of an A/V signal. These include:

- Desired quality
- Distance to transmit
- Compatibility between locations
- Type of signal
- Purpose of signal
- Access control

Anything that is transmitted must originate from a source and terminate at its destination. It is important, therefore, to understand what signals need to go where and how they need to be connected. Do they plug into a wall plate that allows a computer to attach to them to display on your projector, or do they go straight through to the equipment itself? This is pertinent with respect to the clinical monitor and the delivery of signal to or from the monitor. The location and type of connectors is paramount. In one case, a specific monitor may require a VGA feed (HD-15 or DB-15 connector), an audio feed (1/8″ stereo mini phone plug), and a USB connector (the flat/wide connector). In the case of USB, there is an inherent distance limitation that requires the use of special USB repeater amplifier cables for distances up to 80 feet and super extenders that essentially convert the USB signal to a format that is compatible with common network cable for much greater lengths.

Different methods of transmission exist because they solve different sets of problems in conveying A/V signals. The transmission of video is more complex than the transmission of audio given the vastly greater amount of data with video that is transferred. Analog video is also more susceptible to distance limitations than analog audio. There are several options:

- Transmission of analog format A/V signals over "analog" cables – This requires the use of relatively expensive and properly shielded UL-listed A/V cabling but does not require any format conversion at either end. For example: one coaxial cable for the composite video, or a single cable for S-Video, or a single cable composed of three coaxials for component video, plus a pair of coaxial cables for the typical left and right sound channels. The greatest advantage of this method is that these cables generally plug directly into both the source and display units. The disadvantage is that all analog A/V signals degrade over distance and are susceptible to noise pickup. Different analog signal formats have different allowable distances for different levels of quality (read: price) cables before amplifiers are needed. As analog amplifiers are usually indiscriminate in what they boost, eventually, this approach to sending A/V signals reaches a self-limiting distance. Analog A/V cable distribution systems are inherently point-to-point, with all the security advantages and scaling disadvantages this implies.

- Transmission of digital format A/V signals over typical computer data cables – If amplification or sharing is required, inherent with digital signal amplification and distribution is a high expectation that the output matches the input. The most common example is known under various names such as Firewire, iLink, or IEEE-1394a or IEEE-1394b (data throughput for "b" is about twice as fast as "a," with "a" already more than 10 times faster than needed for non-high-definition format video signals). This transmission format was originally created as a data storage format for high-quality digital A/V information on tape, but was then repurposed for transmission as well. Unlike analog, the single Firewire cable carries both audio and video information, and is much less susceptible to noise pickup. While Firewire has an official distance limit of about 15 feet between computer devices, unamplified A/V signals over standard quality Firewire cables of 70 feet have proven to deliver A/V signals identical to cables 3 feet in length. Many cameras and camcorders of price and performance suitable for simulation have a Firewire output, and an ever-growing number of digital recorders have a Firewire input. Two disadvantages are that as of today, few end-point sight and sound presentation devices accept Firewire-formatted signals, and high-quality very long-distance IEEE 1394a or b UL-listed in-wall cable is expensive. Digital A/V-specific cable distribution systems can make use of digital distribution amplifiers, thus allowing easy fan-out of identical signals to numerous recipient devices. The fact that such digital video networks are not normally used for office/lab types of data traffic provides a security and reliability advantage.
- Transmission of Internet Protocol (IP) format A/V signals over intranet/internet network wiring and devices – This method involves sending A/V signals that have been converted into IP format over very low-cost Cat 5 twisted-pair copper cables or optical fibers and then reconverting them at the destination to audio and video formats compatible with the presentation devices. There is considerable cost savings in this methodology in cable and IP distribution devices but as of today, the conversion costs are high. Examples of stand-alone products are the Comet DVIP®[1] (bidirectional, uncompressed, 25 MB per sec digital Firewire A/V data plus an additional 10 MB per sec for delivery information overhead) and VBrick® (bidirectional, compressed, up to 9 MB per sec

analog-to-MPEG A/V data plus up to an additional 5 MB per sec for delivery information overhead). The cost of each unit is not inconsequential and may cost more than running high-quality analog cables, even with amplifiers. In contrast, an example of a software-only solution is the DVTS program[2] (same specifications and totally interoperable with Comet DVIP hardware), which is downloadable and usable at no fee but does require a robust computer at each end to perform the IP/Firewire conversion in real time. Fortunately, computational power is inexpensive today, and is becoming ever more inexpensive.

In order to deliver each full frame of video on time, any IP method, irrespective of the type of device at either end, requires a network that has a large transmission capacity and is not congested with other traffic. Consider information traveling over network cabling as packages. Each subsequent package delivers the next piece of information. Any other data may interrupt or delay your packets from getting to their destination in time to create a smooth A/V experience. Technologies exist (under the heading of QoS – Quality of Service) that prioritize certain types of information versus others, and is an inherent feature in the version six upgrade to IP, known as IPv6. However, this does not exist today for video in most commercially available appliances. Creating and using your own private Internet, called an intranet, gives you much more control on what other traffic can exist on the system. *It is not uncommon for a simulation center to have a dedicated intranet for A/V purposes.* A rule of thumb is that any one link in an IP network should not continuously carry more than 1/3 of its maximum rated traffic flow. Thus, for the DVIP or DVTS example, each such device should have its own, unshared 100 MB per sec cable and network switch port. Ten such devices, all functioning simultaneously, would consume all the useful bandwidth of a 1 Gigabyte per sec network.

- Transmission over the public Internet – The difference between *the* Internet and *your* intranet setup is that you are now on a network that is literally shared by the world. Video streaming of large files can be very problematic. There are buffering and streaming technologies that help with this, but often at the cost of quality when real time is required. This method can have a significant time delay and is prone to any bottlenecks or crashes in its path. It is not uncommon for institutions to have internal networks that can carry 10 times more IP traffic than their connection to the public Internet can carry.
- Wireless transmission – This is an emerging standard that typically uses the 2.4 GHz frequency to send information through the air. Units that use higher frequencies (5 GHz) are prone to less interference from other 2.4 GHz–based systems such as most household wireless handsets. Microwave ovens can cause interference,

as they too operate at 2.4 GHz. It is expected that this technology will continue to develop and become much more prevalent. "Spread spectrum" (or frequency hopping) technology can randomly select different frequencies. This was originally a security measure, but is now widely used to avoid congested frequencies.

For those of you who experience attachment file size limits in your e-mail service, imagine trying to send sequential 35 MB files every second, continuously, uninterrupted, for minutes to hours at a time, from each camera and video source in each simulation laboratory to one or more observation rooms down the hall, across the campus, around the world. Long before clinical simulation sought network infrastructure robust enough to send numerous high-quality video streams, other similar needs promoted the development of networks parallel to the Internet, and are known by names such as Internet2, Next Generation Internet, and LambaRail. Their very purpose is to provide end-to-end very high speed, very high quality of service connectivity, and do so at comparably very high prices.

A note about the cable lengths you need: cable distances are *always* greater than line of sight distances. Even after you measure the actual distances intended for them to span, cables exactly that length always seem to be just a wee bit too short. Be very generous in your length estimations. A few loops of "extra" cable at either end allow for easy device repositioning and minimizes the likelihood of any inadvertent pull on the cable causing a connector to become unplugged or worse, a device or two toppling over.

### 73.6.5 After-the-fact Story Making

Another decision that needs to be made when considering audio-visual equipment is how much, if any, editing will be done after a session is complete, in order to create a new story from the collected sights and sounds. Once again, the budget may limit the initiation of this step because editing will require additional equipment and a person skilled in video editing. However, editing may help to decrease the total volume of recordings and improve archiving for future retrieval. If your center needs editing occasionally, do not overlook the possibility of working with a media services unit on your campus or a nearby commercial vendor. At a minimum, it is useful to mark the beginning of the actual session, so that it is easy to find during debriefing. Many centers start recording before the actual arrival of the trainees in the simulation session to ensure that the recording is started.

If the only purpose to editing is to reduce storage costs, another solution is simply encoding video in a more compressed format. This solution is effortless and can be done without much personnel time. The time of encoding is limited by the size of the video and the computing power of your PC. The product however contains all the video footage but only a fraction of the video data. Just do not compress beyond the point where the essential data worth replaying is lost.

### 73.6.6 Unifying Control System

All the equipment and ideas that have been presented thus far need to be integrated in a way that makes them easy to use. The equipment, for the most part, needs to communicate with each other or a central command appliance. Control of many professional grade units is accomplished by RS-232 linkage and the use of programing. This is what sits behind those fancy touch panels by AMX® or Crestron® that you may have encountered in your institution. If you have the budget for such a system, the programing will literally define the utility of your hardware – so hire a competent and experienced programer (they are usually employed by the A/V vendor). Make a list of features you want and the typical sequence you will use them. Invite the vendor and programer to attend sessions so that they understand the application and see the user's needs in action. *Skipping this step introduces significant risk,* especially when the vendor has little experience with the theatrical aspects of simulation center A/V design.

A system that does not have easy control options and is primarily geared for the consumer retail market can be problematic when programing a larger system. There are a variety of reasons for this that extend beyond the scope of this chapter. Suffice to say that as you design your system, pay careful attention to how many different pieces of equipment you purchase, and whether they are easy to use and complementary to each other. Try to use the same equipment brands and models throughout your center unless there is good reason not to. If you opt for off-the-shelf merchandise, then ensure that each piece of equipment has clear instructions laminated and readily accessible. Keep the original versions of all instructions well secured, and only work from copies.

The obvious preference is to invest in a system that is easy to use and controllable from multiple points. This increases cost substantially. At the lower end, invest in products such as DVRs that are easy to use and have the ability to record, broadcast, and play back. Some can even record while playing back. A good programable touch screen remote control is a worthy investment. Typically, it costs US$ 400 to have one of these programed, if you are not up for the challenge. The beauty of these units is that they offer the user what they need and eliminate what they do not need. They also eliminate the need for multiple wireless remotes that can be easily lost.

## 73.7 Educational Review/Feedback Using Audio/Video

There are three types of "feedback" – immediate, postevent, and remote.

### 73.7.1 Immediate

**Immediate** feedback occurs during the event, usually due to a cause and effect event. For example, if the learners use a defibrillator, then they should see something happen on the ECG, even if it is just the spike from the defibrillator. From an A/V standpoint, immediate feedback usually occurs in two ways. Learners in a separate room are receiving immediate feedback continuously as they watch the live broadcast. For the learner in the simulation room, immediate feedback may be useful when the learners are uncertain whether something occurred. For example, when students request the administration of fluid, the order can be confirmed by using the intercom system. That would be immediate feedback. Another common event is when they administer a drug, and do not see the effect of the drug. They often ask if the machine recognized the drug, and a voice via the paging speaker provides a response to the question.

### 73.7.2 Postevent

**Postevent** feedback is the most common type of feedback. Having a recording of a session is very useful, though not absolutely essential to debriefing. However, if there is disagreement about what actually occurred in the course of the scenario and what was or was not done by the participants, the replay provides an objective and undeniable documentation of events. The replay also gives participants the opportunity to view the entire simulation session from a different perspective and encourages them to reflect on their actions and behavior. This is a common use of the A/V system, and should be high on the list of reasons to have an A/V system. We repeat, it is *your* uses that should always drive the design of *your* A/V system.

### 73.7.3 Remote

**Remote** feedback uses all the distant interconnection technology previously described to bring together participants, observers, and instructors for a shared experience, without each and every one of them actually bringing themselves to a common observation/debriefing room.

Policies need to be developed that determine how long these video segments will be kept (or if they will be kept at all) and who will have access to them. All recording should respect privacy and ensure security of recordings. Students and participants may be asked to sign video release forms. This is mandatory if the videos are to be used for anything outside of the session. Regulations also exist with the use of video for research purposes. The use of simulation recordings for medicolegal discovery has yet to be defined. Some centers opt to simply erase or record over each session that is done for educational purposes alone. Thus, storage and who has future access to the recordings are not potential problems.

There are instances when it is important to retain copies of simulation sessions for remote replay and feedback. Here are two examples to explain what this can do for your simulation program and how it influences your A/V needs.

Example 1 – participant feedback or remediation: sometimes learners need some directed feedback about their performance, particularly during unusual or difficult cases. In such cases, a copy of their performance during a simulation session is provided to their faculty mentor. The mentor then reviews the video with the learners, to provide feedback about their performance during the case. Remote replay capability eliminates the requirement that the observers have to return to the simulation center, and occupy scarce simulation facility resources.

Example 2 – research review: one approach for assessing performance in a research project is to have independent observers review the performance. This can be done by recording the session and giving the videos to the reviewers. By coding the sessions and mixing up the tapes, much of the potential for bias is eliminated – the reviewers were unable to distinguish the intervention groups from the control groups. For digitally recorded sessions, the files can be copied to new media (e.g., DVD), to a research area (or server), renamed (which adds a new time stamp), and renamed with coded names or numbers. Restricted access (passwords) can be used to enhance security.

These two examples point out several additional requirements for the A/V recording session. Many centers that still use VHS tapes to record the sessions will need to have a mechanism to duplicate tapes so that the original is always maintained by the center/program. Unfortunately, there is nothing fast about this old technology. Using VHS tapes is convenient because of the ubiquitous nature of VCRs. Recording videos to other formats such as DVD may be more practical, convenient, and of higher quality. Reviewers also are given much more control of the review process because of the ability to have random access through the DVD technology.

Archiving videos have several implications: (i) media such as VHS tape require substantial space; (ii) Certain media degrade faster than others and are susceptible to damage – VHS tapes just sitting on the shelf degrade over time, as do optical disks; (iii) storage media that is not intended as permanent, such as hard drives, can become victim to drive-failure corruption, or accidental erasure. Saving the information to DVD or hard drive still requires security to be maintained but does help to decrease the storage space requirements of VHS tapes and could potentially make retrieval easier. All storage media is subject to damage from fire, water, and accidental or intentional removal.

## 73.8 Lessons Learned

- Eventually, *all* equipment becomes obsolete and finally unusable. Eventually *all* vendors cease supporting their products and finally cease business all together. This is particularly true of niche products. When basing your system on such a product, ensure that the equipment allows you to scale to other products in the future if your original product or vendor is not available anymore.

- If you are in doubt about what to buy and are confused the gaps between your wants, your needs and your budget, first go with a low-end solution and see what your program's needs really are. Live with your center for a while. There are many centers that have started with just a basic camcorder. Although this may seem like a waste of time and money, if it helps define your needs and prevents the purchase of unnecessary equipment, then it is worth it. If your needs are better defined *and* you have the ability to put in remote-controlled cameras instead of fixed cameras, do it. You might not have to redirect the cameras often, but when you do need to redirect them, you really need to redirect them.

- As you design your system, consider not just the capabilities but the end users' ability to use those capabilities. Underutilization of equipment is primarily the result of not having the properly educated personnel to set up and use the system.

- A poor user interface has sunk products. If you are dealing with a custom installation, then ensure that the vendor allows you sufficient time to try all the features out on items such as touch-panels, etc. Also ask for the code (programing) for the system so that other vendors can access it to add to it if necessary.

- Local vendors may be experts at making a classroom or auditorium, but that does not make them expert at outfitting a simulation center. More often than not, they do not understand the application, which leads to misinterpretation of the customer's vision and a system that is less useful than anticipated. This is a significant problem. Take the time and money to hire a simulation A/V consultant who will help with realizing your vision. Clinical simulation is clinical theater; hire an A/V consultant who has a good track record working with local theaters.

- Do not pay your vendor the full amount until they have completed the job to your satisfaction. Your institution likely has terms that define this, but all too often in large institutions, payments slip through.

- If you do not hire an A/V simulation consultant, then at least show your design to others in the simulation community to see if you are in fact getting what you think you are getting.

- When considering quality, remember that you are not building a premium home theater. You will save a lot if you do not demand high-definition recording, etc., but use judgment to purchase a product that delivers good and reliable content when operated and maintained by the typical people in your simulation team, in your institution.

- Depending on your institutional policy, the cost of most A/V systems will necessitate that bids be solicited. The complexity of the A/V systems, particularly the range of available features, means that most simulation center operators are not able to write the specifications. Many institutions have rules against the company that writes the bid specification, also bidding on the contract, as this is considered unfair. So, you might find it useful to hire a separate consulting firm to write the bid specification.

## 73.9 Example Schematic of Audio/Video Signal Paths and Devices

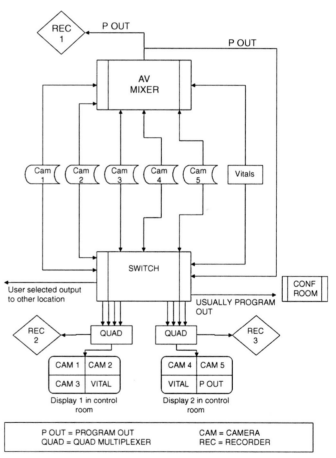

Electronic schematic depicting the use of five cameras (CAM 1–5) plus a vital-signs monitor (Vitals) to produce the following: a mixed image for the learners in the conference room, two quad displays for the operator/controller, recording pathways for the two quad display images to their respective recorders (REC 2 and REC 3), and the mixed image from the mixer (P Out) to a recorder (REC 1). All of the recordings are digital, using real-time encoders. All live broadcast signals are uncompressed. All these systems are controlled via RS-232 protocol. Schematic used with permission from SimHealth Consultants, LLC © 2006.

## Endnotes

1. Comet Digital Video Internet Protocol Device: http://www.comet-can.jp/
2. Digital Video Transport System: http://apps.internet2.edu/dvts.html

# 74

# Simulation Audio/Video Requirements and Working with Audio/Video Installation Professionals

Guillaume Alinier

## 74.1 Deciding What Your A/V System Should Do for You

To determine what you need your audio/video system to do for you, during your next simulation session, or right now as a "gedanken" ("thought") experiment, close your eyes and cover your ears, and then ask yourself what do you have to see and hear to determine what is going on in the session. Note that this process should be applied for all your different sessions, as they each have their own unique sight- and ent sessions, as they each have their own unique sight- and sound-collection requirements. Also, this process should be applied for the different audiences and venues. The sight and sound needs of those in the control room might not be the same as for those in the observation room. Only after you have a clear concept of what you need your audio/video system to do for you should you consider system designs and device specifications. The longer you postpone selecting an item to buy while establishing your requirements, the more likely your system will perform as desired. The more you understand your audio/video system, the less likely it will fail you especially just before a faculty-intensive session or a big presentation before the major donor to your enterprise. Expect your audio/video system to change over time right along with changes in the size and scope of your simulation-based education program. Plan for this change, budget for this change, accept this change.

### 74.1.1  Bleeding Edge or Optimal Edge

There is no upper end to the price you can be charged for audio and video equipment. Why is this? Perhaps, it is because no two people can experience the sights and sounds the same way, hence there is always room for flattery. The good news is that for clinical simulation, we can easily determine for ourselves just what level of performance we need, we want, and we can afford. As an example: given a camera location in the ceiling about 4–5 feet away from where the syringe is going to be used, you need to collect, record, reproduce, and display the image of a 10 ml syringe, see the volume markers and drug label with sufficient resolution to discriminate between ephedrine and epinephrine. As this is an acceptable test method for determining your own eye vision acuity, it should be acceptable for your simulation video system visual resolution. Armed with this functional definition and two labeled syringes, you can now face down any audio/video salesman who wants to unload last year's unsold gear using the latest buzz words, or any administrator who needs extra justification for your unfamiliar purchase.

Once your performance criteria are defined, then comes the functional system design. Both sight and sound share the same key functional blocks, they just use different devices. Determine your functional block requirements by starting from the end of the signal flow, the Presentation. From here, work your way back until you complete Collection:

| | |
|---|---|
| Collection | camera, microphone |
| Connectivity* | cables and connectors |
| Selection | zoom, pan, tilt, orientation, location, direction |
| Signal-processing | white balance, brightness, volume, frequency (frames per second) |
| Recording/Replay | analog or digital formats and the storage media |
| Presentation | video displays, video projectors, loudspeakers, headphones |

*Everything in between all of the other functional blocks, and usually requires twice the length of cable and three times the amount of funding you ever first imagined. A very common mistake made by those new to audio/video system design and installation is to vastly underestimate their connectivity requirements.

### 74.1.2  Device Selection Criteria: Individual Component or Combination

All-in-one camcorders are wonderful for simulation applications in that they package many of the required sight and sound collection and signal-processing capabilities in a single, relatively inexpensive, easily positioned package. The mass retail market for these features results in an endless stream of evermore capable devices at ever lower prices. In the most fundamental sense, a camcorder with a fold-out display and an internal speaker *is* a complete, self-contained audio/video system. That said, there is a threshold of preferred features, and one of the most important is off-camera microphone capability. You usually want the visual collector as far way from the action as possible, while the sound collector as close as possible.

### 74.1.3  Analog or Digital?

Accept analog when you receive it as donated equipment, and use it for pilot projects. Pay for digital and use it for production work. Analog recording devices and their media (tape) have a fatal flaw: serial access. In contrast, almost all reviewers of recorded simulation sessions will need to jump back and forth (random access) to highlight a teaching objective. Only digital recording devices and disk-based media offer the possibility for easy and fast access to desired segments of a recording. Analog recordings have another fatal flaw: degradation by copying. Unlike digital, every analog copy contains less useful information than the previous version. Given that you designed and selected your audio/video system to assure collection of key visual and audible cues, the first analog recording may retain these cues, but with subsequent rerecordings they will become illegible and the whole purpose for making recordings in the first place will be lost. Digital recordings can be exact duplicates of the original. Analog recordings have yet another flaw: degradation by replaying. All analog recording/replay mechanisms require physical contact of a moving flexible storage media (tape) across a solid detection device. The more you replay a tape, and more you use a tape recorder, the more they wear themselves away against each other. Digital disk recording and playback devices are noncontact electromagnetic or optical, and therefore do not degrade with repeated playing. Analog replay devices have very small tolerances for differences in recordings caused by subtle differences in their manufacture and subsequent usage. It is not unheard of a tape recorded in one brand of recorder to not replay in a different brand's device. Similarly with digital recordings, if the session on the disk is not appropriately closed ("finalized"), it may not play in another player. If the format in which the video was saved requires a specific codecs, you might not be able to play it everywhere. And finally, if your file is corrupted because of physical damage or a computer glitch during the transfer, you might not be able to recover any of it.

## 74.1.4 Wide Format High Definition, or Not

Select *or not* until the day when the clinical monitors you use with live patients are in wide format (typically known by its contrast ratio of width to height of 16 × 9 and ubiquitous with all flavors known as "high definition"). Until that time, let others pay the privilege for the newest and flashiest gear. In simulation, as it is currently done, there is no benefit worth the large cost increase for a slightly wider view within one video image. Yes, with wide-screen format, two narrow-format (4 × 3) images can be combined side by side, but the cost of such a combiner, recorder, and display will exceed the costs of two parallel "all-narrow" format devices.

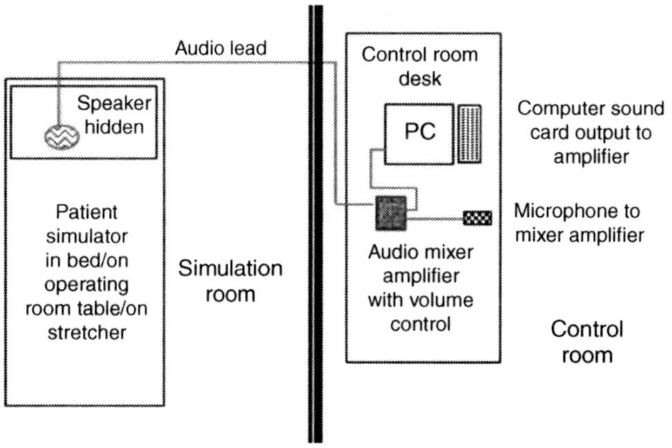

**FIGURE 74.1** Basic audio system for the patient simulator.

# 74.2 Sound Systems

## 74.2.1 Patient Speaking to the Students

Quality vocal communication between your students and their simulated patients are essential. Unfortunately, full-body patient simulator vendors do not take this into account when they sell you their inadequate patient voice system. Fortunately, quality devices, their installation and their use are relatively inexpensive, straightforward, and easy to use. The idea is to collect voices and sounds made in the control area, convey them to the patient, and reproduce them in or very near the patient's head, as shown in Figure 74.1.

The components in the patient's voice audio system are almost identical to that listed above for the overall audio/video system:

| | |
|---|---|
| Collection | public address–type microphone, computer-stored files, sound generator |
| Connectivity | cables and connectors |
| Selection | audio mixer for the various sound sources |
| Signal-processing | audio mixer, preamplifier, amplifier |
| Recording | done by simulator room recording system |
| Presentation | speaker |

Live voices are from a person usually in the control area, and the control area is full of voices that the patient should not convey to the students. The ideal microphone has a *silent* on/off spring loaded switch (normally off!), a very short sound collection range, and a robust frame to survive falls. This class of microphone is called Public Address, or PA microphone.

You may want to have the option of playing prerecorded voices from alternate sources, like a computer or other sound generator. Thus, two or more input sources will have to be connected together. The device for this is called an "audio mixer." In addition, you will require control over the volume, hence the need for an "audio amplifier." These requirements for mixing and amplification are not unique to simulation; two or more channel audio mixer amplifiers are widely available from audio equipment and electronic sound vendors. An amplifier with a power output of 20 Watts is sufficient to drive a patient's voice speaker. There is no need for an expensive, professional singing–quality speaker, but you should make sure that it matches the power output from your audio mixer amplifier so that you do not damage it with the first scream. Alternatively, a number of manufacturers offer small self-powered speakers. These devices have the amplifier, and sometimes as well a preamplifier for a microphone, built inside the frame of the speaker. There is no need for a pair of speakers as our voices are not stereo but mono.

## 74.2.2 Students Speaking to the Patient

The idea is to collect voices and sounds made anywhere in the simulator room, convey them to the control room, and reproduce them for every one in the control area to hear. Given that microphones come in an endless array of types and prices and that no postprocessing can ever recover what a microphone does not collect, then selection of the types, quality (read: price), and number of microphones is an important consideration. If you can only afford one microphone and its attendant signal amplification and processing devices, and your simulator is in a fixed location, then the type known as *Boundary* microphone is an excellent first choice. These are specifically designed to sit flush with a large surface (like a ceiling) and evenly collect spoken voices located over wide distances within the room.

The only warning about its placement is to prevent ventilation air from blowing directly on it, otherwise locate it above the head of the simulated patient (live or plastic). The only drawback in its sole use is that in simulation, all too often, students will converse with their patients with their heads tilted downward, close to their patient's head, and in a quiet voice. If the room is otherwise quiet, a Boundary microphone behind their head and up to 10 feet away will suffice. If the room is noisy, as during a frantic and overpopulated resuscitation scenario, a single Boundary microphone will do a poor job of isolating one speaker's voice. Since every camcorder includes at least one microphone, every additional camcorder adds at least one more microphone to your sound collection system. An additional small, lavaliere sized, microphone, either worn by the live patient or inserted in an ear of the plastic patient is an unobtrusive way to assure that none of the conversation between your students and their patient is too soft to be heard by the people in the control room. As described for the patient's voice, an audio mixer is purpose-designed to mix together a number of different audio sources into one or two channels, and is ideal for blending the signals from two or more microphones.

## 74.3 Audio/Video Options for an Observation Room

### 74.3.1 Live Presentation from the Simulation Room

Audio/video presentation of the live simulation action into an observation room allows additional trainees access to the scenario and freedom to discuss the behaviors without disrupting the students with the simulated patient (Figure 74.2). Such an audio/video link may also overcome physical limitations in simulation laboratory design that did not permit one-way mirrored windows between the simulation room and

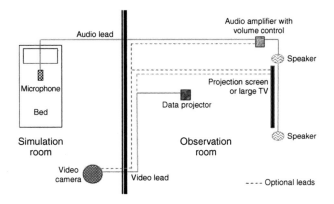

**FIGURE 74.2**   Basic audio/visual installation between a simulation and an observation room.

the observation or control room. In addition, on/off control of access to the sights and sounds from the simulation room can be quick and performed by simple switches that can be located in the simulation control area. The video channels can be from any or all the collected video sources such as cameras and clinical monitors. Likewise, the audio channels can be from any of the microphones, either just those with the simulated patient or include ones in the control area to convey instructors' comments. The listeners in the observation room should have complete control over the volume of the delivered sounds.

## 74.4 Clinical Monitor Image Capture

The more difficult video collection challenge in simulation is the information displayed by the clinical monitor (or patient monitor, or vital-signs monitor). While the installation and use of the required devices is simple, selecting the appropriate devices to produce a satisfactory image is not. The technical cause of this difficulty is the difference in signal format used in the two different visual display industries: computer and television. Until the day when they share the identical format, and it is the only format in use, you will need to use some kind of translation method, and something is always lost in translation. Your method selection difficulty arises from the fact that unlike a few small cheap syringes that you can carry anywhere to test video systems, in this case, the "image" to be collected is generated by a large, heavy, and expensive clinical monitor. This logistical hurdle will reduce your opportunities to try out as many possible solutions compared to the syringe test. Fortunately, your definition for "satisfactory" is just as easily stated: you need to collect, record, reproduce, and display the clinical monitor image with sufficient visual fidelity so that your intended students, at *their* level of competence, will arrive at the same diagnostic decisions when watching the reproduced image that they would when watching the original. ECG pattern recognition is the most challenging visual skill requiring the highest resolution image and is highly subjective in the interpretive minds of the viewers. A simpler test is the accuracy in reading the smallest numbers on the display. For example, if the reported ST segment numbers are clearly distinguished, then the quality of your clinical monitor capture and reproduction system is satisfactory.

There are two basic ways to *collect* the computer-format display signal used by clinical monitors and *convert* it into the television format used by your video system: the direct video camera approach and the scan or media converter approach. In the former, aim a video camera at the clinical monitor display (or a repeater showing the same information). The video camera does the collecting of the photons from the clinical monitor and automatically creates a television-format output signal. If the clinical monitor display is a CRT (not a flat panel), then the collected image will include vertical scrolling bands and barrel distortion. Also, once the camera and presenting display have been mounted and aligned, any vibration of either device, or an object (like a head) coming in between the two will ruin the collected image. Repositioning the orientation or location of the monitor will also remove the monitor's image from view. In contrast, the scan converter is exempt from the camera's line of sight problems as it is hardwired to the clinical monitor's video signal. Note that for simulations using clinical monitors that do not provide easy access to this signal, like most portable monitors, the camera approach is your only option. But in this case, you have to figure out how to keep a camera pointed at a *portable* monitor.

If you can access the clinical monitor's video signal, there is very little downside with using a scan converter. Despite the very large number of different scan converters offered for an equally wide range in price, the first half of the selection process is straightforward. Since the scan converter is a translation device, determine what is at both the input and output sides and use this to select only those scan converters that perform translation between these two. Electronic signal formats *and* physical connector formats must be taken into account, on both ends. The second half of the selection process, making a specific purchase decision, is not straightforward when there is more than one scan converter that meets your format translation needs. The challenge comes from the fact that unlike for almost all other aspects of audio/video choices where the more expensive device is usually the better performer, this simple relationship does not apply to scan converters. Your best approach is to start with the least expensive scan converter choice, test it with your clinical monitor and video system, and then exchange it for the next costlier one. Do this until you run over budget, run out of devices to test, or finally find one that passes your visual acuity test. Expect to spend between US$200 and US$500. Each of you will have to resolve this dilemma according to your budget and institutional culture about expenditures. Once the visual information of the clinical monitor is in the same video format as all the other video signals in your system, it can be relayed, recorded, reproduced, and presented exactly like them.

# 74.5 Video Observation and Presentation Options

Create your A/V system as internally complex as necessary so that its *use* is as automatic, as hands-off as possible. For example: have one camera for every essential view. Avoid requiring someone tasked with redirecting *the* camera to chase the action *during* your scenarios. Also, have one recorder for

each camera. Avoid requiring someone tasked with selecting *which* camera feed goes to *the* recorder. These examples are sure ways for essential actions (the very reason for your A/V system) not to be seen and recorded. However, an occational pan/tilt or zoom is an acceptable added workload in the control room (e.g., to check on an IV drip rate) for the simulator operator.

There are different options for observers to watch unobtrusively. These include simply watching through a one-way window from a darker room, or via video, or a combination of both. The one-way window provides both excellent features and limitations. Once a one-way window is installed, it always works and anyone can use it without any training effort as everyone is familiar with how to glean visual information by looking through a window. Using the one-way window is limited by placement restrictions and sight/sound contamination into the clinical simulation room. If the observers are talking too loudly, they might be heard as windows are a less effective sound barrier than a solid wall. If the one-way window is your only option, the scenario participants might obstruct your view of important interventions. As most full body simulation teaching centers around the body's upper half, orienting the patient feet-first toward the one-way window allows the least obstructed view. If the control or observing room is too bright, participants from the simulation platform might see movement inside the adjacent room. Computer and video monitors show through them.

Similarly, you need to carefully think about your requirements in terms of cameras: number, functionality (controllable zoom, tilt, and rotation or not), and positioning (permanent or mobile). This will depend totally on your training focus, budget, operator skill, and what you want the observers to be able to see. If your cameras can be attached to ceiling tiles as shown on Figure 74.3, ask the installation contractor to provide the system with enough leeway on the signal and power cables so that you are able to move the cameras to anywhere else in the ceiling. Should the configuration of your room change, this would allow you to reposition the camera appropriately.

The number of cameras you are using and the number of observers you expect to have at any one time will influence your choice of display size and mounting position. For these reasons, the displays in your control room will be smaller than those in the observation room. The option of combining your video views using "picture-in-picture," "split screen," or security camera types of "quad screen" will reduce the number of required displays, but this option is not cost free. Devices that mix together several video signals are not free (unless someone gives them to you). Every combination of video signals reduces visual resolution of the individual sources. Do not combine to the extent that you remove the very data you need to see. Displays and cameras are inexpensive today, their prices are falling, and are expected to do so for years to come.

**FIGURE 74.3** Networked Sony digital cameras with auto-focus and remote controllable zoom, tilt, and rotation in the new HICESC simulation center, University of Hertfordshire, U.K. Some cameras have a microphone attached to them and others have the microphone positioned by the head of the patient bed. The overall A/V system integration diagram can be seen in Appendix 74A.1.

## 74.6 Sound Options

You might also want to use one- or two-way audio to communicate between the control room and the observation room without being heard by the simulation participants. As many sound devices have two-channel stereo, and the sounds from the simulation area have little important stereo information, then one channel can be devoted to sounds from the simulation area and the other channel devoted to sounds from the control area. In addition to live teaching, this split audio is useful for discrete voice annotation on recordings made during research studies.

Invariably, individuals in the control area and supporting actors in the simulation area will want to talk with each other during a scenario without the students overhearing them. One solution is the wireless, two-way headset device. Many simulation programs make great use of them for guiding the "on-stage" performance of a novice support actor by an "off-stage" simulation expert. The drawback of the very presence of devices on the most important part of our social anatomy (the head) marks the wearers as even more unfamiliar, further reducing the fidelity of the entire simulated situation. Furthermore, some users cannot disguise the fact that their attention is diverted elsewhere, and thus destroying the key reason for having live support actors in the simulation in the first place: interaction with real humans. An alternative communication device is the telephone: it is so common a feature in most clinical settings that its presence is not distracting, everyone already knows how to use it, everyone is already conditioned to respond to it, everyone already expects bad news to come through it, and last but not least, voice over a telephone can simulate any absent or distant part of the world.

## 74.7 Audio/Video Control Options

The famous KISS principle, Keep It Simple, Stupid, is never more apropos than when controlling your audio/video system during a simulation session. All your attention should be on the students, not your cameras, recorders, displays, and speakers. You can create a very sophisticated system with many options, but the *control* of this technological wonder should be about as complicated as starting a car, putting it into gear, and driving away. The ideal car will just disappear, leaving you with a sense of conveyance in amongst the others on the road. If and when the car gains your attention, you are no longer focused on the world around you. In every one of your decisions about how to design, build, and use your audio/video system, simplicity of actual use is first and foremost. Complex designs are fine, like the complex automatic transmission of a car, but only to the extent that this complexity is hidden from the user and provides an overall net reduction in user effort during simulations.

Recent integration of video within computer networks has provided novel solutions for your consideration. Today, cameras with resolution sufficient for most simulation uses are available with built-in Ethernet capability (Figure 74.3). Some even include microphones. When connected to a network that has sufficient throughput capabilities as required for your video needs, this camera functions as just another data source on the net (see Appendix 74A.1). It also offers many of the same sharing features that are familiar to users of the other, established uses of computer data networks. The downside of a computer/network-based audio/video system is:

- You need immediate access to network competent personnel whenever you want to be sure that your audio/video system is functioning – just in case,
- You need to assure that the rules about who can control your networked system are well understood and followed,
- You need to assure that you have absolute control of where and when your content is being shared,
- You need to assure that no toxic network programs infect your audio/video network.

## 74.8 Control Room

Figuratively, the control room is like the place from where the puppeteer is controlling the marionettes. It is the place where the operators hide their control of the patient simulator from the students, and where instructors observe and discuss with minimal interference. This makes scenarios more realistic as participants do not feel directly observed. This helps them suspend disbelief more easily and encourages them to interact more effectively with the patient simulator. In fixed simulation sites, ideally, the control room is a dedicated room physically separated from the clinical simulation room by a solid, sight-and-sound-blocking wall. However, not every simulation program can afford such a permanent room, nor can they be built in every real clinical area that is temporarily used for a simulated patient experience. In these instances, a semipermanent if not totally portable blind blocks operational activities from the students' view. An example is illustrated in Figure 74.4, and requires minimal alterations to a room. This blind can be built for minimal cost and requires the following elements:

- Office partitions with tinted glazed panels (glass for semipermanent, plastic for portable)
- Thick dark curtains
- Curtain rods and hooks.

Any bright lights within the control boundary should be disabled in order to make the control area darker than the simulation room. This prevents trainees seeing through the

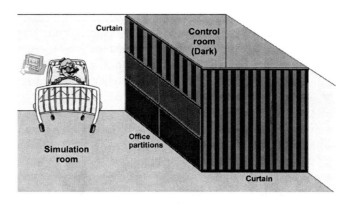

**FIGURE 74.4**  Control room partitioned off within a simulation room.

tinted glass. The curtains can simply be fitted on a rod attached to a suspended ceiling. The floor space required should allow you to fit a desk and chairs to comfortably seat 2–4 people. The installation of such a control room is very practical and can be easily removed should the space need to be used for another purpose in the future.

## 74.9 Working with A/V System Design and Installation Professionals

If you are using contractors who already have performed simulation center or theatrical installations, do not make any assumptions. It is preferable to contact several A/V professionals to obtain different quotes and also to see what technological solutions they can propose to meet your needs. Your needs will certainly differ from anyone else's and you do not want the contractor to miss anything. Before you ever meet a contactor, first have a very detailed set of functional requirements. Then, be able to clearly explain to them, using plans of your center, what you want to be able to achieve. They can't deliver any better than you can convey. It can also be useful for them to see your current facilities so they can have a feel for what you are doing (experiential-based teaching is what we do). It is often easier to go through the explanation of your requirements as if you were actually running a session with participants and observers and explain to the contractor what you are expecting yourself, the participants, and the observers to see or hear at each stage of your session and after the session (storage and retrieval). Ultimately, with the quote, the contractor should give you his "technical translation" of your "educational-simulation talk." It is at that stage that you should check if everything you need has been addressed.

If you are planning a new center from scratch, I would strongly advise you to bring the A/V installation contractor onboard from the start of the project so that the project can be coordinated with the work from the electrical and mechanical engineering contractors. This will ultimately allow you some cost savings and place the cameras and microphones exactly where you want them to be. This may well be the capability of repositioning them over a wide range of places on the walls and across the ceiling. It is only after working alongside architects, building contractors, and electrical and mechanical designers that I have come to realize how busy and cluttered a suspended ceiling void can be with ventilation ducts, outlets, and wires. If things are not planned to fit around the A/V installation, you might be forced to compromise.

# References

Example sources of Audio/Video information usable for simulation purposes (all accessed 12 Oct 2006):

| | |
|---|---|
| MarkerTek, A/V Sales | www.markertek.com |
| Scotia UK | www.smots.org |
| streaming media | www.emedialive.com |
| event DV | www.eventdv.net |
| Digital Content Producer | www.digitalcontentproducer.com |
| Videomaker | www.videomaker.com |
| Vision Systems Design | www.vision-systems.com |
| VideoSystems | www.videosystems.com |
| Digital Video | www.dv.com |
| Government Video | www.governmentvideo.com |

# Rehearsing is the Basis of All Learning

*"Any ideas, plan, or purpose may be placed in the mind through repetition of thought."*

Napoleon Hill

"Play: creative activity, initiated by the young, sustained by enjoyment, modeled upon activities of the old, yet without the irreversible allocation of real resources or the productive purpose that sustain the latter's efforts." Play *is* rehearsal for life. Rehearsing is the basis of all learning. Play is nothing if not an activity that we don't need external persuasion to engage in. All of our students already know how to play. It is through play we learned languages, developed social structures and explored concepts like morality. Now imagine trying to teach any one of these without play, without repetition. Given that all our students learned their earlier lessons through play, and obviously did so quite well, and given that all their patients also will have learned their life lessons through play, would not we in our roles as clinical educators be negligent in fulfilling our responsibilities by dismissing play as a way to help our students become better clinicians?

Just like thoughts, any skills or attitudes can be placed in the hands or heart through repetition of action and their consequences. Education is purposeful modification of behavior, and behaviors come from thoughts, abilities, and priorities. Thus, education is all about changing knowledge, skills, and attitudes. Our learning needs are complex like our nutritional needs are complex: provided by different food sources, digested in different manners, incorporated in different concentrations in different parts of ourselves, used for different purposes. Our abilities, our behaviors, and our performance are dependent upon our ability to assure that every meal includes all essential ingredients.

During each and every simulation session, our students are ingesting from the banquet of experience that we place before them – some more so than others. Every student has a different perception of this feast, depending upon different capabilities and passions that they bring. Thus, each student takes away a different subset of new knowledge, skills, and attitudes. Our task is to craft their experiences so that both the magnitude and direction of their changes lead them to become better clinicians. Our charge is to do so within finite budgets of time, passion, wisdom, and money. Games are goal-oriented play. A game can be a very powerful, effective, and economical tool to help us fulfill this charge.

Anthony Brand not only embraces play and describes the fundamentals of play as they apply to our needs, but also has created a game specifically designed for clinical educators and clinical learners to "play" together. Not only can you incorporate his game in your teachings, but you too can emulate his passion and energy and create your own clinical game or two. Guillaume Alinier offers both a big-picture view of play and games as well as many clear and concise fundamental "rules" of play and games applicable to clinical education. Ronnie Glavin recognizes that our students are not infants (despite some of their immature antics) and describes how we can adapt those concepts of play and games we learned as children to support our efforts with adult learners. Diane Seibert describes the collective findings of professional educators of education and offers pragmatic recommendations on how we can stand on their shoulders to extend our own capabilities. Kristina Stillsmoking presents numerous tools and techniques forged in the heat of hand-to-hand clinical teaching. Hal Doerr and W. Bosseau Murray conclude with a three-pointed structural metaphor upon which they arrange both the theory and practice of simulation-based learning.

# 75

# An Approach for Professional Development: Triad Gaming Techniques in Simulation

Anthony Brand

## 75.1 Play?

"Oh you mean play!" – and yet inherent within the concepts of simulation techniques lies an element of play or gaming. The history and range of simulation approaches are large and include all manner of attempts to model real situations. Here, in addition to presenting a broad overview of techniques and approaches, this topic focuses upon the uses of an enhanced type of role-play in the development of (clinical) skills and attributes.

"Oh you mean play!" – a not untypical reaction when I attempt to describe what I have found to be a highly effective simulation technique. However, when used in an appropriate context with well-briefed participants, role-play simulation using the triad approach of three people produces deep and sustained learning experiences.

## 75.2 Gaming

Simulation covers a wide range of approaches and methods; from those grounded in abstract computational (gaming) techniques to highly people-based situational role playing. These approaches may be seen as delineating extreme ends of a spectrum. At one end taking *roulette* to be emblematic of gaming and associated with casinos in Monte Carlo, we find that this has been refined into a well-established computational approach in the physical sciences using simulation and called the Monte Carlo [1] method. Some will recall the suite of Strategy/Wargames that was developed and introduced in the 1980s by Avalon Hill. These attempted to achieve a high level of authenticity including, where appropriate, the fog of war. Later, the hexagonal squared boards and cardboard counters were replaced with computer-based versions.

While computational modeling is now most frequently linked with digital technology, readers should be aware that earlier analog approaches, including pencil and paper mathematical modeling, existed and produced a high level of valid and reliable data. In the last century, various ingenious devices such as a hydraulic model of the British economy were developed. An example of this apparatus, devised by Bill Philips [2], is on display in the Science Museum, London. It continued to be used as a teaching aid at the London School of Economics until 1992! One presumes that it became redundant with increased confidence in software algorithms.

Even so, electrical analog apparatus has been used to represent heat flows through buildings and construction elements in other areas of education, see Muncey [3] for example. Currently, wind tunnels are still in use, with aircraft and automobile physical models in various forms to investigate dynamic interactions and validate solid–fluid computational models.

Moving to the other end of the spectrum described here, what might be described as role-play, also has a well-established tradition in simulation approaches. Visualization and (guided) imaging is linked with the work of Assagioli [4] and explored further by Ferrucci [5] and Whitmore [6]. Gestalt techniques are described in the work of Perls [7] and Journaling by Progoff [8]. Drawing from what some might seem as distant versions of simulation approaches, we find that Burnard [9] advocates experiential learning activities in teaching interpersonal skills. When introducing the uses of role-play, Burnard establishes the scene with a number of players, which includes the principal actor, supporting actors, an audience, and a facilitator. The work concludes with:

(1) the principal actor self-reports on her performance;
(2) the supporting actors offer the principal actor feedback on her performance;
(3) the audience offers the principal actor feedback on her performance;
(4) these three events are repeated for all the other actors.

Burnard (1989) p. 20

It is possible using the form of triads that are most frequently formed by students or course participants to reconstruct this potentially valuable approach without the need of a cast of thousands.

## 75.3  Triads

Elsewhere in this publication, simulation approaches in a clinical setting using virtual environments and/or sophisticated mannequins, potentially in combination, are presented. These may be regarded as the high-cost and highly technical gaming approaches. However, using a triad of three people (students or course participants) in *all* roles is a low-cost method. When used, for example, in developing clinical communication skills, the three roles are typically patient, practitioner, and observer. This simulation technique is based upon already well-established approaches used in clinical skills training.

Curriculum enhancements in health care and related disciplines have placed increasing emphasis upon the skills and attributes of trainee clinicians, and these are best developed in a context, which simulates in a realistic fashion, encounters between practitioner and patient. This is hardly a new area of training, and, typically, has been addressed by Kurtz, Silverman, and Draper [10]. The Calgary-Cambridge Observation Guide forms a central component of their book, and it has progressively become a core approach in clinical training. Kurtz, Silverman, and Draper present a continuum of teaching methodologies ranging from didactic to experiential and include simulated patients and role-play (Figure 75.1).

They describe a number of challenges associated with the use of simulated patients, the right half of the continuum, and these include expense (if professional actors are used), selection, training, and hidden agendas. Role-play, however, is seen as cheap, requires little training, is always available, and allows the clinical learner to adopt the patient role. Acknowledgment is given to the problems of participants being self-conscious and not being able to shed their clinical knowledge when acting in the patient role. Nonetheless, role-play taken together with Calgary-Cambridge Observation Guide appears to offer a powerful approach to initiating, enhancing, and evaluating communication skills. An overview of the potential approaches for the observation of learners interacting with patients is described by Kurtz, Silverman, and Draper and includes:

- Audio and video recordings
- Real patients
- Simulated patients
- Role-play

Kurtz, Silverman and Draper (1998) pp. 55–56

However, there is a missing dimension in what has been described so far – the use and applicability of a *student* as observer. In the model so far described by Kurtz, Silverman, and Draper, the teacher/facilitator is the observer and potentially engages in a combination of formative and summative assessments/feedback.

In the triad approach described here, the number of participants in the encounter has been increased to three through the addition of an observer. My first introduction to this approach was as a student in a Transpersonal Counseling course. While a superficial examination of the triad approach may be viewed as an artificial and unrealistic play, I can testify to a far deeper

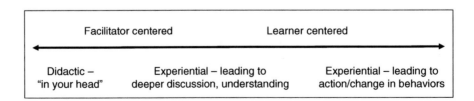

**FIGURE 75.1**   The Continuum of Teaching Methodologies (from Kurtz, Silverman and Draper (1998, p. 54)).

impact being achieved. This has been corroborated in my use as a trainer with a wide range of health care practitioners.

# 75.4 The Approach and Ground Rules

Let us start with the premise that the simulation will cover some form of clinical encounter with people I shall call clinician and client – substitute your own terms as applicable. Then, as indicated above, a third person is introduced – the observer. At first sight, this would appear unrealistic and unrepresentative of real situations but it does, in some ways, resemble the fly on the wall documentary. Yet, this is the essence of formal education: a curriculum orchestrated by the teacher. Ironically, the first and most important ground rule for the observer is that they must not participate in any way in the interactions between clinician and client, and this will be amplified and illustrated in the scenarios provided here. The distancing of the observer may even extend to being out of direct sight of the other two. There are sound reasons for stipulating this golden rule – mainly since the observer will be required to give feedback, in a neutral fashion, about features and aspects of the interactions observed to the other two at a later time. So, a typical session with three students might look as follows:

(1) Role-play between clinician and client;
(2) After the role-play session clinician and client complete a dialog about the encounter from their individual perspectives;
(3) The observer provides independent feedback about the encounter and the dialog.

The independent voice of the observer is a crucial feature associated with the success of this approach. Without this voice, the other participants would lack the benefit of constructive feedback, which may indeed challenge self-perceptions and support deeper situational awareness. Another twist exists – the observer role is found to be a highly privileged one in which this person gains as much, if not more, than the others from the encounter. These features can be further amplified in the following microscenarios.

First, the less than ideal: in which the three independent roles are poorly played. It is in instances such as these where the criticisms of "it just being play" and "being unrealistic" are generated. Even so, it is possible that some practical benefits may still accrue for the participants. The worst-case scenario is when the observer fails to conform to some key ground rules – the first being to maintain complete silence while observing the encounter between therapist and client, even with regard to body language. Inevitably, there will be many aspects of the encounter, which will tempt the observer to interject and offer comments. This would be fatal to the value and purpose of

the game play. By resisting premature feedback, the observer is uniquely placed to help construct an independent reality of the encounter.

This leads on to the next ground rule in which the observer should resist taking sides. A neutral position/voice is best in providing and establishing a feedback dialog. Avoid confrontational comments such as "you did that badly," "you should have done it this way," etc. As will be illustrated in the next scenario, there exist more progressive and enabling approaches, which will allow the student as clinician to realistically own the good and poor aspects of the encounter. While the participants may come to some mutual agreement about the order in which each speaks, this rather unstructured approach may lead to confusing and conflicting feedback. A more robust and structured approach is outlined in the following scenario.

Next, the ideal: here, all three participants have sufficient skill to engage fully in the requirements of each role and in particular, the observer is silent throughout. Experience indicates that the following order of unpacking the encounter has significant benefits and is the most productive:

(1) Clinician
(2) Client
(3) Observer

The clinician is invited to speak first, and addresses the questions "how was it for you; what went well; what went less well; what might you do differently in future?" Sometimes, the clinician is overly self-critical and negative about the performance, but this can be picked up and corrected later by the client and/or observer.

Then follows feedback from the perspective of the client about the encounter. This often contains some less than positive comments that, however, may in the longer term be ameliorated or challenged by the observer.

So, now we arrive at the key reason why the observer has been silent up to this point. On the basis of what has been viewed in the encounter and hearing the self-evaluation of the clinician and feedback from the client, the observer can establish a neutral, nonconfrontational, and mediated dialog. While this will mainly be focused upon the clinician, it does not rule out the client. The style of dialog used by the observer should be reflexive in tone and open in nature. This will help to establish a context in which the student as clinician is best placed to arrive at an accurate evaluation about current skills and attributes. The aim is to achieve a realistic review, which supports and encourages continuing development.

The observer also has the opportunity to reflect back to the client about the feedback provided. Again, a reflexive tone is appropriate; "when you said xxxx, what were your immediate feelings?" and so forth.

## 75.5 Some Observations and Analysis

Returning to the approaches described by Kurtz, Silverman, and Draper, a minor but important distinction needs to be made between role-play in which there is acting to a script, or to a closely defined scenario. Here, the accusation of the process departing from reality starts to appear, with some justification.

Burnard identifies a number of limitations to role-play, which can be summarized as: role-play is never real; it takes place in a safe environment, whereas in real life, it is neither safe or conducted with supportive colleagues; it may appeal only to those who are extroverted; the student as client may be anxious to please, and finally he questions the transfer of knowledge to real settings – Burnard (1989) pp: 20–21. However, he continues in drawing a distinction between role-play and skills rehearsal, and this is most certainly recognized in the triad approach described here.

> It is important to differentiate between role-play and skills rehearsal. Role-play, as we have seen, involves inviting people to play out roles other than their own. Skills rehearsal, on the other hand, is where skills such as counseling and group facilitation are practiced but with the person remaining as who she is at the present time.
>
> Burnard (1989) p. 21

Not withstanding these real concerns, the experiential nature of the encounter has the ability to support deeper learning so long as we avoid the obvious pitfall:

> We had the experience but missed the meaning.
> Eliot [11] (1963) p. 208

In the triad as discussed here, the student as observer is a crucial friend in grounding the experience of the others. Even so, this may not close the loop, as pointed out by Johnson [12], when discussing dream analysis. He adds an additional stage beyond those of discourse and interpretation found in regular analysis:

(1) Making associations
(2) Connecting dream images to inner dynamics
(3) Interpreting
(4) Doing rituals to make the dream concrete
Johnson (1986) p. 51

The approach described here meets many of the cost challenges identified by Kurtz, Silverman, and Draper. With all of the positions being played by students or course participants,

you the session instructor assume much more of a role of facilitator. The most significant responsibility for you is in training the participants to engage effectively in the role positions. Subsequently, facilitation may simply be limited to that of being a time keeper and enabler who smoothly move the participants through the processes described above. The work in progress and developmental nature of the engagements need to be continuously stressed by the Tutor.

## 75.6 Conclusion

Gaming simulation using the triad approach has demonstrated significant advantages in specific areas of professional development. It is likely to be most effective when the participants, though in role, are playing themselves. The potential low cost is an additional benefit, as is the likelihood of it being a low risk but highly effective technique. At the heart of a successful encounter is the concept of experiential learning, see Kolb [13], and reflective practice, see Schön [14].

## References

1. Righini-Brand, A. A Study of the Thermodynamic Properties and Electrical Resistance of the Vanadium-hydrogen System using Ultra-high Vacuum Techniques. PhD Thesis University of London, 1972.
2. Further details of this may be found at: http://www.ingenious.org.uk/Read/TheEnvironment/Predictingtheweather/Justbecauseyoucanmodelit/ http://www.ingenious.org.uk/See/?s=S1&ObjectID={4B441589-6E8D-FD97-D32C-D79849763124}&source=Search&target=SeeMedium both accessed 8 Sept 2007.
3. Muncey, R. W. R. *Heat Transfer Calculations for Buildings.* Applied Science Publishers, 1979.
4. Assagioli, R. *Psychosynthesis.* Crucible, 1990.
5. Ferrucci, P. *What We May Be.* Turnstone, 1982.
6. Whitmore, D. *Psychosynthesis Counselling in Action.* Sage, 1991.
7. Perls, F. S. *Gestalt Therapy.* Verbatim, Bantam Books, 1971.
8. Progoff, I *At a Journal Workshop.* Dialogue House Library, 1975.

9. Burnard, P. *Teaching Interpersonal Skills.* Chapman and Hall, 1989.

10. Kurtz, S., Silverman, J., and Draper, J., *Teaching Learning Communication Skills in Medicine.* Radcliffe Medical Press, 1998.

11. Eliot, T. S. *Collected Poems.* Faber and Faber, 1963.

12. Johnson, R. *Inner Work.* Harper Collins, 1986.

13. Kolb, D. A. *Experiential Learning.* Prentice Hall, 1984.

14. Schön, D. A. *The Reflective Practitioner.* Basic Books, 1983.

# 76

# Learning Through Play: Simulation Scenario = Obstacle Course + Treasure Hunt

Guillaume Alinier

## 76.1 Playing as Learning

Playing is a developmental activity for children of all ages as they learn to interact with their peers and pretend they are in different roles and activities. They will, for example, role-play the preparation of tea and dinner for their friends, or might pretend to fight. All these simple games have their place in childrens' skills development. As strange as it might seem, the same principle still applies to adult learning, but is normally given a different name, otherwise it would not be part of a job, and we would not be paid for taking part in these activities. A reason for giving such activities a different name, is that it is normally more structured, requires preparation, and encourages reflection. For example, the police and armed forces are regularly training in simulated circumstances. Some strategic board games used socially are sometimes used more seriously by military officers to assess offensive or defensive techniques. Another very popular medium of playing for children and teenagers includes computer games. The very same software or games are sometimes used by adults for more intentional behavior or skills changes such as tactical, observation, leadership, or decision-making training. Such interactive computer-based games are a form of simulation and are often categorized as screen-based simulators. An example of such a game includes flight simulators that can help would-be pilots become familiar with the instrumentation and flight characteristics of a range of airplanes. Similar applications are now becoming available for the training of health care professionals. They are now rapidly developing to the extent that they allow several team members to interact with each other (even if they are in different locations) and with one or more model-based computer-generated entities.

An obstacle course challenges *and* rewards by requiring participants to overcome difficulties near the limits of their maximal performance, and offers the opportunity to return for improvement. Simulation allows clinical instructors to "schedule disasters", to challenge their students and to do so safely for *all* involved. Treasure hunts consist of cues and clues requiring both detection and deduction if the participants ever expect to find the reward. Simulation allows clinical instructors to intentionally craft and present clinical "triggers" that they wish their students to gain familiarity with sensing and interpreting.

## 76.2 Creating the Right Atmosphere

Learning can take place under all kinds of circumstances, but is probably achieved to a greater degree when the environment and atmosphere are conducive to it. Clinical simulation might not be a familiar territory for your participants and their instructors, and it will certainly prove a daunting and nerve-racking experience in their first instance. It is important to reassure them to ease off the pressure, while at the same time reminding them that it should be taken very seriously. I do emphasize the fact that we are expecting them to unintentionally make mistakes and hopefully learn from them, so that in real clinical practice they recognize the initial signs of an impending error and prevent its occurrence. The pressure on candidates and their anxiety should be caused by the actual simulated patient case they are dealing with, and not unintentionally added to by the team of faculty or their peers observing. In order to diminish the distractive stress caused by the presence of faculty and peers, a supportive atmosphere should be created, and the candidates should be left to freely manage the patient. Some simple steps to adopt are to keep the nonparticipative candidates and the faculty away from the simulation platform during scenarios, but still allowing them to observe by some other means. Hence, no prompts can be given unintentionally and the observers' reactions will not bother the scenario participants. Although I have to admit that on occasion we have been able to hear a burst of loud laughs from the observers who were two doors away from our simulation room. Such laughter is a good sign: the students are responding both mentally and emotionally – a requirement for forming memories.

## 76.3 The Simulation Teaching Philosophy

Once "practicing skills" have been accomplished, then comes "putting skills in practice" as if it was involving a real patient. Clinical simulation is about realistic scenario-based training where the participants are leading the patient care and treatment, with minimal prompts from faculty. Faculty should remove themselves from the simulated treatment area, and turn over full responsibility to the students, unless they are meant to take part as actors. Trainees or faculty not involved in the patient case should not remain observing directly within the same room, but observe through audio/video links or one-way windows.

The advantage is that in this case no harm can be made to a patient and we can, and should, let participants experience the effect and "own" the consequences of their mistakes in terms of treatment provided or not. The faculty is primarily facilitating the briefing and debriefing during the session from notes taken about the participants' actions during the scenario. This is done to ensure that all important points are raised during the debriefing with the participants. For the same reasons that they were hidden observers during the scenario, faculty should place responsibility upon the participants for their own review of their own behaviors in the scenario. Debriefings should start with a leading, open-ended question from the faculty, one designed to guide the participants' focus toward a review of that scenario's key teaching objectives. The ideal scenario will contain appropriate cues, actions, and reactions to convey that session's teaching objectives. During debriefings, faculty should consider every issue that they have to highlight as an indication that either their design of the scenario needs improvement or their scenario is not matched to the participant's current level of abilities. The different points discussed in this paragraph are summarized in Table 76.1 and presented as the 20 steps to producing a successful simulation session.

## 76.4 Reflection

Learning is enhanced by postexperience reflection. After having taken part in an educational activity, trainees should be

**TABLE 76.1**    Sequence to play the game known as Clinical Simulation

| Activities | Aim |
| --- | --- |
| 1. Welcome the participants | Create a friendly and supportive ("good") atmosphere for the rest of the session. |
| 2. Introduction of faculty | |
| 3. Pre-session evaluation (Optional) | An opportunity to hand out a questionnaire to find out what participants think simulation is about before you start giving explanations.* |
| 4. Explanation of session plan | Inform the participants of the succession of activities forming part of the simulation session. Emphasize if the session purpose is for their learning or your testing. |
| 5. Ice-breaking activity (if participants do not know each others) (Optional) | This will help in creating a more relaxed atmosphere for the participants and should make the communication between participants easier during the scenarios. |
| 6. Explanation of the basic simulation rules, the philosophy | Participants need to know what behavior is expected from them as scenario participants or observers, and what input the faculty will have during the session and scenarios. Tell the participants that they are expected not to discuss the scenarios with other people and respect the confidentiality about the session and the participant's actions: this briefing should never be forgotten. Remind participants that the scenarios are to be taken seriously. |

**TABLE 76.1** (Continued)

| Activities | Aim |
|---|---|
| 7.** Introduction to the patient simulator and simulation environment | This should help participants in familiarizing themselves with a type of patient they are probably not used to care for, as well as operate in an environment that will differ from their usual clinical area. |
| 8.** Basic "warm-up" scenario for the whole group. | This prepares participants for the following scenarios. It helps them relax and perform more realistically in the following scenarios. Valuable teaching points can still be made from that scenario in the discussion following it. |
| 9. Clarification discussion | This helps participants and faculty in clarifying any points that might have been omitted or misinterpreted during the briefing. As a group, verbalize and hand out role-playing cards to all participants. |
| 10. Separation of the participants into groups for the following scenarios | Create appropriate groups with the desired mix of skills. |
| 11. Briefing of the first scenario | Give participants the information they require to start the scenario. This might include patient age, gender, and presenting condition. Participants will be able to find out more details as they question the patient or a relative (actor). |
| 12. Scenario execution | As a faculty, you might intervene with the scenario, but let participants remain in charge. You might be involved as an actor, following a script or assist the participants under their specific guidance. It is important to take notes of issues that need to be raised and discussed during the debriefing. |
| 13. Ending the scenario | Find an appropriate moment to stop the scenario when required. Reconvene all participants and observers in the debriefing room. |
| 14. Facilitate the debriefing by asking the participants to highlight the positive and negative aspects of their performance. Involve the observers as they will have a different perspective | As a facilitator, you should insist that your students monopolize the discussion. Every time your students hear you give them a correction is one less opportunity for them to take responsibility for their own growth and development as a professional. Every time you find yourself needing to comment on errors is one less opportunity for letting the students' experience do the teaching for you. Encourage participants and observers to express their own opinions. The discussion should be conducted in a positive manner to encourage the participants to "change their behavior" next time (rather than pushing them onto the defensive by negative comments). Once all participants and observers have covered all the points they had identified, cross check with your own list of notes taken during the scenario and preidentified for that given scenario. |
| 15. Summarize the important learning points of the scenario | This is to emphasize the important points all the participants should take away. |
| 16. Repeat points 11 to 16 for each of the subsequent (following) scenarios | |
| 17. Conduct an overall session debriefing | You should ask each participant to highlight one important point they will take away from the session ("take home message"). |
| 18. Post-session evaluation | This is an opportunity when you can evaluate the participants' perception of your simulation session. Inform the participants that they might be contacted later in follow-up to determine if the skills learned were applied in their clinical practice. |
| 19. End of session | Close the session on a positive and supportive note. |

* Much time can be saved by sending background information to participants prior to the day of their session. See Chapter 63.

** 7 and 8 are sometimes augmented by the faculty performing a demonstration scenario for the participants.

**The six components of the Reflective Simulation Framework (RSF)**

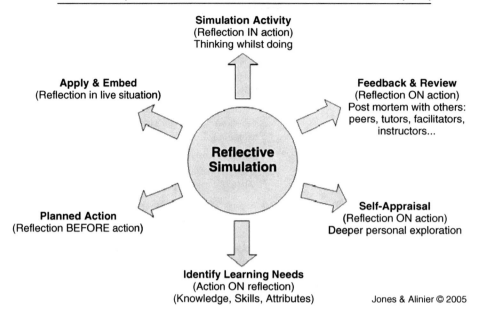

**FIGURE 76.1**   Diagram of the Reflective Simulation Framework [1].

encouraged to reflect on the actions they made during scenarios, and the outcomes of their actions on the patient or efficiency (functioning) of the care team. Reflection is now commonly recognized in health care professional benchmark statements, and seen as a desirable and necessary process for both clinical practice and continuing professional development. It can be described as a process through which students can purposefully revisit, analyze, evaluate, and learn from experiences, whether it occurred in a simulation center or in the real clinical environment. Having people start reflecting might be more easily achieved in the simulation center rather than in the clinical area where events could be perpetually following each other without allowing time for reflection. Figure 76.1 shows our Reflective Simulation Framework [1] designed to structure and guide the students' experiences in ways that foster deep learning about critical incident management. The underlying pedagogies are grounded in the works of Dewey [2] and Kolb [3]. The Reflective Simulation Framework comprises six dimensions, and is an iterative learner-centered model that can be used flexibly to explore the simulated experience in order to enhance learning and practice and crucially act as a basis for multiple feedback systems.

As the use of simulation-based learning has increased considerably, a real need has been identified to provide students and instructors with a more concrete approach to engaging with reflective learning. We propose that unlike other approaches to learning, the Reflective Simulation Framework can be used as an advanced organizer to help accelerate the learning process not just after the event but also prior to and during simulation experiences. We believe that the Reflective

Simulation Framework model should be handed out to students alongside other introductory materials as part of the overall orientation to simulation. (It might also be sent to participants prior to the session to engender some familiarity with the concepts.)

The definitions, features, and use of Reflective Simulation Framework when used in combination present a creative and flexible approach to reflective simulation learning and teaching in structured and guided ways. The visual representation of reflective processes promotes reflection as a conscious formal activity, with purposeful outcomes for personal development and enhanced professional practice.

Reflection should also be an important aspect to the simulation faculty, which is why it is often recommended for faculty to meet and discuss (debrief) after a session. As busy clinicians, time is often not devoted for reflection, but it is really worthwhile and helps the faculty team operate in a more coordinated way. The postsession evaluation questionnaire suggested as step 19 of Table 76.1 can also provide you with some valuable information about how the session went from a participant's point of view. The comments should be considered and reflected upon, so that you can improve the learning experience of the participants or give them, if requested, more information or explanations about particular aspects of the simulation session. Some comments might be specific to particular facilitators and the way they directed a scenario or conducted the debriefing. Although that is very useful to help facilitators develop their expertise, such "personal" feedback should be used with caution and the source should remain confidential. The facilitator could be offended, or the particular group of participants could be picked on at a later stage such as during an examination. The way to avoid personal "feuds" it is to provide overall feedback about the evaluation forms only periodically, hence over a much larger number of participants.

## 76.5 Conclusion

As we have all experienced in our youth, learning does not have to be an unpleasant process. Hard work and exhausting, yes, but unpleasant, no. Facilitated in an appropriate manner, that is, respecting the simulation teaching philosophy and creating an atmosphere conducive to learning, simulation-based clinical training is also very well perceived. Both candidates and faculty enjoy simulation for the critical challenges it can expose them to, while still operating in a safe environment and without putting lives at risk. Being observed, being encouraged to reflect through the debriefing, and receiving feedback presents them with a very particular experiential learning opportunity that ensures they adopt the best practices. During and after simulation-based training, reflection should be encouraged and related to actual clinical practice. The proposed Reflective Simulation Framework is valid not only for trainees and clinicians undertaking professional development activities, but also the faculty team facilitating the simulation session.

The 20 steps to conducting an effective simulation session provide only a small list of primary issues to consider while planning a session. The preparation stage for both parties cannot be left until the last day before a session as it would certainly be detrimental to the quality of the scenarios, hence not maximizing the learning outcomes for the participants. For the faculty team, group reflection can be used to consider the participants' feedback, improve the running of the session, modify or develop scenarios, develop new props, or even consider research questions.

## References

1. Jones, I. and Alinier, G. Reflective Simulation: Enhancing the student' learning experiences through structure and guidance. *Simulation in Healthcare* 1(2), 98 (2006).
2. Dewey, J. *How We Think*. DC Heath & Co. Boston MA, (1933).
3. Kolb, D. *Experiential Learning as the Science of Learning and Development*. Prentice Hall, Englewood Cliffs, NJ, (1984).

# Adult Education
# Methods and Processes

Ronnie J. Glavin

## 77.1 Adult Learning

It will come as no surprise that a field as complex as adult learning will probably not be adequately explained by a single theory [1]. It is very tempting to write a chapter that goes into the background disciplines of epistemology, psychology, sociology, etc., that inform and underpin the theory of adult learning and that have influenced adult education. However, I shall resist that temptation and shall look at the outcomes and published works of some of the people who have had a great influence on adult learning, and look at how this can be translated into good practice when producing simulation-based clinical teaching programs.

## 77.2 Andragogy – The Art and Science of Helping Adults Learn

The key name in this field is Malcolm Knowles who described five assumptions based on his work with adult learners [2].

1. As people mature, their self-concept moves from that of a dependent personality toward one of a self-directing human being. Adult learners are capable of determining their own learning needs and of finding means to meet them.

2. An adult learner accumulates a growing reservoir of experience, which is a rich resource for learning. This experience can be brought to bear on new learning, and enhance the new learning significantly. It can also provide an effective context for the acquisition of new knowledge and skills.

3. The readiness of adult learners to learn more is closely related to the developmental tasks of their social roles. Adult learners value learning that integrates with the demands placed upon them in their everyday life.

4. There is a change in time perspective as people mature – from future application of knowledge to immediacy of application. Thus adult learners are more problem-centered than subject-centered in learning. Generally, adult learners value learning that can be applied to authentic problems that they encounter in everyday life.

5. Adult learners are motivated to learn by internal factors rather than external ones. The internal desire to succeed, the satisfaction of learning, and the presence of personal goals have a greater effect on motivation than external incentives and rewards.

Knowles drew seven principles from these assumptions. These principles are:

1. An effective learning climate should be established. Adult learners should be comfortable, both physically and emotionally. They should feel "safe" and unpressured to express themselves without judgment or ridicule.
2. Adult learners and instructors should be involved in mutual planning of methods and curricular directions. Involvement will help assure that collaboration occurs (between the learners and the instructors) with regard to the content and learning process. It also will increase the relevance to the learners' needs, as they will have provided some input into the curriculum/syllabus.
3. Adult learners should be involved in diagnosing and establishing their own learning needs. Once again, this will help to ensure meaningfulness and will trigger learners' internal (intrinsic) motivation. It will also promote self-assessment and reflection, and effective integration of learning.
4. Adult learners should be encouraged to formulate their own learning objectives. The rationale for this is the same as (3) above. Learners are thus encouraged to take control of their own learning.
5. Adult learners should be encouraged to identify resources and to devise strategies for using them to accomplish their objectives. This principle connects adults' learning needs to practical resources for meeting their objectives, and also provides motivation for using such resources for a specific and focussed purpose.
6. Adult learners should be helped to carry out their learning plans. One of the key elements of motivation is expectancy of success. Learners will become discouraged, and lose their motivation, if a learning task is too difficult. Also, too much pressure without support can lead to a decrement in learning.
7. Involve adult learners in evaluating their learning. This is an essential step in a self-directed learning process. It requires a process of critical reflection on experience.

I shall now look at each of these principles and their application to simulation-based clinical learning. Where necessary, I shall bring in the work of others where I feel that it is relevant.

## 77.3 Establish an Effective Learning Climate

This principle is especially important in simulation-based training because the potential for destruction of "self-image" is considerable. Professionals' image of who they are is connected with the roles and responsibilities they have as part of their job. The self-actualization, that has been a potent motivator, can become a source of vulnerability. There is no hiding place in that simulated clinical environment, and we have had several experienced clinicians comment on how apprehensive they felt. One individual confided that he could not sleep the night before his visit because of fears that "I will let myself down." If participants' perception of their abilities is brought crashing down during a scenario, then the psychological fallout will prevent any positive educational change from taking place. "So tread softly, for you tread upon my dreams" [3].

So, what can you do? We have adopted the following approaches:

a. Appropriate design of the course matched to the intended participants. Any notices, advertisements, requests, or publicity material can be explicit as to the target audience, especially in terms of level of experience or current stage of training (e.g., for first year residents in Anesthesiology; for all clinical career grade staff who currently work in the Emergency Room, etc.).
b. Selection and recruitment of faculty. This is the last place where teaching by humiliation should occur. If would-be faculty do not share this approach, then they do not become faculty. As chefs on television say "First, choose good ingredients. If your basic materials are poor, no amount of subsequent preparation will convert them into good ingredients."
c. Appropriate recruitment of participants. We target the groups we wish to attend. Occasionally, on industry-sponsored courses, we have had participants who were inappropriate for what we had planned. We had to change arrangements, including scenarios, to ensure that the environment retained its psychological safety.
d. Find out the details of your learners' health care backgrounds, not just their job titles. We do this in the general introduction. Some learners will have a wealth of experience in a related field; for example, in the United Kingdom, we currently have International Medical Graduates who are in first year residency posts, but may have several years of experience in their own country.
e. Briefing; every minute spent setting the scene and laying out the boundaries before the simulation experience is worth one hour of tears and handkerchiefs afterward.
f. Have a contingency plan for modifying the scenario in such a way that the learning outcome will not be compromised, but neither will the "hot seater."

g. Have an "emergency exit" or a "fire escape." We always have a "deus ex machina" (i.e., the "machinations of a god" that tells them what to do, or that the instructor, playing a god, causes to happen) – a faculty member who can enter the scenario, sometimes unannounced, to ensure that no adverse events affect the patient or the "hot seater." We always like to have a faculty member on the floor to help prevent a run-away event getting out of control. We also have a telephone in the clinical area and use this to give some advice; for example, a more senior clinician may make an unsolicited enquiry to determine how things are going and may help prevent the hot-seater coming "off the rails."

h. Be prepared to follow up participants who have had upsetting experiences. We have not yet had to put anyone on suicide watch, but we have had individuals who have been upset, and we now have some common sense strategies to help them restore their self-esteem.

We take this follow-up very seriously, as do most of the staff that we know who work in other simulation centers.

## 77.4 Involve Adult Learners in Mutual Planning

This may seem a strange concept for planning of a simulation-based course where not knowing the exact nature of the scenarios is the usual practice. How can the learner be involved with the "mutual planning," and still not know what the scenario will be? However, we believe that when designing a course, it is important to have a *representative* of the target group feeding information into the design and content of the course. "This course is not for our benefit, we present this course for the learners' benefit, and our role is to help you make the best use of the simulation center for that goal."

## 77.5 Involve Adult Learners in Diagnosing Their Own Learning Needs

This operates at two levels. Involvement in the *design* of the course as described above. This involvement may require some help in identifying those learning needs. This may be a casual conversation relating to which areas are difficult, or may involve a formal needs assessment exercise with the target group (see Chapters 7, 8, 12, 13, 14, 15, and 71 on curriculum and simulation). It also operates at the level of helping the individual participant *identify strengths* and *weaknesses* – this will be dealt with in much more detail later. These can help the participant focus on future learning needs.

## 77.6 Encourage Adult Learners to Formulate Their Own Learning Objectives

As with the above category, this also operates at the same two levels. When designing a course, the method at our center, the Scottish Clinical Simulation Center in Glasgow, Scotland, is to enlist a representative faculty that will include individuals responsible for the curriculum, individuals who teach on the course in the real clinical environment, and representatives from the learners. We begin by working out the learning outcomes that we want the course participants to achieve. These outcomes are then translated into scenarios, which in turn are tested – usually with the faculty acting as participants – so that we have a deliverable course that sets out to do what it is supposed to do. The value of having the clinical teachers is that the learning outcomes of the simulator-based course are reinforced in the real clinical environment.

Individual participants will have their own learning needs, and although a scenario may have been designed to promote a particular set of learning outcomes, one always has to take into account the participants' concerns. The areas they wish to concentrate on at debriefing may not be the ones intended by the course designer. It would be a wasted opportunity to ignore such concerns, thus being flexible, sensitive, and responsive to the needs of the participants is yet another of the skills required of all clinical simulation staff and faculty.

## 77.7 Encourage Adult Learners to Identify Resources and to Devise Strategies for Using These Resources

This section overlaps with the chapters on simulation and the curriculum. Simulation is an educational resource. I feel that one of the strengths that I can bring to the table as a simulation center professional is that I know what can be achieved and what is unlikely to be achieved. When designing simulation-based courses, it is very helpful to review the curricular documents of the target group. This allows one to see the bigger picture and identify those learning outcomes for which the different forms of simulation may be helpful. Such planning also helps integrate simulation into an overall package, a package in which one knows what the learners should be bringing with them in terms of abilities, skills, etc. This principle (of encouraging learners to participate) relates to the first principle – effective learning climate. Participants are less likely to become frantic if they have completed the necessary preparation for the simulation-based part of the course.

At the level of the individual participant, what can be done to help remedy the weaknesses that may have been identified outside the learning outcomes of the specific session? Usually, there is not sufficient time during the simulation sessions to address such issues. This is another reason why integrating the simulation-based course into a larger educational package is helpful, because contacts with the course directors have been made, and if specific activities are required, then arrangements can be put in place. Example: a trainee on a course for obstetricians decides to administer norepinephrine by infusion to a patient suffering from massive hemorrhage. It becomes clear in the scenario that the trainee's understanding of vasoactive and inotropic drugs is very patchy. However, this is not a primary learning outcome for the course, nor is it seen by the other participants as a topic worth reviewing in more depth on this course. The instructor then has the challenge of helping that learner address those needs – what resources are appropriate for that trainee? Having the course directors also reemphasize these lessons revealed in the Simulation Laboratory helps ensure that these lessons become ingrained.

## 77.8 Help Adult Learners to Carry Out Their Learning Plans

Simulation-based courses, as described above, are designed to help learners achieve the learning outcomes. This principle states that a key element of motivation is expectancy of success. This relates to the first principle – an effective and safe learning environment. The challenge for the simulation-based course designer is to pitch the scenarios at an appropriate level, while being consistent with achieving the learning outcomes. Scenarios should be stretching yet achievable; that is,

the learner should have the basic abilities to manage the scenario but should have those abilities stretched. So far in this chapter, I have written of learning outcomes and objectives as if they were discrete entities. Education is transformative, and it is worth reviewing some of the writing on this, especially in the context of successful learning. Education is about changing the learner, hence the use of the term transformative.

There are several different theories, but I shall confine discussion to the following:

- The novice-expert continuum of the Dreyfus Brothers
- The zone of proximal development of Vygotsky
- Experiential learning as described by Kolb

I have selected these three because of their relevance to simulation-based clinical training. Aspects of their work overlap. I shall describe all three, and then consider some practical implications of the learning theories to this principle.

### 77.8.1 Dreyfus: Novice-Expert Continuum

This work arose out of activity in the field of Artificial Intelligence in California in the 1960s and 70s. Hubert Dreyfus is a philosopher, and Stuart Dreyfus is an engineer. What their continuum (Table 77.1) describes is the way in which people handle and use knowledge in the journey from novice to expert [4]. Novices behave differently from experts, not necessarily because they have less factual information but because that knowledge has no or little grounding in the real world of professional practice. The perspective of the professional world changes; novices have to deliberately and consciously process the data stream, whereas experts recognize patterns and variations, as well as subtleties of those patterns.

**TABLE 77.1**  Novice-expert continuum (Dreyfus Brothers)

| Stage | Characteristics |
|---|---|
| Novice | Context-free rules, serial processing of information.<br>To the novice, the professional world is black and white – there is a correct way to do something and an incorrect way to do something. |
| Advanced Beginner | Can recognize meaningful components in a clinical case, based on their own experience and the instruction of a more experienced individual. An advanced beginner, like a novice, cannot yet see the bigger picture, but can identify "aspects of the case" as a result of the above. |
| Competent | Learner can now envision longer-term goals or plans, and is not obsessed with the here and now of the clinical case, unlike the novice and advanced beginner. Planning is still a conscious, deliberate process, and so lacks speed and flexibility. |
| Proficient | Learner now comprehends the case as a whole. Learner uses maxims and principles rather than rules ("well generally I do it this way but because patient X has condition Y, I'm going to do it differently"). The learner is now processing information in a parallel fashion, and can focus on the important findings, unlike the novice and advanced beginner. |
| Expert | This is characterized by intuition. When faced with the common professional circumstances and their variants, the experts may not even be conscious of their own thought processes in arriving at a solution. Overall performance is now more fluid, flexible, and fast. |

### 77.8.2 Vygotsky: Zone of Proximal Development

Vygotsky [5] asked himself a central question "How do humans, in their short life trajectory, advance so far beyond their initial biological endowment and in such diverse directions?" He was interested in intellectual development. His *zone of proximal development* can be thought of as the difference between a child's capacity to solve problems on its own, and the capacity to solve them with assistance. The term proximal refers to the range of possible problems that can be put within the grasp of the child by someone with more experience providing "scaffolding" or a support structure to help the child solve the problem. All of the ability to solve the problem does not come from the acquisition of theoretical knowledge (e.g., reading textbooks), but by *social interaction* and having *more experienced people frame the problem*. This may be a very informal process: "I have this patient with symptoms X & Y and signs A & B and I'm not sure what to do – do you have any ideas?" The term scaffolding implies that once the strengthening in the learner's understanding has taken place, the support can be removed without any loss of stability.

Vygotsky's work on the Zone of Proximal Development was mainly focused on children, and Knowles argues that how children learn and how adults learn are different, but as with the novice-expert continuum, we can regard Adult novices as sharing some of the features of children. The learners become more able to solve problems because the "scaffolding" shapes their perspective and transforms their understanding.

### 77.8.3  Kolb: Experiential Learning Cycle

The key to the work by David Kolb [6] is to link episodes of experience with abstract concepts (Figure 77.1).

The cycle can be entered anywhere, but I shall start at the box labeled concrete experience. This implies that something has happened or is ongoing. In order to make sense of it,

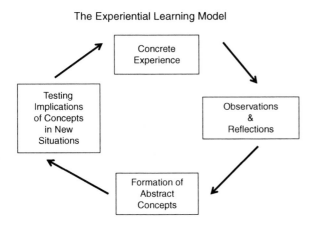

The Experiential Learning Model

**FIGURE 77.1**  Kolb's four-stage model of learning brought about by integrating practical experience into a framework of knowledge.

we have to have to observe and reflect (next box). In other words, we have to make sense of the phenomenon we have just experienced or are experiencing. This will involve the application of concepts (next box) – for example, if we have just seen a patient who has a heart rate of 120 bpm and a blood pressure of 90/60 mmHg, we would probably explain that in terms of cardiovascular shock. This assumes that the concept of cardiovascular shock means something to us.

So, part of the reflective process involves not only the observations, but also bringing into consciousness the abstract concepts and generalizations. We may lack the concepts to explain the phenomena (which takes us back to the "scaffolding" associated with Vygotsky's zone of proximal development).

Or, we may be faced with a discrepancy between what our concepts and generalizations lead us to expect and the phenomenon we are observing; this is known as cognitive dissonance, and can be a very powerful motivating factor in learning. The dissonance may refer to our expectations to manage a problem and the reality that we face.

The next phase (fourth box) is to generate one or more hypotheses – the shock in the example above is due to hypovolemia. We then test that hypothesis by designing an experiment – give fluids and observe what happens (we are back to the "concrete experience"). The conduct of the experiment becomes the first phase in the second cycle. If the patient's condition deteriorates – the patient is now acutely breathless – then we have to go through the cycle again, adding our observations to the previous ones, and using our concepts to make sense of it all. We may hypothesize that this is not hypovolemic shock, but cardiogenic shock and use inotropes and vasodilatation to treat the patient (concrete experience of third execution of cycle).

The essence of the model is that it is transformative – our experiences and situations influence our understanding of the concepts and also influence those concepts (see Chapter 64, Figure 64.15). As we develop these concepts and understanding, so our ability to make sense of the concrete experience improves.

In summary, these three educational models, as I have described them, share the features of our understanding and perception of events and ability to solve problems, as well as manage challenges in our professional domain by being transformed by experiences in that domain. We may look upon experiences as random events depending upon the vagaries of clinical practice, or we may be able to shape the experiences that our learners encounter.

## 77.9 Implications for Simulation-based Clinical Teaching

What are the implications for simulation-based teaching, especially in the context of helping learners carry out their learning plans?

- The obvious feature of clinical simulation is that it allows us to construct and deliver, within an intentional schedule, a particular experience – we can create a clinical scenario (the concrete experience of Kolb's model) that should encourage the learner to focus on particular concepts that are related to the learning outcomes. The choice of those learning outcomes has been commented on earlier.

- The framing of that scenario, the context in which that scenario is set, is one of the great strengths of simulation-based clinical training. An Objective Structured Clinical Examination (OSCE) type of examination tests a practical skill in no particular context – examine this patient's respiratory system, obtain consent from this patient for the following diagnostic procedure, etc., are examples of the type of skill tested. In Objective Structured Clinical Examinations, used widely throughout the world in summative high-stakes assessments, candidates are presented with number of different patient stations, typically ranging from 12 to 20, that they attend sequentially. In a robotic simulation-based setting, we often provide a richer source of information and usually link that with a dynamic set of events where the patient's condition changes, and the learner has to respond to those changes. At the level of the novice, the context does not matter very much, because the rules used by the learner are context free. We may be looking at a relatively simple set of learning outcomes – can the learner recognize a particular clinical case by eliciting some simple information, and can the learner then apply the appropriate rule for dealing with that specific case? An obvious example would be a patient suffering from acute crushing chest pain whose condition deteriorates into an arrhythmia incompatible with a cardiac output. Does the learner notice the onset of cardiac arrest? Does the learner identify the heart rhythm? Does the learner carry out the appropriate management? At the novice level, we make it more likely that the learner will succeed by developing these skills before performing in this scenario. However, if we were presenting a course for more experienced practitioners, then our context would become important because the information we provide in the setting will influence the perceptions of the learners. Remember, at the level of proficient or expert, the practitioners are looking at the larger picture and rely on pattern recognition, often at an intuitive level. If our cues are not consistent with those patterns, then we should not be surprised if the perceptions of the participants do not match what we think should be happening. If the participant does not act in the way we expect, we should review the scenario first of all, and not assume that any deficit is on the part of the (experienced) participant.

- As any simulation center professional knows, the actual concrete experience is only part of the learning process.

Observation and reflection, as usually conducted during debriefing, are of equal, if not more, importance. I shall return to reflection in the next section.

- In terms of Kolb's learning cycle, the phase of testing hypotheses can be a difficult and unsafe phase for a supervisor to allow a learner to conduct in the world of real patients. Not so in simulation, but we must make clear at the prebriefing whether participants can try things they would not normally do in the real world or not. If we criticize their performance subsequently during debriefing, then our learning climate ceases to be effective. Novices are unlikely to try new approaches because they work to rules, but proficient or expert practitioners may decide to try something different to see if it works. This can be a challenge to those producing the scenario, but if our target group consists of experienced professionals, then it is not only no more than we should expect, but is the special domain that is unique to simulation, its greatest added value, and should be embraced whenever the topic and student are ready for this advanced "testing out an idea" phase.

- The social input as described by Vygotsky is a phenomenon that we not only observe at debriefings, but also should encourage. Assumption 2 referred to adults having some prior experience and, as we know, sometimes those who are just a little bit ahead can better explain how to make sense of a new phenomenon. The experiences of your participants are a valuable asset, and I like to spend some time during introductions finding out what people have done, so that I can help the group make use of the prior experience of trainees.

## 77.10 Adult Learners Evaluating Their Own Learning

At its most simple level, self-evaluation of one's own learning carries on from some of the concepts in the previous section. If learning can be thought of as an episode of concrete experience, then the learners are asked to reflect upon themselves and explore what has changed. This leads us into the role of reflection, and also builds on from the previous section. One can help learners carry out their learning plans by encouraging reflection.

The complex nature of professional work, with its "messy" indeterminate problems, lends itself to reflective practice. The work of Donald Schön has been very influential here. Schön [7] contends that formal theoretical preparation can only take us so far in understanding or making sense of the complex nature of professional practice. Schön describes "zones of mastery" around areas of competence. This accords with the models described in the previous section. This is described as "knowing-in-action" and, as with the expert phase, is often intuitive and operates at a subconscious level. However, during professional practice, we encounter events that we do not expect – our normal strategies do not appear to be effective. This triggers two types of reflection, "reflection-in-action" and "reflection-on-action."

*Reflection-in-action* consists of three activities:

- Reframing and reworking the problem from different perspectives
- Establishing where the problem fits into already existing knowledge and expertise
- Understanding the elements and implications present in the problem, its solution, and consequences.

*Reflection-on-action* takes place after the event, is a process of thinking back on what happened, what may have contributed to the unexpected, as well as how this may affect future practice. Through these unexpected events, we come to make sense of the underlying phenomena, and so expand our zones of mastery. Reflection is key to transformation of the learner.

Phenomena require a vocabulary and language of concepts to be understood and appreciated. This applies as equally to professional practice as to modern jazz music appreciation or wine tasting. Such vocabulary can be part of the "scaffolding" referred to by Vygotsky. Simulation offers unparalleled opportunities for reflection:

- Participants have the opportunity to see themselves in action in a way that is almost impossible in the real professional world (the same applies to their instructors).
- Participants have the time and the opportunity to carry out these observations free from any other clinical duty or distraction. What they may require, depending on their stage of training, is help in making sense of what they observe. This is the key to effective debriefing. We begin by asking the learners to:
  - explore the events that took place,
  - make sense of these events,
  - make an assessment of how these events and their subsequent reflection has had an impact upon their understanding of their professional world,
  - describe how their perception of that world has changed,
  - state what impact this will have for their future practice.

Reflection does not stop at the end of debriefing, and a few questions ("take home messages") can help continue that process of reflection-on-action after the simulator course has finished. Never let students leave without giving them a thought or two to ponder. Then, after they have left, ask the teachers to review that session, what behaviors can be changed to do a better teaching job next time.

Another aspect of adult learning that falls within this principle is learning style preferences. I shall not say much about them apart from the fact that they are preferences, and as such, can be modified. The importance comes from the fact that people who are strong in one type of learning style (e.g., reading) may not extract as much from simulation-based training as people with different learning styles (e.g., manipulating). This may be an area to explore if one has a participant on a course that seems to be getting very little out of it. A few questions relating to what types of educational activities they normally seek out may help the learners evaluate their own approach to learning. This equally applies to instructors, who usually teach in the same style that they best like to learn.

## 77.11 Conclusion

Learning in Adulthood, which happens to be the title of Merriam and Caffarella's book [1], is a vast and fascinating topic. I have given a very brief and personal overview of some of the themes that I think are pertinent to the world of simulation. I hope that the examples I have given will put some of the theory into a context and help you make sense of what seems to work in your center and what doesn't. Remember, reflection is not confined to the learners on a course; as faculty, we can improve by reflecting on our own courses and our own performances, and making sense of the phenomenon that is simulation-based training.

## References

The below references will give many hours of interesting reading. In addition to these, I have cited two further references that are very relevant. The article by Roger Kneebone gives an account of some different educational styles in the context of surgical simulation and the book by Patricia Benner is an application of the Dreyfus brothers' model to nurse education.

Kneebone, R. Simulation in surgical training: educational issues and practical implications. *Medical Education*, **37**, 267–277 (2003).

Benner, P. *From novice to expert: Excellence and power in clinical nursing practice.* Commemorative Edition. Prentice Hall Health, Upper Saddle River, NJ (2001).

1. Merriam, S. B., Caffarella, R. S. *Learning in adulthood: A comprehensive guide.* Jossey-Bass, San Francisco, CA, 1991.
2. Knowles, M. *The adult learner: A neglected species.* 4th ed. Gulf Publishing Company, Houston, TX, 1990.

3. Yeats, W. B. He wishes for the cloths of heaven. In: *The collected poems of WB Yeats.* Wordsworth Editions, Hertfordshire, England, 2001.

4. Dreyfus, H. L., Dreyfus, S. E. *Mind over machine: The power of human intuition and expertise in the era of the computer.* Macmillan Inc., New York, NY, 1986.

5. Vygotsky, L. S. *Thought and Language.* The M.I.T. Press, Cambridge, MA, 1962.

6. Kolb, D. A. *Experiential Learning: Experience as the source of learning and development.* Prentice Hall, Englewood Cliffs, NJ, 1984.

7. Schön, D. A. *Educating the reflective practitioner: Toward a new design for teaching and learning in the professions.* Jossey-Bass, San Francisco, CA, 1987.

# 78

# Creating Effective, Interesting, and Engaging Learning Environments

Diane C. Seibert

## 78.1 Education: It's Not Just the Content

Creating effective, interesting, and engaging learning environments is a challenge for even the most seasoned educator. At the conclusion of a given course, many educators will sit down and brainstorm ways to make the course just "a little better" the next time. When asked, most highly skilled and well-respected educators will admit that they're not worried about the content; that usually doesn't need to be changed too much. What they're working on are ways to improve content delivery and their goal is to design a course that's so *interesting* that students learn because they simply can't help themselves. This chapter deals with the challenges surrounding delivery issues and addresses just a few of the factors that educators must consider when constructing a course. How do instructional strategies support learning goals? What *are*

instructional strategies, anyway?! What do adult learners want and need? How is teaching adults different from teaching less mature learners? How do you develop evaluation mechanisms that maximize learning while assisting in the measurement of learning? This chapter provides some general suggestions about what to consider when designing a course; what you need to know about the learner, the content, and the environment to make your learners say "that was an awesome course; you really need to take it."

## 78.2 Educational Frameworks

Why does a book on Simulation need a section on educational theory and frameworks? First, because simulation is an interactive activity that takes full advantage of the stimulus/response interaction between students and the simulated

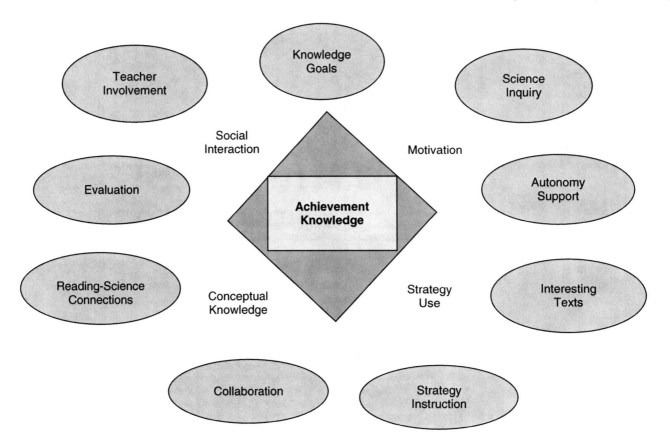

**FIGURE 78.1**   Wigfield & Guthrie Learning Engagement model. www.readingonline.org/articles/handbook/guthrie

participant. Second, humans must have an innate capacity to learn via stimulus/response interaction because this forms the basis for all learning from birth until formal schooling begins. Third, the power of the learning "process" demands at least as much respect and understanding as the content itself. Fourth (but you can probably think of more), because we never really outgrow the ability to learn via stimulus/response interaction, wise educators will make full use of this human predisposition to learn in this way.

Many clinical educators are drawn to teaching because they are expert clinicians; they enjoy teaching and mentoring in the clinical setting and want to pass along their accumulated knowledge and expertise. While it's rare to find a clinician with a formal background in education, most expert clinicians have an innate understanding of the behaviors that clinicians must acquire to become safe and competent. The lack of formal preparation for the educator role creates challenges for clinical educators, however, primarily because they teach the same way they were taught, while trying to avoid repeating the more offensive mistakes some of their own instructors made.

Educational frameworks can be a useful tool to refer to when developing a new course or improving an existing one. Most of them are relatively simple to understand and have been shown to increase instructional effectiveness [1] (Figure 78.1). The key component in many of the frameworks is figuring out a way to increase the learners' motivation to learn (highly motivated learners will go the extra mile to learn something; they may read more, talk about the material with their peers, ask you more questions, etc.). Research has shown that motivation plays a key role by helping to maintain interest and engagement in the learning process. The higher the students' motivation to learn, the more strategies they use, the more complex their understanding becomes, and the more they internalize the knowledge [1].

## 78.3 Elements of Engaging Learning

A few simple definitions go a long way in assisting faculty in using the models effectively. Learning engagement elements are commonly referred to include teacher involvement, autonomy support, collaboration, and real world interaction (active involvement).

### 78.3.1  Teacher Involvement

Students who feel that their teachers know them, care about their progress, and hold realistic, positive goals for them have been shown to be more motivated to learn.

### 78.3.2  Autonomy Support

When teachers offer meaningful learning options while taking into consideration the students' levels of knowledge and capabilities, learning improves. For example, suppose an undergraduate English teacher wants his students to explore a particular ethical dilemma. This year, rather than assigning one book for the class to read, he decided to see what would happen if he allowed the students to choose one of three books containing similar ethical issues. As the course ended, he was convinced both that his learning objectives had been met and that offering choices made an important contribution to the student learning. The classroom discussions were richer because each book presented the ethical dilemma from slightly different perspective. He was also shocked to realize that by the end of the course, many of the students had read all three books because the discussions were so stimulating.

### 78.3.3  Collaboration

Offer students an opportunity to work together in constructing new knowledge. Research has shown that when students work together, they create a unique learning environment that fosters discovery, increases communication, and improves motivation to learn, all of which enhance knowledge acquisition [1]. Since all clinical care is a collaboration of at least one clinician and one patient, why not include collaboration as a formal topic in the curriculum?

### 78.3.4  Real World Interaction

Active involvement provides "hands-on" learning, which is intrinsically motivating (it is also one of the key features in Simulation). For example, suppose you wanted a group of students to learn more about diagnosing a cardiac murmur. You could assign them to read about murmurs, you could show them a video animation and you could discuss heart murmurs in class, but this is often inadequate for students to gain a comprehensive understanding of murmurs. Research has shown (and clinicians intuitively know) that learning improves if students are offered an opportunity to interview and examine an individual with a heart murmur and if you offer this clinical interaction before giving an exam, test scores will be higher [2].

### 78.3.5  Strategy Instruction

Teaching someone *how* to approach a new content domain, in some cases, can be more important than teaching them the content itself. Strategy instruction is a concept that is often neglected in adult education, because it is often considered something that only inexperienced learners need help with (asking a novice reader to sound out a new word, or search for familiar words in a sentence and then trying to understand what the new word means in context, for example). Many graduate faculty quickly come to appreciate that this is not the case, however, because many learners who performed brilliantly in their undergraduate programs find themselves struggling in clinical and/or graduate school. Teaching them some simple strategies to link complex concepts (rather than rote memorization) can be a very effective instructional tool.

### 78.3.6  Reward and Praise

Almost everyone appreciates being rewarded for doing a good job and public recognition for achievement is highly motivating; even for adults!

### 78.3.7  Interesting Texts

This concept comprises far more than just textbooks; it refers to all types of learning materials; web sites, interactive disks, Simulation activities, etc. Finding instructional material that is well written, accurate, and engaging can be a real challenge, as most faculty are painfully aware. Because it's rare to find the "perfect" resource, many faculty either use a variety of materials or they finally give up and write the book themselves (like the one you have in your hands)!

## 78.4  The Adult Learner

Now that the stage is set for thinking about strategic inclusion of educational elements, it's time to consider the individual learner. Who is it you're trying to teach? Are they 8 years old and in the 3rd grade? Are they 15-year-old High School Sophomores? Are they new recruits at a military boot camp? Are they 21-year-old college graduates or are they returning to school after a successful career in another discipline? Are they military corpsmen being prepared for performing medical procedures in the far forward front without direct medical supervision? How much background knowledge do they bring to your classroom? If you are working with adult students, as many of you will be, you have your work cut out for you because they are highly unlikely to be the "blank slate" that you might assume them to be. Adult learners come full of knowledge, most of gleaned from their individual experiences. You know that to optimize their learning, you need to imbed some motivational strategies, but how do you motivate such a diverse group? First you need to realize what they already know (teacher involvement), then offer them some choices so they can start from where they are (autonomy support), provide them with opportunities to collaborate, and offer hands-on learning opportunities, all the while not getting in their way.

We'll start by providing a little more background on what adult learners need when they find themselves in a formal learning environment. The term that most educators use when they are discussing educational strategies is "Pedagogy," which the Cambridge Dictionary defines as "the study of the methods and activities of teaching." http://dictionary.cambridge.org/. Another term that is becoming more commonly used when referring to instructional strategies that should be considered when teaching adults is "Andragogy," a set of assumptions about how adults learn.

Andragogy was popularized by Dr Malcolm Knowles, a British researcher and educator in the early 1970s. Knowles [3] defines adulthood as "the point at which an individual becomes self-directed" and posits that adults will "resent and resist situations in which they feel others are imposing their wills on them" (p. 65). The six principles of Andragogy help to clarify the reasons adults react differently in learning situations than do their less mature counterparts (see Chapter 2).

### 78.4.1  The Learner's Need to Know

Learning never stops, but the reasons why people seek to learn new things changes over time. During the first two decades, people build knowledge and construct frameworks upon which to place new information. After this framework is in place, however, they begin seeking specific information to add to their existent structure. This difference in approach is critical as adults need to be shown the connection between what they're learning, what they think they need to know, and how this learning will advance them toward their goal. Adult educators must clearly state these connections whenever possible, so that the learner perceives the information to be important, relevant, and worth the effort.

### 78.4.2  Learner Self-concept

Adults enter the learning environment because they are seeking specific information and naturally expect to have some control over what and how they learn. Yet, when first offered the autonomy they want, however, many struggle initially, because they have been socialized to learn in a dependent manner and few have the necessary skills to function well in a self-directed learning environment. The exception would be the individual who was home schooled or trained in an environment employing the principles of autonomous learning. Because adults become stressed if they are thrust headfirst into an autonomous learning situation, transition support should be provided in entry-level courses. For example, the first few weeks of the semester could be taught traditionally (faculty talks, students listen), later, students could be asked to develop content, and

toward the end, students could be asked to decide upon a relevant topic and choose the best way to present the information to their peers (written, video, presentation, poster, etc.). One way to keep students motivated is to design learning activities that support skipping and branching. Skipping empowers and maintains motivation by allowing adults with varying degrees of experience and knowledge to test out, or skip over areas in which they can demonstrate competence. Branching allows students to travel down learning paths that might not have been strictly planned by the teacher. This allows an interested learner to explore topics they (and the faculty) might not have anticipated. Adult learners may also resist and resent the fact that faculty have power over their lives, and may revert to more child-like behavior in response to the power inequity. Teachers are continually challenged to maintain professional boundaries while walking that fine line between mentor and colleague.

### 78.4.3  Previous Experience

Faculty should always remain very aware of the fact that adults are not "blank slates," and that most educational experiences are designed to enhance existing knowledge and not reshape individuals for conformity purposes (like military boot camp). People seeking advanced degrees in health care, for example, may have had years of experience in a different role within the health care setting. Adults are eager to apply what they already know to the new situation, and should be offered opportunities to use their existent knowledge whenever possible. Educational activities like case studies, group activities, and simulation are highly effective strategies for harnessing prior knowledge. One note of caution, however; along with a lifetime of experience comes a lifetime of perceptions, biases, stereotypes, and bad habits. Use of reflective learning activities like journaling (keeping a diary) will help to ferret out some of these barriers to fully embracing new information and new behaviors.

### 78.4.4  Readiness to Learn

Adults will become ready to learn when they can see the link between what they're currently learning and their future goals. Faculty need to lay a breadcrumb trail for their students by structuring the content so that the path opens up just a little more as each new skill is acquired. Adults can then see where they're going, and will be ready to tackle the next new chunk of information.

### 78.4.5  Orientation to Learning

Adults are problem centered and task focused. They want to see how what they're learning applies to a problem, to their new role and to their life, and they want to apply their new knowledge immediately; preferably, in a real world setting. Simulation is a perfect vehicle for demonstrating how well they've acquired the requisite knowledge and skills, and clearly supports an adult's need to see the link between learning and future goals. This can be frightening and discouraging, however, if the task is too hard, or they're not ready for it. Learning challenges should be structured to push them along the edge of their knowledge, but not be so difficult that they perceive themselves as "failures." Only from the outer edges can the adult learners expand their domain of knowledge, understanding, and competency.

### 78.4.6 Motivation

Internal priorities, e.g., "I need to learn this so I can do that" are more important to most adults than external rewards, e.g., "I need to get an A in this class." Faculty need to be aware of how motivation plays a role in knowledge acquisition, and should deliberately create educational environments that support internal motivation. Some strategies to consider: (i) asking students for their input when designing classroom activities, not during an end-of-course evaluation; (ii) consider creating content modules, allowing students to attain specific predetermined goals; (iii) ask students to develop material for future cohorts of students, challenging them to think about how to improve the learning environment; (iv) ask students to help decide the sequence that a course will be presented, i.e., topic prioritization. It's very important, particularly in clinical education, that teachers remember that for students, learning is a means to an end, not the end in and of itself. Students rarely lose sight of the overall objective; becoming a competent provider, but all too often teachers are so focused on what's going on in the classroom (or simulation) activity that they forget about the overarching goal.

# 78.5 Kolb Decision-making Model

Adults are active learners. Sitting in a classroom, or worse, standing about a patient simulator and listening to an "expert" talk about what they know is not engaging (children have a tough time with this too), and adults often tune out and go home to learn the content themselves. Adults retain information better if they are actively involved in the learning process, and simulation is a perfect vehicle for adult learning because it provides the link between knowing and doing. David Kolb's Learning Cycle Model, is very applicable to simulation and adult learning. In developing his model, Kolb [4] adapted the works of highly respected learning theorists like John Dewey, Kurt Lewin, Jean Piaged, and J.P. Guilford, and proposes that for adults, learning occurs through experiences, not simply rote memorization. Perhaps, the worst problem is where the learners and their instructors become so/too engrossed in making decisions within simulation and thus short-circuit the learning process as they

oscillate between Active Experimentation and Concrete Experience and do not spend time in Reflective Observation or Abstract Conceptualization. Kolb's model describes the experiential learning process as a cycle consisting of four stages (any of which can be the starting point for the learning experience).

### 78.5.1 Active Experimentation

The learning process begins when the learner acts upon his/her environment and observes what happens.

### 78.5.2 Concrete Experiences

On the basis of their observation, the learner can now anticipate the result if the same action is taken again under similar circumstances.

### 78.5.3 Reflective Observation

From their understanding of what happened in steps 1 and 2 above, the learner begins to consider what might happen under different circumstances. They make extrapolations about their experiences and theorize about "what if." This process can be facilitated in a group setting, and in the simulation environment, it is often accomplished through the use of a debrief or "out-brief."

### 78.5.4 Abstract Conceptualization

This is the ultimate outcome for educators; students who have internalized their knowledge and can apply it across multiple simulation scenarios. Students have truly mastered a concept when they demonstrate this ability.

# 78.6 Putting It All Together

As you may have noticed, there's quite a bit of overlap in the three learning models. For example, Real World Interaction described by Guthrie & Wigfield is very similar to both Kolb's Concrete Experiences and Knowles Orientation to Learning principles, and Autonomy support was viewed by both Guthrie & Wigfield and Knowles as very important in supporting adult learning. None of these elements were intended to function in isolation; they were intended to be combined with other elements to meet specific learning objectives. Here's what a simulation experience might look like if we combine several of these concepts in developing a new simulation experience. As faculty, our overarching goal is to improve the subjective experience for our students, while also achieving (and possibly exceeding!) our learning objectives. From Guthrie & Wigfield, we'll use Real World Interaction and Autonomy Support, from

Knowles, we'll add Learners Need to Know and Prior Experience, and we'll incorporate all four of Kolb's elements as they might emerge as the course unfolds.

At first glance, it would appear that adding *real world interaction* is a no-brainer, because that's what simulation *is*! As teachers begin to use simulation, however, they quickly gain an appreciation for how powerful a learning tool the "real world" can be. The tighter the coupling between the teacher's objectives, the simulation activity, and the learners' personal learning goals are the more engaged students are and the more "real" the simulation is to them. More importantly, if the learning environment has been structured to challenge but not completely overwhelm the student, they invariably stretch to meet the challenge, taking risks that they may not have taken or might not have been allowed to take, in a real patient situation.

As you develop your simulation, start by thinking about what level of student you are trying to engage. If your students are very early in their education, you might begin by creating a formative encounter; for example, the simulator might be hooked up to a heart rate monitor to demonstrate what happens when a particular intravenous drug is administered. As student confidence and knowledge increase, however, you might add challenges like beeping pagers, people moving in and out of the room (staff changing shift), full bladders, sidebar conversations, or competing demands (i.e., sponge counts). A distinct advantage of simulation is that it offers students an opportunity to watch other people making good (and bad) decisions. Although simulation is usually more simplistic than a real world challenge might be, this is not a bad thing. By shining the "light of guided learning" on one critical learning objective at a time, students have an opportunity to achieve mastery and then move on to more complex concepts or to scenarios combining several critical concepts. Dynamic experiences provide students with strategies for managing challenges in ways that static displays never will.

*Autonomy support* is an important concept and is employed when offering students choices. This is often a very scary concept for teachers. It may be difficult to imagine how learning outcomes could be equivalent if student experiences are not all the same, but this is going on every day in simulation centers using standardized patients. For example, 12 students may enter the Simulation Center prepared to take a history and conduct a physical examination on a patient with a musculoskeletal complaint. The students don't know *what* M/S complaint they'll be assigned, though, so they have to prepare for any and all possibilities. They may be asked to evaluate someone with back pain, hip pain, or ankle pain; all they've been told is that they will have an orthopedic complaint to manage. Learning often accelerates *because* of the uncertainty. In developing your simulation session, you might tell the students that they will be asked to manage an ENT (ear, nose, and throat) case, but that it will be either a pediatric *or* an Adult case. You may offer them an opportunity to choose which case they prefer to use in a formative session – with a gentle

reminder to pick the one that might be more challenging for them, because during a summative simulation examination, they will be randomized to one of several cases.

Learners in clinical fields are often frustrated by the volumes of seemingly useless information they are asked to memorize and often ask themselves why they need to know so much minutiae. Medical school students may feel like they're spending 2 years learning things that don't seem relevant to clinical practice. How will knowing everything there is to know about a chloride channel help them provide better care for a patient? This question can be answered quickly using simulation, and the key is early introduction with targeted learning scenarios. The importance of a chloride channel can easily be demonstrated to an entire biochemistry class using a mannequin, a monitor, and a few selected IV medications. This is a very easy "portable" simulation experience that supports the *learners' need to know*.

How can an instructor determine what *previous experiences* individual students have had and then appropriately balance the course content to meet everyone's needs? It may not be possible to completely satisfy all learners within a given group, but it's certainly possible to meet most of them halfway; but only if you ask the question. Many adult learners are delighted to have their prior experience acknowledged even if it's not acted upon. One way to assess prior knowledge is through the use of a pretest. Most faculty are quite comfortable with a pretest providing information about the cohort, but they're not quite as comfortable allowing students to advance in the course based on the results of that exam! Teachers may have to prepare at least two course plans (possibly more) to allow students to advance from different starting points. They then have to figure out how to keep everyone actively engaged while working at different levels (similar to the Montessori method of teaching children). This can be easily integrated into a simulation environment; the basic scenario can be maintained, but the challenges posed to the students can be varied depending upon the level of the learner.

The debrief supports *reflective observation* and is a critical component of learning, particularly in a clinical simulation environment. This is the time for faculty and students to review encounter recordings, compare their performance on various tasks, check to see if the scenario objectives were met, and examine outcomes. No simulation encounter is truly complete without a debrief, no matter what form they take. The simulation exposes students to the lesson and the debrief closes the loop, helping students integrate the new knowledge with existent information. Group discussions are particularly useful if many of the students felt they did not perform well because it allows the cohort to normalize the experience while they reflect upon the experience.

*Abstract conceptualization* is achieved when learners are able to apply what they've learned. Simulation is unique in that it can be used to build mastery and can also be used to demonstrate (to faculty and the student themselves) that they really do "get it." The ability to reason abstractly is critical in health care, because making intuitive leaps of thought is how advances in science and medicine are often made. An excellent way to anchor this new knowledge is to offer students an opportunity to repeat the simulation after they participate in the debrief. Those students that are interested in repeating the scenario and "getting it right" have found the experience to be both rewarding and empowering.

## 78.7 Conclusion

Adult learning principles should be applied whenever adult students are involved in an educational experience. After the initial shock wears off (choices, empowerment, responsibility for learning), learning engagement increases and learning outcomes improve [2]. Simulation is a perfect environment in which to weave adult learning principles into the curriculum because it already embodies several of the engagement elements.

Brainstorming with faculty colleagues about how to operationalize some of these elements results in rich, active, and engaging learning experiences for adult learners.

## References

1. Guthrie, J. T., and Wigfield, A. Engagement and motivation in reading. In: M. Kamil, D. Pearson, P. Mosenthat, and R. Barr (Eds.), *Handbook of reading research III.* pp. 406–422. Longman, New York, 2000.
2. Seibert, D., Guthrie, J., and Adamo, G. Improving learning outcomes: integration of standardized patients and telemedicine technology. *Nursing Education Perspectives.* 25(5), 232–237 (2004).
3. Knowles, M. S., Holton, E. F., and Swanson, R. A. The Adult Learner, *The Definitive Classic in Adult Education.* Elsevier, Burlington, MA, 2005.
4. Kolb, D.A. LSI Learning-style inventory. McBer & Company, Training Resources Group, Boston, MA (1985).

# 79

# Adult Learning: Practical Hands-on Methods for Teaching a Hands-on Subject

Kristina Lee Stillsmoking

## 79.1 Self-directed

In the past, we taught our students by standing in front of them with "content" that was rotely placed (hopefully) into their memories. The next content was piled onto the previous learning until they achieved mastery. For example, I learned to spell through sight-memorizing each word. Needless to say, I had trouble with words I had never seen before and trying to sound out words was a true puzzle.

The same can be said for the sight-memory model that does not work well with the adult learner. Yes, you can spend hours reading books and articles on this very topic, and I hope to condense it into a quick, readable, relevant introduction.

Over the years, educators of adults realized that the adult learner needed a different learning model than do children. What they determined was that an adult learner must be self-directed and find a commonality or application to their need to learn. An increase in self-esteem or pleasure will motivate curiosity in a topic the adult learner wishes to learn more about. The "What's in it for me" is the bottom line of their thinking process. How can they *use* the information presented in a meaningful and productive way in their present day life experience? "*Problem-solving centered*" rather than "*subject centered*" is more applicable to them. They create similarities and connections between past experiences and the present information. The ability to have control and decision-making power over their learning experience is a must. Thus, the more self-directed the experience, the higher the comfort level and the more positive is the learning outcome.

If you, the educator, do not accommodate these specific qualities and characteristics, at best you will lose the attention of the adult learner students, and at worst, the adult learner students may be come resentful and resist outright the learning experience. Therefore, how can you, the educator, integrate learning of content with real-life experiences, which the student can grab hold of now and apply later? You, the educator, should deliver content interactively by adding real-life stories and examples that the adult learner students can relate to, engage in dynamic group sharing of how the teaching topics influence them, as well as integrate the content into their own personal lives (using your real-life examples of integrating knowledge into clinical practice).

## 79.2 Types of Adult Learners

There are various definitions for the four adult learner types, depending on your source, but basically they are very similar in nature, and no one style inherently better than another. Your teaching process must address these four different learning types in order to meet all the types of learners that may be in your class.

> **Kinetic** – the hands-on, fidgety, can't sit still for 10 minutes, let alone an hour, have to touch and manipulate;
> **Audio** – have to hear, read out, listen to stories;
> **Visual** – seeing is believing, the page highlighter, the movie/PowerPoint watcher, user and creator of drawings and maps of all forms;
> **Intellectual** – reflecting and making connections, have to think through problems.

Not all learners fit neatly into one of these four categories. Some of your learners may be global learners where they see the whole picture of the process as opposed to the linear learner who has to understand each step first before the steps can be put together for the global insight. How we learn and remember (Table 79.1) and senses employed (Table 79.2) also vary.

Using the sense of smell is becoming a new avenue for use in simulation training, as we are becoming involved with a behavioral health simulation that plans on using smells to enhance the realism of their the sessions to improve learning and retention.

**TABLE 79.1**   Actions and percentages of how we learn and remember, with examples

| % | Route | Example |
| --- | --- | --- |
| 10 | Reading | Journal |
| 20 | Hearing | Audio recording |
| 30 | Seeing | Reading |
| 50 | Hearing and seeing | Video debriefing |
| 80 | Writing and/or speaking | Preparing for, and lecturing |
| 90 | Doing | Simulation |

**TABLE 79.2**   Senses and percentages involved in learning, with examples

| % | Route | Example |
| --- | --- | --- |
| 1 | Taste | Eating an apple pie |
| 2 | Touch | Blood pressure palpation |
| 3 | Smell | Ketone smell during diabetic keto-acidosis scenario; Smoke during disaster training |
| 11 | Hearing | Heart and lung sounds |
| 83 | Sight | Clinical monitor |

## 79.3 Content Delivery

Now that you appreciate your adult learners, how do you conduct a class that will meet their learning styles and keep them engaged? It is best to start with information the learner may be familiar with and then move to new content. If an unplanned, spontaneous re-direction of interest occurs, it is acceptable to revise your agenda.

### 79.3.1 Student Goals

I start each class by greeting each student at the door and ask them to write down the number one learning experience they want to achieve. I place a flip chart paper on a wall near the door (make sure you don't have pen bleed-through onto the wall). Just before I start, I quickly look over these requests and make sure they are incorporated into my class.

Thus, I am giving the student a sense of control and responsibility over what they hope to learn. Even more important, they have to take responsibility for knowing what they are there for, and what their purpose is.

### 79.3.2 Breaks

Ten minutes in every hour is a good rule of thumb. If the speaker has gone over the allotted time or we need to adjust the timing of the agenda, I always ask for a group consensus.

### 79.3.3 Story Telling or Sharing Experiences

This is vital to a great learning experience, but don't just tell 500 stories as your content. Engage your students into telling an experience that may be related to another student's needs.

### 79.3.4 Humor

Be sure to use it when the students look like they are going to die on the vine or to push hard on a particular point. Of course, its use must be appropriate to the content, class makeup, and personality. The use of cartoons in visual presentations works well.

### 79.3.5 Exercises/Games

Money spent on great training exercise/game books is very helpful, and keeping a "library" of different types is a good idea. There are different categories that you can match to your needs in the presentation. Interjecting appropriate exercises/games during the course of a class will help to emphasize a point(s). I teach a communication and critical thinking course incorporating a bridge building exercise. This exercise divides the students into two three-member teams that must build a half of a suspension bridge across a 4-mile river. The three include a leader and two nonsighted and nonspeaking/nonhearing members. They must all decide the design and building plan using only straws and straight pins. Two masking tape lines on the floor represent the river edges. No speaking is allowed by any member of either of the two teams, and only 15 minutes is allowed to complete the project. Once the 15 minutes is over, I conduct a debriefing to discuss how communication and critical thinking was used in the exercise.

### 79.3.6 Role-playing/Case Studies

Can be very touchy for students who aren't comfortable being in front of a group and pretending to be someone else. On the other hand, you may have the class actor who is more than willing to help. By using this method, you can actually play out your content and apply it to real life.

### 79.3.7 Demonstration

In our all-day Simulation Facilitators Course, we work the old fashioned adage "see one, do one, teach one," and it appeals to all types of learners. After our discussion of many of the topics already discussed above, we give the students a simple scenario that they must program into the mannequin's software. One or two students are given a laptop computer (with the simulator's software installed) to complete this task. This gives them hands-on experience of programing, running, and evaluating the scenarios' parameters. Very powerful tool!

### 79.3.8 Handouts

Keep them simple with short bullet points of concise information. Allow space for the student to add comments.

### 79.3.9 Sleepy Time

If you are lucky like me, I always get the afternoon session of a presentation so the blood is out of the brain and in the stomach. Heads are bobbing and whiplash is happening everywhere. This is a tough time to teach and be remembered. Now is a great time to walk around the room, and put on an active demonstration or exercise.

### 79.3.10 Class Champion

There's always a couple of students who pick up on the process quickly and will make great helpers when you get to the demonstrations, exercises, or games. Encourage them to excel and see how the other students begin to pick up and learn quicker as they can relate more readily to a fellow student.

### 79.3.11 PowerPoint Presentation

Keep it simple silly! A rule of thumb I picked up on years ago in figuring time of presentation, allow approximately 2 minutes per slide. Thus, if you have a 50-minute class and you have 40 slides, there's going to be a problem finishing on time.

Keep your slide simple, and remember they are outlines of the content not the whole content rolled into one slide. Transitions and fancy animations have no place in any presentation unless they clearly support your teaching objectives and the students' learning objectives. Don't forget the size of your classroom, which will determine your font size and color of background. Again there are wonderful, beautiful backgrounds in bright colors, but will your student be able to read the content, let alone focus on the presentation? You can intersperse exercises and games during a presentation. See the example above under the section Exercises and Games.

### 79.3.12 Body Language

Maintain good eye contact with your students and keep your hands from fidgeting, out of your pockets and avoid "ahs," "ok?", "got it?" etc., when speaking. Present your material with an enthusiastic and friendly persona and be willing to genuinely help your students.

## 79.4 Program the Student to Succeed

I have seen other instructors (I use the term lightly) who believe in "throwing the students to the wolves" when learning a new process, or stated another way, "let them fumble around first." Some instructors need proof of incompetence in others so that they can feel accomplished – the dark side of Adult Learners. I personally don't hold to that tenet, as it creates a negative environment for the student, especially if they have never been exposed to the behavior they are to master. For example, how can I expect my students to correctly use a biphasic defibrillator if they have never seen one until I throw them into a learning environment? Give them a demonstration on how the heart attack code should be handled correctly, provide time for them to practice, and finally allow them to incorporate the experience of into the mock code scenario. You must see, hear, and touch the defibrillator before you can use it.

**Feedback** – is very important as it empowers the student and gives the instructor ideas for improvements. Be sure you provide positive reinforcement throughout the class. Make sure you set the ground rules that comments to anyone in attendance must be constructive (goes all ways: instructor to student, student to student, student to instructor, instructor to instructor).

**Environment** – depending on the content being taught and the teaching environment, make it as comfortable, quiet, and nonthreatening as possible. Keep the background noises to a minimum, provide scribble paper, and later in the day have hard candy on each table. Act as a role model of respect for everyone, encourage fun, relaxation, and invite questions and comments.

**Evaluation** – be sure you include this at the end of the day with questions that provide information that is helpful in revisions as necessary. Remember no matter how great a facilitator of learning you are, there will always be one or two nay sayers in the group. As a society, we will always remember the negative experiences and forget the positive ones, even when there are more positive ones. On the other hand, the negative experience may instigate performance improvement. Just like the lessons that can be learned from a less than successful performance during a simulation session, less than glowing evaluations of the simulation session are fertilizer for improving the instructors' side of the simulation performance.

## 79.5 Conclusion

Make the learning experience a memorable time for your students so that what and how they learned will be internalized into their memory and practice, memorable for how much more they knew when they left your simulation shop, memorable for how many of their fundamental motives for placing their adult selves in the subservient role of learner were satisfied, memorable for justifying their trust in you to help them become better clinicians. I still meet up with former students who tell me they will never forget the content of a course I taught them because I addressed their goals and mentored them into being a better clinician. I accomplished this improvement because I met their learning needs, made it fun with humor and provided hands-on experiences that enabled them to attain their goals they had set for themselves as adult learners.

The views/ideas expressed in this article are those of the author and do not reflect the official policy or position of the U.S. Department of the Army, U.S. Department of Defence, or the United States Government.

# 80

# How to Build a Successful Simulation Strategy: The Simulation Learning Pyramid

Hal Doerr
W. Bosseau Murray

## 80.1 Learning via Simulation

One area that needs to be examined by anyone attempting simulation is very basic, i.e., how do you create those key components necessary for learning via simulation? In this chapter, I (HD) will provide the reader the necessary components to build their own "Simulation Learning Pyramid," thereby enabling them to develop and create their own successful simulation-based clinical teaching strategy.

In other chapters, you will find many different ideas for creating an analog environment, such as how to create an operating room, an emergency room, a SICU, or a MICU. All of these ideas are important to suspend disbelief. None of these are new concepts. However, it is quite important, as we have found over the years, that the more realistic the environment is that we provide for our learners, the easier it is for that participant to suspend disbelief. We, as instructors, must create an environment where the quality of input we provide helps the learner accept their simulated patient *is* real, and also, we must remember to incorporate the teaching objectives.

## 80.2 High-fidelity Environment Analog Training: A NASA Example

I wanted to take that one step further and look at the possibility of providing H–E–A–T (High-fidelity Environment Analog Training) to our simulations. This is a concept that was developed for simulation training at NASA for both Astronauts and Flight Surgeons. The term flight surgeon is a military term that describes any physician who has received Aerospace Medical training, and does not necessarily imply any specific surgical training. The Aerospace Medical training involves the study of the physiology of flight, critical care air transport, and those operational issues specific to the health of pilots. The NASA flight surgeons have all studied Aerospace Medicine and have clinical backgrounds in subspecialties such as Internal Medicine, Emergency Medicine, Radiology, and Urology. What this group of doctors do have in common is a strong working knowledge of both the astronauts' physiology in microgravity and the many operational constraints of the shuttle and international space station. In the past, the clinical simulations at NASA were carried out by having astronauts read their symptoms from an index card, which provided the clinical symptoms and information necessary for the flight surgeon sitting at the console at mission control to make a diagnosis.

I wanted to create an environment where we could more effectively suspend disbelief by adding High-fidelity Environment Analog Training. What we have done at NASA is create a ground-based (earth bound) simulation center that mimics all of the communication loops and all of the other components important to mission control to provide an analog environment. This allows for integrated simulations incorporating various personnel in the team such as the flight director, biomedical engineers, flight surgeons, and astronauts. We also have a second laboratory, which is onboard the parabolic flight program aircraft, so that we can actually train astronauts in clinical care using a simulator in a zero-gravity environment (Figure 80.1).

Gravity reduction is accomplished by repeatedly diving the aircraft at a 45 degree angle for 10,000 feet and then pulling up. As the plane floats over the top of the parabola, you have about 30 seconds of reduced gravity. The flight profiles can produce zero Gravity, 1/6th or 1/3rd of earth's gravity to simulate the gravity present in orbit, on the Moon, or on Mars, respectively. During the pull-up portion of the flight, humans and equipment experience about twice the earth's normal gravity. NASA created such flights to enable research, testing, and teaching of medical protocols and skill sets in a reduced gravity environment.

Medical Education Technologies, Inc., Sarasota Florida, U.S.A., designed and built a "one-off" human patient simulator that could withstand the stress of this 2G to 0G environment. Extensive changes were needed to enable the simulator to function with the presence of both an absence of gravity as well as the phase with almost twice the normal gravity. It was also

**FIGURE 80.1** NASA test aircraft used to create a brief zero-gravity environment for the occupants while flying a series of climbs and dives.

reengineered to meet the stress of flight and the researchers' rough treatment. Furthermore, electrical and gas supplies had to comply with specific aviation standards. This became a very extensive project with a great deal of development to be able to spend time in this special environment to assure the functionality of this unique simulator. After the simulated environment and the "flight-certified" simulator had been prepared, we then used the principles of the "Simulation Learning Pyramid" to plan and develop the operations of the center.

# 80.3 Simulation Learning Pyramid

The Simulation Learning Pyramid involves four basic components: the "**Simulator Plan**" at the base, the next layers being the "**Simulation**" and the "**Debriefing**," and at the top of the pyramid is "**Transference**." By sequentially following and combining these four components, we can promote a successful Simulation Learning Pyramid (Figure 80.2).

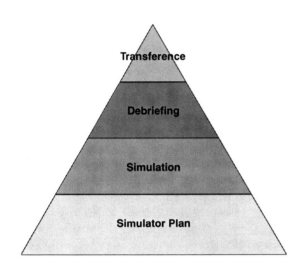

**FIGURE 80.2**  Simulation Learning Pyramid, illustrating how the three phases of simulation support educational transference.

## 80.3.1 Simulation Plan

Let us begin with a brief overview of the "simulation plan." This stage involves many different aspects such as setting goals, i.e., what is it we hope the trainees will learn through this simulation? Equally important, we have to decide what is our plan "B," in case the participants veer far from the objectives in the stem case provided. How are these goals and objectives going to be demonstrated in the clinical presentation provided? Selecting an appropriate patient is a most critical step at this point. Try to choose patients you have treated in the past so that you are familiar with all aspects of their clinical presentation. Choosing patients you are familiar with also

provides the instructor with the ammunition necessary when your students challenge your clinical presentation or clinical interventions. They cannot say: "This will never happen in real life!" My ultimate goal here is to plan both the simulation and the debriefing, so that the information that I am expecting my participants to learn will be set firmly in their memory so that they can take the new knowledge, transform, and utilize this information in their clinical practices.

What are our objectives with this case? First, we want the participants to begin with the ABCs, i.e., airway, breathing, and circulation. In this case, the patient is breathing too fast and has a saturation of 72%, with a low blood pressure. We would hope the trainees' first order of therapy would be to establish an airway and control the patient's breathing (i.e., first steps of ABCs). Typically, the airway is established, and bilateral breath sounds are auscultated and confirmed. This is where learners, in general, will be stumped due to the lack of improvement of the patient's oxygen saturation. They will try any number of interventions trying to improve the patient's saturation. We can suggest to them a hypoxemia algorithm to implement. Algorithm development and implementation is an important part of crisis resource management principles. We can now advance the caregiver to the next level of the algorithm for hypoxemia: first listen over the mouth (or airway: Laryngeal Mask Airway, Endotracheal Tube, etc.) to see if the cuff on the endotracheal tube has ruptured, or possibly the endotracheal tube has been displaced out of the trachea and into the esophagus. Next, listen over both lung apices and then over both the far lateral lung bases. If breath sounds are equal and symmetric at all levels, the source of hypoxemia is not a problem of ventilatory volume (i.e., not respiratory in nature). We must then look to the heart and circulation as the source of the hypoxia. The students now gain a fresh perspective (not lungs, consider heart) and moves forward in the treatment of this patient.

Once we have defined the objective, selected a patient, and created a simulator plan, the next step is to produce the simulation: presenting the patient and the relevant clinical information (history, clinical examination, and relevant laboratory results and X-rays). As a concrete example, I will present a real patient seen in my practice, and presented here with a fictitious name. Eileen Short, a 36-year-old female, was involved in a powerful explosion in a laboratory. She presented with a traumatic below-the-knee amputation to her right leg. She also had glass shrapnel that had penetrated and lacerated her right thigh, abdomen, and chest. She had intact facial hair, and no visible burns above the level of her pelvis. She presented with hypotension, tachycardia, and hypoxemia. Of the three symptoms we will cover today, the key element in this simulation is that of hypoxemia. On arrival to the emergency room, her vital signs were: heart rate 180 beats per minute, blood pressure 70 mmHg systolic by palpation, saturation of 72%, and a respiratory rate of 36 breaths per minute. She was unconscious following the explosion, with a Glasgow Coma Store (GCS) of 3.

In this scenario, the patient provides a simulation framework where multiple levels of caregivers can participate as well as learn and practice their roles. Because we have multiple levels of success built into this case, we can now utilize this patient for not just experienced anesthesiologists and emergency room personnel, but also for primary and secondary responders. This case will also work for firefighters, EMTs, paramedics, nursing staff, and respiratory therapists.

### 80.3.2 Present the Simulation

When we have new learners, I find it important to first set the ground rules for the simulation. For example, "I expect you to treat this like a real patient." Why is that important? Because we have to guide the learners through their initial stage of disbelief, and that this is a real patient for whose outcome they take responsibility. There are many different methods of making this happen, however, I believe that setting the ground rules, talking to the students ahead of time, then setting up a familiarization session with them, so that they understand what the simulator can do, where breath sounds are, where heart sounds are, what pulses they can feel, what clinical signs they can look for, and what the mannequin is capable of and is not capable of. Prepare them for success.

We would now produce the simulation, allowing participants to follow any therapeutic path and sequence that they would like, giving them any information that they request. Following the conclusion of the simulation, the students are then asked to go to a debriefing area for the next portion of our pyramid.

### 80.3.3 Debriefing

Following the simulation, we begin the debriefing. There are multiple different debriefing techniques that will be discussed later (see section 80.7).

### 80.3.4 Transference

Once the debriefing is complete and the students return to their clinical settings, we hope that the information learned in this session will translate into successful clinical skills. The next time the students are in an emergency department or in the midst of a clinical emergency, they will be able to draw on the information they learned in the simulation center and transfer this information to the critical clinical setting, thereby improving their performance and improving their chances of success. Of importance is our desire for transference to real events, i.e., that their confidence in their own competence will be transferred to a successful management of a difficult clinical case. All the knowledge and ability in the world is useless if real world challenges are not successfully managed. We will

now examine how the adult learning principles as espoused by Knowles, and the experiential learning theory by Kolb, fit into the learning pyramid.

## 80.4 Adult Learning Principles (Knowles) as Applied to the Learning Pyramid

There are three points from adult learning principles that we have to consider in the Learning Pyramid:

- Adult learners bring experience that they expect to be used, and respected;
- Adult learners expect to influence "how" they are educated;
- Adult learners expect to influence "how" learning will be evaluated.

The adult learning principles, as formulated by Malcolm Knowles [1], include the concept that adult learners are autonomous and self-directed; they bring a wealth of experience into the learning experience. The experience and expertise they bring with them stems from their training and experience. This expertise can be used in the design phase of the simulation experience, as well as during the debriefing.

The students would like to see how the simulation experience could be applied to their own specific circumstances and/or environments. The simulation should not present an abstract environment (e.g., unrealistic events), it has to be very concrete, one in which they can see where the simulations will transfer directly to their own clinical milieu, and see how this will benefit them in their daily practice. Knowles also encourages active participation. Therefore, simulation should not be a "lecture," but a practical, hands-on experience. Simulation is a very active process where enough latitude should be built into any simulation to allow the student to branch off from the stem question. If the student starts to make a mistake, the trainer should be flexible enough to allow this to happen. Active participation implies allowance for such self-discovery (especially through errors) to happen. As mentioned, this should not be a one-way lecture; this should be a multifaceted interaction.

Adult learners expect their responses and actions to have clinical effects. Adult learners expect their responses "to be acted upon" and have effects within the simulation. There is no better training arena than simulation to do just that, i.e., demonstrate outcomes after action (good or bad). As the trainees respond to events, the simulation should respond to whatever pharmacologic or physical interactions that are applied to that particular patient. A concrete example of this principle in action is well demonstrated with the diversity of responses and treatment interventions in Eileen Short (our simulated patient described earlier). Depending on the teaching objectives, I can have had students treat this particular patient for the following conditions: hemorrhage, Lupus, sickle cell disease, and pneumothorax.

Knowles [1] describes adult learning in four critical elements. These are: motivation, reinforcement, retention, and transference. All these four elements can be mapped to the Learning Pyramid, as will be shown.

## 80.4.1 Motivation

We have to set the tone and the level of the simulation so that we can demonstrate concern and interest for their learning. We are concerned that they learn the critical thought processes, algorithms, and begin to understand the critical thinking that goes into clinical decision making.

We have to provide controlled stress during the learning process. If it is "too easy," the trainees are bored, feel insulted, and do not become engaged. If we apply challenges appropriately, the learners actually learn and retain more. Once we add too much tension to the simulation, we find that people begin to lose interest, shut off, and are no longer able to function clinically – as they do not yet possess the skills needed. The whole idea of simulation is for the students to increase their performance through decreasing their stress from unfamiliarity. However, how do we assess the amount of tension that we are placing on our students? This is where experience (clinical and simulation) is essential, because we need to watch our students, listen to their words, and their tone of voice, as well as observe their body language. Are they beginning to show signs of stress such as sweating, speaking rapidly, pressured speech, or fixating on objects such as a monitor, thereby demonstrating that they are no longer relaxed enough to have their learning experience optimized?

Knowles also mentioned, under motivation, that the appropriate level of difficulty should be set. At our particular program, we train a very wide variety of adult learners, and these include emergency medical technicians, paramedics, nurses, respiratory therapists, technicians, residents at all levels, faculty, flight surgeons, and astronauts. To accommodate all these different groups, we can write scenarios appropriate to the level of each of those groups, which represents a great deal of writing, programing, and scripting. Alternatively, when we develop a new patient/scenario, we design the scenario much like one would design examination questions: there are easy as well as difficult parts to a given examination and scenario. This is an effective way to write simulations that contain multiple levels of success. This "multidifficulty" concept is especially important when we are developing integrated, multidisciplinary simulations. An example would be an operating room team, including a surgeon, an anesthesiologist, a scrub nurse, a circulating nurse, and a nurses' assistant. Each of these caregivers can respond and intervene at different levels, as well as be able to contribute to the success of the team.

The elements of motivation can be mapped to each of the components of the Learning Pyramid:

- **Planning:** We build motivation into the planning base of the pyramid. For instance, the fact that the scenario selected during the planning is a crisis that the trainees would most likely encounter, and may motivate them to want to know how to treat this problem.
- **Executing:** Maintain sufficient tension during the scenario, but give them multiple opportunities to succeed, using their prior experience in combination with lessons learned within the ongoing simulation.
- **Debriefing:** Demonstrate to them that you are concerned about their education by helping them develop solutions.
- **Transference:** Then talk about their prior problems and how they can use the skills learned in the simulation in future crises.

By presenting 'realistic' challenges combined with having fun and excitement during the simulations, trainees become emotionally involved and engaged, and are thereby reinvigorated to learn.

## 80.4.2 Reinforcement

Conveying desired behaviors is key to learning in a clinical simulation. To reinforce desired behaviors, it is important to refer to all positive behaviors during the debriefing. When I see a good attitude, or excitement about learning, complimenting that behavior assists in creating an attitude that I would like to see within the simulation center. Similarly, when we see undesirable behaviors and attitudes, we apply negative reinforcement. Frequently, we will see these undesirable behaviors in more experienced learners who complain such as, "this doesn't seem realistic to me, so it is of no value to me," "this isn't the way the physiology would have presented itself," or "this isn't the way that I would have treated this patient."

There are many new and old approaches to handle this sort of "negative" person. I have found it to be quite effective to determine why this person is displaying these unacceptable behaviors. Generally speaking, it is because I have not, as an instructor, engaged them appropriately at that time. This is probably due to the fact that I am giving them a case that is not appropriate (i.e., too simplistic), for their level of learning. This is where turning up the heat, increasing the challenges, and increasing their perception that there are immediate consequences to their therapy, is a good idea. This patient could go into shock![1] The consequences of their actions are going to be viewed poorly by others, and this in turn increases their stress. I have found frequently that we can strongly motivate them to learn by **planning and running** an appropriately difficult, realistic, and applicable/relevant patient and scenario where during the **debriefing**, one can point out the value of the **transference**.

### 80.4.3 Retention

Learners must see a meaning or purpose to the material, they must have the ability to interpret and apply the material, and they must have the ability to assign the correct degree of importance to the material. While running our simulation, we have to look for the following items:

- Did they grasp the concept that I was trying to demonstrate to them in this hypoxic patient? (Eileen Short – see previous description of this simulated patient.)
- Do they have the tools, skills, and experience to manage the simulation?

While debriefing the simulation, it is important to help the trainees:

- To see how to interpret and apply the concepts that are being presented to them,
- To set the correct degree of importance,
- To think through a clinical emergency.

We need to be able, as educators, to simplify crises into digestible components that can be understood by any learner, find the commonalities in these different types of clinical emergencies, and move forward. For example, I go back to airway, breathing, circulation, as the 1, 2, and 3 of what is important in a case. We have to demonstrate to the learners how to set priorities and implement them. Knowles stated that retention equals the degree of original learning and the amount of practice. This is where simulation allows learning followed by practice to maximize retention.

### 80.4.4 Transference

Transference refers to the ability to take the lessons learned from these sessions and apply them to a new or a novel problem. Transference can be either positive or negative. Positive transference would be described as the use of behaviors that had been deliberately taught in the session and will now be applied by the trainee within the clinical arena. Negative transference would be, for example, learners who refrain from prior bad habits and poor attitudes after the training session. They do not continue to do what they were told not to do. This equals a positive outcome. For instance, trainees are told not to use large dosages of epinephrine for treatment of a low blood pressure, and they consequently do not administer a large, ACLS-type dose (1 mg epinephrine) in a patient with a

beating heart, but instead select 50–100 ug per dose. There are multiple different ways to help to create "transference:"

- We establish positive rewards;
- We look at how we can help them to take the behaviors that they learned within the simulation center to their clinical practice (i.e., a positive means of transference);
- For a learner who makes mistakes and does not self-correct during the simulation, the debriefing is a critical time to enable transference of correct treatments and behaviors.

The debriefing offers the ideal time to let them self-discover what is incorrect, then help them understand why it is correct. When they walk out of the simulation, they have a strong image in their mind of what they are not going to do the next time this particular event occurs. So, negative transference can be a very powerful and effective means of obtaining what we are looking for, i.e., transference of correct therapies and behaviors to the clinical arena. The transference as envisaged by Knowles is the transference to clinical practice that is the ultimate aim of the Learning Pyramid.

## 80.5 Experiential Learning Theory as Applicable to the Learning Pyramid

Kolb [2] proposed a four-staged cyclical learning process:

- Concrete experience
- Reflective observation
- Abstract conceptualization
- Active experimentation

### 80.5.1 Concrete Experience

We have to provide a concrete experience (i.e., clinical events as experienced within clinical simulation). The experience should include lessons to be learned and that can be integrated into our practice after the debriefing.

### 80.5.2 Reflective Observation

Reflective observation occurs during the debriefing phase, where we have to analyze the outcome. We treated a patient, we used critical communication skills we displayed during the scenario, and a definitive outcome was observed. Now, we have to analyze these steps so that we can learn. Without reflection and debriefing, we cannot recognize what worked and what did not work; therefore, we will not change.

### 80.5.3 Abstract Conceptualization

Abstract conceptualization simply means reviewing and understanding the concept. The debriefing helps to transform the concrete experience into a higher and more generically applicable ("abstract") concept. Was this the correct way to determine if the problem existed? Was it the correct treatment? Did we assign and prioritize our tasks in the correct order? How can we take the points we learned in this, say, hypoxic crisis, and apply the principles in, say, a cardiovascular crisis? We first need to form an abstract concept of the points learned during the hypoxic crisis before we can transform the abstract concepts to practical applications for active experimentation in the new (cardiovascular) crisis.

### 80.5.4 Active Experimentation

After the abstract conceptualization phase follows active experimentation. An example of active experimentation to finding solutions can be seen in our scenarios created for NASA. We needed to develop a concrete experience in learning airway management as applicable in the weightless state in outer space. In our particular case, we need to provide for a circumstance of a patient becoming apneic, let's say that the patient had a heart attack, and this happens in a zero-gravity setting, how does one who is floating now intubate a patient who is also floating? When trying to intubate a patient, instructors developing the learning session find that there are many logistical problems. For instance, when an astronaut pushes the laryngoscope against the simulator, both move (but in opposite directions) in the weightless environment (Figure 80.3). New techniques are needed to solve such environmentally specific challenges. From there, the instructors have to go to reflective observation and examine the outcome. Were we effective? Were we successful in establishing the airway? How do we fix (fasten, stabilize) the intubator relative to the victim? In our initial studies, we found out that we were not successful – we could not keep the intubator and victim in close proximity. So then, we have to go to our next phase, abstract conceptualization, reviewing our concepts and ask, do we have the correct concept, the correct treatment protocol, the correct priorities in obtaining that airway in a zero-gravity concept? We then have to develop different techniques for intubation in a zero-gravity setting.

So, this is where active experimentation comes in. We (as instructors) try to find solutions by practicing different concepts. Once we have developed a new idea, e.g., using Velcro to fix the victim and the intubator to the same structure (bed), we now go back to concrete experience. We integrate this new concept into our practice. We have different astronauts and flight surgeons actually test the new methodology. We then return to reflective observation, perform an outcome analysis, and determine whether the new concept is practical and functional. We find out that the learner (who is the instructor in the first cycle, and the trainee with the second cycle) can begin at any point in this cycle; there is no limit to the number of cycles you can make to solve and practice therapies for a given problem.

As we noted in the previous discussion, Kolb's experiential theory also includes the following basic principles:

- The learner can begin at any point in this particular cycle;
- There is no limit to the number of cycles;
- The learner can practice any number of options including making mistakes;
- Without reflection, trainees will continue to repeat their mistakes.

Kolb asserts that "no reflection equals no learning." These concepts are built into the Learning Pyramid: repeatedly running, debriefing until success, and then transference.

## 80.6 Applying Learning Principles to Simulator Session Design

In this section, we present multiple practical examples and implications of applying the aforementioned Learning Principles (Learning Pyramid, Knowles and Kolb) to the design and production of simulator sessions.

Let us say that I am planning to cover challenging cases of hypoxia for a new group of residents coming through my simulation center over the next 2 months. So, I formally identify and write down my goals and objectives. I select from patients that I have had in the past, and determine which events, circumstances, and aspects of those particular patients could be presented in a manner that would be an effective way of attaining my goals and objectives. The next stage is to program and script these scenarios. I present them to test audiences of colleagues knowledgeable in both simulation-based teaching and the clinical concepts of this particular scenario, but naïve to the scenario itself. We are now in our active experimentation phase. We are then going to see how the scenario works, i.e., going into our concrete experience. This will be followed by the reflective observation stage and outcome analysis to analyze if the scenario works (i.e., did it achieve the goals and objectives?). What if it didn't seem to be effective during the test, because the audience didn't learn what I was hoping they would have learned? Therefore, I need to change the scenario. This is where I need to review the concept of understanding (i.e., the abstract conceptualization), and I may need to return to the active experimentation phase. You can start at any one point in the cycle, and there is no limit to the number of cycles you can make in developing or experiencing a given learning case. What is important is that you reflect on the results that

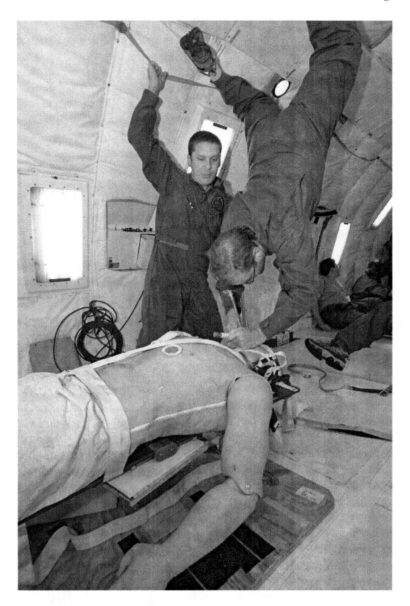

**FIGURE 80.3**  Intubation in zero-gravity. Without the usual connection (provided by gravity) between clinician and patient, forces applied through the laryngoscope tend to flip the clinician over the patient.

you have, look at how your students are doing, and make certain that the necessary changes to achieve the objectives are implemented. The scenario can then become a very exciting, and most of all, engaging simulation.

### 80.6.1 "Fly as You Simulate – Simulate as You Fly"

"Fly as you simulate – Simulate as you fly" is an expression adapted from the aviation industry. This means that, when you are a pilot training in a flight simulator, you are expected to function as if you are in a real aircraft. You should fly that particular "airplane" as well as you possibly can in that simulated setting. Then when you are in your real commercial aircraft, you should fly just as you did in the simulated aircraft. The military has a similar saying: "Fight as you train, train as you fight."

### 80.6.2 No Torpedoes

You can make the most difficult simulated cases ever imagined in the history of clinical education, but you must ask yourself if this is an effective way for your students to learn? You can also continually challenge the trainees by changing the playing field, adding yet another organ system failure, to the point where they cannot possibly succeed. If this is not in your objectives, why do it? What do we accomplish when we force failure on a student? By making that person feel embarrassed or look bad in that particular circumstance, are you accomplishing the goals that you set out to do? The goal that I would suggest is: in our center, we never "torpedo" a learner, because we want them to always experience some (legitimately earned) success, even if they are not completely successful, we want them to learn new material, reinforcing old material and good habits. We want to contribute to the excitement they have about their education.

This same principle of "no torpedoes" applies to instructors new to simulation. Many are apprehensive about being embarrassed, and appropriately so. If your goal is for this novice to embrace simulation-based education, then first tell them that the theme of your simulation program is to make every instructor look brilliant in front of their students, and then follow up these words with actions. So, no torpedoes against the faculty, either.

## 80.6.3  Death of a Simulated Patient

Are you going to allow your simulated patient to die? There are two answers to this important question: 1) No and 2) Yes, but . . .

**No**, never as a "punishment". Generally speaking, we do not allow simulated patients to die. What if the patient is being treated by the caregivers and they are not following correct protocols, which in real life would likely lead to death? In such a case, there are several options:

1. Stop the simulation before death and say: "The simulation is over, now let us sit down, and begin our debriefing process." We can utilize one of two techniques of debriefing: being an instructor and instructing them why this particular outcome was bound to happen; or lead them to self-discovery through a facilitatory debriefing (see later sections for more on these two types of debriefing modes).
2. Stop the simulation before death, point out the clinical symptoms and signs needed to make the correct diagnosis, and initiative the correct therapy, and then either continue from that point on, or start from the beginning.
3. Place the simulator in a holding state or pause mode ("statis" mode) before death, until they the make the diagnosis and start the correct therapy. Then, continue with the progress of the simulation.
4. Before death, help them find the correct diagnosis and therapy (using actors, or suggestions from the instructors, etc.) and continue to work toward the objectives for the session. For instance, if the objective of the session is diagnosis and therapy of pneumothorax, the trainees might miss the diagnosis. If the patient is "allowed" to die, most of the simulator session time might be spent on an advanced resuscitation (ACLS) protocol, which is *not* the stated objective.

**Yes**, but only in special circumstances when the "death" is a well-planned educational objective. For instance, we do allow patients to die, if we are doing simulations based on concepts of death and dying, and learning to give "bad news" to the patient's family. We would then request a psychologist to come in and assist us in talking about death and dying. This is a very, very important and sensitive subject. For any caregiver in medicine, it is important to help them obtain and develop the tools that they need to be able to psychologically deal with the outcome that we all dread.

An additional observation about death and the simulated patient: start with the same unhealthy cardiovascular state as our previously mentioned patient, Eileen Short, in a patient that is very old, frail, and in the Operating Room for septic bowl surgery. Enter the student, a third year resident, and within 10 seconds "kills" their patient by promptly treating the very high heart rate with a beta blocker. The student focused only on half the problem, i.e., the heart rate component, in complete oblivion of the other half, the very low blood pressure. Perhaps promptly hitting him in the face with a large wet fish in the form of asystole is a *very* appropriate teaching approach. But, to prevent this horrendous event from just being child abuse, convert it into a valuable learning experience about resuscitating before anesthetizing by having all the instructors immediately enter the simulation room as one asks the famous questions:

> What did you determine about your patient?
> What did you do about it?
> What happened next?
> Why?
> What could you have done differently?
> Want to try again?

If some students are so oblivious to their lethality, then perhaps a death or two (simulated) is just what they need to awaken them to the responsibility that goes with such power.

Immediately repeating scenarios allows students opportunities to produce a different (hopefully better) outcome given the same starting conditions. It is *not* 'cheating' that they 'now know what will happen'. By repetition of a purpose-driven scenario, students will have first, found the boundaries of *their* competence by direct personal experience, second, *safely* extended themselves beyond into *their* own regions of incompetence, and third, stayed there long enough, under guidance, to develop familiarity, to create new competence. This is *the* fundamental value that only simulation can add to our world of clinical learning.

### 80.6.4  How to Handle Negative Behaviors

Are you going to, the moment someone starts disrupting your class, embarrass them in front of their peers? Are you going to take them out in private to discuss their negative behavior with them? These are both viable means of taking care of negative behaviors. Preempting negative behaviors is even more useful. For instance, by setting up the expectations that you have prior to the beginning of the simulation. Most of our trainees respond best to private criticism and public praise for those behaviors we actually want them to exhibit. This approach applies equally to our fellow instructors who need to develop and exhibit better teaching behaviors, just as it does to our students who need to develop and exhibit better learning and clinical behaviors.

### 80.6.5  How Much Stress

It takes a fair amount of experience to figure out how much stress that learners need for them to perform at their best. So, by monitoring them very closely during the simulation, we can see signs of stress. For instance, we can listen to the tone of their voices, listen to how well they are describing what is going on, etc. Are they standing up straight? Are they showing body language positions demonstrating stress? Are they staring or focusing on a single object such as a monitor for any great length of time? All of these stress indicators need to be assessed during the simulation.

When is a participant allowed to support the person in charge (the "hot seater") at the head of the patient? The person in the "hot seat" is under the most stress, and, therefore, may not be thinking as clearly as the person who has just walked into the room. This is an important learning point – ask for help when you are under stress. Frequently, members of the simulation team will want to assist the hot seater because the hot seater might perhaps not be picking up on a critical vital sign. In some cases, we do not encourage this spontaneous help and advice from team members. We request that support is given only when the hot seater managing the patient requests help. This forces the hot seater to realize that help is often available, and should be asked for. In these scenarios, all other participants are requested to be as competent as they know how, but not to initiate or volunteer any help for the person in charge. In other simulations, we ask the helpers to offer help to the hot seater. This would be the case where we want operating room personnel such as nurses, to learn to become involved in an anesthesia crisis. In one scenario, we employ an actor anesthesiologist who ignores all the signs of malignant hyperthermia, and continues to blame the tachycardia, increased carbon dioxide levels, and high temperature on other causes. The nurses are then "forced" to be assertive, give advice and ask questions until the diagnosis of MH is declared and the proper treatment initiated.

### 80.6.6  Never Discuss Performance Outside the Simulation (Confidentiality, Security)

We request that members of the simulation team do not discuss trainees' performance outside the simulation. The reason this is so very important is that we want to create a simulation center environment where trainees can practice various treatment options (especially "mistakes") without the fear of reprisal or ridicule. We can create a learning environment where the students understand that they are welcome to make mistakes, and this is in fact the best place to make mistakes and learn from them. Trainees may "freeze" and fail to make decisions. Trainees may "panic" and try "shotgun" therapy, i.e., institute multiple different modalities and drugs without waiting to see if any of them have any effect. The idea would be to rather experience and remedy these "mistakes" in simulation, prior to these behaviors happening in the real world.

### 80.6.7  Rules of Engagement

Just as we set up rules prior to a familiarization session in a simulator center, we also set up rules of engagement ("code of conduct") for debriefings. Debriefings are essential to learning and changing behaviors (near the peak of the Learning Pyramid). The rules for debriefing include highlighting the learning objectives, rather than getting sidetracked with controversies about various therapies.

### 80.6.8  Privacy is Maintained

Privacy should be maintained. There are no grades within our simulation center. Trainees' performances are not passed on to the residency director, the clinical competency committee, or to the chair of the department. This provides a safe and nurturing *learning* environment. The video recordings are considered for research purposes only and, therefore, are not discoverable. Trainees are informed that clinical remedial knowledge is their own responsibility. The simulation session can highlight deficient understanding and attitudes, but simulation is not an efficient way of conveying factual knowledge.

### 80.6.9  Have a Backup Plan and be Open to Changes in the Simulation

We always have a backup plan in case a participant's behavior or actions are unexpected. For example, during a transport scenario of an intubated patient, an excited trainee might give excessively large tidal volumes. We might then decide to add a pneumothorax to the objectives, especially if the patient has a known lung disease (e.g., chronic obstructive pulmonary disease). This is an expansion of the objectives, which might include "safe ventilation during transport." The pneumothorax is *not* used to punish the trainee, as this would constitute a "torpedo." The pneumothorax is just another complication of overzealous ventilation, which is now demonstrated (as a practical hands-on crisis) during the scenario, rather than just mentioned (as a passive discussion) during the debriefing. (Note also that the pneumothorax does not progress to a cardiac arrest.) So, have a backup plan in mind, be flexible, and this will help you out with managing your debriefing. Always be open to changes in the simulation, but only when this supports your objectives.

### 80.6.10 Simulation Planning

The several dozen crisis events encountered in the operating room [3, 4] can be encapsulated into five basic emergencies: hypertension, hypotension, tachycardia, bradycardia, and hypoxemia. This is obviously a very simplified system, however, learners do very well when you can simplify diagnoses and therapies into very general concepts. A clinical emergency is typically a combination (or permutation) of the five different crisis events. Trainees need the knowledge (drugs and doses) to manage these events. These drugs and dosages can be learned from reading, lectures, and web sites. However, trainees do not function in isolation; they work in teams, and they need to learn efficient team interactions in simulation scenarios using the principles of Crisis Resource Management. Working in teams cannot be learnt from books. This is where simulation is needed. Team training has been encapsulated by Dr. Gaba in the Crisis Resource Management principles. Relevant principles of Crisis Resource Management include critical thinking, critical communications, and critical algorithm development. When we look at our simulator planning objectives, we present cases necessitating trainees to use CRM skills and behaviors. We want to find out how well they perform, and determine their ability to perform under pressure. We look at how effective they are at dealing with a task-saturated environment, and how they interact with other participants. The video replay and debriefing gives them a window (or a mirror) to view themselves as they change their behaviors and transfer the skills to the real clinical environment.

## 80.7 Debriefing – Rules of Engagement

Debriefing is not a new concept. When hunters came back from their hunts, they sat and discussed and diagramed their kills: how effective they were and who could have done better. Modern debriefing really took hold in the United States in the business world back in the 1960s and 1970s to improve the planning of meetings and congresses. At that time, they initiated the concept of "learning and awareness facilitators:" learning becomes a dialog, not a rote process [5]. Today, we

have a number of resources to help us capture and then "relive" our simulation experiences:

- Video recording and replay,
- An instructor or simulation professional taking key notes,
- Digital and robotic simulators recording vital signs, actions, and the exact time the students performed the interventions.

Let us first start with a "How not to do it example" from the Editors:

> I have one clinical instructor who is passionate about teaching, has boundless energy, is very knowledgeable, and has a terrible manner. "So, what did you do wrong?" is his opening question to the students at the beginning of each debriefing session. His follow-up statement is, "I told you not to do that before the start of the scenario." I am at my wits end to find a way to teach those who are unaware that they are in need of change, other than to use them as a reminder that I make sure each and every instructor new to my simulation center gets a thorough "How to use simulation in a non-punitive way of teaching" course, followed up by regular discussions about debriefings with all clinical instructors.

Now we can continue with a "How to do it."

### 80.7.1 Begin with Positive Reward

A key element in debriefing that we must always begin with is a positive reward. We always emphasize constructive criticism, and not complaints. Once your simulation program has established a policy and a reputation for constructive comments, then the absence of them can be a more powerful commentary on poor behavior than would be to express complaints. So, what we would like to hear from the instructor is "what did they do well?" We would like to set a tone of acceptance. We want the learners to feel as if they did something positive, even if the outcome was not what they had hoped to have. Such a positive comment starts off the debriefing with positive reinforcement. Only use negative reinforcement when necessary in the face of bad behaviors that we wish to extinguish. As actions speak louder than words, the very best way is for the instructors to keep their voices quite and let their simulated patient's condition do the 'negative reinforcement'.

### 80.7.2 Constructive Criticism

All comments must be based on constructive criticism. For example, if a student is looking at another student's performance and says, "Boy, I can't believe you did that. That patient was exhibiting such obvious clinical signs, and you just missed it completely." That is not constructive criticism. Constructive criticism is: "How did you arrive at your decision? How are you going to ensure that you and your team gather all the necessary information next time? I have been in a similar crisis before and I was too hasty in my decision. I did not consult all my team members. How do you think such an action (asking for help) might have helped in this particular case?" (Note that this is an example of Plus-Delta described in Chapter 57 – Scenario Design and Execution.)

### 80.7.3  Leave the Participant Ready to Learn

The idea is that when you finish with these simulation sessions and your debriefing, you want the trainees to be so interested that want to learn more on their own. We do this through positive reinforcement, constructive criticism, and having a strong understanding of adult learning principles. Leave the participant ready and eager to learn. Make sure you give them a condensed review list, i.e., only one or two key items that could actually digest (not an encyclopedic list).

### 80.7.4  Roles of Debriefers and Types of Debriefing

The debriefers' role fall into one of three categories; either a **facilitator** or an **instructor** or a combination of the two, which we term a **facilitative instructor**.

**Facilitator role:** To begin with, a **facilitator** is someone who is going to encourage *self-discovery* amongst the other participants. The facilitator, as a content-neutral party, enables groups to work more effectively in learning or in dialog, and becomes just the guide to that learning or dialogue[2]. A facilitator leader helps a group understand organizational dynamics and thereby creates a process to enable the group members to more fully utilize their potentials. A facilitative individual can be a team player and would assist others in this process.

The organizational psychologist Kaner [5] describes facilitation as requiring "contract neutrality," that is, where the facilitator does not have a position or a stake in the outcome. However, the students' instructors will never be accepted completely as neutral, since the instructors do have a stake, i.e., we want our students to learn, and we are the ultimate judges of how well they learn. Even if students are told that their behaviors during the simulation sessions are not graded *per se*, the students know that they must pass grading hurdles to progress; their simulation instructors define both the hurdles as well as what is acceptable as successful passage over those hurdles. This inherent "interest in the outcome" makes purely neutral facilitation more difficult, but not impossible. However, the value gained in establishing and using contract neutrality throughout the simulation experience is worth the attempt at *simulating* such an environment.

**Instructor role:** Sometimes, our participants have difficulty in grasping the *content* we are trying to teach them in a particular session: therefore, the facilitator might have to become an instructor. For example, if we have a group of clinical students who are obviously not familiar enough with how a certain drug interacts with a specific organ system, it may be necessary to give some instruction at that time, and that would be the appropriate way to go. If it is very clear from the participants that they did not understand the physiology, it is very difficult to be a straight facilitator, because at this point we really do need to make certain that they understand the basic physiologic principles, for instance the Starling curve, or how gas exchange happens inside the lungs.

Whenever clinical content issues become the focus of discussion within a simulation, then the curriculum is in need of review. Students should not be overwhelmed and stumped by simulation scenarios that require factual knowledge they have not yet acquired. If a scenario reveals that certain facts and concepts that had been presented previously had not yet been internalized, then either the deficient student needs to revisit the books or the instructors need to address the deficiencies in their factual presentations prior to the simulation[3].

**Combined facilitator and instructor role:** Sometimes, the trainees do not fully understand the disease process, and it is necessary to become an instructor. For instance, it may be necessary to discuss with them the pathophysiology of heart and lung disease, and pharmacologic interventions. This is more common, generally speaking, with more inexperienced groups. Sometimes, however, we have a group that does understand what is going on, and they capture the essence of the simulation. They might only need guidance and/or remediation with one or two minor clinical points. This may be a time where a facilitator can lead this group to self-actualization or self-discovery. There might actually be much benefit if this clinical teaching point can be turned into a CRM team principle. For instance, during the session, the leader administered an incorrect drug or dose. During the team debriefing, some of the team members might actually have the necessary information. This illustrates the principle of the leader asking for help and the team member to suggest help.

Over years of providing these discussion groups, I have found that you seldom have a pure instructor or pure facilitator role; these two roles most often coexist. As the two roles start to slip together, the pure instructor and the pure facilitator start to blend and merge: this is what I would like to call a **facilitative instructor.**

### 80.7.5  Debriefing Techniques

As described, we can debrief in a number of different ways. Each way is best suited for some purposes, and not one way is perfect for all. There is a strong temptation to be very confrontational, where we try to pin the person down as to what they were doing, what they were thinking, how they reacted, why they reacted that way. Mostly, this is ineffective. Debriefing techniques should be based on the principles (philosophy) and time-tested methods from the beginning of education:

Listen to the Participants – allow sufficient time for their self-direction, self-discovery. Allow the participants to fashion their own debriefing because it may be that there are key elements that they found quite important, that you might not have thought of.

Keep their attention. Always bring their focus back to the main points of the debriefing. You can maintain their attention and cooperation by listening to the participants, thereby requiring that they participate in the process.

Positive reinforcement should begin each session. Therefore, prior to beginning every debriefing, find something positive that each participant did very well. For instance, it can be a critical communication, a demonstrated thought process, followed by an action such as demonstrating a skill set, or establishing an airway. Ideally, each positive example could be taken directly from the explicitly stated goals for that day's simulation session syllabus.

Summarize each *individual's* performance after the debriefing, noting where there is room for improvement. This is a very effective time to give direction toward their individual improvement.

Give direction for *group* improvement – facilitate group direction for change. If the group performs effectively, reward them immediately, by giving them positive reinforcement. If there is still room for this group to work in a more effective team-oriented manner, this point can be used by the facilitator as the direction for improvement by giving other examples of more effective group functioning as a team.

### 80.7.6  Content-neutral or Content-specific Debriefings

Sometimes, debriefing can be **facilitator driven** where we allow the participants to come to their own self-discovery. One of the most effective ways is to video record the simulation, with timing marks at points of interest, so that we know when certain key issues and critical communications occurred. The video recording is then played back from those points to stimulate the discussion so that we can obtain our objectives.

Sometimes, debriefing can be **instructor driven** where the instructor is not only going to describe all of the aspects of physiology and pharmacology, but also describe all of those behaviors that were noticed during that simulation. The instructor also helps lead the group to understand why critical communication, critical thinking, and critical algorithms are important to clinical practice.

Another possible option for debriefing is a form of an **action debriefing,** which can be described as play acting. Two people who are well informed on what the outcome should be will put on a play. One person might verbally attack the other person's

manner of treatment as well as the communications. The idea is to see how trainees react to such a confrontation. Highlighting such a confrontation by neutral play actors can be valuable when we do integrated multidisciplinary simulations, where we want to see how nurses and physicians would respond to an argument or a heated discussion.

The last type of debriefing is an example of a **critical event action debriefing**, which, for instance, may follow any of the natural disasters that have happened over the past number of years. Rescuers are debriefed following a critical action or a critical event, by talking about it and going through the different aspects the critical event. This leads them to self-discovery of what problems this crisis had created for them, and how they can best deal with such crises in the future. Such debriefing has been successful, both in the military and the civilian world. Using these principles, we can provide a simulated crisis and use "critical event action debriefing" to help trainees understand their actions and behaviors, and self-actualize changes in their future behaviors.

## 80.8  Conclusion

This chapter showed the basic components of a clinical simulation, how to follow the principles of crew resource management, including critical thinking, critical algorithm development, and critical communications. For instance, trainees are taught to identify the common elements of most clinical emergencies. These elements can be thought of in a simplified manner as five elements, i.e., hypotension, hypertension, tachycardia, bradycardia, and hypoxemia. Each of these elements can be dealt with separately or as a unit, depending on circumstances.

Keep in mind that for your adult learner students, and your students on their way to becoming adult learners, the principles of adult learning are an invaluable foundation on which to build your scenarios. Once the plan has been created, stick to the plan, and then carry out the simulation, moving up the simulation pyramid to the debriefing.

Using the principles outlined in this chapter, at the end of our debriefing, transference should occur, and what the trainees had experienced in the simulator, should transfer to the clinical arena. By applying H–E–A–T (High-fidelity Environment Analog Training), we have completed the simulation learning pyramid.

## Endnotes

1. There is a temptation is to "punish" the "negative" trainee and cause the patient to "die." However, this is seldom justifiable as an "on the spur of the moment" decision. Also see Section 80.6.2 "No Torpedoes".

2. This is where including nonclinician facilitators have an advantage – as they have no clinical expertise, there is minimal tendency for them to become sidetracked in esoteric clinical issues and they can therefore fully focus on the CRM principles.

3. This is why some Simulation Centers reserve CRM sessions for higher-level trainees. Such centers typically spend about half the debriefing time on Crisis Resource Management issues and half the time of clinical issues (disease, drugs, dosages, etc. – also known as Crisis Management). If the students do not have the background of the drugs and dosages, they do not gain full value from such a CRM session. On the other hand, some centers only focus on purely CRM issues. They can therefore include medical students as participants in multidisciplinary CRM sessions, as the medical students can readily grasp the CRM concepts of communication, team work, avoiding fixations, etc. Even though these medical students do not have a full understanding of the clinical issues, their feedback from participating in focused CRM sessions indicates that they do understand, grasp, and can internalize the CRM principles.

# References

1. Malcolm Shepherd Knowles (Editor). *Andragogy in Action. Applying modern principles of adult education*, A collection of chapters examining different aspects of Knowles' formulation. Jossey Bass, San Francisco. 1984.
2. Kolb, D. A. (with J. Osland and I. Rubin). *Organizational Behavior: An Experiential Approach to Human Behavior in Organizations* 6e, Prentice Hall, Englewood Cliffs, NJ, 1995.
3. Gaba, D., Fish, K., and Howard, S. *Crisis Management in Anesthesiology*. Churchill Livingston, Philadelphia. 1994.
4. Schwid H. – Anesthesia Simulator – http://www.anesoft.com/Products/as.asp Accessed 1 Oct 2007.
5. Kaner, S., Lind, L., Toldi, C., Fisk, S., and Berger, D. *Facilitator's Guide to Participatory Decision Making*. New society Publishers, Gabriola Island, BC. 2001.

# Further Readings

Brew, J. M. Informal Education. *Adventures and reflections.* Faber, London, 1946.

Brockett, R. G. and Hiemstra, R. Self-Direction in Adult Learning. *Perspectives on theory, research and practice.* Routledge, London, 1991.

Brookfield, S. B. 'Self directed learning' in YMCA George Williams College ICE301 *Adult and Community Education Unit 2: Approaching adult education.* YMCA George Williams College, London, 1994.

Candy, P. C. Self-direction for Lifelong Learning. *A comprehensive guide to theory and practice.* Jossey-Bass, San Francisco, 1991.

Carlson, R. 'Malcolm Knowles: Apostle of Andragogy', *Vitae Scholasticae*, 8, 1. http://www.nl.edu/ace/Resources/Knowles.html. 1989.

Davenport 'Is there any way out of the andragogy mess?' In: Thorpe, M., Edwards, R. and Hanson, A. (eds.) *Culture and Processes of Adult Learning*, Routledge, London, 1993 (First published 1987).

Hewitt, D. and Mather, K. *Adult Education: A dynamic for democracy.* Appleton-Century-Crofts, East Norwalk, Con. 1937.

Jarvis, P. 'Malcolm Knowles' In: Jarvis, P. (ed.), *Twentieth Century Thinkers in Adult Education.* Croom Helm, London, 1987a.

Kett, J. F. The Pursuit of Knowledge Under Difficulties. *From self-improvement to adult education in America. 1750–1990,* Stanford University Press, Stanford, Ca. 1994.

Knowles, M. S. Informal Adult Education. Guide for educators based on the writer's experience as a programme organizer in the YMCA. Association Press, New York, 1950.

Knowles, M. S. and Knowles, H. F. *How to Develop Better Leaders*, Association Press, New York, 1955.

Knowles, M. S. and Knowles, H. F. *Introduction to Group Dynamics.* Association Press, Chicago/Cambridge Books, New York, Revised edition 1972, 1959.

Knowles, M. S. *A History of the Adult Education Movement in the U.S.A.* A revised edition was published in 1977. Krieger, New York, 1962.

Knowles, M. S. Self-Directed Learning. *A guide for learners and teachers.* Programmatic guide that is rather objective oriented. Sections on the learner, the teacher and learning resources. Prentice Hall/Cambridge, Englewood Cliffs, 1975, p. 135.

Knowles, M. S. The Modern Practice of Adult Education. *Andragogy versus pedagogy.* Prentice Hall/Cambridge, Englewood Cliffs, 1980, p. 400.

Knowles, M. S. *The Making of an Adult Educator. An autobiographical journey,* 211 + xxii pages. A rather quirky series of reflections on becoming an adult educator; episodes that changed his life; landmarks and heroes in adult education; how his thinking changed; questions he frequently got asked etc. Jossey-Bass, San Francisco, 1989.

Knowles, M. S. *The Adult Learner. A neglected species* (4e). human resource development (HRD). The section on andragogy has some reflection on the debates concerning andragogy. Extensive appendices which includes planning checklists, policy statements and some articles by Knowles – creating lifelong learning communities, from teacher to facilitator etc. Gulf Publishing, Houston, 1990.

Kolb, D. A. and Fry, R. 'Toward an applied theory of experiential learning'. In: C. Cooper (ed.) *Theories of Group Process.* John Wiley, London, 1975.

Kolb, A. and Kolb, D. A. *Experiential Learning Theory Bibliography. 1971–2001*, McBer and Co, Boston, MA.2001. http://trgmcber.haygroup.com/Products/learning/bibliography.htm

Layton, E. (ed.) Roads to Citizenship. *Suggestions for various methods of informal education in citizenship.* Oxford University Press, London, 1940.

Lindeman, E. C. *The Meaning of Adult Education.* 1989th ed. University of Oklahoma, Norman, 1926.

Merriam, S. B. and Caffarella, R. S. Learning in Adulthood. *A comprehensive guide.* Jossey-Bass, San Francisco, 1991.

Quinones, M. and Ehrenstein, A. Training for a Rapidly Changing Work Place. American Psychological Association, Washington, DC, 1997.

Tennant, M. *Psychology and Adult Learning.* Routledge, London, 1996.

Yeaxlee, B. Lifelong Education. *A sketch of the range and significance of the adult education movement.* Cassell and Company, London, 1929.

# Expect the Unexpected

*"Experience is the name everyone gives to their mistakes."*

Oscar Wilde

*"Experience teaches only the teachable."*

Aldous Huxley

If battles are won by the side that makes the fewest mistakes, and if experiences are in the mistakes, and if lessons come from experience, then is loosing a war the best way to learn how to fight? There are many forms of combat, and waging a campaign in presenting a simulation-based course in a far-off land is one of the more challenging contests that clinical educators can undertake. Producing a simulation event at a conference violates many "rules" that we have reinforced throughout this book:

- Know your audience's initial conditions and their numbers;
- Know what your audience wants to gain by their attendance;
- Know your environment;
- Know your resources;
- Know your limitations.

At "away" venues, you not only won't know, but can't know what you should know well before any simulation event; thus, almost everything you have learned from all your experiences within your own well-protected simulation environment "at home" either don't apply or are in opposition to what you ought to know when away. The amount and sources of novelty will overwhelm all but the simplest plans for the most basic of goals. At home, every simulation event should integrate within and support a larger, well-defined teaching program. Away, a realizable goal is more like offering a "free sample," a taste of what is possible, rather than what we might consider typical teaching objectives. Time during the event is the most precious ingredient. Every aspect of your plan of action, from the grand overview to the smallest scenario detail must be chosen like the words in an abstract; every letter, word, concept must have a clear reason to be included, as there is no spare room for unused baggage, no extra time for do-overs.

This intentional reduction of scope is a source of disappointment to many workshop attendees. They make the effort to show up out of a real need. The challenge for the workshop producer is that there is very little in the way of effective pre-sorting and pre-prioritizing all the disparate needs of all the attendees. Often, the result is that no one, or at best, only a very small fraction of the participants are satisfied. This frustration includes the presenters. Expectation management is just as crucial for the producers as it is for the audience.

Kristina Stillsmoking and W. Bosseau Murray have had their share of "experiences" when producing simulation courses at a conference, learned from them, and produced this Logistics document. Let the well-known statement that "amateurs talk tactics while professionals talk logistics" be your guide, for without managing logistics, there will be only failure.

# 81

# Managing a Simulation Session at a Congress, Away from Home Base

Kristina Lee Stillsmoking
W. Bosseau Murray

# 81.1 Management and Preparation

Program management is the glue keeping your simulation program functional, if not also effective and efficient. Often, program management is seen as a high-level administrative function, only possibly performed by the highest paid individuals in the simulation group – the clinicians. However, clinical personnel do not have the time, nor hardly ever the inclination or training for management thought, planning, and action. As someone must fulfill the role and execute the duties of program management, it might as well be the simulation operations professionals. While the appearance of every simulation program may seem unique, they all share a common core of management tasks, functions, and responsibilities because they all share common social cultures. Providing a simulation experience at a congress ("away from home") magnifies the challenges and need for impeccable program management, as well as ensuring that the session builds upon shared common social cultures.

You have before you an idea/goal/project that you have decided to undertake at a congress, because you want to, or your supervisor strongly suggested it. Some requirement or incentive, such as administration, faculty, student, or personal need was the guiding force for your decision and subsequent involvement. So, how do you carry out this simulation project "away from the home base" from inception through to completion?

Perhaps this chapter has way more than you need to know! Maybe, maybe not. You can pick and choose elements that will fit the goals of your project or presentation. Having put on two weekend conferences, several management training sessions, and even simulation sessions in other countries [3], we know how to create order and prevent the development of chaos. Many of the points made in this article are applicable even when you do a "road trip" to another site within your institution, for instance, taking the simulator to the chronic pain clinic or endoscopy suite for CRM training in areas with a low incidence of crisis events. As always, attention to details will keep you from falling on your face.

As the "Queen" of organization, my (KLS) style is to put things into labeled three ring binders with lots of tab sheets. I even go so far as to make an index of the tabs, novel idea. Keeping a good calendar of events goes hand in hand with these binders. I have three calendars to keep me going. Of course, the trick to that many calendars is taking the time to look at them. As you put information into your binders, be sure you put a note or two as what the information is for.

When you have meetings, use a recording tool ("system") that will allow you to go back and check on projects pending. Many of the meetings I go to have minutes about what was said, but there is not a real good record of who is responsible for the follow-up ("action points"). Plus, someone has to pay attention to detail to make sure that the project is reviewed at the next meeting. These pending projects never get completed for this exact reason. Have you ever written down someone's phone number, no name then later you find the number and have no idea whose number it is? Point taken, right?

Attention to detail is the key to success along with a good team who supports the project with shared goals and enthusiasm.

When you look at the big picture, it may be overwhelming, so just take one day or piece of a project at a time. Don't forget to breathe. Or maybe you're an type A+ personality like me who can handle as many as two major projects with a few minor ones thrown in.

Remain positive and adaptable, share reward from successes with your staff, thrive on creativity, and delegate when possible.

# 81.2 Educational Issues in General (Needed for Planning)

It is useful to work through the intellectual exercise of answering each of these questions to assist with your planning process. Not all questions will apply to every session, but going through them might help you avoid missing a critical planning item.

## 81.2.1 Who is the Target Audience/ Learner/Customer?

- Students, faculty, administration, owners, community, others?
- Level of training – high school students, medical students, experienced clinicians, etc.
- Homogenous (all same specialty/subspecialty, e.g., anesthesia) or not homogenous (e.g., trauma course – nurse, paramedic, General Practitioner, trauma surgeon, anesthesiologist, etc.).
- How large is the group, e.g., a small group of 4–5 individuals, with a hands-on session ±15–20 attendees with an interactive debriefing (after a live scenario or video recorded scenario), or larger groups – really a lecture format (none, or only minimal audience participation).

## 81.2.2 What are the Target Audiences' Learning Objectives?

You might have to adjust your original "grandiose" learning objectives to fit the audience and the available time. This usually means to simplify and be less ambitious. It also might be depending on the time available for planning, as well as for producing ("staging") the session.

- **Crisis Management** – for example, is how to treat an arrhythmia: drug, dose of drug, defibrillate or not, e.g., A.C.L.S., and better for single-discipline groups and very junior students – they don't yet have the knowledge to treat the crisis and therefore can't/don't concentrate on the higher-level skills. Crisis Management debriefing might occur with or without video (for instance, rerunning the scenario as you debrief "over" the simulator).
- **Crisis *Resource* Management** – is higher-level "administration" – team work, leadership issues, communication, avoiding fixations, etc., and better for multidisciplinary groups and more senior students. Video debriefing is typically used.

**Contact Hours, Continuing Education, and Continuing Medical Education Considerations** – Might have to reassess and/or rephrase the learning objectives to satisfy the requirements for:

a. Contact Hours (CH) for nursing
b. Continuing Education (CE) – medical technicians and therapists
c. Continuing Medical Education (CME) hours for physicians

Apply early for approval, don't wait until the last hour.

## 81.2.3 Who are the Required Subject Matter Experts?

Supervisors, peers, outside people, etc., might need to be contacted during the design and development phase, as well as asked for their input regarding the points to highlight during debriefing.

## 81.2.4 Who are the Required Administrative Leaders and Supporters?

**How much funding do they provide for the training session?**

*Cost Factors involved: A very careful listing of all potential costs needs to be developed, and matched with resources of funding and donations to cover the costs.*

a. Cost versus benefits outcome
b. Staffing hours/costs
   i. Hosts
   ii. Speakers
c. Supplies/equipment needed/cost
   i. LCD projector
   ii. Microphone
   iii. Handouts
   iv. Candy/chocolate chip cookies/coffee
   v. Scribble paper and pens

d. Budget/origin
   i. Where's it coming from?
   ii. Who has access to it?
e. What hotel is close to presentation site?
   i. Will the presentation be in a hotel?
   ii. Will a discount be given for attendees?
   iii. Set/reserve a block of rooms for speakers
f. Keep a list of travel arrangements, arrival and departure dates/times of speaker's pickup and return to airport arrangements
g. Will there be an honorarium given to presenters? How will this be handled?
   i. At presentation
   ii. Mailed
   iii. Will you give presenters a free room and food? For how many nights?
   iv. Will you give presenters free registration to the conference (presentation, congress)?
h. Where will the project/class be held? What are the cost implications?
   i. Hotel
   ii. Conference room
   iii. Other
   iv. Setup of rooms/equipment for each presenter
i. Will there be food and/or hotel accommodations as needed?
   i. What food sources are available?
   ii. Menu or Samples from caterer
   iii. Bring in your own
   iv. Use of a local restaurant (self-pay); might take excessively long if the local cafeteria happens to be too busy.
j. Marketing
   i. What media will you use to get the message out?
     a. E-mail
     b. Flyers, in and out of house
     c. Brochures
     d. Journal advertisement
   ii. How will fees be collected and payments be made? Who will "keep the books?"

## 81.2.5 What is the Time Line to Completing this Project?

a. Preliminary start date – first meeting of team to decide roles and responsibilities for elements of the project.
b. Overview dates/meetings with team – subsequent progress report/problem-solving meetings.
c. Preliminary due and end dates – dates on which all pieces of the project are due to project manager for review and feedback: planning outline, fleshed out outline, semifinal project, final project.
d. Completion date – finalization of project and presentation to target audience/customer.

### 81.2.6 Will There be any Media Coverage?

a. Lists of contacts of local newspapers, TV, and radio stations
b. Will there be a news conference before or after presentation?
c. Who will speak at the news conference?

### 81.2.7 Agenda Planned

a. Topics to be presented
b. Ideas for presenters and contact information

### 81.2.8 Data Collection

a. Data may be needed for CH, CE, or CME purposes, research purposes, and/or improvement of the course
b. Evaluation of project, facilities, and instructors
c. Use data to improve next similar project
d. Ideas for a different project
e. Share data with speakers

### 81.2.9 How Much Time is Available for Presenting the Session?

What day (weekday, weekend, evening) and length of simulation session?

a. Half day
b. Full day, more than 1 day for multiple sessions
c. How many times will the session be repeated during the congress?
   - 1½–2 hours is probably the minimum to adequately address the CRM issues in a hands-on scenario plus a debriefing.
   - Is it possible (sufficient time) to run another (a second) hands-on scenario with a second debriefing?
   - A longer period of 2–3 hours will allow the attendees to experience a second scenario to practice the newly acquired skills.

### 81.2.10 Sequence: Hands-on Session Before or After the CRM Lecture?

There are various points of view in the simulation community about giving information about the CRM principles before or after the hands-on session.

- CRM lecture first:

  Advantages: enables the participants to practice and implement the CRM principles they have just heard about.

  Disadvantages: the participants typically believe that they will be able to manage the crisis without any

problem, and therefore, they tend to "ignore" ("pooh-pooh") the CRM principles given in a lecture prior to the hands-on session.

- Hands-on session first:

  Advantages: the trainees experience how easy it is to miss a diagnosis or therapeutic modality in the chaos of the crisis. They are turned into "adult learners" who have just experienced a "needs analysis." They now believe how easy it is for chaos to develop, and they are eager to learn how to avoid the chaos and perform better. Furthermore, running the hands-on scenario prior to the CRM lecture makes for dramatic debriefing scenes, which seem to make a longer lasting impression.

  Disadvantages: if there is only time for a single session, the trainees often feel that there is no opportunity for them to "redeem" themselves by applying the newly learnt CRM concepts.

### 81.2.11 Is it Really Necessary to Have a Live Hands-on Scenario as Part of the Session?

A series of historical events (e.g., a case report) or previously taped video recordings can be used for illustration and debriefing – this can be developed by "actors" going through a scenario (and deliberately making the "mistakes" (behaviors) that will be used in the debriefing), or it can be segments of recordings from actual CRM sessions (requires written permission of participants and/or obscuring of facial/recognition features). Consider using appropriately chaotic and other scenarios from commercial video recordings of commercial "entertainment," e.g., E.R., M.A.S.H., Rescue 911, etc. Most in the simulation community believe that it is *very* important for trainees to have a hands-on session to gain the most value, as it turns them into adult learners. However, if the objective is to demonstrate CRM principles and training methods to a very large group, such use of video replay is probably an efficient way of getting the message across, and getting individuals interested in participating in a small group, hands-on session at a later stage.

Some instructors believe that if the remotely situated "observers" feel close to, and involved with the hot seaters (e.g., peers, colleagues, fellow residents or nurses, etc.), then they do get some value from observing their peers during a live CRM session (either watching over a live video link or actually being in the same room). Setting the scene such that the observers are requested and entitled to give advice and help to the hot seaters places the onus on the observers, such that if a diagnosis is not found, or a therapy not instituted, they (the remote observers) are part of such "a less than desirable action" [1–3].

### 81.2.12 Mini-scenarios

Several Simulation Laboratory groups have produced (highly successful) multiple brief 10-min scenarios as part of a workshop where verbal debriefing (CM and CRM) was done in a one-on-one format in 2–3 min (without a video recording). (This is useful when there is neither time, money, nor space for a comprehensive CRM lecture *and* video debriefing.)

### 81.2.13 Role of Hot Seater

The hot seaters may play the roles of themselves (with the skills they possess in real life) or be asked to act as junior personnel with limited knowledge and skill sets (e.g., they cannot or are not allowed to intubate). The participants can also be used as actors, e.g., act as surgeons, scrub nurses, secretaries, etc., to enlarge the group of hot seaters and minimize the number of actors required [3]. Sometimes, however, the trainee actors may go overboard, e.g., act as a really abusive or aggressive surgeon (possibly disrupting the learning issues), and the director needs to have planned a mechanism to neutralize or tone down such excessive behavior.

### 81.2.14 Real-life Events

Try to base the cases in the sessions on real-life events, or combine elements from multiple but separate real-life events into the session. Emphasize to the participants that the scenarios *are* based on real-life events; this preempts them from thinking: "This would never happen to me in real life."

## 81.3 Example: A Real Life Case

A subject, riding on a motorcycle, hit a deer with his leg, and the animal crashed into his chest. He continued to ride with the deer on his lap, until he could slow down, and park the motorcycle against the side of a bridge. After being transported to the hospital, he presented with hypotension, tachycardia, and hypoxia. The differential diagnosis included:

- Hypovolemic shock due to a femur fracture
- Fat embolism due to a long bone fracture
- Lung pathology: pulmonary contusion, hemo- and pneumothorax, tension penumothorax
- Cardiac pathology: cardiac contusion, pericardial tamponnade, myocardial infarct

We mostly use this real-life case to show *one* pathology at a time – compare and contrast the several diagnoses. We use this case for more advanced trainees with two or more

simultaneous pathologies (of varying severity) to help with their critical thinking and diagnostic skills.

## 81.4 Issues Related to the Congress

The type, size, site, and facilities at the congress also need to be taken into consideration in planning the session. A decision also has to be made if the objectives would include an opportunity for all the attendees at the congress to experience the training session (only possible for smaller meetings). If this is not possible, it is important to plan how to select participants.

### 81.4.1 Group Size of Hot Seat Participants

The desire is to get as many participants involved as possible without diluting the experience. The larger group may be required to be actively involved – e.g., discuss events and be able to send a helper [2]. (See Section 81.2.11 – live scenarios versus video recorded scenarios; see Section 81.4.2 – larger group may be observers only.) Typical sessions have one responder with a second helper.

a. Increase group size by sequential entry of participants (this is best for a mix of attendee types) [4].

   This system has the following structure: Send in two *first* responders (nurses, junior residents).
   Send in one or two more *second* responders (often, the primary care physicians that the nurses would typically call for help such as surgeons, senior residents).
   Send in final helper(s) (consultant, cardiologist, anesthesiologist). This system enables up to six participants to be in the "hot seat" and also demonstrates the pitfalls that might occur during the handover of leadership.

b. Have a larger group watching the events (in the same room or via video link).

### 81.4.2 Avoid Embarrassing Participants

a. Hot seaters may be told beforehand which "mistakes" to make (which behaviors to emulate to enable points to be illustrated during the debriefing).
b. The audience may be told that hot seaters were asked to make "mistakes," even though we did not actually ask them to make that specific "mistake."
c. The audience may be asked to participate (and send a helper – see entry (b) in Section 81.3.1); therefore, they become part of the "mistake."
d. During the scenario hot seaters may be "given" the diagnosis if they miss it. This enables the participants to concentrate on the team and group interaction (CRM) aspects, especially during the debriefing.

e. One might manage the debriefing such that the discussion totally avoids all issues of medical management (crisis management – drugs and dosages) and focuses totally on CRM issues. (This might be advisable at international meetings where drugs (names and dosages), equipment, and technology, as well as support systems may vary greatly.)

f. Language and cultural issues. (Soft-spoken or "unintelligible" conversations need to be conveyed to the audience, e.g., an actor can ask: "Did you say X mg of drugs?" or say: "I am giving X mg of drug Y, as you asked.")

g. How to select "hot seat" volunteers.
   - Extroverts who "think aloud" and talk loudly and clearly
   - Stable personalities who can "take it," when they inevitably will make a "mistake"
   - Clear voice so that audience can hear events

h. Burnout of instructors and actors, loss of enthusiasm (when asked to repeat the sessions too many times).

i. Timing issues.
   - Late arrivals – because many clinicians, especially at congresses are perpetually late, we often plan to informally introduce the simulator to the participants as they arrive. This gives them something to do, and this introduction is then given to 1–2 participants at a time, enabling more hands-on participation.
   - Need time for setting up of the simulator between sessions – gives the instructors time to fully prepare the participants for the next session.

j. Booking (scheduling) issues.
   - Advertising and enrollment instructions should be placed in the congress materials
   - Admission tickets to avoid having to turn away participants
   - Overbooking – determine your policy before, not during
   - Admission fee (if relevant) – should be collected prior to the session to avoid delays in starting the session

k. Scenario and debriefing in separate rooms.
   - Frees up simulator for other users/uses
   - Enables setup time for the next session
   - Gives the participants some recuperation time after an intense session before the debriefing session

# 81.5 Issues Related to Venue/ Environment

Each of these points is important in planning the session. For instance, a second room for debriefing has cost implications in a hotel or conference center. Certain essential elements of the high-fidelity simulated environment can be (and need to be) recreated, while it may not be possible to recreate other elements. For instance, the size of the room or the environment (OR, ICU, emergency department, etc.) might not be quite as realistic.

## 81.5.1 Where will the Project/Class be Held?

- Hotel
- Conference room
- Other
- Setup of rooms/equipment for each presenter

## 81.5.2 Observers in the Same Room

The advantages include cost (not a second room, no need for a video link).

- Seating should be planned so that all can see (low chairs in front, high "bar" stools in the back).
- Size of room: large enough to contain all participants and observers.
- Limit number of observers to avoid too much noise.
- Noise in room – depending on the objectives, observers to be asked to comment during events or remain silent For instance, laughter, spontaneous "helpful" but distracting (from realism) comments might be encouraged of discouraged.
- Noise from rest of Congress (prefer a separate room rather than mobile wall divider).
- Control of entry into room/area needs to be planned – latecomers need to be accommodated while passersby need to be directed to continue passing by.
- A separate individual is needed to be dedicated to watch and control observers (minimize rowdiness).

## 81.5.3 Audio/Video Equipment

- At least two cameras should be available in case one fails (bad tape, loses power supply, dead battery, knocked out of position, wires disconnected).
- Video control center should be in a safe place.
- A mobile photographer can move to different positions and get close-up, but the overall "picture/scene" is lost if only a single camera is used. Some centers use the debriefer as the photographer to enable capturing points and behaviors that are to be highlighted in the debriefing.
- Note that there are varying video, CD-ROM, and DVD standards in different machines and in different countries.
- Sound recorder on video camera is often too far away; consider "patching in" a second microphone.
- Ideal if hot seaters have lapel ("lavalier") microphones.
- Avoid noisy environments (Congress, observers).

## 81.5.4 Wireless Communication

- Ideal to have headphones and microphones for "actors" to be in contact with scenario director.
- Signals need to be agreed upon, and rehearsed (if wireless communication is not available).

## 81.5.5 Telephone/Intercom

- This is useful for hot seaters to call for help, and obtain laboratory information, but might be costly or difficult to arrange.
- An actor might be instructed to "answer" a fake telephone and give information (e.g., laboratory results) to the "hot seaters."
- Bring your own inert telephone set(s).

## 81.5.6 Lack of Control Room and One-way Glass Partition

The equipment, personnel, and controller(s) need to be placed in unobtrusive locations (e.g., personnel to manage and control the computer, video, sound, etc., as well as unobtrusive behaviors of personnel in the roles of director, nonactive actors, responders to telephone or intercom, etc.) (see Chapters 23–27).

Unobtrusive manipulation of simulator equipment:

- Changing of gas tanks if empty or near-empty.
- Attention to connections (gas, electrical), which may become undone.
- Resetting and refurbishing components of simulator need to be planned.

## 81.5.7 Waiting Area for Second Responder Needs to be Identified

Many sessions require the second responder or helpers to be naïve to the progress of events up to the point when they are called in to help. A site needs to be identified for them to be waiting.

- Most often "out of earshot" of events.
- A person to call, or bring in the second responder needs to be planned.

## 81.5.8 Storage of Props and Equipment

Items needed for scenarios need to be stored in readily accessible places, for use during a scenario, as well as for setting up the next scenario.

## 81.5.9 Projection and Presentation Systems

Systems are needed to project the lecture, video debriefing, and other educational content. Projection of live events to a remote area for observers requires a hardwire connection (analog or digital), or an Internet-type connection (digital).

A large monitor display or projector and screen is necessary for all observers to see vital signs, anesthesia machine data, waveforms, etc.

- A video camera aimed at the monitor or a scan converter of the monitor for projection of the vital signs to a larger screen is needed for larger groups.
- "Actors" may be asked to continuously announce the vital signs aloud so that the remote participants are aware of any changes that they could not see clearly (however, this might give hints to "hot seaters" if they did not notice the change in vital signs, which may, or may not, be desirable).

Whiteboards/flip chart paper/easels might be needed for debriefing.

## 81.5.10 Handouts Need to be Prepared in Appropriate Quantities

Pathology, blood and other laboratory results, etc., need to be prepared in advance; preprepared sheets with results may be used.

a. Might use hard copy printouts of Power Point presentations for handouts.
b. There is a need to make copies ahead of time (copying in the business center of a hotel is expensive).
c. Floor map of break out rooms, if used, needs to be available to participants.

## 81.5.11 Pre- and Post-questionnaires Need to be Available

The questionnaires are typically designed to explore if the learning objectives have been achieved. Changes in participants' beliefs and perceptions might be explored in the pre- and post-session questionnaire format. This requires matching of pre- and post-session questionnaires with some form of identifier. There may also be rules and regulations regarding CH, CE, and CME requirements to be fulfilled in the questionnaire (e.g., was the presentation free of commercial bias and conflicts of interest?). Questionnaires may also be used to obtain data to evaluate the session, the instructors, and the facilities for future improvements. Release for future use of photographic and video material might also be sought. However, given the nature of CRM sessions, this is not usually encouraged as confidentiality encourages more open discussion, and is believed to improve the learning experience (see next Section 81.5.12).

### 81.5.12 Confidentiality (Volunteers, Performance, Scenario, etc.)

Confidentiality (volunteers, performance, scenario, etc.) forms need to be prepared. Confidentiality of scenario, at least for duration of Congress, needs to be emphasized. The CRM principles should, however, be widely discussed (see comment in Section 81.5.11 regarding photographic release documentation).

### 81.5.13 Knowledge of Scenario

There are several methods to "keep the scenario alive" even though session attendees might discuss the scenario with future CRM session participants.

- Use a different scenario each time (lots of work, and the new scenario needs to be tested and practiced every day).
- Change the demographics and history of the patient (e.g., a healthy male patient, motorcycle accident with a femur fracture, and placement of an internal "rod" leading to a fat embolism, can become an elderly female patient with a hip fracture and renal or cardiac or liver failure who develops the same embolism – fat, or air – very similar cardiovascular presentation).
- A single patient can have one of several complications develop, e.g., an elderly paraplegic patient with coronary artery disease (cardiogenic shock), a 4-day-old CVP (sepsis with septic/distributive shock, or perforated atrium with simple hypovolemic shock, or cardiac tamponnade), with intraabdominal pathology (acute abdomen with dehydration) can have low BP, tachycardia, shock, and loss of consciousness due to one of many causes – every day and every CRM session might have a different cause of shock.
- In many ways, even if the crisis is obvious or fully known (cardiac arrest, loss of electricity, oxygen failure, etc.), it does not matter how the medical crisis is managed, as most of the discussion centers on CRM issues and team interaction. As a general CRM principle, the participants should have the knowledge and skills to diagnose and treat the underlying crisis; otherwise, they continue to focus on the "unfairness" of the *difficult* scenario rather than concentrate on the real educational objectives, i.e., the CRM principles. Therefore, one tends to use relatively simple diagnostic problems, and mostly only a single pathology at a time.

### 81.5.14 Familiarity with All Loaned Clinical Equipment

Instructors need to be familiar with all loaned clinical equipment such as anesthetic equipment, anesthesia/machine, ventilator, and complex vital sign monitors.

## 81.6 Standard CRM Issues in Need of Modification

### 81.6.1 Equipment Needs to be Available

Equipment needs to be available (intense planning and listing – see Appendix 81A.1). Every single piece of imaginable (and unimaginable) equipment needs to be available (with spares) (e.g., a dropped laryngoscope and broken bulb make intubation a problem; lack of a 14 g IV catheter for chest decompression; a perforated endotracheal tube cuff; lack of lubricant).

### 81.6.2 Props Need to be Checked

- Epidural catheter; X-ray and X-ray viewing box; patient's records and chart (ECG, lab reports); IV supplies.
- Props specific for scenario (e.g., wig, radio for background noise).
- (See example of a list in the Appendix 81A.1.)

### 81.6.3 Power Supplies and Batteries

Extension cords, double adapters, screwdrivers (to replace batteries).

### 81.6.4 Roles of "Actors" – Multiple Roles for a Given Actor Versus a Single Role for Each Actor

- Often, only a limited number of actors are available.
- Director may have to fulfill multiple roles, and needs to clearly indicate (dress, announcement) which role is now being played.
- Simulator professional (Computer operator) may double as video and sound operator.
- "Actors" might be mostly from different backgrounds at a multi-institutional simulation session at a Congress – need to practice the sessions.
- Actors may have their own agenda and objectives (as well as pet peeves and/or hobby horses) – all need to agree on joint educational objectives for this specific session.
- All actors accept the Director as being "in overall charge."

### 81.6.5 Changes "On the Fly"

No matter how much thought goes into the planning, participants will come up with a new concept/idea/treatment plan, which might necessitate changes during the session.

Only one person should be identified (typically the session Director) who can change the scenario and sequence of events (e.g., a participant intubates the patient before the patient can give a critical piece of history, or intubates an awake patient without giving any medications at all).

### 81.6.6 Drug Recognition System

For a structured CRM session with clearly defined objectives, a large audience, and "unknown" participants, it might be prudent to disconnect the drug recognition system or override the pharmacological model response to detected drugs. Drug administration errors/wrong drug, wrong dose, wrong timing, wrong response, etc., might take the scenario in a totally unpredictable (and unmanageable?) direction.

For instance, should the patient develop ischemia, ventricular tachycardia, and cardiac arrest with epinephrine (adrenaline), an embolism scenario may turn into a cardiac resuscitation scenario, where every country has different sequences, drugs, responses, etc.

A patient who died under the trainees' care is embarrassing to all concerned, and more time may be spent smoothing ruffled feathers than on CRM principles. This said, however, under special circumstances, this (i.e., a dead patient) might be instructive to highlight and emphasize CRM principles.

### 81.6.7 Fluids and their Messes

Hotels and conference centers don't like bloody messes in their carpeted convention areas, even if it is *only* fake blood. Make adjustments to your home scenarios in how you present body fluids. You are a guest, so please act accordingly.

## 81.7 Developing the CRM Session

### 81.7.1 Time Requirements and Planning for a Single Hands-on Crisis Event

30 min
  a. Prequestionnaire.
  b. Familiarization with the simulator and equipment. (Note: need door keeper to introduce/explain events to latecomers.)
30 min
  c. Setting of scene. Rules of engagement/roles of volunteers.

Crisis event – with the simulator – video recorded.

15 min
  d. Lecture on CRM principles with many examples.
15 min
  e. Typically, a nonmedical video illustrating CRM principles allows participants to become familiar with:
     – using the CRM terminology
     – functioning/discussing with a group of strangers
     – being supported and encouraged by the facilitator.
The circus scene from "A Bug's Life" and aircraft accidents in a flight simulator are commonly used in our center. Video (7 min) (15 min, with interruptions for discussion). Attendees practice to use CRM terminology.

25 min
  f. Own video debriefing.
5 min
  g. Post-questionnaire.

Note: While it is possible to fit in a second crisis event in 120 min, it requires all trainees to arrive promptly on time, and no dawdling in between cases – it typically feels hurried, and is probably not advisable – rather add some more time (30–60 min) for normal human hesitancy and indecisiveness.

### 81.7.2 The Sequence of Sections of the Training Course

The major question is whether to give the CRM lecture after or before the main crisis scenario.
  **CRM Lecture After:**

  a. Pre-questionnaire/data collection/confidentiality form.
  b. Familiarization with the simulator.
  c. Crisis event.
  d. Lecture on CRM.
  e. Practice session using CRM terminology (aircraft video, other video, case discussion).
  f. Own video debriefing.
  g. Postquestionnaire.
  h. Individual debriefing/discussion (specific simulator issues, relook at simulator). (Specific medical issues, controversies of diagnosis and treatment.)

  **CRM Lecture Before:**

Switch c and d (i.e., the CRM lecture is given first and participants try to implement the just-presented CRM principles in the scenario).

### 81.7.3 Pre-questionnaire/Data Collection/ Confidentiality Form

- This may be collected at this stage (after completion of the questionnaire, but before the crisis event), or kept by each observer/participant until the end and handed in as a unit.
- Can be anonymous, with demographic questions only (if kept as a unit, can do before and after statistics) or can be with participant's name.
- Can be filled in at time of booking the session (prerequisite).
- Ask if would be willing to be a volunteer "hot seater" at the time of booking the session.

### 81.7.4 Familiarization with the Simulator

- This may be set as a prerequisite to the session (in trainee's own time).
- This may be done as an optional event in the 15–20 min prior to the official start time.
- This may be done as participants drift in – this is a very useful use of the initial 20–30 minutes of the session, as very few congress participants actually arrive on time (very late participants might not have an opportunity for prior familiarization).
- This may not be done at all (large groups, each having to place a stethoscope on every speaker of the simulator, feel all the pulses, etc., may just take too much time).
- May be done as a group introduction (video to show eyes, chest movement, etc.).
- May be done at the end of the session for observers who missed it earlier.
- Essential for hot seaters to be familiar with functioning of simulator.
- Essential for the group to have an idea of the environment in which the scenario takes place (e.g., a patient's home, an emergency room, a rural hospital, an academic hospital with a trauma center).
- Essential for the hot seaters to understand if they are themselves with their capabilities, or are acting as someone else (e.g., a junior clinician who is expected to not "know it all, and make deliberate mistakes").

### 81.7.5 Crisis Event

- Time should be allowed to set up the simulator and the scenario for the crisis event (after the familiarization).
- This is an appropriate time to select volunteers for hot seaters and set the background (give past medical history, progress of therapy of the patient up to this point).
- Unless the induction of anesthesia is essential to the scenario (e.g., difficult mask/intubation), it is advisable to arrange for the trainees to "take over" an anesthetic. Much time can be "wasted" by "suspicious" hot seaters checking every detail of the anesthesia machine, piping, oxygen pressure, vital signs, etc.
- The crisis event can be quite short (5–7 min) – this greatly enhances (smoothes/eases) the video debriefing, as less time is spent fast forwarding (and back) to find specific points to illustrate.
    (Even 5–7 min feels like an eternity to the hot seaters; it also avoids boredom in the observers.)
    (Some units employ 20–30 min, or even longer, scenarios with great success; however, this is more typical in day-long sessions that occur in the home-based simulation laboratory.)
- The crisis event should be designed (and tested) to "force" behaviors that make for good video debriefing

(e.g., a piece of failed equipment might get all the participants to work on it, while nobody is watching the patient (e.g., an empty oxygen cylinder, a stuck key or locked resuscitation cart/trolley)). An actor can also ask the hot seater to help with a task (thereby creating a distraction). Multiple equipment problems might occur (be given) simultaneously, forcing attention away from the patient.

### 81.7.6 Lecture on CRM Principles

- This should be kept brief.
- A handout with the CRM principles in a large font is a good idea.
- Practical, real-life examples should be used when discussing CRM points (rather than generic principles).

For example:

- When the lecturer says: "The leader should stand back mentally and physically and should direct personnel to perform specific functions," this statement/description is too "generic" and needs to be illustrated by the lecturer crossing arms, stepping back, and giving examples of orders such as: "Jane, bring four units of O-negative blood as quickly as possible."
- Examples should be appropriate for the learner group, e.g., some hospitals/practitioners do not deal with trauma/pediatrics/obstetrics, and trauma examples (hypovolemia due to a ruptured spleen after a motor vehicle crash) are not appropriate, e.g., for a group of internal medicine practitoners or pediatricians, one could use bleeding esophageal varices as the cause of hypovolemia in a CRM session.
- It is difficult to find an example to illustrate each CRM point for each subspeciality, but every effort should be made to achieve this goal. Asking the audience (and recording the responses) is an excellent way of building up examples based on real-life incidents (it also involves the audience).
- Giving examples where the instructor/lecturer was involved or made errors (could have done it differently) is an excellent introduction/method (gives permission) to asking the hot seaters for self-debriefing (i.e., ask them what they will do differently next time).

### 81.7.7 Practice Session Using CRM Terminology in a Non-medical Crisis

- Using a nonmedical video (e.g., aircraft accident) is a nonthreatening method of obtaining audience participation (and prepares them for self-debriefing).
- Asking each participant (in a small group) to mention (use) a CRM principle in the nonmedical video engages

them (might not be possible for each participant to contribute in a larger group, e.g., 25+ observers).

- The participants have the handout with the CRM principles available as an immediate memory aid and a take-home training aid.

### 81.7.8 Debriefing

- Should be done in a supportive, nonpunitive fashion.
- It is advisable for would-be debriefers to undergo a specific training course, as well as practice debriefing under guidance of an experienced instructor to gain expertise in this difficult (and "sensitive") skill.
- Concentrating on CRM issues for larger groups with non-hands-on observers (and avoiding medical issues) helps to avoid and minimize self-deprecation based on perceived incorrect medical management. (Note: many experienced home-based units do mix CM and CRM, e.g., 40% CM and 60% CRM.)
- Terms such as "errors," "wrong," etc., should be avoided in favor of questions such as: "What could you have done differently?" or "What would you do differently next time, now that you have heard about the CRM principles."
- Questions are often better than statements (e.g., "Who is the leader at this stage?" or "Did you tell anybody what you were thinking/planning/doing/preparing?").

### 81.7.9 Post-questionnaire

- Structured questions (pre- and post-) give statistically comparable and reliable data (e.g., anchored Likert scales), and can enable comparison of the pre- with the post-questionnaire.
- Free-form comments are useful to refine the session during a series of sessions at a Congress (and help in planning future sessions).

## 81.8 Conclusion

While many might consider a road show or a simulation session outside the strict confines of the simulation center too daunting, it can be done successfully with meticulous planning and attention to detail. Such "away" sessions greatly expand the number of potential trainees, and are an excellent method to bring the advantages of simulation to a much larger audience who might otherwise not have an opportunity to be introduced to this valuable mode of learning.

The views/ideas expressed in this article are those of the author and do not reflect the official policy or position of the US Department of the Army, US Department of Defense, or the United States Government.

## References

1. Murray, W. B., Henry, J., Jackson, L., Murray, C. and Lamoreaux, R.B. Leadership training: A new application of CRM and distance education in a large group format at a medical simulation facility. *Journal of Education in Peri-operative Medicine* IV(2), 1–14 (2002) http://jepmadmin.org/VolumeIV/Issue2/Issue2v4.htm
2. Elser, C. S., Murray, W. B., Schneider, A. J. L., et al. Simulated crisis in obstetric anesthesia: design and evaluation of distance education presentation. *Journal of Education in Perioperative Medicine* III(2) (2001). http://jepmadmin.org/VolumeIII/Issue2/Issue2v3.htm
3. Lutz, A., and Murray, W. B. The Perceived Value of a Human Simulator and Crisis Resource Management (CRM) Training in the Third World. *J. Clin. Anesth.* 15(4), 308–309 (2003).
4. Murray, W. B., Proctor, L., Henry, J., et al. Increasing the "hot seats" for crisis resource management (CRM) training: planned sequential participant entry. *Anesth. Educ.* 17(2), 6 (1999).

# Borrow Success

*"Don't worry about people stealing an idea. If it's original, you will have to ram it down their throats."*

Howard Aiken

If simulation is nothing but applied symbology, and if symbology is using the essence of one to represent that of another, then the acme or pinnacle of a lesson on simulation is to use the familiar symbols of one domain to convey the concepts of another. Mary Katherine Krause and Margaret Faut-Callahan use the fundamentals of owning and operating a restaurant as a way to present the fundamentals of owning and operating a simulation facility. All of us have patronized restaurants; few of us have actually owned and operated one. Most of us have experienced simulation in one form or another; some of us have actually created our own simulation session, few have created and operated our own simulation facility. Here, the veil of professionalism is pulled back to expose the real work required to produce a quality experience that we are all familiar with and would take pride being responsible for. A quality dining experience starts a long time before the patrons arrive, requires a lot of hard work to set up, and even more to sustain. So too do clinical simulation experiences. The best chefs make the impossible seem ordinary, and make the ordinary invisible. So too do clinical simulators. A restaurant's reputation is only as good as its last serving. So too is that of the simulation program. For those of you who wish to learn more about what a simulation program has to offer to your clinical education program, learn from with those institutions and individuals that have created successful ones. For those of you who wish to learn more about owning and operating the facilities required for a simulation program, perhaps you should study restaurants and chefs that have created successful ones.

# 82

# An Innovative Way to Think about Simulation Laboratory Core Administrative Functions: Comparing Managing a Simulation Laboratory to a Restaurant

Mary Katherine Krause
Margaret Faut-Callahan

## 82.1 Academic Entrepreneurs

There are some aspects about operating a simulation facility that are simple and some that are complex. One might compare operating a simulation program to operating a restaurant. It is unlike any other educational endeavor that most academicians have ever ventured into. In fact, it is quite entrepreneurial. The Webster's Dictionary definition of an entrepreneur is very fitting under this circumstance: "One who launches or manages a business venture, often assuming risks." Typically, academicians do not manage brands or negotiate the sale of educational products and services. Clinical teaching techniques classically do not include using reality simulation, let alone constructing real educational programs on a foundation of synthetic

patients. Persuading teachers to utilize a different methodology requires a business-like approach to behavior modification. It takes an individual with a keen business sense to lead and manage a successful clinical simulation enterprise.

## 82.2 Successful Simulation Program from Successful Business Practices

However, as daunting and high tech as clinical simulation can be, producing a simulation program is still all about putting on a good show day in and day out, with customers' experiences being only as good as their last experience. With your simulation program's reputation on the line every day, the mantra touted in the restaurant industry, "Proper Preparation Prevents Poor Performance," holds true for you as you prepare to open your own simulation program as well as how you plan to use it. The purpose of this chapter is to provide insights into the core *administrative functions* of a simulation program and to describe features that may need to be customized to fit an individual laboratory.

What makes one choose a certain restaurant over another? The location? The convenience? The ambience? The type of food it serves? The quality of food it serves? The variety in the menu? The ability to modify the menu and the dining experience to accommodate your needs? The price? The speed of service? The comforts in having everybody know your name? It can be any one of these things on any given day. At the end of the day, operating a restaurant and operating a simulation laboratory all comes down to the ingredients (your simulators), the menu (your scenarios), the ambience and music (the clinical realism), and the service (the instructors) provided. How well you prepare for the fundamentals will determine your success in the long term.

While comparing the operation of a simulation laboratory to operating a restaurant may seem overly simplistic, both Ram Charan, in "*What the CEO Wants You to Know*" [1], and Jim Collins, in "*Good to Great*" [2], tout the importance of simplicity. Ram Charan explains that what CEOs know that most of us do not is how to use the same common sense that successful street vendors use to perceive what's really changing in the outside world and to cut through the complexity of their business. Jim Collins explains that what separates those who make the biggest impact from all the others is that they took a complex world and simplified it.

## 82.3 What are the Questions You Ask?

What is the vision for your new simulation program? Who are the stakeholders? Who are the decision makers? Will it be a single-purpose, single-discipline simulation laboratory (e.g., a hamburger joint) or will it be full-service operation (e.g., a family diner)? Have you considered your target market? Will your customer base be from a multiservice, single institution (e.g., a museum restaurant); a large, multiservice, multi-institutional citywide area (e.g., a local restaurant chain); or an even larger area covering a region or state(s) (e.g., national chain of multiple types of restaurants)? How do you plan to fund the start-up and maintenance of this new venture? Philanthropy? Grants? (See Appendix 82A.1) Hospital or university funds? How much money do you think you will need? How do you ensure that you have thought about and covered all financial aspects? (See Appendix 82A.2.)

Is this a large or small simulation center relative to your customer base? Are you retrofitting existing space or starting from scratch? Have you allocated funds for marketing and public relations (e.g., press release, open house, brochures, web page creation and maintenance, and personnel time as tour guides, etc.)? Have you obtained permission from users to photograph them and to use their photos on promotional materials? (See Appendix 82A.3.) Have you factored the warranty and annual maintenance contract costs into the total cost of maintaining your simulation equipment and functional spares all essential items to assure no session is ever canceled due to equipment failures? What about replacement parts for your task trainers? What about upgrading when new models become available? What about up front and ongoing training costs for your simulation professionals? What about the income from patient care lost during simulation training and use by clinical faculty?

How do you plan to attract instructors and their students to even try simulation for the first time? One common misperception in starting a simulation laboratory is that "if you build it, they will come." Not so for either a restaurant or a simulation laboratory. In order for diners to be motivated to change what they are currently doing (e.g., preparing their own meal at home or eating at their 'favorite' restaurant, etc.), they must think that their dining experience will be better by coming to your restaurant. Likewise, in order for instructors to be motivated to change what they are currently doing (e.g., teaching through textbooks and lectures, or having trainees practice on patients or each other, etc.), they must think that the training experience will be better by coming to your simulation facility. In the same way that parents (instructors) as customers make the ultimate decision in where the family is going for dinner, the family won't come back if the children (trainees) are not satisfied.

So what's the draw? What's the hook? How are you going to convince instructors to change their curricula to include simulation-based training as a component of their course(s)? How are you going to relieve their anxiety in having to learn something new, risk embarrassment at no longer being the expert, and to help them find the time to do so? And once you convince them to come for a demonstration, how are you going to convince them to expend their time and energy to use simulation with their students?

## 82.4 Who are the People You Include?

In a restaurant, there may be an owner, a general manager, a host, and a chef, in addition to waiters, bussers, bartenders, dishwashers, and cooks. In a small or start-up simulation program, the simulation coordinator plays many of these roles, from host and general manager, to cook and even dishwasher.

As in a restaurant, the host monitors the customers coming in and out of the restaurant and their satisfaction levels. The simulation coordinator does the same and is the first and the last contact that trainers and trainees have in the simulation lab. The person filling this role is the *face* of your program. Choose wisely, support fully, and pay well.

The host needs to be able to accommodate one person dining alone as well as small and large groups dining together. Likewise, the simulation coordinator actively shuffles the rotation schedule to accommodate one person being evaluated, such as a single resident working on a specific skill, small multidisciplinary groups utilizing standardized patients to practice role-plays, such as "delivering bad news," and large multidisciplinary groups enrolled in local metropolitan area chemical bioterrorism workshops, for instance. The host needs to be able to accommodate customers expecting preferential treatment. Likewise, the simulation coordinator maneuvers through the various personalities and behavior styles of a diverse set of instructors who may ultimately represent every clinical discipline in your institution. The host needs to be able to accommodate children needing to be kept occupied during the dining experience with high chairs, crayons, and paper. The simulation coordinator does the same, using airway and other task trainers, screen-based teaching programs, and audio/visual technology to occupy the "down time" of larger groups of users rotating through the simulation laboratory at one time.

As part general manager, the simulation coordinator must be in control, ready for each day, and have the right attitude to always strive to do better. Likewise, the simulator coordinator must monitor inventory, ensuring that proper supplies are available, are in working order, and are up to regulatory code. In a simulation facility, IVs and oxygen take the place of the plates, linens, and utensils in a restaurant; the simulator computer and audio/video system take the place of the oven, deep-fryer, and exhaust system. Both a simulator laboratory and a restaurant must maintain the standards of regulatory agencies such as the departments of public health, occupational safety and health, and transportation, etc. Both simulation labs and restaurants have regulated and nonregulated waste, requiring special handling and collection. In a simulation facility, specialized disposal of needles, bandages, and fake blood take the place of specialized disposal of used cooking oil, grease, and ashes (from a wood-burning stove), etc.

As the dishwasher, the simulation coordinator is responsible for cleaning supplies, maintaining equipment, and even mopping up "blood" and the messy results of "hacks" to enhance a simulator's capabilities. Because of the many full- and part-time roles that a simulator coordinator plays, methods for reducing job burnout and ensuring career progression are important in order to keep one of your most important assets to your simulation laboratory satisfied. These include offering the simulation coordinator "train-the-trainer" opportunities to train or teach, attending simulator society meetings to meet other simulation laboratory coordinators, other personnel to provide "backup" or "after-hours" operational support; training a cadre of skilled faculty instructors in order to make better use of the simulation coordinator's time; and inviting the simulation coordinator to participate in other departmental staff meetings, parties, and special events in order to enhance the stature of simulation within the institution. Methods for ensuring career progression of your simulation coordinator include offering tuition reimbursement, and a more flexible schedule with backup support in order to pursue an advanced degree.

The director(s) and administrative liaison(s) to your simulation laboratory also play the role of owner in terms of having ultimate responsibility for legal contracts, operational performance, and return on investment. They also serve as general manager in terms of determining organizational structure, managing human resources, facilities layout, vendor selection, negotiating agreements, managing the bottom line, marketing, communications, and customer satisfaction.

Simulation clinical instructors not only play the role of customer, but in some programs, they play the role of chef, as well as creating and selecting the scenarios (menus) to be provided for each simulation rotation (seating). While menus are determined on the basis of the type of food the customers want from a restaurant (e.g., fast-food versus gourmet, etc.) or niche of the restaurant (e.g., seafood, steak, pizza, ethnic, etc.), simulation scenarios are based upon the teaching objectives of the instructors and the learning objectives of the students. These goals in turn determine the type of simulator equipment available (e.g., infant, pediatric, or adult human simulators or laparoscopic, endoscopic, or endovascular specialty simulators, i.e., part-task trainers or robotic full-human simulators, etc.), the conditions that are simulated (e.g., cardiac arrest, asthma attack, or preeclampsia, etc.), and the procedures practiced (e.g., defibrillation, intubation, or paracentesis, etc.). Like the waiter or server in a restaurant, simulation laboratory instructors must service the customers, trainees in this case, by listening, prompting, and modifying the menu (scenarios) and dining experience (simulation rotation), to satisfy customers' dining (learning) needs.

Team meetings in successful restaurants occur on a daily basis. Likewise, team meetings for a simulation laboratory should also occur regularly, at least monthly among staff, instructors, and representatives from user group disciplines. Being a good director of a successful simulation laboratory entails being a good team leader, including understanding that all stakeholders have a place where they add the most value and helping them to utilize their talents in order to maximize the team's potential. John Maxwell [3] calls this "the law of the niche," and suggests that in order to help trainees reach their potential and maximize their effectiveness, they must be stretched out of their "comfort zones," which leads to fulfillment, but never out of their "gift zones," which leads to frustration. Good simulation laboratory directors must also possess the discipline of execution, "meshing strategy with reality, aligning people with goals, and achieving the results promised" [4, 5].

## 82.5 What are the Services You Provide?

In the same way that a restaurant must be prepared to provide anywhere from one to four or more courses as part of the dining experience, a simulation laboratory must be prepared to provide one or more scenarios as part of a simulation experience. Thus, the simulation session "*du jour*" could range anywhere from a simple airway management training scenario to a full advanced cardiac life support or crisis management training scenario. Restaurants must be prepared for picky eaters, persons with allergies, or persons on a restricted diet who require modifications to the recipe or add-ons to the meal. Likewise, the chef (instructor) may need to tweak a recipe (specific simulation scenario) to get just the right taste (to respond to a particular trainees' unique caregiving actions).

While restaurants vary in terms of the volume of customers throughout the day, month, and year, simulation center volume will ebb and flow due to instructor availability, university class schedule, and time of day, etc. While a restaurant may need to create flexible staffing plans to adjust to customer volume (e.g., sending one server home early one week and another home early another week), a simulation laboratory may need to create a flexible staffing plan (e.g., hiring part-time staff from other areas of the clinical center or negotiating "trades" of resources and talents, etc.).

## 82.6 What Spaces Do You Need?

In a restaurant, there are the parts of the physical structure that customers see, such as the dining area(s) or bar and the parts of the physical structure that are hidden from view, such as the kitchen. Likewise, simulation customers see the simulation room (dining room) or conference and lecture room(s) (banquet rooms), but rarely the control room (kitchen), where simulation scenario prep and computer programing (meal prep and the staging of the meal) occurs. The true test of a simulation experience is realism in the minds of the students. If boxes of supplies and spare equipment are lining the simulation room that has been prepared ("mocked up") to simulate an OR, ER, or an ICU, then realism is compromised. Adequate space for storage, cleanup, and preparatory work should be allocated so it is hidden from view.

The physical layout of a restaurant (simulation laboratory) is based on the flow of the operations; the design of the restaurant is based on what the restaurant is serving; and the design of the kitchen (control room) is based on what is being served (scenarios). And while the strength of a restaurant (simulation laboratory) lies in its food (simulators) and service, the entire dining experience is evaluated on the basis of ambience (e.g., a fire place, mood lighting, music and service tone in a restaurant; fairness and useful feedback in a simulation laboratory, etc.), layout (e.g., different-sized tables and/or booths in

a restaurant in the place of small and large conference rooms or multipurpose areas in a simulation laboratory, etc.), and special features (e.g., an "open kitchen" to observe the cooking process or outdoor seating in a restaurant in the place of one-way windows and observation rooms in a simulation laboratory, etc.).

## 82.7 Conclusion

And while all of this may seem like a lot to think about, just remember that operating a simulation laboratory has many simple tasks (e.g., scheduling simulation rotations, ordering supplies, and maintaining records, etc.), but the complex human-centered nature of simulation requires highly skilled and motivated individuals who work well as a team, are capable of motivating interdisciplinary groups, and are committed to making the simulation experience the best that it can be for their customers so they'll keep coming back for more! Bon Appetite! *(Special thanks to Augustus Solomon, General Manager, Francesca's Fiore in Forest Park, IL)*

## References

### Suggested Reading

Martin E. Dorf, Restaurants That Work – Case Studies of the Best in the Industry. Whitney Library of Design, an imprint of Watson-Guptill Publications/New York (1992). Excerpts from the flyleaf: "[The book] presents a complete rundown of how successful restaurateurs, teaming up with architects and designers, ply their craft. It answers the questions: How does a freshman restaurateur without any prior experience decide what type of restaurant to open? Where should it be located? Who is the potential customer and competition? What type of food should be offered, and at what price? What should the restaurant look like? The book features invaluable information on building codes, utility requirements, and construction cost analysis, as well as a special appendix on handicapped access laws."

1. Charan, R. *What the CEO Wants You to Know: How Your Company Really Works.* Crown Publishing Group, New York, 2001.
2. Collins, J. *Good to Great.* Harper Collins, New York, 2001.
3. Maxwell, J. *The 17 Indisputable Laws of Teamwork.* TN. Maxwell Motivation, Inc., Nashville, 2001.
4. Welch, J. *Jack: Straight from the Gut.* Warner Books, New York, 2001.
5. Bossidy, L. and Charan, R. *Execution: The Discipline of Getting Things Done.* Crown Publishing Group, New York, 2002.

# Index

CPSIA information can be obtained at www.ICGtesting.com
Printed in the USA
BVOW050050170112

280648BV00007B/1/P